Mathematics
of Physics
and Modern Engineering

I. S. SOKOLNIKOFF
Professor of Mathematics
University of California, Los Angeles

R. M. REDHEFFER
Professor of Mathematics
University of California, Los Angeles

Second Edition

McGRAW-HILL BOOK COMPANY
NEW YORK ST. LOUIS SAN FRANCISCO
TORONTO LONDON SYDNEY

**MATHEMATICS OF PHYSICS
AND MODERN ENGINEERING**

Library of Congress Catalog Card Number 65-28518
59625   1 2 3 4 5 6 7 8 9 0BN7 3 2 1 0 6 9 8 7 6

Preface to
the second edition

THIS NEW EDITION of our book differs materially from its predecessor in content, but the fundamental objective—to provide a sound and inspiring introduction to mathematics and its uses—remains the same.

The increasing use of mathematics in applications has been accompanied by a notable improvement in the mathematical preparation of those seeking careers in science and engineering. For this reason, some introductory material in the first edition has been replaced by new topics, and emphasis on the cultivation of critical mathematical thinking has been strengthened. Aside from its special importance in applications, the subject of infinite series admirably lends itself to the introduction of a variety of concepts that underlie the structure of mathematical analysis. Accordingly, it is treated in the first chapter in this edition. Some familiarity with this chapter is desirable, but not essential, for the study of the other chapters. Indeed, the sole prerequisite for the study of any chapter in this book is a sound first course in calculus.

Chapters 2 and 3, devoted to ordinary differential equations and the Laplace transformation, include much material on linear and nonlinear equations that did not appear in the earlier edition. Although the need for manipulative skill is stressed, the subject is presented as a connected body of knowledge, rather than as a series of isolated techniques. Major revisions and additions have also been made in the presentation of the material on functions of several variables and in the coordinate-free formulation of the basic theorems of vector field theory.

The number of problems and illustrative exercises has been greatly increased, without providing extensive and often useless lists of problems of essentially the same order of difficulty. Our problems fall into three broad categories: (1) those that promote the assimilation of material presented in the preceding sections, (2) those that prepare the student for new concepts to be introduced later, and (3) those that extend the results of the text to new situations and assist the student in making useful generalizations. Throughout the book we have stressed the relation of science to mathematics and of mathematics to science.

The book has ample substance for six consecutive semester courses meeting three hours a week. An increasing number of universities is introducing curricula for students of engineering and physical sciences that require a course in mathematics in every undergraduate year. The topics we have chosen are in substantial agreement with those recommended by the distinguished Committee on Applied Mathematics of the Study Session, sponsored by the National Science Foundation. These recommendations appear in "Curricula in Solid Mechanics," edited by H. Liebowitz and J. M. Allen, Prentice-Hall, Inc., Englewood Cliffs, N. J., 1961.

The contents of this book include what we believe should be the minimum mathematical equipment of a scientific engineer. It may not be out of place to note that the mathematical preparation of physicists and engineers in Russia exceeds the minimum

v

laid down here.   While the curricula of only a few leading American engineering colleges provide now for more than one year of mathematics beyond calculus, their number will continue to increase with the realization that the time allotted to mathematics is a sound capital investment yielding excellent returns both in the time gained in professional studies and in the depth of penetration.

I. S. SOKOLNIKOFF
R. M. REDHEFFER

From the preface to
the first edition

THE RAPIDLY DECREASING time lag between scientific discoveries and applications imposes ever-increasing demands on the mathematical equipment of scientists and engineers. Although the mathematical preparation of engineering students has been strengthened materially in the past thirty years, the introduction of courses beyond the traditional "terminal course" in calculus has been largely confined to a few leading institutions. The reluctance to broaden significantly the program of instruction in mathematics can be attributed in part to the crowded engineering curricula, in part to the failure to sense the central position of mathematics in sciences and technology, and in part to the scarcity of suitable staffs and instructional media. The broadening, however, is inevitable, for it is now generally recognized that no professional engineer can keep abreast of scientific developments without substantially extending his mathematical horizons.

This book, in common with its predecessor written by the senior author some twenty-five years ago, has as its main aim a sound extension of such horizons. The authors not only have been guided by their subjective appraisal of the live present-day needs of the engineering profession but have also taken into account the views of the leaders of engineering thought as expressed in numerous conferences and symposia on engineering education sponsored by the National Science Foundation, the American Society of Engineering Education, and its predecessor the Society for the Promotion of Engineering Education. . . .

A preoccupation with the logic of mathematics and the overemphasis of a convention called rigor are among the best known means for stifling interest in mathematics as a crutch to common sense. On the other hand, a presentation which puts applications above the medium making applications possible is sterile, because it gives no inkling of the supreme importance of generalizations and abstractions in applications. The authors have tried to strike a balance which would make this book both a sound and an inspiring introduction to applied mathematics. . . .

The choice of topics is based on the authors' estimate of the frequency with which the subjects treated occur in applications. The illustrative material, examples, and problems have been chosen more for their value in emphasizing the underlying principles than as a collection of instances of dramatic uses of mathematics in specific situations confronting practicing engineers.

Although the book is written so as to require little, if any, outside help, the reader is cautioned that no amount of exposition can serve as a substitute for concentration in following the course of the argument in a serious discipline. In order to facilitate the understanding of the principles and to cultivate the art of formulating physical problems in the language of mathematics, numerous illustrative examples are worked out in detail. The authors believe with Newton that *exempla non minus doceunt quam precaepta.*

I. S. SOKOLNIKOFF
R. M. REDHEFFER

To ensure *flexibility and to minimize irksome references, the contents of this book are arranged in ten virtually independent chapters.* Each chapter, in turn, is subdivided into relatively independent functional parts. The earlier parts of each chapter are written in an expansive style so as to make them easily accessible to students with some knowledge of beginning calculus. But, toward the end of the chapter, the tempo quickens, and concluding parts include much that is of interest to advanced students.

This organization allows the presentation of several topics at an earlier stage than is appropriate for the book as a whole. Thus, experience has shown that Secs. 1 to 13 of the chapter on infinite series can be assimilated by students having no prior acquaintance with the subject. (Naturally the pace must be leisurely, since the actual content, in both theory and applications, goes beyond that of the usual introductory treatment.) To make the analysis self-contained, an unobtrusive review of theorems on limits has been added, in the form of examples and solved problems.

The subject of linear differential equations in Chap. 3, Secs. 1 to 11, can also be taken up, without haste, at an early stage. This discussion stresses the *general* concepts of linearity, superposition, and stability. We believe that an early acquaintance with these ideas will prove useful to the student throughout his subsequent studies. Indeed, the concept of linear operator occurs repeatedly in the most varied contexts and is used in this book as a unifying device behind the diversity of special methods.

Among other chapters whose early parts require little preparation (when studied at a leisurely pace) we mention Chap. 9 on probability and Chap. 10 on numerical methods. These chapters have been rewritten so that the earlier parts are less demanding than they were in the first edition, although the total content of both chapters has been increased.

If the early sections of selected chapters are used in the manner suggested, the instructor may find it advisable to review these sections when later sections are taken up in a more advanced course. Such a review not only consolidates the material but facilitates acclimatization of students transferring from other institutions.

The steady increase of pace within each chapter has enabled us to include a number of relatively advanced topics. There are conflicting and often prejudiced currents of thought as to what should and what should not be proved in a book of this kind. In our opinion, logical development of the more important results should be carried at least far enough so that the student can see why the results are true. Quite often, the easiest means of doing this is to give a proof.

In line with these thoughts, we have not yielded to the temptation to quote outside sources more often than seems necessary. The mere giving of references would have been easier for us as authors, but experience indicates that many students have no time to look up the needed material. Hence, the net effect of overreliance on higher authority is that major results are accepted without any logical basis.

On the other hand, although the need for logical structure is much the same in different disciplines, the appropriate emphasis is dictated by the special interests and needs of our readers. When faced with a choice between addressing our colleagues and addressing the student, we have always decided in favor of the latter. The result is that, in not a few cases, the use of small and large print is exactly opposite to that which it might be in another setting.

If the instructor is willing to accept this shift of emphasis, he will find that the book contains analytical developments of considerable mathematical content. Of many examples that could be given, we mention estimation by the method of differential inequalities in Chap. 2, the proof of existence and uniqueness for the Laplace transform in Chap. 3, and the convergence criteria for improper multiple integrals in Chap. 6.

We feel that there are some advantages in the judicious inclusion of this sort of material, even when it may go beyond the student's immediate needs. One advantage is that the more elementary topics are appreciated as part of a larger whole. The resulting *broadening of horizons* is a major goal of this book, as it was of its predecessors. Another advantage is that a moderate amount of advanced material greatly increases the usefulness of the book as a reference volume. This is not *the* consideration, but it is a consideration, because our experience has shown that many users of the first edition retained the book after their course work was completed.

The physical arrangement of material in this book is dictated largely by the requirement of flexibility. Thus, most of the discussion of numerical methods is found in a single chapter, Chap. 10, so that it can be used in a course on numerical analysis. Such a course would be difficult to organize if the subject matter were scattered throughout the book. On the other hand, those who wish to include numerical methods in their study of differential equations (for example) will find that it is easy to do so, simply by referring to relevant sections of Chap. 10.

The discussion of series solutions for differential equations has been moved from the chapter on infinite series to that on differential equations, because this is the point at which most users of the book take up such material. However, the dominating technical difficulties involve the theory of series rather than differential equations. We have therefore included a problem directing the student to review relevant sections of Chap. 1. This method of cross reference is used on other occasions. For example, when the Laplace transform is applied to partial differential equations in Chap. 7, a problem outlines a review of appropriate sections on the Laplace transform in Chap. 3.

To minimize the blank spaces in the problem lists, many of our numbered problems consist of several parts which are set across the page. The technical difficulty of the parts ordinarily increases as one proceeds horizontally; conceptual difficulty grows vertically, that is, with increasing problem number. The economical arrangement of problem lists has made it possible to include some important topics that would otherwise have to be excluded for want of space.

Our experience has shown that several independent courses for students of applied mathematics can be based on the material contained in this book. Thus Chaps. 2 and 3 have ample substance for a solid course in ordinary differential equations. Sections 12 to 19 of Chap. 3 can serve as a text for a course in the Laplace transform. Other courses, bearing labels similar to our chapter headings, can easily be organized, because the selection of material for them requires little, if any, skipping from chapter to chapter.

We have included a large number of carefully worked out illustrative examples designed to minimize the student's dependence on the instructor and to foster the habit of independent study.

Contents

Infinite series 1

The general theory

Power series and Taylor's formula

Fourier series

Infinite series

VIRTUALLY EVERY FUNCTION encountered in applications of mathematics can be represented as an infinite series. Many nonelementary integrals can be written by inspection as a power series, and such series also give a simple, systematic method of solving differential equations. Another use of power series is in the study of functions of a complex variable $z = x + iy$; thus, from the series for $\sin x$ one can ascertain the appropriate definition and the important properties of $\sin z$. A type of series known as a *Fourier series* arises when one studies the response of a linear system to a periodic input in circuit analysis, in transmission-line problems, or in the theory of mechanical systems. Fourier series and their generalizations are also useful for solving the boundary-value problems of mathematical physics.

THE GENERAL THEORY

1. Sequences and Series.   Let there be given a succession of numbers

$$a_1, a_2, a_3, a_4, \ldots, a_n, \ldots \tag{1-1}$$

An ordered set of numbers such as this is commonly denoted by $\{a_n\}$ and is called a *sequence*. Thus,

$$1, 3, 5, 7, \ldots, 2n - 1, \ldots \tag{1-2}$$

is a sequence represented by the symbol $\{2n - 1\}$. When the number of terms in the sequence is unlimited, the sequence is called an *infinite sequence;* otherwise it is a finite sequence. The symbol ... terminating expressions (1-1) and (1-2) indicates that both are infinite sequences.

Starting with the sequence (1-1), we can form the expression

$$a_1 + a_2 + a_3 + \cdots + a_n.$$

Such an expression is called a *series* of n terms. An *infinite series* is an expression

$$a_1 + a_2 + a_3 + \cdots + a_n + \cdots \tag{1-3}$$

having infinitely many terms.

Although the plus signs denote addition, it is sometimes convenient to make a distinction between the *series* and the *sum of the series*. Thus,

$$1 + 3 + 5 \quad \text{and} \quad 3 + 3 + 3$$

each have the same sum, 9, but they are different series because they are formed from the different sequences

$$1, 3, 5 \qquad \text{and} \qquad 3, 3, 3$$

respectively. The sum, 9, can be thought to be the numerical value of the series. In an equation such as $1 + 3 + 5 = 9$, or

$$a_1 + a_2 + \cdots + a_n = s_n$$

it is always the numerical value that is intended. Similar remarks apply to infinite series, except that the meaning of the word "sum" then requires further explanation. This matter is discussed next.

There is a substantial theory of infinite series, called the *formal theory*, which does not require that the series have any numerical value as the sum assigned to it. But the series commonly arising in applications have a numerical value, which is obtained by considering the sequence of *partial sums:*

$$a_1, \qquad a_1 + a_2, \qquad a_1 + a_2 + a_3, \qquad a_1 + a_2 + a_3 + a_4, \qquad \ldots \ .$$

The nth partial sum is $s_n = a_1 + a_2 + a_3 + \cdots + a_n$.

For example, the infinite series

$$\frac{1}{1 \cdot 2} + \frac{1}{2 \cdot 3} + \frac{1}{3 \cdot 4} + \frac{1}{4 \cdot 5} + \cdots + \frac{1}{n(n+1)} + \cdots \tag{1-4}$$

has partial sums

$$s_1 = \tfrac{1}{2} \qquad s_2 = \tfrac{1}{2} + \tfrac{1}{6} = \tfrac{2}{3} \qquad s_3 = \tfrac{1}{2} + \tfrac{1}{6} + \tfrac{1}{12} = \tfrac{3}{4}$$

and so on. The nth partial sum is $s_n = n/(n+1)$, as we see by writing

$$s_n = \left(\frac{1}{1} - \frac{1}{2}\right) + \left(\frac{1}{2} - \frac{1}{3}\right) + \left(\frac{1}{3} - \frac{1}{4}\right) + \cdots + \left(\frac{1}{n} - \frac{1}{n+1}\right). \tag{1-5}$$

The inner terms cancel; hence $s_n = 1 - 1/(n+1) = n/(n+1)$. A series such as (1-5) is called a *telescoping series.*

The formula $s_n = n/(n+1)$ obtained for the series (1-4) enables one to add any number of terms with the greatest ease. For example, taking $n = 9$ we get $s_n = \tfrac{9}{10}$. This means that the sum of the first nine terms is 0.9. Taking $n = 99$ we get 0.99 for the sum of the first 99 terms. The sum of the first 999 terms is 0.999, and so on. The partial sums s_n become and remain arbitrarily close to 1, if n is large enough; in other words, $\lim_{n \to \infty} s_n = 1$. Hence, the numerical value assigned to the series (1-4) is 1.

This example is typical of the general case. If the partial sums s_n satisfy

$$\lim_{n \to \infty} s_n = s \tag{1-6}$$

the series (1-3) is said to *converge* to the *sum s*, and we write

$$s = a_1 + a_2 + a_3 + \cdots + a_n + \cdots.$$

If the limit of s_n does not exist, the series is said to *diverge*, and no numerical value is assigned to the series. The precise meaning of the statement (1-6) is that for any pre-assigned positive number ϵ, however small, one can find a number N such that

$$|s - s_n| < \epsilon \qquad \text{for all } n > N. \tag{1-7}$$

For example, applying the definition to the series (1-4) with $s = 1$ we are led to the condition

$$|1 - s_n| = \left| 1 - \frac{n}{n+1} \right| = \left| \frac{1}{n+1} \right| < \epsilon \qquad \text{for all } n > N.$$

This is equivalent to $n + 1 > (1/\epsilon)$ for $n > N$; hence the choice $N = (1/\epsilon) - 1$ fulfills the requirement of the definition. If $\epsilon = 0.1$, $N = 9$ will meet the requirement; if $\epsilon = 0.01$, $N = 99$ will do so; and so on.

The number ϵ in (1-7) can be thought of as a measure of error made in approximating the sum s by the sum of the first n terms. The actual error in the approximation is

$$r_n \equiv s - s_n$$

and the condition (1-7) demands that $|r_n| < \epsilon$ for all sufficiently large values of n. We shall call r_n the *remainder of the series* (1-3) *after n terms*.

The limit of s_n may fail to exist either when s_n increases indefinitely with n or when the partial sums s_n oscillate without approaching a limit as $n \to \infty$. Thus, the series

$$1 + 1 + 1 + 1 + \cdots$$

diverges because its nth partial sum $s_n = n$ increases with n without limit, while the series

$$1 - 1 + 1 - 1 + \cdots$$

diverges because its partial sums $s_1 = 1$, $s_2 = 0$, $s_3 = 1, \ldots$. oscillate.

As another example, consider the *harmonic series*

$$1 + \tfrac{1}{2} + \tfrac{1}{3} + \tfrac{1}{4} + \tfrac{1}{5} + \tfrac{1}{6} + \tfrac{1}{7} + \tfrac{1}{8} + \cdots + 1/n + \cdots.$$

The terms of this series can be grouped as follows:

$$1 + \tfrac{1}{2} + (\tfrac{1}{3} + \tfrac{1}{4}) + (\tfrac{1}{5} + \tfrac{1}{6} + \tfrac{1}{7} + \tfrac{1}{8}) + (\tfrac{1}{9} + \cdots + \tfrac{1}{16}) + \cdots.$$

Each term of the foregoing series is at least as large as the corresponding term of

$$\tfrac{1}{2} + \tfrac{1}{2} + (\tfrac{1}{4} + \tfrac{1}{4}) + (\tfrac{1}{8} + \tfrac{1}{8} + \tfrac{1}{8} + \tfrac{1}{8}) + (\tfrac{1}{16} + \cdots + \tfrac{1}{16}) + \cdots.$$

The latter series, however, reduces to

$$\tfrac{1}{2} + \tfrac{1}{2} + \tfrac{1}{2} + \tfrac{1}{2} + \tfrac{1}{2} + \cdots$$

which is divergent. Hence the original series is also divergent. This discussion illustrates the idea of comparison, which is fundamental in the study of series.

Example 1.   Establish the divergence of $\tfrac{1}{2} + \tfrac{1}{4} + \tfrac{1}{6} + \tfrac{1}{8} + \cdots$.

Here the nth partial sum can be written in the form

$$s_n = \frac{1}{2}\left(1 + \frac{1}{2} + \frac{1}{3} + \frac{1}{4} + \cdots + \frac{1}{n} \right) = \frac{1}{2} S_n$$

where S_n is the nth partial sum of the harmonic series. By the foregoing discussion, S_n increases without limit as $n \to \infty$. Hence the same is true of s_n, and therefore the series diverges.

Example 2.   Show that any given sequence $\{s_n\}$ can be represented as the partial sums of a suitable series, $a_1 + a_2 + a_3 + \cdots + a_n + \cdots$.

Since $s_n = s_1 + (s_2 - s_1) + (s_3 - s_2) + \cdots + (s_n - s_{n-1})$, the stated result is obvious, with

$$a_1 = s_1 \qquad a_n = s_n - s_{n-1} \qquad \text{for } n \geq 2.$$

Example 3.   (Review of the limit concept.) If $\lim s_n = s$ and $\lim t_n = t$, show that $\lim (s_n + t_n) = s + t$, and also $\lim (cs_n) = cs$ for every constant c.

Here, as elsewhere in this chapter, "lim" is used as an abbreviation for $\lim_{n \to \infty}$. To prove the statement, let ϵ be any given positive number. Since $\tfrac{1}{2}\epsilon$ is also positive we can find N_1 and N_2 so that

$$|s_n - s| < \tfrac{1}{2}\epsilon \qquad \text{and} \qquad |t_n - t| < \tfrac{1}{2}\epsilon$$

for $n > N_1$ and $n > N_2$, respectively. If $n > N = \max (N_1, N_2)$, both relations hold, and

$$|(s_n + t_n) - (s + t)| = |(s_n - s) + (t_n - t)| \leq |s_n - s| + |t_n - t| \leq \tfrac{1}{2}\epsilon + \tfrac{1}{2}\epsilon.$$

This establishes the first result. The second is evident when $c = 0$, and when $c \neq 0$ it follows by choosing N so that $|s - s_n| < |c|^{-1}\epsilon$ for all $n > N$.

PROBLEMS

1. Represent the sum of the series $3 + 0.1 + 0.04 + 0.001 + 0.0005 + 0.00009$ as a decimal.

2. Add the two series $1 + 2 + 3 + 4 + 5 + 6 + 7$ and $7 + 6 + 5 + 4 + 3 + 2 + 1$ term by term, and thus deduce the value of the sum of each series.

3. Using the method of Prob. 2, show that $2(1 + 2 + 3 + \cdots + n) = n(n + 1)$.

4. Find the sum of the first n terms of the sequence $a, a + d, a + 2d, a + 3d, \ldots$, where a and d are constant. *Hint:* Separate the part of the sum involving a from that involving d, and use the result of Prob. 3.

5. By means of the identity $2k - 1 = k^2 - (k - 1)^2$, show that $1 + 3 + 5 + \cdots + (2n - 1) = n^2$. *Hint:* The given identity with $k = 1, k = 2, k = 3, \ldots$ shows that the series equals

$$(1^2 - 0^2) + (2^2 - 1^2) + (3^2 - 2^2) + \cdots + [n^2 - (n - 1)^2].$$

6. By use of the identity $3k(k - 1) = k^3 - (k - 1)^3 - 1$, show that

$$1 \cdot 2 + 2 \cdot 3 + 3 \cdot 4 + \cdots + (n - 1)n = \tfrac{1}{3}n(n^2 - 1).$$

7. Show that the series $1 + 1 + 1 + \cdots$ and $(-1) + (-1) + (-1) + \cdots$ are both divergent but that the series $(1 - 1) + (1 - 1) + (1 - 1) + \cdots$ obtained by adding these series term by term is convergent. Give a similar example in which the terms are not constant.

8. By comparison with the harmonic series, show that each of the following series is divergent:

$$\tfrac{1}{4} + \tfrac{1}{5} + \tfrac{1}{6} + \tfrac{1}{7} + \cdots, \qquad \tfrac{1}{3} + \tfrac{1}{6} + \tfrac{1}{9} + \tfrac{1}{12} + \cdots, \qquad \tfrac{1}{9} + \tfrac{1}{12} + \tfrac{1}{15} + \tfrac{1}{18} + \cdots.$$

9. If $a_n > 0.0001$ for each n, show that $a_1 + a_2 + a_3 + \cdots$ is divergent. *Hint:* The nth partial sum exceeds $0.0001n$.

10. If $\lim s_n = s$ and $\lim t_n = t$, prove that $\lim (\alpha s_n + \beta t_n) = \alpha s + \beta t$ for any constants α and β. *Hint:* Apply the second result of Example 3 to the sequences $\{\alpha s_n\}$ and $\{\beta t_n\}$. Then apply the first result of Example 3 to the sequence $\{(\alpha s_n) + (\beta t_n)\}$.

11. If $\{a_n\}$ is a sequence and c is constant, $c\{a_n\}$ is defined to be the sequence with the general term ca_n; that is, $c\{a_n\} \equiv \{ca_n\}$. Compute $5\{a_n\}$, where $\{a_n\}$ is the particular sequence $1, 3, 2, 4, 3, 6$.

12. If $\{a_n\}$ and $\{b_n\}$ are two sequences having the same number of terms, the sum of these two sequences is defined to be the sequence $\{c_n\}$, where $c_n = a_n + b_n$; that is, $\{a_n\} + \{b_n\} = \{a_n + b_n\}$. Compute the sum of the two sequences $1, 3, 5, 7, 9$ and $-1, 0, 1, -1, 0$.

13. Two sequences are said to be equal if they consist of the same terms in the same order; that is, $\{a_n\} = \{b_n\}$ means that $a_n = b_n$ for each relevant n. Using the definitions given in Probs. 11 and 12, show that $1\{a_n\} = \{a_n\}$ and $0\{a_n\} = \{0\}$ for every sequence $\{a_n\}$. (Problems 11 to 13 form the starting point for the *formal theory* of sequences and series, which does not assume convergence.)

2. Sigma Notation. Linearity. Infinite series are often written in the condensed notation

$$\sum_{k=1}^{\infty} a_k = a_1 + a_2 + a_3 + \cdots + a_n + \cdots. \tag{2-1}$$

The symbol Σ, read *sigma*, is the Greek letter S suggesting the English word "sum."

However, depending on context, we shall use (2-1) to represent either the series or the sum of the series. Finite sums or series are expressed similarly, with the limits $(1,\infty)$ replaced by appropriate values. For example,

$$\sum_{k=m}^{n} b_k = b_m + b_{m+1} + \cdots + b_n \qquad n \geq m. \qquad (2\text{-}2)$$

The limits of summation are often omitted, if they need not be emphasized or are clear from the context. Whenever the limits are omitted in Secs. 2 to 9 of this chapter, the summation range is from 1 to ∞.

Although the left side of (2-2) involves k, the value of the expression is independent of k, as is evident from the right side. Thus,

$$\sum_{k=m}^{n} a_k = \sum_{j=m}^{n} a_j = \sum_{p=m}^{n} a_p.$$

The roles of k, j, or p above are analogous to the roles of x, t, and s in the definite integrals

$$\int_a^b f(x)\,dx = \int_a^b f(t)\,dt = \int_a^b f(s)\,ds.$$

Variables such as these, which can be changed at will without any effect, are called *dummy variables*.

As an example of an equation in sigma notation we show that

$$\sum_{k=1}^{n} (\alpha a_k + \beta b_k) = \alpha \sum_{k=1}^{n} a_k + \beta \sum_{k=1}^{n} b_k \qquad (2\text{-}3)$$

for any constants α and β. Indeed, the left side of (2-3) is

$$(\alpha a_1 + \beta b_1) + (\alpha a_2 + \beta b_2) + \cdots + (\alpha a_n + \beta b_n).$$

Collecting the terms involving α and then those involving β, we get

$$\alpha(a_1 + a_2 + \cdots + a_n) + \beta(b_1 + b_2 + \cdots + b_n)$$

and this agrees with the right side of (2-3).

Equation (2-3) indicates that the operator denoted by $\sum_{k=1}^{n}$ is linear. In general, an operator $\mathbf{T}$ is said to be *linear* if the equation

$$\mathbf{T}(\alpha u + \beta v) = \alpha \mathbf{T}u + \beta \mathbf{T}v \qquad (2\text{-}4)$$

holds for all constants α, β and for all u and v such that $\mathbf{T}u$ and $\mathbf{T}v$ are defined. The condition (2-4) can be deduced from the two simpler conditions

$$\mathbf{T}(u + v) = \mathbf{T}u + \mathbf{T}v \qquad \text{and} \qquad \mathbf{T}(cu) = c\mathbf{T}u. \qquad (2\text{-}5)$$

That is, if (2-5) holds whenever $\mathbf{T}u$ and $\mathbf{T}v$ are defined, and for every constant c, then $\mathbf{T}$ is linear.

The concept of linearity pervades several broad fields of modern mathematics, and linear operators are constantly encountered in the applications of mathematics to science and technology. A familiar example of a linear operation is the operation of taking an integral. Thus, by elementary calculus

$$\int_a^b [\alpha u(x) + \beta v(x)]\,dx = \alpha \int_a^b u(x)\,dx + \beta \int_a^b v(x)\,dx \qquad (2\text{-}6)$$

[compare (2-3)]. It follows from Example 3 of Sec. 1 that the operation symbolized by "lim" is also linear; that is,

$$\lim (\alpha s_n + \beta t_n) = \alpha \lim s_n + \beta \lim t_n \qquad (2\text{-}7)$$

whenever the right-hand side makes sense.

As suggested by these examples, the expressions u and v in the general definition of linearity can be of quite diverse sorts. In (2-6) u and v are considered to be integrable functions, and $\mathbf{T}$ is the operation of taking an integral. But if (2-3) is considered in the context of (2-4), u and v are sequences, and $\mathbf{T}$ is the operator that converts a sequence into the sum of the corresponding series.

Upon taking the limit in (2-3) as $n \to \infty$, using (2-7), we get

$$\sum_{k=1}^{\infty} (\alpha a_k + \beta b_k) = \alpha \sum_{k=1}^{\infty} a_k + \beta \sum_{k=1}^{\infty} b_k \qquad (2\text{-}8)$$

if the right-hand series are convergent. This result is important enough to state as a theorem:

THEOREM I. *The operator denoted by Σ is linear, that is,*

$$\Sigma(\alpha a_n + \beta b_n) = \alpha \Sigma a_n + \beta \Sigma b_n$$

whenever the series Σa_n and Σb_n are convergent.

Upon choosing $\alpha = c$ and $\beta = 0$ in Theorem I we get

$$\Sigma c a_n = c \Sigma a_n. \qquad (2\text{-}9)$$

That is, *a convergent series can be multiplied term by term by any constant.* Upon choosing $\alpha = 1$ and $\beta = 1$ or $\beta = -1$ we get

$$\Sigma(a_k \pm b_k) = \Sigma a_k \pm \Sigma b_k. \qquad (2\text{-}10)$$

That is, *two convergent series can be added or subtracted term by term.*

As an illustration, if

$$s = 1 + x + x^2 + \cdots + x^n + \cdots$$

then (2-9) gives $xs = x + x^2 + \cdots + x^n + \cdots$. (The term x^n here arises from the term x^{n-1} in the original series.) Since the result can be written

$$xs = 0 + x + x^2 + \cdots + x^n + \cdots$$

(2-10) gives $s - xs = 1$. This shows that, if the series converges, it must converge to the value $1/(1 - x)$. In the next section it is found that the series does converge, provided $|x| < 1$.

Another obvious but important property is used so often that we state it as a theorem:

THEOREM II. *If finitely many terms of an infinite series are altered, the convergence is not affected (though, of course, the value of the sum may be affected).*

To prove this we denote the original terms by a_k and the new terms by $a_k + b_k$, where all but a finite number[1] of b_k's are zero. The result is then a consequence of (2-10). It should be noted that this argument not only establishes convergence but shows that the new value of the sum can be found by obvious arithmetical calculation. For instance, if the seventh term of a convergent series is increased by 2.4 the sum is also increased by 2.4, and similarly in other cases.

Example 1. Establish the divergence of $\frac{1}{12} + \frac{1}{16} + \frac{1}{20} + \frac{1}{24} + \cdots$.

Multiplying by 4 we get the series $\frac{1}{3} + \frac{1}{4} + \frac{1}{5} + \frac{1}{6} + \cdots$, which is obviously divergent since it differs from the harmonic series $\Sigma(1/n)$ only in that it lacks the first two terms. Hence the given series is also divergent.

---

[1] Any finite series $b_1 + \cdots + b_n$ can be regarded as an infinite series with all terms beyond the nth equal to zero. If we so regard it, the definition of convergence given in Sec. 1 makes the finite series converge to its ordinary sum.

This use of (2-9) to establish divergence is readily justified, even though (2-9) applies to convergent series only. Thus, assume that the original series converges. The foregoing analysis shows, then, that $\Sigma(1/n)$ would have to converge, and that is a contradiction.

Example 2.   Evaluate $\displaystyle\sum_{j=3}^{5} (2^j - j^2)$.

The notation means: Set $j = 3$, then $j = 4$, then $j = 5$, and add the resulting expressions. The value is therefore

$$(2^3 - 3^2) + (2^4 - 4^2) + (2^5 - 5^2) = -1 + 0 + 7 = 6.$$

Example 3.   Display the series $\Sigma(2^j - j^2)$ without using Σ notation.

According to our conventions, the summation is from 1 to ∞. Hence,

$$\Sigma(2^j - j^2) = 1 + 0 - 1 + 0 + 7 + \cdots + (2^n - n^2) + \cdots. \qquad (2\text{-}11)$$

In contrast to the preceding series, this one does not have a sum.

Example 3 shows that there is a certain ambiguity when an infinite series is indicated by giving only the first few terms; that is, from knowledge of the first four terms in (2-11) one could hardly be expected to guess that the fifth term is 7. Nevertheless, we shall often describe infinite series by giving only the first few terms, because in the illustrations we have chosen there is no difficulty in surmising what general term is intended.

PROBLEMS

1. Write the following series in full, without using Σ notation:

(a) Σk;   *(b)* $\Sigma 2^{-k}$;   *(c)* Σk^{-2};   *(d)* $\Sigma(2k + 1)^{-1}$;   *(e)* $\Sigma(0.001 + 2^{-k})$;   *(f)* $\Sigma(k + 1)(k + 2)^{-1}$.

2. Write the following series in condensed form, using Σ notation:

(a) $1 + \frac{1}{2} + \frac{1}{3} + \frac{1}{4} + \cdots$;   *(b)* $1 + \frac{1}{4} + \frac{1}{9} + \frac{1}{16} + \cdots$;

(c) $\frac{2}{3} + \frac{3}{4} + \frac{4}{5} + \frac{5}{6} + \cdots$;   *(d)* $\log \frac{2}{3} + \log \frac{3}{4} + \log \frac{4}{5} + \log \frac{5}{6} + \cdots$;

(e) $1 + 1 + 1 + 1 + \cdots$;   *(f)* $\sin 1 + \sin \frac{1}{4} + \sin \frac{1}{9} + \sin \frac{1}{16} + \cdots$.

3. Compute the third partial sum, s_3, for each series in Prob. 1.

4. Some of the series in Probs. 1 and 2 are divergent because the general term a_k exceeds a positive constant independent of k (cf. Prob. 9 of Sec. 1). Which ones are they?

5. Prove that the series $2^{-1} + 2^{-2} + 2^{-3} + \cdots + 2^{-n} + \cdots$ is convergent. *Hint:* Verify that the nth partial sum s_n satisfies $2s_n - s_n = 1 - 2^{-n}$.

6. In Prob. 5, interpret the sum of the series and the calculation of s_n in binary notation. [Partial answer: $s_n = 0.111 \cdots 11$ (with n 1's) and $2s_n = 1.11 \cdots 110$.]

7. Show that the general linearity property (2-4) follows from the special properties (2-5). *Hint:* The choice $c = \alpha$ shows that $\mathbf{T}(\alpha u)$ is meaningful and $\mathbf{T}(\alpha u) = \alpha \mathbf{T} u$; similarly for $\mathbf{T}(\beta v)$. Now apply the first relation (2-5) to $(\alpha u) + (\beta v)$.

8. (Abel's transformation.)   If $s_n = a_1 + a_2 + \cdots + a_n$, show that

$$\sum_{k=1}^{n} a_k b_k = s_1(b_1 - b_2) + s_2(b_2 - b_3) + \cdots + s_{n-1}(b_{n-1} - b_n) + s_n b_n.$$

Hint: Since $a_n = s_n - s_{n-1}$, the series is $s_1 b_1 + (s_2 - s_1)b_2 + \cdots + (s_n - s_{n-1})b_n$. Collect the two terms containing s_1, then those containing s_2, and so on.

REVIEW OF THE LIMIT CONCEPT

9. Compare the meaning of the following three statements, and thus show that they are equivalent to each other: $\lim a_n = a$, $\lim (a_n - a) = 0$, $\lim |a_n - a| = 0$.

10. If $a_n = a$ for all large n, prove that $\lim a_n = a$. *Solution:* Let $\epsilon > 0$ be given. If $a_n = a$ for $n > N$, then $|a_n - a| = 0$ for $n > N$; hence $|a_n - a| < \epsilon$ for $n > N$.

11. If $\lim a_n = a$ and $\lim a_n = b$, prove that $a = b$. *Solution:* Suppose $|a - b| = 2\epsilon > 0$. For sufficiently large n there is a contradiction,

$$|a - b| = |(a - a_n) + (a_n - b)| \leq |a - a_n| + |a_n - b| < \epsilon + \epsilon = |a - b|.$$

3. A Fundamental Principle.   Frequently it is possible to infer the existence of a limit without knowing its value.   For example, consider the series

$$0.1 + 0.01 + 0.001 + \cdots$$

whose partial sums are $s_1 = 0.1$, $s_2 = 0.1 + 0.01 = 0.11$, $s_3 = 0.1 + 0.01 + 0.001 = 0.111$, and so on.   Each partial sum, being a decimal, is less than 1.   On the other hand, the s_n increase with n.   If the successive values of s_n are plotted as points on a straight line (Fig. 1), the points move to the right but never progress as far as the point 1.   It is intuitively clear that there must be some point s, at the left of 1, that the numbers s_n approach as a limit.   In this case the numerical value of the limit was not ascertained, but its existence has been established with the aid of a fundamental principle:

PRINCIPLE OF MONOTONE CONVERGENCE.   If a sequence $\{s_n\}$ satisfies $s_n \leq s_{n+1} \leq M$ for each n, where M is some constant, then $\lim\limits_{n \to \infty} s_n$ exists.

In other words: *Every bounded increasing sequence has a limit.*   Considering $-s_n$ instead of s_n gives a corresponding statement for decreasing sequences.

From the geometric interpretation of the fundamental principle it appears that when an increasing sequence of partial sums s_n has a limit, the difference between the successive values of s_n must tend to zero as $n \to \infty$.   Since $s_n - s_{n-1} = a_n$, the foregoing statement is equivalent to the assertion that $\lim a_n = 0$.   This can be established from the definition of convergence without appeal to the fundamental principle and without the assumption that s_n is increasing.

Indeed, since

$$a_n = s_n - s_{n-1} \tag{3-1}$$

and since the series converges by hypothesis, we have $\lim s_n = \lim s_{n-1} = s$ as $n \to \infty$. Hence (3-1) shows that

$$\lim a_n = \lim s_n - \lim s_{n-1} = 0. \tag{3-2}$$

We state the result as a theorem:

THEOREM.   If a series converges, the general term must approach zero; hence, if the general term does not approach zero, the series diverges.

The reader is cautioned that the converse of this theorem is *not* true.   For instance, the harmonic series was found to diverge even though the general term $1/n$ approaches zero.

There is a more elaborate version of this theorem which has a converse.   By writing the sums in full we find a relation analogous to (3-1):

FIGURE 1

$$a_m + a_{m+1} + \cdots + a_n = s_n - s_{m-1} \qquad n \geq m > 1. \tag{3-3}$$

If the infinite series converges, so that $\lim s_n = s$, both the sums on the right of (3-3) become arbitrarily close to s, provided m and n are chosen large enough. Hence the right-hand side becomes arbitrarily small in magnitude, and we are led to the following: *If Σa_k converges, then for any $\epsilon > 0$ there is an N such that*

$$|a_m + a_{m+1} + \cdots + a_n| < \epsilon \tag{3-4}$$

whenever $n \geq m \geq N$. This statement admits a converse. *If for each positive ϵ there is an N such that (3-4) holds whenever $n \geq m \geq N$, then Σa_n converges.* Proof of the converse depends on the *Cauchy convergence criterion*, which is discussed next.

The principle of monotone convergence is only one of several equivalent principles, all of which serve the same purpose. Among the most important of these is:

CAUCHY CONVERGENCE CRITERION. Let $\{s_n\}$ *be a sequence with the following property: For each positive ϵ there is an N such that $|s_n - s_m| < \epsilon$ whenever $n \geq m \geq N$. Then s_n has a limit.*

The Cauchy criterion is useful in advanced analysis, because it can be applied to many sequences $\{s_n\}$ in which s_n are not real numbers. If the s_n are real numbers (as they are here) it is possible to show that the principle of monotone convergence and the Cauchy criterion imply each other.

Since the principle of monotone convergence seems simpler and more plausible than the Cauchy criterion, the analysis in this book uses the former principle rather than the latter. For the benefit of interested readers, a short proof of the Cauchy criterion is given at the end of Sec. 7.

Example 1.   A certain series has partial sums $s_n = r^n$, where r is a constant such that $0 \leq r < 1$. By use of the fundamental principle, show that the series converges to zero.

We have to show that $\lim s_n = 0$ as $n \to \infty$, or, in other words,

$$\lim_{n \to \infty} r^n = 0 \qquad \text{for } 0 \leq r < 1. \tag{3-5}$$

Since $r \geq 0$, it is evident that $s_n \geq 0$; hence the sequence s_n is *bounded from below.* Also $r^{n+1} = rr^n$, or, in other words,

$$s_{n+1} = rs_n. \tag{3-6}$$

Since $r < 1$, this shows that $s_{n+1} < s_n$, so that the sequence s_n is *decreasing.* Hence the limit of s_n exists by the fundamental principle. If we write $s = \lim s_n$ and take the limit as $n \to \infty$ in (3-6), there results

$$s = \lim s_{n+1} = \lim (rs_n) = r \lim s_n = rs.$$

From $s = rs$ it follows that $s = 0$, since $r \neq 1$, and this gives (3-5).

Example 2.   Show that the *geometric series*

$$1 + x + x^2 + x^3 + \cdots + x^n + \cdots$$

converges to $1/(1 - x)$ when $|x| < 1$ but diverges when $|x| \geq 1$.

We have to decide whether the partial sums

$$s_n = 1 + x + x^2 + \cdots + x^{n-2} + x^{n-1} \tag{3-7}$$

tend to a limit. If the foregoing equation is multiplied by x, there results

$$xs_n = x + x^2 + \cdots + x^{n-1} + x^n \tag{3-8}$$

and subtracting (3-8) from (3-7) yields $s_n - xs_n = 1 - x^n$. Solving for s_n we get

$$s_n = \frac{1 - x^n}{1 - x} \qquad \text{for } x \neq 1. \tag{3-9}$$

If $|x| < 1$, then $\lim\limits_{n \to \infty} |x|^n = 0$ by (3-5); hence (3-9) gives

$$\lim_{n \to \infty} s_n = \frac{1 - 0}{1 - x} = \frac{1}{1 - x}.$$

This establishes the required convergence when $|x| < 1$. On the other hand, if $|x| \geq 1$, the general term does not approach zero and the series diverges by the theorem given above. The value x is called the *ratio* for the series, since x equals the ratio of two successive terms. We have shown that the geometric series converges if, and only if, the ratio is less than 1 in magnitude.

PROBLEMS

1. Write the following series in full, without using Σ notation:

(a) $\displaystyle\sum_{k=1}^{\infty} \frac{1}{2k}$;   (b) $\displaystyle\sum_{i=2}^{\infty} \left(-\frac{43}{44}\right)^i$;   (c) $\displaystyle\sum_{n=1}^{\infty} \frac{n^2 + 1}{2n + 3}$;   (d) $\displaystyle\sum_{j=7}^{\infty} \frac{3}{j}$;   (e) $\displaystyle\sum_{m=1}^{\infty} (\cos x)^{2m}$.

2. Write the following series in condensed form, using Σ notation:

(a) $(\frac{2}{3})^2 + (\frac{2}{3})^3 + (\frac{2}{3})^4 + (\frac{2}{3})^5 + \cdots$;   (b) $\frac{1}{6} + \frac{1}{9} + \frac{1}{12} + \frac{1}{15} + \cdots$;

(c) $\dfrac{1}{1,000} + \dfrac{1}{1,002} + \dfrac{1}{1,004} + \dfrac{1}{1,006} + \cdots$;   (d) $0.2 - 0.02 + 0.002 - \cdots$;

(e) $\dfrac{1}{1} + \dfrac{1}{1 \cdot 2} + \dfrac{1}{1 \cdot 2 \cdot 3} + \dfrac{1}{1 \cdot 2 \cdot 3 \cdot 4} + \cdots$;   (f) $\frac{1}{20} + \frac{1}{30} + \frac{1}{40} + \frac{1}{50} + \cdots$.

3. Some of the series in Probs. 1 and 2 are divergent because the general term does not approach zero. Which ones are they?

4. Some of the series in Probs. 1 and 2 are convergent because they are geometric series with ratio less than 1 in magnitude (or multiples of such a series). Which ones are they?

5. Some of the series in Probs. 1 and 2 are divergent because $\Sigma(1/n)$ is divergent. Which ones are they?

6. Show that $(1 - 1) + (1 - 1) + (1 - 1) + \cdots$ converges but would diverge if the parentheses were dropped.

7. If Σa_n diverges and Σb_n converges, prove that $\Sigma(a_n - b_n)$ diverges. *Hint:* Note that $a_n = (a_n - b_n) + b_n$, and use linearity.

8. Using the result of Prob. 7 or Theorem I of Sec. 2, test for convergence:

$$\sum \left[\frac{1}{n} - \left(\frac{7}{8}\right)^n\right], \qquad \sum \left[\left(\frac{3}{4}\right)^n - \left(\frac{5}{6}\right)^n\right], \qquad \sum \left[\frac{1}{5n} - (0.999)^n\right], \qquad \sum (2^n - 2^{-n}).$$

REVIEW OF THE LIMIT CONCEPT

9. If $\lim A_n = \lim B_n = 0$, prove that $\lim (A_n B_n) = 0$. *Solution:* Given $\epsilon > 0$, choose N_1 so that $|A_n| < \epsilon$ for $n > N_1$. Also choose N_2 so that $|B_n| < 1$ for $n > N_2$. For $n > N = \max (N_1, N_2)$ we then have $|A_n B_n| < \epsilon$.

10. If $\lim a_n = a$ and $\lim b_n = b$, prove that $\lim (a_n b_n) = ab$. *Outline of solution:* Let $A_n = a_n - a$ and $B_n = b_n - b$. Then, by elementary algebra,

$$a_n b_n - ab = A_n B_n + b a_n + a b_n - 2ab.$$

Since $\lim (A_n B_n) = 0$ by Sec. 2, Prob. 9, and the above Prob. 9, linearity gives

$$\lim (a_n b_n - ab) = 0 + ba + ab - 2ab = 0.$$

11. The equation $\lim a_n = a$ is often written $a_n \to a$. In this notation, write the conclusion of the preceding problem and also the linearity property (2-7).

12 INFINITE SERIES CHAP. 1

4. Improper Integrals and the Integral Test. In calculus definite integrals are defined, at first, only for a finite interval $a \le x \le b$. Extension to an infinite interval is then made by a passage to the limit; thus

$$\int_a^\infty f(x)\,dx = \lim_{b\to\infty} \int_a^b f(x)\,dx. \tag{4-1}$$

The integral on the left of (4-1) is called an *improper integral*. If the limit at the right exists, the improper integral is said to *converge* (to the value of the limit) and to *diverge* if the limit does not exist. The definition is analogous to the corresponding definition

$$\sum_{k=1}^\infty a_k = \lim_{n\to\infty} \sum_{k=1}^n a_k$$

for infinite series.

An example of a divergent improper integral is

$$\int_1^\infty \frac{1}{x}\,dx = \lim_{b\to\infty} \int_1^b \frac{dx}{x} = \lim_{b\to\infty} (\log x \,|_1^b) = \lim_{b\to\infty} \log b = \infty. \tag{4-2}$$

On the other hand, if p is constant and $p \ne 1$, then

$$\int_1^\infty \frac{1}{x^p}\,dx = \lim_{b\to\infty} \int_1^b x^{-p}\,dx = \lim_{b\to\infty}\left(\frac{x^{1-p}}{1-p}\Big|_1^b\right) = \lim_{b\to\infty} \frac{b^{1-p}-1}{1-p}. \tag{4-3}$$

The question of convergence now depends on the behavior of b^{1-p} as $b \to \infty$. If the exponent $1-p$ is positive, $b^{1-p} \to \infty$ and the integral (4-3), like (4-2), is divergent. But if $1-p$ is negative, then $p-1 > 0$ and hence

$$b^{1-p} = \frac{1}{b^{p-1}} \to 0 \qquad \text{as } b \to \infty.$$

In this case the integral (4-3) converges to the value $1/(p-1)$.

The result of this discussion can be summarized as follows:

THEOREM I. *The improper integral $\int_1^\infty \frac{1}{x^p}\,dx$ converges if, and only if, the constant $p > 1$.*

Theorem I suggests the following analogous result for infinite series:

THEOREM II. *The infinite series $\sum_{k=1}^\infty \frac{1}{k^p}$ converges if, and only if, the constant $p > 1$.*

It will be seen that Theorem II is valid; in fact, there is a close connection between infinite series and improper integrals which will now be discussed.

Suppose the terms of an infinite series Σa_k are positive and decreasing; that is, $a_n > a_{n+1} > 0$ for each positive integer n. In this case there is a continuous decreasing function $f(x)$ such that[1]

$$a_n = f(n) \qquad n = 1, 2, 3, \ldots. \tag{4-4}$$

Each term a_n of the series may be thought of as representing the area of a rectangle of base unity and height $f(n)$ (see Fig. 2). The sum of the areas of the first n circumscribed rectangles is greater than the area under the curve from 1 to $n+1$, so that

$$a_1 + a_2 + \cdots + a_n > \int_1^{n+1} f(x)\,dx. \tag{4-5}$$

This shows that, *if the integral $\int_1^\infty f(x)\,dx$ diverges, the sum Σa_k also diverges.*

---

[1] For instance, let the graph of $y = f(x)$ consist of straight-line segments joining the points (n, a_n) and $(n+1, a_{n+1})$.

On the other hand, the sum of the areas of the inscribed rectangles is less than the area under the curve, so that

$$a_2 + a_3 + \cdots + a_n < \int_1^n f(x)\,dx. \quad (4\text{-}6)$$

If the integral converges, we have [since $f(x) > 0$]

$$\int_1^n f(x)\,dx < \int_1^\infty f(x)\,dx \equiv M$$

so that the partial sums are bounded independently of n:

$$s_n = a_1 + a_2 + \cdots + a_n < M + a_1.$$

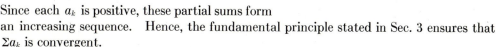

FIGURE 2

Since each a_k is positive, these partial sums form an increasing sequence. Hence, the fundamental principle stated in Sec. 3 ensures that Σa_k is convergent.

The result of this discussion is summarized as follows:

THEOREM III.   For $x \geq 1$, let $f(x)$ be positive, continuous, and decreasing.   Then the series $\sum\limits_{n=1}^{\infty} f(n)$ and the integral $\int_1^\infty f(x)\,dx$ both converge or both diverge.   In either case the partial sums are bounded as follows:

$$\int_1^{n+1} f(x)\,dx < \sum_{k=1}^{n} f(k) < \int_1^n f(x)\,dx + f(1). \qquad (4\text{-}7)$$

Choosing $f(x) = x^{-p}$ in Theorem III, we see that Theorem II is a consequence of Theorem I.   The test for convergence contained in Theorem III is commonly called *the Cauchy integral test*, though it was first discovered by Maclaurin.   The result (4-7) is especially useful because it enables us to estimate the value of the sum.

Example 1.   Show that the series

$$\frac{1}{1+1^2} + \frac{1}{1+2^2} + \frac{1}{1+3^2} + \frac{1}{1+4^2} + \cdots + \frac{1}{1+n^2} + \cdots$$

converges to a value which is between 0.7 and 1.3.

Here we choose $f(x) = 1/(1 + x^2)$.   Since

$$\int_1^b \frac{1}{1+x^2}\,dx = \tan^{-1} x \Big|_1^b \to \frac{\pi}{2} - \frac{\pi}{4} = \frac{\pi}{4} \qquad \text{as } b \to \infty$$

the integral is convergent, and so the series is convergent.   Moreover

$$0.79 \doteq \frac{\pi}{4} < \sum_{k=1}^{\infty} \frac{1}{1+k^2} < \frac{\pi}{4} + \frac{1}{2} \doteq 1.29$$

by letting $n \to \infty$ in (4-7) and noting that $f(1) = \tfrac{1}{2}$.   The next example shows how the accuracy in such an estimate may be improved to any extent desired.

Example 2.   Compute the sum of the following series within ± 0.01:

$$s = 1 + \frac{1}{4} + \frac{1}{9} + \frac{1}{16} + \frac{1}{25} + \frac{1}{36} + \cdots + \frac{1}{n^2} + \cdots.$$

It is easily verified that the first six terms give the sum 1.491.   To estimate the remainder we have, from (4-7) on taking $f(x) = 1/(x + 6)^2$,

$$\int_1^\infty \frac{1}{(x+6)^2}\,dx < \sum_1^\infty \frac{1}{(n+6)^2} < \int_1^\infty \frac{1}{(x+6)^2}\,dx + \frac{1}{49}. \tag{4-8}$$

The two limits in (4-8) are 0.143 and 0.163, as the reader can verify.   Hence

$$1.634 = 1.491 + 0.143 < s < 1.491 + 0.163 = 1.654. \tag{4-9}$$

It is interesting to see how many terms are needed to get the same accuracy by direct computation.   The remainder after n terms is given by (4-7) as

$$\sum_{k=n+1}^\infty \frac{1}{k^2} \doteq \int_{n+1}^\infty \frac{1}{x^2}\,dx = \frac{1}{n+1}.$$

To make this as small as the uncertainty interval $1.654 - 1.634$ obtained in (4-9), we must have $1/(n+1) \le 0.02$, or $n \ge 49$.   Thus, direct summation of the series requires almost 50 terms for the accuracy that we obtained by adding 6 terms only.

PROBLEMS

1. Test the following integrals for convergence, and evaluate if convergent:

$$\int_1^\infty \frac{dx}{1+x}, \quad \int_1^\infty e^{-x}\,dx, \quad \int_1^\infty x^{-4}\,dx, \quad \int_2^\infty \frac{dx}{x(\log x)^2}, \quad \int_2^\infty \frac{dx}{x \log x}.$$

2. Test the following series for convergence:

$$\sum \frac{1}{(n+1)^{3/2}}, \quad \sum n^{-1.01}, \quad \sum_{n=2}^\infty \frac{1}{n(\log n)}, \quad \sum_{n=2}^\infty \frac{1}{n(\log n)^{1.01}}, \quad \sum \frac{2n}{1+n^2}.$$

3. (a) For what values of the constant c does $\int_1^\infty e^{cx}\,dx$ converge?   (b) Using the result (a), discuss the convergence of Σe^{cn}.   (c) Show that the series in part b is a geometric series, and also show that your results are consistent with those of Sec. 3.

4. About how many terms of $\Sigma n^{-0.9}$ are needed to make the sum of those terms exceed 1,000?

5. Estimate the value of $\Sigma n^{-1.1}$ by direct use of Theorem III and also by adding the first five terms and using Theorem III to estimate the remainder.   In both cases, find approximately how many terms of the original series you would have to add to get comparable accuracy.

6. (*Review.*) (a) Form the difference of the divergent series $\Sigma(2n)^{-1}$ and $\Sigma(2n-1)^{-1}$, and show that the resulting series converges.   (b) Consider the same question for the two series Σn^{-1} and $\Sigma(2n-1)^{-1}$.

7. (*Review.*) (a) By (3-9), show that the partial sums of the series

$$1 + \frac{1}{2} + \frac{1}{2^2} + \frac{1}{2^3} + \cdots + \frac{1}{2^n} + \cdots$$

are all less than 2.   (b) Given $n! = 1 \cdot 2 \cdot 3 \cdots n$, show that the partial sums of the series

$$1 + \frac{1}{2!} + \frac{1}{3!} + \frac{1}{4!} + \cdots + \frac{1}{n!} + \cdots$$

are all less than 2.   *Hint:* Compare the partial sums with those of the series in part a.   (c) Deduce, by the fundamental principle, that the series in part b converges.   (d) Show that the remainder r_n after n terms of the series in part b satisfies

$$r_n < \frac{n+2}{(n+1)(n+1)!}.$$

REVIEW OF THE LIMIT CONCEPT

8. A sequence $\{B_n\}$ is said to be *bounded* if there is a constant B such that $|B_n| < B$ for all n. Prove: If $\lim A_n = 0$ and $|B_n|$ is bounded, then $\lim A_n B_n = 0$. *Solution:* Given $\epsilon > 0$, choose N so that $|A_n| < B^{-1}\epsilon$ for $n > N$. Then $|A_n B_n| < \epsilon$ for $n > N$.

9. If $\lim b_n = b > 0$, prove that $b_n > \frac{1}{2}b$ for all sufficiently large n. *Hint:* Choose N so that $|b_n - b| < \frac{1}{2}b$ for $n > N$, and note that $b_n = (b_n - b) + b$.

10. If $\lim a_n = a$ and $\lim b_n = b > 0$, prove that $\lim (a_n/b_n) = a/b$. *Hint:* Problem 9 gives $b_n \neq 0$ for large n; hence by elementary algebra

$$\frac{a_n}{b_n} - \frac{a}{b} = A_n B_n \qquad \text{where } A_n = ba_n - ab_n \qquad \text{and} \qquad B_n = (bb_n)^{-1}.$$

Use the result of Probs. 8 and 9, noting that $\lim A_n = ba - ab = 0$ by linearity.

---

5. Comparison Term by Term. One way to test a series of positive terms for convergence is to compare that series with another whose convergence is known. Let Σa_n and Σb_n be two series with positive terms such that $a_n \leq b_n$ and Σb_n converges. The inequality

$$s_n \equiv \sum_1^n a_n \leq \sum_1^n b_n \leq \sum_1^\infty b_n$$

shows that the partial sums s_n are bounded, and since s_n is increasing, the limit exists by the fundamental principle. It is left for the reader to verify also that, if $a_n \geq b_n \geq 0$ and Σb_n diverges, then Σa_n diverges.

This discussion establishes the following result, known as the *comparison test:*

THEOREM I. *If $0 \leq a_n \leq b_n$, then the convergence of Σa_n follows from the convergence of Σb_n. And if $a_n \geq b_n \geq 0$, then the divergence of Σa_n follows from the divergence of Σb_n.*

Since the first few terms of a series do not affect the convergence, we need the hypothesis not for all n but only for n sufficiently large (see Sec. 2, Theorem II). Similar remarks apply to every convergence test, and we make constant use of this fact in the sequel.

For example, suppose we want to establish the convergence of $\Sigma(9/n^n)$. Although the inequality

$$\frac{9}{n^n} < \frac{1}{2^n} \tag{5-1}$$

is not valid for all n, it is valid when n is sufficiently large. Hence the series converges by comparison with the geometric series. Another example is given by the series

$$\sum_{n=2}^\infty \frac{1}{100 \log n}. \tag{5-2}$$

Although it is not true that

$$\frac{1}{100 \log n} > \frac{1}{n}$$

for all n, this is true for all sufficiently large n, and hence the series (5-2) diverges by comparison with the harmonic series.

We shall now use the comparison test to establish the following general theorem:

THEOREM II.   Let $a_n \leq b_n \leq c_n$ hold for all sufficiently large n.   Then, if Σa_n and Σc_n converge, the same is true of Σb_n.

It should be emphasized that the terms need not be positive.   For proof, subtract a_n from each term, so that the given inequality $a_n \leq b_n \leq c_n$ appears as

$$0 \leq b_n - a_n \leq c_n - a_n.$$

Since Σc_n and Σa_n are convergent, the same is true of $\Sigma(c_n - a_n)$ by linearity (Sec. 2, Theorem I).   The comparison test now shows that $\Sigma(b_n - a_n)$ converges, and by applying the principle of linearity again, we see that the series with general term $b_n = (b_n - a_n) + a_n$ converges too.   This completes the proof.

Theorem II has many important applications.   As one application, we show that *the convergence of a positive series is determined by the asymptotic behavior of the terms as $n \to \infty$*.   It is customary to write $a_n \sim b_n$ (read "a_n is asymptotic to b_n") if

$$\lim_{n \to \infty} \frac{a_n}{b_n} = 1.$$

For example, $n + 1 \sim n$ because $n + 1 = n(1 + 1/n)$ and $(1 + 1/n) \to 1$ as $n \to \infty$.   Similarly, $5n^2 + 3n + 4 \sim 5n^2$, but it is not the case that $2/n \sim 1/n$ even though the difference between these quantities tends to zero.   The essential meaning of asymptotic equality, $a_n \sim b_n$, is that a_n approximates b_n within a small percentage error, when n is large.

Since it is possible to have $a_n \sim b_n$ with b_n much simpler than a_n, the following result is often useful:

THEOREM III.   If $a_n \sim b_n$ and $b_n > 0$, the series Σa_n and Σb_n are both convergent or both divergent.

The hypothesis $\lim (a_n/b_n) = 1$ shows that $\frac{1}{2} \leq a_n/b_n \leq 2$ whenever n is sufficiently large.   Also, taking reciprocals, $2 \geq b_n/a_n \geq \frac{1}{2}$.   These inequalities give, respectively,

$$\tfrac{1}{2}b_n \leq a_n \leq 2b_n \qquad \text{and} \qquad 2a_n \geq b_n \geq \tfrac{1}{2}a_n.$$

Hence, the conclusion follows from Theorem II.

As an illustration, consider the series $\Sigma(n^2 + 5n + 3)^{-\frac{1}{2}}$.   Inasmuch as

$$n^2 + 5n + 3 = n^2\left(1 + \frac{5}{n} + \frac{3}{n^2}\right) \sim n^2 \tag{5-3}$$

we have $(n^2 + 5n + 3)^{-\frac{1}{2}} \sim (n^2)^{-\frac{1}{2}} = 1/n$.   Since $\Sigma(1/n)$ diverges, the given series diverges.

As another illustration, consider the series

$$\sum \left(\frac{n^4 + 4n^3 + 1}{7n^7 + 5n^4 + 8n}\right)^{\frac{2}{5}}.$$

Since $n^4 + 4n^3 + 1 \sim n^4$, and since $7n^7 + 5n^4 + 8n \sim 7n^7$, the general term is asymptotic to

$$\left(\frac{n^4}{7n^7}\right)^{\frac{2}{5}} = \frac{1}{7^{\frac{2}{5}}}\frac{1}{n^{\frac{6}{5}}}.$$

The series with general term $1/n^{\frac{6}{5}}$ converges by Theorem II of Sec. 4, and hence the given series also converges.

These examples illustrate two properties of the relation "$\sim$", which are now set forth explicitly.   First, we show that any polynomial is asymptotic to its leading term.   Indeed, if $a \neq 0$ and $m \geq 1$, then, as $n \to \infty$,

$$\frac{an^m + bn^{m-1} + \cdots + rn + s}{an^m} = 1 + \frac{b}{an} + \cdots + \frac{r}{an^{m-1}} + \frac{s}{an^m} \to 1.$$

This shows that $an^m + bn^{m-1} + \cdots + rn + s \sim an^m$, as stated.

Second, if $a_n \sim b_n$ and $c_n \sim d_n$, it follows that

$$a_n^\alpha c_n^\beta \sim b_n^\alpha d_n^\beta$$

for any constants α and β. To establish this, consider the ratio

$$\frac{a_n^\alpha c_n^\beta}{b_n^\alpha d_n^\beta} = \left(\frac{a_n}{b_n}\right)^\alpha \left(\frac{c_n}{d_n}\right)^\beta \to 1^\alpha 1^\beta = 1.$$

Example 1.   Does $\Sigma n^{-\log n}$ converge? For all large n we have $\log n > 2$ (since $\log n \to \infty$). Hence

$$\frac{1}{n^{\log n}} < \frac{1}{n^2}$$

for all large n, and the series converges by comparison with the convergent series $\Sigma(1/n^2)$.

Example 2.   For what values of the constant c does $\Sigma(5n^2 + 2n \sin n + 7)^c$ converge? Since $|\sin n|$ is bounded, the expression in parentheses is asymptotic to the leading term, $5n^2$; hence the general term $\sim 5^c n^{2c}$. The series converges if $2c < -1$; otherwise it diverges.

PROBLEMS

1. Test the following series for convergence by comparing with the series $\Sigma(1/n^p)$:

$$\sum \frac{4}{\sqrt{n}}, \qquad \sum \frac{1}{2n\sqrt{n+1}}, \qquad \sum \frac{\cos^2 nx}{(2n+1)^2}, \qquad \sum \frac{n}{(2n+1)^2}, \qquad \sum \frac{n}{(2n+1)^3}.$$

2. Test the following series for convergence by using Theorem III:

$$\sum \frac{n^2+1}{n^3+1}, \qquad \sum \frac{n^4+n^2}{3n^6+n}, \qquad \sum \left(\frac{1}{n} + \frac{1}{n^2}\right), \qquad \sum \frac{3^n+4^n}{4^n+5^n}, \qquad \sum \frac{n+1}{n^4+4}, \qquad \sum \frac{n+\cos^2 nx}{n^3+3n+1}.$$

3. Test the following series for convergence by any method:

$$\sum e^{-n^2}, \qquad \sum \frac{1}{n \log(n+1)}, \qquad \sum \frac{1}{n\sqrt{n}}, \qquad \sum \frac{n^4}{n^6+3}, \qquad \sum \frac{n}{n^2+\log n}, \qquad \sum \frac{n^2}{e^n}.$$

4. Test for convergence each series in Sec. 2, Prob. 1.

5. Test for convergence each series in Sec. 2, Prob. 2.   *Hint:* $\log(1+x) \sim x$ and $\sin x \sim x$ as $x \to 0$.

6. (a) If $a_n \sim b_n$ and $b_n \sim c_n$, show that $a_n \sim c_n$.   (b) If $a_n \sim b_n$ and $c_n \sim d_n$, is it necessary that $a_n + c_n \sim b_n + d_n$? Prove your answer by an example.   (c) Find a_n and b_n such that $a_n \sim b_n$ but $a_n - b_n \to \infty$.   (d) Find a_n and b_n such that $a_n/b_n \to \infty$ but $a_n - b_n \to 0$.

REVIEW OF THE LIMIT CONCEPT

7. If $\lim b_n = b$ exists, prove that the sequence $\{b_n\}$ is bounded.   *Solution:* Choose N so that $|b_n - b| < 1$ for $n > N$. Then a bound for $|b_n|$ is given by the largest of the numbers

$$|b_1|, |b_2|, \ldots, |b_N|, |b| + 1.$$

8. By writing $a_n b_n = (a_n - a)b_n + ab_n$ and using linearity, prove that $\lim a_n b_n = ab$ whenever $\lim a_n = a$ and $\lim b_n = b$.   *Hint:* Use Prob. 7 above and Prob. 8 of Sec. 4.

---

6. Comparison of Ratios.   It often happens that the general term of an infinite series is complicated whereas the ratio of two successive terms is simple. For example, consider the series

$$\sum \frac{x^{2n}}{n!} \tag{6-1}$$

where $n!$ (read "n factorial") is the product of the first n integers:

$$n! = 1 \cdot 2 \cdot 3 \cdot 4 \cdots n.$$

Since $(n + 1)! = (n + 1)n!$, the general term a_n of (6-1) satisfies

$$\frac{a_{n+1}}{a_n} = \frac{x^{2n+2}}{(n+1)!} \frac{n!}{x^{2n}} = \frac{x^2}{n+1} \cdot \tag{6-2}$$

The following theorem enables us to deduce convergence by considering this ratio rather than the general term itself:

THEOREM I.   Let Σa_n and Σb_n be two series with positive terms.   If

$$\frac{a_{n+1}}{a_n} \leq \frac{b_{n+1}}{b_n} \qquad n = 1, 2, 3, \ldots \tag{6-3}$$

then the convergence of Σa_n follows from the convergence of Σb_n, and the divergence of Σb_n follows from the divergence of Σa_n.

The proof is simple.   Since

$$a_n = a_1 \frac{a_2}{a_1} \frac{a_3}{a_2} \cdots \frac{a_n}{a_{n-1}} \leq a_1 \frac{b_2}{b_1} \frac{b_3}{b_2} \cdots \frac{b_n}{b_{n-1}} = \frac{a_1}{b_1} b_n$$

the convergence of Σb_n implies that of Σa_n by the comparison test (Theorem I in Sec. 5). The second statement is logically equivalent to the first.

If we take $b_n = r^n$ in Theorem I, then Σb_n converges whenever $r < 1$.   Also

$$\frac{b_{n+1}}{b_n} = \frac{r^{n+1}}{r^n} = r.$$

Hence the theorem shows that Σa_n converges if there is a fixed number $r < 1$ such that

$$\frac{a_{n+1}}{a_n} < r \qquad n = 1, 2, 3, \ldots . \tag{6-4}$$

Since the condition (6-4) is needed only for large n, the series Σa_n also converges whenever

$$\lim \frac{a_{n+1}}{a_n} = r < 1. \tag{6-5}$$

The test based on (6-4) and (6-5) is termed the *ratio test*.   To illustrate the ratio test, we consider the series (6-1).   By (6-2) we have

$$\lim \frac{a_{n+1}}{a_n} = \lim \frac{x^2}{n+1} = 0$$

and hence (6-5) holds for all x.   Thus the series (6-1) converges for all x.

The ratio test is useful but very crude.   It cannot even establish the convergence of a series such as Σn^{-100}, which is rapidly convergent.   To obtain a better test one may use the series $\Sigma(1/n^p)$ for Σb_n rather than the geometric series.   In this case

$$\frac{b_{n+1}}{b_n} = \frac{1}{(n+1)^p} n^p = \left(\frac{n}{n+1}\right)^p = \left(1 + \frac{1}{n}\right)^{-p}.$$

By the binomial theorem (see Sec. 9)

$$\left(1 + \frac{1}{n}\right)^{-p} = 1 - \frac{p}{n} + \frac{p(p+1)}{2n^2} - \cdots$$

and hence

$$\frac{b_{n+1}}{b_n} - 1 \sim -\frac{p}{n}.$$

Since Σb_n converges if $p > 1$ and diverges if $p \leq 1$, we are led to the result stated in part b of Theorem II. The result (6-5) is stated in part a.

THEOREM II. *Let Σa_n be a series of positive terms, and let r and p be constant.* (a) *If $a_{n+1}/a_n \sim r$, then Σa_n converges when $r < 1$ and diverges when $r > 1$.* (b) *If*

$$a_{n+1}/a_n - 1 \sim -p/n$$

then Σa_n converges when $p > 1$ and diverges when $p < 1$.

Example 1. Does $\Sigma(n^2/2^n)$ converge? With $a_n = n^2/2^n$ we have

$$\frac{a_{n+1}}{a_n} = \frac{(n+1)^2}{2^{n+1}} \frac{2^n}{n^2} = \left(\frac{n+1}{n}\right)^2 \frac{1}{2} \sim \frac{1}{2} < 1.$$

Hence the series converges by the ratio test, Theorem IIa.

Example 2. Apply the ratio test to the harmonic series. With $a_n = 1/n$ we have

$$\frac{a_{n+1}}{a_n} = \frac{n}{n+1} \sim 1.$$

Since this is the case $r = 1$, the test gives no information. Moreover,

$$\frac{a_{n+1}}{a_n} - 1 = \frac{n}{n+1} - 1 = -\frac{1}{n+1} \sim -\frac{1}{n}.$$

Since this is the case $p = 1$, the more refined test of Theorem IIb also gives no information.[1]

Example 3. For what values of the constant c does the following converge?

$$\frac{c}{1!} + \frac{c(c+1)}{2!} + \frac{c(c+1)(c+2)}{3!} + \cdots.$$

For sufficiently large n the terms are of constant sign, and multiplication by -1, if necessary, yields a series with positive terms. Hence, Theorem II is applicable. We have

$$\frac{a_{n+1}}{a_n} = \frac{c(c+1)(c+2)\cdots(c+n)}{(n+1)!} \frac{n!}{c(c+1)\cdots(c+n-1)} = \frac{c+n}{n+1}.$$

Since

$$\frac{c+n}{n+1} - 1 = \frac{c-1}{n+1} \sim \frac{c-1}{n} = -\frac{1-c}{n}$$

the series is convergent if $1 - c > 1$ and divergent if $1 - c < 1$. Hence, it is convergent when $c < 0$ and divergent when $c > 0$. In this example Theorem IIa gives no information but Theorem IIb solves the problem completely.

PROBLEMS

1. Determine the convergence by using the ratio test, Theorem IIa:

$$\sum \frac{n}{2^n}, \quad \sum \frac{n!}{n^2}, \quad \sum \frac{n^2}{n!}, \quad \sum \frac{x^{2n}}{2^n}, \quad \sum \frac{(x^2+1)^n}{n!}, \quad \sum \frac{n^{10}}{(1.01)^n}, \quad \sum \frac{2^n n!}{n^n}.$$

2. Show that Theorem IIa gives no information, and test for convergence by Theorem IIb:

$$\sum \frac{1}{n^2}, \quad \sum \frac{n!}{c(c+1)(c+2)\cdots(c+n)}, \quad \sum \frac{(2n)!}{4^n(n!)^2}.$$

3. Test for convergence by any method:

$$\sum \frac{n!}{1\cdot3\cdot5\cdots(2n+1)}, \quad \sum \frac{\log n}{n^2}, \quad \sum \frac{2^n+3^n}{3^n+4^n}, \quad \sum \frac{2n-1}{2n+1}, \quad \sum \frac{(n+\log n)^2}{(n^2+1)^2}.$$

[1] More general tests can be found in I. S. Sokolnikoff, "Advanced Calculus," chap. 7, McGraw-Hill Book Company, New York, 1939.

4. Give an example of a divergent series Σa_n such that all the terms satisfy $a_n > 0$ and $a_{n+1}/a_n < 1$. Does this contradict the remarks made in connection with (6-4)?

5. If x is constant, prove that $\lim (x^n/n!) = 0$. *Hint:* The series $\Sigma(|x|^n/n!)$ converges by the ratio test.

REVIEW OF THE LIMIT CONCEPT

6. The notation "$\lim a_n = \infty$" is sometimes used to mean the following: Given any positive constant A, however large, there is an N such that $a_n > A$ for all $n > N$. If $a_n = 0.00001n$, prove that $\lim a_n = \infty$. *Hint:* Choose $N = 100,000A$.

7. If $\lim a_n = \infty$, prove that $\lim c a_n = \infty$, where c is any positive constant. *Hint:* See Prob. 6.

8. Referring to Prob. 6, prove: If $\lim a_n = \infty$ and b_n is any sequence bounded below, then $\lim (a_n + b_n) = \infty$. *Hint:* Let $b_n \geq -B$. Given $A > 0$, choose N so that $a_n > A + B$, for all $n > N$.

---

7. Absolute Convergence.   Most of the preceding tests for convergence apply to series with positive terms. We shall now see how these tests can be used to establish convergence even when the signs of the terms change infinitely often.

DEFINITION.   *A series Σa_n is said to be **absolutely convergent** if the series of absolute values $\Sigma|a_n|$ is convergent.*

This terminology is justified by the following theorem, which states that an absolutely convergent series is convergent:

THEOREM I.   *If $\Sigma|a_n|$ converges, then Σa_n converges.*

For proof, note that $-|a_n| \leq a_n \leq |a_n|$, and apply Theorem II of Sec. 5.

For example, the series $\Sigma[(-1)^n/n^2]$ is absolutely convergent, since

$$\sum \left| \frac{(-1)^n}{n^2} \right| = \sum \frac{1}{n^2}$$

converges. On the other hand, $\Sigma[(-1)^n/n]$ is not absolutely convergent, because the series of absolute values is the harmonic series, $\Sigma(1/n)$.

A series whose terms are alternately positive and negative is called an *alternating series*. There is a simple test due to Leibniz that establishes the convergence of many such series even when the series does not converge absolutely.

THEOREM II.   *Suppose the alternating series $\Sigma(-1)^{n+1}a_n$ is such that $a_n > a_{n+1} > 0$ and $\lim a_n = 0$. Then the series converges, and the remainder after n terms has a value that is between zero and the first term not taken.*

For example, if the sum of the series is approximated by the first five terms

$$s \doteq s_5 = a_1 - a_2 + a_3 - a_4 + a_5$$

the error in that approximation is between zero and $-a_6$:

$$0 > s - s_5 > -a_6.$$

The value given by s_5 is *too large*, because s_5 ends with a *positive* term, $+a_5$. The value s_6 is too small, since s_6 ends with a negative term, and so on.

To prove the theorem, we have

$$s_{2n} = (a_1 - a_2) + (a_3 - a_4) + \cdots + (a_{2n-1} - a_{2n})$$
$$= a_1 - (a_2 - a_3) - \cdots - (a_{2n-2} - a_{2n-1}) - a_{2n}$$

and hence s_{2n} is positive but less than a_1 for all n. Also $s_2 < s_4 < s_6 \cdots$ so that these sums tend to a limit by the fundamental principle (Sec. 3). Since $s_{2n+1} = s_{2n} + a_{2n+1}$ and $\lim a_{2n+1} = 0$, it follows that the partial sums of odd order tend to the same limit; hence the series converges. The proof of the second statement is left as an exercise for the reader. Actually Theorem II becomes rather obvious when the partial sums s_n are plotted on the x axis.

Since the choice $a_n = 1/n$ satisfies the requirements of Theorem II, the alternating harmonic series

$$s = 1 - \frac{1}{2} + \frac{1}{3} - \frac{1}{4} + \frac{1}{5} - \frac{1}{6} + \cdots + \frac{(-1)^{n+1}}{n} + \cdots \tag{7-1}$$

is convergent. If the sum is approximated by the first two terms, Theorem II says that the error is between 0 and $\frac{1}{3}$; that is, $0 < s - \frac{1}{2} < \frac{1}{3}$, or

$$\tfrac{1}{2} < s < \tfrac{5}{6}. \tag{7-2}$$

Inasmuch as the series of absolute values diverges, we could not establish the convergence by use of Theorem I. A series such as this, which converges but not absolutely, is said to be *conditionally convergent*.

By rearranging the order of terms in a conditionally convergent series, one can make the resulting series converge to any desired value. In illustration of this fact we shall rearrange the series (7-1) in such a way that the new sum is π, though (7-2) shows that the original sum is not π.

The terms of (7-1) are obtained by choosing alternately from the series

$$1 + \tfrac{1}{3} + \tfrac{1}{5} + \tfrac{1}{7} + \cdots \tag{7-3}$$

and from the series

$$-\tfrac{1}{2} - \tfrac{1}{4} - \tfrac{1}{6} - \tfrac{1}{8} - \cdots \tag{7-4}$$

both of which are *divergent*. To form a series that converges to π, first pick out, in order, as many positive terms (7-3) as are needed to make the sum just greater than π. Then pick out, in order, enough negative terms (7-4) so that the sum of all terms so far chosen will be just less than π. Then choose more positive terms until the total sum is just greater than π, and so on. The process is possible because the series (7-3) and (7-4) are divergent; the resulting series converges to π because the error is less than the last term taken.

To get a physical interpretation of this result, suppose we place unit positive charges P at the points

$$x = 1, \ -\sqrt{2}, \ \sqrt{3}, \ -\sqrt{4}, \ \sqrt{5}, \ -\sqrt{6}, \ldots$$

and attempt to find the force on a unit negative charge N located at the origin (see Fig. 3). By Coulomb's law two opposite unit charges a distance $\sqrt{n}$ apart experience an attraction of magnitude $1/n$. Since the attraction of charges at the left of N exerts a force toward the left whereas attraction of the other charges exerts a force toward the right, the total force on N is given formally by the series (7-1). Now, the fact that this series is conditionally convergent makes the force depend not only on the final configuration of charges but also on the manner in which the charges were introduced. If we obtained the final configuration by putting 10 charges at the left, then 1 at the right, then 10 more at the left, and 1 again at the right, and so on, the net force will be directed toward the left. But if we had a preponderance of charges at the right while setting up the final configuration, the final force would be directed toward the right.

FIGURE 3

The foregoing behavior is perhaps not very surprising.   What is surprising is that a rearrangement such as this will always give the *same* value provided the series in question is *absolutely convergent*.   For example, let the configuration consist of unit positive charges P at the points $x = 1, -2, 3, -4, 5, -6, \ldots$, so that the force is given by the absolutely convergent series

$$1 - \frac{1}{2^2} + \frac{1}{3^2} - \frac{1}{4^2} + \cdots + \frac{(-1)^{n+1}}{n^2} + \cdots.$$

In this case, as we shall show, the force does not depend on the way in which the final configuration was reached.

The preceding examples may assist the reader to appreciate the following theorem, which describes what is perhaps the most important property of absolute convergence.

THEOREM III.   The terms of an absolutely convergent series can be rearranged in any manner without altering the value of the sum.

We establish this result first for series of positive terms.   Let Σp_k be such a series and $\Sigma p_k'$ a rearrangement.   For every n we have

$$\sum_{k=1}^{n} p_k' \le \sum_{k=1}^{\infty} p_k$$

inasmuch as each term p_k' is to be found among the terms of Σp_k.   Hence $\Sigma p_k'$ converges (by the fundamental principle), and also

$$\Sigma p_k' \le \Sigma p_k.$$

In just the same way we find $\Sigma p_k \le \Sigma p_k'$, and hence $\Sigma p_k' = \Sigma p_k$.

To obtain the result for an arbitrary but absolutely convergent series Σa_k, denote the rearrangement by $\Sigma a_k'$ and observe that

$$a_k = (a_k + |a_k|) - |a_k| \qquad a_k' = (a_k' + |a_k'|) - |a_k'|. \tag{7-5}$$

By the result for positive series we have $\Sigma |a_k'| = \Sigma |a_k|$, as well as

$$\Sigma (a_k' + |a_k'|) = \Sigma (a_k + |a_k|).$$

Hence (7-5) gives $\Sigma a_k' = \Sigma a_k$ when we recall the linearity property (2-10).

By methods quite similar to the foregoing[1] one can establish the following, which expresses a third fundamental property of absolutely convergent series:

THEOREM IV.   If $\Sigma a_k = a$ and $\Sigma b_k = b$ are absolutely convergent, then these series can be multiplied like finite sums and the product series will converge to ab.   Moreover, the product series is absolutely convergent and hence may be rearranged in any manner.

For example,

$$ab = a_1 b_1 + (a_2 b_1 + a_1 b_2) + (a_3 b_1 + a_2 b_2 + a_1 b_3) + \cdots.$$

Example.   Consider the series $\Sigma (x^n/\sqrt{n})$.   With $a_n = x^n/\sqrt{n}$ we have

$$\frac{|a_{n+1}|}{|a_n|} = \left| \frac{x^{n+1}}{\sqrt{n+1}} \frac{\sqrt{n}}{x^n} \right| = |x| \sqrt{\frac{n}{n+1}} \sim |x|.$$

Hence by Theorem II of Sec. 6, the series converges absolutely if $|x| < 1$ and diverges if $|x| > 1$. To see what happens when $x = \pm 1$, we substitute these values into the original series, obtaining

[1] The proof is given in full in Sokolnikoff, *op. cit.*, pp. 242–244. For the general theory of rearrangement see Paul Johnson and R. M. Redheffer, Scrambled Series, *Am. Math. Monthly*, in press (1966).

$$\sum \frac{(-1)^n}{\sqrt{n}} \quad \text{and} \quad \sum \frac{1}{\sqrt{n}}$$

for $x = -1$ and $x = +1$, respectively. The first series is conditionally convergent (by Theorem II), and the second is divergent. Hence the series converges absolutely when $|x| < 1$, it converges conditionally when $x = -1$, and it diverges for all other values of x.

PROBLEMS

1. Classify the following series as absolutely convergent, conditionally convergent, or divergent:

$$\sum \frac{(-1)^n}{\sqrt{n}}, \quad \sum (-1)^n \frac{2n+3}{2n}, \quad \sum \frac{(-1)^n}{n^2}, \quad \sum \frac{n}{2^n}, \quad \frac{1}{3} - \frac{1\cdot3}{3\cdot6} + \frac{1\cdot3\cdot5}{3\cdot6\cdot9} - \frac{1\cdot3\cdot5\cdot7}{3\cdot6\cdot9\cdot12} + \cdots.$$

2. Determine the values of x for which the following series are absolutely convergent, conditionally convergent, or divergent:

$$\sum (-1)^n \frac{x^n}{n}, \quad \sum (-1)^n \frac{x^{2n}}{(2n)!}, \quad \sum (-1)^n x^n, \quad \sum \frac{1}{nx^n}, \quad \sum \frac{1}{n}\left(\frac{2x}{x+4}\right)^n, \quad \sum n!x^n, \quad \sum \frac{(x-2)^n}{\log(n+1)}.$$

3. Approximately how many terms of the series $\sum\limits_{1}^{\infty} [(-1)^n/n^4]$ are needed to give the sum within 10^{-8}? Evaluate the sum to two places of decimals.

4. Show that Theorem II (Leibniz' test) cannot be applied to the following series, and determine which of these series converge conditionally, converge absolutely, or diverge.

(a) $1 - 3^{-1} + 2^{-1} - 3^{-3} + 2^{-2} - 3^{-5} + 2^{-3} - 3^{-7} + 2^{-4} - 3^{-9} + \cdots$;

(b) $3^{-1} - 3^{-1} + 5^{-1} - 3^{-3} + 7^{-1} - 3^{-5} + 9^{-1} - 3^{-7} + \cdots$;

(c) $3^{-1} - 1 + 7^{-1} - 5^{-1} + \cdots + (4n-1)^{-1} - (4n-3)^{-1} + \cdots$.

Hint: See Sec. 3, Probs. 7 and 8.

5. Using Theorem IV, write $(\Sigma 2^{1-n})^2$ as a single infinite series, and show that the latter converges.

6. How many terms of $\Sigma(-1)^{n-1}n^{-1}$ suffice to approximate the sum within 10^{-3}? Within 10^{-4}?

7. (*Review.*) Determine the values of x for which the following series converge, find the sum when convergent, and sketch the graph of the sum as a function of x:

$$x + \Sigma(x^n - x^{n+1}), \quad x^{1/3} + \Sigma(x^{1/(2n+1)} - x^{1/(2n-1)}).$$

8. (*Review.*) (a) By direct use of the definition of limit show that $0.111111\cdots = \frac{1}{9}$. *Hint:* If $s_1 = 0.1,\ s_2 = 0.11,\ s_3 = 0.111,\ \ldots$, then $|s_1 - \frac{1}{9}| = \frac{1}{90},\ |s_2 - \frac{1}{9}| = \frac{1}{900}$, and so on. (b) With s_n as in part a, and with $\epsilon > 0$, how large must you choose N to make

$$|s_n - \tfrac{1}{9}| < \epsilon \quad \text{for all } n > N?$$

(c) If $s = 0.111111\cdots$, evaluate s by considering $10s - s$ and also by the formula for the sum of a geometric series.

THE CAUCHY CRITERION

9. A sequence $\{s_n\}$ is called a *Cauchy sequence* if it satisfies the following condition: For each $\epsilon > 0$ there is an N such that

$$|s_n - s_m| < \epsilon \quad \text{whenever } m \geq N \text{ and } n \geq N.$$

A *subsequence* is a sequence $s_{n_1}, s_{n_2}, s_{n_3}, \ldots$, where $\{n_k\}$ is an increasing sequence of positive integers. Prove that every Cauchy sequence has a convergent subsequence. *Solution:* Choose n_0 so large that $|s_{n_0} - s_n| < 1$ for all $n > n_0$. Then choose $n_1 > n_0$ so large that $|s_{n_1} - s_n| < \frac{1}{2}$ for all $n > n_1$. Similarly, choose n_k so that n_k exceeds the previously chosen $n_0, n_1, n_2, \ldots$ and so that

$$|s_{n_k} - s_n| < 2^{-k} \quad \text{for all } n > n_k.$$

By the comparison test, the series

$$(s_{n_0} - s_{n_1}) + (s_{n_1} - s_{n_2}) + (s_{n_2} - s_{n_3}) + \cdots$$

converges absolutely; hence it converges. Since the partial sums telescope, this shows that $\lim_{k\to\infty} s_{n_k} = s$ exists.

10. Prove that every Cauchy sequence has a limit. *Solution:* Given $\epsilon > 0$, choose N so large that $|s_m - s_n| < \frac{1}{2}\epsilon$ whenever $m > N$ and $n > N$. Also choose $m = n_k > N$ so large that $|s_m - s| < \frac{1}{2}\epsilon$, where $\{s_{n_k}\}$ is the convergent subsequence obtained in the preceding problem. For $n > N$ we then have

$$|s_n - s| = |(s_n - s_m) + (s_m - s)| \leq |s_n - s_m| + |s_m - s| < \epsilon.$$

---

8. Uniform Convergence.   If a finite number of functions that are all continuous in an interval[1] $[a,b]$ are added together, the sum is also a continuous function in $[a,b]$. The question arises as to whether or not this property will be retained in the case of an infinite series of continuous functions.   Moreover, it is frequently desirable to obtain the derivative (or integral) of a function $f(x)$ by means of term-by-term differentiation (or integration) of an infinite series that defines $f(x)$.   Unfortunately such operations are not always valid, and many important investigations have led to erroneous results because of the improper handling of infinite series.   The analysis of these questions is based on a property known as *uniform convergence*, which was first defined by the mathematical physicist Sir George Stokes.

To illustrate the main idea we consider the particular sequence of functions shown graphically in Fig. 4. Each function $s_n(x)$ is given for $0 \leq x \leq 2$, and the graph of $y = s_n(x)$ consists of three straight-line segments joining the four points $(0,0)$, $(1/2n,1)$, $(1/n,0)$ and $(2,0)$. For fixed n the curve $y = s_n(x)$ has a triangular hump with its apex at $(1/2n,1)$ but, except for this hump, $y = 0$.   As n increases, the hump moves farther and farther to the left.

If x is fixed and $0 < x \leq 2$, then $\lim s_n(x) = 0$ as $n \to \infty$, because eventually the hump is wholly to the left of x.   The same condition holds for $x = 0$, since in this case $s_n(x) = 0$ for all n.   Therefore the sequence has the remarkable property that

$$\lim_{n\to\infty} s_n(x) = 0 \qquad 0 \leq x \leq 2$$

even though the maximum value of each function is

$$\max_{0 \leq x \leq 2} s_n(x) = 1 \qquad n = 1, 2, 3, \ldots.$$

The sequence is convergent but not *uniformly convergent*. This terminology is suggested by the fact that the difference between $s_n(x)$ and its limit can be made small for each fixed x, by suitable choice of n, but it cannot be made uniformly small for all x simultaneously.

The general case is similar to the special example·

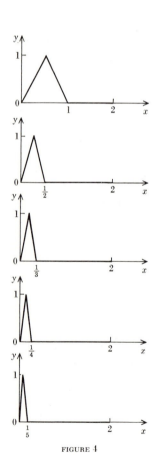

FIGURE 4

[1] We use $[a,b]$ to indicate the closed interval $a \leq x \leq b$.

Let $\{s_n(x)\}$ be any sequence of functions, defined in an interval $[a,b]$ and such that the maximum error

$$\max_{a \le x \le b} |s_n(x) - s(x)| \tag{8-1}$$

tends to 0 as $n \to \infty$. The sequence is then said to converge *uniformly* to the limit $s(x)$. If the maximum is denoted by ϵ_n, the foregoing condition gives

$$|s_n(x) - s(x)| \le \epsilon_n \quad \text{for } a \le x \le b \quad \text{and} \quad \lim \epsilon_n = 0. \tag{8-2}$$

The formulation (8-2) can be used even when the functions $s_n(x)$ happen to be of the kind that do not attain their maximum. That is, if (8-2) holds for some sequence of constants ϵ_n, the sequence converges uniformly to $s(x)$. Upon recalling the meaning of the statement "$\lim \epsilon_n = 0$" occurring in (8-2) we are led to the following:

DEFINITION 1.   A sequence $\{s_n(x)\}$ converges uniformly to $s(x)$ in a given interval $[a,b]$ if for each positive ϵ there is a number N, **independent of x,** *such that*

$$|s_n(x) - s(x)| < \epsilon \quad \text{for all } n > N \quad a \le x \le b.$$

It is the words in boldface that give the whole distinction between ordinary convergence and uniform convergence.

For example, if the apex of the triangles in Fig. 4 had been chosen as the point $(1/2n,1/n)$, instead of $(1/2n,1)$, the convergence would be uniform. In this case, the maximum difference between $s_n(x)$ and the limit 0 would be $1/n$, which tends to 0.

A geometric interpretation of uniform convergence can be obtained by considering the graphs of $y = s(x)$ and of the nth approximating curves $y = s_n(x)$. The condition in the definition is equivalent to

$$s(x) - \epsilon < s_n(x) < s(x) + \epsilon$$

which means that the graph of $y = s_n(x)$ lies in a strip of width 2ϵ centered on the graph of $y = s(x)$ (see Fig. 5). No matter how narrow the strip may be, this condition must hold for all sufficiently large n; otherwise the convergence is not uniform.

Since the value of an infinite series is defined to be the limit of the sequence of partial sums, there is no difficulty in extending the concept of uniform convergence to series. Let $\Sigma u_n(x)$ be a series of functions defined in a given interval $[a,b]$, with partial sums

$$s_n(x) = u_1(x) + u_2(x) + \cdots + u_n(x). \tag{8-3}$$

If the sequence of partial sums converges uniformly to a function $s(x)$, the series $\Sigma u_n(x)$ is said to be *uniformly convergent;* otherwise the series is not uniformly convergent. This concept can be further clarified by considering the remainder $r_n(x)$.

We assume that the series $\Sigma u_n(x)$ converges at each point of the interval $[a,b]$, so that

$$s(x) = \Sigma u_n(x) \quad a \le x \le b.$$

The partial sums are given by (8-3), and the remainder after n terms is

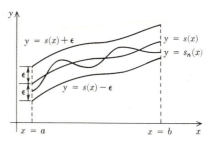

FIGURE 5

$$r_n(x) = s(x) - s_n(x) = u_{n+1}(x) + u_{n+2}(x) + \cdots. \tag{8-4}$$

Since the series converges to $s(x)$, $\lim s_n(x) = s(x)$ as $n \to \infty$; hence

$$\lim r_n(x) = 0. \tag{8-5}$$

The statement embodied in (8-5) means that for any preassigned positive number ϵ, however small, one can find a number N such that

$$|r_n(x)| < \epsilon \qquad \text{for all } n > N.$$

In general, the magnitude of N depends not only on the choice of ϵ but also on the value of x. But if it is possible to find a single, fixed N, for any preassigned positive ϵ, which will serve for all values of x in the interval, the series is *uniformly convergent*.

We summarize the result of this discussion by the following:

DEFINITION 2. *The series $\Sigma u_n(x)$ is uniformly convergent in the interval* $[a,b]$ *if for each* $\epsilon > 0$ *there is a number N,* **independent of x,** *such that the remainder $r_n(x)$ satisfies* $|r_n(x)| < \epsilon$ *for all* $n > N$ *and all x in* $[a,b]$.

Since $r_n(x) = s(x) - s_n(x)$, the requirement of Definition 2 is precisely the same as requiring uniform convergence of the sequence of partial sums.

Example 1.   Consider the series

$$x + (x - 1)x + (x - 1)x^2 + \cdots + (x - 1)x^{n-1} + \cdots.$$

Since

$$s_n(x) = x + (x - 1)x + (x - 1)x^2 + \cdots + (x - 1)x^{n-1} = x^n$$

it is evident that

$$\lim_{n \to \infty} s_n(x) \equiv \lim_{n \to \infty} x^n = 0 \qquad \text{if } 0 \le x < 1.$$

Thus, $s(x) = 0$ for all values of x in the interval $0 \le x < 1$, and therefore

$$|r_n(x)| = |s_n(x) - s(x)| = |x^n - 0| = x^n.$$

Hence, the requirement that $|r_n(x)| < \epsilon$, for an arbitrary ϵ, will be satisfied only if $x^n < \epsilon$. This inequality leads to the condition

$$n \log x < \log \epsilon.$$

Since $\log x$ is negative for x between 0 and 1, it follows that it is necessary to have

$$n > \frac{\log \epsilon}{\log x}$$

which clearly shows the dependence of N on both ϵ and x. In fact, if $\epsilon = 0.01$ and $x = 0.1$, n must be greater than $\log 0.01/\log 0.1 = -2/(-1) = 2$, so that N can be chosen as any number greater than 2. If $\epsilon = 0.01$ and $x = 0.5$, N must be chosen larger than $\log 0.01/\log 0.5$, which is greater than 6. Since the values of $\log x$ approach zero as x approaches 1, the ratio $\log \epsilon/\log x$ will increase indefinitely and it will be impossible to find a single value of N which will serve for $\epsilon = 0.01$ and for all values of x in $0 \le x < 1$. The series is not uniformly convergent for $0 \le x < 1$. This is also obvious from the graphs of x^n.

Example 2.   Consider $\sum_{0}^{\infty} x^n$ on the interval $-\tfrac{1}{2} \le x \le \tfrac{1}{2}$.

According to the result of Sec. 3, Example 2, the sum, partial sum, and remainder are, respectively,

$$s(x) = \frac{1}{1 - x} \qquad s_n(x) = \frac{1 - x^n}{1 - x} \qquad r_n(x) = \frac{x^n}{1 - x}. \tag{8-6}$$

The condition $|r_n(x)| < \epsilon$ gives $|x^n| < \epsilon(1 - x)$ or, upon taking the logarithm and solving for n,

$$n > \frac{\log \epsilon(1 - x)}{\log |x|}. \tag{8-7}$$

Again it appears that the choice of N depends on both x and ϵ, but in this case it is possible to choose an N that will serve for all values of x in $[-\frac{1}{2}, \frac{1}{2}]$. Given a small ϵ, the ratio $\log \epsilon(1 - x)/\log |x|$ assumes its maximum value when $x = +\frac{1}{2}$. Hence if N is chosen so that

$$N > \frac{\log \epsilon/2}{\log \frac{1}{2}} = 1 - \frac{\log \epsilon}{\log 2}$$

the inequality (8-7) will be satisfied for all $n \geq N$.

Upon recalling the conditions for uniform convergence, we see that the series Σx^n converges uniformly for $-\frac{1}{2} \leq x \leq \frac{1}{2}$. However, the series does not converge uniformly in the interval $(-1,1)$, for, in this interval, the ratio appearing in (8-7) will increase indefinitely as x approaches the values ± 1.

--- **PROBLEMS**

1. If $\{s_n(x)\}$ is the sequence of functions depicted in Fig. 4, show that

$$\lim_{n \to \infty} \int_0^2 s_n(x)\, dx = \int_0^2 \lim_{n \to \infty} s_n(x)\, dx.$$

Hint: The integral equals the area of the triangle.

2. Let $\{s_n(x)\}$ be a sequence similar to that in Fig. 4, except that the apex of the triangle is at the point $(1/2n, n)$ rather than $(1/2n, 1)$. Show that $\lim_{n \to \infty} s_n(x) = 0$ for each fixed x, $0 \leq x \leq 2$, but that the maximum tends to infinity. Also show that the conclusion of Prob. 1 does not hold.

3. Let $\{s_n(x)\}$ be a sequence similar to that in Fig. 4, except that the apex of the triangle is at $(1/2n, a_n)$, where $\{a_n\}$ is a given sequence of numbers. Give a necessary and sufficient condition on $\{a_n\}$ such that (a) $\lim_{n \to \infty} s_n(x) = 0$ for $0 \leq x \leq 2$; (b) the convergence in part a is uniform; (c) the limit of the integral of $s_n(x)$ equals the integral of the limit.

4. The partial sums of a series are $s_n(x) = x^n$. Show that the series is uniformly convergent in the interval $[0, \frac{1}{2}]$.

5. By using the definition of uniform convergence, show that

$$\frac{1}{x + 1} - \frac{1}{(x + 1)(x + 2)} - \cdots - \frac{1}{(x + n - 1)(x + n)} - \cdots$$

is uniformly convergent in the interval $0 \leq x \leq 1$. *Hint:* Rewrite the series to show that $s_n(x) = 1/(x + n)$ and therefore $s_n(x) - s(x) = 1/(x + n)$. See Sec. 1.

6. Plot the sequence $s_n(x) = nx/(1 + nx)$ versus x for $0 \leq x \leq 1$ and for $n = 10, 100, 1,000$. Does $\lim s_n(x) = s(x)$ exist for every x? Is the convergence uniform on $0 < x \leq 1$? Is $s(x)$ continuous? Does $\lim \int_\alpha^\beta s_n(x)\, dx = \int_\alpha^\beta s(x)\, dx$ for all α, β on $[0,1]$?

7. If $s_n(x) = 2nxe^{-nx^2}$, $0 \leq x \leq 1$, show that

$$\lim_{n \to \infty} \int_0^1 s_n(x)\, dx - \int_0^1 \lim_{n \to \infty} s_n(x)\, dx = 1.$$

Is the convergence $s_n(x) \to s(x)$ uniform?

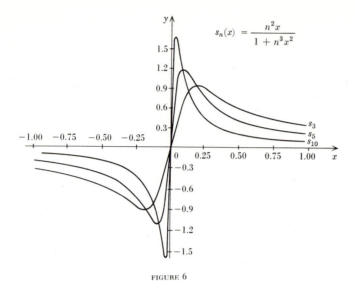

FIGURE 6

8. The partial sums of a certain series are $s_n(x) = n^2x(1 + n^3x^2)^{-1}$.   Show that the series converges to 0 for all x but that the convergence is not uniform in any interval containing $x = 0$ (Fig. 6). Does the area under the curve $y = s_n(x)$ from $x = 0$ to $x = x_0$ tend to 0 as $n \to \infty$?   Does the slope of the curve $y = s_n(x)$ tend to 0 for all x?

9. As in Prob. 8, discuss $s_n(x) = n^2x(1 + n^2x^2 \log n)^{-1}$.

---

9. Theorems about Uniform Convergence.

Generally speaking, any test for convergence becomes a test for uniform convergence provided its conditions are satisfied uniformly, that is, independently of x.   For instance, the ratio test takes the form: If there is a number r *independent of* x such that for all large n

$$\left| \frac{u_{n+1}(x)}{u_n(x)} \right| \leq r < 1$$

then $\Sigma u_n(x)$ converges uniformly.   Similarly, the comparison test takes the form: If $\Sigma v_n(x)$ is a *uniformly convergent* series such that $|u_n(x)| \leq v_n(x)$, then $\Sigma u_n(x)$ converges uniformly.   The simplest example of a uniformly convergent series $\Sigma v_n(x)$ is a series of constants.   Choosing such a series in the comparison test, we are led to the *Weierstrass M test:*

THEOREM I.   If there is a convergent series of constants, ΣM_n, such that $|u_n(x)| \leq M_n$ for all values of x on $[a,b]$, then the series $\Sigma u_n(x)$ is uniformly (and absolutely) convergent on $[a,b]$.

The proof is simple.   Since ΣM_n is convergent, for any prescribed $\epsilon > 0$ there is an N such that

$$M_{n+1} + M_{n+2} + M_{n+3} + \cdots < \epsilon \qquad \text{for all } n > N.$$

By the ordinary comparison test, $\Sigma u_n(x)$ converges for each x, so that $r_n(x)$ is well defined.   We have, also,

$$|r_n(x)| = |u_{n+1}(x) + u_{n+2}(x) + \cdots| \leq |u_{n+1}(x)| + |u_{n+2}(x)| + \cdots$$
$$\leq M_{n+1} + M_{n+2} + \cdots < \epsilon$$

for all $n > N$. Since N does not depend on x, this establishes the theorem. The other tests for uniform convergence mentioned above are established similarly.

The fact that the Weierstrass test establishes the absolute convergence, as well as the uniform convergence, of a series means that it is applicable only to series that converge absolutely. There are other tests that are not so restricted, but these tests are more complex. It should be emphasized that a series may converge uniformly but not absolutely, and vice versa.

To illustrate the use of the M test, consider the series $\Sigma(\sin nx)/n^2$. Since $|\sin nx| \leq 1$ for all values of x, the convergent series $\Sigma(1/n^2)$ will serve as an M series. It follows that $\Sigma(\sin nx)/n^2$ is uniformly and absolutely convergent on every interval, no matter how large.

For another example, consider the geometric series Σx^n. In any interval $[-a,a]$ with $0 < a < 1$ the series of positive constants Σa^n could be used as an M series, since $|x^n| \leq a^n$ on the given interval and since Σa^n converges.

The importance of uniform convergence rests upon the following properties:

THEOREM II. Let $\Sigma u_k(x)$ be a series such that each $u_k(x)$ is a continuous function of x in the interval $[a,b]$. If the series is uniformly convergent in $[a,b]$, then the sum of the series is also a continuous function of x in $[a,b]$.

THEOREM III. If a series of continuous functions $\Sigma u_n(x)$ converges uniformly to $s(x)$ in $[a,b]$, then

$$\int_\alpha^\beta s(x)\, dx = \int_\alpha^\beta u_1(x)\, dx + \int_\alpha^\beta u_2(x)\, dx + \cdots + \int_\alpha^\beta u_n(x)\, dx + \cdots,$$

where $a \leq \alpha \leq b$ and $a \leq \beta \leq b$. Moreover, the convergence is uniform with respect to α and β.

THEOREM IV. Let $\Sigma u_k(x)$ be a series of differentiable functions that converges to $s(x)$ in $[a,b]$. If the series $\Sigma u_k'(x)$ converges uniformly in $[a,b]$, then it converges to $s'(x)$.

The proof is not difficult and serves well to illustrate the idea of uniform convergence (see words in boldface below). In Theorem II, if x and $x + h$ are on $[a,b]$, we have

$$s(x) = s_n(x) + r_n(x)$$
$$s(x + h) = s_n(x + h) + r_n(x + h)$$

and hence
$$s(x + h) - s(x) = s_n(x + h) - s_n(x) + r_n(x + h) - r_n(x). \tag{9-1}$$

Given $\epsilon > 0$, pick n so that $|r_n(t)| < \epsilon$ **for all t** on $[a,b]$. Now, $s_n(x)$ is a finite sum of continuous functions and hence continuous. Therefore

$$|s_n(x + h) - s_n(x)| < \epsilon$$

whenever $|h|$ is sufficiently small. From (9-1) it follows that

$$|s(x + h) - s(x)| \leq |s_n(x + h) - s_n(x)| + |r_n(x + h)| + |r_n(x)|$$
$$< \qquad\qquad \epsilon \qquad\qquad + \qquad \epsilon \qquad + \quad \epsilon \;.$$

This shows that $|s(x + h) - s(x)|$ becomes arbitrarily small provided $|h|$ is sufficiently small, and hence $s(x)$ is continuous.

For Theorem III, note that $s(x)$ and $r_n(x)$ are continuous by Theorem II. Hence

$$\int_\alpha^\beta s(x)\, dx = \int_\alpha^\beta s_n(x)\, dx + \int_\alpha^\beta r_n(x)\, dx.$$

If we choose n so large that $|r_n(x)| < \epsilon$ **for all x** on $[a,b]$, then

$$\left| \int_\alpha^\beta s(x)\, dx - \int_\alpha^\beta s_n(x)\, dx \right| \le \left| \int_\alpha^\beta \epsilon\, dx \right| = |\beta - \alpha|\, \epsilon \le (b - a)\epsilon.$$

Since the finite sum $s_n(x)$ can be integrated term by term and since $(b - a)\epsilon$ is arbitrarily small independently of α and β, the desired result follows. Theorem IV follows from Theorem III when $u_k'(x)$ is continuous;[1] we simply write the differentiated series and integrate term by term.

It should be noted that some of the properties of uniformly convergent sequences become rather obvious when uniform convergence is interpreted graphically. For example, the conclusion of Theorem III is

$$\int_\alpha^\beta \lim_{n \to \infty} s_n(x)\, dx = \lim_{n \to \infty} \int_\alpha^\beta s_n(x)\, dx \tag{9-2}$$

and the truth of (9-2) is strongly suggested by considering appropriate areas in Fig. 5 (Sec. 8).

We now give an example that shows how the theory of uniform convergence is used in the solution of differential equations.

Example. By means of infinite series, find a function y such that $y = 1$ at $x = 0$, and $(1 + x)y' = py$ for $|x| < 1$, where p is any given constant.

Let us seek a solution in the form

$$y = a_0 + a_1 x + a_2 x^2 + \cdots + a_n x^n + \cdots. \tag{9-3}$$

If the coefficients are chosen in such a way that the series converges and can be differentiated term by term, then

$$\begin{aligned} y' &= a_1 + 2a_2 x + 3a_3 x^2 + \cdots + (n+1)a_{n+1}x^n + \cdots \\ xy' &= \quad\quad\ a_1 x + 2a_2 x^2 + \cdots + \quad\quad n a_n x^n + \cdots \\ py &= pa_0 + pa_1 x + pa_2 x^2 + \cdots + \quad\quad pa_n x^n + \cdots. \end{aligned} \tag{9-4}$$

Substitution into the prescribed equation $y' + xy' = py$ gives

$$a_1 + (2a_2 + a_1)x + (3a_3 + 2a_2)x^2 + (4a_4 + 3a_3)x^3 + \cdots = pa_0 + pa_1 x + pa_2 x^2 + pa_3 x^3 + \cdots$$

as the condition that the a_n's must satisfy.

This relation is an equality between the *sums* of the respective series. But it will certainly be satisfied if the *series themselves* agree, that is, if corresponding terms are equal. Equating coefficients of 1, x, x^2, x^3, . . . , we get

$$a_1 = pa_0, \qquad 2a_2 + a_1 = pa_1, \qquad 3a_3 + 2a_2 = pa_2, \qquad 4a_4 + 3a_3 = pa_3,$$

and so on. All these special equations are contained in the general *recurrence relation*

$$(n + 1)a_{n+1} + na_n = pa_n \tag{9-5}$$

which follows by equating the coefficients of x^n [see (9-4)]. Writing (9-5) as

$$a_{n+1} = a_n \frac{p - n}{n + 1} \qquad n = 0, 1, 2, 3, \ldots \tag{9-6}$$

we see that each coefficient is determined in terms of its predecessor.

Since $y = 1$ at $x = 0$ we must have $a_0 = 1$. Then $a_1 = a_0 p = p$,

$$a_2 = a_1 \frac{p - 1}{2} = \frac{p(p - 1)}{2!}, \qquad a_3 = a_2 \frac{p - 2}{3} = \frac{p(p - 1)(p - 2)}{3!}, \qquad \cdots$$

and the solution appears as

$$y = 1 + px + \frac{p(p - 1)}{2!}x^2 + \frac{p(p - 1)(p - 2)}{3!}x^3 + \frac{p(p - 1)(p - 2)(p - 3)}{4!}x^4 + \cdots.$$

Now that we have determined the a_n's we can justify the analysis as follows: The ratio of two successive terms of the series for y' in (9-4) satisfies

[1] A proof free of this restriction is given in K. Knopp, "Theory and Application of Infinite Series," p. 343, Blackie & Son, Ltd., Glasgow, 1928.

$$\left| \frac{(n+1)a_{n+1}x^n}{na_n x^{n-1}} \right| = \frac{n+1}{n} \left| \frac{a_{n+1}}{a_n} \right| |x| = \frac{n+1}{n} \left| \frac{p-n}{n+1} \right| |x|$$

by use of (9-6). This has the limit $|x|$ as $n \to \infty$. Hence the series is uniformly convergent in any interval $|x| \le r$, with $r < 1$, and the differentiation is justified by Theorem III. [The fact that (9-3) converges uniformly follows from the ratio test or from the fact that the terms are smaller in magnitude than those of the series for y'.] This shows that the function y that we obtained actually satisfies the differential equation.

It is easily verified that the function $\tilde{y} = (1 + x)^p$ also satisfies the differential equation, and we can prove that this function agrees with the series solution y, as follows: Let

$$z = (1 + x)^{-p}y.$$

Then $z' = (1 + x)^{-p}y' - p(1 + x)^{-p-1}y = (1 + x)^{-p-1}[(1 + x)y' - py]$. Because of the differential equation which y satisfies the result is 0 for $|x| < 1$. Hence z is constant for $|x| < 1$, and the fact that $z = 1$ at $x = 0$ shows that $z \equiv 1$. This gives $y = (1 + x)^p$, as desired. Thus we have established the *binomial theorem*

$$(1 + x)^p = 1 + \sum_{n=1}^{\infty} \frac{p(p-1)(p-2) \cdots (p-n+1)}{n!} x^n$$

for $|x| < 1$ and for all real values of p.

The series (9-3) is called a *power series*. Because of their great practical importance, power series are discussed in the following eight sections. Additional examples of the use of power series in solving differential equations can be found in Chap. 3.

PROBLEMS

1. Show that Σn^{-x} converges uniformly in the interval $r \le x < \infty$, if $r > 1$.

2. Show that $\Sigma(x^{2n} - x^{2n-2})$ converges for $-1 \le x \le 1$ but that the convergence is not uniform in this interval.

3. Show that $\Sigma 2^{-n} \sin 2^n \pi x$ converges uniformly on every finite interval, and discuss the series obtained by differentiating this series term by term.

4. Show that the function $f(x) = \Sigma(n^4 + x^4)^{-1}$ is continuous for all values of x.

5. Test the following series for uniform convergence:

$$\sum \frac{x^n}{n!}, \quad \sum \frac{x^n}{n}, \quad \sum \frac{x^n}{n^2}, \quad \sum \frac{\cos nx}{n^3}, \quad \sum (10x)^n, \quad \sum n(\sin x)^n, \quad \sum \frac{1}{1 + x^{2n}}.$$

6. Test for uniform convergence the series obtained by term-by-term differentiation of the seven series given in Prob. 5.

7. Show that Σa_n converges absolutely if $\lim \sqrt[n]{|a_n|} = r < 1$. *Hint:* Choose r' so that $r < r' < 1$. Then $\sqrt[n]{|a_n|} < r'$ for sufficiently large n, and hence $|a_n| < (r')^n$.

8. If $\lim \sqrt[n]{|a_n|} = r < 1$, show that $\Sigma a_n x^n$ converges uniformly for $|x| \le 1$. *Hint:* Referring to Prob. 7, use $M_n = (r')^n$ in the Weierstrass M test.

9. (*Review.*) Assuming $f(x)$ positive, continuous, and decreasing, let

$$u_n = f(1) + f(2) + \cdots + f(n) - \int_1^n f(x) \, dx.$$

(*a*) Interpret u_n geometrically as in Sec. 4, and similarly interpret $u_{n+1} - u_n$. Thus show that u_n is positive and decreasing and hence that $\lim u_n$ exists as $n \to \infty$. (*b*) Applying part *a* with $f(x) = 1/x$, deduce the existence of *Euler's constant*

$$\gamma = \lim_{n \to \infty} \left[1 + \frac{1}{2} + \frac{1}{3} + \cdots + \frac{1}{n} - \log n \right] \doteq 0.5772157.$$

(The question whether γ is irrational has been an open problem for over two centuries.)   (*c*) About how many terms of the harmonic series are needed to make the sum of those terms larger than 1,000?

POWER SERIES AND TAYLOR'S FORMULA

10. Properties of Power Series.   One of the most useful types of infinite series is the power series

$$a_0 + a_1 x + a_2 x^2 + \cdots + a_n x^n + \cdots \tag{10-1}$$

which gets its name from the fact that the terms are arranged in ascending powers of the variable x.   To account for the initial term a_0 we introduce the following conventions:

$$\sum = \sum_0^\infty \qquad x^0 = 1 \qquad 0! = 1.$$

Thus, (10-1) can be written compactly as $\Sigma a_n x^n$.

The present convention for Σ differs from that in Secs. 1 to 9 but is appropriate in the discussion of power series.   The equation $x^0 = 1$ is automatic for $x \neq 0$, and $0! = 1$ is suggested by taking $n = 1$ in the general relation $n! = n(n-1)!$.

For many power series the region of convergence is easily determined by the ratio test.   As an illustration, consider the three series

$$\Sigma x^n n!, \qquad \Sigma x^n (n!)^{-1}, \qquad \Sigma x^n. \tag{10-2}$$

For the first of these the ratio of two successive terms leads to

$$\left| \frac{x^n n!}{x^{n-1}(n-1)!} \right| = |xn| = |x|\, n \to \infty \qquad \text{for } x \neq 0$$

and hence the series converges only for $x = 0$.   In just the same way it is found that the second series gives a ratio $|x|/n$, which approaches zero.   Hence the second series converges for all x.   The third series is the geometric series, which, as we know, converges for $|x| < 1$.

It is a remarkable fact that every power series, without exception, behaves like one of these three examples.   The series converges for $x = 0$ only, or it converges for all x, or there is a number r such that the series converges whenever $|x| < r$ but diverges whenever $|x| > r$.   The number r is called the *radius of convergence*, and the interval $|x| < r$ is called the *interval of convergence*.   The fact that every power series has an interval of convergence can be deduced from the following theorem (cf. Sec. 16 for further discussion):

THEOREM I.   If $\Sigma a_n x^n$ converges for a particular value $x = x_0$, then the series converges absolutely whenever $|x| < |x_0|$ and uniformly in the interval $|x| \leq |x_1|$ for each fixed x_1 such that $|x_1| < |x_0|$.   And if it diverges for $x = x_0$, then it diverges for all x such that $|x| > |x_0|$.

To establish Theorem I, observe that $\lim a_n x_0^n = 0$, since $\Sigma a_n x_0^n$ converges (theorem of Sec. 3).   Hence $|a_n x_0^n| < 1$ for all sufficiently large n, or

$$|a_n| < \frac{1}{|x_0|^n} \qquad \text{for all } n > N, \text{ say.} \tag{10-3}$$

This shows that $\Sigma |a_n|\, |x|^n$ converges by comparison with the geometric series

$$\sum \frac{1}{|x_0|^n} |x|^n = \sum \left(\frac{|x|}{|x_0|}\right)^n$$

provided $|x| < |x_0|$. The statement concerning uniform convergence is established by the same calculation, since $\Sigma(|x_1|/|x_0|)^n$ serves as an M series for the Weierstrass M test. Finally, the statement concerning divergence follows, by contradiction, from the result on convergence; that is, if the series converged for x, it would have to converge for x_0, since $|x_0| < |x|$, and this is contrary to the hypothesis.

The uniform convergence mentioned in Theorem I shows that a power series represents a continuous function for all values of x interior to its interval of convergence (see Theorem II in Sec. 9). For instance, $\Sigma x^n = 1/(1 - x)$ is continuous for $|x| < 1$, though not at $x = 1$. We shall soon see that such functions not only are continuous but have derivatives of all orders and the derivatives can be found by term-by-term differentiation of the series.

As an illustration of this fact, consider the geometric series Σx^n mentioned above. Term-by-term differentiation yields the series $\Sigma n x^{n-1}$. Because of the coefficient n, which tends to infinity, one might expect the latter series to have a smaller interval of convergence than the former. Actually, however, the intervals are the same. Since

$$\left|\frac{n x^{n-1}}{(n-1)x^{n-2}}\right| = \left|\frac{n}{n-1} x\right| = \frac{n}{n-1}|x| \rightarrow |x| \qquad \text{as } n \rightarrow \infty$$

the ratio test shows that the differentiated series, like the original series, has the interval of convergence $|x| < 1$. A similar result is found if we differentiate repeatedly. Each differentiation multiplies the ratio by $n/(n-1)$. Inasmuch as $n/(n-1) \rightarrow 1$, this factor does not change the *limit* of the ratio and hence does not change the interval of convergence.

For many power series the ratio $|a_{n+1}/a_n|$ has no limit as $n \rightarrow \infty$, and the foregoing analysis does not apply. However, suppose the series (10-1) converges for some value $x = x_0 \neq 0$, so that, as before, we have the estimate (10-3). If $|x| < |x_0|$, the differentiated series $\Sigma n a_n x^{n-1}$ converges by comparison with

$$\sum n \frac{1}{|x_0|^n} |x|^n = \sum n \left(\frac{|x|}{|x_0|}\right)^n.$$

(Note that the latter series was shown to be convergent in the previous paragraph.) The same calculation establishes *uniform* convergence of the derivative series if $|x| \leq |x_1| < |x_0|$, since

$$\sum n \left(\frac{|x_1|}{|x_0|}\right)^n$$

serves as an M series for the Weierstrass M test. Hence, the result of the differentiation is actually the derivative of the original series $\Sigma a_n x^n$ (see Theorem IV of Sec. 9).

The foregoing argument is practically identical with that used to prove Theorem I. A third use of the same method establishes the corresponding result for the integrated series $\Sigma a_n x^{n+1}/(n + 1)$. In this case the comparison series are, respectively,

$$\sum \frac{1}{n+1} \left(\frac{|x|}{|x_0|}\right)^n \qquad \text{and} \qquad \sum \frac{1}{n+1} \left(\frac{|x_1|}{|x_0|}\right)^n.$$

Summarizing this discussion we can state the following, which is perhaps the most important and useful result in the whole theory of power series:

THEOREM II. A power series may be differentiated (or integrated) term by term in any interval interior to its interval of convergence. The resulting series has the same interval of convergence as the original series and represents the derivative (or integral) of the function to which the original series converges.

Consider, for example, the geometric series

$$(1 - x)^{-1} = 1 + x + x^2 + \cdots + x^n + \cdots \qquad |x| < 1. \tag{10-4}$$

Differentiating term by term we obtain

$$(1 - x)^{-2} = 1 + 2x + 3x^2 + \cdots + nx^{n-1} + \cdots \qquad |x| < 1. \tag{10-5}$$

Differentiating again gives an expansion for $(1 - x)^{-3}$, and so on. Since the series (10-4) converges for $|x| < 1$, Theorem II shows *without further discussion* that all these other expansions are also valid for $|x| < 1$.

On the other hand, if the series (10-4) is integrated term by term from zero to x, there results an expansion

$$-\log (1 - x) = x + \frac{x^2}{2} + \frac{x^3}{3} + \cdots + \frac{x^n}{n} + \cdots \qquad |x| < 1 \tag{10-6}$$

which can be used for numerical computation of the logarithm.

Equations (10-4) to (10-6) give power-series representations for the functions on the left. It will now be established that such representations are always unique.

THEOREM III. If two power series converge to the same sum throughout an interval, then corresponding coefficients are equal.

For proof, assume that $\Sigma a_n x^n = \Sigma b_n x^n$ so that (by the linearity of Σ)

$$0 = (a_0 - b_0) + (a_1 - b_1)x + (a_2 - b_2)x^2 + \cdots + (a_n - b_n)x^n + \cdots.$$

The choice $x = 0$ yields $a_0 = b_0$. Differentiating with respect to x yields

$$0 = (a_1 - b_1) + 2(a_2 - b_2)x + \cdots + n(a_n - b_n)x^{n-1} + \cdots$$

and if we now set $x = 0$, we get $a_1 = b_1$. Upon differentiating again and setting $x = 0$, we get $a_2 = b_2$, and so on.

This process not only shows that the coefficients are uniquely determined but yields a simple formula for their values. Let

$$f(x) = a_0 + a_1 x + a_2 x^2 + \cdots + a_n x^n + \cdots \qquad \text{for } |x| < x_0.$$

Upon differentiating n times we get

$$f^{(n)}(x) = 0 + 0 + 0 + \cdots + 0 + n!a_n + \cdots$$

where the second group of terms "$+ \cdots$" involves x, x^2, or higher powers. These terms disappear when we set $x = 0$, and hence $f^{(n)}(0) = n!a_n$, or

$$a_n = \frac{f^{(n)}(0)}{n!}. \tag{10-7}$$

In the following section we shall be led to the same formula (10-7), though by an entirely different method.

The algebraic properties described in Sec. 2 for series in general give corresponding properties for power series: Two power series may be added term by term, a power series may be multiplied by a constant, and so on. Since power series converge absolutely in the interval of convergence, Theorem IV of Sec. 7 yields the following additional property:

THEOREM IV. Two power series can be multiplied like polynomials for values x which are interior to both intervals of convergence; that is,

$$(\Sigma a_n x^n)(\Sigma b_n x^n) = \Sigma c_n x^n$$

where $c_n = a_0 b_n + a_1 b_{n-1} + a_2 b_{n-2} + \cdots + a_n b_0.$

So far, nothing has been said about the behavior of power series at the ends of the interval of convergence. As a matter of fact, all behaviors are possible. For example, each of the series

$$\sum_1^\infty \frac{x^n}{n^2}, \quad \sum_1^\infty x^n, \quad \sum_1^\infty \frac{x^n}{n}, \quad \sum_1^\infty \frac{(-1)^n x^n}{n}$$

has $|x| < 1$ as the interval of convergence. However, the first series converges at $x = 1$ and -1, the second diverges at $x = 1$ and -1, the third converges at $x = -1$ but diverges at $x = 1$, and the fourth diverges at $x = -1$ but converges at $x = 1$.

For applications, the most important theorem concerning the behavior at the ends of the convergence interval is Abel's theorem[1] on continuity of power series, which reads as follows:

ABEL'S THEOREM. Suppose the power series $\Sigma a_n x^n$ converges for $x = x_0$, where x_0 may be an end point of the interval of convergence. Then

$$\lim_{x \to x_0} \Sigma a_n x^n = \Sigma a_n x_0{}^n$$

provided $x \to x_0$ through values interior to the interval of convergence.

To illustrate the theorem, let $x \to -1$ through values greater than -1 in the series (10-6). The limit of the left side is $-\log 2$, since the logarithm is continuous, and the limit of the right side is $\sum_1^\infty \frac{(-1)^n}{n}$ by virtue of Abel's theorem. Hence,

$$\log 2 = 1 - \frac{1}{2} + \frac{1}{3} - \frac{1}{4} + \cdots + \frac{(-1)^{n+1}}{n} + \cdots.$$

As another example of Abel's theorem, let $x \to 1$ in Theorem IV to obtain the following: *If*

$$c_n = a_0 b_n + a_1 b_{n-1} + \cdots + a_n b_0$$

then $(\Sigma a_n)(\Sigma b_n) = \Sigma c_n$, provided each series is convergent. Hence, with the particular arrangement of the product series which is given by Σc_n, we do not need *absolute* convergence as in Theorem IV of Sec. 7.

── **PROBLEMS**

1. Find the interval of convergence, and determine the behavior at the end points of the interval:

(a) $\quad \sum (-1)^n n x^{n-1}, \quad \sum \frac{x^{2n}}{2^n(n+1)}, \quad \sum \frac{x^{2n+1}}{(2n+1)!}, \quad \sum \frac{n^3 x^{2n}}{9^n}, \quad \sum \frac{x^{4n}}{3^n};$

(b) $\quad \sum \frac{(x+1)^n}{(n+1)2^n}, \quad \sum \frac{(x+1)^n}{(n+1)^2}, \quad \sum \frac{(x-2)^{2n}}{(n+1)(-1)^n}, \quad \sum n^n(x-1)^n, \quad \sum \frac{n}{n+1}\left(\frac{2+x}{2}\right)^n.$

Hint: In the first case of part *b*, let $t = x + 1$ and note that the interval $-2 \le t < 2$ is the same as $-3 \le t - 1 < 1$. The other cases are similar.

2. Show that the radius of convergence of $\Sigma a_n x^n$ is given by $r = \lim_{n \to \infty} |a_n/a_{n+1}|$ whenever this limit exists.

[1] A proof is given in Sokolnikoff, *op. cit.*, pp. 278–279. See also Sec. 2, Prob. 8.

3. (a) By letting $x = -t^2$ in (10-4), obtain the expansion

$$\frac{1}{1 + t^2} = 1 - t^2 + t^4 - t^6 + \cdots + (-1)^n t^{2n} + \cdots.$$

(b) By integrating from zero to x, obtain an expansion for $\tan^{-1} x$.   (c) Using your result (b), show that

$$\frac{\pi}{4} = 1 - \frac{1}{3} + \frac{1}{5} - \frac{1}{7} + \cdots.$$

4. (a) Show that the series $y = \Sigma x^n/n!$ satisfies $y' = y$.   (b) Deduce an expansion for e^x.   For what values of x is this expansion valid?   (c) Obtain series expansions for e and $1/e$ by taking $x = \pm 1$ in part b.   (d) Using your series, compute e and $1/e$ to three significant figures, and check your work by finding the product $e(1/e)$.

5. Using results given in the text, express the following integrals as power series:

$$\int_0^x \frac{dt}{1 + t^4}, \quad \int_0^x \log (1 + t^5) \, dt, \quad \int_0^x \frac{t^6 \, dt}{(1 - t^3)^2}, \quad \int_0^x \left(\frac{t^7}{1 - t}\right)^2 dt, \quad \int_0^x \frac{t^2 \, dt}{1 + t^8}.$$

Hint: In the third case, for example, let $x = t^3$ in (10-5), multiply through by t^6, and finally integrate term by term.

6. By multiplication of series, obtain the expansion of $(1 + x + x^2 + \cdots + x^n + \cdots)^2$.   In particular, compute the coefficients of 1, x, x^2, x^3, and x^n in the product series.

7. For $-3 < x < 3$ and $0 < x < 2$, respectively, show that

$$\frac{3x}{9 + x^2} = \sum (-1)^n \left(\frac{x}{3}\right)^{2n+1}, \qquad \frac{1}{x} = \sum (-1)^n (x - 1)^n.$$

---

11. Taylor's Formula.   The usefulness of power series is greatly increased by the *Taylor formula*, which yields the power-series expansion for an arbitrary function $f(x)$ together with an expression for the remainder after n terms.   Let $f(x)$ be a function with a continuous nth derivative throughout the interval $[a,b]$.   Taylor's formula is obtained by integrating this nth derivative n times in succession between the limits a and x, where x is any point on $[a,b]$.   Thus,

$$\int_a^x f^{(n)}(x) \, dx = f^{(n-1)}(x)\Big|_a^x = f^{(n-1)}(x) - f^{(n-1)}(a)$$

$$\int_a^x \int_a^x f^{(n)}(x) \, (dx)^2 = \int_a^x f^{(n-1)}(x) \, dx - \int_a^x f^{(n-1)}(a) \, dx$$

$$= f^{(n-2)}(x) - f^{(n-2)}(a) - (x - a)f^{(n-1)}(a)$$

$$\cdot \quad \cdot \quad \cdot \quad \cdot \quad \cdot \quad \cdot \quad \cdot \quad \cdot \quad \cdot \quad \cdot \quad \cdot \quad \cdot \quad \cdot \quad \cdot$$

$$\int_a^x \cdots \int_a^x f^{(n)}(x) \, (dx)^n$$

$$= f(x) - f(a) - (x - a)f'(a) - \frac{(x - a)^2}{2!} f''(a) - \cdots - \frac{(x - a)^{n-1}}{(n - 1)!} f^{(n-1)}(a).$$

In the repeated integrals it is understood that the upper limit x of the first integral becomes the variable of integration in the next integral, and so on.

Solving for $f(x)$ gives

$$f(x) = f(a) + (x - a)f'(a) + \cdots + \frac{(x - a)^{n-1}}{(n - 1)!} f^{(n-1)}(a) + R_n \qquad (11\text{-}1)$$

where

$$R_n = \int_a^x \cdots \int_a^x f^{(n)}(x) \, (dx)^n. \qquad (11\text{-}2)$$

The formula given by (11-1) is known as Taylor's formula, and the particular form of R_n given in (11-2) is called the *integral form* of the remainder after n terms. The *Lagrangian form* of the remainder, which is often more useful, is

$$R_n = \frac{(x-a)^n}{n!} f^{(n)}(\xi) \qquad a \le \xi \le x. \tag{11-3}$$

To derive this from the form (11-2), let M be the maximum and m the minimum of $f^{(n)}(t)$ for $a \le t \le x$. Then the integral (11-2) clearly lies between

$$\int_a^x \cdots \int_a^x M \, (dx)^n \qquad \text{and} \qquad \int_a^x \cdots \int_a^x m \, (dx)^n.$$

Upon carrying out the integration we find that these bounds are

$$\frac{(x-a)^n}{n!} M \qquad \text{and} \qquad \frac{(x-a)^n}{n!} m$$

respectively. Since the continuous function $f^{(n)}(t)$ assumes all values between its maximum M and minimum m, there must be a number $t = \xi$ such that (11-3) holds. We have written the inequalities for the case $a < x$; in any case, ξ is between a and x.

The discussion uses the fact that a continuous function defined on a bounded closed interval assumes its maximum and minimum and also assumes all values between these. Both properties follow from the fundamental principle of Sec. 3, as is proved in nearly all books on advanced calculus.[1]

In general, the remainder R_n depends on x, as is obvious from the representation (11-2). It may happen, however, that $f(x)$ has derivatives of all orders and that the remainder R_n approaches zero as $n \to \infty$ for each value x on $[a,b]$. In this case we obtain a representation for $f(x)$ as an infinite series

$$f(x) = \sum_{n=0}^{\infty} \frac{f^{(n)}(a)(x-a)^n}{n!} \tag{11-4}$$

and R_n now gives the error that arises when the series is approximated by its nth partial sum. The series in (11-4) is called *the Taylor series for $f(x)$ about the point $x = a$*. The special case

$$f(x) = \sum_{n=0}^{\infty} \frac{f^{(n)}(0)x^n}{n!} \tag{11-5}$$

is often called *Maclaurin's series*, though Taylor's work preceded Maclaurin's.

To illustrate the use of Taylor's formula, let $f(x) = e^x$. Then $f'(x) = e^x$, $f''(x) = e^x$, ..., and hence $f^{(n)}(0) = 1$. Equation (11-5) suggests that

$$e^x = 1 + x + \frac{x^2}{2!} + \frac{x^3}{3!} + \cdots + \frac{x^n}{n!} + \cdots \tag{11-6}$$

and, indeed, by the ratio test this series converges for all x. However, to show that it converges to e^x we must consider the remainder, which takes the form

$$R_n = \frac{x^n}{n!} e^{\xi} \qquad 0 \le \xi \le x \tag{11-7}$$

when we use (11-3). Since this approaches zero[2] as $n \to \infty$, the series does converge to e^x.

[1] For a simple proof based on decimal notation see R. M. Redheffer, What! Another Note on the Fundamental Theorem of Algebra?, *Am. Math. Monthly*, February, 1964.
[2] The fact that (11-6) converges shows that $x^n/n! \to 0$. (Cf. Sec. 6, Prob. 5.)

As another example, we find the expansion of cos x in powers of $x - (\pi/2)$. The values of f, f', f'', f''' are, respectively,

$$\cos x, \; -\sin x, \; -\cos x, \; \sin x. \tag{11-8}$$

Since the next term is $f^{\mathrm{iv}} = \cos x$, the next four derivatives repeat the sequence (11-8), the next four repeat again, and so on. Evaluating at $x = \pi/2$ we get, respectively,

$$0, -1, 0, 1; \quad 0, -1, 0, 1; \quad 0, -1, 0, 1; \quad \cdots$$

and hence Eq. (11-4) suggests the expansion

$$\cos x = -\left(x - \frac{\pi}{2}\right) + \frac{1}{3!}\left(x - \frac{\pi}{2}\right)^3 - \frac{1}{5!}\left(x - \frac{\pi}{2}\right)^5 + \cdots. \tag{11-9}$$

To determine if the series converges to the function on the left, we consider the remainder after n terms. Now, (11-8) gives $f^{(n)}(x) = \pm\sin x$ or $\pm\cos x$, so that (11-3) implies $|R_n| \le |x - \pi/2|^n/n!$. Since $\lim R_n = 0$, the expansion (11-9) is valid.

Upon setting $x = \pi/2 - t$ and noting that $\cos(\pi/2 - t) = \sin t$, we get an expansion for $\sin t$

$$\sin t = t - \frac{t^3}{3!} + \frac{t^5}{5!} - \frac{t^7}{7!} + \cdots + \frac{(-1)^n t^{2n+1}}{(2n+1)!} + \cdots \tag{11-10}$$

which is consistent with (11-5). It is left as an exercise for the reader to obtain a similar expansion for the cosine by use of (11-5) and (11-3):

$$\cos x = 1 - \frac{x^2}{2!} + \frac{x^4}{4!} - \frac{x^6}{6!} + \cdots + \frac{(-1)^n x^{2n}}{(2n)!} + \cdots. \tag{11-11}$$

In these examples the fact that the series converges *to the function* was established by direct examination of R_n. Such examination is necessary even when the series is found to be convergent by other means. For example, if we define

$$f(x) = e^{-1/x^2} \quad f(0) = 0$$

it can be shown that the Taylor series about $x = 0$ converges for all x but converges to $f(x)$ only when $x = 0$. The trouble with this function is that it does not admit *any* power-series expansion valid over an interval containing $x = 0$, and we have the following:

THEOREM. *Suppose a function $f(x)$ admits a series representation in powers of $x - a$, so that $f(x) = \Sigma a_n(x - a)^n$ for some interval $|x - a| < \epsilon$. Then the Taylor series generated by $f(x)$ coincides with the given expansion [and hence the Taylor series converges to $f(x)$].*

For proof, differentiate n times and set $x = a$, just as in the discussion of (10-7). It will be found that $a_n = f^{(n)}(a)/n!$, and hence the given series is identical with the Taylor series.

The fact that the series now considered are in powers of $x - a$ rather than x causes no trouble. By a translation of axes, $\bar{x} = x - a$, these series become power series of the type considered in the preceding section and hence are subject to the theorems of that section.

This theorem shows that a valid power-series expansion obtained by any method whatever must coincide with the Taylor series. For instance, to find the Taylor series for $\sin x^2$ about $x = 0$, we set $t = x^2$ in (11-10). This is far simpler than direct use of Taylor's formula, as the reader can verify.

Example. Obtain the expansion of

$$f(x) = \frac{1}{(x-2)(x-3)}$$

in powers of $x - 1$. With $t = x - 1$, the given function becomes

$$\frac{1}{(t-1)(t-2)} = \frac{-1}{t-1} + \frac{1}{t-2} = \frac{1}{1-t} - \frac{1}{2}\frac{1}{1-\frac{1}{2}t}$$

$$= \sum t^n - \frac{1}{2}\sum \left(\frac{1}{2}t\right)^n = \sum \left(1 - \frac{1}{2}\frac{1}{2^n}\right)t^n \qquad (11\text{-}12)$$

when we use partial fractions and the known formula for the sum of a geometric series. Upon recalling that $t = x - 1$, we get the required result

$$\frac{1}{(x-2)(x-3)} = \sum_{n=0}^{\infty} \left(1 - \frac{1}{2^{n+1}}\right)(x-1)^n. \qquad (11\text{-}13)$$

Since the two geometric series (11-12) converge for $|t| < 1$ and $|t| < 2$, respectively, the expansion (11-13) is valid for $|x - 1| < 1$. By the above theorem, this expansion coincides with the Taylor series.

PROBLEMS

1. For the following functions, find the Taylor series about the point $x = 0$ and also about the point $x = 1$, when such series exist:

$$(x-1)^3, \qquad 2+x^2, \qquad x^4, \qquad (x+2)^{-1}, \qquad e^{2x}, \qquad \sin \pi x, \qquad \cos(x-1).$$

2. (a) Expand e^x about the point $x = a$ by writing $e^x = e^a e^{x-a}$ and using (11-6). (b) Expand $\log x$ about $x = 1$ by writing $\log x = \log[1 - (1 - x)]$ and using (10-6). (c) Obtain the general Taylor series from Maclaurin's. *Hint:* If $g(t) = f(a + t)$, then

$$f(a+t) = g(t) = \sum \frac{g^{(n)}(0)t^n}{n!} = \sum \frac{f^{(n)}(a)t^n}{n!}.$$

Now let $t = x - a$.

3. Expand the following fractions about the point $x = 1$:

$$\frac{1}{x}, \qquad \frac{1}{x^2-4}, \qquad \frac{1}{x(x^2-4)}, \qquad \frac{x+2}{(x+3)(x+4)}, \qquad \frac{x^2+1}{(x^2-4)(x+3)}.$$

4. Show that the Taylor series for $\sin x$ in powers of $x - a$ converges to $\sin x$ for every value of x and a, and find the expansion of $\sin x$ in powers of $x - \pi/6$.

5. Obtain the Maclaurin series for $\cos x$ by differentiating the series for $\sin x$.

6. By means of the known series for e^u, $\sin u$, and $\log(1 - u)$, find Taylor's expansions for

$$\sin 3x, \qquad e^{-x^2}, \qquad \sin x^2, \qquad e^x + e^{-x}, \qquad e^x - e^{-x}, \qquad x^{-2}\log(1+x^4), \qquad e^x \log(1-x).$$

7. What is the Taylor series for $(1 + x)^p$ if p is constant? Find the interval of convergence, and discuss the absolute convergence at the end points of the interval. *Hint:* Use Theorem IIb of Sec. 6. Analysis of the remainder R_n is difficult and may be omitted. A proof that the series converges to $(1 + x)^p$ was given in Sec. 9.

8. Express $f(x) = x/\sqrt{1+x}$ as a power series in $x/(1 + x)$. *Hint:* Let $t = x/(1 + x)$, and express $f(x)$ in terms of t.

----

12. The Expression of Integrals as Infinite Series. Many difficult integrals can be represented as power series. For example, if we let $x = t^2$ in the series (11-11) for $\cos x$, we get

$$\cos t^2 = 1 - \frac{t^4}{2!} + \frac{t^8}{4!} + \cdots + \frac{(-1)^n t^{4n}}{(2n)!} + \cdots$$

and hence, integrating term by term,

$$\int_0^x \cos t^2 \, dt = x - \frac{x^5}{5 \cdot 2!} + \cdots + \frac{(-1)^n x^{4n+1}}{(4n+1)(2n)!} + \cdots. \tag{12-1}$$

This integral is called the *Fresnel cosine integral;* it is important in the theory of diffraction. Although the Fresnel integral is not expressible in terms of elementary functions in closed form, the expansion (12-1) is valid for all x and gives a representation which is entirely adequate for many purposes.

Sometimes one may obtain a power series involving a parameter rather than the variable of integration, as in the example in Sec. 9. To illustrate this possibility we shall express the arc length of an ellipse as a power series in the eccentricity k. If the equation of the ellipse is given in parametric form as

$$x = a \sin \theta \qquad y = b \cos \theta \qquad a \geq b$$

the arc s satisfies

$$ds^2 = dx^2 + dy^2 = (a^2 \cos^2 \theta + b^2 \sin^2 \theta) \, d\theta^2.$$

Upon noting that $\cos^2 \theta = 1 - \sin^2 \theta$, we obtain

$$ds = a\sqrt{1 - k^2 \sin^2 \theta} \, d\theta$$

where $k = (a^2 - b^2)^{1/2}/a$ is the eccentricity. Hence, the arc length from $\theta = 0$ to $\theta = \phi$ is

$$s = a \int_0^\phi \sqrt{1 - k^2 \sin^2 \theta} \, d\theta \equiv aE(k, \phi).$$

The integral $E(k, \phi)$ defined by this equation is called the *elliptic integral of the second kind.* Although $E(k, \phi)$ is not elementary, it may be expressed as a power series.

By the binomial theorem (Sec. 9),

$$(1 - k^2 \sin^2 \theta)^{1/2} = 1 - \tfrac{1}{2}k^2 \sin^2 \theta - \tfrac{1}{8}k^4 \sin^4 \theta - \cdots \tag{12-2}$$

for $k^2 < 1$, which is the case when $b \neq 0$. Since

$$1 + \tfrac{1}{2}k^2 + \tfrac{1}{8}k^4 + \cdots$$

serves as an M series, (12-2) is uniformly convergent and term-by-term integration is permissible. Hence we obtain the desired expression

$$E(k, \phi) = \phi - \frac{k^2}{2} \int_0^\phi \sin^2 \theta \, d\theta - \frac{k^4}{2 \cdot 4} \int_0^\phi \sin^4 \theta \, d\theta - \cdots$$
$$- \frac{1 \cdot 3 \cdot 5 \cdots (2n-3)}{2 \cdot 4 \cdot 6 \cdots 2n} k^{2n} \int_0^\phi \sin^{2n} \theta \, d\theta - \cdots.$$

In a similar manner it can be shown that the *elliptic integral of the first kind,*

$$F(k, \phi) \equiv \int_0^\phi \frac{d\theta}{\sqrt{1 - k^2 \sin^2 \theta}} \qquad k^2 < 1$$

has the expansion

$$F(k, \phi) = \phi + \frac{k^2}{2} \int_0^\phi \sin^2 \theta \, d\theta + \frac{1 \cdot 3}{2 \cdot 4} k^4 \int_0^\phi \sin^4 \theta \, d\theta + \cdots$$
$$+ \frac{1 \cdot 3 \cdot 5 \cdots (2n-1)}{2 \cdot 4 \cdot 6 \cdots 2n} k^{2n} \int_0^\phi \sin^{2n} \theta \, d\theta + \cdots.$$

The *elliptic integral of the third kind* is

$$\Pi(n, k, \phi) = \int_0^\phi \frac{d\theta}{(1 + n \sin^2 \theta)\sqrt{1 - k^2 \sin^2 \theta}}$$

and this, too, can be expressed as a series by expanding the radical.

Any integral of the form

$$\int (a \sin x + b \cos x + c)^{\pm 1/2} \, dx$$

or of the form

$$\int R(x, \sqrt{ax^4 + bx^3 + cx^2 + dx + e})\, dx, \qquad R(x,y) = \text{rational function}$$

is expressible in terms of the elliptic integrals[1] together with elementary functions. For this reason elliptic integrals have great practical importance and have been extensively tabulated.

PROBLEMS

1. Expand the following integrals as power series:

$$\int_0^x e^{-t^2}\, dt, \qquad \int_0^x \frac{e^t - e^{-t}}{t}\, dt, \qquad \int_0^x \sin(t^2)\, dt, \qquad \int_0^x \frac{\sin t}{t}\, dt, \qquad \int_0^x \frac{1 - \cos x}{x^2}\, dx.$$

2. Express $\int_0^\pi e^{x \sin t}\, dt$ as a power series in x. *Hint:* By Wallis' formula,

$$\int_0^\pi \sin^n t\, dt = \frac{(n-1)(n-3)\cdots 2 \text{ or } 1}{n(n-2)\cdots 2 \text{ or } 1}\, \alpha,$$

where $\alpha = 2$ if n is odd and $\alpha = \pi$ if n is even.

3. If $x \geq 0$ and p is constant, express each of the integrals

$$\int_0^x t^{p-1} e^{-t}\, dt, \qquad \int_0^x t^p \log(1+t)\, dt, \qquad \int_0^x t^{2p} \cos(t^2)\, t\, dt, \qquad \int_0^x \frac{t^p\, dt}{\sqrt{1 - t^4}}$$

as a series in powers of x. For what values of x and p is your expansion valid?

4. The *beta function* is defined by

$$B(p,q) = \int_0^1 x^{p-1}(1-x)^{q-1}\, dx.$$

Express this as a series by using the binomial theorem for $(1-x)^{q-1}$ and integrating term by term. For what values of p and q is the resulting series absolutely convergent? (See Theorem II*b* in Sec. 6. Although the range of integration includes the value $x = 1$, which is an *end point* of the convergence interval, the integration is easily justified by Abel's theorem, Sec. 10.)

13. Approximation by Means of Taylor's Formula. If a function $f(x)$ has a convergent Taylor series, the partial sums of that series can be used to approximate the function. In this way, calculations of great intrinsic complexity are reduced to calculations involving polynomials. The method is especially important because Taylor's formula not only gives a polynomial approximation but gives a means of estimating the error. Thus, the remainder R_n in (11-2) and (11-3) is precisely the difference between $f(x)$ and the nth partial sum of its Taylor series.

To illustrate the use of Taylor's series for numerical computation, let us find sin 10° within $\pm 10^{-7}$. The value $10° = \pi/18$ radian is closer to zero than to any other value of x for which $\sin x$ and its derivatives are easily found, and hence the expansion is taken about the point $x = 0$. To estimate the number of terms required, (11-3) gives

$$|R_n| = \left| \frac{f^{(n)}(\xi)}{n!}\, x^n \right| \leq \frac{x^n}{n!} = \left(\frac{\pi}{18}\right)^n \frac{1}{n!} \doteq \frac{(0.175)^n}{n!} \tag{13-1}$$

when $a = 0$ and when we set $x = \pi/18 = 0.175$ and recall that sin x together with its derivatives is less than 1 in magnitude. The successive bounds for R_n as given by

[1] See, for example, P. Franklin, "Methods of Advanced Calculus," chap. 7, McGraw-Hill Book Company, New York, 1944.

(13-1) may be computed recursively; indeed, the nth value is obtained by applying a factor $(0.175/n)$ to the preceding one. For $n = 1$ the bound (13-1) is 0.175 and the next few are as follows:

Value of n	2	3	4	5	6
Bound for $|R_n|$	1.5×10^{-2}	8.8×10^{-4}	3.9×10^{-5}	1.4×10^{-6}	4.0×10^{-8}

From a list such as this the n sufficient for any prescribed accuracy can be determined at once. In particular, an accuracy of $\pm 10^{-7}$ is found if we take $n = 6$. Thus

$$\sin 10° = \frac{\pi}{18} - \left(\frac{\pi}{18}\right)^3 \frac{1}{3!} + \left(\frac{\pi}{18}\right)^5 \frac{1}{5!} + R_6$$

where $|R_6| \leq 4.0 \times 10^{-8}$; more explicitly, $0 \geq R_6 \geq -4.0 \times 10^{-8}$, since $f^{(6)}(\xi) = -\sin \xi \leq 0$. Inasmuch as the next term of the series is zero, *the first six terms have the same sum as the first seven terms.* Hence the error is also equal to R_7, where

$$0 \geq R_7 \geq -1.0 \times 10^{-9}. \tag{13-2}$$

An improvement of accuracy such as this is to be expected whenever the series is terminated just before one or more terms with zero coefficients.

In modern computing practice an automatic computing machine is so programmed that it keeps track of the remainder, which can often be estimated recursively as in this example. The machine is then instructed to take as many terms as are needed to make the remainder less than some preassigned amount. This process was illustrated in the foregoing calculation, where the value $n = 6$ was chosen, not at random, but by consideration of the desired accuracy.

The reader may have noticed that the series for $\sin (\pi/18)$ is an alternating series with terms decreasing in magnitude. Hence, the estimate for the error (13-2) could have been found by Theorem II of Sec. 7. Taylor's formula, however, has the merit of applying to general power series, whether alternating or not.

Many important approximations are obtained by using the first few terms of the Taylor-series expansion instead of the function itself. For example, the formula

$$k = y''[1 + (y')^2]^{-3/2}$$

for curvature of the curve whose equation is $y = f(x)$ yields

$$k = y'' \left[1 - \frac{3}{2}(y')^2 + \frac{1}{2!}\frac{3}{2} \cdot \frac{5}{2}(y')^4 - \cdots \right]$$

when we use the binomial theorem. The first-term approximation $k \cong y''$ is sufficient for most applications.

As another example, in railroad surveying it is frequently useful to know the difference between the length of a circular arc and the length of the corresponding chord. Let r be the radius of curvature of the arc AB (Fig. 7), and let α be the angle intercepted by the arc. Then, if s is the length of the arc AB and c is the length of the chord AB, $s = r\alpha$ and $c = 2r \sin \frac{1}{2}\alpha$. Since

$$\sin x = x - \frac{x^3}{3!} + \frac{x^5}{5!} \cos \xi$$

where $0 \leq \xi \leq x$, the error in using only the first two terms of the expansion is certainly less than $x^5/5!$. Then,

$$c = 2r \sin \frac{\alpha}{2} = 2r\left(\frac{\alpha}{2} - \frac{\alpha^3}{8 \cdot 6}\right)$$

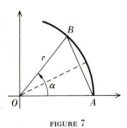

FIGURE 7

with an error less than

$$2r\left(\frac{\alpha^5}{32\cdot120}\right) = \frac{r\alpha^5}{1,920}.$$

Therefore, $s - c = \alpha^3 r/24$ with an error that is less than $r\alpha^5/1,920$.

Example 1.   For the nonelementary integral $\int_0^t e^{u^2}\,du$, obtain a polynomial approximation valid within ±0.00001 when $0 \le x \le \frac{1}{2}$.

According to (11-6) and (11-7),

$$e^x = 1 + x + \frac{x^2}{2!} + \cdots + \frac{x^{n-1}}{(n-1)!} + \frac{x^n}{n!}e^\xi \qquad 0 \le \xi \le x.$$

If we set $x = u^2$, this becomes

$$e^{u^2} = 1 + u^2 + \frac{u^4}{2!} + \cdots + \frac{u^{2n-2}}{(n-1)!} + \frac{u^{2n}}{n!}e^\xi \qquad 0 \le \xi \le u^2$$

and integrating from 0 to x yields

$$\int_0^x e^{u^2}\,du = x + \frac{x^3}{3} + \frac{x^5}{5\cdot2!} + \cdots + \frac{x^{2n-1}}{(2n-1)(n-1)!} + \int_0^x \frac{u^{2n}}{n!}e^\xi\,du.$$

To estimate the integral on the right we note that

$$e^\xi \le e^{u^2} \le e^{x^2}$$

since $\xi \le u^2$ and $u \le x$.   Hence

$$\int_0^x u^{2n}e^\xi\,du \le e^{x^2}\int_0^x u^{2n}\,du = e^{x^2}\frac{x^{2n+1}}{2n+1}.$$

It follows that if we write

$$\int_0^x e^{u^2}\,du = \sum_{n=1}^\infty \frac{x^{2n-1}}{(2n-1)(n-1)!}$$

the error R_{2n} after the term x^{2n-1} satisfies

$$0 \le R_{2n} \le \frac{e^{x^2}x^{2n+1}}{n!(2n+1)}.$$

For an approximation valid within ±0.00001 when $0 \le x \le \frac{1}{2}$, we choose n large enough to make

$$\frac{e^{1/4}(\frac{1}{2})^{2n+1}}{n!(2n+1)} \le 0.00002.$$

Since $e^{1/4} < 1.3$, the above condition is satisfied when

$$n!(2n+1)2^{2n+1} \ge 65,000.$$

By trial we find that $n = 4$ suffices.   This choice of n yields an approximation which is *too small* by 0.00002 at most.   If 0.00001 is added to the approximation, we get

$$\int_0^x e^{u^2}\,du = x + \frac{x^3}{3} + \frac{x^5}{10} + \frac{x^7}{42} + 0.00001 \tag{13-3}$$

within ±0.00001 when $0 \le x \le \frac{1}{2}$.

Example 2.   Obtain a polynomial approximation for $\int_0^x e^{\sin x}\,dx$ valid near $x = 0$.   Keeping terms as far as x^3 and no terms beyond x^3, we have

$$e^{\sin x} = 1 + \sin x + \frac{(\sin x)^2}{2!} + \frac{(\sin x)^3}{3!} + \cdots$$

$$= 1 + \left(x - \frac{x^3}{6}\right) + \frac{1}{2}\left(x - \frac{x^3}{6}\right)^2 + \frac{1}{6}\left(x - \frac{x^3}{6}\right)^3 + \cdots$$

$$= 1 + x + \frac{1}{2}x^2 + 0\cdot x^3 + \cdots.$$

Hence
$$\int_0^x e^{\sin x}\, dx = x + \frac{x^2}{2} + \frac{x^3}{6} + \cdots$$

where the terms omitted involve x^5 or higher powers.

This calculation of the series for $e^{\sin x}$ illustrates a principle which is often useful. Let $f(y) = \Sigma b_n y^n$ and $y = \Sigma a_n x^n$ be power series with nonzero radii of convergence. If $y = 0$ when $x = 0$, the power series for $f(y)$ as a function of x also has a nonzero radius of convergence. This series may be found by substituting the series for y into the series for $f(y)$ and collecting terms.[1] By uniqueness, the series so obtained is the Taylor series.

── **PROBLEMS**

1. It is desired to approximate a function $f(x)$ by a polynomial $p(x)$,
$$p(x) = a_0 + a_1 x + \cdots + a_n x^n$$
in such a way that at the origin $p(x)$ has the same value and the same first n derivatives as $f(x)$. (a) How should the coefficients be determined? *Hint:* $a_0 = p(0) = f(0)$, $a_1 = p'(0) = f'(0)$, $2a_2 = p''(0) = f''(0), \ldots$. (b) If the coefficients are determined as in part a, what relation does $p(x)$ have to the Maclaurin series for $f(x)$?

2. For the following functions, obtain a polynomial approximation valid near $x = 0$ by finding the first three nonzero terms of the Maclaurin series:

$$\tan x, \qquad e^{\tan x}, \qquad \sec x, \qquad \frac{e^x}{1 + e^x}, \qquad \frac{\sin x}{e^x - 1}, \qquad \sqrt{\cos x}, \qquad \cos\sqrt{(x^2 + x)}, \qquad \sin(\sin x).$$

3. (a) By means of series, compute $x - 10\sin(x/10)$ to three significant figures when $x = 1.000$. (b) Attempt the same calculation by using a table of $\sin x$ (note that x is in radians). How many significant figures for $\sin x$ are needed?

4. If $y = 10(\tan x - x)/x^3$, (a) use series to evaluate y near $x = 0$. In particular, what is the limit of y as $x \to 0$? (b) Plot y versus x for $0 < x \le 0.2$. (c) Discuss the construction of such a graph by use of a table of $\tan x$, without series.

5. By use of series, compute to three places of decimals:

(a) $e^{1.1} = ee^{0.1} = 2.7183e^{0.1}$;   (b) $\cos 10° = \cos(\pi/18)$;

(c) $\sin 33° = \sin(30° + 3°)$;   (d) $\sqrt[5]{35} = 2(1 + \tfrac{3}{32})^{1/5}$.

6. Evaluate by series the first three integrals to three places of decimals and the last to two places:

$$\int_0^1 \sin(x^2)\, dx, \qquad \int_0^{1/5} \frac{\sin x\, dx}{\sqrt{1 - x^2}}, \qquad \int_0^{0.1} \frac{\log(1 - z)}{z}\, dz,$$

$$\int_0^1 (2 - \cos x)^{-1/2}\, dx = \int_0^1 \left(1 + 2\sin^2\frac{x}{2}\right)^{-1/2} dx.$$

7. Determine the magnitude of α if the error in the approximation $\sin\alpha \cong \alpha$ is not to exceed 1 per cent. *Hint:* $(\alpha - \sin\alpha)/\alpha = 0.01$ and $\sin\alpha = \alpha - (\alpha^3/3!) + (\alpha^5/5!) - \cdots$.

8. Discuss the percentage error in the approximation (13-3) as $x \to 0$. How would the percentage error behave if the term 0.00001 had not been added? (This shows that it may be better not to alter the Taylor series even when such alteration reduces the absolute error.)

9. As in Example 1, obtain a polynomial approximation for the *Fresnel sine integral* $\int_0^x \sin(t^2)\, dt$ which is valid within ±0.00001 for $0 \le x \le \frac{1}{2}$.

10. Evaluate the following integrals to three places of decimals:

$$\int_0^1 \frac{\sin x}{\sqrt{x}}\, dx, \qquad \int_0^{1/2} \frac{\sin x}{x}\, dx, \qquad \int_0^1 e^{-x^2}\, dx, \qquad \int_0^1 \cos\sqrt{x}\, dx.$$

──

[1] For proof, see Knopp, *op. cit.*, p. 180, or Johnson and Redheffer, *loc. cit.*

14. The Gamma Function. Since many power series involve the factorial function $n!$ it is useful to have a sensible interpretation for $n!$ when n is not an integer. This is given by the *gamma function*, which is defined by

$$\Gamma(p+1) = \int_0^\infty t^p e^{-t}\, dt \qquad p \geq 0. \tag{14-1}$$

We shall find that study of the improper integral (14-1) involves several of the methods that have been introduced in this chapter.

The gamma function was discovered by the celebrated Swiss mathematician L. Euler. Because of its connection with $p!$ the function $\Gamma(p+1)$ is often called the *factorial function* and is written $p!$ or $\Pi(p)$. We can use the notation $p!$ as soon as we have established that (14-1) gives the appropriate value when p is an integer. For comparison with other notations,

$$p! \equiv \Pi(p) \equiv \Gamma(p+1).$$

If $p \geq 1$, integration by parts in (14-1) gives

$$\int_0^\infty t^p e^{-t}\, dt = -t^p e^{-t}\Big|_0^\infty + p \int_0^\infty t^{p-1} e^{-t}\, dt.$$

Since the integrated term drops out, the foregoing relation simplifies to

$$\Gamma(p+1) = p\Gamma(p) \tag{14-2}$$

when we recall (14-1). This calculation is justified by the interpretation of improper integrals given in Sec. 4. That is, write b in place of ∞, carry out the partial integration, and then let $b \to \infty$. The discussion of convergence presents no difficulty (Probs. 1 and 2).

Equation (14-1) does not give $\Gamma(p+1)$ for $p \leq -1$, because the behavior of t^p at $t = 0$ makes the integral diverge. But we can write (14-2) in the form

$$\Gamma(p) = p^{-1}\Gamma(p+1) \tag{14-3}$$

and thus extend the definition of $\Gamma(p)$ as follows:

If any number p between 0 and 1 is used on the left side of (14-3), the right side gives the value of $\Gamma(p)$, for when $p \geq 0$, $\Gamma(p+1)$ is determined by (14-1). If the recursion formula (14-3) is used again, the values for $-1 < p < 0$ can be found from those for $0 < p < 1$; that is, $p+1$ in (14-3) ranges over the interval (0,1) if p ranges over the interval $(-1,0)$. Similarly, when we know $\Gamma(p)$ for $-1 < p < 0$, we can find $\Gamma(p)$ for $-2 < p < -1$, and so on.

Inasmuch as (14-1) gives

$$\Gamma(1) = \int_0^\infty t^0 e^{-t}\, dt = -e^{-t}\Big|_0^\infty = 1 \quad \text{(14-4)}$$

the method fails for $p = 0$. Thus

$$\lim_{p\to 0} \Gamma(p) = \lim_{p\to 0} p^{-1}\Gamma(p+1) = +\infty \text{ or } -\infty$$

according as $p \to 0$ through positive values or through negative values. Similar behavior is found for all negative integers; hence the graph of $\Gamma(p)$ has the appearance shown in Fig. 8. However, by use of (14-1) and (14-2) it is easily verified that $\Gamma(p)$ never vanishes; hence, if we agree that $1/\Gamma(p) = 0$ for p a negative integer, it will follow that the function $1/\Gamma(p)$ is well behaved for every

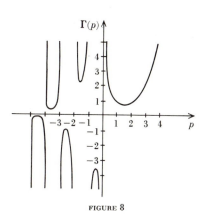

FIGURE 8

value of p without exception. The relation of $\Gamma(p)$ and $1/\Gamma(p)$ is quite analogous to the relation of $\csc p$ and $\sin p$, respectively.

Equations (14-2) and (14-4) give $\Gamma(2) = \Gamma(1) = 1$ and, in succession,

$$\Gamma(3) = 2\Gamma(2) = 2\cdot 1$$
$$\Gamma(4) = 3\Gamma(3) = 3\cdot 2\cdot 1$$
$$\cdots\cdots\cdots\cdots\cdots\cdots$$
$$\Gamma(n+1) = n\Gamma(n) = n(n-1)\cdots 3\cdot 2\cdot 1.$$

Hence the definition

$$p! = \Gamma(p+1) \tag{14-5}$$

furnishes the desired generalization; it gives a meaning to $p!$ for all p except $p = -1$, $-2, -3, \ldots$, and it gives the familiar value when p is a positive integer. The properties

$$p! = p(p-1)! \quad \text{or} \quad p\frac{1}{p!} = \frac{1}{(p-1)!} \tag{14-6}$$

are ensured by (14-2). The former fails when p is zero or a negative integer, but the latter holds for all p, without exception.

An important approximation known as *Stirling's formula* states that

$$p! \sim \sqrt{2\pi p}\, p^p e^{-p} \quad p \to \infty. \tag{14-7}$$

This is true when $p \to \infty$ through arbitrary real values (and even for complex values, provided p is not allowed to become too close to the negative real axis). However, the formula is usually needed only when $p = n$, an integer, and we give the proof for that case.

Equation (14-7) with $p = n$ is equivalent to

$$n!n^{-(n+\frac{1}{2})}e^n \to \sqrt{2\pi} \quad n \to \infty \tag{14-8}$$

and this in turn is equivalent to $L_n \to \log\sqrt{2\pi}$, where L_n is the logarithm of the expression on the left of (14-8):

$$L_n = \log n! - (n+\tfrac{1}{2})\log n + n.$$

Since $n! = n(n-1)!$, a short calculation gives

$$L_n - L_{n-1} = 1 + (n-\tfrac{1}{2})[\log(n-1) - \log n]. \tag{14-9}$$

To estimate the expression in brackets we apply Taylor's formula

$$f(x-1) = f(x) - f'(x) + \tfrac{1}{2}f''(\xi) \quad x-1 < \xi < x$$

with $f(x) = \log x$ and $x = n$. The result is

$$\log(n-1) - \log n = -\frac{1}{n} - \frac{1}{2\xi^2} \quad n-1 < \xi < n.$$

Substitution into (14-9) gives, after rearrangement,

$$L_n - L_{n-1} = \frac{1}{4\xi^2}\left[1 + \frac{2(\xi+n)(\xi-n)}{n}\right].$$

Since $\xi \sim n$ as $n \to \infty$, and $|\xi - n| < 1$, the absolute value of the expression is less than n^{-2} for large n, and therefore the series

$$(L_2 - L_1) + (L_3 - L_2) + (L_4 - L_3) + \cdots + (L_n - L_{n-1}) + \cdots$$

is absolutely convergent. We conclude that the nth partial sum $L_n - L_1$ has a limit as $n \to \infty$, and this shows that $\lim L_n = L$ exists. In other words, the logarithm of the expression on the left of (14-8) has a limit L, so that the expression itself has a limit e^L. With $E = e^L$ our result is

$$\lim \left[n!n^{-(n+\frac{1}{2})}e^n \right] = E. \tag{14-10}$$

To show that $E = \sqrt{2\pi}$ we write (14-10) with n replaced by $2n$ and also with the expression in brackets replaced by its square. These two equations are

$$\lim (2n)!(2n)^{-(2n+\frac{1}{2})}e^{2n} = E \qquad \lim (n!)^2 n^{-(2n+1)}e^{2n} = E^2.$$

Dividing the second equation by the first gives

$$E = \lim \frac{(2^n n!)^2}{(2n)!} \sqrt{\frac{2}{n}}$$

after slight simplification. If we write the factors in full and cancel the even factors of $(2n)!$ with $2^n n!$ we get

$$E = \lim \frac{2 \cdot 4 \cdot 6 \cdots 2n}{1 \cdot 3 \cdot 5 \cdots (2n-1)} \sqrt{\frac{2}{n}}.$$

The term $2/n$ can be replaced by $4/(2n+1)$ without changing the value of the limit, and comparing with Wallis' product

$$\frac{\pi}{2} = \frac{2 \cdot 2 \cdot 4 \cdot 4 \cdot 6 \cdot 6 \cdots}{1 \cdot 3 \cdot 3 \cdot 5 \cdot 5 \cdot 7 \cdots} \tag{14-11}$$

we conclude that $E^2 = 2\pi$. For completeness, the well-known formula (14-11) is derived in Prob. 6.

PROBLEMS

1. If $p \geq 0$ and $\delta > 0$, find the maximum of $t^p e^{-\delta t}$, and thus show that this function is bounded for $t \geq 0$. Deduce that $t^p e^{-2\delta t}$ is exponentially small as $t \to \infty$.

2. Applying Prob. 1 with $\delta = \frac{1}{2}$, show that the integral defining $\Gamma(p+1)$ converges for $p \geq 0$.

3. If $p < 0$ we define

$$\int_0^1 t^p e^{-t}\, dt = \lim_{\delta \to 0+} \int_\delta^1 t^p e^{-t}\, dt.$$

(a) Show that this improper integral diverges for $p \leq -1$ but converges for $p > -1$. (b) Thus show that the integral (14-1) is meaningful for $p > -1$. *Hint:* Break the interval of integration at the point $t = 1$.

4. (a) Justify the integration by parts leading to (14-2) for $p \geq 1$. (b) Using Prob. 3, consider the same question for $p > 0$.

5. For what values of the constant c do the following converge:

$$\sum \frac{n^n e^{cn}}{n!}, \qquad \sum \frac{n^{n+c}e^{-n}}{n!}, \qquad \sum \frac{n^n e^{-n}}{(n+c)!}?$$

6. If $J_n = \int_0^\pi (\sin t)^n\, dt$, obtain Wallis' product from the inequality $J_{2n-1} \geq J_{2n} \geq J_{2n+1}$. *Hint:* Derive Wallis' formula (Sec. 12, Prob. 2) by partial integration.

7. By letting $e^{-t} = s$, $t = s^2$, or $t^p = s$ and $qp = 1$ in (14-1), show that

$$\Gamma(p+1) = \int_0^1 \left(\log \frac{1}{s}\right)^p ds = 2 \int_0^\infty s^{2p+1}e^{-s^2}\, ds = \int_0^\infty e^{-s^q}\, ds.$$

8. Obtain an upper and lower bound for

$$\log n! = \log 1 + \log 2 + \log 3 + \cdots + \log n$$

by considering areas connected with the curve $y = \log x$, as in Sec. 4. Show that the average of the upper and the lower bound gives

$$(n + \tfrac{1}{2}) \log n - n + \text{const}$$

as the principal term for $n \to \infty$. (This problem motivates the form of the approximation established in the text, which otherwise might seem rather mysterious.)

---

15. Complex Numbers.   The equation $z^2 + 1 = 0$ cannot be solved by means of real numbers because the rule of signs does not allow the square of a real number to be negative. But if one adjoins a symbol i to the real numbers, which satisfies the equation

$$i^2 = -1 \tag{15-1}$$

by definition, one can construct the *complex numbers* $z = a + bi$, where a and b are real. The latter satisfy the algebraic laws obeyed by real numbers, and they include the real numbers as a special case. Moreover, complex numbers enable us to solve not only the equation $z^2 + 1 = 0$ but *every* polynomial equation.

The discussion of Secs. 15 to 17 extends the theory of infinite series to allow a complex variable z. In some series, such as

$$\Sigma a_n \sin nz$$

it is not obvious how to define the terms for complex z. But in a power series the general term

$$a_n z^n = a_n zzz \cdots z$$

involves multiplication alone and so is just as meaningful for complex numbers as for real numbers. This is one of the reasons for the predominant role played by power series in developing the theory of functions of a complex variable.

Other reasons are given in Chap. 8, which contains a discussion of complex numbers from a somewhat different point of view. (See also Prob. 7.) The importance of complex functions in the theory of vibration is indicated in Chap. 3.

Since we want to keep the familiar laws of algebra, it is easy to see how addition and multiplication of complex numbers ought to be defined. Indeed, if a, b, and i in $a + ib$ are to be treated like any other numbers of elementary algebra, then

$$(a + ib) + (c + id) = (a + c) + i(b + d). \tag{15-2}$$

This equation is now taken as the definition of addition. In the same way we are led to define multiplication by

$$(a + ib)(c + id) = (ac - bd) + i(bc + ad) \tag{15-3}$$

since elementary algebra would give the product

$$ac + ibc + iad + i^2 bd = ac + i^2 bd + i(bc + ad)$$

and (15-1) asserts that $i^2 = -1$. Finally, we agree that $a + ib = a + bi$.

It is easy to verify that the definitions (15-2) and (15-3) preserve the familiar rules of algebra (including those rules that were not considered in framing the definitions). For example, complex numbers z_k satisfy

$$z_1 + z_2 = z_2 + z_1 \qquad (z_1 + z_2) + z_3 = z_1 + (z_2 + z_3)$$
$$z_1 z_2 = z_2 z_1 \qquad\qquad (z_1 z_2) z_3 = z_1 (z_2 z_3)$$
$$z_1(z_2 + z_3) = z_1 z_2 + z_1 z_3.$$

Also there is a zero, and there is a unit:

$$z + (0 + 0i) = z \qquad z(1 + 0i) = z \qquad \text{for all } z.$$

Moreover, the complex number $a + 0i$ is found to be equivalent in every respect, except notation, to the real number a. Hence in this sense the complex numbers contain the reals as a special case, and we have a right to consider that

$$a + 0i = a. \tag{15-4}$$

Using (15-4) we write 0 and 1 for the zero and unit element of our algebra. Subtraction is defined by considering the equation $(a + ib) + z = 0$; it will be found that

$$-(a + bi) = (-a) + (-b)i = (-1)(a + ib).$$

Division is defined by considering the equation $(a + ib)z = 1$. Multiplying through by $a - ib$ (the *conjugate* of $a + ib$) we get

$$(a - ib)(a + ib)z = a - ib \qquad \text{or} \qquad (a^2 + b^2)z = a - ib.$$

This suggests that

$$(a + ib)^{-1} = (a^2 + b^2)^{-1}(a - ib)$$

provided a or b is not 0. A proof is easily given by (15-3).

The general tenor of this discussion is that the algebra of complex numbers agrees with the algebra of real numbers and we need not hesitate to apply the familiar rules. There is one new feature, however. When we say that $a + ib$ is a symbol for a complex number, we assume, naturally, that different symbols represent different numbers. In other words,

$$\text{if } a + ib = a' + ib' \text{ for } a,\ a',\ b,\ b' \text{ real, then } a = a' \text{ and } b = b'. \tag{15-5}$$

This important relation may be taken as the definition of *equality*. Unlike the algebraic properties described hitherto, (15-5) is true for complex numbers only; it does not hold if i is replaced by a real number.

The following alternative analysis of (15-5) shows the role of the equation $i^2 = -1$ and also shows why $a,\ a',\ b,\ b'$ must be assumed real. If $a + ib = a' + ib'$, then

$$a - a' = i(b' - b).$$

Squaring gives $(a - a')^2 = -(b - b')^2$ because $i^2 = -1$, and hence

$$(a - a')^2 + (b - b')^2 = 0.$$

Since the square of a real number is positive unless the number is zero, the latter equation implies $a - a' = 0$ and $b - b' = 0$.

Complex numbers $z = x + iy$, where x and y are real numbers, are represented graphically in the *z plane* by introducing two perpendicular axes, one for x and one for y (Fig. 9). The x axis is called the *real axis*, and x is the *real part* of $x + iy$; the y axis is the imaginary axis, and y is the imaginary part of $x + iy$. (The imaginary part could also be defined as iy, but the definition as y proves more convenient.) We write $x = \operatorname{Re} z$ and $y = \operatorname{Im} z$, so that

$$z = \operatorname{Re} z + i \operatorname{Im} z \qquad \bar{z} = \operatorname{Re} z - i \operatorname{Im} z$$

where $\bar{z}$ is the conjugate of z.

The *absolute value* of z is the distance from the representative point to the origin; it is denoted by

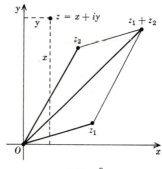

FIGURE 9

$|z|$, as in the case of real numbers.   Evidently, the points satisfying $|z| = r$ lie on a circle of radius r centered at the origin.   The interior of this circle consists of the points $|z| < r$. When $z = x + iy$, then

$$|z| = \sqrt{x^2 + y^2} = \sqrt{z\bar{z}}. \tag{15-6}$$

A short calculation based on (15-6) gives

$$|z_1 z_2| = |z_1|\,|z_2| \qquad \left|\frac{z_1}{z_2}\right| = \frac{|z_1|}{|z_2|} \qquad \text{if } z_2 \neq 0$$

so that the absolute value of a product is the product of the absolute values, and similarly for quotients.

Since real and imaginary parts are added separately, computation of $z_1 + z_2$ can be effected as shown in Fig. 9.   Inasmuch as the sum of two sides of any triangle is greater than the third, the figure gives the important *triangle inequality*

$$|z_1 + z_2| \leq |z_1| + |z_2|. \tag{15-7}$$

One can think of $z = x + iy$ as a vector with components x and y.   The method of adding vectors by adding components agrees with the definition (15-2), and hence the construction is simply the parallelogram rule familiar from mechanics.

If $s_n = u_n + iv_n$ and $s = u + iv$, with u_n, v_n, u, and v real, we define $\lim s_n = s$ to mean that simultaneously

$$\lim u_n = u \qquad \lim v_n = v. \tag{15-8}$$

This shows that the theory of limits for complex numbers can be based on the corresponding theory for real numbers.

Example.   Show that $|z_1 + z_2 + z_3| \leq |z_1| + |z_2| + |z_3|$.   Applying (15-7) to the two numbers $(z_1 + z_2)$ and z_3, we get $|(z_1 + z_2) + z_3| \leq |z_1 + z_2| + |z_3|$.   The desired result now follows by applying (15-7) again.   Evidently, a similar inequality holds for any number of complex numbers.

─── **PROBLEMS**

1. Multiplying the equation $(a + ib)z = c + id$ by the conjugate of $a + ib$, that is, by $a - ib$, obtain $(a^2 + b^2)z = (ac + bd) + i(ad - bc)$.   Thus find z when a or b is not zero.

2. (a) Show that conjugation obeys the rules $\overline{z_1 + z_2} = \bar{z}_1 + \bar{z}_2$ and $\overline{z_1 z_2} = \bar{z}_1\bar{z}_2$.   (b) Deduce $|z_1 z_2| = |z_1|\,|z_2|$ by using part a together with

$$|z_1 z_2|^2 = z_1 z_2\overline{z_1 z_2} = z_1 z_2 \bar{z}_1 \bar{z}_2 = z_1 \bar{z}_1 z_2 \bar{z}_2.$$

(c) Obtain a corresponding result for quotients.   *Hint:* If $zz_2 = z_1$ then $|z|\,|z_2| = |z_1|$ by (b).

3. If $F(z)$ is a polynomial with real coefficients, deduce from Prob. 2a that $F(\bar{z}) = \overline{F(z)}$.   Thus show that the complex roots of the equation $F(z) = 0$ occur in conjugate pairs.   *Hint:* If $F(z) = 0$ then $F(\bar{z}) = \overline{F(z)} = \bar{0} = 0$.

4. Prove that $|\text{Re } z| \leq |z|$, and hence obtain the triangle inequality without any reference to the geometry of triangles.   *Hint:*

$$(z_1 + z_2)(\bar{z}_1 + \bar{z}_2) = |z_1|^2 + 2\,\text{Re } z_1\bar{z}_2 + |z_2|^2$$

and
$$|\text{Re } z_1\bar{z}_2| \leq |z_1\bar{z}_2| = |z_1|\,|z_2|.$$

5. Show that $\lim s_n = s$ if, and only if, $\lim |s_n - s| = 0$.

6. Sketch the set of points in the complex plane denoted by (a) $|z| = 1$; (b) $|z| < 2$; (c) $|z| \geq 1$; (d) $|z - 2| < 1$; (e) $|z + 1| = 1$; (f) $|z - 5| > 4$.   *Hint:* $|z - a|$ is the distance from the representative point for z to that for a.

7. The definition of a complex number as a symbol $a + ib$ is not quite satisfactory, because it leaves unanswered the question: Is the "+" in the symbol "$a + ib$" the same as the "+" occurring elsewhere in the algebraic development? A somewhat different approach is free of this defect and shows that no distinction need be made in the two uses of "+." We define complex numbers as pairs of real numbers, in which $(a,b) = (c,d)$ means $a = c$ and $b = d$. Furthermore, by definition,

$$(a,b) + (c,d) = (a + c, b + d), \qquad (a,b)(c,d) = (ac - bd, bc + ad).$$

(a) Show that $(0,0)$ acts as the zero and $(1,0)$ as the unit. (b) Obtain $-(a,b)$ by considering $z + (a,b) = (0,0)$, where $z = (x,y)$. (c) Show that $(a,0) + (c,0) = (a + c, 0)$ and $(a,0)(c,0) = (ac,0)$ [and hence for algebraic purposes the complex number $(a,0)$ can be identified with the real number a]. (d) If -1 is identified with $(-1,0)$, show that $-z = (-1)z$ for $z = (x,y)$. (e) Show that the special complex number $i = (0,1)$ satisfies $i^2 = (-1,0) = -1$. (f) Show that $(a,b) = (a,0) + (0,1)(b,0) = a + ib$, where the first equality results from the laws of addition and multiplication and the second from certain abbreviations.

---

16. Complex Series. Convergence of infinite series of complex numbers is defined by considering the limit of partial sums, just as for real series. By (15-8) the complex series converges if, and only if, the two real series obtained by considering real and imaginary parts are both convergent. In other words,

$$\Sigma(p_n + iq_n) = p + iq$$

if, and only if, $\Sigma p_n = p$ and $\Sigma q_n = q$. Because of this correspondence of real and complex series, most of the results presented hitherto in this chapter apply with little change to the complex case, and the proof also involves nothing new.

As an illustration, let us show that *the general term of a convergent series approaches zero.* If a_n is complex, we have

$$a_n = \sum_{k=1}^{n} a_k - \sum_{k=1}^{n-1} a_k$$

and taking the limit as $n \to \infty$ yields $\lim a_n = 0$, exactly as in the proof for real series. Alternatively, one can use the *result* for real series. Namely, if $a_k = p_k + iq_k$, the convergence of Σa_k implies the convergence of Σp_k and of Σq_k. Hence, by the theorem for real series, $\lim p_n = 0$ and $\lim q_n = 0$. Consequently, $\lim a_n = 0$.

As a second illustration, we show that *an absolutely convergent series is convergent.* With $a_k = p_k + iq_k$, we have

$$|p_k| = \sqrt{p_k^2} \le \sqrt{p_k^2 + q_k^2} = |a_k|.$$

Hence, if $\Sigma|a_k|$ converges, $\Sigma|p_k|$ converges by the comparison test for real series, and then Σp_k converges, because it is known that for *real* series absolute convergence implies convergence. Similarly, Σq_k converges, and hence $\Sigma(p_k + iq_k)$ converges.

As a third example, the reader may obtain the analog of Theorem I in Sec. 10, for complex series. That is, if $\Sigma a_n z^n$ converges for $z = z_0$, the series converges absolutely for all z such that $|z| < |z_0|$ and uniformly for all z such that $|z| \le |z_1| < |z_0|$. It will be found that the proof is the same, word for word, as the proof in the case of real series. The symbol for absolute value, however, has the meaning assigned in (15-6).

For many series $\Sigma u_n(z)$, the set of points z at which the series converges gives a complicated region in the z plane. It is a remarkable fact that for a power series the region of convergence is always a circle centered at the origin. The circle is called the *circle of convergence,* and the radius of the circle is the *radius of convergence.* We agree to take the radius as zero if the series converges for $z = 0$ only and as infinity if the series converges for all z. At points on the boundary of the circle the series may either converge or diverge, just as in the case of the interval of convergence for real series.

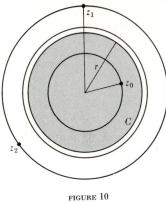

FIGURE 10

For proof that the region is a circle, let $\Sigma a_n z^n$ be a power series which converges for some value $z = z_0 \neq 0$ but diverges for some other value $z = z_1$. As we have already noted, the fact that the series converges for $z = z_0$ makes the series converge throughout the circle $|z| < |z_0|$. On the other hand, the series obviously does not converge throughout any circle containing the point z_1. We let C be *the largest circle* $|z| = r$ such that the series converges at every interior point of C. The radius r of C is at least equal to $|z_0|$ but does not exceed $|z_1|$ (see Fig. 10).

To show that C is the circle of convergence, all we have to do is establish that the series diverges at every point exterior to C. Let z_2 be an exterior point, so that $|z_2| > r$. If the series converges at z_2, it would have to converge throughout the circle $|z| < |z_2|$. But this contradicts the fact that C is the *largest* circle throughout which the series converges, and hence the proof is complete.

The fact that a largest circle exists is clear from the geometry. An analytical proof may be given, if desired, by constructing circles with successively larger radii and using the fundamental principle (Sec. 3). Alternatively, consider the set of values, r, such that the series converges for $|z| < r$. The foregoing discussion shows that if any number r_0 belongs to this set then the interval $0 \le r \le r_0$ also belongs to the set. Hence the set itself is an interval, $0 \le r < r_1$ or $0 \le r \le r_1$. The right-hand end point r_1 of the interval gives the radius of convergence of the series.

The concept of "circle of convergence" makes it possible to extend the definition of many important functions from real to complex values of z. For example, since the series

$$-\log (1 - x) = x + \frac{x^2}{2} + \frac{x^3}{3} + \cdots + \frac{x^n}{n} + \cdots$$

converges for $|x| < 1$, the circle of convergence for the complex series

$$z + \frac{z^2}{2} + \frac{z^3}{3} + \cdots + \frac{z^n}{n} + \cdots$$

contains the interval $-1 < x < 1$ of the real axis. This shows *without further discussion* that the complex series converges for $|z| < 1$. The latter series is the definition of $-\log (1 - z)$. Similarly, e^z, $\sin z$, and $\cos z$ are defined by

$$e^z = \sum \frac{z^n}{n!} \qquad \sin z = \sum \frac{(-1)^n z^{2n+1}}{(2n + 1)!} \qquad \cos z = \sum \frac{(-1)^n z^{2n}}{(2n)!}.$$

As we have already seen, each of these series converges whenever z is a real number, $z = x$. Hence the circle of convergence contains the whole real axis, and these complex series therefore converge for all z.

Example. The series

$$\frac{1}{1 + x^2} = 1 - x^2 + x^4 - x^6 + \cdots + (-1)^n x^{2n} + \cdots$$

diverges for $|x| \ge 1$. If $1/(1 + x^2)$ is regarded as a function of the *real* variable x, there appears to be no reason why the series should diverge when $|x| \ge 1$, for $1/(1 + x^2)$ has derivatives of all orders at every value of x. But when we regard x as a *complex* variable, the divergence is explained by the fact that the denominator $1 + x^2$ vanishes at $x = \pm i$.   Clearly, if the circle of convergence cannot contain the points $\pm i$, the radius of convergence cannot exceed 1.

PROBLEMS

1. Verify that the series $\sum_1^\infty z^n/n^2$ converges absolutely at every boundary point of its circle of convergence whereas $\sum_1^\infty n^2 z^n$ converges at no boundary point of its circle of convergence.

2. If $f(z) = e^{-1/z^2}$ and $f(0) = 0$,
(a) How does $f(z)$ behave when $z = x$ and $x \to 0$ through real values?
(b) How does $f(z)$ behave when $z = iy$ and $y \to 0$ through real values?
(c) Could this function have a power-series expansion valid in some circle containing the origin?

3. Determine the circle of convergence for each of the following series:
$$\Sigma n z^n, \quad \Sigma z^n \log (n + 1), \quad \Sigma n^{-n} z^n, \quad \Sigma e^n z^n, \quad \Sigma e^{-n^2} z^n.$$

4. Referring to Prob. 6 in Sec. 15, describe the region of convergence:
$$\Sigma 2^n (z - 3)^n, \quad \Sigma n!(z + 4)^n, \quad \Sigma (n^2 + n + 1)(z + 1)^n.$$

5. Describe the region of the (x,y) plane in which the following series converge, and show, in particular, that none of the regions is a circle:
$$\Sigma n^2 (x - 2)^n, \quad \Sigma (x^2 + y)^n, \quad \Sigma (x^2 + 2y^2)^n, \quad \Sigma (x^2 - y^2)^n.$$

6. The series $\Sigma (x + y)^n$ can be written
$$1 + (x + y) + (x^2 + 2xy + y^2) + \cdots + (x^n + n x^{n-1} y + \cdots + n x y^{n-1} + y^n) + \cdots.$$

Show that the latter series converges in a region of infinite area but that it would converge at most in the square $|x| < 1$, $|y| < 1$ if the parentheses were dropped.   [A similar remark applies to the series $\Sigma a_n z^n = \Sigma a_n (x + iy)^n$.]   *Hint:* Convergence is possible only if the general term tends to 0.

17. Discussion and Applications.   Many familiar formulas can be extended at once to complex variables.   To establish that
$$\sin^2 z + \cos^2 z = 1 \tag{17-1}$$
for example, we reason as follows: It is known that (17-1) holds when z is a real variable x; hence
$$\left[\sum \frac{(-1)^n x^{2n+1}}{(2n + 1)!} \right]^2 + \left[\sum \frac{(-1)^n x^{2n}}{(2n)!} \right]^2 - 1 = 0 \tag{17-2}$$
when x is real.   The left side of (17-2) is a power series, as we see by imagining that the terms are collected.   Since the power series is zero for an interval of x values, every coefficient must be zero (Sec. 10, Theorem III).   Hence the power series is also zero when x is replaced by a complex variable z, and this establishes (17-1).

The same method can be used to prove formulas involving two variables; for example, from
$$e^{x_1 + x_2} = e^{x_1} e^{x_2} \qquad \text{for } x_1 \text{ and } x_2 \text{ real}$$
it follows *without detailed calculation* that
$$e^{z_1 + z_2} = e^{z_1} e^{z_2} \qquad \text{for } z_1 \text{ and } z_2 \text{ complex}.$$

The systematic development of this idea leads to a branch of mathematical analysis known as the theory of *analytic continuation*.

Upon using iz instead of z in the series for e^z, we get

$$e^{iz} = \sum \frac{(iz)^n}{n!} = 1 - \frac{z^2}{2!} + \frac{z^4}{4!} - \frac{z^6}{6!} + \cdots + i\left(z - \frac{z^3}{3!} + \frac{z^5}{5!} - \frac{z^7}{7!} + \cdots\right)$$

when we write the sum in full, noting that

$$i^2 = -1, \quad i^3 = -i, \quad i^4 = 1, \quad i^5 = ii^4 = i, \quad \ldots$$

The series representations for $\cos z$ and $\sin z$ now give *Euler's formula*,

$$e^{iz} = \cos z + i \sin z \tag{17-3}$$

which expresses the exponential function in terms of the trigonometric functions. On the other hand, (17-3) also leads to

$$\cos z = \frac{e^{iz} + e^{-iz}}{2} \qquad \sin z = \frac{e^{iz} - e^{-iz}}{2i} \tag{17-4}$$

as the reader can verify. These equations are constantly used in the study of periodic phenomena, for example, in network analysis and synthesis, in physical optics, and in electromagnetic theory. The calculations are ordinarily carried out by complex exponentials, and then the appropriate trigonometric form is obtained by taking real or imaginary parts.

It is a familiar fact that $|\sin x| \leq 1$ and $|\cos x| \leq 1$ for real x. But if we choose $z = iy$ in (17-4) we get

$$\cos iy = \frac{e^{-y} + e^y}{2} \qquad \sin iy = \frac{e^{-y} - e^y}{2i}.$$

This shows that $\cos iy$ and $\sin iy$ grow with exponential rapidity on the imaginary axis. Similarly, the function e^z is large for $z = x \to \infty$, but (17-3) shows that

$$|e^{iy}| = 1 \qquad -\infty < y < \infty.$$

We use these estimates of growth in the following example.

Example 1. Suppose $c_i(t)$ are real or complex-valued polynomials in the real variable t such that

$$c_0(t) + c_1(t)e^t + c_2(t)e^{2it} \equiv 0$$

in some small interval $a < t < b$. Prove that each polynomial $c_i \equiv 0$.

Although the equation is given only for real t and in a small interval, the principle of analytic continuation shows that it holds for all t, real or complex. If we multiply by e^{-t} and let $t \to \infty$ through real values we conclude that $\lim c_1(t) = 0$ as $t \to \infty$. Since the polynomial c_1 is asymptotic to its leading term (Sec. 5), this shows that $c_1 \equiv 0$. Similarly, letting $t = iy$, where $y \to -\infty$, we get $c_2 = 0$. The original equation now gives $c_0 = 0$.

As we shall see in Chap. 3, results of this type have great importance in the theory of linear differential equations.

Example 2. Obtain the trigonometric identity

$$\cos u + \cos 2u + \cdots + \cos nu = \frac{\sin (n + \frac{1}{2})u}{2 \sin \frac{1}{2}u} - \frac{1}{2} \tag{17-5}$$

for $u \neq 0, \pm 2\pi, \pm 4\pi, \ldots$, by considering the exponential sum

$$s = e^{iu} + e^{2iu} + \cdots + e^{niu}.$$

The series s is a partial sum of a geometric series with ratio $r = e^{iu}$, and so

$$s = e^{iu}\frac{e^{inu} - 1}{e^{iu} - 1} \tag{17-6}$$

by (3-9). If the numerator and denominator in (17-6) are multiplied by $e^{-i\frac{1}{2}u}$, we get

$$s = e^{i\frac{1}{2}u} \frac{e^{inu} - 1}{e^{i\frac{1}{2}u} - e^{-i\frac{1}{2}u}} = \frac{e^{i(n+\frac{1}{2})u} - e^{i\frac{1}{2}u}}{2i \sin \frac{1}{2}u}$$

upon using (17-4). By (17-3) this yields another expression for the exponential sum s,

$$s = \frac{\cos (n + \frac{1}{2})u + i \sin (n + \frac{1}{2})u - \cos \frac{1}{2}u - i \sin \frac{1}{2}u}{2i \sin \frac{1}{2}u}$$

which leads to (17-5) when real parts are equated.

Example 3.   Show that $\int_0^{2\pi} e^{ikx}\, dx = 0$ if k is a nonzero integer, and deduce

$$\int_0^{2\pi} \cos nx \cos mx\, dx = \begin{cases} 0 & \text{if } m \neq n \\ \pi & \text{if } m = n \end{cases}$$

whenever m and n are positive integers.
   If k is a nonzero integer, (17-3) gives

$$\int_0^{2\pi} e^{ikx}\, dx = \int_0^{2\pi} (\cos kx + i \sin kx)\, dx = 0 \qquad (17\text{-}7)$$

which is the first result. By (17-4),

$$4 \int_0^{2\pi} \cos nx \cos mx\, dx = \int_0^{2\pi} (e^{inx} + e^{-inx})(e^{imx} + e^{-imx})\, dx$$

$$= \int_0^{2\pi} \left[e^{i(m+n)x} + e^{i(m-n)x} + e^{i(n-m)x} + e^{-i(m+n)x} \right] dx.$$

Each term is of the form e^{ikx} with k an integer, and hence we get zero unless $m = n$. If $m = n$, the two middle terms give 2, so that the integral is 4π.

──────────────────────────────── **PROBLEMS**

1. By using the series definition for e^z, show that $(d/dt)e^{ct} = ce^{ct}$ when c is a complex constant. (Since we have not defined the derivative with respect to a complex variable, assume that t is real.)

2. Sum the series $\sin x + \sin 2x + \cdots + \sin nx$.

3. Evaluate $\int e^{ax} \cos bx\, dx$ and $\int e^{ax} \sin bx\, dx$ by considering $\int e^{(a+ib)x}\, dx$ and equating real and imaginary parts. *Hint:* $\dfrac{e^{(a+ib)x}}{a + ib} = \dfrac{e^{ax}e^{ibx}(a - ib)}{a^2 + b^2}$.

4. Evaluate $\int_0^{2\pi} (2 \cos x)^4\, dx$ by using the formula $2 \cos x = e^{ix} + e^{-ix}$ and (17-7).

5. Show that every complex number z can be written in the form $z = re^{i\theta}$, where $r \geq 0$ and $0 \leq \theta < 2\pi$. *Hint:* If $z = x + iy$, introduce polar coordinates (r,θ), so that $x = r \cos \theta$ and $y = r \sin \theta$.

6. (*a*) For $0 \leq r < 1$, obtain the expansion

$$\frac{1}{1 - re^{i\theta}} = \Sigma r^n (\cos n\theta + i \sin n\theta)$$

by letting $z = re^{i\theta}$ in the series for $1/(1 - z)$.
   (*b*) Separate $1/(1 - re^{i\theta})$ into real and imaginary parts, by noting that

$$\frac{1 - re^{-i\theta}}{1 - re^{-i\theta}} \frac{1}{1 - re^{i\theta}} = \frac{1 - r \cos \theta + ir \sin \theta}{1 - 2r \cos \theta + r^2}.$$

(c) From parts a and b, deduce that

$$\frac{1 - r \cos \theta}{1 - 2r \cos \theta + r^2} = \sum_{n=0}^{\infty} r^n \cos n\theta \qquad 0 \leq r < 1$$

$$\frac{r \sin \theta}{1 - 2r \cos \theta + r^2} = \sum_{n=0}^{\infty} r^n \sin n\theta \qquad 0 \leq r < 1.$$

[The first series of part c is an example of a *Fourier cosine series*, and the second is a *Fourier sine series*. The study of such series by real-variable methods forms the topic of the next nine sections.]

FOURIER SERIES

18. The Euler-Fourier Formulas.   Trigonometric series of the form

$$f(x) = \tfrac{1}{2}a_0 + \sum_{n=1}^{\infty} (a_n \cos nx + b_n \sin nx) \tag{18-1}$$

are required in the treatment of many physical problems, for example, in the theory of sound, heat conduction, electromagnetic waves, electric circuits, and mechanical vibrations.   An important advantage of the series (18-1) is that it can represent discontinuous functions, whereas a Taylor series represents only functions that have derivatives of all orders.

We take the point of view that $f(x)$ in (18-1) is known on $(-\pi,\pi)$ and that the coefficients a_n and b_n are to be found.

It is convenient to assume, temporarily, that the series is uniformly convergent, so that it can be integrated term by term from $-\pi$ to π.   Since

$$\int_{-\pi}^{\pi} \cos nx \, dx = \int_{-\pi}^{\pi} \sin nx \, dx = 0 \qquad \text{for } n = 1, 2, \ldots$$

the calculation yields

$$\int_{-\pi}^{\pi} f(x) \, dx = a_0 \pi. \tag{18-2}$$

The coefficient a_n is determined similarly.   Thus, if we multiply (18-1) by $\cos nx$, there results

$$f(x) \cos nx = \tfrac{1}{2}a_0 \cos nx + \cdots + a_n \cos^2 nx + \cdots \tag{18-3}$$

where the terms not written involve products of the form $\sin mx \cos nx$ or of the form $\cos mx \cos nx$ with $m \neq n$.   It is easily verified[1] that, for integral values of m and n,

$$\int_{-\pi}^{\pi} \sin mx \cos nx \, dx = 0 \qquad \text{in general}$$

$$\int_{-\pi}^{\pi} \cos mx \cos nx \, dx = 0 \qquad \text{when } m \neq \pm n$$

and hence integration of (18-3) yields

$$\int_{-\pi}^{\pi} f(x) \cos nx \, dx = a_n \int_{-\pi}^{\pi} \cos^2 nx \, dx = a_n \pi.$$

Therefore, $$a_n = \frac{1}{\pi} \int_{-\pi}^{\pi} f(x) \cos nx \, dx. \tag{18-4}$$

By (18-2), this result is also valid for $n = 0$.   (That is the reason for writing the constant term as $\tfrac{1}{2}a_0$ rather than a_0.)

[1] See Example 3 of Sec. 17.

Similarly, multiplying (18-1) by $\sin nx$ and integrating yield

$$b_n = \frac{1}{\pi} \int_{-\pi}^{\pi} f(x) \sin nx \, dx. \tag{18-4a}$$

The formulas (18-4) are called the *Euler-Fourier formulas*, and the series (18-1) which results when a_n and b_n are determined by the Euler-Fourier formulas is called the *Fourier series* of $f(x)$. More specifically, a Fourier series is a trigonometric series in which the coefficients are given, for some absolutely integrable function[1] $f(x)$, by (18-4).

The distinction between a *convergent trigonometric series* and a *Fourier series* is important in the modern development of the subject and is a genuine distinction. For instance, it is known that the trigonometric series

$$\sum_{n=1}^{\infty} \frac{\sin nx}{\log (1 + n)}$$

is convergent for every value of x without exception, and yet this series is not a Fourier series. In other words, there is no absolutely integrable function $f(x)$ such that

$$\int_{-\pi}^{\pi} f(x) \cos nx \, dx = 0 \qquad \int_{-\pi}^{\pi} f(x) \sin nx \, dx = \frac{\pi}{\log (1 + n)}.$$

On the other hand, a series may be a Fourier series for some function $f(x)$ and yet diverge. Although such functions are not considered in this book, they often arise in practice, for example, in the theory of Brownian motion, in problems of filtering and noise, or in analyzing the ground return to a radar system. Even when divergent, the Fourier series represents the main features of $f(x)$, and for this reason Fourier series are an indispensable aid in problems of the sort just mentioned.

Treatises devoted to Fourier series commonly replace the sign of equality in (18-1) by $\sim$, $\cong$, or some similar symbol to indicate that the series on the right is the Fourier series of the function on the left. We shall continue to use the equality sign because the series obtained in this book do, in fact, converge for every value of x.

To illustrate the calculation of a Fourier series, let $f(x) = x$. By Eqs. (18-4)

$$a_n = \frac{1}{\pi} \int_{-\pi}^{\pi} x \cos nx \, dx = 0$$

$$b_n = \frac{1}{\pi} \int_{-\pi}^{\pi} x \sin nx \, dx = -\frac{2}{n} \cos n\pi = \frac{2}{n} (-1)^{n+1}$$

and substituting in (18-1) suggests that

$$x = 2 \left(\sin x - \frac{\sin 2x}{2} + \frac{\sin 3x}{3} - \cdots \right). \tag{18-5}$$

In Sec. 25 it is shown that the series (18-5) converges to x for $-\pi < x < \pi$. To discuss the convergence outside this interval, we introduce the notion of periodicity. A function $f(x)$ is said to be *periodic* if $f(x + p) = f(x)$ for all values of x, where p is a nonzero constant. Any number p with this property is a *period* of $f(x)$; for instance, $\sin x$ has the periods 2π, -2π, 4π, $\ldots$.

Now, each term of the series (18-5) has period 2π, and hence the sum also has period 2π. The graph of the sum therefore has the appearance shown in Fig. 11. Evidently, the sum is equal to x only on the interval $-\pi < x < \pi$, and not on the whole interval $-\infty < x < \infty$.

It remains to describe what happens at the points $x = \pm\pi$, $\pm3\pi$, $\ldots$, where the sum of the series exhibits an abrupt jump from $-\pi$ to $+\pi$. Upon setting $x = \pm\pi$,

[1] This means that $|f(x)|$, as well as $f(x)$, is integrable.

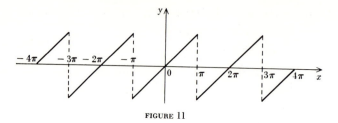

$\pm 3\pi$, ... in (18-5), we see that every term is zero. Hence the sum is zero, and this fact is indicated in the figure by placing a dot at the points in question.

The term $a_n \cos nx + b_n \sin nx$ in (18-1) is sometimes called the nth *harmonic* (from analogy with the theory of musical instruments). The first four harmonics of the series (18-5) are

$$2 \sin x, \qquad -\sin 2x, \qquad \tfrac{2}{3} \sin 3x, \qquad -\tfrac{1}{2} \sin 4x.$$

These and the next two harmonics are plotted as the numbered curves in Fig. 12. The sum of the first four harmonics is

$$y = 2 \sin x - \sin 2x + \tfrac{2}{3} \sin 3x - \tfrac{1}{2} \sin 4x.$$

Since this is a partial sum of the Fourier series, it may be expected to approximate the function x. The closeness of the approximation is indicated by the upper curves in Fig. 12, which show this partial sum of four terms together with the sums of six and

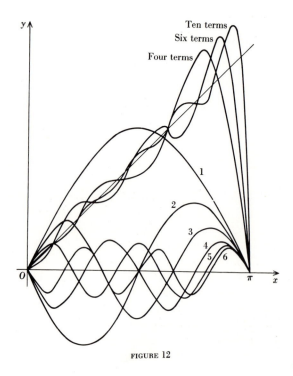

ten terms. As the number of terms increases, the approximating curves approach $y = x$ for each fixed x on $-\pi < x < \pi$ but not for $x = \pm\pi$.

Example 1.   Find the Fourier series of the function defined by

$$f(x) = 0 \quad \text{if } -\pi \leq x < 0; \quad f(x) = \pi \quad \text{if } 0 \leq x < \pi.$$

By (18-4) we have

$$a_0 = \frac{1}{\pi}\left(\int_{-\pi}^{0} 0 \, dx + \int_{0}^{\pi} \pi \, dx\right) = \pi$$

$$a_n = \frac{1}{\pi}\int_{0}^{\pi} \pi \cos nx \, dx = 0 \qquad n \geq 1$$

$$b_n = \frac{1}{\pi}\int_{0}^{\pi} \pi \sin nx \, dx = \frac{1}{n}(1 - \cos n\pi).$$

The factor $(1 - \cos n\pi)$ assumes the following values as n increases:

$n =$	1	2	3	4	5	$\cdots$
$(1 - \cos n\pi) =$	2	0	2	0	2	$\cdots$

Determining b_n by use of this table, we obtain the required Fourier series

$$\frac{\pi}{2} + 2\left(\frac{\sin x}{1} + \frac{\sin 3x}{3} + \frac{\sin 5x}{5} + \cdots\right). \tag{18-6}$$

The successive partial sums are

$$y = \frac{\pi}{2}, \qquad y = \frac{\pi}{2} + 2\sin x, \qquad y = \frac{\pi}{2} + 2\left(\sin x + \frac{\sin 3x}{3}\right), \qquad \cdots$$

The first four of these are plotted, together with the graph of $y = f(x)$, in Fig. 13.

Example 2.   Find the Fourier series for the function defined by

$$f(x) = -x \qquad -\pi \leq x \leq 0; \qquad f(x) = \pi - x \qquad 0 < x < \pi.$$

This function equals the function considered in Example 1 minus that considered in (18-5); the Fourier series can therefore be obtained by subtracting the series (18-6) and (18-5) term by term. The result is

$$\frac{\pi}{2} + 2\left(\frac{\sin 2x}{2} + \frac{\sin 4x}{4} + \frac{\sin 6x}{6} + \cdots\right).$$

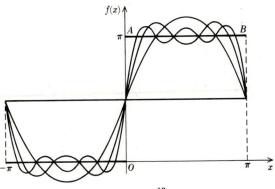

FIGURE 13

Justification of this procedure depends on the fact that the operation of forming the Fourier coefficients is a linear operation; that is, the coefficients for $f + g$ can be obtained by adding the coefficients for f and for g, and if c is any constant, the coefficient for cf can be obtained by multiplying the coefficients for f by c. Further discussion of this important linearity property is given in the problems.

PROBLEMS

1. Evaluate $\int_{-\pi}^{\pi} \cos mx \cos nx \, dx$ for integral m and n by use of the identity

$$2 \cos A \cos B = \cos (A + B) + \cos (A - B).$$

2. Find the Fourier series for $f(x)$ if

$$f(x) = \pi \qquad -\pi < x \le \tfrac{1}{2}\pi; \qquad f(x) = 0 \qquad \tfrac{1}{2}\pi < x \le \pi.$$

3. Find the Fourier series for the function defined by

$$f(x) = 0 \qquad -\pi < x < 0; \qquad f(x) = \sin x \qquad 0 \le x \le \pi.$$

4. Do Prob. 3, with $\sin x$ replaced by $\cos x$.

5. Let $c(f)$ denote the nth Fourier coefficient of a function f, and let $s(f)$ denote the nth Fourier sine coefficient. If α and β are any constants, and f and g are integrable, prove that

$$c(\alpha f + \beta g) = \alpha c(f) + \beta c(g), \qquad s(\alpha f + \beta g) = \alpha s(f) + \beta s(g).$$

(In other words, the operation of forming the Fourier coefficient of a function is a linear operation.) *Hint:* Substitute $\alpha f + \beta g$ for f in (18-4) and recall that integration is linear (Sec. 2).

6. As in Example 2, find the Fourier series for the function defined by

$$f(x) = x \qquad -\pi < x \le 0; \qquad f(x) = x + \pi \qquad 0 < x < \pi.$$

7. As in Example 2, find the Fourier series for the function

$$f(x) = -\tfrac{1}{2}x \qquad -\pi < x < 0; \qquad f(x) = \tfrac{1}{2} - \tfrac{1}{2}x \qquad 0 < x < \pi.$$

8. Show that the sawtooth wave of Fig. 11 can be represented as the sum of a square wave of period 2π and a sawtooth wave of period π. *Hint:* Do not use Fourier series.

19. Remarks on Convergence. The foregoing examples illustrate certain features which are characteristic of Fourier series in general and which will now be discussed from a general standpoint. Each term of the series (18-1) has period 2π, and hence, if $f(x)$ is to be represented by the sum, $f(x)$ must also have period 2π. Whenever we consider a series such as (18-1), we shall suppose that $f(x)$ is given for $-\pi \le x < \pi$ and that outside this interval $f(x)$ is determined by the periodicity condition

$$f(x + 2\pi) = f(x).$$

Of course, any interval $a \le x < a + 2\pi$ would serve equally well.

The term *simple discontinuity* is used to describe the situation that arises when the function $f(x)$ suffers a finite jump at a point $x = x_0$ (see Fig. 14). Analytically, this means that the two limiting values of $f(x)$, as x approaches x_0 from the right-hand and the left-hand sides, exist but are unequal; that is,

$$\lim_{\epsilon \to 0} f(x_0 + \epsilon) \ne \lim_{\epsilon \to 0} f(x_0 - \epsilon) \qquad \epsilon > 0.$$

In order to economize on space, these right-hand and left-hand limits are written as $f(x_0+)$ and $f(x_0-)$, respectively, so that the foregoing inequality can be written as

$$f(x_0+) \ne f(x_0-).$$

A function $f(x)$ is said to be *bounded* if the inequality

$$|f(x)| \leq M$$

holds for some constant M and for all x under consideration. For example, $\sin x$ is bounded, but the function

$$f(x) = x^{-1} \quad \text{for } x \neq 0 \quad f(0) = 0$$

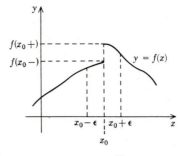

FIGURE 14

is not, even though the latter is well defined for every value of x. It can be shown that, if a bounded function has only a finite number of maxima and minima and only a finite number of discontinuities, all its discontinuities are simple. That is, $f(x+)$ and $f(x-)$ exist at every value of x.

The function illustrated in Fig. 11 satisfies these conditions in every finite interval. On the other hand, the function $\sin(1/x)$ has infinitely many maxima near $x = 0$, and the discontinuity at $x = 0$ is not simple. The function defined by

$$f(x) = x^2 \sin \frac{1}{x} \quad x \neq 0 \quad f(0) = 0$$

also has infinitely many maxima near $x = 0$, although it is continuous and differentiable for every value of x. The behavior of these two functions is illustrated graphically in Figs. 15 and 16.

With these preliminaries, we can state the following theorem, which establishes the convergence of Fourier series for a very large class of functions:

DIRICHLET'S THEOREM.   For $-\pi \leq x < \pi$, suppose $f(x)$ is defined, is bounded, has only a finite number of maxima and minima, and has only a finite number of discontinuities. Let $f(x)$ be defined for other values of x by the periodicity condition $f(x + 2\pi) = f(x)$. Then the Fourier series for $f(x)$ converges to

$$\tfrac{1}{2}[f(x+) + f(x-)]$$

at every value of x [and hence it converges to $f(x)$ at points where $f(x)$ is continuous].

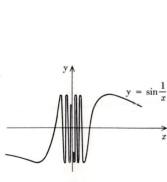

FIGURE 15

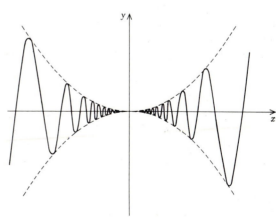

FIGURE 16

The conditions imposed on $f(x)$ are called *Dirichlet conditions*, after the mathematician Dirichlet who discovered the theorem. In Sec. 25 we establish the conclusion under slightly more restrictive conditions which are sufficient, however, for almost all applications.[1]

Example. Consider the Fourier series of the periodic function defined by

$$f(x) = -\pi \qquad -\pi < x < 0; \qquad f(x) = x \qquad 0 < x < \pi.$$

The integral (18-4) may be expressed as an integration from $-\pi$ to 0, followed by integration from 0 to π. If the appropriate formula for $f(x)$ is used in these two intervals, we get

$$a_n = \frac{1}{\pi}\left(\int_{-\pi}^0 -\pi \cos nx\, dx + \int_0^\pi x \cos nx\, dx\right) = \frac{1}{\pi}\frac{\cos n\pi - 1}{n^2}.$$

The integration assumes that $n \neq 0$; if $n = 0$, we get $a_0 = -\pi/2$, as the reader can verify. Similarly,

$$b_n = \frac{1}{\pi}\left(\int_{-\pi} -\pi \sin nx\, dx + \int_0^\pi x \sin nx\, dx\right) = \frac{1}{n}(1 - 2\cos n\pi).$$

Therefore the Fourier series is

$$f(x) = -\frac{\pi}{4} - \frac{2}{\pi}\cos x - \frac{2}{\pi}\frac{\cos 3x}{3^2} - \frac{2}{\pi}\frac{\cos 5x}{5^2} - \cdots$$

$$+ 3\sin x - \frac{\sin 2x}{2} + \frac{3\sin 3x}{3} - \frac{\sin 4x}{4} + \frac{3\sin 5x}{5} - \cdots. \quad (19\text{-}1)$$

By Dirichlet's theorem, equality actually holds at all points of continuity, since $f(x)$ has been defined to be periodic (Fig. 17). At the points of discontinuity $x = 0$ and $x = \pi$, the series converges to

$$\frac{f(0+) + f(0-)}{2} = -\frac{\pi}{2} \quad \text{and} \quad \frac{f(\pi+) + f(\pi-)}{2} = 0 \quad (19\text{-}2)$$

respectively. Either condition leads to the interesting expansion

$$\frac{\pi^2}{8} = \frac{1}{1^2} + \frac{1}{3^2} + \frac{1}{5^2} + \frac{1}{7^2} + \cdots$$

as is seen by making corresponding substitutions in (19-1).

PROBLEMS

1. On the interval $-4\pi \leq x \leq 4\pi$, sketch the graph of the function to which each Fourier series converges in Probs. 2, 3, 4, 6, and 7 of Sec. 18. (Do not recompute the Fourier series.)

2. If $f(x) = -x$ for $-\pi < x < 0$, and $f(x) = 0$ for $0 < x < \pi$, show that the corresponding Fourier series is

$$\frac{\pi}{4} - \frac{2}{\pi}\sum_{n=1}^\infty \frac{\cos(2n-1)x}{(2n-1)^2} + \sum_{n=1}^\infty \frac{(-1)^n \sin nx}{n}.$$

Sketch the graph of the sum of this series on $-4\pi \leq x \leq 4\pi$.

3. Deduce from the expansion of $f(x) = x + x^2$

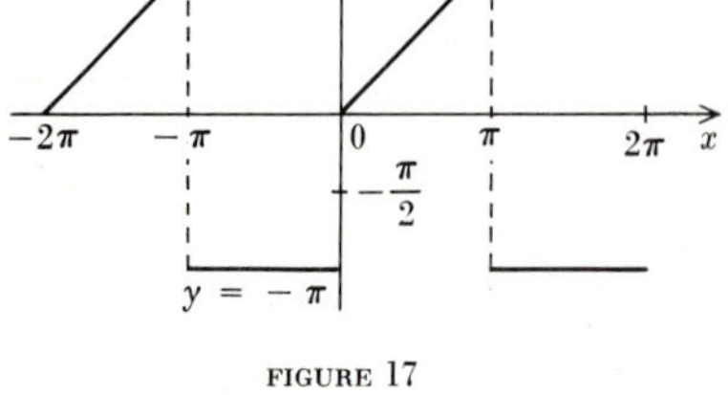

FIGURE 17

[1] For a proof of Dirichlet's theorem in full generality, see E. C. Titchmarsh, "The Theory of Functions," 2d ed., pp. 406–407, Oxford University Press, London, 1939, or E. T. Whittaker and G. N. Watson, "A Course of Modern Analysis," p. 175, Cambridge University Press, London, 1940.

in Fourier series in the interval $(-\pi,\pi)$ that

$$\sum_{n=1}^{\infty} \frac{1}{n^2} = \frac{\pi^2}{6}.$$

4. Let a be constant, with $0 < a < \pi$. Find the Fourier series for the function $f_a(x)$ defined in $-\pi < x \le \pi$ by

$$f_a(x) = 1 \quad 0 < x < a; \quad f_a(x) = 0 \quad \text{elsewhere.}$$

Sketch the graph of the sum of the series for $-4\pi \le x \le 4\pi$.

5. Let a, b, c, d be constant, with $0 < a < b < \pi$. Find the Fourier series of the function $f(x)$ which is equal to c on the interval $(0,a)$, equal to d on (a,b), and to 0 on the rest of $(-\pi,\pi)$. Sketch the graph for several values of c and d. *Hint:* If $f_a(x)$ is the function in Prob. 4, note that $f(x) = (c - d)f_a(x) + df_b(x)$. Use Prob. 5 of the preceding section, taking $\alpha = c - d$, $\beta = d$.

6. Show that, if two functions satisfy the conditions of Dirichlet, their sum need not do so. *Hint:* Try $f(x) = x^2 \sin (1/x) + 2x$, $g(x) = -2x$.

---

20. Even and Odd Functions. For many functions the Fourier sine or cosine coefficients can be determined by inspection, and this possibility is now to be investigated. A function $f(x)$ is said to be *even* if

$$f(-x) \equiv f(x) \tag{20-1}$$

and the function $f(x)$ is *odd* if

$$f(-x) \equiv -f(x). \tag{20-2}$$

For example, x^2 and $\cos x$ are even, whereas x and $\sin x$ are odd. The graph of an even function is symmetric about the y axis, as shown in Fig. 18, and the graph of an odd function is skew symmetric (Fig. 19). By inspection of the figures it is evident that

$$\int_{-a}^{a} f(x)\, dx = 2 \int_{0}^{a} f(x)\, dx \quad \text{if } f(x) \text{ is even} \tag{20-3}$$

$$\int_{-a}^{a} f(x)\, dx = 0 \quad \text{if } f(x) \text{ is odd} \tag{20-4}$$

as the integrals represent the signed areas under the curves.[1] For example,

$$\int_{-a}^{a} \sin nx\, dx = 0$$

since $\sin nx$ is an odd function.

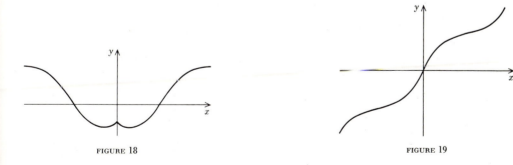

FIGURE 18   FIGURE 19

---

[1] An analytic proof of (20-3) and (20-4) may be based on (20-1) and (20-2).

Products of even and odd functions obey the rules

(even)(even) = even   (even)(odd) = odd   (odd)(odd) = even

which correspond to the familiar rules

$$(+1)(+1) = +1 \qquad (+1)(-1) = -1 \qquad (-1)(-1) = +1.$$

for proof, let $F(x) = f(x)g(x)$, where $f(x)$ and $g(x)$ are even. Then

$$F(-x) = f(-x)g(-x) = f(x)g(x) = F(x)$$

which shows that the product $f(x)g(x)$ is even. The other two relations are verified similarly. As an example, the product $\cos nx \sin mx$ is odd because $\cos nx$ is even and $\sin mx$ is odd. Hence, (20-4) gives

$$\int_{-a}^{a} \cos mx \sin nx \, dx = 0$$

without detailed calculation.

The application of these results is facilitated by the following theorem:

THEOREM. *If $f(x)$ defined in the interval $-\pi < x < \pi$ is even, the Fourier series has cosine terms only and the coefficients are given by*

$$a_n = \frac{2}{\pi} \int_0^{\pi} f(x) \cos nx \, dx \qquad b_n = 0. \tag{20-5}$$

If $f(x)$ is odd, the series has sine terms only and the coefficients are given by

$$b_n = \frac{2}{\pi} \int_0^{\pi} f(x) \sin nx \, dx \qquad a_n = 0. \tag{20-6}$$

To see this, let $f(x)$ be even. Then $f(x) \cos nx$ is the product of an even function times an even function and so is even. Therefore,

$$a_n = \frac{1}{\pi} \int_{-\pi}^{\pi} f(x) \cos nx \, dx = \frac{2}{\pi} \int_0^{\pi} f(x) \cos nx \, dx$$

by (20-3). On the other hand, $f(x) \sin nx$ is an even function times an odd function and so is odd. By (20-4),

$$b_n = \int_{-\pi}^{\pi} f(x) \sin nx \, dx = 0.$$

The result (20-6) is established similarly.

For example, let $f(x) = x$ for $-\pi < x < \pi$ (Fig. 20). Since this function is odd, the Fourier series reduces to a sine series, and we need not bother to write or calculate the cosine terms. It was found in Sec. 18 that

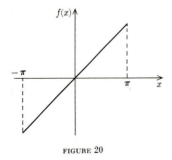

FIGURE 20

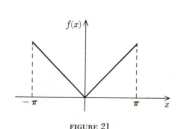

FIGURE 21

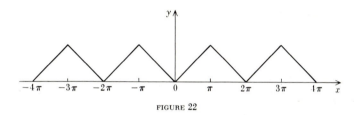

FIGURE 22

$$x = 2 \left(\sin x - \frac{\sin 2x}{2} + \frac{\sin 3x}{3} - \cdots \right). \tag{20-7}$$

As another example, let $f(x) = |x|$ for $-\pi < x < \pi$ (Fig. 21). In this case $f(x)$ is even, so that $b_n = 0$ and

$$a_n = \frac{1}{\pi} \int_{-\pi}^{\pi} |x| \cos nx \, dx = \frac{2}{\pi} \int_0^{\pi} x \cos nx \, dx$$

by use of (20-5). The integration gives $a_0 = \pi$ and

$$a_n = \frac{2}{n^2 \pi} [(-1)^n - 1].$$

Hence, for $-\pi < x < \pi$,

$$|x| = \frac{\pi}{2} - \frac{4}{\pi} \left(\frac{\cos x}{1} + \frac{\cos 3x}{3^2} + \frac{\cos 5x}{5^2} + \cdots \right). \tag{20-8}$$

The function to which this series converges is illustrated in Fig. 22, and the sum of the series (20-7) is presented graphically in Fig. 11.

Since $|x| = x$ for $x \geq 0$, the two series (20-7) and (20-8) converge to the same function x when $0 \leq x < \pi$. The first expression (20-7) is called the *Fourier sine series* for x, and (20-8) is the *Fourier cosine series*. Any function $f(x)$ defined in $(0,\pi)$ which satisfies the Dirichlet conditions can be expanded in a sine series and in a cosine series on $0 < x < \pi$. To obtain a sine series, we extend $f(x)$ over the interval $-\pi < x < 0$ in such a way that the extended function is odd. That is, we define

$$F(x) = f(x) \qquad \text{on } 0 < x < \pi$$
$$F(x) = -f(|x|) \qquad \text{on } -\pi < x < 0.$$

The Fourier series for $F(x)$ consists of sine terms only, since $F(x)$ is odd. And the coefficients are given by (20-6) because $F(x) = f(x)$ on the interval $0 < x < \pi$. Similarly, if it is desired to obtain a cosine series for $f(x)$ on $0 < x < \pi$, the coefficients are given by (20-5).

Example.   Obtain a cosine series and also a sine series for $\sin x$.

For the cosine series, (20-5) gives $b_n = 0$ and, after a short calculation,

$$a_n = \frac{2}{\pi} \int_0^{\pi} \sin x \cos nx \, dx = \frac{2(1 + \cos n\pi)}{\pi(1 - n^2)} \qquad n \neq 1.$$

For $n = 1$ the result of the integration is zero; hence

$$\sin x = \frac{2}{\pi} - \frac{4}{\pi} \left(\frac{\cos 2x}{2^2 - 1} + \frac{\cos 4x}{4^2 - 1} + \frac{\cos 6x}{6^2 - 1} + \cdots \right)$$

when $0 < x < \pi$. Since the sum of the series is an even function, it converges to $|\sin x|$ rather than $\sin x$ when $-\pi \leq x \leq 0$. This shows, by periodicity, that the series converges to $|\sin x|$ for all values of x.

To obtain a sine series, (20-6) gives $a_n = 0$ and

$$b_n = \frac{2}{\pi} \int_0^\pi \sin x \sin nx \, dx = \begin{cases} 0 & \text{for } n \geq 2 \\ 1 & \text{for } n = 1. \end{cases}$$

Hence the Fourier sine series for $\sin x$ is $\sin x$, just as one would expect. That this is not a coincidence is shown by the following:

UNIQUENESS THEOREM. If two trigonometric series of the form (18-1) converge to the same sum for all values of x, then corresponding coefficients are equal.[1]

PROBLEMS

1. Classify the following functions as even, odd, or neither:

$$x^2, \quad x \sin x, \quad x^3 \cos nx, \quad x^4, \quad e^x, \quad g(x^2), \quad xg(x^2), \quad (\sin x)^2, \quad \log \frac{1+x}{1-x}.$$

2. Prove that any function defined on $(-\pi, \pi)$ can be represented as the sum of an even function and an odd function. *Hint:* $f(x) = \frac{1}{2}[f(x) + f(-x)] + \frac{1}{2}[f(x) - f(-x)]$.

3. By taking $f(x) = \pi/4$ in (20-6), show that

$$\frac{\pi}{4} = \sin x + \frac{\sin 3x}{3} + \frac{\sin 5x}{5} + \cdots \qquad 0 < x < \pi.$$

4. A function is defined by $f(x) = \pi$ for $0 < x < \pi/2$ and $f(x) = 0$ elsewhere in $(-\pi, \pi)$. Find the Fourier series, the Fourier sine series, and the Fourier cosine series. In each case, sketch the graph of the sum of the series for $-4\pi \leq x < 4\pi$.

5. By taking $f(x) = x^2$ in (20-5), show that

$$x^2 = \frac{\pi^2}{3} + 4 \sum_{n=1}^{\infty} (-1)^n \frac{\cos nx}{n^2}$$

for $-\pi < x < \pi$, and deduce that

$$\frac{\pi^2}{12} = \frac{1}{1^2} - \frac{1}{2^2} + \frac{1}{3^2} - \frac{1}{4^2} + \cdots.$$

6. Show that if $f(x) = x$ for $0 < x < \pi/2$ and $f(x) = \pi - x$ for $\pi/2 < x < \pi$ then

$$f(x) = \frac{\pi}{4} - \frac{2}{\pi} \left(\frac{\cos 2x}{1^2} + \frac{\cos 6x}{3^2} + \frac{\cos 10x}{5^2} + \cdots \right).$$

7. Show that for $-\pi \leq x \leq \pi$

$$\cos \alpha x = \frac{\sin \pi\alpha}{\pi\alpha} + \sum_{n=1}^{\infty} (-1)^n \frac{2\alpha \sin \pi\alpha}{\pi(\alpha^2 - n^2)} \cos nx$$

when α is not an integer. Deduce

$$\cot \pi\alpha = \frac{1}{\pi} \left(\frac{1}{\alpha} - \sum_{n=1}^{\infty} \frac{2\alpha}{n^2 - \alpha^2} \right).$$

8. Let $f(x) = 1/2\delta$ for $|x| < \delta$, where δ is a small positive constant, and let $f(x) = 0$ elsewhere in $(-\pi, \pi)$. Obtain the Fourier series

$$\pi f(x) = \frac{1}{2} + \sum_{n=1}^{\infty} \frac{\sin n\delta}{n\delta} \cos nx.$$

Plot the graph of the sum of the series, and discuss its behavior as $\delta \to 0$.

---

[1] This theorem is due chiefly to Riemann. It is much deeper than the analogous statement for power series, and the proof would be out of place here. See Titchmarsh, *op. cit.*, pp. 427–432.

9. (a) Show that $f(x)$ is even on a given interval $(-a,a)$ if, and only if, $2f(x) \equiv f(x) + f(-x)$ on this interval, and thus show that the derivative of an even differentiable function is odd.   Illustrate graphically.   (b) Using part a, show that an even polynomial $p(x)$ can contain only even powers of x.

10. Formulate and solve an analog of Prob. 9 for odd functions.

11. By uniqueness, obtain Fourier cosine series for $\sin^2 x$ and $\cos^2 x$ valid on $(-\pi,\pi)$.

---

21. Extension of the Interval.   The methods developed up to this point restrict the interval of expansion to $(-\pi,\pi)$.   In many problems it is desired to develop $f(x)$ in a Fourier series that will be valid over a wider interval.   By letting the length of the interval increase indefinitely one may expect to get an expansion valid for all x.

To obtain an expansion valid on the interval $(-l,l)$, change the variable from x to lz/π.   If $f(x)$ satisfies the Dirichlet conditions on $(-l,l)$, the function $f(lz/\pi)$ can be developed in a Fourier series in z,

$$f\left(\frac{lz}{\pi}\right) = \frac{a_0}{2} + \sum_{n=1}^{\infty} a_n \cos nz + \sum_{n=1}^{\infty} b_n \sin nz \qquad (21\text{-}1)$$

for $-\pi \leq z < \pi$.   Since $z = \pi x/l$, the series (21-1) becomes

$$f(x) = \frac{a_0}{2} + \sum_{n=1}^{\infty} a_n \cos \frac{n\pi x}{l} + \sum_{n=1}^{\infty} b_n \sin \frac{n\pi x}{l}. \qquad (21\text{-}2)$$

By applying (18-4) to the series (21-1), we see that

$$a_n = \frac{1}{\pi} \int_{-\pi}^{\pi} f\left(\frac{lz}{\pi}\right) \cos nz \, dz = \frac{1}{l} \int_{-l}^{l} f(x) \cos \frac{n\pi x}{l} \, dx$$

and

$$b_n = \frac{1}{\pi} \int_{-\pi}^{\pi} f\left(\frac{lz}{\pi}\right) \sin nz \, dz = \frac{1}{l} \int_{-l}^{l} f(x) \sin \frac{n\pi x}{l} \, dx.$$

As an illustration we develop $f(x)$ in Fourier series in the interval $(-2,2)$ if $f(x) = 0$ for $-2 < x < 0$ and $f(x) = 1$ for $0 < x < 2$.   Here

$$a_0 = \tfrac{1}{2}\left(\int_{-2}^{0} 0 \cdot dx + \int_{0}^{2} 1 \cdot dx \right) = 1$$

$$a_n = \tfrac{1}{2}\left(\int_{-2}^{0} 0 \cdot \cos \frac{n\pi x}{2} \, dx + \int_{0}^{2} 1 \cdot \cos \frac{n\pi x}{2} \, dx \right) = \frac{1}{n\pi} \sin \frac{n\pi x}{2} \Big|_{0}^{2} = 0$$

$$b_n = \tfrac{1}{2}\left(\int_{-2}^{0} 0 \cdot \sin \frac{n\pi x}{2} \, dx + \int_{0}^{2} 1 \cdot \sin \frac{n\pi x}{2} \, dx \right) = \frac{1}{n\pi} (1 - \cos n\pi).$$

Therefore,
$$f(x) = \frac{1}{2} + \frac{2}{\pi}\left(\sin \frac{\pi x}{2} + \frac{1}{3} \sin \frac{3\pi x}{2} + \frac{1}{5} \sin \frac{5\pi x}{2} + \cdots \right).$$

If n is any integer, then

$$\cos \frac{n\pi(x + 2l)}{l} = \cos \left(\frac{n\pi x}{l} + 2n\pi \right) = \cos \frac{n\pi x}{l}$$

and similarly for sines.   Hence, each term of the series (21-2) has period $2l$, and therefore the sum also has period $2l$.   For this reason the sum cannot represent an arbitrary function on $(-\infty,\infty)$; it represents periodic functions only.

Subject to the Dirichlet conditions, however, the function can be chosen arbitrarily on the interval $(-l,l)$, and it is natural to inquire if a representation for arbitrary functions on $(-\infty,\infty)$ might be obtained by letting $l \to \infty$. We shall see that such a representation is possible. The process leads to the *Fourier integral theorem*, which has many practical applications.

Assume that $f(x)$ satisfies the Dirichlet conditions in every interval $(-l,l)$ (no matter how large) and that the integral

$$M = \int_{-\infty}^{\infty} |f(x)|\, dx$$

converges. As we have just seen, $f(x)$ is given by (21-2), where[1]

$$a_n = \frac{1}{l}\int_{-l}^{l} f(t) \cos \frac{n\pi t}{l}\, dt \qquad b_n = \frac{1}{l}\int_{-l}^{l} f(t) \sin \frac{n\pi t}{l}\, dt.$$

Substituting these values of the coefficients into (21-2) gives

$$f(x) = \frac{1}{2l}\int_{-l}^{l} f(t)\, dt + \frac{1}{l}\sum_{n=1}^{\infty} \int_{-l}^{l} f(t) \cos \frac{n\pi(t-x)}{l}\, dt \qquad (21\text{-}3)$$

when we recall that

$$\cos\frac{n\pi t}{l}\cos\frac{n\pi x}{l} + \sin\frac{n\pi t}{l}\sin\frac{n\pi x}{l} = \cos\frac{n\pi(t-x)}{l}.$$

Since $\int_{-\infty}^{\infty} |f(x)|\, dx$ is assumed to be convergent,

$$\left|\frac{1}{2l}\int_{-l}^{l} f(t)\, dt\right| \le \frac{1}{2l}\int_{-l}^{l} |f(t)|\, dt \le \frac{M}{2l}$$

which obviously tends to zero as l is allowed to increase indefinitely. Also, if the interval $(-l,l)$ is made large enough, the quantity π/l, which appears in the integrands of the sum, can be made as small as desired. Therefore, the sum in (21-3) can be written as

$$\frac{1}{\pi}\sum_{n=1}^{\infty}\left[\int_{-l}^{l} f(t) \cos n\,\Delta\omega(t-x)\, dt\right]\Delta\omega \qquad (21\text{-}4)$$

where $\Delta\omega = \pi/l$ is small.

This sum suggests the definition of the definite integral of the function

$$F(\omega) = \int_{-l}^{l} f(t) \cos \omega(t-x)\, dt$$

in which the values of the function $F(\omega)$ are calculated at the points $n\Delta\omega$. For large values of l

$$\int_{-l}^{l} f(t) \cos \omega(t-x)\, dt$$

differs little from

$$\int_{-\infty}^{\infty} f(t) \cos \omega(t-x)\, dt$$

and it appears plausible that, as l increases indefinitely, the sum (21-4) will approach the limit

$$\frac{1}{\pi}\int_{0}^{\infty} d\omega \int_{-\infty}^{\infty} f(t) \cos \omega(t-x)\, dt.$$

If such is the case, (21-3) can be written as

----

[1] We use t as the variable of integration to avoid confusion with the x in (21-2). If $f(x)$ is discontinuous at $x = x_0$, the left side of (21-2) means $\frac{1}{2}[f(x_0+) + f(x_0-)]$.

$$f(x) = \frac{1}{\pi} \int_0^\infty d\omega \int_{-\infty}^\infty f(t) \cos \omega(t - x)\, dt. \tag{21-5}$$

The foregoing discussion is heuristic and cannot be regarded as a rigorous proof. However, the validity of formula (21-5) can be established rigorously[1] if the function $f(x)$ satisfies the conditions enunciated above. The integral (21-5) bears the name of the *Fourier integral*.

Formula (21-5) assumes a simpler form if $f(x)$ is an even or an odd function. Expanding the integrand of (21-5) gives

$$\frac{1}{\pi} \int_0^\infty d\omega \left[\int_{-\infty}^\infty f(t) \cos \omega t \cos \omega x\, dt + \int_{-\infty}^\infty f(t) \sin \omega t \sin \omega x\, dt \right]$$

for the right-hand member. If $f(t)$ is odd, then $f(t) \cos \omega t$ is an odd function times an even function and hence is odd. Similarly, $f(t) \sin \omega t$ is even when $f(t)$ is odd. Upon applying (20-4) to the first integral in the foregoing expression and (20-3) to the second integral, we see that

$$f(x) = \frac{2}{\pi} \int_0^\infty d\omega \int_0^\infty f(t) \sin \omega t \sin \omega x\, dt \tag{21-6}$$

when $f(x)$ is odd. A similar argument shows that, if $f(x)$ is even, then

$$f(x) = \frac{2}{\pi} \int_0^\infty d\omega \int_0^\infty f(t) \cos \omega t \cos \omega x\, dt. \tag{21-7}$$

If $f(x)$ is defined only in the interval $(0,\infty)$, both (21-6) and (21-7) may be used, since $f(x)$ may be thought to be defined in $(-\infty,0)$ so as to make it either odd or even. This corresponds to the fact that a function given on $(0,\pi)$ may be expanded in either a sine series or a cosine series.

Since the Fourier series converges to $\frac{1}{2}[f(x+) + f(x-)]$ at points of discontinuity, the Fourier integral does also. In particular, for an odd function the integral converges to zero at $x = 0$; this fact is verified by setting $x = 0$ in (21-6).

Example.   By (21-7) obtain the formula

$$\int_0^\infty \frac{\sin \omega \cos \omega x}{\omega}\, d\omega = \begin{cases} \pi/2 & \text{if } 0 \le x < 1 \\ \pi/4 & \text{if } x = 1 \\ 0 & \text{if } x > 1. \end{cases}$$

We choose $f(x) = 1$ for $0 \le x < 1$ and $f(x) = 0$ for $x > 1$. Then

$$\int_0^\infty f(t) \cos \omega t\, dt = \int_0^1 \cos \omega t\, dt = \frac{\sin \omega}{\omega} \qquad \omega \ne 0.$$

Substitution into (21-7) gives

$$\int_0^\infty \frac{\sin \omega}{\omega} \cos \omega x\, d\omega = \frac{\pi}{2} f(x)$$

after multiplying by $\pi/2$. Upon recalling the definition of $f(x)$, we see that the desired result is obtained for $0 \le x < 1$ and for $x > 1$. The fact that the integral is $\pi/4$ when $x = 1$ follows from

$$\frac{1}{2} = \frac{f(1-) + f(1+)}{2}.$$

[1] See H. S. Carslaw, "Fourier's Series and Integrals," pp. 283–294, The Macmillan Company, New York, 1921, or Titchmarsh, *op. cit.*, p. 433.

PROBLEMS

1. If $f(x)$ is an odd function on $(-l,l)$, show that the Fourier series takes the form

$$f(x) = \sum_{n=1}^{\infty} b_n \sin \frac{n\pi x}{l} \qquad b_n = \frac{2}{l} \int_0^l f(x) \sin \frac{n\pi x}{l} \, dx.$$

Similarly, if $f(x)$ is even, show that

$$f(x) = \frac{a_0}{2} + \sum_{n=1}^{\infty} a_n \cos \frac{n\pi x}{l} \qquad a_n = \frac{2}{l} \int_0^l f(x) \cos \frac{n\pi x}{l} \, dx.$$

2. Expand the function defined by $f(x) = 1$ on $(0,2)$ and $f(x) = -1$ on $(-2,0)$.

3. Expand $f(x) = |x|$ in the interval $(-1,1)$.

4. Expand $f(x) = \cos \pi x$ in the interval $(-1,1)$.

5. Find the expansion in the series of cosines, if

$$f(x) = 1 \qquad 0 < x < \pi; \qquad f(x) = 0 \qquad \pi < x < 2\pi.$$

Hint: Regard $f(x)$ as being an even function.

6. Expand the function defined by

$$f(x) = \tfrac{1}{4} - x \qquad 0 < x < \tfrac{1}{2}; \qquad f(x) = x - \tfrac{3}{4} \qquad \tfrac{1}{2} < x < 1.$$

7. Show that

$$\tfrac{1}{2}l - x = \frac{l}{\pi} \sum_{n=1}^{\infty} \frac{1}{n} \sin \frac{2n\pi x}{l} \qquad 0 < x < l.$$

8. Taking $f(x) = e^{-\beta x}$ in (21-6) and (21-7), show that if $\beta > 0$

$$\int_0^\infty \frac{\alpha \sin \alpha x}{\alpha^2 + \beta^2} \, d\alpha = \frac{\pi}{2} e^{-\beta x} \qquad \int_0^\infty \frac{\cos \alpha x}{\alpha^2 + \beta^2} \, d\alpha = \frac{\pi}{2\beta} e^{-\beta x}.$$

9. An *integral equation* is an equation in which an unknown function appears under an integral sign. If $F(t)$ is known and $f(x)$ is to be found, the *integral equation of Fourier* is

$$\int_0^\infty f(x) \cos xt \, dx = F(t).$$

(*a*) Using (21-7), show that a solution is given by

$$f(x) = \frac{2}{\pi} \int_0^\infty F(t) \cos xt \, dt.$$

(*b*) Solve a similar integral equation by use of (21-6).

---

22. Complex Fourier Series. Fourier Transform.   The Fourier series

$$f(x) = \frac{a_0}{2} + \sum_{n=1}^{\infty} (a_n \cos nx + b_n \sin nx) \tag{22-1}$$

with $\qquad a_n = \frac{1}{\pi} \int_{-\pi}^{\pi} f(t) \cos nt \, dt \qquad b_n = \frac{1}{\pi} \int_{-\pi}^{\pi} f(t) \sin nt \, dt$

can be written, with the aid of the Euler formula (Sec. 17)

$$e^{iu} = \cos u + i \sin u \tag{22-2}$$

in an equivalent form, namely,

$$f(x) = \sum_{n=-\infty}^{\infty} c_n e^{inx}. \qquad (22\text{-}3)$$

The coefficients c_n are defined by the equation

$$c_n = \frac{1}{2\pi} \int_{-\pi}^{\pi} f(t) e^{-int}\, dt \qquad (22\text{-}4)$$

and the limit is interpreted by taking the sum from $-N$ to N, letting $N \to \infty$. Thus, the index n runs through all positive and negative integral values including 0.

Proof of this equivalence is developed in Prob. 1. But if the series (22-3) is uniformly convergent we can obtain (22-4) more simply as follows: Replace x by t and the dummy index n by m, so that

$$f(t) = \sum_{m=-\infty}^{\infty} c_m e^{imt}.$$

Multiplying by e^{-int} gives

$$f(t) e^{-int} = \sum_{m=-\infty}^{\infty} c_m e^{i(m-n)t}.$$

If we now integrate from $-\pi$ to π, the terms with $m \neq n$ integrate to 0, by Sec. 17, Example 3, and the term with $m = n$ gives $2\pi c_n$. The result is (22-4).

To illustrate the use of (22-4), consider the function $f(x) = e^{\alpha x}$ on $(-\pi, \pi)$. Here,

$$2\pi c_n = \int_{-\pi}^{\pi} e^{\alpha t} e^{-int}\, dt = \int_{-\pi}^{\pi} e^{(\alpha-in)t}\, dt = \frac{e^{(\alpha-in)t}}{\alpha-in} \Big|_{-\pi}^{\pi} = \frac{e^{\alpha\pi} e^{-in\pi} - e^{-\alpha\pi} e^{in\pi}}{\alpha-in}.$$

Since (22-2) gives $e^{\pm in\pi} = \cos(\pm n\pi) = (-1)^n$, we obtain

$$c_n = \frac{e^{\alpha\pi} - e^{-\alpha\pi}}{2\pi} \frac{(-1)^n}{\alpha-in} = \frac{\sinh \pi\alpha}{\pi} \frac{(-1)^n}{\alpha^2+n^2} (\alpha + in)$$

and hence, by (22-3),

$$e^{\alpha x} = \frac{\sinh \pi\alpha}{\pi} \sum_{n=-\infty}^{\infty} \frac{(-1)^n}{\alpha^2+n^2} (\alpha + in) e^{inx}. \qquad (22\text{-}5)$$

The methods of Sec. 21 yield

$$f(x) = \sum_{n=-\infty}^{\infty} c_n e^{in\pi x/l} \qquad \text{with } c_n = \frac{1}{2l} \int_{-l}^{l} f(t) e^{-in\pi t/l}\, dt \qquad (22\text{-}6)$$

for the expansion on an arbitrary interval $(-l, l)$. Upon letting $l \to \infty$, we obtain the Fourier integral theorem in the form

$$f(x) = \lim_{A\to\infty} \frac{1}{2\pi} \int_{-A}^{A} d\omega \int_{-\infty}^{\infty} f(t) e^{i\omega(x-t)}\, dt \qquad (22\text{-}7)$$

when $f(x)$ satisfies the conditions postulated for (21-5).

If

$$g(\omega) = \frac{1}{\sqrt{2\pi}} \int_{-\infty}^{\infty} e^{-i\omega t} f(t)\, dt \qquad (22\text{-}8)$$

then (22-7) gives, after renaming some of the variables,

$$f(t) = \lim_{A\to\infty} \frac{1}{\sqrt{2\pi}} \int_{-A}^{A} e^{i\omega t} g(\omega)\, d\omega. \qquad (22\text{-}9)$$

The transform $\mathbf{T}$ defined by

$$\mathbf{T}(f) = \frac{1}{\sqrt{2\pi}} \int_{-\infty}^{\infty} e^{-i\omega t} f(t)\, dt$$

is called the *Fourier transform;* it is one of the most powerful tools in the whole repertoire of modern analysis.

Although the formulas (22-8) and (22-9) are similar, the conditions on the functions f and g are quite different. A more symmetric theory can be based on a type of convergence known as *mean convergence.* Since an independent discussion of mean convergence is given in Sec. 24, with proofs, we shall be brief in the purely descriptive treatment given here.

Let $g_A(t)$ be an integrable function of t, on each finite interval, for each value of the parameter A. It is said that $g_A(t)$ *converges in mean* to $g(t)$, and we write

$$g(t) = \operatorname*{l.i.m.}_{A\to\infty} g_A(t) \tag{22-10}$$

if it is true that

$$\lim_{A\to\infty} \int_{-\infty}^{\infty} |g(t) - g_A(t)|^2 \, dt = 0. \tag{22-11}$$

The initials l.i.m. in (22-10) stand for "limit in mean." As an illustration,

$$g(t) = \operatorname*{l.i.m.}_{A\to\infty} \int_{-A}^{A} e^{-i\omega t} f(t) \, dt \tag{22-12}$$

means that (22-11) holds with $g_A(t)$ replaced by the integral on the right of (22-12). One can write $g(t)$ in (22-12) as an integral from $-\infty$ to ∞ if it is stated that the equation holds in the sense of mean convergence.

We now state the following important theorem:[1]

PLANCHEREL'S THEOREM. *Let $f(t)$ and $g(\omega)$ be integrable on every finite interval, and suppose*

$$\int_{-\infty}^{\infty} |f(t)|^2 \, dt \qquad or \qquad \int_{-\infty}^{\infty} |g(\omega)|^2 \, d\omega \tag{22-13}$$

is finite. Then if either of the equations

$$g(\omega) = \frac{1}{\sqrt{2\pi}} \int_{-\infty}^{\infty} f(t) e^{-i\omega t} \, dt, \qquad f(t) = \frac{1}{\sqrt{2\pi}} \int_{-\infty}^{\infty} g(\omega) e^{i\omega t} \, d\omega$$

holds in the sense of mean convergence, so does the other, and the two integrals (22-13) are equal.

This particular wording is valid for the ordinary Riemann integral. If the Lebesgue integral (Appendix B) is used, the initial hypothesis "Let $f(t)$ and $g(\omega)$ be integrable . . ." can be replaced by "Let $f(t)$ or $g(\omega)$ be integrable . . . ." The integrability of the other function is then an automatic consequence.

The statement that the two integrals (22-13) are equal is called *Parseval's equality.* It holds in the sense of ordinary convergence.

Example 1. Expand $\cos \beta x$ in complex Fourier series, where β is any constant not an integer. By the Euler formula,

$$\cos \beta x = \tfrac{1}{2}(e^{i\beta x} + e^{-i\beta x}). \tag{22-14}$$

The Fourier series for $e^{i\beta x}$ can be obtained by setting $\alpha = i\beta$ in (22-5). Since $\sinh \pi i\beta = i \sin \pi\beta$, the result is

$$e^{i\beta x} = \frac{\sin \pi\beta}{\pi} \sum_{n=-\infty}^{\infty} \frac{(-1)^n}{\beta^2 - n^2} (\beta + n) e^{inx}.$$

[1] Titchmarsh, *op. cit.,* pp. 436–439.

Replacing β by $-\beta$ gives the series for $e^{-i\beta x}$, and taking half the sum of these two series, we get

$$\cos \beta x = \frac{\sin \pi \beta}{\pi} \sum_{n=-\infty}^{\infty} (-1)^n \frac{\beta}{\beta^2 - n^2} e^{inx} \qquad -\pi < x < \pi$$

by (22-14). This calculation is justified because the formula (22-4) depends linearly on the function f, as was the case for real series.

Example 2.   Discuss the Fourier transform for the function defined by

$$f(t) = 1 \qquad \text{for } |t| < a \qquad f(t) = 0 \qquad \text{for } |t| \geq a$$

where a is constant. Here

$$\sqrt{2\pi} \, \mathbf{T}f = \int_{-a}^{a} e^{-i\omega t} f(t) \, dt = \frac{e^{-i\omega t}}{-i\omega} \bigg|_{-a}^{a}$$

where the first equality follows because $f(t) = 0$ for $|t| > a$, and the second because $f(t) = 1$ for $|t| < a$. Hence,

$$\mathbf{T}f = \sqrt{\frac{2}{\pi}} \frac{\sin \omega a}{\omega}.$$

The reader familiar with circuit theory will recognize this as a circuit response to a rectangular pulse, and the reader familiar with physical optics will recognize that we have computed the diffraction pattern for a uniformly illuminated aperture.

The Fourier inversion formula gives

$$\frac{1}{\pi} \int_{-\infty}^{\infty} \frac{\sin \omega a}{\omega} e^{i\omega t} \, d\omega = \begin{cases} 1 \text{ for } |t| < a \\ \frac{1}{2} \text{ for } t = a \\ 0 \text{ for } |t| > a \end{cases}$$

and Parseval's equality is

$$a = \frac{1}{\pi} \int_{-\infty}^{\infty} \frac{\sin^2 \omega a}{\omega^2} \, d\omega.$$

This can be interpreted physically as stating that the energy in the signal equals the energy in the spectrum.

It may be noted that neither $\omega^{-1} \sin \omega$ nor $\omega^{-2} \sin^2 \omega$ has an indefinite integral that is an elementary function, so that these results cannot be obtained by the usual methods of elementary calculus.

PROBLEMS

1. (*a*) By using (22-2) in (22-4), show that

$$2c_n = a_n - ib_n, \qquad 2c_0 = a_0, \qquad 2c_{-n} = a_n + ib_n.$$

(*b*) From these relations, obtain

$$a_n = c_n + c_{-n}, \qquad 2c_0 = a_0, \qquad b_n = i(c_n - c_{-n}).$$

(*c*) Demonstrate equivalence of the real and complex Fourier series. *Hint:* Write the latter as

$$c_0 + \sum_{n=1}^{\infty} (c_n e^{inx} + c_{-n} e^{-inx}).$$

2. By applying Prob. 1*b* to (22-5), obtain the real Fourier series for $e^{\alpha x}$.

3. By setting $x = 0$ in (22-5), obtain the expansion

$$\frac{\pi}{\pi \sinh \alpha \pi} = \sum_{n=-\infty}^{\infty} \frac{(-1)^n}{\alpha^2 + n^2}.$$

4. Deduce the complex Fourier series for $\sin \beta x$ from (22-5).

5. Deduce the real Fourier series for $\sin \beta x$ and $\cos \beta x$ from the result of Prob. 4 and Example 1, respectively.

6. In the (x,y) plane, plot the points

$$(-\pi,0), \qquad (-\delta,0), \qquad (0,\pi/\delta), \qquad (\delta,0), \qquad (\pi,0)$$

where δ is a small positive constant. If the graph of $y = f(x)$ consists of straight lines joining these points, in order, find the complex Fourier series of $f(x)$, and discuss as $\delta \to 0$.

7. Obtain complex series for selected problems in Secs. 18 to 21, and use Prob. 1 to verify agreement with the given answers.

8. Compute the complex Fourier transform Tf, obtain sine and cosine transform pairs by separation into real and imaginary parts, and write the Parseval equality for the following functions $f(t)$:

(a) $f(t) = 0, -\infty < t < 0, f(t) = e^{-t}, t > 0$;
(b) $f(t) = 1, 0 < t < 1, f(t) = 0$ elsewhere;
(c) $f(t) = 1 - |t|$ for $|t| < 1, f(t) = 0$ elsewhere.

---

23. Orthogonal Functions.   A sequence of functions $\theta_n(x)$ is said to be *orthogonal* on the interval (a,b) if

$$\int_a^b \theta_m(x)\theta_n(x)\,dx \begin{cases} = 0 & \text{for } m \neq n \\ \neq 0 & \text{for } m = n. \end{cases} \tag{23-1}$$

For example, the sequence

$$\theta_1(x) = \sin x, \qquad \theta_2(x) = \sin 2x, \qquad \ldots, \qquad \theta_n(x) = \sin nx, \qquad \ldots$$

is orthogonal on $(0,\pi)$ because

$$\int_0^\pi \theta_m(x)\theta_n(x)\,dx = \int_0^\pi \sin mx \sin nx\,dx = \begin{cases} 0 & \text{for } m \neq n \\ \pi/2 & \text{for } m = n. \end{cases}$$

The sequence

$$1, \sin x, \cos x, \sin 2x, \cos 2x, \ldots \tag{23-2}$$

is orthogonal on $(0,2\pi)$, though not on $(0,\pi)$.

In the foregoing sections the functions (23-2) were used to form Fourier series. Actually, one can form series analogous to Fourier series by means of any orthogonal set. These generalized Fourier series are an indispensable aid in electromagnetic theory, acoustics, heat flow, and many other branches of mathematical physics. (Cf. Chap. 7.)

The formula for Fourier coefficients is especially simple if the integral (23-1) has the value 1 for $m = n$. The functions $\theta_n(x)$ are then said to be *normalized*, and $\{\theta_n(x)\}$ is called an *orthonormal* set. If

$$\int_a^b [\theta_n(x)]^2\,dx = A_n$$

in (23-1), it is easily seen that the functions

$$\phi_n(x) = (A_n)^{-\frac12}\theta_n(x)$$

are orthonormal; in other words,

$$\int_a^b \phi_m(x)\phi_n(x)\,dx \begin{cases} = 0 & \text{for } m \neq n \\ = 1 & \text{for } m = n. \end{cases} \tag{23-3}$$

For example, since

$$\int_0^{2\pi} 1\,dx = 2\pi \qquad \int_0^{2\pi} \sin^2 nx\,dx = \pi \qquad \int_0^{2\pi} \cos^2 nx\,dx = \pi$$

for $n \geq 1$, the orthonormal set corresponding to the orthogonal set (23-2) is

$$(2\pi)^{-\frac{1}{2}}, \ \pi^{-\frac{1}{2}} \sin x, \ \pi^{-\frac{1}{2}} \cos x, \ \dots, \ \pi^{-\frac{1}{2}} \sin nx, \ \pi^{-\frac{1}{2}} \cos nx, \ \dots .$$

The product of two different functions in this set gives zero but the square of each function gives 1, when integrated from zero to 2π.

Let $\{\phi_n(x)\}$ be an orthonormal set of functions on (a,b), and suppose that another function $f(x)$ is to be expanded in the form

$$f(x) = c_1\phi_1(x) + c_2\phi_2(x) + \cdots + c_n\phi_n(x) + \cdots . \tag{23-4}$$

To determine the coefficients c_n we multiply by $\phi_n(x)$, getting

$$f(x)\phi_n(x) = c_1\phi_1(x)\phi_n(x) + \cdots + c_n[\phi_n(x)]^2 + \cdots .$$

Here, the terms not written involve products $\phi_n(x)\phi_m(x)$ with $m \neq n$. If we formally integrate from a to b, these terms disappear, and hence

$$\int_a^b f(x)\phi_n(x) \ dx = \int_a^b c_n[\phi_n(x)]^2 \ dx = c_n. \tag{23-5}$$

According to Theorem III of Sec. 9, the term-by-term integration is justified when the series is uniformly convergent and the functions are continuous. The foregoing procedure shows that if $f(x)$ has an expansion of the desired type the coefficients c_n must be given by (23-5). In the following section (23-5) is obtained in a different manner, which does not assume uniform convergence.

The formula (23-5) is called the *Euler-Fourier formula*, the coefficients c_n are called the *Fourier coefficients* of $f(x)$ with respect to $\{\phi_n(x)\}$, and the resulting series (23-4) is called the *Fourier series* of $f(x)$ with respect to $\{\phi_n(x)\}$. The reader can verify that the foregoing results applied to the sequence (23-2) yield the ordinary Fourier series, as described in the foregoing sections.

Orthogonal sets of functions are obtained in practice by solving differential equations, as illustrated in the following example. Additional discussion can be found in Sec. 22 of Chap. 3.

Example. By use of the differential equation $y'' = -\lambda y$, show that the sequence (23-2) is orthogonal on $(-\pi,\pi)$.

Let $y_m = \sin mx$ or $\cos mx$ and $y_n = \sin nx$ or $\cos nx$, so that

$$y_m'' = -m^2 y_m \qquad y_n'' = -n^2 y_n.$$

If the first equation is multiplied by y_n, the second by y_m, and the results subtracted, we get

$$y_n y_m'' - y_n'' y_m = (n^2 - m^2) y_m y_n.$$

The left side is recognized as the derivative of $y_n y_m' - y_n' y_m$; hence, integrating from $-\pi$ to π,

$$(y_n y_m' - y_m y_n') \Big|_{-\pi}^{\pi} = (n^2 - m^2) \int_{-\pi}^{\pi} y_m y_n \ dx. \tag{23-6}$$

When m and n are integers, $y_n y_m' - y_m y_n'$ is periodic; it has the same value at $-\pi$ as at π and so the left side of (23-6) is zero. This yields the desired orthogonality except in the case $m = n$. If $m = n$, however, the relevant integral can be evaluated by inspection:

$$\int_{-\pi}^{\pi} \cos nx \sin nx \ dx = \int_{-\pi}^{\pi} \tfrac{1}{2} \sin 2nx \ dx = 0.$$

PROBLEMS

1. If $\{\phi_n\}$ is orthonormal on the interval $(0,1)$, and if $a > 0$, show that the set of functions $\psi_n(x) = a^{-\frac{1}{2}} \phi_n(x/a)$ is orthonormal on the interval $(0,a)$.

2. The distance between two functions is sometimes defined by

$$d(f,g) = \left\{ \int_a^b [f(x) - g(x)]^2 \ dx \right\}^{\frac{1}{2}}.$$

Show that the distance between any two distinct members of an orthonormal set is $\sqrt{2}$.   [Assume that the set is orthonormal on (a,b).]

3. If $f(x) = c_1\phi_1(x) + c_2\phi_2(x) + c_3\phi_3(x)$, where the c_k are constant and $\{\phi_k\}$ is an orthonormal set on (a,b), show that

$$\int_a^b [f(x)]^2\, dx = c_1{}^2 + c_2{}^2 + c_3{}^2.$$

4. If $\{\phi_n\}$ is orthonormal on (a,b), the nth *reproducing kernel* is

$$k_n(x,y) = \phi_1(x)\phi_1(y) + \phi_2(x)\phi_2(y) + \cdots + \phi_n(x)\phi_n(y).$$

Show that the equation

$$\int_a^b k_n(x,y)f(y)\, dy = f(x)$$

holds for all functions of the form

$$f(x) = c_1\phi_1(x) + c_2\phi_2(x) + \cdots + c_n\phi_n(x).$$

Hint: Verify the equation for $f = \phi_k$ and use linearity.

---

24. The Mean Convergence of Fourier Series.   If we try to approximate a function $f(x)$ by another function $p_n(x)$, the quantity

$$|f(x) - p_n(x)| \qquad \text{or} \qquad [f(x) - p_n(x)]^2 \tag{24-1}$$

gives a measure of the error in the approximation.   The sequence $\{p_n(x)\}$ converges to $f(x)$ whenever the expressions (24-1) approach zero as $n \to \infty$.

These measures of the error are appropriate for discussing convergence at any fixed point x.   But it is often useful to have a measure of error which applies simultaneously to a whole interval of x values, $a \le x \le b$.   Such a measure is easily found if we integrate (24-1) from a to b:

$$\int_a^b |f(x) - p_n(x)|\, dx \qquad \text{or} \qquad \int_a^b [f(x) - p_n(x)]^2\, dx. \tag{24-2}$$

These expressions are called the *mean error* and *mean-square error*, respectively.   If either expression (24-2) approaches zero as $n \to \infty$, the sequence $p_n(x)$ is said to converge *in mean* to $f(x)$ and the term *mean convergence* is used.   The terminology is appropriate because, if the integrals (24-2) are multiplied by $1/(b - a)$, the result is precisely the mean value of the corresponding expressions (24-1).

Even though (24-2) involves an integration that is not present in (24-1), for Fourier series it is much easier to discuss the mean-square error and the corresponding mean convergence than the ordinary convergence.   In the following discussion we use f and ϕ_n as abbreviations for $f(x)$ and $\phi_n(x)$, respectively, and assume that f and ϕ_n are integrable on $a \le x \le b$.   If the integrals are improper, the convergence of $\int_a^b f^2\, dx$ and $\int_a^b \phi_n{}^2\, dx$ is required.

Let $\{\phi_n(x)\}$ be a set of orthonormal functions on $a \le x \le b$, so that, as in the preceding section,

$$\int_a^b \phi_n(x)\phi_m(x)\, dx = \begin{cases} 0 & \text{for } m \ne n \\ 1 & \text{for } m = n. \end{cases} \tag{24-3}$$

We seek to approximate $f(x)$ by a linear combination of $\phi_n(x)$,

$$p_n(x) = a_1\phi_1(x) + a_2\phi_2(x) + \cdots + a_n\phi_n(x)$$

in such a way that the mean-square error (24-2) is minimum:

$$E \equiv \int_a^b [f - (a_1\phi_1 + \cdots + a_n\phi_n)]^2 \, dx = \min. \tag{24-4}$$

Upon expanding the term in brackets we see that (24-4) yields

$$\int_a^b f^2 \, dx - 2 \int_a^b (a_1\phi_1 + \cdots + a_n\phi_n) f \, dx + \int_a^b (a_1\phi_1 + \cdots + a_n\phi_n)^2 \, dx. \tag{24-5}$$

If the Fourier coefficients of f relative to ϕ_k are denoted by

$$c_k = \int_a^b f\phi_k \, dx$$

the second integral (24-5) is

$$\int_a^b (a_1\phi_1 + \cdots + a_n\phi_n) f \, dx = a_1c_1 + a_2c_2 + \cdots + a_nc_n.$$

The third integral (24-5) can be written

$$\int_a^b (a_1\phi_1 + \cdots + a_n\phi_n)(a_1\phi_1 + \cdots + a_n\phi_n) \, dx$$

$$= \int_a^b (a_1{}^2\phi_1{}^2 + a_2{}^2\phi_2{}^2 + \cdots + a_n{}^2\phi_n{}^2 + \cdots) \, dx$$

$$= a_1{}^2 + \cdots + a_n{}^2$$

where the second group of terms "$+ \cdots$" involves cross products $\phi_i\phi_j$ with $i \neq j$. By (24-3) these terms integrate to zero, and the expression reduces to the value indicated.

Hence, (24-5) yields

$$E \equiv \int_a^b f^2 \, dx - 2 \sum_{k=1}^n a_kc_k + \sum_{k=1}^n a_k{}^2 \tag{24-6}$$

for the mean-square error in the approximation. Inasmuch as

$$-2a_kc_k + a_k{}^2 \equiv -c_k{}^2 + (a_k - c_k)^2$$

the error E in (24-6) is also equal to

$$E = \int_a^b f^2 \, dx - \sum_{k=1}^n c_k{}^2 + \sum_{k=1}^n (a_k - c_k)^2 \tag{24-7}$$

and we have established a theorem of central importance:

THEOREM. *If $\{\phi_n\}$ is a set of orthonormal functions, the mean-square error (24-4) can be written in the form (24-7), where c_k are the Fourier coefficients of f relative to ϕ_k.*

From the two expressions (24-4) and (24-7), one obtains a number of interesting and significant theorems with the greatest ease. In the first place, the terms $(a_k - c_k)^2$ in (24-7) are positive unless $a_k = c_k$, in which case they are zero. Hence the choice of a_k that makes E minimum is obviously $a_k = c_k$, and we have the following:

COROLLARY 1. *The partial sums of the Fourier series*

$$c_1\phi_1 + \cdots + c_n\phi_n, \qquad c_k = \int_a^b f\phi_k \, dx$$

give a smaller mean-square error $\int_a^b (f - p_n)^2 \, dx$ than is given by any other linear combination

$$p_n = a_1\phi_1 + \cdots + a_n\phi_n.$$

Upon setting $a_k = c_k$ in (24-7), we see that the minimum value of the error is

$$\min E = \int_a^b f^2 \, dx - \sum_{k=1}^n c_k^2. \tag{24-8}$$

Now, the expression (24-4) shows that $E \geq 0$, because the integrand in (24-4), being a square, is not negative. Since $E \geq 0$ for all choices of a_k, it is clear that the minimum of E (which arises when $a_k = c_k$) is also ≥ 0. The expression (24-8) yields, then,

$$\int_a^b f^2 \, dx - \sum_{k=1}^n c_k^2 \geq 0 \qquad \text{or} \qquad \sum_{k=1}^n c_k^2 \leq \int_a^b f^2 \, dx.$$

Upon letting $n \to \infty$ we obtain, by the principle of monotone convergence:

COROLLARY 2. *If* $c_k = \int_a^b f \phi_k \, dx$ *are the Fourier coefficients of f relative to the orthonormal set ϕ_n, then the series Σc_k^2 converges and satisfies the Bessel inequality*

$$\sum_{k=1}^\infty c_k^2 \leq \int_a^b [f(x)]^2 \, dx. \tag{24-9}$$

Because the general term of a convergent series must approach zero (theorem of Sec. 3) we deduce the following from Corollary 2:

COROLLARY 3. *The Fourier coefficients* $c_n = \int_a^b f \phi_n \, dx$ *tend to zero as $n \to \infty$.*

For applications it is important to know whether or not the mean-square error approaches zero as $n \to \infty$. Evidently the error approaches zero, for some choice of the a_k's, only if the minimum error (24-8) does so. Letting $n \to \infty$ in (24-8) we get the *Parseval equality*

$$\int_a^b f^2 \, dx - \sum_{k=1}^\infty c_k^2 = 0$$

as the condition for zero error:

COROLLARY 4. *If f is approximated by the partial sums of its Fourier series, the mean-square error approaches zero as $n \to \infty$ if, and only if, Bessel's inequality (24-9) becomes Parseval's equality*

$$\sum_{n=1}^\infty c_n^2 = \int_a^b [f(x)]^2 \, dx. \tag{24-10}$$

In other words, the Fourier series converges to f in the mean-square sense if, and only if, (24-10) holds. If this happens for every choice of f, the set $\{\phi_n(x)\}$ is said to be *closed*. A closed set, then, is a set that can be used for mean-square approximation of arbitrary functions. It can be shown that the set of trigonometric functions $\{\cos nx, \sin nx\}$ is closed on $0 \leq x \leq 2\pi$, though the proof is too long for inclusion here.[1] A set $\{\phi_n(x)\}$ is said to be *complete* if there is no nontrivial function $f(x)$ which is orthogonal to all of the ϕ_n's. That is, the set is complete if

$$c_k = \int_a^b f(x) \phi_k(x) \, dx = 0 \qquad \text{for } k = 1, 2, 3, \ldots \tag{24-11}$$

[1] See Titchmarsh, *op. cit.*, p. 414.

implies that

$$\int_a^b [f(x)]^2 \, dx = 0. \qquad (24\text{-}12)$$

Now, whenever (24-10) holds, (24-11) yields (24-12) at once.   Hence we have:

COROLLARY 5.   Every closed set $\{\phi_n(x)\}$ is complete.

The converse is also true: *Every complete set is closed.* This converse, however, requires a more general integral than that of Riemann.   The generalized integral is known as the *Lebesgue* integral; it was first constructed to deal with this very problem.   A brief description of the Lebesgue integral is given in Appendix B.

The notions of *closure* and *completeness* have simple analogs in the elementary theory of vectors.   Thus, a set of vectors $\mathbf{V}_1$, $\mathbf{V}_2$, $\mathbf{V}_3$ is said to be *closed* if every vector $\mathbf{V}$ can be written in the form

$$\mathbf{V} = c_1\mathbf{V}_1 + c_2\mathbf{V}_2 + c_3\mathbf{V}_3$$

for some choice of the constants c_k.   The set of vectors $\mathbf{V}_1$, $\mathbf{V}_2$, $\mathbf{V}_3$ is said to be *complete* if there is no nontrivial vector orthogonal to all of them; that is, the set is complete if the condition

$$\mathbf{V} \cdot \mathbf{V}_k = 0 \qquad \text{for } k = 1, 2, 3$$

implies $\mathbf{V} \cdot \mathbf{V} = 0$.   In this setting, it is obvious that closure and completeness are equivalent, for both conditions simply state that the three vectors $\mathbf{V}_1$, $\mathbf{V}_2$, $\mathbf{V}_3$ are not coplanar.   These matters are taken up more fully in Chap. 4.

PROBLEMS

1. (a) Show that Parseval's equality takes the form

$$\frac{1}{\pi} \int_0^{2\pi} [f(x)]^2 \, dx = \tfrac{1}{2}a_0^2 + \sum_{n=1}^{\infty} (a_n^2 + b_n^2)$$

when $\phi_n(x)$ are the trigonometric functions on $(0,2\pi)$.   (b) Specialize to sine and cosine series.

2. It is desired to approximate 1 by

$$p(x) = a_1 \sin x + a_2 \sin 2x + a_3 \sin 3x$$

in such a way that $\int_0^{\pi} [1 - p(x)]^2 \, dx$ is minimum.   How should the coefficients a_i be determined?

3. Obtain the formula $a_k = c_k$ from (24-4) by using the fact that $\partial E/\partial a_k = 0$ at the minimum value of E.

4. Apply Parseval's equality (Prob. 1) to selected Fourier series in the examples and problems of Secs. 18 to 20.   You need not recompute the Fourier coefficients.

5. For complex Fourier series on $(-\pi,\pi)$ it can be shown that Parseval's equality has the form

$$\sum_{n=-\infty}^{\infty} |c_n|^2 = \frac{1}{2\pi} \int_{-\pi}^{\pi} |f(x)|^2 \, dx$$

where both f and c_n may be complex, and the absolute value has the meaning assigned in Sec. 15. Without recomputing the Fourier coefficients, apply this equality to selected examples and problems in Sec. 22.

---

25. The Pointwise Convergence of Fourier Series.   We shall now obtain an explicit formula for the difference between a function and the nth partial sum of its trigonometric Fourier series. The formula will enable us to establish the convergence for a class of functions which includes all the examples given in this book.

If $f(x)$ is a bounded integrable function of period 2π, the nth partial sum of its Fourier series is

$$s_n(x) = \tfrac{1}{2}a_0 + \sum_{k=1}^{n} (a_k \cos kx + b_k \sin kx) \tag{25-1}$$

where the coefficients are given by

$$a_k = \frac{1}{\pi} \int_{-\pi}^{\pi} f(t) \cos kt \, dt \qquad b_k = \frac{1}{\pi} \int_{-\pi}^{\pi} f(t) \sin kt \, dt. \tag{25-2}$$

Substituting (25-2) into the series (25-1) we get

$$s_n(x) = \frac{1}{\pi} \int_{-\pi}^{\pi} f(t) \left[\tfrac{1}{2} + \sum_{k=1}^{n} (\cos kt \cos kx + \sin kt \sin kx) \right] dt$$

$$= \frac{1}{\pi} \int_{-\pi}^{\pi} f(t) \left[\tfrac{1}{2} + \sum_{k=1}^{n} \cos k(t - x) \right] dt.$$

If we define the *Dirichlet kernel* by

$$D_n(u) = \tfrac{1}{2} + \sum_{k=1}^{n} \cos ku \tag{25-3}$$

the foregoing result takes the simpler form

$$s_n(x) = \frac{1}{\pi} \int_{-\pi}^{\pi} f(t) D_n(t - x) \, dt. \tag{25-4}$$

Setting $t - x = u$ in (25-4) yields

$$s_n(x) = \frac{1}{\pi} \int_{-\pi-x}^{\pi-x} f(x + u) D_n(u) \, du. \tag{25-5}$$

Now, $D_n(u)$ has period 2π by inspection of (25-3), and $f(x)$ also has period 2π. Hence, the integral of $f(u + x)D_n(u)$ over any interval of length 2π is the same as the integral over any other interval of length 2π, and (25-5) may be replaced by

$$s_n(x) = \frac{1}{\pi} \int_{-\pi}^{\pi} f(x + u) D_n(u) \, du. \tag{25-6}$$

Since $D_n(-u) = D_n(u)$ by (25-3), we may replace u by $-u$ in (25-6) to obtain the alternative form

$$s_n(x) = \frac{1}{\pi} \int_{-\pi}^{\pi} f(x - u) D_n(u) \, du. \tag{25-7}$$

The sum of (25-6) and (25-7) yields

$$2s_n(x) = \frac{1}{\pi} \int_{-\pi}^{\pi} [f(x + u) + f(x - u)] D_n(u) \, du.$$

Since the integrand is an even function of u, the integral from 0 to π is half the integral from $-\pi$ to π, and we have thus established that

$$s_n(x) = \frac{1}{\pi} \int_{0}^{\pi} [f(x + u) + f(x - u)] D_n(u) \, du. \tag{25-8}$$

To introduce $f(x)$ into our considerations, we observe that

$$\frac{1}{2} = \frac{1}{\pi} \int_{0}^{\pi} D_n(u) \, du \tag{25-9}$$

since the terms involving $\cos ku$ in (25-3) integrate to zero. If (25-9) is multiplied by $2f(x)$ (which is constant with respect to the integration variable u), we get

$$f(x) = \frac{1}{\pi} \int_0^\pi 2f(x) D_n(u)\ du. \tag{25-10}$$

Subtracting (25-10) from (25-8) gives the fundamental formula

$$s_n(x) - f(x) = \frac{1}{\pi} \int_0^\pi [f(x + u) - 2f(x) + f(x - u)] D_n(u)\ du \tag{25-11}$$

which will now be used to study the convergence of $s_n(x)$ to $f(x)$.

We shall say that $f(x)$ is *piecewise smooth* if the graph of $f(x)$ consists of a finite number of curves on each of which $f'(x)$ exists. We suppose also that the derivative exists at the end points of these curves, in the sense

$$\lim_{u \to 0+} \frac{f(x + u) - f(x+)}{u} \qquad \text{or} \qquad \lim_{u \to 0+} \frac{f(x - u) - f(x-)}{-u} \tag{25-12}$$

where "$u \to 0+$" means $u \to 0$ through positive values. Such a function may have finitely many discontinuities. However, since the Fourier coefficients of $f(x)$ are not altered if $f(x)$ is redefined at a finite number of points, we can assume that

$$f(x) = \frac{f(x+) + f(x-)}{2} \tag{25-13}$$

at every point x, whether $f(x)$ is continuous at x or not.

These preliminaries lead to the following theorem:

THEOREM. *If $f(x)$ is periodic of period 2π, is piecewise smooth, and is defined at points of discontinuity by (25-13), then the Fourier series for $f(x)$ converges to $f(x)$ at every value of x.*

To establish this theorem we recall that the series (25-3) was summed in Sec. 17, Example 2. The result (17-5) yields

$$D_n(u) = \frac{\sin (n + \frac{1}{2})u}{2 \sin \frac{1}{2}u}. \tag{25-14}$$

If we substitute this into (25-11) and replace $2f(x)$ by $f(x+) + f(x-)$ in accordance with (25-13), we get

$$s_n(x) - f(x) = \frac{1}{\pi} \int_0^\pi \frac{f(x + u) - f(x+) + f(x - u) - f(x-)}{2 \sin \frac{1}{2}u} \sin (n + \frac{1}{2})u\ du.$$

Now, the expression

$$\frac{f(x + u) - f(x+)}{2 \sin \frac{1}{2}u} = \frac{f(x + u) - f(x+)}{u} \frac{(u/2)}{\sin (u/2)} \tag{25-15}$$

has a limit as $u \to 0+$, since

$$\lim \frac{u/2}{\sin (u/2)} = 1 \qquad \text{as } u \to 0$$

and since the limits (25-12) exist by hypothesis. If we define the value of (25-15) at $u = 0$ to be this limit, the expression is continuous for $u \geq 0$ as long as the points

$$[x, f(x+)] \qquad \text{and} \qquad [x + u, f(x + u)]$$

are on the same smooth curve belonging to the graph of $f(x)$.

On the other hand, for $u \neq 0$ the function (25-15) is just as well behaved as the numerator $f(x + u) - f(x+)$, since $\sin \frac{1}{2}u$ does not vanish. This shows that the graph of (25-15) consists of a finite number of continuous curves, which have finite limits as one approaches their end points and hence are bounded.

It follows from Corollary 3 of the preceding section that

$$\lim_{n \to \infty} \int_0^\pi \frac{f(x + u) - f(x+)}{2 \sin \frac{1}{2}u} \sin (n + \frac{1}{2}) u \, du = 0.$$

In just the same way it is found that

$$\lim_{n \to \infty} \int_0^\pi \frac{f(x - u) - f(x-)}{2 \sin \frac{1}{2}u} \sin (n + \frac{1}{2}) u \, du = 0$$

and hence the integral representing $s_n(x) - f(x)$ tends to zero as $n \to \infty$. This shows that

$$\lim_{n \to \infty} s_n(x) = f(x)$$

and completes the proof.

Example. Let $f(x)$ be a piecewise-smooth function such that

$$\int_0^1 f(x)x^k \, dx = 0 \qquad \text{for } k = 0, 1, 2, 3, \dots . \tag{25-16}$$

Prove that $f(x) = 0$ at every point of continuity.

We extend the definition of $f(x)$ so that $f(x) = 0$ on $(-\pi,0)$ and on $(1,\pi)$. By the Taylor series for $\cos nx$,

$$\int_{-\pi}^\pi f(x) \cos nx \, dx = \int_{-\pi}^\pi f(x) \left[1 - \frac{(nx)^2}{2!} + \frac{(nx)^4}{4!} - \cdots \right] dx.$$

Term-by-term integration is justified because the series converges uniformly, and the result of the integration is 0 by the hypothesis (25-16). Since a similar calculation applies to $\sin nx$, we conclude that all the Fourier coefficients of $f(x)$ are 0. But, by the above theorem, the Fourier series converges to $f(x)$ at all points of continuity. Hence $f(x) = 0$ at such points.

PROBLEMS

1. If $f(x)$ is piecewise smooth on $[-\pi,\pi]$, show that the Fourier coefficients satisfy $|a_n| \le M_0/n$ and $|b_n| \le M_0/n$, where M_0 is a suitable constant. *Hint:* Let the points x_k divide $[-\pi,\pi]$ into a finite number of intervals on each of which $f'(x)$ is continuous. The Fourier coefficients are obtained by adding integrals of the type

$$\int_{x_k}^{x_{k+1}} f(x) \cos nx \, dx \qquad \text{or} \qquad \int_{x_k}^{x_{k+1}} f(x) \sin nx \, dx$$

and these can be integrated by parts.

2. If $f'(x)$ is continuous and of period 2π, and if $f''(x)$ is piecewise smooth, show that the Fourier coefficients satisfy $|a_n| \le M/n^2$, $|b_n| \le M/n^2$ for a suitable constant M. *Hint:* By partial integration, express the coefficients in terms of $f'(x)$, and then apply Prob. 1.

3. Let $f(x)$ be periodic and satisfy the conditions of Prob. 2. Then the Fourier series of $f(x)$ converges uniformly to $f(x)$ in every interval, and the remainder after n terms does not exceed $2M/n$.

Hint: Since $|\cos kx| \le 1$ and $|\sin kx| \le 1$, the remainder does not exceed $2M \sum_{n+1}^\infty k^{-2}$, and this can be estimated by an integral. The fact that the series converges to $f(x)$ is assured by the above theorem.

4. Suppose $f(x)$ can be approximated within ϵ by a partial sum of its Fourier series on a given finite interval, where $\epsilon > 0$. Prove that $f(x)$ can be approximated within 2ϵ by a polynomial on the same interval. *Hint:* Apply Taylor's theorem to each term of the given partial sum.

26. The Integration and Differentiation of Fourier Series.

If $f(x)$ is piecewise continuous[1] on $[-\pi,\pi]$, the function

$$F(x) = \int_{-\pi}^{x} f(t)\, dt \qquad (26\text{-}1)$$

is continuous and piecewise smooth (Sec. 25).   Moreover, $F(x)$ remains continuous when defined to have period 2π, provided $F(-\pi) = F(\pi)$.   Since $F(-\pi) = 0$, the latter condition reduces to

$$F(\pi) = \int_{-\pi}^{\pi} f(t)\, dt = \pi a_0 = 0 \qquad (26\text{-}2)$$

where $\tfrac{1}{2}a_0$ is the first Fourier coefficient of $f(x)$.   Applying the theorem of the preceding section, we can now deduce that the Fourier series for the periodic function $F(x)$ converges to $F(x)$ at every value of x.

This result holds when $f(x)$ and $|f(x)|$ are only assumed integrable, without being piecewise continuous.   Indeed, one can always write $f(x) = P(x) - N(x)$, where $P(x)$ is positive and $-N(x)$ is negative.   The equation

$$F(x) = \int_{-\pi}^{x} P(t)\, dt - \int_{-\pi}^{x} N(t)\, dt$$

expresses $F(x)$ as the difference of two increasing continuous functions.   Since such functions satisfy the Dirichlet conditions, the desired result can be deduced from Dirichlet's theorem as quoted in Sec. 19.

We shall show next that the Fourier series for $F(x)$ is obtained by integrating the series for $f(x)$.   If $n \geq 1$, the Fourier cosine coefficient A_n of $F(x)$ satisfies

$$\pi A_n = \int_{-\pi}^{\pi} F(x) \cos nx\, dx = F(x) \left.\frac{\sin nx}{n}\right|_{-\pi}^{\pi} - \int_{-\pi}^{\pi} \frac{\sin nx}{n} F'(x)\, dx$$

when we integrate by parts.   Since $F(-\pi) = F(\pi) = 0$, the integrated part drops out, and since $F'(x) = f(x)$, the expression becomes

$$\pi A_n = -\frac{1}{n} \int_{-\pi}^{\pi} \sin nx\, f(x)\, dx = -\pi \frac{b_n}{n}.$$

In the same way $B_n = a_n/n$, and also

$$A_0 = -\frac{1}{\pi} \int_{-\pi}^{\pi} x f(x)\, dx. \qquad (26\text{-}3)$$

These considerations establish the following remarkable theorem:

THEOREM I.   Let $f(x)$ be a function of period 2π which has a Fourier series

$$\Sigma(a_n \cos nx + b_n \sin nx). \qquad (26\text{-}4)$$

Then, with A_0 given by (26-3),

$$\int_{-\pi}^{x} f(t)\, dt = \tfrac{1}{2}A_0 + \Sigma\left(\frac{a_n}{n} \sin nx - \frac{b_n}{n} \cos nx\right) \qquad (26\text{-}5)$$

and this equation holds for all x, even if the Fourier series (26-4) does not converge.   Moreover, the series (26-5) is actually the Fourier series of the function on the left.

In case $a_0 \neq 0$, so that the Fourier series for $f(x)$ is

$$\tfrac{1}{2}a_0 + \Sigma(a_n \cos nx + b_n \sin nx)$$

---

[1] This means that the interval $[-\pi,\pi]$ can be divided by points $x_1, x_2, \ldots, x_n$ into a finite number of intervals on each of which $f(x)$ is continuous.   Also $f(x)$ must have a limit as $x \to x_k+$ and as $x \to x_k-$.

we apply Theorem I to $f(x) - \frac{1}{2}a_0$.   Inasmuch as

$$\int_\alpha^\beta f(x)\,dx = \int_{-\pi}^\beta f(x)\,dx - \int_{-\pi}^\alpha f(x)\,dx = F(\beta) - F(\alpha)$$

for all α and β, the reader can deduce, by Theorem I, that

$$\int_\alpha^\beta f(x)\,dx = \int_\alpha^\beta (\tfrac{1}{2}a_0)\,dx + \Sigma \int_\alpha^\beta (a_n \cos nx + b_n \sin nx)\,dx. \tag{26-6}$$

This result is summarized as follows:

THEOREM II.   Any Fourier series (whether convergent or not) can be integrated term by term between any limits.   The integrated series converges to the integral of the periodic function corresponding to the original series.

For example, according to (18-5) the Fourier series for $\frac{1}{2}x$ is

$$\tfrac{1}{2}x = \sin x - \frac{\sin 2x}{2} + \frac{\sin 3x}{3} + \cdots + (-1)^{n-1}\frac{\sin nx}{n} + \cdots. \tag{26-7}$$

If we integrate from α to x, by Theorem II we get

$$\tfrac{1}{4}(x^2 - \alpha^2) = \sum_{n=1}^\infty (-1)^n \frac{\cos nx - \cos n\alpha}{n^2}.$$

Treating α as constant, we see that

$$\tfrac{1}{4}x^2 = C + \sum_{n=1}^\infty (-1)^n \frac{\cos nx}{n^2} \tag{26-8}$$

where C is constant.   Since C is the first Fourier coefficient of $\frac{1}{4}x^2$,

$$C = \frac{1}{2\pi}\int_{-\pi}^\pi \tfrac{1}{4}x^2\,dx = \frac{\pi^2}{12}. \tag{26-9}$$

Alternatively, because $a_0 = 0$ in (26-7), we can use (26-3) to obtain

$$A_0 = -\frac{1}{\pi}\int_{-\pi}^\pi \tfrac{1}{2}x^2\,dx = -\frac{\pi^2}{3}$$

and hence, by (26-5),

$$F(x) = \tfrac{1}{4}(x^2 - \pi^2) = -\frac{\pi^2}{6} + \sum_{n=1}^\infty (-1)^n \frac{\cos nx}{n^2}.$$

The consistency of this result with (26-8) and (26-9) is easily verified.

Although Fourier series can always be integrated, as we have just seen, the differentiation of Fourier series requires caution.   For example, the series (26-7) converges for all x, and yet the series

$$\cos x - \cos 2x + \cos 3x - \cos 4x + \cdots$$

obtained by differentiating (26-7) diverges for all x.   The trouble is that the function $\frac{1}{2}x$ (when made periodic) is discontinuous at the points $\pm\pi$, $\pm3\pi$, $\pm5\pi$, $\ldots$.

This example is quite typical of the general situation, which can be described as follows: *There is not much hope of being able to differentiate a Fourier series, unless the* **periodic** *function generating the series is continuous at* **every** *value of x.*   On the other hand, when this condition is fulfilled, one usually can differentiate, as is shown by the following theorem:

THEOREM III.   Let $f(x)$ be a continuous function of period 2π.   If $f'(x)$ is piecewise continuous then the Fourier series for $f'(x)$ can be obtained by differentiating the series for $f(x)$.

Repeated application of the theorem gives the corresponding result for higher derivatives.   For instance, the series for $f''(x)$ can be found by differentiating the series for $f(x)$ twice, provided $f''(x)$ satisfies the appropriate conditions.

Theorem III is established by applying Theorem I to $f'(x)$.   Since $f'(x)$ is integrable it has a Fourier series, and the constant term a_0 satisfies

$$\pi a_0 = \int_{-\pi}^{\pi} f'(x)\, dx = f(\pi) - f(-\pi) = 0.$$

Thus, the series for $f'(x)$ has the form (26-4), namely,

$$\Sigma(a_n \cos nx + b_n \sin nx). \tag{26-10}$$

It follows from Theorem I that the Fourier series for the function

$$\int_{-\pi}^{x} f'(t)\, dt = f(x) - f(-\pi)$$

has the form (26-5), and so the series for $f(x)$ is

$$f(-\pi) + \tfrac{1}{2}A_0 + \Sigma\left(\frac{a_n}{n} \sin nx - \frac{b_n}{n} \cos nx\right). \tag{26-11}$$

By inspection, we see that differentiating (26-11) gives (26-10).   In other words, the Fourier series for $f'(x)$ can be found by differentiating the series for $f(x)$; this is the main assertion in Theorem III.

Since the differentiated series is a Fourier series, its convergence can be tested by the usual methods.   In particular, if $f'(x)$ satisfies the Dirichlet conditions, then the Fourier series for $f'(x)$ converges to $f'(x)$ at every point where $f'(x)$ is continuous.

Example.   Polynomials.   If $p(x)$ is an even polynomial on $(-\pi,\pi)$, the Fourier sine coefficients are 0, and the nth cosine coefficient is

$$a_n = \frac{2}{\pi n}(-1)^n \left[\frac{p'(\pi)}{n} - \frac{p'''(\pi)}{n^3} + \frac{p^v(\pi)}{n^5} - \frac{p^{vii}(\pi)}{n^7} + \cdots\right] \tag{26-12}$$

for $n \geq 1$.   If $q(x)$ is an odd polynomial on $(-\pi,\pi)$, the Fourier cosine coefficients are 0, and the nth sine coefficient is

$$b_n = -\frac{2}{\pi n}(-1)^n \left[q(\pi) - \frac{q''(\pi)}{n^2} + \frac{q^{iv}(\pi)}{n^4} - \frac{q^{vi}(\pi)}{n^6} + \cdots\right]. \tag{26-13}$$

To prove (26-13), integration by parts gives

$$b_n = \frac{1}{\pi} \int_{-\pi}^{\pi} q(x) \sin nx\, dx = -\frac{\cos nx}{\pi n} q(x) \Big|_{-\pi}^{\pi} + \frac{1}{\pi} \int_{-\pi}^{\pi} \frac{\cos nx}{n} q'(x)\, dx.$$

Since q is odd, $q(-\pi) = -q(\pi)$, and the result reduces to

$$b_n = -\frac{2}{\pi n}(-1)^n q(\pi) + \frac{1}{\pi} \int_{-\pi}^{\pi} \frac{\cos nx}{n} q'(x)\, dx.$$

If we integrate by parts again, the integrated term vanishes for $n \geq 1$, and

$$b_n = -\frac{2}{\pi n}(-1)^n q(\pi) - \int_{-\pi}^{\pi} \frac{\sin nx}{n^2} q''(x)\, dx. \tag{26-14}$$

Using $b_n(f)$ to denote the Fourier sine coefficient of the arbitrary function f, we write (26-14) in the form

$$b_n(q) = -\frac{2}{\pi n}(-1)^n q(\pi) - \frac{1}{n^2} b_n(q''). \tag{26-15}$$

But q'', like q, is an odd polynomial. Hence, applying (26-15) to q'',

$$b_n(q'') = -\frac{2}{\pi n}(-1)^n q''(\pi) - \frac{1}{n^2} b_n(q^{iv}).$$

Similarly, $b_n(q^{iv})$ can be expressed in terms of $b_n(q^{vi})$, and so on. Since all derivatives of sufficiently high order are zero, the process terminates after a finite number of steps, and successive substitution gives (26-13).

Equation (26-12) can be established in the same way, but a more efficient method is as follows: Since p is even, p' is odd; hence (26-13) applies to p'. Integrating the resulting Fourier series term by term gives the series for p. This establishes (26-12).

If $f(x)$ is an arbitrary polynomial, then $f = p + q$, where p consists of the terms of even degree and q consists of the terms of odd degree. The expansion of f is evidently

$$f(x) = \tfrac{1}{2}a_0 + \sum_{n=1}^{\infty}(a_n \cos nx + b_n \sin nx)$$

where a_n and b_n are given by (26-12) and (26-13), respectively, and

$$a_0 = \frac{1}{\pi}\int_{-\pi}^{\pi} p(x)\, dx.$$

--- **PROBLEMS**

1. By integrating the series (26-8) from 0 to x, deduce that

$$x(x^2 - \pi^2) = 12 \sum_{n=1}^{\infty}(-1)^n \frac{\sin nx}{n^3}.$$

2. By integrating the series in Prob. 1 from $-\pi$ to x, deduce that

$$\frac{1}{48}(x^2 - \pi^2)^2 = \frac{\pi^4}{90} - \sum_{n=1}^{\infty}(-1)^n \frac{\cos nx}{n^4}.$$

3. Show that the following is not a Fourier series:

$$\sum_{n=1}^{\infty} \frac{\sin nx}{\log(1+n)}.$$

Hint: If it is a Fourier series, the integrated series must converge for all x.

4. Obtain the result of Probs. 1 and 2 from (26-13) and (26-12).

5. Obtain the Fourier series for x^5 and x^6 by inspection.

6. The results of a certain experiment are approximated by the polynomial $f(x) = 2 + 3x - x^2 + x^3 + 5x^4$ on $(-\pi,\pi)$. What is the Fourier series of this polynomial? *Hint:* $p(x) = 2 - x^2 + 5x^4$, $q(x) = 3x + x^3$.

7. According to (26-13) the Fourier coefficients for an odd polynomial q have the order of magnitude $1/n$, in general, but they have the order of magnitude $1/n^3$ or less if $q(\pi) = 0$. Explain, by reference to the graph of the periodic function which equals $q(x)$ on $(-\pi,\pi)$. Similarly discuss (26-12).

Nonlinear differential equations 2

Methods of solution

Applications

Existence, uniqueness, and stability

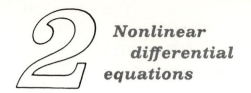

Nonlinear differential equations

THE POWER and effectiveness of mathematical methods in the study of natural sciences stem, to a large extent, from the unambiguous language of mathematics with the aid of which the laws governing natural phenomena can be formulated. Many natural laws, especially those concerned with rates of change, can be phrased as equations involving derivatives or differentials. For example, when a verbal statement of Newton's second law of motion is translated into mathematical symbols, there results an equation relating time derivatives of displacements to forces. Such equations provide a complete qualitative and quantitative characterization of the behavior of mechanical systems under the action of forces. Several broad types of equations studied in Secs. 1 to 13 of this chapter characterize physical situations of great diversity and practical interest. In the concluding six sections we present general methods that are used when simple explicit solutions are not available. The analysis provides a mathematical foundation for developments in Chap. 3. General methods are tending more and more to dominate the field and are essential for a deeper understanding of differential equations. However, this material is conceptually more difficult, and despite its importance it can be omitted on first reading or used chiefly for reference.

METHODS OF SOLUTION

1. Introduction. In this section we present some general concepts and terminology that will be useful throughout this chapter and Chap. 3. A *differential equation* is an equation involving some of the derivatives of a function. Differential equations are divided into two classes, *ordinary* and *partial*. The former contain only one independent variable and derivatives with respect to it. The latter contain more than one independent variable. The order of the highest derivative contained in a differential equation is the *order* of the equation. For example, the equations

$$\left(\frac{d^2y}{dx^2}\right)^4 = \frac{dy}{dx}, \qquad \frac{\partial^3y}{\partial x^3} = \left(\frac{\partial^2y}{\partial x\,\partial t}\right)^4 + xt\sin(yt) \tag{1-1}$$

are an ordinary differential equation of order 2 and a partial differential equation of order 3, respectively.

As explained more fully at the end of this section, differentiation with respect to the independent variable is often denoted by primes. Thus,

$$y' = \frac{dy}{dx}, \qquad y'' = \frac{d^2y}{dx^2}, \qquad \cdots, \qquad y^{(n)} = \frac{d^ny}{dx^n}$$

when x is the independent variable and similarly, with x replaced by t, when t is the independent variable. In this notation the first equation (1-1) is $(y'')^4 = y'$.

A function $y = \phi(x)$ is a *solution* of the differential equation

$$F(x,y,y') = 0 \tag{1-2}$$

in a given interval $a < x < b$ if, on the substitution $y = \phi(x)$, $y' = \phi'(x)$ in (1-2), the left-hand member vanishes identically. Thus, $y = e^{2x}$ is a solution of $y' - 2y = 0$ on every interval because the substitution $y = e^{2x}$, $y' = 2e^{2x}$ into the equation gives an identity, $0 = 0$. A similar definition applies to equations of higher order; for example, $y'' + y = 0$ has a solution $y = \sin x$, as is verified by substitution.

Many differential equations can be solved with ease. For example, given the equation $y' = 6x^2$, integration immediately produces the family of solutions

$$y = 2x^3 + c \qquad c = \text{const.} \tag{1-3}$$

It is obvious that each of these functions satisfies $y' = 6x^2$ and, conversely, if y is any solution of $y' = 6x^2$, then

$$(y - 2x^3)' = y' - 6x^2 = 6x^2 - 6x^2 = 0.$$

Hence $y - 2x^3 = c$, where c is constant,[1] and we conclude that y belongs to the family (1-3). The family is the *complete solution*, because every solution can be obtained by suitable choice of the constant c.

If we want a solution through a given point (x_0, y_0), substituting $x = x_0$ and $y = y_0$ into (1-3) gives

$$y_0 = 2x_0{}^3 + c \qquad \text{or} \qquad c = y_0 - 2x_0{}^3. \tag{1-4}$$

This value of c in (1-3) produces the desired solution,

$$y - y_0 = 2x^3 - 2x_0{}^3. \tag{1-5}$$

The same result can be obtained by writing the differential equation $dy/dx = 6x^2$ in the form

$$dy = 6x^2 \, dx \tag{1-6}$$

and integrating between corresponding limits; that is, the equation

$$\int_{y_0}^{y} dy = \int_{x_0}^{x} 6x^2 \, dx$$

is equivalent to (1-5), as the reader can verify.

This discussion illustrates several properties that hold for differential equations in general. Usually a first-order equation $y' = f(x,y)$ has a family of solutions $y = \phi(x,c)$ containing a single arbitrary constant, and knowledge of y at a single value of x suffices to determine the solution uniquely. A second-order equation $y'' = f(x,y,y')$ usually has a family of solutions containing two arbitrary constants, and two conditions are generally needed to specify a particular solution uniquely. (For example, one can prescribe both y and y' at a particular value x_0.) A third-order equation $y''' = f(x,y,y',y'')$ usually requires three conditions, and so on.

In the present example we found that there is one, and only one, solution through any given point (x_0, y_0) of the (x,y) plane. This behavior is also typical for first-order equations $y' = f(x,y)$, though sometimes the result holds only in a restricted region rather

---

[1] It is a familiar theorem that if the derivative of a function is zero on an interval then the function is constant there.

than in the whole plane.[1]   A family of solutions having this property is said to be a *general solution* of the differential equation.   More precisely, a family of solutions $y = \phi(x,c)$ of the equation $y' = f(x,y)$ is a *general solution* in a given region R, provided that the constant c can be determined so as to satisfy the condition $y_0 = \phi(x_0,c)$ at the arbitrarily specified point (x_0,y_0) of R.   The resulting curve $y = \phi(x,c)$ is called an *integral curve* of the differential equation and is said to satisfy the *initial condition* $y = y_0$ at $x = x_0$.

Thus, $y = 2x^3 + c$ describes a general solution of $y' = 6x^2$ because the constant c can be chosen so as to satisfy the arbitrary initial condition $y(x_0) = y_0$.   The resulting integral curve is given by (1-5).

Similar considerations apply to equations of higher order.   For example, a family of solutions $y = \phi(x,c_1,c_2)$ containing two constants is a *general solution* of a second-order equation

$$y'' = f(x,y,y')$$

if the constants c_1 and c_2 can be chosen so as to satisfy arbitrary initial conditions

$$y(x_0) = y_0 \qquad y'(x_0) = y_1.$$

Here specification of the region can involve y_1 as well as x and y.

We shall further illustrate concepts and terminology by discussing the equation

$$y' = f(x) \qquad a < x < b. \tag{1-7}$$

It is evident that there can be at most one solution through any given point (x_0,y_0) because if y and $\tilde{y}$ are two solutions the equation gives $(y - \tilde{y})' = 0$.   Hence $y - \tilde{y}$ is constant, and the fact that $y = \tilde{y}$ at x_0 makes $y \equiv \tilde{y}$.   A statement that there is at most one solution satisfying given conditions is called a *uniqueness theorem* because it asserts that the solution, when it exists at all, is unique.   The foregoing remarks establish a uniqueness theorem for the equation $y' = f(x)$.

If $y = F(x)$ is a given solution of the equation, then $y = F(x) + c$ is also a solution for every choice of the constant c.   The solution satisfying the initial condition $y = y_0$ at $x = x_0$ is obtained by choosing c so that

$$y_0 = F(x_0) + c. \tag{1-8}$$

We conclude that the family $y = F(x) + c$ is a *general* solution in the region

$$a < x < b \qquad -\infty < y_0 < \infty.$$

The first condition expresses the fact that x is on the interval (a,b), and the second condition indicates that y_0 is unrestricted.   In agreement with the uniqueness theorem, c is uniquely determined by (1-8).

If we write the equation as $dy = f(x)\,dx$ and integrate between corresponding limits, the result is

$$y - y_0 = \int_{x_0}^{x} f(x)\,dx \tag{1-9}$$

when $f(x)$ is continuous.   This formula gives the solution satisfying the initial condition $y = y_0$ at $x = x_0$.   Indeed, $y(x_0) = y_0$ because the integral is 0 when $x = x_0$, and $y' = f(x)$ by the fundamental theorem of calculus.   The statement that an equation has at least one solution satisfying specified conditions is an *existence theorem*.   Formula (1-9) establishes an existence theorem for the equation $y' = f(x)$.

[1] An equation $y' = f(x,y)$ has one, and only one, solution through each point of the region in which f and $\partial f/\partial y$ are continuous. This theorem is carefully formulated in Sec. 14 and proved in Sec. 18. As a matter of personal taste, we postpone such general considerations until a number of specific equations have been examined and solved.   The reader who wishes to do so can take up Sec. 14 at an earlier point in his study of differential equations, and the same applies to the geometric discussion of Sec. 7.

We conclude by mentioning some matters of notation. A differential equation can be written either as an equality between functions, such as

$$y' + py = f \tag{1-10}$$

or as an equality between functional values, such as

$$y'(x) + p(x)y(x) = f(x) \qquad a < x < b. \tag{1-11}$$

When the interval $a < x < b$ is not given, one generally assumes that the equation holds over the largest interval for which the various terms are meaningful.

In certain kinds of analysis the *function* and the *value of the function* must be carefully distinguished by the notation. In differential equations and some other subjects, however, it is convenient to use a more flexible notation that shows the general status of the functions being considered.

Thus, one can write (1-10) and (1-11) in any of the forms

$$y' + py = f(x) \qquad y' + p(x)y = f \qquad y' + p(x)y = f(x) \tag{1-12}$$

to emphasize that p is constant, or that f is constant, or that both may be nonconstant. Only a very little experience with differential equations is needed to recognize that y' in (1-12) is an abbreviation for $y'(x)$, and y for $y(x)$. Hence no purpose is served by emphasizing the variable, and it is common practice not to do so.

In this book we often use notation analogous to (1-12) in the initial statement of a problem and the shorter notation (1-10) in the subsequent discussion.

Example 1. The equation $x\,dx + y\,dy = 0$ can be written $\frac{1}{2}d(x^2 + y^2) = 0$, giving

$$x^2 + y^2 = c \qquad c = \text{const.} \tag{1-13}$$

For $c > 0$ this equation describes a circle centered at the origin.

In the neighborhood of any point (x,y) of the circle we can express y as a differentiable function of x, or x as a differentiable function of y, using one of the relations

$$y = \sqrt{c - x^2} \qquad x = \sqrt{c - y^2} \qquad x = -\sqrt{c - y^2} \qquad y = -\sqrt{c - x^2}. \tag{1-14}$$

Each expression gives $x\,dx + y\,dy = 0$ on an appropriate interval, with x or y the independent variable, as the case may be. In this sense, the circle (1-13) actually satisfies the differential equation. The equation $x^2 + y^2 = c$ is called an *implicit solution*, in contrast to the explicit solutions (1-14).

Strictly speaking, the three differential equations

$$\frac{dy}{dx} = -\frac{x}{y} \qquad \frac{dx}{dy} = -\frac{y}{x} \qquad x\,dx + y\,dy = 0$$

are not equivalent because the first implies that y is a differentiable function of x, the second implies that x is a differentiable function of y, and the third allows solutions of either type. Nevertheless, in many practical problems a subdivision into cases as suggested by (1-14) is very artificial, and the distinction between explicit and implicit solutions is not emphasized in the examples and problems of this book.

Example 2. A particle of constant mass $m > 0$ moves in a straight line subject to a known force $mf(t)$ at time t. Discuss conditions needed for unique determination of the motion.[1]

Let $s = s(t)$ denote the position of the particle at time t, relative to some point chosen as the origin. By definition, the *velocity* is $v = s'(t)$ and the *acceleration* is $a = v'(t)$, where the primes denote differentiation with respect to the time t. Although systematic discussion of Newton's law is postponed to Sec. 11, the reader is doubtless familiar with the statement that

$$(\text{mass})(\text{acceleration}) = \text{force}.$$

This gives the differential equation $mv' = mf(t)$, or $v' = f(t)$.

For unique determination of the velocity we must know the value at some particular time.

---

[1] This example and similar ones given later indicate the relevance of differential equations in fields other than mathematics. The mathematical theory itself is independent of applications.

For example, if the initial velocity $v(0) = v_0$ is known, then $v(t)$ is uniquely determined. After v is found, the equation $s' = v(t)$ gives s, provided the initial displacement s_0 is known. We conclude that the motion is determined if both the initial displacement and velocity are given.

Since the displacement satisfies the second-order equation $s'' = f(t)$, the fact that two conditions are needed agrees with the general discussion of the text and provides a physical interpretation.

When $f(t)$ is continuous the equations $v' = f(t)$, $s' = v(t)$ can be solved by use of (1-9), with appropriate changes of notation. The result is

$$v(t) = v_0 + \int_0^t f(\tau)\, d\tau \qquad s(t) = s_0 + \int_0^t v(\tau)\, d\tau. \tag{1-15}$$

These equations give a complete description of the motion.

PROBLEMS

Here and elsewhere, the letter c introduced without explanation denotes a constant.

1. Find the orders of $y'' + y = \sin t$, $\quad y'' = f(y,y')$, $\quad xy^3 + (y'')^2 = 1$, $\quad y' = x^2$.

2. (a) If $y' = 2x$, show that $(y - x^2)' = 0$, and conclude that $y = x^2 + c$. (b) Show that the equation $y' = 2x$ has one, and only one, solution through each of the points $(0,0)$, $(1,1)$, $(0,1)$, $(1,-1)$, and find these solutions. (c) Sketch the solutions $y = x^2 + c$ for several choices of the constant c. By reference to the differential equation, explain why the result is a family of parallel curves. (d) Is $y = x^2 + c$ a general solution in the whole (x,y) plane?

3. If $y'' = 0$, show that $y = cx + c_1$, where c and c_1 are constant. Is the converse true? *Hint:* $y' = c$ gives $(y - cx)' = 0$.

4. Referring to Prob. 3, show that the equation $y'' = 0$ has one, and only one, solution through any given point, with prescribed slope, and it also has one, and only one, solution through any two points, in general. What are the exceptional cases?

5. (a) Obtain a solution, $y = u$, of the equation $y'' = 2x$. (b) If v is any other solution, show that $(v - u)'' = 0$, and deduce $v = u + cx + c_1$ by reference to Prob. 3. (c) Show that the equation $y'' = 2x$ has a unique solution such that $y(0) = 0$, $y'(0) = 1$, and find it.

6. Verify that the given expression is a solution of the differential equation immediately below it:

$$y = ce^x, \qquad y = c_1 \sin x + c_2 \cos x, \qquad xy = \int f(x)\, dx, \qquad 2e^y = e^x + ce^{-x}.$$
$$y' = y \qquad y'' + y = 0 \qquad\qquad xy' + y = f(x) \qquad y' + 1 = e^{x-y}.$$

7. Find implicit solutions by inspection:

$$d(x^2 y) = 0, \qquad d(xy) = 0, \qquad x\, dy + y\, dx = 0, \qquad d(x^2 + 3y^2) = 0, \qquad 2x\, dx + 6y\, dy = 0.$$

8. The force on a particle of mass m near the surface of the earth is $-mg$, where g is the acceleration of gravity. Obtain the familiar formulas

$$v = v_0 - gt \qquad s = s_0 + v_0 t - \tfrac{1}{2}gt^2$$

for freely falling bodies, by letting $f(t) = -g$ in (1-15). [The value of g in the cgs system is approximately 980 cm per sec², and in the fps system 32.2 ft per sec². It should be emphasized that (1-15) has far greater scope than the special case considered in this problem.]

9. A stone is thrown vertically upward with velocity 8 fps at time $t = 0$. Write an expression for the position and velocity at time t and also for the velocity as a function of distance s. Find the time at which the velocity is zero, and show that the height is then maximum. Show that the maximum height agrees with that obtained by equating kinetic and potential energy, that is, with $mgh = mv_0^2/2$. [Cf. Prob. 8.]

10. An iceboat weighing 500 lb is driven by a wind that exerts a force of 25 lb. Five pounds of this force is expended in overcoming frictional resistance. What speed will this boat acquire at

the end of 30 sec if it starts from rest? *Hint:* The
force producing the motion is $F = 25 - 5 = 20$.
Hence, $500\, dv/dt = 20g$.

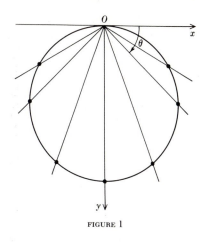

11. A particle of mass m slides down an inclined
plane, making an angle θ with the horizontal. If
the initial velocity is zero and friction may be
neglected, the component of force in the direction
of motion is $F = mg \sin \theta$. What are the velocity
of the particle and the distance traveled during
the time t? Find the speed as a function of the
vertical distance fallen, and verify that the same
result would be given by equating energies as in
Prob. 9. At any given instant, show that the
locus of particles obtained for various θ's is a circle[1]
(Fig. 1).

FIGURE 1

12. A body is set sliding down an inclined plane
with an initial velocity of v_0 fps. If the angle made
by the plane with the horizontal is θ and the coefficient of friction is μ, show that the distance
traveled in t sec is

$$s = \tfrac{1}{2}g(\sin \theta - \mu \cos \theta)t^2 + v_0 t.$$

13. Write the equations of Prob. 1 as equalities between functional values and also as equalities
between functions.

---

2. Exponential Growth.
Some interesting conclusions concerning the uses of differential equations are suggested by the example

$$\frac{dy}{dx} = ky \qquad k = \text{const} \tag{2-1}$$

and we shall discuss this equation in detail. The equation is called the equation of
exponential growth (or decay) because its solutions are exponential functions.

Indeed, on any interval on which $y \neq 0$ Eq. (2-1) can be written as

$$\frac{dy}{y} = k\, dx. \tag{2-2}$$

Then[2] $\log |y| = kx + c_1$, where c_1 is constant, and so $|y| = e^{kx}e^{c_1}$. Distinguishing the
cases $y > 0$ and $y < 0$ gives

$$y = ce^{kx} \tag{2-3}$$

where $c = \pm e^{c_1}$ is constant. The result is verified by substitution.

The foregoing method has the disadvantage that it requires the assumption $y \neq 0$,
and it does not readily guarantee that (2-3) contains all solutions. We now give another
method that is free of these defects.

Writing (2-1) as $y' - ky = 0$, multiply by e^{-kx}. The result is equivalent to

$$(ye^{-kx})' = 0 \tag{2-4}$$

as the reader will verify. Hence $ye^{-kx} = c$, where c is constant, and this gives (2-3).

---

[1] The result of this problem was known to the Italian physicist Galileo before Newton's laws were
available.
[2] Recall that $|y| = y$ for $y \geq 0$, and $|y| = -y$ for $-y > 0$. In the latter case the formula $\int y^{-1}\, dy =
\log(-y) + c$ is easily verified by differentiation.

It follows that (2-3) is the *complete solution,* in the sense that every solution is obtained by suitable choice of the constant c.

To indicate the connection with general theory, we remark that (2-2) illustrates the method of *separation of variables,* and (2-4) illustrates the method of *integrating factors.* As shown in Secs. 3 and 5, both methods apply to broad classes of differential equations.

Because of the presence of the arbitrary constant c, the differential equation does not determine y uniquely, but an extra condition is needed. For example, if the solution curve y versus x is to contain a prescribed point (x_0, y_0), substitution into (2-3) gives

$$ce^{kx_0} = y_0 \qquad \text{or} \qquad c = y_0 e^{-kx_0}.$$

The solution is determined uniquely as

$$y = y_0 e^{k(x - x_0)}. \tag{2-5}$$

In agreement with the usual behavior of first-order equations as described in Sec. 1, there is one, and only one, solution through the given point (x_0, y_0). Since this is obtained by specializing the constant c, we conclude that $y = ce^{kx}$ is a general solution.

In the following examples and problems it is found that the equation $y' = ky$ applies in such diverse fields as mechanics, circuit theory, biology, and optics. This discussion illustrates a particularly striking and characteristic feature of the subject of differential equations; namely, *one and the same differential equation can describe the behavior of several different physical systems.* The study of differential equations thus leads to precise, quantitative analogies between seemingly unrelated processes. These analogies form the basis for the science of analog computation and allow the transfer of techniques from one field to another. (For example, the electrical concept of impedance can be applied to the study of mechanics because the underlying differential equations have the same form in both disciplines.)

When differential equations are applied to physical problems it is frequently necessary to assess the effect of small changes in the data. To illustrate this statement, let x denote the time t, and let the initial condition in (2-1) be given at $t = 0$. The equation and its solution (2-5) are then

$$y' = ky \qquad y = y_0 e^{kt}.$$

Now $y \to \infty$ as $t \to \infty$ if k and y_0 are positive, but $y = y_0$ if $k = 0$, and $y \to 0$ as $t \to \infty$ if k is negative. It should be emphasized that the magnitude of k need not be large; for example, the three respective behaviors ∞, y_0; 0 are found for

$$k = 0.0000001 \qquad k = 0 \qquad k = -0.0000001.$$

Thus, in general, small changes (or perturbations) in the coefficients of a differential equation may completely alter the nature of its solutions. This remark has an important bearing on the problem of constructing differential equations that purport to represent the behavior of physical systems. In physical problems, the coefficients in a differential equation are usually related to physical quantities. Such quantities are determined from measurements which are subject to experimental errors. For this reason, it is important to know what effect small variations in the coefficients of a given equation have on the character of its solutions. When small changes in the coefficients result in small changes in the solutions, the solutions are termed *stable.*

Another form of the stability problem occurs in the study of the dependence of solutions on small changes in the initial values. In practice one ordinarily seeks particular solutions that satisfy specified initial data. The initial data are generally determined either experimentally or from a specific assumption that certain physical conditions hold. (For example, one may assume that the deflection of a beam at a given point is zero.)   If the initial conditions are altered slightly, is it true that the solution of a given equation will not be affected by a great deal?   The fact that solutions of differential equations need not be continuous functions of initial conditions is also clear from the above discussion.

Thus, if $k > 0$, the solution for $y_0 > 0$ tends to ∞, the solution for $y_0 = 0$ is 0, and the solution for $y_0 < 0$ tends to $-\infty$.   Here again, the magnitude of y_0 can be arbitrarily small without invalidating the conclusion.

The chief purpose of these remarks is to point out the existence of the stability problem and the need for caution when a differential equation is used without explicit knowledge of its solutions.   In most of the following applications explicit solutions are available, and stability can be studied, if desired, as in the foregoing discussion.   An introduction to the general theory is given in the final sections of this chapter.

Before deriving (2-1) in various contexts, it is convenient to introduce the concept of *asymptotic equality*. We write $a \sim b$ and read "a is asymptotic to b," if a and b are two quantities whose ratio tends to 1 in the course of a given limiting process.   Thus,

$$a \sim b \qquad \text{means} \qquad \lim \frac{a}{b} = 1.$$

For example, the asymptotic equality

$$\sin \tfrac{1}{2}\Delta\theta \sim \tfrac{1}{2}\Delta\theta \qquad \Delta\theta \to 0 \tag{2-6}$$

follows by using the Taylor series for the sine or by applying l'Hospital's rule to $(\sin x)/x$, where $x = \frac{1}{2}\Delta\theta$.   As suggested by Prob. 1, asymptotic equality has several of the properties of ordinary equality.

Example 1.   Certain biological organisms grow at a rate proportional to their size at a given time.   Show that, in such cases, the law of growth is wholly determined by the size at two different times, and find the law.

If S denotes the size at time t, the rate of growth is dS/dt.   Hence the statement about growth is equivalent to the equation $S' = kS$, where k is constant.   By (2-3) with $x = t$ and $y = S$ the solution is

$$S = ce^{kt} \qquad c, k \text{ const.} \tag{2-7}$$

Without loss of generality, we assume that the two values of size are given at time $t = 0$ and time $t = t_1$.   Thus, $S_0 = c$, $S_1 = ce^{kt_1}$, where S_0 and S_1 are known.   By division, $e^{kt_1} = S_1/S_0$; raising to power $1/t_1$ gives e^k; and by (2-7)

$$S = c(e^k)^t = S_0 \left(\frac{S_1}{S_0} \right)^{t/t_1}.$$

Example 2.   A capacitor of capacity C is discharged through a resistance R, as shown in Fig. 2. If Q is the charge on the capacitor at time t, V is the voltage across the capacitor, and I is the current in the circuit, show that

$$\frac{dQ}{dt} = kQ \qquad \frac{dV}{dt} = kV \qquad \frac{dI}{dt} = kI \qquad \text{where } k = -\frac{1}{RC}. \tag{2-8}$$

We have $I = dQ/dt$ by the definition of current and $Q = CV$ by the definition of capacity. The voltage across the resistor is IR by Ohm's law, and, since the voltage drop around the circuit must be 0,

$$IR + C^{-1}Q = 0.$$

The first relation (2-8) is obtained by substituting $I = dQ/dt$, and the others follow from the first.

Example 3.   A flexible rope of negligible weight is wrapped around a capstan as shown in Fig. 3.   How does the tension depend on the angle of wrap?

FIGURE 2

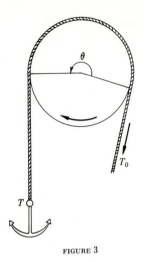

FIGURE 3

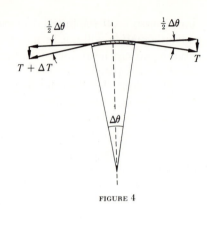

FIGURE 4

Figure 4 shows a small piece of rope with ends at positions specified by the angles θ and $\theta + \Delta\theta$. The statement that the rope is flexible means that the forces due to tension are directed tangentially as shown. If these forces have magnitudes T and $T + \Delta T$ at the two ends, the corresponding component of force in a direction normal to the small piece at its center is

$$\Delta F = (T + \Delta T) \sin \tfrac{1}{2} \Delta\theta + T \sin \tfrac{1}{2} \Delta\theta. \qquad (2\text{-}9)$$

If the short piece of rope were straight, the frictional force would be directed tangentially and would have magnitude $\mu \, \Delta F$, where μ is the coefficient of static or sliding friction, as the case may be. It is an experimental fact that a similar relation for the curved piece holds as an asymptotic equality, namely,

$$\Delta T \sim \mu \, \Delta F \qquad \Delta\theta \to 0. \qquad (2\text{-}10)$$

This together with (2-9) shows that $\Delta T \to 0$ and hence, by (2-9) and (2-6),

$$\Delta F \sim T \, \Delta\theta. \qquad (2\text{-}11)$$

Multiplying (2-11) by μ gives $\mu \, \Delta F \sim \mu T \, \Delta\theta$, so that (2-10) gives

$$\Delta T \sim \mu \, \Delta F \sim \mu T \, \Delta\theta.$$

By the result of Prob. 1b we get $\Delta T \sim \mu T \, \Delta\theta$, or

$$1 = \lim \frac{\Delta T}{\mu T \, \Delta\theta} = \frac{1}{\mu T} \lim \frac{\Delta T}{\Delta\theta} = \frac{1}{\mu T} \frac{dT}{d\theta}.$$

Thus $dT/d\theta = \mu T$. The solution $T = T_0 e^{\mu\theta}$ indicates that the tension is independent of the radius of the capstan but depends only on the number of turns. This surprising behavior results from the assumption that the rope is perfectly flexible.

Example 4. In a colony of bacteria each bacterium divides into two after a time interval, on an average, of length τ. If there are n bacteria at time $t = 0$ and m at time $t = 1$, with n large, estimate the value of τ.

The hypothesis suggests that $dN/dt = kN$, approximately, with improved accuracy as the number of bacteria, N, increases. Also $t = 0$ corresponds to $N = n$, $t = 1$ corresponds to $N = m$, and since the number doubles in the interval τ, we conclude that $t = \tau$ corresponds to $N = 2n$. By the method of integrating between corresponding limits,

$$\int_n^m \frac{dN}{N} = \int_0^1 k \, dt \qquad \text{and} \qquad \int_n^{2n} \frac{dN}{N} = \int_0^\tau k \, dt.$$

These equations give $\log m - \log n = k$ and $\log 2 = k\tau$, respectively, so that

$$\tau = \frac{\log 2}{\log m - \log n}.$$

An obvious objection to this discussion is that an integer-valued function, N, cannot have a derivative unless it is constant. This objection, however, is superficial. One can approximate N within 1 by a smooth function, and an error of one bacterium is not significant in the legitimate uses of the result.

A more serious difficulty is that the underlying principles are statistical. The hypothesis indicates what a large number of bacteria will probably do, and not what they must do. By means of the theory of probability (Chap. 9) it can be shown that the differential equation is reliable when the sample is large but that the use of a small sample can lead to discrepancies. See also Prob. 10.

── **PROBLEMS**

1. (*a*) Read the discussion of asymptotic equality in Chap. 1, Sec. 5. (*b*) If $a \sim b$ and $b \sim c$, prove that $a \sim c$.

2. (*A law of chemical change.*) The rate of decomposition of a certain chemical substance is proportional to the amount of the substance still unchanged. If the amount of the substance at the end of t hr is x and x_0 is the initial amount, show that $x = x_0 e^{-kt}$, where k is the constant of proportionality. Find k if x changes from 1,000 to 500 g in 2 hr.

3. (*Newton's law of cooling.*) The rate at which a body is cooling is proportional to the difference in the temperatures of the body and the surrounding medium. It is known that the temperature of a body fell from 120 to 70°C in 1 hr when it was placed in air at 20°C. How long will it take the body to cool to 40°C? 30°C? 20°C? *Hint:* Let $y = T - 20°$.

4. (*Lambert's law of absorption.*) The percentage of incident light absorbed in passing through a thin layer of material is proportional to the thickness of the material. If 1 in. of material reduces the light to half its intensity, how much additional material is needed to reduce the intensity to one-eighth of its initial value? Obtain the answer by inspection, and check by solving an appropriate differential equation.

5. (*Viscous damping.*) A torpedo moving in still water is retarded with a force proportional to the velocity. Find the speed at the end of t sec and the distance traveled in t sec if the initial speed is 30 mph. *Hint:* (Mass)(acceleration) = force.

6. (*Law of radioactive decay.*) Radioactive substances decay at a rate proportional to the amount present, the time required for decay of half the amount initially present being the *half-life*. Obtain an expression for the half-life of a radioactive substance if the amounts A_0 and A are present at times t_0 and $t_0 + \tau$, respectively. Check by comparison with Example 4. Does the result depend on t_0?

7. If 3 g of a radioactive substance is present at time $t = 1$ year and 1 g at $t = 4$ years, how much was present initially, and what is the half-life? (For terminology see Prob. 6.)

8. A curve $y = f(x)$ in cartesian coordinates is such that the projection of the tangent on the x axis has the constant length $1/k$ as shown in Fig. 5. A second curve $r = g(\theta)$ in polar coordinates cuts the radius at the constant angle $\psi = \tan^{-1}(1/k)$, as shown in Fig. 6. (*a*) Obtain the differential equations

$$\frac{dy}{dx} = ky \qquad \text{and} \qquad \frac{dr}{d\theta} = kr$$

in the two cases, respectively, and show, conversely, that positive solutions of these equations have the stated geometric properties. (*b*) Discuss the application of the results of part *a* to Examples 1 to 4 and Probs. 2 to 6.

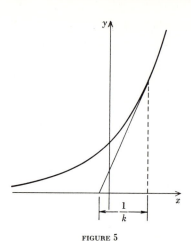

FIGURE 5

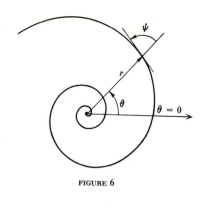

FIGURE 6

9. A sailor raising a 2,000-lb anchor with the aid of a capstan finds that a tension of 100 lb is just sufficient to prevent the rope from slipping. Suppose the rope starts to slip, so that the coefficient of friction μ changes from its former static-friction value of 0.4 to a sliding-friction value of 0.2. How much tension must the sailor exert to stop the slipping?

10. According to accepted physical principles, the probability of radioactive change for any given atom in the time interval Δt is asymptotic to $k\,\Delta t$ and does not depend on the state of other atoms. (The proportionality constant depends, however, on the element being considered.) If A measures the number of atoms in a sample of a radioactive element at time t, justify the approximate equality $\Delta A \doteq -kA\,\Delta t$, and discuss the validity of the conclusion $dA/dt = -kA$. (Because of the large numbers involved, theoretical objections have less relevance here than they did in Example 4. A millionth of a microgram contains about 10^{13} atoms. In our view the statistical basis underlying certain applications has been sufficiently emphasized by this problem and Example 4, and we do not discuss the subject again.)

---

3. Separation of Variables.

Many differential equations have the form

$$f(x)\,dx = g(y)\,dy$$

in which x alone occurs on one side of the equation and y alone on the other. The variables are said to be *separated*, and an equation that can be reduced to this form by reversible algebraic operations is said to be a *separable equation*. When f and g are continuous, a solution containing an arbitrary constant is readily obtained by integration.

As an illustration, the equation

$$dy + e^x y\,dx = e^x y^2\,dx \tag{3-1}$$

is separable, because upon rearrangement it becomes $dy = e^x(y^2 - y)\,dx$, or

$$\frac{dy}{y^2 - y} = e^x\,dx \qquad y \neq 0 \qquad y \neq 1.$$

Near a point where $y \neq 0$ and $y \neq 1$, integration gives

$$\log\left|\frac{y - 1}{y}\right| = e^x + c$$

and the initial condition $y(x_0) = y_0$ is satisfied by appropriate choice of c. For example, the solution through $(0,\frac{1}{2})$ has $c = \log 1 - e^0 = -1$. The process breaks down if $y_0 = 0$ or 1, but a solution for this case is $y \equiv 0$ or $y \equiv 1$, respectively, as is clear from (3-1).

By using definite integrals rather than indefinite integrals the solution through (x_0,y_0) can be obtained without introduction of an arbitrary constant. A statement sufficiently general for applications is:

UNIQUENESS AND EXISTENCE FOR THE SEPARABLE EQUATION. In a given region R of the (x,y) plane, let $f(x)$ and $g(y)$ be continuous, and suppose $f(x)$ and $g(y)$ do not vanish simultaneously. Then the equation

$$f(x)\,dx = g(y)\,dy$$

has one, and only one, solution through each point (x_0,y_0) of R, and the solution is given implicitly by

$$\int_{x_0}^{x} f(t)\,dt = \int_{y_0}^{y} g(t)\,dt.$$

Although the proof is not presented here, the main idea is to show that y is determined uniquely as a differentiable function of x at points where $g(y) \neq 0$, and x is determined uniquely as a differentiable function of y at points where $f(x) \neq 0$. The theorem then follows from the fact that one or the other condition holds at each point of the region considered. As shown in Sec. 7, the hypothesis that $f(x)$ and $g(y)$ do not vanish simultaneously is essential.

There are several broad classes of equations that are not separable as they stand but can be reduced to separable form by a simple change of variable. For example, the equation

$$\frac{dy}{dx} = F\left(\frac{y}{x}\right) \tag{3-2}$$

suggests the substitution $u = y/x$. Then $y = xu$, and

$$\frac{dy}{dx} = x\frac{du}{dx} + u = F(u)$$

where the second equality follows from (3-2). Rearrangement gives

$$\frac{du}{F(u) - u} = \frac{dx}{x} \tag{3-3}$$

which is separated. Hence u can be found by integration, and then $y = xu$ satisfies (3-2).

Equation (3-2) requires $x \neq 0$ and (3-3) also requires $u \neq F(u)$. A value u_0 such that $u_0 = F(u_0)$ is called a *fixed point* for F. If u_0 is a fixed point, (3-2) has an obvious solution $y = u_0 x$.

As another illustration, if a and b are constant, the equation

$$y' = F(a + bx + y) \tag{3-4}$$

suggests the substitution $u = a + bx + y$. Then

$$u' = y' + b = F(u) + b$$

where the second equality follows from (3-4). Separating variables gives

$$\frac{du}{F(u) + b} = dx. \tag{3-5}$$

Thus we get u, and $y = u - a - bx$. The analysis here requires $F(u) + b \neq 0$. If $F(u_0) = -b$ the original equation has a solution $y = u_0 - a - bx$.

Since both (3-2) and (3-4) were reduced to separable form, the existence and uniqueness theorem for separable equations can be used, if desired, to get a similar theorem for these other equations.

Example 1.   Solve $y^2 + x^2 y' = xyy'$.   Here

$$\frac{dy}{dx} = \frac{y^2}{xy - x^2} = \frac{(y/x)^2}{y/x - 1}$$

so that $F(u) = u^2/(u-1)$ in (3-2).   We compute $F(u) - u$ and substitute in (3-3), with the result

$$\frac{u-1}{u}\, du = \frac{dx}{x}.$$

Integration yields $u - \log|u| = \log|x| + c$, and since $u = y/x$, the final answer is

$$\frac{y}{x} - \log|y| = c.$$

Although the analysis requires $y \neq 0$, it is evident that $y \equiv 0$ satisfies the original equation.

The result can be simplified, as shown next.   If we write $c = \log|c_1|$, as we may for all c, the solution appears as

$$\frac{y}{x} = \log|y| + \log|c_1| = \log|yc_1|.$$

Since we can choose the sign of c_1 so that $c_1 y$ is positive, there is no real loss of generality in dropping the absolute value.   Hence, writing c instead of c_1,

$$y = x \log cy.$$

This form is not only simpler but gives added insight.   If multiplied by c, it becomes $cy = cx \log cy$; hence, all integral curves, except the curve $y = 0$, are geometrically similar to the single curve obtained for $c = 1$.   A corresponding remark applies to the general equation (3-2), since the latter is unchanged when (x,y) is replaced by (cx,cy), $c \neq 0$.

Example 2.   Solve $dy - dx = x\, dx + y\, dx$.   Rewriting gives $y' = x + y + 1$, which has the form (3-4) with $F(u) = u$ and $b = 1$.   The transformed equation (3-5) is

$$\frac{du}{u+1} = dx \qquad \text{where } u = x + y + 1.$$

Integration gives $u + 1 = ce^x$; hence $y = ce^x - x - 2$.

Example 3.   By the preceding example, solve $y' = (1 + x + y)^{-1}$.   If we interchange x and y then dy/dx becomes dx/dy, so that y' becomes $1/y'$, and the resulting equation is the same as that in Example 2.   Interchanging x and y in the solution of Example 2 gives the solution of the present problem: $x = ce^y - y - 2$.

PROBLEMS[1]

1. Solve (a), and in (b) get the solution through $(0,1)$:

(a) $(1 + x^2)\, dy = dx$, $\sqrt{1 - x^2}\, dy = \sqrt{1 - y^2}\, dx$, $y' = xy^2 - x$, $y' \sin y = \sin^2 x$;

(b) $dy = y^2\, dx$, $(1 + x)y' = 1 + y$, $\sin x \cos^2 y = y' \cos^2 x$, $\sqrt{1 + x}\, dy = (1 + y^2)\, dx$.

2. Solve by letting $y = ux$, and in (b) find the solution through $(1,1)$:

(a) $(x^2 + y^2)y' + 2xy = 0$, $xy' - y = \sqrt{x^2 - y^2}$, $(y - xy')\cos(y/x) = x$;

(b) $(x + y)y' = x - y$, $x^2 y\, dx = (x^3 - y^3)\, dy$, $x^2 y' = xy - y^2$, $x^3 y' = 2xy^2 - y^3$.

[1] The problems in Secs. 3 to 6 are intended chiefly for the acquisition of important manipulative skills, and we have included only as many as we consider sufficient for this purpose.   About 900 additional problems with answers can be found in J. L. Brenner, "Problems in Differential Equations," W. H. Freeman and Company, San Francisco, 1963.

3. A function $f(x,y)$ is *homogeneous of degree n* if $f(tx,ty) = t^n f(x,y)$ holds for all x, y, and $t \neq 0$ such that the right side is defined. Verify that the following functions are homogeneous and find the degrees: $x^3 + x^2y + y^3$, $x^2 \sin(x/y) + xy$, $x(\log y - \log x)$, $(x^5 + y^5)^5$, $(x^3y + y^3x)e^{y/x}$.

4. (*a*) By picking $t = 1/x$ in Prob. 3, show that a homogeneous function of degree n can be written in the form $f(x,y) = x^n \phi(y/x)$ when $x \neq 0$. (*b*) If $P(x,y)$ and $Q(x,y)$ are homogeneous functions of degree n, and if $xQ(x,y) \neq 0$ holds at all relevant points (x,y), show that the differential equation $P\,dx + Q\,dy = 0$ can be written $y' = \phi(y/x)$ and hence solved by setting $u = y/x$. (The importance of this observation is that it facilitates recognition of the corresponding class of differential equations.)

5. Solve by separating variables or by letting $y = ux$:

(*a*) $\sinh x\, dy + \cosh y\, dx = 0$, $xy' = x + y$, $(x - \sqrt{xy})y' = y$, $\sqrt{xy} + y = xy'$;

(*b*) $x^2 y'(1 - y) = y^2$, $xy' = y + xe^{y/x}$, $y'(1 - \log y) = \tan x \sec^2 x$, $y' \sin x = \sin y$.

6. Solve by an appropriate substitution:

$$y' = 1 + \cot(2 + y - x), \qquad y' = (x + y + 3)^{-2}, \qquad xyy' = x^2 + y^2, \qquad e^x(y' + 1) = e^y.$$

7. A man and a parachute weighing w lb fall from rest under the force of gravity. If the air resistance is proportional to the square of the speed v, and if the limiting speed is v_0, find the speed as a function of the time t and as a function of the distance fallen s. *Hint:* $(w/g)(dv/dt) = w - kv^2$.

8. Water is flowing out through a circular hole in the side of a cylindrical tank 2 ft in diameter. The velocity of the water in the jet has the value it would attain by a free fall through a distance equal to the head. How long will it take the water to fall from a height of 25 ft to a height of 9 ft above the orifice if the stream of water is 1 in. in diameter?

9. (*Review.*) Water is flowing from a 2-in. horizontal pipe running full. Find the discharge in cubic feet per second if the jet of water strikes the ground 4 ft beyond the end of the pipe when the pipe is 2 ft above the ground. *Hint:* The horizontal and vertical components of the motion of a water particle can be treated independently. Neglect air resistance.

---

4. Exact Differential Equations. We now consider an important class of differential equations

$$P(x,y)\,dx + Q(x,y)\,dy = 0 \tag{4-1}$$

under the hypothesis that $P(x,y)$ and $Q(x,y)$ have continuous partial derivatives at every point under consideration.[1] The expression on the left of (4-1) is said to be *exact* if it coincides with the differential dF of some function $F(x,y)$, that is, if

$$P(x,y)\,dx + Q(x,y)\,dy = \frac{\partial F}{\partial x}\,dx + \frac{\partial F}{\partial y}\,dy \equiv dF \tag{4-2}$$

where x and y are independent variables. In these circumstances Eq. (4-1) is simply $dF = 0$, and its solution, therefore, is

$$F(x,y) = c. \tag{4-3}$$

The truth of this is strongly suggested by the notation, but further discussion is needed because x and y are not independent when (4-1) holds. Thus, y is a function of x, or x of y, or perhaps x and y are functions of a parameter t. The equivalence of (4-1) with the equation $dF = 0$ actually follows from the chain rule established in Chap. 5, Sec. 5. The chain rule permits going in both directions; if $x(t)$ and $y(t)$ are differentiable functions satisfying $F(x,y) = c$ then (4-1) holds. and, conversely, if (4-1) holds on a curve $x = x(t)$, $y = y(t)$, with x and y differentiable, $F[x(t),y(t)]$ is constant.

---

[1] The discussion depends on a few theorems concerning functions of two variables presented in Chap. 5, and we refer to these as they are used. However, the reader meeting the subject for the first time is not expected to know these theorems, nor is such knowledge needed for effective use of the results obtained.

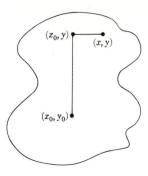

FIGURE 7

Since the variables in (4-2) are independent, we can let $dx = 0$ or $dy = 0$ and deduce the representation

$$P(x,y) = \frac{\partial F}{\partial x} \qquad Q(x,y) = \frac{\partial F}{\partial y} \qquad (4\text{-}4)$$

for the coefficients P and Q.   It is proved in books on advanced calculus that

$$\frac{\partial^2 F}{\partial x\, \partial y} = \frac{\partial^2 F}{\partial y\, \partial x}$$

if these derivatives are continuous; hence differentiation of (4-4) gives a necessary condition

$$\frac{\partial P}{\partial y} = \frac{\partial Q}{\partial x} \qquad (4\text{-}5)$$

for $P\, dx + Q\, dy$ to be exact.   The condition is also sufficient, in general, as indicated next.

We suppose that the region has the properties suggested by Fig. 7; namely, the segment joining (x_0,y_0) to (x_0,y) is required to be in the region, and so is the segment joining (x_0,y) to (x,y).   Here (x_0,y_0) is a suitably chosen fixed point and (x,y) is an arbitrary point.

On integrating the first of Eqs. (4-4) with respect to x we get

$$F(x,y) = \int_{x_0}^{x} P(x,y)\, dx + f(y) \qquad (4\text{-}6)$$

where $f(y)$ is an arbitrary function of y, since y appearing in the integral is treated as constant.   The expression (4-6) clearly satisfies $\partial F/\partial x = P(x,y)$ for every choice of $f(y)$. To make $\partial F/\partial y = Q(x,y)$ we require

$$\frac{\partial}{\partial y} \int_{x_0}^{x} P(x,y)\, dx + f'(y) = Q(x,y). \qquad (4\text{-}7)$$

It is proved in books on advanced calculus that the order of integration and differentiation can be interchanged, under our present assumptions, and hence the first term in (4-7) is

$$\int_{x_0}^{x} \frac{\partial}{\partial y} P(x,y)\, dx = \int_{x_0}^{x} \frac{\partial}{\partial x} Q(x,y)\, dx = Q(x,y) - Q(x_0,y).$$

Here the first equality follows from (4-5) and the second from the fundamental theorem of calculus.   Substitution into (4-7) gives $f'(y) = Q(x_0,y)$, integration gives $f(y)$, and hence by (4-6)

$$F(x,y) = \int_{x_0}^{x} P(x,y)\, dx + \int_{y_0}^{y} Q(x_0,y)\, dy. \qquad (4\text{-}8)$$

The same theorems used to justify these steps justify their reversal, so that $F(x,y)$ satisfies (4-4).   In other words, $\partial F/\partial x = P$ and $\partial F/\partial y = Q$.

For example, the equation $(2xy + 1)\, dx + (x^2 + 4y)\, dy = 0$ is exact since each derivative in (4-5) equals $2x$.   Substitution into (4-8) with $x_0 = y_0 = 0$ gives

$$F(x,y) = \int_{0}^{x} (2xy + 1)\, dx + \int_{0}^{y} 4y\, dy = x^2 y + x + 2y^2$$

and hence the solution is $x^2 y + x + 2y^2 = c$.

The integral (4-8) is an example of a *line integral*, and the rectilinear path shown in Fig. 7 is the *path of integration*.   Other choices of the path lead to other formulas. For example, the path shown in Fig. 8 produces the formula

$$F(x,y) = \int_{x_0}^{x} P(x,y_0)\, dx + \int_{y_0}^{y} Q(x,y)\, dy. \quad (4\text{-}9)$$

The proof that $dF = P\, dx + Q\, dy$ is practically identical with the proof just given, except that it starts with the second relation (4-4) rather than with the first. Naturally, the region must be restricted so that the path of integration does not go outside it. In practice it is advisable to try both formulas (4-9) and (4-8) and use the one that leads to s'mpler integrals.

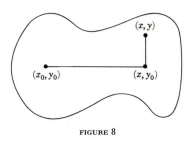

FIGURE 8

Another result is obtained by using a straight-line path from (x_0,y_0) to (x,y). If $x_0 = y_0 = 0$ the formula is

$$F(x,y) = \int_{0}^{1} [xP(tx,ty) + yQ(tx,ty)]\, dt. \quad (4\text{-}10)$$

Here the condition is that the segment joining (x,y) to the origin be in the region whenever (x,y) is in the region. A region having this property is said to be *star-shaped* with respect to the origin.

We do not give the proof of (4-10) because direct verification is rather long and contributes nothing to the main ideas. Actually all three formulas (4-8) to (4-10) can be obtained by inspection as soon as the reader is familiar with the theory of line integrals (Chap. 5, Secs. 19 to 23).

To illustrate the use of (4-10), we consider the exact equation

$$[6(x + y)^5 + e^x]\, dx + [6(x + y)^5 + e^y]\, dy = 0.$$

We compute $P(xt,yt)$ by writing xt in place of x and yt in place of y, thus:

$$P(xt,yt) = 6(xt + yt)^5 + e^{xt} = 6t^5(x + y)^5 + e^{xt}.$$

The formula for $Q(xt,yt)$ is similar, and substitution gives

$$F(x,y) = \int_{0}^{1} [6t^5x(x + y)^5 + xe^{xt} + 6t^5y(x + y)^5 + ye^{yt}]\, dt$$

$$= (x + y)^6 + e^x + e^y - 2.$$

Equation (4-10) has an interesting application to homogeneous functions. A function $f(x,y)$ is said to be *homogeneous of degree n* if

$$f(tx,ty) = t^n f(x,y) \quad (4\text{-}11)$$

holds for all x,y and $t \neq 0$ for which the right-hand side is defined. (The condition is required here only for $t > 0$.) If P and Q are homogeneous of degree $n > -1$, substitution of

$$P(tx,ty) = t^n P(x,y) \qquad Q(tx,ty) = t^n Q(x,y)$$

into (4-10) produces the formula

$$F(x,y) = \int_{0}^{1} [xt^n P(x,y) + yt^n Q(x,y)]\, dt = \frac{xP(x,y) + yQ(x,y)}{n + 1}. \quad (4\text{-}12)$$

It follows that an exact equation $P(x,y)\, dx + Q(x,y)\, dy = 0$ has a solution

$$xP(x,y) + yQ(x,y) = c$$

if P and Q are homogeneous functions of the same degree $n > -1$ with continuous derivatives in a neighborhood of the origin.

Example. Solve $[x^2(x^3 + y^3)^{1/3} + e^x]\, dx + [y^2(x^3 + y^3)^{1/3} + y^5]\, dy = 0$. The left-hand member is the sum of the two differentials

$$(x^3 + y^3)^{1/3}(x^2\, dx + y^2\, dy) \qquad \text{and} \qquad (e^x\, dx + y^5\, dy)$$

both of which are exact. The coefficients in the first differential are homogeneous of degree $n = 3$, as the reader can verify, and hence by (4-12)

$$(x^3 + y^3)^{1/3}(x^2\,dx + y^2\,dy) = d\,\frac{(x^3 + y^3)^{4/3}}{4}.$$

Similarly,

$$e^x\,dx + y^5\,dy = d(e^x + \tfrac{1}{6}y^6)$$

as is clear by inspection. The sum of these two differentials duplicates the left side of the given differential equation, so that the solution is

$$\tfrac{1}{4}(x^3 + y^3)^{4/3} + e^x + \tfrac{1}{6}y^6 = c.$$

The method depends on the fact that

$$d(F_1 + F_2) = dF_1 + dF_2$$

whenever the right-hand side makes sense. It is also a consequence of the definition that $d(cF) = c\,dF$ for any constant c. Any operator d satisfying these two conditions is said to be *linear*. The linearity of the differential is an important aid in the solution of differential equations.

PROBLEMS

1. Integrate the following equations if they are exact:
(a) $e^x\,dx = dy - dx$, $4x\,dx = e^{x/y}(dy - 2y^{-1}\,dx)$, $3x^2y - y^3 = (3y^2x - x^3)y'$;
(b) $x\,dx + y\,dy = 0$, $(y\,dx + dy)e^x = 0$, $(3x^2y + y^3)\,dx + (x^3 + 3y^2x)\,dy = 0$;
(c) $(y\cos xy + 2x)\,dx + x\cos xy\,dy = 0$, $(y^2 + 2xy + 1)\,dx + (2xy + x^2)\,dy = 0$;
(d) $3x^2y\,dx = (3x^2y^2 - x^3)\,dy$, $xy(2x\,dy + y\,dx)\sin xy^2 = \cos xy^2\,dx$.

2. Show that the following are exact, integrate by inspection, and check by differentiation:

$$4x^2(x^3 + y^3)^{1/3}\,dx + 4y^2(x^3 + y^3)^{1/3} + y^3\,dy = 0, \qquad (x^2 + y^2)^2(x\,dx + y\,dy) + x^5\,dx = 0.$$

3. Show that the equation $(y^2 + xy + 1)\,dx + (x^2 + xy + 1)\,dy = 0$ becomes exact when multiplied by e^{xy}, and thus solve it. (The function e^{xy} is an *integrating factor* for this equation, as discussed in the following section.)

4. Referring to Prob. 3, show that the following equations have the integrating factors $(x + y)^{-2}$ and e^{x^2y}, respectively, and solve:

$$y\,dx - x\,dy = (dx + dy)(x + y), \qquad x^3\,dy + (2x^2y + 1)\,dx = 0.$$

5. Show that a separated equation with continuous coefficients is exact, and compare the solutions given by (4-8) and (4-9) with that in the theorem of the preceding section.

6. What condition must the constant coefficients a, b, c, A, B, C satisfy if

$$(ax^2 + 2bxy + cy^2)\,dx + (Ax^2 + 2Bxy + Cy^2)\,dy$$

is an exact differential dF, and what is F in that case? *Hint:* The first question is answered by (4-5) and the second by (4-12).

7. Formulate and solve an analog of Prob. 6 for polynomials of degree 1 and also for polynomials of degree 3.

8. (*Review of separable equations.*) Find the curves in the xy plane which satisfy the following conditions: (a) The tangents pass through the origin; (b) the normals pass through the origin; (c) the segment of tangent between a point on the curve and the x axis has unit length; (d) the area bounded by the curve, the x axis, and the ordinate equals the ordinate; (e) the area equals the length of the curve from $(0,1)$ to (x,y). *Hint:* In part e, solve algebraically for y' in terms of y and then separate variables.

5. Integrating Factors. In the preceding section, we found that the general differential equation

$$P(x,y)\, dx + Q(x,y)\, dy = 0$$

can be solved with ease when it is exact. Equations encountered in practice usually are not exact but become exact when multiplied by a suitable function $\mu(x,y)$. Such a function is called an *integrating factor*.

Thus, in order to solve $x\, dy - y\, dx = 0$, which is not exact as it stands, multiply through by $1/xy$. The resulting equation

$$\frac{dy}{y} - \frac{dx}{x} = 0$$

is exact and can be solved by inspection. Here the integrating factor is $\mu(x,y) = 1/xy$. Another integrating factor is $1/x^2$, and a third is $1/y^2$. These three factors reduce the original equation to

$$d\left(\log\frac{y}{x}\right) = 0 \qquad d\left(\frac{y}{x}\right) = 0 \qquad d\left(-\frac{x}{y}\right) = 0$$

respectively.

All three results are essentially equivalent, but the first breaks down when $x = 0$ or $y = 0$, the second when $x = 0$, and the third when $y = 0$. The singular behavior is introduced by the integrating factor. Some integrating factors also introduce extraneous solutions that make $\mu(x,y) = 0$ but do not satisfy the differential equation. Any suspected solution should be checked and eliminated, if need be, by substitution.

A systematic method of finding integrating factors can be based on the theory of Sec. 4. The expression

$$\mu(P\, dx + Q\, dy)$$

is exact if the coefficients μP and μQ are continuously differentiable and satisfy

$$\frac{\partial}{\partial y}\mu P = \frac{\partial}{\partial x}\mu Q. \tag{5-1}$$

This follows from the criterion (4-5), the role of P and Q being now taken by μP and μQ.

As an illustration, we show that the *linear equation* (cf. Prob. 7)

$$\frac{dy}{dx} + r(x)y = f(x) \tag{5-2}$$

has an integrating factor $\mu(x)$ depending on x alone, when r and f are continuous. If (5-2) is multiplied by dx, the coefficients

$$P(x,y) = r(x)y - f(x) \qquad Q(x,y) = 1$$

are read off by inspection, and substitution into (5-1) gives

$$\frac{\partial}{\partial y}\mu(ry - f) = \frac{\partial\mu}{\partial x}$$

or $\mu r = \mu'$. This separable equation for μ has the solution

$$\mu(x) = e^{R(x)} \qquad \text{where } R(x) = \int r(x)\, dx. \tag{5-3}$$

Multiplying (5-2) by $\mu(x)$, we get an equation that can be written

$$\frac{d}{dx}(e^R y) = e^R f \tag{5-4}$$

and from this, y is determined with ease.

For example, consider the equation

$$(x + 1)y' + 2y = 12(x + 1)^4 \qquad x \neq -1. \tag{5-5}$$

Dividing by $x + 1$ shows that this equation is linear with

$$r(x) = 2(x + 1)^{-1} \qquad f(x) = 12(x + 1)^3.$$

The integral of $r(x)$ is $2 \log |x + 1|$, and so

$$e^{R(x)} = e^{2 \log |x+1|} = (e^{\log |x+1|})^2 = (x + 1)^2.$$

Equation (5-4) is

$$\frac{d}{dx} [(x + 1)^2 y] = 12(x + 1)^5$$

and thus we get the solution $(x + 1)^2 y = 2(x + 1)^6 + c$. The same result would be obtained if we multiply (5-5) by the integrating factor $x + 1$. The advantage of reducing the equation to the form (5-2) is that the procedure is systematic.

As a general rule, construction of an integrating factor automatically leads to an existence and uniqueness theorem. For example, since e^R and e^{-R} do not vanish, Eq. (5-4) is *strictly equivalent* to (5-2) on any interval in which the relevant functions are continuous. Upon applying (1-9) to (5-4) we obtain:

EXISTENCE AND UNIQUENESS FOR THE LINEAR EQUATION. *On a given interval (a,b), let $f(x)$ and $r(x)$ be continuous, and let $a < x_0 < b$. Then the equation $y' + ry = f$ has one, and only one, solution such that $y(x_0) = y_0$, and the solution is given by*

$$e^{R(x)} y(x) = \int_{x_0}^x e^{R(t)} f(t) \, dt + y_0 \qquad \text{where } R(x) = \int_{x_0}^x r(s) \, ds.$$

The *Bernoulli equation*

$$y' + r(x)y = y^n f(x) \qquad n = \text{const} \tag{5-6}$$

is a generalization of the linear equation, to which it reduces when $n = 0$. This equation admits an integrating factor of the form $y^m \mu(x)$, where m is constant. Values of m and $\mu(x)$ are determined by substitution into (5-1), with the result

$$\mu(x,y) = y^{-n} e^{(1-n)R} \qquad R = \int r(x) \, dx. \tag{5-7}$$

Multiplying (5-6) by $(1 - n)\mu(x,y)$ yields an equation equivalent to

$$\frac{d}{dx} [y^{1-n} e^{(1-n)R}] = (1 - n)e^{(1-n)R} f \qquad \text{for } y \neq 0 \tag{5-8}$$

which can be solved by integration. We omit the derivation of (5-7) because the final result (5-8) is easy to verify directly. In any region in which $y \neq 0$ Eq. (5-8) produces an existence and uniqueness theorem for the Bernoulli equation similar to that for the linear equation.

Example 1.   Solve $y' + y = xy^3$. Here $n = 3$, $f(x) = x$, and

$$R(x) = \int r(x) \, dx = \int dx = x$$

in (5-8), with the result

$$\frac{d}{dx} (y^{-2} e^{-2x}) = -2e^{-2x} x \qquad y \neq 0.$$

Integration gives the solution $y^{-2} = x + \frac{1}{2} + ce^{2x}$. There is also a solution $y \equiv 0$, corresponding to $c = \infty$.

Example 2.   Solve $(x - y^3) \, dy = y \, dx$. If x and y are interchanged the result is the linear equation $y' - y/x = -x^2$, for which

$$R(x) = \int \left(-\frac{1}{x}\right) dx = -\log |x| \qquad e^{R(x)} = \pm\frac{1}{x} \qquad x \neq 0.$$

Equation (5-4) gives the solution $2y = cx - x^3$; hence a solution to the original problem is $2x = cy - y^3$. The same is obtained by the integrating factor $1/y^2$, as the reader can verify. There is also a solution $y \equiv 0$, corresponding to $c = \infty$.

Example 3.   (*Parabolic Mirror.*)   Find a mirror such that light from a point source at the origin O is reflected in a beam parallel to the x axis.

Let the ray of light OP strike the mirror at P and be reflected along PR (Fig. 9). If PQ is the tangent at P and α, β, ϕ, and θ are the angles indicated, $\alpha = \beta$ by the optical law of reflection and $\alpha = \phi$ by geometry. Hence $\beta = \phi$. The equation

$$\tan \theta = \tan (\beta + \phi) = \tan 2\phi = \frac{2 \tan \phi}{1 - \tan^2 \phi}$$

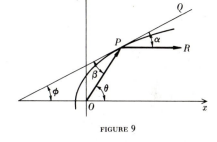

FIGURE 9

gives

$$\frac{y}{x} = \frac{2y'}{1 - (y')^2}$$

since $y' = \tan \phi$. Solution of this quadratic equation for y' gives

$$yy' = -x \pm \sqrt{x^2 + y^2}.$$

The equation admits an integrating factor $\mu(x,y) = (x^2 + y^2)^{-\frac{1}{2}}$ as we see by rearranging in the form

$$\frac{x\, dx + y\, dy}{\pm\sqrt{x^2 + y^2}} = dx.$$

The left side is an exact differential, which gives, on integrating,

$$\pm\sqrt{x^2 + y^2} = x + c$$

or, on squaring, $y^2 = 2cx + c^2$. The curves form a family of parabolas with the focus at the origin.

PROBLEMS

1. The following problems give a few of the integrable combinations that commonly occur in practice. Verify the equations by differentiating:

(a) $d(xy) = x\, dy + y\, dx$, $d\left(\log \dfrac{y}{x}\right) = \dfrac{x\, dy - y\, dx}{xy}$, $d\left(\tan^{-1} \dfrac{y}{x}\right) = \dfrac{x\, dy - y\, dx}{x^2 + y^2}$;

(b) $d\left(\dfrac{x}{y}\right) = -\dfrac{x\, dy - y\, dx}{y^2}$, $d\left(\dfrac{y}{x}\right) = \dfrac{x\, dy - y\, dx}{x^2}$, $d\sqrt{x^2 + y^2} = \dfrac{x\, dx + y\, dy}{\sqrt{x^2 + y^2}}$.

2. Solve the following equations by finding a suitable integrating factor:

(a) $x\, dy + x^2\, dx = y\, dx$, $(xy^2 + y)\, dx = (x^2 y - x)\, dy$, $x\, dx + y\, dy = \sqrt{x^2 + y^2}\, d(xy)$;

(b) $x\, dy + 3y\, dx = xy\, dy$, $(x^2 + y^2 + 2x)\, dy = 2y\, dx$, $y\, dx - x\, dy = d(x^2 y^2)$;

(c) $x\, dy - y\, dx = xy\, dy$, $(x^2 - y^2)\, dy = 2xy\, dx$, $x\, dy - y\, dx = d(x^2 + y^2)^{\frac{1}{2}}$.

3. Solve these linear equations, and in part b find a solution through $(0, -1)$:

(a) $(1 + x^2)y' = x - xy$, $(x^2 + 1)y' + 2xy = x^2$, $xy' + y = x^2 \sin x$, $y' + e^x y = e^{x+1}$;

(b) $y' + xy = x$, $y' = e^{-x^2} - 2xy$, $y' + y \cos x = \cos^3 x$, $y' + 4x^3 y = x^7$.

4. Solve the following examples of Bernoulli's equation:

(a) $y^3 y' + x^{-1} y^4 = x$, $y' + y = xy^3$, $xy' + y = x^3 y^6$, $y' + x^{-2} y^2 = x^{-1} y$;

(b) $d(xy) = y^2 \log x\, dx$, $y' + xy = x^3 y^3$, $y^3 + 3y^2 y' = 4$, $y - y' = 3y^3 e^{-2x}$.

5. Solve by interchanging x and y, or otherwise:

$$dx = (yx^3 - x)\, dy, \qquad y\, dx = (x + y^3)\, dy, \qquad y + xy' = e^y y', \qquad 1 + xy' \tan y = y'.$$

6. Solve by any method:

(a) $y' = y + \cos x - \sin x$, $dx = y(1 - x)\, dy$, $xyy' = y^2 - x\sqrt{x^2 - y^2}$;

(b) $x^2(1 + 4y^2)\, dx + 3x^3y\, dy = 0$, $y^{-1} \sin^{-1} x\, dx = (e^y - 1)\, dy$, $y' + xy = y$.

7. Show that the operator $\mathbf{T}$ defined by $\mathbf{T}y = y' + r(x)y$ is *linear*, that is,

$$\mathbf{T}(\alpha u + \beta v) = \alpha \mathbf{T}u + \beta \mathbf{T}v$$

for all constants α, β and all functions u, v for which $\mathbf{T}u$ and $\mathbf{T}v$ are meaningful.   [The equation $\mathbf{T}y = f(x)$ in (5-2) is called "linear" not only because it is linear in y and y' but because it is associated with the linear operator $\mathbf{T}$.   For further discussion see Secs. 9 and 16 of this chapter; also see Chap. 3.]

8. Let $P(x,y)\, dx + Q(x,y)\, dy = 0$ have a solution $F(x,y) = c$ in a given region, so that the solution curves $x = x(t)$, $y = y(t)$ satisfy

$$\frac{\partial F}{\partial x} x' + \frac{\partial F}{\partial y} y' = 0 \qquad \text{and} \qquad Px' + Qy' = 0.$$

Assuming that certain relevant quantities do not vanish, deduce that the coefficients of x' and y' in the two equations are proportional, so that we can define simultaneously

$$\mu(x,y) = \frac{\partial F/\partial x}{P(x,y)} \qquad \mu(x,y) = \frac{\partial F/\partial y}{Q(x,y)}.$$

Hence $\mu(P\, dx + Q\, dy) = dF$ or, in other words, μ is an integrating factor.   (The method is interesting as an existence proof, but it presupposes knowledge of the solution.)

---

6. The Derivative as New Variable.   This section contains several applications of the substitution $p = y'$, in which the derivative is introduced as a new variable.   The physical meaning of the substitution depends on circumstances.   In problems of dynamics the independent variable is the time t, and $p = dy/dt$ is the velocity.   In problems about the geometry of curves, $p = y'$ is the slope at the given point; in problems of growth, p is the rate of increase; and so on.

As our first illustration, we consider equations that can be solved algebraically for p as a function of x and y.   We assume y to be continuously differentiable.

Example 1.   Solve $2p^2 - (2y^2 + x)p + xy^2 = 0$, where $p = dy/dx$.   Factoring gives

$$(p - y^2)(2p - x) = 0$$

so that, at each x, either $p = y^2$ or $p = \frac{1}{2}x$.   The assumption that p is continuous ensures that one or the other of these relations actually holds throughout an interval.   Hence, with $p = dy/dx$, they can be regarded as differential equations and solved in the usual way.

From $dy/dx = y^2$ and from $dy/dx = \frac{1}{2}x$ we obtain, respectively,

$$xy + 1 + c_1 y = 0 \qquad \text{and} \qquad 4y - x^2 + c_2 = 0.$$

Additional solutions can be constructed by joining curves of one family to curves of the other at points of the parabola $y^2 = \frac{1}{2}x$, where the two values of the slope p agree.   It is possible to show that these additional solutions are included if the result is written in the form

$$(xy + 1 + c_1 y)(4y - x^2 + c_2) = 0$$

though this aspect of the subject is not emphasized here.   (Cf. Sec. 7.)

Another method is to solve for y as a function of x and p, and then differentiate with respect to x, using $dy/dx = p$. We shall apply this method to the *d'Alembert-Lagrange equation*,

$$y = xf(p) + g(p) \qquad \text{where } p = y'. \tag{6-1}$$

It is assumed that f and g are differentiable functions of p and that p is a differentiable function of x, so that the chain rule applies.

Differentiating with respect to x yields

$$p = xf'(p)\frac{dp}{dx} + f(p) + g'(p)\frac{dp}{dx}$$

or, if $p \neq f(p)$,

$$\frac{dx}{dp} - \frac{f'(p)x}{p - f(p)} = \frac{g'(p)}{p - f(p)}. \tag{6-2}$$

This equation is linear in x, that is, of the form $dx/dp + M(p)x = N(p)$, and can be solved by the method of the preceding section. Its solution for x as a function of p, together with (6-1), yields a solution of the original equation in parametric form, with p as parameter.

If $p_0 = f(p_0)$, so that p_0 is a fixed point, the equation has an obvious solution $y = p_0 x + g(p_0)$. Further discussion of the condition $p = f(p)$ is given in Prob. 2.

Example 2. Solve $p^5 + py = 1$, where $p = dy/dx$. Since it is impractical to solve this equation in the form $p = f(y)$ (which would have led to a separable equation) we solve it for y and obtain

$$y = p^{-1} - p^4. \tag{6-3}$$

Differentiating with respect to x leads to

$$p\,dx = -p^{-2}\,dp - 4p^3\,dp$$

since $dy = p\,dx$ by definition of p. Dividing by p and integrating gives

$$x = \tfrac{1}{2}p^{-2} - \tfrac{4}{3}p^3 + c$$

which, together with (6-3), describes the desired solution in parametric form. There is little advantage in eliminating the parameter p even when it is possible to do so. Plotting the solution curves as p varies, one obtains not only the locus of points (x,y) but also the slope p at each point.

The substitution $p = y'$ can be used to reduce many higher-order equations to equations of the first order. For example, since the derivative of the first derivative is the second derivative, $y'' = p'$ whenever y'' exists. The second-order equation $y'' = f(x)$ is equivalent to $p' = f(x)$; an integration gives p, and another integration gives y.

The same method applies to the vastly more general equation

$$F(x,y',y'') = 0. \tag{6-4}$$

Upon substituting $p = y'$ we get $F(x,p,p') = 0$. This first-order equation is solved for p, and an integration gives y.

Example 3. Find solutions of $y'' \sin y' = \sin x$ that satisfy $y'(1) = 1$. Being free of y, the equation has the form (6-4), and it becomes $p' \sin p = \sin x$ when we set $p = y'$. Solving this separable equation yields

$$-\cos p = -\cos x + c.$$

The condition $p = 1$ at $x = 1$ gives first $c = 0$, and then $p = x$. Writing dy/dx for p in the equation $p = x$ gives, on integration, the final answer

$$y = \tfrac{1}{2}x^2 + c.$$

The result contains an arbitrary constant because two conditions are required, in general, to determine the solution of a second-order equation; the single condition $y'(1) = 1$ is not enough.

In the present example, we can prescribe the value of y at any point. For instance, the solution satisfying $y(3) = 5$ is obtained when $c = \frac{1}{2}$.

Another class of second-order equations can be solved by means of the relation

$$y'' = \frac{dp}{dx} = \frac{dp}{dy}\frac{dy}{dx} = \frac{dp}{dy}p. \tag{6-5}$$

This holds if $y = \phi(x)$, where ϕ is any twice-differentiable function such that the equation $y = \phi(x)$ can be inverted[1] to express x as a differentiable function of y. According to (6-5) the second-order equation

$$F(y,y',y'') = 0 \qquad ' = \frac{d}{dx} \tag{6-6}$$

is reduced to a first-order equation

$$F(y,p,pp') = 0 \qquad ' = \frac{d}{dy}. \tag{6-7}$$

After p is found we get y from the separable equation $dy/dx = p(y)$.

The solution of (6-7) for p is said to give a *first integral* of the original equation (6-6). Similarly, determination of p from (6-4) gives a *first integral* of that equation. This terminology is suggested by the fact that a single integration is usually involved in the calculation of p, and a second integration is then needed to get y.

Example 4. Solve $yy'' + y^2 = 2(y')^2$. The substitution $y' = p$, $y'' = p\,dp/dy$ gives

$$yp\frac{dp}{dy} + y^2 = 2p^2 \qquad \text{or} \qquad \frac{dp}{dy} = 2\frac{p}{y} - \frac{y}{p}$$

when $yp \neq 0$. Since the right-hand side has the form $F(p/y)$ we set $p = vy$ and solve as in Sec. 3. The resulting first integral is

$$p = \pm y\sqrt{1 + c^2y^2}.$$

Writing $p = dy/dx$ and separating variables gives the final answer,

$$cy = \operatorname{csch}(x + c_1).$$

(The solution $y \equiv 0$, excluded in the above calculation, corresponds to $c = \infty$.)

To verify the result we can observe that the differential equation is unaltered by the substitutions $y \to cy$, $x \to x + c_1$. Hence, it suffices to check the single solution $y = \operatorname{csch} x$.

A final example, taken from the theory of pursuit, involves several of the methods that were introduced in this discussion.

Example 5. A boat A moves along the y axis with constant speed a. Find the path of a second boat B which moves in the left-hand half of the xy plane with constant speed b and always points directly at A (see Fig. 10).

At a time t min after A is at $(0,0)$, we shall have A at $(0,at)$ and B at (x,y), say. Since the line AB is tangent to the path of B, the slope of this line equals the slope of the path, so that

$$\frac{y - at}{x - 0} = \frac{dy}{dx}$$

or $xy' - y = -at$. To eliminate t, we differentiate and obtain

$$(xy' - y)' = xy'' = -a\frac{dt}{dx}. \tag{6-8}$$

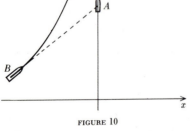

FIGURE 10

[1] It is shown in books on calculus that the inversion is possible if $\phi'(x) \neq 0$ on the interval in question. See Chap. 5, Sec. 9.

Since $ds/dt = b$, where s is an arc on the trajectory, we have

$$\frac{dt}{dx} = \frac{dt}{ds}\frac{ds}{dx} = \frac{1}{b}\sqrt{1 + y'^2}. \tag{6-9}$$

With r defined as a/b, substituting (6-9) in (6-8) yields

$$xy'' = -r\sqrt{1 + y'^2}.$$

This is reduced to a separable equation by letting $p = y'$, as in the discussion of (6-4). The solution is

$$p = \sinh\left(r \log \frac{c}{x}\right) = \frac{1}{2}\left[\left(\frac{c}{x}\right)^r - \left(\frac{x}{c}\right)^r\right] \qquad r = \frac{a}{b}$$

and y is found by integration.

PROBLEMS

1. Solve (a) as in Example 1, (b) as in Example 2, and (c) by either method:

(a) $p^2 - 2py = 3y^2$, $p^2 + 1 = 2p$, $p^2 + p(yx^{-1} - xy^{-1}) = 1$, $p^3 + (xy)^3 = py(y + px^3)$;

(b) $p^4 = p^2y + 2$, $p^3 + p = e^y$, $y = xp + p^2$, $p^2(e^p + 1) = y + 1$, $y = xp^2 + 1$;

(c) $py = 1 + p^2x$, $p^2 + y^2 = 1$, $p(p + 2x - y) = 2xy$, $p^2 + xp = e^x(x + p)$, $x = 2y - p^3$.

2. The *Clairaut equation* is $y = xp + g(p)$. It is a special case of the d'Alembert-Lagrange equation (6-1). Derive the solution $y = xc + g(c)$ and verify it. Are there other solutions?

3. Solve (a) as in Example 3, (b) as in Example 4, and (c) by either method:

(a) $(1 - x^2)y'' = xy'$, $xy'' = x^2 + y'$, $xy'' + 2y' = 3x$, $xy'' + 1 = e^{-y'}$, $xy'' + x(y')^2 = y'$;

(b) $y'' = 0$, $y'' = yy'$, $y'' = y'$, $y'' + 4y = 0$, $y'' = (y')^2$, $y'' = e^y y'$, $2y'' = \cos y$;

(c) $x^2y'' = 1 - x$, $yy'' = (y')^2$, $xy'' = 4x - 2y'$, $y'' = 1 + (y')^2$, $x^2e^y(xy'' + 3) = 1$.

4. Obtain first integrals for the following equations, given $y'(1) = m$. Show that certain choices of m simplify the determination of y, and find y in those cases:

$$y'y'' = 5x^4, \qquad 2y'y''x + (y')^2 = 4x^3, \qquad e^{y'+x}(y'' + 1) = 4x, \qquad (xy')(xy')' = 2x^2.$$

5. Obtain first integrals for the following equations such that $y' = m$ when $y = 0$. Show that certain choices of m simplify the determination of y, and find y in those cases:

$$y'' - y = 1, \qquad y'' = e^y, \qquad y'' = y^3, \qquad y'' + (y')^2 = yy', \qquad y'' + (y')^2 + y = 0.$$

6. If $p = y'$ the plane with cartesian coordinates (y, p) is called the *phase plane*, or *phase space*. Discuss the phase-plane graph of p versus y for (a) $y'' + y = 0$; (b) selected examples in Prob. 3b; and (c) selected examples in Prob. 5.

REVIEW PROBLEMS

7. Solve the following equations by any method:

(a) $d(xy) + y^3\,dy = 0$, $y(1 + x^2)^{-1}\,dx + \tan^{-1} x\,dy = 0$, $(1 + x^2)\,dy = (1 + y^2)\,dx$;

(b) $\sin 2y\,dx + 2x\cos 2y\,dy = 0$, $e^xy' = e^x + e^y$, $dx + 2x\,dy = y\,dy$, $1 = (yx^3 - x)y'$;

(c) $(x^2 + y^2)\,dx = xy\,dy$, $y' = 2y + e^{3x}$, $y^2\,dx = (xy - x^3e^y)\,dy$, $y' + 3y = xy^3$;

(d) $x(y')^2 + 1 = (1 + x)y'$, $x(1 - y') = 1$, $xy' = (x^2 + y^2)^{\frac{1}{2}}$, $y' = (x + y)^2$.

8. Let $x = u - h$, $y = v - k$, where h and k are suitably chosen constants, and thus reduce the equation

$$\frac{dy}{dx} = \frac{x - y - 2}{x + y + 6} \qquad \text{to} \qquad \frac{dv}{du} = \frac{u - v}{u + v}.$$

Solve this equation as in Sec. 3, and deduce $x^2 - 2xy - y^2 - 4x - 12y = c$ for the solution of the original equation. *Hint:* It will be found desirable to have $h - k + 2 = 0$, $h + k - 6 = 0$.

9. As in the previous problem, solve $(x - y + 1)\,dy = (1 - x - y)\,dx$.

10. As in Prob. 8, reduce to the form $du/dv = F(u/v)$ but do not solve:

$$y' = \frac{x+y-1}{2x+y+2}, \qquad y' = \frac{3x+y+6}{2x+y+7}, \qquad y' = \sin\frac{x-y+2}{x+y+3}, \qquad y' = \left(\frac{x+2y+3}{2x-y-5}\right)^3.$$

11. Find the path of a small boat in a wide river with uniform current if the boat has constant speed relative to the water and always heads toward a fixed point on the bank.

12. Solve Example 5 completely under the assumption that A is at $(0,0)$ and B is at $(x_0,0)$, at time $t = 0$. Distinguish the cases $r = 1$ and $r \neq 1$. If $r < 1$, at what point and when does B overtake A? If $r = 1$, how close can B get to A?

---

7. Geometric Interpretation. Envelopes. The foregoing methods of solution are called *solution by quadrature* because they give the solution by a succession of integrations and elementary operations. In the following sections we present a variety of nontrivial applications, each of which leads to an equation of the type considered in Secs. 3 to 6. Hence, the equation is solvable by quadrature.

Nevertheless it must not be concluded that all equations are solvable by quadrature. An example is the *Riccati equation*

$$y' = p(x) + q(x)y + r(x)y^2$$

for general coefficients p, q, r. This equation occurs in the theory of neutron transport, in electromagnetic theory, and in optics. Other equations not solvable by quadrature occur in the theory of nonlinear oscillations, in control theory, and in problems of mathematical physics.

Here we consider geometric methods of obtaining information about the differential equation without actually solving it. The method allows one to make qualitative statements about the solution even when the solution is not available, and it gives insight into the problem of uniqueness.

The function $f(x,y)$ in the general first-order equation

$$\frac{dy}{dx} = f(x,y) \tag{7-1}$$

gives the slope of the solution curve at each point (x,y). Hence the solution curves are increasing functions of x in regions of the xy plane in which $f(x,y)$ is positive and decreasing in regions where $f(x,y)$ is negative. For continuous $f(x,y)$ the boundary between these regions is part or all of the curve

$$f(x,y) = 0. \tag{7-2}$$

Equation (7-2) gives the locus of the critical points, and their character (maximum, minimum, neither) is shown by the sign of $f(x,y)$ at neighboring points. The inflection points and sense of concavity are similarly found from

$$y'' = f_x + f_y y' = f_x + f_y f \tag{7-3}$$

where $f_x \equiv \partial f/\partial x$ and $f_y \equiv \partial f/\partial y$. Equation (7-3) is a special case of the chain rule (Chap. 5, Sec. 5).

As an illustration, we shall discuss the integral curves of

$$y' = xy - 1$$

without solving the equation. The hyperbola $xy = 1$ is the locus where $y' = 0$. If $xy > 1$, then $y' > 0$ and the solution curves are increasing, but if $xy < 1$, they are decreasing. Hence, $xy = 1$

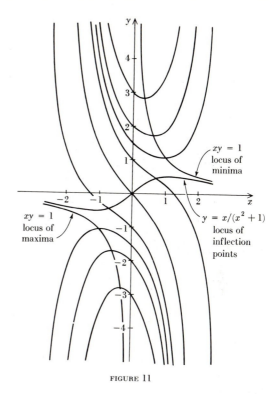

$xy = 1$
locus of
minima

$xy = 1$
locus of
maxima

$y = x/(x^2 + 1)$
locus of
inflection
points

FIGURE 11

gives a locus of minima in the first quadrant, maxima in the third quadrant. Since $y' = -1$ when $x = 0$ or $y = 0$, all integral curves intersect the axes at an angle of 135°. From

$$y'' = xy' + y = x(xy - 1) + y = y(x^2 + 1) - x$$

the curve is concave up if $y > x/(x^2 + 1)$ and concave down when this inequality is reversed. The curves have the appearance shown in Fig. 11.

The method of this example can be extended to give more detailed information. At any point where $f(x,y) = c$ the solution curves have slope c; hence they approximate a straight line of slope c. These straight lines determine a direction at each point (x,y). The resulting set of directions, one at each point, is called a *direction field*. From this viewpoint a first-order equation $y' = f(x,y)$ is equivalent to a direction field in the (x,y) plane. An integral curve is a curve whose direction at each of its points coincides with the direction of the field at that point.

We can get an idea of the appearance of the direction field by computing $f(x,y)$ at a large number of points and drawing short lines with this value as slope, as in Figs. 12 to 14. The three figures correspond, respectively, to the three equations

$$\frac{dy}{dx} = \frac{y}{x} \qquad \frac{dy}{dx} = -\frac{x}{y} \qquad \frac{dy}{dx} = \frac{2xy}{x^2 - y^2}. \tag{7-4}$$

For economy of space the direction field is shown only in the first quadrant; extension to other quadrants is made by symmetry.

When a direction field is needed for numerical analysis it is often convenient to introduce the curves, called *isoclines*, on which the field direction is constant. For the

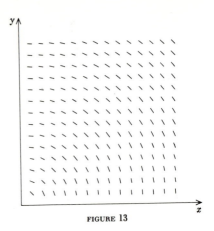

FIGURE 12   FIGURE 13

differential equation $y' = f(x,y)$ the isocline corresponding to slope c of the field is given by

$$f(x,y) = c. \tag{7-5}$$

Figure 15 illustrates the method of isoclines as applied to the equation

$$\frac{dy}{dx} = -\frac{8x}{3y} - 1.$$

The integral curves are spirals, four of which are shown, and the isoclines are straight lines through the origin.   On the isocline making an angle θ with the x axis $x/y = \cot \theta$; hence the integral curves have the slope

$$m = -\tfrac{8}{3} \cot \theta - 1.$$

This formula was used to compute the slopes of the short segments in Fig. 15.   By sliding a triangle along a fixed ruler a number of short segments were obtained from a *single* computation of m.   That is the advantage of the method of isoclines.

The reader may find it instructive to compare m as given by the formula with the slopes of the short segments in Fig. 15 and to verify by inspection that the spirals have the expected relation to the direction field.

Another use of the geometric interpretation is to clarify some cases in which the solution is not unique.   Let a family of integral curves

$$\phi(x,y,c) = 0$$

be a general solution of a first-order equation

$$F(x,y,y') = 0$$

in some region of the (x,y) plane.   An *envelope* of the family is a fixed curve C such that every member of the family is tangent to C and such that C is tangent, at each of its points, to some member of the family (see Fig. 16).   At a point (x,y) on the envelope, the values x, y, y' for the envelope are the same as for the integral curve, and hence these values x, y, y' satisfy the above equation.   Thus *an envelope of a family of solutions is again a solution*.   At a point on the envelope one would expect the uniqueness to fail, for there are at least two solutions: the envelope and the member of the family through the point in question.

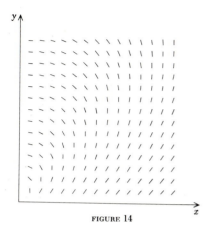

FIGURE 14

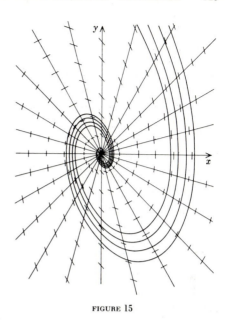

FIGURE 15

According to familiar results of calculus the envelope is obtained by eliminating c between the equations[1]

$$\phi(x,y,c) = 0 \qquad \text{and} \qquad \phi_c(x,y,c) = 0. \tag{7-6}$$

In treatises on differential equations it is shown that the envelope can also be obtained under suitable hypotheses, by eliminating p between the equations

$$F(x,y,p) = 0 \qquad \text{and} \qquad F_p(x,y,p) = 0. \tag{7-7}$$

The advantage of the latter procedure is that it does not require knowledge of a family of solutions.

It should be stressed that neither (7-6) nor (7-7) always yields a solution, and the results must be checked by substitution into the differential equation. If a solution is actually obtained by either process, it is often called a *singular solution*.[2]

Example. Discuss the differential equation

$$(y')^2 = 4y$$

and show, in particular, that there are infinitely many solutions through each given point (x_0, y_0), with $y_0 \geq 0$.

Separation of variables gives the family of parabolas

$$y = (x - c)^2 \tag{7-8}$$

illustrated in Fig. 16. It is geometrically evident

[1] For a statement of conditions under which this holds, see Philip Franklin, "A Treatise on Advanced Calculus," p. 527, John Wiley & Sons, Inc., New York, 1940. A discussion of (7-7) can be found in E. B. Wilson, "Advanced Calculus," p. 136, Ginn and Company, Boston, 1912, or A. E. Taylor, "Advanced Calculus," p. 394, Ginn and Company, Boston, 1955.
[2] For a precise definition, see Sec. 14, Prob. 4.

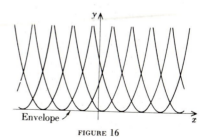

Envelope

FIGURE 16

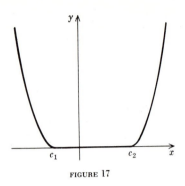

FIGURE 17

that the family has $y = 0$ as envelope, and this conclusion is confirmed by eliminating c between (7-8) and the equation

$$\phi_c = 2(x - c) = 0. \qquad (7\text{-}9)$$

The envelope is also obtained by eliminating p between the original equation

$$p^2 = 4y$$

and the equation $2p = 0$, which results from differentiation with respect to p.

Additional solutions can be constructed by following one of the parabolas to the envelope, then following the envelope, and continuing on another parabola. A graph of such a solution is shown in Fig. 17. Its equation is

$$y = (x - c_1)^2 \qquad y = 0 \qquad y = (x - c_2)^2 \qquad (7\text{-}10)$$

on the intervals $x < c_1$, $c_1 \le x \le c_2$, $c_2 < x$, respectively. Since the parabolas are tangent to the line $y = 0$ it is clear that (7-10) defines a continuously differentiable curve and that the curve satisfies the differential equation at every one of its points. (An analytic proof is easily given, though we shall not do so.) To get a family of solutions containing a prescribed point (x_0,y_0) in the upper half plane, choose c_1 so that the curve contains (x_0,y_0), and let c_2 remain arbitrary.

The common equation of all solutions can be written in the form

$$y[y - (x - c)^2] = 0. \qquad (7\text{-}11)$$

Any differentiable function $y = f(x)$ satisfying (7-11) is a solution of the differential equation, and conversely. Although the envelope is not a member of the family (7-8), it is a member of (7-11). A simpler family, which also contains the envelope, is

$$y = 0 \quad \text{for } x < c \qquad y = (x - c)^2 \quad \text{for } x \ge c.$$

The envelope corresponds to $c = \infty$.

PROBLEMS

1. (a) For each of the equations $x\, dy - y\, dx = 0$, $x\, dx + y\, dy = 0$, sketch the integral curves and verify agreement with the general behavior of the direction field as shown in Figs. 12 and 13. (b) How many solutions are there through the origin? Through points (x_0,y_0) other than the origin? (c) How would the answer to part b change if only explicit solutions of the form $y = f(x)$ are allowed? Of the form $x = g(y)$? *Hint:* A differentiable curve $y = f(x)$ cannot have a vertical tangent.

2. For each of the equations

$$dy = x\, dx, \qquad x\, dx = y\, dy, \qquad x\, dy + y\, dx = 0, \qquad x\, dy = (y - x)\, dx,$$

(a) sketch the locus $y' = 0$ in the xy plane. (b) Indicate the regions in which y is increasing; decreasing. (c) When is y concave up? Down? (d) At what slope do the solutions cross the axes? (e) Sketch the locus where the solutions have slope 1, -1, 2, -2, 5, -5. (f) Sketch the solutions as well as you can. (g) Verify your work by solving the equation.

3. (a) Can two equations have the same isoclines but different general solutions? Explain. (b) Can the isoclines coincide with the solutions? *Hint:* Both questions are answered by Figs. 12 to 14 and the accompanying equations (7-4).

4. In what regions of the xy plane are solutions of $y' = \sin(x^2 + y^2)$ increasing? Sketch the locus where the slope y' is maximum; zero; minimum. Thus try to sketch the solutions. Show also that the curvature of the solutions is $2x$ at all points of the circle $x^2 + y^2 = 2\pi$.

5. (*a*) Show that $y - c = (x - c)^2$ represents a family of congruent parabolas with the vertex on the line $y = x$, and sketch. (*b*) By differentiating with respect to c, obtain the envelope $y = x - \frac{1}{4}$. (*c*) By direct computation, verify that the parabolas and the envelope have the same slope at corresponding points. (*d*) Obtain a first-order differential equation for the family, and (*e*) verify that $y = x - \frac{1}{4}$ is a singular solution of this equation.

6. Obtain the equation $y + c^2 = 2cx$ for the tangent line to the parabola $y = x^2$ at $x = c$. Sketch the parabola and a number of its tangents, and verify by (7-6) that the parabola is the envelope of its tangents. Also obtain a differential equation for the family of tangents, and verify that $y = x^2$ is a singular solution of this equation. Can $y = x^2$ be obtained by (7-7)? (If the tangents are regarded as light rays, the envelope $y = x^2$ is an example of a *caustic curve*. One of the main uses of singular solutions is in geometrical optics.)

7. Show that the equation $xy' = 2y$ has a two-parameter family of solutions defined by

$$y = c_1 x^2 \quad \text{for } x < 0 \qquad y = c_2 x^2 \quad \text{for } x \geq 0$$

and sketch some of these solutions. Conclude that there are infinitely many distinct solutions through any point (x_0, y_0), provided $x_0 \neq 0$. How many solutions are there if $x_0 = 0$? Sketch the direction field for this equation, and explain how the nonuniqueness could have been predicted without solving the equation.

8. Discuss the equation $2y' = 3y^{1/3}$ analytically and geometrically. In particular, verify that there is one, and only one, solution through each point (x_0, y_0), if $y_0 \neq 0$, but infinitely many if $y_0 = 0$.

9. Obtain an algebraic equation for the family of circles with radius a and centers on the x axis (Fig. 18). Thus obtain the differential equation

$$y^2 (dy)^2 = (a^2 - y^2)(dx)^2.$$

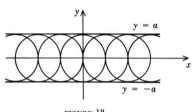

FIGURE 18

Get the envelope by (7-6) and also by (7-7), and verify that it satisfies the differential equation. Sketch some of the odd-looking solutions that are obtained by going back and forth between the circles and the envelope.

10. How are the curves in Prob. 9 restricted if only explicit solutions of the form $y = f(x)$ are allowed? Of the form $x = g(y)$? What if it is also required that the equation be explicitly solved for y', for example,

$$y' = \sqrt{a^2 y^{-2} - 1}$$

with the positive square root?

APPLICATIONS

8. Orthogonal Trajectories. In a variety of practical investigations it is desirable to find a family of curves that intersect the curves of a given family at right angles. For example, when a steady current flows in a homogeneous conducting sheet, the lines of equal potential are perpendicular to the lines of current flow. Similarly, heat flows in a direction at right angles to the curves of constant temperature, and corresponding statements can be made for hydrodynamics and electromagnetic theory. The polar and cartesian coordinate systems are familiar examples of orthogonal families of curves.

It is convenient to assume that each point of the region considered lies on just one curve in each of the two families, and, of course, that the curves meet there at right

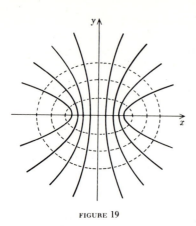

FIGURE 19

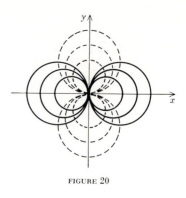

FIGURE 20

angles.   Then the curves of either family are said to be *orthogonal trajectories* of the curves of the other family.   For example, the solid curves in Fig. 19 are orthogonal trajectories of the dotted curves in a region that excludes two points of the horizontal axis.

We suppose that the given family of curves can be described by a differential equation of the form

$$y' = f(x,y). \tag{8-1}$$

By definition, the orthogonal family cuts the curves of the given family at right angles. At any point (x,y) of the given region, the slope of the orthogonal curve is the negative reciprocal of the slope y' in (8-1); hence the orthogonal curves must satisfy

$$y' = -[f(x,y)]^{-1}. \tag{8-2}$$

The question of existence and uniqueness of the orthogonal family hinges on the corresponding questions for the differential equation (8-2).

Although for logical clarity we assumed that (8-1) is solved for y', in practice the equation need not be reduced to that form.   The essence of the matter is that dy/dx be replaced by $-dx/dy$ in the differential equation of the given family, regardless of the particular form that equation may have.   This way of looking at the subject makes it easy to extend the method to polar coordinates.

Thus, if a curve is given in polar coordinates, the tangent of the angle α made by the radius vector and the curve is equal to $r\,d\theta/dr$, by a familiar formula of calculus. Since increasing α by 90° changes $\tan\alpha$ into $-\cot\alpha$, we get the differential equation for the orthogonal trajectories, in general, if we replace $r\,d\theta/dr$ by $-dr/(r\,d\theta)$ in the appropriate differential equation.

Example.   Consider the family $x^2 + y^2 = cx$.   Differentiation gives

$$2x + 2yy' = c \tag{8-3}$$

and, substituting this value of c into the given equation, we get

$$2xyy' = y^2 - x^2 \tag{8-4}$$

after slight simplification.   The equation of the orthogonal trajectories is

$$-2xy(y')^{-1} = y^2 - x^2$$

or, equivalently, $2xy\,dx + (y^2 - x^2)\,dy = 0$.   This equation can be solved by letting $y = vx$, as in Sec. 3, or by the integrating factor y^{-2}.   Either procedure gives

$$x^2 + y^2 - cy = 0$$

which represents a family of circles tangent to the x axis (see Fig. 20).

There is one feature of the analysis that should be carefully observed. If (8-3) is used instead of (8-4), the result of replacing y' by $-1/y'$ is

$$2x - 2y(y')^{-1} = c. \tag{8-5}$$

This equation is valid, but it requires that c *not be treated as constant*. Indeed, although $c = x^{-1}(x^2 + y^2)$ is constant on each curve of the given family, it is not constant on the orthogonal trajectories and so it is not constant in the differential equation (8-5). That is the reason why the differential equation must be obtained in a form free of c, for the purposes in view here.

PROBLEMS

1. Sketch the following families of curves, obtain a differential equation free of the parameter c, find the orthogonal trajectories, and add them to your sketch:

(a) $x^2 + y^2 = c^2$, $xy = c$, $y = cx^n$, $9x^2 + 4y^2 = c^2$, $x^3 - 3xy^2 = c$, $y = \log(cx)$;
(b) $y = ce^x$, $y^2 = x + c$, $y = ce^{x^2}$, $r = c$, $r = e^{c\theta}$, $r = c(1 - \cos\theta)$, $r = c\sin\theta$.

2. Show that the orthogonal trajectories of the solutions of $x\,dx + y\,dy = 0$ are the isoclines of the original equation.

3. (a) Obtain the equation $yy' + y'^2 + x = 0$ for the orthogonal trajectories of the family $y = cx + 1/c$. (b) Show that $y = 2\sqrt{x}$ is the envelope of the family. (c) At points of the curve $y = 2\sqrt{x}$, find the slope of the solutions of $yy' + (y')^2 + x = 0$ in terms of x. Then find the slope of the curve $y = 2\sqrt{x}$ in terms of x. How are these two slopes related? Why? (d) Sketch the family, the envelope, and the orthogonal trajectories in a single diagram.

4. A cylindrical tumbler containing liquid is rotated with a constant angular velocity about the axis of the tumbler. Show that the surface of the liquid assumes the shape of a paraboloid of revolution. *Hint:* The resultant force acting on a particle of the liquid is directed normally to the surface. This resultant is compounded of the force of gravity and the centrifugal force, since pressure at the free boundary is zero.

5. If a and b are constant and λ a parameter, show that the family of curves

$$\frac{x^2}{a^2 + \lambda} + \frac{y^2}{b^2 + \lambda} = 1$$

satisfies an equation, free of λ, which is unaltered when y' is replaced by $-1/y'$. What does this indicate concerning the orthogonal trajectories?

6. The family of curves intersecting a given family of curves at a fixed angle α is called the *isogonal family*. Show that if the differential equation of the given family is $F(x,y,y') = 0$ then the differential equation of the isogonal family is

$$F\left(x,y,\frac{y' - m}{1 + my'}\right) = 0$$

in general, where $m = \tan\alpha$. What if $\alpha = \pi/2$?

7. Referring to Prob. 6, find curves intersecting the following families at an angle of 45° and sketch:
(a) $xy = c$; (b) $x^2 + y^2 = 2cx$.

8. The wind blows with speed a parallel to the xy plane in the positive direction of the x axis, while sound is emitted from a fixed source in the xy plane. What are the lines along which the sound is propagated in the xy plane?   *Hint:* If the speed of sound in still air is v_0, the lines of propagation are the orthogonal trajectories of the family of circles

$$(x - at)^2 + y^2 = (v_0t)^2$$

where t is the time, reckoned from the instant of departure of the sound wave from the source. Eliminate t and obtain two equations of the Lagrange type.

---

9. Diffusion and Linear Equations.   Problems involving chemical reactions, mixtures, and solvents often lead to differential equations of the type considered in the foregoing pages. The basic technique in deriving these equations is to keep track of the income and outgo of the various ingredients, as given by physical or chemical principles. Since the technique is much the same in different problems, it suffices to discuss a single typical example.

Suppose that a tank contains G gal of water and that brine containing w lb of salt per gallon flows into the tank and out again at a constant rate r gpm, starting at time $t = 0$. At the same time a piece of rock salt is dropped into the tank, where it dissolves at a constant rate of q lb per min. The mixture being kept uniform by stirring, it is required to find the amount of salt present at any time $t \geq 0$.

The differential equation is obtained by writing the equation of continuity

increase equals income minus outgo

for the amount of salt. We call this amount $y = y(t)$ at time t. In the time interval from t to $t + \Delta t$ the number of gallons entering the tank is $r \, \Delta t$, since the rate of flow is r. Now each gallon contains w lb of salt. Hence the $r \, \Delta t$ gal contains

$$wr \, \Delta t$$

pounds of salt, and this, then, represents income due to the inflowing brine.

The income due to the dissolving salt is

$$q \, \Delta t$$

by the definition of q.

It remains to compute the amount of salt lost in the mixture leaving the system. The number of *gallons* leaving is $r \, \Delta t$, the concentration of the mixture in pounds per gallon is y/G at time t, and hence the number of *pounds* leaving is

$$G^{-1}r\bar{y} \, \Delta t.$$

Here $\bar{y}$ denotes the mean value of y over the interval $(t, t + \Delta t)$. We assume y to be continuous, so that $\bar{y} \rightarrow y$ as $\Delta t \rightarrow 0$.

The total change Δy in the time interval Δt is the sum of the first two expressions obtained minus the third, so that

$$\Delta y = wr \, \Delta t + q \, \Delta t + G^{-1}r\bar{y} \, \Delta t.$$

Dividing by Δt and letting $\Delta t \rightarrow 0$ give the differential equation

$$\frac{dy}{dt} = wr + q - G^{-1}ry. \qquad (9\text{-}1)$$

The initial condition is $y(0) = 0$, since the tank contained water initially.

This linear equation with constant coefficients is readily solved by the method of Sec. 5. The solution

$$ry = (wr + q)G(1 - e^{-tr/G}) \tag{9-2}$$

gives a qualitative and quantitative description of the system under discussion. As a check, we note that the behavior for t small and t large is, respectively,

$$y \doteq (wr + q)t \qquad ry \doteq G(wr + q).$$

The first equation expresses the fact that not much salt is lost from the system in the initial stages, and the second expresses the fact that income of salt must balance outgo in the steady state.

Equation (9-1) is a linear equation with constant coefficients. Linear equations are encountered not only in the theory of diffusion but in chemistry, in circuit analysis, in the theory of population growth, and in many other branches of science. Any theorem about the equation is, at the same time, a theorem about each of the underlying physical, chemical, or biological systems. We proceed to discuss the general first-order linear equation, which contains (9-1) as a special case.

Example. The number of individuals y in a certain population (of molecules, people, or bacteria) is governed by the linear equation

$$y' = f(t) - p(t)y \qquad t \geq 0$$

where t is time. Discuss the meaning of the coefficients f and p, and derive some general properties of the system.

When f and p are positive, $f(t)$ measures the strength of an external influence that promotes growth, while $p(t)y$ represents an inhibitory term, proportional to the size of the population y. Analogous interpretations are possible when f or p is negative and also when they change sign. (The latter behavior could be due to seasonal variations of environment, for example.)

If y_1 and y_2 are two different solutions of the equation, corresponding to two different behaviors of the system, it is easily verified that the difference $y = y_1 - y_2$ satisfies $y' = -p(t)y$. Assuming $p(t)$ continuous, we can solve this separable equation and obtain

$$y_1(t) - y_2(t) = ce^{-P(t)} \qquad \text{where } P(t) = \int_0^t p(\tau)\, d\tau. \tag{9-3}$$

The constant c equals the initial value, $y_1(0) - y_2(0)$.

It follows from (9-3) that $\lim (y_1 - y_2) = 0$ as $t \to \infty$ if, and only if,

$$\int_0^\infty p(\tau)\, d\tau \equiv \lim_{t \to \infty} \int_0^t p(\tau)\, d\tau = \infty. \tag{9-4}$$

This is the condition for the population to be *asymptotically stable* with respect to changes of initial conditions. Borrowing from the terminology of circuit theory, we can say that the population has a well-defined steady-state behavior if (9-4) holds; the effect of the initial conditions is then a transient. But when the integral converges, the initial conditions influence the infinitely remote future.

If $y_1(t) = y_2(t)$ at any particular time $t = t_0$ Eq. (9-3) indicates that $c = 0$, and hence[1] $y_1(t) \equiv y_2(t)$. This shows that the behavior of the system is determined by its state at any particular time t_0. Two solutions are either distinct at every value of t, or they coincide at every value of t—they cannot agree at some times and fail to agree at others.

In conclusion we note that any three distinct solutions y_1, y_2, y_3 satisfy the remarkable identity

$$\frac{y_1(t) - y_2(t)}{y_1(t) - y_3(t)} = \text{const}$$

which does not involve the coefficients $p(t)$ and $f(t)$. This is a consequence of (9-3).

[1] This proof of uniqueness is independent of that in Sec. 5.

PROBLEMS

1. (a) If $w = 1$, $G = 2$, $q = 3$, and $r = 4$ in the problem considered in the text, show that the concentration of salt at the end of 4 min is $\frac{7}{4}(1 - e^{-8})$ lb per gal. (b) Discuss the application of the example to (9-1), and verify by (9-2).

2. How would the discussion in the text change if the rock salt had been added at time $t = t_0$ instead of time $t = 0$?

3. How would the discussion change if the rock salt dissolved at a rate proportional to the amount undissolved, rather than at the constant rate q? *Hint:* If A is this amount, $dA/dt = -kA$. From this, find A at time t, and, from that, find $q = -dA/dt$.

4. Let A be the amount of a substance at the beginning of a chemical reaction, and let y be the amount of the substance entered in the reaction after t sec. Assuming that the rate of change of the substance is proportional to the amount remaining, deduce that $dy/dt = c(A - y)$, where c is a constant depending on the reaction. Show that $y = A(1 - e^{-ct})$.

5. Let a solution contain two substances whose amounts expressed in gram molecules, at the beginning of a reaction, are A and B. If an equal amount x of both substances has changed at the time t, and if the rate of change is jointly proportional to the amounts of the substances remaining, obtain the equation $dy/dt = k(A - y)(B - y)$. Solve, assuming that $y = 0$ when $t = 0$.

6. Formulate the appropriate differential equation if the rate at which a substance dissolves is jointly proportional to the amount present and to the difference between the actual concentration and the saturate concentration.

7. The current I in the RL circuit of Fig. 21 satisfies

$$L\frac{dI}{dt} + RI = E_0(t)$$

where $E_0(t)$ is the applied voltage and L and R are positive constants. (a) Solve this equation, assuming that $I = 0$ at $t = 0$. (b) Evaluate explicitly when $E_0(t) = E_0$ is constant, and sketch the solution $I(t)$ versus t. (c) Show that the constants L, R, and E_0 can be chosen so that the equation in this problem exactly duplicates Eq. (9-1). (By uniqueness the solutions also agree; hence the circuit could be used for analog computation.)

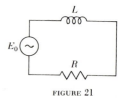

FIGURE 21

8. A radioactive substance A decomposes into a new substance B, which in turn decomposes into a third substance C. Set up a differential equation for the amount of B at time t. *Outline of solution:* The rate of increase of B is equal to the rate at which B is formed from A minus the rate at which B decomposes. Thus, denoting the amounts by A and B,

$$\frac{dB}{dt} = -\frac{dA}{dt} - k_1 B.$$

By Sec. 2, $A = A_0 e^{-kt}$, and substitution gives a linear equation for B.

9. A belt weighing w lb per ft passes over a pulley of radius r whose plane of rotation is vertical. If θ denotes the angle of lap measured from the horizontal and $\Delta s = r\,\Delta\theta$, obtain the formulas

$$\mu\,\Delta F \sim \Delta T - w\,\Delta s\cos\theta \qquad \Delta F \sim T\,\Delta\theta + w\,\Delta s\sin\theta$$

in the notation of Sec. 2, Example 3. Hence the tension T satisfies

$$\frac{dT}{d\theta} = \mu T + wr(\mu\sin\theta + \cos\theta).$$

Discuss this linear equation from the point of view of the example in this section.

10. The Hanging Chain. Let it be required to find the curve assumed by a flexible chain in equilibrium under gravity (Fig. 22). With s as arc from the point $x = 0$, let the weight density of the chain be $w(s)$ lb per ft and let the loading function be $f(x)$ lb per ft. The equation of the curve $y = y(x)$ will be obtained from the fact that the portion of chain between 0 and x is in static equilibrium.

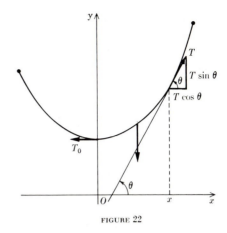

FIGURE 22

Equating horizontal forces gives

$$T_0 = T \cos \theta \tag{10-1}$$

where T_0 is the tension at the lowest point and θ the angle made by the tangent to the curve with the horizontal. Similarly, equating vertical forces gives

$$\int_0^s w(s)\, ds + \int_0^x f(x)\, dx = T \sin \theta \tag{10-2}$$

since the weight of the chain-plus-load must be balanced by the vertical component of T. Both (10-1) and (10-2) require that the function $y(x)$ be differentiable (so that θ is well defined), and they use the fact that the tension is tangential, for a flexible chain.

From (10-1) we have $T = T_0/\cos \theta$, so that

$$T \sin \theta = T_0 \tan \theta = T_0 y' \tag{10-3}$$

the latter equation resulting from the definition of y' as slope. Substitution in (10-2) gives

$$\int_0^s w(s)\, ds + \int_0^x f(x)\, dx = T_0 y'. \tag{10-4}$$

When w and f are continuous, (10-4) may be differentiated with respect to x, a procedure that leads to the differential equation for the curve

$$w(s) \frac{ds}{dx} + f(x) = T_0 y'' \tag{10-5}$$

in view of the fact that

$$\frac{d}{dx} \int_0^s w(s)\, ds = \left[\frac{d}{ds} \int_0^s w(s)\, ds \right] \frac{ds}{dx}.$$

Example. Show that a uniform flexible chain acted upon by gravitational forces alone assumes the shape of a catenary, and find the tension in terms of the height y.

Here $f(x) = 0$, $w(s) = w_0$, a constant. Since $ds/dx = \sqrt{1 + y'^2}$, Eq. (10-5) gives

$$w_0 \sqrt{1 + y'^2} = T_0 y''. \tag{10-6}$$

This is a second-order equation, which can be reduced to one of first order by the method of Sec. 6. With $p = dy/dx$ we have $y'' = p\, dp/dy$ and so (10-6) becomes

$$c\sqrt{1 + p^2} = p \frac{dp}{dy} \qquad c = \frac{w_0}{T_0}.$$

This equation is separable, the solution being

$$cy = \sqrt{1 + p^2} \tag{10-7}$$

when the axis is so chosen that $cy = 1$ when $p = 0$. Equations (10-1) and (10-7) now give

$T = T_0 \sec \theta = T_0\sqrt{1 + \tan^2 \theta} = T_0\sqrt{1 + p^2} = T_0 cy = w_0 y$, which gives the tension. Writing $p = dy/dx$ in (10-7) and separating variables yield

$$y = \frac{T_0}{w_0} \cosh \frac{w_0 x}{T_0}$$

as the reader will verify.

PROBLEMS

1. A flexible weightless cable supports a uniform roadway weighing w_0 lb per ft. The tensions at the highest and lowest points are T and T_0, the roadway is $2a$ ft long, the sag is b, and the length of the cable is $2s$. If the cable is symmetric about the y axis and has its lowest point at the origin, show that the equation of the curve is

$$y = \frac{w_0 x^2}{2T_0}$$

and thus obtain the relations $T_0 = w_0 a^2/2b$, $T = w_0(a/b)\sqrt{a^2/4 + b^2}$,

$$s = \int_0^a \sqrt{1 + \left(\frac{2bx}{a^2}\right)^2}\, dx \qquad s - a \sim \frac{2b^2}{3a}$$

as $b \to 0$. *Hint:* $\sqrt{1 + u} - 1 \sim u/2$ as $u \to 0$.

2. One end of a flexible uniform telephone wire is b ft above the lowest point, at a distance a ft from it measured horizontally and at a distance s ft from it along the wire. If $u = aw_0/T_0$, in the notation of this section, show that u satisfies the transcendental equations $(\cosh u - 1)/u = b/a$, $(\sinh u)/u = s/a$ and hence by division the nontranscendental equation $\tanh (u/2) = b/s$. Also find the relations $T_0 = w_0 a/u = w_0 s \operatorname{csch} u$, $T = w_0 a (\cosh u)/u = w_0 s \coth u$ for T_0 and for T, the tension at the highest point. The reader familiar with infinite series will obtain simplified expressions by expansion of the hyperbolic functions when u is small, that is, when the tension is large.

---

11. Newton's Laws of Motion.
In this section and the next we discuss the laws of motion and gravitation formulated by the English mathematician and natural philosopher Sir Isaac Newton. Born in 1642 on Christmas Day, Newton was one of the greatest mathematical physicists of all time and was the first to be in possession of a really powerful mathematical technique. Before age 25 he had formulated the laws that bear his name, carried out decisive experiments in optics, and developed a substantial part of the differential and integral calculus. Applying mathematics with astonishing skill to the study of Nature, Newton attacked questions as difficult as the three-body problem (which still has its puzzles today) and obtained results in other fields that are now derived by potential theory or by the calculus of variations. Some of his assertions in the theory of equations resisted proof for nearly two centuries.

Newton's second law of motion states that the rate of change of momentum is equal to the impressed force. In symbols,

$$\frac{d}{dt}(mv) = F \tag{11-1}$$

where F is the force, m is the mass of the moving particle, and v is its velocity. The extension of (11-1) to continuous mass distributions and to motion in a curved path is discussed presently, but first we assume that the particle moves in a straight line, its distance from some fixed point on the line being s. The force can depend on the time t, on the displacement s, and, in the case of motion through a resisting medium, on the velocity v. Also the mass may be variable in some problems, for example, those concerned with rocket flight or high-speed electrons.

Since

$$\frac{d(mv)}{dt} = \frac{d(mv)}{ds}\frac{ds}{dt} = \frac{d(mv)}{ds} v \qquad (11\text{-}2)$$

Eq. (11-1) may be put in the form

$$v\frac{d(mv)}{ds} = F \qquad (11\text{-}3)$$

which gives an alternative statement of Newton's law. Multiplying (11-3) by m leads to $(mv)d(mv)/ds = Fm$, which may be written

$$\frac{d}{ds}\frac{1}{2}(mv)^2 = Fm. \qquad (11\text{-}4)$$

This gives still another formulation.

In case m and F are known in terms of s only, $m = m(s)$, $F = F(s)$, then (11-4) may be solved completely:

$$\tfrac{1}{2}(mv)^2 - \tfrac{1}{2}(m_0v_0)^2 = \int_{s_0}^{s} F(s)m(s)\,ds. \qquad (11\text{-}5)$$

If the mass is constant, (11-5) becomes

$$\tfrac{1}{2}mv^2 - \tfrac{1}{2}mv_0^2 = \int_{s_0}^{s} F(s)\,ds \qquad (11\text{-}6)$$

since $m(s)$ may be factored out of the integral. Then (11-6) is the *law of conservation of energy*, for the left side of (11-6) is the change in kinetic energy while the right side represents the work done when the particle moves from s_0 to s. The steps leading to (11-6) are reversible if $F(s)$ is continuous, and hence *Newton's law is equivalent to the principle of conservation of energy, when the mass is constant and the force is a continuous function of position only.*

For several particles of masses m_i moving on the same straight line, addition gives $(d/dt)\Sigma m_iv_i = \Sigma F_i$, or

$$\frac{d}{dt} m\bar{v} = F \qquad \text{where } \bar{v} = \frac{\Sigma m_iv_i}{\Sigma m_i}. \qquad (11\text{-}7)$$

Here $m = \Sigma m_i$ is the total mass, $\bar{v}$ is the *mean velocity*, and $F = \Sigma F_i$ is the total force. However, when the internal forces cancel in pairs by Newton's law of equal and opposite reaction, F is also the total *external* force acting on the system.

If the masses are constant then $\bar{v} = d\bar{s}/dt$, where $\bar{s}$ gives the position of the center of mass; $m\bar{s} = \Sigma m_is_i$. In that case the center of mass follows (11-1). Extension to continuous mass distributions is made analogously, the equations being defined as the limiting form of those obtained for a set of approximating discrete distributions.

As an illustration, we consider a simple distributed-mass problem that leads to an important equation in the theory of rockets. Suppose a gun containing a bullet moves with positive velocity v on a straight horizontal frictionless track and points in a direction opposite to that of the motion. The mass of bullet-plus-gun is m, and that of the bullet is $-\Delta m$, where Δm is negative. If the bullet is fired with velocity c relative to the gun, it is to be established that

$$\Delta(mv) = (v - c + \eta)\,\Delta m \qquad \text{where } |\eta| \leq |\Delta v|. \qquad (11\text{-}8)$$

Here $\Delta(mv)$ denotes the change in momentum of the gun system; that is, it denotes the momentum of the gun after firing, minus the momentum of bullet-plus-gun before firing.

By Newton's law, the momentum of bullet-plus-gun is constant since there is (we assume) no force acting on this system as a whole. Hence,

$$mv = (m + \Delta m)(v + \Delta v) + (-\Delta m)v_b \tag{11-9}$$

where $v + \Delta v$ is the new velocity of the gun and v_b of the bullet. The precise value of v_b hinges on the interpretation of the constant c; but, whatever interpretation we use, it is evident that

$$v + \Delta v - c \leq v_b \leq v - c.$$

Since (11-9) is equivalent to $\Delta(mv) = \Delta m\, v_b$, the desired conclusion (11-8) follows.

To see the relevance of this result to rocket problems, suppose a rocket fires some of its mass backward at a variable rate r kg per sec and at a variable speed c m per sec relative to the rocket. If m and v denote the mass and velocity of the rocket at time t, we can apply (11-8), divide by Δt, and pass to the limit. Since $\eta \to 0$ when v is continuous, the result is

$$\frac{d(mv)}{dt} = (v - c)\frac{dm}{dt}$$

or $d(mv) = (v - c)\, dm$. This gives the differential in momentum due to the rocket motor. The differential $d(mv)$ due to external forces is $F\, dt$ by (11-1), and addition produces the useful formula

$$m\, dv + c\, dm = F\, dt \tag{11-10}$$

since $d(mv) = m\, dv + v\, dm$. It should be emphasized that m, v, c, F, and the rate of discharge $r = -dm/dt$ can all depend on t.

So far we have assumed that the motion is in a straight line. If a particle is moving in space, so that its position at time t is given by the cartesian coordinates $x = x(t)$, $y = y(t)$, $z = z(t)$, the appropriate law of motion is obtained by applying (11-1) to each coordinate separately. In particular, when the mass is constant the equations are

$$mx'' = F_x \qquad my'' = F_y \qquad mz'' = F_z \tag{11-11}$$

where the primes denote differentiation with respect to t and where F_x, F_y, F_z are forces along the three axes, respectively, at the point $[x(t), y(t), z(t)]$.

Applying (11-6) to each equation (11-11) with $s = x$, y, or z and adding give

$$\tfrac{1}{2}mv^2 - \tfrac{1}{2}mv_0^2 = \int_{x_0}^{x} F_x\, dx + \int_{y_0}^{y} F_y\, dy + \int_{z_0}^{z} F_z\, dz \tag{11-12}$$

where $v = [(x')^2 + (y')^2 + (z')^2]^{1/2}$ is the *speed* at time t and v_0 is the speed at time t_0. The precise meaning of the first integral in (11-12) is

$$\int_{t_0} F_x[x(\tau), y(\tau), z(\tau)]x'(\tau)\, d\tau$$

and the other integrals are interpreted similarly. Equation (11-12) is a principle of *conservation of energy* for motion in a curved path.

The reader familiar with vector analysis will recognize that (11-11) is equivalent to the single equation $m\mathbf{R}'' = \mathbf{F}$, where $\mathbf{R}$ is the position vector and $\mathbf{F}$ the force vector. Besides its brevity, this formulation has the distinct advantage that it is independent of the coordinate system. The proper setting for a discussion of (11-12) is also given by vector analysis. The sum on the right represents a *line integral*

$$\int \mathbf{F} \cdot d\mathbf{R}$$

of the tangential component of $\mathbf{F}$ along the path and hence it measures the work done on the particle by $\mathbf{F}$. These matters are discussed in Chap. 4, Sec. 8, and Chap. 6, Sec. 3.

Example 1.   Assuming conservation of energy for motion in a curved path, obtain an expression giving the period of a simple pendulum.

Let P denote the position of a pendulum bob suspended from O, and let θ be the angle made by OP with the position of equilibrium OQ, as shown in Fig. 23. The work required to change θ to any other value α is the work required to raise the bob through a vertical distance $a \cos \theta - a \cos \alpha$, if a is the pendulum length. With α chosen as the angle for maximum displacement, so that $v = 0$ at $\theta = \alpha$, conservation of energy gives

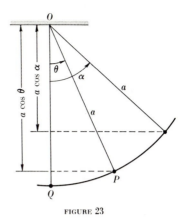

FIGURE 23

$$\tfrac{1}{2}mv^2 = \tfrac{1}{2}m\left(\frac{ds}{dt}\right)^2 = \tfrac{1}{2}ma^2\left(\frac{d\theta}{dt}\right)^2$$

$$= mga(\cos \theta - \cos \alpha)$$

where m is the mass of the bob. Separating variables gives t in terms of θ and hence, implicitly, θ in terms of t. In particular, the time required for θ to increase from 0 to α is

$$\frac{T}{4} = \sqrt{\frac{a}{2g}} \int_0^\alpha \frac{d\theta}{\sqrt{\cos \theta - \cos \alpha}} \tag{11-13}$$

so that the period T depends on the amplitude α. This dependence leads to the *circular error* in pendulum clocks, as discussed in the problems at the end of this section.

Example 2. Starting with velocity v_0, an electron is accelerated for a distance s by a constant electric field of magnitude E. What is the terminal velocity?

This problem arises in the theory of high-speed electron accelerators; hence relativistic effects cannot be neglected. If c is the velocity of light, a fundamental equation of Einstein gives the mass m of the electron in terms of its rest mass m_0 and its velocity v by

$$m = \frac{m_0}{\sqrt{1 - v^2/c^2}}. \tag{11-14}$$

If we write
$$\frac{v}{c} = \sin \theta \qquad \sqrt{1 - \frac{v^2}{c^2}} = \cos \theta \tag{11-15}$$

as we may for $v < c$, then (11-14) gives $m = m_0 \sec \theta$ and $mv = cm_0 \tan \theta$. Substituting in (11-3) with $F = Ee$, where e is the charge on the electron, gives

$$\frac{d}{ds}(cm_0 \tan \theta) = \frac{F}{v} = \frac{eE}{c} \csc \theta.$$

Hence $\sec^2 \theta (d\theta/ds) = (eE/m_0c^2) \csc \theta$, and by integration

$$\sec \theta = \sec \theta_0 + \frac{seE}{m_0c^2} \tag{11-16}$$

where θ_0 refers to the initial value. For numerical calculation it is more efficient to use (11-16) with trigonometric tables than to obtain v explicitly by (11-15).

PROBLEMS

1. A brick is set moving in a straight line over ice with an initial velocity v_0. If the coefficient of friction between the brick and the ice is μ, how long will it be before the brick stops? By integrating $ds = v\,dt$, obtain the total distance traveled, s_0, and verify that the work μmgs_0 done by frictional forces equals the loss in kinetic energy. (This problem illustrates the fact that conservation of energy applies to dissipative systems as well as to the so-called "conservative" systems.)

2. If $F = F(t)$ is a continuous function of t, show that Newton's law admits the first integral

$$mv - m_0v_0 = \int_{t_0}^t F(t)\,dt.$$

3. A particle moving in a straight line is subject to a force which is directed toward the origin and has magnitude proportional to the distance from the particle to the origin.   Obtain the differential equation and solution

$$s'' + \omega^2 s = 0 \qquad s = A \cos(\omega t - \phi)$$

where ω, A, and ϕ are constant.   (This is the equation for simple harmonic motion, an important type of periodic motion that arises in many mechanical and electrical systems.   For detailed discussion, see Chap. 3, Sec. 1.)

4. The equation of a cycloid is $x = \theta + \sin\theta$, $y = 1 - \cos\theta$.   Show that the arc s from the lowest point satisfies $s^2 = 8y$, and deduce the equation

$$4v^2 = g(s_0{}^2 - s^2)$$

for a particle sliding without friction down the curve.   By differentiation, show that $4s'' + gs = 0$, so that the motion is simple harmonic (Prob. 3).   Note that the period is independent of the starting point, and the time to reach the bottom is also independent of the starting point.   In Chap. 3, Sec. 18, it is shown that the cycloid is the only curve with this property.

5. A rocket fires some of its mass backward at a constant rate r kg per sec and at a constant speed c m per sec relative to the rocket.   If the rocket is subject to the constant force F, show that

$$v - v_0 = \left(c + \frac{F}{r}\right) \log \frac{m_0}{m}$$

where v_0 is the initial velocity, m_0 the initial mass, and v and m are the velocity and mass at time t.   If $F = 0$, what is the speed when half the mass is used up?   *Hint:* Let $m = m_0 - rt$ in (11-10), and separate variables.

6. Suppose that, instead of having F constant, the rocket of Prob. 5 is subject to a retarding force of magnitude $mg + kv$, where g and k are constant.   From (11-10), get a linear equation for v as a function of t, show that $(m_0 - rt)^{-k/r}$ is an integrating factor, and find v and s at time t.

7. A projectile is fired, with an initial velocity v_0, at an angle α with the horizontal.   Find the equation of the path under the assumption that the force of gravity is the only force acting on the projectile.   For what α is the range maximum?   Describe the region that is within the range of the gun.   *Hint:* Find the envelope of the trajectories when v_0 is fixed but α varies.

8. If $F = F(v)$ and $m = m(v)$ in Newton's law, obtain the following formulas under suitable hypotheses of monotony and continuity:

$$t - t_0 = \int_{v_0}^{v} \frac{m(v) + vm'(v)}{F(v)} \, dv \qquad s - s_0 = \int_{v_0}^{v} \frac{m(v) + vm'(v)}{F(v)} v \, dv.$$

Hint: Use $d(mv) = m(v)\,dv + vm'(v)\,dv$ in (11-1) and (11-3).

9. In a microwave electron accelerator the field is $E \sin \omega t$.   If an electron starts with velocity v_0, find the maximum possible terminal velocity.   *Hint:* The maximum occurs when the time for passage is exactly π/ω, for an electron starting at time $t = 0$.   Use the result of Prob. 8.

10. A particle starts from rest and falls through a resistive medium, so that the force in the direction of motion is $F = mg - f(v)$.   If $f(v)$ is continuous and $f(0) < mg$, show that the limiting velocity as $t \to \infty$ is the first positive root, $\bar{v}$, of the equation $f(v) = mg$, in general.   *Hint:* Use the first relation of Prob. 8.   The validity of the result hinges on the divergence of the integral as $v \to \bar{v}$.

11. A particle thrown upward with positive velocity v_0 in a resistive medium is subject to the force $F = -mg - f(v)$, where $f(v)$ is continuous and positive.   Show that the time for maximum height and the maximum height are, respectively,

$$t = \int_0^{v_0} \frac{m}{mg + f(v)} \, dv \qquad s = \int_0^{v_0} \frac{mv}{mg + f(v)} \, dv.$$

Hint: In Prob. 8 a plot of t versus v is monotone decreasing, so that $v = 0$ for just one value of t.

12. Obtain another expression for the period T of the simple pendulum in the text by applying $\cos \psi = 1 - 2 \sin^2 \tfrac{1}{2}\psi$ with $\psi = \theta$ and $\psi = \alpha$.   Then introduce a new variable ϕ defined by

$$\sin \tfrac{1}{2}\theta = k \sin \phi \qquad \text{where } k = \sin \tfrac{1}{2}\alpha.$$

and get the formula

$$T = 4\sqrt{\frac{a}{g}} \int_0^{\pi/2} \frac{d\phi}{\sqrt{1 - k^2 \sin^2 \phi}}.$$

This is an elliptic integral of the first kind (Chap. 1, Sec. 12).

13. For $k < 1$, expand the result of Prob. 12 in the form

$$T = 2\pi \sqrt{\frac{a}{g}} \left[1 + \left(\frac{1}{2}\right)^2 k^2 + \left(\frac{1\cdot3}{2\cdot4}\right)^2 k^4 + \left(\frac{1\cdot3\cdot5}{2\cdot4\cdot6}\right)^2 k^6 + \cdots \right].$$

Neglecting terms of order α^4, obtain the approximate formula

$$T_\alpha = T_0(1 + \tfrac{1}{16}\alpha^2) \qquad \alpha \text{ in radians}$$

for dependence of the period on α. Plot the circular error in seconds per week versus α for $0 \le \alpha \le 3°$ if the pendulum has a period of exactly 1 sec when $\alpha \to 0$.

---

12. Newton's Law of Gravitation. Kepler's Laws.

Another law of Newton is the law of gravitation, to which he was led in his profound analysis of planetary motion. This law states that two mass points attract each other with a force proportional to the product of their masses and inversely proportional to the square of the distance between them. If the magnitude of the force of attraction is denoted by F, the masses of the two bodies by m_1 and m_2, and the distance between them by r, then

$$F = \frac{\gamma m_1 m_2}{r^2} \tag{12-1}$$

where γ is a proportionality constant, called the *gravitational constant*. In the cgs system the value of γ is 6.664×10^{-8}.

It can be established that a uniform spherical shell attracts a particle at an external point as if the whole mass of the shell were collected at the center (see Chap. 6, Sec. 14). By integration, the same is true for a solid sphere, provided the density is a function of the radius only.

A second integration gives a corresponding result for two spheres. Hence, (12-1) holds not only for two mass points but for two mutually external spheres in which the density depends only on the distance to the center. This fact enormously simplifies the use of Newton's law in astronomy, because for most astronomical purposes one can treat the sun and planets as spheres.

As a simple illustration of Newton's law, consider the motion of a particle falling without resistance toward the earth. The force on a particle of mass m at the earth's surface is mg, and since the preceding paragraph indicates that the force has the form kr^2, we conclude that

$$F = mg\left(\frac{r_e}{r}\right)^2.$$

Here r_e is the earth's mean radius and $r \ge r_e$ is the distance from the particle to the center of the earth. The value of g depends slightly on location, but this dependence is usually neglected.

Conservation of energy gives

$$\tfrac{1}{2}mv^2 - \tfrac{1}{2}mv_0^2 = -\int_{r_0}^r mg\left(\frac{r_e}{r}\right)^2 dr = mgr_e^2\left(\frac{1}{r} - \frac{1}{r_0}\right)$$

after carrying out the integration. If the particle starts from rest at a very great distance, the velocity with which it strikes the earth is

$$v_e = \sqrt{2gr_e}$$

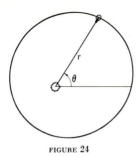

FIGURE 24

as we see by setting $v_0 = 0$, $r_0 = \infty$, $r = r_e$. This terminal velocity is also the minimum velocity of escape for a particle which leaves the earth never to return. Since r_e is approximately 4,000 miles, v_e is nearly 7 miles per sec.

There is a close connection between Newton's law of gravitation and three laws of planetary motion discovered empirically by the German mathematical astronomer and mystic Johannes Kepler (1571–1630). Because of its importance in astronomy, in the theory of satellites, and in the history of science, this connection is now developed in detail.

Kepler's first law states that *the planets move in ellipses with the sun at one focus*. The second law states that *the radius from the planet to the sun sweeps out equal areas in equal times*, in every part of the orbit. The third law states that *the square of the time for a single traverse of the orbit is proportional to the cube of the planet's mean distance to the sun*. The term "mean distance" can be variously defined, but in astronomy it denotes the average of the maximum and minimum distances.

Without essential loss of generality the following discussion is carried out in two dimensions. The (x,y) plane is the plane of the orbit, the origin is at the center of the sun, and the point (x,y) denotes the center of the planet (Fig. 24). To get Kepler's laws it must be assumed that the influence of other planets is negligible. However, we shall not assume initially that the force is given by (12-1) or that the orbit is an ellipse. Instead, the subject is developed in the general context of plane motion.

Let the components of force on the planet be denoted by mf_x and mf_y, where m is the mass of the planet. Newton's law of motion gives

$$x'' = f_x \qquad y'' = f_y$$

after division by the common factor m. If the second equation is multiplied by the complex unit[1] $i = \sqrt{-1}$ and added to the first, the result is

$$w'' = f \tag{12-2}$$

where $w = x + iy$, $f = f_x + if_y$.

Since both Newton's and Kepler's laws refer to the radius from the planet to the sun, it is convenient to introduce polar coordinates (r,θ) as shown in Fig. 24. Then $x = r \cos \theta$, $y = r \sin \theta$, and so

$$w = r(\cos \theta + i \sin \theta) = re^{i\theta}.$$

The second equality follows from the Euler de Moivre formula of Chap. 1, Sec. 17 (cf. also Prob. 2). By differentiation

$$w' = (r' + ir\theta')e^{i\theta} \qquad w'' = [(r' + ir\theta')' + i\theta'(r' + ir\theta')]e^{i\theta}. \tag{12-3}$$

The identity $(r\theta')' + r'\theta' = r^{-1}(r^2\theta')'$ gives another form for w'', with the result

$$[r'' - r(\theta')^2 + ir^{-1}(r^2\theta')']e^{i\theta} = f \tag{12-4}$$

upon substitution into (12-2). This is the general polar equation for plane motion subject to the arbitrary force function, $f = f(r,\theta)$.

[1] Familiarity with complex numbers is required here and in Sec. 16 but not in the rest of this chapter. The use of complex numbers in the exposition of Kepler's laws was suggested to us by R. P. Agnew's highly individual and interesting book, "Differential Equations," McGraw-Hill Book Company, New York, 1960. Further applications of complex numbers in the study of two-dimensional problems can be found in Chap. 8.

According to Newton's law the force is directed toward the origin. Analytically this means that f has the form $f(r,\theta) = F(r,\theta)e^{i\theta}$, where F is real. In that case $(r^2\theta')' = 0$ in (12-4), so that

$$r^2\theta' = C \qquad C = \text{const.} \tag{12-5}$$

The result (12-5) can be written $\frac{1}{2}r^2\,d\theta = \frac{1}{2}C\,dt$, and by integration

$$\int_{\theta_0}^{\theta} \tfrac{1}{2}r^2\,d\theta = \tfrac{1}{2}C(t - t_0). \tag{12-6}$$

The left side represents the area swept out by the radius vector, and the right side is proportional to the elapsed time. Hence (12-6) is the analytic statement of Kepler's second law. Conversely, if Kepler's law holds then $r^2\theta' = C$, hence $(r^2\theta')' = 0$, and f in (12-4) is a real multiple of $e^{i\theta}$. Thus, Kepler's second law is equivalent to the statement that the force is always directed along a line joining the planet to the sun. A force field of this kind is said to be *central*, because the lines of action of all forces in it pass through a single point.

From now on we assume that the force field is central, so that $\theta' = C/r^2$ by (12-5). This formula suggests that we take $s = 1/r$ as a new variable, and the substitution $s = 1/r$ is further suggested by the fact that Newton's law involves $1/r^2$. Then $\theta' = Cs^2$ and, by the chain rule,

$$\frac{dr}{dt} = -\frac{1}{s^2}\frac{ds}{dt} = -\frac{1}{s^2}\frac{ds}{d\theta}\frac{d\theta}{dt} = -C\frac{ds}{d\theta}.$$

Substitution of this and of $r\theta' = Cs$ into (12-3) produces the formula

$$w' = C\left(-\frac{ds}{d\theta} + is\right)e^{i\theta} \qquad s = \frac{1}{r}. \tag{12-7}$$

We can find w'' by differentiating with respect to θ and multiplying by $d\theta/dt$, that is, by Cs^2. The result is

$$w'' = -C^2s^2\left(\frac{d^2s}{d\theta^2} + s\right)e^{i\theta} = f \tag{12-8}$$

where the second equality follows from Newton's law $w'' = f$. This is the differential equation for the orbit, s versus θ, of any particle in a central force field.

According to Kepler's first law, which is to be discussed next, the orbit is an ellipse with the origin at one focus. Since $s = 1/r$, the polar equation of the ellipse has the form

$$s = \alpha - \beta \cos(\theta - \theta_0) \tag{12-9}$$

where α, β, and θ_0 are constant. Substitution into (12-8) yields

$$f = -C^2s^2\alpha e^{i\theta} \tag{12-10}$$

so that the magnitude of f is proportional to s^2, as stated in Newton's law. Conversely, if f is given by (12-10), then (12-8) gives

$$\frac{d^2s}{d\theta^2} + s = \alpha.$$

The quantity $\tilde{s} = s - \alpha$ therefore satisfies a similar equation with $\alpha = 0$. This is the equation for simple harmonic motion (Sec. 11, Prob. 3), and we conclude that s has the form (12-9). Therefore the orbit is a conic, and, since for a planet the orbit is bounded, it must be an ellipse.

Kepler's third law, which we have not yet considered, depends on the fact that

$$f = -ks^2e^{i\theta} \tag{12-11}$$

where the constant k depends only on the mass of the sun, not on the mass of the planets. (Recall that the mass m canceled out in the derivation of $w'' = f$.) Comparing (12-10) and (12-11) gives

$$\alpha C^2 = k.$$

On the other hand, the time T for a complete revolution satisfies

$$\tfrac{1}{2}CT = \int_0^{2\pi} \tfrac{1}{2}r^2 \, d\theta = A$$

by (12-6), where A is the area of the ellipse. It follows that the square of the time is

$$T^2 = \frac{4A^2}{C^2} = \frac{4A^2\alpha}{k}. \tag{12-12}$$

To interpret this result we introduce the notation $2a$, $2b$, $2c$, for the major axis, minor axis, and distance between foci, respectively. According to (12-9) the extreme values of s are $\alpha + \beta$ and $\alpha - \beta$, so that the extreme values of r are the reciprocals. By inspection of a figure,

$$2a = \frac{1}{\alpha - \beta} + \frac{1}{\alpha + \beta} \qquad 2c = 2a - \frac{2}{\alpha + \beta}.$$

Since elementary analytic geometry gives $b^2 = a^2 - c^2$ and $A = \pi ab$, it is a simple matter to express the area A in terms of α and β. Substitution into (12-12) yields

$$kT^2 = (2\pi)^2 a^3 \tag{12-13}$$

where the semimajor axis a is the mean of the maximum and minimum distances. This is Kepler's third law. Conversely, if Kepler's third law holds independently of the masses of the various planets, we conclude that k is independent of the mass. The force itself is therefore proportional to the mass, as stated by Newton.

The discussion of Newton's and Kepler's laws affords an interesting contrast between theoretical and empirical modes of procedure. Kepler's remarkable investigations, sustained by deep faith in the mathematical harmony of Nature, were extended over a period of 30 years, and one of his many notebooks contains over 800 pages of calculations in small handwriting. Nevertheless his work reduces to three statements of fact about the solar system, while Newton's applies to an enormous diversity of problems throughout the universe.

--- **PROBLEMS**

1. (a) Review the discussion of complex numbers and the Euler–de Moivre formula in Chap. 1, Secs. 15 to 17.   (b) Complex-valued functions are differentiated by differentiating real and imaginary parts separately, as in the text. Show that the familiar rules for differentiating sums and products hold for complex-valued functions.

2. If $e^{i\theta}$ is defined so that its derivative is $ie^{i\theta}$, show that the function

$$y = e^{i\theta}(\cos \theta - i \sin \theta)$$

satisfies $dy/d\theta = 0$. Conclude that y is constant, and by setting $\theta = 0$, get $y \equiv 1$. Multiply by $\cos \theta + i \sin \theta$ to obtain

$$e^{i\theta} = \cos \theta + i \sin \theta.$$

Conversely, if $e^{i\theta}$ is defined by this formula then the derivative of $e^{i\theta}$ is $ie^{i\theta}$.

3. By equating gravitational and centrifugal force, obtain the equation

$$T = 2\pi \frac{a}{r_e} \sqrt{\frac{a}{g}}$$

for the period of a satellite in a circular orbit of radius a. Also verify that this result is consistent with Kepler's third law. *Hint:* $mv^2/a = mk/a^2$ and $k = gr_e^2$. In this and the following problems, the satellite is assumed so high that air resistance can be neglected and is revolving around the earth rather than the sun. Thus, k has a different value from that in the text.

4. Referring to a dictionary for the relevant constants, apply the formula of Prob. 3 to the moon.

5. (*a*) If a particle moves in a central force field, as in the text, show that its speed v satisfies

$$\left(\frac{v}{C}\right)^2 = \left(\frac{ds}{d\theta}\right)^2 + s^2.$$

Hint: $v = |w'|$ in (12-7). (*b*) If the orbit is described by $s = \alpha - \beta \cos \theta$ where α and β are positive constants, show that

$$v = C(\alpha^2 + \beta^2 - 2\alpha\beta \cos \theta)^{1/2} \qquad \alpha C^2 = k$$

and interpret geometrically by the law of cosines.

6. Show that the principle of conservation of energy gives

$$\frac{v^2}{2} - \frac{k}{r} = \text{const}$$

for the speed v of a satellite, and verify consistency with the result of Prob. 5*b*. *Hint:* The energy of the orbit is the work needed to raise the satellite to the distance r, plus the kinetic energy $\frac{1}{2}mv^2$.

7. The point of an orbit at which a satellite is farthest from the earth is the *apogee*, and the point nearest the earth is the *perigee*. (*a*) If $s = \alpha - \beta \cos \theta$, with $\alpha > \beta > 0$, use Prob. 5*b* to compute the values of r and v at the apogee and perigee in terms of α and β. (*b*) Suppose a satellite is launched so that it has speed v_0 at height r_0 above the earth's center. Show that the mean radius of the orbit, a, satisfies

$$\frac{1}{a} + \frac{1}{g}\left(\frac{v_0}{r_e}\right)^2 = \frac{2}{r_0}.$$

Hence the time for return depends only on r_0 and v_0, not on the direction of launching. *Hint:* Use Prob. 6 and part *a*. Assume, naturally, that the satellite does not hit the earth. What does this mean about the value of r at perigee?

8. The text neglects to discuss the sign of α and of $\alpha - \beta$. Supply the missing discussion and show, in particular, that $\alpha < 0$ corresponds to a force of repulsion, while $\alpha \le |\beta|$ makes the orbit unbounded.

9. If the earth were a uniform sphere the gravitational force on a particle of mass m at a distance $r \le r_e$ from the center would be directed toward the center and have the magnitude

$$|F| = mg \frac{r}{r_e}$$

by Prob. 10, and this is now assumed. Imagine a straight tube extending from a point P on the earth's surface through the earth to another point Q on the surface. A particle is released in the tube at some point R, so that it slides back and forth without friction. Show that the period of oscillation is independent of P, Q, and R and is the same as the period of a satellite in an orbit of radius r_e.

10. It is known that a uniform spherical shell exerts no force on a particle which is in the hollow space enclosed by the shell (Chap. 6, Sec. 14). Show that the force on a particle of mass m at distance r from the center of a sphere is

$$\frac{\gamma m}{r^2} \int_0^r 4\pi u^2 \rho(u) \, du$$

when the density of the material forming the sphere is a continuous function $\rho(u)$ of the distance

u to the center. The special case $\rho(u) = 0$ for $u > r_0$ gives the result for a particle outside the sphere, and $\rho = $ const gives the formula of Prob. 9.

11. Obtain the differential equation

$$\frac{d}{dr}\left(\frac{r^2}{\rho}\frac{dp}{dr}\right) + 4\pi\gamma r^2\rho = 0$$

relating the density ρ and pressure p in the interior of a spherical star if each is a function of the distance to the center only: $\rho = \rho(r)$, $p = p(r)$. Assume $\rho(r)$ continuous. *Outline of solution:* Consider a column of material of unit cross section extending along a radius from r to r_s, the radius of the star. The pressure at the base of this column equals the total downward force on the column due to gravitation. The differential force on an element of the column from $r = q$ to $r = q + dq$ is given by

$$dF = \gamma\frac{\rho(q)\,dq}{q^2}\left[\int_0^q 4\pi u^2\rho(u)\,du\right] \equiv \phi(q)\,dq$$

in accordance with Prob. 10, and the total force is given by integration:

$$p(r) = F = \int_r^{r_s}\phi(q)\,dq = -\int_{r_s}^r\phi(q)\,dq.$$

Differentiate with respect to r, multiply by $r^2/\rho(r)$, and differentiate again. (This equation, due to Chandrasekhar, shows how Newton's law is used in astrophysics.)

---

13. The Elastic Curve. Linear Approximation.   Consider a horizontal elastic beam under the action of vertical loads. It is assumed that all the forces acting on the beam lie in a plane containing the central axis of the beam. Choose the x axis along the central axis of the beam in the undeformed state and the positive y axis down (Fig. 25). Under the action of external forces F_i the beam is bent and its central axis deformed. The deformed central axis, shown in the figure by the dashed line, is known as the *elastic curve*, and it is an important problem in the theory of elasticity to determine its shape.

A beam made of elastic material that obeys Hooke's law is known to deform in such a way that the curvature K of the elastic curve is proportional to the bending moment M. In fact,

$$K = \frac{y''}{[1 + (y')^2]^{3/2}} = \frac{M}{EI} \tag{13-1}$$

where E is Young's modulus, I is the moment of inertia of the cross section of the beam about a horizontal line passing through the centroid of the section and lying in the plane of the cross section, and y is the ordinate of the elastic curve. The important relation (13-1) bears the name *Bernoulli-Euler law*, after Euler and another Swiss mathematician, James Bernoulli.

We assume EI constant in the sequel. If M is a known function of x, the equation can be reduced to a first-order equation by the substitution $p = y'$. But the left side leads to an integral that one would ordinarily evaluate by letting $p = \tan\theta$, and hence one might as well use this substitution in the first place. Then

$$y' = \tan\theta \qquad y'' = (\sec^2\theta)\theta'$$

and (13-1) becomes $(\cos\theta)\theta' = M/EI$. The result of integration is

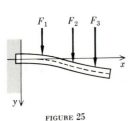

FIGURE 25

$$\sin \theta = \sin \theta_0 + \frac{1}{EI} \int_0^x M(\xi)\, d\xi \qquad y = \int_0^x \tan \theta\, dx \qquad (13\text{-}2)$$

where θ_0 is the value of θ at $x = 0$.

Conditions at the ends of the beam are called *boundary conditions* because they are imposed at the boundary of the interval $(0,l)$ in which the differential equation holds. The boundary condition $y(0) = \theta_0 = 0$ means that the left-hand end of the beam is clamped horizontally as shown in Fig. 25. The two-point boundary condition $y(0) = y(l) = 0$ means that the ends are both supported, and so on.

Solution of boundary-value problems is vastly simpler when (13-1) is replaced by the *linear approximation* $y'' = M/EI$ and we shall now inquire whether the approximation is justified. As an indication of the importance of the question it can be said that nearly all books on beam theory and strength of materials assume $y'' = M/EI$. A justification can be provided by rigorous solution of the problem of bending and flexure of beams in the linear theory of elasticity.[1] Here we give a somewhat different approach, which applies, in principle, to broad classes of nonlinear problems.

It is convenient to assume that M depends on x only and that $|M/EI| \leq \delta$, where δ is a known constant. Then (13-2) gives

$$|\sin \theta - \sin \theta_0| \leq \delta x \qquad 0 \leq x \leq l. \qquad (13\text{-}3)$$

A result such as this is called an *a priori estimate*, because it is obtained without prior knowledge of the solution.

For simplicity we assume the boundary condition $\theta_0 = 0$, so that the beam is clamped as shown in Fig. 25. Then, by (13-3)

$$|\sin \theta| \leq \delta x \qquad (13\text{-}4)$$

which shows that θ is small if δ is small. We conclude that $y' = \tan \theta$ is also small; hence Eq. (13-1) is approximated by the simpler equation

$$y'' = \frac{M}{EI}. \qquad (13\text{-}5)$$

But for applications, this is not enough. We must prove that the *solutions* of (13-1) are approximated by the *solutions* of (13-5), and we need a specific bound for the error. Such a bound is obtained as follows:

Let (13-1) be written in the form

$$y'' = (1 + \eta)\frac{M}{EI} \qquad \text{where } \eta = [1 + (y')^2]^{3/2} - 1. \qquad (13\text{-}6)$$

Since $y' = \tan \theta$ we have $1 + (y')^2 = \sec^2 \theta = (1 - \sin^2 \theta)^{-1}$, and (13-4) gives

$$0 \leq \eta(x) \leq [1 - (\delta x)^2]^{-3/2} - 1. \qquad (13\text{-}7)$$

If $\tilde{y}$ satisfies $\tilde{y}'' = M/EI$ and y satisfies (13-6) then $u = y - \tilde{y}$ satisfies

$$u'' = (1 + \eta)\frac{M}{EI} - \frac{M}{EI} = \eta \frac{M}{EI}. \qquad (13\text{-}8)$$

The boundary conditions give $u(0) = u'(0) = 0$, and u is obtained by integrating twice from 0 to x. The estimate $|M/EI| \leq \delta$ thus leads to

$$|u| \leq \int_0^x \int_0^x \delta\eta(x)(dx)^2.$$

[1] See I. S. Sokolnikoff, "Mathematical Theory of Elasticity," 2d ed., pp. 102, 106, 212, McGraw-Hill Book Company, New York, 1956.

Upon using (13-7) and evaluating the elementary integrals we get the final result,

$$|y(x) - \bar{y}(x)| \leq \tfrac{1}{2}\delta^3 x^4 (1 + \cos\theta)^{-2} \qquad \text{if } \delta x < 1. \tag{13-9}$$

The error is always less than $\tfrac{1}{2}\delta^3 x^4$, and for small δ it is about $\tfrac{1}{8}\delta^3 x^4$, since then $\cos\theta \doteq 1$. On the other hand, y usually has the order of magnitude δx^2, so that the fractional error has the order of magnitude $\delta^2 x^2$. This is small if δ is small, and, in that sense, the linear approximation is justified.

We conclude by remarking that (13-8) gives $u \geq 0$ if $M \geq 0$, and hence $y \geq \bar{y}$. This shows that the linear approximation underestimates the true deflection, and, indeed,

$$\bar{y} \leq y \leq \bar{y} + \tfrac{1}{2}\delta^3 x^4 \qquad \text{if } M \geq 0 \text{ and } \delta x < 1. \tag{13-10}$$

Example 1. Consider a cantilever beam of length l, built in at the end $x = 0$ and carrying in addition to a distributed load $w(x)$ lb per ft a concentrated load W lb and a couple L ft-lb applied at the end $x = l$ (Fig. 26). Assume the deflection small.

The resultant moment in a cross section x ft from the end $x = 0$, produced by the loads acting to the right of that section, is

$$M(x) = \int_x^l (\xi - x)w(\xi)\,d\xi + W(l - x) + L. \tag{13-11}$$

If $w(x) = 0$ and $L = 0$, this formula yields $M = W(l - x)$; hence, from (13-5), the approximate differential equation of the central line of a cantilever beam subjected to the end load W is

$$y'' = \frac{W}{EI}(l - x).$$

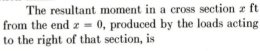

FIGURE 26

On integrating this equation we get

$$y = \frac{W}{EI}\left(\frac{x^2}{2} - \frac{x^3}{6}\right) + c_1 x + c_2.$$

The integration constants c_1 and c_2 can be evaluated from the conditions $y(0) = 0$, $y'(0) = 0$, stating that the displacement and the slope of the central line vanish at the built-in end. It is readily checked that these conditions lead to

$$y = \frac{W}{2EI}\left(lx^2 - \frac{x^3}{3}\right)$$

so that the displacement d at the free end is $d = Wl^3/3EI$.

If we assume that $M(x) = W(l - x)$, as before, but that the linear approximation is not used, we get $\delta = Wl/EI$ for the maximum of M/EI. Equation (13-10) with $x = l$ shows that the displacement d of the free end satisfies

$$\tfrac{1}{3}\delta l^2 < d < \tfrac{1}{3}\delta l^2[1 + \tfrac{1}{2}(\delta l)^2]$$

if $\delta l < 1$. The approximate solution $\tfrac{1}{3}\delta l^2$ obtained from the linear theory is satisfactory if the quantity $\delta l = Wl^2/EI$ is small compared with unity.

Since the deflection is small we did not distinguish between the length of the beam and the x coordinate, l, of the right-hand end. We now give a problem in which this distinction is essential.

Example 2. A thin weightless rod with a concentrated weight W at its end is pushed through a snug-fitting horizontal hole until the weight is at a distance l to the right of the hole. After the system has stopped oscillating, how large can l be?

Here again $M = W(l - x)$ and $\theta_0 = 0$, but the linear theory gives no clue to the solution of this problem; it does not even predict $l < \infty$. This behavior is characteristic of nonlinear problems. A linear approximation can be very useful when the variation is small, and wholly useless in other cases.

Substituting $M = W(l - x)$ into (13-2) with $\theta_0 = 0$ gives

$$\sin \theta = \frac{W}{EI} \int_0^x (l - \xi)\, d\xi = \frac{W}{EI} x(l - \tfrac{1}{2}x).$$

The maximum occurs at $x = l$ and must, of course, give $\sin \theta \leq 1$. Thus,

$$\frac{Wl^2}{2EI} \leq 1.$$

The resulting value

$$l = \sqrt{\frac{2EI}{W}}$$

is approached as the length of the bar is increased indefinitely, but it is not attained.

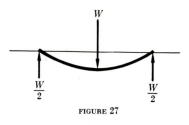

FIGURE 27

PROBLEMS

1. A beam of length l is freely supported at its ends and is loaded in the center by a concentrated vertical load W, which is large in comparison with the weight of the beam (see Fig. 27). By symmetry, the behavior of this beam is the same as that of a cantilever beam of length $l/2$ loaded by a concentrated load of magnitude $W/2$ at its free end. Verify this equivalence by direct computation of the elastic curve, using the linear approximation. *Hint:*

$$M = \begin{cases} \dfrac{-W}{2} x & 0 < x \leq \dfrac{l}{2} \\[2ex] \dfrac{-W}{2}(l - x) & \dfrac{l}{2} \leq x \leq l. \end{cases}$$

2. A uniform unloaded beam of length l weighs w lb per ft. Find the maximum deflection when it is used as a cantilever beam and also when it is freely supported at each end. *Hint:* Since the reaction at the end $x = l$ is $R = W/2$, the moment in the cross section at a distance x from the end $x = 0$ is

$$M = w \int_x^l (\xi - x)\, d\xi - \frac{wl}{2}(l - x).$$

Use the linear approximation.

3. Considering Example 2 without the assumption $\theta_0 = 0$, plot the maximum l versus the variable $s = \sin \theta_0$. Also find the maximum height attained by the bar, given θ_0 and l.

4. (*a*) Assuming appropriate conditions, reduce (13-1) to a first-order equation when $M = M(x,y')$ and when $M = M(y,y')$. (*b*) By letting $y' = \tan \theta$ in the latter result for $M = M(y)$, deduce

$$\cos \theta = \cos \theta_0 - \frac{1}{EI} \int_0^y M(\eta)\, d\eta \qquad x = \int_0^y \cot \theta\, dy.$$

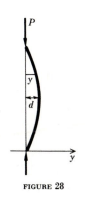

FIGURE 28

5. If $|M| \leq \delta EI$ in the case $M = M(y)$ of Prob. 4, deduce

$$|\cos \theta - \cos \theta_0| \leq \delta y.$$

6. Discuss the approximate solution of $y''[1 + (y')^2]^{-3/2} = M(x,y)$ by drawing circles in the (x,y) plane, and test your method when $M(x,y) = 1 - x$, $y(0) = y'(0) = 0$. *Hint:* K is the reciprocal of the radius of curvature.

7. If a long column is subjected to an axial load P, as shown in Fig. 28, the moment is $M = Py$. Obtain a first integral of (13-1) by Prob. 4*b*, and show that the length of the central line is given by the formula

$$s = 2\sqrt{\frac{EI}{P}} F\left(k, \frac{\pi}{2}\right)$$

where d is the maximum deflection, $k^2 = d^2P/4EI$, and $F(k,\pi/2)$ is the elliptic integral of the first kind (Sec. 11, Prob. 12).

8. Examine the improvement in the error estimate of Example 1 when, instead of the crude value $|M(x)| \leq lW$, the exact value $M(x) = W(x - l)$ is used in the analysis.

9. (a) Let $O(s)$ denote "terms of the order of s as $s \to 0$" and similarly in other cases, such as $O(\delta x)$. Assuming $\theta_0 = 0$, show that (13-4) gives $\sin \theta = \theta + O(\delta^3 x^3)$ and conclude that $\tan \theta = \sin \theta + O(\delta^3 x^3)$. Hence, by (13-2),

$$y = \int_0^x \sin \theta \, dx + \int_0^x O(\delta^3 x^3) \, dx = \tilde{y} + O(\delta^3 x^4).$$

(b) The reader sufficiently skilled in analysis will discuss the problem of getting a specific error estimate, such as (13-10), in this manner.

EXISTENCE, UNIQUENESS, AND STABILITY

14. A General Existence Theorem.   In this section we state a theorem concerning existence of solutions for a broad class of first-order equations

$$y' = f(x,y) \qquad y(x_0) = y_0 \tag{14-1}$$

and we give a method of approximating the solution to any desired degree of accuracy. It is assumed that $f(x,y)$ is continuous in a given region R containing (x_0,y_0) in its interior and satisfies an inequality

$$|f(x,y)| \leq m \qquad m = \text{const.}$$

The term "solution" means an explicit solution $y = \phi(x)$ whose graph is defined on an interval and lies in R. Solutions through the given interior point (x_0,y_0) are allowed to extend to the right and left until they meet the boundary of R, but no farther.

For present purposes a *region R* is simply a set of points, though usually the concept is more restricted (cf. Chap. 5). A *neighborhood* of a point (x_0,y_0) is a circular disk centered at (x_0,y_0), that is, a set of points (x,y) satisfying

$$(x - x_0)^2 + (y - y_0)^2 < r^2$$

where r is some positive constant. A point (x_0,y_0) is an *interior* point of R if it has a neighborhood that is wholly contained in R, and (x_0,y_0) is a *boundary* point of R if every neighborhood of (x_0,y_0) has points belonging to R as well as points not belonging to R. The *boundary* of R is the set of all boundary points and the *interior* of R is the set of all interior points. It is said that R is *bounded* if R is contained in some circular disk, that is, in some neighborhood.

A precise definition of continuity for functions of two or more variables is given in Chap. 5. The main property used here is that, if $f(x,y)$ is continuous, and $y = \phi(x)$ is also continuous, then $f[x,\phi(x)]$ is continuous, hence integrable.

Under the foregoing assumptions one can get an idea of the shape of a solution curve $y = \phi(x)$ in the following way:

Let us choose a point (x_0,y_0) and compute $y' = f(x_0,y_0)$. The number $f(x_0,y_0)$ determines a direction of the curve at (x_0,y_0). Now, let (x_1,y_1) be a point near (x_0,y_0) in the direction specified by $f(x_0,y_0)$. Then $y' = f(x_1,y_1)$ determines a new direction at (x_1,y_1). Upon proceeding a short distance in this new direction, we select a new point (x_2,y_2) and at this point determine a new slope $y' = f(x_2,y_2)$. As this process is continued, a curve is built up consisting of short line segments (Fig. 29). If the successive points (x_0,y_0),

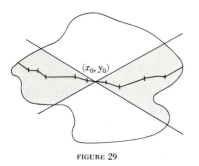

FIGURE 29

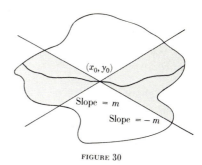

FIGURE 30

(x_1,y_1), (x_2,y_2), . . . are chosen near one another, the polygonal curve can be made to approximate a smooth curve $y = \phi(x)$ which is a solution[1] of $y' = f(x,y)$ associated with the initial point (x_0,y_0).

In view of the condition $|f(x,y)| \le m$ all the straight lines used in this process have slopes between $-m$ and m; hence the polygonal curve lies between the two lines of slopes $\pm m$ through (x_0,y_0) as shown in Fig. 29. It is natural to surmise that every solution through (x_0,y_0) must have the same property. This surmise is capable of proof, without use of the polygonal approximation and without assuming that the solution is unique.

Indeed, if $|f(x,y)| \le m$ and $y' = f(x,y)$, then $|y'| \le m$ and hence

$$|y - y_0| = \left| \int_{x_0}^{x} y' \, dx \right| \le \left| \int_{x_0}^{x} |y'| \, dx \right| \le m|x - x_0| \qquad (14\text{-}2)$$

which is equivalent to the result stated. The set of interior points of R satisfying $|y - y_0| \le m|x - x_0|$ is referred to as the *shaded region* (Fig. 30).

Besides the condition $|f(x,y)| \le m$ we shall need

$$|f(x,y) - f(x,\bar{y})| \le k|y - \bar{y}| \qquad k = \text{const.} \qquad (14\text{-}3)$$

This states that for every pair of points (x,y) and $(x,\bar{y})$ in R, $f(x,y)$ does not change too rapidly when y changes. The inequality (14-3) is called a *Lipschitz condition* after the German mathematician R. Lipschitz. It is useful not only in the theory of differential equations but in Fourier series, in the calculus of variations, and in several other subjects.

Under suitable hypothesis the mean-value theorem gives

$$f(x,y) - f(x,\bar{y}) = f_y(x,\eta)(y - \bar{y})$$

where η is between y and $\bar{y}$. This suggests that (14-3) can be replaced by the simpler but more restrictive condition[2]

$$|f_y(x,y)| \le k \qquad (14\text{-}4)$$

for all (x,y) interior to R.

We can now state the following theorem, established in Sec. 18:

---

[1] Although proof of uniqueness requires a supplementary assumption similar to (14-3) below, it was discovered by the Italian mathematician G. Peano that *some* sequence of these polygonal curves actually converges to *some* solution under the sole hypothesis that f is continuous. See E. A. Coddington and N. Levinson, "Theory of Ordinary Differential Equations," chap. 1, McGraw-Hill Book Company, New York, 1955.

[2] It is shown in books on advanced calculus that (14-4) is automatic, for some k, if $f_y(x,y)$ is continuous in the region consisting of R together with its boundary. A similar remark applies to the condition $|f(x,y)| \le m$.

THEOREM.   In a bounded region R, let $f(x,y)$ be continuous and satisfy $|f(x,y)| \leq m$ as well as (14-3) or (14-4).   Then the equation $y' = f(x,y)$ has one, and only one, solution $y = \phi(x)$ containing the interior point (x_0, y_0).   The solution lies in the shaded region and can be extended to the right and left of x_0 until it meets the boundary of R.

It should be observed that the theorem not only establishes existence of the solution in the neighborhood of (x_0, y_0) but gives an estimate for the interval over which the solution can be extended.   This interval is at least $a \leq x \leq b$, where a and b are chosen so that the region bounded by the lines

$$x = a \qquad x = b \qquad y - y_0 = m(x - x_0) \qquad y - y_0 = -m(x - x_0)$$

lies wholly interior to R (Fig. 31).   (If R is unbounded, the solution may be extensible infinitely far in one or both directions.)

This theorem is due chiefly to the great French mathematician Augustin Louis Cauchy (1789–1857).   Cauchy not only inaugurated the modern theory of differential equations but is founder of the theory of functions of a complex variable and cofounder, with Navier, of the theory of elasticity.

Cauchy based his proof of this theorem on the construction of Fig. 29, and the approximating polygonal curve is called *Cauchy's polygon*.   However, instead of developing that procedure here, we prefer to use a method of successive approximation due to another French mathematician, Emile Picard.

In applying Picard's method to the general problem

$$y' = f(x,y) \qquad y(0) = y_0$$

one forms a sequence of functions $y_1, y_2, y_3, y_4, \ldots$ satisfying

$$y_2' = f(x,y_0), \qquad y_2' = f(x,y_1), \qquad y_3' = f(x,y_2), \qquad \ldots.$$

In each case the initial condition is $y = y_0$ at $x = x_0$.   The differential equations are readily solved in succession, since the right-hand members are known from the preceding step.   Indeed, the equation

$$y_{n+1}' = f(x,y_n) \qquad y_{n+1}(x_0) = y_0$$

has the solution

$$y_{n+1} = y_0 + \int_{x_0}^{x} f(x,y_n)\, dx \tag{14-5}$$

as noted in Sec. 1.   We shall presently prove that all these functions y_n are defined for $a \leq x \leq b$ and tend to a solution y on this interval.   First, however, we give some concrete applications of Picard's method and of the existence theorem.

Example 1.   Apply Picard's method to the problem

$$y' = 1 + xy \qquad y(0) = 6.$$

Picard's method consists in taking the initial value $y_0 = 6$ as an approximate solution and obtaining a better approximation by solving

$$y' = 1 + xy_0 \qquad y(0) = 6.$$

When $y_0 = 6$ the solution of this equation is

$$y_1 = 6 + \int_{0}^{x} (1 + 6x)\, dx = 6 + x + 3x^2.$$

A better approximation, $y = y_2$, is now obtained from

$$y' = 1 + xy_1 \qquad y(0) = 6.$$

Substituting the value of y_1 and solving give

$$y_2 = 6 + x + 3x^2 + \tfrac{1}{3}x^3 + \tfrac{3}{4}x^4.$$

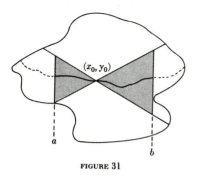

FIGURE 31

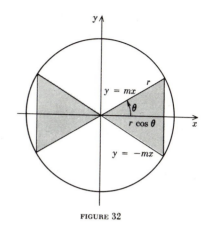

FIGURE 32

Similarly, y_3 is determined from $y' = 1 + xy_2$, and so on.

Example 2.   Estimate the interval of existence for the solution of

$$y' = x^2 + y^2 \qquad y(0) = 0.$$

Here $|y'| \leq m$ in a circle of radius $r = \sqrt{m}$. By Fig. 32 the solution can be extended to the right and left of the initial value over the distance $r \cos \theta$, where $\tan \theta = m$. This gives the estimate

$$|x - x_0| \leq r \cos \theta = \frac{r}{\sqrt{1 + m^2}} \tag{14-6}$$

for the interval, whenever $|f(x,y)| \leq m$ in a circle of radius r centered at (x_0,y_0). In the present case $r = \sqrt{m}$ and $x_0 = 0$, so that the interval is

$$|x| \leq \sqrt{\frac{m}{1 + m^2}}.$$

The maximum occurs when $m = 1$. Putting $m = 1$, we conclude that the solution exists at least over an interval of length $\sqrt{2}$ centered at the origin.

PROBLEMS

1. Verify that the solutions of $2xy\, dx = (x^2 - y^2)\, dy$ represents a family of circles tangent to the x axis at the origin (cf. Figs. 14 and 20). Thus show that there are infinitely many explicit solutions $y = \phi(x)$ through certain points (x_0,y_0) if the equation is considered in the whole (x,y) plane, but there is at most one solution through each point if the region is the plane minus the origin. Discuss the connection of this result with the general existence theorem. *Hint:* Since the circles are mutually tangent at the origin you can proceed from one circle to another without violating differentiability.

2. For the equation $y' = (1 - x^2)^{-1}$, verify that the lines $x = -1$ and $x = 1$ divide the plane into three regions, in each of which the above theorem is applicable. Sketch the solutions in each region and note that there are really three distinct families of curves. Verify that routine use of the methods of Sec. 1 produces a single equation for the solutions, which is valid in each of the three regions. Similarly discuss the equation $y' = (1 - y^2)$.

3. As in Example 2, get an estimate for the region of existence of y if $y' = 1 + y^2$ and $y(0) = 0$. Compare with the exact result obtained by separation of variables.

4. For purposes of this problem a *singular point* of a differential equation is a point such that in every sufficiently small neighborhood of the point there are two or more distinct solutions whose

graph contains the point.   A *singular solution* is a locus of singular points whose equation, $y = \phi(x)$, happens to give a solution.   (*a*) With reference to the theorem in this section, explain why the singular points can be found, in general, by examining the locus where $f_y(x,y)$ does not exist. (*b*) The reader familiar with the implicit-function theorem (Chap. 5, Sec. 9) will explain why singular points for $F(x,y,y') = 0$ are among the points, in general, where $F_p(x,y,p) = 0$.   *Hint:* When $F_p \neq 0$ the equation can be solved for y' and the theorem applies.

5. If $p = y'$, in what regions of the (x,y) plane do the following equations of form $F(x,y,y') = 0$ lead to two distinct differential equations of form $y' = f(x,y)$?   Solve by the method of Sec. 6, Example 1, and sketch both sets of integral curves, including the singular solutions, if any.   Check whether the latter can be found by the equation $\phi_c(x,y,c) = 0$ of Sec. 7, or by the locus of points where $f_y(x,y)$ does not exist, or by the equation $F_p(x,y,p) = 0$:

$$x(p^2 + 4) = 2yp, \qquad y(1 + p^2) = 2xp, \qquad 2p(p - x) = 2y - x^2, \qquad xp^2 + x = 2yp.$$

Hint: In the third equation, set $u^2 = 4y - x^2$.

6. Show that the theorem in this section gives the existence theorem of Sec. 5 (apart from the specific formula) but does not directly give the theorem of Sec. 3, even if attention is confined to explicit solutions $y = \phi(x)$.

7. Reexamine selected first-order equations in the examples and problems of Secs. 3 to 6, with a view to determining *all* solutions, and *only* solutions, in appropriate regions.   For purposes of this problem, no solution must be introduced or lost by algebraic manipulations, and the distinction between explicit solutions $x = \phi(y)$ or $y = \phi(x)$ is to be carefully observed.   *Hint:* In general, the locus of points where $f_y(x,y)$ fails to exist divides the plane into regions, in each of which the theorem of this section applies.

8. On an interval containing x_0, suppose

$$u' = f(x,u) \qquad v' = g(x,v) \qquad w' = h(x,w).$$

If $f(x,y) < g(x,y) < h(x,y)$ holds throughout the relevant region R of the (x,y) plane, and if $u(x_0) = v(x_0) = w(x_0)$, show that[1]

$$u(x) < v(x) < w(x)$$

for $x > x_0$ and that the inequalities are reversed for $x < x_0$.   Interpret graphically.   *Hint:* Since $u'(x_0) < v'(x_0)$, $u(x) < v(x)$ for $x > x_0$ and x near x_0.   If the conclusion $u(x) < v(x)$ fails for some $x > x_0$, consider the *first point* x_1 beyond x_0 at which $u(x_1) = v(x_1)$.   It will be found that the condition $u'(x_1) < v'(x_1)$ leads to a contradiction.   The other inequalities follow by symmetry.

Existence of a *first point* x_1 can be deduced from the fundamental principle quoted in Chap. 1, Sec. 3, but is taken for granted here.

9. If $y' = \sin xy$ and $y(0) = 1$, show that

$$e^{x^2/\pi} < y < e^{x^2/2} \qquad 0 < x < 0.8.$$

Hint: The graph of $\sin u$ indicates that $2u/\pi < \sin u < u$ holds for $0 < u < \pi/2$.   Use Prob. 8.

10. Apply the result of Prob. 8 to the inequality $|y'| \leq m$.   *Hint:* $|y'| < M$ holds for every $M > m$.   After estimating $|y - y_0|$, let $M \to m$.   This proof does not require that y' be continuous or even integrable.

11. (*a*) Discuss the uniqueness of solutions of $y' = |y|^m$ near the origin, where m is constant. (*b*) Let $g(y)$ be continuous for $|y| \leq 1$ and positive, except that $g(0) = 0$.   Show that the equation $y' = g(y)$ has at least two distinct solutions through the origin if

$$\int_0^1 \frac{ds}{g(s)}$$

converges.   [It is not difficult to show that solution (*b*) is uniquely determined on (0,1) if the integral diverges.   More generally, the American mathematician William Fogg Osgood proved in

---

[1] This useful method of comparison is due in the main to the Russian mathematician S. A. Chaplygin. It can be extended to equations of higher order and to partial differential equations.

1898 that uniqueness holds if $k|y - \tilde{y}|$ in (14-3) is replaced by $g(|y - \tilde{y}|)$, where g is any function of the above type with a divergent integral.   Many other conditions giving uniqueness have been found since then.   For *existence* there is required only the continuity of f, as has already been stated.]

12. Obtain the first three Picard approximations for selected problems in Sec. 5, using (*a*) $y(0) = 0$, (*b*) $y(1) = 2$, or (*c*) $y(x_0) = y_0$.

---

15. Systems of Equations.   Phase Space.   As the reader will recall from Sec. 6, a second-order equation

$$y'' = f(x,y,y')$$

can often be simplified by the substitution $y' = p$.   The result

$$y' = p \qquad p' = f(x,y,p)$$

is an example of a *first-order system* in two unknowns, y and p.   In a like manner the third-order equation

$$y''' = f(x,y,y',y'')$$

can be reduced to a first-order system in the three unknowns $y = y_1$, $y' = y_2$, $y'' = y_3$. The result is

$$y_1' = y_2 \qquad y_2' = y_3 \qquad y_3' = f(x,y_1,y_2,y_3).$$

This is a special case of the system

$$
\begin{aligned}
y_1' &= f_1(x,y_1,y_2,y_3) \\
y_2' &= f_2(x,y_1,y_2,y_3) \\
y_3' &= f_3(x,y_1,y_2,y_3).
\end{aligned}
\tag{15-1}
$$

A reduction of the nth-order equation to a system of n first-order equations is of some practical importance in numerical integration of equations on differential analyzers and electronic calculators.   Such computing devices are usually so designed that it is simpler to calculate n first derivatives than one derivative of order n.   The reduction has also numerous advantages in theoretical considerations.

Systems of differential equations appear naturally in problems involving dynamical systems with several degrees of freedom.   Thus, the motion of a particle constrained to move on a surface can be described by two positional coordinates[1] (x,y).   These coordinates satisfy equations of the form

$$x'' = F(x,y,x',y',t) \qquad y'' = G(x,y,x',y',t).$$

This pair of second-order equations can be reduced to a system of four first-order equations in the unknowns x, y, x', y'.

Analysis of systems is greatly simplified by use of an appropriate notation.   We denote the unknown functions collectively by a single boldface letter $\mathbf{y}$; for example, if there are three functions y_1, y_2, y_3 then

$$\mathbf{y} = (y_1,y_2,y_3).$$

When y_1, y_2, y_3 are real numbers, an expression $\mathbf{y}$ of this sort is called a *vector*, and the individual values y_1, y_2, and y_3 are the *coordinates* of $\mathbf{y}$.   When y_1, y_2, and y_3 are real-valued functions rather than numbers, $\mathbf{y}$ is properly called a *vector-valued function*, since its value, $\mathbf{y}(x)$, is a vector for each value of x.   However, now that the distinction has

---

[1] If a particle moves on a sphere, for example, x and y may be taken as the latitude and longitude, respectively.

been pointed out, we shall use[1] the term "vector" also in the sense of "vector-valued function."

The vector $\mathbf{y}$ in the foregoing discussion is said to be of dimension 3 because it has three coordinates. In general, a vector with n coordinates has *dimension n*. Two vectors of the same dimension are *equal* if corresponding coordinates agree. For example,

$$(y_1,y_2,y_3) = (\tilde{y}_1,\tilde{y}_2,\tilde{y}_3) \quad \text{means} \quad y_1 = \tilde{y}_1 \quad y_2 = \tilde{y}_2 \quad y_3 = \tilde{y}_3.$$

A vector is differentiated by differentiating its coordinates. As an illustration, if

$$\mathbf{y} = (y_1,y_2,y_3) \quad \text{then} \quad \mathbf{y}' = (y_1',y_2',y_3').$$

A function of the arguments y_1, y_2, y_3 is also a function of the single argument $\mathbf{y}$. For instance, the three functions on the right of the system (15-1) can be denoted, respectively, by

$$f_1(x,\mathbf{y}) \qquad f_2(x,\mathbf{y}) \qquad f_3(x,\mathbf{y}).$$

In terms of the vector $\mathbf{f} = (f_1,f_2,f_3)$, the system can therefore be written in the compact form

$$\mathbf{y}' = \mathbf{f}(x,\mathbf{y}).$$

It should be noted that the equivalence of this equation with (15-1) requires not only the appropriate definition of $\mathbf{f}$ and $\mathbf{y}$ but also the definitions of vector differentiation and vector equality.

Similar considerations apply to systems in any number of unknowns. If we set

$$\mathbf{f} = (f_1,f_2,f_3, \ldots ,f_n)$$

where $f_k = f_k(x,y_1,y_2, \ldots ,y_n)$, a general first-order system is $\mathbf{y}' = \mathbf{f}(x,\mathbf{y})$. It is left for the reader to write this in expanded form similar to (15-1) and to verify that the nth-order equation

$$y^{(n)} = f[x,y,y',y'', \ldots ,y^{(n-1)}]$$

can be reduced to a system of this type by taking

$$y_1 = y, \qquad y_2 = y', \qquad y_3 = y'', \qquad \ldots, \qquad y_n = y^{(n-1)}$$

as new unknowns.

An important aid in visualizing the solutions of systems is the *phase space*, in which the solutions $\mathbf{y}$ are represented without showing the explicit dependence on t. We shall explain the main ideas by interpreting the system

$$\frac{dx}{dt} = f(x,y) \qquad \frac{dy}{dt} = g(x,y) \tag{15-2}$$

in the (x,y) plane.

The equation $x = \phi(t)$, $y = \eta(t)$ of a solution can be thought to give the position of a moving point in the (x,y) plane at time t. The *orbit* or *trajectory* is the path followed, that is, the graph of the equation $\psi(x,y) = 0$ obtained by eliminating t. If $f(x,y) \neq 0$ or $g(x,y) \neq 0$ the orbit satisfies

$$\frac{dy}{dx} = \frac{g(x,y)}{f(x,y)} \qquad \text{or} \qquad \frac{dx}{dy} = \frac{f(x,y)}{g(x,y)} \tag{15-3}$$

as is seen when one equation (15-2) is divided by the other.

The velocity of the moving point is described by the vector (x',y'), since dx/dt represents the velocity in the x direction, and dy/dt represents the velocity in the y

---

[1] This practice is quite common in applications. It is justified, in part, by the fact that algebraic rules governing vector-valued functions have the same form as those governing vectors. In this chapter we are really using vector notation, rather than vectors. The invariance needed for full characterization of the concept of *vector* is not relevant here (cf. Chaps. 4 and 6).

direction.   Although we have not introduced a pictorial representation of vectors, the reader is doubtless familiar with the fact that the slope of the vector (x',y') is y'/x'.   This agrees with the slope dy/dx of the orbit as given by (15-3); hence the velocity vector is tangent to the orbit (cf. Chap. 4, Sec. 7).

Upon attaching the velocity vector (x',y') or

$$[f(x,y),\ g(x,y)]$$

to each point (x,y), we obtain an interpretation of the system as a vector field, corresponding to the direction field introduced in Sec. 7.   If the point $[x(t),y(t)]$ is thought to be a particle in a fluid flowing in the (x,y) plane, the field describes the velocity field, and the orbits are the streamlines, that is, the paths of the fluid particles.

At a point (x_0,y_0) where $f(x_0,y_0) = g(x_0,y_0) = 0$, neither of the orbital equations (15-3) is meaningful, and such a point is said to be a *singularity* for the differential system.   Since the system then admits $x = x_0$ and $y = y_0$ as solution, which does not depend on the time t, the singular point is also called a *point of equilibrium.*

Roughly speaking, an equilibrium point (x_0,y_0) is a point of stable equilibrium if the fluid particles initially close to (x_0,y_0) remain close throughout their subsequent history. The point is *asymptotically stable* if all particles sufficiently near to (x_0,y_0) tend to (x_0,y_0) as limit when $t \to \infty$.

To illustrate these ideas we shall examine the system

$$x' = y, \qquad y' = 2x + y \tag{15-4}$$

in some detail.   The orbits satisfy

$$\frac{dy}{dx} = \frac{2x + y}{y} = 2\frac{x}{y} + 1 \tag{15-5}$$

which can be solved by setting $y = ux$ as in Sec. 3.   For qualitative description, however, it is easier to use the geometric method of Sec. 7.   If $x = 0$ then $dy/dx = 1$, and hence the curves intersect the y axis at an angle of 45°.   Similarly, they intersect the x axis at right angles.   The locus where $dy/dx = 0$ is the line $y = -2x$.   Above this line the solutions of (15-5) are increasing, and below it they are decreasing for $y > 0$.   Since dy/dx is constant when x/y is constant, the isoclines are straight lines through the origin, and one can obtain detailed information without difficulty. The orbits are shown in Fig. 33.

As noted in Sec. 3, an equation $dy/dx = f(y/x)$ may have solutions of the form $y = mx$, corresponding to fixed points m of the function f.   The substitution of $y = mx$ into (15-5) yields a quadratic for m, of which the roots are $m = -1$ and $m = 2$.   Hence, the two lines $y = -x$ and $y = 2x$ are solutions.   These straight-line solutions are a valuable aid in sketching the other solutions.

So far we have discussed only the orbits.   It remains to describe the motion of a point $[x(t),y(t)]$ on an orbit as t varies from $-\infty$ to ∞.   Since $x' = y$ by (15-4) it is clear that x increases with t in the half plane $y > 0$, and x decreases with t when $y < 0$.   This leads to the direction of flow specified by the arrows in Fig. 33.   As a check, we note that $y' > 0$ if $2x + y > 0$, and $y' < 0$ if $2x + y < 0$.   The dividing line is the line $y = -2x$ obtained previously as locus of the extrema. Considering the half plane in which y increases with t, we get the same arrows as those specified by consideration of x'.

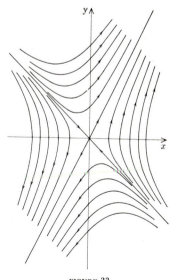

FIGURE 33
Saddle. Roots of opposite sign.

As is clearly indicated by the flow pattern, a particle initially near but not at the origin will eventually move away, and the origin is said to be an *unstable singularity*, or a point of *unstable equilibrium*.

A discussion similar to that in this example can be carried out for any system of the form

$$x' = ax + by \qquad y' = cx + dy \qquad a, b, c, d \text{ const.}$$

It is found that the general character of the solution depends on the roots of the quadratic equation

$$\lambda^2 - (a + d)\lambda + (ad - bc) = 0 \qquad\qquad (15\text{-}6)$$

and we shall summarize the results here. A proof can be given along the lines suggested in Prob. 6 or by the method of Sec. 16.

If the roots of (15-6) are real and of opposite sign, the phase trajectories have the general appearance shown in Fig. 33, and the origin is called a *saddle point*. A saddle point is always unstable.

When the roots are real and of the same sign, the trajectories have the general appearance shown in Fig. 34, and the origin is called a *node*. If both roots are negative, the point $[x(t), y(t)]$ moves toward the origin as indicated by the arrows in the figure, and the node is stable. If both roots are positive, the direction of the arrows must be reversed, and the node is unstable.

When the roots are complex the trajectories spiral around the origin as shown in Fig. 35, and the origin is called a *focus*. If the real part of the roots is negative, the point spirals toward the origin as indicated by the arrows in the figure, and the focus is stable. If the real part is positive the direction of the arrows must be reversed, and the focus is unstable.

The conditions on the roots in the above discussion embrace all cases whose general character is not altered by small perturbations of the coefficients a, b, c, d. The various degenerate cases, which are sensitive to arbitrarily small changes in a, b, c, d, are taken up in Probs. 3 and 4.

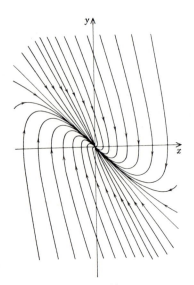

FIGURE 34
Node. Roots of same sign.

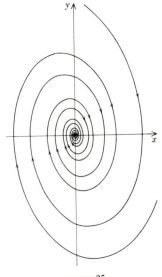

FIGURE 35
Focus. Roots complex.

PROBLEMS

1. With $y_1 = y, y_2 = y'$, write selected second-order equations in Sec. 6, Prob. 3a, b, c, as first-order systems.

2. For each of the following systems: (a) Sketch the orbits in the phase plane by geometric methods. (b) Check by finding the equation of the orbits. (c) Describe the motion of a point on a representative orbit as t increases from $-\infty$ to ∞. (d) If t is replaced by $-t$, what is the effect on the original system, on the orbits, and on the answer to part c? (This shows that the kinematic interpretation as a moving point is more informative than the static interpretation as an orbit.)

$$\begin{cases} x' = y \\ y' = 0 \end{cases} \quad \begin{cases} x' = x \\ y' = y \end{cases} \quad \begin{cases} x' = y \\ y' = -x \end{cases} \quad \begin{cases} x' = -x \\ y' = y \end{cases} \quad \begin{cases} x' = x \\ y' = 2y \end{cases} \quad \begin{cases} x' = x \\ y' = x + y \end{cases} \quad \begin{cases} x' = x + y \\ y' = -x + y. \end{cases}$$

Hint: In the last case, introduce polar coordinates $r = \sqrt{x^2 + y^2}$, $\theta = \tan^{-1}(y/x)$ and compute r' and θ' by use of the given differential equations.

3. (a) If a root $\lambda = 0$ in (15-6), show that $x'd = y'b$, and conclude that the orbits are parallel straight lines unless $a = b = c = d = 0$, when they reduce to points. (b) Show that for the system $x' = ax$, $y' = ay$, with $a \neq 0$, the root λ is double and the orbits are radial lines. (c) Discuss the stability in parts a and b. (The configuration in part b is called a *singular node*. All other cases of nonzero double roots lead to a *degenerate node*, for which the configuration resembles Fig. 34. The two straight-line solutions, however, are coincident.)

4. If the roots are purely imaginary in (15-6), show that $d = -a$, so that the equation of the orbits can be written

$$(ax + by)\,dy = (cx - ay)\,dx.$$

Observe that this is exact and obtain, on integrating,

$$cx^2 - 2axy - by^2 = \text{const.}$$

Since $ad > bc$ for imaginary roots, this represents a family of concentric ellipses. (The resulting singularity is called a *center*. It is stable but not asymptotically stable.)

5. Discuss conditions for a solution $y = mx$ or $x = my$, and show that in either case m must satisfy a quadratic with the same discriminant $D = (a - d)^2 + 4bc$ as has the quadratic (15-6).

6. Referring to Fig. 15, discuss the orbits and stability for each of the systems

$$\begin{cases} x' = ax \\ y' = -y \end{cases} \quad \begin{cases} x' = -ax \\ y' = -y \end{cases} \quad \begin{cases} x' = -ax + y \\ y' = -x - ay \end{cases}$$

where a is a positive constant. (In the three cases leading to Figs. 33 to 35 it can be shown that a suitable linear change of variables

$$t = KT \qquad x = AX + BY \qquad y = CX + DY$$

reduces the general system to one of the forms considered here.)

7. A point moving in a straight line is subject to a force depending only on the displacement s and velocity s', so that by Newton's law the differential equation has the form

$$s'' = g(s,s').$$

Show that the variables $x = s$, $y = s'$ satisfy (15-2) with $f(x,y) = y$.

8. In Prob. 7, let $g(x,y) = g(y)$, a function of y only. Show that the trajectories in phase space have the form

$$x = G(y) + c \qquad c = \text{const,}$$

where $G(y)$ is any integral of $y[g(y)]^{-1}$. Hence, the orbits form a family of parallel curves. Sketch for $g(y) = y^b$, where $b = 0, 1, 2, 3$.

9. In Prob. 7, let $g(x,y) = g(x)$ be a function of x only. Show that the trajectories in phase space have the form

$$\tfrac{1}{2}y^2 = G(x) + c \qquad c = \text{const,}$$

where $G(x)$ is any integral of $g(x)$. Sketch for $G(x) = \frac{1}{2}x^2, -\frac{1}{2}x^2, -x^4$. (A closed trajectory in phase space usually corresponds to a periodic motion. For an example, see Prob. 10.)

10. With a suitable choice of units the pendulum problem of Sec. 11, Example 1, leads to the equation of Prob. 9 with $G(x) = \cos x$. Discuss the interpretation in phase space, distinguishing the cases in which the pendulum is oscillating, is balanced, and is performing complete revolutions. *Hint:* Consider $c < -1, c = -1, -1 < c < 1, c = 1, 1 < c$, and note that $y^2 \geq 0$, since y is real.

---

16. An Introduction to Linear Systems. The usefulness of vector notation is greatly increased by some additional definitions introduced now. The zero vector is

$$\mathbf{0} = (0,0, \ldots, 0).$$

Two vectors are added or subtracted by adding or subtracting corresponding coordinates; for example,

$$(y_1,y_2,y_3) + (\bar{y}_1,\bar{y}_2,\bar{y}_3) = (y_1 + \bar{y}_1,\, y_2 + \bar{y}_2,\, y_3 + \bar{y}_3).$$

Although we shall not multiply two vectors, we do multiply a vector by a number, using the definition

$$c(y_1,y_2, \ldots,y_n) = (cy_1,cy_2, \ldots,cy_n).$$

Similarly, a vector is integrated by integrating each coordinate, as illustrated for $n = 3$ by

$$\int_{x_0}^{x} \mathbf{y}\, dx = \left(\int_{x_0}^{x} y_1\, dx,\, \int_{x_0}^{x} y_2\, dx,\, \int_{x_0}^{x} y_3\, dx \right).$$

We now use vector methods to discuss the linear system

$$\frac{dy_1}{dt} = a_{11}y_1 + a_{12}y_2 + \cdots + a_{1n}y_n + f_1(t)$$

$$\frac{dy_2}{dt} = a_{21}y_1 + a_{22}y_2 + \cdots + a_{2n}y_n + f_2(t) \qquad (16\text{-}1)$$

$$. \quad . \quad . \quad . \quad . \quad . \quad . \quad . \quad . \quad . \quad . \quad .$$

$$\frac{dy_n}{dt} = a_{n1}y_1 + a_{n2}y_2 + \cdots + a_{nn}y_n + f_n(t)$$

where the coefficients a_{ij} are constant and $f_i(t)$ are continuous functions on a given interval $a < t < b$. The system is called *linear* partly because it is linear in the unknown functions y_i and their derivatives, and partly because it is associated with a linear operator $\mathbf{T}$, as explained in the example at the end of this section.

Linear systems play a prominent role in the theory of oscillation of electrical and mechanical systems and in several branches of mathematical physics. Indeed, the possibility of reducing an nth-order equation to a first-order system shows that the whole of Chap. 3 of this book is actually concerned with linear systems.

For the modest objectives in view just now we assume that all the functions f_i are 0. The system is then said to be *homogeneous*. When $n = 1$ the linear homogeneous system with constant coefficients reduces to the form $y' = \lambda y$, which has a solution

$$y = ke^{\lambda t} \qquad k, \lambda \text{ const}$$

by the results of Sec. 2. In the general case we therefore seek a solution

$$\mathbf{y} = \mathbf{k}e^{\lambda t} \qquad \mathbf{k}, \lambda \text{ const} \qquad (16\text{-}2)$$

or, which is the same thing,[1]

---

[1] The equivalence of (16-2) and (16-3) follows from the rules of vector algebra given previously.

$$y_1(t) = k_1 e^{\lambda t}, \qquad y_2(t) = k_2 e^{\lambda t}, \qquad \ldots, \qquad y_n(t) = k_n e^{\lambda t}, \qquad (16\text{-}3)$$

where the constants k_i and λ are to be determined so that Eqs. (16-1) are satisfied identically with $f_i = 0$.

The substitution from (16-3) in (16-1) yields

$$\lambda k_1 e^{\lambda t} = (a_{11}k_1 + a_{12}k_2 + \cdots + a_{1n}k_n)e^{\lambda t}$$
$$\lambda k_2 e^{\lambda t} = (a_{21}k_1 + a_{22}k_2 + \cdots + a_{2n}k_n)e^{\lambda t}$$
$$\cdot \quad \cdot \quad \cdot \quad \cdot \quad \cdot \quad \cdot \quad \cdot \quad \cdot \quad \cdot \quad \cdot$$
$$\lambda k_n e^{\lambda t} = (a_{n1}k_1 + a_{n2}k_2 + \cdots + a_{nn}k_n)e^{\lambda t}.$$

On dividing each equation by $e^{\lambda t}$ and transposing all terms to one side, we get the system

$$
\begin{aligned}
(a_{11} - \lambda)k_1 + \quad & a_{12}k_2 + \cdots + & a_{1n}k_n = 0 \\
a_{21}k_1 + (a_{22} - \lambda)k_2 + \cdots + & & a_{2n}k_n = 0 \\
\cdot \quad \cdot \quad \cdot \quad \cdot \quad \cdot \quad \cdot \quad & \cdot \quad \cdot \quad \cdot \quad \cdot \\
a_{n1}k_1 + \quad & a_{n2}k_2 + \cdots + & (a_{nn} - \lambda)k_n = 0.
\end{aligned}
\qquad (16\text{-}4)
$$

This system is a system of linear homogeneous algebraic equations for the unknown k's. It has an obvious solution

$$k_1 = k_2 = \cdots = k_n = 0$$

corresponding to the trivial solution

$$y_1 = y_2 = \cdots = y_n = 0.$$

Since we are interested in solutions (16-3) which are not all zero, we must seek values of the k's which are not all zero. Now, a system of equations (16-4) will have such solutions for the k_i if, and only if, its determinant[1]

$$D = \begin{vmatrix} a_{11} - \lambda & a_{12} & \cdots & a_{1n} \\ a_{21} & a_{22} - \lambda & \cdots & a_{2n} \\ \cdot & \cdot \quad \cdot \quad \cdot \quad \cdot \quad \cdot & & \cdot \\ a_{n1} & a_{n2} & \cdots & a_{nn} - \lambda \end{vmatrix} = 0. \qquad (16\text{-}5)$$

The equation $D = 0$ is called the *characteristic equation* for the system (16-1). On expanding the determinant, we see that (16-5) is an algebraic equation of degree n in λ, and thus it has n real or complex roots:

$$\lambda = \lambda_1, \qquad \lambda = \lambda_2, \qquad \ldots, \qquad \lambda = \lambda_n.$$

These numbers λ_i are called *characteristic roots*, and the associated nonzero vectors $\mathbf{k}_i$ are called *characteristic vectors*. (The vectors $\mathbf{k}_i$ must be carefully distinguished from the numbers k_i which are coordinates of a single vector $\mathbf{k}$.)

Corresponding to each characteristic root λ_i is a solution of the form (16-2), so that the totality of solutions obtained by this process is

$$\mathbf{k}_1 e^{\lambda_1 t}, \; \mathbf{k}_2 e^{\lambda_2 t}, \ldots, \; \mathbf{k}_n e^{\lambda_n t}.$$

By Prob. 9*b* the function

$$\mathbf{y} = c_1 \mathbf{k}_1 e^{\lambda_1 t} + c_2 \mathbf{k}_2 e^{\lambda_2 t} + \cdots + c_n \mathbf{k}_n e^{\lambda_n t}. \qquad (16\text{-}6)$$

is also a solution, for every choice of the constants c_i.

As an illustration, we solve the system

$$y_1' = 2y_1 + 3y_2 \qquad y_2' = 2y_1 + y_2.$$

The characteristic equation (16-5) now reads

$$\begin{vmatrix} 2 - \lambda & 3 \\ 2 & 1 - \lambda \end{vmatrix} = 0 \qquad \text{or} \qquad \lambda^2 - 3\lambda - 4 = 0.$$

[1] See Appendix A.

The roots are $\lambda_1 = -1$, $\lambda_2 = 4$. Corresponding to $\lambda_1 = -1$ we have a solution

$$y = (k_1, k_2)e^{-t} \quad \text{or} \quad y_1 = k_1 e^{-t} \quad y_2 = k_2 e^{-t}.$$

To determine the constants k_1 and k_2 we form the system (16-4), namely,

$$(2 - \lambda)k_1 + 3k_2 = 0 \qquad 2k_1 + (1 - \lambda)k_2 = 0 \tag{16-7}$$

set $\lambda = -1$, and solve for the k_i's. The result is $k_1 = -k_2$, and thus one of the k_i's can be chosen at will. If we take $k_1 = 1$ we get the solution

$$y = (1, -1)e^{-t} \quad \text{or} \quad y_1 = e^{-t} \quad y_2 = -e^{-t}$$

corresponding to $\lambda = -1$.

Similarly, the choice $\lambda = 4$ in (16-7) gives the solution

$$y = (1, \tfrac{2}{3})e^{4t} \quad \text{or} \quad y_1 = e^{4t} \quad y_2 = \tfrac{2}{3}e^{4t}$$

as the reader can verify. Additional solutions are

$$\mathbf{y} = c_1(1, -1)e^{-t} + c_2(1, \tfrac{2}{3})e^{4t}$$

where c_1 and c_2 are constant, or, in coordinate form,

$$y_1 = c_1 e^{-t} + c_2 e^{4t} \qquad y_2 = -c_1 e^{-t} + \tfrac{2}{3}c_2 e^{4t}.$$

If the roots λ_i are all unequal it is not difficult to show that (16-6) represents a *general solution* on the interval $-\infty < t < \infty$; that is, a suitable choice of c_i gives

$$\mathbf{y}(t_0) = \mathbf{y}_0$$

for any given choice of t_0 and y_0. Furthermore, the solution is *complete*, in the sense that every solution, without exception, is representable in the form (16-6). These matters are discussed more fully in the following section and in Chap. 3. However, the net result is that the *linear homogeneous system with constant coefficients is completely solved by the formula* (16-6), *provided the characteristic roots λ_i are all unequal*. A similar statement holds when there are multiple roots λ but involves additional solutions in the form of polynomials in t multiplied by $e^{\lambda t}$.

Granting the truth of this statement, we can deduce an important conclusion regarding stability of the solutions. Let $\mathbf{y} = \mathbf{u}(t)$ be the solution of the inhomogeneous system (16-1) satisfying $\mathbf{u}(0) = \mathbf{u}_0$, and let $\mathbf{y} = \mathbf{v}(t)$ be another solution satisfying $\mathbf{v}(0) = \mathbf{v}_0$. The solution $\mathbf{u}$ is said to be *asymptotically stable* if[1]

$$\lim_{t \to \infty} [\mathbf{u}(t) - \mathbf{v}(t)] = \mathbf{0} \tag{16-8}$$

holds for every choice of $\mathbf{v}_0$. Now by Prob. 9a the function $\mathbf{y} = \mathbf{u} - \mathbf{v}$ satisfies the homogeneous system, that is, the system with $f_i = 0$. Hence, $\mathbf{y}$ can be represented in the form (16-6) (or in the corresponding form for multiple roots).

If $\lambda = \sigma + i\omega$ the Euler-de Moivre formula gives

$$|e^{\lambda t}| = |e^{\sigma t} e^{i\omega t}| = e^{\sigma t}$$

and hence $e^{\lambda t} \to 0$ as $t \to \infty$ if, and only if, $\sigma < 0$. A similar result holds for $p(t)e^{\lambda t}$, where $p(t)$ is any polynomial. Applying this result to each term of (16-6), we get the following:

THEOREM.   *The solutions of a linear system with constant coefficients are asymptotically stable if, and only if, all the characteristic roots have negative real parts.*

Although for simplicity we defined stability by (16-8), ordinarily the limit is required to be uniform in $\mathbf{v}_0$ when $|\mathbf{v}_0|$ ranges over a given finite interval. This stronger kind of stability can also

[1] The meaning of (16-8) is that a corresponding limit holds for each coordinate separately.

be established by (16-6), when we examine the dependence of the constants c_i on the initial conditions.

We conclude this discussion of linear equations by showing how the concept of *linearity* is formulated in vector notation. Several applications are given in Prob. 9.

Example. For purposes of this example the coordinates of a vector **y** are written one above the other, rather than side by side. Let **T** be an operator which operates on vectors **y** according to the rule

$$\mathbf{T}\begin{pmatrix} y_1 \\ y_2 \end{pmatrix} = \begin{pmatrix} y_1' - a_{11}y_1 - a_{12}y_2 \\ y_2' - a_{21}y_1 - a_{22}y_2 \end{pmatrix} \tag{16-9}$$

If **f** is a two-dimensional vector, what is the coordinate form of the equation **Ty = f**?

By definition of vector equality, the equation **Ty = f** means that corresponding coordinates agree. Hence, the coordinate form is

$$y_1' - a_{11}y_1 - a_{12}y_2 = f_1$$
$$y_2' - a_{21}y_1 - a_{22}y_2 = f_2.$$

This is the system (16-1) for $n = 2$.

The system for arbitrary n can also be written in the form

$$\mathbf{Ty} = \mathbf{f} \tag{16-10}$$

where **T** is defined by an equation similar to (16-9). Even when the coefficients $a_{ij} = a_{ij}(t)$ are not constant, it is found that **T** is *linear;* that is,

$$\mathbf{T}(\alpha\mathbf{u} + \beta\mathbf{v}) = \alpha\mathbf{Tu} + \beta\mathbf{Tv}$$

for all constants α and β and all functions **u** and **v** for which the right-hand side is meaningful. As shown in Prob. 9, this notation leads to a remarkably simple proof of fundamental facts concerning linear systems.

PROBLEMS

1. (a) Obtain the general solution: $y_1' = y_1 + y_2$, $y_2' = 4y_1 + y_2$. (b) Apply the method of this section to the equations in Sec. 15, Prob. 2.

2. For the system $x' = ax + by$, $y' = cx + dy$ considered in Sec. 15, show that the characteristic equation is (15-6), and verify that the criterion for stability of the above theorem agrees with the discussion of Sec. 15.

3. Solve Prob. 6 of Sec. 15 by the method of this section and compare your results with Figs. 33 to 35.

4. A system of second-order equations

$$y_i'' = a_{i1}y_1 + a_{i2}y_2 + \cdots + a_{in}y_n \qquad i = 1, 2, \ldots, n$$

where the a_{ij} are constant, is frequently encountered in dynamics. (a) Write this system in expanded form similar to (15-1). (b) Show by assuming a solution in the form $y = k \cos (\lambda t + \alpha)$ that one is led to the following characteristic equation for λ:

$$\begin{vmatrix} a_{11} + \lambda^2 & a_{12} & \cdots & a_{1n} \\ a_{21} & a_{22} + \lambda^2 & \cdots & a_{2n} \\ \cdots\cdots\cdots\cdots\cdots\cdots\cdots & & & \\ a_{n1} & a_{n2} & \cdots & a_{nn} + \lambda^2 \end{vmatrix} = 0.$$

The constants k_i are determined from the system of linear equations analogous to (16-4), and the constant α remains arbitrary.

5. Reduce the system of n second-order linear equations with constant coefficients,

$$\frac{d^2 y_i}{dt^2} = \sum_{j=1}^{n} a_{ij} y_j + \sum_{j=1}^{n} b_{ij} \frac{dy_j}{dt} \qquad i = 1, 2, \ldots, n$$

to a system of $2n$ first-order equations.

6. If $\lambda = \sigma + i\omega$ we define $e^{\lambda t} = e^{\sigma t} e^{i\omega t} = e^{\sigma t}(\cos \omega t + i \sin \omega t)$ (compare Sec. 12). Show that $|e^{\lambda t}| = |e^{\sigma t}|$, and that $(e^{\lambda t})' = \lambda e^{\lambda t}$. [These properties were used in the discussion of (16-2).]

7. Show that three-dimensional vectors obey the laws: $a\mathbf{0} = \mathbf{0}$, $0\mathbf{u} = \mathbf{0}$, $1\mathbf{u} = \mathbf{u}$, $\mathbf{u} + \mathbf{0} = \mathbf{u}$, $\mathbf{u} + \mathbf{v} = \mathbf{v} + \mathbf{u}$, $a(\mathbf{u} + \mathbf{v}) = a\mathbf{u} + a\mathbf{v}$, $(a + b)\mathbf{u} = a\mathbf{u} + b\mathbf{u}$, $\mathbf{u} + (\mathbf{v} + \mathbf{w}) = (\mathbf{u} + \mathbf{v}) + \mathbf{w}$, $a(b\mathbf{u}) = (ab)\mathbf{u}$. (Similar results for n-dimensional vectors were used in the text.)

8. Show that the operator $\mathbf{T}$ in the example is linear. *Hint:* Substitute $\mathbf{y} = \alpha\mathbf{u} + \beta\mathbf{v}$ and separate the terms involving α from those involving β. The proof for arbitrary n is similar.

9. In this problem, $\mathbf{T}$ is a linear operator, as in (16-10). (*a*) Show that the difference of two solutions of $\mathbf{Ty} = \mathbf{f}$ satisfies the homogeneous equation $\mathbf{Ty} = \mathbf{0}$. *Solution:* If $\mathbf{Tu} = \mathbf{f}$ and $\mathbf{Tv} = \mathbf{f}$ then

$$\mathbf{T(u - v)} = \mathbf{Tu} - \mathbf{Tv} = \mathbf{f} - \mathbf{f} = \mathbf{0}.$$

Parts *b*, *c*, and *d* below are equally easy.

(*b*) (Superposition principle.) If $\mathbf{u}$ and $\mathbf{v}$ satisfy the homogeneous equation $\mathbf{Ty} = \mathbf{0}$, show that $\alpha\mathbf{u} + \beta\mathbf{v}$ also satisfies it for every choice of the constants α and β.

(*c*) (Principle of the complementary function.) If $\mathbf{u}$ is a particular solution of $\mathbf{Ty} = \mathbf{f}$, and $\mathbf{v}$ is any solution of the associated homogeneous equation, show that $\mathbf{y} = \mathbf{u} + \mathbf{v}$ satisfies $\mathbf{Ty} = \mathbf{f}$.

(*d*) (Generalized superposition principle.) If $\mathbf{Tu} = \mathbf{f}$ and $\mathbf{Tv} = \mathbf{g}$, show that the function $\mathbf{y} = \alpha\mathbf{u} + \beta\mathbf{v}$ satisfies $\mathbf{Ty} = \alpha\mathbf{f} + \beta\mathbf{g}$, where α and β are any constants.

---

17. Remarks on Nonlinear Systems.[1]

When Picard's method is applied to the vector equation

$$\mathbf{y}' = \mathbf{f}(x, \mathbf{y}) \qquad \mathbf{y}(x_0) = \mathbf{y}_0$$

the successive approximations $\mathbf{y}_1, \mathbf{y}_2, \mathbf{y}_3, \ldots$ are determined from

$$\mathbf{y}_1' = \mathbf{f}(x, \mathbf{y}_0), \qquad \mathbf{y}_2' = \mathbf{f}(x, \mathbf{y}_1), \qquad \mathbf{y}_3' = \mathbf{f}(x, \mathbf{y}_2), \qquad \ldots$$

with the initial condition $\mathbf{y}(x_0) = \mathbf{y}_0$ in each case. (The vectors $\mathbf{y}_k$ here must be carefully distinguished from the numerical functions y_k which are the coordinates of a single vector $\mathbf{y}$.) In view of our definition of vector addition and integration, the solution of the $(n + 1)$st Picard equation can be written

$$\mathbf{y}_{n+1} = \mathbf{y}_0 + \int_{x_0}^{x} \mathbf{f}(x, \mathbf{y}_n)\, dx. \tag{17-1}$$

This follows by applying the results of Sec. 1 to each coordinate.

As an illustration, consider the system

$$y_1' = 2xy_1 + y_2 \qquad y_2' = y_1 y_2 + 1 \qquad y_1(0) = 1 \qquad y_2(0) = 0. \tag{17-2}$$

The initial approximation is given by the initial conditions as

$$\mathbf{y}_0 = (1, 0).$$

Substitution of $y_1 = 1$ and $y_2 = 0$ into the right-hand members of (17-2) gives

$$y_1' = 2x \qquad y_2' = 1.$$

---

[1] The remainder of this chapter presupposes a higher degree of mathematical maturity than is required elsewhere in this book and can be omitted without loss of continuity.

Solving these with the initial condition $y_1(0) = 1$, $y_2(0) = 0$ we get

$$y_1 = x^2 + 1 \qquad y_2 = x.$$

Hence the first Picard approximation is

$$\mathbf{y}_1 = (x^2 + 1, x).$$

These values of y_1 and y_2 are substituted into the right-hand members of the differential equations to give

$$y_1' = 2x^3 + 3x \qquad y_2' = x^3 + x + 1.$$

Solving these equations with the initial conditions $y_1(0) = 1$, $y_2(0) = 0$ we get the second Picard approximation,

$$\mathbf{y}_2 = (\tfrac{1}{2}x^4 + \tfrac{3}{2}x^2 + 1, \tfrac{1}{4}x^4 + \tfrac{1}{2}x^2 + x). \tag{17-3}$$

The process can be continued, though it becomes increasingly complex.

To state conditions under which the process converges we need inequalities involving vectors. It is not possible, in general, to introduce between vectors a relation "$<$" that has the familiar properties of this relation as applied to numbers. However, if $\mathbf{y}$ is a vector with coordinates y_k we define

$$|\mathbf{y}| = [(y_1)^2 + (y_2)^2 + \cdots + (y_n)^2]^{1/2} \tag{17-4}$$

and refer to $|\mathbf{y}|$ as the *magnitude* or *length* of $\mathbf{y}$. The terminology is suggested by the fact that, for $n = 3$, $|\mathbf{y}|$ is the ordinary Euclidean length of the line joining the point (y_1,y_2,y_3) to the origin. This particular measure of magnitude is the one that the reader is most likely to encounter in further study of vectors. Other measures, which are sometimes more useful in numerical analysis, are discussed in the problems following Sec. 18.

Although geometric visualization fails when the dimension is higher than 3, it is often desirable to use geometric terminology. A vector $\mathbf{Y}$ is said to represent (or even to be) a *point* in a space of a corresponding number of dimensions. A *neighborhood* of a point $\mathbf{Y}_0$ is a set of points $\mathbf{Y}$ satisfying

$$|\mathbf{Y} - \mathbf{Y}_0| < r$$

for some positive constant r, and the concepts of bounded region, interior point, and boundary point are defined in terms of neighborhoods as in Sec. 14. The definitions of limit and continuity for vector-valued functions also parallel the definitions for $n = 1$, the role formerly taken by the absolute value now being taken by the magnitude. These matters are discussed in Chap. 5 (cf. also Sec. 18, Prob. 1).

Without further ado we state an existence theorem for the general first-order system

$$\mathbf{y}' = \mathbf{f}(x,\mathbf{y}) \qquad \mathbf{y}(x_0) = \mathbf{y}_0$$

in a given region R specified by allowable values of x, $y_1, y_2, \ldots, y_n$. It is assumed that $\mathbf{y}$ and $\mathbf{f}$ have the same dimension, n, and that

$$|\mathbf{f}(x,\mathbf{y})| \le m \qquad |\mathbf{f}(x,\mathbf{y}_0)| \le m_0 \qquad |\mathbf{f}(x,\mathbf{y}) - f(x,\bar{\mathbf{y}})| \le k|\mathbf{y} - \bar{\mathbf{y}}| \tag{17-5}$$

holds throughout R for some constants m, m_0, and k. Here the symbol $|\mathbf{f}|$ denotes the magnitude of $\mathbf{f}$, as explained in the foregoing paragraphs. However, since the magnitude reduces to the ordinary absolute value when $n = 1$, the same symbol can be, and is, used in both cases. Thus, $|x - x_0|$ denotes the absolute value of $x - x_0$, which is the same as the magnitude of the one-dimensional vector $(x - x_0)$.

EXISTENCE THEOREM. *In a bounded region R let $\mathbf{f}(x,\mathbf{y})$ be continuous and satisfy (17-5), and let $(x_0,\mathbf{y}_0)$ be an interior point. Then the problem*

$$\mathbf{y}' = \mathbf{f}(x,\mathbf{y}) \qquad \mathbf{y}(x_0) = \mathbf{y}_0$$

has one, and only one, solution.   This solution can be extended to the right and left of x_0 until it reaches the boundary of R, and if $\mathbf{y}_n$ is the nth Picard approximation, we have

$$|\mathbf{y} - \mathbf{y}_0| \leq m|x - x_0| \qquad |\mathbf{y} - \mathbf{y}_n| \leq \frac{m_0}{k} \frac{h^{n+1}}{(n+1)!} e^h$$

where $h = k|x - x_0|$.

The last statement holds over the common interval of definition of $\mathbf{y}$ and $\mathbf{y}_n$ and, in particular, it holds over the interval $a \leq x \leq b$ in the higher-dimensional analog of Fig. 31.   If $\mathbf{f}(x,\mathbf{y})$ is continuous and satisfies the Lipschitz condition

$$|\mathbf{f}(x,\mathbf{y}) - \mathbf{f}(x,\bar{\mathbf{y}})| \leq k|\mathbf{y} - \bar{\mathbf{y}}| \qquad \text{for } a \leq x \leq b \qquad |y| < \infty \tag{17-6}$$

the solution can be extended over the interval $a < x < b$ even if $|\mathbf{f}(x,\mathbf{y})| \leq m$ is not assumed.

The scope of this theorem is vastly greater than the scope of the theorem in Sec. 14, even though the proofs are similar (Sec. 18).   For example, we can establish existence and uniqueness for the nth-order equation

$$y^{(n)} = f(x,y,y',y'', \ldots, y^{(n-1)}) \tag{17-7}$$

by taking $y_1 = y$, $y_2 = y'$, ..., $y_n = y^{(n-1)}$.   The initial condition $\mathbf{y}(x_0) = \mathbf{y}_0$ has the form[1]

$$y(x_0) = y_0{}^0, \qquad y'(x_0) = y_1{}^0, \qquad \ldots, \qquad y^{(n-1)}(x_0) = y_{n-1}{}^0, \tag{17-8}$$

where $y_0{}^0, y_1{}^0, \ldots, y_{n-1}{}^0$ are coordinates of the prescribed initial vector $\mathbf{y}_0$.   Since $y_1' = y_2$, $y'' = y_3$, and so on, the appropriate vector function in $\mathbf{y}' = \mathbf{f}(x,\mathbf{y})$

$$\mathbf{f}(x,\mathbf{y}) = [y_2,y_3, \ldots, y_n, f(x,\mathbf{y})] \tag{17-9}$$

where $\mathbf{f}(x,\mathbf{y})$ is given by (17-7) as $f(x,\mathbf{y}) = f(x,y_1,y_2, \ldots, y_n)$.

If $f(x,\mathbf{y})$ satisfies

$$|f(x,\mathbf{y})| \leq M \qquad |f(x,\mathbf{y}) - f(x,\bar{\mathbf{y}})| \leq K|\mathbf{y} - \bar{\mathbf{y}}| \tag{17-10}$$

where M and K are constant, a brief calculation based on (17-9) gives

$$|\mathbf{f}(x,\mathbf{y})|^2 \leq |\mathbf{y}|^2 + M^2 \qquad |\mathbf{f}(x,\mathbf{y}) - \mathbf{f}(x,\bar{\mathbf{y}})|^2 \leq |\mathbf{y} - \bar{\mathbf{y}}|^2 + K|\mathbf{y} - \bar{\mathbf{y}}|^2.$$

Hence, (17-5) holds with

$$m = \sqrt{N^2 + M^2} \qquad k = \sqrt{1 + K^2}$$

where N is a bound for $|\mathbf{y}|$ in R.   We conclude that *the problem* (17-7) *has one, and only one, solution satisfying the initial condition* (17-8), *if $f(x,\mathbf{y})$ is continuous and satisfies* (17-10).   The extent of the solution is described in the main theorem.

In conclusion, we remark that stability of the solutions with respect to small changes of initial conditions or of the function $\mathbf{f}$ is established in Sec. 19.   However, this analysis applies only to a finite interval.   Stability on an infinite interval can be established if the system is sufficiently well approximated by a linear system satisfying the criterion of the theorem in Sec. 16.   For $n = 2$ it is even possible to show, under a suitable hypothesis, that the solutions of the nonlinear system have the general behavior illustrated in Figs. 33 to 35, near the origin, if the solutions of the approximating linear system have this behavior.[2]

---

[1] We use $y_k{}^0$ to avoid confusion with the functions y_k that occur elsewhere in this discussion.

[2] For a precise statement and proof see Coddington and Levinson, *op. cit.*, chap. 13.   Further discussion of stability on an infinite interval, with interesting examples, can be found in L. S. Pontryagin, "Ordinary Differential Equations," chap. 5, Addison-Wesley Publishing Company, Inc., Reading, Mass., 1962, and Joseph LaSalle and Solomon Lefschetz, "Stability by Liapunov's Direct Method with Applications," Academic Press Inc., New York, 1961.

Example 1.   If the functions $a_{ij}(x)$ and $f_i(x)$ are continuous for $a \leq x \leq b$ and if $a < x_0 < b$, show that the n equations

$$y'_1 = a_{i1}y_1 + a_{i2}y_2 + \cdots + a_{in}y_n + f_i \qquad i = 1, 2, \ldots, n$$

have one, and only one, solution $\mathbf{y} = (y_1, y_2, \ldots, y_n)$ such that

$$y_1(x_0) = y_1{}^0, \qquad y_2(x_0) = y_2{}^0, \qquad \ldots, \qquad y_n(x_0) = y_n{}^0.$$

Also, this solution is valid on the whole interval $a < x < b$.

When the system is written $\mathbf{y}' = \mathbf{f}(x,\mathbf{y})$, the right-hand member of the ith equation is $f_i(x,\mathbf{y})$, and we obtain by subtraction

$$f_i(x,\mathbf{y}) - f_i(x,\tilde{\mathbf{y}}) = a_{i1}(y_1 - \tilde{y}_1) + a_{i2}(y_2 - \tilde{y}_2) + \cdots + a_{in}(y_n - \tilde{y}_n).$$

Since the functions $a_{ij}(x)$ are continuous on the closed interval $a \leq x \leq b$ they are bounded, so that $|a_{ij}| \leq A$ for some constant A.   Also

$$|y_i - \tilde{y}_i| \leq |\mathbf{y} - \tilde{\mathbf{y}}|$$

as is clear by inspection.   Hence

$$|f_i(x,\mathbf{y}) - f_i(x,\mathbf{y})| \leq nA|\mathbf{y} - \tilde{\mathbf{y}}| \tag{17-11}$$

for $i = 1, 2, \ldots, n$.   If we square these relations and add them together, we get an inequality of the type (17-6), and the result follows from this.

Using the Schwarz inequality of Sec. 18, Prob. 2, it is possible to show that k and m_0 in (17-5) can be any constants such that

$$\Sigma\Sigma a_{ij}{}^2 \leq k^2 \qquad (\Sigma\Sigma a_{ij}{}^2 + |\mathbf{f}|^2)(1 + |\mathbf{y}_0|^2) \leq m_0{}^2$$

where $\mathbf{y}_0 = \mathbf{y}(x_0)$.   For numerical estimation these bounds are much sharper than those given by the previous method.

Example 2.   If the functions $p_i(x)$ and $f(x)$ are continuous for $a \leq x \leq b$, and if $a < x_0 < b$, show that the problem

$$y^{(n)} + p_{n-1}(x)y^{(n-1)} + \cdots + p_1(x)y' + p_0(x)y = f(x)$$

has one, and only one, solution satisfying the initial conditions

$$y(x_0) = y_0, \qquad y'(x_0) = y_1, \qquad \ldots, \qquad y^{(n-1)}(x_0) = y_{n-1}.$$

Also, this solution is valid over the whole interval $a < x < b$.

When the equation is reduced to a first-order system as in the discussion of (17-7) it will be found that the conclusion is a special case of the result of Example 1.

Examples 1 and 2 provide a mathematical foundation for the theory of linear systems and linear equations, as presented in the following chapter.

PROBLEMS

1. Obtain the first three Picard approximations y_n for the equation

$$y' = x^2 + y^2 \qquad y(0) = 0$$

in the intersection of the unit circle and the strip $|x| \leq \frac{1}{2}\sqrt{2}$ (cf. Sec. 14, Example 2).   By an uncritical use of the theorem of this section, get the error estimate

$$|y - y_n| \leq \frac{|x|}{2} \frac{(\sqrt{2}|x|)^n}{(n+1)!} e^{\sqrt{2}|x|}.$$

2. In the preceding problem, obtain the stronger estimate

$$|y - y_n| \leq |x|^3 \frac{X^n}{(n+1)!} e^x \qquad \text{where } X = \tfrac{4}{3}x^4.$$

Hint: The graphs of y and y_n lie in the shaded region.   Do not pick k or m_0 larger than necessary for the given x.

3. In this problem, p, q, and r' are continuous functions on an interval containing x_0 in its interior, and $r(x) \neq 0$ on this interval. (a) Show that the substitution $y = -z'/rz$ transforms the Riccati equation

$$y' = p + qy + ry^2$$

into an equivalent linear equation

$$rz'' - (qr + r')z' + pr^2z = 0.$$

Hint: $z' = -rzy$, $z'' = -r'zy - rz'y - rzy' = ?$ (b) If $z = u$ and $z = v$ satisfy the linear equation with the respective initial conditions

$$u(x_0) = 1 \qquad u'(x_0) = 0 \qquad v(x_0) = 0 \qquad v'(x_0) = -r(x_0)$$

show that the solution of the Riccati equation satisfying $y(x_0) = c$ is

$$y = -\frac{1}{r}\frac{u' + cv'}{u + cv}.$$

Hint: Use (a) with $z = u + cv$, and uniqueness. (c) Study the dependence of y on the initial value c.

4. A family of solutions $\mathbf{y} = \mathbf{u}(x,\mathbf{c})$ is said to be a *general solution* of the system $\mathbf{y}' = \mathbf{f}(x,\mathbf{y})$ in a given region R if the constant vector $\mathbf{c}$ can be chosen so as to satisfy the initial condition

$$\mathbf{u}(x_0,\mathbf{c}) = \mathbf{y}_0$$

at any arbitrarily specified point $(x_0,\mathbf{y}_0)$ interior to R. If the hypothesis of the existence theorem holds, show that a general solution is a complete solution, in the sense that each particular solution $\mathbf{y} = \mathbf{v}$ is a member of the family. *Hint:* Choose $\mathbf{c}$ so that

$$\mathbf{u}(x_0,\mathbf{c}) = \mathbf{v}(x_0)$$

at some suitable value x_0. Then $\mathbf{u}(x,\mathbf{c}) \equiv \mathbf{v}(x)$, by uniqueness. (Under the conditions of this problem every general solution represents the same family of functions, and one is justified in speaking of *the general solution* rather than *a general solution*.)

5. Suppose the homogeneous system associated with the system of Example 1 has a pair of solutions

$$\mathbf{y}_1 = (y_1^{(1)}, y_1^{(2)}), \qquad \mathbf{y}_2 = (y_2^{(1)}, y_2^{(2)})$$

such that the determinant

$$\begin{vmatrix} y_1^{(1)} & y_1^{(2)} \\ y_2^{(1)} & y_2^{(2)} \end{vmatrix}$$

is not zero at any point of the interval $a < x < b$. Show that the general solution of the inhomogeneous system is given by

$$\mathbf{y} = c_1\mathbf{y}_1 + c_2\mathbf{y}_2 + \mathbf{u}$$

where $\mathbf{u}$ is any particular solution of the inhomogeneous system. By Prob. 4, conclude that every solution can be represented in this form. *Hint:* The fact that the given expression is a solution follows from Sec. 16, Prob. 9. To show that the solution is general, let $\mathbf{y}_0$ be an arbitrary vector and choose the constants c_i so as to satisfy $\mathbf{y}(x_0) = \mathbf{y}_0$.

6. Show that the determinant considered in Prob. 5 is identically zero if it is zero at one point x_0, $a < x_0 < b$. *Hint:* By Appendix A it is possible to find constants c_i, not both 0, so that

$$c_1\mathbf{y}_1(x_0) + c_2\mathbf{y}_2(x_0) = \mathbf{0}.$$

Since the solution $\mathbf{y} = \mathbf{0}$ satisfies the same initial condition, uniqueness gives

$$c_1\mathbf{y}_1(x) + c_2\mathbf{y}_2(x) = \mathbf{0}.$$

7. Generalize Probs. 5 and 6 to systems of dimension n. (The nth-order determinant formed by the coordinates of the solution vectors $\mathbf{y}_1$, $\mathbf{y}_2$, $\ldots$, $\mathbf{y}_n$ is called the *Wronskian*. Further discussion of the Wronskian is given in Chap. 3.)

8. In a region R, suppose the equation $\mathbf{y}' = \mathbf{f}(\mathbf{y})$ satisfies the hypothesis of the existence theorem and that the solutions are defined for $-\infty < t < \infty$. (a) If $\mathbf{y} = \mathbf{u}(t)$ is a solution, show that

$y = \mathbf{u}(t + t_0)$ is also a solution for every choice of the constant t_0. (b) Show that if two orbits in phase space intersect they coincide. *Hint:* If $\mathbf{u}(t_1) = \mathbf{v}(t_2)$ then $\mathbf{u}(t + t_1) \equiv \mathbf{v}(t + t_2)$ since the latter functions are solutions and have the same value at $t = 0$. [The system $\mathbf{y}' = \mathbf{f}(\mathbf{y})$ is called *autonomous* because the variable t does not occur explicitly on the right-hand side. Phase-plane orbits of the general system $\mathbf{y}' = \mathbf{f}(t,\mathbf{y})$ can intersect without coinciding.]

---

18. Proof of Convergence.   We now establish the theorem of Sec. 14.   It is assumed that $f(x,y)$ is continuous in the bounded region R and satisfies the three inequalities

$$|f(x,y)| \leq m \qquad |f(x,y_0)| \leq m_0 \qquad |f(x,y) - f(x,\bar{y})| \leq k|y - \bar{y}|$$

where m, m_0, and k are constant.   Evidently $m_0 \leq m$, and so m_0 could be replaced by m. We introduce m_0, nevertheless, because it gives a better estimate for the error $|y - y_n|$ and has some theoretical advantages (cf. Example 2 below).

The Picard approximations $y_1, y_2, y_3, \ldots, y_n, \ldots$ satisfy

$$y_1' = f(x,y_0), \qquad y_2' = f(x,y_1), \qquad y_3' = f(x,y_2), \qquad \ldots$$

with the initial condition $y_j(x_0) = y_0$.   We begin by showing that all these curves $y = y_n(x)$ are defined for $a \leq x \leq b$ and lie in the shaded region of Fig. 31.

As the reader will recall, the *shaded region* is defined by

$$|y - y_0| \leq m|x - x_0|.$$

The fact that the graph of $y = y_0$ lies in the shaded region for $a \leq x \leq b$ is thus evident by inspection.   If it is assumed that the graph of $y = y_n(x)$ is defined for $a \leq x \leq b$ and lies in the shaded region, then $|f(x,y_n)| \leq m$ for $a \leq x \leq b$; hence $|y_{n+1}'| \leq m$, and we conclude as in (14-2) that the curve $y = y_{n+1}(x)$ lies in the shaded region too.

The next step in the proof is to estimate the difference of two successive approximations.   If the equations

$$y_{n+1}' = f(x,y_n) \qquad \text{and} \qquad y_n' = f(x,y_{n-1})$$

are subtracted we get

$$|y_{n+1}' - y_n'| \leq |f(x,y_n) - f(x,y_{n-1})| \leq k|y_n - y_{n-1}|$$

upon taking absolute values and using the Lipschitz condition.   In terms of the differences

$$w_n = y_n - y_{n-1}$$

the result is $|w_{n+1}'| \leq k|w_n|$, valid for $n \geq 1$.   We shall use this inequality to obtain the following estimate for $n \geq 1$, $a \leq x \leq b$:

$$|w_n(x)| \leq \frac{m_0}{k} \frac{h^n}{n!} \qquad \text{where } h = k|x - x_0|. \tag{18-1}$$

The definitions of y_1 and of m_0 give

$$|y_1 - y_0| = \left| \int_{x_0}^{x} f(x,y_0)\, dx \right| \leq \left| \int_{x_0}^{x} m_0\, dx \right| = m_0|x - x_0|.$$

This agrees with (18-1) for $n = 1$.   Also, since $w_n(x_0) = 0$ we have

$$|w_{n+1}| = \left| \int_{x_0}^{x} w_{n+1}'\, dx \right| \leq \left| \int_{x_0}^{x} |w_{n+1}'|\, dx \right| \leq \left| \int_{x_0}^{x} k|w_n|\, dx \right|$$

where the last inequality follows from $|w_{n+1}'| \leq k|w_n|$.   If we substitute the estimate (18-1) into the foregoing inequality the result is

$$|w_{n+1}| \leq \int_{x_0}^{x} \frac{m_0 k^n (x - x_0)^n}{n!}\, dx = \frac{m_0 k^n |x - x_0|^{n+1}}{(n + 1)!}$$

for $x \geq x_0$, and by symmetry the same estimate holds for $x < x_0$. This shows that (18-1) holds for $n + 1$ if it holds for n, and the proof follows by mathematical induction.

The relation (18-1) indicates that the series

$$w_1 + w_2 + \cdots + w_n + \cdots \tag{18-2}$$

is uniformly convergent on $a \leq x \leq b$; we use the largest value of h on this interval and apply the Weierstrass M test (Chap. 1, Sec. 9). Since the nth partial sum of the series (18-2) is

$$(y_1 - y_0) + (y_2 - y_1) + \cdots + (y_n - y_{n-1}) = y_n - y_0$$

we see that $y = \lim y_n$ exists as a uniform limit on the interval $a \leq x \leq b$. The estimate

$$|f(x,y) - f(x,y_n)| \leq k|y - y_n|$$

shows that

$$\lim f(x,y_n) = f(x,y)$$

also holds as a uniform limit on $a \leq x \leq b$. Since $y'_{n+1} = f(x,y_n)$ we conclude that the sequence of derivatives converges uniformly. The derivatives are continuous, since $f(x,y_n)$ is continuous, and Theorem IV of Chap. 1, Sec. 9, therefore gives

$$y' = \lim y'_n = \lim f(x,y_{n-1}) = f(x,y).$$

This shows that $y' = f(x,y)$ for $a \leq x \leq b$, establishing existence of the solution.

Uniqueness follows from more general theorems presented in Sec. 19. However, for the reader's convenience, another proof is given here.

Let $\bar{y}$ be another solution, so that $\bar{y}' = f(x,\bar{y})$. By subtraction,

$$|y' - \bar{y}'| = |f(x,y) - f(x,\bar{y})| \leq k|y - \bar{y}|. \tag{18-3}$$

Since both $y = y(x)$ and $y = \bar{y}(x)$ lie in the shaded region, we have

$$|y - \bar{y}| = |(y - y_0) - (\bar{y} - y_0)| \leq 2m|x - x_0|.$$

We can therefore apply (18-3) repeatedly as in the derivation of (18-1). At the nth step the result is the bound (18-1) with $m_0 = 2m$, and since this tends to 0 as $n \to \infty$, we conclude that $|y - \bar{y}| = 0$.

So far, the solution has been constructed only for the interval $a \leq x \leq b$ of Fig. 31. If the solution curve intersects the line $x = b$ at the interior point (x_0^*, y_0^*), where $x_0^* = b$, we can repeat the foregoing procedure, using this point as the initial point instead of (x_0, y_0). The result so obtained is extended again, and so on. Evidently, a similar procedure applies to $x = a$.

To show that the method actually gets us to the boundary of R, let $x = b_0$ be the *largest value* of x to which the solution can be extended in this fashion. If the curve on the interval $x_0 \leq x < b_0$ does not come arbitrarily near to the boundary as $x \to b_0$, the distance from each point of the curve to the boundary is larger than some positive constant δ. If we extend the solution sufficiently near to the line $x = b$, it will be found that the foregoing method enables us to extend the solution, in one step, beyond $x = b_0$ (cf. Sec. 14, Example 1). This contradiction shows that $\delta > 0$ is not possible.

The fact that a *largest value* $x = b_0$ exists is geometrically plausible and can be proved by means of the fundamental principle quoted in Chap. 1, Sec. 3. It is not difficult to show also that $\lim y(x)$ exists as $x \to b_0$ through values less than b_0. In other words, the point $[x, y(x)]$ actually tends to a definite point on the boundary of R and we say, briefly, that the solution meets the boundary. [Existence of this limit hinges on the fact that $|f(x,y)| \leq m$. If we assume only that $f(x,y)$ is continuous at the interior points of R the solution can still be extended arbitrarily near the boundary but might oscillate without having a limit.]

To grasp the essentials of Picard's method the reader may find it helpful to read the foregoing proof again, skipping everything in fine print.

Example 1.   If y_n denotes the nth Picard approximation over the interval $a \leq x \leq b$, as in the text, show the difference between y_n and the true solution y satisfies

$$|y - y_n| \leq \frac{m_0}{k} \frac{h^{n+1}}{(n+1)!} e^h \qquad \text{where } h = k|x - x_0|. \qquad (18\text{-}4)$$

Since $y_n - y_0$ is the sum of the first n w_k's, while $y - y_0$ is the sum of all the w_k's, we conclude by subtraction that $y - y_n$ is the sum of all the w_k's beyond w_n.   Hence, by (18-1)

$$|y - y_n| \leq \frac{m_0}{k} \left[\frac{h^{n+1}}{(n+1)!} + \frac{h^{n+2}}{(n+2)!} + \frac{h^{n+3}}{(n+3)!} + \cdots \right].$$

Upon factoring out the first term we obtain a series whose terms are smaller than those in the series for e^h, and (18-4) follows.

Naturally, the conclusion holds only on the common interval of definition of y and y_n.   Thus, it holds at least for $a \leq x \leq b$ in Fig. 31, but it may not hold for the whole range of definition of y, because y_n may not be defined over this whole range.

Example 2.   In the strip R defined by $a \leq x \leq b$, $-\infty < y < \infty$, let $f(x,y)$ be continuous and satisfy $|f_y(x,y)| \leq k$, where k is constant.   Show that the equation $y' = f(x,y)$ has a unique solution $y = \phi(x)$ through each point (x_0, y_0) of R and that the solution is valid on the whole interval $a < x < b$.

If $|f(x,y)|$ is bounded in the strip, the result is a special case of the theorem of Sec. 14.   Thus, the only points of the shaded region which are boundary points of R occur at $x = a$ and $x = b$; hence the solution is extensible from $x = a$ to $x = b$.

Since $|f(x,y)|$ may not be bounded the present result is not a special case but is established by reexamining the proof.   The reason for introducing the shaded region is to be sure that the approximating curves $y = y_n(x)$ do not go out of R before reaching the lines $x = a$ and $x = b$ of Fig. 31.   In the present example each of these curves is automatically defined for $a \leq x \leq b$, by the recursion formula (14-5).   Furthermore, although $|f(x,y)|$ may not be bounded, $|f(x,y_0)|$ is bounded, since it is continuous for $a \leq x \leq b$.   Hence we can introduce m_0, though we cannot introduce m, and the remainder of the proof proceeds as before.   (The discussion is somewhat simpler, because the problem of extending the solution beyond $x = a$ and $x = b$ does not arise here.)

Example 3.   Extend the foregoing results to systems, and thus obtain the existence theorem of Sec. 17.

The extension depends on the two properties

$$|\mathbf{y} + \bar{\mathbf{y}}| \leq |\mathbf{y}| + |\bar{\mathbf{y}}| \qquad \left| \int_{x_0}^{x} \mathbf{y} \, dx \right| \leq \int_{x_0}^{x} |\mathbf{y}| \, dx \qquad (18\text{-}5)$$

which are established in Probs. 2 and 3 (cf. also Prob. 1).   To give the proof we merely replace y, f, and y_n by $\mathbf{y}$, $\mathbf{f}$, and $\mathbf{y}_n$, respectively, in the foregoing argument.   The symbol of absolute value has the meaning assigned to it in (17-4), but the discussion is virtually unchanged because of (18-5).   Analogs of the theorems of Chap. 1, Sec. 9, for vector-valued functions are easily obtained by repeating the proofs of these theorems in vector notation or by applying the theorems in their present form to each coordinate.   The estimate for $|\mathbf{y} - \mathbf{y}_n|$ follows as in Example 1, and the remarks accompanying (17-6) as in Example 2.

PROBLEMS ON VECTORS

1. (*a*) Show that $|w_k| \leq |\mathbf{w}|$ holds for any vector $\mathbf{w} = (w_1, w_2, \ldots, w_n)$.   (*b*) A sequence of vectors $\{\mathbf{y}_m\}$ is said to have limit $\mathbf{y}$ if $|\mathbf{y}_m - \mathbf{y}|$ tends to 0.   Applying part a with $\mathbf{w} = \mathbf{y}_m - \mathbf{y}$, show that $\lim \mathbf{y}_m = \mathbf{y}$ holds if, and only if, the corresponding limit holds for each coordinate.

2. (*a*) If $\mathbf{a} = (a_1, a_2, \ldots, a_n)$ and $\mathbf{b} = (b_1, b_2, \ldots, b_n)$, show that the two inequalities

$$|\mathbf{a} + \mathbf{b}| \leq |\mathbf{a}| + |\mathbf{b}| \qquad \sum_{i=1}^{n} a_i b_i \leq |\mathbf{a}| \, |\mathbf{b}|$$

are equivalent, in the sense that either follows from the other. *Hint:* The first inequality, when squared, is equivalent to

$$\Sigma(a_i + b_i)^2 \leq |\mathbf{a}|^2 + 2|\mathbf{a}|\,|\mathbf{b}| + |\mathbf{b}|^2.$$

(*b*) Prove the second inequality by summing the equations

$$(|\mathbf{a}|b_i - |\mathbf{b}|a_i)^2 = |\mathbf{a}|^2 b_i{}^2 - 2|\mathbf{a}|\,|\mathbf{b}|a_i b_i + |\mathbf{b}|^2 a_i{}^2$$

from $i = 1$ to $i = n$, noting that the left side is nonnegative. [The inequality $|\mathbf{a} + \mathbf{b}| \leq |\mathbf{a}| + |\mathbf{b}|$ is called the *triangle inequality* because when $n = 2$ or $n = 3$ it can be interpreted as stating that one side of a triangle does not exceed the sum of the other two sides. The inequality $\Sigma a_i b_i \leq |\mathbf{a}|\,|\mathbf{b}|$ is often called the *Schwarz inequality*, though it is due to Cauchy. (Cf. Chap. 4, Sec. 14.)]

3. (*a*) By induction, extend the triangle inequality (Prob. 2) to any number of vectors. (*b*) Show that continuous vector-valued functions **f** satisfy

$$\left| \int_a^b \mathbf{f}(x)\,dx \right| \leq \int_a^b |\mathbf{f}(x)|\,dx.$$

Hint: Approximate each integral by a sum, using the same points of subdivision in both cases. Continuity of **f** means that each coordinate is continuous, $a \leq x \leq b$.

4. Instead of the definition of the text, suppose we define

$$|\mathbf{a}| = |a_1| + |a_2| + \cdots + |a_n|.$$

Show that $|\mathbf{a} + \mathbf{b}| \leq |\mathbf{a}| + |\mathbf{b}|$ holds with this new definition, the conclusions of Probs. 1 and 3 also hold, and hence the proof of the existence theorem follows as before. (Note that the content of the theorem is different, however.)

5. As in Prob. 4, justify the definition

$$|\mathbf{a}| = \max\,(|a_1|, |a_2|, \ldots, |a_n|).$$

6. Let $w = |\mathbf{w}|$, where **w** is differentiable and $|\mathbf{w}|$ is defined as in the text. Show that $|w'| \leq |\mathbf{w}'|$ at points where $w > 0$. *Hint:* Differentiate the relation $w^2 = w_1{}^2 + w_2{}^2 + \cdots + w_n{}^2$, and apply the Schwarz inequality (Prob. 2).

---

19. Differential Inequalities and Comparison.   The stability theorem of Sec. 16 was easy to prove because we had an explicit formula for the error, $|\mathbf{u} - \mathbf{v}|$. However, the central problem of error estimation and stability concerns equations for which explicit solutions are not available. In discussing this problem we shall illustrate the method of differential inequalities developed, chiefly in this century, by Chaplygin in Russia, Perron and Collatz in Germany, Nagumo in Japan, and mathematicians of several other nations.

Without loss of generality we assume that the initial condition is given at $t = 0$ and that the functions listed below are defined and continuous on an interval $0 \leq t < t_0$. Extension to the case $-t_0 < t \leq 0$ can be made by taking $-t$ as the new variable, but in many applications t is the time and the conclusions are desired, in fact, only for $t > 0$.

We use the letters

$$\epsilon,\ \bar{\epsilon},\ p,\ P,\ U,\ V,\ u,\ v,\ w,\ y,\ \mathbf{u},\ \mathbf{v},\ \mathbf{w}$$

as abbreviations for $\epsilon(t)$, $\bar{\epsilon}(t)$, $U(t)$, and so on. However, if not further qualified, any relation containing these letters is to hold on the whole interval $0 < t < t_0$, rather than just for a single t.

A *differential inequality* is the relation obtained from a differential equation when the equality sign in it is replaced by a sign of inequality. Sometimes the initial condition as well as the equation is replaced by an inequality; for example,

$$U' - pU \leq \epsilon \qquad U(0) \leq \delta \tag{19-1}$$

is a differential inequality and initial condition associated with the linear differential equation and initial condition

$$V' - pV = \epsilon \qquad V(0) = \delta. \qquad (19\text{-}2)$$

By Sec. 5 the solution of the latter is $V = (\delta + \bar{\epsilon})e^P$, where

$$P(t) = \int_0^t p(\tau)\, d\tau \qquad \bar{\epsilon}(t) = \int_0^t e^{-P(\tau)}\epsilon(\tau)\, d\tau. \qquad (19\text{-}3)$$

This can be used to solve the inequality (19-1). With $W = U - V$ it is found that (19-2) and (19-1) give $(e^{-P}W)' \leq 0$, and hence $W \leq 0$, or $U \leq V$. In other words,

$$U(0) \leq \delta \qquad \text{and} \qquad U' - pU \leq \epsilon \qquad \text{implies} \qquad U \leq (\delta + \bar{\epsilon})e^P. \qquad (19\text{-}4)$$

By Prob. 1 the inequality $U' - pU \leq \epsilon$ is actually needed only at points where $U > V$, and we use this observation in the sequel.

If u is an approximate solution of a differential equation and boundary condition, and v is an exact solution, the error in the initial value of u is

$$u(0) - v(0).$$

To get an equally convenient expression for the error in the differential equation, we introduce the notation

$$\mathbf{T}u = u' - f(t,u) \qquad \mathbf{T}v = v' - f(t,v).$$

If the desired differential equation is $\mathbf{T}v = g(t)$, the amount by which u fails to satisfy it is represented by

$$\mathbf{T}u - \mathbf{T}v.$$

Our problem, then, is to estimate the *solution error* $u - v$ in terms of the *initial error* $u(0) - v(0)$ and the *equation error* $\mathbf{T}u - \mathbf{T}v$. We shall establish the following:

THEOREM I. Let $\mathbf{T}u = u' - f(t,u)$, *and suppose* f *admits the estimate*

$$\frac{f(t,u) - f(t,v)}{u - v} \leq p \qquad \text{for} \qquad u \neq v \qquad 0 \leq t < t_0. \qquad (19\text{-}5)$$

Suppose further that $\delta + \bar{\epsilon} \geq 0$ *in* (19-3). *Then*

$$u(0) - v(0) \leq \delta \qquad \text{and} \qquad \mathbf{T}u - \mathbf{T}v \leq \epsilon \qquad \text{implies} \qquad u - v \leq (\delta + \bar{\epsilon})e^P.$$

It should be observed that neither u nor v need be an exact solution of the underlying differential equation, so that the result is independent of existence theorems.

To prove the theorem, define $U = u - v$ and note that $u(0) - v(0) \leq \delta$ gives $U(0) \leq \delta$. Also $\mathbf{T}u - \mathbf{T}v \leq \epsilon$ gives

$$U' \leq \epsilon + f(t,u) - f(t,v) \leq \epsilon + pU$$

for $U > 0$, since (19-5) can then be multiplied through by $u - v = U$. In particular, the inequality $U' \leq \epsilon + pU$ holds whenever U exceeds the desired upper bound V given by (19-2). Hence, the conclusion follows from (19-4).

By the mean-value theorem, (19-5) holds if

$$f_y(t,y) \leq p \qquad \text{for all } y \text{ between } u \text{ and } v. \qquad (19\text{-}6)$$

This requirement is less general than that of the theorem but is convenient in applications.

As an illustration, suppose the exact solution $y = v$ of the problem

$$\mathbf{T}y \equiv y' - e^{-t} - 2t^3 (\cos y - y) = 0 \qquad y(0) = -1 \qquad (19\text{-}7)$$

is approximated by $u = -e^{-t}$. (This would be the exact solution if the term containing t^3 were absent.) It is easily seen that (19-6) holds with $p = 0$. Also $\mathbf{T}u = 2t^3(u - \cos u)$; hence

$$\mathbf{T}u - \mathbf{T}v \leq 0 \qquad \mathbf{T}v - \mathbf{T}u = 4t^3. \qquad (19\text{-}8)$$

These elementary estimates, which could be improved, follow from the fact that $-1 < u < 0$. Since $p = \delta = 0$ the first result (19-8) in Theorem I gives $u - v \leq 0$. To use the second result we take $\epsilon = 4t^3$ and apply Theorem I with u and v interchanged. This gives $v - u \leq t^4$, and so the exact solution v satisfies

$$-e^{-t} \leq v \leq -e^{-t} + t^4 \qquad t > 0.$$

For many purposes it is sufficient to replace $p(t)$ by its maximum, so that $p = k$, a constant. If ϵ is also constant the values of P and $\bar{\epsilon}$ are easily computed explicitly, and we are led to the following:

THEOREM II. Let $\mathbf{T}u = u' - f(t,u)$, *let* ϵ, δ, *and* k *be constant, and let* (19-5) *or* (19-6) *hold with* $p = k$. *Then*

$$|u(0) - v(0)| \leq \delta \qquad and \qquad |\mathbf{T}u - \mathbf{T}v| \leq \epsilon \qquad implies \qquad |u - v| \leq \delta e^{kt} + \epsilon \frac{e^{kt} - 1}{k}.$$

If $k = 0$ the right-hand side is replaced by its limit, $\delta + \epsilon t$. Theorem II is established by applying Theorem I as it stands and with u and v interchanged, using the same ϵ both times.

The result can be used to prove stability of the solution on any finite interval $0 \leq t < t_0$. Roughly speaking, we must show that the solution error is small whenever the initial error and the equation errors are both small. In a strict formulation, the solution v is *stable* if for every positive constant η there is a positive constant δ, independent of u, such that

$$|u(0) - v(0)| \leq \delta \qquad and \qquad |\mathbf{T}u - \mathbf{T}v| \leq \delta \qquad implies \qquad |u - v| \leq \eta.$$

By inspection of the error bound in Theorem II we see that an exact solution, v, is stable provided (19-5) or (19-6) holds for this v and all relevant u. The choice $\epsilon = \delta = 0$ gives uniqueness, $u = v$.

There is no difficulty in extending the results to vector equations if the one-sided inequality (19-5) is replaced by an appropriate two-sided inequality. In the notation of Sec. 17 we shall establish:

THEOREM III. Let $\mathbf{Tu} = \mathbf{u}' - \mathbf{f}(t,\mathbf{u})$ *and let*

$$|\mathbf{f}(t,\mathbf{u}) - \mathbf{f}(t,\mathbf{v})| \leq p|\mathbf{u} - \mathbf{v}|. \tag{19-9}$$

Then $|\mathbf{u}(0) - \mathbf{v}(0)| \leq \delta$ *and* $|\mathbf{Tu} - \mathbf{Tv}| \leq \epsilon$ *implies* $|\mathbf{u} - \mathbf{v}| \leq (\delta + \bar{\epsilon})e^P$.

Continuity and boundedness of $\mathbf{f}$ are not needed here, though both are needed in the existence theorem.

For proof, set $\mathbf{w} = \mathbf{u} - \mathbf{v}$. The condition $|\mathbf{Tu} - \mathbf{Tv}| \leq \epsilon$ gives

$$|\mathbf{w}' - [\mathbf{f}(t,\mathbf{u}) - \mathbf{f}(t,\mathbf{v})]| \leq \epsilon$$

or, in an obvious notation, $|\mathbf{w}' - \Delta\mathbf{f}| \leq \epsilon$. Since $\mathbf{w}' = (\mathbf{w}' - \Delta\mathbf{f}) + \Delta\mathbf{f}$ we deduce that

$$|\mathbf{w}'| \leq \epsilon + |\Delta\mathbf{f}| \leq \epsilon + p|\mathbf{u} - \mathbf{v}| = \epsilon + p|\mathbf{w}|$$

where the second inequality follows from the hypothesis (19-9). According to the result of Sec. 18, Prob. 6, we have $w' \leq |\mathbf{w}'|$, where $w = |\mathbf{w}| > 0$. Hence $w = |\mathbf{u} - \mathbf{v}|$ satisfies $w' \leq \epsilon + pw$ whenever $w > 0$, and the desired estimate follows from (19-4) with $U = w$.

If $p = k$ is constant, the error bound is evaluated as in Theorem II. Since the initial point could be taken as x_0 rather than 0 we get a result of the form

$$|\mathbf{u}(x_0) - \mathbf{v}(x_0)| \leq \delta \qquad and \qquad |\mathbf{Tu} - \mathbf{Tv}| \leq \epsilon \qquad implies \qquad |\mathbf{u} - \mathbf{v}| \leq \delta e^h + \epsilon \frac{e^h - 1}{k}$$

where $h = k|x - x_0|$. (The estimate for $x < x_0$ follows by symmetry.) This holds on any interval containing x_0 throughout which

$$|\mathbf{f}(x,\mathbf{u}) - \mathbf{f}(x,\mathbf{v})| \le k|\mathbf{u} - \mathbf{v}|.$$

In particular, it holds under the hypothesis (17-5). Since the error bound tends to 0 as $\delta \to 0$ and $\epsilon \to 0$ we conclude that *the solution given by the existence theorem of Sec. 17 is stable.*

PROBLEMS

1. If $\phi(0) \le 0$, $\phi(t)$ is continuous for $0 \le t < t_0$, and $\phi'(t) \le 0$ at all points where $\phi(t) > 0$, show that $\phi(t) \le 0$ for $0 \le t < t_0$. Thus establish the strengthened form of (19-4) by taking $\phi = e^{-p}W$. *Hint:* If $\phi(t_2) > 0$ at some point t_2, consider the *last point* t_1 on the interval $0 \le t \le t_2$ at which $\phi(t_1) \le 0$, and apply Rolle's theorem on the interval (t_1,t_2). Existence of a *last point* t_1 follows from the fundamental principle of Chap. 1, Sec. 3.

2. (a) Applying Theorem II to (19-7) with $\epsilon = 4t_0^3$, obtain

$$|u - v| \le 4t_0^3 t \qquad \text{and} \qquad |u - v| \le 4t^4 \qquad 0 < t < t_0.$$

If $t_0 = 1$, show that the respective errors given by the text and by the above are ± 0.00005, ± 0.4, and ± 0.0004 at $t = 0.1$. (b) Most treatises on differential equations require $|f_y| \le k$ instead of our one-sided inequality $f_y \le k$. Show that this condition in (19-7) gives $|u - v| < 54$ for $t = 1$, instead of the relation $0 < v - u < 1$ obtained in the text. *Hint:* $\epsilon = k = 4$ in Theorem II.

3. Let $\mathbf{T}u = u' - f(t,u)$ with $f_y(t,y) \le$ const as in Theorem II. Show that

$$u(0) \le v(0) \qquad \text{and} \qquad \mathbf{T}u \le \mathbf{T}v \qquad \text{implies} \qquad u \le v.$$

An operator $\mathbf{T}$ with this property is said to be *monotone in the sense of Collatz*. *Hint:* Use Theorem II with $\delta = \epsilon = 0$.

4. Let $\mathbf{T}y = y' - f(t,y)$, where $f(t,s)$ is nondecreasing as a function of s for each t. Using Theorem I with $p = 0$, show that

$$|u(0) - v(0)| \le \delta \qquad \text{and} \qquad |\mathbf{T}u - \mathbf{T}v| \le \epsilon \qquad \text{implies} \qquad |u - v| \le \int_0^t \epsilon(t)\, dt.$$

5. If the hypothesis of Theorem II holds on $0 < t < \infty$ with $k < 0$, show that

$$|u - v| \le \max\left(\delta, \frac{\epsilon}{|k|}\right) \qquad 0 < t < \infty$$

and hence that stability holds on the whole interval $0 < t < \infty$.

6. On a given interval $a < x < b$, let an operator $\mathbf{T}$ be defined by

$$\mathbf{T}u = u - f(x,u',u'').$$

Suppose that $f[x,u'(x),s]$ or $f[x,v'(x),s]$ is a nondecreasing function of s for each x, and that

$$|u(a) - v(a)| \le \delta \qquad |u(b) - v(b)| \le \delta \qquad |\mathbf{T}u - \mathbf{T}v| \le \epsilon$$

where δ and ϵ are constant. Prove that $|u - v| \le \max(\epsilon,\delta)$. *Outline of solution:* If $u(x) - v(x)$ exceeds $\max(\epsilon,\delta)$ at some x, it assumes a maximum at a point $x = \xi$ between a and b. The conditions

$$u(\xi) - v(\xi) > \epsilon \qquad u'(\xi) = v'(\xi) \qquad u''(\xi) \le v''(\xi)$$

lead at once to a contradiction. We assume $u - v$ continuous, $a \le x \le b$.

Linear differential equations and Laplace transform

165

3
Linear differential equations and Laplace transform

LINEAR DIFFERENTIAL EQUATIONS occupy a prominent place in the study of the response of elastic structures to impressed forces and in the analysis of electrical circuits and servomechanisms. They also appear in numerous boundary-value problems in the theory of diffusion and heat flow, in quantum mechanics and fluid mechanics, and in electromagnetic theory.

The first part of this chapter treats equations with constant coefficients, emphasizing the theory of oscillation and its interpretation in phase space. The important concepts of linearity, superposition, stability, and steady-state behavior are introduced here. The second part gives a comprehensive account of the Laplace transform, which is the basis for much of the modern theory of linear system and circuit analysis. The third part contains an introduction to equations with variable coefficients, including series solutions, orthogonality, and a general discussion of linear dependence. Throughout this chapter, the development is simplified by systematic use of the concept of a linear operator. Apart from their use in differential equations, linear operators are important in several other areas of mathematics and mathematical physics.

EQUATIONS WITH CONSTANT COEFFICIENTS

1. Mechanical and Electrical Systems. We begin our study with a simple mechanical system which is a prototype of more general systems that appear in the analysis of vibrations of elastic structures.[1]

If a mass m is applied to one end of a suspended spring whose other end is fixed, it will produce the elongation s, which, according to Hooke's law, is proportional to the applied force $F = mg$, g being the gravitational acceleration. Thus,

$$F = ks = mg$$

where k is the stiffness of the spring.

If an additional force is applied to produce an extension y, after which this additional force is removed, the spring will start to oscillate (Fig. 1). The problem is to determine the position of the end point of the spring at any subsequent time.

The forces acting on the mass m are the force of gravity mg downward, which will be taken as the positive direction for the displacement y, and the tension T in the spring, which acts in the direction opposite to the force of gravity. Since T is the tension

---

[1] This discussion and similar ones given later are intended to show the use of linear equations in applications and to give a concrete physical model for the benefit of readers who prefer such a model. It should be emphasized that the theory of linear differential equations is in no way dependent on any physical principles, either mechanical or electrical.

166

when the elongation is $s + y$, Hooke's law states that $T = k(s + y)$. The resultant downward force due to the combined action of the spring and gravity is, therefore,

$$F = mg - k(s + y) = -ky$$

when it is recalled that $mg = ks$.

It is convenient to denote differentiation by a prime, so that

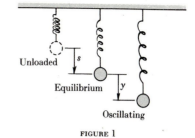

$$y' \equiv \frac{dy}{dt}, \qquad y'' \equiv \frac{d^2y}{dt^2}, \qquad \dots, \qquad y^{(n)} \equiv \frac{d^ny}{dt^n}.$$

The velocity is then y' and the acceleration y''. By Newton's second law,

$$(\text{mass})(\text{acceleration}) = (\text{force})$$

and so $my'' = -ky$, or

$$my'' + ky = 0.$$

This is the differential equation for free oscillations of the undamped mass-spring system.

We now suppose that the mass is in a resisting medium in which the damping force is proportional[1] to the velocity y'. Since the resisting medium opposes the motion, the damping acts in a direction opposite to the velocity. Hence the resistive force is $-ry'$, where r is a positive constant. Newton's law now gives $my'' = -ky - ry'$, or

$$my'' + ry' + ky = 0.$$

This is the equation for free oscillations of the damped mass-spring system.

Next, let the point of support vibrate in accordance with some law that gives the displacement of the top of the spring as a function of time, say $x = x(t)$, where x is measured positively downward. Since the elongation of the spring is diminished by the amount x, the term ky in the foregoing discussion must be replaced by $k(y - x)$. The result is

$$my'' + ry' + ky = kx(t). \tag{1-1}$$

This equation describes forced oscillations of the damped mass-spring system. The equation for an undamped system is obtained by setting $r = 0$, and the equation for free oscillations is obtained by setting $x = 0$. Hence, (1-1) summarizes all the preceding analysis.

Although detailed study of (1-1) occupies several sections of this chapter, the special case

$$my'' + ky = 0 \tag{1-2}$$

can be solved easily. The key to the solution is to introduce the velocity $v = y'$ as a new variable. By the definition of y'' and the chain rule,

$$y'' = \frac{dv}{dt} = \frac{dv}{dy}\frac{dy}{dt} = v\frac{dv}{dy}.$$

Substituting into (1-2) gives $mv\, dv + ky\, dy = 0$ after multiplication by dy, and by integration we get

$$\tfrac{1}{2}mv^2 + \tfrac{1}{2}ky^2 = E \qquad E = \text{const.} \tag{1-3}$$

[1] This kind of damping is called *viscous damping*.

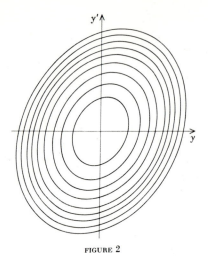

FIGURE 2

This important result expresses the principle of conservation of energy. The term $\frac{1}{2}mv^2$ represents the kinetic energy in the moving mass, and the term $\frac{1}{2}ky^2$ gives the work

$$\int_0^y F\, dy = \int_0^y ky\, dy$$

required to stretch the spring by an amount y. Hence, $\frac{1}{2}ky^2$ represents the potential energy. When there is no damping and no forcing, one would expect the total energy E to remain constant; this expectation is confirmed by the analysis. Similar behavior is found for broad classes of nondissipative systems.

According to (1-3) the curve given by $y = y(t)$, $v = v(t)$ traces an ellipse in the (y,v) plane (Fig. 2). The ellipse is called the *energy ellipse*, and the (y,y') plane is called the *phase space*. The size of the ellipse is determined by the initial displacement and velocity, and the point at the common center of all the ellipses corresponds to the equilibrium position, $y = v = 0$. When the moving mass is perturbed by an additional impulse, the orbit in phase space moves to a *new energy level*, that is, to a new ellipse.

Solving (1-3) for v gives

$$v = \pm\omega\sqrt{A^2 - y^2} \qquad \text{where } \omega^2 = \frac{k}{m} \qquad A^2 = 2\frac{E}{k}. \tag{1-4}$$

Since $v = dy/dt$, Eq. (1-4) with the minus sign is equivalent to

$$-\frac{dy}{\sqrt{A^2 - y^2}} = \omega\, dt$$

and this integrates to give $\cos^{-1}(y/A) = \omega t - \phi$, where ϕ is constant. Thus

$$y = A\cos(\omega t - \phi) \qquad A \text{ and } \phi \text{ const} \qquad \omega = \sqrt{\frac{k}{m}}. \tag{1-5}$$

Expansion of $\cos(\omega t - \phi)$ in (1-5) gives another form for y:

$$y = c_1 \cos\omega t + c_2 \sin\omega t \tag{1-6}$$

where $c_1 = A\cos\phi$ and $c_2 = A\sin\phi$ are also constant. It is easily shown that each $y(t)$ in (1-5) or (1-6) satisfies $y'' + \omega^2 y = 0$, and we shall presently prove that the equation has no other solutions. Thus, ignoring the plus sign in (1-4) involves no loss of generality.

Motion governed by $y = A\cos(\omega t - \phi)$ as in (1-5) is called *simple harmonic motion*, with *amplitude A* and *phase ϕ*. (Here the term "phase" has no relation to the expression "phase space" introduced previously.) Since

$$\cos\left[\omega\left(t + \frac{2\pi}{\omega}\right) - \phi\right] = \cos(\omega t - \phi)$$

the simple harmonic motion is periodic,[1] with period $T = 2\pi/\omega$. This is the time required for one complete oscillation. The *frequency*

$$f = \frac{1}{T} = \frac{\omega}{2\pi}$$

---

[1] A function f has period T if $f(t + T) \equiv f(t)$, where T is a nonzero constant.

is the number of oscillations per second, and ω is the *angular frequency* in radians per second.

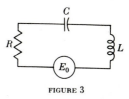

FIGURE 3

We conclude this discussion by comparing the mass-spring system with a corresponding electrical problem. It will be seen that a striking analogy exists between the mechanical and electrical systems. This analogy permits one to replace a study of complicated mechanical systems by the analysis of performance of mathematically equivalent electrical circuits.

Let a source of emf be connected in series with a capacitor, a coil, and a resistor, as shown in Fig. 3. If the current is I, the voltage drop across the resistance R is IR, by Ohm's law, and the drop across an inductance L is

$$L\frac{dI}{dt} \equiv LI'$$

by the law of Faraday. The voltage equation is therefore

$$LI' + RI + V = E_0(t) \tag{1-7}$$

where $E_0(t)$ is the voltage across the source and V is the voltage across the capacitor.

It is known that the charge Q on a capacitor plate satisfies $Q = CV$, where the constant of proportionality, C, is the *capacitance*. Hence $I = Q' = CV'$ (since $I \equiv dQ/dt$ by definition). Substitution into (1-7) gives

$$LV'' + RV' + SV = SE_0(t) \qquad S = C^{-1} \tag{1-8}$$

after division by C. The constant $S = 1/C$ is called the *elastance*.

Upon comparing (1-8) with (1-1) we see an exact correspondence between the variables

$$V, L, R, S$$

of the present problem and the variables

$$y, m, r, k$$

of the mechanical problem. The impressed displacement $x(t)$ of the top of the spring corresponds to the impressed voltage $E_0(t)$ of the source.

Mathematical correspondences of this kind form the basis for analog computers, in which mechanical systems are simulated by electrical circuits. The advantage is that electrical components are easier to manufacture and to interconnect than are mechanical components, and a voltage is more easily measured than is the displacement of a moving object. Another advantage is that the whole technology of transmission-line theory, with its concepts of impedance, transfer functions, and so on, can be applied to mechanical systems.[1]

Example. A mass of 50 g stretches a spring from 10 to 30 cm when hung on its end. The mass is raised 3 cm from its equilibrium position and released with a downward speed of 14 cm per sec. Neglecting frictional losses, describe the subsequent motion.

Since $g = 980$ cm per sec^2 in the cgs system, the 50-g mass has a weight of $(50)(980)$ dynes, and by Hooke's law $20k = (50)(980)$ or $k = 2,450$ dynes per cm. Hence $\omega = \sqrt{k/m} = 7$, and (1-6) gives

$$y = c_1 \cos 7t + c_2 \sin 7t.$$

The initial displacement is $y(0) = -3$, if y is measured positively downward, so that $c_1 = -3$. Differentiating and using $y'(0) = 14$ we get $7c_2 = 14$, with the result

[1] For a comprehensive discussion of electromechanical analogs see J. E. Alexander and J. M. Bailey, "Systems Engineering Mathematics," Prentice-Hall, Inc., Englewood Cliffs, N.J., 1962.

$$y = -3 \cos 7t + 2 \sin 7t \qquad t \geq 0.$$

It should be noted that two initial conditions determine the motion: the initial displacement $y(0)$ and the velocity $y'(0)$. Two conditions are needed because the underlying differential equation is of the second order. We shall presently find that a third-order equation would require three conditions, and so on.

── **PROBLEMS**

In these and subsequent problems involving springs it is assumed that the spring is weightless and does not become slack or exceed its elastic limit. If nothing is said about resistance, the resistance is to be neglected.

1. A mass M is hung on the end of a spring of stiffness k and allowed to come to equilibrium. The total length of the spring is now y_0. If an additional mass m increases the length to y_1, show that $mg = k(y_1 - y_0)$ independently of the mass M.

2. A force of 1,000 dynes stretches a spring 1 cm. A mass of 100 g is suspended at the end of the spring and allowed to come to equilibrium. (*a*) Find the differential equation of motion and the frequency of vibration. (*b*) As in the example, obtain displacement versus time if the mass is pulled down 2 cm from its equilibrium position and then released. (*c*) Similarly, solve if the mass is projected down from its equilibrium position with speed 10 cm per sec.

3. Two equal masses are suspended from the end of a spring of stiffness k. After equilibrium is reached, one of the masses falls off. Describe the motion of the remaining mass. *Hint:* Measure y from the equilibrium position of a single mass, as in the text, and see Prob. 1.

4. Using (1-6), find y, given that $y'' + y = 0$, $y(0) = y_0$, $y'(0) = y_1$.

5. Determine ω so that $V = A \cos \omega t$ satisfies $LV'' + C^{-1}V = 0$, where ω, A, L, and C are nonzero constants.

6. If $m = k = 1$, show that the energy ellipse is a circle. Draw concentric circles corresponding to $E = 0, 1, 2, 3, 4, 5$.

7. A lossless mass-spring system is observed to oscillate with angular frequency ω. When a known mass m_1 is added, the frequency drops to a new value, ω_1. Express m and k of the original system in terms of ω, ω_1, m_1.

8. A truck being lowered by a crane sinks 6 in. from the moment the tires just touch the ground. Estimate the frequency of free oscillations when the truck is on the road. (Regard the truck as a lossless mass-spring system, with mass m and stiffness constant k. Since 6 in. = ½ ft, Hooke's law gives $\frac{1}{2}k = mg$, where $g = 32.2$ ft per sec².)

9. Solve (1-4) with the plus sign and show that the same family of solutions (1-6) is obtained.

10. Review the discussion of terminology and notation given in Chap. 2, Sec. 1.

──

2. Linear Operators. The equation obtained for mechanical and electrical vibrations in Sec. 1 has the form

$$y'' + p_1 y' + p_0 y = f(t) \tag{2-1}$$

where p_1 and p_0 are constant. This can be written $\mathbf{T}y = f(t)$, where $\mathbf{T}$ is the operator defined by

$$\mathbf{T}y = y'' + p_1 y' + p_0 y. \tag{2-2}$$

It should be emphasized that (2-2) describes the action of $\mathbf{T}$, not only on y but on any function to which $\mathbf{T}$ may be applied. For example,

$$\mathbf{T}u = u'' + p_1 u' + p_0 u \qquad \mathbf{T}v = v'' + p_1 v' + p_0 v \tag{2-3}$$

and so on. The sole condition on u, v, or y is that they be twice differentiable, and it is said briefly that the *domain* of $\mathbf{T}$ is the class of twice-differentiable functions.

We shall presently show that $\mathbf{T}$ satisfies

$$\mathbf{T}(u + v) = \mathbf{T}u + \mathbf{T}v \qquad \text{and} \qquad \mathbf{T}(cu) = c\mathbf{T}u \tag{2-4}$$

for any constant c and for any functions u and v such that $\mathbf{T}u$ and $\mathbf{T}v$ make sense. An operator of this kind is said to be *linear*. The two conditions (2-4) give the single condition

$$\mathbf{T}(\alpha u + \beta v) = \alpha \mathbf{T}u + \beta \mathbf{T}v \qquad \alpha, \beta \text{ const} \tag{2-5}$$

as shown in Chap. 1, Sec. 2, Prob. 7. This is the property needed in applications, and we regard it as another definition of linearity; that is, if (2-5) holds for all constants α and β and for all functions u and v such that $\mathbf{T}u$ and $\mathbf{T}v$ make sense, $\mathbf{T}$ is a linear operator.

A familiar example of a linear operation is the operation of taking a derivative. Thus, by elementary calculus,

$$\frac{d}{dx}(\alpha u + \beta v) = \alpha \frac{d}{dx} u + \beta \frac{d}{dx} v$$

or $(\alpha u + \beta v)' = \alpha u' + \beta v'$. Differentiating again we get

$$(\alpha u + \beta v)'' = \alpha u'' + \beta v''$$

and so on. This shows that the operator d^2/dx^2 associated with the second derivative is also linear, and similarly for higher derivatives.

It is easy to verify that the operator $\mathbf{T}$ in (2-2) is linear. Setting $y = \alpha u + \beta v$ gives

$$\mathbf{T}y = (\alpha u + \beta v)'' + p_1(\alpha u + \beta v)' + p_0(\alpha u + \beta v)$$

$$= (\alpha u'' + \beta v'') + p_1(\alpha u' + \beta v') + p_0(\alpha u + \beta v).$$

Collecting the terms involving α and then those involving β, we get

$$\alpha(u'' + p_1 u' + p_0 u) + \beta(v'' + p_1 v' + p_0 v).$$

This is $\alpha \mathbf{T}u + \beta \mathbf{T}v$, by (2-3); hence the proof of linearity is complete.

Problems involving linear operators are greatly simplified by use of the following fundamental theorem:

PRINCIPLE OF SUPERPOSITION. Let $\mathbf{T}$ be any linear operator. Then if $y = u$ and $y = v$ are both solutions of $\mathbf{T}y = 0$, the same is true of $y = c_1 u + c_2 v$, for any constants c_1 and c_2.

The proof is immediate, because by linearity $\mathbf{T}(c_1 u + c_2 v) = c_1 \mathbf{T}u + c_2 \mathbf{T}v$, and $\mathbf{T}u$ and $\mathbf{T}v$ are 0 by hypothesis. As an illustration, since $y'' + \omega^2 y = 0$ is associated with a linear operator, the solution

$$c_1 \cos \omega t + c_2 \sin \omega t$$

can be obtained by superposition of the two particular solutions $\cos \omega t$, $\sin \omega t$.

A simple method of generating solutions u and v is based on the fact that the exponential function e^{st} has the property of reproducing itself when differentiated:

$$y = e^{st}, \qquad y' = se^{st}, \qquad y'' = s^2 e^{st}, \qquad \ldots \ldots \tag{2-6}$$

Substitution of e^{st} into a linear differential equation with constant coefficients therefore produces an algebraic equation in the unknown coefficient s.

Thus, the differential equation

$$y'' + p_1 y' + p_0 y = 0 \tag{2-7}$$

leads to the algebraic equation

$$s^2 + p_1 s + p_0 = 0 \tag{2-8}$$

when we use (2-6) and divide through by e^{st}. If (2-8) has distinct roots s_1 and s_2, we get two solutions

$$e^{s_1 t} \quad \text{and} \quad e^{s_2 t}.$$

Superposition now gives the solution

$$y = c_1 e^{s_1 t} + c_2 e^{s_2 t}$$

containing the arbitrary constants c_1 and c_2. Equation (2-8) is called the *characteristic equation* of (2-7).

Example. The displacement y in a certain overdamped mass-spring system satisfies $y'' + 3y' + 2y = 0$. If the initial displacement is 5 and the initial velocity is 3, find the value of y at any subsequent time.

The function e^{st} satisfies the given differential equation if

$$s^2 + 3s + 2 = 0 \quad \text{or} \quad (s+1)(s+2) = 0.$$

The two roots $s = -1$ and $s = -2$ give the two solutions e^{-t} and e^{-2t}; hence

$$y = c_1 e^{-t} + c_2 e^{-2t}$$

is also a solution for every choice of the constants c_1 and c_2. Since the initial displacement is 5, setting $t = 0$ gives

$$c_1 + c_2 = 5$$

and, since the initial velocity is 3, differentiating and setting $t = 0$ give

$$c_1(-1)e^{-t} + c_2(-2)e^{-2t} = -c_1 - 2c_2 = 3.$$

By addition we get $-c_2 = 8$, and thus $c_1 = 5 - c_2 = 13$. The solution is

$$y = 13e^{-t} - 8e^{-2t}.$$

PROBLEMS

1. Obtain two solutions of $y'' - 3y' + 2y = 0$ by trying $y = e^{st}$. Thus obtain a family of solutions containing two arbitrary constants, and, from this, find a particular solution satisfying $y(0) = 1$, $y'(0) = 1$.

2. Obtain a family of solutions containing two arbitrary constants:

$$y'' + 7y' + 12y = 0, \qquad y'' + 9y' + 18y = 0, \qquad y'' + 8y' + 15y = 0, \qquad y'' = y.$$

3. Obtain a solution satisfying $y(0) = 4$, $y'(0) = -1$:

$$y'' + 7y' + 10y = 0, \qquad y'' + 8y' + 12y = 0, \qquad y'' + 9y' + 14y = 0, \qquad y'' = 9y.$$

4. By trying $y = e^{st}$, obtain a nonzero solution of $y' + p_0 y = 0$, where p_0 is constant. Thus obtain a solution assuming a prescribed value y_0 at $t = t_0$.

5. Describe an appropriate domain of functions y for each of the following operators, and state whether the operator is linear on that domain: (a) $\mathbf{T}y = 2y''$; (b) $\mathbf{T}y = y^2$; (c) $\mathbf{T}y = y' + 1$; (d) $\mathbf{T}y = \int_0^1 ty(t)\,dt$; (e) $\mathbf{T}y = y^{(n)}$; (f) $\mathbf{T}y = 3y^{(iv)} + y''$.

6. Prove that the operator defined by $\mathbf{T}y = y''' + p_2(t)y'' + p_1(t)y' + p_0(t)y$ is linear on the class of functions y that are three times differentiable on a given interval, if the $p_i(t)$ are any functions defined on that interval.

7. (*Review.*) Let $4y'' + 9y = 0$, $-\infty < t < \infty$. (a) Sketch y' versus y in phase space, if $y(0) = 3$, $y'(0) = 0$. (b) What other initial conditions $y(0) = y_0$, $y'(0) = y_1$ would lead to the same graph in phase space?

8. (*Uniqueness deduced from the energy integral.*) In this problem, $\mathbf{T}y = y'' + y$ and $2E = y^2 + (y')^2$. (*a*) Show that $E' = y'\mathbf{T}y$ and hence E is constant on any interval on which $\mathbf{T}y = 0$. (*b*) If $\mathbf{T}u = \mathbf{T}v = 0$ on an interval, and $u(t_0) = v(t_0)$, $u'(t_0) = v'(t_0)$ at a point t_0 of the interval, deduce that $u = v$. *Hint:* By linearity, $y = u - v$ satisfies $\mathbf{T}y = 0$, and the initial conditions give $E(t_0) = 0$. Hence $E = 0$, and since y is real, this makes $y = 0$.

---

3. General Solutions. The Wronskian.

Let us suppose that by some means we have obtained two solutions $y = u(t)$ and $y = v(t)$ of a linear second-order equation

$$y'' + p_1 y' + p_0 y = 0. \tag{3-1}$$

Then by the superposition principle

$$y(t) = c_1 u(t) + c_2 v(t) \tag{3-2}$$

is a solution for every choice of the constants c_i. We say that (3-2) is a *general solution* of (3-1) provided that for a suitable choice of the constants c_i the solution satisfies the initial conditions,

$$y(t_0) = y_0 \qquad y'(t_0) = y_1 \tag{3-3}$$

for *arbitrary* t_0, y_0, and y_1. To determine the restrictions on $u(t)$ and $v(t)$ ensuring that the solution is, indeed, general, we insert (3-2) in (3-3) and obtain two linear algebraic equations,

$$\begin{aligned} c_1 u(t_0) + c_2 v(t_0) &= y_0 \\ c_1 u'(t_0) + c_2 v'(t_0) &= y_1 \end{aligned} \tag{3-4}$$

for c_1 and c_2. The system (3-4) can be solved for c_1 and c_2 (for arbitrarily specified t_0, y_0, and y_1) if, and only if, the determinant

$$W(u,v) \equiv \begin{vmatrix} u(t) & v(t) \\ u'(t) & v'(t) \end{vmatrix} \neq 0 \tag{3-5}$$

for every $t = t_0$ in the interval. If $W(u,v) = 0$ for some value $t = t_0$, the constants c_i cannot be determined at t_0 for every choice of y_0 and y_1 and the solution is not general. The determinant $W(u,v)$ is called the *Wronskian* after the Polish mathematician G. Wronski, who deduced the criterion (3-5). The Wronskian is denoted by $W(u,v)$ or by $W(t)$, according as the dependence on the functions u, v or on the variable t is emphasized. Thus, $W(t) = W[u(t),v(t)]$.

If (3-5) holds over a given interval $a < t < b$, the functions u and v are called *independent* solutions over that interval.

As an illustration, recall that the equation $y'' + \omega^2 y = 0$ has a solution

$$y = c_1 \cos \omega t + c_2 \sin \omega t \tag{3-6}$$

obtained by superposition of the two solutions $u = \cos \omega t$ and $v = \sin \omega t$. In this case

$$W(u,v) = \begin{vmatrix} \cos \omega t & \sin \omega t \\ -\omega \sin \omega t & \omega \cos \omega t \end{vmatrix} = \omega \tag{3-7}$$

and hence (3-6) is a general solution for $\omega \neq 0$. This means that (3-6) is general enough to satisfy an arbitrarily specified initial displacement

$$c_1 \cos \omega t_0 + c_2 \sin \omega t_0 = y_0$$

and an arbitrarily specified initial velocity

$$-c_1 \omega \sin \omega t_0 + c_2 \omega \cos \omega t_0 = y_1$$

at any given initial time t_0. Equation (3-7) also indicates that the functions $\sin \omega t$ and $\cos \omega t$ are independent over every interval when $\omega \neq 0$. As is clear from this example, a general solution is in reality a *family* of solutions yielding one solution for each choice of the constants c_i.

The foregoing analysis suggests an interesting conclusion concerning the form of possible solutions of $y'' + \omega^2 y = 0$. Let y be a solution on some interval $a < t < b$, and let t_0 be a given point of this interval. When $\omega \neq 0$ the constants c_1 and c_2 can be chosen in such a way that the function

$$Y = c_1 \cos \omega t + c_2 \sin \omega t$$

satisfies $Y(t_0) = y(t_0)$ and $Y'(t_0) = y'(t_0)$, no matter what the values $y_0 = y(t_0)$ and $y_1 = y'(t_0)$ may be. With this choice of c_1 and c_2, the functions y and Y have the same value and the same derivative at t_0. But according to the fundamental uniqueness theorem (Chap. 2, Sec. 17) there can be *at most one* solution with a given value and derivative at t_0. Therefore the two solutions coincide, $y \equiv Y$, and this shows that y admits the representation

$$y = c_1 \cos \omega t + c_2 \sin \omega t.$$

In short, if $\omega \neq 0$, every solution of $y'' + \omega^2 y = 0$ can be obtained by superposition of the two solutions $\cos \omega t$ and $\sin \omega t$.

A similar result is proved in the same way for the general linear second-order equation, $\mathbf{T}y = 0$. If y_1 and y_2 are two solutions such that $W(y_1,y_2) \neq 0$ holds at a given point t_0, we can determine constants so that the function

$$Y = c_1 y_1 + c_2 y_2$$

agrees with y, in value and derivative, at t_0. If $\mathbf{T}$ satisfies the conditions needed for the uniqueness theorem we conclude $y \equiv Y$; hence, y can be obtained by superposition of the two solutions y_1 and y_2. These matters are discussed more fully in Sec. 24.

In the preceding section we found that (3-1) has solutions of the form

$$y = c_1 e^{s_1 t} + c_2 e^{s_2 t} \tag{3-8}$$

where s_1 and s_2 satisfy the characteristic equation. Since

$$W(e^{s_1 t}, e^{s_2 t}) = \begin{vmatrix} e^{s_1 t} & e^{s_2 t} \\ s_1 e^{s_1 t} & s_2 e^{s_2 t} \end{vmatrix} = e^{s_1 t} e^{s_2 t} \begin{vmatrix} 1 & 1 \\ s_1 & s_2 \end{vmatrix} \tag{3-9}$$

the solution (3-8) is a general solution if $s_1 \neq s_2$. But if $s_1 = s_2$ the solution is not general because it reduces to a multiple of the single solution, $e^{s_1 t}$. This fact is reflected in the vanishing of the Wronskian.

We can get an idea of the nature of the general solution when $s_1 = s_2$ as follows: By the superposition principle the expression

$$c(e^{s_1 t} - e^{s_2 t})$$

is a solution for every choice of the constant c. When $s_1 \neq s_2$ we choose $c = (s_1 - s_2)^{-1}$ to obtain the solution

$$\frac{e^{s_1 t} - e^{s_2 t}}{s_1 - s_2}.$$

As $s_2 \to s_1$ the limit is the derivative of e^{st} with respect to s at $s = s_1$. (This follows from the *definition* of derivative, without calculation.) Thus the process suggests that

$$\frac{d}{ds} e^{st} = t e^{st} = t e^{s_1 t} \qquad \text{at } s = s_1 \tag{3-10}$$

will be a solution when $s_1 = s_2$. Although the method by which we obtained (3-10) does not in itself prove that (3-10) is a solution, that fact is easily verified by direct substitution.

The result of this discussion is that (3-1) has the solution

$$y = c_1 e^{s_1 t} + c_2 t e^{s_1 t} \qquad s_1 = s_2 \tag{3-11}$$

when the characteristic equation has equal roots.   Proof that (3-11) is a general solution is facilitated by the following theorem, which is established presently.

THEOREM. $W(uy_1, uy_2) = u\, W(y_1, y_2)$, *for any differentiable functions u, y_1, and y_2.*

Applying this with $u = e^{s_1 t}$, $y_1 = 1$, and $y_2 = t$, we find that the Wronskian associated with (3-11) is

$$W(e^{s_1 t}, t e^{s_1 t}) = e^{2 s_1 t} W(1,t) = e^{2 s_1 t} \begin{vmatrix} 1 & t \\ 0 & 1 \end{vmatrix} = e^{2 s_1 t}.$$

Since $e^{s_1 t} \neq 0$, this shows that (3-11) is a general solution.

To prove this theorem, recall that $(uy)' = uy' + u'y$. Using this and a familiar property of determinants (Appendix A), we get

$$W(uy_1, uy_2) = \begin{vmatrix} uy_1 & uy_2 \\ uy_1' + u'y_1 & uy_2' + u'y_2 \end{vmatrix} = \begin{vmatrix} uy_1 & uy_2 \\ uy_1' & uy_2' \end{vmatrix} + \begin{vmatrix} uy_1 & uy_2 \\ u'y_1 & u'y_2 \end{vmatrix}.$$

The second determinant on the right vanishes because the rows are proportional, and the first is $u^2 W(y_1, y_2)$.

Example 1.   Prove that the functions $e^{-\sigma t} \cos \omega t$ and $e^{-\sigma t} \sin \omega t$ are independent, if σ and ω are any constants with $\omega \neq 0$.

By the above theorem the Wronskian is $e^{-2\sigma t} W(\cos \omega t, \sin \omega t)$, and the latter expression is nonzero by (3-7).

Example 2.   Find two independent solutions of $y'' + 2y' + y = 0$, and thus obtain a general solution.

The characteristic equation in this case is $s^2 + 2s + 1 = 0$, or $(s + 1)^2 = 0$.   Corresponding to the double root $s = -1$ are the two independent solutions $y_1 = e^{-t}$, $y_2 = t e^{-t}$.   Accordingly, a general solution is $y = (c_1 + c_2 t) e^{-t}$.

PROBLEMS

1. Prove that $y = c_1 e^{2t} + c_2 t e^{2t}$ is a general solution of $y'' - 4y' + 4y = 0$, and find the solution for which $y(0) = 1$, $y'(0) = 4$.   Also find the solution for which $y(0) = y'(0) = 0$.

2. Find a general solution: (a) $y'' + 6y' + 9y = 0$, $y'' + 25y = 10y'$, $y'' = 0$; (b) $y'' = 2y' - y$, $4y'' + 25y = 20y'$, $3y'' = 2y' + 8y$, $y'' + y' = y$.

3. Compute $W(1,t)$, $W(t^a, t^b)$, $W(e^t, e^{-t})$, $W(\sinh t, \cosh t)$.

4. How does $W(1, e^{rt})$ behave as $r \to 0$?

5. If u and v are differentiable and $u \neq 0$, show that

$$\frac{W(u,v)}{u^2} = \frac{d}{dt} \frac{v}{u}.$$

If $W(u,v) \equiv 0$, what do you conclude about the ratio v/u?

6. Assuming u_1, u_2, v_1, and v_2 differentiable, prove

$$W(u_1 v_1, u_2 v_2) = u_1 u_2 W(v_1, v_2) + v_1 v_2 W(u_1, u_2).$$

7. If u'' and v'' exist, show that $(d/dt)W(u,v)$ is obtained by differentiating the elements of the bottom row.

8. Given $\mathbf{T}y = y'' + p_1 y' + p_0 y$, where p_1 and p_0 are constant, prove the identity $u\mathbf{T}v - v\mathbf{T}u = W' + p_1 W$, where $W = W(u,v)$. When $u\mathbf{T}v = v\mathbf{T}u$, deduce $W(x) = W(x_0)e^{-p_1(x-x_0)}$. [Hence $W(x) \neq 0$, or $W(x) \equiv 0$. This applies, in particular, if $\mathbf{T}u = \mathbf{T}v = 0$.]

9. (*Review.*) By substitution into the equation, determine necessary and sufficient conditions on the three constants c_1, c_2, c_3 such that $y = c_1 \cos(c_2 t + c_3)$ satisfies $y'' + \omega^2 y = 0$ for all t, where ω is constant.

---

4. Complex Solutions.   As seen in Sec. 1, the displacement y for an undamped mass-spring system satisfies

$$y'' + \omega^2 y = 0 \qquad \omega = \text{const.} \tag{4-1}$$

If we try to solve this equation by setting $y = e^{st}$, we get $y'' = s^2 e^{st}$; hence

$$s^2 + \omega^2 = 0 \tag{4-2}$$

is the condition for s. This equation has roots $s = i\omega$ and $s = -i\omega$, where i is the imaginary unit, $i = \sqrt{-1}$. It is natural to inquire whether the corresponding complex solutions

$$e^{i\omega t} \qquad \text{and} \qquad e^{-i\omega t}$$

can be given an interpretation that yields the real solutions

$$\cos \omega t \qquad \text{and} \qquad \sin \omega t$$

obtained in Sec. 1. We shall find that such an interpretation is possible and has the utmost importance in the study of vibration.

To get an idea of the meaning that should be attached to $e^{i\omega t}$, recall that the family

$$c_1 \cos \omega t + c_2 \sin \omega t \tag{4-3}$$

contains all solutions of (4-1). (This statement was established for real solutions in Sec. 3, and its extension to complex solutions presents no difficulty; see Example 2.)

Since we want to define $e^{i\omega t}$ so that it is also a solution, $e^{i\omega t}$ must necessarily be a member of the family (4-3). In other words,

$$e^{i\omega t} = c_1 \cos \omega t + c_2 \sin \omega t \tag{4-4}$$

for suitably chosen constants c_1 and c_2. If (4-4) can be differentiated by the usual rules we get

$$i\omega e^{i\omega t} = -c_1 \omega \sin \omega t + c_2 \omega \cos \omega t. \tag{4-5}$$

Since $e^0 = 1$ the choice $t = 0$ in (4-4) gives $c_1 = 1$, and the choice $t = 0$ in (4-5) gives $c_2 = i$. Substituting these values into (4-4) produces the *Euler-de Moivre formula*,

$$e^{i\omega t} = \cos \omega t + i \sin \omega t. \tag{4-6}$$

Thus, we have obtained an interpretation of e^z when $z = i\omega t$ is pure imaginary. Extension to arbitrary complex numbers $z = a + ib$ is made by defining

$$e^{a+ib} = e^a e^{ib} = e^a(\cos b + i \sin b). \tag{4-7}$$

Note that the second equality (4-7) follows from (4-6) with $\omega t = b$.

The steps leading to (4-7) depend on an explicit agreement concerning the differentiation of complex functions and concerning the appropriate properties of e^z. If $w = u + iv$ is a complex function, that is, if u and v are real-valued functions of the real variable t, the derivative is defined by $w' = u' + iv'$. It is easy to show that the familiar rules of calculus then remain valid for complex functions of a real variable t. The discussion of e^z was motivated by a desire to preserve the familiar properties

$$e^{p+q} = e^p e^q, \qquad \frac{d}{dt} e^{rt} = r e^{rt} \tag{4-8}$$

for complex constants p, q, r. What we showed is that, if the desired extension is possible at all, it must be given by (4-7). The argument does not in itself prove that (4-7) satisfies (4-8), but this fact is easily verified by direct computation. Formula (4-6) and its generalization (4-7) are so useful that the reader may find it advisable to write them a few times, until he knows them by memory.

In stating an equation such as $w = u + iv$, without further explanation, we always mean that u and v are real. Then u is called the *real part* and v the *imaginary part* of w. (The imaginary part could be defined as iv, but the definition v proves more convenient.) The real part is denoted by the symbol Re and the imaginary part by Im, so that

$$w = \operatorname{Re} w + i \operatorname{Im} w.$$

In this notation the Euler-de Moivre formula is equivalent to the two equations

$$\operatorname{Re}(e^{i\omega t}) = \cos \omega t \qquad \operatorname{Im}(e^{i\omega t}) = \sin \omega t. \tag{4-9}$$

When a linear operator is applied to a complex function $u + iv$ it is assumed that $\mathbf{T}u$ and $\mathbf{T}v$ are meaningful and, by linearity,

$$\mathbf{T}(u + iv) = \mathbf{T}u + i\mathbf{T}v. \tag{4-10}$$

However, the decomposition on the right of (4-10) may not necessarily be into real and imaginary parts, because $\mathbf{T}u$ or $\mathbf{T}v$ might not be real. If $\mathbf{T}u$ is real whenever u is a real function for which $\mathbf{T}u$ makes sense, the operator $\mathbf{T}$ is said to be *real*. As an illustration, the operator defined by

$$\mathbf{T}w = w'' + p_1 w' + p_0 w$$

is real if the constants p_1 and p_0 are real.

Much of the usefulness of complex solutions depends on the following:

THEOREM. Let $\mathbf{T}$ be a real linear operator, and let w be a complex function such that $\mathbf{T}w = 0$. Then the real and imaginary parts of w satisfy this same equation separately.

If $w = u + iv$, it is to be shown that $\mathbf{T}u = 0$ and $\mathbf{T}v = 0$. But $\mathbf{T}w = 0$ gives $\mathbf{T}u + i\mathbf{T}v = 0$ by (4-10), and since $\mathbf{T}u$ and $\mathbf{T}v$ are real, the desired result follows at once.

For example, $w = e^{i\omega t}$ satisfies the real linear equation $w'' + \omega^2 w = 0$; hence the real and imaginary parts satisfy the equation also. By inspection of (4-9) we get the solutions $\cos \omega t$ and $\sin \omega t$ introduced at the beginning of this discussion.

Example 1. Setting $y = e^{st}$, obtain a real general solution of

$$y'' + 4y' - 13y = 0.$$

The suggested substitution leads to $s^2 + 4s - 13 = 0$, which has the roots

$$s = \frac{-4 \pm \sqrt{16 - 52}}{2} = -2 \pm 3i.$$

Corresponding to the root $-2 + 3i$ is the solution

$$e^{st} = e^{(-2+3i)t} = e^{-2t}(\cos 3t + i \sin 3t) \tag{4-11}$$

where the right-hand form follows from (4-7). Since the equation in this example has real coefficients, the real and imaginary parts

$$e^{-2t}\cos 3t \qquad e^{-2t}\sin 3t$$

of the complex solution (4-11) are also solutions. Forming a linear combination we get

$$y = c_1 e^{-2t}\cos 3t + c_2 e^{-2t}\sin 3t.$$

By Example 1 of Sec. 3, this solution is general. It is left for the reader to verify that the same family is obtained from the second root, $-2 - 3i$.

Example 2. Prove that every complex solution of the equation $w'' + w = 0$ has the form $c_1\cos t + c_2\sin t$, where c_1 and c_2 are suitably chosen complex constants.

If $w = u + iv$ then u and v satisfy the given equation separately. Since these functions are real, the result of the preceding section shows that

$$u = a_1\cos t + a_2\sin t \qquad v = b_1\cos t + b_2\sin t$$

for suitable real constants a_1, a_2, b_1, b_2. The equation $w = u + iv$ now gives the desired representation, with

$$c_1 = a_1 + ib_1 \qquad \text{and} \qquad c_2 = a_2 + ib_2.$$

PROBLEMS

1. Read the discussion of complex numbers in Chap. 1, Sec. 15, and do Probs. 1 to 3 of that section. Also read the infinite-series derivation of the Euler formula given in Chap. 1, Sec. 17.

2. If $w_1 = u_1 + iv_1$ and $w_2 = u_2 + iv_2$ are differentiable, deduce $(w_1 w_2)' = w_1 w_2' + w_2 w_1'$ from the corresponding result for real functions.

3. For the equation $y'' - 10y' + 26y = 0$, obtain the complex solution $y = e^{(5+i)t} = e^{5t}e^{it}$ by trying $y = e^{st}$. Thus get two independent real solutions.

4. Obtain a general solution in real form:

(a) $y'' + 3y' = 54y, \quad y'' + 6y = 5y', \quad y'' + 6y' + 10y = 0;$
(b) $y'' + y = 2y', \quad y'' - 4y = 0, \quad y'' + 10y' + 29y = 0;$
(c) $y'' + 4y = 0, \quad y'' + 4y = 4y', \quad y'' + 3y' + 4y = 0;$
(d) $y'' + 5y = 4y', \quad y'' + 50y = 2y', \quad y'' + 6y' + 35y = 0.$

5. If ω, c_1, and c_2 are real and $C = c_1 - ic_2$, prove that $c_1\cos\omega t + c_2\sin\omega t = \text{Re}\,(Ce^{i\omega t})$.

6. (a) Find a general solution of $y'' + 4y' + a^2 y = 0$, where a is a real constant. (b) For $a^2 = 3, 4, 5$, respectively, get the particular solutions

$$y = e^{-t}, \qquad y = te^{-2t}, \qquad y = e^{-2t}\cos t.$$

(c) Sketch each solution of part b as a function of t, $0 \le t \le 4\pi$.

7. (a) Find a general solution of $y'' + 2by' + 2y = 0$, where b is a real constant. (b) For $b^2 = 1, 2, 3$, respectively, get the particular solutions

$$y = e^{-bt}\cos t, \qquad y = e^{-bt}t, \qquad y = e^{-bt}e^t.$$

8. (*Review.*) If the position of a particle at time t is $y = A\cos(\omega t - \phi)$, where A, ω, and ϕ are constant, discuss the graphs of y versus y', of y versus y'', and of y' versus y''. Verify agreement with the results of Sec. 1.

5. Free Damped Oscillations.

The equation obtained for free oscillations in Sec. 1 can be written in the form

$$y'' + 2by' + a^2 y = 0 \tag{5-1}$$

where b and a are positive constants. In the case of a spring,

$$2b = \frac{r}{m} \qquad a^2 = \frac{k}{m} \qquad (5\text{-}2)$$

and y is the displacement, whereas if (5-1) refers to an electrical circuit, then

$$2b = \frac{R}{L} \qquad a^2 = \frac{1}{LC} \qquad (5\text{-}3)$$

and y denotes the voltage across the capacitor.

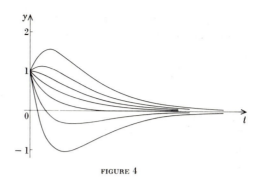

FIGURE 4

The characteristic polynomial $s^2 + 2bs + a^2$ associated with (5-1) has roots

$$s = -b \pm \beta \qquad \text{where } \beta = \sqrt{b^2 - a^2}. \qquad (5\text{-}4)$$

It will prove instructive to interpret the physical significance of the solution in the three cases $b > a$, $b = a$, and $b < a$, for which β is real and nonzero, or $\beta = 0$, or $\beta = i\omega$ is pure imaginary, respectively.

Whenever $\beta \neq 0$ the roots are distinct, and a general solution is therefore

$$y = c_1 e^{(-b+\beta)t} + c_2 e^{(-b-\beta)t}. \qquad (5\text{-}5)$$

If $b > a$ then β is real, and the equation $\beta^2 = b^2 - a^2$ shows that $|\beta| < b$. Hence, both of the quantities $-(b - \beta)$ and $-(b + \beta)$ in the exponents of (5-5) are negative. This is the *overdamped* case. The retarding force is so great that no vibration can occur, and the solution decays exponentially as $t \to \infty$. Figure 4 shows the behavior of several displacement curves with a common initial value but various initial velocities. The latter is the slope at the left-hand point of the curves. In certain cases there is almost half an oscillation, but more than half an oscillation is not possible.

The constants c_1 and c_2 in (5-5) depend on the initial displacement y_0 and the initial velocity y_1. Setting $t = 0$ in (5-5) gives

$$c_1 + c_2 = y_0$$

and differentiating and setting $t = 0$ in (5-5) give

$$c_1 - c_2 = (y_1 + by_0)\beta^{-1}$$

when we use $c_1 + c_2 = y_0$. The value of c_1 is now obtained by adding these equations, and c_2 by subtracting them. For example, the particular solution $y = u$ corresponding to $y_0 = 1$, $y_1 = 0$ is found to be

$$u(t) = e^{-bt}\left(\cosh \beta t + b\,\frac{\sinh \beta t}{\beta}\right) \qquad (5\text{-}6)$$

while the solution $y = v$ corresponding to $y_0 = 0$, $y_1 = 1$ is

$$v(t) = e^{-bt}\,\frac{\sinh \beta t}{\beta}. \qquad (5\text{-}7)$$

It is a remarkable fact that the general solution y can be expressed in terms of u and v, as

$$y(t) = y_0 u(t) + y_1 v(t). \qquad (5\text{-}8)$$

Indeed, the function y in (5-8) satisfies the differential equation by the superposition principle, and it satisfies the initial conditions

$$y(0) = y_0 u(0) + y_1 v(0) = y_0 \qquad y'(0) = y_0 u'(0) + y_1 v'(0) = y_1$$

by virtue of the initial conditions for u and v. Since there is only one solution with these properties, (5-8) is justified.

When $a = b$, the two roots of the characteristic equation are also equal. According to the results of Sec. 3, a general solution is

$$y = e^{-bt}(c_1 + c_2 t) \qquad \text{for } a = b. \tag{5-9}$$

This is the *critically damped*[1] case, so called because the motion becomes oscillatory if the damping is reduced by an arbitrarily small amount. The general character of the motion is similar to that in the overdamped case; hence we do not give a detailed analysis here. (See Probs. 1 and 2.)

The most interesting case occurs when $b < a$, because then the roots of the characteristic equation are complex, and the motion is *underdamped*. If ω is defined by

$$\omega^2 = a^2 - b^2 \qquad \omega > 0$$

comparison with $\beta^2 = b^2 - a^2$ in (5-4) shows that $\beta = \pm i\omega$. All the preceding calculations are carried out as before, with this value of β. In particular, since the familiar definition of cosh and sinh in terms of exponentials gives

$$\cosh i\omega t = \cos \omega t \qquad \frac{\sinh i\omega t}{i\omega} = \frac{\sin \omega t}{\omega}$$

the substitution $\beta = \pm i\omega$ in (5-6) and (5-7) produces the solutions

$$u(t) = e^{-bt}\left(\cos \omega t + b \frac{\sin \omega t}{\omega}\right) \qquad v(t) = e^{-bt}\frac{\sin \omega t}{\omega} \tag{5-10}$$

for which $(y_0, y_1) = (1,0)$ or $(0,1)$, respectively. The general solution (5-8) is now

$$y = e^{-bt}(c_1 \cos \omega t + c_2 \sin \omega t) \tag{5-11}$$

where $c_1 = y_0$ and $\omega c_2 = y_0 b + y_1$.

Graphical analysis of (5-11) is facilitated by the fact that

$$c_1 \cos \omega t + c_2 \sin \omega t \equiv A \cos (\omega t - \phi) \tag{5-12}$$

where $A \cos \phi = c_1$ and $A \sin \phi = c_2$, as in Sec. 1. By a short calculation

$$A = \sqrt{c_1{}^2 + c_2{}^2} \qquad \phi = \tan^{-1}\frac{c_2}{c_1} \qquad -\frac{\pi}{2} \le \phi \le \frac{\pi}{2}$$

and substitution of (5-12) into (5-11) gives

$$y = A e^{-bt} \cos (\omega t - \phi). \tag{5-13}$$

The motion consists of a cosine factor, with period $2\pi/\omega$, multiplied by the damping factor e^{-bt}. Since the cosine oscillates between -1 and 1, the graph of y versus t oscillates between the curves

$$y = -A e^{-bt} \qquad \text{and} \qquad y = A e^{-bt}.$$

It touches the lower curve at the values $\omega t = \phi + \pi,\ \phi + 3\pi,\ \phi + 5\pi,\ \dots$, where the cosine is -1, and it touches the upper curve at the values halfway between these (Fig. 5).

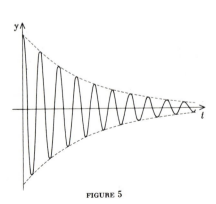

FIGURE 5

[1] The term *deadbeat* is often used.

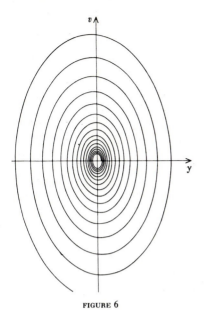

FIGURE 6

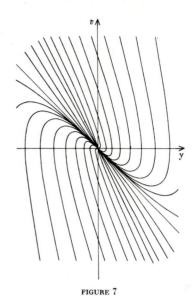

FIGURE 7

Since $\omega^2 = a^2 - b^2$ it is clear that $\omega \to a$ as $b \to 0$, and this agrees with the value $\omega = \sqrt{k/m}$ obtained for undamped oscillation in Sec. 1. If we set $a = \omega_0$ to emphasize that a is the angular frequency corresponding to $b = 0$, our equation becomes

$$\omega^2 + b^2 = \omega_0^2. \tag{5-14}$$

The point (ω, b) traces part of a circle of radius ω_0 as b varies. Evidently $\omega < \omega_0$ and so the effect of the damping is to slow down the oscillations. When divided by ω_0^2, Eq. (5-14) takes the interesting form

$$\left(\frac{\omega}{\omega_0}\right)^2 + \left(\frac{b}{a}\right)^2 = 1$$

where b/a is the ratio of the actual damping b to the critical damping a.

Differentiating the equation $y'' + 2by' + a^2y = 0$ with respect to t shows that this same equation is satisfied by the velocity $v = y'$. Hence, v admits a representation like that previously found for y. Thus, for $b < a$ the point (y,v) in phase space is described by

$$y = Ae^{-bt} \cos(\omega t - \phi) \qquad v = A_1 e^{-bt} \cos(\omega t - \phi_1) \tag{5-15}$$

where A, ϕ, A_1, and ϕ_1 are constant.

It is not difficult to show that the curve

$$y = A \cos(\omega t - \phi) \qquad v = A_1 \cos(\omega t - \phi_1)$$

is an ellipse, in general, which is traversed once each time t increases by the amount $T = 2\pi/\omega$. Hence the curve (5-15) has much the same behavior, except that the coordinates are multiplied by the steadily decreasing factor e^{-bt}. The curve (5-15) is therefore a spiral (Fig. 6). Increasing t by the amount T takes us once around the spiral; we return to a point on the same ray through the origin, but nearer the origin by a factor e^{-bT}.

If the vibrating system is suddenly disturbed by a new impulse the point (y,v) moves to a new spiral, along which it again proceeds toward the origin. The corresponding behavior for the undamped case is illustrated in Fig. 2, and for the overdamped case, the curves have the general appearance shown in Fig. 7. A point of the type shown in Fig. 2 is called a *center*, that in Fig. 6

is a *focus*, and that in Fig. 7 is a *node*. When $b = 0$ the origin is said to be a *stable* point for the motion, while for $b > 0$ the origin is said to be *asymptotically stable*. The reason for this terminology will be clear from the following discussion.

─── **PROBLEMS**

1. Show that the constants in the deadbeat case (5-9) are given in terms of initial values by $c_1 = y_0$, $c_2 = y_1 + by_0$, and thus obtain the formula

$$y = y_0e^{-bt}(1 + bt) + y_1e^{-bt}t \qquad a = b.$$

2. Show that the result of Prob. 1 can be found by letting $\beta \to 0$ in the overdamped case and also by letting $\omega \to 0$ in the underdamped case. [Use (5-6), (5-7), and (5-10).]

3. The force of 98,000 dynes extends a spring 2 cm. A mass of 200 g is suspended at the end, and the spring is pulled down 10 cm and released. Find the position of the mass at any instant t if the resistance of the medium is neglected.

4. Solve Prob. 3 under the assumption that the spring is viscously damped. It is given that the resistance is 2,000 dynes for a velocity of 1 cm per sec. What must the resistance be in order that the motion be deadbeat?

5. A capacitor of capacity 4 μf is charged so that the potential difference of the plates is 100 volts. The capacitor is then discharged through a coil of resistance 500 ohms and inductance 0.5 henry. Find the potential difference at any later time t. How large must the resistance be in order that the discharge just fails to be oscillatory? Determine the potential difference for this case. *Hint:* $LV'' + RV' + C^{-1}V = 0$.

6. Solve Prob. 5 if $R = 100$ ohms, $C = 0.5$ μf, and $L = 0.001$ henry.

7. A simple pendulum of length l is oscillating through a small angle θ in a medium in which the resistance is proportional to the velocity. Show that the differential equation of the motion is

$$\frac{d^2\theta}{dt^2} + 2k\frac{d\theta}{dt} + \frac{g}{l}\theta = 0.$$

Discuss the motion, and show that the period of the periodic factor is $2\pi/\sqrt{\omega^2 - k^2}$, where $\omega^2 = g/l$.

8. Observe that (5-11) is the real form of the solution appropriate to the root $s = -b + i\omega$, and derive the constants c_1 and c_2 directly from the initial conditions $y(0) = y_0$, $y'(0) = y_1$.

9. Let $y(t) = Ae^{-bt}\cos(\omega t - \phi)$. If $T = 2\pi/\omega$ is the period of the cosine factor, show that

$$\frac{y(t + T)}{y(t)} = e^{-bT}$$

independently of the value of t. Show that successive maxima y_k, y_{k+1} of the curve y versus t are spaced by a time interval T and hence that the foregoing equation also gives y_{k+1}/y_k. You need not compute the maximum values y_k. [The quantity $\delta = \log(y_k/y_{k+1}) = 2\pi b/\omega$ is called the *logarithmic decrement*. It is useful for determining parameters of the motion from a plot of y versus t.]

10. Show that the convergence of $Ae^{-bt}\cos(\omega t - \phi)$ to $A\cos(\omega_0 t - \phi)$ as $b \to 0$ is uniform on any fixed interval $0 \le t \le t_0$ but is not uniform on $0 \le t < \infty$ even if b is restricted to positive values during the limiting process.

11. A body suspended from the end of a viscously damped spring makes 60 oscillations per minute, and after 1 min its amplitude is halved. Find the differential equation of the motion.

6. Nonhomogeneous Equations.

The equation for free oscillations considered in the preceding section has the form $\mathbf{T}y = 0$, where $\mathbf{T}$ is a linear operator. Such an equation is said to be *homogeneous*, while an equation of the form $\mathbf{T}y = f \not\equiv 0$ is *nonhomogeneous*. As an illustration,

$$y'' + y = 4e^{-t} \tag{6-1}$$

is a nonhomogeneous equation, with $\mathbf{T}y = y'' + y$ and with $f(t) = 4e^{-t}$.

The function $2e^{-t}$ is a solution of (6-1) because, upon substitution, it reduces (6-1) to an identity in t. A single solution such as this is called a *particular solution*, or sometimes a *particular integral*, to distinguish it from the general solution which contains arbitrary constants. We shall now show that the general solution can be obtained from any particular solution by adding to it the general solution of the homogeneous equation. The latter is called the *complementary function*, because it completes the given particular solution.

For example, a general solution of (6-1) is

$$y = c_1 \cos t + c_2 \sin t + 2e^{-t}$$

obtained by adding the general solution of $y'' + y = 0$ to the particular solution $2e^{-t}$ of (6-1). Here the complementary function is $c_1 \cos t + c_2 \sin t$. The basis of this method is given by the following theorem:

PRINCIPLE OF THE COMPLEMENTARY FUNCTION. Let u be a particular solution of $\mathbf{T}y = f$, where $\mathbf{T}$ is any linear operator, and let v satisfy the homogeneous equation $\mathbf{T}y = 0$. Then $y = u + v$ satisfies $\mathbf{T}y = f$, and every solution of $\mathbf{T}y = f$ can be obtained in this way.

The fact that $\mathbf{T}(u + v) = f$ is evident, because $\mathbf{T}(u + v) = \mathbf{T}u + \mathbf{T}v$ by linearity, and $\mathbf{T}u = f$ and $\mathbf{T}v = 0$ by hypothesis. The precise meaning of the second assertion is that, if $\mathbf{T}y = f$, then y can be written in the form $y = u + v$, where $\mathbf{T}v = 0$. The truth of this follows from the identity

$$y = u + (y - u)$$

giving $v = y - u$. Then $\mathbf{T}v = \mathbf{T}(y - u) = \mathbf{T}y - \mathbf{T}u = f - f = 0$, as desired.

For example, if y_1 and y_2 are independent solutions of

$$\mathbf{T}y \equiv y'' + p_1 y' + p_0 y = 0 \tag{6-2}$$

let $v = c_1 y_1 + c_2 y_2$, and let u be any particular solution of

$$y'' + p_1 y' + p_0 y = f(t). \tag{6-3}$$

The theorem asserts, then, that the function

$$y = u + c_1 y_1 + c_2 y_2 \tag{6-4}$$

satisfies (6-3). Although u was only a particular solution, the function y in (6-4) is actually a general solution. Proof of this is virtually identical with that given for the homogeneous equation in Sec. 3 and so is not repeated here.

We shall now explain a simple and powerful technique for obtaining particular solutions, known as the *method of undetermined coefficients*. As an illustration, let it be required to find a solution of

$$y'' + 2y = 6 + 8t. \tag{6-5}$$

Since the right-hand member is a linear function of t we try a linear function

$$y = a_0 + a_1 t.$$

The coefficients a_0 and a_1 are, at present, undetermined. But the substitution into (6-5) gives $2(a_0 + a_1t) = 6 + 8t$, which holds for $a_0 = 3$ and $a_1 = 4$. Thus, $y = 3 + 4t$ satisfies (6-5).

The trial solution $y = a_0 + a_1t$ fails for the equation

$$y'' + 2y' = 6 + 8t$$

because this choice makes both y'' and y' constant. To make $y'' + 2y'$ equal an expression of *first* degree in t we must clearly start with y of *second* degree in t. We therefore try

$$y = t(a_0 + a_1t).$$

The substitution now gives $2a_1 + 2a_0 + 4a_1t = 6 + 8t$, and this is satisfied when $a_1 = 2$ and $a_0 = 1$.

If our equation is $y'' = 6 + 8t$ the differentiation reduces the degree of y by 2. Hence to get the *first*-degree expression $6 + 8t$ we must start with an expression of *third* degree in t. The trial solution

$$y = t^2(a_0 + a_1t)$$

gives $y'' = 2a_0 + 6a_1t$, which equals $6 + 8t$ if $a_0 = 3$ and $a_1 = \frac{4}{3}$.

The method also succeeds when the right-hand member involves exponential functions. Thus the right-hand member of

$$y'' + 3y' + 2y = 2e^t \tag{6-6}$$

suggests that it probably has a solution of the form $y = ae^t$, since the differentiation of exponentials yields exponentials. Accordingly we take $y = ae^t$ as a trial solution. The coefficient a is at present undetermined; but substitution into (6-6) gives

$$ae^t + 3ae^t + 2ae^t = 2e^t$$

and hence, dividing by e^t, we get $6a = 2$, or $a = \frac{1}{3}$. Thus, $y = \frac{1}{3}e^t$ is a solution of (6-6).

If a general solution is desired, we merely add the complementary function. The characteristic equation of the associated homogeneous equation

$$y'' + 3y' + 2y = 0 \tag{6-7}$$

is $s^2 + 3s + 2 = 0$, or $(s + 1)(s + 2) = 0$. Hence (6-7) has solutions e^{-t} and e^{-2t}, and a general solution of (6-6) is $y = c_1e^{-t} + c_2e^{-2t} + \frac{1}{3}e^t$.

If we attempt to solve

$$y'' + 3y' + 2y = 2e^{-t} \tag{6-8}$$

by taking the trial solution $y = ae^{-t}$, we get a nonsensical result, $0 = 2e^{-t}$. The reason that the trial solution ae^{-t} is not suitable is the following: The homogeneous equation (6-7) associated with (6-8) has a solution $y = ae^{-t}$, as we have just seen, and substitution of it in (6-8) naturally makes the left-hand member vanish. In this case we take the trial solution in the form $y = ate^{-t}$. Substitution into (6-8) gives

$$a(te^{-t} - 2e^{-t}) + 3a(-te^{-t} + e^{-t}) + 2ate^{-t} = 2e^{-t}$$

which reduces to $ae^{-t} = 2e^{-t}$. Hence $a = 2$, and a solution is $y = 2te^{-t}$.

It should be noticed that, in each of these examples, the fact that the trial function is a solution is automatically verified in the course of the calculation. This is a characteristic feature of the method of undetermined coefficients; one is therefore justified in using the method even before any general theory has been presented.[1] Hence, we simply state what the procedure is. Let the equation be

$$y'' + p_1y' + p_0y = (A_0 + A_1t + \cdots + A_nt^n)e^{rt} \tag{6-9}$$

---

[1] See Sec. 19, Example 2.

where the p's, the A's, and r denote constants.   The facts are as follows: *If r is not a root of the characteristic equation*

$$s^2 + p_1 s + p_0 = 0 \tag{6-10}$$

then Eq. (6-9) has a solution of the form

$$y = (a_0 + a_1 t + \cdots + a_n t^n)e^{rt}. \tag{6-11}$$

If r is a simple root of (6-10) the trial solution (6-11) must be multiplied by t, and if the root is double, by t^2.   It should be noted that the exceptional forms containing the factor t or t^2 are obtained by differentiating the general form (6-11) with respect to the exponent r.

Example.   Solve the initial-value problem

$$y'' + 3y' = 1 - 9t^2 \qquad y(0) = 0 \qquad y'(0) = 1. \tag{6-12}$$

Since $e^0 = 1$, this corresponds to the case $r = 0$ in (6-11).   The value 0 is a root of the characteristic equation $s^2 + 3s = 0$, and hence we seek a solution in the form

$$y = t(a_0 + a_1 t + a_2 t^2).$$

We compute $y' = a_0 + 2a_1 t + 3a_2 t^2$ and $y'' = 2a_1 + 6a_2 t$, and substitute into the given differential equation.   The result is

$$(2a_1 + 6a_2 t) + 3(a_0 + 2a_1 t + 3a_2 t^2) = 1 + 0t - 9t^2 \tag{6-13}$$

where the term $0t \equiv 0$ has been introduced to emphasize that the coefficient of t in $1 - t^2$ is 0. Since (6-13) is an identity in t we can put $t = 0$ and show that the constant term on the left agrees with the constant term on the right; that is, $2a_1 + 3a_0 = 1$.   Similarly, the coefficients of t and t^2 on the left must agree with those on the right, giving $6a_2 + 6a_1 = 0$ and $9a_2 = -9$, respectively.   Solving these three equations for a_i, starting with the last, we get

$$a_2 = -1 \qquad a_1 = 1 \qquad a_0 = -\tfrac{1}{3}$$

so that a particular solution is $y = t(-\tfrac{1}{3} + t - t^2)$.

Since the characteristic equation $s(s + 3) = 0$ has roots $s = 0$ and $s = -3$, a general solution is

$$y = c_1 + c_2 e^{-3t} + t(-\tfrac{1}{3} + t - t^2). \tag{6-14}$$

The initial conditions $y(0) = 0$ and $y'(0) = 1$ demand that

$$c_1 + c_2 = 0 \qquad \text{and} \qquad -3c_2 - \tfrac{1}{3} = 1.$$

Thus $c_2 = -\tfrac{4}{9}$, $c_1 = \tfrac{4}{9}$, and the final answer is (6-14) with these values of c_i.   By Chap. 2, Sec. 17, the solution is unique.

PROBLEMS

1. Obtain a particular solution:

(a) $y'' + y = 12t^2$, $y'' + y' = 12t^2$, $y'' = 12t^2$;

(b) $y'' + 3y' = 10e^{2t}$, $y'' - 3y' + 2y = 12e^{2t}$, $y'' - 4y' + 4y = 12e^{2t}$.

2. Obtain a general solution:

(a) $y'' - 5y' + 6y = e^{4t}$, $y'' + 2y' + y = t$, $y'' + 5y' + 6y = e^t$;

(b) $y'' - 2y' + y = t$, $y'' - y = 5t - 2$, $y'' - y = e^{2t}(t - 1)$;

(c) $y'' - 2y' + y = te^t$, $y'' - 6y' + 9y = e^{3t}$, $y'' + 9y' = 3$;

(d) $y'' + 9y = (t - 1)^2$, $y'' - y = e^t$, $y'' + y = t^3 + t$;

(e) $y'' - 5y' + 6y = t^3 e^{2t}$, $y'' - 2y' + y = e^t(t - 1)$, $y'' = t^3 e^{-t}$;

(f) $y'' - 5y' + 6y = 3t^2$, $y'' - 5y' = 3t^2$, $y'' + 3y' = e^{5t}$.

3. Obtain the solution satisfying $y(0) = y'(0) = 0$:

$$y'' + y' = 1 + 2t, \qquad y'' + y' = 0, \qquad y'' + 4y' + 3y = 1, \qquad y'' - 2y' = t + e^{2t}.$$

4. Solve each equation subject to the given initial conditions:

(a) $y'' + y' = 1 + 2t$, $y(0) = 0$, $y'(0) = -2$; $\quad y'' + 4y' + 3y = t$, $y(0) = 1$, $y'(0) = 0$;

(b) $y'' + 4y' + 3y = t$, $y(0) = -\frac{4}{9}$, $y'(0) = \frac{3}{9}$; $\quad y'' + y' = 6t^2$, $y(0) = 0$, $y'(0) = 1$.

---

7. Further Discussion of Nonhomogeneous Equations.   The method of the foregoing section applies to equations of the form

$$\mathbf{T}y = Ae^{\sigma t}\cos \omega t \qquad \mathbf{T}y = Ae^{\sigma t}\sin \omega t \tag{7-1}$$

where $\mathbf{T}y = y'' + p_1 y' + p_0 y$, and where the various constants are real.   Since

$$e^{\sigma t}e^{i\omega t} = e^{\sigma t}(\cos \omega t + i \sin \omega t) \tag{7-2}$$

the right-hand members of (7-1) are the real and imaginary parts of $Ae^{(\sigma + i\omega)t}$, respectively. Instead of (7-1), we therefore consider the equation

$$\mathbf{T}y \equiv y'' + p_1 y' + p_0 y = Ae^{(\sigma + i\omega)t} \tag{7-3}$$

and obtain its solution $y = u + iv$.   Then the real part u of such a solution satisfies the first equation (7-1), and the imaginary part v satisfies the second.   Proof of this depends on the following:

PRINCIPLE OF EQUATING REAL PARTS.   Let $w = u + iv$ be a complex solution of $\mathbf{T}w = f$, where $\mathbf{T}$ is any real linear operator.   Then the real part of w satisfies $\mathbf{T}u = \operatorname{Re} f$, and the imaginary part satisfies $\mathbf{T}v = \operatorname{Im} f$.

By linearity, $\mathbf{T}(u + iv) = \mathbf{T}u + i\mathbf{T}v$; this equals f by hypothesis.   Hence the real part of f is $\mathbf{T}u$, and the imaginary part is $\mathbf{T}v$, as stated in the conclusion of the theorem.

As an illustration, let it be required to solve

$$y'' + y = 3 \sin 2t. \tag{7-4}$$

Since $e^{2it} = \cos 2t + i \sin 2t$, we consider, instead of (7-4), the equation

$$y'' + y = 3(\cos 2t + i \sin 2t) = 3e^{2it}. \tag{7-5}$$

The imaginary part of a solution of (7-5) satisfies (7-4).   Equation (7-5) has the form (6-9) with $r = 2i$, and since neither of the roots of the characteristic equation $s^2 + 1 = 0$ is equal to $2i$, we take the trial solution $y = ae^{2it}$.   Then $y' = 2iae^{2it}$, $y'' = (2i)^2 ae^{2it} = -4ae^{2it}$, and substitution in (7-5) yields

$$-4ae^{2it} + ae^{2it} = 3e^{2it}.$$

Thus $a = -1$, and consequently $y = -e^{2it}$ satisfies (7-5).   The imaginary part of $-e^{2it}$ is $-\sin 2t$, giving a solution of (7-4).

The methods of this and the preceding section can be extended to equations in which the right-hand member is a sum of several functions of the types considered in those sections.   The basis for this extension is the following:

GENERAL PRINCIPLE OF SUPERPOSITION.   Let y_1 satisfy the equation $\mathbf{T}y_1 = f_1$ and let y_2 satisfy $\mathbf{T}y_2 = f_2$, where $\mathbf{T}$ is any linear operator.   Then, for any constants c_1 and c_2, the function $y = c_1 y_1 + c_2 y_2$ satisfies $\mathbf{T}y = c_1 f_1 + c_2 f_2$.

The proof is immediate, since $\mathbf{T}(c_1 y_1 + c_2 y_2) = c_1 \mathbf{T}y_1 + c_2 \mathbf{T}y_2$ by linearity, and $\mathbf{T}y_1 = f_1$ and $\mathbf{T}y_2 = f_2$ by hypothesis.   Upon choosing $f_1 = f_2 = 0$ we get the principle of superposition for homogeneous equations as stated in Sec. 2. The choice $f_1 = f$, $f_2 = 0$ gives the theorem of Sec. 6.

For example, to find a solution of $y'' + y = 3 \sin 2t + 1 + 2e^t$, consider the three equations

$$y'' + y = 3 \sin 2t \qquad y'' + y = 1 \qquad y'' + y = 2e^t.$$

A particular integral of the first is $y = -\sin 2t$ by the discussion of (7-4), and solutions of the second and third are $y = 1$ and $y = e^t$, respectively, as is clear by inspection. Hence, an integral of the given equation is

$$y = -\sin 2t + 1 + e^t.$$

We conclude this discussion with a result of great importance in the analysis of forced oscillations. A particular solution u_0 of the equation $\mathbf{T}y = f$ is said to be *asymptotically stable* if every other solution u satisfies

$$\lim_{t \to \infty} (u_0 - u) = 0.$$

Since u_0 can correspond to one set of initial conditions and u to another, the statement "$u_0 - u \to 0$" means that the effect of initial conditions becomes negligible after a time. Thus all the different solutions have much the same behavior as $t \to \infty$, and any one of them can be taken as a description of the steady state. Conditions justifying this concept of the steady-state solution are described in the following:

PRINCIPLE OF ASYMPTOTIC STABILITY. Let $y = u_0$ be any particular solution of the equation $\mathbf{T}y = f(t)$, where $\mathbf{T}$ is linear. Then u_0 is asymptotically stable if, and only if, every solution of the homogeneous equation $\mathbf{T}y = 0$ tends to 0 as $t \to \infty$.

Indeed, let u be any solution of the equation $\mathbf{T}u = f$. By linearity $\mathbf{T}(u_0 - u) = \mathbf{T}u_0 - \mathbf{T}u = f - f = 0$, and hence $u_0 - u$ satisfies the homogeneous equation. If *every* solution of the homogeneous equation tends to 0, clearly $u_0 - u$ does so too; this shows that u_0 is stable.

On the other hand, if some solution v of the homogeneous equation does not tend to 0, we could pick $u = u_0 + v$. Then u satisfies $\mathbf{T}u = f$, but $u_0 - u = -v$ does not tend to 0. This shows that u_0 is not stable and completes the proof.

For operators of the kind that we have been considering, the homogeneous equation $\mathbf{T}y = 0$ has solutions

$$y = c_1 e^{s_1 t} + c_2 e^{s_2 t} \qquad \text{or} \qquad y = (c_1 + c_2 t) e^{s_1 t} \tag{7-6}$$

where the exponents are roots of the characteristic equation,

$$s^2 + p_1 s + p_0 = 0. \tag{7-7}$$

The functions (7-6) tend to 0 as $t \to \infty$ (for every choice of the constants c_i) if, and only if, the exponents have negative real parts. In this case all solutions of

$$y'' + p_1 y' + p_0 y = f(t) \tag{7-8}$$

are asymptotically stable, and in the contrary case no solution of (7-8) is asymptotically stable. The condition of stability is usually satisfied in practice because most physical systems have a certain amount of dissipation (cf. discussion of the case $b > 0$ in Sec. 5).

Complex numbers are often plotted in the plane by assigning the point (x,y) to the number $z = x + iy$. The condition that the real part, x, be negative then means that the representative point lies in the left half plane. Hence, *if the roots $s = \sigma + i\omega$ of the characteristic equation lie in the left half plane the system is stable*, and otherwise not. Entirely similar considerations apply to higher-order equations, as will be seen presently.

Example 1.   Describe the steady-state behavior for solutions of

$$y'' + 2y' + y = e^{-t} \cos t. \tag{7-9}$$

Since $e^{-t} \cos t$ is the real part of $e^{-t}(\cos t + i \sin t) = e^{-t}e^{it}$, we consider, instead of (7-9), the equation

$$y'' + 2y' + y = e^{t(-1+i)}. \tag{7-10}$$

The roots of the characteristic equation $s^2 + 2s + 1 = 0$ are both -1, and since -1 is different from the exponent $(-1 + i)$ occurring in (7-10), we take the trial solution $y = ae^{(-1+i)t}$. Substitution into (7-10) gives

$$[(-1 + i)^2 + 2(-1 + i) + 1]ae^{(-1+i)t} = e^{(-1+i)t}$$

and hence, after simplification, $a = -1$. The solution $-e^{(-1+i)t} = -e^{-t}e^{it}$ has a real part $-e^{-t} \cos t$, and this satisfies (7-9). The solution is asymptotically stable, because the roots -1, -1 of the characteristic equation have negative real parts. Thus, $y = -e^{-t} \cos t$ describes the steady state.

Example 2.   Solve $y'' + \omega^2 y = \cos \omega t$, where ω is a positive constant. Since $e^{i\omega t} = \cos \omega t + i \sin \omega t$, we consider the equation

$$y'' + \omega^2 y = e^{i\omega t}. \tag{7-11}$$

The exponent $i\omega$ in (7-11) agrees with a root $s = i\omega$ of the characteristic equation, $s^2 + \omega^2 = 0$; hence a suitable trial solution is not $ae^{i\omega t}$, but $ate^{i\omega t}$. For $y = ate^{i\omega t}$, differentiation gives

$$y' = i\omega a te^{i\omega t} + ae^{i\omega t} \qquad y'' = -\omega^2 a te^{i\omega t} + 2i\omega ae^{i\omega t}$$

and, substituting into (7-11), we get $2i\omega ae^{i\omega t} = e^{i\omega t}$. Thus $2i\omega a = 1$ or $2\omega a = -i$. Therefore the function

$$y = ate^{i\omega t} = -\frac{i}{2\omega} t(\cos \omega t + i \sin \omega t) \tag{7-12}$$

satisfies (7-11). The real part of y in (7-12) satisfies the given equation; adding the complementary function gives the final answer,

$$y = \frac{t}{2\omega} \sin \omega t + c_1 \cos \omega t + c_2 \sin \omega t.$$

PROBLEMS

1. (*a*) For the equation $y'' + 4y' + 3y = 130e^{2it}$, show that the trial solution $y = ae^{2it}$ leads to $(-1 + 8i)a = 130$. Multiplying by $(-1 - 8i)$, obtain a and hence

$$y = (-2 - 16i)(\cos 2t + i \sin 2t).$$

(*b*) By inspection of your result (*a*), get solutions of

$$y'' + 4y' + 3y = 130 \cos 2t \qquad \text{and} \qquad y'' + 4y' + 3y = 130 \sin 2t.$$

These solutions describe the steady-state behavior. Why?

2. Obtain a general solution in real form:

(*a*) $y'' - 3y' + 2y = \cos 2t$, $y'' + 4y = \cos 3t$;
(*b*) $2y'' + 2 \cos t + 6 \sin t = y + y'$, $y'' + 5y' + 6y = 3e^{-2t} + e^{3t}$;
(*c*) $y'' + 2y' + 5y = e^t \sin 2t$, $y'' - y' - 6y = e^{3t} \cos 3t$;
(*d*) $y'' - 25y = e^{5t} + t^2 - 4t$, $y'' + y = 3 \sin 2t - 9 \cos 3t$;
(*e*) $y'' - 2y' + 10y = e^t + \sin 3t$, $y'' + y = (1 + 2 \sin t) \sin t$.

3. Obtain the solution satisfying $y(0) = y'(0) = 0$ in parts *a* and *b*:

(*a*) $y'' - y = \sin t$, $y'' + 2y' + 5y = 0$, $y'' + y = t \sin t$;
(*b*) $y'' - 2y' = e^{-t} \cos t$, $y'' + y = 1 + \cos t$, $y'' - y' = \cos 3t$.

4. Solve each equation, using the initial conditions below it:

$$y'' + y = -\sin 2t, \qquad y'' - y' = 2(1-t), \qquad y'' + 4y = \sin t,$$
$$y(\pi) = y'(\pi) = 1, \qquad y(0) = y'(0) = 1, \qquad y(0) = 0, y'(0) = 1.$$

5. If p_1 and p_0 are real, prove that both roots of $s^2 + p_1 s + p_0 = 0$ have negative real parts when, and only when, p_1 and p_0 are positive. *Hint:* Recall that $-p_1$ is the sum of the roots, and p_0 is the product. Using this result, test the following equations for stability and obtain a representative of the steady-state solution if stable:

(a) $y'' + 4y' + 3y = \cos \omega t, \quad y'' + 7y' = \cos \omega t, \quad y'' + 4y' + 13y = t \cos t;$
(b) $4y'' + 8y' + 13y = \sin \omega t, \quad y'' = t \sin \omega t, \quad 9y'' + 6y' + 2y = \sin \omega t.$

6. Solve $y' = e^{(a+ib)t}$ by trying $y = Ae^{(a+ib)t}$, and thus show that

$$(a^2 + b^2) \int e^{at} \cos bt \, dt = e^{at}(a \cos bt + b \sin bt) + c.$$

7. As in Prob. 6, consider $y' = te^{(a+ib)t}$.

8. If a, b, and ω are real constants, obtain a particular solution:

$$y'' + 2y' + a^2 y = \cos \omega t, \qquad y'' + 2by' + y = \cos \omega t \qquad \text{for } b \neq 0, \qquad \text{same for } b = 0.$$

---

8. Forced Oscillations and Resonance.

The equation obtained for forced oscillations in Sec. 1 has the form

$$y'' + 2by' + a^2 y = a^2 f(t) \tag{8-1}$$

where the function $f(t)$ is known. In the mechanical problem, $f(t)$ is the displacement $x(t)$ of the top of the spring, and in the electrical circuit, $f(t)$ is the impressed voltage $E_0(t)$.

A particularly important choice of $f(t)$ is a *sinusoidal input*,

$$f(t) = A_0 \cos \omega t \qquad \text{or} \qquad f(t) = A_0 \sin \omega t \tag{8-2}$$

ω and A_0 being positive constants. Part of the importance of this choice stems from the fact that the output of an a-c generator is, as a rule, sinusoidal. But more fundamental is the fact that an *arbitrary* periodic input $f(t)$ can generally be expanded in Fourier series (Chap. 1, Secs. 18 to 26). Thus, if f has period T and satisfies mild smoothness conditions, $f(t)$ can be approximated by sums of the form

$$\tfrac{1}{2}a_0 + a_1 \cos \omega_1 t + b_1 \sin \omega_1 t + \cdots + a_n \cos \omega_n t + b_n \sin \omega_n t \tag{8-3}$$

where $T\omega_k = 2\pi k$, $k = 1, 2, 3, \ldots$. If, now, we determine the response to the particular functions (8-2), the general superposition principle gives the response to $f(t)$ in (8-3).

We can get an idea of the effect of a sinusoidal input by considering the undamped equation

$$y'' + a^2 y = A_0 \cos \omega t \tag{8-4}$$

for which $b = 0$. According to Sec. 5, the constant a equals the natural frequency ω_0 of the system, that is, the frequency of the free oscillations. If the impressed frequency $\omega \neq \omega_0$, it is easily seen that (8-4) has a solution $A \cos \omega t$, where A is constant. But if $\omega = \omega_0$, all functions $A \cos \omega t$ or $A \sin \omega t$ make the left member vanish, and a solution is

$$y = \frac{A_0}{2\omega} t \sin \omega t$$

by Example 2 of Sec. 7. The oscillatory factor $\sin \omega t$ has the same frequency as the impressed oscillation, but the amplitude increases beyond all bounds as $t \to \infty$. This phenomenon, known as *resonance*, is of profound importance in numerous engineering and physical situations.

The failure of the Tacoma bridge was explained by some authorities on the basis of forced vibrations, and there are instances of the collapse of buildings induced by the rhythmic swaying of dancing couples. The failure of propeller shafts is often attributed to forced torsional vibrations, and accidents have been caused by resonance in connection with power steering in automobiles. See also Joshua 6:5.

The behavior for $b = 0$ suggests that the forced oscillations may also be large for $b > 0$, when b is small and the impressed frequency is close to the natural frequency. Such an expectation proves to be well founded, as seen next.

Corresponding to (8-1) and (8-2) is the complex equation

$$y'' + 2by' + a^2y = a^2A_0e^{i\omega t}. \tag{8-5}$$

The trial solution $y = Ae^{i\omega t}$ in (8-5) leads to

$$A(-\omega^2 + 2bi\omega + a^2)e^{i\omega t} = a^2A_0e^{i\omega t}$$

and hence a particular integral is

$$y = Ae^{i\omega t} \quad \text{where } A = \frac{a^2}{a^2 - \omega^2 + 2bi\omega} A_0. \tag{8-6}$$

From the discussion of free oscillations in Sec. 5 and the principle of stability in Sec. 7, it is clear that the part of the general solution of (8-5) that is due to free vibrations decays exponentially with time. Thus, the function (8-6) describes the steady-state behavior.

To get a physical interpretation, the denominator is written in *polar form* $Re^{i\phi}$. The equation

$$a^2 - \omega^2 + 2ib\omega = Re^{i\phi} = R(\cos\phi + i\sin\phi)$$

holds if $a^2 - \omega^2 = R\cos\phi$ and $2b\omega = R\sin\phi$. Hence,

$$R = \sqrt{(a^2 - \omega^2)^2 + 4b^2\omega^2} \quad \tan\phi = \frac{2b\omega}{a^2 - \omega^2} \quad -\frac{\pi}{2} \le \phi \le \frac{\pi}{2}. \tag{8-7}$$

The result of substituting the form $Re^{i\phi}$ into (8-6) is

$$y = A_0\frac{a^2}{\sqrt{(a^2 - \omega^2)^2 + 4b^2\omega^2}}e^{i(\omega t - \phi)}. \tag{8-8}$$

Since $\quad e^{i(\omega t - \phi)} = \cos(\omega t - \phi) + i\sin(\omega t - \phi)$

the real and imaginary parts of the solution (8-8) can be read by inspection. The real part satisfies (8-1) with $f(t) = A_0\cos\omega t$, and the imaginary part satisfies (8-1) with $f(t) = A_0\sin\omega t$.

According to (8-8), the steady-state response has the same frequency ω as the impressed oscillation, but it does not have the same phase. When $\omega < a = \omega_0$ the angle ϕ in (8-7) is between 0 and $\pi/2$, so that there is a phase lead; the opposite behavior is found when $\omega > a$. As $\omega \to 0$ it is evident that $\phi \to 0$; hence the system tends to follow a slow oscillation in phase. This agrees with one's intuitive expectation. But there is a discontinuity at $\omega = a$, which could hardly be foreseen intuitively. Namely, as $\omega \to a$ through values less than a the phase tends to $\pi/2$, corresponding to a 90° lead, whereas if $\omega \to a$ through larger values than a, the limiting phase denotes a 90° lag.

If we write $a = \omega_0$ to emphasize that a is the natural frequency of the undamped system, the term multiplying A_0 in (8-8) is

$$G = \frac{\omega_0^2}{\sqrt{(\omega_0^2 - \omega^2)^2 + 4b^2\omega^2}} \quad \text{where } a = \omega_0. \tag{8-9}$$

Since the input amplitude is A_0 and the response amplitude is GA_0, the coefficient G is a measure of the *gain* of the system. When ω is large, G is approximately $\omega_0{}^2/\omega^2$, and this is small. Thus, a rapid impressed oscillation does not produce much response. But if $\omega = \omega_0$ then $G = \omega_0/(2b)$, and this can be dangerously large when the damping term b is close to 0.

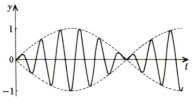

FIGURE 8

Example.   (*Beats.*)   If ω and ω_0 are unequal positive constants, solve

$$y'' + \omega_0{}^2 y = \cos \omega t \qquad y(0) = y'(0) = 0$$

and discuss the behavior when ω is nearly equal to ω_0.

The trial solution $y = A \cos \omega t$ gives $A(-\omega^2 + \omega_0{}^2) = 1$ after division by $\cos \omega t$. Thus, a particular solution is $y = (\omega_0{}^2 - \omega^2)^{-1} \cos \omega t$. Adding the complementary function, we get the general solution

$$y = c_1 \cos \omega_0 t + c_2 \sin \omega_0 t + (\omega_0{}^2 - \omega^2)^{-1} \cos \omega t.$$

The condition $y' = 0$ at $t = 0$ gives $c_2 = 0$ by inspection, and the condition $y = 0$ at $t = 0$ gives $c_1 = -(\omega_0{}^2 - \omega^2)^{-1}$. Thus,

$$y = (\omega_0{}^2 - \omega^2)^{-1}(\cos \omega t - \cos \omega_0 t).$$

By use of a trigonometric identity, the result can be written

$$y = 2(\omega_0{}^2 - \omega^2)^{-1} \sin \tfrac{1}{2}(\omega_0 + \omega)t \sin \tfrac{1}{2}(\omega_0 - \omega)t. \tag{8-10}$$

Since ω is nearly equal to ω_0, the second sine term is slowly varying in comparison with the first, and the curve has the general appearance shown in Fig. 8. The low-frequency factor $\sin \tfrac{1}{2}(\omega_0 - \omega)t$ induces an *amplitude modulation* on the high-frequency factor, $\sin \tfrac{1}{2}(\omega + \omega_0)t$. The resulting modulated oscillation is an example of the phenomenon of *beats*.

PROBLEMS

1. Show that $G^2 > 1$ if, and only if, $\omega^2 + 4b^2 < 2\omega_0{}^2$. This is the condition for amplification, that is, for the steady-state response to have a larger amplitude than the input.

2. When a and b are fixed, show that the frequency giving maximum gain is $\omega^2 = a^2 - 2b^2$ if this expression is positive, and the maximum gain is

$$G_{\max} = \frac{a^2}{2b\sqrt{a^2 - b^2}} \qquad \text{for } a^2 \geq 2b^2.$$

But if there is so much damping that $a^2 < 2b^2$ then the maximum occurs at $\omega = 0$ and has the value $G = 1$. *Hint:* G is maximum when the expression under the radical is minimum.

3. If $\delta = \omega - \omega_0$ and $|\delta|$ is small, obtain the approximate equation

$$4G^2 \doteq \frac{\omega_0{}^2}{\delta^2 + b^2}.$$

Sketch the corresponding graph of G^2 versus δ for fixed $b > 0$, and show that G^2 assumes half its maximum value when $\delta = b$. Hence, if G^2 is plotted versus ω, the full width at half maximum is approximately $2b$. (This interpretation of loss by means of frequency response is useful in the analysis of microwave and acoustic resonators.)

4. For $y'' + 2by' + y = \cos \omega t$, obtain a solution $y = G \cos (\omega t - \phi)$ by the method of the text, and check by reference to results of the text. Then make a careful plot of G versus ω for $b = 10$, 1, 0.1, 0.01 (cf. Probs. 2 and 3). Also plot ϕ versus ω for these values of b.

5. The equation considered in Prob. 4 has $a = A_0 = 1$. Reduce the general case to this one by the substitution $y(t) = A_0 Y(at)$. What are the new values of b and ω?

6. Solve $y'' = \cos \omega t$, $y(0) = y'(0) = 0$, and make a graph of y versus t showing how the solution builds up from its value at $t = 0$. Is the solution stable?

7. A package falling from an airplane is subjected to a retarding force due to air resistance, which is proportional to the speed. Obtain a differential equation of the form $my'' = mg - ry'$, where y is the distance measured downward from the height of release. Find a solution of the form $y = at$, and show that its derivative gives the limiting speed as $t \to \infty$. Are your results consistent with those of the text for $a = \omega \to 0$?

8. (a) Show that the equation $Y' + pY = g(t)$ can be written $(e^{pt}Y)' = e^{pt}g(t)$, hence solved by inspection. (b) Let e^{rt} be a solution of the homogeneous equation associated with (8-1), so that $r^2 + 2br + a^2 = 0$. Show that the substitution $y = e^{rt}v$ reduces (8-1) to the form

$$Y' + (2r + 2b)Y = a^2 e^{-rt}f(t) \qquad \text{where } Y = v'.$$

By part a, get Y; by integration, get v; and then $y = e^{rt}v$. [Note that this method solves the problem of forced oscillations with an arbitrary force function $f(t)$.]

9. Let the generator voltage in the RLC series circuit of Sec. 1 be the real part of $E_0 e^{i\omega t}$, where E_0 and ω are constant, so that the complex voltage equation is

$$LV'' + RV' + SV = SE_0 e^{i\omega t} \qquad S = C^{-1}.$$

(a) Show that the corresponding complex current equation is

$$LI'' + RI' + SI = i\omega E_0 e^{i\omega t}.$$

(b) For either equation, show that the natural frequency is $\omega_0 = (LC)^{-\frac{1}{2}}$ and the condition for critical damping is $CR^2 = 4L$. Also express the steady-state voltage gain in terms of R, L, C, and ω.

(c) Show that the current equation has the solution $I = ZE_0 e^{i\omega t}$, where the *complex impedance* $Z = (R + i\omega L - iS/\omega)^{-1}$. Assuming $L > 0$, what conditions on R and C ensure that this solution represents the steady state?

---

9. Higher-order Equations. The foregoing methods are easily extended to linear nth-order equations

$$y^{(n)} + p_{n-1}y^{(n-1)} + \cdots + p_1y' + p_0y = f(t) \qquad (9\text{-}1)$$

with constant coefficients. If the left member of (9-1) is denoted by $\mathbf{T}y$, a discussion similar to that of Sec. 2 shows that the operator $\mathbf{T}$ is linear. Hence the superposition principle, the principle of equating real parts, and all the other general principles established for linear operators in the foregoing discussion apply without change to (9-1).

A family of solutions of (9-1) is called a *general solution* provided the constants in it can be determined so as to satisfy the arbitrarily specified initial conditions

$$y(t_0) = y_0, \qquad y'(t_0) = y_1, \qquad \ldots, \qquad y^{(n-1)}(t_0) = y_{n-1},$$

where t_0 is an arbitrary point on the interval in which (9-1) is expected to hold. Just as was the case for $n = 2$, a general solution of (9-1) can be obtained from any particular solution by adding to it a general solution of the associated homogeneous equation:

$$y^{(n)} + p_{n-1}y^{(n-1)} + \cdots + p_1y' + p_0y = 0. \qquad (9\text{-}2)$$

We do not have to repeat the proof of this, because the principle of the complementary function, from which the result follows, is valid for all linear operators (Sec. 6).

The homogeneous equation (9-2) can be solved by the substitution $y = e^{st}$, exactly as was the case for $n = 2$. It is found that $y = e^{st}$ satisfies (9-2) if the constant coefficient s is a root of the *characteristic equation*

$$P(s) \equiv s^n + p_{n-1}s^{n-1} + \cdots + p_1 s + p_0 = 0. \tag{9-3}$$

If this equation has roots $s_1, s_2, \ldots, s_n$ then (9-2) has solutions

$$e^{s_1 t}, \qquad e^{s_2 t}, \qquad \ldots, \qquad e^{s_n t}.$$

By the superposition principle the function

$$y = c_1 e^{s_1 t} + c_2 e^{s_2 t} + \cdots + c_n e^{s_n t} \tag{9-4}$$

is also a solution of (9-2) for every choice of the constants c_i. The polynomial $P(s)$ in (9-3) is called the *characteristic polynomial* of (9-1).

As was the case for $n = 2$, the functions corresponding to a multiple root s are found by differentiating e^{st} with respect to the parameter s. Thus, if s is a root of multiplicity m, the corresponding solutions are

$$e^{st}, \qquad te^{st}, \qquad t^2 e^{st}, \qquad \ldots, \qquad t^{m-1}e^{st}$$

and superposition leads to a family of solutions:

$$y = (c_0 + c_1 t + \cdots + c_{m-1}t^{m-1})e^{st}.$$

A similar family of solutions is obtained for each multiple root, giving a correspondingly modified form of (9-4). A simple proof that these expressions are, in fact, solutions is given in Sec. 19, Example 2. The fact that the procedure produces a general solution is indicated by the results of Sec. 24.

As an illustration, consider the fourth-order equation

$$y^{(iv)} - 2y''' + 2y'' - 2y' + y = 0. \tag{9-5}$$

The characteristic equation is $s^4 - 2s^3 + 2s^2 - 2s + 1 = 0$, or, upon factoring,

$$(s^2 + 1)(s - 1)^2 = 0.$$

There are two simple roots $s_1 = i$, $s_2 = -i$, and a double root $s_3 = s_4 = 1$. Solutions corresponding to these roots are

$$e^{it}, \qquad e^{-it}, \qquad e^t, \qquad te^t$$

and a general solution is

$$y = c_1 e^{it} + c_2 e^{-it} + c_3 e^t + c_4 te^t.$$

Since the coefficients in (9-5) are real, the real and imaginary parts of e^{it} are also solutions, by the theorem of Sec. 4. The corresponding real form is

$$y = C_1 \cos t + C_2 \sin t + (c_3 + c_4 t)e^t.$$

In conclusion we remark that calculation of particular solutions by the method of undetermined coefficients for functions $f(t)$ of the type considered in Secs. 6 and 8 follows, with obvious minor modifications, the pattern of those sections. Without further ado we illustrate the procedure.

Example 1. Find a solution of $y''' + y'' + 2y' = t^2 + 3t + 1$.

On recalling the discussion of Sec. 6 we take the trial solution

$$y = t(a_0 + a_1 t + a_2 t^2).$$

Substitution into the given equation gives

$$(6a_2) + (2a_1 + 6a_2 t) + 2(a_0 + 2a_1 t + 3a_2 t^2) = t^2 + 3t + 1$$

or, after rearrangement,

$$(2a_0 + 2a_1 + 6a_2) + (6a_2 + 4a_1)t + 6a_2 t^2 = t^2 + 3t + 1.$$

Hence
$$6a_2 = 1$$
$$6a_2 + 4a_1 = 3$$
$$2a_0 + 2a_1 + 6a_2 = 1$$

and we conclude that
$$a_2 = \tfrac{1}{6} \qquad a_1 = \tfrac{1}{2} \qquad a_0 = -\tfrac{1}{2}.$$

Accordingly, $y = t(-\tfrac{1}{2} + \tfrac{1}{2}t + \tfrac{1}{6}t^2)$ is a solution.

Example 2. Describe the steady-state solution of the third-order equation

$$y''' + 3y'' + 4y' + 2y = 20 \cos t. \tag{9-6}$$

Here the characteristic equation can be factored in the form $(s^2 + 2s + 2)(s + 1) = 0$; hence all the roots have negative real parts. According to the principle of asymptotic stability (Sec. 7) the steady-state solution is adequately described by the behavior of any particular solution.

To obtain a particular solution, consider, instead of (9-6), the equation

$$y''' + 3y'' + 4y' + 2y = 20e^{it}. \tag{9-7}$$

The trial solution $y = ae^{it}$ leads to

$$ae^{it}(-i - 3 + 4i + 2) = 20e^{it}$$

or $a(-1 + 3i) = 20$. Multiplying by $(-1 - 3i)$, the conjugate of the coefficient of a, we get the equivalent form

$$a(-1 + 3i)(-1 - 3i) = 20(-1 - 3i) \qquad \text{or} \qquad a(1 + 9) = 20(-1 - 3i).$$

Hence $a = 2(-1 - 3i)$, and the function

$$y = (-2 - 6i)e^{it} = (-2 - 6i)(\cos t + i \sin t) \tag{9-8}$$

satisfies the complex equation (9-7). Since $20 \cos t$ is the real part of $20e^{it}$, the real part of the complex solution (9-8) satisfies (9-6). Hence our steady-state solution is

$$y = -2 \cos t + 6 \sin t.$$

All other solutions differ from this one by a term that tends to zero with exponential rapidity as $t \to \infty$.

PROBLEMS

1. What is the general solution of $y^{(iv)} = 0$? Of $y^{(iv)} = \cos t + 24$?

2. Consider $y''' - 3y'' + 2y' = f(t)$. Obtain the general solution when $f(t) \equiv 0$, by use of the characteristic equation $s(s^2 - 3s + 2) = 0$. Get a particular solution when $f(t) = 4$ by the trial solution $y = at$, and a particular solution when $f(t) = 60e^{5t}$ by the trial solution $y = ae^{5t}$. Thus get the general solution of

$$y''' - 3y'' + 2y' = 4 + 60e^{5t}.$$

3. Obtain a solution of $y^{(iv)} + y' = e^{i\omega t}$ by taking $y = ae^{i\omega t}$. Thus get general solutions of $y^{(iv)} + y' = \cos \omega t$ and $y^{(iv)} + y' = \sin \omega t$.

4. Find a solution of $y''' - y' = 1$ that satisfies $y(0) = y'(0) = y''(0) = 2$.

5. Find the solution of $y''' + 2y'' - y' - 2y = 2e^{-3t} + 4t^2$ that satisfies $y(0) = y'(0) = y''(0) = 0$.

6. Find a solution of $y^{(iv)} + 6y''' + 8y'' = 6y' + 9y$.

7. Show that the characteristic polynomial of $y''' + 8y'' + 19y' + 12y = \cos \omega t$ is $(s^2 + 5s + 4)(s + 3)$, and obtain a steady-state solution. Is the solution asymptotically stable?

8. If $\mathbf{T}y = y''' + p_2 y'' + p_1 y' + p_0 y$, show that $\mathbf{T}(ae^{rt}) = (ae^{rt})P(r)$, where a and r are constant and $P(s)$ is the characteristic polynomial associated with $\mathbf{T}$. Thus obtain a particular solution of $\mathbf{T}y = Ae^{rt}$ where A is constant and $P(r) \neq 0$.

10. The Euler Column. Rotating Shaft.

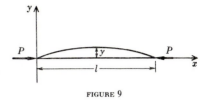

FIGURE 9

In all the applications discussed hitherto the variable of differentiation has been the time t, and the arbitrary constants in our solutions have been determined by giving the value and some of the derivatives at a single time, $t = t_0$. We now present some applications in which the independent variable is a space variable x, and the extra conditions are specified not just at one point but at two. Problems of this type, known as *boundary-value problems*, have great importance in several branches of mathematical physics.

It is known from experiments that a long rectilinear rod subjected to the action of axial compressive forces is compressed and retains its initial shape as long as the compressive forces do not exceed a certain critical value. Upon gradual increase of the compressive load P, a value of $P = P_1$ is reached when the rod buckles suddenly and becomes curved. The deflections of rods so compressed become extremely sensitive to minute changes of the load and increase rapidly with the increase in P. A detailed analysis of this *instability* or *buckling* phenomenon depends on rather delicate considerations in the nonlinear theory of elasticity. However, if the argument of Euler is followed, it is possible to deduce the magnitude of the critical load P_1 from linear differential equations governing small deflections of loaded rods.

Thus, consider a rod of uniform cross section and length l, compressed by the forces P applied to its ends (Fig. 9). Initially this rod is straight, but after the critical load P_1 is reached, it becomes curved, and we denote the deflection of its central line by y.

It is known from the Bernoulli-Euler law (Chap. 2, Sec. 13) that for small deflections

$$\frac{d^2y}{dx^2} = \frac{M}{EI}$$

where, in our case, the bending moment $M = -Py$. Thus

$$y'' + k^2y = 0 \tag{10-1}$$

where $k^2 \equiv P/EI$ and the primes denote differentiation with respect to x. Equation (10-1) must be solved subject to the end conditions

$$y(0) = 0 \qquad y(l) = 0 \tag{10-2}$$

since the ends of the rod remain on the x axis.

The *boundary-value problem* characterized by Eqs. (10-1) and (10-2) is quite different from the initial-value problems considered heretofore. In the initial-value problems one seeks solutions of differential equations satisfying specified conditions at *one* point only, while in the boundary-value problem stated above the solution y must satisfy conditions (10-2) assigned at *two* points $x = 0$ and $x = l$. It is not obvious that a solution of a differential equation satisfying specified conditions at two points exists in general. We shall see, however, that for suitable choices of the parameter k Eq. (10-1) does have solutions, other than the trivial solution $y = 0$, which vanish at the end points $x = 0$, $x = l$.

The general solution of (10-1) is $y = c_1 \cos kx + c_2 \sin kx$ and, on imposing the conditions (10-2), we get two equations

$$0 = c_1 \cos k0 + c_2 \sin k0 \qquad 0 = c_1 \cos kl + c_2 \sin kl.$$

These demand that

$$c_1 = 0 \qquad c_2 \sin kl = 0. \tag{10-3}$$

The choice $c_1 = c_2 = 0$ gives $y = 0$, corresponding to the rectilinear shape of the rod. If the rod does not remain straight, $c_2 \neq 0$, and we conclude from (10-3) that $\sin kl = 0$, so that

$$k = \frac{n\pi}{l} \qquad n = 0, 1, 2, \ldots. \tag{10-4}$$

The choice $n = 0$ again gives $y = 0$. If $n = 1$, then $k = \pi/l$, and on recalling the definition of k, we see that the corresponding value of P is

$$P_1 = EI\frac{\pi^2}{l^2}. \tag{10-5}$$

This is the *critical*, or the *Euler, load.*
   The shape of the central line of the rod, in this case, is

$$y = c_2 \sin \frac{\pi x}{l}.$$

The choice of $n = 2, 3, \ldots$ in (10-4) gives other "critical loads" $P_2, P_3, \ldots$ and the corresponding solutions

$$y = c_2 \sin \frac{n\pi x}{l}.$$

The maximum deflection c_2 is not determined in this analysis.
   Another interesting problem, essentially of the same sort, arises in the study of rotating shafts. It has been noted that, when a long shaft supported by bearings at $x = 0$ and $x = l$ is allowed to rotate, its initially rectilinear shape is preserved only if the speed of rotation ω does not exceed a certain critical value ω_1. On approaching the speed ω_1 the shaft starts pulsating and its shape changes. On further increase of the speed another critical value ω_2 is reached when the shaft starts beating and its shape changes again, and so on. This phenomenon can, in part, be explained by calculations similar to those used in determining the Euler load.
   Let us suppose that the shaft is rotating with the angular speed ω. An element of length dx of the shaft experiences the centrifugal force

$$F\,dx = \rho\,dx\,\omega^2 y$$

where ρ is the density per unit length of the shaft and y is the deflection at the point x. Thus,

$$F = \rho\omega^2 y \tag{10-6}$$

is the force per unit length of the shaft distributed along its length. It is shown in books on strength of materials that when the forces F acting on a rod are normal to its axis then $F = M'' \equiv d^2M/dx^2$, where the bending moment M is given by the Bernoulli-Euler law:

$$M = EIy''. \tag{10-7}$$

Thus, $F = (EIy'')''$ and if the *flexural rigidity EI* is constant, we get

$$\frac{d^4y}{dx^4} = \frac{F}{EI}. \tag{10-8}$$

The substitution for F from (10-6) gives the desired equation for the rotating shaft:

$$\frac{d^4y}{dx^4} - k^4 y = 0 \qquad \text{where } k^4 \equiv \frac{\rho\omega^2}{EI}. \tag{10-9}$$

Since the roots of the characteristic equation $s^4 - k^4 = 0$ are $s = \pm k$, $s = \pm ki$, the general solution of (10-9) is

$$y = c_1 e^{kx} + c_2 e^{-kx} + c_3 \cos kx + c_4 \sin kx. \tag{10-10}$$

If at the points of support $x = 0$, $x = l$ the deflection y and the moment M are zero, then [see (10-7)]

$$y(0) = 0 \qquad y''(0) = 0; \qquad y(l) = 0 \qquad y''(l) = 0. \tag{10-11}$$

The substitution from (10-10) into the boundary conditions (10-11) yields four equations:

$$\begin{aligned}
c_1 + c_2 + c_3 &= 0 \qquad\qquad c_1 + c_2 - c_3 = 0 \\
c_1 e^{kl} + c_2 e^{-kl} + c_3 \cos kl + c_4 \sin kl &= 0 \\
c_1 e^{kl} + c_2 e^{-kl} - c_3 \cos kl - c_4 \sin kl &= 0.
\end{aligned} \tag{10-12}$$

The solution $c_1 = c_2 = c_3 = c_4 = 0$, yielding $y = 0$, corresponds to the straight shaft. The system (10-12) also has nonzero solutions for certain values of k. From the first two equations (10-12) we find

$$c_1 = -c_2 \qquad c_3 = 0$$

and the substitution of these values in the two remaining equations gives

$$c_1 = c_2 = c_3 = 0 \qquad c_4 \sin kl = 0.$$

Thus, $\sin kl = 0$ unless $c_4 = 0$, and hence

$$k = \frac{n\pi}{l} \qquad n = 1, 2, 3, \ldots.$$

Using the value of k in (10-9) for $n = 1$ gives the first critical speed

$$\omega_1 = \frac{\pi^2}{l^2} \sqrt{\frac{EI}{\rho}}.$$

The critical speeds $\omega_2, \omega_3, \ldots$ are determined by taking $n = 2, 3, \ldots.$

PROBLEMS

1. (a) Solve $y'' + 3y' = 0$, $y(0) = y(3) = 0$; $y'' + \pi^2 y = 0$, $y(0) = y(1) = 0$;
(b) consider $y'' + y = 2e^x$, $y(0) = y(\pi) = 0$; $y'' + y = 1$, $y(0) = y(2\pi) = 0$.

2. A vibrating mass-spring system satisfies $my'' + ky = 0$, where k and m are positive constants. It is observed that $y = 0$ at $t = 0$ and also that $y = 0$ at a later time T. What are the possible values of the ratio k/m?

3. By a suitable choice of units, the equation for small static deflection of a uniformly loaded beam can be brought into the form $y^{(iv)} = 24$, $0 < x < 1$. The boundary conditions are $y(0) = y'(0) = y(1) = y'(1) = 0$ if the ends are clamped, $y(0) = y''(0) = y(1) = y''(1) = 0$ if the ends are supported by a pivot. (a) Show that, in either case, y has the form

$$y = x^4 + ax^3 + bx^2 - (1 + a + b)x$$

where a and b are constant. (b) Find a and b in the two cases and thus prove that the maximum deflection when the ends are clamped is one-fifth as much as it is when the ends are pivoted.

4. Discuss Prob. 3 when one end is clamped and the other is pivoted.

5. When a beam lies on an elastic foundation, in addition to the transverse external load $F(x)$ there is a restoring force $R = -a^2 y$ proportional to the deflection y. The equation of the axis of the beam then has the form

$$EI y^{(iv)} + a^2 y = F(x).$$

Solve this equation for $F(x) = p$, a constant, by assuming that the ends of the beam are hinged so that

$$y(0) = y''(0) = y(l) = y''(l) = 0.$$

6. The differential equation of the buckling of an elastically supported beam under an axial load P has the form

$$\frac{d^4y}{dx^4} + \frac{P}{EI}\frac{d^2y}{dx^2} + \frac{k}{EI}y = 0$$

where EI is the flexural rigidity and k is the modulus of the foundation.   Solve this equation.

---

11. Differential Operators.   We now return to the general nth-order equation

$$y^{(n)} + p_{n-1}y^{(n-1)} + \cdots + p_1y' + p_0y = f(t) \tag{11-1}$$

where the coefficients are constant and the superscripts denote differentiation with respect to t.   Study of this equation is simplified by use of another notation, namely,

$$Dy = y', \qquad D^2y = y'', \qquad D^3y = y''', \qquad \ldots, \qquad D^my = y^{(m)}, \qquad \ldots.$$

Here $D = d/dt$ is the operator that transforms any differentiable function of t into its derivative.   [For instance, $D(\sin t) = \cos t$.]

Since the $(m + 1)$st derivative is obtained by differentiating the mth derivative, it is evident that

$$D^{m+1}y = D(D^my) \qquad m = 1, 2, 3, \ldots$$

whenever either side of this equation is defined.   Thus $D^{m+1} = DD^m$, just as if D were a number.   Further analogy to the algebra of numbers is given by the properties

$$D(u + v) = Du + Dv, \qquad D(cu) = c\,Du \qquad c \text{ constant} \tag{11-2}$$

which express the fact that differentiation is a linear operation.

We agree that

$$(D^3 + p_2D^2 + p_1D + p_0)y = D^3y + p_2D^2y + p_1Dy + p_0y \tag{11-3}$$

and similarly for polynomials of higher degree.   Hence, (11-1) can be expressed in the form $(D^n + p_{n-1}D^{n-1} + \cdots + p_1D + p_0)y = f(t)$, or $P(D)y = f(t)$, where

$$P(D) = D^n + p_{n-1}D^{n-1} + \cdots + p_1D + p_0. \tag{11-4}$$

Although (11-3) resembles the linearity property (11-2), the two statements have a different mathematical status.   Equation (11-3) gives the definition of the expression on the left, while (11-2) is a theorem about differentiation.

The symbol $(D + s_1)(D + s_2)y$ is interpreted to mean that $D + s_1$ operates on $(D + s_2)y$, that is, on $y' + s_2y$.   By a short calculation,

$$(D + s_1)(D + s_2)y = [D^2 + (s_1 + s_2)D + s_1s_2]y$$

provided s_1 and s_2 are constant.   From the structure of the right-hand member it follows that

$$(D + s_1)(D + s_2)y = (D + s_2)(D + s_1)y$$

and a similar relation is obtained, by use of this one, for any number of factors.

The substance of these remarks is that the differential operator $F(D)$ in (11-4) behaves as if it were an algebraic polynomial.   This polynomial is exactly the same as the characteristic polynomial

$$P(s) = s^n + p_{n-1}s^{n-1} + \cdots + p_1s + p_0$$

except that s is replaced by D.   Hence, we can get the characteristic polynomial by writing s in place of D in $P(D)$.

For example, consider $D(D + 1)^3y = 0$.   Since $P(D) = D(D + 1)^3$, the characteristic polynomial is $P(s) = s(s + 1)^3$.   The characteristic equation is $s(s + 1)^3 = 0$, which gives the solution $y = c_1 + c_2e^{-t} + c_3te^{-t} + c_4t^2e^{-t}$.

It is evident that

$$e^{st} = e^{st}   \qquad De^{st} = se^{st}   \qquad D^2e^{st} = s^2e^{st}   \qquad D^3e^{st} = s^3e^{st}$$

and so on.   If we multiply the first equation by p_0, the second by p_1, the third by p_2, and so on, and then add, the result is

$$(p_0 + p_1D + p_2D^2 + \cdots + D^n)e^{st} = (p_0 + p_1s + p_2s^2 + \cdots + s^n)e^{st}.$$

This is equivalent to

$$P(D)e^{st} = P(s)e^{st}. \tag{11-5}$$

To illustrate the use of (11-5), consider the equation

$$D(D^2 + 3D + 1)^2y = 45e^{it}. \tag{11-6}$$

The trial solution $y = ae^{it}$ gives

$$D(D^2 + 3D + 1)^2ae^{it} = i(i^2 + 3i + 1)^2ae^{it} = -9iae^{it}.$$

This reproduces the right-hand member of (11-6) if $-9ia = 45$, or $a = 5i$.   Thus, $y = 5ie^{it}$ satisfies (11-6).   Since the coefficients are real, we deduce that $y = -5 \sin t$ and $y = 5 \cos t$ satisfy $D(D^2 + 3D + 1)^2y = 45 \cos t$ and $D(D^2 + 3D + 1)^2y = 45 \sin t$, respectively.

The method of this example leads to the following theorem:

THEOREM.   *Suppose that all the roots of the characteristic polynomial $P(s)$ have negative real parts and that ω and C are constant, with ω real.   Then the steady-state response to an input $f(t) = Ce^{i\omega t}$ in (11-1) is $y = [P(i\omega)]^{-1}Ce^{i\omega t}$.*

For proof, the trial solution $y = ae^{i\omega t}$ gives

$$P(D)ae^{i\omega t} = P(i\omega)ae^{i\omega t} = Ce^{i\omega t}$$

by (11-5) with $s = i\omega$; hence $y = [P(i\omega)]^{-1}Ce^{i\omega t}$ is a solution.   But the principle of asymptotic stability shows that *any* particular solution gives an adequate description of the steady state.   This completes the proof.

The function $H(s) = 1/P(s)$ is often called the *system function*, or the *response function*, and the complex variable $s = \sigma + i\omega$ is the *complex frequency*.   If the system function is written in polar form

$$\frac{1}{P(s)} \equiv H(s) = Ge^{-i\delta}$$

then G determines the amplitude and δ the phase of the steady-state output relative to the input, under the hypothesis of the above theorem.

This statement is evident from the complex form, since the input and output in the theorem are, respectively, $Re^{i(\omega t - \phi)}$ and $GRe^{i(\omega t - \phi - \delta)}$, where $C = Re^{-i\phi}$ is the polar form of C.   When the coefficients p_k are real the same conclusion follows by comparing the corresponding real expressions $R \cos (\omega t - \phi)$ and $GR \cos (\omega t - \phi - \delta)$.

For a general linear system the response function is the function by which a given input must be multiplied to get the output, or response, of the system.   The physical interpretation depends on circumstances.   If the input is current and the output is voltage, the response function is an impedance.   But if the input is voltage and the output is current, the response function is an admittance.   When both input and output are measured in the same units the response function is dimensionless and is ordinarily a transmission coefficient or reflection coefficient.[1]

Example 1.   The motion of an oscillatory system is governed by the equation

$$(0.1D^2 + D + 10)^2 y = 150 \cos(\omega t - \phi)$$

where ω and ϕ are real constants. What value of the frequency ω maximizes the amplitude of the steady-state response, and what is the maximum amplitude?

The leading coefficient in $P(D)$ is not 1, but Eq. (11-5) and the conclusions dependent thereon are still valid, as is clear from the derivation. We have

$$|P(i\omega)| = |(10 - 0.1\,\omega^2) + i\omega|^2 = (10 - 0.1\,\omega^2)^2 + \omega^2$$

and differentiating with respect to ω^2 shows that the maximum of $|P(i\omega)|$ occurs at $\omega^2 = 50$. This frequency gives the maximum gain, $G = 1/|P(i\omega)| = \frac{1}{75}$. The resulting response has amplitude $\frac{150}{75} = 2$.

Example 2.   Solve $x' + 2x - 2y = t$, $y' - 3x + y = e^t$.

The equations can be written in operator notation as

$$(D + 2)x - 2y = t \qquad -3x + (D + 1)y = e^t. \tag{11-7}$$

Operate on the second of these with $\frac{1}{3}(D + 2)$ to obtain

$$-(D + 2)x + \tfrac{1}{3}(D + 2)(D + 1)y = \tfrac{1}{3}(D + 2)e^t = e^t$$

and add to the first equation. The result after simplification is

$$(D^2 + 3D - 4)y = 3e^t + 3t.$$

This equation can be solved for y as in Sec. 6, to get

$$y = \tfrac{3}{5}te^t - \tfrac{3}{4}t - \tfrac{9}{16} + c_1 e^t + c_2 e^{-4t}.$$

The second equation in (11-7) now gives $x = \frac{1}{3}(D + 1)y - \frac{1}{3}e^t$, or

$$x = \tfrac{2}{5}te^t - \tfrac{1}{4}t - \tfrac{7}{16} + (\tfrac{2}{3}c_1 - \tfrac{2}{15})e^t - c_2 e^{-4t}.$$

The equations in this example constitute a *linear system* with constant coefficients. Although the method of solution is general, we prefer to base systematic study of systems on the *Laplace transform*, introduced in the next section.

PROBLEMS

1. Solve $(2D^2 - D - 1)y = 0$, $(D^2 - 1)y = 0$, $(D^2 + D)y = 2y$, $(D - 3)^2 y = 0$.

2. Find a general solution:

(a) $(D - 5)(2D + 3)Dy = 0$, $(D^2 + 1)(D^2 + 2D + 5)y = 0$;
(b) $(D^3 + 3D^2 + 3D + 1)y = 0$, $(D^3 + 8)y = 0$;
(c) $(D^3 - 2D^2 + D)y = 0$, $(D^4 + 3D^3 + 3D^2 + D)y = 0$;
(d) $(D^4 - k^4)y = 0$, $(D^3 - 3D^2 + 4)y = 0$;
(e) $(D^3 - D^2 + 4D)y = 4t + e^t$, $(D^4 + 1)y = 2 \cos t$;
(f) $(D - 1)(D - 2)^2 y = t^2$, $(D + 1)(D - 1)(D - 2)y = 1 - e^t$;
(g) $(D^4 + 2D^3 + D^2)y = 0$, $(D^3 + D^2 + D + 1)y = te^t$;
(h) $(D^4 + D^3)y = \cos 4t$, $(D^3 + 6D + 7)y = -24e^t \cos 2t$;
(i) $(D^4 - 2D^3 + D^2)y = t^3$, $(D^3 - 4D^2 + 5D - 2)y = 2t + 3$.

3. The following equations are stable, by Sec. 7, Prob. 5. Describe the steady-state behavior:

(a) $(D + 1)(D^2 + 3D + 2)y = 6 \cos 2t$, $(D + 2)(D^2 + 5D + 6)y = 5 \sin 3t$;
(b) $(D^2 + 10D + 29)(D + 4)y = \cos t$, $(D^2 + 6D + 10)(D + 2)y = \sin t$.

4. Find the solution in Example 2 satisfying $x(0) = y(0) = 0$.

[1] See E. A. Guillemin, "Theory of Linear Physical Systems," John Wiley & Sons, Inc., New York, 1963.

5. Write the following systems in operator form, and solve:

(*a*) $x' = y, y' = -x$; $x' = y, y' = x$; $x' = 3x - 2y, y' = 2x - y$;

(*b*) $x'' = y, y'' = x$; $x'' = y, y' = x$; $y' + x' = 2y, y' - x' = 2x$.

6. Solve $(D + 1)x + (2D + 1)y = e^t$, $(D - 1)x + (D + 1)y = 1$.

7. Solve the first equation with $y(0) = y'(0) = y''(0) = 1$, and the second with $y(0) = y'(0) = y''(0) = 0$: $(D^3 - D)y = 6 - 3t^2$, $(D^3 + 2D^2 + 2D + 1)y = t$.

THE LAPLACE TRANSFORM

12. General Principles. We now present a systematic and elegant procedure that is widely used in circuit analysis and in the study of feedback and control. The theory takes its form from a symbolic method developed by the English engineer Oliver Heaviside. It enables one to solve many problems without going to the trouble of finding the general solution and then evaluating the arbitrary constants. The procedure can be extended to systems of equations, to partial differential equations, and to integral equations, and it often yields results more readily than other techniques.

The basis of this method is the transformation defined by

$$F(s) = \int_0^\infty f(t)e^{-st}\,dt = \mathbf{L}f. \tag{12-1}$$

The function $F(s)$ is the *Laplace transform* of $f(t)$, and the operator $\mathbf{L}$ that transforms f into F is the *Laplace transform operator*. For example, if $f(t) = e^{at}$, where a is a real or complex constant,[1] the transform is

$$\int_0^\infty e^{at}e^{-st}\,dt = \int_0^\infty e^{-(s-a)t}\,dt = \frac{e^{-(s-a)}}{-(s-a)}\bigg|_0^\infty = \frac{1}{s-a} \tag{12-2}$$

provided $s > \operatorname{Re} a$. When $s \leq \operatorname{Re} a$ the integral diverges, and $F(s)$ is not defined.

The formula (12-1) represents a superposition of exponential functions, e^{-st}, where the superposition is over t and s is a parameter. Similar superpositions of exponential functions have already been noted in connection with linear differential equations; for example, the expression

$$c_1 e^{s_1 t} + c_2 e^{s_2 t} + \cdots + c_n e^{s_n t}$$

is associated with the homogeneous equation. In this case, t is a parameter and the superposition is over s.

Since the argument st of e^{-st} must be a pure number, s has the units of $1/t$. Hence, if t is time, s is *frequency* (cf. also Sec. 11).

It should be emphasized that (12-1) describes the action of $\mathbf{L}$, not only on f but on any function to which $\mathbf{L}$ can be applied. Thus,

$$\mathbf{L}u = \int_0^\infty u(t)e^{-st}\,dt \qquad \mathbf{L}v = \int_0^\infty v(t)e^{-st}\,dt \tag{12-3}$$

and so on. Using (12-3) we show that the operator $\mathbf{L}$ is linear, that is,

$$\mathbf{L}(\alpha u + \beta v) = \alpha \mathbf{L}u + \beta \mathbf{L}v \tag{12-4}$$

for any constants α and β and any functions u and v such that $\mathbf{L}u$ and $\mathbf{L}v$ make sense. Indeed, the left side of (12-4) is

---

[1] Complex-valued functions are integrated by integrating the real and imaginary parts. The familiar rules of calculus continue to hold for such functions, so that (12-2) follows from the formula $(e^{rt})' = re^{rt}$, with $r = -(s - a)$. See Sec. 4.

$$\int_0^\infty [\alpha u(t) + \beta v(t)]e^{-st}\, dt = \alpha \int_0^\infty u(t)e^{-st}\, dt + \beta \int_0^\infty v(t)e^{-st}\, dt$$

and this equals the right side of (12-4) by virtue of (12-3).

The chief purpose of the foregoing remarks is to explain what a Laplace transform is. We now outline the method by which the transform is used to solve linear differential equations

$$y^{(n)} + p_{n-1}y^{(n-1)} + \cdots + p_1 y' + p_0 y = f(t) \tag{12-5}$$

where the coefficients are constant and f is suitably restricted.

If the Laplace transform of both sides of (12-5) is taken, it is shown in the following section that the resulting equation is

$$P(s)\mathbf{L}y = F(s) + P_0(s) \tag{12-6}$$

where $P(s)$ is the characteristic polynomial associated with (12-5) and $P_0(s)$ is a polynomial that depends on the initial conditions. The function $F(s) = \mathbf{L}f$ is ordinarily available from a table of transforms, that is, from a table of corresponding pairs $F(s)$ and $f(t)$. Equation (12-6) then gives $\mathbf{L}y$, and the table of transforms (see inside back cover) now gives the solution y. The table serves much the same purpose as does a double-entry dictionary in the study of a foreign language.

For success of the method it is essential that f have a Laplace transform, and this matter is discussed next. We say that f is *admissible* if the following hold for $t \geq 0$:

$$f(t) \text{ is sectionally continuous}[1] \text{ on every finite interval}$$
and $$|f(t)| \leq Me^{at} \qquad \text{for some choice of constants } M \text{ and } a. \tag{12-7}$$

Under these conditions the integral converges for $s > a$, just as in the foregoing example. In fact,

$$\int_0^\tau |f(t)|e^{-st}\, dt \leq \int_0^\tau Me^{at}e^{-st}\, dt \leq M\int_0^\infty e^{-(s-a)t}\, dt.$$

Since the latter integral has the finite value (12-2), the integral on the left remains bounded as $\tau \to \infty$. This establishes not only the convergence but the *absolute convergence* of the integral defining $\mathbf{L}(f)$. The convergence is uniform if $s \geq \sigma_0 > a$, where σ_0 is fixed; hence the operations we shall carry out later are justified.

The foregoing calculation shows that the integral converges absolutely if s is complex, provided $\operatorname{Re} s \geq \sigma_0 > a$. For example, the proof of (12-2) is exactly the same when s is complex and $\operatorname{Re} s > \operatorname{Re} a$. If s is chosen to be pure imaginary and $f(t)$ is defined as 0 for $t < 0$, the Laplace transform is essentially the same as the Fourier transform $\mathbf{T}f$ introduced in Chap. 1, Sec. 22.

For functions $F(s)$ commonly encountered in applications, the extension to complex values of s presents no difficulty. However the extension is not emphasized here because, as shown presently, knowledge of $F(s)$ for real s suffices to determine $f(t)$.

The class of admissible functions is large enough for most practical purposes. All polynomials are admissible, the exponential function is admissible, and any periodic function that is sectionally continuous on its period is also admissible. Also, if $f_1(t)$ and $f_2(t)$ are admissible, the same is true of the three functions

$$f_1(t) + f_2(t) \qquad f_1(t)f_2(t) \qquad \int_0^t f_1(\tau)\, d\tau. \tag{12-8}$$

For proof, let f_1 and f_2 satisfy (12-7) with the constants M_1, a_1 and M_2, a_2, respectively. Then $f_1 + f_2$ clearly satisfies (12-7) with the constants

$$M = M_1 + M_2 \qquad a = \max(a_1, a_2).$$

---

[1] This means that the interval can be divided by points $t_1, t_2, \ldots, t_n$ into a finite number of intervals on each of which $f(t)$ is continuous. Also $f(t)$ must have a limit as $t \to t_k+$ and $t \to t_k-$. The following discussion uses a comparison test for integrals which can be verified in the same way as the corresponding test for series in Chap. 1, Sec. 5.

Proof that the other expressions (12-8) are admissible is equally simple and is left to the reader. By repeated use of (12-8) one can form a great variety of admissible functions, thus justifying the statement made at the beginning of the preceding paragraph.

Although the derivative of an admissible function need not be admissible (Prob. 8), a condition of this kind holds for solutions of (12-5). In Sec. 19, we establish the following:

EXISTENCE THEOREM FOR THE TRANSFORM. If f(t) is admissible, and y is any solution of the differential equation (12-5) then $y, y', y'', \ldots, y^{(n)}$ are also admissible; hence these functions, as well as f, have Laplace transforms.

The last step in the proposed solution of (12-5) requires that we recover y from its transform $\mathbf{L}y$. Many operations on y do not have the property of determining y uniquely. (For example, knowledge of the derivative Dy determines y only to within an additive constant.) But the Laplace transform determines y uniquely, as stated in the following theorem:

UNIQUENESS THEOREM FOR THE TRANSFORM. Let y and ȳ be admissible. If $\mathbf{L}y = \mathbf{L}\bar{y}$ for all large s, then $y = \bar{y}$ at every point where both functions are continuous.

The proof is given in Sec. 16. Since alteration of y at a single point does not affect $\mathbf{L}y$, the ambiguity at points of discontinuity is inherent in the problem. On the other hand, the ambiguity has no practical importance, because physically meaningful solutions of (12-5) are, as a rule, continuous.

The operation by which we recover f from $\mathbf{L}f = F(s)$ is called *inverse Laplace transformation* and is denoted by $\mathbf{L}^{-1}$. According to the theorem, $\mathbf{L}^{-1}F(s) = f(t)$ is substantially unique for $t > 0$, even if $F(s)$ is known only for large s. Hence, whenever we compute a Laplace transform in the sequel, or state an equality between transforms, it is understood that s is real and sufficiently large to justify the calculation.

Example 1.   Let $f(t) = t^b$. The change of variable $x = st$ yields

$$\int_0^\infty t^b e^{-st}\, dt = \int_0^\infty \left(\frac{x}{s}\right)^b e^{-x} \frac{dx}{s} = \frac{1}{s^{b+1}} \int_0^\infty x^b e^{-x}\, dx.$$

According to Chap. 1, Sec. 14, the latter integral is convergent for $b > -1$ and represents the generalized factorial $b!$. Hence

$$\mathbf{L}(t^b) = b! s^{-(b+1)} \qquad \text{for } b > -1.$$

Example 2.   The choice $a = i\omega$ in (12-2) yields

$$\mathbf{L}(e^{i\omega t}) = \mathbf{L}(\cos \omega t + i \sin \omega t) = \frac{1}{s - i\omega}. \tag{12-9}$$

Upon equating real and imaginary parts we get

$$\mathbf{L}(\cos \omega t) = \frac{s}{s^2 + \omega^2} \qquad \mathbf{L}(\sin \omega t) = \frac{\omega}{s^2 + \omega^2}$$

for all real s and ω. Differentiation with respect to ω gives

$$\mathbf{L}(t \sin \omega t) = \frac{2\omega s}{(s^2 + \omega^2)^2} \qquad \mathbf{L}(t \cos \omega t) = \frac{s^2 - \omega^2}{(s^2 + \omega^2)^2}. \tag{12-10}$$

Proceeding in this fashion one can construct a table of transforms, such as is given on the inside back cover of this book. Although the foregoing derivation assumes ω is real, the formulas are also valid for ω complex. This can be established by expressing $\cos \omega t$ and $\sin \omega t$ as a sum of exponentials and applying (12-9) to each term.[1]

[1] It also follows from the principle of analytic continuation (Chap. 1, Sec. 17).

PROBLEMS

1. Compute $\mathbf{L}1$ and $\mathbf{L}t$ directly from the definition, and check by the table of transforms.

2. Using (12-2) with $a = b$ and $a = -b$, find $\mathbf{L}(\sinh bt)$ and $\mathbf{L}(\cosh bt)$.   Check by letting $\omega = ib$ in entries $3a$ and $4a$ of the transform table (see also entry 7).

3. Using the table of transforms and linearity, find transforms of

$$3 + 2t - 4t^2, \qquad 2\cos 3t - 4\sin 2t, \qquad e^t(1 + \sin t), \qquad \sin^2 t \equiv \tfrac{1}{2}(1 - \cos 2t).$$

4. Using the table of transforms and linearity, find functions $f(t)$ whose transforms are

$$\frac{1}{s-2} + \frac{2}{s+3} + \frac{3}{s+4}, \qquad \frac{5}{s^2+4} + \frac{6}{s^2-4}, \qquad \frac{3s+1}{s^2+16}, \qquad \frac{1}{s+2} + \frac{4s-5}{(s+3)^2+4}.$$

5. By (12-1), find $\mathbf{L}[f(t)]$ if $f(t)$ is 0 for $t > 1$ and is defined as follows for $0 \le t \le 1$:

$$f(t) = 5, \qquad f(t) = e^{-t}, \qquad f(t) = 1 - 3t, \qquad f(t) = \sin \pi t, \qquad f(t) = te^{-t}.$$

6. In this problem a, b, c, A, B, C are constant, with a, b, c nonnegative.   The *unit step* $H_a(t)$ is defined by $H_a(t) = 0$ for $t < a$, $H_a(t) = 1$ for $t \ge a$.   (*a*) By (12-1), find $\mathbf{L}[H_a(t)]$, and thus find $\mathbf{L}y$, where $y = AH_a(t) + BH_b(t)$.   (*b*) Sketch the graph of y in part a for several well-chosen values of the constants.   (*c*) Similarly consider $y = AH_a(t) + BH_b(t) + CH_c(t)$.

7. As in Chap. 1, Sec. 14, Prob. 3, it is not difficult to prove that $\mathbf{L}f$ converges when f is piecewise continuous on every finite interval for $t > 0$, and $|f(t)| \le Me^{at}$ for $t > 1$, and $|f(t)| \le Mt^b$ for $0 < t \le 1$, where $b > -1$ and M, a, and b are constant.   If we say that a function satisfying these conditions is admissible, show that the product of two admissible functions might not be admissible.   [Try $f_1(t) = f_2(t) = t^{-\frac{2}{3}}$.   That is the reason why the extended definition of admissibility is not used in this text.]

8. Show that $\sin(e^{t^2})$ is admissible but that its derivative is not.

9. If f is admissible, use the equation following (12-7) to show that $F(s) \to 0$ as $s \to \infty$.   [The same is true when it is known only that $\mathbf{L}f$ converges for some value $s = s_0$.   Hence, if $F(s)$ does not approach 0 as $s \to \infty$, $F(s)$ cannot be the Laplace transform of any function f.]

10. Given a power series $H(x) = \Sigma a_n x^n$, convergent for small $|x|$, write $a_n = f(n)$, $x = e^{-s}$, and verify that

$$\sum_{n=0}^{\infty} a_n x^n = \sum_{n=0}^{\infty} f(n)e^{-ns} = \sum_{t=0}^{\infty} f(t)e^{-st} = F(s)$$

for sufficiently large s, where $F(s)$ is defined by this equation.   In this sense, the Laplace transform is an integral analog of a power series.   [The functions $H(x)$ or $F(s)$ are called *generating functions* for the sequence $\{a_n\}$.   The French mathematician Pierre Simon de Laplace, contemporary of Napoleon, developed his method of generating functions to solve certain *difference equations* encountered in probability theory.   However, the transform method of solving differential equations is due to the great Swiss mathematician Leonhard Euler.   He introduced what is now called the Laplace transform for this purpose before Laplace was born.]

---

13. The Transformed Equation.

If $f(t)$ is a continuous function with an admissible derivative $f'(t)$, integration by parts gives

$$\int_0^\tau e^{-st}f'(t)\,dt = e^{-st}f(t)\Big|_0^\tau + \int_0^\tau se^{-st}f(t)\,dt.$$

Since f is the integral of f', (12-8) indicates that f is admissible.   When s is sufficiently large this shows that $e^{-s\tau}f(\tau) \to 0$ as $\tau \to \infty$, and hence the integrated term reduces to

$-f(0)$. Upon recalling the definition of the Laplace transform, we obtain the following fundamental formula:

$$\mathbf{L}[f'(t)] = s\mathbf{L}[f(t)] - f(0). \tag{13-1}$$

If f is not continuous at 0 it is clear from the derivation that $f(0)$ must be interpreted to be the limit of $f(t)$ as $t \to 0$ through positive values.

The choice $f = y$ in (13-1) gives

$$\mathbf{L}y' = s\mathbf{L}y - y(0). \tag{13-2}$$

If y' is continuous and y'' is admissible, the choice $f = y'$ gives

$$\mathbf{L}y'' = s\mathbf{L}y' - y'(0) = s[s\mathbf{L}y - y(0)] - y'(0)$$

in view of (13-2). Hence,

$$\mathbf{L}y'' = s^2\mathbf{L}y - sy(0) - y'(0). \tag{13-3}$$

Repetition of this process gives

$$\mathbf{L}y^{(m)} = s^m\mathbf{L}y - s^{m-1}y_0 - s^{m-2}y_1 - \cdots - sy_{m-2} - y_{m-1} \tag{13-4}$$

where $y_0 = y(0)$, $y_1 = y'(0)$, . . . , $y_{m-1} = y^{(m-1)}(0)$. It is assumed that $y^{(m)}(t)$ is admissible and that all the lower derivatives are continuous for $t > 0$. The values at 0 are limits as $t \to 0$ from the right.

For example, consider the initial-value problem

$$y'' + 4y' + 13y = 0 \qquad y(0) = 5 \qquad y'(0) = -22. \tag{13-5}$$

The transform of this equation is $\mathbf{L}y'' + 4\mathbf{L}y' + 13\mathbf{L}y = 0$, or

$$(s^2\mathbf{L}y - 5s + 22) + 4(s\mathbf{L}y - 5) + 13\mathbf{L}y = 0$$

by (13-3) and (13-2). Solving for $\mathbf{L}y$ we get

$$\mathbf{L}y = \frac{5s - 2}{s^2 + 4s + 13} = \frac{5s - 2}{(s+2)^2 + 9} = \frac{5(s+2) - 12}{(s+2)^2 + 9}$$

where the second and third expressions were found by completing the square in the denominator and then rewriting the numerator. The objective is to get a form matching entry 2b of the table. Taking $a = 5$, $b = -12$, $c = 2$, and $\omega = 3$ in entry 2b, we get

$$y = 5e^{-2t}\cos 3t - 4e^{-2t}\sin 3t.$$

We now transform the general equation with constant coefficients, which can be written

$$y^{(n)} + p_{n-1}y^{(n-1)} + \cdots + p_1y' + p_0y = f(t) \tag{13-6}$$

or, in the notation of Sec. 11, $P(D)y = f(t)$. The solution satisfying

$$y(0) = 0, \qquad y'(0) = 0, \qquad \ldots, \qquad y^{(n-1)}(0) = 0 \tag{13-7}$$

and defined for $t \geq 0$ is called the *rest solution* because the initial conditions (13-7) mean that the system is initially at rest.

For the rest solution, (13-4) gives $\mathbf{L}y^{(m)} = s^m\mathbf{L}y$ by inspection, when $m \leq n$. In other words,

$$\mathbf{L}y' = s\mathbf{L}y, \qquad \mathbf{L}y'' = s^2\mathbf{L}y, \qquad \ldots, \qquad \mathbf{L}y^{(n)} = s^n\mathbf{L}y.$$

The transform of (13-6) is therefore

$$s^n\mathbf{L}y + p_{n-1}s^{n-1}\mathbf{L}y + \cdots + p_1s\mathbf{L}y + p_0\mathbf{L}y = \mathbf{L}f.$$

Since $\mathbf{L}y$ can be factored out of the left member, the result is

$$P(s)\mathbf{L}y = F(s) \qquad \text{rest solution} \tag{13-8}$$

where $F(s) = \mathbf{L}f$ and $P(s) = s^n + p_{n-1}s^{n-1} + \cdots + p_1 s + p_0$ is the characteristic polynomial.

It is not difficult to take account of general initial conditions

$$y(0) = y_0, \qquad y'(0) = y_1, \qquad \ldots, \qquad y^{(n-1)}(0) = y_{n-1}. \tag{13-9}$$

According to (13-4) the term $\mathbf{L}y^{(n)}$ now contains an extra polynomial of degree $n-1$, and the other terms have extra polynomials of lower degree. Hence the procedure gives

$$P(s)\mathbf{L}y - P_0(s) = F(s) \qquad \text{general case} \tag{13-10}$$

where $P_0(s)$ is a polynomial, of degree $n-1$, depending on the initial conditions.

Carrying out the details, we get

$$P_0(s) = s^{n-1}y_0 + s^{n-2}(y_1 + p_{n-1}y_0) + \cdots + (y_{n-1} + p_{n-1}y_{n-2} + \cdots + p_2 y_1 + p_1 y_0)$$

but it is usually better to repeat the derivation of (13-10) in any particular example.

The relation (13-10) gives the important formula

$$\mathbf{L}y = \frac{F(s)}{P(s)} + \frac{P_0(s)}{P(s)} \tag{13-11}$$

for the transform of the solution. To get a physical interpretation we consider two simpler problems. The first problem is

$$P(D)u = f(t) \qquad u(0) = u'(0) = \cdots = u^{(n-1)}(0) = 0$$

which corresponds to forced oscillations with null initial conditions. The second problem is

$$P(D)v = 0, \qquad v(0) = y_0, \qquad v'(0) = y_1, \qquad \ldots, \qquad v^{(n-1)}(0) = y_{n-1}$$

which corresponds to free oscillations with arbitrary initial conditions. According to (13-8), and (13-10) with $F = 0$, the solution of the two problems satisfy, respectively,

$$\mathbf{L}u = \frac{F(s)}{P(s)} \qquad \mathbf{L}v = \frac{P_0(s)}{P(s)}.$$

The general formula (13-11) indicates that $\mathbf{L}y = \mathbf{L}u + \mathbf{L}v$; hence, by the uniqueness theorem,

$$y = u + v. \tag{13-12}$$

Thus, y is a superposition of the two simpler functions, u and v. For dissipative systems, u is a representative of the *steady state*, as discussed in Sec. 7, while v represents a *transient* due to the initial conditions. The separation of steady-state and transient effects embodied in (13-11) and (13-12) is one of many advantages of the transform method.

Most problems at the end of this section have been chosen so that it is easy to recover the solution y from its transform $\mathbf{L}y$. (A systematic discussion is given in the next section.) It should be recorded, however, that this final step is by no means a primary concern in technological applications. On the contrary, one generally works with the transform $\mathbf{L}y$ throughout most of the process of system development, and only at the end does one bother to get y. The main part of the discussion proceeds in the *frequency domain*, that is, in the domain of the complex variable s, rather than in the time domain.

This technique is possible because many important properties of the system response can be read directly from $\mathbf{L}y$ in (13-11). For example, the points where $P_0(s)/P(s)$ is infinite (the *poles* of the function) are zeros of the characteristic polynomial, $P(s)$. According to the discussion of Sec. 7 the system is stable if, and only if, these zeros are all in the left half of the complex s plane. If some zero is pure imaginary, say $s_1 = i\omega$, we know from Sec. 4 that the free oscillations have a term $e^{i\omega t}$, with corresponding real form $\sin \omega t$ or $\cos \omega t$. This indicates that the system contains an element with no dissipation—perhaps an RLC circuit with resistance $R = 0$ or a mass-spring component with no viscosity.

Deeper reasons for working in the frequency domain rest on the fact that transcendental operations on $f(t)$ often correspond to algebraic operations on $F(s)$. An example of this is given in (13-8), which shows that the relation between $\mathbf{L}y$ and $\mathbf{L}f$ is far simpler than that between y and f. Indeed, if we consider the input to be $F(s)$ rather than $f(t)$, and the output to be $\mathbf{L}y$, the effect of the system on rest solutions is completely specified by multiplying by the system function $1/P(s)$. The concept of the *system function* is thus extended to the arbitrary forcing term $f(t)$, while the development of Sec. 11 was valid for sinusoidal functions only.

Analysis and synthesis of linear systems in the frequency domain is a broad and sophisticated area of technology, with methods and vocabulary of its own.[1]

Example. Find $\mathbf{L}y$, if $y^{(iv)} + y'' = \sin \omega t$, $y(0) = y'(0) = 0$, $y''(0) = y'''(0) = 1$. The initial conditions and (13-4) give

$$s^4 \mathbf{L}y - s - 1 + s^2 \mathbf{L}y = \mathbf{L} \sin (\omega t) = \omega(s^2 + \omega^2)^{-1}$$

and hence, solving for $\mathbf{L}y$,

$$\mathbf{L}y = \frac{\omega}{(s^2 + \omega^2)(s^2 + 1)s^2} + \frac{s+1}{(s^2 + 1)s^2}.$$

The result can be interpreted, in part, by noting that $s^2(s^2 + 1)$ is the characteristic polynomial. The presence of the factor s^2 indicates that $y = t$ satisfies the homogeneous equation and hence the solution is not stable. For $\omega = 1$ the roots $s = i\omega$ and $s = i$ coincide, and there is resonance.

PROBLEMS

1. If $y' + y = 0$, $y(0) = 1$, show that $\mathbf{L}y = (s + 1)^{-1}$ and thus get $y = e^{-t}$.

2. As in Prob. 1, solve $y' + py = 0$, $y(0) = y_0$, where p is constant.

3. Obtain $\mathbf{L}y$ if $y'' = 2$, $y(0) = y'(0) = 1$. Thus get y, and verify that y satisfies both the equation and the initial conditions.

4. By means of the Laplace transform, obtain rest solutions for each equation, and verify that it has the expected properties:

$$y'' = 24, \qquad y''' = 24, \qquad y^{(iv)} = 24, \qquad y'' = 24t, \qquad y^{(v)} = 0, \qquad y^{(v)} = 24t^2.$$

5. The static deflection of a certain cantilever beam satisfies $y^{(iv)} = 0$, $y(0) = y'(0) = 0$. Find a family of solutions containing two arbitrary constants.

6. Find a function $f(t)$ whose transform is

$$\frac{2s - 5}{3s^2 + 12s + 8} = \frac{2s - 5}{3(s + 2)^2 - 4} = \frac{1}{3}\frac{2(s + 2) - 9}{(s + 2)^2 - \frac{4}{3}}.$$

(Use entry 2b of the table of transforms with $\omega = 2i/\sqrt{3}$. See also entry 7.)

7. (a) If $y'' + 2y' + 2y = 0$, $y(0) = 1$, $y'(0) = -1$, show that $(s^2 + 2s + 2)\mathbf{L}y = s + 1$, and find y by the method of Prob. 6. (b) Solve $y'' + y = 0$, $y(0) = y'(0) = 1$.

8. (a) If $y'' + 2y' = 0$, $y(0) = 1$, $y'(0) = -\frac{1}{2}$, show that $(s^2 + 2s)\mathbf{L}y = s + \frac{3}{2}$, and find y by the method of Prob. 6. (b) Solve $y'' + 3y' + y = 0$, $y(0) = 1$, $y'(0) = -3$.

9. Solve the four homogeneous equations in Prob. 3 of Sec. 2 by the Laplace transform, using the initial conditions given there. Verify agreement with the book's answers.

---

[1] See, for example, R. N. Clark, "Introduction to Automatic Control Systems," John Wiley & Sons, Inc., New York, 1962, and J. G. Truxal, "Automatic Feedback Control System Synthesis," McGraw-Hill Book Company, New York, 1955.

10. Solve the four homogeneous equations in Prob. 1 of Sec. 11, subject (a) to the initial conditions $y(0) = 0$, $y'(0) = 1$, and (b) to the conditions $y(0) = 1$, $y'(0) = 0$. Verify that your solutions agree with the general solutions given in the answers, and also verify that $y(0)$ has the expected value.[1]

11. (a)–(d) Solve selected examples from the twelve homogeneous equations in Prob. 4 of Sec. 4, using arbitrary initial conditions $y(0) = y_0$, $y'(0) = y_1$.

12. Transform $y'' + 2by' + a^2y = f(t)$ subject to $y(0) = y_0$, $y'(0) = y_1$, and express $\mathbf{L}y$ as a term depending on f plus a term depending on the initial conditions.

13. Find $\mathbf{L}y$ for the rest solution of each of the four equations in Prob. 3 of Sec. 11.

14. Find $\mathbf{L}y$, given that $y(0) = y_0$, $y'(0) = y_1$, $y''(0) = y_2$, and

$$y''' + p_2y'' + p_1y' + p_0y = A \cos \omega t.$$

15. Let $f(t)$ be continuous for $-\infty < t < \infty$ except at a point t_0, where the left-hand limit $f(t_0-)$ and the right-hand limit $f(t_0+)$ exist but are unequal. Under suitable additional hypotheses, show that

$$\int_{-\infty}^{\infty} e^{-st}f'(t) \, dt = s \int_{-\infty}^{\infty} e^{-st}f(t) \, dt - [f(t_0+) - f(t_0-)]e^{-st_0}.$$

In the special case in which $t_0 = 0$ and $f(t) = 0$ for $t < 0$, observe that the result is equivalent to $\mathbf{L}[f'(t)] = s\mathbf{L}[f(t)] - f(0+)$; in other words, the term $f(0)$ in (13-1) is explained by the discontinuity of the extended function at $t = 0$. (*Hint:* Express the integral as an integral from $-\infty$ to t_0 plus an integral from t_0 to ∞, and integrate each of these by parts. A similar result is valid when there are several discontinuities.)

---

14. Partial Fractions.   In many important cases $F(s)$ is a *rational function*, that is, a quotient of two polynomials.[2] Hence $\mathbf{L}y$ as given by the analysis of the preceding section is also rational, and we can determine y by expanding $\mathbf{L}y$ in partial fractions.

The reader is presumably familiar with partial fractions, from their use for integrating rational functions in elementary calculus. Nevertheless, a brief discussion is presented here.

If $s_1, s_2, \ldots, s_n$ are unequal real or complex numbers and if $Q(s)$ is any polynomial of degree less than n, there are constants a_k such that the *partial-fraction expansion*

$$\frac{Q(s)}{(s - s_1)(s - s_2) \cdots (s - s_n)} \equiv \frac{a_1}{s - s_1} + \frac{a_2}{s - s_2} + \cdots + \frac{a_n}{s - s_n}$$

holds for $s \neq s_k$. The constants a_k can be determined by clearing of fractions and equating coefficients of corresponding powers of s. But a more efficient procedure is to multiply through by $s - s_k$ and let $s \to s_k$.

In illustration of this last remark, consider

$$\frac{Q(s)}{(s - s_1)(s - s_2)(s - s_3)} \equiv \frac{a_1}{s - s_1} + \frac{a_2}{s - s_2} + \frac{a_3}{s - s_3} \tag{14-1}$$

where $Q(s)$ is a polynomial of degree 2 or less. Multiplying through by $s - s_1$ we get

$$\frac{Q(s)}{(s - s_2)(s - s_3)} = a_1 + \left(\frac{a_2}{s - s_2} + \frac{a_3}{s - s_3}\right)(s - s_1).$$

---

[1] This method of partial verification is to be used in Prob. 11 and in analogous cases throughout Secs. 13 to 19.

[2] The corresponding class of functions $f(t)$ is identical with that considered in the method of undetermined coefficients (Secs. 6 to 9).

When $s \to s_1$ the terms involving a_2 and a_3 disappear, and the result is

$$a_1 = \frac{Q(s_1)}{(s_1 - s_2)(s_1 - s_3)}.$$

Formulas for a_2 and a_3 are obtained similarly (or by cyclic permutation of the subscripts).

The mathematical justification is as follows: Clearing of fractions in (14-1), we get

$$Q(s) = a_1(s - s_2)(s - s_3) + a_2(s - s_1)(s - s_3) + a_3(s - s_1)(s - s_2). \tag{14-2}$$

The above choice of a_1, a_2, and a_3 makes this an equality for $s = s_1$, s_2, and s_3. Hence, the two *second-degree* polynomials (14-2) agree for *three* distinct values of s. It follows that the two polynomials are identically equal, and the desired identity is established. Proof in the case of n roots is entirely similar and is left to the reader. It will be found that two polynomials of degree $n - 1$ agree at the n values s_1, s_2, ..., s_n and hence are identically equal.

We have assumed the s_k to be all unequal, so that there are no multiple roots. If s_1 is a root of multiplicity m the term $a_1(s - s_1)^{-1}$ in the partial-fraction expansion must be replaced by

$$\frac{a_{11}}{s - s_1} + \frac{a_{12}}{(s - s_1)^2} + \cdots + \frac{a_{1m}}{(s - s_1)^m} \tag{14-3}$$

where the a_{1k} are constant. Naturally, a similar result applies to the other roots.

The proof follows by differentiating the original identity with respect to the root in question.[1] For example, if we differentiate (14-1) with respect to s_1 (remembering that the constants a_k depend on s_1) we get

$$\frac{Q(s)}{(s - s_1)^2(s - s_2)(s - s_3)} = \frac{a_1'}{s - s_1} + \frac{a_1}{(s - s_1)^2} + \frac{a_2'}{s - s_2} + \frac{a_3'}{s - s_3}$$

where the primes denote differentiation with respect to s_1. Differentiating again gives the expansion for a triple root, and so on. Similarly, one can differentiate with respect to s_2 and s_3 or with respect to s_k in the general expansion for n roots. The result is an algebraic identity, valid for all real values of the variables s, s_1, s_2, ..., s_n. By the principle of analytic continuation (Chap. 1, Sec. 17) the identity remains valid when these variables are complex.

We conclude by remarking that it is sometimes convenient to use quadratic factors of the denominator, instead of the linear factors $s - s_k$ considered heretofore. The appropriate form is obtained by grouping pairs of terms in the basic expansion.

For example, corresponding to the quadratic factor $(s - s_1)(s - s_2)$ one gets the expression

$$\frac{a_1}{s - s_1} + \frac{a_2}{s - s_2} \equiv \frac{(a_1 + a_2)s - a_1 s_2 - a_2 s_1}{(s - s_1)(s - s_2)}.$$

The numerator is now a linear function, instead of a constant. (Similarly, if we consider cubic factors as denominators in the partial-fraction expansion, the numerators would turn out to be quadratic.) Repeated factors are treated in a manner analogous to (14-3). For example, the part of the expansion arising from a triple factor $(s^2 + \alpha s + \beta)^3$ in the denominator is of the form

$$\frac{a_{11} + b_{11}s}{s^2 + \alpha s + \beta} + \frac{a_{12} + b_{12}s}{(s^2 + \alpha s + \beta)^2} + \frac{a_{13} + b_{13}s}{(s^2 + \alpha s + \beta)^3}.$$

This is obtained by twice differentiating the original expansion corresponding to the simple factor $s^2 + \alpha s + \beta$. The differentiation is with respect to β.

---

[1] Compare discussion of the corresponding problem in Secs. 3 and 6.

Example 1.   Solve $y''' - y' = \sin t$, given $y(0) = 2$, $y'(0) = 0$, $y''(0) = 1$.   The transformed equation is

$$s^3 \mathbf{L}y - 2s^2 - 1 - (s\mathbf{L}y - 2) = \mathbf{L} \sin t = (s^2 + 1)^{-1}.$$

Solving for $\mathbf{L}y$,

$$\mathbf{L}y = \frac{2s^2 - 1}{s^3 - 3} + \frac{1}{(s^3 - s)(s^2 + 1)} = \frac{2s^3 + s}{(s^2 - 1)(s^2 + 1)}.$$

To determine y we factor the denominator and use partial fractions; thus,

$$\frac{2s^3 + s}{(s - 1)(s + 1)(s^2 + 1)} = \frac{a}{s - 1} + \frac{b}{s + 1} + \frac{c + ds}{s^2 + 1}.$$

Multiplying through by $s - 1$ and then letting $s \to 1$ give

$$a = \frac{2 + 1}{(1 + 1)(1 + 1)} = \frac{3}{4}.$$

Similarly, $b = \frac{3}{4}$.   To get c and d we could (as always) clear of fractions.   But it is simpler to assign to s any two values that have not already been used.   The choice $s = 0$ shows that $c = 0$, and multiplying by s and letting $s \to \infty$, we get $2 = a + b + d$.   Thus $d = \frac{1}{2}$, and

$$\mathbf{L}y = \frac{3}{4}\frac{1}{s - 1} + \frac{3}{4}\frac{1}{s + 1} + \frac{1}{2}\frac{s}{s^2 + 1}.$$

Entries 1a and 4a of the table of transforms now give the solution

$$y = \frac{3}{4}e^t + \frac{3}{4}e^{-t} + \frac{1}{2}\cos t.$$

Example 2.   Obtain the rest solution of $(D + 1)^3(D + 2)y = 6$.   Since $\mathbf{L}6 = 6s^{-1}$ the transformed equation is

$$\mathbf{L}y = \frac{6}{(s + 1)^3(s + 2)s} \equiv \frac{a}{(s + 1)^3} + \frac{b}{(s + 1)^2} + \frac{c}{s + 1} + \frac{d}{s + 2} + \frac{e}{s}.$$

Multiplying by $(s + 1)^3$ and letting $s \to -1$ give $a = -6$.   Similarly, $d = 3$ and $e = 3$.   Multiplying by s and letting $s \to \infty$ give $0 = c + d + e$, and hence $c = -6$.

To get b we assign to s any value other than the values $-1, -2, 0, \infty$ that have already been used.   The choice $s = 1$ gives

$$\frac{1}{4} = \frac{1}{8}a + \frac{1}{4}b + \frac{1}{2}c + \frac{1}{3}d + e$$

and hence $b = 0$.   By the table of transforms,

$$y = -3t^2e^{-t} - 6e^{-t} + 3e^{-2t} + 3. \tag{14-4}$$

This function satisfies the given conditions at $t = 0$, and it satisfies the differential equation, with $f(t) = 6$, for all t.   However, a solution obtained by the Laplace transform may be physically meaningful only for $t > 0$.   Thus, if the input $f(t) = 6$ was obtained by closing a switch at time $t = 0$, then, in reality, $f(t) = 0$ for $t < 0$ and $f(t) = 6$ for $t > 0$.   In that case, y is given by (14-4) for $t > 0$, and by $y = 0$ for $t < 0$.   Similar remarks apply to the foregoing example and to all solutions obtained in the sequel.

PROBLEMS

1. If y satisfies $y'' - 3y' + 2y = 4$, $y(0) = 2$, $y'(0) = 3$, show that

$$\mathbf{L}(y) = \frac{2s^2 - 3s + 4}{s(s - 1)(s - 2)}.$$

Deduce that $y = 2 - 3e^t + 3e^{2t}$.

2. Solve by the Laplace transform $y'' + 4y = \sin t$, $y(0) = 1$, $y'(0) = 0$.

3. Find $\mathbf{L}(y)$ and solve: $y''' + y'' = e^t + t + 1$, $y(0) = y'(0) = y''(0) = 0$.

4. Find y if $y = 0$ for $t < 0$, $y'' + y = 1$ for $t > 0$, and $y'(t)$ is continuous for all t.   Does the differential equation hold at $t = 0$?

5. Obtain the four rest solutions in Prob. 3 of Sec. 6.

6. Solve the seven second-order initial-value problems in Probs. 4 of Secs. 6 and 7.

7. (a)–(f) Find rest solutions for selected examples among the eighteen second-order nonhomogeneous equations of Prob. 2 in Sec. 6.

8. (a), (b) Obtain the six rest solutions in Prob. 3 of Sec. 7.

9. Solve Probs. 4 and 5 of Sec. 9, and Prob. 7 of Sec. 11 (four third-order initial-value problems).

10. Let $P(s)$ and $Q(s)$ be polynomials, as in the text, and let $P(s)$ have a root $s - s_1$ of multiplicity m. Thus,

$$\frac{Q(s)}{P(s)} \equiv \frac{Q(s)}{(s - s_1)^m R(s)} \equiv \frac{G(s)}{(s - s_1)^m}$$

where $R(s_1) \neq 0$ and $G(s)$ is defined by the equation. By expanding $G(s)$ in Taylor series

$$G(s) = G(s_1) + G'(s_1) \frac{s - s_1}{1!} + G''(s_1) \frac{(s - s_1)^2}{2!} + \cdots$$

about the point $s = s_1$, show that the part of the partial-fraction expansion of $P(s)/Q(s)$ associated with the multiple root s_1 is

$$\frac{G(s_1)}{(s - s_1)^m} + \frac{G'(s_1)}{1!(s - s_1)^{m-1}} + \frac{G''(s_1)}{2!(s - s_1)^{m-2}} + \cdots + \frac{G^{(m-1)}(s_1)}{(m - 1)!(s - s_1)}.$$

11. Apply the formula of Prob. 10 to the function

$$\frac{P(s)}{Q(s)} = \frac{6}{(s + 1)^3(s + 2)s}$$

and verify agreement with Example 2. [At the triple root $s_1 = -1$, $m = 3$, and $G(s) = 6(s + 2)^{-1}s^{-1}$.]

APPLICATIONS

12. (a)–(d) By the Laplace transform, solve Probs. 3 to 6 of Sec. 5.

13. The switch in an RC series circuit is closed at time $t = 0$, so that the input voltage suddenly assumes the value $E_0 \sin \omega t$, where E_0 and ω are positive constants. Given that the differential equation is

$$RCV' + V = E_0 \sin \omega t \qquad t > 0$$

find V by the Laplace transform, and make a sketch showing how V builds up from its initial value 0. Similarly discuss $RCV' + V = E_0 \cos \omega t$.

14. The voltage across the capacitor in an LC series circuit satisfies

$$LCV'' + V = E_0 \sin \omega t \qquad \text{or} \qquad LCV'' + V = E_0 \cos \omega t$$

for $t > 0$, and $V(t) = 0$ for $t \leq 0$. Discuss as in the preceding problem, distinguishing the cases $\omega = (LC)^{-\frac{1}{2}}$ and $\omega \neq (LC)^{-\frac{1}{2}}$.

15. (a) If a, b, and A denote positive constants, show that the rest solution of the equation $y'' + 2by' + a^2y = a^2A$ satisfies

$$y = A - Ae^{-bt} \left(\cos \omega t + b \frac{\sin \omega t}{\omega} \right) \qquad y' = a^2Ae^{-bt} \frac{\sin \omega t}{\omega}$$

for $\omega \neq 0$. Show that the value for $\omega = 0$ equals the limit as $\omega \to 0$, and rewrite in terms of hyperbolic functions when $\omega = i\beta$ is pure imaginary. (b) Referring to Sec. 1, interpret y as displacement in a mass-spring system and as voltage in a series RLC circuit. Also interpret y' in terms of velocity, momentum, charge, and current. (c) Sketch y and y' versus t for the three cases $\omega > 0$, $\omega = 0$, $\omega = i\beta$.

15. Systems.   As was noted in Chap. 2, any differential equation can be reduced to a system of first-order equations.   For example, the equation

$$y'' + 2by' + a^2y = f(t) \tag{15-1}$$

can be written in the form

$$\begin{aligned} y_1' &= y_2 \\ y_2' &= -a^2y_1 - 2by_2 + f(t) \end{aligned} \tag{15-2}$$

where $y_1 = y$ and $y_2 = y'$.   This is a first-order system in two unknowns with constant coefficients.   A more general system of the same sort is

$$\begin{aligned} y_1' &= a_{11}y_1 + a_{12}y_2 + f_1(t) \\ y_2' &= a_{21}y_1 + a_{22}y_2 + f_2(t) \end{aligned} \tag{15-3}$$

where $f_1(t)$ and $f_2(t)$ are forcing functions and a_{ij} are constant.

In Sec. 6 it was found that appropriate initial conditions for (15-1) involve the specification of both $y(0)$ and $y'(0)$.   This corresponds to specifying $y_1(0)$ and $y_2(0)$ in (15-2). The latter conditions are also appropriate for the general system (15-3); hence (15-3) can be transformed by means of the equations

$$\mathbf{L}y_1' = s\mathbf{L}y_1 - y_1(0) \qquad \mathbf{L}y_2' = s\mathbf{L}y_2 - y_2(0). \tag{15-4}$$

Indeed, if the various functions are admissible, the transform of (15-3) is

$$s\mathbf{L}y_1 - y_1(0) = a_{11}\mathbf{L}y_1 + a_{12}\mathbf{L}y_2 + \mathbf{L}f_1$$

$$s\mathbf{L}y_2 - y_2(0) = a_{21}\mathbf{L}y_1 + a_{22}\mathbf{L}y_2 + \mathbf{L}f_2$$

by (13-1).   Collecting terms in $\mathbf{L}y_1$ and $\mathbf{L}y_2$ on one side of the equations and the remaining terms on the other, we get

$$\begin{aligned} (a_{11} - s)\mathbf{L}y_1 + a_{12}\mathbf{L}y_2 &= -\mathbf{L}f_1 - y_1(0) \\ a_{21}\mathbf{L}y_1 + (a_{22} - s)\mathbf{L}y_2 &= -\mathbf{L}f_2 - y_2(0). \end{aligned} \tag{15-5}$$

This is a set of simultaneous linear equations in the two unknowns $\mathbf{L}y_1$ and $\mathbf{L}y_2$. The coefficient determinant is a quadratic polynomial, $P(s)$, which is called the *characteristic polynomial*.   Thus,

$$P(s) = \begin{vmatrix} a_{11} - s & a_{12} \\ a_{21} & a_{22} - s \end{vmatrix}. \tag{15-6}$$

It is convenient to abbreviate the right-hand members (15-5) by means of

$$\tilde{F}_1(s) = F_1(s) + y_1(0) \qquad \tilde{F}_2(s) = F_2(s) + y_2(0) \tag{15-7}$$

where $F_1(s) = \mathbf{L}f_1$ and $F_2(s) = \mathbf{L}f_2$.   The solution of (15-5) by Cramer's rule is then

$$\mathbf{L}y_1 = \frac{-1}{P(s)}\begin{vmatrix} \tilde{F}_1(s) & a_{12} \\ \tilde{F}_2(s) & a_{22} - s \end{vmatrix} \qquad \mathbf{L}y_2 = \frac{-1}{P(s)}\begin{vmatrix} a_{11} - s & \tilde{F}_1(s) \\ a_{21} & \tilde{F}_2(s) \end{vmatrix} \tag{15-8}$$

(cf. Appendix A) if s is not a root of $P(s) = 0$.   When $F_1(s)$ and $F_2(s)$ are rational we can get y_i from $\mathbf{L}y_i$ by partial fractions, as in the foregoing section.

To illustrate the procedure in a numerical case, consider

$$y_1' + 2y_2' + y_1 - y_2 = 25 \qquad 2y_1' + y_2 = 25e^t \tag{15-9}$$

with the initial conditions $y_1(0) = 0$, $y_2(0) = 25$.   These equations are not in *standard form;* that is, they are not solved for y_1' and y_2'.   Nevertheless, we can find $\mathbf{L}y_1$ and $\mathbf{L}y_2$ by the technique used in deriving (15-8).

The transform of (15-9) leads to

$$sLy_1 + 2(sLy_2 - 25) + Ly_1 - Ly_2 = 25s^{-1}$$

$$2sLy_1 + Ly_2 = 25(s - 1)^{-1}.$$

Solving for Ly_1 we get

$$Ly_1 = \frac{25}{4s(s - 1)^2(s + \frac{1}{4})} = \frac{25}{s} - \frac{9}{s - 1} + \frac{5}{(s - 1)^2} - \frac{16}{s + \frac{1}{4}}$$

and, according to entries $1a$ and $1b$ of the table of transforms,

$$y_1 = 25 - 9e^t + 5te^t - 16e^{-t/4}.$$

It should be noticed that this method enables us to find y_1 without finding y_2, and the initial conditions are satisfied automatically.

Naturally one can also solve for Ly_2, expand in partial fractions, and get y_2, Or, more simply, one can get y_2 from the second equation (15-9). Either method gives

$$y_2 = 33e^t - 10te^t - 8e^{-t/4}.$$

Returning to the general equation (15-8), replace $\bar{F}_1$ and $\bar{F}_2$ by their values (15-7). The result is

$$Ly_1 = -\frac{1}{P(s)}\begin{vmatrix} F_1(s) & a_{12} \\ F_2(s) & a_{22} - s \end{vmatrix} - \frac{1}{P(s)}\begin{vmatrix} y_1(0) & a_{12} \\ y_2(0) & a_{22} - s \end{vmatrix} \tag{15-10}$$

and similarly for Ly_2. This formula separates the effect of the forcing functions from the effect of the initial conditions in a most elegant manner. The first term involves F_1 and F_2 but not the initial conditions, while the second term involves $y_1(0)$ and $y_2(0)$ but not the forcing functions.

The second term of (15-10) is a fraction whose numerator is a linear function and whose denominator is the characteristic polynomial, $P(s)$. If $P(s)$ has the unequal roots s_1 and s_2 the partial-fraction expansion is of the form

$$\frac{A_1}{s - s_1} + \frac{A_2}{s - s_2}$$

and the corresponding function in the time domain is

$$A_1e^{s_1t} + A_2e^{s_2t}. \tag{15-11}$$

We conclude that the system is *asymptotically stable* if both roots s_1 and s_2 have negative real parts. In that case the term involving the initial conditions tends exponentially to 0 as $t \to \infty$, and one can speak of the steady-state behavior. The latter is described by any particular solution, for example, by the *rest solution*, for which all initial values are 0.

Similar considerations apply to systems in more than two unknowns. Thus, if a system analogous to (15-3) is written for three functions y_1, y_2, y_3, the coefficient determinant of the transformed system is found to be

$$P(s) = \begin{vmatrix} a_{11} - s & a_{12} & a_{13} \\ a_{21} & a_{22} - s & a_{23} \\ a_{31} & a_{32} & a_{33} - s \end{vmatrix} \tag{15-12}$$

This is called the *characteristic polynomial*, and its roots s_i are the *characteristic roots* or the *natural complex frequencies*. The condition for stability is that all roots of $P(s)$ be in the left half plane.

It is also possible to extend the method to systems of higher-order equations. The transform of each equation is obtained as in Sec. 13, and the result is a set of linear equations for Ly_i. If the forcing terms f_i have rational transforms F_i, the method of partial fractions succeeds again. We illustrate these remarks by discussing a second-order system.

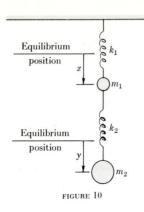

Equilibrium position

k_1

x

m_1

Equilibrium position

k_2

y

m_2

FIGURE 10

Example. Let masses m_1 and m_2 be suspended in tandem from springs of stiffness coefficients k_1 and k_2 as shown in Fig. 10. If x and y denote displacements of the masses m_1 and m_2 from their equilibrium positions, a procedure similar to that in Sec. 1 gives the two equations

$$m_1 x'' = k_2(y - x) - k_1 x \qquad m_2 y'' = -k_2(y - x)$$

when friction is negligible. Evaluate $\mathbf{L}x$ and $\mathbf{L}y$ under the hypothesis that

$$x(0) = 0 \qquad x'(0) = v_1 \qquad y(0) = 0 \qquad y'(0) = v_2. \quad (15\text{-}13)$$

Dividing the equations by m_1 or m_2, respectively, gives

$$x'' = c(y - x) - ax \qquad y'' = -b(y - x) \quad (15\text{-}14)$$

where $a = k_1/m_1$, $b = k_2/m_2$, and $c = k_2/m_1$. If we recall (13-3), the transform of (15-14) subject to (15-13) is found to be

$$(s^2 + a + c)\mathbf{L}x \qquad -c\mathbf{L}y = v_1$$
$$-b\mathbf{L}x + (s^2 + b)\mathbf{L}y = v_2$$

after slight rearrangement. Hence, by Cramer's rule,

$$\mathbf{L}x = \frac{1}{P(s)}\begin{vmatrix} v_1 & -c \\ v_2 & s^2 + b \end{vmatrix} \qquad \mathbf{L}y = \frac{1}{P(s)}\begin{vmatrix} s^2 + a + c & v_1 \\ -b & v_2 \end{vmatrix}$$

where $P(s)$ is the coefficient determinant,

$$P(s) = \begin{vmatrix} s^2 + a + c & -c \\ -b & s^2 + b \end{vmatrix} \equiv s^4 + s^2(a + b + c) + ab.$$

The equation $P(s) = 0$ is a quadratic in $z = s^2$ which has a positive discriminant and therefore has two unequal real roots. The roots z_1, z_2 must be negative, because the coefficients are positive. Hence $s = \pm i\omega_1$, $s = \pm i\omega_2$, where ω_1 and ω_2 are real and distinct. The resulting partial-fraction expansion

$$\frac{A_1 + B_1 s}{s^2 + \omega_1^2} + \frac{A_2 + B_2 s}{s^2 + \omega_2^2}$$

indicates that the motion of each mass is a combination of two simple harmonic motions of frequencies ω_1 and ω_2. These are the *natural frequencies* of the system.

The roots of $P(s) = 0$ are on the imaginary axis because we assumed that frictional forces could be neglected. If each mass is subject to a small amount of dissipation the roots would be in the left half plane, and the system would be asymptotically stable.

PROBLEMS

1. (a)–(e) Given the initial conditions $x(0) = 1$, $y(0) = 0$, find $\mathbf{L}x$ and $\mathbf{L}y$ in each of the following five systems. Obtain x and y from the table of transforms, and get a partial check by verifying that your x and y satisfy the expected initial conditions:

$$\begin{cases} x' = -5x + 3y \\ y' = -4x + 2y \end{cases} \begin{cases} x' = -5x + y \\ y' = -x - 3y \end{cases} \begin{cases} x' = -6x + 2y \\ y' = 3x - 7y \end{cases} \begin{cases} x' = x + 2y \\ y' = -2x - 3y \end{cases} \begin{cases} x' = -x + y \\ y' = -x - y \end{cases}$$

2. (a)–(e) Solve selected systems in Prob. 1, using $x(0) = 0$, $y(0) = 1$.

3. Explain how to satisfy arbitrary initial conditions $x(0) = x_0$, $y(0) = y_0$ in Prob. 1 by superposition of the results in Probs. 1 and 2.

4. (a)–(e) Find $\mathbf{L}x$ and $\mathbf{L}y$ in Prob. 1 with x' and y' replaced by x'' and y'', respectively, in each case. Use initial conditions

$$x(0) = 1, \qquad x'(0) = 0, \qquad y(0) = 0, \qquad y'(0) = 1.$$

5. Find Ly, Lz, y, and z and check by substitution:

$$y' + 3y + z' + 2z = e^{-2t}, \qquad 2y' + 2y + z' + z = 1, \qquad y(0) = z(0) = 0.$$

6. Find y, given $y' + z' = z' + w' = w' + y' = y$, $y(0) = z(0) = w(0) = 1$.

7. (a), (b) Solve some of the six systems in Prob. 5 of Sec. 11, using $x(0) = 1$ and null initial conditions otherwise. Verify agreement with the general answer given in the text, and verify that your solutions satisfy the expected initial conditions.

APPLICATIONS

8. Write $y'' + 2by' + a^2y = 0$ as a system in $u = y$ and $v = y'$, where a and b are positive constants. Solve by the Laplace transform subject to $u(0) = y_0$, $v(0) = y_1$. Interpret physically in two ways by reference to Sec. 5, and verify agreement with the results of Sec. 5 in the three cases $b > a$, $b = a$, and $b < a$.

9. Obtain the rest solution of the system

$$y_1' = y_2, \qquad y_2' + 2by_2 + a^2y_1 = a^2A \cos \omega t,$$

where a, b, A, and ω are constants. Interpret physically in two ways, and check by Sec. 8.

10. Find the equation of the path of a particle whose coordinates satisfy

$$mx'' + Hey' = Ee \qquad my'' - Hex' = 0$$

where m, e, H, and E are constant. Assume $x = y = x' = y' = 0$ at $t = 0$. This system occurs in the determination of the ratio of charge to mass of an electron.

11. The currents I_1 and I_2 in the two coupled circuits shown in Fig. 11 satisfy the following differential equations:

$$M\frac{d^2I_1}{dt^2} + L_2\frac{d^2I_2}{dt^2} + R_2\frac{dI_2}{dt} + \frac{I_2}{C_2} = 0$$

$$M\frac{d^2I_2}{dt^2} + L_1\frac{d^2I_1}{dt^2} + R_1\frac{dI_1}{dt} + \frac{I_1}{C_1} = 0.$$

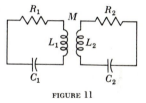

FIGURE 11

Find the Laplace transform of I_1 and I_2 in terms of the circuit constants and the initial values and discuss the solution when R_1 and R_2 are 0.

---

16. Additional Properties of the Transform.[1] The usefulness of the Laplace transform is greatly increased by some general properties that are derived now. As our first result, we note that the obvious equation

$$\int_0^\infty e^{-(s+c)t}f(t)\,dt = \int_0^\infty e^{-st}e^{-ct}f(t)\,dt$$

is equivalent to

$$F(s + c) = \mathbf{L}[e^{-ct}f(t)] \qquad \text{where } F(s) = \mathbf{L}[f(t)]. \tag{16-1}$$

This holds for any constant c, provided f is admissible and s is large enough. The advantage of (16-1) is that it produces new formulas from old ones and thus extends the scope of a table of transforms. For instance, when applied to (12-10) the formula gives

---

[1] The advanced topics in Secs. 16 to 19 are essential for a deeper understanding of Laplace transformation but can be omitted on first reading without loss of continuity.

$$\mathbf{L}(te^{-ct}\cos\omega t) = \frac{(s+c)^2 - \omega^2}{[(s+c)^2 + \omega^2]^2} \qquad \mathbf{L}\left(te^{-ct}\frac{\sin\omega t}{\omega}\right) = \frac{2(s+c)}{[(s+c)^2 + \omega^2]^2}.$$

By completing the square, any quadratic denominator $s^2 + as + b$ can be put into the form of the expressions in brackets.

A property related to (16-1) is

$$\mathbf{L}[f(t-c)] = e^{-cs}\mathbf{L}[f(t)] \qquad \text{where } c \geq 0 \text{ and } f(t) = 0 \text{ for } t \leq 0. \tag{16-2}$$

Thus, for admissible f the substitution $x = t - c$ gives

$$\int_0^\infty f(t-c)e^{-st}\,dt = \int_{-c}^\infty f(x)e^{-s(x+c)}\,dx = e^{-sc}\int_{-c}^\infty f(x)e^{-sx}\,dx.$$

The limits $(-c,\infty)$ can be changed to $(0,\infty)$ if $f(t) = 0$ on the interval $(-c,0)$, and (16-2) follows.

Equation (16-2) can be used to obtain the transform of an admissible periodic function $f(t)$ of period $T > 0$. Since the Laplace transform involves the values of $f(t)$ only for $t \geq 0$, we redefine f in such a way that $f(t) = 0$ for $t < 0$. It is also convenient to let

$$f_0(t) = f(t) \qquad \text{for } 0 \leq t < T \qquad \text{and} \qquad f_0(t) = 0 \text{ elsewhere.} \tag{16-3}$$

We can then state the interesting identity

$$f(t) = f_0(t) + f(t - T) \qquad \text{for } t \geq 0. \tag{16-4}$$

Indeed, the result holds for $0 \leq t < T$ because $f(t - T) = 0$ in that case. It holds for $t \geq T$ because $f_0(t) = 0$ for $t \geq T$, and because $f(t) = f(t - T)$ by periodicity. A graphical interpretation is given in Fig. 12.

Upon taking the Laplace transform of (16-4), using (16-2) with $c = T$, we get

$$\mathbf{L}f = \mathbf{L}f_0 + e^{-sT}\mathbf{L}f.$$

Hence, solving for $\mathbf{L}f$, and recalling (16-3),

$$\mathbf{L}f = (1 - e^{-sT})^{-1}\mathbf{L}f_0 \qquad \text{where } \mathbf{L}f_0 = \int_0^T f(t)e^{-st}\,dt. \tag{16-5}$$

This is the desired formula.

As an illustration, for the *square wave* shown in Fig. 13, we have

$$f_0(t) = 1 \qquad \text{for } 0 < t < a \qquad \text{and } f_0(t) = 0 \text{ elsewhere.}$$

The period is $T = b$, and so by (16-5)

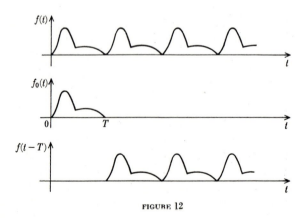

FIGURE 12

$$\mathbf{L}f = (1 - e^{-bs})^{-1} \int_0^a e^{-st}\,dt = \frac{1}{s}\frac{1 - e^{-as}}{1 - e^{-bs}}.$$

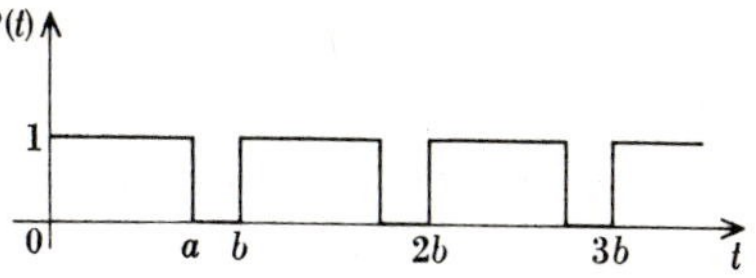

Another important property of the Laplace transform is

FIGURE 13

$$\mathbf{L}\int_0^t f(\tau)\,d\tau = s^{-1}\mathbf{L}[f(t)] \qquad (16\text{-}6)$$

where f is admissible. This follows by applying $\mathbf{L}f_1' = s\mathbf{L}f_1 - f_1(0)$ to

$$f_1(t) = \int_0^t f(\tau)\,d\tau$$

noting that $f_1(0) = 0$ and that $f_1'(t) = f(t)$ at points of continuity.

To illustrate the use of (16-6), recall that the voltage V across a capacitor is related to the charge Q by $V = SQ$, where $S = 1/C$ is the elastance. Since the current $I = dQ/dt$, integration gives

$$V = SQ = S\int_0^t I\,dt + SQ_0$$

where Q_0 is the charge on the capacitor initially, that is, at time $t = 0$. In many practical problems $Q_0 = 0$, and this condition is now assumed.

If the capacitor is connected in series with a resistance R, an inductance L, and source of voltage E as in Sec. 1, the voltage equation is

$$L\frac{dI}{dt} + RI + S\int_0^t I\,dt = E$$

and its transform with $I(0) = 0$ is

$$(sL + R + s^{-1}S)\mathbf{L}I = \mathbf{L}E. \qquad (16\text{-}7)$$

It should be noted that the condition $Q_0 = 0$ is automatically built into the analysis.

For a circuit with n independent loops a calculation of the same sort can be carried out for each loop. The result after transformation is a system of the form

$$\sum_{j=1}^{n} (sL_{ij} + R_{ij} + s^{-1}S_{ij})\mathbf{L}I_j = \mathbf{L}E_i \qquad (16\text{-}8)$$

where I_j is the current and E_j the impressed voltage associated with the jth loop. The resistances R_{ij} and elastances S_{ij} are 0 for $i \neq j$ but the inductances L_{ij} need not have this property because of the possibility of mutual inductance between branches.

The reader familiar with matrices (Chap. 4) will recognize that the system (16-8) can be written in the form (16-7), where L, R, and S are matrices, with R and S diagonal, and where I and E denote the current and voltage vectors, respectively. The matrix equation is the *equilibrium equation for the circuit on a loop basis*. It gives the mathematical foundation for the general study of linear lumped-parameter systems.

The properties established here enable us to prove the uniqueness theorem of Sec. 12, namely, *if y and $\bar{y}$ are admissible functions such that $\mathbf{L}y = \mathbf{L}\bar{y}$ for large s, then $y = \bar{y}$ at points of continuity*. Indeed, the linearity property gives $\mathbf{L}(y - \bar{y}) = 0$ for large s, or $\mathbf{L}u = 0$ where $u = y - \bar{y}$. If we now define $v(t) = e^{-ct}u(t)$, with c a sufficiently large constant, the shift property (16-1) indicates that $\mathbf{L}v = 0$ for all $s \geq 0$, and not just for large s.

To get a smoother function one can integrate $v(t)$ from 0 to t. The Laplace transform of the integrated function vanishes for $s > 0$ by (16-6), since $\mathbf{L}v = 0$. If we multiply by e^{-t} and integrate again, the transform of the resulting function $V(t)$ also vanishes for $s > 0$. The substitution $x = e^{-t}$ gives

$$\mathbf{L}V = \int_0^\infty e^{-st} V(t)\, dt = \int_0^1 x^{s-1} V(-\log x)\, dx.$$

Since $\mathbf{L}V = 0$ for $s > 0$ we certainly have $\mathbf{L}V = 0$ for $s = 1, 2, 3, 4, \ldots$, and the example in Chap. 1, Sec. 25, then gives[1] $V \equiv 0$. By differentiation $v = 0$ at points where v is continuous; hence $y = \bar{y}$ at points where $y - \bar{y}$ is continuous.

We conclude with a brief discussion of the table of transforms, which summarizes these and other properties of the Laplace transform. Entries 8a and 8b constitute the linearity property mentioned in Sec. 12, and 9a and 9b are (16-2) and (16-1), respectively. For 10a we let $\tau = ct$ to obtain

$$\int_0^\infty f(ct) e^{-st}\, dt = \int_0^\infty f(\tau) e^{-(s/c)\tau} d\left(\frac{\tau}{c}\right) = \frac{1}{c} F\left(\frac{s}{c}\right)$$

provided the constant $c > 0$. Writing $1/c$ instead of c in 10a gives 10b, again for $c > 0$. The result 11a was established in Sec. 13, and 11b follows by differentiating (12-1). Statement 12a is (16-6), and 12b follows by integrating (12-1), assuming absolute convergence. Proof of the *convolution theorem*, entry 13, is postponed to Sec. 18.

Example 1. Find y, given that $y(0) = y'(0) = y''(0) = 0$ and

$$(D^2 + 4D + 13)^2 y = 0 \qquad y'''(0) = 8.$$

By the initial conditions the transform of $D^4 y$ is $s^4 \mathbf{L}y - 8$, and the transform of $D^m y$ is $s^m \mathbf{L}y$ for $m \le 3$. A moment's consideration shows that the transformed equation must be

$$(s^2 + 4s + 13)^2 \mathbf{L}y = 8.$$

Upon solving for $\mathbf{L}y$ and completing the square we get

$$\mathbf{L}y = \frac{8}{[(s + 2)^2 + 9]^2}$$

and entry 3b of the table of transforms, followed by 9b, gives

$$y = \tfrac{1}{27} (\sin 3t - 3t \cos 3t) e^{-2t}.$$

Example 2. A switch is closed at time $t_0 = 2\pi$ so that a voltage $V(t)$ suddenly changes from its original value 0 to the value $2 \cos t$. At time $t_0 + \pi/4$ the switch is opened again, and the voltage $V(t)$ drops back to 0. By use of the table of transforms, find $\mathbf{L}V$.

Let $H(t)$ denote the *Heaviside unit function*, defined by

$$H(t) = 0 \quad \text{for } t < 0 \qquad H(0) = \tfrac{1}{2} \qquad H(t) = 1 \quad \text{for } t > 0.$$

The function $H(t - 2\pi)$ vanishes for $t < 2\pi$ and equals 1 for $t > 2\pi$; hence

$$H(t - 2\pi)\, 2 \cos t$$

describes the effect of closing the switch at time $t = 2\pi$. Applying similar considerations to $H(t - 2\pi - \pi/4)\, 2 \cos t$, we find that

$$V(t) = H(t - 2\pi)\, 2 \cos t - H(t - \tfrac{9}{4}\pi)\, 2 \cos t \qquad (16\text{-}9)$$

holds, except perhaps at $t = 2\pi$ and $t = \tfrac{9}{4}\pi$, where the physics of the problem is not clear.

The first term of (16-9) equals $H(t - 2\pi)\, 2 \cos (t - 2\pi)$. Since

$$\mathbf{L}[H(t)\, 2 \cos t] = \mathbf{L}(2 \cos t) = \frac{2s}{s^2 + 1}$$

entry 9a of the table gives

$$\mathbf{L}[H(t - 2\pi) \cos (t - 2\pi)] = e^{-2\pi s} \frac{2s}{s^2 + 1}. \qquad (16\text{-}10)$$

---

[1] Because $V(-\log x)$ has a continuous derivative. This is the reason for introducing V in preference to v.

To use a similar method for the second term of (16-9) we must express $2 \cos t$ as a function of the variable $\tau = t - \frac{3}{4}\pi$. Substitution gives

$$2 \cos t = 2 \cos (\tau + \frac{3}{4}\pi) = \sqrt{2} \cos \tau - \sqrt{2} \sin \tau.$$

The second term in (16-9) is therefore

$$\sqrt{2}\, H(t - \frac{3}{4}\pi)[\cos (t - \frac{3}{4}\pi) - \sin (t - \frac{3}{4}\pi)]$$

and its transform is

$$\sqrt{2}\, e^{-\frac{3}{4}\pi s}(\mathbf{L} \cos t - \mathbf{L} \sin t) = \sqrt{2}\, e^{-\frac{3}{4}\pi s} \frac{s - 1}{s^2 + 1}. \tag{16-11}$$

By superposition $\mathbf{L}V$ is the difference of the two expressions (16-10) and (16-11). It is left for the reader to obtain the same result directly from the definition,

$$\mathbf{L}V = \int_{2\pi}^{\frac{3}{4}\pi} 2 \cos t e^{-st}\, dt.$$

PROBLEMS

1. A *full-rectified* and a *half-rectified* wave are defined, respectively, by

$$f_1(t) = |\sin \omega t| \qquad f_2(t) = \frac{1}{2}(|\sin \omega t| + \sin \omega t).$$

Sketch the graph of each function, and obtain the transforms

$$\mathbf{L}f_1(t) = \frac{\omega}{\omega^2 + s^2} \coth \frac{1}{2} \frac{\pi s}{\omega}$$

$$\mathbf{L}f_2(t) = \frac{\omega}{\omega^2 + s^2} \frac{1}{1 - e^{-\pi s/\omega}}.$$

2. Show that the transform of the *sawtooth wave* of Fig. 14 is

$$\frac{a}{s}\left(\frac{1}{bs} - \frac{1}{e^{bs} - 1} \right).$$

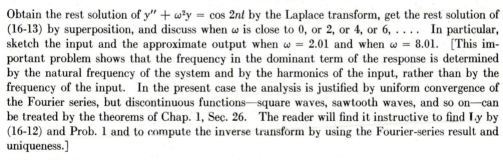

FIGURE 14

3. According to the example in Chap. 1, Sec. 20, the full-rectified wave $|\sin t|$ can be expanded in Fourier series, so that the equation

$$y'' + \omega^2 y = \frac{1}{2}\pi\, |\sin t| \tag{16-12}$$

is equivalent to

$$y'' + \omega^2 y = 1 - \frac{2 \cos 2t}{2^2 - 1} - \frac{2 \cos 4t}{4^2 - 1} - \frac{2 \cos 6t}{6^2 - 1} - \cdots. \tag{16-13}$$

Obtain the rest solution of $y'' + \omega^2 y = \cos 2nt$ by the Laplace transform, get the rest solution of (16-13) by superposition, and discuss when ω is close to 0, or 2, or 4, or 6, ... In particular, sketch the input and the approximate output when $\omega = 2.01$ and when $\omega = 8.01$. [This important problem shows that the frequency in the dominant term of the response is determined by the natural frequency of the system and by the harmonics of the input, rather than by the frequency of the input. In the present case the analysis is justified by uniform convergence of the Fourier series, but discontinuous functions—square waves, sawtooth waves, and so on—can be treated by the theorems of Chap. 1, Sec. 26. The reader will find it instructive to find $\mathbf{L}y$ by (16-12) and Prob. 1 and to compute the inverse transform by using the Fourier-series result and uniqueness.]

4. By entry 12b of the table of transforms, show that

$$\mathbf{L}\left(\frac{\sin \omega t}{\omega t}\right) = \frac{1}{\omega} \cot^{-1} \frac{s}{\omega}, \qquad \mathbf{L}\frac{1 - e^{-\beta t}}{t} = \log \left(1 + \frac{\beta}{s}\right), \qquad \mathbf{L}\left(\frac{1 - \cos \omega t}{t}\right) = \frac{1}{2} \log \left(1 + \frac{\omega^2}{s^2}\right).$$

Also obtain the transforms of the integrals of these functions by means of entry 12a. (Note that the integrals are nonelementary.)

5. Let $f(t) = 1$ for $0 \leq t < 1$, and $f(t) = 0$ elsewhere. Sketch the graph of $f(t)$ for $t \geq 0$, and also the graphs of the functions obtained by integrating once and twice from 0 to t. Find Laplace transforms of all three functions.

6. Discuss the derivation of the first seven entries of the transform table from knowledge of $\mathbf{L}(\sin t)$ and $\mathbf{L}(\cos t)$.

7. Solve $(D^2 + 6D + 7)^2 y = 0$, given $y(0) = y'(0) = y''(0) = 0$, $y'''(0) = 1$.

8. (a)–(d) (*Review.*) Obtain rest solutions for some of the eight second-order inhomogeneous equations in Prob. 2 of Sec. 7.

9. Following the pattern of Prob. 3, discuss $y'' + \omega^2 y = f(t)$, where the periodic function $f(t)$ is given by selected real or complex Fourier series from Secs. 18 to 22 of Chap. 1.

10. Do Prob. 5 of Sec. 12 by the method of Example 2.

---

17. Discontinuities. The Dirac Distribution. Closing a switch in an electrical circuit introduces a discontinuity in the corresponding input function, as we have already seen. A discontinuity may also be produced by a sudden impulse in a mechanical system. The Laplace transform is a most effective means of dealing with such situations, because the transform of many discontinuous functions is just as simple as $\mathbf{L}(e^t)$ or $\mathbf{L}(\sin t)$.

In this section we consider the response of a system to an impulse function which acts over a very short time interval but produces a large effect. The physical situation is typified by a lightning stroke on a transmission line or by a hammer blow on a mechanical system.

To formulate the idea of an impulse, let a be a small positive constant and let $\delta_a(t)$ be the function illustrated in Fig. 15. That is,

$$\delta_a(t) = a^{-1} \qquad \text{for } 0 < t < a$$

and $\delta_a(t) = 0$ elsewhere. The Laplace transform is

$$\mathbf{L}[\delta_a(t)] = \int_0^a a^{-1} e^{-st}\, dt = (sa)^{-1}(1 - e^{-sa}) \equiv h^{-1}(1 - e^{-h})$$

where $h = sa$. By using the Taylor series for e^{-h}, or by applying l'Hospital's rule to the quotient $(1 - e^{-h})/h$, we find that the limit is 1 as $h \to 0$. Thus,

$$\lim_{a \to 0} \mathbf{L}[\delta_a(t)] = 1.$$

It is a conceptual aid to introduce an expression $\delta(t)$, which describes the effect of $\delta_a(t)$ as $a \to 0$, and to say that

$$\mathbf{L}[\delta(t)] = 1. \tag{17-1}$$

We call $\delta(t)$ the *Dirac distribution* or the unit impulse. The legitimacy of this procedure requires discussion, which is given presently. First, however, the use of $\delta(t)$ is illustrated by an example.

Consider a mass-spring system in which the displacement satisfies

$$y'' + y = f(t) \qquad y(0) = 0 \qquad y'(0) = 0$$

where $f(t)$ is the forcing function. To determine the response to a unit impulse at $t = 0$ we replace $f(t)$ by $\delta(t)$; thus $y'' + y = \delta(t)$. The Laplace transform yields

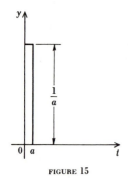

FIGURE 15

$$s^2 \mathbf{L}(y) + \mathbf{L}(y) = \mathbf{L}[\delta(t)] = 1$$

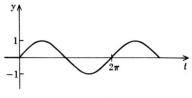

FIGURE 16

when we use (17-1). Hence $\mathbf{L}(y) = 1/(1+s^2)$, or $y = \sin t$ for $t > 0$. The initial conditions require $y = 0$ for $t \le 0$, and the graph has the appearance illustrated in Fig. 16.

The function y is continuous, but it is not differentiable at $t = 0$. Thus, the initial condition $y'(0) = 0$ is *not* satisfied. Indeed, $y'(t) \to 0$ as $t \to 0$ through negative values, but $y'(t) = \cos t \to 1$ as $t \to 0$ through positive values. The unit impulse produces a jump, of magnitude 1, in $y'(t)$.

To clarify the meaning of the result we solve

$$y'' + y = \delta_a(t) \qquad y(0) = y'(0) = 0$$

and then let $a \to 0$. One way of doing this is to write the general solution for $t < 0$, for $0 < t < a$, and for $a < t$. The arbitrary constants are determined by requiring that y and y' be continuous at $t = 0$ and at $t = a$. After a straightforward but tedious calculation one gets

$$y = 0, \qquad y = a^{-1}(1 - \cos t), \qquad y = a^{-1} \sin a \sin t - a^{-1}(1 - \cos a) \cos t$$

on the three intervals, respectively. Since $a^{-1}(1 - \cos a) \to 0$ and $a^{-1} \sin a \to 1$ as $a \to 0$, letting $a \to 0$ gives the same solution, $y = \sin t$, as was obtained in the foregoing discussion.

Although $\delta(t)$ is often called the "Dirac delta function," it is not a function. Indeed, we have already observed that $\mathbf{L}f \to 0$ as $s \to \infty$ for every admissible function f, and δ does not have this property, because $\mathbf{L}\delta = 1$. It is possible to generalize the concept of function and to generalize, correspondingly, the definition of $\mathbf{L}$. The process leads to a branch of mathematics known as the *theory of distributions*. In this theory manipulations with $\delta(t)$ of the type carried out in the foregoing discussion are fully justified.[1]

The basis for the theory is the fact that $\delta(t)$ never occurs alone but only in combination with certain functions, $f(t)$. Instead of engaging in a futile attempt to define $\delta(t)$ for each value of t, we therefore define *the action of $\delta(t)$ on each function $f(t)$*. The process resembles the discussion of the foregoing example, in which the response to $\delta(t)$ was found without investigating $\delta(t)$ itself.

If $f(t)$ is defined at $t = 0$ we agree that the result of applying $\delta(t)$ to $f(t)$ is $f(0)$, and we express this fact in the symbolic notation

$$\int_{-\infty}^{\infty} f(t)\delta(t)\, dt = f(0). \tag{17-2}$$

It is not just $\delta(t)$ but the whole formula on the left that is being defined here. The formula associates the number $f(0)$ with the function f. An operation such as this, which assigns a number to a function, is called a *functional*.

Functionals are often written by use of the notation $\mathbf{N}f$, where $\mathbf{N}$ is the functional and $\mathbf{N}f$ is the value of the functional on f. For example, (17-2) is equivalent to $\mathbf{N}f = f(0)$, where $\mathbf{N}$ is a functional denoted symbolically by the integral together with $\delta(t)$. The left side of (17-2) indicates that the functional is being applied to f.

Another example of a functional is described by the formula

$$\mathbf{N}f = \int_{-\infty}^{\infty} f(t)g(t)\, dt \tag{17-3}$$

[1] See M. J. Lighthill, "Introduction to Fourier Analysis and Generalized Functions," Cambridge University Press, New York, 1958, and A. H. Zemanian, "Distribution Theory and Transform Analysis," McGraw-Hill Book Company, New York, 1965.

where $g(t)$ is a given integrable function; for example, $g(t) = 1$, or $g(t) = e^t$. Here the integral is an ordinary integral, as in elementary calculus.

The functional (17-3) is said to be *represented* by the function $g(t)$. The functional defined by $Nf = f(0)$ cannot be represented by an ordinary function in this way, but it is symbolically represented by the distribution $\delta(t)$. In this sense the concept of distribution is a generalization of the concept of function, and an integral containing a distribution is a generalization of an ordinary integral. The process of definition and generalization is now discussed more fully.

If $\delta(t)$ were a function, and if the integral (17-2) were an ordinary integral, a change of variable would give

$$\int_{-\infty}^{\infty} \delta(t - t_0) f(t)\, dt = \int_{-\infty}^{\infty} \delta(t) f(t + t_0)\, dt. \tag{17-4}$$

According to (17-2) the right-hand side of (17-4) is the value of $f(t - t_0)$ at $t = 0$, namely, $f(t_0)$. This is now taken as the definition of the left-hand member of (17-4). Thus, $\delta(t - t_0)$ represents a functional that assigns the number $f(t_0)$ to the function f.

Similarly, if $\delta(t)$ were a continuously differentiable function vanishing for large $|t|$, an integration by parts would give

$$\int_{-\infty}^{\infty} \delta'(t) f(t)\, dt = -\int_{-\infty}^{\infty} \delta(t) f'(t)\, dt. \tag{17-5}$$

According to (17-2) the right-hand side reduces to $-f'(0)$. This is now taken as the definition of the left-hand side. Hence $\delta'(t)$ represents a functional that assigns the value $-f'(0)$ to the function f. The same line of thought is readily extended to higher derivatives and to the argument $t - t_0$ as in (17-4). The result is

$$\int_{-\infty}^{\infty} \delta^{(n)}(t - t_0) f(t)\, dt = (-1)^n f^{(n)}(t_0) \tag{17-6}$$

provided $f^{(n)}(t)$ exists[1] at $t = t_0$.

Appropriate definitions also allow integration over an interval (a,b) rather than $(-\infty,\infty)$. Let $g(t) = 1$ for $a < t < b$ and let $g(t) = 0$ elsewhere in $(-\infty,\infty)$. If the integral were an ordinary integral, and if $\delta(t)$ were a function, we would have

$$\int_a^b f(t) \delta(t - t_0)\, dt = \int_a^b f(t) g(t) \delta(t - t_0)\, dt = \int_{-\infty}^{\infty} f(t) g(t) \delta(t - t_0)\, dt.$$

Now, according to the discussion of (17-4) the integral on the right is the value of $f(t)g(t)$ at $t = t_0$. This gives $f(t_0)$ if t_0 is on the interval (a,b), since $g(t_0) = 1$ in that case. Otherwise $g(t_0) = 0$, and $f(t_0)g(t_0) = 0$.

In view of these considerations, we adopt the following convention in (17-6) when the interval of integration is (a,b) rather than $(-\infty,\infty)$. If $a < t_0 < b$ we define the result to be the right side of (17-6), and if $t_0 < a$ or $t_0 > b$ we define the result to be 0. Naturally, a similar convention holds for other integral expressions.

When t_0 is an end point, $t_0 = a$ or $t_0 = b$, the appropriate definition depends on circumstances. In the Laplace transform one should regard $t_0 = 0$ as being obtained by letting $t_0 \to 0$ through positive values. This has the effect of treating $t_0 = 0$ as an interior point of the interval of integration. [Cf. Fig. 15, and recall the use of *right-hand* limits in the formula for $\mathbf{L}(y^{(n)})$.]

As the reader may have observed, every definition is formulated so that $\delta(t)$ behaves much like a function vanishing outside the immediate neighborhood of 0. It is no

---

[1] In the theory of distributions $\delta^n(t)$ is defined for more general classes of functions f by use of the so-called *dual space*. See L. Schwartz, "La Théorie de distribution," Hermann & Cie, Paris, 1950. See also B. Friedman, "Principles and Techniques of Applied Mathematics," chap. 3, John Wiley & Sons, Inc., New York, 1956.

wonder, therefore, that manipulation with $\delta(t)$, as if it were such a function, usually leads to correct answers.

Example.   Discuss the formula $\mathbf{L}\delta(t) = 1$, and, more generally,

$$\mathbf{L}[\delta^{(n)}(t)] = s^n \qquad n = 0, 1, 2, 3, \ldots . \tag{17-7}$$

We apply (17-6) with $f(t) = e^{-st}$. Since $f^{(n)}(t) = (-s)^n e^{-st}$ the result is

$$\int_0^\infty \delta^{(n)}(t) e^{-st}\, dt = (-1)^n (-s)^n e^{-st}\Big|_{t=0} = s^n.$$

What is really involved here is a definition: We agree to interpret the Laplace integral in the sense of the preceding discussion, when the integrand involves $\delta(t)$.

PROBLEMS

1. The voltage V of a certain circuit satisfies

$$V'' + 4V' + 3V = E(t)$$

where E is the applied voltage. Find the response of the system to a unit impulse at $t = 0$ if $V = 0$ for $t < 0$.

2. (*a*) Solve the equations $y' = \delta(t)$, $y'' = \delta(t)$, $y''' = \delta(t)$, assuming that $y = 0$ for $t < 0$ and that y and as many derivatives as possible are continuous. (*b*) Show that y, y', and y'' have a jump of value 1 at $t = 0$ in the three cases, respectively.

3. Discuss continuity of solutions of

$$y''' = f(t) \qquad \text{for } t > 0 \qquad y = 0 \qquad \text{for } t \leq 0$$

when $f(t) = t$, 1, $\delta(t)$, $\delta'(t)$, or $\delta''(t)$, respectively.

4. A functional $\mathbf{N}$ is said to be *linear* if $\mathbf{N}(\alpha f + \beta g) = \alpha \mathbf{N}f + \beta \mathbf{N}g$ for all constants α and β and all functions f and g such that $\mathbf{N}f$ and $\mathbf{N}g$ make sense. The functional represented by $\delta(t)$ in (17-2) is linear, because the value of $\alpha f(t) + \beta g(t)$ at $t = 0$ is $\alpha f(0) + \beta g(0)$. Which of the following describe linear functionals?

$$\mathbf{N}f = f(0) + f(3) \qquad \mathbf{N}f = f(2) + \int_0^1 f(t)\, dt \qquad \mathbf{N}f = [f(1)]^2 \qquad \mathbf{N}f = f''(5) + \int_0^\infty e^{-t} f(3t)\, dt.$$

5. We agree that $\alpha(t)\delta(t - t_0) + \beta(t)\delta(t - t_1)$ represents the functional that assigns the value $\alpha(t_0)f(t_0) + \beta(t_1)f(t_1)$ to the function f. Write this as a statement about integrals, and compute

$$\int_{-\infty}^\infty \sin t \left[3\delta\left(t - \frac{\pi}{2}\right) - 4t\delta\left(t + \frac{\pi}{2}\right) \right] dt.$$

[*Ans.:* $3 \sin \frac{1}{2}\pi - 4(-\frac{1}{2}\pi) \sin (-\frac{1}{2}\pi) = 3 - 2\pi$. Other linear expressions, such as

$$\alpha(t)\delta(t - t_0) + \beta(t)\delta'(t - t_1) \qquad \text{or} \qquad \alpha(t)\delta(t) + \beta(t)$$

are interpreted similarly.]

6. Instead of the particular function $\delta_a(t)$, one could consider any positive integrable function $\eta_a(t)$ that vanishes outside the interval $0 < t < a$ and has unit area. (*a*) Sketch such a function. (*b*) Using the inequality

$$e^{-sa}\eta_a(t) \leq e^{-st}\eta_a(t) \leq \eta_a(t) \qquad s > 0$$

show that $\mathbf{L}\eta_a$ lies between e^{-sa} and 1, and hence $\mathbf{L}\eta_a \to 1$ as $a \to 0$.

7. (*a*) Given $y'' + p_1 y' + p_0 y = f(t)$, where f is admissible, show that the effect of replacing $f(t)$ by $f(t) + \alpha\delta(t)$ is exactly the same as the effect of increasing the initial value $y'(0)$ by the constant α. (*b*) Discuss a corresponding assertion for equations of order n.

8. (*a*)–(*i*) (*Review.*) Solve selected higher-order equations among the eighteen in Prob. 2 of Sec. 11, taking $y(0) = 1$ and 0 for the other initial values.

18. The Convolution.   Consider a linear system in which the effect at the present time t of a stimulus $f(\tau)\,d\tau$ at any past time τ is proportional to the stimulus. On physical grounds we assume that the proportionality constant depends only on the elapsed time $t - \tau$ and hence has the form $g(t - \tau)$.   The effect at the present time t is therefore

$$f(\tau)g(t - \tau)\,d\tau.$$

Since the system is linear the response to the whole past history can be obtained by adding these separate effects, and we are led to the integral

$$h(t) = \int_0^t f(\tau)g(t - \tau)\,d\tau. \tag{18-1}$$

The lower limit is 0 because it is assumed that the process started at time $t = 0$; in other words, $f(\tau) = 0$ for $\tau < 0$.

The expression (18-1) is called the *convolution* of f and g.   It gives the response at the present time t as a weighted superposition over the inputs at the times $\tau \leq t$.   The weighting factor $g(t - \tau)$ characterizes the system, and $f(\tau)$ characterizes the past history of the input.   Because of this physical interpretation, the convolution is sometimes called the *superposition integral*.

The importance of the convolution can hardly be overestimated.   It plays a prominent role in the study of heat conduction, wave motion, plastic flow and creep, and other areas of mathematical physics (cf. Chap. 7).   The convolution is also encountered in several branches of mathematics; for example, a large part of the theory of distributions can be based on (18-1) rather than on the concept of a linear functional.[1]

The function h in (18-1) is often denoted by $f * g$, so that $(f * g)(t) = h(t)$ is the value of $f * g$ at the argument t.   A similar notation applies in other cases; for example,

$$(a * b)(t) = \int_0^t a(\tau)b(t - \tau)\,d\tau. \tag{18-2}$$

The $*$ product behaves like ordinary multiplication, in that

$$a * (b + c) = a * b + a * c \qquad a * b = b * a \qquad (a * b) * c = a * (b * c). \tag{18-3}$$

(See Prob. 6.)   We now show that convolution in the time domain actually does correspond to multiplication in the frequency domain.   This is the content of the following:

CONVOLUTION THEOREM.   If f and g are admissible then $\mathbf{L}(f * g) = (\mathbf{L}f)(\mathbf{L}g)$.

To simplify the proof, define $f(t) = 0$ for all negative t.   With a similar convention for $g(t)$ the Laplace transforms can be written[2]

$$\mathbf{L}f = \int_{-\infty}^{\infty} e^{-st}f(t)\,dt \qquad \mathbf{L}g = \int_{-\infty}^{\infty} e^{-st}g(t)\,dt \tag{18-4}$$

and the convolution $h(t)$ is

$$h(t) = \int_{-\infty}^{\infty} f(\tau)g(t - \tau)\,d\tau. \tag{18-5}$$

Indeed, the lower limit $-\infty$ in (18-5) can be replaced by 0 because $f(\tau) = 0$ when $\tau < 0$, and the upper limit can be replaced by t because $g(t - \tau) = 0$ when $t - \tau < 0$.   The convolution theorem now follows by a change in order of integration, as indicated next.

---

[1] This alternative approach is due to Mikusinski.   See A. Erdélyi, "Operational Calculus and Generalized Functions," Holt, Rinehart and Winston, Inc., New York, 1962.
[2] Transforms of the type (18-4) are called *bilateral*, in contrast to the *unilateral* transform (12-1).   An account of the bilateral Laplace transform may be found in B. Van der Pol and H. Bremmer, "Operational Calculus," Cambridge University Press, London, 1950.

It is evident from (18-1) that $h(t) = 0$ for $t < 0$ and so

$$\mathbf{L}h = \int_{-\infty}^{\infty} \left[\int_{-\infty}^{\infty} f(\tau)g(t - \tau)\, d\tau \right] e^{-st}\, dt = \int_{-\infty}^{\infty} \left[\int_{-\infty}^{\infty} g(t - \tau)e^{-st}\, dt \right] f(\tau)\, d\tau.$$

The inversion of order is justified for large s by absolute convergence. If the variable t in the inner integral is replaced by $t = x + \tau$ we get

$$\int_{-\infty}^{\infty} \left[\int_{-\infty}^{\infty} g(x)e^{-s(x+\tau)}\, dx \right] f(\tau)\, d\tau = \left[\int_{-\infty}^{\infty} e^{-s\tau}f(\tau)\, d\tau \right] \left[\int_{-\infty}^{\infty} e^{-sx}g(x)\, dx \right]$$

since $e^{-s(x+\tau)} = e^{-sx}e^{-s\tau}$. The result is $(\mathbf{L}f)(\mathbf{L}g)$.

To illustrate the convolution theorem, consider the problem

$$y'' + \omega^2 y = \omega^2 f(t) \qquad y(0) = y'(0) = 0$$

where ω is constant and f is admissible. The transform gives

$$\mathbf{L}y = (\mathbf{L}f)\omega^2(\omega^2 + s^2)^{-1}.$$

By the table of transforms, $\omega^2(\omega^2 + s^2)^{-1} = \mathbf{L}g$, with $g(t) = \omega \sin \omega t$; hence

$$\mathbf{L}y = (\mathbf{L}f)(\mathbf{L}g) = \mathbf{L}(f * g)$$

in view of the convolution theorem. Uniqueness now gives $y = f * g$, or

$$y = \omega \int_0^t f(\tau) \sin \omega(t - \tau)\, d\tau. \tag{18-6}$$

Here we have a formula for the solution with an *arbitrary* forcing function $f(t)$. Naturally the result contains all the solutions obtained by special devices in the foregoing pages.

The same method applies to the general equation

$$y^{(n)} + p_{n-1}y^{(n-1)} + \cdots + p_1y' + p_0y = f(t) \tag{18-7}$$

or $P(D)y = f(t)$, where the coefficients are constant and $P(s)$ is a characteristic polynomial (see Sec. 9). Thus, if g is a function such that

$$\mathbf{L}g = [P(s)]^{-1} \tag{18-8}$$

the transformed equation for the rest solution of (18-7) is

$$\mathbf{L}y = (\mathbf{L}f)[P(s)]^{-1} = (\mathbf{L}f)(\mathbf{L}g) = \mathbf{L}(f * g)$$

and $y = f * g$ as before. The function g is the rest solution of

$$g^{(n)} + p_{n-1}g^{(n-1)} + \cdots + p_1g' + p_0g = \delta(t)$$

because the latter satisfies $P(s)\mathbf{L}g = 1$ by definition.

Determination of g is accomplished by partial fractions. In particular, if $P(s)$ has simple roots s_k then

$$\frac{1}{P(s)} = \sum \frac{A_k}{s - s_k} = \sum A_k \mathbf{L}(e^{s_k t}) = \mathbf{L}\left(\sum A_k e^{s_k t} \right) \tag{18-9}$$

and, hence, $g(t)$ is the sum on the right. Substitution into $f * g$ gives *Heaviside's expansion theorem* for the rest solution of (18-7):

$$y = \int_0^t f(\tau)[A_1 e^{s_1(t-\tau)} + A_2 e^{s_2(t-\tau)} + \cdots + A_n e^{s_n(t-\tau)}]\, d\tau. \tag{18-10}$$

Since $(f * g)(t)$ involves the values of f and g only on the finite interval $(0,t)$, the assumption that f and g are admissible can be replaced by the weaker assumption that f and g are piecewise continuous on every finite interval. For example, the conclusions hold for $f(t) = e^{t^2}$ or e^{e^t}, though neither of these functions has a Laplace transform.

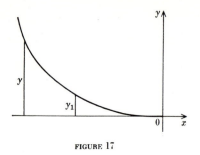

FIGURE 17

To prove this assertion, let us show that $y = f * g$ holds on a given finite interval $0 \le t \le t_0$. If we redefine f and g so that $f(t) = g(t) = 0$ for $t > t_0$, the new functions are admissible, and the foregoing analysis applies. This establishes the conclusion for $0 \le t \le t_0$; since t_0 is arbitrary, the seeming restriction is, in fact, no restriction at all. Many other results can be extended in the same way; hence the method of Laplace transformation has greater scope than appears at first sight.

We conclude with an example that shows how the convolution theorem is used in the solution of integral equations.

Example.   Starting from rest, a particle slides down a frictionless curve under gravity (see Fig. 17). It is required to determine the shape of the curve so that the time of descent will be independent of the starting point. A curve of this sort is called a *tautochrone*.

If the particle starts at a height y, its velocity v when the height is $y_1 = \eta$ can be found by equating potential and kinetic energies. The result is

$$\tfrac{1}{2}mv^2 = mg(y - \eta) \qquad \text{or} \qquad v = (2g)^{1/2}(y - \eta)^{1/2} \tag{18-11}$$

where g is the acceleration of gravity. Denoting the arc along the curve by σ, we see that the time for descent is

$$\int \frac{d\sigma}{v} = \int \frac{1}{v}\frac{d\sigma}{d\eta}\, d\eta = \int \frac{1}{v} f(\eta)\, d\eta$$

where $f(\eta)$ stands for $d\sigma/dy$ at $y = \eta$. Since the time is constant and since v is given by (18-11), the problem reduces to

$$\int_0^y f(\eta)(y - \eta)^{-1/2}\, d\eta = c_0$$

where c_0 is constant. This is an *integral equation* for f.

Taking the Laplace transform gives $L(f)L(y^{-1/2}) = L(c_0)$, or

$$L(f)(-\tfrac{1}{2})!\, s^{-1/2} = c_0 s^{-1}.$$

This gives $L(f) = c_1 s^{-1/2}$, where c_1 is constant, and hence $f(y) = cy^{-1/2}$, where c is constant. Thus we are led to the differential equation

$$f(y) = \frac{d\sigma}{dy} = \left[1 + \left(\frac{dx}{dy}\right)^2\right]^{1/2} = cy^{-1/2}.$$

If we set $y = c^2 \sin^2 \tfrac{1}{2}\phi$, a short calculation yields

$$x = \tfrac{1}{2}c^2(\phi + \sin \phi) \qquad y = \tfrac{1}{2}c^2(1 - \cos \phi)$$

which are the parametric equations of a cycloid.

PROBLEMS

Here f is admissible and α, β, ω, ω_1 are constant.

1. Find the rest solution: $y' + y = f(t)$, $y'' - y = f(t)$, $y'' + 2y' + 2y = f(t)$.

2. Find the rest solution of $y'' + (\alpha + \beta)y' + \alpha\beta y = f(t)$ when $\alpha \ne \beta$.

3. Solve $y'' = f(t)$, and $y'' + 2\alpha y' + \alpha^2 y = f(t)$, by the convolution theorem.

4. By solving $y^{(n)} = f(t)$ in two ways, prove that

$$\int_0^t \int_0^t \cdots \int_0^t f(t)(dt)^n = \int_0^t \frac{(t - \tau)^{n-1}}{(n-1)!} f(\tau)\, d\tau.$$

(On the left, the upper limit of each integral becomes the variable of integration for the next.)

5. Solve $y'' + \omega^2 y = \omega^2 \sin \omega_1 t$ by (18-6), and check by the method of undetermined coefficients.

6. Assuming a, b, c admissible, deduce the algebraic properties (18-3) from the uniqueness theorem for the Laplace transform.   *Hint:*

$$\mathbf{L}(a * b) = (\mathbf{L}a)(\mathbf{L}b) = (\mathbf{L}b)(\mathbf{L}a) = \mathbf{L}(b * a).$$

7. If $f(t) = t^{a-1}$ and $g(t) = t^{b-1}$ in Eq. (18-1), let $\tau = tx$ and thus obtain

$$f * g = t^{a+b-1} \int_0^1 x^{a-1}(1 - x)^{b-1} \, dx.$$

Deduce the *Euler formula for the beta function,*

$$B(a,b) = \int_0^1 x^{a-1}(1 - x)^{b-1} \, dx = \frac{(a - 1)!(b - 1)!}{(a + b - 1)!}.$$

Hint: By the convolution theorem applied to the first result,

$$\mathbf{L}[B(a,b)t^{a+b-1}] = \mathbf{L}(t^{a-1})\mathbf{L}(t^{b-1}).$$

8. By taking $a = b = \frac{1}{2}$ in the preceding problem, and letting $t = \sin^2 \theta$, show that $B(\frac{1}{2},\frac{1}{2}) = \pi$ and hence $(-\frac{1}{2})! = \sqrt{\pi}$.   Referring to Chap. 1, Sec. 14, deduce that

$$\int_0^\infty e^{-s^2} \, ds = \frac{1}{2}\sqrt{\pi}.$$

9. Let $h(t)$ be the rest solution of $P(D)h = H(t)$, where $P(D)$ is as in the text and $H(t)$ is the *Heaviside unit function:*

$$H(t) = 0 \quad \text{for } t < 0 \qquad H(t) = 1 \quad \text{for } t \geq 0.$$

If y is the rest solution of $P(D)y = f(t)$, deduce $\mathbf{L}y = (s\mathbf{L}f)(\mathbf{L}h)$, and hence, by entry 11$a$ of the table of transforms and the convolution theorem,

$$y = \int_0^t f'(\tau)h(t - \tau) \, d\tau + f(0)h(t).$$

[This is the *Heaviside superposition principle.*   It holds for rest solutions if $f(t)$ is continuous and $f'(t)$ is piecewise continuous on every finite interval $0 \leq t \leq t_0$.]

10. Find the steady-state solution of $y'' + 3y' + 2y = f(t)$ by Heaviside's expansion theorem, evaluate explicitly when $f(t) = H(t)$, e^{3t}, and t, and, referring to Prob. 9, obtain the solutions for e^{3t} and for t from that for $H(t)$.

---

19. Mathematical Justification.

In this section we establish the existence of solutions of

$$y^{(n)} + p_{n-1}y^{(n-1)} + \cdots + p_1 y' + p_0 y = f(t) \qquad a < t < b \qquad (19\text{-}1)$$

and also the existence of their Laplace transforms.   For mathematical purposes, a *solution* of (19-1) is defined to be a function $y = y(t)$ such that the equation holds at all points where f is continuous and such that

$$y, y', y'', \ldots, y^{(n-1)} \text{ are continuous for } a < t < b. \qquad (19\text{-}2)$$

At points where $f(t)$ is discontinuous we allow Eq. (19-1) to fail.   The condition (19-2) means that the whole of the bad behavior of f is absorbed, so to speak, by the highest derivative $y^{(n)}$.

Similar considerations apply when f is a distribution or a combination of functions and distributions.   As an illustration, when $f(t) = \delta(t - t_0)$ one would require continuity of all derivatives except $y^{(n)}$ and $y^{(n-1)}$ at t_0.   When $f(t) = \delta'(t - t_0)$ one would require continuity of all except $y^{(n)}$, $y^{(n-1)}$, and $y^{(n-2)}$ at t_0, and so on.   It is possible to extend the following analysis to such cases and also to systems, though this is not done here.

By defining $f(t) = 0$ outside the interval (a,b) one can reduce the general case to the case $-\infty < t < \infty$; hence we take $a = -\infty$, $b = \infty$ in (19-1). For simplicity, f is assumed piecewise continuous on every finite interval.

We begin our discussion with the special case

$$(D - s)y = f(t) \qquad s = \text{const.} \tag{19-3}$$

If (19-3) is multiplied by e^{-st} the resulting equation can be written in the form

$$D(e^{-st}y) = e^{-st}f(t). \tag{19-4}$$

Since y is continuous, we conclude that $e^{-st}y$ is given uniquely by integration, if $y(0)$ is known, and hence

$$y = e^{st} \int_0^t e^{-s\tau}f(\tau)\,d\tau + ce^{st} \tag{19-5}$$

where $c = y(0)$ is constant.

The operator (19-5) that transforms f into the solution y depends both on s and on the constant of integration, c. We denote this operator by $\mathbf{T}(s,c)$, so that

$$\mathbf{T}(s,c)f \equiv e^{st}\int_0^t e^{-s\tau}f(\tau)\,d\tau + ce^{st}. \tag{19-6}$$

The result of our analysis is that y *is a solution of* $(D - s)y = f$ *if, and only if,* $y = \mathbf{T}(s,c)f$ *for some constant* c.

One can solve the equation

$$(D - s_2)(D - s_1)y = f(t) \tag{19-7}$$

by setting $f_1 = (D - s_1)y$. With due account of the definition of *solution* as applied to (19-7), it is found that

$$y = \mathbf{T}(s_1,c_1)f_1 \qquad \text{and} \qquad f_1 = \mathbf{T}(s_2,c_2)f \tag{19-8}$$

where $c_1 = f_1(0)$ and $c_2 = f(0)$. Conversely, every such y is a solution of (19-7) in our sense. The result (19-8) is equivalent to

$$y = \mathbf{T}(s_1,c_1)\mathbf{T}(s_2,c_2)f.$$

In just the same way, y *is a solution of*

$$(D - s_n)(D - s_{n-1}) \cdots (D - s_1)y = f(t) \tag{19-9}$$

if, and only if, y has the form

$$y = \mathbf{T}(s_1,c_1)\mathbf{T}(s_2,c_2) \cdots \mathbf{T}(s_n,c_n)f \tag{19-10}$$

for some choice of the constants $c_1, c_2, \ldots, c_n$. Since Eq. (19-1) can always be put in the form (19-9), the formula represents the complete solution of (19-1).

Equation (19-10) has advantages in numerical work, because the operator $\mathbf{T}(s,c)$ involves integration over a finite range only. Also, $\mathbf{T}(s,c)$ can be programmed once for all, and the iteration (19-10) is a uniform procedure well suited to automation. It should be emphasized that the roots need not be distinct; the values s_k are just repeated as often as necessary. By contrast, the method of undetermined coefficients requires the extra factor t^m, in the case of a multiple root, while the method of Laplace transformation requires a different form for the partial-fraction expansion.

To see the relevance of (19-10) to the theory of the Laplace transform, recall that the sum and product of two admissible functions is admissible, and so is the integral of an admissible function. Since $\mathbf{T}(s,c)f$ is generated by applying these three operations to f and to the admissible functions e^{st} and e^{-st}, we conclude that $\mathbf{T}(s,c)f$ *is admissible whenever f is admissible.* It follows at once from (19-10) that y *is admissible if f is admissible,* and that is the main assertion in the existence theorem of Sec. 12.

To show that the derivatives are admissible we use mathematical induction. For $n = 1$ we have $y' = f - p_0 y$; hence, since f and y are admissible, so is y'. Assume now that the result holds for equations of order $n - 1$. If we set $u = y'$ then (19-1) is an equation of order $n - 1$ in u, with the right-hand member $f - p_0 y$. Since $f - p_0 y$ is admissible, the induction hypothesis indicates that u together with its first $n - 1$ derivatives is admissible. This shows that y together with its first n derivatives is admissible and completes the proof.

It is not difficult to prove that the constants $c_1, c_2, \ldots, c_n$ in (19-10) are uniquely determined by the initial values

$$y(0) = y_0, \qquad y'(0) = y_1, \ldots, y^{(n-1)}(0) = y_{n-1}. \tag{19-11}$$

We conclude that (19-1) *has one, and only one, solution satisfying* (19-11). The same holds if the initial values are prescribed at the arbitrary time t_0, rather than at time $t = 0$, since the change of variable $t \to t - t_0$ reduces that case to the case just considered. Hence (19-10) yields the basic existence and uniqueness theorem for (19-1) without reference to Chap. 2 and without the theory of linear dependence developed in Sec. 24.

The foregoing proof hinges on the possibility of factoring the operator $F(D)$, and this in turn follows from the fundamental theorem of algebra. Since the latter requires complex numbers, we see that the use of complex solutions is not just an engineering convenience but is intimately bound up with the mathematical structure of the problem.

We conclude this discussion with the following useful theorem, which simplifies the determination of particular solutions and allows an efficient treatment of repeated roots:

EXPONENTIAL SHIFT THEOREM.   *If $P(D)$ is a polynomial in the differentiation symbol D, then*

$$P(D)e^{st}u(t) = e^{st}P(D + s)u(t)$$

for any constant s and any function u such that $P(D)u$ is defined.

In other words, the factor e^{st} can be taken out from under the differential operator $P(D)$, provided D is replaced by $D + s$. The statement is readily verified for the special case $P(D) = D^k$, by mathematical induction. That is,

$$D^k e^{st}u(t) = e^{st}(D + s)^k u(t)$$

for $k = 0, 1, 2, 3, \ldots$. Multiplying by $p_k(t)$ we obtain the theorem for $P(D) = p_k(t)D^k$, and it then follows by linearity for a sum of such terms. The coefficients $p_k(t)$ need not be constant.

Several uses of the shift theorem are illustrated in the following examples.

Example 1. To solve $(D + 3)(D - 1)^3 y = 48te^t$ we try $y = u(t)e^t$. By the shift theorem with $s = 1$,

$$(D + 3)(D - 1)^3 ue^t = e^t(D + 4)D^3 u.$$

This satisfies the given equation if $(D + 4)D^3 u = 48t$, or

$$(D + 4)v = 48t \qquad \text{where } D^3 u = v.$$

By inspection $v = 12t - 3$, integrating three times gives u, and then $y = ue^t$. The general solution is therefore

$$y = (\tfrac{1}{2}t^4 - \tfrac{1}{2}t^3)e^t + c_1 e^{-3t} + (c_2 + c_3 t + c_4 t^2)e^t.$$

Example 2. Let $P(s)$ be a polynomial, with constant coefficients, having a root $s = r$ of multiplicity m. If $u(t)$ is any polynomial in t of degree less than m, prove that $y = u(t)e^{rt}$ satisfies $P(D)y = 0$.

The hypothesis implies $P(D) = Q(D)(D - r)^m$, where Q is a polynomial, and hence $y = e^{rt}u(t)$ gives

$$P(D)y = e^{rt}P(D + r)u = e^{rt}Q(D + r)D^m u$$

by the shift theorem.   If u is a polynomial of degree less than m, then $D^m u \equiv 0$, and so $P(D)y = 0$.

This is the main theorem concerning multiple roots for the homogeneous equation.   More generally, if the equation is

$$P(D)y \equiv Q(D)(D - r)^m y = A(t)e^{rt}$$

rather than $P(D)y = 0$, the substitution $y = u(t)e^{rt}$ leads to

$$Q(D + r)v = A(t) \qquad \text{where } v = D^m u.$$

When $A(t)$ is a polynomial it is not hard to show that v can be chosen to be a polynomial of the same degree, and an m-fold integration of that polynomial then gives u.   We conclude that *the given equation has a solution of the form*

$$y = t^m a(t)e^{rt}$$

where $a(t)$ is a polynomial of the same degree as $A(t)$.   This justifies the method of undetermined coefficients as presented in Secs. 6 to 9.

Example 3.   Discuss the resonant solutions, as $t \to \infty$, of

$$(D + 3)(D^2 + 1)^5 y = 128 \cos \omega t. \tag{19-12}$$

For resonance, $\omega = 1$.   Since $\cos t$ is the real part of e^{it} we consider, instead of (19-12), the equation

$$(D + 3)(D + i)^5 (D - i)^5 y = 128 e^{it}. \tag{19-13}$$

The choice $y = u(t)e^{it}$ gives

$$e^{it}(D + 3 + i)(D + 2i)^5 D^5 u = 128 e^{it}$$

by the shift theorem or, writing $v = D^5 u$,

$$(D + 3 + i)(D + 2i)^5 v = 128.$$

We try $v = a$, where a is constant, and get $(3 + i)(2i)^5 a = 128$.   The equation gives $v = -\tfrac{2}{5} - \tfrac{6}{5}i$, integrating five times gives u, and hence

$$y = \left(-\frac{2}{5} - \frac{6}{5}i\right)\frac{t^5}{5!}e^{it}.$$

This satisfies (19-13), and its real part satisfies (19-12) with $\omega = 1$.   Although the problem is not stable, the solution describes the dominant behavior as $t \to \infty$, since the complementary function has terms of order t^4 at most.

PROBLEMS

1. (a) If $(D^2 + 2D + 2)y = e^{-t} \cos t$, show that the choice $y = e^{-t}u(t)$ gives $(D^2 + 1)u = \cos t$ by the shift theorem.   Obtain $u = \tfrac{1}{2}t \sin t$ by the Laplace transform, and thus solve the original equation.   (b) Solve the original equation by direct application of the Laplace transform, without using the shift theorem.

2. Describe the steady-state behavior:

$$(D + 1)(D + 4)^3 y = e^{-4t}, \quad (D^3 + 2D + 7)y = -24e^t \cos 2t, \quad (D^2 + 2D + 2)(D + 1)^5 y = te^{-t}$$

3. Describe the resonant solutions:

$$(D^2 + 1)(D^2 + 4)^2 y = \cos \omega t \qquad (D + 2)(D^2 + 1)^3 y = \sin \omega t.$$

4. Show that the choice $c_1 = c_2 = \cdots = c_n = 0$ in the text gives the rest solution, and thus get the rest solution of $(D^2 + 3D + 2)y = f(t)$.

REVIEW PROBLEMS

5. The current I in an RL circuit satisfies $LI' + RI = V$, where $V = V(t)$ is the applied voltage. At time $t = 0$ a switch is closed, so that V suddenly assumes the value $V_0 + A \sin \omega t$.   (Here L, R, V_0, A, and ω are constants.)   By use of the Laplace transform, find I for $t > 0$.

6. Find the response of the circuit in Prob. 5 to a unit impulse at time $t = 0$, assuming that $V = 0$ for $t < 0$.

7. Find the steady-state solution in Prob. 1 when V is an arbitrary admissible function by (a) the Heaviside expansion theorem and (b) the superposition principle of Sec. 18, Prob. 9.

8. If $\mathbf{L}(y) = F(s)$, use entry 11b of the table of transforms to get

$$\mathbf{L}(ty) = -F', \qquad \mathbf{L}(ty') = -(sF)', \qquad \mathbf{L}(ty'') = -(s^2F)' + y(0).$$

9. A function y satisfies $ty'' + y' + ty = 0$ and has a Laplace transform $\mathbf{L}(y) = F(s)$. By use of Prob. 8, show that

$$F'(1 + s^2) = -sF$$

and thus deduce that $y = cJ_0(t)$, where c is constant. [The equation in this problem has variable coefficients, and $J_0(t)$ is a nonelementary function called the *Bessel function of order* 0. Discussion of these matters forms the topic of the next five sections.]

10. By the Laplace transform, obtain rest solutions (a) in Example 1, (b) in Prob. 2, (c) in Prob. 3, and (d) in Prob. 3 of Sec. 11.

11. Discuss the solution of Example 3 by the Laplace transform. (Cf. entry 9b of the table of transforms.)

12. If a_{ij}, A_i, and ω are constant (a) obtain $\mathbf{L}y_1$ and $\mathbf{L}y_2$ from each of the systems

$$\begin{cases} y_1' = a_{11}y_1 + a_{12}y_2 + A_1 \cos \omega t \\ y_2' = a_{21}y_1 + a_{22}y_2 + A_2 \cos \omega t \end{cases} \qquad \begin{cases} y_1'' = a_{11}y_1 + a_{12}y_2 + A_1 \cos \omega t \\ y_2'' = a_{21}y_1 + a_{22}y_2 + A_2 \cos \omega t \end{cases}$$

and check by reference to Sec. 15. (b) A system is said to be *lossless* if the characteristic roots are pure imaginary. Show that the first system is stable and has real roots if, and only if, the second system is lossless. (c) Show that another system with coefficients b_{ij} instead of a_{ij} has the same characteristic roots as the original system if, and only if,

$$a_{11} + a_{22} = b_{11} + b_{22} \qquad \text{and} \qquad a_{11}a_{22} - a_{12}a_{21} = b_{11}b_{22} - b_{12}b_{21}.$$

(d) Generalize parts a, b, and c to systems in three unknowns y_i.

EQUATIONS WITH VARIABLE COEFFICIENTS

20. Series Solutions. Legendre Functions.   The *general linear equation* has the form

$$y^{(n)} + p_{n-1}(x)y^{(n-1)} + \cdots + p_1(x)y' + p_0(x)y = f(x) \tag{20-1}$$

where the coefficients p_k need not be constant but can depend on x. We write the variable as x and use primes to denote differentiation with respect to x, because these equations are often encountered in boundary-value problems, in which the variable is a space variable rather than the time t. Also the choice of x as variable agrees with the notation used for power series in Chap. 1.

If (20-1) is written in the form $\mathbf{T}y = f$, a discussion virtually identical with that given for constant coefficients in Sec. 2 shows that the operator $\mathbf{T}$ is linear. Hence the general principle of superposition, the principle of equating real parts, the principle of asymptotic stability, and all other theorems stated for general linear operators in Secs. 2 to 8 can be applied without hesitation in the present context.

General linear equations can seldom be solved by means of the Laplace transform, but they can often be solved by the method of undetermined coefficients. As the reader will recall from Secs. 6 to 9, the linear equation with constant p_k's has a polynomial solution y, if f is a polynomial, and this polynomial solution can be found by equating coefficients of corresponding powers of x. A similar result applies to many equations with variable p_k's, except that instead of a polynomial we must choose y to be an infinite series. The process of equating coefficients leads to infinitely many equations, but it turns out that these can be solved recursively, one at a time. We shall illustrate the process as it applies to homogeneous second-order equations

$$y''(x) + p_1(x)y' + p_0(x)y = 0. \tag{20-2}$$

The second-order case is emphasized because it is typical of the general case, and it includes many of the equations encountered in applications. The homogeneous case is emphasized because, as shown in Sec. 23, the second-order inhomogeneous equation can be solved completely if a single nonvanishing solution of the homogeneous equation is known.

Let it be required to solve the differential equation

$$y'' - xy' + y = 0 \tag{20-3}$$

by means of the trial solution

$$y = a_0 + a_1 x + a_2 x^2 + a_3 x^3 + \cdots + a_n x^n + \cdots . \tag{20-4}$$

As in Chap. 1, Secs. 10 to 17, we use Σ to denote a summation from 0 to ∞, so that the foregoing series is $\Sigma a_n x^n$. Then

$$y = \Sigma a_n x^n \qquad y' = \Sigma n a_n x^{n-1} \qquad y'' = \Sigma n(n-1)a_n x^{n-2}$$

and so, writing it in full,

$$
\begin{aligned}
y'' &= 2a_2 + 3\cdot 2a_3 x + 4\cdot 3a_4 x^2 + \cdots + (n+2)(n+1)a_{n+2}x^n + \cdots \\
-xy' &= \qquad\quad - a_1 x - 2a_2 x^2 - \cdots - \qquad\qquad na_n x^n - \cdots \\
y &= a_0 + \quad a_1 x + \quad a_2 x^2 + \cdots + \qquad\qquad a_n x^n + \cdots .
\end{aligned}
$$

According to (20-3) the sum of these three power series is zero, and therefore, by Theorem III of Chap. 1, Sec. 10, the coefficient of x^n in this sum is zero for each n:

$$
\begin{array}{ll}
2a_2 + a_0 = 0 & \text{coefficient of } x^0 \\
3\cdot 2a_3 - a_1 + a_1 = 0 & \text{coefficient of } x^1 \\
\cdot \quad \cdot \quad \cdot \quad \cdot \quad \cdot & \\
(n+2)(n+1)a_{n+2} - na_n + a_n = 0 & \text{coefficient of } x^n.
\end{array}
$$

Hence,

$$a_{n+2} = \frac{n-1}{(n+1)(n+2)}\, a_n. \tag{20-5}$$

This recursion formula gives, in succession,

$$a_2 = -\frac{1}{2}\, a_0 = -\frac{1}{2!}\, a_0$$

$$a_4 = \frac{1}{3\cdot 4}\, a_2 = -\frac{1}{4!}\, a_0$$

$$a_6 = \frac{3}{5\cdot 6}\, a_4 = -\frac{3}{6!}\, a_0, \cdots$$

$$\cdot \quad \cdot \quad \cdot \quad \cdot \quad \cdot \quad \cdot \quad \cdot \quad \cdot \quad \cdot$$

Similarly, $a_3 = 0 \cdot a_1 = 0$, $a_5 = 0$, $a_7 = 0$, and so on. The formal solution is

$$y = a_0 \left(1 - \frac{1}{2!}\, x^2 - \frac{1}{4!}\, x^4 - \frac{1\cdot 3}{6!}\, x^6 - \frac{1\cdot 3\cdot 5}{8!}\, x^8 - \cdots \right) + a_1 x$$

with the two arbitrary constants a_0 and a_1. There should be two constants in the general solution because the equation is of the second order.

The ratio of successive nonzero terms of the infinite series satisfies

$$\left| \frac{a_{2n+2}x^{2n+2}}{a_{2n}x^{2n}} \right| = \frac{2n-1}{(2n+1)(2n+2)} |x^2| \sim \frac{1}{2n} |x^2| \tag{20-6}$$

as is seen by using (20-5) with $2n$ written in place of n. Since the limit of (20-6) is zero, the series converges for all x. Hence the term-by-term differentiation is justified, and the formal solution is valid as an actual solution for $-\infty < x < \infty$.

Not every equation can be solved by means of power series. For example, the equation $2xy' = 3y$ has solutions $y = cx^{3/2}$ which cannot be expanded in series of the form (20-4). The following theorem, which is stated without proof, gives sufficient conditions for existence of series solutions:[1]

THEOREM. *Let $y'' + p_1(x)y' + p_0(x)y = 0$ have coefficients $p_1(x)$ and $p_0(x)$ which can be expanded in a power series for $|x| < r$. Then every solution y can be expanded in a power series for $|x| < r$.*

As an illustration, consider *Legendre's equation*

$$(1 - x^2)y'' - 2xy' + p(p+1)y = 0 \tag{20-7}$$

where p is constant. If the equation is divided by $1 - x^2$, so that the coefficient of y'' is 1 as in the theorem, we find

$$p_1(x) = -\frac{2x}{1-x^2} \qquad p_0(x) = \frac{p(p+1)}{1-x^2}.$$

Since these coefficients have power-series expansions valid for $|x| < 1$, the theorem asserts that the solution y also has a power-series expansion for $|x| < 1$.

To obtain this solution we assume $y = \Sigma a_n x^n$ and get

$$y'' = \Sigma a_{n+2}(n+2)(n+1)x^n$$
$$-x^2y'' = \Sigma - a_n n(n-1)x^n$$
$$-2xy' = \Sigma - 2na_n x^n$$
$$p(p+1)y = \Sigma a_n p(p+1)x^n.$$

By (20-7) the sum of these series is zero. Considering the coefficient of x^n yields

$$a_{n+2}(n+2)(n+1) = a_n[n(n+1) - p(p+1)] \tag{20-8}$$

after slight simplification. For all $n \geq 0$ we have

$$a_{n+2} = -a_n \frac{(p-n)(p+n+1)}{(n+1)(n+2)} \tag{20-9}$$

after factoring the term in brackets in (20-8) and dividing by $(n+2)(n+1)$. The coefficients for even n are determined from a_0, and those for odd n from a_1. Computing the coefficients successively, we get the final result

$$y = a_0 \left[1 - \frac{p(p+1)}{2!} x^2 + \frac{p(p-2)(p+1)(p+3)}{4!} x^4 - \cdots \right]$$
$$+ a_1 \left[x - \frac{(p-1)(p+2)}{3!} x^3 + \frac{(p-1)(p-3)(p+2)(p+4)}{5!} x^5 - \cdots \right].$$

Equation (20-8) shows that the series converge for $|x| < 1$ if we apply the ratio test to the ratio of successive nonzero terms. But if p is a positive integer, $p = m$, either the coefficient of

[1] See E. A. Coddington, "An Introduction to Ordinary Differential Equations," pp. 138–142, Prentice-Hall, Inc., Englewood Cliffs, N.J., 1961. A corresponding result for nonlinear equations can be found in Philip Franklin, "A Treatise on Advanced Calculus," pp. 516–520, John Wiley & Sons, Inc., New York, 1940.

a_0 or the coefficient of a_1 reduces to a polynomial, depending on whether p is even or odd. Choosing a_0 and a_1 so that the polynomials have the value unity when $x = 1$, we get the sequence

$$1, \quad x, \quad \tfrac{3}{2}x^2 - \tfrac{1}{2}, \quad \tfrac{5}{2}x^3 - \tfrac{3}{2}x, \quad \tfrac{35}{8}x^4 - \tfrac{15}{4}x^2 + \tfrac{3}{8}, \quad \cdots$$

for $m = 0, 1, 2, 3, \ldots$. These are the *Legendre polynomials* $P_m(x)$, which arise in several branches of applied mathematics.

Example. If α and β are constant, solve the first-order inhomogeneous equation $y' + xy = \alpha + \beta x$, subject to $y(0) = \beta$. The choice $y = \Sigma a_n x^n$ leads to

$$y' + xy = \Sigma[(n+1)a_{n+1} + a_{n-1}]x^n = \alpha + \beta x \tag{20-10}$$

provided we agree that $a_{-1} = 0$. Equating corresponding coefficients of x^0, x^1, x^2, $\ldots$ in (20-10) gives

$$a_1 = \alpha, \quad 2a_2 + a_0 = \beta, \quad 3a_3 + a_1 = 0, \quad 4a_4 + a_2 = 0, \quad 5a_5 + a_3 = 0, \quad \ldots \tag{20-11}$$

By the first of these relations, the third, the fifth, and so on, we get

$$a_1 = \alpha \qquad a_3 = -\frac{1}{3}\alpha \qquad a_5 = -\frac{1}{5}a_3 = \frac{1}{3 \cdot 5}\alpha$$

and so on. Since $a_0 = y(0)$ the initial condition gives $a_0 = \beta$. Hence the second relation (20-11) gives $a_2 = 0$, the fourth gives $a_4 = 0$, and so on. The solution is, therefore,

$$y = \alpha\left(\frac{x}{1} - \frac{x^3}{1 \cdot 3} + \frac{x^5}{1 \cdot 3 \cdot 5} - \frac{x^7}{1 \cdot 3 \cdot 5 \cdot 7} + \cdots\right) + \beta.$$

PROBLEMS

1. (a) Review Chap. 1, Sec. 2, and do Probs. 1 and 2 of that section. (b) Study the example in Chap. 1, Sec. 9. (c) Read Theorem III and its proof in Chap. 1, Sec. 10. (d) Study the example in Sec. 6 of this chapter. (e) Solve the example of this section by writing series for y', xy, and $y' + xy$ in expanded form, without Σ notation.

2. Given $y(0) = 1$, solve by series: $y' = y$, $y' = 2y$, $y' = y + x$, $y' + y = 1$, $y' = xy$.

3. Solve $y' - y = x^2$ if $y(0) = -2$, and solve $(1 - x)y' = 1 + x - y$ if $y(0) = 0$.

4. Obtain the first three terms of a series solution $y = \Sigma a_n x^n$ for the problem $y' = 2xy$, $y = 1$ when $x = 0$. From these three terms, compute the curvature $k = y''[1 + (y')^2]^{-3/2}$ at $x = 0$. Is your value for curvature exact or only approximate?

5. By considering the equation $y' = (1 - x^2)^{-1/2}$, obtain a series expansion for $\sin^{-1} x$. In particular, show that

$$\frac{\pi}{6} = \frac{1}{2} + \frac{1}{2}\frac{1}{3 \cdot 2^3} + \frac{1 \cdot 3}{2 \cdot 4}\frac{1}{5 \cdot 2^5} + \frac{1 \cdot 3 \cdot 5}{2 \cdot 4 \cdot 6}\frac{1}{7 \cdot 2^7} + \cdots.$$

6. Solve $y'' + xy = 0$ if $y(0) - 1 = y'(0) = 0$; solve $y'' = x^2 y$ if $y(0) = y'(0) = 1$.

7. Solve by series: $y'' + y = 0$, $y'' + y = 1$, $y'' - xy' = 1$, $y'' = xy'$, $y'' = 1 + xy$.

8. (a) For the equation $(2 - x)y'' + (x - 1)y' = y$ what radius of convergence is guaranteed by the theorem of this section? (b) Show that $y = \Sigma a_n x^n$ leads to

$$2(n + 2)(n + 1)a_{n+2} - (n + 1)^2 a_{n+1} + (n - 1)a_n = 0.$$

(c) By taking $a_0 = a_1 = 1$, find a solution satisfying $y = y' = 1$ at $x = 0$. (d) By another choice of a_0 and a_1, find a solution satisfying $y = -1$, $y' = 1$ at $x = 0$. (e) Using parts c and d, express the general solution in terms of elementary functions. Is the radius of convergence larger than guaranteed by the above theorem?

9. (a) Solve by means of power series if $y(0) = y'(0)$:

$$(x^2 - 3x + 2)y'' + (x^2 - 2x - 1)y' + (x - 3)y = 0.$$

(b) Solve $y'' - (x - 2)y = 0$ by assuming $y = \Sigma a_n(x - 2)^n$.

10. (a) For *Hermite's equation* $y'' - 2xy' + 2py = 0$, obtain the two solutions

$$y = 1 - \frac{2p}{2!} x^2 + \frac{2^2 p(p-2)}{4!} x^4 - \frac{2^3 p(p-2)(p-4)}{6!} x^6 + \cdots$$

$$y = x - \frac{2(p-1)}{3!} x^3 + \frac{2^2(p-1)(p-3)}{5!} x^5 - \frac{2^3(p-1)(p-3)(p-5)}{7!} x^7 + \cdots$$

by taking $y(0) = 1$, $y'(0) = 0$, or $y(0) = 0$, $y'(0) = 1$, respectively.   (b) When p is a nonnegative integer m the series terminates.   The resulting polynomials, normalized so that the leading coefficient is 2^m, are the *Hermite polynomials* $H_m(x)$.   Verify that the first five are:

$$1, \qquad 2x, \qquad 4x^2 - 2, \qquad 8x^3 - 12x, \qquad 16x^4 - 48x^2 + 12, \qquad \ldots$$

11. (a) For *Chebychev's equation* $(1 - x^2)y'' - xy' + p^2 y = 0$, obtain the solutions

$$y = 1 - \frac{p^2}{2!} x^2 + \frac{p^2(p^2 - 2^2)}{4!} x^4 - \frac{p^2(p^2 - 2^2)(p^2 - 4^2)}{6!} x^6 + \cdots$$

$$y = x - \frac{p^2 - 1^2}{3!} x^3 + \frac{(p^2 - 1^2)(p^2 - 3^2)}{5!} x^5 - \frac{(p^2 - 1^2)(p^2 - 3^2)(p^2 - 5^2)}{7!} x^7 + \cdots$$

(b) The polynomials obtained when p is a nonnegative integer m, normalized so that the leading coefficient is 2^{m-1}, are the *Chebychev polynomials* $T_m(x)$.   Verify that the first five are

$$1, \qquad x, \qquad 2x^2 - 1, \qquad 4x^3 - 3x, \qquad 8x^4 - 8x^2 + 1, \qquad \ldots$$

12. (a) Show that *Laguerre's equation* $xy'' + (1 - x)y' + py = 0$ has a solution

$$y = 1 - px - \frac{p(1 - p)}{(2!)^2} x^2 - \frac{p(1 - p)(2 - p)}{(3!)^2} x^3 - \cdots$$

even though it does not satisfy the conditions of the theorem in this section.   (b) If p is a nonnegative integer m, the solutions, normalized so that $y(0) = m!$, are the *Laguerre polynomials* $L_m(x)$.   Verify that the first four are 1, $1 - x$, $2 - 4x + x^2$, $6 - 18x + 9x^2 - x^3, \ldots$.

13. It can be shown that the Legendre polynomials $P_n(x)$ satisfy[1]

$$(1 - 2xh + h^2)^{-\frac{1}{2}} = P_0(x) + P_1(x)h + P_2(x)h^2 + \cdots + P_n(x)h^n + \cdots.$$

Verify this equality through the terms in h^4.   *Hint:* Expand $[1 - (2hx - h^2)]^{-\frac{1}{2}}$ by the binomial theorem, and collect powers of h.   The function $(1 - 2xh + h^2)^{-\frac{1}{2}}$ is called the *generating function* of the sequence $\{P_n(x)\}$.

14. For $n = 0, 1, 2, 3$, verify *Rodrigues' formula*[2] for the Legendre polynomials,

$$P_n(x) = \frac{1}{2^n n!} \frac{d^n}{dx^n} (x^2 - 1)^n.$$

15. If y_1 and y_2 both satisfy $y'' + p_1 y' + p_0 y = 0$, express p_1 and p_0 in terms of the y's at any point where the Wronskian $y_1 y_2' - y_2 y_1'$ is not zero.   Thus prove a converse of the theorem of this section, to the effect that p_1 and p_0 have power-series expansions about $x = 0$ if two independent solutions y_1 and y_2 have such expansions.

16. Referring to Prob. 15, find a second-order linear homogeneous equation with solutions x^3 and x^4.

---

21. Generalized Power Series. Bessel's Equation.

An important differential equation was encountered by the German astronomer and mathematician F. W. Bessel in a study of planetary motion.   The *Bessel functions* which arise from the solution of this equation are indispensable in the study of vibration of chains, propagation of electric

[1] E. J. Whittaker and G. N. Watson, "Modern Analysis," pp. 302–303, Cambridge University Press, London, 1952.
[2] *Ibid.*

currents in cylindrical conductors, heat flow in cylinders, vibration of circular membranes, and many other problems of applied mathematics.

Bessel's equation is

$$x^2 y'' + xy' + (x^2 - p^2)y = 0 \tag{21-1}$$

where p is a constant. The theorem of Sec. 20 does not apply to this equation, since

$$p_1(x) = \frac{1}{x} \qquad p_2(x) = 1 - \frac{p^2}{x^2}$$

and these functions cannot be expanded in power series near $x = 0$. For this reason a power-series solution $y = \Sigma a_n x^n$ cannot be expected.

It can be shown, however, that a wide class of equations, including (21-1), have solutions of the form

$$y = x^\rho \sum a_n x^n = \sum_{n=0}^{\infty} a_n x^{n+\rho} \qquad a_0 \neq 0 \tag{21-2}$$

where ρ is constant.[1]

THEOREM I. *Let $xp_1(x)$ and $x^2p_0(x)$ have power-series expansions valid for $|x| < r$.* *Then the equation*

$$y'' + p_1(x)y' + p_0(x)y = 0$$

has a solution of the form (21-2), also valid for $|x| < r$.

The novelty is that ρ is allowed to be any number whereas, if (21-2) is an ordinary power series, ρ must be an integer. Since ρ may be increased at will, the assumption $a_0 \neq 0$ involves no loss of generality. Both Theorem I here and the special case given in Sec. 20 are commonly referred to as "Frobenius' theorem," though they were established earlier by the German mathematician Lazarus Fuchs.

Since Bessel's equation gives

$$xp_1(x) = 1 \qquad x^2p_0(x) = x^2 - p^2$$

Theorem I asserts that the series (21-2), when found, will be valid for all x. To obtain this series, note that

$$x^2 \sum_{n=0}^{\infty} a_n x^{n+\rho} = \sum_{n=0}^{\infty} a_n x^{n+\rho+2} = \sum_{n=2}^{\infty} a_{n-2} x^{n+\rho} \tag{21-3}$$

as is seen by writing it out in full. The limits $(2,\infty)$ on the latter summation may be changed to $(0,\infty)$ if we agree to define

$$a_n = 0 \qquad \text{for all negative } n. \tag{21-4}$$

Hence,
$$x^2 y'' = \Sigma a_n(n + \rho)(n + \rho - 1)x^{n+\rho}$$
$$xy' = \Sigma a_n(n + \rho)x^{n+\rho}$$
$$x^2 y = \Sigma a_{n-2} x^{n+\rho}$$
$$-p^2 y = \Sigma - p^2 a_n x^{n+\rho}.$$

According to Eq. (21-1), which we wish to solve, the sum of the four terms on the left of the above equations is zero. Hence, the same is true for the series on the right. Equating to zero the coefficient of $x^{n+\rho}$ in the sum of these series gives

$$a_n(n + \rho)(n + \rho - 1) + a_n(n + \rho) + a_{n-2} - p^2 a_n = 0$$

or, after simplification,

$$a_n[(n + \rho)^2 - p^2] + a_{n-2} = 0. \tag{21-5}$$

---

[1] A proof is given in Coddington, *op. cit.*, pp. 160–162.

Equation (21-5) is valid for all n. For negative n Eq. (21-5) holds automatically by virtue of (21-4). The first nontrivial case of (21-5) is called the *indicial equation;* it is obtained in the present example by putting $n = 0$ and takes the form

$$a_0(\rho^2 - p^2) + 0 = 0 \qquad \text{indicial equation.} \tag{21-6}$$

This shows that $\rho = p$ or $\rho = -p$, since $a_0 \neq 0$. The other values of a_n are determined from (21-5) in the form

$$a_n = - \frac{1}{(n + \rho)^2 - p^2} a_{n-2}. \tag{21-7}$$

The choice $n = 1$ gives[1] $a_1 = 0$; hence $a_n = 0$ for all odd n. Also

$$a_2 = - \frac{a_0}{(2 + \rho)^2 - p^2} \qquad a_4 = - \frac{a_2}{(4 + \rho)^2 - p^2}$$

and so on. In this way it is easily verified that the series corresponding to $\rho = p$ is

$$y = a_0 x^p \left[1 - \frac{x^2}{2(2p + 2)} + \frac{x^4}{2 \cdot 4(2p + 2)(2p + 4)} - \cdots \right]$$

and that the series for $\rho = -p$ is the same, with $-p$ in place of p.

When p is a nonnegative integer, the expression can be simplified by use of factorials, as follows: We take a factor 2 from each term of the denominator and place it with the x in the numerator, obtaining

$$y = a_0 x^p \left[1 - \frac{(x/2)^2}{1(p + 1)} + \frac{(x/2)^4}{1 \cdot 2(p + 1)(p + 2)} - \cdots \right]. \tag{21-8}$$

If the denominators are now multiplied by $p!$, there results

$$y = a_0 p! x^p \left[\frac{1}{p!} - \frac{(x/2)^2}{1 \cdot (p + 1)!} + \frac{(x/2)^4}{2!(p + 2)!} - \cdots \right]$$

and since $x^p = 2^p (x/2)^p$, this yields $y = a_0 p! 2^p J_p(x)$, where

$$J_p(x) = \sum_{n=0}^{\infty} \frac{(-1)^n (x/2)^{2n+p}}{n!(p + n)!}. \tag{21-9}$$

The function $J_p(x)$ is called *the Bessel function of order p.* The graphs of $J_0(x)$ and $J_1(x)$ are shown in Fig. 18.

The differential equation (21-1) is meaningful even when p is not a positive integer,

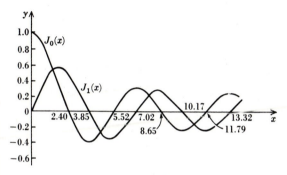

FIGURE 18

---

[1] Provided the denominator in (21-7) does not vanish. See Prob. 9.

and the series solution (before introduction of factorials) is also well defined for general p. It is natural to inquire if we can define $p!$ so that (21-9) is meaningful and satisfies (21-1) for p unrestricted. A glance at (21-8) shows that such an extension may not be possible when p is a negative integer, but there appears to be no difficulty otherwise.

In Sec. 14 of Chap. 1 it was found that the definition of $p!$ actually can be extended in such a way that the equation

$$p \frac{1}{p!} = \frac{1}{(p-1)!}$$

holds for all p. Using this relation one can verify that $J_p(x)$ and $J_{-p}(x)$ are solutions of (21-1), no matter what value p may have.

For most values of p the functions $J_{-p}(x)$ and $J_p(x)$ are independent, in the sense that neither is a constant multiple of the other. If $p = 0$, however, the two roots of the indicial equation are both $\rho = 0$, so that we obtain only the single function $J_0(x)$. Another exceptional case arises when $p = \pm 1, \pm 2, \ldots$. Although the series (21-8) is meaningless when p is a negative integer, the series (21-9) is well defined and satisfies

$$J_{-m}(x) = (-1)^m J_m(x) \qquad m = 0, 1, 2, 3, \ldots. \tag{21-10}$$

To see this, we observe that

$$J_{-m}(x) = \sum_{n=0}^{\infty} \frac{(-1)^n (x/2)^{2n-m}}{n!(n-m)!} = \sum_{n=m}^{\infty} \frac{(-1)^n (x/2)^{2n-m}}{n!(n-m)!} \tag{21-11}$$

since the factor $1/(n-m)!$ is zero when $n < m$. If the sums (21-9) and (21-11) are written in full, (21-10) follows at once. Because of (21-10) the functions $J_{-m}(x)$ and $J_m(x)$ are dependent, so that we obtain only one solution rather than two.

This failure of the method to provide both solutions is not a serious shortcoming, since the second solution can always be found from the first by the method of Sec. 23. Carrying out the calculation in the general case yields the following theorem:

THEOREM II.   When the roots ρ_1 and ρ_2 of the indicial equation are distinct and do not differ by an integer, the method of Theorem I yields two linearly independent solutions. If the roots differ by an integer, a second solution can be found by assuming that

$$y_2 = c y_1(x) \log x + \sum_{n=0}^{\infty} a_n x^{n+\rho_2} \tag{21-12}$$

where $y_1(x)$ is the solution given by Theorem I for the root $\rho = \rho_1$.

It was discovered by Frobenius that the second solution can also be found by differentiating the first solution with respect to the exponent ρ (cf. discussion of the corresponding problem in Sec. 3).

By setting $y_1(x) = J_0(x)$, for example, one can show that a second solution of Bessel's equation for $p = 0$ is

$$K_0(x) = J_0(x) \log x - \sum_{k=1}^{\infty} \frac{(-1)^k (x/2)^{2k}}{(k!)^2} \left(1 + \frac{1}{2} + \cdots + \frac{1}{k}\right).$$

This function is called the Bessel function of the zeroth order and second kind. Thus, the general solution of

$$xy'' + y' + xy = 0$$

is $y = c_1 J_0(x) + c_2 K_0(x)$. Other functions $K_m(x)$ of the second kind are obtained similarly. By considering linear combinations of $J_m(x)$ and $K_m(x)$ we get the *modified Bessel functions* of the first and second kinds, denoted in the literature by $I_m(x)$, $Y_m(x)$, $N_m(x)$, and $H_m(x)$.

Example.   (*The Euler-Cauchy equation.*)   Show that $x^2y'' + 2xy' = 0$ can be solved by a generalized power series having just one term.   The substitution $y = x^\rho$ yields

$$\rho(\rho - 1) + 2\rho = 0 \tag{21-13}$$

after division by x^ρ. (This is actually the indicial equation, in the sense of the foregoing discussion.) Since $\rho = 0$ and $\rho = -1$ satisfy (21-13) we obtain the two independent solutions $y = 1$ and $y = x^{-1}$.

An equation of the form

$$x^n y^{(n)} + p_{n-1}x^{n-1}y^{(n-1)} + \cdots + p_1 xy' + p_0 y = f(x)$$

is usually called *Cauchy's equation*, though it was examined earlier by Euler. The coefficients x^k compensate for the lowering of the exponent caused by the differentiation and so the substitution $y = x^\rho$ into the equation always leads to a polynomial equation for ρ. If ρ is a root of multiplicity m the corresponding family of solutions is

$$x^\rho, \qquad x^\rho \log x, \qquad x^\rho(\log x)^2, \qquad \ldots, \qquad x^\rho(\log x)^{m-1}$$

as is proved by the result of Prob. 11. In agreement with Frobenius' theorem, the same family is obtained by differentiating the original solution, x^ρ, with respect to ρ.

PROBLEMS

1. (*a*) By assuming a solution of the form $y = x^\rho$, solve

$$x^2y'' - 4xy' + 6y = 0, \qquad x^2y'' + 2xy' - n(n+1)y = 0.$$

(*b*) Verify that the same results are obtained by use of a generalized power series, $y = \Sigma a_n x^{n+\rho}$.

2. If $y(0) = 1$, $y'(0) = 0$, solve $xy'' + 2y' + xy = 0$ and $xy'' + y' + xy = 0$.

3. Show that $J_0'(x) = -J_1(x)$ and also that

$$\frac{d}{dx} x^n J_n(x) = x^n J_{n-1}(x) \qquad \frac{d}{dx} x^{-n} J_n(x) = -x^{-n} J_{n+1}(x).$$

Deduce that

$$J_{n-1}(x) - J_{n+1}(x) = 2J_n'(x) \qquad J_{n-1}(x) + J_{n+1}(x) = \frac{2n}{x} J_n(x).$$

4. The *confluent hypergeometric equation* is

$$xy'' + py' = xy' + qy$$

where p and q are constant.

(*a*) According to Theorem I, what range of validity do you expect for a solution of the form $\Sigma a_n x^{n+\rho}$?

(*b*) Assuming that $y = \Sigma a_n x^{n+\rho}$, verify that $xy'' = \Sigma a_{n+1}(n + \rho + 1)(n + \rho)x^{n+\rho}$ and find similar expressions for py', xy', and qy.

(*c*) By considering the coefficient of $x^{n+\rho}$, deduce

$$a_{n+1}(n + \rho + 1)(n + \rho + p) = a_n(n + \rho + q)$$

for all values of n.

5. In Prob. 4, (*a*) show that the roots of the indicial equation are $\rho = 0$ and $\rho = 1 - p$.   *Hint:* The first nontrivial case arises when $n = -1$.

(*b*) When $\rho = 0$, show that

$$a_{n+1} = a_n \frac{n + q}{(n+1)(n+p)}$$

if p is not zero or a negative integer.   Thus get the solution

$$y_1 = a_0\left[1 + \frac{q}{p}\frac{x}{1!} + \frac{q(q+1)}{p(p+1)}\frac{x^2}{2!} + \frac{q(q+1)(q+2)}{p(p+1)(p+2)}\frac{x^3}{3!} + \cdots\right].$$

(*c*) Similarly, obtain the solution corresponding to $\rho = 1 - p$ when p is not a positive integer.

6. For the *hypergeometric equation* of Gauss,

$$x(1 - x)y'' + [c - (a + b + 1)x]y' - aby = 0$$

obtain one solution in the form

$$y = \sum_{n=0}^{\infty} \frac{\Gamma(a + n)\,\Gamma(b + n)}{\Gamma(1 + n)\,\Gamma(c + n)}\, x^n$$

when a and b are not negative integers.

7. For the equation

$$xy'' + (1 - 2p)y' + xy = 0$$

obtain a recursion formula for the coefficients of a series solution, and show that one solution is $y = x^p J_p(x)$.

8. As in Prob. 7, show that $y = x^{1/2} J_p(\lambda x)$ satisfies

$$4x^2 y'' + (4\lambda^2 x^2 - 4p^2 + 1)y = 0.$$

9. (*a*) Given that $\Gamma(\tfrac{1}{2}) = \sqrt{\pi}$, obtain the formulas

$$J_{1/2}(x) = \sqrt{\frac{2}{\pi x}}\, \sin x \qquad J_{-1/2}(x) = \sqrt{\frac{2}{\pi x}}\, \cos x.$$

(*b*) What is the general solution of Bessel's equation with $p = \tfrac{1}{2}$? (This shows that Theorem I may yield the general solution even if $\rho_2 - \rho_1$ is an integer.)

10. The generating function of the sequence $J_n(x)$ is

$$e^{(x/2)\,(h - 1/h)} = \sum_{n=-\infty}^{\infty} J_n(x)h^n.$$

Verify that the coefficient of h^0 in the expansion of the exponential is, in fact, $J_0(x)$. *Hint:* By the series for e^u the exponential is

$$\sum \left(\frac{x}{2}\right)^n \left(h - \frac{1}{h}\right)^n \frac{1}{n!}.$$

Pick out the term independent of h in the binomial expansion of $(h - 1/h)^n$ when n is even, and note that there is no such term when n is odd.

THE EULER–CAUCHY EQUATION

11. (*a*) If $x = e^t$, primes denote differentiation with respect to x, and D denotes d/dt, show that

$$y' = e^{-t}Dy, \qquad y'' = e^{-2t}D(D - 1)y, \qquad y''' = e^{-3t}D(D - 1)(D - 2)y,$$

and so on. *Hint:* $dy/dx = (dy/dt) \div (dx/dt)$. (*b*) Using part *a*, show that the Euler-Cauchy equation can be reduced to a linear equation with constant coefficients, namely,

$$[D(D - 1) \cdots (D - n + 1) + \cdots + p_2 D(D - 1) + p_1 D + p_0]y = f(e^t).$$

12. By setting $x = e^t$ as in Prob. 11, show that the equation

$$x^3 y''' + xy' - y = x \log x \qquad \text{gives} \qquad (D - 1)^3 y = te^t.$$

Verify that the latter has a solution $y = \tfrac{1}{24} t^4 e^t$, and thus solve the original equation.

13. As in Prob. 12, solve: (*a*) $x^2 y'' + 4xy' + 2y = \log x$, $x^2 y'' + y = x^2$;
(*b*) $x^3 y''' - 4x^2 y'' + 5xy' - 2y = 1$, $x^2 y'' - 2xy' + 2y = x \log x$.

---

22. Boundary Values. Orthogonality. Many problems of mathematical physics lead to differential equations of the form

$$\frac{d}{dx}\left[p(x)\frac{dy}{dx}\right] + q(x)y = \lambda r(x)y \qquad \lambda = \text{const} \tag{22-1}$$

with additional conditions specified at the end points of the interval (a,b) in which the equation is to hold. Often, the only possible solution to such a problem is $y = 0$, but for special values of the constant λ it may happen that the problem has nonzero solutions. These special values $\lambda_1, \lambda_2, \lambda_3, \ldots$ are called *characteristic values*, and the corresponding nonzero solutions $y_1, y_2, y_3, \ldots$ are *characteristic functions*.[1]

For example, consider the problem $y'' = \lambda y$ corresponding to $p = r = 1, q = 0$. For boundary conditions we assume $y(0) = y(1) = 0$. If $\lambda = 0$ the general solution is $y = c_1 + c_2 x$, and the boundary conditions give $c_1 = c_2 = 0$, or $y \equiv 0$. If $\lambda > 0$ the sole possibility again is $y \equiv 0$, as the reader can verify. Hence we assume that λ is a negative number, $\lambda = -\omega^2$. In this case the general solution is

$$y = c_1 \cos \omega x + c_2 \sin \omega x.$$

The condition $y(0) = 0$ gives $c_1 = 0$, and the condition $y(1) = 0$ gives $\sin \omega = 0$. The characteristic values λ are

$$-\pi^2, \qquad -(2\pi)^2, \qquad -(3\pi)^2, \qquad -(4\pi)^2, \qquad \ldots$$

and a set of characteristic functions is (cf. Sec. 10)

$$\sin \pi x, \qquad \sin 2\pi x, \qquad \sin 3\pi x, \qquad \sin 4\pi x, \qquad \ldots$$

Characteristic functions satisfy a remarkable identity that is now derived. Writing (22-1) in the compact form

$$(py')' + qy = \lambda r y \qquad a < x < b \tag{22-2}$$

let y_m be a solution when λ has the value λ_m and let y_n be a solution when λ has a different value, λ_n. Thus,

$$(py_m')' + qy_m = \lambda_m r y_m \qquad (py_n')' + qy_n = \lambda_n r y_n.$$

If the first of these equations is multiplied by y_n and the second by y_m we get

$$y_n(py_m')' - y_m(py_n')' = \lambda_m r y_n y_m - \lambda_n r y_m y_n \tag{22-3}$$

upon subtracting the resulting equations. Since

$$[y_n(py_m') - y_m(py_n')]' = y_n(py_m')' + y_n'(py_m') - y_m(py_n')' - y_m'(py_n')$$

and this equals the left side of (22-3), the foregoing result (22-3) can be written

$$\frac{d}{dx}[p(y_n y_m' - y_m y_n')] = (\lambda_m - \lambda_n) r y_m y_n.$$

Integrating from a to b yields the identity

$$p(y_n y_m' - y_m y_n')\Big|_a^b = (\lambda_m - \lambda_n) \int_a^b r y_m y_n \, dx \tag{22-4}$$

when r is continuous on (a,b). In case the integral is improper, both sides are interpreted as limits for $x \to a+$ and $x \to b-$. (See Chap. 1, Secs. 4 and 19.)

If the conditions at a and b are such that the left side of (22-4) is zero, we deduce the *orthogonality equation*

$$\int_a^b r y_m y_n \, dx = 0 \qquad m \neq n. \tag{22-5}$$

For example, this holds if the solutions are required to vanish at a and b, or if their derivatives are required to vanish at a and b, or, more generally, if there are two linear relations

$$\alpha_1 y(a) + \alpha_2 y'(a) = 0 \qquad \beta_1 y(b) + \beta_2 y'(b) = 0. \tag{22-6}$$

Here α_1 and α_2 are constant, not both 0, and similarly for β_1 and β_2.

[1] The German *Eigenwert* and *Eigenfunktion* are also used, as are various hybrids of English and German.

To prove that (22-6) makes the left side of (22-4) vanish, let y_m and y_n both satisfy the first condition (22-6). Thus,

$$\alpha_1 y_m(a) + \alpha_2 y_m'(a) = 0$$
$$\alpha_1 y_n(a) + \alpha_2 y_n'(a) = 0.$$

This is a linear system in the two constants α_1 and α_2. Since α_1 and α_2 are not both 0, the coefficient determinant must vanish; hence, the expression on the left of (22-4) vanishes at the lower limit, $x = a$. Similarly, by the second relation (22-6), it vanishes at $x = b$.

When (22-5) holds, the functions y_n are said to be *orthogonal with respect to the weight r*. If in addition

$$\int_a^b y_n{}^2 r\, dx = 1$$

the set is *orthonormal with respect to the weight r*. For $r = 1$ this agrees with the definition given in Chap. 1, Sec. 23.

The importance of orthogonality is that it greatly facilitates expansion of functions in series of the form

$$f(x) = c_1 y_1(x) + c_2 y_2(x) + \cdots + c_n y_n(x) + \cdots.$$

Indeed, if the y_n are orthonormal, multiplying by $r(x)y_n(x)$ and integrating term by term suggests that

$$c_n = \int_a^b f(x) r(x) y_n(x)\, dx. \tag{22-7}$$

For broad classes of y_m and f it can be shown that the *generalized Fourier series* obtained with this choice of c_k actually converges to the function; in other words, an analog of Dirichlet's theorem holds (cf. Chap. 1, Secs. 19, 24, and 25). The subject leads to a branch of mathematics known as *spectral theory*.[1]

Example 1. Show that the Legendre functions $P_n(x)$ are orthogonal on the interval $(-1,1)$. Legendre's equation (20-7) may be written

$$[(1 - x^2)y']' = \lambda y$$

where λ is constant; $\lambda = -n(n + 1)$ when $y = P_n(x)$. The special case $p = (1 - x^2)$, $q = 0$, $r = 1$ in (22-2) and (22-4) yields

$$(1 - x^2)(P_n P_m' - P_m P_n') \Big|_{-1}^{1} = [-m(m + 1) + n(n + 1)] \int_{-1}^{1} P_m(x)P_n(x)\, dx.$$

Since $(1 - x^2)$ vanishes at ± 1, the left side is zero and the orthogonality follows. It can be shown[2] also that

$$\int_{-1}^{1} [P_n(x)]^2\, dx = \frac{2}{2n + 1}$$

and hence the corresponding orthonormal set is

$$\varphi_n(x) = (n + \tfrac{1}{2})^{1/2} P_n(x).$$

Example 2. Let the sequence $\rho_1, \rho_2, \ldots$ be the distinct positive roots of the equation $J_\mu(x) = 0$, so that $J_\mu(\rho_n) = 0$. If $\mu \geq 0$ the functions

$$\phi_n(x) = \sqrt{2x}\, \frac{J_\mu(\rho_n x)}{J_\mu'(\rho_n)} \tag{22-8}$$

are orthonormal in the interval $(0,1)$.

[1] For an authoritative account in a nonabstract setting see E. A. Coddington and N. Levinson, "Theory of Ordinary Differential Equations," chap. 7, McGraw-Hill Book Company, New York, 1955.

[2] See, for example, Whittaker and Watson, *op. cit.*, p. 305. A discussion of orthogonal functions can be found in Jack Indritz, "Methods in Analysis," chap. 5, The Macmillan Company, New York, 1963. For a development including numerical applications see F. B. Hildebrand, "Introduction to Numerical Analysis," chap. 7, McGraw-Hill Book Company, New York, 1956.

By (21-1) it is found that $y = J_\mu(\rho x)$ satisfies

$$(xy')' - \frac{\mu^2 y}{x} = \lambda xy \qquad \lambda = -\rho^2$$

and hence (22-4) holds with $p = r = x$. If we choose $y_n = J_\mu(\rho x)$ and $y_m = J_\mu(\rho_m x)$, the left side of (22-4) is

$$x\left[J_\mu(\rho x) \frac{d}{dx} J_\mu(\rho_m x) - J_\mu(\rho_m x) \frac{d}{dx} J_\mu(\rho x) \right]\Big|_0^1$$

which reduces to $J_\mu(\rho)\rho_m J_\mu'(\rho_m)$, since $J_\mu(\rho_m) = 0$. It follows that

$$(-\rho_m{}^2 + \rho^2) \int_0^1 x J_\mu(\rho_m x) J_\mu(\rho x)\, dx = \rho_m J_\mu(\rho) J_\mu'(\rho_m). \qquad (22\text{-}9)$$

Since $J_\mu(\rho_n) = 0$, the choice $\rho = \rho_n$ in (22-9) yields the orthogonality

$$\int_0^1 x J_\mu(\rho_m x) J_\mu(\rho_n x)\, dx = 0 \qquad m \neq n. \qquad (22\text{-}10)$$

Moreover, differentiating (22-9) with respect to ρ we get

$$2\rho \int_0^1 x J_\mu(\rho_m x) J_\mu(\rho x)\, dx + (\rho^2 - \rho_m{}^2) \int_0^1 x^2 J_\mu(\rho_m x) J_\mu'(\rho x)\, dx = \rho_m J_\mu'(\rho) J_\mu'(\rho_m)$$

which reduces to

$$2 \int_0^1 x [J_\mu(\rho_m x)]^2\, dx = [J_\mu'(\rho_m)]^2 \qquad (22\text{-}11)$$

when $\rho = \rho_m$. These equations show that the sequence (22-8) is orthonormal on (0,1), as desired.

The fact that the equation $J_\mu(x) = 0$ has infinitely many roots ρ_n is established in treatises on Bessel functions; analysis of such questions for general differential equations constitutes the *Sturm-Liouville theory*.[1]

PROBLEMS

1. Review the discussion of orthogonality in Chap. 1, Secs. 23 and 24.

2. By considering the equation $y'' = \lambda y$, show that the sequence $\{\sin n\pi x/l\}$ is orthogonal on the interval $(0,l)$, and construct the corresponding orthonormal set.

3. If $m + n$ is positive, show that

$$(m^2 - n^2) \int_0^l x^{-1} J_m(x) J_n(x)\, dx = l[J_n(l) J_m'(l) - J_m(l) J_n'(l)].$$

Hint: Bessel's equation (21-1) can be written $(xy')' + xy = \lambda y/x$, where $\lambda = n^2$ when $y = J_n(x)$. To avoid difficulty at $x = 0$, consider $\int_\epsilon^l$ and let $\epsilon \to 0$. The convergence follows from (21-9), since (21-9) gives

$$J_m(x) J_n(x) \sim (\text{const})\, x^{m+n} \qquad \text{as } x \to 0.$$

4. It can be shown that, as $|x| \to \infty$,

$$J_n(x) \sim \sqrt{\frac{2}{\pi x}} \cos\left(x - \frac{\pi}{4} - \frac{n\pi}{2} \right) \qquad J_n'(x) \sim -\sqrt{\frac{2}{\pi x}} \sin\left(x - \frac{\pi}{4} - \frac{n\pi}{2} \right).$$

By letting $l \to \infty$ in Prob. 3, deduce

$$(m^2 - n^2) \int_0^\infty x^{-1} J_m(x) J_n(x)\, dx = \frac{2}{\pi} \sin (m - n) \frac{\pi}{2} \qquad m + n > 0.$$

5. If p_1 is continuous, determine p and q so that

$$p(y'' + p_1 y' + p_0 y) \equiv (py')' + qy.$$

[1] Coddington and Levinson, *loc. cit.*

Thus reduce the general second-order equation $Ty = \lambda r_0 y$ to the form (22-2).   (It will be found that $p = e^P$, where $P = \int p_1 \, dx$.)

6. Referring to Prob. 5 above and to Probs. 10 to 12 of Sec. 20, show that the Hermite, Chebychev, and Laguerre polynomials satisfy, respectively,

$$(e^{-x^2}y')' + 2ne^{-x^2}y = 0, \qquad (\sqrt{1 - x^2}\,y')' + \frac{n^2y}{\sqrt{1 - x^2}} = 0, \qquad (xe^{-x}y')' + ne^{-x}y = 0$$

where n is an integer, $n \geq 0$.   For $m \neq n$, deduce that

$$\int_{-\infty}^{\infty} e^{-x^2}H_n(x)H_m(x) \, dx = \int_{-1}^{1} \frac{T_n(x)\,T_m(x)}{\sqrt{1 - x^2}} \, dx = \int_0^{\infty} e^{-x}L_n(x)L_m(x) \, dx = 0.$$

[In the references cited it is shown that the values of the integrals for $m = n \geq 1$ are $2^n n! \sqrt{\pi}$, $\tfrac{1}{2}\pi$, and $(n!)^2$, respectively.   This holds for $n = 0$, except that the Chebychev polynomial gives π instead of $\tfrac{1}{2}\pi$.]

---

23. Abel's Identity.   Reduction of Order.   We now discuss some simple but important identities connected with the general second-order linear equation $Ty = f$, where T is the operator defined by

$$\mathbf{T}y = y'' + p_1(x)y' + p_0(x)y. \tag{23-1}$$

Since the coefficients are variable our results will apply to Legendre's equation, Bessel's equation, and so on.   It is convenient to assume that f and p_1 are continuous and to define

$$P(x) = \int_{x_0}^{x} p_1(t) \, dt \qquad a < x_0 < b. \tag{23-2}$$

We begin by establishing the following: *Let u and v be any functions such that $u\mathbf{T}v = v\mathbf{T}u$.   Then the Wronskian $W(u,v)$ satisfies*

$$W(x) = W(x_0)e^{-P(x)} \qquad where \qquad W(x) \equiv W[u(x),v(x)]. \tag{23-3}$$

Since $W = uv' - vu'$ we have $W' = uv'' - vu''$, and Eq. (23-1) yields

$$u\mathbf{T}v - v\mathbf{T}u = (uv'' - vu'') + p_1(uv' - vu') = W' + p_1W.$$

The hypothesis $u\mathbf{T}v = v\mathbf{T}u$ gives $W' + p_1W = 0$, or, equivalently, $(e^PW)' = 0$.   This shows that e^PW is constant, and (23-3) follows.   The result (23-3) is known as *Abel's identity*, in honor of the Norwegian mathematician Niels Henrik Abel.

For example, Bessel's equation (21-1) has $p_1(x) = x^{-1}$; hence we can take $P(x) = \log x$ for $x > 0$.   The result of applying Abel's identity to $u = J_p$ and $v = J_{-p}$ is

$$J_p(x)J'_{-p}(x) - J_{-p}(x)J'_p(x) = cx^{-1} \tag{23-4}$$

where c is constant.   To determine c, we replace the Bessel functions by the first term of their series expansions (21-9) and let $x \to 0$.   The value is

$$c = \frac{-2p}{p!(-p)!} = -\frac{2}{\pi}\sin \pi p$$

where the second equality is proved in the theory of the gamma function.[1]

Another useful identity is obtained by setting $y = uv$, where u and v are arbitrary twice-differentiable functions.   Since $y'' = (uv' + vu')'$ the result of substituting in (23-1) is

$$\mathbf{T}(uv) = (uv'' + 2u'v' + vu'') + p_1(uv' + vu') + p_0uv.$$

[1] See Emil Artin, "The Gamma Function," pp. 25–26, Holt, Rinehart and Winston, Inc., New York, 1964.

The terms with v as a factor give $v(u'' + p_1u' + p_0u)$, which is recognized as $v\mathbf{T}u$.   Hence,

$$\mathbf{T}(uv) = uv'' + (2u' + p_1u)v' + v\mathbf{T}u. \tag{23-5}$$

It follows from (23-5) that *the nonhomogeneous equation can be solved completely if a single nonvanishing solution of the homogeneous equation is known.*   For proof, let u be the known solution of $\mathbf{T}u = 0$ and let a solution of $\mathbf{T}y = f$ be sought in the form $y = uv$. By (23-5) the equation $\mathbf{T}y = f$ becomes

$$uv'' + (2u' + p_1u)v' = f. \tag{23-6}$$

This is a first-order equation in the unknown $V = v'$ and can be solved by the method of Chap. 2, Sec. 5.   Indeed, if we multiply by ue^P the result is $(v'u^2e^P)' = ue^Pf$, integration yields

$$v'u^2e^P = \int ue^Pf \, dx + c \tag{23-7}$$

and dividing by u^2e^P and integrating again, we get v.   The steps are reversible since f and p_1 are continuous; hence $y = uv$ satisfies $\mathbf{T}y = f$.

It is not hard to prove that a *general* solution of $\mathbf{T}y = f$ is given by this process, on any interval in which the basic condition $u \neq 0$ is fulfilled.   In particular, if $f = 0$, Eq. (23-7) with $c = 1$ yields

$$v = \int u^{-2}e^{-P} \, dx. \tag{23-8}$$

The resulting solutions $y = u$ and $y = uv$ of the homogeneous equation are independent because their ratio v has a nonvanishing derivative $u^{-2}e^{-P}$.

As another application we show, if p_1 is differentiable, that *the equation* $\mathbf{T}y = f$ *can be transformed into another equation in which the first-derivative term does not occur.*   For proof, take $y = uv$ as before but, instead of having $\mathbf{T}u = 0$, let

$$2u' + p_1u = 0. \tag{23-9}$$

In this case, by (23-5) the equation $\mathbf{T}y = f$ becomes

$$uv'' + v\mathbf{T}u = f \tag{23-10}$$

which lacks the term v'.   We can easily get $u' = -\frac{1}{2}p_1u$ and u'' from (23-9), with the result that

$$\mathbf{T}u = u(p_0 - \tfrac{1}{2}p_1' - \tfrac{1}{4}p_1^2).$$

The value $u = e^{-\frac{1}{2}P}$ is found by separating variables in (23-9), and substitution into (23-10) reduces the latter to

$$v'' + (p_0 - \tfrac{1}{2}p_1' - \tfrac{1}{4}p_1^2)v = e^{\frac{1}{2}P}f. \tag{23-11}$$

In summary, *if v is any solution of* (23-11), *the function*

$$y = e^{-\frac{1}{2}P}v$$

satisfies the general equation $\mathbf{T}y = f$.   The advantage of (23-11) is that it lacks the term v'.

Example 1.   The equation

$$y'' - \frac{4x}{2x - 1}y' + \frac{4}{2x - 1}y = 0 \qquad x > \tfrac{1}{2}$$

has an obvious solution $y = x$.   To get a second solution we compute

$$-P(x) = \int \frac{4x}{2x - 1} \, dx = \int \left[2 + \frac{2}{2x - 1} \right] dx = 2x + \log (2x - 1)$$

so that e^{-P} in (23-8) is $(2x - 1)e^{2x}$.   Thus,

$$v = \int x^{-2}(2x - 1)e^{2x}\,dx = \int 2x^{-1}e^{2x}\,dx - \int x^{-2}e^{2x}\,dx$$

$$= x^{-1}e^{2x} + \int x^{-2}e^{2x}\,dx - \int x^{-2}e^{2x}\,dx = x^{-1}e^{2x}.$$

A second solution, therefore, is $y = xv = e^{2x}$. Although the formula (23-8) requires $u \neq 0$ (in the present case $x \neq 0$) the solution e^{2x} is actually valid without restriction.

Example 2.   In an interval (a,b) containing the origin, suppose $uTv = vTu$, and suppose $v(0) \neq 0$. If $W(u,v) = 0$ at some point of (a,b), prove there is a constant c such that $u = cv$ in a neighborhood of $x = 0$.

If $W(x_0) = 0$, Abel's identity gives $W(x) \equiv 0$. This shows that $(u/v)' = 0$ at all points where $v \neq 0$, and the result follows at once.

The statement has an interesting application to series solutions.   Let

$$u = a_0 + a_1x + a_2x^2 + \cdots \quad \text{and} \quad v = b_0 + b_1x + b_2x^2 + \cdots$$

be series solutions of $Ty = 0$, with $b_0 \neq 0$. If $W(u,v) = 0$ at any point where the solutions are valid we deduce that $u = cv$, and uniqueness of power series then shows that $a_n = cb_n$ for all n. In other words, *any two power-series solutions are independent over their whole interval of definition, or else one is a constant multiple of the other.* A similar result applies to generalized power-series solutions, since the latter can be expressed as ordinary power series about a point $x_1 \neq 0$ sufficiently close to 0.

PROBLEMS

1. The following equations have an obvious solution $y = 1$.   Obtain a second solution by (23-8):

$$xy'' + 2y' = 0, \qquad xy'' + y' = 0, \qquad y'' \sin 2x = 2y' \cos 2x.$$

2. Verify that the following equations have solutions $y = x$, $y = x^{-1} \sin x$, $y = x^2$, respectively, and thus obtain a general solution:

$$y'' = 2\,\frac{xy' - y}{1 - x^2}, \qquad y'' + \frac{2}{x}\,y' + y = 0, \qquad y'' = \frac{2x^2 - 2x - 2}{x^3 + x^2}\,y - \frac{x^2 - 2x - 2}{x^2 + x}\,y'.$$

3. Solve $(2x - x^2)y'' + (x^2 - 2)y' + 2(1 - x)y = 0$.   *Hint:* Try $y = e^x$.

4. Verify that $y'' + (x + x^2)y' + (x^3 + 1)y = 0$ has a solution of the form $y = e^{ax^2}$, find the constant a, and thus solve the equation

$$y'' + (x + x^2)y' + (x^3 + 1)y = f(x)$$

by two integrations.   Partial answer: For $f(x) = 0$ a solution is

$$y = e^{-\frac{1}{2}x^2} \int e^{\frac{1}{2}x^2 - \frac{1}{3}x^3}\,dx.$$

5. Remove the first-derivative term from Bessel's equation (21-1) and thus show that the functions $v = \sqrt{x}\,J_p(x)$, $v = \sqrt{x}\,J_{-p}(x)$ both satisfy

$$x^2v'' + (x^2 - p^2 + \tfrac{1}{4})v = 0.$$

6. Explain how to solve $y'' + p_1y' + p_0y = f$ when $p_0 - \tfrac{1}{2}p_1' - \tfrac{1}{4}p_1^2 \equiv \omega^2$ is constant.   (The method often gives good approximation if ω is slowly varying, rather than constant.)

7. On a given interval (a,b), let $y = c_1u + c_2v$ be a family of solutions of $Ty = 0$.   Suppose the constants c_i can be determined so that y satisfies *arbitrarily specified* initial conditions $y(x_0) = y_0$, $y'(x_0) = y_1$ at a given point x_0.   Show that arbitrary initial conditions can also be satisfied at any other point x_1; in other words, y is a general solution.   [The hypothesis implies $W(x_0) \neq 0$; hence by Abel's identity $W(x_1) \neq 0$.]

8. For $|x| < r$, let $u = a_0 + a_1x + \cdots$ and $v = b_0 + b_1x + \cdots$ be power-series solutions of $Ty = 0$. Prove that $y = c_1u + c_2v$ is a general solution if, and only if, $a_0b_1 \neq a_1b_0$. (See Prob. 7, or Example 2.)

9. Let u be a nonvanishing solution of $\mathbf{T}u = 0$. Obtain a second solution v, with $W(u,v) = 1$ at x_0, by the use of Abel's identity. (Regard $uv' - u'v = e^{-P}$ as a differential equation for the unknown function v. An integrating factor is u^{-2}.)

10. (a) Apart from a constant factor, determine the Wronskian for the Legendre, Hermite, Chebychev, and Laguerre equations. (b) Remove the second term from these equations. [See (20-7) and Probs. 10 to 12 of Sec. 20.]

11. If y_1 and y_2 satisfy $(py')' + qy = 0$, where $p \neq 0$ and p' is continuous, show by Abel's identity that $p(x)W(y_1,y_2)$ is constant. Apply to Prob. 6, Sec. 22.

12. (a) Let $\mathbf{T}y = y''' + p_2y'' + p_1y' + p_0y$, where $p_i = p_i(x)$. If $\mathbf{T}u = 0$ and $u \neq 0$, show that the substitution $y = uv$ reduces $\mathbf{T}y = f$ to a second-order equation in $V = v'$. (b) Similarly discuss reduction of the order of the nth-order equation $\mathbf{T}y = f$.

13. Find the general solution of $x^3y''' - 3x^2y'' + 6xy' - 6y = 0$ which has two linearly independent solutions $y = x$ and $y = x^2$. See Prob. 12.

14. Solve $y'' + p_1(x)y' + p_0(x)y = f(x)$, if one of the following holds on an interval:

$$p_1 + p_0x = 0, \qquad xp_1 = 2 + x^2p_0, \qquad 1 + p_1 + p_0 = 0, \qquad 1 + p_0 = p_1, \qquad p_1x = 1 + p_0x^2 \log x.$$

Hint: Try $y = x^a$, e^{ax}, or $\log (ax)$, and use (23-7).

---

24. Linear Dependence.

The discussion of the Wronskian and independence outlined for second-order equations in Sec. 3 is easily extended to homogeneous equations of the nth order,

$$y^{(n)} + p_{n-1}(x)y^{(n-1)} + \cdots + p_1(x)y' + p_0(x)y = 0. \tag{24-1}$$

If $y_1, y_2, \ldots, y_n$ are solutions of (24-1) the superposition principle shows that

$$y = c_1y_1 + c_2y_2 + \cdots + c_ny_n \tag{24-2}$$

is also a solution for any choice of the constants c_i. Such an expression is called a *general solution* on a given interval (a,b) provided the constants c_i can be determined so as to satisfy the arbitrarily specified initial conditions[1]

$$y(x_0) = \tilde{y}_0, \qquad y'(x_0) = \tilde{y}_1, \qquad \ldots, \qquad y^{(n-1)}(x_0) = \tilde{y}_{n-1} \tag{24-3}$$

at any point x_0 of (a,b). If (24-2) is substituted into (24-3) the result is a system of n equations in the n constants $c_1, c_2, \ldots, c_n$, with coefficient determinant $W(x_0)$, where

$$W(x) \equiv \begin{vmatrix} y_1 & y_2 & \cdots & y_n \\ y_1' & y_2' & \cdots & y_n' \\ \cdot & \cdot & \cdots & \cdot \\ y_1^{(n-1)} & y_2^{(n-1)} & \cdots & y_n^{(n-1)} \end{vmatrix} \tag{24-4}$$

The determinant (24-4) is called the *Wronskian* and is denoted by $W(y_1,y_2,\ldots,y_n)$ or by $W(x)$, according as dependence on the functions y_k or on the variable x is emphasized. Thus,

$$W(x) = W[y_1(x),y_2(x),\ldots,y_n(x)].$$

We conclude that (24-2) *is a general solution if, and only if,* $W(x) \neq 0$ *for all x on (a,b).*

---

[1] We use the notation $\tilde{y}_k$ to avoid confusion between the constants $\tilde{y}_1, \tilde{y}_2, \ldots$ on the right of (24-3) and the functions $y_1, y_2, \ldots$ that occur elsewhere in this discussion.

As an illustration, for $n = 3$ the three equations (24-3) are

$$c_1y_1(x_0) + c_2y_2(x_0) + c_3y_3(x_0) = \bar{y}_0$$
$$c_1y_1'(x_0) + c_2y_2'(x_0) + c_3y_3'(x_0) = \bar{y}_1$$
$$c_1y_1''(x_0) + c_2y_2''(x_0) + c_3y_3''(x_0) = \bar{y}_2$$

with coefficient determinant

$$W(x) = W(y_1,y_2,y_3) = \begin{vmatrix} y_1 & y_2 & y_3 \\ y_1' & y_2' & y_3' \\ y_1'' & y_2'' & y_3'' \end{vmatrix} \tag{24-5}$$

evaluated at x_0. There is a solution for every choice of the constants $\bar{y}_0$, $\bar{y}_1$, $\bar{y}_2$ if, and only if, $W(x_0) \neq 0$. This happens for each x_0 on (a,b) if, and only if, $W(x)$ does not vanish anywhere on (a,b).

The functions $y_1, y_2, \ldots, y_n$ are said to be *linearly dependent* on an interval (a,b) if there are constants $c_1, c_2, \ldots, c_n$, not all zero, such that

$$c_1y_1(x) + c_2y_2(x) + \cdots + c_ny_n(x) \equiv 0 \qquad a < x < b. \tag{24-6}$$

When no such constants exist, the y_k are *linearly independent* on (a,b). The functions now need not satisfy (24-1), but we assume that they have $n-1$ derivatives at least. Then differentiation of (24-6) leads to a system of equations in the constants c_i whose determinant, again, is $W(x)$. For linear dependence this system has a nontrivial solution, c_i. Since a system of homogeneous equations has a nontrivial solution only if the coefficient determinant is zero, we conclude that *the functions y_i can be linearly dependent only if their Wronskian vanishes at every point of (a,b).*

For example, when $n = 3$ the three equations obtained from (24-6) as it stands and by differentiating are

$$c_1y_1 + c_2y_2 + c_3y_3 = 0$$
$$c_1y_1' + c_2y_2' + c_3y_3' = 0$$
$$c_1y_1'' + c_2y_2'' + c_3y_3'' = 0.$$

These equations can have a solution c_i, not all 0, only if the coefficient determinant is zero. The latter is $W(y_1,y_2,y_3)$ by (24-5).

The condition $W(x) \neq 0$ is sufficient but not necessary for linear independence. For example, $y_1 = x^3$ and $y_2 = |x|^3$ are linearly independent on any interval containing the origin, although their Wronskian vanishes for every value of x. These functions satisfy the equation $y'' + p_0(x)y = 0$, where

$$p_0(x) = 6x^{-2} \qquad \text{for } x \neq 0 \qquad p_0(0) = 0.$$

We now show that such an example is not possible if the coefficients of the differential equation are bounded:

THEOREM I. *On a given interval (a,b), let $y_1, y_2, \ldots, y_n$ be solutions of one and the same differential equation (24-1) with bounded coefficients. Then the following statements are equivalent:*

(i) *The functions $y_1, y_2, \ldots, y_n$ are linearly independent.*

(ii) $c_1y_1 + c_2y_2 + \cdots + c_ny_n$ *is a general solution.*

(iii) $W(y_1,y_2,\ldots,y_n) \neq 0$ *holds for at least one x.*

(iv) $W(y_1,y_2,\ldots,y_n) \neq 0$ *holds for every x.*

We show that (i) $\Rightarrow$ (iv) $\Rightarrow$ (ii) $\Rightarrow$ (iii) $\Rightarrow$ (i), where $\Rightarrow$ means "implies." Indeed, if (iv) does not hold then $W(x_0) = 0$ for some x_0 on (a,b). Hence we can determine constants $c_1, c_2, \ldots, c_n$, not all zero, so that the expression

$$y = c_1 y_1 + c_2 y_2 + \cdots + c_n y_n \qquad (24\text{-}7)$$

satisfies zero initial conditions, $y(x_0) = 0, \ldots, y^{(n-1)}(x_0) = 0$. [This follows from the previous discussion of (24-2) to (24-4).] Since the function $y \equiv 0$ also satisfies (24-1) with zero initial conditions, the uniqueness theorem of Chap. 2, Sec. 17, shows that these two solutions coincide; that is, the expression (24-7) equals 0 for all x on (a,b), and this means that the functions $y_1, y_2, \ldots, y_n$ are linearly dependent. Hence (i) $\Rightarrow$ (iv). The remaining implications (iv) $\Rightarrow$ (ii) $\Rightarrow$ (iii) $\Rightarrow$ (i) were established in the foregoing discussion, even without assuming that the coefficients are bounded.

If any one of the four conditions holds, an arbitrary solution y can be written in the form

$$y = c_1 y_1 + c_2 y_2 + \cdots + c_n y_n.$$

This follows from (ii) and uniqueness, as in Sec. 3. We say, briefly, that the family $\{y_1, y_2, \ldots, y_n\}$ forms a *basis* for the set of solutions.

According to Theorem I, the Wronskian of n solutions is either identically zero, or it is never zero. A more detailed result of the same kind is contained in the following:

THEOREM II. *If* $y_1, y_2, \ldots, y_n$ *are any solutions of* (24-1) *on* (a,b), *then their Wronskian* $W(x)$ *satisfies* $W' + p_{n-1}(x)W = 0$ *on* (a,b).

Indeed, if $y_1, y_2, \ldots, y_n$ are functions possessing nth order derivatives, a discussion similar to that in Sec. 3 shows that the derivative of their Wronskian is obtained by differentiating the elements of the bottom row. When the y_k's satisfy (24-1) the resulting nth-order derivatives can be expressed in terms of lower-order derivatives by means of

$$y_k^{(n)} = -p_{n-1} y_k^{(n-1)} - \cdots - p_1 y_k' - p_0 y_k \qquad k = 1, 2, \ldots, n.$$

If we add suitable multiples of the first $n - 1$ rows to the last row the determinant reduces to

$$-p_{n-1} W(y_1, y_2, \ldots, y_n)$$

that is, $W' = -p_{n-1} W$, as stated in Theorem II.

All the essential features of the foregoing argument are contained in the case $n = 3$, and for the reader's convenience we write this case in full. According to the results of Appendix A, the derivative of $W(y_1, y_2, y_3)$ is obtained by differentiating each row separately and adding the resulting determinants, thus:

$$\begin{vmatrix} y_1' & y_2' & y_3' \\ y_1' & y_2' & y_3' \\ y_1'' & y_2'' & y_3'' \end{vmatrix} + \begin{vmatrix} y_1 & y_2 & y_3 \\ y_1'' & y_2'' & y_3'' \\ y_1'' & y_2'' & y_3'' \end{vmatrix} + \begin{vmatrix} y_1 & y_2 & y_3 \\ y_1' & y_2' & y_3' \\ y_1''' & y_2''' & y_3''' \end{vmatrix}.$$

The first two determinants are 0 by inspection, and hence the derivative is the third determinant. If the y_k's satisfy (24-1) with $n = 3$ then

$$y_1''' = -p_2 y_1'' - p_1 y_1' - p_0 y_1$$

and similarly for y_2 and y_3. Substitution gives $W'(x)$ in the form

$$\begin{vmatrix} y_1 & y_2 & y_3 \\ y_1' & y_2' & y_3' \\ -p_2 y_1'' - p_1 y_1' - p_0 y_1 & -p_2 y_2'' - p_1 y_2' - p_0 y_2 & -p_2 y_3'' - p_1 y_3' - p_0 y_3 \end{vmatrix}.$$

If we add p_0 times the first row to the bottom row, and then p_1 times the second row to the bottom row, we see that the terms involving p_0 and p_1 in the bottom row can be dropped. Factoring out $-p_2$ gives $-p_2 W(y_1, y_2, y_3)$, so that $W' = -p_2 W$.

If $p_{n-1}(x)$ is continuous in Theorem II, one can solve for W in the form

$$W(x) = W(x_0) e^{-p(x)} \qquad \text{where } P(x) = \int_{x_0}^{x} p_{n-1}(t) \, dt. \qquad (24\text{-}8)$$

This extends Abel's identity (23-3) to equations of the nth order. Equation (24-8) shows that $W(x) = 0$ if $W(x_0) = 0$ and thus establishes the equivalence of (iii) and (iv) in Theorem I without any hypothesis on the coefficients $p_{n-2}, \ldots, p_0$. By following suggestions in Prob. 8 the reader will find that the continuity of p_{n-1} can also be dispensed with, if $|p_{n-1}|$ is bounded.

Example 1.   Show that the functions $y_1 = x$, $y_2 = x^2$, $y_3 = x^3$ are linearly independent on every interval (a,b), no matter how short.   The Wronskian for this set of functions is

$$W(y_1, y_2, y_3) = \begin{vmatrix} x & x^2 & x^3 \\ 1 & 2x & 3x^2 \\ 0 & 2 & 6x \end{vmatrix} = 2x^3.$$

Since any given interval $a < x < b$ contains a point $x \neq 0$ we have $W(x) \neq 0$ for at least one point of the interval, and the independence follows.

Example 2.   If ω and α are constant, with $\omega \neq 0$ and $\alpha \neq \pm i\omega$, show that the Wronskian of the three functions $y_1 = \sin \omega x$, $y_2 = \cos \omega x$, $y_3 = e^{\alpha x}$ never vanishes.

By the results of Sec. 11 we know that these three functions are solutions of one and the same linear third-order homogeneous equation with constant coefficients. If $W(0) \neq 0$ it will follow from Theorem I that $W(x) \neq 0$.   But

$$W(0) = \begin{vmatrix} 0 & 1 & 1 \\ \omega & 0 & \alpha \\ 0 & -\omega^2 & \alpha^2 \end{vmatrix} = -\omega(\alpha^2 + \omega^2)$$

which is not zero under the given hypothesis.

Example 3.   On a given interval (a,b), suppose

$$c_1 e^{s_1 x} + c_2 e^{s_2 x} + \cdots + c_n e^{s_n x} = 0 \tag{24-9}$$

where the c_i and s_i are constants, with no two s_i equal.   Prove that each $c_i = 0$.

The result is evident for $n = 1$ and is established in general by mathematical induction.   Suppose the desired theorem holds for $n - 1$ functions, so that the equation

$$\gamma_1 e^{\sigma_1 x} + \gamma_2 e^{\sigma_2 x} + \cdots + \gamma_{n-1} e^{\sigma_{n-1} x} = 0 \tag{24-10}$$

implies $\gamma_i = 0$, whenever the γ_i are constants and the σ_i are unequal constants.   If the given equation (24-9) is multiplied by $e^{-s_n x}$ and differentiated with respect to x the result is an equation of the form of (24-10), where

$$\sigma_i = s_i - s_n \qquad \gamma_i = \sigma_i c_i.$$

By the induction hypothesis, $\gamma_i = 0$, and therefore $c_i = 0$.

The same method succeeds when the c_i are polynomials, $c_i(x)$.   [After multiplication by $e^{-s_n x}$ we simply differentiate often enough to get rid of $c_n(x)$, noting that the derivatives of $c(x)e^{sx}$ have the form $\gamma(x)e^{\sigma x}$, where γ is a polynomial of the same degree as c.]   This shows that the functions $c_i(x)e^{s_i x}$ are linearly independent, and so the procedure followed in Secs. 1 to 19 of this chapter actually yields a general solution.

PROBLEMS

1. If $y_1 = x^2 + 2x$, $y_2 = x^3 + x$, $y_3 = 2x^3 - x^2$, find constants c_1 and c_2 such that $y_3 = c_1 y_1 + c_2 y_2$. What does this indicate about the Wronskian?   Verify your conclusion by direct computation.

2. Test for linear dependence by use of the definition of linear dependence and also by computing the Wronskian:

(a) $(x + 1)^2$, $(x - 1)^2$, $3x$; e^{-x}, 1, e^x, $\sinh x$; e^{ix}, $\sin x$, $\cos x$;

(b) $x^2 - 2x + 5$, $3x - 1$, $\sin x$; e^x, xe^x, $x^2 e^x$; 1, $\sin x$, $\cos x$; $1, x, x^2, x^3$.

3. In Sec. 5 it was found that the solution of a certain linear equation $\mathbf{T}y = 0$ satisfying $(y,y') = (c_1,c_2)$ at 0 can be written $y = c_1y_1 + c_2y_2$, where y_1 and y_2 satisfy $\mathbf{T}y = 0$ with initial conditions $(y,y') = (1,0)$ and $(0,1)$, respectively. Generalize this result.

4. Referring to Example 1 and Theorem I, show that x, x^2, x^3 cannot be solutions of one and the same third-order equation (24-1), with bounded coefficients, on any interval containing $x = 0$. Show, however, that $1, x, x^2, x^3$ satisfy a fourth-order equation with constant coefficients.

5. If u and y_k have $n - 1$ derivatives, establish the following relation (a) for $n = 3$ and (b) in general:

$$W(uy_1, uy_2, \ldots, uy_n) = u^n W(y_1, y_2, \ldots, y_n).$$

6. By Prob. 5 and Example 1, find $W(xe^x, x^2e^x, x^3e^x)$.

7. In Theorem II, if $|P_{n-1}(x)| \le m$, where m is constant, show that

$$|W(x)| \ge |W(x_0)|e^{-m|x-x_0|}.$$

8. On an interval (a,b) let $W' + p_{n-1}(x)W = 0$, where p_{n-1} is not necessarily continuous or integrable but $p_{n-1}(x) \le m$ for some constant m. Assuming $W(x_0) \ne 0$, define w by $W(x) = W(x_0)e^w$ near $x = x_0$. Deduce that $w' + p_{n-1} = 0$, hence $w' \ge -m$, and therefore

$$|W(x)| \ge |W(x_0)|e^{-m|x-x_0|} \qquad x_0 \le x < b.$$

Similarly consider $p_{n-1}(x) \ge -m$, and thus obtain the result of Prob. 7.

9. Let y_1 and y_2 be differentiable and suppose $W(y_1,y_2) \ne 0$. Prove that the zeros of y_1 separate those of y_2, in the following sense: If $y_1 = 0$ at x_1 and at another point x_2, then y_2 is not zero at x_1 or x_2 but has a zero between x_1 and x_2. *Hint:* If $y_2 \ne 0$ for $x_1 \le x \le x_2$ then y_1/y_2 has a maximum or minimum, and $(y_1/y_2)' = 0$ there.

---

25. Variation of Parameters. Green's Function.

According to Sec. 23, the equation

$$\mathbf{T}y \equiv y'' + p_1(x)y' + p_0(x)y = f(x) \qquad a < x < b \tag{25-1}$$

is simplified by the substitution $y = y_1v$, where y_1 satisfies the homogeneous equation $\mathbf{T}y_1 = 0$. If two independent solutions y_1 and y_2 are available for the homogeneous equation, the substitution

$$y = v_1y_1 + v_2y_2 \tag{25-2}$$

leads to still greater simplification, as is now shown. The method is similar to the method of undetermined coefficients except that the unknowns are functions v_1 and v_2, rather than constants.

We assume that $f(x)$ is continuous for $a \le x \le b$ and that

$$y = c_1y_1 + c_2y_2$$

is a general solution of $\mathbf{T}y = 0$ on (a,b). The expression (25-2) is obtained when the constant parameters c_1 and c_2 are replaced by the functions v_1 and v_2; for this reason, the procedure based on (25-2) is called the method of *variation of parameters*.

If we substitute $y = v_1y_1 + v_2y_2$ into the equation $\mathbf{T}y = f$ we obtain one condition to be satisfied by the two unknown functions v_1 and v_2. We shall impose a second condition on v_1 and v_2 in a way that would tend to simplify the calculations.

Differentiating the relation $y = v_1y_1 + v_2y_2$ gives

$$y' = (v_1y_1' + v_2y_2') + (v_1'y_1 + v_2'y_2). \tag{25-3}$$

Now, the calculation of y'' will be materially simplified if v_1 and v_2 are chosen so that the expression in the second parentheses in (25-3) vanishes. Accordingly, we set

$$v_1'y_1 + v_2'y_2 = 0 \tag{25-4}$$

and take $$y' = v_1y_1' + v_2y_2'. \tag{25-5}$$

Then $$y'' = v_1y_1'' + v_2y_2'' + v_1'y_1' + v_2'y_2'. \tag{25-6}$$

The substitution from (25-2), (25-5), and (25-6) in the original equation (25-1) yields, on rearrangement,

$$v_1(y_1'' + p_1y_1' + p_2y_1) + v_2(y_2'' + p_1y_2' + p_2y_2) + v_1'y_1' + v_2'y_2' = f(x).$$

But since y_1 and y_2 are known to satisfy $Ty = 0$, the two expressions in parentheses vanish. We thus get

$$v_1'y_1' + v_2'y_2' = f(x). \tag{25-7}$$

The pair of equations (25-4) and (25-7) can be solved for v_1' and v_2' to yield

$$v_1' = \frac{\begin{vmatrix} 0 & y_2 \\ f & y_2' \end{vmatrix}}{\begin{vmatrix} y_1 & y_2 \\ y_1' & y_2' \end{vmatrix}} \qquad v_2' = \frac{\begin{vmatrix} y_1 & 0 \\ y_1' & f \end{vmatrix}}{\begin{vmatrix} y_1 & y_2 \\ y_1' & y_2' \end{vmatrix}}. \tag{25-8}$$

The determinant in the denominator is the Wronskian $W(y_1,y_2)$, and, by Theorem I of the preceding section, it does not vanish anywhere on (a,b). Upon integrating (25-8) we get v_1 and v_2, and $y = v_1y_1 + v_2y_2$ is therefore

$$y = y_1 \int \frac{-fy_2}{W(y_1,y_2)}\, dx + y_2 \int \frac{fy_1}{W(y_1,y_2)}\, dx. \tag{25-9}$$

Since the constants of integration produce an added term of the form $c_1y_1 + c_2y_2$, the result is actually a *general* solution of $Ty = f$.

As an illustration, consider $x^2y'' - 2xy' + 2y = f(x)$, for $x \neq 0$. It is easily checked that a pair of linearly independent solutions of the homogeneous equation is $y_1 = x$, $y_2 = x^2$, for which the Wronskian is

$$W(y_1,y_2) = x^2.$$

To reduce the given equation to the standard form (25-1) we must divide through by x^2. This has the effect of replacing $f(x)$ by $x^{-2}f(x)$. The result of substituting in (25-9) is then

$$y = -x \int x^{-2}f(x)\, dx + x^2 \int x^{-3}f(x)\, dx.$$

For example, when $f(x) = x \log x$ the solution reduces to

$$y = -x[1 + \log x + \tfrac{1}{2}(\log x)^2] + c_1x + c_2x^2$$

which agrees with the discussion of the Euler-Cauchy equation in Sec. 21.

The basic formula (25-9) has several important applications. Let x_0 be any point on the interval (a,b) and let the indefinite integrals be replaced by definite[1] integrals with respect to a dummy variable t from x_0 to x. Since the coefficients $y_1 = y_1(x)$ and $y_2(x)$ are constant as far as the integration is concerned, they can be moved under the integral signs. The result is the formula

$$y(x) = \int_{x_0}^{x} \frac{y_1(t)y_2(x) - y_1(x)y_2(t)}{y_1(t)y_2'(t) - y_1'(t)y_2(t)} f(t)\, dt. \tag{25-10}$$

This gives the solution of $Ty = f$ that satisfies $y(x_0) = y'(x_0) = 0$.

The condition $y'(x_0) = 0$ can be verified by differentiating the integral as explained in Chap. 5, Sec. 14, but a more elementary method is to note that (25-10) yields $v_1(x_0) = v_2(x_0) = 0$. This together with (25-4) gives $(v_1y_1 + v_2y_2)' = 0$ at x_0.

---

[1] The validity of (25-8) is assured by the fundamental theorem of calculus, since f is continuous.

Another special case of (25-9) is useful in the solution of boundary-value problems. Let the first integral in (25-9) be replaced by a definite integral from b to x, and the second by an integral from a to x. The limits of integration (b,x) can be changed to (x,b) if we change the sign; hence

$$y(x) = \int_a^x \frac{y_1(t)y_2(x)}{W(t)} f(t)\, dt + \int_x^b \frac{y_1(x)y_2(t)}{W(t)} f(t)\, dt \qquad (25\text{-}11)$$

where $W(t) = W[y_1(t),y_2(t)]$. If we define

$$G(x,t) = \frac{y_1(t)y_2(x)}{W(t)} \qquad \text{or} \qquad G(x,t) = \frac{y_1(x)y_2(t)}{W(t)} \qquad (25\text{-}12)$$

according as $a < t < x$ or $x < t < b$, Eq. (25-11) is equivalent to

$$y(x) = \int_a^b G(x,t)f(t)\, dt. \qquad (25\text{-}13)$$

Let it be supposed, now, that y_1 and y_2 satisfy boundary conditions at a and b of the form

$$\alpha_1 y_1(a) + \beta_1 y_1'(a) = 0 \qquad \alpha_2 y_2(b) + \beta_2 y_2'(b) = 0. \qquad (25\text{-}14)$$

As $x \to a+$ it is clear that we should use the second formula (25-12), and as $x \to b-$ we should use the first, t being fixed on (a,b). This shows that G satisfies boundary conditions similar to (25-14) at both a and b, when G is regarded as a function of x. By differentiating under the integral in (25-13), we verify that corresponding boundary conditions are also satisfied by y. (See Chap. 5, Sec. 14.) *In other words, the formula* (25-13) *gives the solution of* $\mathbf{T}y = f$ *that satisfies*

$$\alpha_1 y(a) + \beta_1 y'(a) = 0 \qquad \alpha_2 y(b) + \beta_2 y'(b) = 0.$$

It should be noted that the analysis automatically establishes the existence of such a solution, provided functions y_i satisfying (25-14), $\mathbf{T}y = 0$, and the basic condition $W(y_1,y_2) \neq 0$ can be found. The function $G(x,t)$ determined by this analysis is called *Green's function* for the boundary-value problem associated with (25-1) and (25-14).

For example, consider $y'' + y = f(x)$, $y'(0) = 0$, $y(\pi) = 0$. Here suitable solutions of the homogeneous equation $y'' + y = 0$ are

$$y_1 = \cos x \qquad y_2 = \sin x.$$

These satisfy $y_1'(0) = 0$, $y_2(\pi) = 0$, and $W(y_1,y_2) = 1$. By (25-12) we have $G(x,t) = \cos t \sin x$ or $\cos x \sin t$, according as $t < x$ or $x < t$. The solution is given by (25-13), but for numerical evaluation it is preferable to use (25-11). Thus, when $f(x) \equiv 1$ we get

$$y = \int_0^x \cos t \sin x\, dt + \int_x^\pi \cos x \sin t\, dt = 1 + \cos x.$$

It is easily checked that $y'' + y = 1$, $y'(0) = 0$, $y(\pi) = 0$.

The method of variation of parameters is due to the French-Italian mathematician Joseph Louis Lagrange (who also made important contributions to number theory and was one of the founders of analytical mechanics). Lagrange applied his method to the general equation

$$y^{(n)} + p_{n-1}(x)y^{(n-1)} + \cdots + p_1(x)y' + p_0(x)y = f(x) \qquad (25\text{-}15)$$

and we shall briefly indicate how this is done.

Assuming that $c_1 y_1 + c_2 y_2 + \cdots + c_n y_n$ is the general solution of the associated homogeneous equation, we seek an integral of (25-15) in the form

$$y = v_1 y_1 + v_2 y_2 + \cdots + v_n y_n \qquad (25\text{-}16)$$

where the v_i's are unknown functions. To determine them, we form the set of $n - 1$

equations obtained by equating to zero the terms involving the v_i's in the expressions resulting from differentiating (25-16) successively $n-1$ times. The nth equation is obtained by inserting corresponding values of derivatives in (25-15). After straightforward but tedious calculation, the result is the formula

$$y(x) = \sum_{i=1}^{n} y_i(x) \int \frac{W_i(y_1, y_2, \ldots, y_n)}{W(y_1, y_2, \ldots, y_n)} f(x) \, dx \qquad (25\text{-}17)$$

where $W(y_1, y_2, \ldots, y_n)$ is the Wronskian and W_i is the determinant obtained from W by replacing the ith column by $(0, 0, \ldots, 0, 1)$. For $n = 2$, Eq. (25-17) reduces, as it should, to $y_1 v_1 + y_2 v_2$, where the v_i's satisfy (25-8).

Example.   Find an integral of

$$y''' + \frac{1}{x^2} y' - \frac{1}{x^3} y = \frac{1}{x^2} \log x \qquad x > 0. \qquad (25\text{-}18)$$

A set of linearly independent solutions of the corresponding homogeneous equation is known to be

$$y_1 = x \qquad y_2 = x \log x \qquad y_3 = x(\log x)^2. \qquad (25\text{-}19)$$

Accordingly, we take the integral of (25-18) in the form

$$y = v_1 x + v_2 x \log x + v_3 x(\log x)^2. \qquad (25\text{-}20)$$

For the third-order equation the procedure just sketched yields the system of three equations:

$$\begin{aligned} v_1' y_1 + v_2' y_2 + v_3' y_3 &= 0 \\ v_1' y_1' + v_2' y_2' + v_3' y_3' &= 0 \\ v_1' y_1'' + v_2' y_2'' + v_3' y_3'' &= f(x). \end{aligned} \qquad (25\text{-}21)$$

The reader will verify that, on setting $f(x) = (1/x^2) \log x$ and noting (25-19), the system (25-21) yields

$$v_1' = \frac{1}{2x} (\log x)^3 \qquad v_2' = -\frac{1}{x} (\log x)^2 \qquad v_3' = \frac{1}{2x} \log x$$

and we can take

$$v_1 = \tfrac{1}{8}(\log x)^4 \qquad v_2 = -\tfrac{1}{3}(\log x)^3 \qquad v_3 = \tfrac{1}{4}(\log x)^2.$$

Substitution in (25-20) gives finally $y = (x/24)(\log x)^4$.

PROBLEMS

1. Use the method of variation of parameters to find integrals of the following equations with constant coefficients:

(a) $y' + 3y = x^3$, $\quad y'' - 2y' + y = x$, $\quad y'' + y = \cot x$;

(b) $y'' + 5y' + 6y = e^x$, $\quad y''' - 3y' + 2y = 2(\sin x - 2 \cos x)$, $\quad y'' - 2y = 4x^3 e^{x^2}$.

2. Find the solution of $y' + p_0(x)y = f(x)$ by the method of variation of parameters, and verify agreement with (25-17).

3. By the method of variation of parameters, find a particular integral of

$$x^2 y'' - 3xy' - 5y = x^2 \log x$$

where the general solution of the homogeneous equation is $c_1 x^{-1} + c_2 x^5$.

4. Solve $(1 - x)y'' + xy' - y = (1 - x)^2$, given that e^x and x satisfy the associated homogeneous equation.

5. (a) Show that Green's function for $y'' = f(x)$, $y(0) = y(1) = 0$, is

$$G(x,t) = t(x - 1) \quad \text{for } t < x; \qquad G(x,t) = x(t - 1) \quad \text{for } x < t.$$

(b) Using $G(x,t)$, compute y when $f(x) = x^n$, $n \geq 0$, and verify that y has the expected properties.

(c) Apply part b when $f(x)$ has a convergent power-series expansion $\Sigma a_n x^n$.

6. By (25-10), solve $(D - s_1)(D - s_2)y = f(x)$, where s_1 and s_2 are unequal constants and $y(0) = y'(0) = 0$. Show that the result agrees with Heaviside's expansion theorem (Sec. 18).

7. (*a*) Solve $(D - 1)^2 y = f(x)$, $y(x_0) = y'(x_0) = 0$. (*b*) Find Green's function for the boundary conditions $y(0) = y'(1) = 0$. (*c*) Discuss the application of Green's function to the four boundary-value problems of Sec. 10, Prob. 1.

8. Discuss the existence of Green's function for the problem

$$y'' + \omega^2 y = 0 \qquad y(0) = y(1) = 0 \qquad \omega = \text{const.}$$

9. Prove that Green's function is symmetric, that is, $G(x,t) = G(t,x)$ if $p_1(x) \equiv 0$.

10. On any interval not containing the origin, show that a solution of the inhomogeneous Bessel equation $x^2 y'' + xy' + (x^2 - p^2)y = f(x)$ is

$$y = \tfrac{1}{2}\pi \csc \pi p \int_{x_0}^{x} [J_p(x)J_{-p}(t) - J_p(t)J_{-p}(x)]\frac{f(t)}{t}\, dt$$

provided p is not an integer.   *Hint:* See Eq. (23-4).

Algebra and geometry of vectors; matrices

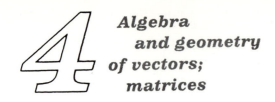

4 Algebra and geometry of vectors; matrices

It is desirable to treat directed quantities like *force* or *velocity* (which are independent of coordinate systems) without reference to a set of coordinate axes. Such a coordinate-free treatment is made possible by the analytical shorthand known as *vector analysis*. The trajectory of a particle, the dynamics of rigid bodies, and the theory of fluid flow are readily studied by vector methods, as are also such topics as the geometry of curves and surfaces. Introduction of coordinates yields a correspondence between vectors and sets of numbers, and this correspondence permits the use of vector methods in the study of linear equations. Such a study leads to the concept of *matrix*, which has proved fruitful in a variety of fields, ranging from circuit analysis to quantum theory.

FUNDAMENTAL OPERATIONS

1. Scalars, Vectors, and Equality. Some quantities appearing in the study of physical phenomena can be completely specified by their magnitude alone. Thus, the mass of a body can be described by the number of grams, the temperature by degrees on some scale, the volume by the number of cubic units, and so on. A quantity that (after a suitable choice of units) can be completely characterized by a single number is called a *scalar*. There are also quantities, called *vectors*, that require for their complete characterization the specification of direction as well as magnitude. An example of a vector quantity is the displacement of translation of a particle. If a particle is displaced from a position P to a new position P' (Fig. 1), the change in position can be represented graphically by the directed line segment PP' whose length equals the amount of the displacement and whose direction is from P to P'. Similarly, a force of magnitude

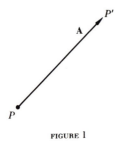

FIGURE 1

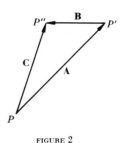

FIGURE 2

258

K dynes can be represented by a line segment whose length is K units and whose direction coincides with that of the force.

The initial point P of a directed line segment representing a vector is called the *origin*, and the representation as an arrow suggests that the terminal point be called the *head* of the vector. In many problems the location of the origin for any given vector is immaterial, and in such problems two vectors are regarded as equal if they have the same length and the same direction. Such vectors, which need not coincide to be equal, are termed *free vectors*. In mechanics, it is sometimes convenient to specify vectors by giving the line of action as well as the length and direction. Equality of these so-called *sliding vectors* means that the lengths, directions, and lines of action coincide. Again, in the treatment of space curves and trajectories one is led to specify the origin of the vector as well as its length and direction. Such vectors are termed *bound vectors*.

To distinguish vectors from scalars, boldface type is used for vectors in this book. The length (or *magnitude*) of the vector $\mathbf{A}$ is denoted by $|\mathbf{A}|$:

$$|\mathbf{A}| = \text{length of } \mathbf{A}. \qquad (1\text{-}1)$$

Equality is denoted by the usual symbol: $\mathbf{A} = \mathbf{B}$. For the most part this chapter deals with free vectors, and hence "$\mathbf{A} = \mathbf{B}$" means that $\mathbf{A}$ and $\mathbf{B}$ have the same length and direction.

2. Addition, Subtraction, and Multiplication by Scalars.

If a particle is displaced from its initial position P to P', so that $\overrightarrow{PP'} = \mathbf{A}$, and subsequently it is displaced to a position P'', so that $\overrightarrow{P'P''} = \mathbf{B}$, the displacement from the original position P to the final position P'' can be accomplished by the single displacement $\overrightarrow{PP''} = \mathbf{C}$. Thus, it is logical to write

$$\mathbf{A} + \mathbf{B} = \mathbf{C}$$

as the definition of vector addition (Fig. 2). In words, if the initial point of the vector $\mathbf{B}$ is placed in coincidence with the terminal point of the vector $\mathbf{A}$, then the vector $\mathbf{C}$, which joins the initial point of $\mathbf{A}$ with the terminal point of $\mathbf{B}$, is called the *sum* of $\mathbf{A}$ and $\mathbf{B}$ and is denoted by $\mathbf{A} + \mathbf{B} = \mathbf{C}$. This is the familiar *parallelogram law of addition* used in physics, and its extension to three or more vectors is obvious. The symbol $+$ behaves like the $+$ of elementary algebra, in that

$$\begin{aligned} \mathbf{A} + \mathbf{B} &= \mathbf{B} + \mathbf{A} \qquad \text{commutative law} \\ \mathbf{A} + (\mathbf{B} + \mathbf{C}) &= (\mathbf{A} + \mathbf{B}) + \mathbf{C} \qquad \text{associative law.} \end{aligned} \qquad (2\text{-}1)$$

A proof is implicit in Figs. 3 and 4. The associative law enables us to omit parentheses, writing $\mathbf{A} + \mathbf{B} + \mathbf{C}$ for $\mathbf{A} + (\mathbf{B} + \mathbf{C})$.

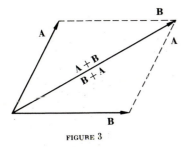

FIGURE 3

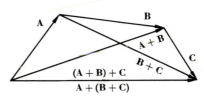

FIGURE 4

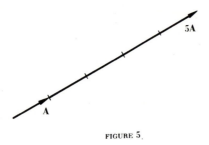

FIGURE 5.

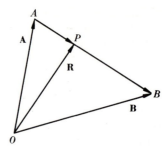

FIGURE 6

It is desirable to give meaning to expressions like **5A**, the product of a scalar and a vector. In agreement with the meaning of multiplication familiar from arithmetic, one defines

$$5\mathbf{A} = \mathbf{A} + \mathbf{A} + \mathbf{A} + \mathbf{A} + \mathbf{A} \tag{2-2}$$

(Fig. 5) and similarly in other cases. By a natural extension of this reasoning, *t***A** *is defined as a vector whose length is* $|t|\,|\mathbf{A}|$ *and whose direction is that of* **A** *if t is positive but opposite to that of* **A** *if t is negative.* One defines **A***t* by the equation

$$\mathbf{A}t = t\mathbf{A}. \tag{2-3}$$

It follows that 1**A** = **A** and also

$$
\begin{aligned}
s(t\mathbf{A}) &= (st)\mathbf{A} && \text{associative law} \\
(s + t)\mathbf{A} &= s\mathbf{A} + t\mathbf{A} && \text{distributive law} \\
t(\mathbf{A} + \mathbf{B}) &= t\mathbf{A} + t\mathbf{B} && \text{distributive law.}
\end{aligned}
\tag{2-4}
$$

A vector of zero length is denoted by **0** and termed the *zero vector.* To introduce the idea of subtraction, one defines −**A** as the solution of the equation **A** + **X** = **0**. Evidently, −**A** is a vector equal in length to **A** but of opposite direction, so that −**A** = (−1)**A**. As in elementary algebra, **B** − **A** is used as an abbreviation for **B** + (−**A**).

Since the laws governing the addition of vectors and multiplication of vectors by scalars are identical with those met in ordinary algebra, one is justified in using the familiar rules of algebra to solve linear equations involving vectors.

Example 1. The point *P* in Fig. 6 divides the segment *AB* in the ratio *m/n*. Express **R** in terms of the vectors **A**, **B** and the scalars *m, n*.

The vector $\mathbf{X} = \overrightarrow{AB}$ satisfies **A** + **X** = **B** by the definition of vector addition, and hence, solving for **X**,

$$\mathbf{X} = \mathbf{B} - \mathbf{A}. \tag{2-5}$$

(This exemplifies the so-called "head-minus-tail" rule for vector subtraction.) The vector $\overrightarrow{AP}$ is $m/(m + n)$ times **X**, by the hypothesis and by the definition of multiplication by scalars. Since $\mathbf{R} = \mathbf{A} + \overrightarrow{AP}$ we have, finally,

$$\mathbf{R} = \mathbf{A} + \frac{m}{m + n}(\mathbf{B} - \mathbf{A}) = \frac{n\mathbf{A} + m\mathbf{B}}{m + n}.$$

Example 2. Prove that the medians of every triangle intersect at a point two-thirds of the way from each vertex to the opposite side.

Let two sides of the triangle be specified by vectors **A** and **B**, as in Fig. 7, so that the third side is **B** − **A** (cf. Example 1). The vector median to the side **B** − **A** is one-half the diagonal of the parallelogram on **A**, **B**; hence this median is (**A** + **B**)/2 (compare the special case *m* = *n*

of Example 1). If the point M in Fig. 7 is two-thirds
of the way from the vertex O to the side $\mathbf{B} - \mathbf{A}$, along
this median, then

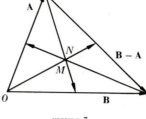

$$OM = \frac{2}{3}\frac{\mathbf{A} + \mathbf{B}}{2} = \frac{\mathbf{A} + \mathbf{B}}{3}. \qquad (2\text{-}6)$$

The vector median to the side $\mathbf{A}$ is $\mathbf{A}/2 - \mathbf{B}$, again
by the head-minus-tail rule. If N is at a point two-
thirds of the way toward the side $\mathbf{A}$ on this median,
then

FIGURE 7

$$ON = \mathbf{B} + \frac{2}{3}\left(\frac{\mathbf{A}}{2} - \mathbf{B}\right) = \frac{\mathbf{A} + \mathbf{B}}{3}.$$

Comparison with (2-6) shows that the two points M, N coincide. That the third median also
has the required behavior follows by interchanging the roles of $\mathbf{A}$ and $\mathbf{B}$.

PROBLEMS

1. Sketch a vector $\mathbf{A}$ of length 1.5 in., parallel to the lower edge of your paper and having an arrow
on its right-hand end. Sketch a second vector $\mathbf{B}$ of length 1 in., making an angle of 30° with $\mathbf{A}$.
Now sketch $2\mathbf{A}$, $3\mathbf{B}$, $\mathbf{A} + \mathbf{B}$, $\mathbf{A} - \mathbf{B}$, $2\mathbf{A} - 3\mathbf{B}$, $(\mathbf{A} + \mathbf{B})/2$.

2. Give a condition on three vectors $\mathbf{A}$, $\mathbf{B}$, $\mathbf{C}$ which ensures that they can form a triangle. Generalize to n vectors $\mathbf{A}$, $\mathbf{B}$, $\mathbf{C}$, . . . , $\mathbf{L}$.

3. Graphically and algebraically, show how to find two vectors $\mathbf{A}$ and $\mathbf{B}$ if their sum $\mathbf{S}$ and difference $\mathbf{D}$ are known.

4. Sketch three vectors $\mathbf{A}$, $\mathbf{B}$, $\mathbf{C}$ issuing from a common point. On your figure, show the vec-
tors $\mathbf{A} - \mathbf{C}$, $\mathbf{B} - \mathbf{A}$, $\mathbf{C} - \mathbf{B}$, and thus illustrate the algebraic identity $(\mathbf{A} - \mathbf{C}) + (\mathbf{B} - \mathbf{A}) +
(\mathbf{C} - \mathbf{B}) = 0$.

5. (*a*) Write a vector of unit length which has the same direction as a given nonzero vector $\mathbf{A}$.
(*b*) Using the result (*a*), write a vector bisecting the angle formed by two nonzero vectors $\mathbf{A}$, $\mathbf{B}$
issuing from a common point.

6. Show that a line from a vertex of a parallelogram to the midpoint of a nonadjacent side trisects
a diagonal.

7. Show that the diagonals of a parallelogram bisect each other.

8. If O is a point in the plane of a triangle ABC and P, Q, and R are the midpoints of the sides
of the triangle, show that $\overrightarrow{OA} + \overrightarrow{OB} + \overrightarrow{OC} = \overrightarrow{OP} + \overrightarrow{OQ} + \overrightarrow{OR}$.

9. (*a*) If $|\mathbf{A}| = 3$, $|\mathbf{B}| = 4$, $|\mathbf{C}| = 5$, and $\mathbf{A} + \mathbf{B} + \mathbf{C} = 0$, what can be said about $\mathbf{A}$, $\mathbf{B}$, and $\mathbf{C}$?
(*b*) If $\mathbf{A}$ and $\mathbf{B}$ are vectors from the origin O to the points A and B, and P is the midpoint of
the vector $\mathbf{B} - \mathbf{A}$, find the vector $\overrightarrow{OP}$.

10. If $\mathbf{B} - \mathbf{A} = \mathbf{C} - \mathbf{D}$ and $\mathbf{A}$, $\mathbf{B}$, $\mathbf{C}$, $\mathbf{D}$ are vectors from the origin to the points A, B, C, D,
show that $ABCD$ is a parallelogram.

---

3. Base Vectors. Any vector $\mathbf{A}$ lying in the plane of two noncollinear vectors $\mathbf{a}$ and $\mathbf{b}$
can be resolved into so-called "components" directed along $\mathbf{a}$ and $\mathbf{b}$. This resolution
is accomplished by constructing the parallelogram whose sides are parallel to $\mathbf{a}$ and $\mathbf{b}$
(Fig. 8). Then one can write

$$\mathbf{A} = x\mathbf{a} + y\mathbf{b}$$

where x and y are the appropriate scalars.

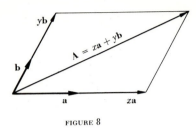

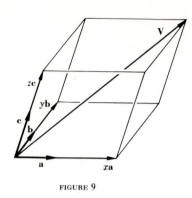

FIGURE 8

FIGURE 9

If three noncoplanar vectors **a**, **b**, and **c** are given, any vector **V** can be expressed uniquely as

$$\mathbf{V} = x\mathbf{a} + y\mathbf{b} + z\mathbf{c} \tag{3-1}$$

where **V** is the diagonal of the parallelepiped whose edges are $x\mathbf{a}$, $y\mathbf{b}$, and $z\mathbf{c}$ (Fig. 9). The vectors **a**, **b**, and **c** are called the *base vectors*, and the scalars x, y, and z the *measure numbers*.

An Important set of base vectors, denoted by **i**, **j**, and **k**, consists of unit vectors directed along the positive directions of the x, y, and z axes, respectively (Fig. 10).   It is assumed that the system of axes is a *right-handed system;* that is, a right-hand screw directed along the positive z axis advances in the positive direction when it is rotated from the positive x axis toward the positive y axis through the smaller (90°) angle. Because **i**, **j**, **k** are mutually orthogonal, the representation

$$\mathbf{A} = x\mathbf{i} + y\mathbf{j} + z\mathbf{k}$$

yields the important formula

$$|\mathbf{A}|^2 = x^2 + y^2 + z^2 \tag{3-2}$$

by use of Pythagoras's theorem (Fig. 11).

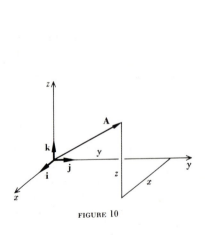

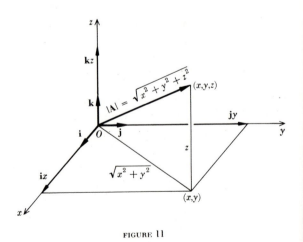

FIGURE 10

FIGURE 11

Example. If $\mathbf{A} = \mathbf{i} + 2\mathbf{j} + 3\mathbf{k}$ and $\mathbf{B} = -\mathbf{j} + 4\mathbf{k}$, compute the length of $2\mathbf{A} - \mathbf{B}$. Since $2\mathbf{A} - \mathbf{B} = 2\mathbf{i} + 5\mathbf{j} + 2\mathbf{k}$, we have

$$|2\mathbf{A} - \mathbf{B}| = (2^2 + 5^2 + 2^2)^{\frac{1}{2}} = \sqrt{33}.$$

PROBLEMS

1. (*a*) In the form $a\mathbf{i} + b\mathbf{j} + c\mathbf{k}$, write two vectors of length 5 parallel to the y axis. (*b*) If $\mathbf{A} = \mathbf{i} + 2\mathbf{j} + 3\mathbf{k}$, $\mathbf{B} = \mathbf{i} + \mathbf{j} + \mathbf{k}$, $\mathbf{C} = \mathbf{i} - \mathbf{k}$, compute $\mathbf{A} + \mathbf{B}$, $(\mathbf{A} + \mathbf{B}) + \mathbf{C}$, $\mathbf{B} + \mathbf{C}$, and $\mathbf{A} + (\mathbf{B} + \mathbf{C})$. What law does this illustrate? (*c*) In part *b*, find $5\mathbf{A}$, $-2\mathbf{A}$, the sum of these vectors, and the vector $3\mathbf{A}$. What law does this illustrate? (*d*) Also find $3\mathbf{A}$, $3\mathbf{B}$, the sum of these vectors, and the vector $3(\mathbf{A} + \mathbf{B})$. (*e*) In part *b*, a certain vector $\mathbf{D}$ is such that $\mathbf{A}$, $\mathbf{B}$, $\mathbf{D}$ can be placed head against tail to form a triangle. What is the z component of $\mathbf{D}$?

2. Sketch the triangle with vertices at the heads of $\mathbf{i} + \mathbf{j} + \mathbf{k}$, $2\mathbf{j} + \mathbf{k}$, and $2\mathbf{i} + \mathbf{j}$, and make the sides into vectors with head against tail. Find the vectors forming the sides of the triangle, and verify that the sum of these vectors is zero.

3. Draw a figure illustrating the inequality $|\mathbf{A} + \mathbf{B}| \le |\mathbf{A}| + |\mathbf{B}|$, and by combining this with (3-2), deduce an algebraic inequality. Can you give a purely algebraic proof?

4. (*a*) Let $\mathbf{A}, \mathbf{B}, \mathbf{C}, \ldots$ be vectors from the center to the vertices of a regular decagon (ten-sided polygon). By choosing a suitable basis $\mathbf{i}$, $\mathbf{j}$ and using symmetry, show that the sum $\mathbf{A} + \mathbf{B} + \mathbf{C} + \cdots$ is zero. (*b*) By picking another basis $\mathbf{i}'$, $\mathbf{j}'$, with $\mathbf{i}'$ making an angle θ with $\mathbf{A}$, deduce the identity $\cos \theta + \cos (\theta + \pi/5) + \cos (\theta + 2\pi/5) + \cdots + \cos (\theta + 9\pi/5) = 0$.

5. Show that the vectors $\mathbf{A} = 3\mathbf{i} - 4\mathbf{j} - 4\mathbf{k}$, $\mathbf{B} = 2\mathbf{i} - \mathbf{j} + \mathbf{k}$, $\mathbf{C} = \mathbf{i} - 3\mathbf{j} - 5\mathbf{k}$ form the sides of a right triangle.

---

4. The Dot Product. The *dot product*[1] of two vectors is defined to be the product of their lengths by the cosine of the angle between them. In symbols,

$$\mathbf{A} \cdot \mathbf{B} = |\mathbf{A}|\,|\mathbf{B}| \cos (\mathbf{A}, \mathbf{B}) \tag{4-1}$$

where $(\mathbf{A}, \mathbf{B})$ is the angle from $\mathbf{A}$ to $\mathbf{B}$. Thus $\mathbf{A} \cdot \mathbf{B}$ is a *scalar*, not a vector. Geometrically,

$$\begin{aligned} \mathbf{A} \cdot \mathbf{B} &= |\mathbf{A}| \times (\text{projection of } \mathbf{B} \text{ on } \mathbf{A}) \\ &= |\mathbf{B}| \times (\text{projection of } \mathbf{A} \text{ on } \mathbf{B}). \end{aligned} \tag{4-2}$$

Evidently $(\mathbf{A}, \mathbf{B})$ can be measured in several ways. However, since $\cos \theta = \cos (-\theta) = \cos (2\pi - \theta)$, these different measures all yield the same value for $\mathbf{A} \cdot \mathbf{B}$. The fact that $\cos \theta = \cos (-\theta)$ also yields

$$\mathbf{A} \cdot \mathbf{B} = \mathbf{B} \cdot \mathbf{A} \qquad \text{commutative law} \tag{4-3}$$

and one easily verifies the additional properties

$$(t\mathbf{A}) \cdot \mathbf{B} = t(\mathbf{A} \cdot \mathbf{B}) \qquad \text{associative law} \tag{4-4}$$

$$\mathbf{A} \cdot (\mathbf{B} + \mathbf{C}) = \mathbf{A} \cdot \mathbf{B} + \mathbf{A} \cdot \mathbf{C} \qquad \text{distributive law.} \tag{4-5}$$

For proof of (4-5), use (4-1) to transform (4-5) into

$$|\mathbf{A}|\,|\mathbf{B} + \mathbf{C}| \cos \psi = |\mathbf{A}|\,|\mathbf{B}| \cos \phi + |\mathbf{A}|\,|\mathbf{C}| \cos \theta \tag{4-6}$$

[1] The terms *scalar product* and *inner product* are often used.

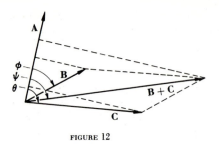

FIGURE 12

where the angles are defined in Fig. 12. Now (4-6) follows from

$$|\mathbf{B} + \mathbf{C}| \cos \psi = |\mathbf{B}| \cos \phi + |\mathbf{C}| \cos \theta \quad (4\text{-}7)$$

and (4-7) is evident from Fig. 12, when the vectors are coplanar and the angles are in the first quadrant. In view of (4-2) the property amounts merely to the assertion that projections are additive, and the extension to arbitrary angles is not difficult.

For the mutually orthogonal unit vectors $\mathbf{i}, \mathbf{j}, \mathbf{k}$ introduced in Sec. 3 we have, by inspection of (4-1),

$$\begin{aligned} \mathbf{i} \cdot \mathbf{i} = \mathbf{j} \cdot \mathbf{j} = \mathbf{k} \cdot \mathbf{k} &= 1 \\ \mathbf{i} \cdot \mathbf{j} = \mathbf{j} \cdot \mathbf{k} = \mathbf{i} \cdot \mathbf{k} &= 0. \end{aligned} \quad (4\text{-}8)$$

Hence, expanding the product by (4-4) and (4-5), we get

$$(x\mathbf{i} + y\mathbf{j} + z\mathbf{k}) \cdot (x_1\mathbf{i} + y_1\mathbf{j} + z_1\mathbf{k}) = xx_1 + yy_1 + zz_1. \quad (4\text{-}9)$$

By (4-1) and (4-9) the dot product gives a simple way to find the angle between two vectors and, in particular, to decide when two vectors are perpendicular. Indeed, if we agree to regard the zero vector as perpendicular to every vector, then from (4-1)

$$\mathbf{A} \cdot \mathbf{B} = 0 \quad \text{if, and only if, } \mathbf{A} \perp \mathbf{B}. \quad (4\text{-}10)$$

The case in which $\mathbf{B}$ is parallel to $\mathbf{A}$ is also worthy of note. In particular when $\mathbf{B} = \mathbf{A}$ we have

$$\mathbf{A} \cdot \mathbf{A} = |\mathbf{A}|^2. \quad (4\text{-}11)$$

Example.   Compute the cosine of the angle between $\mathbf{A}$ and $\mathbf{B}$ if $\mathbf{A} = \mathbf{i} + \mathbf{j} + 2\mathbf{k}$, $\mathbf{B} = -\mathbf{i} + z\mathbf{k}$, and find a value of z for which $\mathbf{A} \perp \mathbf{B}$.

We have $\mathbf{A} \cdot \mathbf{B} = -1 + 0 + 2z = 2z - 1$ and hence, by (4-1),

$$\cos (\mathbf{A}, \mathbf{B}) = \frac{2z - 1}{|\mathbf{A}|\,|\mathbf{B}|} = \frac{2z - 1}{\sqrt{6 + 6z^2}}.$$

The result is zero, and hence the vectors are perpendicular, when $z = \frac{1}{2}$.

PROBLEMS

1. Given $\mathbf{A} = \mathbf{i} + 2\mathbf{j} + 3\mathbf{k}$, $\mathbf{B} = -\mathbf{i} + 2\mathbf{j} + \mathbf{k}$, $\mathbf{C} = 2\mathbf{i} + \mathbf{j}$.   (*a*) Find the dot product of $3\mathbf{i} + 2\mathbf{j} + \mathbf{k}$ with each of these vectors.   (*b*) Find $\mathbf{A} \cdot \mathbf{B}$, $\mathbf{A} \cdot \mathbf{C}$, $\mathbf{B} + \mathbf{C}$, $\mathbf{A} \cdot (\mathbf{B} + \mathbf{C})$. What law is illustrated?   (*c*) Find $2\mathbf{A}$ and $(2\mathbf{A}) \cdot \mathbf{B}$. Compare $\mathbf{A} \cdot \mathbf{B}$ as found in part *b*.   (*d*) Find the angle between $\mathbf{A}$ and $\mathbf{B}$.   (*e*) Find the projection of $\mathbf{A}$ on $\mathbf{C}$.   (*f*) Find a scalar s such that $\mathbf{A} + s\mathbf{B}$ is perpendicular to $\mathbf{A}$.   (*g*) Find a vector of the form $\mathbf{i}x + \mathbf{j}y + \mathbf{k}$ which is perpendicular both to $\mathbf{A}$ and to $\mathbf{B}$.

2. (*a*) Show that $\mathbf{i} + \mathbf{j} + \mathbf{k}$, $\mathbf{i} - \mathbf{k}$, and $\mathbf{i} - 2\mathbf{j} + \mathbf{k}$ are mutually orthogonal.   (*b*) Choose x, y, and z so that $\mathbf{i} + \mathbf{j} + 2\mathbf{k}$, $-\mathbf{i} + z\mathbf{k}$, and $2\mathbf{i} + x\mathbf{j} + y\mathbf{k}$ are mutually orthogonal.

3. (*a*) If $\mathbf{A} \cdot \mathbf{B} = \mathbf{A} \cdot \mathbf{C}$ for some $\mathbf{A} \neq 0$, is it necessary that $\mathbf{B} = \mathbf{C}$?   Illustrate your answer by an example.   (*b*) If $\mathbf{A} \cdot \mathbf{B} = \mathbf{A} \cdot \mathbf{C}$ for every $\mathbf{A}$, is it necessary that $\mathbf{B} = \mathbf{C}$?

4. (*a*) What is the cosine of the angle between the two vectors $-\mathbf{i} + 2\mathbf{j} + 2\mathbf{k}$ and $2\mathbf{i} + \mathbf{j} - 2\mathbf{k}$?   (*b*) Work Prob. 5 of Sec. 3 by using the dot product.

5. (*a*) Show that $\mathbf{A} = \mathbf{i}\cos\alpha + \mathbf{j}\sin\alpha$, $\mathbf{B} = \mathbf{i}\cos\beta + \mathbf{j}\sin\beta$ are unit vectors making angles α and β with the x axis of the xy plane. Form the scalar product $\mathbf{A}\cdot\mathbf{B}$ and thus deduce the formula for $\cos(\alpha - \beta)$. (*b*) The vectors $\mathbf{A}$, $\mathbf{B}$, $\mathbf{C}$ form the sides of a triangle and $\mathbf{A} = \mathbf{B} - \mathbf{C}$. Form $\mathbf{A}\cdot\mathbf{A}$ and deduce the cosine law of trigonometry.

---

5. The Cross Product. Besides the multiplication just considered there is a second kind of multiplication, which yields a product known as the *vector product* or *cross product*. The cross product of $\mathbf{A}$ and $\mathbf{B}$, denoted by $\mathbf{A} \times \mathbf{B}$, is a vector $\mathbf{C}$ which is normal to the plane of $\mathbf{A}$ and $\mathbf{B}$ and is so directed that the vectors $\mathbf{A}$, $\mathbf{B}$, $\mathbf{C}$ form a right-handed system. The length of $\mathbf{C}$ is the product of the length of $\mathbf{A}$ by the length of $\mathbf{B}$ by the sine of the smaller angle between them:

$$|\mathbf{A} \times \mathbf{B}| = |\mathbf{A}|\,|\mathbf{B}|\sin(\mathbf{A},\mathbf{B}). \tag{5-1}$$

The expression (5-1) represents the area of the parallelogram having $\mathbf{A}$, $\mathbf{B}$ as adjacent edges (Fig. 13). The reader is warned, incidentally, that (5-1) does not give $\mathbf{A} \times \mathbf{B}$; it gives the length $|\mathbf{A} \times \mathbf{B}|$ only.

Since rotation from $\mathbf{B}$ to $\mathbf{A}$ is opposite to that from $\mathbf{A}$ to $\mathbf{B}$, we have

$$\mathbf{A} \times \mathbf{B} = -\mathbf{B} \times \mathbf{A} \tag{5-2}$$

so that the commutative law does not hold for vector products. On the other hand, it is the case that

$$(t\mathbf{A}) \times \mathbf{B} = t(\mathbf{A} \times \mathbf{B}) \qquad \text{associative law} \tag{5-3}$$

$$\mathbf{A} \times (\mathbf{B} + \mathbf{C}) = \mathbf{A} \times \mathbf{B} + \mathbf{A} \times \mathbf{C} \qquad \text{distributive law.} \tag{5-4}$$

The proof of Eq. (5-3) is trivial, and (5-4) is readily established if we note that $\mathbf{A} \times \mathbf{V}$ is obtained from the arbitrary vector $\mathbf{V}$ by performing the following three operations O_i illustrated in Fig. 14.

O_1: Project $\mathbf{V}$ on the plane perpendicular to $\mathbf{A}$ to obtain a vector $\mathbf{V}_1 \perp \mathbf{A}$ of magnitude $|\mathbf{V}|\sin(\mathbf{A},\mathbf{V})$.

O_2: Multiply $\mathbf{V}_1$ by $|\mathbf{A}|$ to obtain $\mathbf{V}_2 \perp \mathbf{A}$ of magnitude $|\mathbf{A}|\,|\mathbf{V}|\sin(\mathbf{A},\mathbf{V})$.

O_3: Rotate $\mathbf{V}_2$ about $\mathbf{A}$ through 90° to obtain $\mathbf{V}_3 = \mathbf{A} \times \mathbf{V}$.

It is easily checked that each of these operators is distributive; that is, $O_i(\mathbf{B} + \mathbf{C}) = O_i\mathbf{B} + O_i\mathbf{C}$ for all vectors $\mathbf{B}$ and $\mathbf{C}$. Hence the composite operator $O_3O_2O_1$ is distributive; namely,

$$\begin{aligned}
O_3O_2O_1(\mathbf{B} + \mathbf{C}) &= O_3O_2(O_1\mathbf{B} + O_1\mathbf{C}) & \text{since } O_1 \text{ is distributive} \\
&= O_3(O_2O_1\mathbf{B} + O_2O_1\mathbf{C}) & \text{since } O_2 \text{ is distributive} \\
&= O_3O_2O_1\mathbf{B} + O_3O_2O_1\mathbf{C} & \text{since } O_3 \text{ is distributive.}
\end{aligned}$$

Because $O_3O_2O_1\mathbf{V} = \mathbf{A} \times \mathbf{V}$ for every vector $\mathbf{V}$, the latter equation yields (5-4).

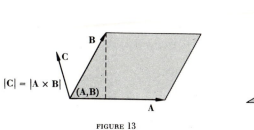

FIGURE 13

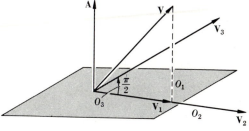

FIGURE 14

The definitions of vector product and of $\mathbf{i}$, $\mathbf{j}$, $\mathbf{k}$ lead to

$$\mathbf{i} \times \mathbf{i} = \mathbf{j} \times \mathbf{j} = \mathbf{k} \times \mathbf{k} = 0$$
$$\mathbf{i} \times \mathbf{j} = -\mathbf{j} \times \mathbf{i} = \mathbf{k} \qquad \mathbf{j} \times \mathbf{k} = -\mathbf{k} \times \mathbf{j} = \mathbf{i} \qquad \mathbf{k} \times \mathbf{i} = -\mathbf{i} \times \mathbf{k} = \mathbf{j}. \tag{5-5}$$

If $\mathbf{A}$ and $\mathbf{B}$ are given by their components as

$$\mathbf{A} = x\mathbf{i} + y\mathbf{j} + z\mathbf{k} \qquad \mathbf{B} = x_1\mathbf{i} + y_1\mathbf{j} + z_1\mathbf{k}$$

expansion by means of (5-3) and (5-4) and simplification by means of (5-5) yield

$$\mathbf{A} \times \mathbf{B} = \mathbf{i}(yz_1 - zy_1) + \mathbf{j}(x_1z - xz_1) + \mathbf{k}(xy_1 - yx_1)$$

which may be written as a determinant[1]

$$\mathbf{A} \times \mathbf{B} = \begin{vmatrix} \mathbf{i} & \mathbf{j} & \mathbf{k} \\ x & y & z \\ x_1 & y_1 & z_1 \end{vmatrix} \equiv \mathbf{i}\begin{vmatrix} y & z \\ y_1 & z_1 \end{vmatrix} - \mathbf{j}\begin{vmatrix} x & z \\ x_1 & z_1 \end{vmatrix} + \mathbf{k}\begin{vmatrix} x & y \\ x_1 & y_1 \end{vmatrix}. \tag{5-6}$$

Example.   Find a vector perpendicular to $\mathbf{i} + 2\mathbf{k}$ and $\mathbf{i} + \mathbf{j} - \mathbf{k}$, and find the area of the triangle with these two vectors as adjacent sides.

Both questions are settled by calculating the cross product.   We have, from (5-6),

$$(\mathbf{i} + 2\mathbf{k}) \times (\mathbf{i} + \mathbf{j} - \mathbf{k}) = \begin{vmatrix} \mathbf{i} & \mathbf{j} & \mathbf{k} \\ 1 & 0 & 2 \\ 1 & 1 & -1 \end{vmatrix}$$

$$= \mathbf{i}\begin{vmatrix} 0 & 2 \\ 1 & -1 \end{vmatrix} - \mathbf{j}\begin{vmatrix} 1 & 2 \\ 1 & -1 \end{vmatrix} + \mathbf{k}\begin{vmatrix} 1 & 0 \\ 1 & 1 \end{vmatrix}$$

$$= -2\mathbf{i} + 3\mathbf{j} + \mathbf{k}.$$

This vector is perpendicular to the given vectors.   The area of the triangle is half the area of the parallelogram and hence the desired area is $\frac{1}{2}\,|-2\mathbf{i} + 3\mathbf{j} + \mathbf{k}| = \frac{1}{2}\sqrt{14}$.

--- PROBLEMS

1. Given $\mathbf{A} = \mathbf{i} + 2\mathbf{j} + \mathbf{k}$, $\mathbf{B} = 3\mathbf{i} + 2\mathbf{j}$, $\mathbf{C} = -\mathbf{i} + \mathbf{j} + 3\mathbf{k}$.   (*a*) Find $\mathbf{A} \times \mathbf{B}$, $\mathbf{A} \times \mathbf{C}$, $\mathbf{A} \times \mathbf{B} + \mathbf{A} \times \mathbf{C}$, $\mathbf{B} + \mathbf{C}$, and $\mathbf{A} \times (\mathbf{B} + \mathbf{C})$.   What law is illustrated?   (*b*) Find a vector perpendicular to $\mathbf{B}$ and $\mathbf{C}$, and verify your answer by use of the dot product.   (*c*) If $\mathbf{A}$, $\mathbf{B}$, $\mathbf{C}$ have their origins at a common point, find a vector perpendicular to the plane in which their heads lie.   (*d*) Find the area of the triangle formed by the heads in part *c*.

2. Show that the cross product for each two of the following vectors is parallel to the third: $\mathbf{i} + \mathbf{j} + \mathbf{k}$, $\mathbf{i} - \mathbf{k}$, $\mathbf{i} - 2\mathbf{j} + \mathbf{k}$.   What does this indicate about the vectors?

3. Give an example of three unequal vectors such that the cross product of any two is perpendicular to the third.

4. If $\mathbf{A} \times \mathbf{B} = 0$ and $\mathbf{A} \cdot \mathbf{B} = 0$, is it necessary that $\mathbf{A} = 0$ or $\mathbf{B} = 0$?

5. In refraction at the plane interface of two homogeneous media, let $\mathbf{A}$, $\mathbf{B}$, $\mathbf{C}$ be unit vectors along the incident, reflected, and refracted rays, respectively, and let $\mathbf{N}$ be the unit normal to the interface.   (*a*) Show that the law of reflection is equivalent to $\mathbf{A} \times \mathbf{N} = \mathbf{B} \times \mathbf{N}$.   (*b*) Show that the law of refraction is equivalent to $n_1\mathbf{A} \times \mathbf{N} = n_2\mathbf{C} \times \mathbf{N}$, where n_1 and n_2 are the indices of refraction.

6. The terminal points of the vectors $\mathbf{A} = 3\mathbf{i} - \mathbf{k}$, $\mathbf{B} = 3\mathbf{i} + 2\mathbf{j}$, and $\mathbf{C} = \mathbf{i} + \mathbf{j} - 2\mathbf{k}$ determine the plane ABC.   Find a unit vector $\mathbf{N}$ perpendicular to this plane, and find the projection of $\mathbf{A}$ on the plane ABC.

7. Are the following pairs of vectors orthogonal?   Parallel?   (*a*) $2\mathbf{i} + 3\mathbf{j} + 6\mathbf{k}$, $3\mathbf{i} - 6\mathbf{j} + \mathbf{k}$; (*b*) $12\mathbf{i} + 4\mathbf{j} - 6\mathbf{k}$, $6\mathbf{i} + 2\mathbf{j} - 3\mathbf{k}$; (*c*) $2\mathbf{i} + 3\mathbf{j} + 6\mathbf{k}$, $6\mathbf{i} + 2\mathbf{j} - 3\mathbf{k}$; (*d*) $30\mathbf{i} - 14\mathbf{j} + 21\mathbf{k}$, $2\mathbf{i} + 3\mathbf{j} + 6\mathbf{k}$; (*e*) $6\mathbf{i} + 2\mathbf{j} - 3\mathbf{k}$, $30\mathbf{i} - 14\mathbf{j} + 21\mathbf{k}$.

[1] The reader unfamiliar with second- or third-order determinants is referred to Appendix A.

8. (*a*) If $\mathbf{C} = \mathbf{A} \times \mathbf{B}$, show that $\mathbf{A} \cdot \mathbf{C} = 0$, $\mathbf{B} \cdot \mathbf{C} = 0$.   (*b*) If $\mathbf{C} = \mathbf{A} \times \mathbf{B}$ and $\mathbf{B} = \mathbf{A} \times \mathbf{C}$, show that $\mathbf{B} = \mathbf{C} = 0$.

9. What is a unit normal vector to the plane passing through the points (*a*) (1,0,0), (0,1,0), (0,0,1); (*b*) (1,1,1), (1,2,3), (2,3,1); (*c*) (1,1,−2), (2,−1,1), (1,3,−1)?

10. Find the areas of the triangles determined by the points in each part of Prob. 9.

---

6. Continued Products.   With the two multiplications previously defined, we can form the products $(\mathbf{A} \cdot \mathbf{B})\mathbf{C}$, $\mathbf{A} \cdot (\mathbf{B} \times \mathbf{C})$, and $\mathbf{A} \times (\mathbf{B} \times \mathbf{C})$; some of the other possible combinations, however, have no meaning. For example, $(\mathbf{A} \cdot \mathbf{B}) \times \mathbf{C}$ is meaningless because the two factors in a cross product must both be vectors.

The first product, $(\mathbf{A} \cdot \mathbf{B})\mathbf{C}$, denotes simply the product of the scalar $\mathbf{A} \cdot \mathbf{B}$ with the vector $\mathbf{C}$ and may be dismissed without further comment. By definition of dot product, the second expression, $\mathbf{A} \cdot (\mathbf{B} \times \mathbf{C})$, called the *scalar triple product*, has the value

$$\mathbf{A} \cdot (\mathbf{B} \times \mathbf{C}) = |\mathbf{A}| \cos \theta |\mathbf{B} \times \mathbf{C}| \qquad (6\text{-}1)$$

where θ is the angle between $\mathbf{A}$ and $\mathbf{B} \times \mathbf{C}$. Since $\mathbf{B} \times \mathbf{C}$ is perpendicular to the face of the parallelepiped containing $\mathbf{B}$ and $\mathbf{C}$ (Fig. 15), and since $|\mathbf{B} \times \mathbf{C}|$ is the area of this face, (6-1) shows that $\mathbf{A} \cdot (\mathbf{B} \times \mathbf{C})$ *represents the signed volume of the parallelepiped having* $\mathbf{A}$, $\mathbf{B}$, $\mathbf{C}$ *as adjacent edges.* Moreover, we have the formula

$$\mathbf{A} \cdot (\mathbf{B} \times \mathbf{C}) = \begin{vmatrix} A_x & A_y & A_z \\ B_x & B_y & B_z \\ C_x & C_y & C_z \end{vmatrix} \qquad \begin{cases} \mathbf{A} = \mathbf{i}A_x + \mathbf{j}A_y + \mathbf{k}A_z \\ \mathbf{B} = \mathbf{i}B_x + \mathbf{j}B_y + \mathbf{k}B_z \\ \mathbf{C} = \mathbf{i}C_x + \mathbf{j}C_y + \mathbf{k}C_z \end{cases} \qquad (6\text{-}2)$$

as will now be seen. The expression (5-6) yields

$$\mathbf{B} \times \mathbf{C} = \begin{vmatrix} \mathbf{i} & \mathbf{j} & \mathbf{k} \\ B_x & B_y & B_z \\ C_x & C_y & C_z \end{vmatrix} = \mathbf{i}P + \mathbf{j}Q + \mathbf{k}R \qquad (6\text{-}3)$$

say, where P, Q, R are certain second-order determinants. Taking the dot product of $\mathbf{i}A_x + \mathbf{j}A_y + \mathbf{k}A_z$ with (6-3) leads to

$$\mathbf{A} \cdot (\mathbf{B} \times \mathbf{C}) = A_xP + A_yQ + A_zR$$

which is the expansion of the determinant (6-2) on elements of the first row.

Since interchanging two rows of a determinant merely changes its sign, (6-2) yields the useful relations

$$\begin{aligned} \mathbf{A} \cdot (\mathbf{B} \times \mathbf{C}) &= \mathbf{B} \cdot (\mathbf{C} \times \mathbf{A}) = \mathbf{C} \cdot (\mathbf{A} \times \mathbf{B}) \\ &= -\mathbf{B} \cdot (\mathbf{A} \times \mathbf{C}) = -\mathbf{A} \cdot (\mathbf{C} \times \mathbf{B}) = -\mathbf{C} \cdot (\mathbf{B} \times \mathbf{A}). \end{aligned} \qquad (6\text{-}4)$$

These results as to magnitude are evident from the volume interpretation, though further discussion is needed to establish the algebraic sign in this way. Because of (6-4) it is customary to write

$$\mathbf{A} \cdot (\mathbf{B} \times \mathbf{C}) = \mathbf{A} \cdot \mathbf{B} \times \mathbf{C} = (\mathbf{ABC}). \qquad (6\text{-}5)$$

To evaluate the *vector triple product* $\mathbf{A} \times (\mathbf{B} \times \mathbf{C})$, let $\mathbf{i}$ be a unit vector parallel to $\mathbf{B}$ and $\mathbf{j}$ a unit vector perpendicular to $\mathbf{i}$ in the plane of $\mathbf{B}$ and $\mathbf{C}$. Thus

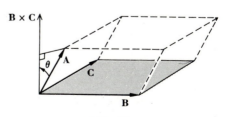

FIGURE 15

$$\mathbf{B} = B_x\mathbf{i} \qquad \mathbf{C} = C_x\mathbf{i} + C_y\mathbf{j} \qquad \mathbf{A} = A_x\mathbf{i} + A_y\mathbf{j} + A_z\mathbf{k} \qquad (6\text{-}6)$$

where $\mathbf{k}$ is a unit vector perpendicular to $\mathbf{i}$ and $\mathbf{j}$, so oriented that the three form a right-handed system. Since $\mathbf{B} \times \mathbf{C} = B_xC_y\mathbf{k}$ by (6-6) and (5-6), we have

$$\begin{aligned}\mathbf{A} \times (\mathbf{B} \times \mathbf{C}) &= -A_xB_xC_y\mathbf{j} + A_yB_xC_y\mathbf{i} \\ &= (A_xC_x + A_yC_y)B_x\mathbf{i} - A_xB_x(C_x\mathbf{i} + C_y\mathbf{j}) \\ &= \mathbf{B}(\mathbf{A}\cdot\mathbf{C}) - \mathbf{C}(\mathbf{A}\cdot\mathbf{B}). \end{aligned} \qquad (6\text{-}7)$$

Example. Establish the identity

$$(\mathbf{A} \times \mathbf{B})\cdot(\mathbf{C} \times \mathbf{D}) = \begin{vmatrix} \mathbf{A}\cdot\mathbf{C} & \mathbf{B}\cdot\mathbf{C} \\ \mathbf{A}\cdot\mathbf{D} & \mathbf{B}\cdot\mathbf{D} \end{vmatrix}. \qquad (6\text{-}8)$$

The expression is the scalar triple product of $\mathbf{A} \times \mathbf{B}$, $\mathbf{C}$, and $\mathbf{D}$. Interchanging the dot and cross, as we may by (6-4), we obtain

$$\begin{aligned}(\mathbf{A} \times \mathbf{B})\cdot\mathbf{C} \times \mathbf{D} &= (\mathbf{A} \times \mathbf{B}) \times \mathbf{C}\cdot\mathbf{D} = [(\mathbf{A}\cdot\mathbf{C})\mathbf{B} - (\mathbf{B}\cdot\mathbf{C})\mathbf{A}]\cdot\mathbf{D} \\ &= (\mathbf{A}\cdot\mathbf{C})(\mathbf{B}\cdot\mathbf{D}) - (\mathbf{B}\cdot\mathbf{C})(\mathbf{A}\cdot\mathbf{D}) \end{aligned} \qquad (6\text{-}9)$$

since $(\mathbf{A} \times \mathbf{B}) \times \mathbf{C} = -\mathbf{C} \times (\mathbf{A} \times \mathbf{B}) = (\mathbf{A}\cdot\mathbf{C})\mathbf{B} - (\mathbf{B}\cdot\mathbf{C})\mathbf{A}$ by (6-7).

PROBLEMS

1. Verify (6-2), (6-7), and (6-8) by direct calculation for the special case $\mathbf{A} = \mathbf{i} + \mathbf{j}$, $\mathbf{B} = -\mathbf{i} + 2\mathbf{k}$, $\mathbf{C} = \mathbf{j} + 2\mathbf{k}$, $\mathbf{D} = \mathbf{i} + \mathbf{j} + \mathbf{k}$.

2. (a) In Prob. 1, find the volume of the parallelepiped having $\mathbf{A}$, $\mathbf{B}$, and $\mathbf{C}$ as adjacent edges. (b) Find x such that the vectors $2\mathbf{i} + \mathbf{j} - 2\mathbf{k}$, $\mathbf{i} + \mathbf{j} + 3\mathbf{k}$, and $x\mathbf{i} + \mathbf{j}$ are coplanar. *Hint:* A certain parallelepiped must have zero volume. (c) State a simple necessary and sufficient condition that three arbitrary vectors $\mathbf{A}$, $\mathbf{B}$, $\mathbf{C}$ be coplanar. (d) Evaluate $(\mathbf{AAB})$ and $(\mathbf{ABA})$, where $\mathbf{A}$, $\mathbf{B}$ are arbitrary.

3. By (6-7), show that $\mathbf{A} \times (\mathbf{B} \times \mathbf{C}) + \mathbf{B} \times (\mathbf{C} \times \mathbf{A}) + \mathbf{C} \times (\mathbf{A} \times \mathbf{B}) = \mathbf{0}$.

4. Show that $(\mathbf{B} \times \mathbf{C}) \times (\mathbf{C} \times \mathbf{A}) = \mathbf{C}(\mathbf{ABC})$, and deduce

$$(\mathbf{A} \times \mathbf{B})\cdot(\mathbf{B} \times \mathbf{C}) \times (\mathbf{C} \times \mathbf{A}) = (\mathbf{ABC})^2.$$

5. The vectors $\mathbf{A}$, $\mathbf{B}$, $\mathbf{C}$ issue from a common point and have their heads in a plane. Show that $(\mathbf{A} \times \mathbf{B}) + (\mathbf{B} \times \mathbf{C}) + (\mathbf{C} \times \mathbf{A})$ is perpendicular to this plane.

---

7. Differentiation. If for each value of a scalar t a vector $\mathbf{R}(t)$ is defined, $\mathbf{R}$ is said to be a vector function of t. In a particular problem t may denote the time and $\mathbf{R}$ the position vector of a moving point relative to some origin O. As in the calculus of scalars, $\mathbf{R}(t)$ is said to be a *continuous vector function* of t at $t = t_0$ provided that

$$\lim_{t \to t_0} \mathbf{R}(t) = \mathbf{R}(t_0). \qquad (7\text{-}1)$$

The precise meaning of (7-1) is that $|\mathbf{R}(t) - \mathbf{R}(t_0)|$ becomes as small as desired whenever t is sufficiently near t_0.

The cartesian components of the vector $\mathbf{R}(t)$ are functions of t, so that one may write

$$\mathbf{R}(t) = \mathbf{i}x(t) + \mathbf{j}y(t) + \mathbf{k}z(t). \qquad (7\text{-}2)$$

It follows from (7-1) that the functions $x(t)$, $y(t)$, $z(t)$ are continuous if, and only if, $\mathbf{R}(t)$ is continuous.

We define the derivative of $\mathbf{R}(t)$ with respect to t by the formula

$$\frac{d\mathbf{R}}{dt} = \lim_{\Delta t \to 0} \frac{\mathbf{R}(t + \Delta t) - \mathbf{R}(t)}{\Delta t} = \lim_{\Delta t \to 0} \frac{\Delta \mathbf{R}}{\Delta t} \tag{7-3}$$

where $\Delta \mathbf{R} \equiv \mathbf{R}(t + \Delta t) - \mathbf{R}(t)$. The substitution of (7-2) in the definition (7-3) leads immediately to the result that $\mathbf{R}$ is differentiable if, and only if, x, y, z are, and in that case

$$\frac{d\mathbf{R}}{dt} = \mathbf{i}\,\frac{dx}{dt} + \mathbf{j}\,\frac{dy}{dt} + \mathbf{k}\,\frac{dz}{dt}. \tag{7-4}$$

As in scalar calculus we shall write $\mathbf{R}'(t)$ for $d\mathbf{R}/dt$, $\mathbf{R}''(t)$ for $d^2\mathbf{R}/dt^2$, and so on.

Products involving vectors are differentiated by the familiar rules of elementary calculus, and the proof of these rules also involves only familiar ideas. For example, the formula

$$\frac{d}{dt}(\mathbf{A} \times \mathbf{B}) = \mathbf{A} \times \frac{d\mathbf{B}}{dt} + \frac{d\mathbf{A}}{dt} \times \mathbf{B} \tag{7-5}$$

follows from

$$\begin{aligned}
\Delta(\mathbf{A} \times \mathbf{B}) &= (\mathbf{A} + \Delta\mathbf{A}) \times (\mathbf{B} + \Delta\mathbf{B}) - \mathbf{A} \times \mathbf{B} \\
&= \mathbf{A} \times \Delta\mathbf{B} + \Delta\mathbf{A} \times \mathbf{B} + \Delta\mathbf{A} \times \Delta\mathbf{B}
\end{aligned}$$

when we divide by Δt and let $\Delta t \to 0$. Of course, the order of the factors in (7-5) must be preserved, since the cross product is not commutative.

A geometric interpretation of the derivative may be obtained as follows: Let the vector $\mathbf{R}(t)$ be regarded as a *bound* vector with its origin at the origin of coordinates. The head of $\mathbf{R}$ then traces a space curve as t varies (see Fig. 16). The vector

$$\Delta\mathbf{R} = \mathbf{R}(t + \Delta t) - \mathbf{R}(t) \tag{7-6}$$

is directed along a secant of the curve, $\Delta\mathbf{R}/\Delta t$ is parallel to this secant, and hence $\lim(\Delta\mathbf{R}/\Delta t)$ is tangent. *Thus, the vector $\mathbf{R}'(t)$ is tangent to the space curve $\mathbf{R} = \mathbf{R}(t)$ whenever $\mathbf{R}'(t)$ exists and $\mathbf{R}'(t) \neq \mathbf{0}$.*

To interpret the magnitude $|\mathbf{R}'(t)|$, let s be the length of the curve from the fixed point given by $t = t_0$ to the variable point given by t. Assuming $\mathbf{R}'(t) \neq 0$, we have $\Delta\mathbf{R} \neq \mathbf{0}$ for small $\Delta t > 0$; hence

$$\frac{\Delta s}{\Delta t} = \frac{\Delta s}{|\Delta\mathbf{R}|} \frac{|\Delta\mathbf{R}|}{\Delta t} = \frac{\Delta s}{|\Delta\mathbf{R}|} \left| \frac{\Delta\mathbf{R}}{\Delta t} \right|. \tag{7-7}$$

Since $|\Delta\mathbf{R}|$ is the length of the chord, and since the ratio $\Delta s/|\Delta\mathbf{R}|$ of arc to chord[1] tends to 1, Eq. (7-7) gives

$$\frac{ds}{dt} = \left| \frac{d\mathbf{R}}{dt} \right| \tag{7-8}$$

when $\Delta t \to 0$. Thus, *the vector $\mathbf{R}'(t)$ has magnitude $|\mathbf{R}'| = ds/dt$, where s is the arc length along the curve.* If $\mathbf{R}'(t)$ is continuous, the

[1] We assume that s increases with t; otherwise a minus sign is needed. The fact that (arc)/(chord) $\to 1$ follows from the familiar interpretation of arc as limit of lengths of inscribed polygons. It is also possible to take (arc)/(chord) $\to 1$ as one of the defining properties of arc and proceed, as in the text, to obtain the formula (7-9).

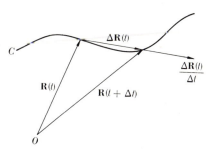

FIGURE 16

arc is given explicitly by

$$s = \int_{t_0}^{t} |\mathbf{R}'(t)|\, dt = \int_{t_0}^{t} \sqrt{(x')^2 + (y')^2 + (z')^2}\, dt. \tag{7-9}$$

Introduction of s as parameter instead of t facilitates the study of space curves (see Sec. 10).

We shall say that a curve C is *smooth* if $\mathbf{R}'(t) \neq \mathbf{0}$ and is continuous at all points of C.

In two dimensions the interpretation of $\mathbf{R}'(t)$ given here agrees with the results of elementary calculus. Let a smooth curve C be represented parametrically by $x = x(t)$, $y = y(t)$, so that the slope is given by

$$\text{slope} = \frac{dy}{dx} = \frac{dy/dt}{dx/dt} \equiv \frac{y'}{x'} \tag{7-10}$$

for $x' \neq 0$. If the same curve is described in the form $\mathbf{R} = \mathbf{i}x + \mathbf{j}y$, we have $\mathbf{R}' = \mathbf{i}x' + \mathbf{j}y'$; hence the slope of the vector $\mathbf{R}'$ is y'/x'. In view of (7-10), *the fact that $\mathbf{R}'$ is tangent to the curve agrees with the fact that dy/dx is the slope of the curve.* The formula $ds/dt = |\mathbf{R}'|$ is also familiar; it states that

$$\frac{ds}{dt} = \sqrt{(x')^2 + (y')^2} = \sqrt{\left(\frac{dx}{dt}\right)^2 + \left(\frac{dy}{dt}\right)^2}$$

which becomes $ds^2 = dx^2 + dy^2$ when squared and multiplied by $(dt)^2$.

Physically, one may regard t as time, so that the head of the bound vector $\mathbf{R}(t)$ gives the position of a moving particle at time t. Since the velocity is defined to be $\mathbf{V} = \mathbf{R}'(t)$, the foregoing result means that *the velocity vector is tangent to the trajectory and has magnitude equal to the speed ds/dt with which the particle is moving.*

Example 1. The position of a particle at time t is determined by the bound vector

$$\mathbf{R}(t) = \mathbf{i}t + \mathbf{j}t^3 + \mathbf{k}\sin t.$$

Find a vector tangent to the orbit at time t, and find the speed of the particle at time $t = 0$.

We have $\mathbf{R}'(t) = \mathbf{i} + 3\mathbf{j}t^2 + \mathbf{k}\cos t$, which is the required tangent vector. At $t = 0$ the velocity is $\mathbf{R}'(0) = \mathbf{i} + \mathbf{k}$, and hence the speed is $ds/dt = |\mathbf{R}'(0)| = \sqrt{2}$.

Example 2. If a differentiable vector $\mathbf{R}(t)$ has constant length, show that $\mathbf{R}'$ is perpendicular to $\mathbf{R}$, and interpret geometrically.

From $\mathbf{R} \cdot \mathbf{R} = $ const, differentiation yields $\mathbf{R} \cdot \mathbf{R}' + \mathbf{R}' \cdot \mathbf{R} = 0$, whence $\mathbf{R}' \cdot \mathbf{R} = 0$. Geometrically, if $\mathbf{R}$ is a bound vector of constant length, its head traces a curve lying on a sphere. The tangent to the curve is tangent to the sphere and so perpendicular to the radius vector. Thus, $\mathbf{R}' \perp \mathbf{R}$.

PROBLEMS

1. If $\mathbf{R}(t) = \mathbf{i}2t + \mathbf{j}3t^2 + \mathbf{k}t^3$, (a) find the derivative $\mathbf{R}'(t)$. (b) At the point $(2,3,1)$ find a tangent to the space curve that is traced by the head of $\mathbf{R}$ when $\mathbf{R}$ is regarded as a bound vector. *Hint:* The point $(2,3,1)$ corresponds to $t = 1$. (c) If $\mathbf{R}(t)$ is a bound vector giving the position of a moving particle at time t, find the velocity and speed of this particle at time $t = 1$.

2. (a) Differentiate the vector $\mathbf{R}(t) = \mathbf{i}t + \mathbf{j}\sin t + \mathbf{k}\cos t$, compute $|\mathbf{R}'(t)|$, and simplify. (b) If $\mathbf{R}(t)$ is a bound vector, find the length of the curve traced by the head of $\mathbf{R}$ as t varies from $t = 0$ to $t = 2$.

3. By writing $\mathbf{A} \cdot \mathbf{B}$ in component form and differentiating, deduce $(\mathbf{A} \cdot \mathbf{B})' = \mathbf{A}' \cdot \mathbf{B} + \mathbf{A} \cdot \mathbf{B}'$.

4. If $\mathbf{R}_0$ and $\mathbf{A}$ are constant, find a vector tangent to the curve described by the bound vector $\mathbf{R} = \mathbf{R}_0 + \mathbf{A}t$.

5. If $\mathbf{R}(t)$ is a bound vector giving the position of a moving particle at time t, the acceleration is defined to be $\mathbf{A} = \mathbf{R}''(t)$. Show that $\mathbf{A}$ is constant if $\mathbf{R}(t) = \mathbf{R}_0 + \mathbf{R}_1 t + \mathbf{R}_2 t^2$, where $\mathbf{R}_0$, $\mathbf{R}_1$, and $\mathbf{R}_2$ are constant vectors. Is the converse true?

6. Show that $(\mathbf{ABC})' = (\mathbf{A'BC}) + (\mathbf{AB'C}) + (\mathbf{ABC'})$, when $\mathbf{A}$, $\mathbf{B}$, $\mathbf{C}$ are differentiable, and write in determinant form.

7. If $\mathbf{R} = \mathbf{A} + f(t)\mathbf{B}$, where $\mathbf{A}$ and $\mathbf{B}$ are constant and f is twice differentiable, then $\mathbf{R}' \times \mathbf{R}'' = \mathbf{0}$. If $\mathbf{R}$ gives the position of the moving point P and t is time, what is the physical meaning of this condition?

8. At time t the vector from the origin to a moving point P is $\mathbf{R} = \mathbf{i}a \cos \omega t + \mathbf{j}b \sin \omega t$, where a, b, ω are constants. (a) Show that P moves on an ellipse. (b) Find the velocity $\mathbf{V}$ of P, and show that $\mathbf{R} \times \mathbf{V} = $ const. (c) Show that the acceleration is directed toward the origin and its magnitude is proportional to the distance of P from the origin (see Prob. 5).

9. If $\mathbf{R}(t)$ is a function of t having second derivatives, compute the derivatives of $\mathbf{R} \cdot \mathbf{R}' \times \mathbf{R}''$ and $\mathbf{R} \times \mathbf{R}'$.

10. Show that the solutions of the differential equations $d^2\mathbf{R}/dt = \pm k^2\mathbf{R}$ are $\mathbf{R} = \mathbf{A}e^{kt} + \mathbf{B}e^{-kt}$, $\mathbf{R} = \mathbf{A} \cos kt + \mathbf{B} \sin kt$, where $\mathbf{A}$ and $\mathbf{B}$ are constant vectors.

APPLICATIONS

8. Mechanics and Dynamics.   The work W done by a constant force $\mathbf{F}$ producing a displacement $\mathbf{S}$ in the direction of $\mathbf{F}$ is $|\mathbf{F}|\ |\mathbf{S}|$. More generally, if $\mathbf{F}$ makes an angle θ with $\mathbf{S}$, the work is $|\mathbf{F}|\ |\mathbf{S}| \cos \theta$, and hence

$$W = \mathbf{F} \cdot \mathbf{S}. \tag{8-1}$$

Because of this equation the dot product plays a central role in certain branches of mechanics.

To illustrate the application of cross products, let the vector $\boldsymbol{\Omega}$ represent the angular velocity of a rotating body; that is, let $\boldsymbol{\Omega}$ be a vector whose magnitude is the angular speed in radians per second and whose direction is parallel to the axis of rotation. The positive sense of $\boldsymbol{\Omega}$ is chosen as that in which a right-handed screw would advance if the screw were rotated in the same direction as the body. Let $\mathbf{R}$ be a vector locating any point P of the body relative to some point O on the axis of rotation. It is required to find the instantaneous velocity $\mathbf{V}$ of the point P. If the distance of P from the axis of rotation is a, then by Fig. 17

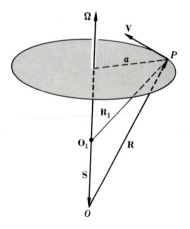

$$|\mathbf{V}| = |\boldsymbol{\Omega}|a = |\boldsymbol{\Omega}|\ |\mathbf{R}|\sin (\mathbf{R},\boldsymbol{\Omega}).$$

Moreover, $\mathbf{V}$ is normal to the plane of $\mathbf{R}$ and $\boldsymbol{\Omega}$ and is so directed that $\boldsymbol{\Omega}$, $\mathbf{R}$, and $\mathbf{V}$ form a right-handed system. Hence,

$$\mathbf{V} = \boldsymbol{\Omega} \times \mathbf{R}. \tag{8-2}$$

The result is independent of the origin O, for if a new origin O_1 is chosen and P is specified by a vector $\mathbf{R}_1$ from O_1, then

$$\mathbf{R}_1 = \mathbf{R} + \mathbf{S}$$

FIGURE 17

where $\mathbf{S}$ is parallel to $\boldsymbol{\Omega}$ (see Fig. 17).   Hence $\boldsymbol{\Omega} \times \mathbf{S} = \mathbf{O}$, and therefore

$$\boldsymbol{\Omega} \times \mathbf{R}_1 = \boldsymbol{\Omega} \times (\mathbf{R} + \mathbf{S}) = \boldsymbol{\Omega} \times \mathbf{R} + \boldsymbol{\Omega} \times \mathbf{S} = \boldsymbol{\Omega} \times \mathbf{R}.$$

Another example from dynamics illustrates the compactness of vector notation. Let O be a fixed point in a rigid body, and let a force $\mathbf{F}$ be applied at a point P of the body, which is located by the bound vector $\mathbf{R}$ whose origin is at O.   The force $\mathbf{F}$ establishes a *torque* or *moment* $\mathbf{T}$ which tends to rotate the body about an axis that passes through O and is normal to the plane of $\mathbf{R}$ and $\mathbf{F}$.   The magnitude of $\mathbf{T}$ is given by

$$|\mathbf{T}| = |\mathbf{R}|\,|\mathbf{F}|\,\sin(\mathbf{R},\mathbf{F}).$$

In addition, $\mathbf{R}$, $\mathbf{F}$, and $\mathbf{T}$ form a right-handed system, so that

$$\mathbf{T} = \mathbf{R} \times \mathbf{F}. \tag{8-3}$$

That the choice of O is immaterial follows as in the discussion of (8-2).   Similarly one shows that $\mathbf{F}$ may slide along its line of action without affecting the result; that is, $\mathbf{F}$ may be regarded as a sliding vector.

To illustrate the use of (8-3), we obtain a formula for the so-called *center of mass* of a system of mass points.   The force on a point of mass m in a gravitational field is given by $m\mathbf{F}$, where m is the mass of the point and $\mathbf{F}$ is a vector specifying the strength of the field at the point in question.   We assume a uniform field, so that $\mathbf{F}$ is independent of position.   From (8-3)

$$(\mathbf{R} - \mathbf{P}) \times m\mathbf{F} \tag{8-4}$$

represents the moment about the point[1] P of the gravitational force on a point of mass m at R. If there are n points of masses $m_1, m_2, \ldots, m_n$ located by the vectors $\mathbf{R}_1, \mathbf{R}_2, \ldots, \mathbf{R}_n$, respectively, the total moment about the point P due to all of them is

$$\Sigma(\mathbf{R}_i - \mathbf{P}) \times m_i\mathbf{F}. \tag{8-5}$$

It is desired to find a single mass point such that its moment (8-4) reproduces the total moment (8-5) for all choices of $\mathbf{F}$ and $\mathbf{P}$.   Equating the moments (8-4) and (8-5) leads to

$$[m\mathbf{P} - \Sigma m_i\mathbf{P} - m\mathbf{R} + \Sigma m_i\mathbf{R}_i] \times \mathbf{F} = \mathbf{0} \tag{8-6}$$

after rearrangement.   Since $\mathbf{F}$ is arbitrary in (8-6), the factor in brackets must vanish, so that

$$\mathbf{P}(m - \Sigma m_i) = m\mathbf{R} - \Sigma m_i\mathbf{R}_i. \tag{8-7}$$

The fact that $\mathbf{P}$ is arbitrary in (8-7) now gives

$$m = \Sigma m_i \qquad m\mathbf{R} = \Sigma m_i\mathbf{R}_i. \tag{8-8}$$

Conversely, (8-8) ensures the validity of (8-7) and hence of (8-6) independently of $\mathbf{F}$ and $\mathbf{P}$.

This discussion was carried out by equating moments only.   Equation (8-8) shows, however, that the total gravitational force is also preserved, since the mass of the point equals the total mass of the collection.

The point R with position vector

$$\mathbf{R} = \frac{m_1\mathbf{R}_1 + m_2\mathbf{R}_2 + m_n\mathbf{R}_n}{m_1 + m_2 + \cdots + m_n} \tag{8-9}$$

determined by (8-8) is called the *center of mass*.   Evidently the collection of points, regarded as a rigid body, would balance about the point R as pivot, for the moment (8-4) is zero when $\mathbf{P} = \mathbf{R}$, and hence the moment (8-5) also vanishes.

Still another example of the use of vectors in mechanics is given by Newton's laws. Relative to an origin O, which is regarded as fixed, let the position of a particle at time t

[1] The vectors $\mathbf{R}$, $\mathbf{P}$, and $\mathbf{R}_i$ are bound position vectors with a common origin for the points R, P, and R_i, respectively.

be specified by the bound vector $\mathbf{R}(t)$. The velocity vector $\mathbf{V}$ is $d\mathbf{R}/dt$, as indicated in Sec. 7, and the *momentum vector* is defined by

$$\mathbf{M} = m\mathbf{V} = m\frac{d\mathbf{R}}{dt} \tag{8-10}$$

where m is the mass of the particle at time t. In this notation Newton's second law of motion takes the simple form

$$\mathbf{F} = \frac{d\mathbf{M}}{dt} \tag{8-11}$$

where $\mathbf{F}$ is the force on the particle at time t. If m is constant the result is

$$\mathbf{F} = m\frac{d^2\mathbf{R}}{dt^2}. \tag{8-12}$$

We shall use (8-10) and (8-12) to derive some interesting properties of the center of mass. Suppose given n particles with masses m_i and positions denoted by $\mathbf{R}_i$ ($i = 1, 2, \ldots, n$), where each m_i is independent of t. The total momentum of the system satisfies

$$\Sigma m_i\frac{d\mathbf{R}_i}{dt} = \frac{d}{dt}\Sigma m_i\mathbf{R}_i = m\frac{d}{dt}\frac{\Sigma m_i\mathbf{R}_i}{m} = m\frac{d\mathbf{R}}{dt} \tag{8-13}$$

where $m = \Sigma m_i$ is the total mass and where $\mathbf{R}$ locates the center of mass [(8-9)]. Thus, *the total momentum of the system equals that of a single particle which has mass m and moves with the same velocity as the center of mass of the system.*

If (8-13) is differentiated with respect to t, there results

$$\Sigma \mathbf{F}_i = m\frac{d^2\mathbf{R}}{dt^2}$$

when we let $\mathbf{F}_i$ be the force on the ith particle and use (8-12). Since internal forces cancel in pairs by Newton's law of equal and opposite reaction, the sum $\Sigma\mathbf{F}_i$ represents the total *external* force acting on the system. Hence *the center of mass has the same acceleration as a particle of mass m acted on by a force equal to the sum of the external forces acting on the system.*

Example 1. Parallel forces $\mathbf{F}$, $-\mathbf{F}$ of equal magnitude but opposite direction constitute a *couple*. Find the total moment, and show that it is the same about every point.

Let $\mathbf{R}$ be a vector from a given point O to a point P on the line of action of $\mathbf{F}$, and $\mathbf{R}_1$ to a point P_1 on the line of action of $-\mathbf{F}$ (Fig. 18). The total torque is

$$\mathbf{R} \times \mathbf{F} + \mathbf{R}_1 \times (-\mathbf{F}) = (\mathbf{R} - \mathbf{R}_1) \times \mathbf{F} = \overrightarrow{(P_1P)} \times \mathbf{F}.$$

Since this is independent of O, the result follows. Notice that $\mathbf{F}$ and $-\mathbf{F}$ must be regarded as sliding vectors (Sec. 1) rather than free vectors, since the line of action is fixed.

Example 2. A system of forces $\mathbf{F}_i$ acting at various points R_i of a rigid body is such that $\Sigma\mathbf{F}_i = \mathbf{0}$. If the total torque about one point is zero, the total torque about every point is zero.

From $\Sigma(\mathbf{R}_0 - \mathbf{R}_i) \times \mathbf{F}_i = 0$, say, we are to deduce $\Sigma(\mathbf{R} - \mathbf{R}_i) \times \mathbf{F}_i = 0$. The two equations may be written

$$\mathbf{R}_0 \times (\Sigma\mathbf{F}_i) = \Sigma\mathbf{R}_i \times \mathbf{F}_i \tag{8-14}$$

$$\mathbf{R} \times (\Sigma\mathbf{F}_i) = \Sigma\mathbf{R}_i \times \mathbf{F}_i. \tag{8-15}$$

Equation (8-14) gives $\Sigma\mathbf{R}_i \times \mathbf{F}_i = \mathbf{0}$, since $\Sigma\mathbf{F}_i = \mathbf{0}$, and (8-15) follows.

Example 3. The moment of the momentum vector $\mathbf{M}$ about a point is called the *angular momentum* of the particle about that point. According to the *principle of angular momentum*, the rate of increase of angular momentum about a point equals the resultant

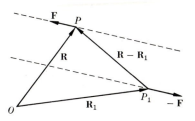

FIGURE 18

torque about that point. Show that this principle is equivalent to Newton's law, $\mathbf{F} = d\mathbf{M}/dt$.

If $\mathbf{A}$ is the angular momentum about the origin, then $\mathbf{A} = \mathbf{R} \times \mathbf{M}$, where $\mathbf{R}$ gives the position of the point. Thus

$$\frac{d\mathbf{A}}{dt} = \mathbf{R} \times \frac{d\mathbf{M}}{dt} + \frac{d\mathbf{R}}{dt} \times \mathbf{M} = \mathbf{R} \times \frac{d\mathbf{M}}{dt} + \mathbf{V} \times (m\mathbf{V}) = \mathbf{R} \times \frac{d\mathbf{M}}{dt}.$$

The principle of angular momentum $d\mathbf{A}/dt = \mathbf{R} \times \mathbf{F}$ is therefore equivalent to

$$\mathbf{R} \times \frac{d\mathbf{M}}{dt} = \mathbf{R} \times \mathbf{F}. \tag{8-16}$$

If this holds for every choice of origin, that is, for every $\mathbf{R}$, then necessarily $d\mathbf{M}/dt = \mathbf{F}$. Conversely, if $d\mathbf{M}/dt = \mathbf{F}$, (8-16) holds for every $\mathbf{R}$.

Example 4.   If P is a point on a disk rotating with constant angular velocity $\boldsymbol{\Omega}$ about an axis perpendicular to the center of the disk, the position of P relative to the xy axes fixed in space is

$$\mathbf{R}(t) = \mathbf{i}R \cos \Omega t + \mathbf{j}R \sin \Omega t \tag{8-17}$$

where R is the distance of P from the center (Fig. 19).

The velocity and acceleration of P are

$$\mathbf{V} = \frac{d\mathbf{R}}{dt} = -\mathbf{i}R\Omega \sin \Omega t + \mathbf{j}R\Omega \cos \Omega t \tag{8-18}$$

$$\mathbf{A} = \frac{d\mathbf{V}}{dt} = -\mathbf{i}R\Omega^2 \cos \Omega t - \mathbf{j}R\Omega^2 \sin \Omega t = -\mathbf{R}(t)\Omega^2. \tag{8-19}$$

Thus the acceleration of P is directed toward the center of the disk and is proportional to the angular speed. This is the so-called *centripetal* acceleration.

Now let a particle initially located at P move along the radius R (say a bead sliding on a wire fixed to the disk) while the disk is rotating with constant angular velocity $\boldsymbol{\Omega}$. If $\mathbf{r}$ is a unit vector along the given radius of the disk and if the particle moves in the direction of $\mathbf{r}$ with constant speed k, then

$$\mathbf{R} = kt\mathbf{r}. \tag{8-20}$$

To compute the velocity of the particle relative to the fixed axes xy, we note from (8-17) that

$$\mathbf{r} = \mathbf{i} \cos \Omega t + \mathbf{j} \sin \Omega t \tag{8-21}$$

and differentiate (8-20). We get

$$\mathbf{V} = \frac{d\mathbf{R}}{dt} = k\mathbf{r} + kt \frac{d\mathbf{r}}{dt} \tag{8-22}$$

and, since $\mathbf{r}$ is a unit vector, $d\mathbf{r}/dt$ is orthogonal to $\mathbf{r}$ by Example 2 of Sec. 7. Equation (8-22) states that $\mathbf{V}$ is made up of two components; one of these is the component $k\mathbf{r}$ directed along the radius $\mathbf{r}$ and the other, $kt \, d\mathbf{r}/dt$, at right angles to the radius in the direction of rotation of the disk. The first of these, obviously, represents the velocity of the particle relative to the disk, and the other is the component due to the rotation of the disk. To obtain the acceleration we differentiate (8-22) and get

$$\mathbf{A} = \frac{d\mathbf{V}}{dt} = 2k \frac{d\mathbf{r}}{dt} + kt \frac{d^2\mathbf{r}}{dt^2}$$

but, from (8-21), $d^2\mathbf{r}/dt = -\Omega^2\mathbf{r}$ so that

$$\mathbf{A} = 2k \frac{d\mathbf{r}}{dt} - k\Omega^2 t\mathbf{r}. \tag{8-23}$$

Accordingly, one component of the acceleration of a moving particle, $-k\Omega^2 t\mathbf{r}$, is directed toward

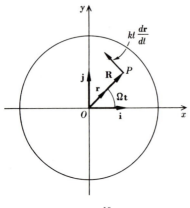

FIGURE 19

the center of the disk and the other, $2k \, dr/dt$, is at right angles to the radius. The first of these $\mathbf{A}_r \equiv -k\Omega^2 l\mathbf{r} = -\Omega^2\mathbf{R}$ is the centripetal acceleration and the other $\mathbf{A}_c = 2k \, dr/dt$ is known as the *compound acceleration of Coriolis*. A particle constrained to slide on a wire thus experiences a force at right angles to the wire, as well as along the wire.

PROBLEMS

1. Given $\mathbf{A} = \mathbf{i} + 2\mathbf{j} + \mathbf{k}$, $\mathbf{B} = \mathbf{i} - \mathbf{k}$, $\mathbf{C} = 2\mathbf{i} + \mathbf{j}$, with $\mathbf{A}, \mathbf{B}$ having their origins at a common point, (a) find the work done by a force $\mathbf{A}$ in a displacement $\mathbf{B}$. (b) Find the work done in a displacement from the head of $\mathbf{A}$ to the head of $\mathbf{B}$ under a force $\mathbf{C}$. (c) Find the work done in the displacement $\mathbf{A}$ subject to simultaneous forces $\mathbf{B}$ and $\mathbf{C}$.

2. In Prob. 1: (a) Find the torque about the origin of $\mathbf{A}$ due to a force $\mathbf{C}$ through the head of $\mathbf{A}$. (b) Find the torque about the head of $\mathbf{A}$ due to a force $\mathbf{C}$ acting through the head of $\mathbf{B}$.

3. In Prob. 1: (a) If the figure formed by $\mathbf{A}$ and $\mathbf{B}$ rotates about $\mathbf{A}$ with angular velocity Ω, find the velocity of the head of $\mathbf{B}$. (b) Find the velocity of the head of $\mathbf{A}$ if the figure formed by $\mathbf{A}$ and $\mathbf{B}$ rotates with angular velocity Ω about an axis parallel to $\mathbf{C}$ through the head of $\mathbf{B}$.

4. Two coordinate systems have a common origin at all times, but the second has a vectorial angular velocity Ω relative to the first. Show that $\mathbf{V}_1 = \mathbf{V}_2 + (\Omega \times \mathbf{R})$, where $\mathbf{V}_1$ and $\mathbf{V}_2$ are the velocity vectors in the first and second systems of a point whose position vector is $\mathbf{R}$ in the first system.

5. Show that the torque due to two couples is the sum of the torques.

6. Three points labeled 1, 2, 3 have masses 1, 2, 3 and positions $2\mathbf{i} + \mathbf{j} + 2\mathbf{k}$, $\mathbf{i} - \mathbf{k}$, $3\mathbf{j}$, respectively. (a) Find the center of mass. (b) Find the total mass of 2, 1, and their center of mass. From this obtain, again, the center of mass for all three.

7. The vectors $\mathbf{A}, \mathbf{B}, \mathbf{C}, \mathbf{D}, \mathbf{E}$ give the positions of the two vertices of a regular pentagon as referred to an origin not necessarily in its plane. Show that this resultant is equal to $5\mathbf{R}$, where $\mathbf{R}$ gives the position of the center. *Hint:* Place a unit mass at each vertex, and find the center of mass in two ways.

8. (a) Show that $\mathbf{F} \cdot \mathbf{V}$ represents the rate at which work is done on a particle moving with velocity $\mathbf{V}$ under a force $\mathbf{F}$. (b) When the mass is constant, show that

$$\frac{d}{dt} \frac{m|\mathbf{V}|^2}{2} = \mathbf{F} \cdot \mathbf{V}$$

so that the rate of increase of kinetic energy equals the rate at which work is done on the particle.

9. A wheel of radius a rolls along a straight line so that its center moves with constant speed V_0. Find the velocity of the point P on the rim relative to the center of the wheel. *Hint:* If $\mathbf{R}$ is the position vector of P and $\mathbf{R}_0$ is the position vector of the center O, the position vector $\mathbf{r}$ of P relative to O is $\mathbf{r} = \mathbf{R} - \mathbf{R}_0$. Write $\mathbf{R}_0 = \mathbf{i}v_0 t + \mathbf{j}a$ and show that $\mathbf{R} = \mathbf{i}a(\omega t - \cos \omega t) + \mathbf{j}a(1 - \sin \omega t)$, where ω is the angular velocity.

10. Refer to Prob. 9 and compute the velocity and acceleration of P along the trajectory of P when $t = \pi a/v_0$, $t = \pi a/2v_0$, $t = 2\pi a/v_0$.

11. Compute the Coriolis acceleration in Example 4 if $\mathbf{R}$ in Eq. (8-20) is replaced by (a) $\mathbf{R} = k\mathbf{r}t^2$ and (b) $\mathbf{R} = k\mathbf{r} \sin t$.

---

9. Lines and Planes. If $\mathbf{R}$ is a bound vector with its origin at the origin of coordinates, the direction numbers x, y, z are the same as the coordinates of the head of $\mathbf{R}$, and one may speak indifferently of "the point $\mathbf{R}$" or "the vector $\mathbf{R}$." This correspondence between vectors and points enables us to use vectors in geometry. Here we consider

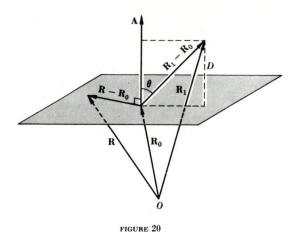

FIGURE 20

the geometry of lines and planes, which is especially simple; the following sections are concerned with general curves and surfaces.

Suppose we have a plane through the point $\mathbf{R}_0$ and perpendicular to the constant vector $\mathbf{A}$. If the point $\mathbf{R}$ is in the plane, then $\mathbf{R} - \mathbf{R}_0$ is perpendicular to $\mathbf{A}$, and conversely (Fig. 20). Hence the equation of the plane is

$$(\mathbf{R} - \mathbf{R}_0)\cdot\mathbf{A} = 0. \tag{9-1}$$

If D is the distance from the point $\mathbf{R}_1$ to the plane, then

$$D = |\mathbf{R}_1 - \mathbf{R}_0|\,|\cos\theta| = \frac{|\mathbf{R}_1 - \mathbf{R}_0|\,|\mathbf{A}|\,|\cos\theta|}{|\mathbf{A}|} = \frac{|(\mathbf{R}_1 - \mathbf{R}_0)\cdot\mathbf{A}|}{|\mathbf{A}|} \tag{9-2}$$

where θ is the angle between $\mathbf{A}$ and $\mathbf{R}_1 - \mathbf{R}_0$.

Next, consider a line through the point $\mathbf{R}_0$ and parallel to a constant vector $\mathbf{A}$. If the point $\mathbf{R}$ is on this line, the vector $\mathbf{R} - \mathbf{R}_0$ is parallel to $\mathbf{A}$, and conversely (Fig. 21). Hence, the equation of the line is

$$(\mathbf{R} - \mathbf{R}_0) \times \mathbf{A} = 0. \tag{9-3}$$

If D is the perpendicular distance from the point $\mathbf{R}_1$ to this line, then

$$D = |\mathbf{R}_1 - \mathbf{R}_0|\,|\sin\theta| = \frac{|\mathbf{R}_1 - \mathbf{R}_0|\,|\mathbf{A}|\,|\sin\theta|}{|\mathbf{A}|} = \frac{|(\mathbf{R}_1 - \mathbf{R}_0) \times \mathbf{A}|}{|\mathbf{A}|}. \tag{9-4}$$

In (9-3) the fact that $\mathbf{R} - \mathbf{R}_0$ is parallel to $\mathbf{A}$ may also be expressed by

$$\mathbf{R} - \mathbf{R}_0 = \mathbf{A}t$$

where t is a scalar. Thus we obtain the equation of the straight line in a parametric form,

$$\mathbf{R} = \mathbf{R}_0 + \mathbf{A}t \qquad -\infty < t < \infty \tag{9-5}$$

which is often more useful than (9-3). It is left to the reader to deduce the cartesian equation by setting

$$\mathbf{R}_0 = x_0\mathbf{i} + y_0\mathbf{j} + z_0\mathbf{k} \qquad \mathbf{A} = a\mathbf{i} + b\mathbf{j} + c\mathbf{k}$$

in (9-5) and equating components. Eliminating t yields the symmetric equation of the straight line

$$\frac{x - x_0}{a} = \frac{y - y_0}{b} = \frac{z - z_0}{c} \qquad (9\text{-}6)$$

which may also be found from (9-3).

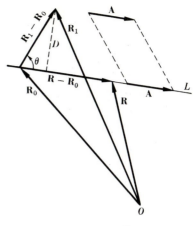

FIGURE 21

Example 1.   Show that every equation of the form

$$ax + by + cz + d = 0 \qquad a, b, c, d \text{ const} \qquad (9\text{-}7)$$

represents a plane with $\mathbf{A} = a\mathbf{i} + b\mathbf{j} + c\mathbf{k}$ as normal, and conversely.

If $\mathbf{R} = \mathbf{i}x + \mathbf{j}y + \mathbf{k}z$ is a general point and $\mathbf{R}_0 = \mathbf{i}x_0 + \mathbf{j}y_0 + \mathbf{k}z_0$ a fixed point on the locus (9-7), writing (9-7) in vector form yields

$$\mathbf{R \cdot A} + d = 0 \qquad \mathbf{R}_0 \cdot \mathbf{A} + d = 0.$$

Subtracting these equations we obtain (9-1), which shows that the locus is a plane. On the other hand, (9-1) itself has the form (9-7), with $d = -\mathbf{R}_0 \cdot \mathbf{A}$; hence the converse is also true.

Example 2.   Find the equation of a line which passes through the point $\mathbf{i} - \mathbf{j}$ and is parallel to the two planes $x + y = 3$, $2x + y + 3z = 4$.

The respective normals to the planes are $\mathbf{i} + \mathbf{j}$ and $2\mathbf{i} + \mathbf{j} + 3\mathbf{k}$, and hence the line of intersection of the planes is parallel to the cross product:

$$(\mathbf{i} + \mathbf{j}) \times (2\mathbf{i} + \mathbf{j} + 3\mathbf{k}) = 3\mathbf{i} - 3\mathbf{j} - \mathbf{k}.$$

Since this vector is parallel to both planes, it gives the direction of the required line, and so the equation is

$$\mathbf{R} = \mathbf{i} - \mathbf{j} + (3\mathbf{i} - 3\mathbf{j} - \mathbf{k})t \qquad -\infty < t < \infty.$$

PROBLEMS

1. (*a*) Find a vector normal to the plane $x + 2y + 3z = 1$. (*b*) Find the angle between this plane and the plane $x + y + z + 2 = 0$. (*c*) What is the distance from the point $3\mathbf{i} + 2\mathbf{j} + \mathbf{k}$ to the plane in part *a*? (*d*) Show that the points $\mathbf{i}$ and $-\mathbf{j} + \mathbf{k}$ lie in the plane in part *a*. (*e*) Find a vector lying in the plane of part *a*. *Hint:* Subtract the vectors of part *d*. (*f*) Verify that the vector of part *e* is normal to the normal found in part *a*.

2. (*a*) Find a vector parallel to the line $\mathbf{R} = \mathbf{i} + \mathbf{k} + (\mathbf{i} + 2\mathbf{j} + 3\mathbf{k})t$. (*b*) If $\mathbf{R} = \mathbf{i}x + \mathbf{j}y + \mathbf{k}z$ in part *a*, find x, y, and z in terms of t. (*c*) In part *a*, find the distance from the point $\mathbf{i} + 2\mathbf{j} + 3\mathbf{k}$ to the given line. (*d*) Show that the line in part *a* intersects the line $\mathbf{R} = 2\mathbf{k} + (3\mathbf{i} + 2\mathbf{j} + \mathbf{k})s$. *Hint:* Equate the two expressions for $\mathbf{R}$, and consider each component. It will be found that all three equations are satisfied by $s = t = \tfrac{1}{2}$. (*e*) Find the intersection point in part *d*. (*f*) Find the point where the line in part *a* intersects the plane $2x - y + 3z = 4$. *Hint:* Substitute the result of part *b* into the equation of the plane, find t, then find $\mathbf{R}$.

3. (*a*) Find the equation of the line common to the two planes $x + 2y + 4z = 1$, $x + y = 3$ in the form $\mathbf{R} = \mathbf{R}_0 + \mathbf{A}t$. *Hint:* Let $z = t$, and solve for x and y in terms of t. (*b*) Find a vector parallel to the intersection of the planes by use of the cross product as in Example 2. (*c*) Verify that your answers to parts *a* and *b* are consistent. (*d*) Find the equation of all planes perpendicular to both planes. (*e*) Write the equation of the line that is parallel to both planes and passes through the point $-3\mathbf{i} + \mathbf{k}$.

4. (*a*) In terms of t, find the square of the distance from the point $\mathbf{i} + 2\mathbf{j} + 3\mathbf{k}$ to a general point on the line $\mathbf{R} = 3\mathbf{i} + 2\mathbf{j} + \mathbf{k} + (\mathbf{i} + \mathbf{j} + \mathbf{k})t$. (*b*) By differentiating, find the t for which the distance is minimum and the minimum value. (*c*) Check by the distance formula.

5. In the form (9-5) obtain the equation of a line perpendicular to the plane $x + y + 3z = 0$ at the origin. At what point does this line intersect the plane $y = 3z + 1$?

6. If the lines $\mathbf{R} = \mathbf{R}_0 + \mathbf{A}t$ and $\mathbf{R} = \mathbf{R}_1 + \mathbf{B}t$ are not parallel, the perpendicular distance between them is

$$D = \frac{|(\mathbf{R}_1 - \mathbf{R}_0) \cdot \mathbf{A} \times \mathbf{B}|}{|\mathbf{A} \times \mathbf{B}|}.$$

Hint: By a suitable figure show that the distance is the length of the projection of $\mathbf{R}_1 - \mathbf{R}_0$ on the common perpendicular to the two lines.

---

10. Normal Lines and Tangent Planes.

If a smooth curve $C\colon x = x(t)$, $y = y(t)$, $z = z(t)$ lies on a smooth surface which has the equation

$$u(x,y,z) = c \tag{10-1}$$

where c is constant, then

$$u[x(t),y(t),z(t)] \equiv c \tag{10-2}$$

identically in t. At a fixed point $\mathbf{R}_0 = \mathbf{i}x_0 + \mathbf{j}y_0 + \mathbf{k}z_0$ on C (Fig. 22) we differentiate (10-2) by the chain rule to obtain

$$\frac{\partial u}{\partial x}\frac{dx}{dt} + \frac{\partial u}{\partial y}\frac{dy}{dt} + \frac{\partial u}{\partial z}\frac{dz}{dt} = 0. \tag{10-3}$$

This may be written as[1]

$$\mathbf{n} \cdot \mathbf{R}'(t) = 0 \tag{10-4}$$

where $\mathbf{R}(t) = \mathbf{i}x(t) + \mathbf{j}y(t) + \mathbf{k}z(t)$ and where

$$\mathbf{n} = \mathbf{i}\frac{\partial u}{\partial x} + \mathbf{j}\frac{\partial u}{\partial y} + \mathbf{k}\frac{\partial u}{\partial z} \qquad \text{at } (x_0,y_0,z_0). \tag{10-5}$$

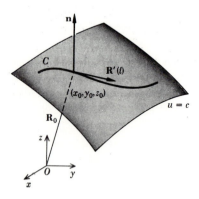

FIGURE 22

Since $\mathbf{R}'(t)$ is tangent to the curve C by Sec. 7, it follows from (10-4) that $\mathbf{n}$ is normal to the curve C. And since this is true for every choice of C, the vector $\mathbf{n}$ must be normal to the surface. The tangent plane is the plane perpendicular to $\mathbf{n}$ at $\mathbf{R}_0$; hence its equation is $\mathbf{n} \cdot (\mathbf{R} - \mathbf{R}_0) = 0$ by Sec. 9.

To illustrate the use of (10-5) we find a normal vector and tangent plane for the ellipsoid

$$x^2 + 2y^2 + 3z^2 = 12$$

at the point $(1,2,-1)$. Since $u = x^2 + 2y^2 + 3z^2$, the partial derivatives are $2x$, $4y$, and $6z$. Evaluating these at $(1,2,-1)$ and substituting in (10-5) give the normal vector

$$\mathbf{n} = 2\mathbf{i} + 8\mathbf{j} - 6\mathbf{k}.$$

[1] The assumptions which underlie this result are clear from the derivation. We assume u differentiable (so that the chain rule holds), and we assume that not all the partial derivatives are zero (otherwise $\mathbf{n} = \mathbf{0}$, and $\mathbf{n}$ does not determine a direction). The analysis shows, then, that $\mathbf{n}$ *is perpendicular to every differentiable curve* $\mathbf{R} = \mathbf{R}(t)$ *that passes through the point* $\mathbf{R}_0$ *and lies in the surface*. It is this property that enables us to consider $\mathbf{n}$ as "normal to the surface."

The tangent plane is perpendicular to **n** and contains the point $(1,2,-1)$. Hence its equation is

$$x + 4y - 3z = 12$$

as the reader can verify.

PROBLEMS

1. By use of (10-5), find a vector normal to the plane $ax + by + cz + d = 0$. Compare Sec. 9, Example 1.

2. At the point $(2,1,3)$ on the surface $xyz = x^2 + 2$, find (*a*) a normal vector, (*b*) an equation for the tangent plane, (*c*) an equation for the normal line.

3. Show that the surfaces $xyz = 1$ and $x^2 + y^2 - 2z^2 = 0$ intersect at right angles at the point $(1,1,1)$; that is, the tangent planes are perpendicular.

4. The two surfaces $x^2 + y^2 + z^2 = 6$ and $2x^2 + 3y^2 + z^2 = 9$ intersect at $(1,1,2)$. Find the angle between the tangent planes at this point.

5. In Prob. 4, find a vector tangent to the curve in which the surfaces intersect. *Hint:* The required vector is perpendicular to both normals.

6. At what point on the surface of the ellipsoid $x^2/a^2 + y^2/b^2 + z^2/c^2 = 1$ does the normal to the ellipsoid make equal angles with the coordinate axes?

7. In Prob. 6, find the equation of the tangent plane to the ellipsoid that cuts the segments of equal lengths from the coordinate axes.

8. Show that the tangent planes to the surface $xyz = a^3$ form tetrahedrons of constant volume with the coordinate planes.

9. Show that the normal at any point to a smooth surface of revolution $z = F(\sqrt{x^2 + y^2})$ intersects the axis of revolution if $F'(r) \neq 0$, $r = \sqrt{x^2 + y^2}$.

10. Find the angle between the normals to the cylinder $x^2 + y^2 = a^2$ and the sphere $(x - a)^2 + y^2 + z^2 = a^2$ at their common point $(a/2, a\sqrt{3}/2, 0)$.

11. Prove that the cone $x^2 + y^2 = z^2$ is tangent to the sphere $x^2 + y^2 + (z - 2)^2 = 2$ at $(0,\pm 1,1)$.

---

11. Frenet's Formulas.   It was shown in Sec. 7 that the vector $\mathbf{R}'(t)$ is tangent to a smooth space curve $\mathbf{R} = \mathbf{R}(t)$ and has the length $|\mathbf{R}'| = ds/dt$, where s is the arc along the curve. If the parameter t is equal to the arc length measured from some point on the curve, so that $t = s$ and

$$\mathbf{R} = \mathbf{R}(s) \tag{11-1}$$

then $ds/dt = 1$. In this case the vector

$$\mathbf{T} = \frac{d\mathbf{R}}{ds} \tag{11-2}$$

is a tangent vector of *unit length*. From $\mathbf{T} \cdot \mathbf{T} = 1$ we deduce that $d\mathbf{T}/ds$ is perpendicular to $\mathbf{T}$ (Sec. 7, Example 2). Hence we may write

$$\frac{d\mathbf{T}}{ds} = \kappa\mathbf{N} \qquad \kappa > 0 \tag{11-3}$$

where **N** is a unit vector perpendicular to **T** and where κ is a scalar multiplier. The vector **N** defined by (11-3) is called the *principal normal*, and the scalar κ is called the *curvature*. The plane of **T** and **N** is termed the *osculating plane*. We define $\kappa = 0$ for a straight line.

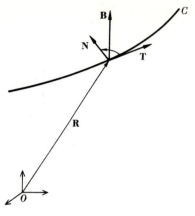

FIGURE 23

If we introduce a third unit vector $\mathbf{B}$ defined by $\mathbf{B} = \mathbf{T} \times \mathbf{N}$, then the system $\mathbf{T}$, $\mathbf{N}$, $\mathbf{B}$ forms a right-handed set of orthogonal unit vectors, analogous to the vectors $\mathbf{i}$, $\mathbf{j}$, $\mathbf{k}$ introduced previously. By Fig. 23,

$$\mathbf{N} \times \mathbf{B} = \mathbf{T} \quad \mathbf{B} \times \mathbf{T} = \mathbf{N} \quad \mathbf{T} \times \mathbf{N} = \mathbf{B}. \quad (11\text{-}4)$$

The vector $\mathbf{B}$ is called the *binormal;* the figure formed by $\mathbf{T}$, $\mathbf{B}$, $\mathbf{N}$ is sometimes referred to as the *trihedral* associated with the curve.

Differentiating the relation $\mathbf{B} = \mathbf{T} \times \mathbf{N}$ and using (11-3) give

$$\mathbf{B}' = \mathbf{T} \times \mathbf{N}' + \mathbf{T}' \times \mathbf{N}$$
$$= \mathbf{T} \times \mathbf{N}' + \kappa \mathbf{N} \times \mathbf{N} = \mathbf{T} \times \mathbf{N}'$$

and hence $\mathbf{B}'$ is perpendicular to $\mathbf{T}$. It is also perpendicular to $\mathbf{B}$, since $\mathbf{B} \cdot \mathbf{B} = 1$, and therefore $\mathbf{B}'$ is parallel to $\mathbf{N}$:

$$\frac{d\mathbf{B}}{ds} = \tau\mathbf{N}. \quad (11\text{-}5)$$

The scalar multiple τ in (11-5) is called the *torsion;* it measures the rate at which the curve twists out of its osculating plane. We define $\tau = 0$ for a straight line. (See also Example 2 below.)

To evaluate $d\mathbf{N}/ds$, recall that $\mathbf{N} = \mathbf{B} \times \mathbf{T}$. Hence

$$\mathbf{N}' = \mathbf{B} \times \mathbf{T}' + \mathbf{B}' \times \mathbf{T} = \kappa\mathbf{B} \times \mathbf{N} + \tau\mathbf{N} \times \mathbf{T} \quad (11\text{-}6)$$

by (11-3) and (11-5). When we use (11-4), Eq. (11-6) reduces to

$$\frac{d\mathbf{N}}{ds} = -\kappa\mathbf{T} - \tau\mathbf{B}. \quad (11\text{-}7)$$

Equations (11-3), (11-5), and (11-7) are known as the *Frenet-Serret formulas;* they are of fundamental importance in the theory of space curves.

By equating the lengths of the two vectors in (11-3) and recalling $|\mathbf{N}| = 1$ we obtain

$$\kappa = |\mathbf{T}'| = |\mathbf{R}''| \qquad ' = \frac{d}{ds}. \quad (11\text{-}8)$$

To get a similar formula for τ we differentiate (11-3), obtaining

$$\mathbf{T}'' = \kappa\mathbf{N}' + \kappa'\mathbf{N} = \kappa(-\kappa\mathbf{T} - \tau\mathbf{B}) + \kappa'\mathbf{N} \quad (11\text{-}9)$$

by (11-7). Hence, by (11-3), (11-9), and (11-4),

$$\mathbf{T}' \times \mathbf{T}'' = \kappa\mathbf{N} \times (-\kappa^2\mathbf{T} - \kappa\tau\mathbf{B} + \kappa'\mathbf{N}) = \kappa^3\mathbf{B} - \kappa^2\tau\mathbf{T}$$

since $\mathbf{N} \times \mathbf{N} = 0$. Taking the dot product with $\mathbf{T}$ yields

$$\mathbf{T} \cdot \mathbf{T}' \times \mathbf{T}'' = -\kappa^2\tau. \quad (11\text{-}10)$$

If we solve (11-10) for τ, express κ^2 in terms of $\mathbf{R}$ by (11-8), and express $\mathbf{T}$ in terms of $\mathbf{R}$ by (11-2), there results

$$\tau = -\frac{\mathbf{R}' \cdot \mathbf{R}'' \times \mathbf{R}'''}{\mathbf{R}'' \cdot \mathbf{R}''} \quad (11\text{-}11)$$

which is the desired formula. When $\mathbf{R} = \mathbf{i}x(s) + \mathbf{j}y(s) + \mathbf{k}z(s)$, Eqs. (11-8) and (11-11) give, respectively,

$$\kappa^2 = (x'')^2 + (y'')^2 + (z'')^2 \qquad \kappa^2\tau = -\begin{vmatrix} x' & y' & z' \\ x'' & y'' & z'' \\ x''' & y''' & z''' \end{vmatrix}. \quad (11\text{-}12)$$

It can be shown that $\kappa(s)$ and $\tau(s)$ determine the curve completely, apart from its position in space.[1]

Since the equation of a smooth curve can always be expressed in terms of its arc as parameter, the foregoing theory suffers no loss of generality by assuming $t = s$. In many physical problems, however, it is more fruitful to take the time t as parameter and this possibility is now to be examined.

Let $\mathbf{R} = \mathbf{R}(t)$ give the position of a moving particle at time t, so that the velocity is $\mathbf{V} = \mathbf{R}'(t)$. With $v = ds/dt$ we have[2]

$$\mathbf{V} = \frac{d\mathbf{R}}{dt} = \frac{d\mathbf{R}}{ds}\frac{ds}{dt} = \mathbf{T}v \tag{11-13}$$

upon using (11-2). Since (11-3) gives

$$\frac{d\mathbf{T}}{dt} = \frac{d\mathbf{T}}{ds}\frac{ds}{dt} = \kappa \mathbf{N}v$$

we get

$$\frac{d\mathbf{V}}{dt} = \mathbf{T}\frac{dv}{dt} + v\frac{d\mathbf{T}}{dt} = \mathbf{T}\frac{dv}{dt} + \kappa v^2 \mathbf{N} \tag{11-14}$$

upon differentiating (11-13). Hence *the acceleration vector $\mathbf{A} = d\mathbf{V}/dt$ lies in the osculating plane, its tangential component has magnitude equal to the linear acceleration dv/dt, and its normal component has magnitude κv^2*. This is a far-reaching generalization of the familiar results

$$\mathbf{A}_{\text{tangential}} = 0 \qquad \mathbf{A}_{\text{normal}} = \frac{v^2}{r}$$

for uniform motion in a circle of radius r.

Taking the cross product of (11-13) and (11-14) with $\mathbf{V}$ replaced by $\mathbf{R}'$ we obtain

$$\mathbf{R}' \times \mathbf{R}'' = \kappa v^3 \mathbf{T} \times \mathbf{N} = \kappa v^3 \mathbf{B}.$$

Hence, the direction of the binormal is given by $\mathbf{R}' \times \mathbf{R}''$ even when the parameter is t rather than s. Since $\mathbf{B}$ is a unit vector, we have

$$\mathbf{B} = \frac{\mathbf{R}' \times \mathbf{R}''}{|\mathbf{R}' \times \mathbf{R}''|} \qquad ' = \frac{d}{dt} \tag{11-15}$$

and similarly, the unit vector $\mathbf{T}$ is obtained from

$$\mathbf{T} = \frac{\mathbf{R}'}{|\mathbf{R}'|} \qquad ' = \frac{d}{dt}. \tag{11-16}$$

Knowing $\mathbf{B}$ and $\mathbf{T}$, we find $\mathbf{N}$ from

$$\mathbf{N} = \mathbf{B} \times \mathbf{T}. \tag{11-17}$$

These formulas enable us to compute the trihedral when the curve is given with an arbitrary parameter t, provided $ds/dt > 0$.

Example 1. Find the equation of the osculating plane at $t = 1$ for the curve $\mathbf{R} = t\mathbf{i} + 2t^2\mathbf{j} + t^3\mathbf{k}$. Differentiation gives

$$\mathbf{R}'(1) = \mathbf{i} + 4\mathbf{j} + 3\mathbf{k} \qquad \mathbf{R}''(1) - 4\mathbf{j} + 6\mathbf{k}.$$

Hence by (11-15) the binormal $\mathbf{B}$ is parallel to

$$(\mathbf{i} + 4\mathbf{j} + 3\mathbf{k}) \times (4\mathbf{j} + 6\mathbf{k}) = 12\mathbf{i} - 6\mathbf{j} + 4\mathbf{k}.$$

[1] See, for example, L. P. Eisenhart, "An Introduction to Differential Geometry," sec. 6, pp. 25–27, Princeton University Press, Princeton, N.J., 1940.

[2] In agreement with the results of Sec. 7, Eq. (11-13) expresses the fact that the velocity is tangent to the orbit and has magnitude equal to the speed.

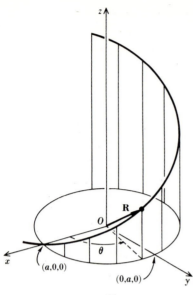

FIGURE 24

The osculating plane is normal to **B** and contains the point

$$\mathbf{R}(1) = \mathbf{i} + 2\mathbf{j} + \mathbf{k}.$$

Hence its equation is $6x - 3y + 2z = 2$, as the reader can verify.

Example 2.   A curve is a plane curve if, and only if, the torsion is zero.

If the curve is a plane curve, not a straight line, the osculating plane is well defined and is the plane of the curve. Hence **B** is constant, and the torsion vanishes by (11-5). Suppose, conversely, that the torsion is zero. Then **B** is constant by (11-5), and therefore, using (11-2),

$$\frac{d}{ds}(\mathbf{B}\cdot\mathbf{R}) = \mathbf{B}\cdot\frac{d\mathbf{R}}{ds} = \mathbf{B}\cdot\mathbf{T} = 0.$$

This gives $\mathbf{B}\cdot\mathbf{R} = \text{const}$, which is the equation of a plane.

Example 3.   Consider the circular helix (Fig. 24) with the equation

$$\mathbf{R} = \mathbf{i}a\cos\theta + \mathbf{j}a\sin\theta + \mathbf{k}p\theta \qquad a,\ p \text{ positive const.} \tag{11-18}$$

Here, the parametric equations are $x = a\cos\theta$, $y = a\sin\theta$, $z = p\theta$.

By (11-8) $\kappa = |\mathbf{R}''|$, where primes denote differentiation with respect to the arc parameter s. From (11-18)

$$d\mathbf{R} = -\mathbf{i}a\sin\theta\,d\theta + \mathbf{j}a\cos\theta\,d\theta + \mathbf{k}p\,d\theta$$

so that $ds^2 = d\mathbf{R}\cdot d\mathbf{R} = (a^2\sin^2\theta + a^2\cos^2\theta + p^2)(d\theta)^2 = (a^2 + p^2)\,d\theta^2$

and therefore $d\theta/ds = 1/\sqrt{a^2 + p^2} \equiv h$, say.   It follows that

$$\frac{d\mathbf{R}}{ds} = \frac{d\mathbf{R}}{d\theta}\frac{d\theta}{ds} = (-\mathbf{i}a\sin\theta + \mathbf{j}a\cos\theta + \mathbf{k}p)h$$

$$\frac{d^2\mathbf{R}}{ds^2} = \frac{d}{d\theta}\frac{d\mathbf{R}}{ds}\frac{d\theta}{ds} = (-\mathbf{i}a\cos\theta - \mathbf{j}a\sin\theta)h^2$$

$$\frac{d^3\mathbf{R}}{ds^3} = \frac{d}{d\theta}\frac{d^2\mathbf{R}}{ds^2}\frac{d\theta}{ds} = (\mathbf{i}a\sin\theta - \mathbf{j}a\cos\theta)h^3.$$

On making use of formula (11-8) we find

$$\kappa^2 = (\mathbf{R}''\cdot\mathbf{R}'') = (a^2\sin^2\theta + a^2\cos^2\theta)h^4 = a^2h^4$$

so that
$$\kappa = \frac{a}{a^2 + p^2}.$$

According to (11-12), the torsion is

$$\tau = -\frac{h^6}{\kappa^2}\begin{vmatrix} -a\sin\theta & a\cos\theta & p \\ -a\cos\theta & -a\sin\theta & 0 \\ a\sin\theta & -a\cos\theta & 0 \end{vmatrix} = \frac{-p}{a^2 + p^2}.$$

If $p = 0$, we get a circle of radius a by inspection of (11-18). In this case $\tau = 0$ because the curve is a plane curve and $\kappa = 1/a$ because the radius is always equal to the constant a. The behavior as $p \to \infty$ may be discussed similarly.

PROBLEMS

1. Given the curve $\mathbf{R}(t) = \mathbf{i}(t^2 - 1) + 2t\mathbf{j} + (t^2 + 1)\mathbf{k}$. (*a*) Find a unit tangent at $t = -1$. (*b*) Find the equation of the normal plane at this point. (*c*) Find the length of the curve from $t = 0$ to $t = 1$.

2. (*a*) If $\mathbf{R}(t)$ in Prob. 1 represents an orbit, find the velocity and acceleration at time t. (*b*) By use of part *a* and Eq. (11-13), find the speed v at time t. (*c*) By use of parts *a* and *b* and (11-14), find the curvature κ and the principal normal $\mathbf{N}$ at time t.

3. If the components of $\mathbf{R}(t)$ are second-degree polynomials in t, then $\mathbf{R} = \mathbf{R}(t)$ is a plane curve. (*a*) Prove this by use of (11-12) and Example 2. (*b*) Find the equation of the plane.

4. Show that (*a*) the tangents to a helix make a fixed angle with the axis of the helix and (*b*) the principal normal is perpendicular to the axis of the helix.

5. (*a*) Given a particle moving according to the law $\mathbf{R}(t) = \mathbf{i}t + \mathbf{j}t^2$, find a unit tangent and a unit normal to the orbit at $t = 1$. (*b*) Find the cartesian components of $\mathbf{V}$ and $\mathbf{A}$ at $t = 1$. (*c*) By use of the dot product and part *a*, find the tangential components V_t and A_t of $\mathbf{V}$ and $\mathbf{A}$ at $t = 1$. (*d*) Find ds/dt as $|\mathbf{R}'(t)|$, and from this find d^2s/dt^2. Compare with part *c*.

6. In Prob. 5, (*a*) show that V_n, the normal component of $\mathbf{V}$, is zero, and find that of $\mathbf{A}$ at $t = 1$ by use of the dot product and Prob. 5*a*. (*b*) By part *a* and $A_n = \kappa|\mathbf{V}|^2$, find the curvature of the orbit at $t = 1$. (*c*) Show that the cartesian equation of the orbit is $y = x^2$, and compute the curvature by $\kappa = y''/(1 + y'^2)^{3/2}$. Compare with part *b*. (*d*) Explain how to find A_n in terms of A_x, A_y, and A_t, and use this to check some of your work.

7. If the curve C is determined by $\mathbf{R} = \mathbf{i}t + \mathbf{j}t^2 + \mathbf{k}t^3$, find the triad of vectors $\mathbf{T}$, $\mathbf{N}$, $\mathbf{B}$ at the point $(1,1,1)$.

8. Find the equations of the tangent line and osculating plane to the curve in Prob. 7 at the point $(2,4,8)$.

9. If the parameter t in Prob. 7 is the time, find the normal and tangential components of the acceleration at $t = 0$ of a particle moving on C.

10. What is the equation of the osculating plane to the curve $\mathbf{R} = \mathbf{i}e^t + \mathbf{j}e^{-t} + \mathbf{k}\sqrt{2}t$ at the point for which $t = 0$?

11. Find the curvature and torsion at any point of the hyperbolic helix $\mathbf{R} = \mathbf{i}a \cosh t + \mathbf{j}a \sinh t + \mathbf{k}at$.

LINEAR VECTOR SPACES AND MATRICES

12. Spaces of Higher Dimensions. There is nothing mysterious about the idea of spaces whose dimensionality is greater than three. In locating objects in the familiar three-dimensional space of our physical intuition, we have found it convenient to introduce a coordinate system and to specify the location of any point in the object by means of three numbers termed the *coordinates of the point*. Thus, if a cartesian system of axes is introduced, we can associate with each point P an ordered triple of labels (x,y,z).

In dealing with the state of gas determined by the pressure p, volume v, and temperature T, it is often useful to visualize the triples of values (p,v,T) as coordinates of points in three-dimensional space, but such a visualization fails when the number of variables characterizing the gas state exceeds three. Thus, the state of gas may (and generally does) depend not only on the pressure, volume, and temperature, but also on the time t. Although a quadruple of values (p,v,T,t) cannot be represented as a point in a fixed coordinate system in three-dimensional space, the geometric visualization is of much lesser importance than the *analytic* apparatus developed for coping with the geometric problems. This apparatus (analytic geometry and vector analysis) makes use of the tools of algebra and analysis which involve operations on ordered sets of quantities such as (p,v,T,l) or $(x_1,x_2, \ldots ,x_n)$, which are valid regardless of the number of variables appearing in the set.

The habits of using the language associated with geometric thinking are so strong, however, that it is natural to continue speaking figuratively of a quadruple of numbers (p,v,T,l) as representing a point in four-dimensional space and more generally to refer to an ordered set of n values $(x_1,x_2, \ldots ,x_n)$ as *a point in n-dimensional space.* The values $x_1, x_2, \ldots , x_n$ may be of quite diverse sorts; the first three, for example, may be associated with cartesian coordinates of some point M in three-dimensional physical space, x_4 may represent the magnitude of electric charge located at M, x_5 may stand for the time of observation, and so on. But whatever meaning we choose to attach to the individual values x_i, we can speak of the *n-tuple* $(x_1,x_2, \ldots ,x_n)$ as representing a point P in n-dimensional space.

In three-dimensional space we found it useful to associate with every pair of points P_1 and P_2 an entity $\overrightarrow{P_1P_2}$ which we called a vector **a**, and we have developed a set of rules for operations with vectors which form the basis for the algebra and calculus of vectors.

Although in the initial formulation of these rules we have been guided by geometric considerations, we have distilled out geometry by giving a set of *algebraic* laws governing operations with vectors. The most important of these laws, discussed in Secs. 2 and 4, are as follows:

$$\mathbf{a} + \mathbf{b} = \mathbf{b} + \mathbf{a} \qquad (\mathbf{a} + \mathbf{b}) + \mathbf{c} = \mathbf{a} + (\mathbf{b} + \mathbf{c}) \qquad \mathbf{a} + 0 = \mathbf{a} \qquad \mathbf{a} - \mathbf{a} = 0$$

$$\alpha(\mathbf{a} + \mathbf{b}) = \alpha\mathbf{a} + \alpha\mathbf{b} \qquad (\alpha + \beta)\mathbf{a} = \alpha\mathbf{a} + \beta\mathbf{a} \qquad (\alpha\beta)\mathbf{a} = \alpha(\beta\mathbf{a})$$

$$\mathbf{a}\cdot\mathbf{b} = \mathbf{b}\cdot\mathbf{a} \qquad \mathbf{a}\cdot(\mathbf{b} + \mathbf{c}) = \mathbf{a}\cdot\mathbf{b} + \mathbf{a}\cdot\mathbf{c} \qquad (\alpha\mathbf{a})\cdot\mathbf{b} = \alpha(\mathbf{a}\cdot\mathbf{b}).$$

Here α, β are real numbers and **a**, **b**, **c** are vectors. The product $\alpha\mathbf{a}$ is a vector, as is the sum $\mathbf{a} + \mathbf{b}$, but the product $\mathbf{a}\cdot\mathbf{b}$ is a real number, not a vector.

We can continue using the suggestive language of three-dimensional vector geometry and say that every pair of points P_1, P_2 in an n-dimensional space determines a vector **a**. However, in devising rules for operating on such vectors we adopt the foregoing algebraic laws, which contain no reference to the dimensionality of space.

The dimensionality of space, we recall, entered only when we made use of these laws in those calculations which involved the *representations of vectors by components in special coordinate systems.* Thus in Sec. 3 we considered a vector in the plane determined by a pair of noncollinear vectors and introduced the notion of base vectors and the so-called "components" of the vector along the base vectors. We also saw that a vector in three-dimensional space can be represented uniquely in terms of its components in the directions of three noncoplanar base vectors. These remarks suggest that the dimensionality of space is in some way connected with the number of base vectors needed to represent a given vector by components. In providing a generalization of the representation of vectors by components in spaces of higher dimensions, we need the notion of linear dependence of a set of vectors which we develop next.

13. The Dimensionality of Space. Linear Vector Spaces.

The concept of linear dependence of a set of vectors $\mathbf{a}_1, \mathbf{a}_2, \ldots , \mathbf{a}_n$ is intimately connected with the idea of dimensionality of space.

FIGURE 25

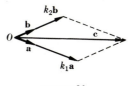

FIGURE 26

DEFINITION. A set of n vectors $\mathbf{a}_1$, $\mathbf{a}_2$, $\ldots$, $\mathbf{a}_n$ *is linearly dependent if there exists a set of numbers* α_1, α_2, $\ldots$, α_n, *not all of which are zero, such that*

$$\alpha_1 \mathbf{a}_1 + \alpha_2 \mathbf{a}_2 + \cdots + \alpha_n \mathbf{a}_n = \mathbf{0}. \tag{13-1}$$

If no such numbers exist, the vectors $\mathbf{a}_1$, $\mathbf{a}_2$, $\ldots$, $\mathbf{a}_n$ *are said to be linearly independent.*[1]

To get at the geometric meaning of this definition, consider two vectors **a** and **b** which are like or oppositely directed (Fig. 25). Then we can find a number $k \neq 0$ such that

$$\mathbf{b} = k\mathbf{a}. \tag{13-2}$$

We can write this equation in symmetric form by setting $k = -\alpha/\beta$, so that (13-2) reads

$$\alpha \mathbf{a} + \beta \mathbf{b} = \mathbf{0}. \tag{13-3}$$

Since neither α nor β is zero, it follows from our definition of linear dependence that two collinear vectors are always linearly dependent. Inasmuch as every vector **b** directed along **a** can be represented in the form (13-2), formula (13-2) serves to define a *one-dimensional linear vector space*. We observe that every two vectors in such a space are linearly dependent.

If we consider two noncollinear vectors **a** and **b** (Fig. 26), every vector **c** in their plane can be represented in the form

$$\mathbf{c} = k_1 \mathbf{a} + k_2 \mathbf{b} \tag{13-4}$$

by a suitable choice of the constants k_1 and k_2. Equation (13-4) can be written as

$$\alpha \mathbf{a} + \beta \mathbf{b} + \gamma \mathbf{c} = \mathbf{0} \tag{13-5}$$

in which not all constants α, β, γ are zero. Formula (13-4) determines every vector **c** in the plane of **a** and **b**, and it thus defines a *two-dimensional linear vector space*, while formula (13-5) ensures that every three vectors in the two-dimensional space are linearly dependent.

If we take three noncoplanar vectors **a**, **b**, **c** (Fig. 27), we can represent every vector **d** in the form

$$\mathbf{d} = k_1 \mathbf{a} + k_2 \mathbf{b} + k_3 \mathbf{c} \tag{13-6}$$

from which follows the relation

$$\alpha \mathbf{a} + \beta \mathbf{b} + \gamma \mathbf{c} + \delta \mathbf{d} = \mathbf{0} \tag{13-7}$$

in which α, β, γ, δ are not all zero.

FIGURE 27

[1] Cf. the definition of linear dependence of a set of functions in Chap. 3, Sec. 24. See also Chap. 2, Secs. 15 and 16.

Equation (13-7) states that in a *three-dimensional linear vector space* defined by (13-6) four vectors are invariably linearly dependent.

The foregoing discussion indicates a relationship between the dimensionality of a vector space with the number of linearly independent vectors required to represent any vector in one-, two-, or three-dimensional vector space.

We generalize this relationship by saying that in an n-dimensional linear vector space every vector $\mathbf{x}$ can be represented in the form

$$\mathbf{x} = k_1\mathbf{a}_1 + k_2\mathbf{a}_2 + \cdots + k_n\mathbf{a}_n \tag{13-8}$$

where $\mathbf{a}_1, \mathbf{a}_2, \ldots, \mathbf{a}_n$ is any set of n linearly independent vectors. It follows from (13-8) that in such a space every set of more than n vectors is linearly dependent.

We shall call a given set of n linearly independent vectors the *base vectors* (or the *basis*) of the n-dimensional linear vector space, and we shall term the numbers $(k_1, k_2, \ldots, k_n)$ the *measure numbers* associated with the basis $\mathbf{a}_1, \mathbf{a}_2, \ldots, \mathbf{a}_n$.

In Sec. 3 we noted that every vector $\mathbf{V}$ in three-dimensional vector space can be represented uniquely by taking as a basis any set of three linearly independent vectors $\mathbf{a}, \mathbf{b}, \mathbf{c}$. But we saw that a special set of mutually orthogonal unit vectors $\mathbf{i}, \mathbf{j}, \mathbf{k}$ when used as a basis greatly simplifies the calculations. This suggests the desirability of representing a vector $\mathbf{x}$ in the n-dimensional space in the form (13-8) in which the base vectors $\mathbf{a}_i$ are the analogs of the unit vectors $\mathbf{i}, \mathbf{j}, \mathbf{k}$. The construction of an analogous set of base vectors requires the extension of the concepts of length and orthogonality to sets of vectors in n-dimensional space. In making these extensions we suppose that the scalar product $\mathbf{a} \cdot \mathbf{b}$ of $\mathbf{a}$ and $\mathbf{b}$ is a real number, and that $\mathbf{a} \cdot \mathbf{a} > 0$ unless $\mathbf{a} = \mathbf{0}$. Further, the operation of scalar multiplication obeys the laws (4-3) and (4-5).

We recall that in three-dimensional space two vectors $\mathbf{a}$ and $\mathbf{b}$ are orthogonal if

$$\mathbf{a} \cdot \mathbf{b} = 0$$

and $\mathbf{a}$ is a unit vector if

$$\mathbf{a} \cdot \mathbf{a} = 1.$$

We extend these definitions to vectors in n-dimensional space and show that, when any set of n linearly independent vectors $\mathbf{a}_1, \mathbf{a}_2, \ldots, \mathbf{a}_n$ is given, one can construct a new set of vectors $\mathbf{e}_1, \mathbf{e}_2, \ldots, \mathbf{e}_n$, such that

$$\begin{aligned} \mathbf{e}_i \cdot \mathbf{e}_j &= 0 \qquad \text{if } i \neq j \\ &= 1 \qquad \text{if } i = j. \end{aligned} \tag{13-9}$$

A set of vectors satisfying the conditions (13-9) is called an *orthonormal set*.

Let the set of vectors $\mathbf{a}_1, \mathbf{a}_2, \ldots, \mathbf{a}_n$ be linearly independent, so that the equation

$$\alpha_1\mathbf{a}_1 + \alpha_2\mathbf{a}_2 + \cdots + \alpha_n\mathbf{a}_n = \mathbf{0} \tag{13-10}$$

can be satisfied only by choosing $\alpha_1 = \alpha_2 = \cdots = \alpha_n = 0$. It follows that $\mathbf{a}_1 \neq \mathbf{0}$, for if it were a zero vector, the choice

$$\alpha_1 = 1, \qquad \alpha_2 = 0, \ldots, \alpha_n = 0$$

would satisfy (13-10) and hence the vectors $\mathbf{a}_i$ would be linearly dependent, thus contradicting our initial assumption.

We shall write

$$\mathbf{a}_i \cdot \mathbf{a}_i \equiv |\mathbf{a}_i|^2$$

and call $|\mathbf{a}_i|$ the *length* of $\mathbf{a}_i$. Now denote the product of $\mathbf{a}_1$ by the reciprocal of its length $|\mathbf{a}_1|$ by $\mathbf{e}_1$, so that

$$\mathbf{e}_1 = \frac{\mathbf{a}_1}{|\mathbf{a}_1|}.$$

Since $\mathbf{e}_1 \cdot \mathbf{e}_1 = 1$, $\mathbf{e}_1$ is a unit vector.   The vectors

$$\mathbf{e}_1, \mathbf{a}_2, \ldots, \mathbf{a}_n$$

are obviously linearly independent.   Consider next the vector

$$\mathbf{e}_2' = \mathbf{a}_2 - (\mathbf{a}_2 \cdot \mathbf{e}_1)\mathbf{e}_1.$$

The scalar product $\mathbf{e}_2' \cdot \mathbf{e}_1$ is

$$\mathbf{e}_2' \cdot \mathbf{e}_1 = \mathbf{a}_2 \cdot \mathbf{e}_1 - (\mathbf{a}_2 \cdot \mathbf{e}_1)\mathbf{e}_1 \cdot \mathbf{e}_1 = 0$$

since $\mathbf{e}_1$ is a unit vector.   Thus $\mathbf{e}_2'$ is orthogonal to $\mathbf{e}_1$ and the vector

$$\mathbf{e}_2 \equiv \frac{\mathbf{e}_2'}{|\mathbf{e}_2'|}$$

is a unit vector orthogonal to $\mathbf{e}_1$.

The set of vectors

$$\mathbf{e}_1, \mathbf{e}_2, \mathbf{a}_3, \ldots, \mathbf{a}_n$$

is linearly independent, and we construct the vector

$$\mathbf{e}_3' = \mathbf{a}_3 - (\mathbf{a}_3 \cdot \mathbf{e}_1)\mathbf{e}_1 - (\mathbf{a}_3 \cdot \mathbf{e}_2)\mathbf{e}_2$$

which is orthogonal to both $\mathbf{e}_1$ and $\mathbf{e}_2$.   The vector

$$\mathbf{e}_3 = \frac{\mathbf{e}_3'}{|\mathbf{e}_3'|}$$

is a unit vector, and the set of vectors

$$\mathbf{e}_1, \mathbf{e}_2, \mathbf{e}_3, \mathbf{a}_4, \ldots, \mathbf{a}_n$$

is a linearly independent set.   We continue the process by forming

$$\mathbf{e}_4' = \mathbf{a}_4 - (\mathbf{a}_4 \cdot \mathbf{e}_1)\mathbf{e}_1 - (\mathbf{a}_4 \cdot \mathbf{e}_2)\mathbf{e}_2 - (\mathbf{a}_4 \cdot \mathbf{e}_3)\mathbf{e}_3$$

which is orthogonal to $\mathbf{e}_1$, $\mathbf{e}_2$, and $\mathbf{e}_3$, and normalize it by dividing it by $|\mathbf{e}_4'|$.   The set of vectors

$$\mathbf{e}_1, \mathbf{e}_2, \mathbf{e}_3, \mathbf{e}_4, \mathbf{a}_5, \ldots, \mathbf{a}_n$$

is linearly independent, and a continuation of the procedure yields after n steps the desired set of orthonormal vectors

$$\mathbf{e}_1, \mathbf{e}_2, \ldots, \mathbf{e}_n.$$

14. Cartesian Reference Frames.   When the base vectors $\mathbf{i}, \mathbf{j}, \mathbf{k}$ of Sec. 3 are oriented along the xyz axes, the coordinates of their terminal points are

$$\mathbf{i}: \quad (1,0,0)$$
$$\mathbf{j}: \quad (0,1,0)$$
$$\mathbf{k}: \quad (0,0,1).$$

By analogy it can be said that when a set of orthonormal base vectors $\mathbf{e}_1, \mathbf{e}_2, \ldots, \mathbf{e}_n$ is oriented along "a cartesian reference frame in n-dimensional Euclidean space" the terminal points of the base vectors have the coordinates

$$\mathbf{e}_1: \quad (1,0,0, \ldots, 0)$$
$$\mathbf{e}_2: \quad (0,1,0, \ldots, 0)$$
$$\mathbf{e}_3: \quad (0,0,1, \ldots, 0)$$
$$\cdots \cdots \cdots \cdots$$
$$\mathbf{e}_n: \quad (0,0,0, \ldots, 1).$$

In this reference frame every vector $\mathbf{x}$ has the representation

$$\mathbf{x} = x_1\mathbf{e}_1 + x_2\mathbf{e}_2 + \cdots + x_n\mathbf{e}_n \tag{14-1}$$

where the x_i are the *components* of $\mathbf{x}$.

On making use of the distributive law of scalar multiplication, we find that

$$\mathbf{x} \cdot \mathbf{x} = x_1^2 + x_2^2 + \cdots + x_n^2 \tag{14-2}$$

since

$$\mathbf{e}_i \cdot \mathbf{e}_j = \delta_{ij} \tag{14-3}$$

where the symbol δ_{ij}, the *Kronecker delta*, means

$$\begin{aligned} \delta_{ij} &= 1 \quad \text{if } i = j \\ &= 0 \quad \text{if } i \neq j. \end{aligned}$$

From (14-2) we conclude that the length $|\mathbf{x}|$ of the vector $\mathbf{x}$ is given by the formula

$$|\mathbf{x}| = \sqrt{x_1^2 + x_2^2 + \cdots + x_n^2}.$$

This is the *formula of Pythagoras in n-dimensional Euclidean space*.

Also, if

$$\mathbf{y} = y_1 \mathbf{e}_1 + y_2 \mathbf{e}_2 + \cdots + y_n \mathbf{e}_n \tag{14-4}$$

then on forming the scalar product $\mathbf{x} \cdot \mathbf{y}$ we find

$$\mathbf{x} \cdot \mathbf{y} = x_1 y_1 + x_2 y_2 + \cdots + x_n y_n \tag{14-5}$$

which has the same structure as formula (4-9).

For the sum of two vectors $\mathbf{x}$, $\mathbf{y}$ with components

$$\begin{aligned} \mathbf{x}: \quad & (x_1, x_2, \ldots, x_n) \\ \mathbf{y}: \quad & (y_1, y_2, \ldots, y_n) \end{aligned}$$

we have the vector $\mathbf{x} + \mathbf{y}$ with components

$$\mathbf{x} + \mathbf{y}: (x_1 + y_1, x_2 + y_2, \ldots, x_n + y_n) \tag{14-6}$$

and for the product of $\mathbf{x}$ by a scalar α,

$$\alpha\mathbf{x}: (\alpha x_1, \alpha x_2, \ldots, \alpha x_n). \tag{14-7}$$

If we have two vectors $\mathbf{x}$ and $\mathbf{y}$ in Euclidean three-dimensional space, we have a useful inequality

$$(\mathbf{x} \cdot \mathbf{y})^2 \leq (\mathbf{x} \cdot \mathbf{x})(\mathbf{y} \cdot \mathbf{y}) \tag{14-8}$$

which follows directly from the fact that

$$\cos^2 \theta = \frac{(\mathbf{x} \cdot \mathbf{y})^2}{(\mathbf{x} \cdot \mathbf{x})(\mathbf{y} \cdot \mathbf{y})} \leq 1.$$

We show next that the formula (14-8), known as the *Cauchy-Schwarz inequality*, is valid in an *n*-dimensional Euclidean space.

Indeed, if $\mathbf{y} \neq \mathbf{0}$,

$$(\mathbf{x} \cdot \mathbf{x})(\mathbf{y} \cdot \mathbf{y}) - (\mathbf{x} \cdot \mathbf{y})^2 = \mathbf{y} \cdot \mathbf{y} \left[\mathbf{x} \cdot \mathbf{x} - 2\frac{(\mathbf{x} \cdot \mathbf{y})^2}{\mathbf{y} \cdot \mathbf{y}} + \frac{(\mathbf{x} \cdot \mathbf{y})^2}{\mathbf{y} \cdot \mathbf{y}} \right] = |\mathbf{y}|^2 \left| \mathbf{x} - \mathbf{y}\frac{\mathbf{x} \cdot \mathbf{y}}{\mathbf{y} \cdot \mathbf{y}} \right|^2 \geq 0$$

which proves the inequality (14-8). We note that the equality sign in (14-8) holds if, and only if, $\mathbf{y} = \mathbf{0}$ or $\mathbf{x} = \alpha\mathbf{y}$ for some scalar α.

The formula (14-8) enables us to establish the result

$$|\mathbf{x} + \mathbf{y}| \leq |\mathbf{x}| + |\mathbf{y}| \tag{14-9}$$

analogous to the "triangle inequality" of Prob. 3 in Sec. 3. We compute

$$\begin{aligned} |\mathbf{x} + \mathbf{y}|^2 = (\mathbf{x} + \mathbf{y}) \cdot (\mathbf{x} + \mathbf{y}) &= \mathbf{x} \cdot \mathbf{x} + \mathbf{y} \cdot \mathbf{y} + 2\mathbf{x} \cdot \mathbf{y} \\ &\leq |\mathbf{x}|^2 + |\mathbf{y}|^2 + 2|\mathbf{x} \cdot \mathbf{y}|. \end{aligned} \tag{14-10}$$

But from (14-8)

$$(\mathbf{x} \cdot \mathbf{x})(\mathbf{y} \cdot \mathbf{y}) \geq |\mathbf{x} \cdot \mathbf{y}|^2$$

so that

$$|\mathbf{x}| \cdot |\mathbf{y}| \geq |\mathbf{x} \cdot \mathbf{y}|. \tag{14-11}$$

The substitution from (14-11) in (14-10) yields

$$|\mathbf{x} + \mathbf{y}|^2 \leq |\mathbf{x}|^2 + |\mathbf{y}|^2 + 2|\mathbf{x}| \cdot |\mathbf{y}| = (|\mathbf{x}| + |\mathbf{y}|)^2$$

and on extracting the square root we get the inequality (14-9).

In quantum mechanics and in several other branches of physics it is necessary to consider ordered sets of *complex* numbers $(x_1, x_2, \ldots, x_n)$. Such sets can be viewed as components of a vector $\mathbf{x}$ in an n-dimensional *complex vector space*. For the definition of addition of two complex vectors $\mathbf{x}$, $\mathbf{y}$ with components

$$\mathbf{x}: \quad (x_1, x_2, \ldots, x_n)$$
$$\mathbf{y}: \quad (y_1, y_2, \ldots, y_n)$$

we can take formula (14-6) and define the multiplication by a scalar α (real or complex) by (14-7). To make the length $|\mathbf{x}|$ of the complex vector $\mathbf{x}$ real, we adopt as the definition of scalar product of $\mathbf{x}$ and $\mathbf{y}$ the formula

$$\mathbf{x} \cdot \mathbf{y} = \bar{x}_1 y_1 + \bar{x}_2 y_2 + \cdots + \bar{x}_n y_n \tag{14-12}$$

in which $\bar{x}_i$ denotes the conjugate of the complex number x_i. This formula specializes to (14-5) when the components of vectors are real, since for real numbers $\bar{x}_i = x_i$. We note from (14-12) that

$$\mathbf{y} \cdot \mathbf{x} = x_1 \bar{y}_1 + x_2 \bar{y}_2 + \cdots + x_n \bar{y}_n$$

so that

$$\mathbf{x} \cdot \mathbf{y} = \overline{\mathbf{y} \cdot \mathbf{x}}$$

since the conjugate of the sum of complex numbers is equal to the sum of their conjugates and the conjugate of the product is the product of the conjugates.

Formula (14-12) yields

$$\mathbf{x} \cdot \mathbf{x} = \bar{x}_1 x_1 + \bar{x}_2 x_2 + \cdots + \bar{x}_n x_n \tag{14-13}$$

so that $|\mathbf{x}| = \sqrt{\mathbf{x} \cdot \mathbf{x}}$ is a real number.

The definition of linear independence of a set of complex vectors is that given in Sec. 13 where the constants α_i are now in the field of complex numbers.

PROBLEMS

1. Show that if one starts with the definition of a vector $\mathbf{x}$ as an n-tuple of n real or complex numbers $(x_1, x_2, \ldots, x_n)$ and uses for the definition of sum and product the formulas

$$\mathbf{x} + \mathbf{y} = (x_1 + y_1, \ldots, x_n + y_n) \qquad k\mathbf{x} = (kx_1, \ldots, kx_n) \qquad \mathbf{x} \cdot \mathbf{y} = \sum_{i=1}^{n} \bar{x}_i y_i$$

then $\quad (\mathbf{x} + \mathbf{y}) \cdot \mathbf{z} = \mathbf{x} \cdot \mathbf{z} + \mathbf{y} \cdot \mathbf{z}, \; \mathbf{x} \cdot (\mathbf{y} + \mathbf{z}) = \mathbf{x} \cdot \mathbf{y} + \mathbf{x} \cdot \mathbf{z}, \; (k\mathbf{x}) \cdot \mathbf{y} = \bar{k}(\mathbf{x} \cdot \mathbf{y}), \; \mathbf{x} \cdot (k\mathbf{y}) = k(\mathbf{x} \cdot \mathbf{y}).$

2. Prove that, if $\mathbf{a}^{(1)}$, $\mathbf{a}^{(2)}$, $\ldots$, $\mathbf{a}^{(n)}$ is a set of n linearly independent vectors in a complex n-dimensional vector space, the only vector $\mathbf{x}$ orthogonal to each of the vectors $\mathbf{a}^{(i)}$ is the zero vector.

3. Prove that a set of mutually orthogonal vectors is always linearly independent.

4. Modify the proof of orthogonalization in Sec. 13 so that it applies to a set of linearly independent complex vectors.

---

15. Summation Convention. Cramer's Rule.   In dealing with expressions involving sums of quantities it is often useful to adopt the following summation convention: *If in some expression a certain summation index occurs twice, we omit writing the summation symbol Σ and agree to sum the terms in the expression for all admissible values of the index.*

Thus in a linear form $\sum_{i=1}^{3} a_i x_i$ the summation index i appears twice under the summation symbol Σ, and we shall write $a_i x_i$ to mean $a_1 x_1 + a_2 x_2 + a_3 x_3$. The symbol $\sum_{i=1}^{3} a_{ii} = a_{11} + a_{22} + a_{33}$ will be written simply as a_{ii}. Again, a double sum

$$\sum_{i=1}^{3} \sum_{j=1}^{3} a_{ij} x_i x_j = a_{11} x_1 x_1 + a_{12} x_1 x_2 + a_{13} x_1 x_3 + a_{21} x_2 x_1 + a_{22} x_2 x_2$$
$$+ a_{23} x_2 x_3 + a_{31} x_3 x_1 + a_{32} x_3 x_2 + a_{33} x_3 x_3$$

which has two repeated summation indices i and j under the summation symbols, will be written as

$$a_{ij} x_i x_j.$$

The range of admissible values of the indices, of course, has to be specified. Thus, the expression

$$a_{ij} x_j \qquad \begin{cases} i = 1, 2, 3 \\ j = 1, 2, 3, 4 \end{cases}$$

represents *three linear forms*

$$a_{11} x_1 + a_{12} x_2 + a_{13} x_3 + a_{14} x_4$$
$$a_{21} x_1 + a_{22} x_2 + a_{23} x_3 + a_{24} x_4$$
$$a_{31} x_1 + a_{32} x_2 + a_{33} x_3 + a_{34} x_4$$

corresponding to the three possible choices $i = 1, 2, 3$ of the *free index i*. The summation index j is often called the *dummy index* because it can be replaced by any other letter having the same range of summation. The dummy index is analogous to the variable of integration in a definite integral, which can also be changed at will. Thus

$$a_{ij} x_i x_j = a_{kr} x_k x_r$$

it being understood that the indices i, j, k, r range over the same sets of values. Unless a statement to the contrary is made, we shall suppose that the indices have the range of values from 1 to n. We shall thus write formulas (14-4) and (14-5), for example, as $\mathbf{y} = y_i \mathbf{e}_i$, $\mathbf{x} \cdot \mathbf{y} = x_i y_i$, and (14-13) as $\mathbf{x} \cdot \mathbf{x} = \bar{x}_i x_i$. We shall make use of this summation notation among other places in writing formulas for the product of determinants and for the expansion of determinants.

We recall that a determinant

$$|a_{ij}| \equiv \begin{vmatrix} a_{11} & a_{12} \cdots a_{1n} \\ a_{21} & a_{22} \cdots a_{2n} \\ \cdot & \cdot \cdot \cdot \cdot \cdot \cdot \\ a_{n1} & a_{n2} \cdots a_{nn} \end{vmatrix} \qquad (15\text{-}1)$$

of order n represents an algebraic sum of $n!$ terms formed from the elements a_{ij} in such a way that one, and only one, element from each row i and each column j appears in each term.[1]

The product of two determinants $|a_{ij}|$ and $|b_{ij}|$, each of the order n, can be written as a single determinant $|c_{ij}|$ of order n in which the element c_{ij} in the ith row and jth column is[2]

$$c_{ij} = a_{ik} b_{jk} = a_{i1} b_{j1} + a_{i2} b_{j2} + \cdots + a_{in} b_{jn}. \qquad (15\text{-}2)$$

---

[1] A discussion of determinants is contained in Appendix A.

[2] Since the number n is *fixed*, the term $a_{in} b_{jn}$ does not represent the sum of terms with respect to n. Here n is *not* a summation index. Cf. Appendix A, Formula (1-10).

Inasmuch as the value of the determinant $|b_{ij}|$ is unchanged when its rows and columns are interchanged, the value of the determinant

$$|c_{ij}| = |a_{ij}| \, |b_{ij}|$$

with the elements (15-2) is the same as that of the determinant $|c_{ij}|$ with the elements

$$c_{ij} = a_{ik}b_{kj} = a_{i1}b_{1j} + a_{i2}b_{2j} + \cdots + a_{in}b_{nj}. \tag{15-3}$$

If the cofactor[1] of the element a_{ij} in the determinant (15-1) is denoted by A_{ij}, we can expand a_{ij} in terms of the cofactors of elements in any row or column of the determinant. A reference to (1-5) in Appendix A will show that the following formulas include the Laplace developments of (15-1):

$$a_{ij}A_{ik} = \delta_{jk}a \tag{15-4}$$

$$a_{ji}A_{ki} = \delta_{jk}a \tag{15-5}$$

where δ_{jk} is the Kronecker delta and a stands for the value of $|a_{ij}|$; for if in (15-4) $k \neq j$, the expression $a_{ij}A_{ik}$ represents the sum of products of the elements in the jth column by the cofactors of the elements in the kth column. The value of such a sum is zero, since it represents the expansion of a determinant with two like columns. If $j = k$, the sum $a_{ij}A_{ik}$ is the sum of products of the elements in the jth column by the cofactors of those elements, yielding the value $a = |a_{ij}|$. Similar statements apply to (15-5) if we replace the word "column" by "row."

Formula (15-4) enables us to give a compact derivation of Cramer's rule for solving a system of n linear equations

$$a_{ij}x_j = b_i \tag{15-6}$$

in n unknowns x_j.

We multiply both members of (15-6) by the cofactors A_{ik} and sum with respect to i. We get

$$a_{ij}A_{ik}x_j = A_{ik}b_i.$$

But by (15-4) this is

$$\delta_{jk}ax_j = A_{ik}b_i.$$

The sum $\delta_{jk}x_j = x_k$, and we conclude that

$$x_k = \frac{A_{ik}b_i}{a} \tag{15-7}$$

whenever $a \neq 0$. The numerator in (15-7) is the determinant obtained by replacing the elements in the kth column of $|a_{ij}|$ by the b_i. The reader finding the foregoing calculations too concise will find a more expansive discussion in Sec. 2 of Appendix A.

PROBLEMS

1. Write the following expressions in full:

(a) $\delta_{ij}a_j$; (b) $\delta_{ij}x_ix_j$; (c) $a_{ij}b_{jk} = \delta_{ik}$; (d) $a_{ijk}x_k$; (e) $(\partial f/\partial x_i)\,dx_i$; (f) $(\partial f_i/\partial x_k)\,dx_k$; (g) δ_{ii}; (h) $a_i = (\partial y_i/\partial x_j)b_j$; (i) $a_{ij}a_{ik} = \delta_{jk}$; (j) $a_{ij}x_ix_j$; (k) $\delta_{ij}\delta_{ik}$; (l) $a_{ik}x_k = b_i$. The symbols δ_{ij} denote the Kronecker deltas.

2. Write out the determinants represented by the expansion $a_{i2}A_{i3}$ and $a_{3i}A_{i4}$, where A_{ij} is the cofactor of the element a_{ij} in $|a_{ij}|$. Also write the determinants represented by $a_{i2}A_{i2}$ and $a_{3i}A_{i3}$.

[1] See Appendix A. We recall that the cofactor of a_{ij} is the signed minor M_{ij} of the element a_{ij}, the sign being $(-1)^{i+j}$.

3. Expand the determinants:

$$(a) \begin{vmatrix} a_{11} & a_{12} & a_{13} & a_{14} \\ 0 & a_{22} & a_{23} & a_{24} \\ 0 & 0 & a_{33} & a_{34} \\ 0 & 0 & 0 & a_{44} \end{vmatrix}; \quad (b) \begin{vmatrix} 1 & x_1 & x_1{}^2 \\ 1 & x_2 & x_2{}^2 \\ 1 & x_3 & x_3{}^2 \end{vmatrix}; \quad (c) \begin{vmatrix} 1 & 1 & 1 \\ x_1 & x_2 & x_3 \\ x_1{}^2 & x_2{}^2 & x_3{}^2 \end{vmatrix}; \quad (d) \begin{vmatrix} a_1 & 0 & 0 \\ a_2 & b_2 & 0 \\ a_3 & b_3 & c_3 \end{vmatrix}.$$

4. Multiply determinant (b) in Prob. 3 by determinants (c) and (d).

---

16. Matrices.   In this section we introduce the concept of a matrix and discuss some rules of operation with matrices which are of value in the study of linear transformations.

An $m \times n$ matrix is an ordered set of mn quantities a_{ij} arranged in a rectangular array of m rows and n columns.   If $m = n$, the array is called a *square matrix* of order n. The quantities a_{ij} are called the *elements* of the matrix.   Thus, a matrix is an array

$$A \equiv \begin{pmatrix} a_{11} & a_{12} \cdots a_{1n} \\ a_{21} & a_{22} \cdots a_{2n} \\ \cdot & \cdot \cdot \cdot \cdot \cdot \cdot \\ a_{m1} & a_{m2} \cdots a_{mn} \end{pmatrix} \tag{16-1}$$

where parentheses are used to enclose the array of elements.   We shall denote matrices by capital letters, or when it is desired to exhibit a typical element of the matrix (16-1), we shall write (a_{ij}).

If the order of the elements in (16-1) is changed, or if any element is changed, a different matrix results.   For example, a triple of values (a_1, a_2, a_3) representing the cartesian coordinates of a point is a 1×3 matrix.   If $a_1 \neq a_2$, the matrix (a_2, a_1, a_3) obviously represents a different point.

Two $m \times n$ matrices $A = (a_{ij})$ and $B = (b_{ij})$ are said to be equal if, and only if, $a_{ij} = b_{ij}$ for each i and j.   That is, $A = B$ only when the elements in like positions of the two arrays are equal.

We define the sum $A + B$ of two $m \times n$ matrices $A = (a_{ij})$, $B = (b_{ij})$ to be the array

$$A + B = (a_{ij} + b_{ij}) \tag{16-2}$$

and their difference $A - B$ to be the array

$$A - B = (a_{ij} - b_{ij}).$$

We shall agree to say that the product of the matrix $A = (a_{ij})$ by a constant k, written kA, is a matrix each of whose elements is multiplied by k.   Thus $kA = (ka_{ij})$.

If we have an $m \times n$ matrix A and an $n \times p$ matrix B, we define the product of A and B, written AB, by the formula

$$AB = (a_{ij}b_{jk}) \tag{16-3}$$

where, as agreed in Sec. 15, the repeated index j is summed from 1 to n.   Thus, the product AB is an $m \times p$ matrix, and we can multiply two matrices only if the number of columns in the first factor is equal to the number of rows in the second.

Example 1.   If

$$A = \begin{pmatrix} 1 & 0 & 2 \\ 2 & -1 & 3 \\ 0 & 5 & 6 \end{pmatrix} \quad \text{and} \quad B = \begin{pmatrix} 2 & -1 & 1 \\ 0 & 1 & 2 \\ 1 & -2 & -1 \end{pmatrix}$$

$$A + B = \begin{pmatrix} 1+2 & 0-1 & 2+1 \\ 2+0 & -1+1 & 3+2 \\ 0+1 & 5-2 & 6-1 \end{pmatrix} = \begin{pmatrix} 3 & -1 & 3 \\ 2 & 0 & 5 \\ 1 & 3 & 5 \end{pmatrix}$$

and

$$AB = \begin{pmatrix} (1)(2) + & (0)(0) + (2)(1) & (1)(-1) + & (0)(1) + (2)(-2) & (1)(1) + & (0)(2) + (2)(-1) \\ (2)(2) + (-1)(0) + (3)(1) & (2)(-1) + (-1)(1) + (3)(-2) & (2)(1) + (-1)(2) + (3)(-1) \\ (0)(2) + & (5)(0) + (6)(1) & (0)(-1) + & (5)(1) + (6)(-2) & (0)(1) + & (5)(2) + (6)(-1) \end{pmatrix}$$

$$= \begin{pmatrix} 4 & -5 & -1 \\ 7 & -9 & -3 \\ 6 & -7 & 4 \end{pmatrix}.$$

Also, if
$$C = \begin{pmatrix} 2 \\ 3 \\ -1 \end{pmatrix}$$

then

$$AC = \begin{pmatrix} 1 & 0 & 2 \\ 2 & -1 & 3 \\ 0 & 5 & 6 \end{pmatrix} \begin{pmatrix} 2 \\ 3 \\ -1 \end{pmatrix} = \begin{pmatrix} (1)(2) + & (0)(3) + (2)(-1) \\ (2)(2) + (-1)(3) + (3)(-1) \\ (0)(2) + & (5)(3) + (6)(-1) \end{pmatrix} = \begin{pmatrix} 0 \\ -2 \\ 9 \end{pmatrix}.$$

We observe that the rule (16-2) for the addition of matrices requires that $A + B = B + A$, but it does not follow from (16-3) that the order of factors in the product AB can be interchanged even when the matrices are square. Indeed, for

$$A = \begin{pmatrix} 0 & 1 \\ 1 & 0 \end{pmatrix} \quad \text{and} \quad B = \begin{pmatrix} -1 & 0 \\ 0 & 1 \end{pmatrix}$$

the rule (16-3) gives

$$AB = \begin{pmatrix} 0 & 1 \\ -1 & 0 \end{pmatrix} \quad \text{while } BA = \begin{pmatrix} 0 & -1 \\ 1 & 0 \end{pmatrix}.$$

Thus, the multiplication of matrices, in general, is not commutative.

However, if we have two square matrices of order n which have zero elements everywhere except possibly on the main diagonal, it follows from (16-3) that

$$\begin{pmatrix} a_1 & 0 & \cdots & 0 \\ 0 & a_2 & \cdots & 0 \\ \cdot & \cdot & \cdot & \cdot \\ 0 & 0 & \cdots & a_n \end{pmatrix} \cdot \begin{pmatrix} b_1 & 0 & \cdots & 0 \\ 0 & b_2 & \cdots & 0 \\ \cdot & \cdot & \cdot & \cdot \\ 0 & 0 & \cdots & b_n \end{pmatrix} = \begin{pmatrix} a_1 b_1 & 0 & \cdots & 0 \\ 0 & a_2 b_2 & \cdots & 0 \\ \cdot & \cdot & \cdot & \cdot \\ 0 & 0 & \cdots & a_n b_n \end{pmatrix}.$$

Such matrices are called *diagonal*.

Thus for two diagonal matrices A and B,

$$AB = BA.$$

A diagonal matrix in which all elements along the main diagonal are equal is called a *scalar matrix*. A particular scalar matrix

$$I = \begin{pmatrix} 1 & 0 \cdots 0 \\ 0 & 1 \cdots 0 \\ \cdot & \cdot \cdot \cdot \\ 0 & 0 \cdots 1 \end{pmatrix} \tag{16-4}$$

is called the *identity* (or *unit*) *matrix*.

We note that if I is the identity matrix and A is any square matrix then[1]

$$IA = AI = A. \tag{16-5}$$

By analogy with the rules of ordinary algebra, we define the zero matrix O to be the matrix such that

$$O + A = A.$$

It follows from (16-2) that all elements of the zero matrix are zeros. We observe that

[1] More generally we can show that if $AX = XA$ for *every* $n \times n$ matrix A then X is a scalar matrix. See Prob. 6.

the product of two matrices may be a zero matrix even when neither of the factors is a zero matrix.   Thus, if

$$A = \begin{pmatrix} 1 & 1 & 0 \\ 0 & 0 & 0 \\ 0 & 1 & 0 \end{pmatrix} \quad \text{and} \quad B = \begin{pmatrix} 0 & 0 & 0 \\ 0 & 0 & 0 \\ 1 & 0 & 0 \end{pmatrix}$$

then

$$AB = \begin{pmatrix} 0 & 0 & 0 \\ 0 & 0 & 0 \\ 0 & 0 & 0 \end{pmatrix}.$$

If the matrix is square, it is possible to form from the elements of the matrix a determinant whose elements have the same arrangement as those of the matrix.   This determinant is called the *determinant of the matrix*.   From any matrix, other matrices can be obtained by striking out a number of rows and columns.   Certain of these matrices will be square matrices, and the determinants of these matrices are called *determinants of the matrix*.   For an $m \times n$ matrix, there are square matrices of orders $1, 2, \ldots, p$, where p is equal to the smaller of the numbers m and n.

Example 2.   The 2×3 matrix

$$A \equiv \begin{pmatrix} a_{11} & a_{12} & a_{13} \\ a_{21} & a_{22} & a_{23} \end{pmatrix}$$

contains the first-order square matrices (a_{11}), (a_{12}), (a_{23}), etc., obtained by striking out any two columns and any one row.   It also contains the second-order square matrices

$$\begin{pmatrix} a_{11} & a_{12} \\ a_{21} & a_{22} \end{pmatrix} \quad \begin{pmatrix} a_{11} & a_{13} \\ a_{21} & a_{23} \end{pmatrix} \quad \begin{pmatrix} a_{12} & a_{13} \\ a_{22} & a_{23} \end{pmatrix}$$

obtained by striking out any column of A.

In many applications, it is useful to employ the notion of the rank of a matrix A. This is defined in terms of the determinants of A.   *A matrix A is said to be of rank r if there is at least one r-rowed determinant of A that is not zero, whereas all determinants of A of order higher than r are zero or nonexistent.*[1]

Example 3.   If

$$A \equiv \begin{pmatrix} 1 & 0 & 1 & 3 \\ 2 & 1 & 0 & -2 \\ -1 & -1 & 1 & 5 \end{pmatrix}$$

the third-order determinants are

$$\begin{vmatrix} 1 & 0 & 1 \\ 2 & 1 & 0 \\ -1 & -1 & 1 \end{vmatrix} = 0 \qquad \begin{vmatrix} 1 & 0 & 3 \\ 2 & 1 & -2 \\ -1 & -1 & 5 \end{vmatrix} = 0$$

$$\begin{vmatrix} 1 & 1 & 3 \\ 2 & 0 & -2 \\ -1 & 1 & 5 \end{vmatrix} = 0 \qquad \begin{vmatrix} 0 & 1 & 3 \\ 1 & 0 & -2 \\ -1 & 1 & 5 \end{vmatrix} = 0.$$

Since

$$\begin{vmatrix} 1 & 0 \\ 2 & 1 \end{vmatrix} \neq 0$$

there is at least one second-order determinant different from zero, whereas all third-order determinants of A are zero.   Therefore, the rank of A is 2.

It should be observed that a matrix is said to have rank zero if all its elements are zero.

If $A = (a_{ij})$ and $B = (b_{ij})$ are two square matrices, then

$$AB = (a_{ik}b_{kj})$$

[1] Cf. Appendix A, Sec. 2.

and the determinant of the matrix AB is

$$|AB| = |a_{ik}b_{kj}|. \qquad (16\text{-}6)$$

We note with reference to (15-3) that the elements in the ith row and jth column of the determinant in (16-6) are precisely those that appear in the product of two determinants $|A| = |a_{ij}|$ and $|B| = |b_{ij}|$. Thus

$$|AB| = |A| \cdot |B| \qquad (16\text{-}7)$$

or, in words, *the determinant $|AB|$ of the product of two matrices A and B is equal to the product of determinants $|A|$ and $|B|$.*

It follows from (16-7) that whenever the product of two matrices is a zero matrix, *then the determinant of at least one of the factors is zero.*

A square matrix whose determinant is zero is called a *singular matrix*.

PROBLEMS

1. Make use of the definitions in this section to establish the following theorems for matrices:

(a) $A + B = B + A$; $\qquad$ (b) $(A + B) + C = A + (B + C)$;

(c) $(A + B)C = AC + BC$; $\qquad$ (d) $C(A + B) = CA + CB$.

2. Verify that the matrices A and B in Example 1 of this section do not commute.

3. Multiply:

(a) $\begin{pmatrix} 1 & 2 & 3 \\ 3 & 1 & 2 \\ 1 & 3 & 2 \end{pmatrix} \cdot \begin{pmatrix} 2 \\ 1 \\ 3 \end{pmatrix}$, $\qquad$ (b) $\begin{pmatrix} 1 & 2 & 3 \\ 3 & 1 & 2 \\ 1 & 3 & 2 \end{pmatrix} \cdot \begin{pmatrix} 2 & 0 & 0 \\ 1 & 0 & 0 \\ 3 & 0 & 0 \end{pmatrix}$.

4. Show that $(AB)C = A(BC)$ on the proper restrictions on the size.

5. Determine the ranks of the matrices:

$$A = \begin{pmatrix} 1 & 2 & 3 \\ 1 & 4 & 2 \\ 2 & 6 & 5 \end{pmatrix}; \quad B = \begin{pmatrix} 1 & 0 & 1 \\ 0 & 0 & 1 \\ 1 & 1 & 1 \end{pmatrix}; \quad C = \begin{pmatrix} 2 & -3 & 4 \\ 1 & 2 & -3 \\ 4 & 1 & -2 \end{pmatrix};$$

$$D = \begin{pmatrix} 2 & -4 & 1 \\ 3 & 1 & -2 \end{pmatrix}; \quad E = \begin{pmatrix} k & 0 & 0 \\ 0 & k & 0 \\ 0 & 0 & k \end{pmatrix}.$$

Is $AB = BA$? Is $AE = EA$? Are these matrices singular?

6. If $AX = XA$ for every matrix A, show that X is a scalar matrix. *Hint:* Let $X = (x_{ij})$; then, since $AX = XA$, $a_{ij}x_{jk} = x_{ij}a_{jk}$ for all choices of a_{ij} and a_{jk}. Now choose $a_{ij} = \delta_{i(p)}\delta_{j(q)}$, where $\delta_{i(p)}$ and $\delta_{j(q)}$ are the Kronecker deltas and p and q have fixed but arbitrary values ranging from 1 to n, and conclude that $x_{ij} = x_{ji} = 0$ if $i \neq j$ and $x_{ii} = x_{jj}$ for each i and j.

---

17. Linear Transformations.   The matrix notation introduced in the preceding section enables us to study effectively properties of linear transformations.

A set of n linear relations

$$y_i = a_{ij}x_j \qquad i, j = 1, 2, \ldots, n \qquad (17\text{-}1)$$

where the a_{ij} are constants, defines a linear transformation of the set of n variables x_i into a new set y_i.

We can regard the quantities $x_1, x_2, \ldots, x_n$ as components (or measure numbers) of some vector $\mathbf{x}$ referred to a set of base vectors $\mathbf{a}_1, \mathbf{a}_2, \ldots, \mathbf{a}_n$ in n-dimensional vector space. The quantities $y_1, y_2, \ldots, y_n$ can be viewed as components of another vector $\mathbf{y}$ referred to the same basis. The relations (17-1) then represent a transformation of the vector $\mathbf{x}$ into another vector $\mathbf{y}$. Since the lengths of $\mathbf{x}$ and $\mathbf{y}$ and their orientations relative to the base vectors $\mathbf{a}_i$ are different in general, we can look upon the transformation (17-1) as representing a deformation of space.

When the components of $\mathbf{x}$ and $\mathbf{y}$ are represented by the column matrices

$$X = \begin{pmatrix} x_1 \\ x_2 \\ \cdot \\ \cdot \\ \cdot \\ x_n \end{pmatrix} \qquad Y = \begin{pmatrix} y_1 \\ y_2 \\ \cdot \\ \cdot \\ \cdot \\ y_n \end{pmatrix}$$

the set of relations (17-1) can be written in the form

$$Y = AX \tag{17-2}$$

where $A = (a_{ij})$ is the matrix of the coefficients in the linear transformation (17-1) and the product AX is computed by the rule (16-3).

If A is a nonsingular matrix, we can solve Eqs. (17-1) for the x_i by Cramer's rule (15-7) and obtain the *inverse* transformation

$$x_i = \frac{A_{ji}}{a} y_j \tag{17-3}$$

where A_{ij} is the cofactor of the element a_{ij} in the determinant $a = |a_{ij}|$ of the matrix A.

The set of equations (17-3) can be written in matrix notation as

$$X = A^{-1}Y$$

where A^{-1} is

$$A^{-1} = \begin{pmatrix} \dfrac{A_{11}}{a} & \dfrac{A_{21}}{a} & \cdots & \dfrac{A_{n1}}{a} \\[2mm] \dfrac{A_{12}}{a} & \dfrac{A_{22}}{a} & \cdots & \dfrac{A_{n2}}{a} \\[2mm] \cdot & \cdot & \cdot & \cdot \\[1mm] \dfrac{A_{1n}}{a} & \dfrac{A_{2n}}{a} & \cdots & \dfrac{A_{nn}}{a} \end{pmatrix}. \tag{17-4}$$

It is natural to call A^{-1} the *inverse matrix* of A. We note that the inverse matrix can be constructed whenever A is nonsingular, that is, whenever the determinant $|A| \equiv a$ does not vanish.

If we form the product of A and A^{-1}

$$AA^{-1} = \left(a_{ik} \frac{A_{jk}}{a} \right) \tag{17-5}$$

and recall[1] that

$$a_{ik}A_{jk} = \delta_{ij}a$$

we can write (17-5) as

$$AA^{-1} = (\delta_{ij}) = I \tag{17-6}$$

where I is the identity matrix.

Since the determinant of the product of two matrices is equal to the product of their determinants, we conclude from (17-6) that

$$|A^{-1}A| = |A^{-1}| \cdot |A| = |I| = 1$$

so that

$$|A^{-1}| = \frac{1}{|A|}. \tag{17-7}$$

---

[1] See (15-5), but note the relation of the subscripts on the A_{ij} to the rows and columns in (17-4).

Multiplying (17-4) by A gives

$$A^{-1}A = I = AA^{-1}. \tag{17-8}$$

In addition to the inverse matrix A^{-1} we shall make frequent use of the matrix

$$A' = \begin{pmatrix} a_{11} & a_{21} & \cdots & a_{n1} \\ a_{12} & a_{22} & \cdots & a_{n2} \\ \cdot & \cdot & \cdots & \cdot \\ a_{1n} & a_{2n} & \cdots & a_{nn} \end{pmatrix} \tag{17-9}$$

obtained by interchanging the rows and columns in the matrix

$$A = \begin{pmatrix} a_{11} & a_{12} & \cdots & a_{1n} \\ a_{21} & a_{22} & \cdots & a_{2n} \\ \cdot & \cdot & \cdots & \cdot \\ a_{n1} & a_{n2} & \cdots & a_{nn} \end{pmatrix}. \tag{17-10}$$

The matrix A' is called the *transpose* of A.

On using the laws of addition and multiplication of matrices it is easy to show that

$$(A + B)' = A' + B' \qquad (kA)' = kA' \qquad (AB)' = B'A'. \tag{17-11}$$

(Note order.)

If we recall the relation (17-8),

$$A^{-1}A = I$$

and form the transpose

$$(A^{-1}A)' = I' = I$$

we get, on making use of (17-11),

$$A'(A^{-1})' = I. \tag{17-12}$$

This shows that $(A^{-1})'$ is the inverse of A' or, in other words,

$$(A^{-1})' = (A')^{-1}. \tag{17-13}$$

The important result embodied in (17-13) states that the *inverse of the transpose of the matrix A is equal to the transpose of its inverse.*

In many calculations it is necessary to compute the inverse of the product of two nonsingular matrices A and B. We can obtain the desired result as follows: Because $(AB)B^{-1}A^{-1} = AIA^{-1} = AA^{-1} = I$, we have

$$(AB)^{-1} = B^{-1}A^{-1}. \tag{17-14}$$

(Note order.)   This result can be extended in an obvious way to more than two matrices, so that, for example,

$$(ABC)^{-1} = C^{-1}B^{-1}A^{-1}.$$

Example 1.   Compute A^{-1} for the matrix

$$A = \begin{pmatrix} 1 & 2 \\ 3 & -1 \end{pmatrix}.$$

Here $\qquad\qquad A_{11} = -1 \qquad A_{12} = -3 \qquad A_{21} = -2 \qquad A_{22} = 1.$

Since $a = |A| = -7$,

$$A^{-1} = \begin{pmatrix} \frac{1}{7} & \frac{2}{7} \\ \frac{3}{7} & -\frac{1}{7} \end{pmatrix}.$$

We note that $|A^{-1}| = -\frac{1}{7} = 1/|A|$.

Example 2.   If A is a nonsingular matrix, show that the matric equations

$$AX = I \qquad \text{and} \qquad XA = I$$

have unique solutions $X = A^{-1}$.

On multiplying both members of the given equations by A^{-1}, we get

$$A^{-1}AX = A^{-1}I \quad \text{and} \quad XAA^{-1} = IA^{-1}.$$

But $$A^{-1}A = AA^{-1} = I \quad \text{and} \quad A^{-1}I = IA^{-1} = A^{-1}$$

so that $$X = A^{-1}.$$

If we have two successive linear transformations

$$y_i = a_{ij}x_j \qquad z_i = b_{ij}y_j \tag{17-15}$$

the direct transformation from the variables x_i to the z_i is obtained by inserting for the y_j in the second set of Eqs. (17-15) from the first set. We thus get

$$z_i = b_{ij}a_{jk}x_k. \tag{17-16}$$

The transformation (17-16) is called the *product* of the transformations in (17-15). If the variables $(x_1,x_2, \ldots ,x_n)$, $(y_1,y_2, \ldots ,y_n)$, and $(z_1,z_2, \ldots ,z_n)$ are interpreted as components of the vectors **x, y, z**, represented by column matrices

$$X = \begin{pmatrix} x_1 \\ x_2 \\ \cdot \\ \cdot \\ \cdot \\ x_n \end{pmatrix} \qquad Y = \begin{pmatrix} y_1 \\ y_2 \\ \cdot \\ \cdot \\ \cdot \\ y_n \end{pmatrix} \qquad Z = \begin{pmatrix} z_1 \\ z_2 \\ \cdot \\ \cdot \\ \cdot \\ z_n \end{pmatrix}$$

we can write Eqs. (17-15) as

$$Y = AX \qquad Z = BY \tag{17-17}$$

and the product transformation (17-16) as

$$Z = BAX. \tag{17-18}$$

Thus, when the variables x_i are subjected to a linear transformation (17-15) with a matrix A and the variables y_i are subjected to a linear transformation with a matrix B, the product transformation has the matrix BA. Since BA in general is not equal to AB, the order in which the transformations are performed is material.

When it is desired to interpret Eqs. (17-15) as transformations on the components of the vectors **x, y, z**, Eqs. (17-17) and (17-18) can be written in the forms

$$\mathbf{y} = A\mathbf{x} \qquad \mathbf{z} = B\mathbf{y} \qquad \mathbf{z} = BA\mathbf{x} \tag{17-19}$$

where **x, y, z** are regarded as the column matrices X, Y, Z, respectively. The matrices in Eqs. (17-19) can be viewed as *operators* transforming a given vector into another vector. Since

$$A(k\mathbf{x}) = kA\mathbf{x} \qquad k \text{ const}$$

and $$A(\mathbf{x} + \mathbf{y}) = A\mathbf{x} + A\mathbf{y}$$

one often speaks of A as a *linear operator*.

PROBLEMS

1. If

$$A = \begin{pmatrix} 1 & 2 & -1 \\ 3 & 0 & 2 \\ 4 & 5 & 0 \end{pmatrix} \qquad \text{and} \qquad B = \begin{pmatrix} 1 & 0 & 0 \\ 2 & 1 & 0 \\ 0 & 1 & 3 \end{pmatrix}$$

find A^{-1} and B^{-1}. Verify that $(AB)' = B'A'$ and $(AB)^{-1} = B^{-1}A^{-1}$.

2. Prove that $(ABC)' = C'B'A'$.

3. Prove that $(A^{-1})^{-1} = A$.

4. Prove that, if A is singular, there exists no matrix B such that $AB = I$.

5. If $y_1 = x_1 \cos \alpha - x_2 \sin \alpha$, $y_2 = x_1 \sin \alpha + x_2 \cos \alpha$, find A^{-1}, A' and show that $A^{-1} = A'$. If $\mathbf{x}$ is a vector with components (x_1, x_2), what is the geometric relation of $\mathbf{x}$ to $\mathbf{y}$? Write the inverse transformation $\mathbf{x} = A^{-1}\mathbf{y}$.

6. If $y_1 = x_1 - x_2$, $y_2 = x_1 + x_2$, what is A^{-1}? Is $A^{-1} = A'$? If $\mathbf{x}$ is a vector with components (x_1, x_2), what is the geometric relation of $\mathbf{x}$ to $\mathbf{y}$?

7. If

$$y_1 = \frac{1}{\sqrt{2}} x_1 + \frac{1}{\sqrt{2}} x_3 \qquad y_2 = x_2 \qquad y_3 = -\frac{1}{\sqrt{2}} x_1 + \frac{1}{\sqrt{2}} x_3$$

compute the matrix A' for the inverse transformation and compare it with the given matrix A. If $\mathbf{x}$ is a vector with components (x_1, x_2, x_3) and $\mathbf{y}$ is a vector with components (y_1, y_2, y_3), what is the geometric relation of $\mathbf{x}$ to $\mathbf{y}$?

8. Let

$$A = \begin{pmatrix} 1 & 2 \\ 2 & 3 \end{pmatrix}$$

and consider the vector $\mathbf{y} = A\mathbf{x}$. Compute $\mathbf{x} = A^{-1}\mathbf{y}$. Is it true that $A' = A^{-1}$?

9. If

$$y_1 = x_1 \cos \alpha + x_2 \sin \alpha \qquad y_2 = -x_1 \sin \alpha + x_2 \cos \alpha$$

and

$$z_1 = y_1 \cos \beta + y_2 \sin \beta \qquad z_2 = -y_1 \sin \beta + y_2 \cos \beta$$

find the product transformation directly and also by computing the product of the matrices as in (17-18). Compute BA and AB, $A^{-1}B^{-1}$, and $(BA)^{-1}$. Also find $(BA)'$ and compare it with $(BA)^{-1}$.

10. If $y_1 = 2x_1 + x_2$, $y_2 = x_1 - x_2$ and $z_1 = y_1 - y_2$, $z_2 = 2y_1 + y_2$ perform the calculations required in Prob. 9.

---

18. Transformation of Base Vectors. In the preceding section we interpreted the set of linear relations

$$y_i = a_{ij}x_j \tag{18-1}$$

as transformations of components $(x_1, x_2, \ldots, x_n)$ of a vector $\mathbf{x}$ into components $(y_1, y_2, \ldots, y_n)$ of another vector $\mathbf{y}$ when the vectors are referred to the same basis $(\mathbf{a}_i)$, so that

$$\mathbf{x} = x_i \mathbf{a}_i \qquad \text{and} \qquad \mathbf{y} = y_i \mathbf{a}_i. \tag{18-2}$$

If we introduce a new system of base vectors $\boldsymbol{\alpha}_i$, obtained from the set $\mathbf{a}_i$ by a linear transformation

$$\boldsymbol{\alpha}_i = b_{ij}\mathbf{a}_j \qquad \text{with } |b_{ij}| \neq 0 \tag{18-3}$$

the vectors $\mathbf{x}$ and $\mathbf{y}$ in the new reference system will have certain representations

$$\mathbf{x} = \xi_i \boldsymbol{\alpha}_i \qquad \mathbf{y} = \eta_i \boldsymbol{\alpha}_i. \tag{18-4}$$

We raise two questions: (1) What is the relation of the components of vectors in the two representations (18-2) and (18-4) when the base vectors are transformed by (18-3)? (2) What is the form of the transformation of the components ξ_i into η_i which corresponds to the deformation of the vector $\mathbf{x}$ characterized by Eqs. (18-1)?

To answer the first question we insert from (18-3) in (18-4) and get

$$\mathbf{x} = b_{ij}\xi_i \mathbf{a}_j \tag{18-5}$$

while a reference to (18-2) shows that

$$\mathbf{x} = \mathbf{a}_i x_i \equiv \mathbf{a}_j x_j. \tag{18-6}$$

From (18-5) and (18-6) we conclude that

$$x_j = b_{ij}\xi_i. \tag{18-7}$$

This formula is the desired relationship connecting the components of $\mathbf{x}$ when it is referred to two different base systems related by (18-3).

We note that in the transformation (18-3) the summation is on the second index j while in (18-7) it is on the first index. In other words, the matrix of coefficients b_{ij} in (18-7) is the transpose of the matrix (b_{ij}) in (18-3).

If we write the matrix in (18-3) as

$$(b_{ij}) = B$$

the set of equations (18-7) can be written as

$$\mathbf{x} = B'\boldsymbol{\xi} \tag{18-8}$$

$\mathbf{x}$ and $\boldsymbol{\xi}$ being the column matrices with components $(x_1, x_2, \ldots, x_n)$ and $(\xi_1, \xi_2, \ldots, \xi_n)$.

On multiplying (18-8) by $(B')^{-1}$ on the left we get the solution for $\boldsymbol{\xi}$ in the form

$$\boldsymbol{\xi} = (B')^{-1}\mathbf{x}. \tag{18-9}$$

Formulas (18-8) and (18-9) give a complete answer to the first question.

The relationship connecting the components $(y_1, y_2, \ldots, y_n)$ with $(\eta_1, \eta_2, \ldots, \eta_n)$ can be represented similarly by

$$\mathbf{y} = B'\boldsymbol{\eta} \quad \text{and} \quad \boldsymbol{\eta} = (B')^{-1}\mathbf{y}. \tag{18-10}$$

We proceed next to the answer of the question concerning the form of the deformation of space (18-1) in the new reference frame $\boldsymbol{\alpha}_i$.

We write Eqs. (18-1) in matrix form as

$$\mathbf{y} = A\mathbf{x} \tag{18-11}$$

substitute for $\mathbf{x}$ from (18-8) and for $\mathbf{y}$ from (18-10), and obtain

$$B'\boldsymbol{\eta} = AB'\boldsymbol{\xi}. \tag{18-12}$$

To solve for $\boldsymbol{\eta}$ we multiply on the left by $(B')^{-1}$ and get

$$\boldsymbol{\eta} = (B')^{-1}AB'\boldsymbol{\xi}. \tag{18-13}$$

Thus the relationship between the components $(\xi_1, \xi_2, \ldots, \xi_n)$ and $(\eta_1, \eta_2, \ldots, \eta_n)$ is determined by the matrix

$$S \equiv (B')^{-1}AB'. \tag{18-14}$$

Since the matrix S characterizes the same deformation of space as the matrix A, the matrices A and S related in the manner of (18-14) are termed *similar*. To avoid carrying primes, we set $B' = C$, and formula (18-14) then assumes the form

$$S = C^{-1}AC \tag{18-15}$$

and (18-13) becomes

$$\boldsymbol{\eta} = S\boldsymbol{\xi}. \tag{18-16}$$

One of the important problems in the theory of linear transformations is to determine a reference frame in which the equations for the deformation of space assume forms which admit of simple interpretations. For example, if it proves possible to find a matrix C such that the matrix S in (18-15) has the diagonal form

$$S = \begin{pmatrix} \lambda_1 & 0 & \cdots & 0 \\ 0 & \lambda_2 & \cdots & 0 \\ \cdot & \cdot & \cdot & \cdot \\ 0 & \cdot & \cdots & \lambda_n \end{pmatrix} \tag{18-17}$$

then Eq. (18-16) shows that

$$\eta_1 = \lambda_1 \xi_1, \qquad \eta_2 = \lambda_2 \xi_2, \qquad \ldots, \qquad \eta_n = \lambda_n \xi_n.$$

In three-dimensional space these correspond to simple elongations (or contractions) of the components of the vector in the directions of base vectors $\boldsymbol{\alpha}_i$ determined by the matrix $C = B'$ [see (18-3)].

Whether or not a matrix C reducing A to the diagonal form S can be found clearly depends on the nature of deformation specified by A. In many problems in dynamics and in the theory of elasticity, the deformation matrix A will be symmetric, and we shall see in Sec. 20 that such matrices can always be diagonalized by finding a suitable matrix C. This fact is of cardinal importance because it enormously simplifies the analysis of many problems.

In the following section we shall study properties of the matrix A in (18-1) for those transformations that leave the length of every vector $\mathbf{x}$ unchanged. In three dimensions such transformations represent rotations and reflections.

19. Orthogonal Transformations.

Let us refer our n-dimensional space to a set of orthonormal base vectors $\mathbf{e}_1, \mathbf{e}_2, \ldots, \mathbf{e}_n$, introduced in Sec. 13. Relative to this basis the vector $\mathbf{x}$ has the representation

$$\mathbf{x} = \mathbf{e}_i x_i$$

and its length $|\mathbf{x}|$ can be computed from the formula

$$|\mathbf{x}|^2 = x_i x_i. \tag{19-1}$$

Let us investigate the structure of the matrix A in the class of transformations

$$y_i = a_{ij} x_j \tag{19-2}$$

which leave the length $|\mathbf{x}|$ of the vector unchanged. Now, the square of the length of the vector $\mathbf{y}$ is

$$|\mathbf{y}|^2 = y_i y_i \tag{19-3}$$

and since we suppose that $|\mathbf{x}| = |\mathbf{y}|$,

$$y_i y_i = x_i x_i. \tag{19-4}$$

We insert in (19-4) from (19-2) and get

$$(a_{ij} x_j)(a_{ik} x_k) = x_i x_i$$

or

$$a_{ij} a_{ik} x_j x_k = \delta_{jk} x_j x_k \tag{19-5}$$

since

$$\delta_{jk} x_j x_k = x_{kk} = x_i x_i.$$

On equating the coefficients of $x_j x_k$ in (19-5), we get the set of restrictive conditions

$$a_{ij} a_{ik} = \delta_{jk} \tag{19-6}$$

on the coefficients a_{ij} if the transformation (19-2) is to leave the length of every vector unchanged.

Equations (19-6), when written for $n = 3$, are

$$
\begin{aligned}
a_{11}{}^2 + a_{21}{}^2 + a_{31}{}^2 &= 1 & a_{12}a_{13} + a_{22}a_{23} + a_{32}a_{33} &= 0 \\
a_{12}{}^2 + a_{22}{}^2 + a_{32}{}^2 &= 1 & a_{13}a_{11} + a_{23}a_{21} + a_{33}a_{31} &= 0 \\
a_{13}{}^2 + a_{23}{}^2 + a_{33}{}^2 &= 1 & a_{11}a_{12} + a_{21}a_{22} + a_{31}a_{32} &= 0.
\end{aligned}
$$

The determinant of the matrix in (19-6) is

$$|a_{ij} a_{ik}| = |\delta_{jk}| = 1 \tag{19-7}$$

and if we recall the rule for multiplication of determinants [cf. (15-2)], we conclude from (19-7) that

$$|a_{ij} a_{ik}| = |a_{ij}| \cdot |a_{ij}| = a^2 = 1 \tag{19-8}$$

where a is the determinant of (a_{ij}). Equation (19-8) states that $a = \pm 1$. In three dimensions the situation when $a = 1$ corresponds to a rotation of space relative to a set of fixed xyz axes determined by the unit vectors **i**, **j**, **k**. The circumstance when $a = -1$ corresponds to a transformation of reflection (say, $x = -x, y = -y, z = -z$) or to a reflection followed by a rotation.

A transformation (19-2) in which the coefficients a_{ij} satisfy (19-6) is called an *orthogonal transformation;* it is called the *transformation of rotation* if $|a_{ij}| = 1$, whatever be the dimensionality of space.

If we denote by A' the transpose of $(a_{ij}) = A$ in (19-2), we can write the *orthogonality condition* (19-6) in matrix form as

$$A'A = I. \tag{19-9}$$

On multiplying this by A^{-1} on the right we get

$$A' = A^{-1}. \tag{19-10}$$

Thus, *in an orthogonal transformation the inverse matrix A^{-1} is equal to the transpose A' of A.*

When Eqs. (19-2) are written in the form

$$\mathbf{y} = A\mathbf{x}$$

we can write their solutions for the x_i as

$$\mathbf{x} = A^{-1}\mathbf{y}. \tag{19-11}$$

We conclube from (19-10) that the solutions of Eqs. (19-2), when the transformation is orthogonal, are

$$x_i = a_{ji}y_j. \tag{19-12}$$

In Sec. 17, we saw that the matrix of the product of two linear transformations is the product of the matrices of the component transformations. Using this fact and the property (19-10) it is easy to show that the *product of two orthogonal transformations is an orthogonal transformation.*

—————————————————————————— **PROBLEMS**

1. Verify that the transformations

(a) $y_1 = x_1 \cos \alpha - x_2 \sin \alpha, \quad y_2 = x_1 \sin \alpha + x_2 \cos \alpha$

and *(b)* $y_1 = \dfrac{1}{\sqrt{2}} x_1 + \dfrac{1}{\sqrt{2}} x_3, \quad y_2 = x_2, \quad y_3 = -\dfrac{1}{\sqrt{2}} x_1 + \dfrac{1}{\sqrt{2}} x_3$

are orthogonal. Do they represent rotations?

2. Discuss the transformation

$$y_1 = \frac{1}{\sqrt{2}} x_1 - \frac{1}{\sqrt{2}} x_2 \quad y_2 = \frac{1}{3\sqrt{2}} x_1 + \frac{1}{3\sqrt{2}} x_2 + \frac{4}{3\sqrt{2}} x_3 \quad y_3 = \tfrac{2}{3}x_1 + \tfrac{2}{3}x_2 - \tfrac{1}{3}x_3.$$

Find the inverse transformation.

3. Prove that the product of any number of orthogonal transformations is an orthogonal transformation.

4. If A is a symmetric matrix (so that $A' = A$) and S is an orthogonal transformation, prove that the matrix $B = S^{-1}AS$ is symmetric. Thus, orthogonal transformations do not destroy the symmetry of A.

5. Let

$$A = \begin{pmatrix} 1 & 1 & 0 \\ 1 & 2 & -1 \\ 0 & -1 & 3 \end{pmatrix} \qquad C = \begin{pmatrix} c_{11} & c_{12} & c_{13} \\ c_{21} & c_{22} & c_{23} \\ c_{31} & c_{32} & c_{33} \end{pmatrix}$$

where C is an orthogonal matrix. Write the set of equations which the c_{ij} must satisfy if $C^{-1}AC = S$, where S is a diagonal matrix.

6. Is the transformation

$$y_1 = 3x_1 - x_2$$
$$y_2 = -2x_1 + x_2$$

orthogonal? Find the inverse transformation. Determine the components of **x**: (x_1, x_2) and **y**: (y_1, y_2) when the base vectors $\mathbf{e}_1$, $\mathbf{e}_2$ are rotated through 45 and 90°.

7. If $y_i = a_{ij}x_j$ is a linear transformation for the components of a complex vector **x**: $(x_1, x_2, \ldots, x_n)$, which preserves the length $|\mathbf{x}|$ of the vector, show that $\bar{a}_{ij}a_{ik} = \delta_{jk}$ or $\bar{A}'A = I$, where $\bar{A}$ is the *conjugate matrix* formed by replacing every element a_{ij} of A by $\bar{a}_{ij}$. Transformations such that $\bar{A}' = A^{-1}$ are called *unitary;* they are of great importance in quantum mechanics.

---

20. The Diagonalization of Matrices. We saw in Sec. 18 that the determination of a nonsingular matrix C such that the given matrix A reduces to the diagonal form S by a similarity transformation $C^{-1}AC$ is equivalent to determining a set of base vectors relative to which the transformation

$$y_i = a_{ij}x_j \tag{20-1}$$

assumes the form

$$\eta_1 = \lambda_1\xi_1, \qquad \eta_2 = \lambda_2\xi_2, \qquad \ldots, \qquad \eta_n = \lambda_n\xi_n. \tag{20-2}$$

We thus seek a solution of the matric equation

$$C^{-1}AC = S \tag{20-3}$$

in which $A = (a_{ij})$ is a given matrix, C the unknown matrix

$$C = \begin{pmatrix} c_{11} & c_{12} & \cdots & c_{1k} & \cdots & c_{1n} \\ c_{21} & c_{22} & \cdots & c_{2k} & \cdots & c_{2n} \\ \cdot & \cdot & \cdot & \cdot & \cdot & \cdot \\ c_{n1} & c_{n2} & \cdots & c_{nk} & \cdots & c_{nn} \end{pmatrix} \tag{20-4}$$

and S is the diagonal matrix,

$$S = \begin{pmatrix} \lambda_1 & 0 & \cdots & 0 \\ 0 & \lambda_2 & \cdots & 0 \\ \cdot & \cdot & \cdot & \cdot \\ 0 & 0 & \cdots & \lambda_n \end{pmatrix}. \tag{20-5}$$

On multiplying (20-3) on the left by C we get an equivalent matric equation

$$AC = CS \tag{20-6}$$

provided that the solution of (20-6) yields a nonsingular matrix C.

Now the matric equation (20-6) is equivalent to a system of linear equations

$$a_{ij}c_{jk} = c_{ik}\lambda_k \qquad \text{no sum on } k, \ k = 1, \ldots, n \tag{20-7}$$

obtained by equating the corresponding elements in the products AC and CS.

For every fixed value of k, the system (20-7) represents a set of n linear *homogeneous* equations for the unknowns $(c_{1k}, c_{2k}, \ldots, c_{nk})$ appearing in the kth column of (20-4). The fact that the system (20-7) is homogeneous can be made plainer by rewriting it in the form

$$(a_{ij} - \delta_{ij}\lambda_k)c_{jk} = 0 \qquad \text{no sum on } k. \tag{20-8}$$

We recall that a system of homogenous equations has solutions other than the obvious solution $c_{1k} = c_{2k} = \cdots = c_{nk} = 0$ if, and only if, its determinant[1]

$$|a_{ij} - \lambda\delta_{ij}| = 0. \tag{20-9}$$

On writing this determinant in full,

$$\begin{vmatrix} a_{11} - \lambda & a_{12} & \cdots & a_{1n} \\ a_{21} & a_{22} - \lambda & \cdots & a_{2n} \\ \cdots & \cdots & \cdots & \cdots \\ a_{n1} & a_{n2} & \cdots & a_{nn} - \lambda \end{vmatrix} = 0 \tag{20-10}$$

we see that (20-10) is an algebraic equation of degree n in λ. Accordingly, there are n roots of this equation, say $\lambda = \lambda_1$, $\lambda = \lambda_2$, ..., $\lambda = \lambda_n$, and corresponding to each root $\lambda = \lambda_k(k = 1, \ldots, n)$, the system (20-8) will have a solution

$$(c_{ik}, c_{2k}, \ldots, c_{nk}). \tag{20-11}$$

The solution (20-11) yields the kth column of the matrix C. If the roots $\lambda_1, \lambda_2, \ldots, \lambda_n$ are all distinct, one can prove that the matrix C will be nonsingular.[2] When the roots λ_k are not distinct, it is impossible, in general, to reduce A by the similarity transformation (20-3) to the diagonal form, because the desired nonsingular matrix C may not exist. In important special cases, however (for example, when A is a real and symmetric matrix), one can construct C such that S has the diagonal form even when some, or even all, roots are equal.

A brief discussion of this is contained in the following section.

As a matter of terminology, Eq. (20-9) is called the *characteristic equation* and its solutions are *characteristic values* of the matrix (a_{ij}). The solutions (20-11) of the system (20-8) corresponding to these characteristic values are called *characteristic vectors*.

Example 1.   Reduce the matrix

$$A = (a_{ij}) = \begin{pmatrix} 1 & -1 \\ -1 & 1 \end{pmatrix} \tag{20-12}$$

to the diagonal form S by the similarity transformation $C^{-1}AC$.

The characteristic equation (20-9) here is

$$\begin{vmatrix} 1 - \lambda & -1 \\ -1 & 1 - \lambda \end{vmatrix} = (1 - \lambda)^2 - 1 = 0.$$

Its solutions are $\lambda_1 = 0$, $\lambda_2 = 2$. The desired matrix C in our case has the form

$$C = \begin{pmatrix} c_{11} & c_{12} \\ c_{21} & c_{22} \end{pmatrix} \tag{20-13}$$

the columns in which satisfy the system of equations (20-8), yielding

$$\begin{aligned} (a_{11} - \lambda_k)c_{1k} + a_{12}c_{2k} = 0 & \quad \text{no sum on } k \\ a_{21}c_{1k} + (a_{22} - \lambda_k)c_{2k} = 0 & \quad k = 1, 2. \end{aligned} \tag{20-14}$$

Since $a_{11} = 1$, $a_{12} = -1$, $a_{21} = -1$, $a_{22} = 1$, we get, on setting $k = 1$ and $\lambda_1 = 0$,

$$\begin{aligned} c_{11} - c_{21} = 0 \\ -c_{11} + c_{21} = 0. \end{aligned} \tag{20-15}$$

---

[1] Appendix A, Sec. 2. Note that this determinantal equation when written in matrix form is $|A - \lambda I| = 0$.

[2] Or in the language of vectors, if we regard each column of C as a vector $\mathbf{c}^{(k)}$: $(c_{1k}, c_{2k}, \ldots, c_{nk})$, the vectors $\mathbf{c}^{(k)}(k = 1, 2, \ldots, n)$ will be linearly independent. A simple proof of this is given in I. S. Sokolnikoff, "Tensor Analysis," 2d ed., pp. 32–33, John Wiley & Sons, Inc., New York, 1964.

A proof for symmetric matrices is given in the following section.

As is always the case with nontrivial homogeneous systems of equations,[1] there are infinitely many solutions of the system (20-15). If we set $c_{11} = a$ (any constant), Eqs. (20-15) give $c_{21} = a$.

Thus the vector $\mathbf{c}^{(1)}$: (c_{11},c_{21}) appearing in the first column of (20-13) has the components $c_{11} = c_{21} = a$. Since any matrix C accomplishing the reduction will do, we can take[2] $a = 1$.

The substitution of $k = 2$ and $\lambda_2 = 2$ in (20-14) yields the system

$$(1 - 2)c_{12} - c_{22} = 0$$
$$-c_{12} + (1 - 2)c_{22} = 0$$

or

$$-c_{12} - c_{22} = 0.$$

Again there are infinitely many solutions, and if we take $c_{12} = a$, then $c_{22} = -a$. We can set $a = 1$ if we wish, so that the elements of the second column in (20-13) are $c_{12} = 1$, $c_{22} = -1$. The desired matrix C, therefore, is

$$C = \begin{pmatrix} 1 & 1 \\ 1 & -1 \end{pmatrix}.$$

The inverse of C is easily found to be

$$C^{-1} = \begin{pmatrix} \tfrac{1}{2} & \tfrac{1}{2} \\ \tfrac{1}{2} & -\tfrac{1}{2} \end{pmatrix}$$

so that $C^{-1}AC$ is

$$\begin{pmatrix} \tfrac{1}{2} & \tfrac{1}{2} \\ \tfrac{1}{2} & -\tfrac{1}{2} \end{pmatrix} \cdot \begin{pmatrix} 1 & -1 \\ -1 & 1 \end{pmatrix} \cdot \begin{pmatrix} 1 & 1 \\ 1 & -1 \end{pmatrix}.$$

On multiplying these matrices we get

$$S = \begin{pmatrix} 0 & 0 \\ 0 & 2 \end{pmatrix} \tag{20-16}$$

as we should, since

$$S = \begin{pmatrix} \lambda_1 & 0 \\ 0 & \lambda_2 \end{pmatrix}$$

as we knew from the start [see (20-5)].

If we interpret A as a matrix operator characterizing the deformation of a vector $\mathbf{x}$ into a vector $\mathbf{y}$ [see (18-11)], the result (20-16) states that in a suitable reference frame the components of $\mathbf{x}$ and $\mathbf{y}$ are related by

$$\eta_1 = 0 \cdot \xi_1 \qquad \eta_2 = 2\xi_2.$$

We thus have a deformation of space corresponding to the twofold elongation in the direction of one of the base vectors. In the notation of Sec. 18, $C = B'$, so that one can actually write out Eqs. (18-3) for the transformation of the base vectors. This, however, is seldom required because the essential matter is to determine the deformation characterized by A rather than a reference frame giving a simple form of the deformation.

Example 2. Determine the characteristic values of the matrix

$$(a_{ij}) = \begin{pmatrix} 1 & -1 & -1 \\ -1 & 1 & -1 \\ -1 & -1 & 1 \end{pmatrix}. \tag{20-17}$$

The characteristic equation this time is

$$\begin{vmatrix} 1 - \lambda & -1 & -1 \\ -1 & 1 - \lambda & -1 \\ -1 & -1 & 1 - \lambda \end{vmatrix} = (1 - \lambda)^3 - 3(1 - \lambda) - 2 = 0.$$

We easily check that the solutions of this cubic are $\lambda_1 = 2$, $\lambda_2 = 2$, $\lambda_3 = -1$. Since we have a double root $\lambda_1 = \lambda_2 = 2$, the solution of the system (20-8) will enable us to determine only two linearly independent columns of the matrix C. The matrix (20-17), however, is *real* and *symmetric*, and one can, in fact, construct the third column of the matrix C such that

[1] See Appendix A, Sec. 2.
[2] Usually one normalizes solutions so that the length of the column vector $\mathbf{c}^{(k)}$ is 1. This would correspond to the choice of $a = 1/\sqrt{2}$, since $c_{11}{}^2 + c_{12}{}^2 = 1$.

$$C^{-1}AC = S$$

where
$$S = \begin{pmatrix} \lambda_1 & 0 & 0 \\ 0 & \lambda_2 & 0 \\ 0 & 0 & \lambda_3 \end{pmatrix} = \begin{pmatrix} 2 & 0 & 0 \\ 0 & 2 & 0 \\ 0 & 0 & -1 \end{pmatrix}.$$

However, the theory presented in this section does not explain how this can be accomplished.

PROBLEMS

1. Diagonalize the matrix

$$A = \begin{pmatrix} 2 & 3 \\ -1 & -2 \end{pmatrix}$$

and determine, in the manner of Example 1, the matrix C. Discuss the meaning of A when viewed as an operator characterizing a deformation of space.

2. Find the roots of the characteristic equation for the matrix

$$A = \begin{pmatrix} 3 & 1 \\ -2 & 2 \end{pmatrix}.$$

3. Find a matrix C reducing

$$A = \begin{pmatrix} -1 & 1 & 2 \\ 0 & -2 & 1 \\ 0 & 0 & -3 \end{pmatrix}$$

to the diagonal form by the transformation $C^{-1}AC$.

4. Diagonalize the matrix

$$\begin{pmatrix} 2 & 4 & -6 \\ 4 & 2 & -6 \\ -6 & -6 & -15 \end{pmatrix}.$$

5. Prove that the roots of characteristic equations of all similar matrices are equal. *Hint:* Write the characteristic equation of $C^{-1}AC$ [cf. (20-9)] in the form $|C^{-1}AC - \lambda I| = 0$. But

$$|C^{-1}AC - \lambda I| = |C^{-1}(A - \lambda I)C| = |A - \lambda I|,$$

since $|C^{-1}| = 1/|C|$.

---

21. Real Symmetric Matrices and Quadratic Forms.   Let the matrix $A = (a_{ij})$ in a linear transformation

$$y_i = a_{ij}x_j \qquad i, j = 1, \ldots, n \tag{21-1}$$

be real and symmetric, so that $A' = A$ (or $a_{ij} = a_{ji}$). We shall indicate that in this case the matrix A can always be reduced by the transformation $C^{-1}AC$ to the diagonal form S. Moreover, C can be chosen as an orthogonal matrix, that is, a matrix such that $C^{-1} = C'$ [cf. Eq. (19-10)].

Linear transformations with real symmetric matrices dominate the study of deformations of elastic media. Real symmetric matrices also occur in the study of quadratic forms

$$Q(x_1, x_2, \ldots, x_n) \equiv a_{ij}x_ix_j \qquad i, j = 1, 2, \ldots, n \tag{21-2}$$

which arise in many problems concerned with vibrations of dynamical systems.

We can always suppose that the coefficients in a quadratic form (21-2) are symmetric because every quadratic form Q can be symmetrized by writing it as

$$Q = \tfrac{1}{2}(a_{ij} + a_{ji})x_ix_j = b_{ij}x_ix_j$$

in which the coefficients

$$b_{ij} = \tfrac{1}{2}(a_{ij} + a_{ji})$$

are obviously symmetric.   Henceforth we shall suppose that our quadratic forms have been symmetrized so that $a_{ij} = a_{ji}$.

It will follow from discussion in this section that the problems of reduction of the transformation (21-1) *with symmetric coefficients* to the form

$$\eta_1 = \lambda_1\xi_1, \qquad \eta_2 = \lambda_2\xi_2, \qquad \ldots, \qquad \eta_n = \lambda_n\xi_n \tag{21-3}$$

and of the quadratic form (21-2) to the form

$$Q = \lambda_1\xi_1{}^2 + \lambda_2\xi_2{}^2 + \cdots + \lambda_n\xi_n{}^2 \tag{21-4}$$

are mathematically identical.

We first note several properties of quadratic forms.   If the variables x_i in (21-2) are subjected to a linear transformation

$$x_i = c_{ij}\xi_j \tag{21-5}$$

the form (21-2) becomes

$$Q = a_{ij}(c_{ik}\xi_k)(c_{jr}\xi_r) = a_{ij}c_{ik}c_{jr}\xi_k\xi_r.$$

We denote the coefficients of $\xi_k\xi_r$ by b_{kr}, so that

$$Q = b_{kr}\xi_k\xi_r \tag{21-6}$$

where

$$b_{kr} = a_{ij}c_{ik}c_{jr}. \tag{21-7}$$

Since i and j are the summation indices and $a_{ij} = a_{ji}$, we see that the value of b_{kr} is not changed by an interchange of k and r.   Thus, we conclude that the symmetry of the coefficients in a quadratic form (21-2) is not destroyed when the variables x_i are changed by a linear transformation (21-5).

If we write (21-7) in the form

$$b_{kr} = c_{ik}(a_{ij}c_{jr})$$

we see that the sum $a_{ij}c_{jr}$ is an element in the ith row and the rth column of the matrix

$$AC \equiv D \qquad \text{or} \qquad (a_{ij}c_{jr}) \equiv (d_{ir}).$$

The product $c_{ik}(a_{ij}c_{jr}) = c_{ik}d_{ir}$ is the element in the kth row and the rth column of the matrix $C'D$.   Thus we can write (21-7) as

$$B = C'AC. \tag{21-8}$$

The result (21-8) can be stated as a theorem.

THEOREM.   When the variables x_i in a quadratic form (21-2) with a matrix A are subjected to a linear transformation (21-5) with a matrix C, the resulting quadratic form has the matrix $C'AC$.

If the linear transformation (21-5) is orthogonal, then $C' = C^{-1}$, and hence (21-8) can be written as

$$B = C^{-1}AC. \tag{21-9}$$

We conclude from (21-9) that the *reduction of a symmetric matrix to the diagonal form by an orthogonal transformation calls for a solution of the matric equation*

$$S = C^{-1}AC. \tag{21-10}$$

This equation is identical with that considered in the preceding section.

When the roots of the characteristic equation

$$|a_{ij} - \lambda\delta_{ij}| = 0 \tag{21-11}$$

are distinct and real, the method of Sec. 20 enables one to compute a matrix C which can be shown to be orthogonal. As a matter of fact, the desired matrix C can always be found whenever the matrix A is real and symmetric. Moreover, it can be shown that the roots of symmetric real matrices are invariably real.

The fact that the columns of C are linearly independent can be established easily when A is symmetric and the λ_k are all unequal. Let $\mathbf{c}$ be a characteristic vector for λ and let $\mathbf{c}'$ be a characteristic vector for a different value, λ'. Then, from (20-7),

$$a_{ij}c_j = \lambda c_i \qquad \text{and} \qquad a_{ij}c_j' = \lambda' c_i'.$$

Multiplying these equations by c_i' and c_i, respectively, gives

$$a_{ij}c_i'c_j = \lambda c_i'c_i \qquad \text{and} \qquad a_{ij}c_j'c_i = \lambda' c_i c_i'$$

after summing on i. Since $a_{ij} = a_{ji}$, the left sides of these equations are equal. Hence by subtraction,

$$0 = \lambda c_i'c_i - \lambda' c_i c_i' = (\lambda - \lambda')c_i c_i'.$$

Since $\lambda \neq \lambda'$ we get $\mathbf{c} \cdot \mathbf{c}' = c_i c_i' = 0$, so that the vectors $\mathbf{c}$ and $\mathbf{c}'$ are orthogonal and thus linearly independent. Taking $\lambda' = \bar{\lambda}$, we deduce similarly that λ is real.[1]

If the roots λ_i are all positive, Eq. (21-4) shows that the quadratic form (21-2) assumes positive values for all nonzero values of the variables x_i. Such quadratic forms are called *positive definite*. They appear in numerous investigations in mathematical physics.

An analog of a symmetric quadratic form (21-2) in which the variables x_i are complex is a *bilinear form*[2]

$$H = a_{ij}\bar{x}_i x_j \tag{21-12}$$

in which $a_{ij} = \bar{a}_{ji}$. Such forms are called *Hermitian*, and their matrices $(a_{ij}) = A$ are *Hermitian matrices*. Since $a_{ij} = \bar{a}_{ji}$, it follows that the elements on the main diagonal of A are necessarily real and that

$$A' = \bar{A}.$$

From the structure of (21-12) it follows that the Hermitian forms assume only real values for arbitrary complex values x_i, for on taking the conjugate of (21-12), we get

$$\bar{H} = \bar{a}_{ij}x_i\bar{x}_j = a_{ji}\bar{x}_j x_i = H$$

which proves that H is real.

Hermitian forms occur in quantum mechanics, and a discussion of the reduction of a quadratic form to a sum of squares (21-4) can be generalized to show that (21-12) can be reduced to the form

$$H = \lambda_1\bar{\xi}_1\xi_1 + \lambda_2\bar{\xi}_2\xi_2 + \cdots + \lambda_n\bar{\xi}_n\xi_n$$

by a linear transformation (21-5) with a *unitary* matrix C defined in Prob. 7 of Sec. 19.

22. Solution of Systems of Linear Equations.

In Sec. 15 we derived Cramer's rule for solving the system of equations

$$a_{ij}x_j = b_i. \tag{22-1}$$

When the number of equations in (22-1) is large, Cramer's rule is inefficient, since it requires evaluating determinants of high orders. For this reason all practical methods of solving the system (22-1) depend on reducing it by some process to an equivalent system whose matrix is sufficiently simple to enable one to compute the unknowns without great effort.

The system (22-1) can be written in matrix notation as

$$A\mathbf{x} = \mathbf{b} \tag{22-2}$$

where $A = (a_{ij})$, $\mathbf{x}$ is the column matrix $(x_1, x_2, \ldots, x_n)$, and $\mathbf{b}$ is the column matrix $(b_1, b_2, \ldots, b_n)$. If A is nonsingular, the solution of (22-2) is

$$\mathbf{x} = A^{-1}\mathbf{b} \tag{22-3}$$

[1] For proofs utilizing the notation of this section, see Sokolnikoff, "Tensor Analysis," pp. 34–39.
[2] Cf. Prob. 7 of Sec. 19.

so that the determination of unknowns hinges on constructing the inverse matrix A^{-1}. The development of effective methods for inverting matrices is a major problem of numerical analysis. One such method depends on a reduction of the system (22-2) to an equivalent system

$$Bx = c \tag{22-4}$$

in which B has the triangular form

$$\begin{pmatrix} 1 & b_{12} & b_{13} & \cdots & b_{1n} \\ 0 & 1 & b_{23} & \cdots & b_{2n} \\ \cdot & \cdot & \cdot & \cdot & \cdot & \cdot \\ 0 & 0 & 0 & \cdots & 1 \end{pmatrix}$$

in which the elements below the main diagonal are all zero. When the system (22-4) is written in full, it has the appearance of Eqs. (4-2) in Chap. 10, whose solutions, as shown in Sec. 4 of Chap. 10, can be obtained quite readily.[1]

Among other methods for solving the system (22-2) is the method of orthogonalization, the essence of which is as follows: Let us seek a matrix C such that the product

$$CA = D \tag{22-5}$$

is an orthogonal matrix. Since D is required to be orthogonal, it follows from (19-9) that

$$DD' = D'D = I \tag{22-6}$$

where D' is the transpose of D.

On multiplying (22-2) on the left by $A'C'C$, we get

$$A'C'CAx = A'C'Cb \tag{22-7}$$

and since

$$A'C' = (CA)' = D'$$

by virtue of (17-11) and (22-5), we can write (22-7) as

$$D'Dx = D'Cb.$$

However, by (22-6), $D'D = I$, so that we finally have

$$x = D'Cb. \tag{22-8}$$

Formula (22-8) gives the solution of the system (22-1) once a matrix C is determined. We do not present the classical procedure for constructing C (known as the Gram-Schmidt method) because of the rather special character of the problem.[2]

---

[1] This is the so-called "Gauss reduction method" discussed in Chap. 10 and in Appendix A.

[2] See, however, Sec. 13.

Functions of several variables

Differentiable functions

Applications of differentiation

Integrals with several variables

5

Functions
of several
variables

THE CONSIDERATIONS of the preceding chapters were confined primarily to functions $y = f(x)$ of a single independent variable x. One does not have to go far to encounter functional relationships depending on two or more independent variables. In courses in analytic geometry and calculus the reader has learned that a functional relationship of the form $z = f(x,y)$ may be represented as a surface, and he has made use of partial derivatives to study some properties of surfaces. In this chapter the familiar concepts underlying the study of real functions of two variables are sharpened and extended to functions of many variables. The bearing of such extensions on the calculation of rates of change and maximum and minimum values of functions of several variables is indicated in numerous problems of practical interest.

The concluding sections of the chapter deal with integrals of functions of several variables. They contain an introduction to the calculus of variations—a subject of great importance in physics and technology. Many situations can be characterized by statements to the effect that certain integrals attain extreme values. The determination of such extremes is in the province of calculus of variations.

DIFFERENTIABLE FUNCTIONS

1. Basic Notions. To define a real function of a real variable x one must specify the set E of admissible values of x. If for each x in E there exists a rule that determines a real value y, it is said that y is a *real function* of x. Functional relationships between y and x are usually denoted by the symbol $y = f(x)$.

In most elementary considerations the set E is an interval on the x axis of an orthogonal cartesian reference frame in which $y = f(x)$ is represented by the ordinate. Thus, if the point set E is an interval $-1 \leq x < 1$ and $y = f(x)$ is given by the formula $y = x^2$, the point (x,y) traces a portion of a parabola as x ranges over the segment $-1 \leq x < 1$ of the x axis. It is quite meaningless to inquire what the value of this particular function is at the point $x = 2$, since $x = 2$ does not belong to the interval $-1 \leq x < 1$. In the first courses in calculus, functions are usually specified by formulas without reference to the set of admissible values of x. It is, then, understood that the set of admissible x's consists of all real values of x for which the formula gives real values to y. If, for example, one states without qualification that $y = x^2$, then the set E is assumed to consist of all real values of x. If $y = 1/\sqrt{1 + x}$ then E is the interval $-1 < x < \infty$, and so on.

Similarly, one must specify the set E of pairs of real values (x,y) in defining a real function of two independent real variables x and y. One can speak of the pair (x,y) as coordinates of a point P in the cartesian xy plane, and if a definite rule associates with each point (x,y) a real number z, it is said that z is a *real function* of x and y on the given set E of pairs of values (x,y). The functional relationship between z and the points (x,y) is symbolized by $z = f(x,y)$, where f stands for the rule that determines z from (x,y). If z is specified by a formula such as

$$z = x^2 + y^2 \tag{1-1}$$

or

$$z = \sqrt{1 - x^2 - y^2} \tag{1-2}$$

without characterizing the set E of admissible pairs of values (x,y), it is assumed that E consists of all real values for which the formula produces real values of z. Equation (1-1), for example, represents the surface of the paraboloid of revolution, and the set E consists of all points (x,y) in the xy plane; Eq. (1-2) represents the surface of a hemisphere, and the admissible points in E lie in a circular disk $x^2 + y^2 \leq 1$. The function $z = 1/\sqrt{1 - x^2 - y^2}$, on the other hand, is defined in a circular disk $x^2 + y^2 < 1$ that does not include its boundary $x^2 + y^2 = 1$.

In a similar way one can specify sets E of triples of real values (x,y,z) and define a function $u = f(x,y,z)$ of the variables x, y, and z. For example, the volume of a rectangular parallelepiped with edges x, y, z is given by the formula

$$u = xyz. \tag{1-3}$$

Since (1-3) is stated to represent the volume, the set E here is the set of all real values $x > 0$, $y > 0$, $z > 0$. When the triples are represented by points in the xyz coordinate system, the set E consists of points in the first octant, excluding the coordinate planes $x = 0$, $y = 0$, $z = 0$. To be sure, formula (1-3) yields real values u when $-\infty < x < +\infty$, $-\infty < y < +\infty$, $-\infty < z < +\infty$, but the set E associated with the formula for the volume consists only of positive values of x, y, and z.

The extension of these concepts to functions of more than three variables is immediate. We specify the set E of n-tuples $(x_1, x_2, \ldots, x_n)$ of real values and speak of a particular n-tuple as being a point P in the n-dimensional space. Such space consists simply of all possible n-tuples of real values $(x_1, x_2, \ldots, x_n)$. If for each point P in the given set E some rule determines a real value u, it is said that u is a real function of points $(x_1, x_2, \ldots, x_n)$ in E, and one writes $u = f(x_1, x_2, \ldots, x_n)$ or, more simply, $u = f(P)$.

The individual numbers $x_1, x_2, \ldots, x_n$ determining P are called coordinates of P. Since the notion of distance between a pair of points plays a useful role in Euclidean geometry, we define the distance D between the pair of points $P(x_1, x_2, \ldots, x_n)$ and $P_0(x_1{}^0, x_2{}^0, \ldots, x_n{}^0)$ in n-dimensional space by the formula[1]

$$D = \sqrt{(x_1 - x_1{}^0)^2 + (x_2 - x_2{}^0)^2 + \cdots + (x_n - x_n{}^0)^2}. \tag{1-4}$$

This is an obvious generalization of the Pythagorean formula for the distance D between two points $P(x,y,z)$ and $P_0(x_0,y_0,z_0)$ in three-dimensional Euclidean space. The space metrized by formula (1-4) is called n-dimensional Euclidean space.

If we fix the point $P_0(x_1{}^0, x_2{}^0, \ldots, x_n{}^0)$ and consider only the points $P(x_1, x_2, \ldots, x_n)$ for which D in (1-4) has a constant value a, we say (by analogy with the similar situation in three-dimensional space) that the points P lie on the n-dimensional sphere of radius a, with the center at $P_0(x_1{}^0, x_2{}^0, \ldots, x_n{}^0)$. It follows from (1-4) that the equation of such a sphere is

$$(x_1 - x_1{}^0)^2 + (x_2 - x_2{}^0)^2 + \cdots + (x_n - x_n{}^0)^2 = a^2.$$

We shall use these definitions in formulating the notions of limits and continuity of functions of several variables.

---

[1] Cf. Chap. 4, Secs. 12 to 14.

2. Limits and Continuity. If $P_0(x_0,y_0)$ is a point in the xy plane, the set of points interior to the circle of radius δ with its center at (x_0,y_0) is called the δ *neighborhood* of P_0. Thus, the points (x,y) of the δ neighborhood of P_0 satisfy the condition $(x - x_0)^2 + (y - y_0)^2 < \delta^2$. If $P_0(x_1^0,x_2^0,\ldots,x_n^0)$ is a point in the n-dimensional Euclidean space, the δ neighborhood of P_0 is the set of points $P(x_1,x_2,\ldots,x_n)$ satisfying the inequality

$$(x_1 - x_1^0)^2 + (x_2 - x_2^0)^2 + \cdots + (x_n - x_n^0)^2 < \delta^2. \tag{2-1}$$

The term *vicinity* is frequently used to mean neighborhood.

A point P of any set of points E is called an *interior point* provided there exists a neighborhood of P that contains only points that belong to E. A point P (which may or may not belong to E) is a *boundary point* of E if *every* neighborhood of P contains some points that belong to E and some that do not belong to E. The totality of boundary points of the set E is called the *boundary* of E. When all boundary points belong to E, the set E is called *closed*.

In this book we shall be concerned mostly with sets of points in three-dimensional Euclidean space, which we shall usually refer to a set of cartesian xyz axes. A set of such points is said to form an *open region R* provided that:[1]

1. Every point of R is an interior point.

2. Every two points of R can be joined by a continuous curve (say a broken line consisting of straight-line segments) all of whose points belong to R.

We observe that an open region contains none of its boundary points. When *all* boundary points of an open region R are adjoined to R, the resulting set of points is said to form a *closed region $\overline{R}$*. We shall frequently use the term *region* to mean an open region. We see that an open region R and the closed region $\overline{R}$ are generalizations of an open interval $a < x < b$, usually denoted by (a,b), and the closed interval $a \le x \le b$, denoted by $[a,b]$.

As an illustration, the set of points satisfying the condition $x^2 + y^2 + z^2 < 1$ is an open region R consisting of interior points of a sphere. The points of the region $\overline{R}$ satisfying the conditions $x^2 + y^2 + z^2 \le 1$ include all points on the boundary of the sphere as well as the interior points. Thus, $\overline{R}$ is a closed region. The weak inequality $x^2 + y^2 + z^2 \ge 1$ defines a closed region which includes points on the boundary of the sphere $x^2 + y^2 + z^2 = 1$ as well as all points exterior to this sphere.

The set of points satisfying the condition $-1 < x \le 1, 1 < y < -1$ is neither open nor closed. It is a region that includes all interior points of the square $|x| < 1, |y| < 1$ and only that part of the boundary of the square for which $x = 1$.

The region R is said to be *bounded* if the distance D [formula (1-4)] of every point of R from the origin $(0,0,\ldots,0)$ is less than some number B. Thus, the region whose points satisfy $x^2 + y^2 + z^2 < 1$ is bounded, but the region defined by $x^2 + y^2 + z^2 > 1$ is unbounded.

An assortment of definitions introduced in this section is essential if we are to make the notions of limits and continuity of functions tolerably precise. Let the functions $f(x,y)$ be determined in some region R of the xy plane. Let $P_0(x_0,y_0)$ be a fixed point in R and let

$$(x_1,y_1), \quad (x_2,y_2), \quad \ldots, \quad (x_n,y_n), \quad \ldots$$

be an arbitrary sequence of points in R such that the sequence of points $\{(x_n,y_n)\}$

[1] The notion of the open region in n-dimensional space can be formulated in a similar way, if we define the concept of a straight-line segment in n-dimensional space. This is not difficult to do by using linear equations to define straight lines.

approaches (x_0,y_0). That is, the distance from (x_n,y_n) to (x_0,y_0) tends to zero as $n \to \infty$. For every such sequence of points, we can construct a sequence

$$f(x_1,y_1), \quad f(x_2,y_2), \quad \ldots, \quad f(x_n,y_n).$$

If this sequence $\{f(x_i,y_i)\}$ has the same limit A for every choice of sequences $\{x_i,y_i\}$ in R, $f(x,y)$ is said to have *the limit A* as $(x,y) \to (x_0,y_0)$. In symbols,

$$\lim_{(x,y)\to(x_0,y_0)} f(x,y) = A. \tag{2-2}$$

Another formulation can be proved equivalent to the formulation just given, though the proof of equivalence is not presented here. This other meaning of (2-2) is that for every positive number ϵ (no matter how small) we can find a neighborhood

$$(x - x_0)^2 + (y - y_0)^2 < \delta$$

of the point (x_0,y_0) such that for all points (x,y) of this neighborhood

$$|f(x,y) - A| < \epsilon.$$

It should be observed that in defining the limit A nothing was said about the value of $f(x,y)$ at the point (x_0,y_0). Now, if $f(x_0,y_0) = A$, so that

$$\lim_{(x,y)\to(x_0,y_0)} f(x,y) = f(x_0,y_0) \tag{2-3}$$

$f(x,y)$ is said to be *continuous* at the point (x_0,y_0). A function that is not continuous at a given point is said to be *discontinuous* at that point. In ordinary parlance, the continuity of $f(x,y)$ at (x_0,y_0) means that the values of $f(x,y)$ can be made to differ from $f(x_0,y_0)$ by as little as desired throughout a sufficiently small neighborhood of the point (x_0,y_0).

We extend the definition of continuity to functions of more than two variables in an obvious way: A function $f(x_1,x_2, \ldots ,x_2)$ is continuous at the point $P_0(x_1{}^0,x_2{}^0, \ldots ,x_n{}^0)$ whenever

$$\lim_{P\to P_0} f(x_1,x_2, \ldots ,x_n) = f(x_1{}^0,x_2{}^0, \ldots ,x_n{}^0).$$

The "point P" here means the set of n real numbers $(x_1,x_2, \ldots ,x_n)$. Clearly,

$$f(x_1,x_2, \ldots ,x_n)$$

cannot be continuous at $(x_1,x_2, \ldots ,x_n)$ if it is not defined at that point.

Whenever $f(x_1,x_2, \ldots ,x_n)$ is continuous at every point P of the given region R, it is said to be *continuous in the region R*. Functions with which we shall deal for the most part will be continuous in some region, open or closed.

We record without proof[1] the following properties of continuous functions which are used in the sequel:

1. If $f(P)$ is continuous at $P = P_0$ and $f(P_0) \neq 0$, there exists a neighborhood of P_0 in which $f(P)$ has the same sign as $f(P_0)$ throughout the neighborhood.

2. If $f(P)$ is continuous in a closed bounded region R, then $f(P)$ is bounded in R; that is, there exists a number $M > 0$ such that $f(P) \leq M$ for all P in R.

3. If $f(P)$ is continuous in a closed bounded region R, then $f(P)$ takes on its maximum and minimum values in R.

[1] These properties are suggested by geometric considerations, but the arithmetic proofs of the properties 2 and 3 depend on the use of a rather sophisticated *Bolzano-Weierstrass theorem*, which asserts that from a given bounded sequence $\{x_n\}$ it is always possible to construct a subsequence that converges to a finite limit. The proof of property 1 for the function $f(x)$ is outlined in Prob. 5.

PROBLEMS

1. Describe the regions of definition and the surfaces defined by the following functions z:

(a) $x - y + z = 1$

(b) $z = y$

(c) $y^2 + z^2 = 25$

(d) $z = 1/(x^2 + y^2)$

(e) $z = 1/x$

(f) $z = \sqrt{1 - (x - 1)^2 - y^2}$.

2. The *contour* or *level* lines for the surface $z = f(x,y)$ are the families of curves determined by the equation $f(x,y) = $ const.   Sketch the level lines for each of the surfaces:

(a) $z = x^2 + y^2$; (b) $z = \log (x^2 + y)$; (c) $z = f(y/x)$; (d) $z = f(x^2 + y^2)$; (e) $z = \sqrt{(y - x^2)/x^2}$.

3. Refer to the definitions of level lines in Prob. 2 and define, in a similar way, *level surfaces* for the function $u = f(x,y,z)$.   Describe the families of level surfaces in each of the following:

(a) $u = x + y + z$ (b) $u = x^2 + y^2 + z^2$ (c) $u = \dfrac{x^2 + y^2 + 2z}{x^2 + y^2}$.

4. Find points of discontinuity for each of the following functions:

(a) $z = \dfrac{xy + 1}{x^2 - y}$ (b) $z = \dfrac{1}{1 - (x^2 + y^2)}$ (c) $z = \dfrac{2xy}{x^2 + y^2}$ if $x \neq 0, y \neq 0$,

$\qquad\qquad\qquad\qquad\qquad\qquad\qquad\qquad\qquad\qquad = 0$ if $x = 0, y = 0$.

5. Prove the theorem: If $\lim\limits_{x \to x_0} f(x) = A \neq 0$, there exists a neighborhood of x_0 throughout which $f(x)$ has the same sign as the sign of A.   *Hint:* The definition of the limit is equivalent to the statement that $A - \epsilon < f(x) < A + \epsilon$, whenever $x_0 - \delta < x < x_0 + \delta$.   Suppose first that $A > 0$, take $\epsilon = A/2$, and conclude that $f(x) > A/2 > 0$.   If $A < 0$, consider $-f(x)$ and observe that $\lim\limits_{x \to x_0} [-f(x)] = -A$.

---

3. Partial Derivatives.   Let $u = f(x,y)$ be a function of two independent variables x, y, and let it be defined at a point (x_0,y_0) and for all values of (x,y) in some neighborhood of (x_0,y_0).   If y is set equal to y_0, then u becomes a function of one variable x, namely,

$$u = f(x,y_0).$$

If this function has a derivative with respect to x, the derivative is called the *partial derivative of $f(x,y)$ with respect to x for $y = y_0$*.   In like manner, if x is assigned a constant value x_0, the derivative with respect to y of the resulting function $f(x_0,y)$ is called the *partial derivative* of $f(x,y)$ *with respect to y for $x = x_0$*.   The customary notations for the partial derivative of $u = f(x,y)$ with respect to x are

$$\frac{\partial u}{\partial x}, \qquad u_x, \qquad f_x, \qquad \text{and} \qquad \frac{\partial f}{\partial x}.$$

The partial derivatives of a function $f(x_1,x_2, \ldots ,x_n)$ of n independent variables are obtained by fixing in it the values of $n - 1$ variables and computing the derivative of the resulting function of a single variable.   Thus,

$$f(x,y) = yx^2 - 2yx \tag{3-1}$$

has the partial derivatives

$$\frac{\partial f}{\partial x} = 2xy - 2y \qquad \frac{\partial f}{\partial y} = x^2 - 2x. \tag{3-2}$$

If $u = f(x,y)$ is a function of two independent variables, it is easy to provide a simple geometric interpretation of partial derivatives u_x and u_y.   The equation $u = f(x,y)$

is the equation of a surface (see Fig. 1). If x is given a fixed value x_0, $u = f(x_0,y)$ is the equation of the curve AB on the surface formed by the intersection of the surface and the plane $x = x_0$. Then

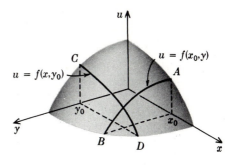

FIGURE 1

$$u_y \equiv \frac{\partial u}{\partial y} = \lim_{\Delta y \to 0} \frac{f(x_0, y_0 + \Delta y) - f(x_0,y_0)}{\Delta y}$$

is the slope at any point of AB. Similarly, if y is assigned the constant value y_0, then $u = f(x,y_0)$ is the equation of the curve CD on the surface and

$$u_x \equiv \frac{\partial u}{\partial x} = \lim_{\Delta x \to 0} \frac{f(x_0 + \Delta x, y_0) - f(x_0,y_0)}{\Delta x}$$

is the slope at any point of CD.

In Chap. 6 we shall see that the partial derivatives u_x, u_y, u_z of $u = f(x,y,z)$ can be interpreted as rectangular components of a certain vector, called the *gradient* of u. This vector provides a measure of the space rate of change of u.

The partial derivatives $f_{x_1}, f_{x_2}, \ldots, f_{x_n}$ of $f(x_1,x_2, \ldots, x_n)$ are functions of x_1, x_2, $\ldots$, x_n, and they may have partial derivatives with respect to some or all of these variables. These derivatives are called *second partial derivatives* of $f(x_1,x_2, \ldots, x_n)$. If there are only two independent variables, $f(x,y)$ may have the second partial derivatives

$$\frac{\partial}{\partial x}\left(\frac{\partial f}{\partial x}\right) \equiv \frac{\partial^2 f}{\partial x^2} \equiv f_{xx} \qquad \frac{\partial}{\partial y}\left(\frac{\partial f}{\partial x}\right) \equiv \frac{\partial^2 f}{\partial y\, \partial x} \equiv f_{xy}$$

$$\frac{\partial}{\partial x}\left(\frac{\partial f}{\partial y}\right) \equiv \frac{\partial^2 f}{\partial x\, \partial y} \equiv f_{yx} \qquad \frac{\partial}{\partial y}\left(\frac{\partial f}{\partial y}\right) \equiv \frac{\partial^2 f}{\partial y^2} \equiv f_{yy}.$$

It should be noted that f_{xy} means that $\partial f/\partial x$ is first found and then $(\partial/\partial y)(\partial f/\partial x)$ is determined, so that the subscripts indicate the order in which the derivatives are computed. In

$$\frac{\partial^2 f}{\partial y\, \partial x} = \frac{\partial}{\partial y}\left(\frac{\partial f}{\partial x}\right)$$

the order is in keeping with the meaning of the symbol, so that the order appears as the reverse of the order in which the derivatives are taken.

For the function $f(x,y)$ in (3-1) we get, on noting (3-2),

$$f_{xy} = \frac{\partial}{\partial y}\left(\frac{\partial f}{\partial x}\right) = \frac{\partial}{\partial y}(2xy - 2y) = 2x - 2$$

$$f_{yx} = \frac{\partial}{\partial x}\left(\frac{\partial f}{\partial y}\right) = \frac{\partial}{\partial x}(x^2 - 2x) = 2x - 2$$

$$f_{yy} = \frac{\partial}{\partial y}(x^2 - 2x) = 0 \qquad f_{xx} = \frac{\partial}{\partial x}(2xy - 2y) = 2y.$$

In this example, $f_{xy} = f_{yx}$; indeed, one rarely meets functions for which the so-called "mixed" derivatives are unequal. In fact, one can prove[1] that $f_{xy} = f_{yx}$ whenever these derivatives are continuous at the point in question.

[1] The proof of this is given in many books on advanced calculus. See, for example, I. S. Sokolnikoff, "Advanced Calculus," sec. 31, McGraw-Hill Book Company, New York, 1939.

The process of defining partial derivatives of higher orders is obvious from the foregoing, and it is possible to establish equalities such as $f_{xyx} = f_{xxy} = f_{yxx}$ and $f_{yxy} = f_{xyy} = f_{yyx}$ whenever these derivatives are continuous at the point in question.

We note in conclusion that, although the notation $\partial u / \partial x$ for the partial derivative u_x suggests a quotient of some quantities analogous to the differentials dy and dx in the notation dy/dx for the derivative of $y = f(x)$, no such interpretation is available for partial derivatives. To stress the point that $\partial u / \partial x$ should never be thought of as a fraction, we give an example.

Example. Consider the equation for an ideal gas $pv = RT$, where p is the pressure, v is the volume, T is the absolute temperature, and R is a physical constant. It should be noted first that the concept of partial derivatives hinges on the agreement as to which variables in a given functional relationship are assumed to be independent. Thus, if we solve the gas equation for p, we obtain

$$p = \frac{RT}{v}.$$

We can then compute

$$\frac{\partial p}{\partial v} = -\frac{RT}{v^2} \quad \text{and} \quad \frac{\partial p}{\partial T} = \frac{R}{v}. \tag{3-3}$$

On the other hand, if we solve for v, we get

$$v = \frac{RT}{p}$$

in which p and T are now regarded as the independent variables, and we can, therefore, compute

$$\frac{\partial v}{\partial T} = \frac{R}{p} \qquad \frac{\partial v}{\partial p} = -\frac{RT}{p^2}. \tag{3-4}$$

We can also solve for T and get

$$T = \frac{pv}{R}$$

in which p and v are to be considered as the independent variables, so that

$$\frac{\partial T}{\partial p} = \frac{v}{R} \qquad \frac{\partial T}{\partial v} = \frac{p}{R}. \tag{3-5}$$

From Eqs. (3-3) to (3-5) we obtain

$$\frac{\partial p}{\partial v} \frac{\partial v}{\partial T} \frac{\partial T}{\partial p} = -\frac{RT}{v^2} \frac{R}{p} \frac{v}{R} = -1 \tag{3-6}$$

since $pv = RT$. But if it were possible to treat the terms in the left-hand member of (3-6) as fractions, we should have obtained $+1$.

PROBLEMS

1. Find $\partial z / \partial x$ and $\partial z / \partial y$ for each of the following functions:

(a) $z = y/x$; (b) $z = x^3 y + \tan^{-1}(y/x)$; (c) $z = \sin xy + x$; (d) $z = e^x \log y$; (e) $z = x^2 y + \sin^{-1} x$.

2. Find $\partial u / \partial x$, $\partial u / \partial y$, and $\partial u / \partial z$ for each of the following functions:

(a) $u = x^2 y + yz - xz^2$ (b) $u = xyz + \log xy$ (c) $u = z \sin^{-1}(x/y)$
(d) $u = (x^2 + y^2 + z^2)^{1/2}$ (e) $u = (x^2 + y^2 + z^2)^{-1/2}$.

3. Verify that $\partial^2 f / \partial x\, \partial y = \partial^2 f / \partial y\, \partial x$ for

(a) $f = \cos xy^2$ (b) $f = \sin^2 x \cos y$ (c) $f = e^{y/x}$.

4. Prove that if

(a) $f(x,y) = \log (x^2 + y^2) + \tan^{-1}\dfrac{y}{x}$ then $\dfrac{\partial^2 f}{\partial x^2} + \dfrac{\partial^2 f}{\partial y^2} = 0$

(b) $f(x,y,z) = (x^2 + y^2 + z^2)^{-\frac{1}{2}}$ then $\dfrac{\partial^2 f}{\partial x^2} + \dfrac{\partial^2 f}{\partial y^2} + \dfrac{\partial^2 f}{\partial z^2} = 0.$

5. Prove that

(a) $f_x + f_y + f_z = 0$ if $f = (x - y)(y - z)(z - x)$

(b) $f_x + f_y + f_z = 1$ if $f = x + \dfrac{x - y}{y - z}.$

6. Find u_{xx}, if: (a) $u = (x^2 + y^2 + z^2)/2$;   (b) $u = (x^2 + y^2 + z^2)^{-\frac{1}{2}}.$

---

4. Total Differentials.

The differential dy of a function $y = f(x)$ at the point x_0 is defined by the formula

$$dy = f'(x_0)\, dx \tag{4-1}$$

where $dx \equiv \Delta x$ is an arbitrary increment of the *independent* variable x. We agree to call an increment of the *independent* variable x the *differential* of x.

Since

$$f'(x_0) = \lim_{\Delta x \to 0} \frac{\Delta y}{\Delta x} = \lim_{\Delta x \to 0} \frac{f(x_0 + \Delta x) - f(x_0)}{\Delta x} \tag{4-2}$$

we can write, on recalling the definition of the limit,

$$\frac{\Delta y}{\Delta x} = f'(x_0) + \epsilon \tag{4-3}$$

where $\lim_{\Delta x \to 0} \epsilon = 0.$ Hence

$$\Delta y = f'(x_0)\, \Delta x + \epsilon\, \Delta x. \tag{4-4}$$

The substitution from (4-1) in (4-4) then yields

$$\Delta y = dy + \epsilon\, \Delta x \qquad \lim \epsilon = 0 \qquad \text{as } \Delta x \to 0. \tag{4-5}$$

Figure 2 illustrates geometrically the relations between Δy, dy, and dx, and formula (4-5) shows that, for small values of Δx, the increment Δy is a good approximation to the differential dy in the sense that

$$\frac{\Delta y - dy}{\Delta x} = \epsilon \tag{4-6}$$

where $\epsilon \to 0$ as $\Delta x \to 0.$

One can construct a similar approximation to the increment Δu for the function $u = f(x,y)$ when x and y are allowed to acquire the respective increments Δx and $\Delta y.$

Indeed, let (x_0,y_0) be a point at which the partial derivatives $f_x(x_0,y_0)$, $f_y(x_0,y_0)$ of $u = f(x,y)$ exist. Now, the increment resulting from the replacement of x_0 by $x_0 + \Delta x$ and of y_0 by $y_0 + \Delta y$ is

$$\Delta u = f(x_0 + \Delta x, y_0 + \Delta y) - f(x_0,y_0) \tag{4-7}$$

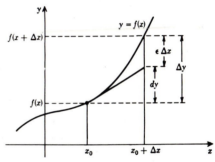

FIGURE 2

and if the differential du is defined by the formula

$$du = f_x(x_0,y_0) \, \Delta x + f_y(x_0,y_0) \, \Delta y \tag{4-8}$$

we can say that du is a good approximation to Δu provided that [cf. (4-5)]

$$\Delta u = du + \epsilon_1 \, \Delta x + \epsilon_2 \, \Delta y \tag{4-9}$$

where $\lim \epsilon_1 = 0$ $\lim \epsilon_2 = 0$ as $\Delta x \to 0$ and $\Delta y \to 0$.

Formula (4-9) shows that the approximation of Δu by du is "good" in the sense that

$$\frac{\Delta u - du}{\sqrt{(\Delta x)^2 + (\Delta y)^2}} = \frac{\epsilon_1 \, \Delta x + \epsilon_2 \, \Delta y}{\sqrt{(\Delta x)^2 + (\Delta y)^2}} \to 0 \tag{4-10}$$

as $\Delta x \to 0$ and $\Delta y \to 0$. If u is a function of a single variable x, the criterion (4-10) reduces to (4-6).

We shall say that the function $u = f(x,y)$, for which $f_x(x_0,y_0)$ and $f_y(x_0,y_0)$ exist, is *differentiable* at the point (x_0,y_0) whenever the condition (4-9) holds, and du defined by (4-8) is then called the *total differential* of $u(x,y)$ at (x_0,y_0).

As in the case of functions of one independent variable, we agree to write the increments Δx and Δy in the independent variables as dx and dy, respectively, so that (4-8) reads

$$du = f_x(x_0,y_0) \, dx + f_y(x_0,y_0) \, dy. \tag{4-11}$$

It follows from (4-9) and (4-8) that $\Delta u \to 0$ when Δx and Δy approach zero and, thus, $u = f(x,y)$ is continuous at the point (x_0,y_0) if it is differentiable at that point.

We shall say that $u = f(x,y)$ is differentiable in a given region if it is differentiable at every point of the region.

Although the formula (4-11) [or (4-8)] has a meaning whenever f_x and f_y exist at a given point, it may not represent the *total differential du*, because the differentiability condition (4-9) may not be satisfied. However, as we shall see next, the formula (4-8) represents the total differential du if $f_x(x,y)$ and $f_y(x,y)$ are continuous at (x_0,y_0). This matter is of great practical importance, since in applications one almost invariably deals with functions whose partial derivatives are continuous. Accordingly, one can conclude at once that $u(x,y)$ is differentiable.

THEOREM. *Let $u = f(x,y)$ have partial derivatives in some neighborhood of the point (x_0,y_0) and let $f_x(x,y)$ and $f_y(x,y)$ be continuous at (x_0,y_0); then the increment Δu can be represented in the form (4-9) and thus $u = f(x,y)$ is differentiable at (x_0,y_0).*

Let $u = f(x,y)$ satisfy the conditions of this theorem, and consider the increment

$$\Delta u = f(x_0 + \Delta x, \, y_0 + \Delta y) - f(x_0,y_0). \tag{4-12}$$

If we add and subtract $f(x_0, \, y_0 + \Delta y)$ in the right-hand member of (4-12), we obtain

$$\Delta u = [f(x_0 + \Delta x, \, y_0 + \Delta y) - f(x_0, \, y_0 + \Delta y)] + [f(x_0, \, y_0 + \Delta y) - f(x_0,y_0)]. \tag{4-13}$$

The expression in the first pair of brackets in (4-13) is the increment in the function $f(x,y)$ when the second variable in it has a fixed value $y_0 + \Delta y$, so that, in effect, we are dealing with the function of one variable x to which the mean-value theorem of differential calculus is applicable. This theorem states that, if $f(x)$ has the derivative $f'(x)$ at every point of the interval $x_0 - \Delta x \le x \le x_0 + \Delta x$, then $f(x_0 + \Delta x) - f(x_0) = f'(\xi) \, \Delta x$, where ξ is an intermediate point in the interval. Accordingly, we can write

$$f(x_0 + \Delta x, \, y_0 + \Delta y) - f(x_0, \, y_0 + \Delta y) = f_x(x_0 + \theta_1 \, \Delta x, \, y_0 + \Delta y) \, \Delta x \tag{4-14}$$

where $0 < \theta_1 < 1$.

Similarly, the application of the mean-value theorem to the expression in the second set of brackets in (4-13), in which x has a fixed value, yields

$$f(x_0, y_0 + \Delta y) - f(x_0, y_0) = f_y(x_0, y_0 + \theta_2 \Delta y) \Delta y \tag{4-15}$$

where $0 < \theta_2 < 1$.

Now if the partial derivatives $f_x(x_0, y_0)$ and $f_y(x_0, y_0)$ are continuous at (x_0, y_0), then

$$\begin{aligned} f_x(x_0 + \theta_1 \Delta x, y_0 + \Delta y) &= f_x(x_0, y_0) + \epsilon_1 \\ f_y(x_0, y_0 + \theta_2 \Delta y) &= f_y(x_0, y_0) + \epsilon_2 \end{aligned} \tag{4-16}$$

where $\lim \epsilon_1 = 0$ and $\lim \epsilon_2 = 0$ as Δx and Δy approach zero. Hence we can write (4-14) and (4-15) in the forms

$$f(x_0 + \Delta x, y_0 + \Delta y) - f(x_0, y_0 + \Delta y) = [f_x(x_0, y_0) + \epsilon_1] \Delta x$$
$$f(x_0, y_0 + \Delta y) - f(x_0, y_0) = [f_y(x_0, y_0) + \epsilon_2] \Delta y$$

so that (4-13) becomes

$$\Delta u = f_x(x_0, y_0) \Delta x + f_y(x_0, y_0) \Delta y + \epsilon_1 \Delta x + \epsilon_2 \Delta y.$$

But this is precisely the formula (4-9) if we note the definition (4-8).

Calculations similar in every respect can be carried out for a function $f(x_1, x_2, \ldots, x_n)$ of n independent variables x_i, and we conclude, as above, that the total differential du is invariably given by the formula

$$\begin{aligned} du &= \frac{\partial f}{\partial x_1} dx_1 + \frac{\partial f}{\partial x_2} dx_2 + \cdots + \frac{\partial f}{\partial x_n} dx_n \\ &\equiv \frac{\partial u}{\partial x_1} dx_1 + \frac{\partial u}{\partial x_2} dx_2 + \cdots + \frac{\partial u}{\partial x_n} dx_n \end{aligned} \tag{4-17}$$

whenever the partial derivatives f_{x_i} are continuous functions.

It should be noted that the total differential du is equal to the sum of n terms involving *independent increments* dx_i; thus, it is a *linear function* in the increments $dx_i \equiv \Delta x_i$. It gives a good approximation to Δu in the sense that the ratio $(\Delta u - du)/\sqrt{(\Delta x_1)^2 + (\Delta x_2)^2 + \cdots + (\Delta x_n)^2}$ tends to zero when $\Delta x_1, \Delta x_2, \ldots, \Delta x_n$ all approach zero. This provides a basis for the use of differentials in approximate calculations and in making estimates of errors.

When a number of small changes are taking place simultaneously in a system, each one proceeds as if it were independent of the others, and the total change is the sum of the effects due to the independent changes. Physically this corresponds to the principle of superposition of effects.

Example 1. Find the total differential of $u = e^x y z^2$. Since u_x, u_y, u_z are obviously continuous functions, formula (4-17) yields

$$du = e^x y z^2 \, dx + e^x z^2 \, dy + 2e^x y z \, dz.$$

Example 2. A metal box without a top has inside dimensions 6 by 4 by 2 ft. If the metal is 0.1 ft thick, find the actual volume of the metal used and compare it with the approximate volume found by using the differential.

The actual volume is ΔV, where

$$\Delta V = 6.2 \times 4.2 \times 2.1 - 6 \times 4 \times 2 = 54.684 - 48 = 6.684 \text{ ft}^3.$$

Since $V = xyz$, where $x = 6$, $y = 4$, $z = 2$,

$$\begin{aligned} dV &= yz \, dx + xz \, dy + xy \, dz \\ &= 8(0.2) + 12(0.2) + 24(0.1) = 6.4 \text{ ft}^3. \end{aligned}$$

Example 3.   Two sides of a triangular piece of land are measured as 100 and 125 ft, and the included angle is measured as 60°.   If the possible errors are 0.2 ft in measuring the sides and 1° in measuring the angle, what is the approximate error in the area?

Since $A = \frac{1}{2}xy \sin \alpha$,

$$dA = \frac{1}{2}(y \sin \alpha\, dx + x \sin \alpha\, dy + xy \cos \alpha\, d\alpha)$$

and the approximate error is therefore

$$dA = \frac{1}{2}\left[125 \left(\frac{\sqrt{3}}{2}\right)(0.2) + 100 \left(\frac{\sqrt{3}}{2}\right)(0.2) + 100(125)\left(\frac{1}{2}\right)\frac{\pi}{180} \right] = 74.0 \text{ ft}^2.$$

PROBLEMS

1. A closed cylindrical tank is 4 ft high and 2 ft in diameter (inside dimensions).   What is the approximate amount of metal in the wall and the ends of the tank if they are 0.2 in. thick?

2. The angle of elevation of the top of a tower is found to be 30°, with a possible error of 0.5°.   The distance to the base of the tower is found to be 1,000 ft, with a possible error of 0.1 ft.   What is the possible error in the height of the tower as computed from these measurements?

3. What is the possible error in the length of the hypotenuse of a right triangle if the legs are found to be 11.5 and 7.8 ft, with a possible error of 0.1 ft in each measurement?

4. The constant C in Boyle's law $pv = C$ is calculated from the measurements of p and v.   If p is found to be 5,000 lb per ft^2 with a possible error of 1 per cent and v is found to be 15 ft^3 with a possible error of 2 per cent, find the approximate possible error in C computed from these measurements.

5. The volume v, pressure p, and absolute temperature T of a perfect gas are connected by the formula $pv = RT$, where R is a constant.   If $T = 500°$, $p = 4,000$ lb per ft^2, and $v = 15.2$ ft^3, find the approximate change in p when T changes to 503° and v to 15.25 ft^3.

6. In estimating the cost of a pile of bricks measured as 6 by 50 by 4 ft, the tape is stretched 1 per cent beyond the estimated length.   If the count is 12 bricks to 1 ft^3 and bricks cost $20 per thousand, find the approximate error in cost.

7. In determining specific gravity by the formula $s = A/(A - W)$, where A is the weight in air and W is the weight in water, A can be read within 0.01 lb and W within 0.02 lb.   Find approximately the maximum error in s if the readings are $A = 1.1$ lb and $W = 0.6$ lb.   Find the maximum relative error $\Delta s/s$.

8. The equation of a perfect gas is $pv = RT$.   At a certain instant a given amount of gas has a volume of 16 ft^3 and is under a pressure of 36 psi.   Assuming $R = 10.71$, find the temperature T. If the volume is increasing at the rate of $\frac{1}{3}$ cfs and the pressure is decreasing at the rate $\frac{1}{8}$ psi per sec, find the rate at which the temperature is changing.

9. The period of a simple pendulum with small oscillations is $T = 2\pi\sqrt{l/g}$.   If T is computed using $l = 8$ ft and $g = 32$ ft per sec per sec, find the approximate error in T if the true values are $l = 8.05$ ft and $g = 32.01$ ft per sec per sec.   Find also the percentage error.

10. The diameter and altitude of a can in the shape of a right-circular cylinder are measured as 4 and 6 in., respectively.   The possible error in each measurement is 0.1 in.   Find approximately the maximum possible error in the values computed for the volume and the lateral surface.

11. We define an approximate relative error e in the differentiable function f by the formula $e = df/f$.   Show that the approximate relative error of the product is equal to the sum of the approximate relative errors of the factors.   *Hint:* $e = d \log f$.

5. Differentiation of Composite Functions.   Chain Rule.

Let $u = f(x,y)$ be differentiable in some region R. If $x = x(t)$ and $y = y(t)$ are differentiable functions of the variable t in an interval $t_1 \leq t \leq t_2$, such that the point (x,y) remains in R, then $u = f(x,y)$ becomes a *composite function* of the variable t. When t acquires an increment Δt, x and y acquire increments Δx and Δy. But $u = f(x,y)$ is differentiable, and, for these increments Δx and Δy, formula (4-9) gives

$$\Delta u = \frac{\partial u}{\partial x} \Delta x + \frac{\partial u}{\partial y} \Delta y + \epsilon_1 \Delta x + \epsilon_2 \Delta y \tag{5-1}$$

where ϵ_1 and ϵ_2 approach zero as $\Delta x \to 0$ and $\Delta y \to 0$. Since $x = x(t)$ and $y = y(t)$ are differentiable functions of t, they are continuous and their increments Δx and Δy approach zero as $\Delta t \to 0$. We conclude that ϵ_1 and ϵ_2 also approach zero as $\Delta t \to 0$. On dividing both members of (5-1) by Δt, we get

$$\frac{\Delta u}{\Delta t} = \frac{\partial u}{\partial x} \frac{\Delta x}{\Delta t} + \frac{\partial u}{\partial y} \frac{\Delta y}{\Delta t} + \epsilon_1 \frac{\Delta x}{\Delta t} + \epsilon_2 \frac{\Delta y}{\Delta t} \tag{5-2}$$

and, upon letting $\Delta t \to 0$, we obtain

$$\frac{du}{dt} = \frac{\partial u}{\partial x} \frac{dx}{dt} + \frac{\partial u}{\partial y} \frac{dy}{dt}. \tag{5-3}$$

If in the foregoing calculations we had taken a differentiable function

$$u = f(x_1, x_2, \ldots, x_n)$$

of n variables x_i in which each variable x_i is a differentiable function $x_i = x_i(t)$ $(i = 1, 2, \ldots, n)$, specified in some interval $t_1 \leq t \leq t_2$, we would have deduced the rule

$$\frac{du}{dt} = \frac{\partial u}{\partial x_1} \frac{dx_1}{dt} + \frac{\partial u}{\partial x_2} \frac{dx_2}{dt} + \cdots + \frac{\partial u}{\partial x_n} \frac{dx_n}{dt}. \tag{5-4}$$

The important rule (5-4) is known as the *chain rule*.

On multiplying the equality (5-4) by dt, we get

$$du = \frac{\partial u}{\partial x_1} dx_1 + \frac{\partial u}{\partial x_2} dx_2 + \cdots + \frac{\partial u}{\partial x_n} dx_n \tag{5-5}$$

which shows that the differential of a *composite* function $u = f(x_1, x_2, \ldots, x_n)$ in which each variable $x_i = x_i(t)$ is a function of the variable t is given by the formula that has *exactly* the same appearance as (4-17) in which the variables x_i are independent.

Let us consider next a differentiable function

$$u = f(x_1, x_2, \ldots, x_n) \tag{5-6}$$

of n variables x_i, where each variable x_i is a function of the variables $t_1, t_2, \ldots, t_m$, say,[1]

$$x_i = x_i(t_1, t_2, \ldots, t_m) \qquad i = 1, 2, \ldots, n. \tag{5-7}$$

If we suppose that the functions in (5-7) have partial derivatives $\partial x_i / \partial t_k$ with respect to each variable t_k $(k = 1, 2, \ldots, m)$, then on fixing in (5-7) all variables except t_k, we obtain a composite function $u = f(x_1, x_2, \ldots, x_n)$ of just one variable t_k. The derivative of u with respect to t_k can then be computed by the chain rule (5-4) and we get

$$\frac{\partial u}{\partial t_k} = \frac{\partial u}{\partial x_1} \frac{\partial x_1}{\partial t_k} + \frac{\partial u}{\partial x_2} \frac{\partial x_2}{\partial t_k} + \cdots + \frac{\partial u}{\partial x_n} \frac{\partial x_n}{\partial t_k} \tag{5-8}$$

where we write $\partial u / \partial t_k$ (instead of du/dt_k) and $\partial x_i / \partial t_k$ (instead of dx_i/dt_k), because all variables in (5-7) except t_k are held fast.

[1] We suppose, as above, that if the region of definition of (5-6) is R then the region of definition of functions in (5-7) is such that the point $(x_1, x_2, \ldots, x_n)$ determined by (5-7) is also in R.

Formula (5-8) provides the rule for computing the partial derivatives of a composite function (5-6) of several variables $t_1, t_2, \ldots, t_k$, and it is valid whenever

$$u = f(x_1, x_2, \ldots, x_n)$$

is differentiable and the derivatives $\partial x_i / \partial t_k$ in (5-7) exist. If we now suppose that all partial derivatives in the right-hand member of (5-8) are continuous, all partial derivatives $\partial u / \partial t_k$ $(k = 1, 2, \ldots, m)$ appearing in the left-hand member of (5-8) will be continuous. From the theorem of the preceding section it would then follow that $u = f(x_1, x_2, \ldots, x_n)$ is a differentiable function of $t_1, t_2, \ldots, t_m$, and the total differential du is again given by (5-5).

The fact that the total differential of a composite function has the same form irrespective of whether the variables involved are independent or not permits one to use the same formulas for calculating differentials as those established for the functions of a single variable. Thus,

$$d(u + v) = du + dv$$

$$d(uv) = \frac{\partial(uv)}{\partial u} du + \frac{\partial(uv)}{\partial v} dv = v\, du + u\, dv$$

and so forth.

Example 1.   If $u = xy + yz + zx$, $x = t$, $y = e^{-t}$, and $z = \cos t$,

$$\frac{du}{dt} = (y + z)\frac{dx}{dt} + (x + z)\frac{dy}{dt} + (x + y)\frac{dz}{dt}$$

$$= (e^{-t} + \cos t)(1) + (t + \cos t)(-e^{-t}) + (t + e^{-t})(-\sin t)$$

$$= e^{-t} + \cos t - te^{-t} - e^{-t}\cos t - t\sin t - e^{-t}\sin t.$$

This example illustrates the fact that this method of computing du/dt is often shorter than the old method in which the values of x, y, and z in terms of t are substituted in the expression for u before the derivative is computed.

Example 2.   If $f(x,y) = x^2 + y^2$, where $x = r \cos \varphi$ and $y = r \sin \varphi$, then

$$\frac{\partial f}{\partial r} = \frac{\partial f}{\partial x}\frac{\partial x}{\partial r} + \frac{\partial f}{\partial y}\frac{\partial y}{\partial r} = 2x \cos \varphi + 2y \sin \varphi = 2r \cos^2 \varphi + 2r \sin^2 \varphi = 2r.$$

$$\frac{\partial f}{\partial \varphi} = \frac{\partial f}{\partial x}\frac{\partial x}{\partial \varphi} + \frac{\partial f}{\partial y}\frac{\partial y}{\partial \varphi} = 2x(-r \sin \varphi) + 2y(r \cos \varphi)$$

$$= -2r^2 \cos \varphi \sin \varphi + 2r^2 \cos \varphi \sin \varphi = 0.$$

Also, $df = 2r\, dr$ or $df = 2x\, dx + 2y\, dy.$

Since $f(x,y) = x^2 + y^2 = r^2$, these results could have been obtained directly.

Example 3.   Let $z = e^{xy}$, where $x = \log(u + v)$ and $y = \tan^{-1}(u/v)$. Then,

$$\frac{\partial z}{\partial x} = ye^{xy} \qquad \frac{\partial z}{\partial y} = xe^{xy} \qquad \frac{\partial x}{\partial u} = \frac{1}{u + v} \qquad \text{and} \qquad \frac{\partial y}{\partial u} = \frac{v}{v^2 + u^2}$$

if we exclude those points where these derivatives cease being continuous.

Hence,

$$\frac{\partial z}{\partial u} = \frac{\partial z}{\partial x}\frac{\partial x}{\partial u} + \frac{\partial z}{\partial y}\frac{\partial y}{\partial u} = \frac{ye^{xy}}{u + v} + \frac{xe^{xy}v}{v^2 + u^2}.$$

Similarly,

$$\frac{\partial z}{\partial v} = \frac{ye^{xy}}{u + v} - \frac{xe^{xy}u}{v^2 + u^2}.$$

The same results can be obtained by noting that

$$dz = ye^{xy}\, dx + xe^{xy}\, dy.$$

But

$$dx = \frac{\partial x}{\partial u} du + \frac{\partial x}{\partial v} dv = \frac{1}{u + v} du + \frac{1}{u + v} dv$$

and
$$dy = \frac{\partial y}{\partial u} du + \frac{\partial y}{\partial v} dv = \frac{v}{v^2 + u^2} du - \frac{u}{v^2 + u^2} dv.$$

Hence,
$$dz = ye^{xy} \frac{du + dv}{u + v} + xe^{xy} \frac{v\,du - u\,dv}{v^2 + u^2}$$

$$= \left(\frac{ye^{xy}}{u + v} + \frac{xe^{xy}v}{v^2 + u^2} \right) du + \left(\frac{ye^{xy}}{u + v} - \frac{xe^{xy}u}{v^2 + u^2} \right) dv.$$

But
$$dz = \frac{\partial z}{\partial u} du + \frac{\partial z}{\partial v} dv$$

and since du and dv are independent differentials, equating the coefficients of du and dv in the two expressions for dz gives

$$\frac{\partial z}{\partial u} = \frac{ye^{xy}}{u + v} + \frac{xe^{xy}v}{v^2 + u^2} \quad \text{and} \quad \frac{\partial z}{\partial v} = \frac{ye^{xy}}{u + v} - \frac{xe^{xy}u}{v^2 + u^2}.$$

Example 4.   A function $f(x_1, x_2, \ldots, x_n)$ of n variables $x_1, x_2, \ldots, x_n$ is said to be *homogeneous of degree m* if the function is multiplied by λ^m when the arguments $x_1, x_2, \ldots, x_n$ are replaced by $\lambda x_1, \lambda x_2, \ldots, \lambda x_n$, respectively.   For example, $f(x,y) = x^2/\sqrt{x^2 + y^2}$ is homogeneous of degree 1, because the substitution of λx for x and λy for y yields $\lambda x^2/\sqrt{x^2 + y^2}$.   Again, $f(x,y) = (1/y) + (\log x - \log y)/x$ is homogeneous of degree -1, whereas $f(x,y,z) = z^2/\sqrt[3]{x^2 + y^2}$ is homogeneous of degree $\frac{4}{3}$.   As is clear from these examples, λ is assumed to be different from zero.

There is an important theorem, due to Euler, concerning homogeneous functions.

EULER'S THEOREM.   If $u = f(x_1, x_2, \ldots, x_n)$ is homogeneous of degree m and has continuous first partial derivatives, then

$$x_1 \frac{\partial f}{\partial x_1} + x_2 \frac{\partial f}{\partial x_2} + \cdots + x_n \frac{\partial f}{\partial x_n} = mf(x_1, x_2, \ldots, x_n).$$

The proof of the theorem follows at once upon substituting

$$x_1' = \lambda x_1, \qquad x_2' = \lambda x_2, \qquad \ldots, \qquad x_n' = \lambda x_n.$$

Then, since $f(x_1, x_2, \ldots, x_n)$ is homogeneous of degree m,

$$f(x_1', x_2', \ldots, x_n') = \lambda^m f(x_1, x_2, \ldots, x_n).$$

Differentiating with respect to λ gives

$$\frac{\partial f}{\partial x_1'} x_1 + \frac{\partial f}{\partial x_2'} x_2 + \cdots + \frac{\partial f}{\partial x_n'} x_n = m\lambda^{m-1} f(x_1, x_2, \ldots, x_n).$$

If λ is set equal to 1, then $x_1 = x_1'$, $x_2 = x_2'$, $\ldots$, $x_n = x_n'$, and the theorem follows.

PROBLEMS

In the following problems, assume that the regions of definition of functions are such that the partial derivatives are continuous.

1. Find du/dt, if (a) $u = e^x \sin yx$ and $x = t^2$, $y = t - 1$, $z = 1/t$; (b) $u = \tan^{-1}(y/x)$ and $x = e^t + e^{-t}$, $y = e^t - e^{-t}$.

2. If a point moves so that $x = t$, $y = t^2$, $z = t^3$, where t is time, find the rate at which it moves away from the origin $(0,0,0)$ when $t = 1$.   *Hint:* Find dr/dt, where $r = \sqrt{x^2 + y^2 + z^2}$.

3. If f is a function of u and v, where $u = \sqrt{x^2 + y^2}$ and $v = \tan^{-1}(y/x)$, find $\partial f/\partial x$, $\partial f/\partial y$, $\sqrt{(\partial f/\partial x)^2 + (\partial f/\partial y)^2}$.

4. If f is a function of u and v, where $u = r \cos \theta$ and $v = r \sin \theta$, find

$$\frac{\partial f}{\partial r} \qquad \frac{\partial f}{\partial \theta} \qquad \sqrt{\left(\frac{\partial f}{\partial r}\right)^2 + \frac{1}{r^2}\left(\frac{\partial f}{\partial \theta}\right)^2}.$$

5. If $x = x' \cos \theta - y' \sin \theta$, $y = x' \sin \theta + y' \cos \theta$, prove that

$$\left(\frac{\partial f}{\partial x}\right)^2 + \left(\frac{\partial f}{\partial y}\right)^2 = \left(\frac{\partial f}{\partial x'}\right)^2 + \left(\frac{\partial f}{\partial y'}\right)^2.$$

6. Find the total differential if $u = x^2 + y^2$, $x = r \cos \theta$, and $y = r \sin \theta$.

7. If $f = e^{xy}$, where $x = \log (u^2 + v^2)^{\frac{1}{2}}$ and $y = \tan^{-1}(u/v)$, find $\partial f/\partial u$ and $\partial f/\partial v$.

8. If $z = (u + v)/(1 - uv)$, $u = y \sin x$, and $v = e^{yx}$, find $\partial z/\partial x$ and $\partial z/\partial y$.

9. Find $\partial z/\partial r$ and $\partial z/\partial s$ if $z = (x - y)/(1 + xy)$, $x = \tan(r - s)$, and $y = e^{rs}$.

10. (a) Find du/dx if $u = x^2 + y^2$ and $y = \tan x$; (b) find $\partial u/\partial r$ and $\partial u/\partial \theta$, if $u = x^2 - 4y^2$ and $x = r \sec \theta$, $y = r \tan \theta$; (c) if $V = f(x,y,z)$, where $x = r \cos \theta$, $y = r \sin \theta$, $z = t$, compute $\partial V/\partial r$, $\partial V/\partial \theta$, $\partial V/\partial t$.

11. Assume that the derivatives of desired orders are continuous and verify that (a) $z = f(x^2 + y^2)$ satisfies $yz_x - xz_y = 0$; (b) $z = xy + xf(y/x)$ satisfies $xz_x + yz_y - z - xy = 0$; (c) $z = f(u,v)$ satisfies $z_t = az_x + bz_y$, if $u = x + at$ and $v = y + bt$; (d) $z = xf(y/x) + yg(y/x)$ satisfies $x^2z_{xx} + 2xyz_{xy} + y^2z_{yy} = 0$.

12. Verify Euler's theorem for each of the following functions:

(a) $f(x,y,z) = x^2y + xy^2 + 2xyz$

(b) $f(x,y) = \sqrt{y^2 - x^2} \sin^{-1}\frac{x}{y}$

(c) $f(x,y) = \frac{1}{y^2} + \frac{\log x - \log y}{x^2}$

(d) $f(x,y,z) = \frac{z^2}{\sqrt{x^2 - y^2}}$

(e) $f(x,y,z) = (x^2 + y^2 + z^2)^{-\frac{1}{2}}$

(f) $f(x,y) = e^{x/y}$

(g) $f(x,y) = \frac{\sqrt{x + y}}{y}$

(h) $f(x,y) = \frac{x^2 + y^2}{x^2 - y^2}.$

---

6. Implicit Functions.   In calculus and analytic geometry, equations of plane curves often appear in the form

$$f(x,y) = 0. \tag{6-1}$$

For example, the equation of a circle of radius a with its center at the origin (Fig. 3a) is

$$x^2 + y^2 - a^2 = 0 \qquad a > 0 \tag{6-2}$$

and the equation of the folium of Descartes (Fig. 3b) is

$$x^3 + y^3 - 3axy = 0 \qquad a > 0. \tag{6-3}$$

It appears from the graph of the circle in Fig. 3a that, corresponding to each value of x in the interval $-a < x < a$, Eq. (6-2) determines two values of y and one value (zero) at $x = \pm a$. These observations also follow from the solution

$$y = \pm\sqrt{a^2 - x^2} \tag{6-4}$$

of Eq. (6-2) for y.

The graph of the folium in Fig. 3b shows that Eq. (6-3) determines at least one value of y in the interval $-\infty < x < \infty$. Although in this case an explicit solution of (6-3) in the form $y = \varphi(x)$, where $\varphi(x)$ is differentiable, can be obtained by solving the cubic, it is not difficult to write the equation in the form (6-1), which cannot be solved for y but which nonetheless determines one or more values of y in a suitable range of

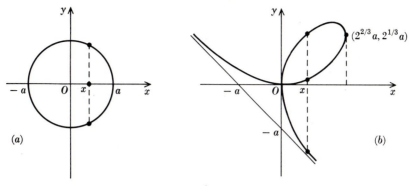

FIGURE 3

values of x. Of course, not every equation of the form (6-1) determines y as a function of x. For example, $x^2 + y^2 + a^2 = 0$ cannot be satisfied by any pair of real values (x,y) if $a \neq 0$. But, if one or more single-valued differentiable functions $y = \varphi(x)$ are determined by Eq. (6-1), it is said that $y = \varphi(x)$ is an *implicit function* specified by Eq. (6-1).

Let us suppose now that Eq. (6-1) is satisfied by a pair of real values (x_0,y_0), so that $f(x_0,y_0) = 0$. Let us further assume that, in some interval about $x = x_0$, Eq. (6-1) defines a single-valued function $y = \varphi(x)$, which has a derivative at $x = x_0$, and that the partial derivatives $f_x(x,y)$, $f_y(x,y)$ are continuous at (x_0,y_0) with $f_y \neq 0$. We shall see that the derivative of $y = \varphi(x)$ at $x = x_0$ can be computed from (6-1) without actually solving this equation for y. Now, the left-hand member of (6-1), which we denote by $u = f(x,y)$, can be viewed as a composite function of x, namely,

$$u = f(x,y) \qquad \text{where } y = \varphi(x). \qquad (6-5)$$

By hypothesis, $y = \varphi(x)$ satisfies Eq. (6-1) in some interval about $x = x_0$, and hence $u \equiv 0$ in that interval. Remembering that $u \equiv 0$ and differentiating (6-5) with respect to x by the chain rule [formula (5-3) with $t = x$], we get an identity,

$$0 = \frac{\partial f}{\partial x} + \frac{\partial f}{\partial y}\frac{dy}{dx}.$$

Thus $$\frac{dy}{dx} = -\frac{\partial f/\partial x}{\partial f/\partial y} \qquad (6-6)$$

provided that $\partial f/\partial y \neq 0$ at (x_0,y_0).

Example 1. We saw that the equation $x^2 + y^2 - a^2 = 0$, $a > 0$, defines two single-valued differentiable functions (6-4). In this case, $f(x,y) = x^2 + y^2 - a^2$, and $\partial f/\partial x = 2x$, $\partial f/\partial y = 2y$ are continuous functions for all values of x and y. Formula (6-6) then yields, for each function in (6-4), $dy/dx = -x/y$, except when $y = 0$. But $y = 0$ when $x = \pm a$, and at the points $(a,0)$, $(-a,0)$ the derivatives of (6-4) cease to exist.

Example 2. If we apply formula (6-6) to Eq. (6-3), defining the folium of Descartes, we find, on setting $f(x,y) = x^3 + y^3 - 3axy$, that $\partial f/\partial x = 3x^2 - 3ay$ and $\partial f/\partial y = 3y^2 - 3ax$. These partial derivatives are continuous for all values of x and y and hence, by (6-6),

$$\frac{dy}{dx} = -\frac{x^2 - ay}{y^2 - ax}$$

except when $y^2 - ax = 0$. It is easy to check that the only points on the folium for which $y^2 - ax = 0$ are $(0,0)$ and $(2^{2/3}a, 2^{1/3}a)$, and at these points dy/dx does not exist.

Example 3.   Let $f(x,y) = 0$ represent the locus of a curve, and let $P(x_0,y_0)$ be a point on the curve. If the curve has a tangent at $P(x_0,y_0)$, the equation of the tangent line is

$$y - y_0 = \left(\frac{dy}{dx}\right)_{x=x_0} (x - x_0).$$

It follows from (6-6) that this equation can be written in the form

$$f_x(x_0,y_0)(x - x_0) + f_y(x_0,y_0)(y - y_0) = 0.$$

PROBLEMS

1. Find the equation of the tangent line to the ellipse $x^2/a^2 + y^2/b^2 = 1$ at the point (x_0,y_0).

2. Find the equation of the tangent line to the ellipse $x = a \cos \theta$, $y = b \sin \theta$ at the point where $\theta = \pi/4$.

3. Find dy/dx at those points of the curve $3x^3y^2 + x \cos y = 0$ where the derivative exists.

7. Differentiation of Implicit Functions.   The considerations of the preceding section can be extended to equations involving more than two variables. Thus, let the equation

$$f(x,y,z) = 0 \tag{7-1}$$

be satisfied at some point (x_0,y_0,z_0) so that $f(x_0,y_0,z_0) = 0$. Let us also assume that Eq. (7-1) determines a real value z at each (x,y) in some neighborhood of the point (x_0,y_0) in the xy plane. We then say that Eq. (7-1) defines an implicit function

$$z = \varphi(x,y) \tag{7-2}$$

in the neighborhood of (x_0,y_0).

If $z = \varphi(x,y)$ is single-valued and the partial derivatives of the function $u = f(x,y,z)$ are continuous at (x_0,y_0,z_0), we can compute the partial derivatives $\partial\varphi/\partial x$ and $\partial\varphi/\partial y$ at (x_0,y_0) directly from (7-1) provided that $\partial f/\partial z \neq 0$ at (x_0,y_0,z_0). Indeed, if we set $y = y_0$ in (7-1), we obtain the equation

$$f(x,y_0,z) = 0 \tag{7-3}$$

that involves only two variables x and z. This equation determines z as an implicit function of x in some neighborhood of $x = x_0$ and hence formula (6-6) is available for the computation of $\partial z/\partial x$. We get, on making use of (6-6),

$$\frac{\partial z}{\partial x} = -\frac{\partial f/\partial x}{\partial f/\partial z} \qquad \text{if } \frac{\partial f}{\partial z} \neq 0. \tag{7-4}$$

Similarly, on fixing the value of x in (7-1), we obtain

$$\frac{\partial z}{\partial y} = -\frac{\partial f/\partial y}{\partial f/\partial z} \qquad \text{if } \frac{\partial f}{\partial z} \neq 0. \tag{7-5}$$

Entirely similar considerations apply to an implicit function $u = \varphi(x_1,x_2, \ldots ,x_n)$ determined by the equation

$$f(x_1,x_2, \ldots ,x_n,u) = 0 \tag{7-6}$$

and we conclude that

$$\frac{\partial u}{\partial x_i} = -\frac{\partial f/\partial x_i}{\partial f/\partial u} \qquad i = 1, 2, \ldots , n \tag{7-7}$$

whenever the partial derivatives in the right-hand member of Eq. (7-7) are continuous and $\partial f/\partial u \neq 0$.

Example 1.   Let $x^2 + 2y^2 - 3xz + 1 = 0$.   Here the partial derivatives of $u = x^2 + 2y^2 - 3xz + 1$ are obviously continuous for all values of x, y, z, and formulas (7-4) and (7-5) give

$$\frac{\partial z}{\partial x} = - \frac{2x - 3z}{-3x} \qquad \frac{\partial z}{\partial y} = - \frac{4y}{-3x}$$

provided that $x \neq 0$.   The given equation, in this case, can be easily solved for $z = \varphi(x,y)$ to yield $z = (x^2 + 2y^2 - 1)/3x$, $x \neq 0$.   The reader may find it instructive to compute $\partial z/\partial x$ and $\partial z/\partial y$ from this solution and compare the result with the expressions obtained from formulas (7-4) and (7-5).

Example 2.   The equation

$$x^2 + y^2 + z^2 - 1 = 0 \tag{7-8}$$

determines two differentiable functions

$$z = \pm\sqrt{1 - x^2 - y^2} \tag{7-9}$$

in the region $x^2 + y^2 < 1$.   This time, formulas (7-4) and (7-5) yield

$$\frac{\partial z}{\partial x} = - \frac{x}{z} \qquad \frac{\partial z}{\partial y} = - \frac{y}{z} \qquad \text{if } z \neq 0. \tag{7-10}$$

But if $z = 0$, Eq. (7-8) states that $x^2 + y^2 = 1$, so that the partial derivatives in (7-10) are not determined along the circle $x^2 + y^2 = 1$.   This is obvious from geometric considerations (see Sec. 3) since (7-8) is the equation of sphere.

Partial derivatives of higher orders (if they exist) can be computed in a similar way by successive differentiations.   We illustrate the formal procedure in the following examples.

Example 3.   Obtain d^2y/dx^2 of an implicit function defined by

$$f(x,y) = 0. \tag{7-11}$$

On differentiating Eq. (7-11) with respect to x, we get

$$f_x(x,y) + f_y(x,y)\frac{dy}{dx} = 0 \tag{7-12}$$

and differentiating Eq. (7-12) again we find

$$f_{xx}(x,y) + 2f_{xy}(x,y)\frac{dy}{dx} + f_{yy}(x,y)\left(\frac{dy}{dx}\right)^2 + f_y(x,y)\frac{d^2y}{dx^2} = 0. \tag{7-13}$$

If $f_y(x,y) \neq 0$, we can solve (7-12) for dy/dx, substitute $dy/dx = -f_x(x,y)/f_y(x,y)$ in (7-13), and solve the resulting equation for dy^2/dx^2.   The result is

$$\frac{d^2y}{dx^2} = - \frac{f_{xx}f_y{}^2 - 2f_{xy}f_xf_y + f_{yy}f_x{}^2}{f_y{}^3}. \tag{7-14}$$

Formula (7-14) is meaningful when the derivatives are continuous and $f_y \neq 0$.   The formal procedure indicated above can be continued to obtain derivatives of higher orders.

Example 4.   Let it be required to find the derivatives of second order of the function z defined implicitly by the equation

$$\frac{x^2}{a^2} + \frac{y^2}{b^2} + \frac{z^2}{c^2} = 1.$$

Differentiating this equation with respect to x and y gives

$$\frac{2x}{a^2} + \frac{2z}{c^2}\frac{\partial z}{\partial x} = 0 \qquad \frac{2y}{b^2} + \frac{2z}{c^2}\frac{\partial z}{\partial y} = 0. \tag{7-15}$$

Differentiating the first of Eqs. (7-15) with respect to x and y, one obtains

$$\frac{2}{a^2} + \frac{2}{c^2}\left(\frac{\partial z}{\partial x}\right)^2 + \frac{2z}{c^2}\frac{\partial^2 z}{\partial x^2} = 0 \qquad \frac{2}{c^2}\frac{\partial z}{\partial x}\frac{\partial z}{\partial y} + \frac{2z}{c^2}\frac{\partial^2 z}{\partial x\,\partial y} = 0.$$

Solving for $\partial^2 z/\partial x^2$ and $\partial^2 z/\partial x\,\partial y$ and making use of (7-15), one obtains

$$\frac{\partial^2 z}{\partial x^2} = -\frac{c^2}{a^4}\frac{a^2 z^2 + c^2 x^2}{z^3} \qquad \frac{\partial^2 z}{\partial x\,\partial y} = -\frac{c^4}{a^2 b^2}\frac{xy}{z^3}.$$

In a similar way the differentiation of the second of Eqs. (7-15) with respect to y yields

$$\frac{\partial^2 z}{\partial y^2} = -\frac{c^2}{b^4}\frac{b^2 z^2 + c^2 y^2}{z^3}.$$

These calculations are meaningful at all points of the given ellipsoid for which $z \neq 0$ (compare with Example 2).

Example 5.   Let $f(x,y,z) = 0$ define a surface $z = f(x,y)$ having a tangent plane (that is not perpendicular to the xy plane) at a point $P_0(x_0,y_0,z_0)$ on the surface.   From calculus, the equation of such tangent plane at P_0 is

$$z - z_0 = \left(\frac{\partial z}{\partial x}\right)(x - x_0) + \left(\frac{\partial z}{\partial y}\right)(y - y_0) \tag{7-16}$$

where partial derivatives are computed at (x_0,y_0).

It follows at once from (7-4) and (7-5) that (7-16) can be cast in the form

$$\left(\frac{\partial f}{\partial x}\right)_{P_0}(x - x_0) + \left(\frac{\partial f}{\partial y}\right)_{P_0}(y - y_0) + \left(\frac{\partial f}{\partial z}\right)_{P_0}(z - z_0) = 0 \tag{7-17}$$

which does not require solving the given equation for z.

PROBLEMS

1. Find z_x and z_y if (a) $x^3 y - \sin z + z^2 = 0$; (b) $xz^2 - yz^2 + xy^2 z - 5 = 0$; (c) $xz^2 - yz + 3xy = 0$.

2. Find z_x, z_y, z_{xx}, z_{xy}, and z_{yy} at $(1,1,1)$, if $x^2 - y^2 + z^2 = 1$.

3. Refer to Example 5 and find the equation of the tangent plane to the surface $x^2 - y^2 + z^2 = 1$ at the point $(1,1,1)$.

4. The procedure indicated in Example 3 enables one to compute the coefficients in the Taylor expansion

$$y = \varphi(x_0) + \varphi'(x_0)(x - x_0) + \cdots + \frac{\varphi^{(n)}(x_0)}{n!}(x - x_0)^n + \cdots$$

of the function $y = \varphi(x)$ defined by $f(x,y) = 0$.   Use this procedure to show that

$$y = 1 - 2(x - 2) - \frac{5}{2!}(x - 2)^2 - \frac{30}{3!}(x - 2)^3 + \cdots$$

is the solution of $x^2 + y^2 - 5 = 0$ in the neighborhood of the point $(2,1)$.

5. If $z = \frac{1}{2}(x^2 + y^2 + z^2)$, find z_{xx}.

---

8. Change of Variables.   Implicit Functions Defined by Systems of Equations.
The main purpose of this section is to develop manipulative skill in calculating the derivatives of implicit functions and to indicate the formal modes of attack on the problem.   The continuity of the functions and their partial derivatives is assumed throughout this section and will not be referred to again.

Let

$$w = f(u,v) \tag{8-1}$$

denote a function of two independent variables u and v, and suppose that u and v are connected with some other variables x and y by means of the relations

$$x = x(u,v) \qquad y = y(u,v) \tag{8-2}$$

which can be solved for x and y to yield

$$u = u(x,y) \qquad v = v(x,y). \tag{8-3}$$

The expressions (8-3), when substituted for u and v in (8-1), yield a function of x and y,

$$w = F(x,y). \tag{8-4}$$

The partial derivatives of w with respect to x and y can be calculated from (8-4) directly, but frequently it is impracticable to obtain the solution (8-3), and we consider an indirect mode of calculation. By the rule for the differentiation of composite functions,

$$\frac{\partial w}{\partial u} = \frac{\partial w}{\partial x}\frac{\partial x}{\partial u} + \frac{\partial w}{\partial y}\frac{\partial y}{\partial u}$$
$$\frac{\partial w}{\partial v} = \frac{\partial w}{\partial x}\frac{\partial x}{\partial v} + \frac{\partial w}{\partial y}\frac{\partial y}{\partial v}. \tag{8-5}$$

The partial derivatives $\partial x/\partial u$, $\partial y/\partial u$, $\partial x/\partial v$, and $\partial y/\partial v$ can be calculated from (8-2), and hence they may be regarded as known functions of u and v. The partial derivatives in the left-hand members of (8-5) are also known functions of u and v, since they can be calculated from (8-1).

Hence, Eqs. (8-5) may be regarded as linear equations for the determination of $\partial w/\partial x$ and $\partial w/\partial y$. Assuming that the *Jacobian* $J(u,v)$ defined by

$$J(u,v) \equiv \begin{vmatrix} \dfrac{\partial x}{\partial u} & \dfrac{\partial y}{\partial u} \\[2mm] \dfrac{\partial x}{\partial v} & \dfrac{\partial y}{\partial v} \end{vmatrix}$$

is not zero and solving by Cramer's rule give

$$\frac{\partial w}{\partial x} = \frac{\begin{vmatrix} \dfrac{\partial w}{\partial u} & \dfrac{\partial y}{\partial u} \\[2mm] \dfrac{\partial w}{\partial v} & \dfrac{\partial y}{\partial v} \end{vmatrix}}{J(u,v)} \qquad \frac{\partial w}{\partial y} = \frac{\begin{vmatrix} \dfrac{\partial x}{\partial u} & \dfrac{\partial w}{\partial u} \\[2mm] \dfrac{\partial x}{\partial v} & \dfrac{\partial w}{\partial v} \end{vmatrix}}{J(u,v)}.$$

The resulting expressions for $\partial w/\partial x$ and $\partial w/\partial y$ are known functions of u and v and thus can be treated exactly like (8-1) if it is desirable to calculate the derivatives of higher orders.

As an example, consider the function $w(r,\theta)$, and let it be required to calculate the partial derivatives of w with respect to x and y, where $x = r\cos\theta$ and $y = r\sin\theta$. Now

$$\frac{\partial w}{\partial r} = \frac{\partial w}{\partial x}\frac{\partial x}{\partial r} + \frac{\partial w}{\partial y}\frac{\partial y}{\partial r} = \frac{\partial w}{\partial x}\cos\theta + \frac{\partial w}{\partial y}\sin\theta$$

$$\frac{\partial w}{\partial \theta} = \frac{\partial w}{\partial x}\frac{\partial x}{\partial \theta} + \frac{\partial w}{\partial y}\frac{\partial y}{\partial \theta} = -\frac{\partial w}{\partial x}r\sin\theta + \frac{\partial w}{\partial y}r\cos\theta.$$

Solving these equations for $\partial w/\partial x$ and $\partial w/\partial y$ in terms of $\partial w/\partial r$ and $\partial w/\partial \theta$ gives

$$\frac{\partial w}{\partial x} = \cos\theta\,\frac{\partial w}{\partial r} - \frac{\sin\theta}{r}\frac{\partial w}{\partial \theta}$$

$$\frac{\partial w}{\partial y} = \sin\theta\,\frac{\partial w}{\partial r} + \frac{\cos\theta}{r}\frac{\partial w}{\partial \theta}.$$

The Jacobian J is, in this case,

$$\begin{vmatrix} \cos\theta & \sin\theta \\ -r\sin\theta & r\cos\theta \end{vmatrix} = r$$

which does not vanish unless $r = 0$.

As a somewhat more complicated instance of implicit differentiation, consider a pair of equations

$$F(x,y,u,v) = 0 \qquad G(x,y,u,v) = 0 \tag{8-6}$$

and let it be supposed that they can be solved for u and v in terms of x and y to yield

$$u = u(x,y) \qquad v = v(x,y). \tag{8-7}$$

The partial derivatives of u and v with respect to x and y can be obtained in the following manner. Considering x and y as the independent variables and differentiating Eqs. (8-6) with respect to x and y give

$$\frac{\partial F}{\partial x} + \frac{\partial F}{\partial u}\frac{\partial u}{\partial x} + \frac{\partial F}{\partial v}\frac{\partial v}{\partial x} = 0 \qquad \frac{\partial F}{\partial y} + \frac{\partial F}{\partial u}\frac{\partial u}{\partial y} + \frac{\partial F}{\partial v}\frac{\partial v}{\partial y} = 0$$

$$\frac{\partial G}{\partial x} + \frac{\partial G}{\partial u}\frac{\partial u}{\partial x} + \frac{\partial G}{\partial v}\frac{\partial v}{\partial x} = 0 \qquad \frac{\partial G}{\partial y} + \frac{\partial G}{\partial u}\frac{\partial u}{\partial y} + \frac{\partial G}{\partial v}\frac{\partial v}{\partial y} = 0. \tag{8-8}$$

Equations (8-8) are linear in $\partial u/\partial x$, $\partial u/\partial y$, $\partial v/\partial x$, and $\partial v/\partial y$. If

$$J(u,v) \equiv \begin{vmatrix} \dfrac{\partial F}{\partial u} & \dfrac{\partial F}{\partial v} \\[2mm] \dfrac{\partial G}{\partial u} & \dfrac{\partial G}{\partial v} \end{vmatrix} \neq 0$$

the partial derivatives in question can be determined from (8-8) by Cramer's rule.

A special case of Eqs. (8-6) is useful in applications. Let

$$x = f(u,v) \qquad y = g(u,v).$$

Differentiating these equations with respect to x and remembering that x and y are independent variables, one obtains

$$1 = \frac{\partial f}{\partial u}\frac{\partial u}{\partial x} + \frac{\partial f}{\partial v}\frac{\partial v}{\partial x}$$

$$0 = \frac{\partial g}{\partial u}\frac{\partial u}{\partial x} + \frac{\partial g}{\partial v}\frac{\partial v}{\partial x}. \tag{8-9}$$

These equations can be solved for $\partial u/\partial x$ and $\partial v/\partial x$ if

$$J(u,v) \equiv \begin{vmatrix} \dfrac{\partial f}{\partial u} & \dfrac{\partial f}{\partial v} \\[2mm] \dfrac{\partial g}{\partial u} & \dfrac{\partial g}{\partial v} \end{vmatrix} \neq 0.$$

Example 1. Let $u^2 - v^2 + 2x = 0$, and $uv - y = 0$.

Differentiating with respect to x,

$$2\left(u\frac{\partial u}{\partial x} - v\frac{\partial v}{\partial x} + 1\right) = 0 \qquad v\frac{\partial u}{\partial x} + u\frac{\partial v}{\partial x} = 0.$$

Hence

$$\frac{\partial u}{\partial x} = -\frac{u}{u^2 + v^2} \qquad \frac{\partial v}{\partial x} = \frac{v}{u^2 + v^2}.$$

Differentiating the first of these results with respect to x gives

$$\frac{\partial^2 u}{\partial x^2} = \frac{-(\partial u/\partial x)(u^2 + v^2) + 2[u(\partial u/\partial x) + v(\partial v/\partial x)]u}{(u^2 + v^2)^2}$$

$$= \frac{u(u^2 + v^2) - 2u(u^2 - v^2)}{(u^2 + v^2)^3} = \frac{u(3v^2 - u^2)}{(u^2 + v^2)^3}.$$

One obtains similarly $\partial^2 v/\partial x^2$, $\partial^2 u/\partial x\, \partial y$, and higher derivatives.

Example 2.   Let $x = u + v$, $y = 3u + 2v$.   (a)

Differentiating with respect to x,

$$1 = \frac{\partial u}{\partial x} + \frac{\partial v}{\partial x} \qquad 0 = 3\frac{\partial u}{\partial x} + 2\frac{\partial v}{\partial x}$$

so that

$$\frac{\partial u}{\partial x} = -2 \qquad \frac{\partial v}{\partial x} = 3.$$

It is easily checked that

$$\frac{\partial u}{\partial y} = 1 \qquad \frac{\partial v}{\partial y} = -1.$$

Equations (a) can be solved for u and v in terms of x and y, and the result is

$$u = -2x + y \qquad v = 3x - y.$$

Regarding u and v as the independent variables and differentiating these equations with respect to u, one finds

$$1 = -2\frac{\partial x}{\partial u} + \frac{\partial y}{\partial u} \qquad 0 = 3\frac{\partial x}{\partial u} - \frac{\partial y}{\partial u}.$$

Hence,

$$\frac{\partial x}{\partial u} = 1 \qquad \frac{\partial y}{\partial u} = 3.$$

This agrees with the result obtained by direct differentiation of (a), as of course it should. Note that $\partial u/\partial x$ and $\partial x/\partial u$ are not reciprocals.

Example 3.   If $w = uv$ and

$$u^2 + v + x = 0 \qquad v^2 - u - y = 0 \qquad\qquad (b)$$

one can obtain $\partial w/\partial x$ as follows: Differentiation of w with respect to x gives

$$\frac{\partial w}{\partial x} = u\frac{\partial v}{\partial x} + v\frac{\partial u}{\partial x}.$$

The values of $\partial u/\partial x$ and $\partial v/\partial x$ can be calculated from (b) as was done in Example 1. The reader will check that

$$\frac{\partial w}{\partial x} = -\frac{u + 2v^2}{1 + 4uv} \qquad \frac{\partial w}{\partial y} = \frac{2u^2 - v}{1 + 4uv}.$$

PROBLEMS

1. If $u^2 + v^2 + y^2 - 2x = 0$, $u^3 + v^3 - x^3 + 3y = 0$, find $\partial u/\partial x$, $\partial v/\partial x$, $\partial u/\partial y$, and $\partial v/\partial y$.

2. Find $\partial w/\partial x$ and $\partial w/\partial y$ if $w = u/v$, and $x = u + v$, $y = 3u + 2v$.

3. Show that if $f(x,y,z) = 0$ then $(\partial z/\partial x)(\partial x/\partial z) = 1$ and $(\partial x/\partial y)(\partial y/\partial z)(\partial z/\partial x) = -1$. Note that, in general, $\partial z/\partial x$ and $\partial x/\partial z$ are not reciprocals.

4. If $x = x(u,v)$, $y = y(u,v)$ with $\partial x/\partial u = \partial y/\partial v$, and $\partial x/\partial v = -\partial y/\partial u$, show that

$$\frac{\partial^2 w}{\partial u^2} + \frac{\partial^2 w}{\partial v^2} = \left(\frac{\partial^2 w}{\partial x^2} + \frac{\partial^2 w}{\partial y^2}\right)\left[\left(\frac{\partial x}{\partial u}\right)^2 + \left(\frac{\partial x}{\partial v}\right)^2\right].$$

5. Show that the expressions

$$V_1 = \left(\frac{\partial z}{\partial x}\right)^2 + \left(\frac{\partial z}{\partial y}\right)^2 \qquad \text{and} \qquad V_2 = \frac{\partial^2 z}{\partial x^2} + \frac{\partial^2 z}{\partial y^2}$$

upon change of variable by means of $x = r \cos \theta$, $y = r \sin \theta$, become

$$V_1 = \left(\frac{\partial z}{\partial r}\right)^2 + \frac{1}{r^2}\left(\frac{\partial z}{\partial \theta}\right)^2 \quad \text{and} \quad V_2 = \frac{\partial^2 z}{\partial r^2} + \frac{1}{r^2}\frac{\partial^2 z}{\partial \theta^2} + \frac{1}{r}\frac{\partial z}{\partial r}.$$

6. Show that

$$\frac{\partial^2 V}{\partial t^2} = c^2 \frac{\partial^2 V}{\partial x^2}$$

if $V = f(x + ct) + g(x - ct)$, where f and g are any functions possessing continuous second derivatives.

7. Show that

$$\frac{\partial^2 V}{\partial x^2} + \frac{\partial^2 V}{\partial y^2} = e^{-2r}\left(\frac{\partial^2 V}{\partial r^2} + \frac{\partial^2 V}{\partial \theta^2}\right)$$

if $x = e^r \cos \theta$, $y = e^r \sin \theta$.

8. Find $\partial u/\partial x$ if $u^2 - v^2 - x^3 + 3y = 0$, $u + v - y^2 - 2x = 0$.

9. Prove that

$$\frac{\partial u}{\partial x}\frac{\partial y}{\partial u} + \frac{\partial v}{\partial x}\frac{\partial y}{\partial v} = 0$$

if $F(x,y,u,v) = 0$ and $G(x,y,u,v) = 0$.

10. If $V_1(x,y,z)$ and $V_2(x,y,z)$ satisfy the equation

$$\nabla^2 V \equiv \frac{\partial^2 V}{\partial x^2} + \frac{\partial^2 V}{\partial y^2} + \frac{\partial^2 V}{\partial z^2} = 0$$

show that $U \equiv V_1(x,y,z) + (x^2 + y^2 + z^2)V_2(x,y,z)$

satisfies the equation

$$\nabla^2 \nabla^2 U = 0$$

where $$\nabla^2 \equiv \frac{\partial^2}{\partial x^2} + \frac{\partial^2}{\partial y^2} + \frac{\partial^2}{\partial z^2}.$$

11. To indicate explicitly the variables entering in the Jacobian

$$J(u,v) = \begin{vmatrix} \dfrac{\partial x}{\partial u} & \dfrac{\partial y}{\partial u} \\ \dfrac{\partial x}{\partial v} & \dfrac{\partial y}{\partial v} \end{vmatrix}$$

one frequently writes $J(u,v) = J\left(\frac{x,y}{u,v}\right)$. The Jacobian

$$J(x,y) = \begin{vmatrix} \dfrac{\partial u}{\partial x} & \dfrac{\partial v}{\partial x} \\ \dfrac{\partial u}{\partial y} & \dfrac{\partial v}{\partial y} \end{vmatrix}$$

of the transformation (8-3) is written as $J(x,y) = J\left(\frac{u,v}{x,y}\right)$. Prove that

(a) $J\left(\frac{u,v}{x,y}\right)J\left(\frac{x,y}{u,v}\right) = 1$ (b) $J\left(\frac{u,v}{x,y}\right)J\left(\frac{x,y}{\xi,\eta}\right) = J\left(\frac{u,v}{\xi,\eta}\right)$

where $u = u(x,y)$, $v = v(x,y)$, $x = x(\xi,\eta)$, and $y = y(\xi,\eta)$. *Hint:* Write the Jacobians and multiply.

9. Remarks on the Existence of Implicit Functions.   In our discussion of implicit functions we considered the equation

$$f(x_1, x_2, \ldots, x_n, u) = 0 \tag{9-1}$$

which is such that, for the point $P_0(x_1^0, x_2^0, \ldots, x_n^0, u^0)$, $f(x_1^0, x_2^0, \ldots, x_n^0, u^0) = 0$. We further assumed that in some neighborhood of $(x_1^0, x_2^0, \ldots, x_n^0)$ Eq. (9-1) defines a single-valued function

$$u = \varphi(x_1, x_2, \ldots, x_n) \tag{9-2}$$

with continuous partial derivatives $\partial u / \partial x_i$, which is such that $\varphi(x_1^0, x_2^0, \ldots, x_n^0) = u^0$. We saw that if $f(x_1, x_2, \ldots, x_n, u)$ has continuous partial derivatives in the neighborhood of P_0 then it is possible to compute $\partial u / \partial x_i$ $(i = 1, 2, \ldots, n)$ directly from (9-1) by formula (7-7), provided that $\partial f / \partial u \neq 0$ at P_0. The partial derivatives $\partial u / \partial x_i$ so obtained are continuous in some neighborhood of $(x_1^0, x_2^0, \ldots, x_n^0)$.

The question now arises whether the assumption that the function $f(x_1, x_2, \ldots, x_n, u)$ has continuous derivatives at P_0 and that $\partial f / \partial u \neq 0$ at P_0 ensures the existence of a single-valued solution (9-2) with continuous partial derivatives in the neighborhood of $(x_1^0, x_2^0, \ldots, x_n^0)$. The answer to this question is in the affirmative, and it forms the substance of the Fundamental Existence Theorem on implicit functions.

We do not present the proof of this theorem here[1] and give only a precise (but rather long) statement of the general theorem which includes the special case of Eq. (9-1).

THEOREM.   Let a system of n equations

$$F_i(x_1, x_2, \ldots, x_m, u_1, u_2, \ldots, u_n) = 0 \qquad i = 1, 2, \ldots, n \tag{9-3}$$

in $m + n$ variables $x_1, x_2, \ldots, x_m, u_1, u_2, \ldots, u_n$ have a real solution $x_1^0, x_2^0, \ldots, x_m^0$, $u_1^0, u_2^0, \ldots, u_n^0$, so that

$$F_i(x_1^0, x_2^0, \ldots, x_m^0, u_1^0, u_2^0, \ldots, u_n^0) = 0 \qquad i = 1, 2, \ldots, n. \tag{9-4}$$

If the functions

$$F_i(x_1, x_2, \ldots, x_m, u_1, u_2, \ldots, u_n)$$

regarded as functions of $m + n$ independent variables, are continuous and have continuous first partial derivatives in some region R of the $(m + n)$-dimensional space enclosing the point P^0, whose coordinates are $(x_1^0, x_2^0, \ldots, x_m^0, u_1^0, u_2^0, \ldots, u_n^0)$, and if the Jacobian

$$\begin{vmatrix} \dfrac{\partial F_1}{\partial u_1} & \dfrac{\partial F_1}{\partial u_2} & \cdots & \dfrac{\partial F_1}{\partial u_n} \\[2mm] \dfrac{\partial F_2}{\partial u_1} & \dfrac{\partial F_2}{\partial u_2} & \cdots & \dfrac{\partial F_2}{\partial u_n} \\[2mm] \cdot & \cdot & \cdots & \cdot \\[2mm] \dfrac{\partial F_n}{\partial u_1} & \dfrac{\partial F_n}{\partial u_2} & \cdots & \dfrac{\partial F_n}{\partial u_n} \end{vmatrix}$$

is different from zero at the point P^0, then the set of Eqs. (9-3) can be solved as

$$u_i = f_i(x_1, x_2, \ldots, x_m) \qquad i = 1, 2, \ldots, n \tag{9-5}$$

in the vicinity of the point $(x_1^0, x_2^0, \ldots, x_m^0)$ in such a way that

$$u_i^0 = f_i(x_1^0, x_2^0, \ldots, x_m^0) \qquad i = 1, 2, \ldots, n.$$

[1] See, for example, Sokolnikoff, "Advanced Calculus," chap. 12; W. Maak, "Modern Calculus," chap. 9, Holt, Rinehart and Winston, Inc., New York, 1963.

Moreover, the set of solutions (9-5) is unique and the functions u_i are continuous together with their first partial derivatives in some neighborhood of the point $(x_1{}^0, x_2{}^0, \ldots, x_m{}^0)$.

APPLICATIONS OF DIFFERENTIATION

10. Taylor's Formula for Functions of Several Variables.   Taylor's formula for functions of one variable can be generalized to functions of several variables having continuous derivatives of order n in a given region. We outline the procedure only for functions of two variables, since it has all the features of the more general case.

Let $f(x,y)$ have continuous derivatives of order n in some region containing the point $x = a$, $y = b$, and let $x = a + \alpha$, $y = b + \beta$ be a nearby point in this region. The equations of a straight-line segment joining these points can be taken in the form

$$x = a + \alpha t \qquad y = b + \beta t \qquad 0 \le t \le 1 \tag{10-1}$$

and, along this segment,

$$f(x,y) = f(a + \alpha t, b + \beta t) \equiv F(t) \tag{10-2}$$

is a composite function of t having continuous derivatives of order n. Accordingly, $F(t)$ can be represented by the Maclaurin formula,

$$F(t) = F(0) + F'(0)t + \frac{F''(0)}{2!} t^2 + \cdots + \frac{F^{(n)}(\theta t)}{n!} t^n \qquad 0 < \theta < 1. \tag{10-3}$$

The successive derivatives of $F(t)$ can be computed from (10-2) by the rule (5-3) for differentiating composite functions.

Thus,

$$F'(t) = f_x(x,y) \frac{dx}{dt} + f_y(x,y) \frac{dy}{dt}$$

$$= f_x(x,y)\alpha + f_y(x,y)\beta$$

where in the last step we recalled (10-1).

Similarly, $F''(t) = [f_{xx}(x,y)\alpha + f_{yx}(x,y)\beta] \dfrac{dx}{dt} + [f_{xy}(x,y)\alpha + f_{yy}(x,y)\beta] \dfrac{dy}{dt}$

$$= f_{xx}(x,y)\alpha^2 + 2f_{xy}(x,y)\alpha\beta + f_{yy}(x,y)\beta^2$$

and

$$F^{(n)}(t) = \frac{\partial^n f}{\partial x^n} \alpha^n + C_{n,1} \frac{\partial^n f}{\partial x^{n-1} \partial y} \alpha^{n-1}\beta + \cdots + C_{n,n-1} \frac{\partial^n f}{\partial x\, \partial y^{n-1}} \alpha\beta^{n-1} + \frac{\partial^n f}{\partial y^n} \beta^n$$

where the $C_{n,r} \equiv n!/[r!(n-r)!]$ are the binomial coefficients.

But it is clear from (10-1) that $t = 0$ corresponds to the point $x = a$, $y = b$, so that

$$F(0) = f(a,b) \qquad F'(0) = f_x(a,b)\alpha + f_y(a,b)\beta$$
$$F''(0) = f_{xx}(a,b)\alpha^2 + 2f_{xy}(a,b)\alpha\beta + f_{yy}(a,b)\beta^2$$

and so on.

The substitution of these values in (10-3) gives

$$f(x,y) = f(a,b) + [f_x(a,b)\alpha + f_y(a,b)\beta]t$$

$$+ \frac{1}{2!} [f_{xx}(a,b)\alpha^2 + 2f_{xy}(a,b)\alpha\beta + f_{yy}(a,b)\beta^2]t^2 + \cdots + R_n \tag{10-4}$$

where $$R_n = \frac{F^{(n)}(\theta t)}{n!} t^n \qquad 0 < \theta < 1.$$

But, from (10-1), $\alpha t = x - a$, $\beta t = y - b$, so that (10-4) takes the form

$$f(x,y) = f(a,b) + f_x(a,b)(x - a) + f_y(a,b)(y - b)$$

$$+ \frac{1}{2!} [f_{xx}(a,b)(x - a)^2 + 2f_{xy}(a,b)(x - a)(y - b) + f_{yy}(a,b)(y - b)^2]$$

$$+ \cdots + R_n \tag{10-5}$$

which is the desired Taylor's formula.

Example.   Obtain the Taylor expansion of $\tan^{-1}(y/x)$ about $(1,1)$ up to the third-degree terms:

$$f(x,y) = \tan^{-1}\frac{y}{x} \qquad f(1,1) = \tan^{-1}1 = \frac{\pi}{4}$$

$$f_x(x,y) = -\frac{y}{x^2 + y^2} \qquad f_x(1,1) = -\tfrac{1}{2}$$

$$f_y(x,y) = \frac{x}{x^2 + y^2} \qquad f_y(1,1) = \tfrac{1}{2}$$

$$f_{xx}(x,y) = \frac{2xy}{(x^2 + y^2)^2} \qquad f_{xx}(1,1) = \tfrac{1}{2}$$

$$f_{xy}(x,y) = \frac{y^2 - x^2}{(x^2 + y^2)^2} \qquad f_{xy}(1,1) = 0$$

$$f_{yy}(x,y) = \frac{-2xy}{(x^2 + y^2)^2} \qquad f_{yy}(1,1) = -\tfrac{1}{2}.$$

Then, from (10-5),

$$\tan^{-1}\frac{y}{x} = \frac{\pi}{4} - \tfrac{1}{2}(x - 1) + \tfrac{1}{2}(y - 1) + \frac{1}{2!}[\tfrac{1}{2}(x - 1)^2 - \tfrac{1}{2}(y - 1)^2] + \cdots.$$

PROBLEMS

1. Obtain the Taylor expansion for $xy^2 + \cos xy$ about $(1,\pi/2)$ up to the third-degree terms.

2. Expand by Taylor's formula $f(x,y) = e^{xy}$ at $(1,1)$, obtaining three terms.

3. Expand by Taylor's formula $e^x \cos y$ at $(0,0)$ up to the fourth-degree terms.

4. Show that for small values of x and y $e^x \sin y = y + xy$ (approx) and $e^x \log(1 + y) = y + xy - \frac{y^2}{2}$ (approx).

5. Expand by Taylor's formula $f(x,y) = x^3y + x^2y + 1$ about $(0,1)$.

6. Expand by Taylor's formula $\sqrt{1 - x^2 - y^2}$ about $(0,0)$ up to the third-degree terms, and verify that your result agrees with the binomial expansion of $[1 - (x^2 + y^2)]^{1/2}$.

7. If $|x|$ and $|y|$ are small compared with 1, use Taylor's formula up to and including the terms of second degree to approximate the following functions:

(a) $\tan^{-1}\dfrac{1 + x}{1 - y}$ $\left[Ans.: \dfrac{\pi}{4} + \tfrac{1}{2}(x + y) - \tfrac{1}{4}(x^2 - y^2).\right]$

(b) $\dfrac{\cos x}{\cos y}$ $[Ans.: 1 - \tfrac{1}{2}(x^2 - y^2).]$

8. Expand $f(x,y) = y^x$ in the neighborhood of $(1,1)$ up to and including the terms of second degree.
[*Ans.:* $1 + (y - 1) + (x - 1)(y - 1).$]

11. Maxima and Minima of Functions of Several Variables.   A function $f(x,y)$ defined in a region R is said to have a *relative maximum* at a point (a,b) if

$$\Delta f \equiv f(a + h, b + k) - f(a,b) \leq 0 \qquad\qquad (11\text{-}1)$$

for all values of h and k in a neighborhood of (a,b).   It is said to have a *relative minimum* at (a,b) if

$$\Delta f \equiv f(a + h, b + k) - f(a,b) \geq 0 \qquad\qquad (11\text{-}2)$$

for all values (h,k) in a neighborhood of (a,b).

The requirement that the inequalities (11-1) and (11-2) hold for all values of (h,k) in a neighborhood of (a,b) implies that we are concerned here only with the interior and not the boundary points of the region.   A function may attain a maximum or a minimum value on the boundary of the region, but the behavior of functions on the boundary requires a separate investigation, the nature of which will be clear from the sequel. The greatest and least values assumed by $f(x,y)$ in the closed region are called, respectively, the *absolute maximum* and the *absolute minimum*.   In the following discussion we dispense with the adjective "relative," and we shall refer to relative maxima and minima simply as *maxima* and *minima*.

Let it be assumed that $f(x,y)$ attains a maximum (or minimum) at some interior point (a,b).   Then the function $f(x,b)$ of the variable x must attain a maximum (or minimum) at $x = a$.   From the study of functions of one variable it follows that the derivative of $f(x,b)$, if it exists, must vanish at $x = a$.   The derivative may cease to exist at the critical points when the behavior of the function is like that shown in Fig. 4 in the neighborhood of $x = a_1$, $x = a_2$, and $x = a_3$.   Thus, a *necessary* condition for a maximum (or minimum) of $f(x,b)$ at $x = a$ is that $f_x = 0$, *if this derivative exists at $x = a$*.

A similar consideration of the function $f(a,y)$ leads to the conclusion that $f_y = 0$ at $y = b$, whenever this derivative exists.

The coordinates (a,b) of a differentiable function $f(x,y)$ thus satisfy the pair of equations

$$f_x = 0 \qquad f_y = 0$$

at any point (a,b), where $f(x,y)$ attains a maximum or minimum.

This discussion can obviously be extended to functions of any number of variables to yield a theorem.

THEOREM I.   A function $f(x_1, x_2, \ldots, x_n)$ of n independent variables x_i attains a maximum or a minimum in the interior of the region only for those values of the variables x_i for which $f_{x_1}, f_{x_2}, \ldots, f_{x_n}$ either vanish simultaneously or some of them cease to exist.

We emphasize that the conditions stated in this theorem are necessary but not sufficient for maxima or minima.   We recall a similar situation in the study of maxima and minima of a function $f(x)$ of one variable, when the curve $y = f(x)$ has the point of inflection with a horizontal tangent.

The points of the region at which all first derivatives of $f(x_1, x_2, \ldots, x_n)$ vanish simultaneously or cease to exist are called the *critical points* of f.   Whether a critical point P_0 is associated with a maximum, minimum, or neither depends on the sign of $\Delta f \equiv f(P) - f(P_0)$ for all points P in the neighborhood of P_0.   Thus, if (a,b) is a critical point of $f(x,y)$ with con-

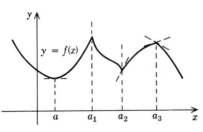

FIGURE 4

tinuous second derivatives, Taylor's formula (10-5), with $x = a + h$, $y = b + k$, yields

$$\Delta f = f(a + h, b + k) - f(a,b) = R_2$$

where R_2 is the remainder after two terms in (10-5). Accordingly, the sign of Δf agrees with that of R_2, and we see that the character of the critical point determines the behavior of the second derivatives of $f(x,y)$. We record the following theorem:[1]

THEOREM II. Let $f(x,y)$ have continuous second derivatives in an open region containing the point (a,b). If $f_x(a,b) = 0$, $f_y(a,b) = 0$ and $D \equiv f_{xy}^2(a,b) - f_{xx}(a,b)f_{yy}(a,b) < 0$, then

(i) $f(x,y)$ has a relative maximum at (a,b) if $f_{xx}(a,b) < 0$;
(ii) $f(x,y)$ has a relative minimum at (a,b) if $f_{xx}(a,b) > 0$.

If $f_x(a,b) = 0$, $f_y(a,b) = 0$ and $D > 0$, then $f(x,y)$ has neither maximum nor minimum at (a,b) and the point (a,b) is called a saddle point of $f(x,y)$.

We observe that Theorem II provides no information about the nature of the critical point if $D = 0$ at that point. In this case the remainders of order higher than R_2 in Taylor's formula must be investigated. But in many practical problems the character of the critical point can be discerned from geometric or physical considerations. Thus, the function $f(x,y) = x^4 + y^4$ obviously has a minimum at the critical point $(0,0)$ and $f(x,y) = -(x^4 + y^4)$ has a maximum at $(0,0)$, but for either of these functions $D = 0$ at $(0,0)$.

Example 1.   Determine the extreme values of

$$f(x,y) = \sin x + \sin y + \sin (x + y) \tag{a}$$

where $0 \leq x \leq 2\pi$, $0 \leq y \leq 2\pi$.

Note that it is required to determine the maximum and minimum values of $f(x,y)$ in the *closed* square shown in Fig. 5.

Now, the critical points in the *interior* of the square must satisfy simultaneously

$$f_x = \cos x + \cos (x + y) = 0 \qquad f_y \equiv \cos y + \cos (x + y) = 0. \tag{b}$$

We conclude from (b) that the critical points of $f(x,y)$ are those points for which $\cos x = \cos y$, that is, $x = y \pm 2n\pi$ $(n = 0, 1, 2, \ldots)$. But the points in question lie within the square in Fig. 5 and hence $y = x$. On setting $y = x$ in (b) we get $\cos x + \cos 2x = 0$, which is the same as

$$2 \cos^2 x + \cos x - 1 = 0 \tag{c}$$

since $\cos 2x = 2 \cos^2 x - 1$. The solutions of the quadratic equation (c) are $\cos x = -1$, $\cos x = \frac{1}{2}$, and therefore $x = \pi$, $x = \pi/3$, $x = 5\pi/3$. Accordingly, there are three critical points in the interior of the square,

$$(\pi,\pi), \qquad \left(\frac{\pi}{3},\frac{\pi}{3}\right), \qquad \left(\frac{5\pi}{3},\frac{5\pi}{3}\right). \tag{d}$$

To determine the nature of these points we compute the value of D in Theorem II and find, after simple calculations, that $D = 0$ at (π,π), so that the nature of the critical point (π,π) is not determined by Theorem II. We find similarly that, at the point $(\pi/3,\pi/3)$, $D = \frac{3}{4} - 3 < 0$ and, since $f_{xx}(\pi/3,\pi/3) = -2\sqrt{3}/2 < 0$, we conclude that this point yields a maximum $f(\pi/3,\pi/3) = 3\sqrt{3}/2$. Finally, at

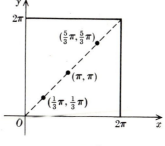

FIGURE 5

[1] The proof of this theorem, although not difficult, is rather long. See, for example, A. E. Taylor, "Advanced Calculus," p. 232, Ginn and Company, Boston, 1955, or Sokolnikoff, "Advanced Calculus," p. 322.

$(5\pi/3,5\pi/3)$, $D = \frac{3}{4} - 3 < 0$, $f_{xx}(5\pi/3,5\pi/3) = \sqrt{3} > 0$, and hence this point yields a minimum $f(5\pi/3,5\pi/3) = -3\sqrt{3}/2$. Inasmuch as there are no other critical points within the square, we conclude that the value $f(\pi,\pi) = 0$ is neither maximum nor minimum.

It remains to investigate the behavior of $f(x,y)$ on the boundary of the square. Now, along the boundary that coincides with the x axis, $y = 0$, and $f(x,0) = \sin x + \sin 0 + \sin (x + 0) = 2 \sin x$. Thus, the extreme values of $f(x,0)$ are 2 and -2. It is obvious from the form of $f(x,y)$ that on the remaining sides of the square $f(x,y)$ also varies between 2 and -2, and we conclude that the extreme values of $f(x,y)$ in the given closed region are $3\sqrt{3}/2$ and $-3\sqrt{3}/2$.

Example 2.   Investigate the maxima and minima of the surface

$$\frac{x^2}{a^2} - \frac{y^2}{b^2} = 2cz.$$

Now, $$\frac{\partial z}{\partial x} = \frac{1}{c}\frac{x}{a^2} \qquad \frac{\partial z}{\partial y} = -\frac{1}{c}\frac{y}{b^2}$$

which vanish when $x = y = 0$. But

$$\frac{\partial^2 z}{\partial x^2} = \frac{1}{a^2 c} \qquad \frac{\partial^2 z}{\partial y^2} = -\frac{1}{b^2 c} \qquad \frac{\partial^2 z}{\partial x\,\partial y} = 0.$$

Hence, $D = 1/a^2 b^2 c^2$, and consequently, there is no maximum or minimum at $x = y = 0$. The surface under consideration is a saddle-shaped surface called a *hyperbolic paraboloid*. The points for which the first partial derivatives vanish and $D > 0$ are sometimes called *minimax*. The reason for this odd name appears from a consideration of the shape of the hyperbolic paraboloid near the origin of the coordinate system. The reader will benefit from sketching it in the vicinity of $(0,0,0)$.

─── **PROBLEMS**

1. Show that the function $f(x,y) = \sin x + \sin y + \sin (x + y)$, considered in Example 1, takes on a maximum at $(\pi/3,\pi/3)$ and a minimum at $(0,0)$ if $0 \le x \le \pi/2$, $0 \le y \le \pi/2$.

2. Show that, of all rectangular parallelepipeds with the given volume V, the cube has the least surface.

3. Show that, of all rectangular parallelepipeds with the given surface area A, the cube has the greatest volume.

4. If a positive number a is written in the form $a = x_1 + x_2 + x_3$, show that the product $y = x_1 x_2 x_3$ is a maximum when $x_1 = x_2 = x_3$.

5. If a positive number $a = x_1 x_2 x_3 x_4$ is represented as the product of positive factors x_i, the least value of a is attained when $x_1 = x_2 = x_3 = x_4 = \sqrt[4]{a}$. Show this.

6. If $x_1 x_2 x_3 \cdots x_n = 1$, where the factors x_i are positive, show that $x_1 + x_2 + \cdots + x_n \ge n$.

7. Show that, of all triangles inscribed in a given circle, the equilateral triangle has the greatest area. *Hint:* Let the central angles subtending the sides of the triangle be α, β, γ. Note that $\alpha + \beta + \gamma = 2\pi$, and show that the area of the triangle is proportional to $f(\alpha,\beta) = \sin \alpha + \sin \beta - \sin (\alpha + \beta)$.

8. Find the volume of the largest rectangular parallelepiped that can be inscribed in the ellipsoid

$$\frac{x^2}{a^2} + \frac{y^2}{b^2} + \frac{z^2}{c^2} = 1.$$

9. Find the dimensions of the largest rectangular parallelepiped that has three faces in the coordinate planes and one vertex in the plane

$$\frac{x}{a} + \frac{y}{b} + \frac{z}{c} = 1.$$

10. A pentagonal frame is composed of a rectangle surmounted by an isosceles triangle. What are the dimensions for maximum area of the pentagon if the perimeter is given as P?

11. A floating anchorage is designed with a body in the form of a right-circular cylinder with equal ends that are right-circular cones. If the volume is given, find the dimensions giving the minimum surface area.

12. Given n points P_i whose coordinates are (x_i, y_i, z_i) $(i = 1, 2, \ldots, n)$. Show that the coordinates of the point $P(x, y, z)$, such that the sum of the squares of the distances from P to the P_i is a minimum, are given by

$$\left(\frac{1}{n} \sum_{i=1}^{n} x_i, \frac{1}{n} \sum_{i=1}^{n} y_i, \frac{1}{n} \sum_{i=1}^{n} z_i \right)$$

---

12. Constrained Maxima and Minima. The discussion in the preceding section was confined to the calculation of the maximum and minimum values of functions of several independent variables. In a large number of investigations, it is required that the maximum and minimum values of a function $f(x_1, x_2, \ldots, x_n)$ be found when the variables x_i are connected by some functional relationships, so that the x_i are no longer independent. Such problems are called problems in *constrained* maxima to distinguish them from the problems in *free* maxima discussed in Sec. 11. As in Sec. 11, our concern here is with the relative maxima and minima at the interior points of the region. We assume that $f(x_1, x_2, \ldots, x_n)$ has continuous derivatives in the given region R and that the variables x_i satisfy one or more relations of the type $\varphi_j(x_1, x_2, \ldots, x_n) = 0$ $(j = 1, 2, \ldots, m < n)$, where the functions φ_j are also continuously differentiable in the interior of R.

To avoid circumlocution, we shall speak of the maximum or minimum values as the *extreme* values. Thus, let us consider the problem of finding the extreme values of the function

$$u = f(x, y, z) \tag{12-1}$$

in which the variables x, y, z are constrained by the relation

$$\varphi(x, y, z) = 0. \tag{12-2}$$

We naturally suppose that the point (x_0, y_0, z_0) at which the function $u = f(x, y, z)$ takes on its extreme value satisfies the constraining relation (12-2). This problem can be solved by the procedure of Sec. 11 as follows: Suppose that the constraining relation (12-2) is solved for one of the variables, say z, to yield a continuously differentiable function

$$z = \Phi(x, y). \tag{12-3}$$

This surely will be possible in some neighborhood of the point (x_0, y_0), if $\partial \varphi / \partial z \neq 0$ at (x_0, y_0, z_0). If one substitutes z from (12-3) in (12-1), there results the function

$$u = f[x, y, \Phi(x, y)] \equiv F(x, y) \tag{12-4}$$

of two independent variables x, y to which the considerations of Sec. 11 apply.

However, either the solution (12-3) may be difficult to obtain or the function $F(x, y)$ in (12-4) may be so unwieldy that the simultaneous equations $F_x(x, y) = 0$, $F_y(x, y) = 0$ are unpleasant to deal with. In this event an ingenious method devised by the great French analyst Lagrange often leads to a manageable and symmetric system of equations for the determination of extreme values. The central idea of the method hinges on the following observation: In Sec. 11, we saw that a necessary condition for a relative extremum of the differentiable function $f(x_1, x_2, \ldots, x_n)$ of n independent variables is the simultaneous vanishing of all partial derivatives f_{x_i} at the point where the extremum is attained.   Inasmuch as the total differential of f is

$$df = f_{x_1} \, dx_1 + f_{x_2} \, dx_2 + \cdots + f_{x_n} \, dx_n$$

it is clear that $df = 0$ whenever each $f_{x_i} = 0$.   Conversely, if $df = 0$, the partial derivatives f_{x_i} vanish, since the dx_i are independent.   But it is also true that the vanishing of the total differential is a necessary condition for an extremum of $f(x_1, x_2, \ldots, x_n)$ even when the variables x_i are dependent because of the invariant character of df established in Sec. 5.   We can thus state a theorem:

THEOREM.   A necessary condition for an extremum of a continuously differentiable function $f(x_1, x_2, \ldots, x_n)$ is the vanishing of its total differential at the maximum and minimum points of the function.

We proceed now to a discussion of the *method of Lagrange multipliers* for determining the extreme values of the function in (12-1) subject to the equation of constraint (12-2).

By the theorem just stated, the differential of (12-1) vanishes at the critical points so that

$$\frac{\partial f}{\partial x} \, dx + \frac{\partial f}{\partial y} \, dy + \frac{\partial f}{\partial z} \, dz = 0. \tag{12-5}$$

Also, since $\varphi(x,y,z) = 0$, its total differential vanishes and we can write

$$\frac{\partial \varphi}{\partial x} \, dx + \frac{\partial \varphi}{\partial y} \, dy + \frac{\partial \varphi}{\partial z} \, dz = 0. \tag{12-6}$$

Let Eq. (12-6) be multiplied by some parameter λ and then added to (12-5).   The result is

$$\left(\frac{\partial f}{\partial x} + \lambda \frac{\partial \varphi}{\partial x} \right) dx + \left(\frac{\partial f}{\partial y} + \lambda \frac{\partial \varphi}{\partial y} \right) dy + \left(\frac{\partial f}{\partial z} + \lambda \frac{\partial \varphi}{\partial z} \right) dz = 0. \tag{12-7}$$

If we regard x and y as independent variables, and suppose that $\partial\varphi/\partial z \neq 0$ at the point where the extremum is attained, we can find a λ such that at this point

$$\frac{\partial f}{\partial z} + \lambda \frac{\partial \varphi}{\partial z} = 0. \tag{12-8}$$

With this choice of λ, Eq. (12-7) reduces to

$$\left(\frac{\partial f}{\partial x} + \lambda \frac{\partial \varphi}{\partial x} \right) dx + \left(\frac{\partial f}{\partial y} + \lambda \frac{\partial \varphi}{\partial y} \right) dy = 0.$$

But since dx and dy are independent increments, we conclude from this equation that

$$\frac{\partial f}{\partial x} + \lambda \frac{\partial \varphi}{\partial x} = 0 \qquad \frac{\partial f}{\partial y} + \lambda \frac{\partial \varphi}{\partial y} = 0. \tag{12-9}$$

The system of three equations (12-8) and (12-9) contains four unknowns x, y, z, λ, and we must adjoin to it the fourth equation (12-2) to obtain the complete system for the determination of the unknowns.

If $\partial\varphi/\partial z = 0$ at the point where the extremum is attained, but $\partial\varphi/\partial y \neq 0$, the roles of z and y in the foregoing discussion are interchanged.   Clearly, the method will fail to yield the desired value of λ when φ_x, φ_y, and φ_z vanish simultaneously at the point where $f(x,y,z)$ has an extremum.

Before proceeding to extend the Lagrange method to the study of extreme values of functions with several constraining conditions, we consider some instructive examples.

Example 1.   Find the maximum and the minimum distances from the origin to the curve

$$5x^2 + 6xy + 5y^2 - 8 = 0.$$

The problem here is to determine the extreme values of

$$f(x,y) = x^2 + y^2$$

subject to the condition

$$\varphi(x,y) \equiv 5x^2 + 6xy + 5y^2 - 8 = 0.$$

Equations (12-9) and (12-2) in this case read

$$2x + \lambda(10x + 6y) = 0$$
$$2y + \lambda(6x + 10y) = 0$$
$$5x^2 + 6xy + 5y^2 - 8 = 0.$$

Multiplying the first of these equations by y and the second by x and then subtracting give

$$6\lambda(y^2 - x^2) = 0$$

so that $y = \pm x$. Substituting these values of y in the third equation gives two equations for the determination of x, namely,

$$2x^2 = 1 \quad \text{and} \quad x^2 = 2.$$

The first of these gives $f \equiv x^2 + y^2 = 1$, and the second gives $f \equiv x^2 + y^2 = 4$. Obviously, the first value is a minimum, whereas the second is a maximum. The given curve is an ellipse of semiaxes 2 and 1 whose major axis makes an angle of 45° with the x axis.

Example 2.   Find the dimensions of the rectangular box, without a top, of maximum capacity whose surface is 108 in.²

    The function to be maximized is

$$f(x,y,z) \equiv xyz$$

subject to the condition

$$xy + 2xz + 2yz = 108. \tag{12-10}$$

Equations (12-8) and (12-9) yield

$$yz + \lambda(y + 2z) = 0$$
$$xz + \lambda(x + 2z) = 0 \tag{12-11}$$
$$xy + \lambda(2x + 2y) = 0.$$

To solve these equations, we multiply the first by x, the second by y, and the last by z, and add. There results

$$\lambda(2xy + 4xz + 4yz) + 3xyz = 0$$

or

$$\lambda(xy + 2xz + 2yz) + \tfrac{3}{2}xyz = 0.$$

Substituting from (12-10) gives

$$108\lambda + \tfrac{3}{2}xyz = 0$$

or

$$\lambda = -\frac{xyz}{72}.$$

Substituting this value of λ in (12-11) and dividing out common factors give

$$1 - \frac{x}{72}(y + 2z) = 0 \qquad 1 - \frac{y}{72}(x + 2z) = 0$$
$$1 - \frac{z}{72}(2x + 2y) = 0.$$

From the first two of these equations, it is evident that $x = y$. The substitution of $x = y$ in the third equation gives $z = 18/y$. Substituting for y and z in the first equation yields $x = 6$. Thus, $x = 6$, $y = 6$, and $z = 3$ give the desired dimensions.

Example 3.   Find the shortest distance from the origin to the curve $y = (x - 1)^{3/2}$ in Fig. 6. We apply the procedure employed in Example 1 to minimize

$$f(x,y) = x^2 + y^2 \tag{12-12}$$

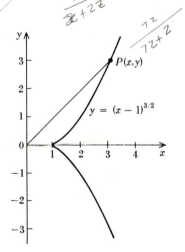

FIGURE 6

subject to the constraining condition

$$\varphi(x,y) \equiv y^2 - (x - 1)^3 = 0. \tag{12-13}$$

Equations (12-9) now yield

$$\begin{aligned} 2x - 3\lambda(x - 1)^2 &= 0 \\ 2y + 2\lambda y &= 0 \end{aligned} \tag{12-14}$$

which must be solved together with (12-13). The system (12-13) and (12-14) has no solutions for x, y, and λ. This becomes obvious on noting that the minimum is attained at $x = 1$, $y = 0$, and if we insert these values in (12-14), the first of the resulting equations yields a nonsensical result $2 = 0$ while the second is true for all values of λ. The reason that the Lagrange method this time has failed to give the solution is simple. The method depends on the assumption that not *both* φ_x and φ_y vanish at the point where the extremum is attained. In our case $\varphi_x(1,0) = 0$ and $\varphi_y(1,0) = 0$. The moral of this example is that the Lagrange method yields the solution of the problem only when the system of Eqs. (12-8) and (12-9) can be solved for λ.

PROBLEMS

1. Work Probs. 2, 3, 4, 5, 6, 8, and 9 of Sec. 11, by using Lagrange multipliers.

2. Prove that the point of intersection of the medians of a triangle possesses the property that the sum of the squares of its distances from the vertices is a minimum.

3. Find the maximum and the minimum of the sum of the angles made by a line from the origin with (*a*) the coordinate axes of a cartesian system and (*b*) the coordinate planes.

4. Find the maximum distance from the origin to the folium of Descartes $x^3 + y^3 - 3axy = 0$, $x \geq 0$, $y \geq 0$.

5. Find the shortest distance from the origin to the plane $ax + by + cz = d$.

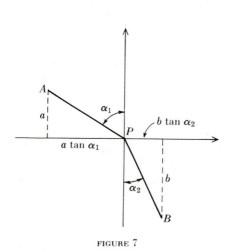

FIGURE 7

6. Find the points on the ellipse $x^2/4 + y^2/9 = 1$ that are at the greatest and least distances from the line $3x - y - 9 = 0$.

7. The Fermat principle in optics states that the path APB (Fig. 7) taken by a ray of light in passing across the plane separating two different optical media is such that the travel time t is a minimum. Use this principle to deduce the law of refraction $\sin \alpha_1/\sin \alpha_2 = v_1/v_2$, where α_1 is the angle of incidence, α_2 is the angle of refraction, and v_1 and v_2 are the speeds of light in the two media. *Hint:* Refer to Fig. 7 and show that the travel time $t = a/(v_1 \cos \alpha_1) + b/(v_2 \cos \alpha_2)$ must be minimized subject to the condition $a \tan \alpha_1 + b \tan \alpha_2 = $ const.

8. Three material points with masses m_1, m_2, m_3 are located at $P_1(x_1,y_1)$, $P_2(x_2,y_2)$, $P_3(x_3,y_3)$, respectively. Find the point $P(x,y)$ such that the sum of the squares of the distances of P from P_1, P_2, and P_3 is a minimum.

9. Heat loss Q in a wire with resistance R carrying steady current I is $Q = I^2R$. How must you branch the current into three wires with the resistances R_1, R_2, and R_3 if the heat loss is to be a minimum? *Hint:* $Q = \Sigma I_i^2 R_i$ and $I = \Sigma I_i$.

10. Find the sides of the triangle that has the least perimeter for a given area A. Note that $A = \sqrt{s(s - a)(s - b)(s - c)}$, where $s = \frac{1}{2}(a + b + c)$, a, b, and c being the sides of the triangle.

13. Lagrange Multipliers.

We now extend the considerations of Sec. 12 to cases where the extremum of a continuously differentiable function $f(x_1, x_2, \ldots, x_n)$ is sought under several conditions of constraint.

We consider first the function

$$w = f(x,y,u,v) \tag{13-1}$$

in which the variables are constrained by two relations

$$\varphi_1(x,y,u,v) = 0 \qquad \varphi_2(x,y,u,v) = 0. \tag{13-2}$$

We assume that in the neighborhood of the point $P_0(x_0,y_0,u_0,w_0)$, where $f(x,y,u,v)$ has an extremum, the functions f, φ_1, and φ_2 have continuous first partial derivatives and that $\varphi_1(x_0,y_0,u_0,v_0) = 0$, $\varphi_2(x_0,y_0,u_0,w_0) = 0$. We further suppose that the Jacobian

$$J(u,v) = \begin{vmatrix} \dfrac{\partial \varphi_1}{\partial u} & \dfrac{\partial \varphi_1}{\partial v} \\[2mm] \dfrac{\partial \varphi_2}{\partial u} & \dfrac{\partial \varphi_2}{\partial v} \end{vmatrix} \neq 0 \qquad \text{at } P_0 \tag{13-3}$$

so that Eqs. (13-2) can be solved (see Sec. 9) for u and v as functions of x, y in the neighborhood of (x_0,y_0).

If w takes on the extreme values for certain values of (x,y,u,v), then for such values

$$\frac{\partial f}{\partial x}\, dx + \frac{\partial f}{\partial y}\, dy + \frac{\partial f}{\partial u}\, du + \frac{\partial f}{\partial v}\, dv = 0 \tag{13-4}$$

by the theorem in the preceding section. Also, (13-2) yields two equations:

$$\frac{\partial \varphi_1}{\partial x}\, dx + \frac{\partial \varphi_1}{\partial y}\, dy + \frac{\partial \varphi_1}{\partial u}\, du + \frac{\partial \varphi_1}{\partial v}\, dv = 0$$

$$\frac{\partial \varphi_2}{\partial x}\, dx + \frac{\partial \varphi_2}{\partial y}\, dy + \frac{\partial \varphi_2}{\partial u}\, du + \frac{\partial \varphi_2}{\partial v}\, dv = 0.$$

We multiply the first of these by λ_1 and the second by λ_2, add the results to (13-4), and obtain

$$\left(\frac{\partial f}{\partial x} + \lambda_1 \frac{\partial \varphi_1}{\partial x} + \lambda_2 \frac{\partial \varphi_2}{\partial x} \right) dx + \left(\frac{\partial f}{\partial y} + \lambda_1 \frac{\partial \varphi_1}{\partial y} + \lambda_2 \frac{\partial \varphi_2}{\partial y} \right) dy$$

$$+ \left(\frac{\partial f}{\partial u} + \lambda_1 \frac{\partial \varphi_1}{\partial u} + \lambda_2 \frac{\partial \varphi_2}{\partial u} \right) du + \left(\frac{\partial f}{\partial v} + \lambda_1 \frac{\partial \varphi_1}{\partial v} + \lambda_2 \frac{\partial \varphi_2}{\partial v} \right) dv = 0. \tag{13-5}$$

Now, since at the points where the extreme values are attained (13-3) holds, we can find the values of λ_1 and λ_2 such that

$$\frac{\partial f}{\partial u} + \lambda_1 \frac{\partial \varphi_1}{\partial u} + \lambda_2 \frac{\partial \varphi_2}{\partial u} = 0$$

$$\frac{\partial f}{\partial v} + \lambda_1 \frac{\partial \varphi_1}{\partial v} + \lambda_2 \frac{\partial \varphi_2}{\partial v} = 0 \tag{13-6}$$

and, accordingly, (13-5) reduces to the sum of two terms involving arbitrary differentials dx and dy. The fact that they are arbitrary enables us to conclude that

$$\frac{\partial f}{\partial x} + \lambda_1 \frac{\partial \varphi_1}{\partial x} + \lambda_2 \frac{\partial \varphi_2}{\partial x} = 0$$

$$\frac{\partial f}{\partial y} + \lambda_1 \frac{\partial \varphi_1}{\partial y} + \lambda_2 \frac{\partial \varphi_2}{\partial y} = 0. \tag{13-7}$$

The system of six equations (13-6), (13-7), and (13-2) serves to determine the parameters λ_1, λ_2 and the point (x,y,u,v) at which the extremum is attained.

The foregoing procedure may be extended to cover the case of more than two constraining conditions to obtain the following rule:

RULE.   *In order to determine the extreme values of a continuously differentiable function*

$$f(x_1, x_2, \ldots, x_n) \tag{13-8}$$

whose variables are subjected to m continuously differentiable constraining relations

$$\varphi_i(x_1, x_2, \ldots, x_n) = 0 \qquad i = 1, 2, \ldots, m \tag{13-9}$$

form the function

$$F = f + \sum_{i=1}^{m} \lambda_i \varphi_i$$

and determine the parameters λ_i and the values of $x_1, x_2, \ldots, x_n$ from the n equations

$$\frac{\partial F}{\partial x_j} = 0 \qquad j = 1, 2, \ldots, n \tag{13-10}$$

and the m equations (13-9).

It should be carefully noted that the applicability of this rule to specific problems depends on the possibility of determining the multipliers λ_i.   The existence of the λ_i was established above only under the hypothesis that the appropriate Jacobian $J \neq 0$, which ensures the existence of solutions of the system of Eqs. (13-9) for some m variables x_i in the neighborhood of the extremum (see Sec. 9).

Example.   As an illustration, consider the problem of determining the maximum and the minimum distances from the origin to the curve of intersection of the ellipsoid

$$\frac{x^2}{a^2} + \frac{y^2}{b^2} + \frac{z^2}{c^2} = 1$$

with the plane

$$Ax + By + Cz = 0.$$

The square of the distance from the origin to any point (x,y,z) is

$$f = x^2 + y^2 + z^2$$

and it is necessary to find the extreme values of this function when the point (x,y,z) is common to the ellipsoid and the plane.   The constraining relations are, therefore,

$$\varphi_1 \equiv \frac{x^2}{a^2} + \frac{y^2}{b^2} + \frac{z^2}{c^2} - 1 = 0 \tag{a}$$

and

$$\varphi_2 \equiv Ax + By + Cz = 0. \tag{b}$$

The function $F = f + \lambda_1 \varphi_1 + \lambda_2 \varphi_2$ is, in this case,

$$F = x^2 + y^2 + z^2 + \lambda_1 \left(\frac{x^2}{a^2} + \frac{y^2}{b^2} + \frac{z^2}{c^2} - 1 \right) + 2\lambda_2(Ax + By + Cz)$$

where the factor of 2 is introduced in the last term for convenience.   Equations (13-10) then become

$$x + \lambda_1 \frac{x}{a^2} + \lambda_2 A = 0 \qquad y + \lambda_1 \frac{y}{b^2} + \lambda_2 B = 0 \qquad z + \lambda_1 \frac{z}{c^2} + \lambda_2 C = 0. \tag{c}$$

These equations, together with (a) and (b), give five equations for the determination of the five unknowns x, y, z, λ_1, and λ_2.   If the first, second, and third of Eqs. (c) are multiplied by x, y, and z, respectively, and then added, there results

$$x^2 + y^2 + z^2 + \lambda_1\left(\frac{x^2}{a^2} + \frac{y^2}{b^2} + \frac{z^2}{c^2}\right) + \lambda_2(Ax + By + Cz) = 0.$$

Making use of (a) and (b), it is evident that

$$\lambda_1 = -(x^2 + y^2 + z^2) = -f.$$

Setting this value of λ_1 in the first of Eqs. (c) gives

$$x\left(1 - \frac{f}{a^2}\right) + \lambda_2 A = 0 \quad \text{or} \quad x = -\frac{\lambda_2 A a^2}{a^2 - f}.$$

We find similarly from the remaining equations in (c)

$$y = -\frac{\lambda_2 B b^2}{b^2 - f} \quad \text{and} \quad z = -\frac{\lambda_2 C c^2}{c^2 - f}.$$

When these values of x, y, and z are substituted in (b), we obtain

$$\frac{A^2 a^2}{a^2 - f} + \frac{B^2 b^2}{b^2 - f} + \frac{C^2 c^2}{c^2 - f} = 0$$

from which f can be readily determined by solving the quadratic equation in f.

PROBLEMS

1. Find the points on the ellipse $x^2/a^2 + y^2/b^2 = 1$ at which the normal line to the ellipse is at the greatest distance from the origin.

2. What are the dimensions of the rectangular parallelepiped of maximum volume if the sum of the lengths of its edges is s?

3. Show that the cylinder with the greatest surface that can be inscribed in a sphere of radius a has the altitude $h = a\sqrt{2 - 2/\sqrt{5}}$ and the radius of the base $r = (a/2)\sqrt{2 + 2\sqrt{5}}$.

4. If x, y, z are positive, show that $\frac{1}{3}(x + y + z) \geq \sqrt[3]{xyz}$. What theorem concerning the arithmetic and geometric means is suggested by this problem? Can you prove it? *Hint:* Maximize $u = xyz$ subject to $x + y + z = a$.

5. Find the triangle of minimum perimeter that can be inscribed in a given triangle.

INTEGRALS WITH SEVERAL VARIABLES

14. Differentiation under the Integral Sign. The fundamental theorem of integral calculus states that whenever $f(x)$ is a continuous function in the closed interval (a,b) and $F(x)$ is any function such that $F'(x) = f(x)$, then

$$\int_{u_0}^{u_1} f(x)\, dx = F(u_1) - F(u_0) \tag{14-1}$$

for any two points u_0 and u_1 in the interval. If u_0 and u_1 are differentiable functions of another variable α, so that

$$u_0 = u_0(\alpha) \qquad u_1 = u_1(\alpha) \qquad \alpha_0 \leq \alpha \leq \alpha_1$$

the right-hand member in (14-1) is a function of α and the chain rule gives

$$\frac{dF(u_1)}{d\alpha} = F'(u_1)\frac{du_1}{d\alpha} = f(u_1)\frac{du_1}{d\alpha}.$$

Since a similar result holds for $F(u_0)$, differentiation of (14-1) yields the important formula

$$\frac{d}{d\alpha} \int_{u_0(\alpha)}^{u_1(\alpha)} f(x) \, dx = f(u_1) \frac{du_1}{d\alpha} - f(u_0) \frac{du_0}{d\alpha}. \tag{14-2}$$

If the variable α in (14-1) occurs under the integral sign, so that the integral takes the form

$$\varphi(\alpha) = \int_{u_0}^{u_1} f(x,\alpha) \, dx \tag{14-3}$$

we can compute the derivative of $\varphi(\alpha)$ by calculating the limit of the difference quotient $\Delta\varphi/\Delta\alpha$ as $\Delta\alpha \to 0$ whenever $f(x,\alpha)$ is continuous and has a continuous partial derivative $f_\alpha(x,\alpha)$ in the rectangle $u_0 \le x \le u_1$, $\alpha_0 \le \alpha \le \alpha_1$. This calculation is simple when the limits u_0, u_1 are constant. Indeed, in this case (14-1) gives

$$\Delta\varphi = \varphi(\alpha + \Delta\alpha) - \varphi(\alpha) = \int_{u_0}^{u_1} f(x, \, \alpha + \Delta\alpha) \, dx - \int_{u_0}^{u_1} f(x,\alpha) \, dx$$

$$= \int_{u_0}^{u_1} [f(x, \, \alpha + \Delta\alpha) - f(x,\alpha)] \, dx.$$

Dividing by $\Delta\alpha$ and taking the limit as $\Delta\alpha \to 0$ give

$$\varphi'(\alpha) \equiv \lim_{\Delta\alpha \to 0} \frac{\varphi(\alpha + \Delta\alpha) - \varphi(\alpha)}{\Delta\alpha} = \lim_{\Delta\alpha \to 0} \int_{u_0}^{u_1} \frac{f(x, \, \alpha + \Delta\alpha) - f(x,\alpha)}{\Delta\alpha} \, dx \tag{14-4}$$

provided the limit on the right exists.

If we knew that

$$\lim_{\Delta\alpha \to 0} \int_{u_0}^{u_1} \frac{f(x, \, \alpha + \Delta\alpha) - f(x,\alpha)}{\Delta\alpha} \, dx = \int_{u_0}^{u_1} \lim_{\Delta\alpha \to 0} \frac{f(x, \, \alpha + \Delta\alpha) - f(x)}{\Delta\alpha} \, dx \tag{14-5}$$

then the right-hand member of (14-4) would give

$$\int_{u_0}^{u_1} \frac{\partial f}{\partial \alpha} \, dx$$

by the definition of partial derivative. We could then conclude that

$$\frac{d}{d\alpha} \int_{u_0}^{u_1} f(x,\alpha) \, dx = \int_{u_0}^{u_1} f_\alpha(x,\alpha) \, dx \qquad u_0, \, u_1 \text{ const.} \tag{14-6}$$

Interchanging an integral with a limit operation as in (14-5) is not valid in general,[1] but the equality of (14-5) can, in fact, be justified when $f_\alpha(x,\alpha)$ is continuous, and hence (14-6) holds in that case.

Equation (14-2) requires that the integrand be independent of α, while (14-6) assumes that the limits of integration are independent of α.

When the limits and also the integrand depend on α, it can be shown[2] that the correct formula is given by addition of (14-2) and (14-6), namely,

$$\frac{d}{d\alpha} \int_{u_0(\alpha)}^{u_1(\alpha)} f(x,\alpha) \, dx = f(u_1,\alpha) \frac{du_1}{d\alpha} - f(u_0,\alpha) \frac{du_0}{d\alpha} + \int_{u_0(\alpha)}^{u_1(\alpha)} f_\alpha(x,\alpha) \, dx \tag{14-7}$$

provided that $u_0(\alpha)$ and $u_1(\alpha)$ are differentiable and $f(x,\alpha)$ and $f_\alpha(x,\alpha)$ are continuous. The formula (14-7), known as *Leibniz's formula*, will now be illustrated by several examples.

[1] The reader can verify that

$$\lim_{\alpha \to 0} \int_0^1 \frac{1}{\log \alpha} \frac{2x}{x^2 + \alpha^2} \, dx \ne \int_0^1 \lim_{\alpha \to 0} \frac{1}{\log \alpha} \frac{2x}{x^2 + \alpha^2} \, dx.$$

[2] See Sokolnikoff, "Advanced Calculus," pp. 121–122.

Example 1.   Evaluate $\dfrac{d}{d\alpha} \displaystyle\int_0^1 \log (x^2 + \alpha^2)\, dx$.   Inasmuch as the limits are constants, (14-6) yields, when $\alpha \neq 0$,

$$\frac{d}{d\alpha} \int_0^1 \log (x^2 + \alpha^2)\, dx = \int_0^1 \frac{2\alpha}{x^2 + \alpha^2}\, dx.$$

The resulting integral is easily evaluated by the fundamental theorem of integral calculus, since

$$\frac{d}{dx}\left(2 \tan^{-1} \frac{x}{\alpha}\right) = \frac{2\alpha}{x^2 + \alpha^2}.$$

We thus obtain

$$\frac{d}{d\alpha} \int_0^1 \log (x^2 + \alpha^2)\, dx = 2 \tan^{-1} \frac{x}{\alpha}\Big|_0^1 = 2 \tan^{-1} \frac{1}{\alpha}.$$

Example 2.   If $\varphi(x) = \int_0^{x^{1/3}} x^2\, dx$, find $\varphi'(x)$ first by evaluating the integral and then differentiating, and also by the Leibniz rule.   To avoid confusing the parameter x appearing in the limits of the integral with the variable of integration x, we write

$$\varphi(x) = \int_0^{x^{1/3}} t^2\, dt = \tfrac{1}{3} t^3 \Big|_0^{x^{1/3}} = \tfrac{1}{3} x.$$

Hence $\varphi'(x) = \tfrac{1}{3}$.   On the other hand, the application of the rule (14-2) yields

$$\varphi'(x) = \frac{d}{dx} \int_0^{x^{1/3}} t^2\, dt = (x^{1/3})^2 \frac{dx^{1/3}}{dx} = \frac{1}{3}$$

thus checking the result previously obtained.

Example 3.   Find $d\varphi/d\alpha$ if $\varphi(\alpha) = \int_{-\alpha^2}^{2\alpha} e^{-x^2/\alpha^2}\, dx$.   Since the integrand and the limits in this integral are functions of α, we use formula (14-7).   Then

$$\frac{d\varphi}{d\alpha} = \int_{-\alpha^2}^{2\alpha} \frac{2x^2}{\alpha^3} e^{-x^2/\alpha^2}\, dx + e^{-4}(2) - e^{-\alpha^2}(-2\alpha)$$

$$= \int_{-\alpha^2}^{2\alpha} \frac{2x^2}{\alpha^3} e^{-x^2/\alpha^2}\, dx + 2e^{-4} + 2\alpha e^{-\alpha^2}.$$

The integral appearing in this expression cannot be evaluated in a closed form in terms of elementary functions, but it can be readily computed in infinite series (see Chap. 1, Sec. 12).

Example 4.   Formula (14-6) can sometimes be used to evaluate definite integrals.   Thus consider

$$\varphi(\alpha) = \int_0^1 \frac{x^\alpha - 1}{\log x}\, dx \qquad \alpha \geq 0. \tag{14-8}$$

Differentiating under the integral sign, we get

$$\varphi'(\alpha) = \int_0^1 \frac{x^\alpha \log x}{\log x}\, dx = \int_0^1 x^\alpha\, dx.$$

The evaluation of the integral is easy, and we find

$$\varphi'(\alpha) = \frac{x^{\alpha+1}}{\alpha + 1}\Big|_0^1 = \frac{1}{\alpha + 1}.$$

Integrating again we get

$$\varphi(\alpha) = \log (\alpha + 1) + c. \tag{14-9}$$

To evaluate the constant c, we note that, for $\alpha = 0$, (14-8) gives $\varphi(0) = 0$ while (14-9) for $\alpha = 0$ requires that $\varphi(0) = \log 1 + c$.   Hence $c = 0$.   We finally have

$$\int_0^1 \frac{x^\alpha - 1}{\log x}\, dx = \log (\alpha + 1).$$

PROBLEMS

---

1. Find $d\varphi/d\alpha$ if $\varphi(\alpha) = \int_0^{\pi/2} \sin \alpha x \, dx$ by using the Leibniz formula, and check your result by direct calculation.

2. Find $d\varphi/d\alpha$ if $\varphi(\alpha) = \int_0^\pi (1 - \alpha \cos x)^2 \, dx$.   **3.** Find $d\varphi/d\alpha$ if $\varphi(\alpha) = \int_0^{\alpha^2} \tan^{-1} \frac{x}{\alpha^2} \, dx$.

4. Find $d\varphi/d\alpha$ if $\varphi(\alpha) = \int_0^\alpha \tan (x - \alpha) \, dx$.

5. Find $d\varphi/dx$ if $\varphi(x) = \int_0^{x^2} \sqrt{x} \, dx$.

6. Show in the manner of Example 4 that

$$\varphi(\alpha) = \int_0^\pi \log (1 + \alpha \cos x) \, dx = \pi \log \frac{1 + \sqrt{1 - \alpha^2}}{2} \qquad \text{if } \alpha^2 < 1.$$

7. Differentiate under the integral sign, and thus evaluate $\int_0^\pi \dfrac{dx}{(\alpha - \cos x)^2}$ by using

$$\int_0^\pi \frac{dx}{\alpha - \cos x} = \frac{\pi}{(\alpha^2 - 1)^{1/2}} \qquad \text{if } \alpha^2 > 1.$$

8. Show that

$$\int_0^\pi \log (1 - 2\alpha \cos x + \alpha^2) \, dx = \begin{cases} 0 & \text{if } \alpha^2 < 1 \\ \pi \log \alpha^2 & \text{if } \alpha^2 > 1. \end{cases}$$

9. Verify that

$$y = \frac{1}{k} \int_0^x f(\alpha) \sin k(x - \alpha) \, d\alpha$$

is a solution of the differential equation

$$\frac{d^2 y}{dx^2} + k^2 y = f(x)$$

where k is a constant.

10. If $b > 0$ and $\alpha > 0$, $\int_0^b \dfrac{dx}{1 + \alpha x} = \dfrac{1}{\alpha} \log (1 + \alpha b)$.   Differentiate this integral with respect to α and show that

$$\int_0^b \frac{x \, dx}{(1 + \alpha x)^2} = \frac{1}{\alpha^2} \log (1 + \alpha b) - \frac{b}{\alpha(1 + \alpha b)}.$$

11. Find the curvature at $x = 1$ of the curve $y = y(x)$ defined by the integral

$$y(x) = \int_\pi^{2\pi} \frac{\sin \alpha x}{\alpha} \, d\alpha.$$

---

15. The Calculus of Variations.   Physical laws can often be deduced from concise mathematical principles to the effect that certain integrals attain extreme values.   Thus, the Fermat principle of optics asserts that the actual path traversed by the light particle is such that the integral representing the travel time between two points in every medium is a minimum.   Also a considerable part of mechanics can be deduced from the principle of minimum potential energy, stating that the equilibrium configuration of a mechanical system corresponds to the minimum value of a certain integral related to the work done on the system by the forces acting on it.   For example, the shape assumed by a flexible chain fixed between two fixed points is such that its center of gravity is as low as possible.

To say that the center of gravity is as low as possible is equivalent to saying that the potential energy of the system is as small as possible.

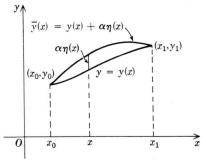

The problems concerned with the determination of extreme values of integrals whose integrands contain unknown functions belong to the *calculus of variations*. The simplest of such problems concerns the determination of an unknown function $y = y(x)$ for which the integral

FIGURE 8

$$I = \int_{x_0}^{x_1} F(x,y,y')\, dx \qquad (15\text{-}1)$$

between two fixed points $P_0(x_0,y_0)$ and $P_1(x_1,y_1)$ is a minimum.   The integrable function F of the variables x, y, and $y' \equiv dy/dx$ is assumed to be known.

If we imagine that the points P_0 and P_1 in the xy plane are joined by a sufficiently smooth curve $y = f(x)$, the substitution of $y = f(x)$ and $y' = f'(x)$ in the integrand of (15-1) yields the integral $I(f)$ whose value, ordinarily, depends on the choice of the curve $y = f(x)$.   We ask the question: What is the equation of the curve $y = y(x)$ joining P_0 and P_1 which makes the value of the integral (15-1) a minimum?   To be certain that this question makes sense, it is necessary to impose some restrictions on the integrand in (15-1) and to specify how the curves that enter in competition for the minimum value of I are to be chosen.

We shall suppose that $F(x,y,y')$, viewed as a function of its arguments x, y, and y', has continuous partial derivatives of the second order, and we assume that there is a curve $y = y(x)$ with continuously turning tangent that minimizes the integral.   We then choose the competing family of curves as follows: Let $y = \eta(x)$ be any function with continuous second derivatives which vanishes at the end points of the interval (x_0,x_1). Then

$$\eta(x_0) = 0 \qquad \eta(x_1) = 0. \qquad (15\text{-}2)$$

If α is a small parameter,

$$\bar{y}(x) = y(x) + \alpha\eta(x) \qquad (15\text{-}3)$$

represents a family of curves passing through (x_0,y_0) and (x_1,y_1), since the *minimizing curve* $y = y(x)$ passes through these points and $\eta(x_0) = \eta(x_1) = 0$.   The situation here is that indicated in Fig. 8.   The vertical deviation of a curve in the family (15-3) from the minimizing curve is $\alpha\eta(x)$; it is called the *variation of $y(x)$*.

Now if we substitute $\bar{y}$ and $\bar{y}'$ from (15-3) for y and y' in the integral (15-1), we get a function of α:

$$I(\alpha) = \int_{x_0}^{x_1} F[x, y(x) + \alpha\eta(x), y'(x) + \alpha\eta'(x)]\, dx. \qquad (15\text{-}4)$$

For $\alpha = 0$, Eq. (15-3) yields $\bar{y}(x) = y(x)$, and since $y = y(x)$ minimizes the integral, we conclude that $I(\alpha)$ must have a minimum for $\alpha = 0$.   A necessary condition for this is

$$\left.\frac{dI}{d\alpha}\right|_{\alpha=0} = 0. \qquad (15\text{-}5)$$

We can compute the derivative of $I(\alpha)$ by differentiating (15-4) under the integral sign and get

$$I'(\alpha) = \int_{x_0}^{x_1} \frac{\partial}{\partial\alpha} F(x,Y,Y')\, dx \qquad (15\text{-}6)$$

where we have set

$$Y \equiv y(x) + \alpha\eta(x) \qquad Y' \equiv y'(x) + \alpha\eta'(x). \tag{15-7}$$

But by the rule for the differentiation of composite functions,

$$\frac{\partial F(x,Y,Y')}{\partial \alpha} = \frac{\partial F}{\partial Y}\frac{\partial Y}{\partial \alpha} + \frac{\partial F}{\partial Y'}\frac{\partial Y'}{\partial \alpha} = \frac{\partial F}{\partial Y}\eta(x) + \frac{\partial F}{\partial Y'}\eta'(x)$$

so that (15-6) can be written as

$$I'(\alpha) = \int_{x_0}^{x_1}\left[\frac{\partial F}{\partial Y}\eta(x) + \frac{\partial F}{\partial Y'}\eta'(x)\right]dx. \tag{15-8}$$

Since $I'(0) = 0$ by (15-5), we get, on setting $\alpha = 0$ in (15-8),

$$\int_{x_0}^{x_1}\left[\frac{\partial F}{\partial y}\eta(x) + \frac{\partial F}{\partial y'}\eta'(x)\right]dx = 0 \tag{15-9}$$

because for $\alpha = 0$ it is evident from (15-7) that $Y = y(x)$, $Y' = y'(x)$.

The second term in the integral (15-9) can be integrated by parts to yield

$$\int_{x_0}^{x_1}\frac{\partial F}{\partial y'}\eta'(x)\,dx = \frac{\partial F}{\partial y'}\eta(x)\Big|_{x_0}^{x_1} - \int_{x_0}^{x_1}\eta(x)\frac{d}{dx}\left(\frac{\partial F}{\partial y'}\right)dx = -\int_{x_0}^{x_1}\eta(x)\frac{d}{dx}\left(\frac{\partial F}{\partial y'}\right)dx$$

since the integrated part drops out because of (15-2). Accordingly, we can write (15-9) as

$$\int_{x_0}^{x_1}\eta(x)\left[\frac{\partial F}{\partial y} - \frac{d}{dx}\left(\frac{\partial F}{\partial y'}\right)\right]dx = 0. \tag{15-10}$$

But $\eta(x)$ is an arbitrary function vanishing at the end points of the interval. Since the integral (15-10) must vanish for *every* choice of $\eta(x)$, it is easy to conclude that[1]

$$\frac{\partial F}{\partial y} - \frac{d}{dx}\left(\frac{\partial F}{\partial y'}\right) = 0. \tag{15-11}$$

This equation is called the *Euler equation*. On carrying out the differentiation indicated in (15-11), we get the second-order ordinary differential equation

$$\frac{\partial F}{\partial y} - \frac{\partial^2 F}{\partial x\,\partial y'} - \frac{\partial^2 F}{\partial y\,\partial y'}y' - \frac{\partial^2 F}{\partial y'^2}y'' = 0 \tag{15-12}$$

for the determination of the minimizing function $y(x)$.

The general solution of (15-12) contains two arbitrary constants which must be chosen so that the curve $y = y(x)$ passes through (x_0,y_0) and (x_1,y_1).

It should be noted that the solution of Euler's equation (15-11) may not yield the minimizing curve because the condition (15-5) is necessary but not sufficient for a minimum. Ordinarily one must verify whether or not this solution yields the curve that actually minimizes the integral, but frequently geometrical or physical considerations enable one to tell whether the curve so obtained makes the integral a maximum, a minimum, or neither.

Similar calculations when performed on the integral

$$I(y) = \int_{x_0}^{x_1} F(x,y,y',y'',\ldots,y^{(n)})\,dx$$

[1] The proof is by contradiction. Assume that the function in the brackets of (15-10) is not zero at some point $x = \xi$ of the interval (x_0,x_1). Then, since it is a continuous function, there will be a subinterval l about $x = \xi$ throughout which $F_y - (d/dx)F_{y'}$ has the same sign as at $x = \xi$. Choose $\eta(x)$ so that it has the same sign as $F_y - (d/dx)F_{y'}$ in l and vanishes outside this subinterval. For such a choice of $\eta(x)$, the integrand in (15-10) will be positive, and thus the integral will fail to vanish as demanded by (15-10).

yield the Euler equation

$$F_y - \frac{d}{dx} F_{y'} + \frac{d^2}{dx^2} F_{y''} - \cdots - (-1)^n \frac{d^n}{dx^n} F_{y^{(n)}} = 0$$

where the subscripts are used to denote partial derivatives.

Example. What is the equation of the curve $y = y(x)$ for which the area of the surface of revolution got by revolving the curve about the x axis is a minimum?

The integral to be minimized in this problem is

$$I = 2\pi \int_{x_0}^{x_1} y \, ds = 2\pi \int_{x_0}^{x_1} y\sqrt{1 + y'^2} \, dx. \tag{15-13}$$

It has the form (15-1) with

$$F(x,y,y') = 2\pi y\sqrt{1 + y'^2}. \tag{15-14}$$

The substitution from (15-14) in the Euler equation (15-11), after simple calculations, yields

$$\sqrt{1 + y'^2} - \frac{d}{dx} \frac{yy'}{\sqrt{1 + y'^2}} = 0$$

or

$$yy'' - y'^2 - 1 = 0$$

This second-order equation is easily solved by setting $y' = p$, $y'' = p \, dp/dy$ (cf. Chap. 2, Sec. 6). The result is

$$y = c_1 \cosh \frac{x - c_2}{c_1} \tag{15-15}$$

so that the desired curve is a catenary. The integration constants c_1 and c_2 in the general solution (15-15) must be determined so that the curve passes through given points (x_0,y_0), (x_1,y_1).

Frequently one seeks a maximum or minimum value of the integral (15-1) subject to the condition that another integral

$$J = \int_{x_0}^{x_1} G(x,y,y') \, dx \tag{15-16}$$

have a known constant value. A physical problem of this sort occurs in finding the shape of the chain that minimizes the potential energy while the length of the chain is given. This is one of the so-called "isoperimetric" problems of the calculus of variations.[1]

It is natural to attempt to solve the problem $I = \min$ subject to the condition $J = \text{const}$ by the method of Lagrange multipliers. We construct the integral

$$I + \lambda J = \int_{x_0}^{x_1} [F(x,y,y') + \lambda G(x,y,y')] \, dx \tag{15-17}$$

and consider the free extremum of the integral (15-17). The corresponding Euler equation (15-11) is

$$\frac{\partial(F + \lambda G)}{\partial y} - \frac{d}{dx} \frac{\partial(F + \lambda G)}{\partial y'} = 0 \tag{15-18}$$

and, on carrying out the indicated differentiation[2] in (15-18), we get the second-order ordinary differential equation containing the parameter λ. The general solution of this equation, in addition to λ, will contain two arbitrary integration constants. The integration constants and the parameter λ must then be determined so that the curve $y = y(x)$ passes through the given end points and satisfies the constraining condition (15-16).

The justification of this procedure is based on an argument similar to that used

[1] Isoperimetric because the length (or the perimeter) of the curve is given.
[2] Compare Eq. (15-12).

above, where instead of the one-parameter family of the neighboring curves (15-3) one constructs a suitable two-parameter family.[1]

The sole object of this section is to give a glimpse into a topic that is of steadily growing importance in physics and engineering.[2]

── **PROBLEMS**

1. Show that the curve of minimum length joining a pair of given points in the plane is a straight line.   *Hint:* Minimize

$$\int_{x_0}^{x_1} \sqrt{1 + y'^2}\, dx.$$

2. Solve Prob. 1 by taking

$$\int ds = \int \sqrt{1 + r^2 \left(\frac{d\theta}{dr}\right)^2}\, dr.$$

3. When a bead slides from rest along any smooth curve C from the point P to a point Q on C, the speed v of the bead is $v = \sqrt{2gh}$, where h is the vertical distance from P to Q.   Hence the travel time from P to Q is $t = \int_P^Q \frac{ds}{v}$.   Choose P at the origin, and show that the curve for which the travel time is a minimum is a cycloid.   See Prob. 6.

4. Consider the integral $I = \int_{x_0}^{x_1} \frac{\sqrt{1 + y'^2}}{y}\, dx$, and show that the general solution of the associated Euler equation is $y^2 + (x - c_1)^2 = c_2$.   See Prob. 6 below and Prob. 4 in Chap. 2, Sec. 11.

5. Obtain Euler's equation for the integral

$$I(y) = \int_{x_0}^{x_1} [p(x)(y')^2 + q(x)y^2 + 2f(x)y]\, dx.$$

Special cases of this integral arise in the study of deflection of bars and strings.

6. Show that the Euler equation (15-11) can be written in the form $(d/dx)(F - y'F_{y'}) = F_x$, and conclude that if $F(x,y,y')$ does not contain x explicitly then the first integral of (15-11) is $F - y'F_{y'} = \text{const.}$

──

16. A Review of Double Integrals.   Change of Variables.   Let $f(x,y)$ be a piece-wise-continuous function defined in some closed bounded region R in the xy plane. The double integral $\int_R f(x,y)\, dA$ is defined, we recall, by forming the sum $I_n = \sum_{i=1}^{n} f(\bar{x}_i, \bar{y}_i)\, \Delta A_i$, in which the ΔA_i are the areas of the subregions ΔR_i into which R is subdivided in an arbitrary manner and $(\bar{x}_i, \bar{y}_i)$ is a point arbitrarily chosen in ΔR_i.   The limit of the sum I_n, when the number of subregions is allowed to increase indefinitely by making the linear dimensions of ΔR_i approach zero, is called the *double integral* $\int_R f(x,y)\, dA$.

When R is bounded by two continuous curves $y = g_1(x)$, $y = g_2(x)$, where $g_1(x) \leq g_2(x)$ and $a \leq x \leq b$ (Fig. 9a), the value of $\int_R f(x,y)\, dA$ can be determined by repeated evaluations of two ordinary integrals.   Thus,

[1] See G. A. Bliss, "Calculus of Variations," Carus Monograph, The Open Court Publishing Co., LaSalle, Ill., 1925.
[2] A readable account of this subject is given by L. A. Pars, "An Introduction to the Calculus of Variations," John Wiley & Sons, Inc., New York, 1962.

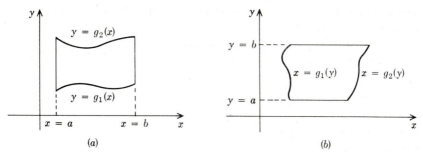

FIGURE 9

$$\int_R f(x,y)\ dA\ =\ \int_{x=a}^{x=b} dx \int_{y=g_1(x)}^{y=g_2(x)} f(x,y)\ dy \tag{16-1}$$

where in integrating with respect to y the variable x is treated as a fixed parameter.

When R is bounded by continuous curves $x = g_1(y)$, $x = g_2(y)$, where $g_1(y) \leq g_2(y)$, $a \leq y \leq b$ (Fig. 9b), the labels x and y in the repeated integrals in (16-1) are interchanged. Because of the additive property of integrals, formulas of the type of (16-1) can be used to evaluate double integrals over regions that can be subdivided into a finite number of subregions of the types depicted in Fig. 9.

The evaluation of double integrals can sometimes be simplified by a change of variables. Thus, if we set

$$u = u(x,y) \qquad v = v(x,y) \tag{16-2}$$

where $u(x,y)$ and $v(x,y)$ have continuous partial derivatives in R, the transformation (16-2) carries every point (x,y) of R into some point (u,v) in a certain region T in the uv plane (Fig. 10). If we further assume that the transformation (16-2) has a single-valued continuously differentiable inverse

$$x = x(u,v) \qquad y = y(u,v) \tag{16-3}$$

the correspondence between the points of the regions T and R will be one-to-one. The Fundamental Theorem for implicit functions given in Sec. 9 then ensures that the solution (16-3) of Eqs. (16-2) exists in some neighborhood of the point $P(u,v)$, whenever the Jacobian

$$J(u,v)\ =\ \begin{vmatrix} \dfrac{\partial x}{\partial u} & \dfrac{\partial x}{\partial v} \\[2mm] \dfrac{\partial y}{\partial u} & \dfrac{\partial y}{\partial v} \end{vmatrix} \neq 0 \qquad \text{at } P. \tag{16-4}$$

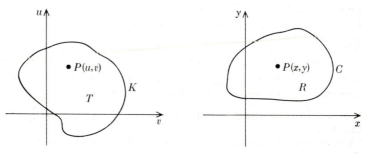

FIGURE 10

When the condition (16-4) is satisfied (except possibly at a finite number of points of T), it can also be shown that[1]

$$\iint\limits_{R} f(x,y)\,dy\,dx = \iint\limits_{T} f[x(u,v),y(u,v)]|J(u,v)|\,du\,dv. \tag{16-5}$$

An important special case of (16-3) is the transformation

$$x = \rho\cos\theta \qquad y = \rho\sin\theta \tag{16-6}$$

relating the polar coordinates (ρ,θ) to the cartesian coordinates (x,y). On substituting from (16-6) in (16-4) with $\rho = u$, $\theta = v$, we find $J(u,v) = \rho$, and thus (16-5) yields

$$\iint\limits_{R} f(x,y)\,dy\,dx = \iint\limits_{T} f(\rho\cos\theta,\ \rho\sin\theta)\rho\,d\rho\,d\theta \tag{16-7}$$

where T is the map of the region R by (16-6). When the boundary of T is described by simple equations in the $\rho\theta$ plane, the evaluation of (16-7) may prove simpler than the corresponding evaluation in the xy plane. Although the Jacobian of (16-6) vanishes when $\rho = 0$, formula (16-7) is valid even when $\rho = 0$ is a point in T because the value of the integral is not changed when its integrand is redefined at a finite number of points of the region.

As an illustration of the use of (16-5) we consider a few examples.

Example 1.   The volume V of one octant of the sphere $x^2 + y^2 + z^2 = a^2$ can be computed in cartesian coordinates by evaluating

$$V = \iint\limits_{R} f(x,y)\,dy\,dx = \int_{x=0}^{x=a} dx \int_{y=0}^{y=\sqrt{a^2-x^2}} \sqrt{a^2 - x^2 - y^2}\,dy.$$

Here R is one quadrant of the disk $x^2 + y^2 \le a^2$. The evaluation of this integral in the xy plane is a bit unpleasant. But the map T of R by (16-6) is the rectangle $0 \le \rho \le a$, $0 \le \theta \le \pi/2$ in the cartesian $\rho\theta$ plane. Accordingly, we can express V, by using (16-7), as

$$V = \int_{\theta=0}^{\theta=\pi/2} d\theta \int_{\rho=0}^{\rho=a} \sqrt{a^2 - \rho^2}\,\rho\,d\rho$$

which is easy to evaluate.

Example 2.   The moment of inertia I of the area of the circle $x^2 + y^2 - ax = 0$ about the x axis (Fig. 11) can be computed by a laborious evaluation of the integral

$$I = \iint\limits_{R} y^2\,dA = \int_{x=0}^{x=a} dx \int_{y=-\sqrt{ax-x^2}}^{y=\sqrt{ax-x^2}} y^2\,dy.$$

But the boundary of the given circle in polar coordinates is $\rho = a\cos\theta$, and by using (16-7) we get at once

$$I = \int_{\theta=0}^{\theta=\pi} d\theta \int_{\rho=0}^{\rho=a\cos\theta} \rho^2 \sin^2\theta\,\rho\,d\rho = \int_0^\pi \frac{a^4\cos^4\theta\sin^2\theta}{4}\,d\theta = \frac{\pi a^4}{64}.$$

Example 3.   A linear transformation

$$x = au + bv \qquad y = cu + dv \tag{16-8}$$

for which the Jacobian $J(u,v) = ad - bc \neq 0$, is often useful in evaluating double integrals over a closed region R bounded by straight lines. Since (16-8) is a linear transformation, a polygon in the xy plane maps into some polygon in the uv plane. When the region in the xy plane is a parallelogram, the constants in (16-8) can be chosen so that the map of the parallelogram is a

---

[1] The proof of this formula under slightly more stringent restrictions than those imposed here on the transformation (16-3) follows from Sec. 21. A less restrictive proof will be found in T. M. Apostol, "Mathematical Analysis," chap. 10, Addison-Wesley Publishing Company, Inc., Reading, Mass., 1957, and in several other books on advanced calculus.

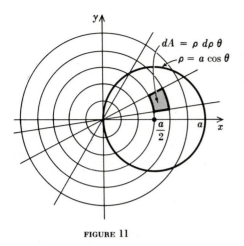

FIGURE 11

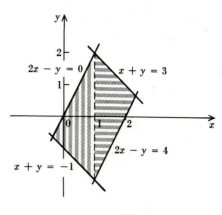

FIGURE 12

rectangle.   The transformation (16-8) may then prove useful in evaluating repeated integrals in (16-5) over the rectangular region T.

As an illustration, consider the integral

$$V = \iint_R (x + y)\, dA \tag{16-9}$$

where R is a parallelogram bounded by the lines $x + y = -1$, $x + y = 3$, $2x - y = 0$, $2x - y = 4$ (Fig. 12).

The region R can be subdivided into two shaded subregions of the type shown in Fig. 9, and the integral (16-9) expressed as the sum of two double integrals over these subregions.   It is simpler, however, to proceed as follows: Let us set

$$u = x + y \qquad v = 2x - y \tag{16-10}$$

or

$$x = \frac{u + v}{3} \qquad y = \frac{2u - v}{3}.$$

Under the transformation (16-10) the parallelogram in Fig. 12 maps into a rectangle T bounded by the lines $u = -1$, $u = 3$, $v = 0$, $v = 4$.   It is easily found that $J(u,v) = -\frac{1}{3}$, and we get from (16-5)

$$V = \iint_T u|-\tfrac{1}{3}|\, du\, dv = \tfrac{1}{3} \int_{-1}^{3} u\, du \int_0^4 dv = \tfrac{16}{3}.$$

─── **PROBLEMS**

1. Evaluate $\iint_R e^{-(x^2+y^2)}\, dy\, dx$, where R is the region bounded by the circle $x^2 + y^2 = a^2$.   Use polar coordinates.

2. Find the area outside the cardioid $\rho = a(1 + \cos\theta)$ and inside the circle $\rho = 3a \cos\theta$.

3. Find the coordinates of the center of gravity of the area between the circles $\rho = 2\sin\theta$ and $\rho = 4\sin\theta$.

4. Evaluate $\iint_R e^{(x+y)/(x-y)}\, dy\, dx$, where R is the triangular region bounded by $x = 0$, $y = 0$, $x + y = 1$.   *Hint:* Set $u = x + y$, $v = x - y$.   [*Ans.:* $\frac{1}{4}(e - e^{-1})$.]

5. Evaluate $\iint_R \sqrt{1 - x^2/a^2 - y^2/b^2}\, dy\, dx$, where R is the region bounded by the ellipse $x^2/a^2 + $

$y^2/b^2 = 1$.   (*Ans.:* $2\pi ab/3$.)   *Hint:* Set $x = a\rho \cos\theta$, $y = b\rho \sin\theta$, and note that the straight lines $\theta = $ const, $\rho = $ const in the cartesian $\rho\theta$ plane correspond to the families of radial lines and ellipses in the xy plane.

6. Use the transformation in Prob. 5 to evaluate the integral $\iint\limits_R xy\,dy\,dx$, where R is one quadrant of the ellipse $x^2/a^2 + y^2/b^2 = 1$, $x \ge 0$, $y \ge 0$.   (*Ans.:* $a^2b^2/8$.)

7. Let $I = \int_R f(\sqrt{x^2+y^2})\,dA = \int_0^2 dx \int_0^x f(\sqrt{x^2+y^2})\,dy$, where f is continuous.   Show that in polar coordinates $I = \int_0^{\pi/4} d\theta \int_0^{2\sec\theta} \rho f(\rho)\,d\rho$.

8. If R is the triangle bounded by $y = x$, $y = -x$, $y = 1$, show that

$$\iint\limits_R f(x,y)\,dA = \int_{\pi/4}^{3\pi/4} d\theta \int_0^{\csc\theta} \rho f(\rho\cos\theta, \rho\sin\theta)\,d\rho$$

if (ρ,θ) are polar coordinates and f is continuous.

9. If $u = x + y$, $v = x - y$, and $f(x,y)$ is continuous, show that

$$\int_0^1 dx \int_0^1 f(x,y)\,dy = \frac{1}{2}\left[\int_0^1 du \int_{-u}^u f\left(\frac{u+v}{2}, \frac{u-v}{2}\right)dv + \int_1^2 du \int_{u-2}^{2-u} f\left(\frac{u+v}{2}, \frac{u-v}{2}\right)dv\right].$$

Hint: Show that in the uv plane the region of integration is bounded by the lines $u = v$, $u + v = 2$, $u - v = 2$, and $u = -v$.

10. Obtain the element of area $dA = |J(u,v)|\,du\,dv$ if (*a*) $x = u + a$, $y = v + b$; (*b*) $x = au$, $y = bv$; (*c*) $x = u\cos\alpha - v\sin\alpha$, $y = u\sin\alpha + v\cos\alpha$, where a, b, and α are constants.   Interpret your results geometrically by examining the maps of rectangular regions.

---

17. Change of Variables in Triple Integrals.   The extension of the considerations of Sec. 16 to triple integrals is direct.   Thus, let R be a closed region (Fig. 13) bounded by a pair of surfaces $z = \varphi_1(x,y)$, $z = \varphi_2(x,y)$, and by the cylinder whose projection on the xy plane is a two-dimensional region R' of the type considered in Fig. 9a.   We suppose that $z = \varphi_1(x,y)$ and $z = \varphi_2(x,y)$ are continuous and $\varphi_1(x,y) \le \varphi_2(x,y)$ in R'.   Then the triple integral $\int_R f(x,y,z)\,d\tau$, where $d\tau$ is the volume element and $f(x,y,z)$ is piecewise continuous in R, can be evaluated by repeated integrations:

$$\int_R f(x,y,z)\,d\tau = \int_{R'} dA \left[\int_{z=\varphi_1(x,y)}^{z=\varphi_2(x,y)} f(x,y,z)dz\right]$$
$$= \int_{x=a}^{x=b} dx \int_{y=g_1(x)}^{y=g_2(x)} dy \int_{z=\varphi_1(x,y)}^{z=\varphi_2(x,y)} f(x,y,z)\,dz. \tag{17-1}$$

When the variables in this triple integral are changed by the transformation

$$x = x(u,v,w) \qquad y = y(u,v,w) \qquad z = z(u,v,w) \tag{17-2}$$

in which the functions in (17-2) are continuously differentiable and

$$J(u,v,w) = \begin{vmatrix} \dfrac{\partial x}{\partial u} & \dfrac{\partial x}{\partial v} & \dfrac{\partial x}{\partial w} \\[2mm] \dfrac{\partial y}{\partial u} & \dfrac{\partial y}{\partial v} & \dfrac{\partial y}{\partial w} \\[2mm] \dfrac{\partial z}{\partial u} & \dfrac{\partial z}{\partial v} & \dfrac{\partial z}{\partial w} \end{vmatrix} \ne 0 \tag{17-3}$$

in the region T into which R is mapped by (17-2), then [cf. (16-5)]

$$\int_R f(x,y,z)\, d\tau = \iiint_T f[x(u,v,w), y(u,v,w),$$

$$z(u,v,w)] |J(u,v,w)|\, du\, dv\, dw. \quad (17\text{-}4)$$

We illustrate the use of (17-4) by two examples.

Example 1.   The transformation from cylindrical coordinates (ρ,θ,z) to cartesian coordinates (x,y,z) is

$$x = \rho \cos\theta \qquad y = \rho \sin\theta \qquad z = z. \quad (17\text{-}5)$$

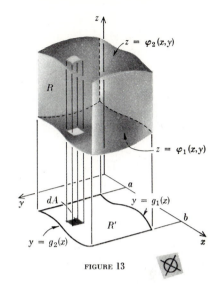

FIGURE 13

If we suppose that $\rho > 0$, $0 \le \theta < 2\pi$, and z is unrestricted, then (17-5) maps a region R, in the space of the variables (x,y,z), into a region T in the $\rho\theta z$ space, in a one-to-one manner. It is easy to check that the Jacobian (17-3) with $u = \rho$, $v = \theta$, $w = z$ is $J(\rho,\theta,z) = \rho$. Accordingly,

$$\iiint_R f(x,y,z)\, dx\, dy\, dz = \iiint_T f(\rho \cos\theta,\, \rho \sin\theta,\, z)\rho\, d\rho\, d\theta\, dz. \quad (17\text{-}6)$$

To illustrate the use of (17-6), let I_x represent the moment of inertia about the x axis of the solid cylinder $x^2 + y^2 \le a^2$ bounded by the planes $z = 0$, $z = b$. Then

$$I_x = \sigma \int_R (y^2 + z^2)\, d\tau \quad (17\text{-}7)$$

if the density σ of the solid is constant.

On making use of (17-6) we find

$$I_x = \sigma \int_0^a \rho\, d\rho \int_0^{2\pi} d\theta \int_0^b (\rho^2 \sin^2\theta + z^2)\, dz$$

which on integration gives $I_x = \sigma \pi a^2 b(3a^2 + 4b^2)/12$. The evaluation of (17-7) in cartesian coordinates is considerably more involved.

Example 2.   The transformation from spherical to cartesian coordinates is

$$x = \rho \sin\theta \cos\varphi \qquad y = \rho \sin\theta \sin\varphi \qquad z = \rho \cos\theta \quad (17\text{-}8)$$

and its inverse is single-valued if $\rho > 0$, $0 < \theta < \pi$, $0 \le \varphi < 2\pi$. The Jacobian (17-3) with $\rho = u$, $\theta = v$ $w = \varphi$ is $J(\rho,\theta,\varphi) = \rho^2 \sin\theta$, and (17-4) when expressed in spherical coordinates reads

$$\int_R f(x,y,z)\, d\tau = \iiint_T f(\rho \sin\theta \cos\varphi,\, \rho \sin\theta \sin\varphi,\, \rho \cos\theta)\rho^2 \sin\theta\, d\rho\, d\theta\, d\varphi \quad (17\text{-}9)$$

where T is the map of R by (17-8).

As an illustration, let us find the x coordinate of the center of gravity of the part of the solid sphere $x^2 + y^2 + z^2 = a^2$ in the first octant. We recall that the x coordinate of the center of gravity of a uniform solid is

$$\bar{x} = \frac{\displaystyle\int_R x\, d\tau}{\displaystyle\int_R d\tau}.$$

The value of the denominator in this expression can be written at once, since the solid under

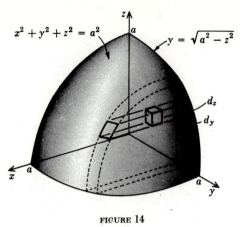

$$x^2 + y^2 + z^2 = a^2$$

$$y = \sqrt{a^2 - z^2}$$

FIGURE 14

consideration is one octant of a sphere of radius a. Thus $\int_R d\tau = (\tfrac{1}{8})4\pi a^3/3 = \pi a^3/6$. In cartesian coordinates, the numerator can be expressed as (Fig. 14)

$$\int_R x \, d\tau =$$

$$\int_{z=0}^{z=a} \int_{y=0}^{y=\sqrt{a^2-z^2}} \int_{x=0}^{x=\sqrt{a^2-y^2-z^2}} x \, dx \, dy \, dz.$$

Although the evaluation of this integral is not difficult, it is simpler to use the spherical coordinates. We find, from (17-9),

$$\int_R x \, d\tau =$$

$$\int_0^{\pi/2} d\varphi \int_0^{\pi/2} d\theta \int_0^a \rho \sin\theta \cos\varphi \rho^2 \sin\theta \, d\rho = \frac{\pi a^4}{16}.$$

Thus, $\bar{x} = (\pi a^4/16)/(\pi a^3/6) = 3a/8$.

PROBLEMS

1. Use cylindrical coordinates, defined by (17-5), to compute the moment of inertia of the volume of the right-circular cylinder of height h and radius a about the axis of the cylinder. Also, evaluate the integral in cartesian coordinates. (*Ans.:* $\tfrac{1}{2}\pi a^4/h$.)

2. Use cylindrical coordinates to find the volume enclosed by the circular cylinder $\rho = 2a\cos\theta$, the cone $z = \rho$, and the plane $z = 0$. (*Ans.:* $32a^3/9$.)

3. Use spherical coordinates to find the center of mass of the solid hemisphere $x^2 + y^2 + z^2 = a^2$, $z \geq 0$, if the density at any point is proportional to the distance of the point from the origin. (*Ans.:* $\bar{x} = 0$, $\bar{y} = 0$, $\bar{z} = 2a/5$.)

4. Show that the attraction of a homogeneous sphere at a point exterior to the sphere is the same as though all the mass of the sphere were concentrated at the center of the sphere. Assume the inverse-square law of force. See Chap. 6, Sec. 14.

5. The Newtonian potential V, due to a body T, at a point P is defined by the equation $V(P) = \int_T \dfrac{dm}{r}$, where dm is the element of mass of the body and r is the distance from the point P to the element of mass dm. Show that the potential of a homogeneous spherical shell of inner radius b and outer radius a is

$$V = 2\pi\sigma(a^2 - b^2) \qquad \text{if } r < b$$

and

$$V = \tfrac{4}{3}\pi\sigma \frac{a^3 - b^3}{r} \qquad \text{if } r > a$$

where σ is the density.

6. Find the Newtonian potential on the axis of a homogeneous circular cylinder of radius a.

7. Show that the force of attraction of a right-circular cone upon a point at its vertex is $2\pi\sigma h(1 - \cos\alpha)$, where h is the altitude of the cone and 2α is the angle at the vertex.

8. Show that the force of attraction of a homogeneous right-circular cylinder upon a point on its axis is

$$2\pi\sigma[h + \sqrt{R^2 + a^2} - \sqrt{(R + h)^2 + a^2}]$$

where h is altitude, a is radius, and R is the distance from the point to one base of the cylinder.

18. Surface Integrals. A surface is usually defined as a locus of points determined by the equation

$$z = f(x,y) \qquad (18\text{-}1)$$

where $f(x,y)$ is a continuous function specified ·in some closed region of the xy plane. This definition, however, is too broad to permit one to formulate a meaningful concept of the surface area. Since most surfaces encountered in applications are two-sided and piecewise smooth, we confine our considerations to such surfaces only.[1]

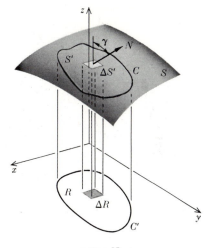

FIGURE 15

The surface defined by (18-1) is called *smooth* if it has continuous partial derivatives $\partial z/\partial x$ and $\partial z/\partial y$ at each of its points. This implies that a smooth surface has a continuously turning tangent plane and hence a well-defined normal at each of its points.[2]

It is intuitively clear that a small element of a smooth surface is nearly flat, so that a neighborhood of any point on it is well approximated by a portion of the tangent plane. This observation suggests a procedure for constructing a meaningful definition of the area of a smooth surface.

Thus, let S' be a smooth portion of the surface S bounded by a closed curve C (Fig. 15). We shall suppose that S' is such that every line parallel to some coordinate axis (say the z axis) cuts S' in just one point. If the projection C' of C on the xy plane encloses the region R, we can subdivide R into n small subregions ΔR_i by the families of straight lines parallel to the x and y axes. The planes through these lines, normal to the region R, cut from S' small regions $\Delta S_i'$ of areas $\Delta\sigma_i$. Let ΔA_i be the area of ΔR_i. The projection of $\Delta\sigma_i$ on the xy plane is, approximately,

$$\Delta A_i = \cos\gamma_i\,\Delta\sigma_i$$

where $\cos\alpha_i$, $\cos\beta_i$, and $\cos\gamma_i$ are the direction cosines of the normal N to S at a point (x_i,y_i,z_i) of $\Delta S_i'$. Since

$$\cos\alpha_i : \cos\beta_i : \cos\gamma_i = (z_x)_i : (z_y)_i : -1$$

where $z_x \equiv \partial z/\partial x$, $z_y \equiv \partial z/\partial y$, and

$$\cos^2\alpha_i + \cos^2\beta_i + \cos^2\gamma_i = 1$$

we have

$$\cos\gamma_i = \frac{-1}{\pm\sqrt{(z_x)_i{}^2 + (z_y)_i{}^2 + 1}}.$$

Using the positive value for $\cos\gamma_i$, which amounts to the choice of the positive direction of N, we can write

$$\Delta\sigma_i \doteq \sec\gamma_i\,\Delta A_i = \sqrt{(z_x)_i{}^2 + (z_y)_i{}^2 + 1}\;\Delta A_i.$$

---

[1] See also remarks about admissible surfaces in Chap. 6, Sec. 5.

[2] We recall that the equation of the tangent plane to (18-1) at a point $P_0(x_0,y_0,z_0)$ is

$$z - z_0 = \left(\frac{\partial z}{\partial x}\right)_P(x - x_0) + \left(\frac{\partial z}{\partial y}\right)_P(y - y_0)$$

so that the direction of the normal at P is determined by the ratios $(\partial z/\partial x)_P : (\partial z/\partial y)_P : -1$ [cf. Chap. 5, Eq. (7-16)].

The surface area of S' can then be approximated by the sum

$$\sum_{i=1}^{n} \Delta \sigma_i = \sum_{i=1}^{n} \sqrt{(z_x)_i{}^2 + (z_y)_i{}^2 + 1} \, \Delta A_i$$

and we define the area σ of S' by the double integral

$$\sigma = \int_R \sqrt{(z_x)^2 + (z_y)^2 + 1} \, dA \equiv \int_R \sec \gamma \, dA. \tag{18-2}$$

The integral (18-2) can be evaluated by repeated integrations for regions of the type depicted in Fig. 9 to yield

$$\sigma = \iint\limits_R \sqrt{(z_x)^2 + (z_y)^2 + 1} \, dy \, dx.$$

By considering the projections R' and R'' of S' on the other coordinate planes, we deduce similar formulas:

$$\sigma = \int_{R'} \sec \alpha \, dA \qquad \sigma = \int_{R''} \sec \beta \, dA.$$

To obtain the surface area of a piecewise-smooth surface we need merely to add the areas of its smooth pieces.

The surface integral of a continuous function $\varphi(x,y,z)$ specified on the surface S' is defined as follows: Let S' be subdivided into subregions $\Delta S_i'$ of areas $\Delta \sigma_i$ and form the sum

$$\sum_{i=1}^{n} \varphi(x_i, y_i, z_i) \, \Delta \sigma_i \tag{18-3}$$

where (x_i, y_i, z_i) is some point in $\Delta S_i'$. The limit of the sum (18-3) as $n \to \infty$ in such a way that the greatest linear dimensions of the $\Delta S_i'$ tend to zero is the surface integral of $\varphi(x,y,z)$ over S'. It is denoted by the symbol

$$\int_{S'} \varphi(x,y,z) \, d\sigma. \tag{18-4}$$

The integral (18-4) is usually evaluated by repeated integrations. Thus, if

$$d\sigma = \sec \gamma \, dA = \sqrt{(z_x)^2 + (z_y)^2 + 1} \, dx \, dy$$

then

$$\int_{S'} \varphi(x,y,z) \, d\sigma = \iint\limits_R \varphi[x,y,f(x,y)] \sqrt{(z_x)^2 + (z_y)^2 + 1} \, dx \, dy$$

where $z = f(x,y)$ is the equation of S' and R is the projection of S' on the xy plane.

Example 1.   Find the surface area of the sphere $x^2 + y^2 + z^2 = a^2$ cut off by the cylinder $x^2 - ax + y^2 = 0$ (Fig. 16).

From symmetry it is clear that it will suffice to determine the surface area in the first octant and multiply the result by 4.   Now

$$\sigma = \int_R \sqrt{(z_x)^2 + (z_y)^2 + 1} \, dy \, dx$$

and, since $z = \sqrt{a^2 - x^2 - y^2}$,

$$z_x \equiv \frac{\partial z}{\partial x} = \frac{-x}{\sqrt{a^2 - x^2 - y^2}}$$

$$z_y \equiv \frac{\partial z}{\partial y} = \frac{-y}{\sqrt{a^2 - x^2 - y^2}}.$$

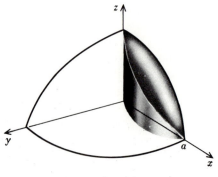

FIGURE 16

Thus, the integral becomes

$$\sigma = 4 \int_0^a \int_0^{\sqrt{ax-x^2}} \sqrt{\frac{x^2+y^2}{a^2-x^2-y^2}+1}\, dy\, dx = 4 \int_0^a \int_0^{\sqrt{ax-x^2}} \frac{a\, dy\, dx}{\sqrt{a^2-x^2-y^2}}.$$

It is simpler to evaluate this integral by transforming to cylindrical coordinates.   The equation of the cylinder becomes $r = a \cos \theta$, and that of the sphere

$$z = \sqrt{a^2-x^2-y^2} = \sqrt{a^2-r^2}.$$

Thus, $$\sigma = 4 \int_0^{\pi/2} \int_0^{a\cos\theta} \frac{ar\, dr\, d\theta}{\sqrt{a^2-r^2}} = 4a^2 \left(\frac{\pi}{2} - 1 \right).$$

Example 2.   Find the z coordinate of the center of gravity of the surface area of one octant of the sphere $x^2 + y^2 + z^2 = a^2$.   Now,

$$\bar{z} = \frac{\int_{S'} z\, d\sigma}{\int_{S'} d\sigma} = \frac{\int_0^a \int_0^{\sqrt{a^2-y^2}} z \sqrt{(z_x)^2+(z_y)^2+1}\, dx\, dy}{4\pi a^2/8} = \frac{2 \int_0^a \int_0^{\sqrt{a^2-y^2}} a\, dx\, dy}{\pi a^2} = \frac{a}{2}.$$

PROBLEMS

1. Find, by the method of this section, the area of the surface of the sphere $x^2 + y^2 + z^2 = a^2$ that lies in the first octant.

2. Find the area of the surface of the sphere $x^2 + y^2 + z^2 = a^2$ cut off by the cylinder $x^2 - ax + y^2 = 0$.

3. Find the volume bounded by the cylinder and the sphere of Prob. 2.

4. Find the area of the surface of the cylinder $x^2 + y^2 = a^2$ cut off by the cylinder $y^2 + z^2 = a^2$.

5. Find the coordinates of the center of gravity of the area of the portion of the surface of the sphere cut off by the right-circular cone whose vertex is at the center of the sphere.

6. If a sphere is inscribed in a right-circular cylinder, the surfaces of the sphere and the cylinder intercepted by a pair of planes perpendicular to the axis of the cylinder are equal in area.   Prove it.

7. Find the surface area in the first octant of the part of the cone $y^2 = x^2 + z^2$ bounded by the plane $y + z = a$.   *Hint:* Integrate over the projection of the surface on the yz plane.

------

19. Line Integrals in the Plane.[1]   A curve C represented by $y = f(x)$, $a \leq x \leq b$, is said to be *continuous* if $f(x)$ is continuous in $[a,b]$ and to be *continuously differentiable*, or *smooth*, if $f'(x)$ is continuous in $[a,b]$.   When C is specified by the parametric equations $y = y(t)$, $x = x(t)$, $t_0 \leq t \leq t_1$, then C is continuous if $y(t)$ and $x(t)$ are continuous in $[t_0, t_1]$ and C is smooth if the derivatives $y'(t)$, $x'(t)$ are continuous and do not vanish for the same value of t in the interval $[t_0, t_1]$.   It is further required that a smooth curve will not intersect itself.   That is, the equations $x(t_2) = x(t_3)$, $y(t_2) = y(t_3)$ can be satisfied only if $t_2 = t_3$.   Such a curve has two definite end points corresponding to $t = t_0$ and $t = t_1$ and is called a *simple arc*.   If the end points coincide, so that $x(t_0) = x(t_1)$, $y(t_0) = y(t_1)$, the curve is called a *simple closed curve*.

   A curve consisting of a finite number of simple arcs joined end to end is called a *piecewise-smooth curve*, or *path*.   A piecewise-smooth curve may intersect itself at one or

------

[1] For a discussion of line integrals in space, see Chap. 6.

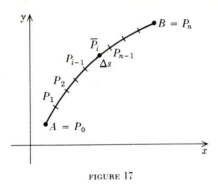

FIGURE 17

more points.   It is closed if its end points coincide.   All curves considered in the remainder of this chapter are either smooth or piecewise smooth.

The familiar "limit-of-the-sum" definition of the definite integral $\int_a^b f(x)\ dx$, $a \leq x \leq b$, can be generalized to define the line integral $\int_C f(x,y)\ ds$ of a function $f(x,y)$ specified along a rectifiable curve C.   The term "rectifiable" means that the curve has a definite length.   We recall that the length l of a piecewise-smooth curve C is determined by the formula $l = \int_a^b \sqrt{1 + [y'(x)]^2}\ dx$, when the equation of C is given in the form $y = y(x)$, $a \leq x \leq b$.

Let A and B be the end points of a piecewise-smooth curve C (Fig. 17).   We subdivide C into n segments by the points $P_1, P_2, \ldots, P_{n-1}$ so chosen that the lengths of arcs $\widehat{AP_1}, \widehat{AP_2}, \ldots, \widehat{AP_i}, \ldots$ increase.   If the length of $\widehat{AP_i}$ is denoted by s_i, then $s_1 < s_2 < \cdots < s_n$, where $s_n = l$ is the length of C.   We denote the length of arc $\widehat{P_{i-1}P_i}$ by Δs_i, so that $\Delta s_i = s_i - s_{i-1}$, and form the sum

$$I_n = \sum_{i=1}^{n} f(\overline{x}_i, \overline{y}_i)\ \Delta s_i \tag{19-1}$$

where $P_i(\overline{x}_i, \overline{y}_i)$ is a point arbitrarily chosen on each arc $\widehat{P_{i-1}P_i}$.   Now, if for all possible modes of subdividing C into segments Δs_i the sum (19-1) has the limit I as all the $\Delta s_i \to 0$, we write

$$\lim \sum_{i=1}^{n} f(\overline{x}_i, \overline{y}_i)\ \Delta s_i = \int_C f(x,y)\ ds \tag{19-2}$$

where the limit $I \equiv \int_C f(x,y)\ ds$ is called a *line integral* of $f(x,y)$ along C.   The curve C is often spoken of as the *path of integration.*

As is the case with definite integrals, the limit (19-2) can be shown to exist if $f(x,y)$ is piecewise continuous along C.   Further, if $f(x,y)$ is continuous on C, the argument used to establish the mean-value theorem for definite integrals gives this time the mean-value theorem

$$\int_C f(x,y)\ ds = f(\xi,\eta)l \tag{19-3}$$

where (ξ,η) is some point on C between A and B and l is the length of C.

It follows from the definition (19-2) that

$$\int_C [\alpha f(x,y) + \beta g(x,y)]\ ds = \alpha \int_C f(x,y)\ ds + \beta \int_C g(x,y)\ ds$$

for any constants α and β, whenever the integrals in this formula exist.   In other words, the operation of forming line integrals is *linear.*   (Cf. Chap. 1, Sec. 2.)

If in the formulation of (19-2) we proceed from the point B to the point A, so that $\Delta s_i = s_{i-1} - s_i < 0$, we obtain

$$\int_{\substack{C \\ A \text{ to } B}} f(x,y)\ ds = - \int_{\substack{C \\ B \text{ to } A}} f(x,y)\ ds. \tag{19-4}$$

This is a generalization of the familiar result $\int_a^b f(x)\ dx = - \int_b^a f(x)\ dx$.

The line integral

$$I = \int_C f(x,y)\, ds \qquad (19\text{-}5)$$

can be evaluated by reducing it to the familiar definite integral when the equation of C is given in the form

$$y = y(x) \qquad a \leq x \leq b. \qquad (19\text{-}6)$$

Indeed, the value of the integrand $f(x,y)$ along C is $f[x,y(x)]$, and since $ds = \sqrt{1 + [y'(x)]^2}\, dx$, (19-5) suggests that

$$I = \int_{x=a}^{x=b} f[x,y(x)]\sqrt{1 + [y'(x)]^2}\, dx \qquad (19\text{-}7)$$

which is a definite integral of the type $\int_a^b F(x)\, dx$. This surmise can be established by showing that the difference between the value of the definite integral (19-7) and the sum (19-1) tends to zero when $\max |\Delta s_i| \to 0$. Similar remarks apply to the formulation of (19-9) and (19-12).

In many applications it is more convenient to specify C in the parametric form

$$x = x(s) \qquad y = y(s) \qquad 0 \leq s \leq l \qquad (19\text{-}8)$$

where x and y are continuously differentiable functions of the parameter s representing the length of C measured from the point A to an *arbitrary* point P on C (Fig. 17).

The substitution from (19-8) in (19-5) then gives the definite integral

$$I = \int_{s=0}^{s=l} f[x(s),y(s)]\, ds. \qquad (19\text{-}9)$$

If, on the other hand, C is specified by two continuously differentiable functions

$$x = x(t) \qquad y = y(t) \qquad t_0 \leq t \leq t_1 \qquad (19\text{-}10)$$

where the parameter t is such that the point $P[x(t),y(t)]$ on C traces out the curve from A to B when t increases from t_0 to t_1, then ds in (19-5) has the form

$$ds = \sqrt{[x'(t)]^2 + [y'(t)]^2}\, dt \qquad (19\text{-}11)$$

and (19-5) yields the definite integral

$$I = \int_{t_0}^{t_1} f[x(t),y(t)]\sqrt{[x'(t)]^2 + [y'(t)]^2}\, dt. \qquad (19\text{-}12)$$

As an illustration of the evaluation of line integrals, consider

$$I = \int_C (x - y^2)\, ds \qquad (a)$$

where C is a segment of straight line $y = 2x$, $0 \leq x \leq 1$. We obtain, on using (19-7),

$$I = \int_C (x - y^2)\, ds = \int_0^1 (x - 4x^2)\sqrt{1 + 2^2}\, dx = \sqrt{5}\left[\frac{x^2}{2} - \frac{4x^3}{3}\right]_0^1 = -5\frac{\sqrt{5}}{6}.$$

If the curve C in (a) is an arc of the circle

$$x = a \cos t \qquad y = a \sin t \qquad 0 \leq t \leq \frac{\pi}{2} \qquad (b)$$

we find from (19-11) that $ds = \sqrt{(-a \sin t)^2 + (a \cos t)^2}\, dt = a\, dt$ and, on making the substitution from (b) in (a), we get

$$I = \int_C (x - y^2)\, ds = \int_0^{\pi/2} (a \cos t - a^2 \sin^2 t)a\, dt = a^2 \sin t \Big|_0^{\pi/2} - a^3\left(\frac{t}{2} - \frac{\sin 2t}{4}\right)\Big|_0^{\pi/2} = a^2 - \frac{\pi a^3}{4}.$$

In many applications the integrand $f(x,y)$ in (19-5) has one of the forms

$$f(x,y) = M(x,y)\frac{dx}{ds} \qquad f(x,y) = N(x,y)\frac{dy}{ds} \tag{19-13}$$

where $dx/ds = \cos\alpha$ and $dy/ds = \sin\alpha$, α being the angle made by the tangent to C with the x axis (Fig. 18). We then write

$$\int_C f(x,y)\,ds = \int_C M(x,y)\frac{dx}{ds}\,ds \equiv \int_C M(x,y)\,dx$$

$$\int_C f(x,y)\,ds = \int_C N(x,y)\frac{dy}{ds}\,ds \equiv \int_C N(x,y)\,dy.$$

On adding these line integrals we get the line integral

$$I = \int_C \left[M(x,y)\frac{dx}{ds} + N(x,y)\frac{dy}{ds} \right] ds \equiv \int_C [M(x,y)\,dx + N(x,y)\,dy] \tag{19-14}$$

which can be expressed as a definite integral when the equations of C are specified in one of the forms considered above.

Example 1.   Evaluate the line integral

$$I = \int_C (M\,dx + N\,dy) \equiv \int_C \left(\frac{y}{x^2+y^2}\,dx - \frac{x}{x^2+y^2}\,dy \right) \tag{a}$$

where C is the semicircle

$$x = \cos t \qquad y = \sin t \qquad 0 \le t \le \pi. \tag{b}$$

Since along C

$$dx = -\sin t\,dt \qquad dy = \cos t\,dt \tag{c}$$

the substitution from (b) and (c) in (a) gives the integral

$$I = \int_0^\pi (-\sin^2 t - \cos^2 t)\,dt = -\pi.$$

Example 2.   Evaluate the line integral

$$I = \int_C [x^2 y\,dx - (x+y)\,dy] \tag{d}$$

where C is the segment of a straight line joining the point $(0,0)$ with $(2,1)$ (Fig. 19). We can write the equation of this segment as $y = x/2$, $0 \le x \le 2$. But along this segment $dy = \frac{1}{2}\,dx$, and the substitution in (d) gives a definite integral

$$I = \int_0^2 \left[x^2 \left(\frac{x}{2} \right) dx - \left(x + \frac{x}{2} \right)(\tfrac{1}{2}\,dx) \right] = \tfrac{1}{2}.$$

If the path C in (d) is the segment of the parabola $x = 2y^2$ joining the same points $(0,0),(2,1)$ (Fig. 19) we have $dx = 4y\,dy$, and the integral (d), when evaluated over the parabola, is

$$I = \int_0^1 [(2y^2)^2 y(4y\,dy) - (2y^2 + y)\,dy] = {}^{47}\!/_{42}.$$

The reader will verify this result by expressing the line integral (d) as a definite integral evaluated along the x axis.

Example 3.   Evaluate the line integral

$$I = \int_C (xy^2\,dx - x^2 y\,dy) \tag{e}$$

where C is the broken line joining the points $P(0,0)$, $Q(1,1)$, $R(2,1)$ in that order (Fig. 20).

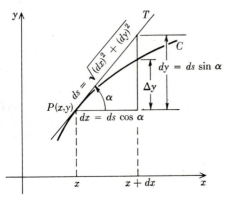

FIGURE 18

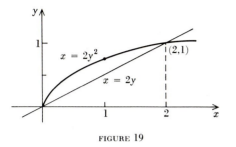

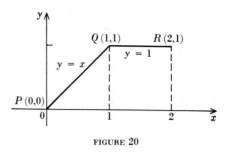

FIGURE 19

FIGURE 20

We write the equation of the segment PQ as $y = x$, $0 \le x \le 1$, and that of QR as $y = 1$, $1 \le x \le 2$. Then along PQ, $dy = dx$ and along QR, $dy = 0$. Accordingly the integral (e) when evaluated along PQ is

$$I_1 = \int_0^1 [x(x^2)\, dx - x^2(x)\, dx] = 0$$

and its value along the path QR is

$$I_2 = \int_1^2 [x(1)\, dx - x^2(1)0] = \tfrac{3}{2}.$$

Thus $I = I_1 + I_2 = \tfrac{3}{2}$.

PROBLEMS

1. Evaluate $\int_C (x^2 + y^2)\, ds$, where C is an arc of the circle $x = a \cos (s/a)$, $y = a \sin (s/a)$, $0 \le s \le \pi a$. (*Ans.:* πa^3.)

2. Show that (a) $\int_C (x - y^2)\, dx = \tfrac{1}{6}$, (b) $\int_C 2xy\, dy = \tfrac{2}{3}$, (c) $\int_C [(x - y^2)\, dx + 2xy\, dy] = \tfrac{5}{6}$, if C is the segment of the straight line $y = x$, $0 \le x \le 1$.

3. Show that $\int_C [(2x^2 + 4xy)\, dx + (2x^2 - y^2)\, dy] = \tfrac{41}{3}$, if C is the segment of the parabola $y = t^2$, $x = t$, $1 \le t \le 2$.

4. Show that $\int_C (x^2 + y^2)\, dy = -\tfrac{2}{3}$, if the path C is the broken line joining the points $(0,0)$, $(1,1)$, $(1,0)$ in that order.

5. Find the values of the line integral $\int_C [\sqrt{y}\, dx + (x - y)\, dy]$ along the following paths C joining the points $(0,0)$ and $(1,1)$: (a) $x = t$, $y = t$; (b) $x = t^2$, $y = t$; (c) $x = t$, $y = t^2$; (d) $x = t$, $y = t^3$. Sketch the paths.

6. Find the values of the line integral $\int_C [(x^2 + y^2)\, dx + 2xy\, dy]$ along the following paths C joining the points $(0,0)$, $(1,1)$: (a) $y = x$; (b) $y = x^2$; (c) $x = y^2$; (d) $x = y^3$; (e) $y = x^3$.

7. Find the value of $\int_C (x\, dy + y\, dx)$ along the following paths C: (a) the arc of the circle $x^2 + y^2 = a^2$ joining the points $(-a,0)$, $(a,0)$ in that order; (b) the path formed by the straight lines joining the points $(-a,0)$, $(-a,a)$, (a,a), $(a,0)$ in that order; (c) the straight line joining the points $(-a,0)$, $(a,0)$ in that order.

8. Find the values of $\int_C (x^2\, dx + y^2\, dy)$ along the following paths: (a) the circular arc $x = \sin t$, $y = \cos t$ joining the points $(1,0)$ and $(0,1)$ in that order; (b) the straight-line segment joining $(1,0)$ and $(0,1)$ in that order; (c) the segment of the parabola $y = 1 - x^2$ joining the points $(1,0)$ and $(0,1)$ in that order.

9. Find the values of the line integral

$$\int_C (y \cos x\, dx + \sin x\, dy)$$

if the path C is (a) the straight-line segment joining the points $(0,0)$, $(\pi/2,\pi/2)$; (b) the arc of the circle $x^2 + y^2 = \pi^2/4$ joining the points $(0,0)$, $(\pi/2,\pi/2)$.

10. Evaluate $\int_C (y \sin x - x \cos y)$ if the path C is the straight-line segment joining the point $(0,0)$ with an arbitrary point (x,y).

---

20. Line Integrals Independent of the Path.   The considerations of this chapter that have preceded the discussion of integrals of functions of several variables were largely concerned with the properties of functions in the neighborhood of a point.   The concept of an integral, on the other hand, requires the characterization of functions throughout certain regions.   To make the fundamental theorem of integral calculus (14-1) available for the evaluation of multiple integrals in Secs. 16 to 18, it proved necessary to impose certain restrictions, not only on the character of functions but also on the types of regions in which such formulas as (16-1) and (17-1) are meaningful.   The properties of line integrals, as we shall see, crucially depend on the nature of regions in which the integrands are specified, and we now describe several types of regions with which we shall deal.

Throughout the remainder of this chapter we shall be concerned only with closed connected regions whose boundaries may consist of one or more piecewise-smooth curves (Fig. 21c).   A region R is said to be *connected* if every two points belonging to R can be joined by a piecewise-smooth curve that lies wholly in R.   Thus, a configuration consisting of two shaded disjoint circular regions in Fig. 21a is not connected, but the shaded region bounded by two concentric circles C_0 and C_1 in Fig. 21b is connected, and so is the shaded region bounded by several closed curves in Fig. 21c.

As stated in Sec. 19, a nonintersecting closed curve consisting of a finite number of arcs joined end to end is called a *simple closed curve*, and the region R is said to be simply connected if *every* simple closed curve drawn in its interior can be shrunk by continuous deformation to a point without crossing the boundaries of R.   Thus, the interior of a square or an ellipse, or the interior of a plane region bounded by one simple closed curve, is simply connected.   The interior of the ring in Fig. 21b is not simply connected, because a closed curve K (shown by the dotted line) cannot be shrunk to a point without crossing the boundary C_1.   A region that is not simply connected is called *multiply connected*.

Let $u(x,y)$ be a differentiable (and hence single-valued) function defined in some simply or multiply connected region R, and consider its differential

$$du = \frac{\partial u}{\partial x} \, dx + \frac{\partial u}{\partial y} \, dy.$$

If we write

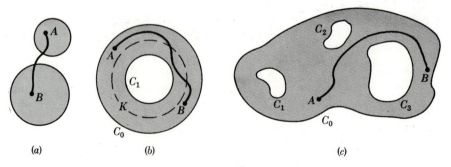

(a)   (b)   (c)

FIGURE 21

$$\frac{\partial u}{\partial x} \equiv M(x,y) \qquad \frac{\partial u}{\partial y} = N(x,y) \tag{20-1}$$

and assume that $M(x,y)$ and $N(x,y)$ are continuous in R, we can form the line integral

$$I = \int_C [M(x,y)\ dx + N(x,y)\ dy] \tag{20-2}$$

over an arbitrary path C joining any two given points A and B in R (Fig. 21c). Let the equations of such a path be

$$x = x(t) \qquad y = y(t) \qquad t_0 \le t \le t \tag{20-3}$$

where the parameter $t = t_0$ gives the point A and $t = t_1$ gives the point B. On substituting from (20-3) in (20-2) we obtain a definite integral, which by virtue of (20-1) is

$$I = \int_{t_0}^{t_1} du[x(t),y(t)] = u_B - u_A \tag{20-4}$$

where u_B and u_A are the values of $u(x,y)$ at B and A, respectively.

Equation (20-4) states that the value I of (20-2) when its integrand $M\ dx + N\ dy$ is a total differential of some function $u(x,y)$ is equal to the difference of the values of $u(x,y)$ at the points A and B. Consequently, I is the same for all piecewise-smooth paths joining A and B, and the integral (20-2) is said to be *independent of the path*. When the path C is closed, the points A and B coincide and hence the value of the integral is zero for every closed path. We can thus state the following:

THEOREM. *If the integrand in the line integral $\int_C (M\ dx + N\ dy)$ is the total differential $du = M\ dx + N\ dy$ of some function $u(x,y)$ with continuous partial derivatives $u_x(x,y)$, $u_y(x,y)$ in the given region R, then the line integral vanishes for every piecewise-smooth closed path C in R.*

To say that the line integral vanishes for every closed path in R is equivalent to saying that the integral is independent of the path connecting a pair of given points in R. It is customary then to write (20-2) as

$$I = \int_{(x_0,y_0)}^{(x_1,y_1)} (M\ dx + N\ dy)$$

where no reference is made to the path C and where only the end points of the path are indicated in the limits of the integral.

If $A(x_0,y_0)$ is a fixed point and $B(x,y)$ is an arbitrary point in R, it follows from (20-4) that

$$\int_{(x_0,y_0)}^{(x,y)} (M\ dx + N\ dy) = u(x,y) - u(x_0,y_0). \tag{20-5}$$

Thus the integral in (20-5) with the variable upper limit defines uniquely the function $u(x,y)$ whenever the integrand

$$M\ dx + N\ dy \equiv \frac{\partial u}{\partial x}\ dx + \frac{\partial u}{\partial y}\ dy \tag{20-6}$$

and $\partial u/\partial x$ and $\partial u/\partial y$ are continuous in R. In this case

$$\frac{\partial u}{\partial x} = M(x,y) \qquad \frac{\partial u}{\partial y} = N(x,y)$$

and, if $u(x,y)$ also has continuous mixed derivatives

$$\frac{\partial^2 u}{\partial y\ \partial x} = \frac{\partial M}{\partial y} \qquad \frac{\partial^2 u}{\partial x\ \partial y} = \frac{\partial N}{\partial x}$$

at all points of R, we conclude that

$$\frac{\partial M}{\partial y} = \frac{\partial N}{\partial x} \tag{20-7}$$

is a necessary condition for (20-6) to be the total differential of some function $u(x,y)$. However, the condition (20-7) does not ensure that the expression $M(x,y)\,dx + N(x,y)\,dy$ is, in fact, the total differential of some function $u(x,y)$ unless, as we shall see in Sec. 22, the integral (20-5) is defined in a simply connected region.

––– PROBLEMS

1. The following integrals are independent of the path. Show by choosing any convenient path that (a) $\int_{(0,1)}^{(1,2)} [(x^2 + y^2)\,dx + 2xy\,dy] = 13\frac{2}{3}$; (b) $\int_{(0,0)}^{(\pi/2,\pi/2)} (y \cos x + \sin x\,dy) = \frac{\pi}{2}$; (c) $\int_{(1,1)}^{(2,3)} [(x + 1)\,dx + (y + 1)\,dy] = 17\frac{1}{2}$.

2. Demonstrate by choosing two different paths joining the points $(0,0)$, $(1,1)$ that the following line integrals are not independent of the path: (a) $\int_C [(x^2 - y^2)\,dx + 2xy\,dy]$; (b) $\int_C [(x - y^2)\,dx + 2xy\,dy]$; (c) $\int_C (x^2 - y^2)\,dx$.

––

21. Green's Theorem in the Plane.[1]   We establish next an important theorem that relates the line integral evaluated over a simple closed curve with a double integral defined over the interior of the plane region bounded by the curve.

Consider first a region R of the type shown in Fig. 22 to which the fundamental theorem of integral calculus can be applied to evaluate the double integral

$$\iint_R \frac{\partial M(x,y)}{\partial y}\,dx\,dy. \tag{21-1}$$

We suppose that the boundary of R in Fig. 22 is piecewise smooth with $g_1(x) \le g_2(x)$, $a \le x \le b$, and that $\partial M / \partial y$ is continuous in R.   We can then write (cf. Sec. 16)

$$\iint_R \frac{\partial M}{\partial y}\,dx\,dy = \int_a^b dx \int_{y=g_1(x)}^{y=g_2(x)} \frac{\partial M}{\partial y}\,dy.$$

But by the fundamental theorem of integral calculus (14-1),

$$\int_{y=g_1(x)}^{y=g_2(x)} \frac{\partial M(x,y)}{\partial y}\,dy = M(x,y)\bigg|_{y=g_1(x)}^{y=g_2(x)} = M[x,g_2(x)] - M[x,g_1(x)].$$

Thus,

$$\iint_R \frac{\partial M}{\partial y}\,dx\,dy = \int_a^b M[x,g_2(x)]\,dx - \int_a^b M[x,g_1(x)]\,dx$$

$$= -\int_b^a M[x,g_2(x)]\,dx - \int_a^b M[x,g_1(x)]\,dx. \tag{21-2}$$

The definite integrals in the right-hand member of (21-2) are equivalent to line integrals over the curves $\widehat{SQ}$ and $\widehat{PT}$, since

$$\int_{\widehat{SQ}} M(x,y)\,dx = \int_b^a M[x,g_2(x)]\,dx$$

$$\int_{\widehat{PT}} M(x,y)\,dx = \int_a^b M[x,g_1(x)]\,dx. \tag{21-3}$$

[1] This theorem was deduced by an English mathematical physicist George Green (1793–1841) and, independently of Green, by a Russian mathematician M. Ostrogradsky (1801–1861). In some books it is called the Green-Ostrogradsky theorem.

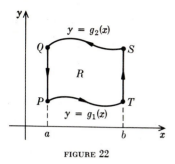

FIGURE 22

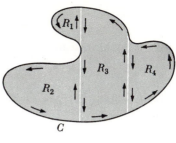

FIGURE 23

If we add to the sum of the line integrals in the left-hand members of (21-3) two line integrals

$$\int_{\widehat{QP}} M(x,y)\,dx \qquad \int_{\widehat{TS}} M(x,y)\,dx$$

which vanish over the rectilinear paths QP and TS (since $dx = 0$ over these paths) we obtain the line integral $\int_C M(x,y)\,dx$ over the path $C = \widehat{PT} + \widehat{TS} + \widehat{SQ} + \widehat{QP}$ that forms the boundary of R. We conclude that

$$\iint_R \frac{\partial M}{\partial y}\,dx\,dy = -\int_C M(x,y)\,dx \tag{21-4}$$

where the line integral is evaluated by describing C counterclockwise. The result (21-4) is Green's theorem in the simplest case.

Formula (21-4) remains valid if the region R can be subdivided into a finite number of subregions R_i (Fig. 23), each of which is of the type shown in Fig. 22, because, on forming the sum of the integrals like (21-4) over the subregions R_i, the line integrals over the subdividing lines (shown by the white lines in Fig. 23) are traversed twice in opposite directions. Such integrals cancel out in pairs, and the sum of the remaining line integrals is equal to the line integral over the boundary C.

Similar consideration can be applied to the integral $\iint_R (\partial N/\partial x)\,dx\,dy$, if $\partial N/\partial x$ is continuous in a region R of the type shown in Fig. 9b. We find

$$\iint_R \frac{\partial N}{\partial x}\,dx\,dy = \int_C N(x,y)\,dy \tag{21-5}$$

where the path C is also traversed counterclockwise. The formula (21-5) is obviously valid for regions that can be subdivided into several subregions of the type shown in Fig. 9b.

When the region R satisfies the requirements of both formulas (21-4) and (21-5), we can combine these formulas and obtain *Green's theorem*:

$$\iint_R \left(\frac{\partial M}{\partial y} - \frac{\partial N}{\partial x} \right) dy\,dx = -\int_C [M(x,y)\,dx + N(x,y)\,dy]. \tag{21-6}$$

Formula (21-6) is valid if $\partial M/\partial y$ and $\partial N/\partial x$ are continuous in the special simply connected regions considered above. A rather careful analysis, which we do not present here, shows that (21-6) remains valid in every bounded simply connected plane region R with a simple closed curve C as its boundary.

An interesting formula for the areas bounded by such curves follows at once from (21-6). If we set $M(x,y) = -y$ and $N(x,y) = x$, we obtain

$$\iint\limits_{R} 2dy\,dx = \int_{C} (-y\,dx + x\,dy).$$

But the double integral is equal to twice the area A of the region bounded by C. Thus,

$$A = \tfrac{1}{2}\int_{C} (-y\,dx + x\,dy). \tag{21-7}$$

This formula sometimes enables one to compute easily the areas of regions bounded by plane curves.

For example, if C is an ellipse, $x = a\cos\theta$, $y = b\sin\theta$, $0 \le \theta \le 2\pi$, we get at once from (21-7)

$$A = \tfrac{1}{2}\int_{0}^{2\pi} ab(\sin^2\theta + \cos^2\theta)\,d\theta = \tfrac{1}{2}ab\int_{0}^{2\pi} d\theta = \pi ab.$$

Formula (21-6) can be used to justify formula (16-5) for the change of variables in double integrals. Thus, if the variables x, y are changed by the transformation

$$x = x(u,v) \qquad y = y(u,v) \tag{16-3}$$

considered in Sec. 16, we get, on setting $M = 0$, $N = x$ in (21-6),

$$\iint\limits_{R} dy\,dx = \int_{C} x\,dy. \tag{21-8}$$

But it follows from (16-3) that along C

$$dy = \frac{\partial y}{\partial u}\,du + \frac{\partial y}{\partial v}\,dv$$

so that (21-8) can be written as

$$\iint\limits_{R} dy\,dx = \int_{K} x(u,v)\left(\frac{\partial y}{\partial u}\,du + \frac{\partial y}{\partial v}\,dv\right) \tag{21-9}$$

where K is the boundary of the region T into which R is mapped by the transformation (16-3). If we transform by Green's theorem (21-6) the line integral in (21-9) into a double integral taken over the region T by setting

$$M(u,v) = x(u,v)\,\frac{\partial y(u,v)}{\partial u} \qquad N(u,v) = x(u,v)\,\frac{\partial y(u,v)}{\partial v}$$

we get

$$\iint\limits_{R} dy\,dx = -\iint\limits_{T}\left[\frac{\partial}{\partial v}\left(x\,\frac{\partial y}{\partial u}\right) - \frac{\partial}{\partial u}\left(x\,\frac{\partial y}{\partial v}\right)\right] du\,dv$$

$$= -\iint\limits_{T}\left(\frac{\partial x}{\partial v}\,\frac{\partial y}{\partial u} - \frac{\partial x}{\partial u}\,\frac{\partial y}{\partial v}\right) du\,dv = \iint\limits_{T} |J(u,v)|\,du\,dv. \tag{21-10}$$

In obtaining this result we tacitly assumed that $\partial^2 y/\partial u\,\partial v = \partial^2 y/\partial v\,\partial u$, which is certainly true if these mixed partial derivatives are continuous. Also, the absolute value of the Jacobian was introduced since the area of R is intrinsically positive. Formula (16-5) follows from (21-10) on recalling the definition of a double integral.

Example. Use Green's theorem (21-6) to evaluate $I = \int_{C} (x^2 y\,dx + y^3\,dy)$, where C is a simple closed path formed by the curves $y^3 = x^2$ and $y = x$ (Fig. 24). Here $M(x,y) = x^2 y$, $N(x,y) = y^3$, so that $\partial M/\partial y - \partial N/\partial x = x^2$. Hence, by (21-6),

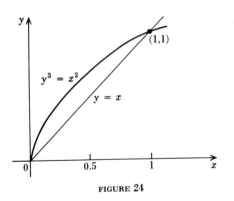

FIGURE 24

$$I = -\iint_R x^2\, dx\, dy = -\int_0^1 dy \int_{y^{3/2}}^y x^2\, dx = -\tfrac{1}{44}.$$

The reader will check this result by evaluating the line integral directly.

PROBLEMS

1. Use formula (21-7) to find the areas bounded by:

(a) $y^2 = 9x$ and $y = 3x$;
(b) $y = 0$, $x = 0$, and $x + y = a$, $a > 0$;
(c) $x = a \cos^3 \theta$, $y = a \sin^3 \theta$, $0 \le \theta < 2\pi$. (*Ans.: $3\pi a^2/8$.*)

2. Evaluate the following line integrals directly and by Green's theorem:

(a) $\int_C [(xy - x^2)\, dx + x^2 y]\, dy$, where C is the closed path formed by $y = 0$, $x = 1$, $y = x$;
(b) $\int_C (x^2 y\, dx + y\, dy)$, where C is the closed path formed by $y^2 = x$ and $y = x$;
(c) $\int_C (y^{-1}\, dx + x^{-1}\, dy)$, where C is the closed path formed by $y = x$, $y = 1$, and $x = 4$;
(d) $\int_C [(xy + x + y)\, dx + (xy + x - y)\, dy]$, where C is the triangle formed by $x = 0$, $y = 0$, $x = 1$. What is the value of this integral if C is the ellipse $x^2/a^2 + y^2/b^2 = 1$? (*Ans.: $-\tfrac{1}{6}$.*)
(e) $\int_C [(2x - y)\, dx + (x + 3y)\, dy]$, where C is the ellipse $x^2/4 + y^2 = 1$. (*Ans.: 0.*)

3. Evaluate the line integrals in Prob. 2 of Sec. 20 by Green's theorem if C is the circle $x^2 + y^2 = 1$. Verify your calculations by evaluating the line integrals directly.

---

22. Consequences of Green's Theorem. We have seen in Sec. 21 that, if $M(x,y)$, $N(x,y)$, $\partial M/\partial y$, and $\partial N/\partial x$ are continuous in a bounded simply connected region R with a simple closed curve C as its boundary, then

$$\iint_R \left(\frac{\partial M}{\partial y} - \frac{\partial N}{\partial x} \right) dy\, dx = -\int_C (M\, dx + N\, dy) \tag{22-1}$$

where the path C is described counterclockwise.

It is obvious from (22-1) that, if $\partial M/\partial y - \partial N/\partial x = 0$ at all points of R, the line integral $\int_C (M\, dx + N\, dy)$ evaluated over *every* simple closed curve C' contained in R is equal to zero. Conversely, if the line integral in (22-1) vanishes for every such curve C' enclosing a subregion R' of R, then

$$\iint_{R'} \left(\frac{\partial M}{\partial y} - \frac{\partial N}{\partial x} \right) dy\, dx = 0. \tag{22-2}$$

This enables us to prove that

$$\frac{\partial M}{\partial y} = \frac{\partial N}{\partial x} \qquad \text{at every point of } R. \tag{22-3}$$

Suppose that

$$\frac{\partial M}{\partial y} - \frac{\partial N}{\partial x} \neq 0 \tag{22-4}$$

at some point P of R and let, for definiteness, the difference in (22-4) be positive at P. Inasmuch as $\partial M/\partial y - \partial N/\partial x$ is continuous in R, there is a region R' including P throughout which the integrand in (22-2) is positive. (See concluding paragraphs of Sec. 2.)

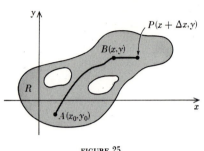

FIGURE 25

But if the integrand in a double integral is positive, the double integral over R' is also positive, and this contradicts the statement (22-2). Thus the assumption (22-4) is not tenable and (22-3) must be true. We have established a theorem:

THEOREM I. Let $M(x,y)$, $N(x,y)$, $\partial M/\partial y$, and $\partial N/\partial x$ be continuous in a bounded closed simply connected region R with a piecewise-smooth boundary curve C. Then a necessary and sufficient condition for the line integral $\int_{C'} (M\,dx + N\,dy)$ to vanish for every simple closed path C' in R is that $\partial M/\partial y = \partial N/\partial x$ at all points of R.

It follows from considerations of Sec. 20 that the vanishing of the line integral $\int_C (M\,dx + N\,dy)$ over every simple closed path C in any connected region R is equivalent to the statement that the integral is independent of the path. Accordingly, the integral $\int_{(x_0,y_0)}^{(x,y)} [M(x,y)\,dx + N(x,y)\,dy]$ defines at every point (x,y) of R a single-valued function

$$u(x,y) = \int_{(x_0,y_0)}^{(x,y)} [M(x,y)\,dx + N(x,y)\,dy]. \tag{22-5}$$

Now, by definition,

$$\frac{\partial u}{\partial x} = \lim_{\Delta x \to 0} \frac{u(x + \Delta x,\, y) - u(x,y)}{\Delta x} \tag{22-6}$$

in which $u(x + \Delta x,\, y)$ is determined by the integral

$$u(x + \Delta x,\, y) = \int_{(x_0,y_0)}^{(x+\Delta x,\,y)} [M(x,y)\,dx + N(x,y)\,dy]. \tag{22-7}$$

This integral can be evaluated over any suitable path joining the points (x_0,y_0) and $(x + \Delta x,\, y)$. It suits our purpose to choose as our path (Fig. 25) any convenient curve $\overarc{AB}$ joining $A(x_0,y_0)$ with $B(x,y)$ and the straight-line segment BP parallel to the x axis connecting $B(x,y)$ with $P(x + \Delta x,\, y)$. Then, on forming (22-6), with the aid of (22-5) and (22-7), we get

$$\frac{\partial u}{\partial x} = \lim_{\Delta x \to 0} \frac{1}{\Delta x} \left[\int_{\overarc{AP}} [M(x,y)\,dx + N(x,y)\,dy] - \int_{\overarc{AB}} [M(x,y)\,dx + N(x,y)\,dy] \right]$$

$$= \lim_{\Delta x \to 0} \frac{1}{\Delta x} \int_{BP} [M(x,y)\,dx + N(x,y)\,dy].$$

But, along BP, $y = \text{const}$, $dy = 0$; hence

$$\frac{\partial u}{\partial x} = \lim_{\Delta x \to 0} \frac{1}{\Delta x} \int_x^{x+\Delta x} M(x,y)\,dx. \tag{22-8}$$

On applying the mean-value theorem for definite integrals to (22-8), we get

$$\frac{\partial u}{\partial x} = \lim_{\Delta x \to 0} \left[\frac{1}{\Delta x} M(\xi,y)\,\Delta x \right] = M(x,y). \tag{22-9}$$

We get in a similar way

$$\frac{\partial u}{\partial y} = N(x,y). \tag{22-10}$$

Accordingly, by Sec. 4, the expression

$$M \, dx + N \, dy = \frac{\partial u}{\partial x} \, dx + \frac{\partial u}{\partial y} \, dy \tag{22-11}$$

is the total differential du of the function defined by (22-5). It should be observed that the results (22-9) to (22-11) are valid in simply and multiply connected regions, since they have been deduced on the assumption that the integral (22-5) is independent of the path in *any* given connected region.

On noting the result (22-11) and referring to Theorem I, we see that we have established the following:

THEOREM II. If $M(x,y)$ and $N(x,y)$ satisfy the conditions of Theorem I in a simply connected region R, then a necessary and sufficient condition for the expression $M(x,y) \, dx + N(x,y) \, dy$ to be the total differential of some function $u(x,y)$ is that $\partial M/\partial y = \partial N/\partial x$ at all points of R. The function $u(x,y)$ is then determined by (22-5).

We shall see in Sec. 23 that Theorem II may not hold if the region R is multiply connected, because formula (22-1) was obtained on the assumption that R is a simply connected region.

─── **PROBLEMS**

1. Use Green's theorem to show that

$$\int_C \left[(x^2 + xy) \, dx + (y^2 + x^2) \, dy \right] = 0$$

if C is the boundary of the square $x = \pm 1, y = \pm 1$. Note that $\partial(x^2 + xy)/\partial y \neq \partial(y^2 + x^2)/\partial x$. Does this contradict Theorem I of this section? Evaluate this integral when C is the boundary of the square $x = 0, y = 0, x = 1, y = 1$.

2. Show that the following integrals defined in the entire xy plane are independent of the path, and find $u(x,y)$ by formula (22-5) and by integrating Eqs. (22-9) and (22-10):

(a) $\displaystyle\int_{(0,0)}^{(x,y)} \left[(x^2 + y^2) \, dx + 2xy \, dy \right]$

(b) $\displaystyle\int_{(0,0)}^{(x,y)} \left[y \cos x \, dx + \sin x \, dy \right]$

(c) $\displaystyle\int_{(-1,-1)}^{(x,y)} \left[(x + 1) \, dx + (y + 1) \, dy \right]$

(d) $\displaystyle\int_{(0,0)}^{(x,y)} (y^2 \, dx + 2xy \, dy)$.

3. If a simply connected region R does not include the point $(0,0)$, show that the line integral

$$u(x,y) = \int_{(x_0,y_0)}^{(x,y)} \left(\frac{x}{x^2 + y^2} \, dx + \frac{y}{x^2 + y^2} \, dy \right)$$

is independent of the path and that $u = \frac{1}{2} \log (x^2 + y^2)$. Can you apply Green's theorem to evaluate this line integral over a circular path $x^2 + y^2 = 1$? Evaluate the integral over this circle.

───

23. Green's Theorem for Multiply Connected Regions.   Formula (22-1) can be extended to a bounded multiply connected region having several simple closed curves as its boundaries (Fig. 27). As in Sec. 22, we suppose that $M(x,y)$, $N(x,y)$, $\partial M/\partial y$, and $\partial N/\partial x$ are continuous in the closed region R.

We begin with a doubly connected region, that is, a region bounded externally by a simple closed curve C_0 and internally by a simple closed curve C_1 (Fig. 26). We introduce a "cut" C joining some point P_0 on C_0 with a point P_1 on C_1, where C is a nonintersecting

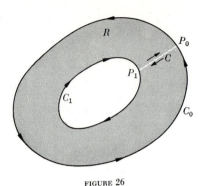

FIGURE 26

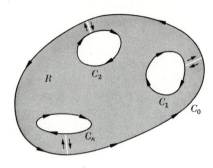

FIGURE 27

piecewise-smooth curve. The cut C can be visualized as a slit in the region with the boundary $C + C_0 + C_1$ of the slit region. The slit region is then simply connected, and if we apply formula (22-1) to it, we get

$$-\iint\limits_{R} \left(\frac{\partial M}{\partial y} - \frac{\partial N}{\partial x}\right) dy\, dx = \oint_{C_0} (M\, dx + N\, dy) + \int_{P_0}^{P_1} (M\, dx + N\, dy)$$

$$+ \oint_{C_1} (M\, dx + N\, dy) + \int_{P_1}^{P_0} (M\, dx + N\, dy). \quad (23\text{-}1)$$

The arrows on the integrals in (23-1) refer to the direction of integration along C_0 and C_1 as shown in Fig. 26, and the integrals $\int_{P_0}^{P_1}$ and $\int_{P_1}^{P_0}$ are evaluated along C in the direction indicated by the limits. Inasmuch as $\int_{P_0}^{P_1} = -\int_{P_1}^{P_0}$, Eq. (23-1) reduces to

$$-\iint\limits_{R} \left(\frac{\partial M}{\partial y} - \frac{\partial N}{\partial x}\right) dy\, dx = \oint_{C_0} (M\, dx + N\, dy) + \oint_{C_1} (M\, dx + N\, dy). \quad (23\text{-}2)$$

An obvious extension of this result to the region R bounded externally by C_0 and internally by n contours C_i (Fig. 27) yields

$$-\iint\limits_{R} \left(\frac{\partial M}{\partial y} - \frac{\partial N}{\partial x}\right) dy\, dx = \oint_{C_0} (M\, dx + N\, dy) + \sum_{i=1}^{n} \oint_{C_i} (M\, dx + N\, dy). \quad (23\text{-}3)$$

An important result follows directly from formula (23-3) if it is supposed that continuously differentiable functions M and N are such that

$$\frac{\partial M}{\partial y} = \frac{\partial N}{\partial x} \quad (23\text{-}4)$$

in the region R. If (23-4) holds in R, the double integral in (23-3) vanishes and we get

$$\oint_{C_0} (M\, dx + N\, dy) = -\sum_{i=1}^{n} \oint_{C_i} (M\, dx + N\, dy)$$

$$= \sum_{i=1}^{n} \oint_{C_i} (M\, dx + N\, dy).$$

Thus, the line integral over the exterior contour C_0 taken in the counterclockwise direction is equal to the sum of the line integrals over the interior contours C_i taken in the same direction.   In particular, if there is only one interior contour C_1 (Fig. 26), we conclude that

$$\int_{C_0} (M\, dx + N\, dy) = \int_{C_1} (M\, dx + N\, dy). \qquad (23\text{-}5)$$

These integrals need not vanish.   If, however, continuously differentiable functions M and N are also defined in the region interior to C_1 and satisfy the condition (23-4) in that region, the value of the integral $\int_{C_1} (M\, dx + N\, dy)$ is zero, inasmuch as the integral on the left in (23-5) vanishes by theorem (22-1).

Example.   Consider the function $u = \tan^{-1}(y/x)$ in the region R bounded by the concentric circles $x^2 + y^2 = a^2$ and $x^2 + y^2 = b^2$ (Fig. 28).   The partial derivatives of u,

$$\frac{\partial u}{\partial x} = \frac{-y}{x^2 + y^2} \equiv M \qquad \frac{\partial u}{\partial y} = \frac{x}{x^2 + y^2} \equiv N \qquad (23\text{-}6)$$

are obviously continuously differentiable in R (since $b > 0$) and hence Green's theorem [formula (23-2)] is applicable to R.   It is easy to check that $\partial M/\partial y = \partial N/\partial x$ at all points of the region R; hence by (23-2)

$$\oint_{C_0} (M\, dx + N\, dy) = \oint_{C_1} (M\, dx + N\, dy) \qquad (23\text{-}7)$$

where C_0 and C_1 are labeled in Fig. 28.   To evaluate (23-7) we can take equations of C_0 in the form $x = a \cos \theta$, $y = a \sin \theta$, $0 \leq \theta \leq 2\pi$, and on using (23-6) we find

$$\int_{C_0} (M\, dx + N\, dy) = \int_{C_0} \frac{-y\, dx + x\, dy}{x^2 + y^2} = \int_0^{2\pi} d\theta = 2\pi. \qquad (23\text{-}8)$$

This result shows that the line integral in (23-8) is not independent of the path; if it were, its value over every closed path would be zero, by Sec. 20.

If $P(x,y)$ is an arbitrary point on C_0 and the integral in (23-8) is evaluated along C_0 from the point $P_0(a,0)$ to $P(x,y)$ in a counterclockwise sense, we get

$$u(x,y) = \int_{(a,0)}^{(x,y)} \frac{-y\, dx + x\, dy}{x^2 + y^2} = \int_0^{\theta} d\theta = \theta. \qquad (23\text{-}9)$$

The angle $\theta = \tan^{-1} y/x$ is the polar angle of the point (x,y), and hence $u(x,y) = \tan^{-1}(y/x)$. Now by (23-6)

$$\frac{\partial u}{\partial x} dx + \frac{\partial u}{\partial y} dy = \frac{-y\, dx + x\, dy}{x^2 + y^2} \qquad (23\text{-}10)$$

and even though, in this case, $\partial M/\partial y = \partial N/\partial x$ at all points of R, the expression in (23-10) is not a total differential.   The function $u(x,y) = \tan^{-1}(y/x)$ is multiple-valued, and the total differentials are defined only for single-valued functions.   We also recall that the line integral $\int_C (M\, dx + N\, dy)$ over every simple closed path C invariably vanishes if $M\, dx + N\, dy$ is the total differential.   By introducing a crosscut as in Fig. 26, the reader will show that the value of the integral $\oint_C \dfrac{-y\, dx + x\, dy}{x^2 + y^2}$ is 2π for every simple closed path C enclosing C_1 and that its value is 0 for every simple closed path that does not enclose C_1.

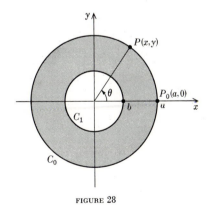

FIGURE 28

PROBLEMS

1. Evaluate $\int_C (-y\,dx + x\,dy)$ by Green's theorem (23-2), if C is the boundary of the ring formed by the circles $x^2 + y^2 = 1$ and $x^2 + y^2 = 4$.   (*Ans.: 6π.*)

2. Evaluate $\oint_C \dfrac{x\,dy + y\,dx}{x^2 + a^2 y^2}$, where C is the ellipse $x^2 + a^2 y = 1$.   (*Ans.: 0.*)

3. Apply Green's theorem (23-2) to evaluate $\int_C \dfrac{x\,dx + y\,dy}{x^2 + y^2}$, where $C = C_0 + C_1$ is the boundary of the region in Fig. 28.   What is the relation of the integrand to the function $u = \frac{1}{2} \log (x^2 + y^2)$?

4. Evaluate $\int_C [x(x^2 - y^2)\,dx - y(x^2 - y^2)\,dy]$, where C is the boundary of the region in Fig. 28. Is the integrand a total differential of some function $u(x,y)$?   If so, what is $u(x,y)$?   [*Ans.: $u = \frac{1}{4}(x^2 - y^2) + \text{const.}$*]

Vector field theory — 6

Scalar and vector fields

Illustrations and applications

Vector field theory

THIS CHAPTER is concerned with scalar and vector functions in Euclidean three-dimensional space. The emphasis throughout is placed on the invariant properties of scalar and vector functions, that is, properties that do not depend on the coordinate representation. Nature has no cognizance of special reference frames, which are merely useful tools in carrying out calculations. The principal object of this chapter is to deduce several transformation theorems involving line, surface, and volume integrals. These theorems, associated with the names of Gauss, Green, Stokes, and Ostrogradsky, are indispensable in the study of fluids, thermodynamics, electrodynamics, and every branch of mechanics of continua. They are also of profound importance in the theoretical development of mathematical analysis.

SCALAR AND VECTOR FIELDS

1. Invariants. Scalar and Vector Functions. In the preceding chapter we studied functions of several variables, specified in cartesian reference frames. The reason for according preference to cartesian systems is that the computations associated with the development of the idea of the rate of change (the derivative of a function), or that of the area (the integral), are relatively simple in cartesian coordinates. On the other hand, the ideas themselves are obviously independent of the chosen reference system, which merely provides a language for description of the ideas. Moreover, in the course of the development of a geometric or physical concept with the aid of the special language provided by this or that reference frame, it is not always clear that the results obtained are universally valid. Accordingly, in formulating the basic concepts and theorems concerned with natural phenomena, it is desirable to phrase them in an invariant form, that is, in a form that makes no use of special reference frames.

A point in space, for example, is an invariant. In cartesian coordinates it may be described by a triple of numbers (x,y,z), but in visualizing a point no scientist thinks of just one (or any) particular triple of numbers. If the coordinate system is changed, the same point is usually characterized by a different triple of numbers, but the change of coordinates does nothing to the point itself. Again, a set of points forming a curve or surface is invariant, although the formulas used to represent curves and surfaces depend on the chosen reference frame. In this chapter we formulate all basic definitions and theorems, not concerned with specific calculations, in an invariant form.

Let R be a region[1] of space at each point P of which a number $u(P)$ is determined.

---

[1] In Sec. 1 we use *region* to mean point set, though usually the concept of region is more restricted.

We then say that $u = u(P)$ is a *scalar point function* in R. The temperature at a point of a body is an example of a scalar point function; the hydrostatic pressure at points on the surface of a body immersed in a fluid is another example.

Besides point functions we can also consider scalar functions of regions, or *region functions*. If a number $u(D)$ is determined for each bounded region D in a suitable class of regions, u is a *scalar region function*. A simple example of a scalar region function is the area $a(D)$; this is defined for the class of plane regions having area. If a mass distribution is given over the region R, the mass $m(D)$ contained in the region D is another example. This is defined for the class of subregions of R having mass. Neither area nor mass can be suitably defined for *all* subregions of a given region R, and one sometimes uses the term *admissible* to specify the regions on which a given region function $u(D)$ may be defined.

All region functions considered in this book are additive. This means that if D is divided into two nonoverlapping parts D_1 and D_2 such that[1] $D = D_1 + D_2$ and D_1 and D_2 are admissible, then D is admissible, and $u(D) = u(D_1) + u(D_2)$. The area function $a(D)$ mentioned above is additive, because, if D_1 and D_2 are disjoint plane regions having area, their union $D_1 + D_2$ also has area, and the area is the sum of the areas of D_1 and D_2. Similarly, the mass function $m(D)$ is additive.

When a vector $\mathbf{v}(P)$ is determined at each point P of R, $\mathbf{v} = \mathbf{v}(P)$ is said to be a *vector point function*. The velocity $\mathbf{v}(P)$ of points in a moving fluid is an example of a vector point function. The definition of a vector region function is similar. Unless otherwise noted, we shall assume that the scalar and vector functions that concern us are single-valued; in other words, they are *functions* in the accepted sense of the term. Scalar and vector functions are sometimes called *fields*, and the study of scalar and vector fields is called *field theory*.

To facilitate calculations with scalar and vector functions it may prove convenient to refer the region of definition to some special coordinate system. If the system is cartesian, we shall write $u(P)$ as $u(x,y,z)$ and $\mathbf{v}(P)$ as $\mathbf{v}(x,y,z)$. The vector $\mathbf{v}(x,y,z)$ at any point can be represented by its components $v_x(x,y,z)$, $v_y(x,y,z)$, $v_z(x,y,z)$ along the appropriate base vectors. It should be noted, however, that the introduction of coordinate systems is a matter of convenience and that $u(P)$ and $\mathbf{v}(P)$ depend only on the choice of P in the field and not on any special reference frame selected to locate P. The fact that scalar and vector point functions are independent of coordinate systems is spoken of as *invariance*, and we shall see that it is possible to associate with $u(P)$ and $\mathbf{v}(P)$ certain new scalar and vector functions that have important invariant significance.

PROBLEMS

1. Give three physical examples of (*a*) scalar and (*b*) vector fields not mentioned in the text.

2. Sketch the vector velocity field associated with a disk rotating with constant angular velocity about an axis through its center perpendicular to the plane of the disk.

3. Sketch the vector field formed by unit tangents to the family of parabolas $y = cx^2$, where c is a parameter.

4. Which of the following is, in general, an additive scalar region function: temperature, volume, heat, velocity, speed, electrostatic charge, intensity of light?

---

[1] We use $D_1 + D_2$ to represent the union of D_1 and D_2, that is, for the set of points belonging to at least one of the sets forming the sum.

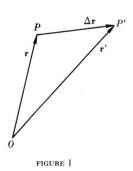

FIGURE 1

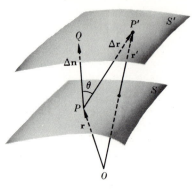

FIGURE 2

2. Continuity of Scalar and Vector Functions. Gradient.

Consider a scalar point function $u = u(P)$ or a vector point function $\mathbf{v}(P)$ in some region R.

It is said that $u(P)$ and $\mathbf{v}(P)$ are *continuous* at P if[1]

$$\lim_{P' \to P} u(P') = u(P) \qquad \text{and} \qquad \lim_{P' \to P} \mathbf{v}(P') = \mathbf{v}(P).$$

Functions continuous at every point of the region are said to be continuous in the region.

Let $u(P)$ be a continuous scalar function in the given region. We select a point O in this region for the origin of position vectors $\mathbf{r} \equiv \overrightarrow{OP}$. If P' is some point in the neighborhood of P, we denote $\overrightarrow{OP'}$ by $\mathbf{r}'$ and write (Fig. 1)

$$\mathbf{r}' \equiv \mathbf{r} + \Delta\mathbf{r}.$$

The difference quotient

$$\frac{u(P') - u(P)}{|\Delta\mathbf{r}|} \equiv \frac{u(P') - u(P)}{\Delta s} \tag{2-1}$$

where $|\Delta\mathbf{r}| = \Delta s$, gives an approximate space rate of change of $u(P)$, and we can study the limit of (2-1) as P' is made to approach P. We choose P' so that $\Delta\mathbf{r}$ remains parallel to a given fixed unit vector $\mathbf{t}$, as $\Delta\mathbf{r} \to 0$. If this limit exists we write

$$\lim_{\Delta s \to 0} \frac{u(P') - u(P)}{\Delta s} = \frac{du}{ds}. \tag{2-2}$$

and call du/ds the *directional* derivative of u in the direction specified by $\mathbf{t}$. A different choice of $\mathbf{t}$ yields different vectors $\Delta\mathbf{r}$ and, in general, a different value for du/ds at P. If all derivatives du/ds at P are continuous, $u(P)$ is *continuously differentiable* at P. Unless a statement to the contrary is made, we assume in the sequel that all functions $u(P)$ are continuously differentiable for each point P of the region R in which $u(P)$ is defined.

A set of points at which $u(P)$ assumes a constant value c determines a *level surface* S, which we denote by $u(P) = c$.

Let us consider a pair of such surfaces S and S' determined by $u = c$ and $u = c + \Delta c$, where Δc is a small change in c (Fig. 2). If P is a point on S and P' on S', the change $\Delta u \equiv u(P') - u(P)$ is Δc, and this is independent of the position of P' on S'. But the average space rate of change

$$\frac{u(P') - u(P)}{|\Delta\mathbf{r}|} \equiv \frac{\Delta u}{\Delta s} \tag{2-3}$$

[1] Cf. Chap. 4, Sec. 7.

clearly depends on the magnitude of $\Delta\mathbf{r}$.   The limit of this ratio as $\Delta\mathbf{r}$ is made to approach zero by making $\Delta c \rightarrow 0$ is the directional derivative (2-2) in the fixed direction determined by $\Delta\mathbf{r} = \mathbf{k}t$.   The greatest space rate of change of u will occur when P' is taken on the normal $\overrightarrow{PQ} \equiv \Delta\mathbf{n}$ to the surface S (Fig. 2), since for this position of P' the denominator $|\Delta\mathbf{r}|$ in (2-3) has its minimum value $|\Delta\mathbf{n}|$.   Indeed,

$$\Delta\mathbf{n} \doteq \Delta\mathbf{r} \cos \theta \tag{2-4}$$

where θ is the angle between the normal $\overrightarrow{PQ}$ to S and $\overrightarrow{PP'}$.

On taking account of (2-4), we conclude that

$$\frac{du}{dn} = \frac{1}{\cos \theta} \frac{du}{ds} \equiv \frac{du}{ds} \sec \theta. \tag{2-5}$$

The derivative du/dn in the direction of the normal to the level surface $u = \text{const}$ is called the *normal derivative of* $u(P)$.

If $\mathbf{n}$ is a unit vector at P, pointing in the direction for which $\Delta u > 0$, we can construct a vector, called the *gradient of u*, namely,

$$\text{grad } u \equiv \mathbf{n} \frac{du}{dn}. \tag{2-6}$$

This vector represents in both the direction and magnitude the *greatest space rate of increase* of $u(P)$, provided, of course, that $du/dn \neq 0$.   The gradient vector (2-6) is clearly independent of the choice of coordinate systems and hence is an invariant.

To get a physical interpretation, let $u(P)$ be the temperature field.   Then the gradient is directed along the normal to the isothermal surface $u = c$, and its direction is that in which the rate of increase of temperature is a maximum.

If we introduce cartesian coordinates and denote $u(P)$ by $u(x,y,z)$, the assumption that $u(P)$ is continuously differentiable in R ensures that the partial derivatives u_x, u_y, u_z are continuous.   We can then write (see Chap. 5, Sec. 4)

$$\Delta u = \frac{\partial u}{\partial x} \Delta x + \frac{\partial u}{\partial y} \Delta y + \frac{\partial u}{\partial z} \Delta z + \epsilon_1 \Delta x + \epsilon_2 \Delta y + \epsilon_3 \Delta z \tag{2-7}$$

where ϵ_1, ϵ_2, ϵ_3 approach zero as Δx, Δy, $\Delta z \rightarrow 0$.   On dividing (2-7) by $\Delta s \equiv \sqrt{(\Delta x)^2 + (\Delta y)^2 + (\Delta z)^2}$ and letting $\Delta s \rightarrow 0$, we get (cf. Chap. 5, Sec. 5)

$$\frac{du}{ds} = \frac{\partial u}{\partial x} \frac{dx}{ds} + \frac{\partial u}{\partial y} \frac{dy}{ds} + \frac{\partial u}{\partial z} \frac{dz}{ds} \tag{2-8}$$

since ϵ_1, ϵ_2, ϵ_3 approach zero as $\Delta s \rightarrow 0$.   Clearly, $dx/ds = \cos(x,s)$, $dy/ds = \cos(y,s)$, $dz/ds = \cos(z,s)$ are the direction cosines of the unit vector $\mathbf{t}$ coinciding with $d\mathbf{r}$ (Fig. 3). Since the position vector $\mathbf{r}$ of the point P is $\mathbf{r} = \mathbf{i}x + \mathbf{j}y + \mathbf{k}z$,

$$\mathbf{t} \equiv \frac{d\mathbf{r}}{ds} = \mathbf{i} \frac{dx}{ds} + \mathbf{j} \frac{dy}{ds} + \mathbf{k} \frac{dz}{ds}. \tag{2-9}$$

We see that (2-8) can be written as the scalar product of the vector

$$\nabla u \equiv \mathbf{i} \frac{\partial u}{\partial x} + \mathbf{j} \frac{\partial u}{\partial y} + \mathbf{k} \frac{\partial u}{\partial z} \tag{2-10}$$

and the unit vector $\mathbf{t}$ in (2-9).   Thus,

$$\frac{du}{ds} = \nabla u \cdot \mathbf{t}. \tag{2-11}$$

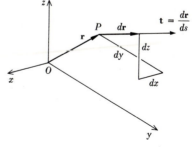

FIGURE 3

Inasmuch as the greatest value of du/ds is assumed when the direction of $\mathbf{t}$ coincides with that of the normal $\mathbf{n}$ to the level surface $u = $ const, we conclude that

$$\nabla u = \operatorname{grad} u \qquad (2\text{-}12)$$

for the right-hand member of (2-11) can be interpreted as the component of the vector ∇u in the direction $\mathbf{t}$ and the maximum component du/ds is obtained when $\mathbf{t}$ is directed along ∇u.

It follows from (2-10) and (2-12) that a formula for calculating grad u in cartesian coordinates is

$$\operatorname{grad} u = \mathbf{i}\,\frac{\partial u}{\partial x} + \mathbf{j}\,\frac{\partial u}{\partial y} + \mathbf{k}\,\frac{\partial u}{\partial z}. \qquad (2\text{-}13)$$

On comparing (2-6) and (2-13), we see that

$$|\operatorname{grad} u| = \left|\frac{du}{dn}\right| = \sqrt{\left(\frac{\partial u}{\partial x}\right)^2 + \left(\frac{\partial u}{\partial y}\right)^2 + \left(\frac{\partial u}{\partial z}\right)^2}.$$

Formula (2-10) suggests a definition of the differential vector operator ∇, called *del* or *nabla*,[1]

$$\nabla \equiv \mathbf{i}\,\frac{\partial}{\partial x} + \mathbf{j}\,\frac{\partial}{\partial y} + \mathbf{k}\,\frac{\partial}{\partial z}$$

analogous to the scalar differential operator $D \equiv d/dx$ introduced in Chap. 3, Sec. 11. The product of ∇ and the scalar $u(x,y,z)$ is interpreted to mean (2-10). The reader will show that

$$\begin{aligned}
\nabla(u + v) &= \nabla u + \nabla v \\
\nabla(uv) &= u\nabla v + v\nabla u
\end{aligned} \qquad (2\text{-}14)$$

whenever u and v are continuously differentiable scalar functions of (x,y,z). A formula for grad u in orthogonal curvilinear coordinates is deduced in Sec. 13.

The directional derivative dv/ds of a vector point function $\mathbf{v}(P)$ is defined by formula (2-2) in which $u(P)$ is replaced by $\mathbf{v}(P)$. When $\mathbf{v}(P)$ is expressed in the form

$$\mathbf{v} = \mathbf{i}v_x + \mathbf{j}v_y + \mathbf{k}v_z \qquad (2\text{-}15)$$

where $\mathbf{i}$, $\mathbf{j}$, $\mathbf{k}$ are the base vectors in the system x,y,z,

$$\frac{d\mathbf{v}}{ds} = \mathbf{i}\,\frac{dv_x}{ds} + \mathbf{j}\,\frac{dv_y}{ds} + \mathbf{k}\,\frac{dv_z}{ds}. \qquad (2\text{-}16)$$

We have already employed a similar formula in Chap. 4, Sec. 7, to calculate the derivatives of the position vector $\mathbf{R} = \mathbf{i}x + \mathbf{j}y + \mathbf{k}z$ with respect to the time parameter t.

Example 1.   Find the directional derivative of $u = xyz^2$ at $(1,0,3)$ in the direction of the vector $\mathbf{i} - \mathbf{j} + \mathbf{k}$. Compute the greatest rate of change of u and the direction of the maximum rate of increase of u.

On substituting $u = xyz^2$ in (2-10), we find that the gradient u is given by

$$\nabla u = \mathbf{i}yz^2 + \mathbf{j}xz^2 + \mathbf{k}2xyz.$$

At $(1,0,3)$

$$\nabla u = \mathbf{i}0 + \mathbf{j}9 + \mathbf{k}0 = 9\mathbf{j}.$$

Thus, the greatest rate of change $|\nabla u| = 9$, and the direction of the maximum rate of change is along the y axis.   Since the unit vector $\mathbf{t}$ in the direction of the vector $\mathbf{i} - \mathbf{j} + \mathbf{k}$ is

$$\mathbf{t} = \frac{1}{\sqrt{3}}\,(\mathbf{i} - \mathbf{j} + \mathbf{k})$$

[1] From Greek ναβλα, meaning "harp."

we find on using (2-11) that the desired directional derivative is

$$\frac{du}{ds} = \nabla u \cdot \mathbf{t} = 9\mathbf{j} \cdot \frac{1}{\sqrt{3}}(\mathbf{i} - \mathbf{j} + \mathbf{k}) = -\frac{9}{\sqrt{3}}.$$

Example 2.   Find the unit normals to the surface $x^2 - y^2 + z^2 = 6$ at (1,2,3).

The surface in this example is a level surface for the function $u = x^2 - y^2 + z^2$. Since the gradient of u is normal to the level surface $u = $ const, we have by (2-12)

$$\operatorname{grad} u = \nabla u = \mathbf{i}2x - \mathbf{j}2y + \mathbf{k}2z$$

which at (1,2,3) has the value

$$\nabla u = \mathbf{i}2 - \mathbf{j}4 + \mathbf{k}6.$$

But this vector is directed along the unit normal $\mathbf{n}$ to $u = x^2 - y^2 + z^2 = 6$ in the direction of increasing u.   Hence

$$\mathbf{n} = \frac{\nabla u}{|\nabla u|} = \frac{1}{\sqrt{56}}(\mathbf{i}2 - \mathbf{j}4 + \mathbf{k}6).$$

The direction of the other unit normal vector is opposite to this.

PROBLEMS

1. Compute the directional derivative of $u = x^2 + y^2 + z^2$ at (1,2,3) in the direction of the line

$$\frac{x}{3} = \frac{y}{4} = \frac{z}{5}.$$

Find the maximum rate of increase of u at (1,2,3); at (0,1,2).

2. Find grad u if (a) $u = (x^2 + y^2 + z^2)^{-\frac{1}{2}}$ and (b) $u = \log(x^2 + y^2 + z^2)$.

3. Find the directional derivative of $u = x^2y - y^2z - xyz$ at $(1,-1,0)$ in the direction of the vector $\mathbf{i} - \mathbf{j} + 2\mathbf{k}$.

4. Find the directional derivative of $u = xyz$ at (1,2,3) in the direction from (1,2,3) to $(1,-1,-3)$.

5. Find the unit normal vector in the direction of the exterior normal to the surface $x^2 + 2y^2 + z^2 = 7$ at $(1,-1,2)$.

6. Find the unit vectors normal to $xyz = 2$ at $(1,-1,-2)$.

7. Show that $\nabla r^n = nr^{n-2}\mathbf{r}$, where $\mathbf{r} = \mathbf{i}x + \mathbf{j}y + \mathbf{k}z$ and $r = |\mathbf{r}|$.

8. Use the result of Prob. 7 to compute the directional derivative of $u = (x^2 + y^2 + z^2)^{\frac{3}{2}}$ at $(-1,1,2)$ in the direction of the vector $\mathbf{i} - 2\mathbf{j} + \mathbf{k}$.

9. Compute the directional derivative of

$$\mathbf{v} = \mathbf{i}(x^2 - y^2) + \mathbf{j}(xyz - 1) + \mathbf{k}z$$

at (1,2,0) in the direction from (1,2,0) to (0,0,0).

10. Find the angle between the gradients of $u = \sqrt{x^2 + y^2}$ and $v = x - 3y + 3xy$ at the point (3,4).

11. Find the directional derivative of $u = (x^2 + y^2 + z^2)^{\frac{1}{2}}$ in the direction of the gradient of u.

12. If $w = F(u,v)$, $u = f(x,y)$, and $v = g(x,y)$ are continuously differentiable, prove that

$$\nabla w = \frac{\partial F}{\partial u}\nabla u + \frac{\partial F}{\partial v}\nabla v.$$

13. Compute $|\nabla u|$ at (1,0,0) and determine the level surfaces for $u = \sin^{-1}(z/\sqrt{x^2 + y^2})$, $x^2 + y^2 \neq 0$.

14. If $u = f(r)$, where $r = (x^2 + y^2 + z^2)^{\frac{1}{2}}$ and $f(r)$ is continuously differentiable, show that $\nabla u = f'(r)(\mathbf{i}x/r + \mathbf{j}y/r + \mathbf{k}z/r)$ and that the level surfaces are spheres.

15. If u and v are continuously differentiable scalars, show that (a) $\nabla(u^n) = nu^{n-1}\nabla u$; (b) $\nabla(u/v) = (v\nabla u - u\nabla v)/v^2$.

16. Show that the directional derivative of $u = y^2/x$ at every point of the ellipse $2x^2 + y^2 = 1$ in the direction of the normal to the ellipse has the value zero.

---

3. Line Integrals in Three-dimensional Space.   In Chap. 5, Sec. 19, we defined the line integral $\int_C f(P)\,ds$, where $f(P)$ was a function $f(x,y)$ specified along a plane curve C. A word-for-word repetition of the formulation of that section yields the definition of the line integral of a piecewise-continuous scalar function $f(P)$ specified over a piecewise-smooth space curve C.

When C is given in vector form as $\mathbf{r} = \mathbf{r}(s)$, s being the arc parameter measured from some point A on C (Fig. 4), the curve C is said to be smooth if the tangent vector $\mathbf{t} = d\mathbf{r}/ds$ is continuous at each point of C. The definitions of *simple arc, simple closed curve, piecewise-smooth curve,* and *path* in space parallel those given for plane curves in Chap. 5, Sec. 19, and are not repeated here.

Unless a statement is made to the contrary, we shall assume that the space curves C considered in this chapter are piecewise smooth.

Let $\mathbf{v} = \mathbf{v}(P)$ be a piecewise-continuous vector function specified along a smooth curve C given by the vector equation $\mathbf{r} = \mathbf{r}(s)$, $0 \leq s \leq l$. Then, the scalar product

$$\mathbf{v}\cdot d\mathbf{r} = \mathbf{v}\cdot\frac{d\mathbf{r}}{ds}\,ds = \mathbf{v}\cdot\mathbf{t}\,ds$$

where $\mathbf{t} = d\mathbf{r}/ds$ is the unit tangent vector to C at P (Fig. 4). If we write $u(P) \equiv \mathbf{v}\cdot\mathbf{t}$, we can form the line integral

$$\int_C u(P)\,ds = \int_C \mathbf{v}\cdot\mathbf{t}\,ds \equiv \int_C \mathbf{v}\cdot d\mathbf{r}. \tag{3-1}$$

This integral has an obvious physical meaning when $\mathbf{v}(P)$ is the force acting on a particle moving along C, since the integrand $\mathbf{v}\cdot d\mathbf{r}$ represents the work done in displacing the particle through a distance $ds = |d\mathbf{r}|$, and the integral (3-1) then defines the work done in moving the particle along C.

As in Chap. 5, Sec. 19, the line integral (3-1) can be evaluated by reducing it to a definite integral when $\mathbf{v}(P)$ and the equation of C are specified in some coordinate system. Thus, if

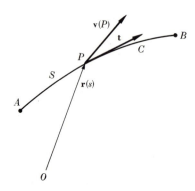

FIGURE 4

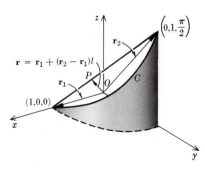

FIGURE 5

$$\mathbf{v}(P) = \mathbf{i}v_x(x,y,z) + \mathbf{j}v_y(x,y,z) + \mathbf{k}v_z(x,y,z) \tag{3-2}$$

where v_x, v_y, v_z are cartesian components of the vector function $\mathbf{v}(P)$, and C is defined by

$$\mathbf{r} = \mathbf{i}x(t) + \mathbf{j}y(t) + \mathbf{k}z(t) \qquad t_0 \le t \le t_1$$

or $\qquad\qquad x = x(t) \qquad y = y(t) \qquad x = z(t) \qquad t_0 \le t \le t_1 \tag{3-3}$

then $d\mathbf{r} = \mathbf{i}\,dx(t) + \mathbf{j}\,dy(t) + \mathbf{k}\,dz(t)$ and (3-1) assumes the form[1]

$$\int_C \mathbf{v}\cdot d\mathbf{r} = \int_C [v_x(x,y,z)\,dx + v_y(x,y,z)\,dy + v_z(x,y,z)\,dz]. \tag{3-4}$$

On substituting in the integrand of (3-4) from (3-3) one obtains a definite integral of the type $\int_{t_0}^{t_1} F(t)\,dt$. We illustrate the evaluation of such integrals by examples.

Example 1. Evaluate the integral $\int_C \mathbf{r}\cdot d\mathbf{r}$ when C is the helical path

$$x = \cos t \qquad y = \sin t \qquad z = t \tag{3-5}$$

joining the points determined by $t = 0$ and $t = \pi/2$ and also when C is the straight line joining these points.

Since $\mathbf{r} = \mathbf{i}x + \mathbf{j}y + \mathbf{k}z$, we get, on using (3-5),

$$\mathbf{r} = \mathbf{i}\cos t + \mathbf{j}\sin t + \mathbf{k}t$$

$$d\mathbf{r} = (-\mathbf{i}\sin t + \mathbf{j}\cos t + \mathbf{k})\,dt.$$

Hence
$$\int_C \mathbf{r}\cdot d\mathbf{r} = \int_0^{\pi/2} t\,dt = \frac{\pi^2}{8}. \tag{3-6}$$

If the path C is a straight line joining the same points $(1,0,0)$ and $(0,1,\pi/2)$, we can write its equation in vector form as (cf. Chap. 4, Sec. 9)

$$\mathbf{r} = \mathbf{r}_1 + (\mathbf{r}_2 - \mathbf{r}_1)t \tag{3-7}$$

where $\mathbf{r}_1$ and $\mathbf{r}_2$ are the position vectors of $(1,0,0)$ and $(0,1,\pi/2)$, respectively (Fig. 5).

The parameter t clearly varies between 0 and 1 since, for $t = 0$, (3-7) yields $\mathbf{r} = \mathbf{r}_1$ and for $t = 1$, $\mathbf{r} = \mathbf{r}_2$. But $\mathbf{r}_1 = \mathbf{i}$, $\mathbf{r}_2 = \mathbf{j} + (\pi/2)\mathbf{k}$, so that (3-7) reduces to

$$\mathbf{r} = \mathbf{i} + \left(\mathbf{j} + \frac{\pi}{2}\mathbf{k} - \mathbf{i}\right)t.$$

Hence
$$\int_C \mathbf{r}\cdot d\mathbf{r} = \int_0^1 \left[-1 + \left(2 + \frac{\pi^2}{4}\right)t\right]dt = \frac{\pi^2}{8}.$$

This is the same value as we got for the helical path. In the following section we shall see why this particular integral is independent of the path.

Example 2. Compute the value of $\int_C \mathbf{v}\cdot d\mathbf{r}$, where $\mathbf{v} = \mathbf{i}y + \mathbf{j}2x$ and C is the straight line joining $(0,1)$ and $(1,0)$. Evaluate also when C is the arc of a circle of radius 1 centered at the origin (Fig. 6).

Since $\mathbf{r} = \mathbf{i}x + \mathbf{j}y$, we have $d\mathbf{r} = \mathbf{i}\,dx + \mathbf{j}\,dy$ and therefore

$$\int_C \mathbf{v}\cdot d\mathbf{r} = \int_C (y\,dx + 2x\,dy). \tag{3-8}$$

To evaluate this integral along the rectilinear path in Fig. 6, we write the equation of the path in the form

$$y = -x + 1 \tag{3-9}$$

and insert (3-9) in (3-8). Since $dy = -dx$, we get

---

[1] This is the usual form in which the line integrals are introduced in the coordinate treatment of line integrals in space. See formula (19-14) in Chap. 5, which is a special case of (3-4) obtained by setting $v_x = M$ and $v_y = N$, $v_z = 0$.

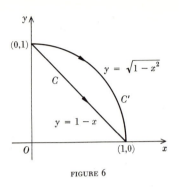

FIGURE 6

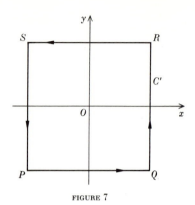

FIGURE 7

$$\int_C \mathbf{v}\cdot d\mathbf{r} = \int_0^1 [(-x+1)\,dx - 2x\,dx] = \int_0^1 (1-3x)\,dx = -\tfrac{1}{2}.$$

The integration here is performed so that the path C is traced from the point $(0,1)$ to $(1,0)$.

To compute the value of the integral (3-8) over the circular path C' joining the same two points, we note that $y = \sqrt{1-x^2}$ along C', $dy = -x\,dx/\sqrt{1-x^2}$, so that

$$\int_{C'} \mathbf{v}\cdot d\mathbf{r} = \int_0^1 \left[\sqrt{1-x^2}\,dx - \frac{2x^2}{\sqrt{1-x^2}}\,dx \right]$$

$$= \int_0^1 \frac{1-3x^2}{\sqrt{1-x^2}}\,dx = -\frac{\pi}{4}.$$

Again, the path C' is traced from $(0,1)$ to $(1,0)$. If the direction of description of C' is reversed, so that the circle is traced from $(1,0)$ to $(0,1)$, the limits in the integral must be interchanged, and we get $+\pi/4$ for the value of the integral.

Example 3.   Evaluate $\int_C \mathbf{v}\cdot d\mathbf{r}$, where $\mathbf{v} = (\mathbf{i}y - \mathbf{j}x)/(x^2 + y^2)$ and C is the circular path $x^2 + y^2 = 1$ described counterclockwise.

This integral can be evaluated as in the preceding example by substituting in the integrand $y = \sqrt{1-x^2}$ for points on the upper half of the circle C and $y = -\sqrt{1-x^2}$ on the lower half. It is simpler, however, to write the equations of the path in parametric form

$$x = \cos\theta \qquad y = \sin\theta \qquad 0 \le \theta \le 2\pi. \tag{3-10}$$

We thus get

$$\mathbf{r} = \mathbf{i}x + \mathbf{j}y = \mathbf{i}\cos\theta + \mathbf{j}\sin\theta$$

$$d\mathbf{r} = (-\mathbf{i}\sin\theta + \mathbf{j}\cos\theta)\,d\theta$$

$$\mathbf{v} = \frac{\mathbf{i}\sin\theta - \mathbf{j}\cos\theta}{\sin^2\theta + \cos^2\theta} = \mathbf{i}\sin\theta - \mathbf{j}\cos\theta.$$

Hence

$$\int_C \mathbf{v}\cdot d\mathbf{r} = \int_0^{2\pi} (-\sin^2\theta - \cos^2\theta)\,d\theta = -2\pi.$$

If the path is traced in the clockwise direction, we get $+2\pi$.

It may prove instructive to evaluate this integral over the square C' formed by the lines $x = \pm 1, y = \pm 1$ (Fig. 7).

The integral over C' is equal to the sum of four integrals evaluated over the paths PQ, QR, RS, SP.

Now along PQ, $y = -1$, $dy = 0$, $\mathbf{r} = \mathbf{i}x - \mathbf{j}$, $d\mathbf{r} = \mathbf{i}\,dx$, and $\mathbf{v} = (-\mathbf{i} - \mathbf{j}x)/(x^2 + 1)$.

Hence

$$\int_{PQ} \mathbf{v}\cdot d\mathbf{r} = \int_{-1}^1 \frac{-dx}{x^2+1} = -\tan^{-1} x \Big|_{-1}^{1} = -\frac{\pi}{2}.$$

Along the path QR, $x = 1$, $\mathbf{r} = \mathbf{i} + \mathbf{j}y$, $d\mathbf{r} = \mathbf{j}\,dy$, $\mathbf{v} = (\mathbf{i}y - \mathbf{j})/(1 + y^2)$, so that

$$\int_{QR} \mathbf{v}\cdot d\mathbf{r} = \int_{-1}^{1} \frac{-dy}{1 + y^2} = -\frac{\pi}{2}.$$

In a similar way we find that

$$\int_{RS} \mathbf{v}\cdot d\mathbf{r} = \int_{SP} \mathbf{v}\cdot d\mathbf{r} = -\frac{\pi}{2}$$

so that the integral

$$\int_{C'} \mathbf{v}\cdot d\mathbf{r} = 4\left(-\frac{\pi}{2}\right) = -2\pi.$$

This time we obtained the same result as we did for the circular path. In Sec. 4 we shall see that this is not an accident and that the value of this integral for every simple closed path enclosing the origin is -2π.

─── **PROBLEMS**

1. Evaluate the integral in Example 2 over the path C consisting of straight-line segments joining the points $(0,1)$, $(0,0)$, $(1,0)$ in that order.

2. Evaluate the integral in Example 1 over the polygonal path joining the points $(1,0,0)$, $(1,1,0)$, $(1,1,\pi/2)$ in that order.

3. Compute the value of the integral $\int_C (xy\,dx - y\,dy + dz)$ over the straight lines joining: (a) $(0,0,0)$ and $(1,1,1)$; (b) $(0,0,1)$ and $(0,1,1)$; (c) $(0,0,0)$ and $(1,2,3)$. Note that this integral has the form $\int_C \mathbf{v}\cdot d\mathbf{r}$.

4. Compute the integral $\int_C \mathbf{v}\cdot d\mathbf{r}$, where $\mathbf{v} = \mathbf{i}x - \mathbf{j}y + \mathbf{k}z$, over the helical path in Example 1. Also evaluate it over the rectilinear path.

5. Compute the work done in displacing a particle of unit mass in a constant gravitational field $\mathbf{F} = -\mathbf{k}g$ along the following paths:

(a) Straight line joining $(0,0,0)$ and $(1,1,1)$;
(b) A polygonal path joining $(0,0,0)$, $(1,1,0)$, $(1,1,1)$ in that order.

6. Evaluate $\int_C [x\,dx + y\,dy + (x + y - 1)\,dz]$, where C is the segment of the straight line joining $(1,1,1)$ and $(2,3,4)$.

7. The force field $\mathbf{F}$ is given by $\mathbf{F} = \mathbf{j}x^2$. Find the work done by the field when a particle of unit mass is displaced along the segment of the parabola $x = 1 - y^2$ between $(1,0)$ and $(0,1)$.

8. Compute the work done by the force $\mathbf{F}$ acting on a particle moving along the upper half of the ellipse $x = a\cos\theta$, $y = b\sin\theta$, if $\mathbf{F}$ is directed toward the center of the ellipse and its magnitude is equal to the distance of the particle from the center.

9. Evaluate $\int_C (xz\,dx + x\,dy - yz\,dz)$ over a polygonal path C joining $(0,0,1)$, $(1,0,0)$, $(0,1,0)$ in that order.

10. Evaluate $\int_C [(y - z)\,dx + (z - x)\,dy + (x - y)\,dz]$ over the circle C formed by the intersection of the sphere $x^2 + y^2 + z^2 = a^2$ and the plane $x + y + z = 0$.

───

4. Line Integrals of Gradients. Independence of the Path. The reader should review the definitions and discussion of line integrals presented in Chap. 5, Sec. 20. As in that section, we shall be concerned here with bounded connected regions of space. Such regions may be bounded by one or more sufficiently smooth closed surfaces (described in Sec. 5), and the definitions of the simply and multiply connected three-

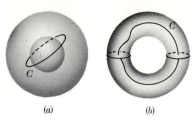

(a)   (b)

FIGURE 8

dimensional regions are identical with those given in Sec. 20 of Chap. 5. Thus, a region R is simply connected if every simple closed curve drawn in the interior of R can be shrunk to a point of R without crossing the bounding surfaces.

It should be noted, however, that the region bounded by two concentric spheres in Fig. 8a is simply connected, but the interior of a torus (an anchor ring) is not.

Let R be simply or multiply connected, and let C be a piecewise-smooth curve in the interior of R.

We consider the line integral $I = \int_C \mathbf{v} \cdot d\mathbf{r}$ in which the vector function $\mathbf{v}$, specified along C, is the gradient ∇u of some continuously differentiable (and hence single-valued) scalar point function $u(P)$ at each point P of R. Then,

$$I = \int_C \mathbf{v} \cdot d\mathbf{r} = \int_C \nabla u \cdot d\mathbf{r}. \tag{4-1}$$

The integrand in (4-1), being the scalar product of two vectors, is an *invariant;* that is, its value is the same whatever coordinate system one chooses to represent ∇u and $d\mathbf{r}$. But, in cartesian coordinates,

$$\mathbf{v} = \nabla u = \mathbf{i}\,\frac{\partial u}{\partial x} + \mathbf{j}\,\frac{\partial u}{\partial y} + \mathbf{k}\,\frac{\partial u}{\partial z} \qquad d\mathbf{r} = \mathbf{i}\,dx + \mathbf{j}\,dy + \mathbf{k}\,dz$$

so that

$$\mathbf{v} \cdot d\mathbf{r} = \nabla u \cdot d\mathbf{r} = \frac{\partial u}{\partial x}\,dx + \frac{\partial u}{\partial y}\,dy + \frac{\partial u}{\partial z}\,dz. \tag{4-2}$$

Since u is assumed to be continuously differentiable, the partial derivatives in (4-2) are continuous at all points of R; hence the right-hand member of (4-2) is the total differential du of the scalar function $u(x,y,z)$. (See Chap. 5, Sec. 4.) Accordingly, we can write (4-1) as

$$I = \int_C \nabla u \cdot d\mathbf{r} = \int_{P_0}^{P_1} du = u(P_1) - u(P_0) \tag{4-3}$$

where P_0 and P_1 are the end points of C. The result in (4-3) states that the value of the line integral $\int_C \nabla u \cdot d\mathbf{r}$ depends only on the difference of the values of $u(P)$ at the end points of the path C, and hence the integral is independent of the choice of the paths C. It should be observed that formula (4-3) is a generalization of the fundamental theorem of integral calculus [formula (14-1) in Chap. 5] and it permits one to write at once the value of the line integral (4-1) whenever $\mathbf{v}(P)$ is known to be the gradient of a single-valued scalar function $u(P)$.

We conclude, as in Chap. 5, Sec. 20, that when C is any closed path in R then $\int_C \nabla u \cdot d\mathbf{r} = 0$.

Thus the statement that a line integral $\int_C \mathbf{v} \cdot d\mathbf{r}$ is independent of the path in a given connected region R is equivalent to the statement $\int_C \mathbf{v} \cdot d\mathbf{r} = 0$ for every closed path C in R.

We can state the following:

THEOREM I.   *The line integral $\int_C \nabla u \cdot d\mathbf{r}$ in which $u(P)$ is a continuously differentiable (and hence single-valued) function in a given region R is independent of the path and so it vanishes for every closed path C in R.*

A trivial generalization of calculations, that have led to formula (22-11) of Chap. 5, enables one to show that, if the line integral[1]

$$\int_{P_0}^{P} \mathbf{v} \cdot d\mathbf{r} = \int_{(x_0,y_0,z_0)}^{(x,y,z)} [v_x(x,y,z) \, dx + v_y(x,y,z) \, dy + v_z(x,y,z) \, dz] = u(x,y,z) \qquad (4\text{-}4)$$

is independent of the path and thus defines a single-valued function $u(x,y,z)$, then

$$\frac{\partial u}{\partial x} = v_x(x,y,z) \qquad \frac{\partial u}{\partial y} = v_y(x,y,z) \qquad \frac{\partial u}{\partial z} = v_z(x,y,z) \qquad (4\text{-}5)$$

whenever v_x, v_y, v_z are continuous in a given connected region R. Since, by (4-5), $\nabla u = \mathbf{i}(\partial u/\partial x) + \mathbf{j}(\partial u/\partial y) + \mathbf{k}(\partial u/\partial z) = \mathbf{v}$, we can state the following:

THEOREM II. If a vector point function $\mathbf{v}(P)$ is continuous in a simply or multiply connected region R, and if the line integral $\int_C \mathbf{v} \cdot d\mathbf{r}$ is independent of the path, then a single-valued scalar $u(P)$ exists such that $\mathbf{v} = \Delta u$ in R.

When $\mathbf{v}(P)$ is specified in cartesian coordinates, $u(P) = u(x,y,z)$ can be determined by evaluating the line integral (4-4) over any convenient path in R, or by constructing, with the aid of (4-5), the function $u(x,y,z)$ such that $du = v_x \, dx + v_y \, dy + v_z \, dz$.

If $v_x(x,y,z)$, $v_y(x,y,z)$, $v_z(x,y,z)$ have continuous first partial derivatives in R, we obtain, by differentiating (4-5),

$$\frac{\partial^2 u}{\partial y \, \partial x} = \frac{\partial v_x}{\partial y} \qquad \frac{\partial^2 u}{\partial x \, \partial y} = \frac{\partial v_y}{\partial x} \qquad (4\text{-}6)$$

and, since the mixed derivatives in (4-6) are continuous, we conclude that

$$\frac{\partial v_x}{\partial y} = \frac{\partial v_y}{\partial x}. \qquad (4\text{-}7)$$

We find in the same way that

$$\frac{\partial v_z}{\partial y} = \frac{\partial v_y}{\partial z} \qquad \frac{\partial v_x}{\partial z} = \frac{\partial v_z}{\partial x}. \qquad (4\text{-}8)$$

We have shown that, if $\mathbf{v}$ is *known* to be the gradient $\nabla u = \mathbf{i}v_x + \mathbf{j}v_y + \mathbf{k}v_z$ of a scalar function $u(x,y,z)$ with continuous second derivatives in R, the cartesian components of $\mathbf{v}$ satisfy[2] Eqs. (4-7) and (4-8).

But as we have already observed in the corresponding case in Chap. 5, Sec. 22, the satisfaction of Eqs. (4-7) and (4-8) does not ensure the existence of a single-valued function $u(x,y,z)$, such that $\mathbf{v} = \nabla u$, unless we impose certain restrictions on the connectivity of R.

—————————————————————— **PROBLEMS**

1. Given that $u = \frac{1}{2}r^2$, where $r^2 = x^2 + y^2 + z^2$, show that the integral $\int_C \mathbf{r} \cdot d\mathbf{r}$, where $\mathbf{r} = \mathbf{i}x + \mathbf{j}y + \mathbf{k}z$, is independent of the path in every region containing the path C. Find the value of this integral when C is the rectilinear path joining $(0,0,0)$ and $(1,1,1)$.

2. If u is known to be single-valued in the xy plane and $du = (y - x^2) \, dx + (x + y^2) \, dy$, find $u(x,y)$.

---

[1] We omit the subscript C on the integral when it is independent of the path and use the end points P_0 and P of the possible paths in the limits of the line integral.

[2] Compare Eq. (20-7) in Chap. 5.

3. If u is single-valued in every region R and $du = yz\,dx + zx\,dy + xy\,dz$, find $u(x,y,z)$ by integrating $\int_C du$ over the rectilinear path joining $(0,0,0)$ with an *arbitrary* fixed point (x_1,y_1,z_1). *Hint:* Write equations of C in the form $x = x_1 t$, $y = y_1 t$, $z = z_1 t$, $0 \le t \le 1$, integrate, and conclude that $u = xyz$.

4. (*a*) If $u = \log (x^2 + y^2)$, show that $\int_C \nabla u \cdot d\mathbf{r} = 0$ if C is the circle $x^2 + y^2 = 1$.

(*b*) If $u = \tan^{-1} (y/x)$, show that $\int_C \nabla u \cdot d\mathbf{r} = 2\pi$. Note that the conditions (4-16) and (4-17) are satisfied at all points of the xy plane except $x = 0$, $y = 0$. See Chap. 5, Sec. 23.

---

5. Regions and Their Boundaries.   The principal results in the remainder of this chapter concern two remarkable invariant formulas. One of these, deduced by Gauss from physical and geometric considerations, relates certain volume integrals to surface integrals; the other, deduced by Stokes, connects certain surface integrals with line integrals.[1]

These formulas are of such profound importance that it is difficult to conceive how the theories of electrodynamics, fluid mechanics, thermodynamics, and elasticity could have been developed without them. The formulas also led mathematicians to inquire more deeply into the meaning of surface and volume integrals. The main upshot of such investigations is that a surface can be an extraordinarily complex set of points and that a satisfactory invariant characterization of surfaces may prove impossible to achieve with the tools of coordinate geometry. Some leading scholars of mathematical analysis even *postulate* the validity of Stokes' formula and then maintain that "surfaces, their boundaries, and surface integrals must be defined in such a way as to assure the validity of the Stokes formula." [2]

In line with these thoughts we shall not obscure the invariant concepts, which enabled Gauss and Stokes to deduce their formulas, by an overmeticulous description of surfaces and boundaries. However, we do mention two properties that are genuinely pertinent to the analysis.

The first of these properties is smoothness. A surface is *smooth* if the unit normal $\mathbf{n}$ is a continuous function of position over the whole surface, including the boundary.[3] This property is relevant because one deals with integrals involving the normal, and these integrals might not exist if the normal behaves too wildly. Since an integral over the whole surface can be obtained by adding integrals over parts of the surface, it is sufficient to assume that the surface can be divided by smooth curves into a finite number of pieces, each of which is smooth. Such a surface is called *piecewise smooth*, or *sectionally smooth*. For example, the surface of a cube is sectionally smooth. The surface of a cone is not sectionally smooth because of the troublesome vertex; but isolated points do not affect the value of an integral, and cones are allowed in our analysis. The cone is thought to be the limit of the sectionally smooth surfaces one gets by truncating the cone.

The second major property is orientability. Intuitively, a surface is *orientable* if it has two sides, which could be distinguished by painting with two different colors. Not

---

[1] Karl Friedrich Gauss (1777–1855), a German mathematician, made a phenomenal contribution to mathematics and physics. He ranks among the great geniuses of all time. George Gabriel Stokes (1819–1903) was born in Ireland but spent most of his life in Cambridge, England. He was an applied mathematician of great attainments, whose principal contributions were to fluid mechanics.

[2] Quotation from Wilhelm Maak, "Modern Calculus," p. 383, Holt, Rinehart and Winston, Inc., New York, 1963.

[3] The definition of a smooth surface given in Chap. 5, Sec. 18, follows directly from the invariant definition given here once the surface is specified by Eq. (18-1) of Chap. 5.

all surfaces are two-sided. A one-sided surface can be formed, for example, by gluing the ends of a long strip in such a way that the upper side of one end of the strip is joined onto the under side of the other end (Fig. 9). If two oppositely directed normals PN and PN' are drawn at any point P of the surface, the normal PN when carried along the path $PABCP$ will coincide with PN'. It may be noted that this surface, which is called a *Möbius band*, has a simple closed curve as its boundary.

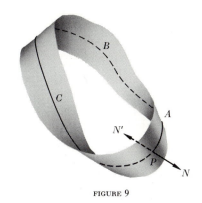

FIGURE 9

For mathematical purposes, a smooth surface is *orientable* if it is possible to define a continuous field of unit normals, $\mathbf{n}(P)$, over the whole surface. The Möbius band is ruled out, because choice of a normal $\mathbf{n}$ at any given point inevitably leads to the opposite choice $-\mathbf{n}$ when an attempt is made to extend the field $\mathbf{n}(P)$.

If a surface is part of the boundary of a solid, the two possible normals can be distinguished as an *interior* normal, which penetrates into the solid, and an *exterior* normal which (if sufficiently short) does not. The concept of orientability is thus extended to sectionally smooth surfaces bounding a solid; namely, take the *exterior* normal field on each of the smooth pieces. A surface with a unique normal field obtained in this way is sometimes said to *inherit* an orientation from the solid that it bounds.

In the following discussion the word "surface" is used to mean "sectionally smooth orientable surface" although an extension to more general cases can be made by a limiting process. The surface is assumed to have a finite area, but it need not always consist of a single connected part. (See Prob. 3.) For want of a better name we shall call regions bounded by closed surfaces *regular* regions.

PROBLEMS

1. Place a piece of notepaper on carbon paper with the carbon side up (so that a pencil mark on one side of the notepaper is reproduced on the other side). Draw a circle with an arrow indicating counterclockwise rotation. Such a circle, together with its image via the carbon paper, represents an oriented loop lying in the surface. What happens if you try to cover a Möbius band with oriented loops of this kind?

2. Experimentally, or otherwise, consider whether a Möbius band might be identical with part of the boundary of an ordinary three-dimensional solid.

3. Describe the exterior normals for the shell-like region bounded by two concentric spheres S_1 and S_2. Thus assign a suitable orientation to the boundary $S = S_1 + S_2$, even though S consists of two separate pieces.

4. (*a*) Show that the field of normals on the curved surface of a cone is continuous but not uniformly continuous. (*b*) Discuss the question of subdividing a cone into smooth pieces in such a way that each of the pieces, together with its boundary, admits a continuous field of normals.

---

6. Divergence. Let T be a regular region bounded by a surface S, and let $\mathbf{n}(P)$ be the exterior normal at the arbitrary point P of S (Fig. 10). Thus, $\mathbf{n}(P)$ is a unit vector, normal to S at points where S is smooth, and directed toward the *exterior* of T. It should

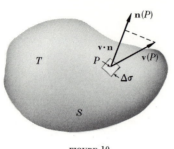

FIGURE 10

be emphasized that S is the *entire* boundary of T; hence S might consist of several disconnected pieces. For example, if T is the region between two concentric spheres, S consists of the inner spherical surface S_1 together with the outer spherical surface S_2. On the inner surface, $\mathbf{n}(P)$ is directed toward the center but on the outer surface, $\mathbf{n}(P)$ is directed away from the center.

If $\mathbf{v}(P)$ is a continuously differentiable vector function defined in T and on S, the *flux* F is the region function

$$F = \int_S \mathbf{v} \cdot \mathbf{n}\, d\sigma \qquad (6\text{-}1)$$

where $d\sigma$ is the element of surface area of S.

This integral has the form[1]

$$\int_S u(P)\, d\sigma \qquad (6\text{-}2)$$

where $u(P) = \mathbf{v} \cdot \mathbf{n}$ is piecewise continuous on S. A surface integral of a scalar function, such as this, is defined by subdividing S into small subregions ΔS_i, bounded by simple closed curves C_i, of areas $\Delta \sigma_i$ and forming the sum $I = \Sigma u(P_i)\, \Delta\sigma_i$, where P_i is interior to ΔS_i. The limit of this sum when the greatest linear dimension of all the subregions ΔS_i tend to 0 for all modes of subdivision of S is the integral (6-1). The situation is more involved than in the similar definition of the line integral, because the elements ΔS_i have more complexity than the corresponding elements Δs_i considered there. However, the existence of the limit can be established for piecewise-continuous functions defined on surfaces admitted by the discussion of Sec. 5.

We observe that F changes sign if $\mathbf{n}$ is chosen as the interior unit normal. When $\mathbf{v}(P)$ is the velocity of particles of a fluid moving in R, the flux F represents the net amount of the fluid leaving the region T per unit time. If the fluid is incompressible and there are no points in T where the fluid is generated or absorbed, the value of the integral (6-1) is zero. The points in T at which the fluid is generated are termed *sources* and those at which it is absorbed are *sinks*. If there are no sources or sinks in T and the fluid is compressible, the density of the fluid in T is changing when the flux $F \neq 0$.

Now let P be an arbitrary point in T, and consider the ratio

$$\frac{1}{\tau} \int_S \mathbf{v} \cdot \mathbf{n}\, d\sigma \qquad (6\text{-}3)$$

which represents the flux of $\mathbf{v}$ over S per unit volume. If we let S shrink toward P in such a way that the maximum diameter of $T \to 0$ (and hence $\tau \to 0$), the limit of the ratio (6-3) is called the *divergence* of $\mathbf{v}$ at P.

We denote divergence of $\mathbf{v}$ by div $\mathbf{v}(P)$ and write $T \to 0$ to indicate that the maximum diameter of T tends to zero. Thus

$$\operatorname{div} \mathbf{v}(P) = \lim_{T \to 0} \frac{1}{\tau} \int_S \mathbf{v} \cdot \mathbf{n}\, d\tau. \qquad (6\text{-}4)$$

This limit provides a measure of the expansion of the fluid at the point P.

In this process one may let $T \to 0$ while staying similar to a fixed solid, or one may let T become arbitrarily thin compared with its length, and so on. If the regions T are restricted as ex-

---

[1] See also the coordinate treatment of surface integrals in Chap. 5, Sec. 18.

plained in Sec. 5, it is possible to show that the limit $L = \text{div } \mathbf{v}$ exists, independently of the shape of T, and that the convergence is uniform in the following sense: *Given any $\epsilon > 0$, there is a $\delta > 0$ such that*

$$\left| \frac{1}{\tau} \int_\sigma \mathbf{v} \cdot \mathbf{n} \, d\sigma - L \right| < \epsilon$$

provided the maximum diameter of T is less than δ. This matter is discussed in Prob. 10 of the next section, although we do not profess to give a complete proof.

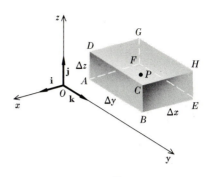

To calculate div $\mathbf{v}$ in cartesian coordinates we consider a region T in the shape of a rectangular parallelepiped with center at $P(x,y,z)$ and with edges Δx, Δy, Δz (Fig. 11). The flux of $\mathbf{v}$ over the surface of this parallelepiped is easily computed. Since $\mathbf{v} = \mathbf{i}v_x + \mathbf{j}v_y + \mathbf{k}v_z$, the normal component $\mathbf{v} \cdot \mathbf{n}$ of $\mathbf{v}$ over the face $ABCD$ is v_x. Hence the outflow over that face is $(v_x)_{x+\frac{1}{2}\Delta x} \, \Delta y \, \Delta z$, where $(v_x)_{x+\frac{1}{2}\Delta x}$ is the mean value of v_x over $ABCD$. Similarly, the outflow over the parallel face $EFGH$ is

$$(-v_x)_{x-\frac{1}{2}\Delta x} \, \Delta y \, \Delta z$$

where the minus sign appears because the exterior normal to $EFGH$ is $-\mathbf{i}$ and hence $\mathbf{v} \cdot \mathbf{n} = -v_x$.

Thus the net outflow over a pair of faces parallel to the yz plane is

$$\left[(v_x)_{x+\frac{1}{2}\Delta x} - (v_x)_{x-\frac{1}{2}\Delta x} \right] \Delta y \, \Delta z \equiv v_x \Big|_{x-\frac{1}{2}\Delta x}^{x+\frac{1}{2}\Delta x} \Delta y \, \Delta z.$$

Proceeding in the same way with the remaining faces, we get for the total outflow

$$\int_\sigma \mathbf{v} \cdot \mathbf{n} \, d\sigma = v_x \Big|_{x-\frac{1}{2}\Delta x}^{x+\frac{1}{2}\Delta x} \Delta y \, \Delta z + v_y \Big|_{y-\frac{1}{2}\Delta y}^{y+\frac{1}{2}\Delta y} \Delta x \, \Delta z + v_z \Big|_{z-\frac{1}{2}\Delta z}^{z+\frac{1}{2}\Delta z} \Delta x \, \Delta y.$$

Dividing by $\tau = \Delta x \, \Delta y \, \Delta z$ and taking the limit give

$$\text{div } \mathbf{v}(P) = \lim \left(\frac{v_x \Big|_{x-\frac{1}{2}\Delta x}^{x+\frac{1}{2}\Delta x}}{\Delta x} + \frac{v_y \Big|_{y-\frac{1}{2}\Delta y}^{y+\frac{1}{2}\Delta y}}{\Delta y} + \frac{v_z \Big|_{z-\frac{1}{2}\Delta z}^{z+\frac{1}{2}\Delta z}}{\Delta z} \right) \qquad (6\text{-}5)$$

as Δx, Δy, and Δz approach zero in any manner. If $\mathbf{v}$ is continuously differentiable, it is not difficult to show that the three limits in (6-5) are the respective partial derivatives, so that we obtain the important formula

$$\text{div } \mathbf{v}(P) = \frac{\partial v_x}{\partial x} + \frac{\partial v_y}{\partial y} + \frac{\partial v_z}{\partial z}. \qquad (6\text{-}6)$$

In terms of the differential operator

$$\nabla = \mathbf{i}\frac{\partial}{\partial x} + \mathbf{j}\frac{\partial}{\partial y} + \mathbf{k}\frac{\partial}{\partial z}$$

introduced in Sec. 2, we can consider a symbolic scalar product

$$\nabla \cdot \mathbf{v} = \left(\mathbf{i}\, \frac{\partial}{\partial x} + \mathbf{j}\, \frac{\partial}{\partial y} + \mathbf{k}\, \frac{\partial}{\partial z} \right) \cdot (\mathbf{i} v_x + \mathbf{j} v_y + \mathbf{k} v_z)$$

$$\equiv \frac{\partial v_x}{\partial x} + \frac{\partial v_y}{\partial y} + \frac{\partial v_z}{\partial z}.$$

On comparing this with (6-6) we see that

$$\operatorname{div} \mathbf{v} = \nabla \cdot \mathbf{v}. \tag{6-7}$$

We can also define the *Laplacian operator* ∇^2 by the formula

$$\nabla^2 = \nabla \cdot \nabla = \left(\mathbf{i}\, \frac{\partial}{\partial x} + \mathbf{j}\, \frac{\partial}{\partial y} + \mathbf{k}\, \frac{\partial}{\partial z} \right) \cdot \left(\mathbf{i}\, \frac{\partial}{\partial x} + \mathbf{j}\, \frac{\partial}{\partial y} + \mathbf{k}\, \frac{\partial}{\partial z} \right)$$

$$= \frac{\partial^2}{\partial x^2} + \frac{\partial^2}{\partial y^2} + \frac{\partial^2}{\partial z^2} \tag{6-8}$$

and observe that, if $\mathbf{v} = \nabla u$, then

$$\operatorname{div} \nabla u = \nabla \cdot \nabla u = \nabla^2 u. \tag{6-9}$$

Furthermore, if the symbol $\nabla \times \mathbf{v}$ is defined by the rule for computing vector products, we get

$$\nabla \times \mathbf{v} = \begin{vmatrix} \mathbf{i} & \mathbf{j} & \mathbf{k} \\ \dfrac{\partial}{\partial x} & \dfrac{\partial}{\partial y} & \dfrac{\partial}{\partial z} \\ v_x & v_y & v_z \end{vmatrix}$$

$$= \mathbf{i} \left(\frac{\partial v_z}{\partial y} - \frac{\partial v_y}{\partial z} \right) + \mathbf{j} \left(\frac{\partial v_x}{\partial z} - \frac{\partial v_z}{\partial x} \right) + \mathbf{k} \left(\frac{\partial v_y}{\partial x} - \frac{\partial v_x}{\partial y} \right). \tag{6-10}$$

It is worth observing that the condition $\nabla \times \mathbf{v} = \mathbf{0}$ requires that each component of the vector $\nabla \times \mathbf{v}$ be zero. We can therefore write Eqs. (4-7) and (4-8) in the compact form $\nabla \times \mathbf{v} = \mathbf{0}$.

In Sec. 13 we give a formula for div $\mathbf{v}$ analogous to (6-6) when the vector $\mathbf{v}$ is referred to an arbitrary orthogonal curvilinear coordinate system. It is important to note that the definition (6-4) is independent of the choice of coordinates, so that div $\mathbf{v}$ is an invariant.

Example 1.   If $\mathbf{v} = \mathbf{i} 3x^2 + \mathbf{j} 5xy^2 + \mathbf{k} xyz^3$, compute div $\mathbf{v}$ at (1,2,3) and $\nabla \times \mathbf{v}$ at (x,y,z).

Since $v_x = 3x^2$, $v_y = 5xy^2$, $v_z = xyz^3$, the substitution in (6-6) yields div $\mathbf{v} = 6x + 10xy + 3xyz^2$. At the point (1,2,3), div $\mathbf{v} = 6 + 20 + 54 = 80$. If $\mathbf{v}$ is interpreted as the velocity vector of fluid particles, we conclude the fluid is expanding at (1,2,3).

To compute $\nabla \times \mathbf{v}$ we use the formula (6-10) and find

$$\nabla \times \mathbf{v} = \mathbf{i}(xz^3 - 0) + \mathbf{j}(0 - yz^3) + \mathbf{k}(5y^2 - 0).$$

Since this vector is not identically zero, we conclude that no scalar function $u(x,y,z)$ exists such that $\mathbf{v} = \nabla u$.

Example 2.   Let it be required to compute the flux of $\mathbf{v} = \mathbf{i} x + \mathbf{j} y + \mathbf{k} z$ over the surface of the cylinder $x^2 + y^2 = a^2$ bounded by the planes $z = 0$, $z = b$. The unit exterior normal over the lateral surface L of the cylinder is $\mathbf{n} = \mathbf{i} x/a + \mathbf{j} y/a$; hence the flux over L is

$$\int_L \mathbf{v} \cdot \mathbf{n} \, d\sigma = \int_L \left(\frac{x^2 + y^2}{a} \right) d\sigma = \int_L a \, d\sigma = 2\pi a^2 b$$

since b is the height of the cylinder. The exterior unit normal $\mathbf{n}$ over the top B of the cylinder bounded by the plane $z = b$ is $\mathbf{n} = \mathbf{k}$; hence $\mathbf{v} \cdot \mathbf{n}|_{z=b} = b$, and $\int_B \mathbf{v} \cdot \mathbf{n} \, d\sigma = \int_B b \, d\sigma = \pi a^2 b$. On the base of the cylinder bounded by the plane $z = 0$, $\mathbf{n} = -\mathbf{k}$; hence $\mathbf{v} \cdot \mathbf{n} = -z$ and, since $z = 0$ on this base, $\mathbf{v} \cdot \mathbf{n}|_{z=0} = 0$. Accordingly, the flux $F = 2\pi a^2 b + \pi a^2 b = 3\pi a^2 b$.

Example 3.   Let **v** be the Newtonian force field $\mathbf{v} = m\mathbf{r}/r^3$, where $m = $ const, and $\mathbf{r} = \mathbf{i}x +$ $\mathbf{j}y + \mathbf{k}z$. The components of this "inverse-square force field" are $v_x = mx/r^3$, $v_y = my/r^3$, $v_z = mz/r^3$, and they are continuously differentiable except when $r = 0$. Accordingly, in every region that excludes the point $(0,0,0)$, we have

$$\frac{\partial v_x}{\partial x} = m\frac{r^2 - 3x^2}{r^5} \qquad \frac{\partial v_y}{\partial y} = m\frac{r^2 - 3y^2}{r^5} \qquad \frac{\partial v_z}{\partial z} = m\frac{r^2 - 3z^2}{r^5}$$

and div $\mathbf{v} = \partial v_x/\partial x + \partial v_y/\partial y + \partial v_z/\partial z = 0$. The flux of **v** over the surface of the sphere S, given by $x^2 + y^2 + z^2 = a^2$, is $\int_S \mathbf{v} \cdot \mathbf{n}\, d\sigma = \int_S \left(\dfrac{m}{a^2}\right) d\sigma = 4\pi m$, since the unit exterior normal to S is $\mathbf{n} = \mathbf{i}x/a + \mathbf{j}y/a + \mathbf{k}z/a = \mathbf{r}/a$. In the language of fluid mechanics and electrostatics, the point $(0,0,0)$ is a source. The Gauss formula, deduced in Sec. 7, shows that the flux of **v** over every closed surface that does not contain $(0,0,0)$ in its interior is zero.

PROBLEMS

1. Find div **v** if (*a*) $\mathbf{v} = \mathbf{i}x + \mathbf{j}y + \mathbf{k}z$; (*b*) $\mathbf{v} = \mathbf{i}(x/r) + \mathbf{j}(y/r) + \mathbf{k}(z/r)$, where

$$r = \sqrt{x^2 + y^2 + z^2} \neq 0$$

(*c*) $\mathbf{v} = \mathbf{i}(z - y) + \mathbf{j}(x - z) + \mathbf{k}(y - x)$.

2. Compute $\nabla^2(1/r)$ and $\nabla^2 r$, where $r = \sqrt{x^2 + y^2 + z^2}$.

3. If u, **u**, and **v** are continuously differentiable, show that (*a*) div $(\mathbf{u} + \mathbf{v}) = $ div $\mathbf{u} + $ div $\mathbf{v}$, (*b*) div $(u\mathbf{v}) = \nabla \cdot (u\mathbf{v}) = \nabla u \cdot \mathbf{v} + u\nabla \cdot \mathbf{v}$; (*c*) div $(\mathbf{u} \times \mathbf{v}) = \nabla \cdot (\mathbf{u} \times \mathbf{v}) = \mathbf{v} \cdot (\nabla \times \mathbf{u}) - \mathbf{u} \cdot (\nabla \times \mathbf{v})$.

4. Show that div $(\mathbf{r} \times \mathbf{a}) = 0$ if $\mathbf{r} = \mathbf{i}x + \mathbf{j}y + \mathbf{k}z$ and **a** is a constant vector.

5. Find div $(u\mathbf{v})$ if $u = x^2 + y^2 + z^2$ and $\mathbf{v} = \mathbf{i}x + \mathbf{j}y + \mathbf{k}z$.   Also find div $(\nabla u \times \mathbf{v})$.

6. Compute div **v** for the central force field $\mathbf{v} = f(r)\mathbf{r}/r$, where $f(r)$ is differentiable and $\mathbf{r} = \mathbf{i}x + \mathbf{j}y + \mathbf{k}z$. What is the flux of **v** over the sphere $x^2 + y^2 + z^2 = a^2$, if (*a*) $f(r) = 1/r^2$ and (*b*) $f(r) = r$?

7. If $u = m/r$ and $v = -\nabla u$, show that div $\nabla u = 0$ if $r = \sqrt{x^2 + y^2 + z^2} \neq 0$. Compare with Example 3.

---

7. The Divergence Theorem.

An important relationship connecting the surface integral (6-1) for the flux of a vector field with the volume integral of its divergence is deduced in this section. The resulting integral transformation theorem, known as the *Gauss, or divergence, theorem*, is fundamental to all developments in mechanics of continuous media.

Let T be a regular region bounded by the surface S, and let $\mathbf{v}(P)$ be a vector function which is continuously differentiable at every point of $T + S$. We subdivide T into k cells ΔT_i in the shape of rectangular boxes and parts of boxes (Fig. 12) and compute the divergence (6-4) at some point P_i of ΔT_i,

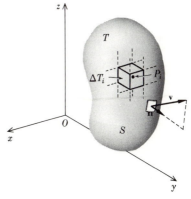

FIGURE 12

$$\text{div } \mathbf{v}(P_i) = \lim_{\Delta \tau_i \to 0} \frac{1}{\Delta \tau_i} \int_{\Delta S_i} (\mathbf{v} \cdot \mathbf{n}) \, d\sigma \tag{7-1}$$

for each cell ΔT_i. [The role of T and S in (6-4) is now taken by ΔT_i and ΔS_i.] On recalling the definition of this limit, we can rewrite (7-1) in the form

$$\int_{\Delta S_i} \mathbf{v} \cdot \mathbf{n} \, d\sigma = (\text{div } \mathbf{v})_i \, \Delta \tau_i + \epsilon_i \, \Delta \tau_i \tag{7-2}$$

where the $\epsilon_i \to 0$ as $\Delta \tau_i \to 0$ and where $(\text{div } \mathbf{v})_i \equiv \text{div } \mathbf{v}(P_i)$. We next form the sum

$$\sum_{i=1}^{k} \int_{\Delta S_i} \mathbf{v} \cdot \mathbf{n} \, d\sigma = \sum_{i=1}^{k} (\text{div } \mathbf{v})_i \, \Delta \tau_i + \sum_{i=1}^{k} \epsilon_i \, \Delta \tau_i \tag{7-3}$$

over all the cells and observe that the surface integrals in (7-3) over the interfaces of adjacent cells vanish, since the exterior normals $\mathbf{n}$ to the common faces of the boxes point in opposite directions. Thus the surviving terms in the sum on the left in (7-3) correspond to surface elements belonging to the surface S, and hence this sum is equal to $\int_S \mathbf{v} \cdot \mathbf{n} \, d\sigma$. The sum $\Sigma(\text{div } \mathbf{v})_i \, \Delta \tau_i$ approximates the volume integral $\int_T \text{div } \mathbf{v} \, d\tau$, and indeed the approximation can be made as close as one wishes by sufficiently decreasing the size (that is, the maximum diameter) of the cells. The sum $\Sigma \epsilon_i \, \Delta \tau_i$ involves products of small quantities ϵ_i and $\Delta \tau_i$ and is estimated as follows: Given $\epsilon > 0$, suppose the subdivision is so fine that $|\epsilon_i| < \epsilon$ for all the cells at once. In this case $|\Sigma \epsilon_i \, \Delta \tau_i| \leq \epsilon \Sigma \, \Delta \tau_i = \epsilon \tau$ where τ is the volume of the region. As $\epsilon \to 0$ the estimate becomes arbitrarily small, and we conclude from (7-3) that

$$\int_S \mathbf{v} \cdot \mathbf{n} \, d\sigma = \int_T \text{div } \mathbf{v} \, d\tau. \tag{7-4}$$

This is the *divergence theorem*.

The argument that has just been outlined is the simplest and most direct method of seeing why the divergence theorem is true. It has the further advantage of being free of coordinates and thus puts the *invariant* nature of the theorem into due prominence. Since the divergence theorem is a sophisticated theorem of modern analysis, one cannot expect that the above discussion constitutes a complete proof. It does, however, embody several of the ideas used in such a proof.

For certain very simple regions the divergence theorem can be proved by partial integration, much as in the discussion of Green's theorem given in Chap. 5, Sec. 21. We do not present the details, because introduction of coordinates obscures the physical meaning, and the needed restrictions on T are actually quite irrelevant to the divergence theorem itself.

After this preliminary step is taken, one can use these simple regions for the cells $\Delta \tau_i$ and apply the result of Prob. 10. This gives the uniform estimate of the ϵ_i, and the argument of the text becomes a valid proof. Additional refinements are needed to get the theorem in the full generality implied by the discussion of Sec. 5, but the main features of the analysis are as we have indicated them.[1]

The divergence theorem (7-4) expresses certain surface integrals as volume integrals, and since it contains no reference to any special coordinate system, the result is true in all coordinate systems.

In particular, if $\mathbf{v}$ and $\mathbf{n}$ are expressed in terms of their cartesian components

$$\mathbf{n} = \mathbf{i} \cos (x,n) + \mathbf{j} \cos (y,n) + \mathbf{k} \cos (z,n)$$

we can write (7-4), on recalling (6-6), as

---

[1] The details can be found in many books. See, for example, O. D. Kellogg, "Foundations of Potential Theory," p. 84, Springer-Verlag OHG, Berlin, 1929.

$$\int_S \left[v_x \cos{(x,n)} + v_y \cos{(y,n)} + v_z \cos{(z,n)} \right] d\sigma = \int_T \left(\frac{\partial v_x}{\partial x} + \frac{\partial v_y}{\partial y} + \frac{\partial v_z}{\partial z} \right) d\tau. \quad (7\text{-}5)$$

Example.   Apply the theorem (7-4) to $\mathbf{v} = \mathbf{i}(x/r) + \mathbf{j}(y/r) + \mathbf{k}(z/r)$, where $r = \sqrt{x^2 + y^2 + z^2}$ and the region T is the sphere $x^2 + y^2 + z^2 \le a^2$.

We readily find that if $r \ne 0$,

$$\frac{\partial v_x}{\partial x} = \frac{r^2 - x^2}{r^3} \qquad \frac{\partial v_y}{\partial y} = \frac{r^2 - y^2}{r^3} \qquad \frac{\partial v_z}{\partial z} = \frac{r^2 - z^2}{r^3}$$

so that by (6-6)

$$\operatorname{div} \mathbf{v} = \frac{3r^2 - r^2}{r^3} = \frac{2}{r} \qquad \text{if } r \ne 0$$

and, since the integrand is discontinuous when $r = 0$, care must be exercised in evaluating this integral.   The fact that this integral exists is most easily shown by transforming it to spherical coordinates.   The value of this integral can be obtained by computing the surface integral in the left-hand member of formula (7-4).   Inasmuch as the normal $\mathbf{n}$ to the sphere is directed along its radius, $\mathbf{n} = \mathbf{i}(x/r) = \mathbf{j}(y/r) + \mathbf{k}(z/r)$, and we find at once $\int_S \mathbf{v} \cdot \mathbf{n} \, d\sigma = \int_S 1 \, d\sigma = 4\pi a^2$.

── **PROBLEMS**

1. Show that $\int_S \mathbf{r} \cdot \mathbf{n} \, d\sigma = 3\tau$, where $\mathbf{r}$ is the position vector $\mathbf{r} = \mathbf{r}(P)$ of points P on the surface S forming the boundary of a simply connected region T of volume τ.   *Hint:* Apply formula (7-4).

2. Compute $\int_S \mathbf{v} \cdot \mathbf{n} \, d\sigma$, where S is the surface of the cylinder $x^2 + y^2 = a^2$ bounded by the planes $z = 0, z = b$ and where $\mathbf{v} = \mathbf{i}x - \mathbf{j}y + \mathbf{k}z$.

3. Find $\int_S \mathbf{r} \cdot \mathbf{n} \, d\sigma$, where $\mathbf{r}$ is the position vector of points on the surface of the ellipsoid $(x^2/a^2) + (y^2/b^2) + (z^2/c^2) = 1$.

4. Find the value of $\int_S \mathbf{v} \cdot \mathbf{n} \, d\sigma$, where $\mathbf{v} = r^2(\mathbf{i}x + \mathbf{j}y + \mathbf{k}z)$, $r^2 = x^2 + y^2 + z^2$, and S is the surface of the sphere $x^2 + y^2 + z^2 = a^2$.   Compute the integral directly and also with the aid of the divergence theorem.

5. If $\mathbf{v} = \nabla u$ and $\nabla^2 u = \rho$, where ρ is a specified continuous scalar point function, show that

$$\int_S \frac{du}{dn} \, d\sigma = \int_T \rho \, d\tau.$$

Hint: Recall that $du/dn = \nabla u \cdot \mathbf{n}$.

6. Assume that all derivatives of required orders are continuous and use the divergence theorem to show that

$$\int_T \operatorname{div} (u \, \nabla v) \, d\tau = \int_S u \, \nabla v \cdot \mathbf{n} \, d\sigma.$$

Show that this equation can be written as

$$\int_T u \nabla^2 v \, d\tau = \int_S u \frac{dv}{dn} \, d\sigma - \int_T \nabla u \cdot \nabla v \, d\tau.$$

This important relation is known as *Green's first identity.*

7. Using Prob. 6, obtain the symmetrical form of Green's identity, namely,

$$\int_T (u \nabla^2 v - v \nabla^2 u) \, d\tau = \int_S \left(u \frac{dv}{dn} - v \frac{du}{dn} \right) d\sigma$$

which is also known as *Green's second identity.*   (It is assumed in this identity that both u and v have continuous second derivatives.)   Green's identities are perhaps the most frequently encountered transformation formulas in mathematical physics.

8. If the twice-differentiable function u satisfies Laplace's equation $\nabla^2 u = 0$ in T, what is the value of $\int_S \frac{du}{dn} \, d\sigma$?   *Hint:* Set $v = 1$ in Green's second identity, Prob. 7.

9. Evaluate the following integrals with the aid of the divergence theorem:

(a) $\int_S [x^3 \cos(x,n) + y^3 \cos(y,n) + z^3 \cos(z,n)] \, d\sigma$, where S is the surface of the sphere $x^2 + y^2 + z^2 = a^2$;

(b) $\int_S \mathbf{r} \cdot \mathbf{n} \, d\sigma$, where $\mathbf{r} = \mathbf{i}x + \mathbf{j}y + \mathbf{k}z$ is the position vector on the surface S of the tetrahedron formed by the planes $x = 0$, $y = 0$, $z = 0$, and $x + y + z = a$;

(c) $\int_S [x^2 \cos(x,n) + y^2 \cos(y,n) + z^2 \cos(z,n)] \, d\sigma$, where S is the surface of the cone $x^2/a^2 + y^2/a^2 = z^2/b^2$, $0 \le z \le b$;

(d) $\int_S [yz \cos(x,n) + xz \cos(y,n) + xy \cos(z,n)] \, d\sigma$, where S is the surface in the first octant bounded by the planes $x = 0$, $y = 0$, $z = b$, and by the cylinder $x^2 + y^2 = a^2$;

(e) $\int_S (x^2 + y^2 + z^2)^{1/2} [\cos(x,n) + \cos(y,n) + \cos(z,n)] \, d\sigma$, where S is the surface of the sphere $x^2 + y^2 + z^2 = a^2$.

10. If $u(P)$ is a continuous function, the *modulus of continuity* is defined by $\omega(\delta) = \max |u(P) - u(P_0)|$ for $|P - P_0| \le \delta$. Let T be a region of diameter $\le \delta$ containing the point P_0, and let $\mathbf{v}(P)$ be continuously differentiable in a region containing T and its boundary S. Assuming that the divergence theorem holds for T, show that

$$\left| \frac{1}{\tau} \int_S \mathbf{v} \cdot \mathbf{n} \, d\sigma - \operatorname{div} \mathbf{v}(P_0) \right| \le \omega(\delta)$$

where $\omega(\delta)$ is the modulus of continuity for the function $u(P) = \operatorname{div} \mathbf{v}(P)$. [This establishes uniformity of the limit defining $\operatorname{div} \mathbf{v}(P_0)$ for a very broad class of regions.]

---

8. Curl of a Vector Field.   In Sec. 3, the line integral $\int_C u(P) \, ds$ was presented in an invariant form (3-1), and in Secs. 6 and 7 we outlined the definitions of the surface integral $\int_S u(P) \, d\sigma$ and the volume integral $\int_T u(P) \, d\tau$ without reference to any special coordinate system.

Formally one is tempted to extend these "limits-of-the-sum" definitions to such symbols as

$$\int_C \mathbf{v}(P) \, ds, \qquad \int_S \mathbf{v}(P) \, d\sigma, \qquad \int_T \mathbf{v}(P) \, d\tau \tag{8-1}$$

in which $\mathbf{v}(P)$ is a vector function and ds, $d\sigma$, and $d\tau$ are the elements of arc length, surface, and volume, respectively. Thus, there is a temptation to define the volume integral $\int_T \mathbf{v}(P) \, d\tau$ by the formula

$$\int_T \mathbf{v}(P) \, d\tau = \lim \sum \mathbf{v}(P_i) \, \Delta\tau_i \tag{8-2}$$

in which it is imagined that the region T is divided into elements of volume $\Delta\tau_i$. A definition such as (8-2) requires forming sums $\Sigma \mathbf{v}(P_i) \, \Delta\tau_i$ of the *bound* vectors $\mathbf{v}(P_i)$ which are determined at different points of the body. There is a question whether the rules for addition of free vectors given in Chap. 4 can be used to provide a sensible meaning for (8-2). Without going into details, we state it as a fact that the definition suggested by (8-2) makes sense in those geometries where the distance between a pair of points is given by the Pythagorean formula.[1]

If $\mathbf{v}(P)$ is expressed in terms of its cartesian components as $\mathbf{v} = \mathbf{i}v_x(x,y,z) + \mathbf{j}v_y(x,y,z) + \mathbf{k}v_z(x,y,z)$, the integrals in (8-1) can be reduced to the evaluation of three ordinary integrals by writing, for example,

[1] Spaces so metrized are called Euclidean, and it is only with such that we are concerned in this book.

$$\int_T \mathbf{v}(P)\, d\tau = \mathbf{i} \int_T v_x\, d\tau + \mathbf{j} \int_T v_y\, d\tau + \mathbf{k} \int_T v_z\, d\tau.$$

No such simple means of evaluating integrals of the type (8-1) are available in curvilinear coordinates because the base vectors in curvilinear coordinate systems usually vary from point to point in space. (See Sec. 12.) This remark may serve to explain why cartesian coordinates are so prominent in calculations involving vectors.

We saw in Sec. 6 that with every continuously differentiable vector function $\mathbf{v}(P)$ one can associate a scalar div $\mathbf{v}(P)$ defined by the formula

$$\operatorname{div} \mathbf{v}(P) = \lim \frac{1}{\tau} \int_S \mathbf{n} \cdot \mathbf{v}\, d\sigma \qquad (8\text{-}3)$$

which has a simple physical meaning.

We show next that $\mathbf{v}(P)$ can also be associated with a vector field called curl $\mathbf{v}$, defined by an analogous formula

$$\operatorname{curl} \mathbf{v}(P) = \lim \frac{1}{\tau} \int_S \mathbf{n} \times \mathbf{v}\, d\sigma. \qquad (8\text{-}4)$$

We shall see that curl $\mathbf{v}(P)$ bears an interesting relation to the concept of *circulation* in the vector field.

Let $\mathbf{v}(P)$ be continuously differentiable in some region that contains in its interior a simple closed curve C bounding a plane region R of area A. At a given point P of R we construct a unit normal $\boldsymbol{\nu}$ so directed that $\boldsymbol{\nu}$ points in the direction of an advancing right-hand screw when C is traversed in the positive sense (Fig. 13). We then construct a right cylinder of small height h with elements parallel to $\boldsymbol{\nu}$ and with base A and denote its surface by S and its volume by τ.

Since $\boldsymbol{\nu}$ is a constant vector, formula (8-4) yields

$$\boldsymbol{\nu} \cdot \operatorname{curl} \mathbf{v} = \lim \frac{1}{\tau} \int_S \boldsymbol{\nu} \cdot \mathbf{n} \times \mathbf{v}\, d\sigma. \qquad (8\text{-}5)$$

But along the bases of the cylinder, $\boldsymbol{\nu}$ is parallel to the normal $\mathbf{n}$, and hence the triple scalar product $\boldsymbol{\nu} \cdot \mathbf{n} \times \mathbf{v}$ vanishes over the bases. Accordingly, the integral in (8-5) need be computed only over the lateral surface of the cylinder. We can thus write

$$\boldsymbol{\nu} \cdot \operatorname{curl} \mathbf{v} = \lim \frac{1}{\tau} \int_C \boldsymbol{\nu} \cdot \mathbf{n} \times \mathbf{v} h\, ds \qquad (8\text{-}6)$$

since $d\sigma = h\, ds$.

But $\boldsymbol{\nu} \cdot \mathbf{n} \times \mathbf{v} = \mathbf{v} \cdot \boldsymbol{\nu} \times \mathbf{n}$ by Chap. 4, Eq. (6-4), and $\boldsymbol{\nu} \times \mathbf{n} = \mathbf{t}$ along C, where $\mathbf{t}$ is the unit tangent vector to C. Thus the integrand in (8-6) can be written

$$\boldsymbol{\nu} \cdot \mathbf{n} \times \mathbf{v} h\, ds = \mathbf{v} \cdot \boldsymbol{\nu} \times \mathbf{n} h\, ds = \mathbf{v} \cdot \mathbf{t} h\, ds = h \mathbf{v} \cdot d\mathbf{r}$$

where $d\mathbf{r}$ is the differential of the position vector $\mathbf{r}$ of a point on C. If we further note that $\tau = hA$, we can rewrite (8-6) as

$$\boldsymbol{\nu} \cdot \operatorname{curl} \mathbf{v} = \lim_{A \to 0} \frac{1}{A} \int_C \mathbf{v} \cdot d\mathbf{r}. \qquad (8\text{-}7)$$

The line integral $\int_C \mathbf{v} \cdot d\mathbf{r}$ is called the *circulation of* $\mathbf{v}$ *along* C. If $\mathbf{v}$ represents the velocity of a fluid, then $\mathbf{v} \cdot d\mathbf{r} = \mathbf{v} \cdot \mathbf{t}\, ds$ takes account of the tangential component of velocity $\mathbf{v}$ and a fluid particle moving with this velocity

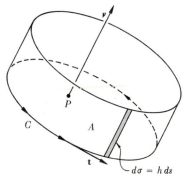

FIGURE 13

circulates along C. A particle moving with velocity $\mathbf{v} \cdot \mathbf{n}$ normal to C, on the other hand, crosses C. That is, it flows either into or out of the region bounded by C. Hence formula (8-7) provides a measure of *the circulation per unit area* at the point P. This formula can be used to compute the cartesian components of the vector curl $\mathbf{v}$ by taking ν successively as the $\mathbf{i}$, $\mathbf{j}$, $\mathbf{k}$ base vectors and by evaluating the limit in the right-hand member in the manner we deduced formula (6-6). It is somewhat simpler, however, to get the formula for curl $\mathbf{v}$ in cartesian coordinates from the definition (8-4) with the aid of the divergence theorem.

Now the components of $\mathbf{n} \times \mathbf{v}$ in cartesian coordinates are $\mathbf{n} \times \mathbf{v} \cdot \mathbf{i}$, $\mathbf{n} \times \mathbf{v} \cdot \mathbf{j}$, $\mathbf{n} \times \mathbf{v} \cdot \mathbf{k}$. Consequently,

$$\begin{aligned}
\mathbf{n} \times \mathbf{v} &= \mathbf{i}(\mathbf{n} \times \mathbf{v} \cdot \mathbf{i}) + \mathbf{j}(\mathbf{n} \times \mathbf{v} \cdot \mathbf{j}) + \mathbf{k}(\mathbf{n} \times \mathbf{v} \cdot \mathbf{k}) \\
&\equiv \mathbf{i}(\mathbf{n} \cdot \mathbf{v} \times \mathbf{i}) + \mathbf{j}(\mathbf{n} \cdot \mathbf{v} \times \mathbf{j}) + \mathbf{k}(\mathbf{n} \cdot \mathbf{v} \times \mathbf{k}).
\end{aligned} \tag{8-8}$$

On inserting from (8-8) in (8-4) we get

$$\operatorname{curl} \mathbf{v} = \mathbf{i} \lim \frac{1}{\tau} \int_S (\mathbf{n} \cdot \mathbf{v} \times \mathbf{i})\, d\sigma + \mathbf{j} \lim \frac{1}{\tau} \int_S (\mathbf{n} \cdot \mathbf{v} \times \mathbf{j})\, d\sigma$$

$$+ \mathbf{k} \lim \frac{1}{\tau} \int_S (\mathbf{n} \cdot \mathbf{v} \times \mathbf{k})\, d\sigma. \tag{8-9}$$

But a comparison of the right-hand member of (8-3) with (8-9) enables us to rewrite (8-9) in the form

$$\operatorname{curl} \mathbf{v} = \mathbf{i} \operatorname{div} (\mathbf{v} \times \mathbf{i}) + \mathbf{j} \operatorname{div} (\mathbf{v} \times \mathbf{j}) + \mathbf{k} \operatorname{div} (\mathbf{v} \times \mathbf{k}). \tag{8-10}$$

On inserting $\mathbf{v} = \mathbf{i} v_x + \mathbf{j} v_y + \mathbf{k} v_z$ in (8-10) we get

$$\operatorname{curl} \mathbf{v} = \mathbf{i} \operatorname{div} (\mathbf{j} v_z - \mathbf{k} v_y) + \mathbf{j} \operatorname{div} (\mathbf{k} v_x - \mathbf{i} v_z) + \mathbf{k} \operatorname{div} (\mathbf{i} v_y - \mathbf{j} v_x)$$

and a simple calculation making use of formula (6-6) yields the desired result:

$$\operatorname{curl} \mathbf{v} = \mathbf{i} \left(\frac{\partial v_z}{\partial y} - \frac{\partial v_y}{\partial z} \right) + \mathbf{j} \left(\frac{\partial v_x}{\partial z} - \frac{\partial v_z}{\partial x} \right) + \mathbf{k} \left(\frac{\partial v_y}{\partial x} - \frac{\partial v_x}{\partial y} \right). \tag{8-11}$$

If we recall the expression for the symbolic vector ∇, introduced in Sec. 2, we can write (8-11) compactly as

$$\operatorname{curl} \mathbf{v} = \begin{vmatrix} \mathbf{i} & \mathbf{j} & \mathbf{k} \\ \dfrac{\partial}{\partial x} & \dfrac{\partial}{\partial y} & \dfrac{\partial}{\partial z} \\ v_x & v_y & v_z \end{vmatrix} \equiv \nabla \times \mathbf{v}. \tag{8-12}$$

An analogous formula for curl $\mathbf{v}$ in orthogonal curvilinear coordinates is given in Sec. 13, and several useful relations involving the use of the curl operator are recorded in Prob. 1.

Example 1.   Compute curl $\mathbf{v}$ if $\mathbf{v} = \mathbf{i} xyz + \mathbf{j} xyz^2 + \mathbf{k} x^3 yz$. The substitution of $v_x = xyz$, $v_y = xyz^2$, $v_z = x^3 yz$ in (8-11) yields

$$\operatorname{curl} \mathbf{v} = \mathbf{i}(x^3 z - 2xyz) + \mathbf{j}(xy - 3x^2 yz) + \mathbf{k}(yz^2 - xz).$$

Example 2.   Deduce with the aid of the divergence theorem the relation

$$\int_T \nabla u\, d\tau = \int_S u\mathbf{n}\, d\sigma \tag{8-13}$$

where u is a continuously differentiable scalar point function in some domain containing a region T.

Now in cartesian coordinates

$$\begin{aligned}
\mathbf{n} &\equiv \mathbf{i} \cos (x,n) + \mathbf{j} \cos (y,n) + \mathbf{k} \cos (z,n) \\
&= \mathbf{i}(\mathbf{n} \cdot \mathbf{i}) + \mathbf{j}(\mathbf{n} \cdot \mathbf{j}) + \mathbf{k}(\mathbf{n} \cdot \mathbf{k})
\end{aligned}$$

and
$$\nabla u = \mathbf{i}\,\frac{\partial u}{\partial x} + \mathbf{j}\,\frac{\partial u}{\partial y} + \mathbf{k}\,\frac{\partial u}{\partial z}$$

so that (8-13) is equivalent to the three equations

$$\int_T \frac{\partial u}{\partial x}\,d\tau = \int_S (\mathbf{i}u)\cdot\mathbf{n}\,d\sigma \qquad \int_T \frac{\partial u}{\partial y}\,d\tau = \int_S (\mathbf{j}u)\cdot\mathbf{n}\,d\sigma \qquad \int_T \frac{\partial u}{\partial z}\,d\tau = \int_S (\mathbf{k}u)\cdot\mathbf{n}\,d\sigma.$$

But these are the special cases of formula (7-4) applied to vectors $\mathbf{v} = \mathbf{i}u$, $\mathbf{v} = \mathbf{j}u$, $\mathbf{v} = \mathbf{k}u$, and thus (8-13) is established.

Formula (8-13) can serve as a basis for a definition of ∇u in the form

$$\nabla u = \lim \frac{1}{\tau}\int_S un\,d\sigma \tag{8-14}$$

analogous to (8-3) and (8-4).

PROBLEMS

1. Show that under suitable hypothesis on continuity of the derivatives:

(a) curl $(\mathbf{A} + \mathbf{B})$ = curl $\mathbf{A}$ + curl $\mathbf{B}$

(b) div curl $\mathbf{A} = \nabla\cdot\nabla \times \mathbf{A} = 0$

(c) curl curl $\mathbf{A} = \nabla$ div $\mathbf{A} - \nabla^2\mathbf{A}$, where $\nabla^2\mathbf{A} \equiv \mathbf{i}\nabla^2 A_x + \mathbf{j}\nabla^2 A_y + \mathbf{k}\nabla^2 A_z$

(d) curl $\nabla u = \nabla \times (\nabla u) = \mathbf{0}$

(e) curl $(u\mathbf{A}) = \nabla \times (u\mathbf{A}) = u\nabla \times \mathbf{A} + \nabla u \times \mathbf{A}$

(f) div $(\mathbf{A} \times \mathbf{B}) = \mathbf{B}\cdot$curl $\mathbf{A} - \mathbf{A}\cdot$curl $\mathbf{B}$

(g) curl $(\mathbf{A} \times \mathbf{B}) = \mathbf{A}\nabla\cdot\mathbf{B} - \mathbf{B}\nabla\cdot\mathbf{A} + (\mathbf{B}\cdot\nabla)\mathbf{A} - (\mathbf{A}\cdot\nabla)\mathbf{B}$, where $(\mathbf{A}\cdot\nabla)\mathbf{B} \equiv \mathbf{C}$ is the vector with components $C_x = A_x\partial B_x/\partial x + A_y\partial B_x/\partial y + A_z\partial B_x/\partial z$, $C_y = A_x\partial B_y/\partial x + A_y\partial B_y/\partial y + A_z\partial B_y/\partial z$, $C_z = A_x\partial B_z/\partial x + A_y\partial B_z/\partial y + A_z\partial B_z/\partial z$.

2. Compute curl $\mathbf{A}$ if $r = \sqrt{x^2 + y^2 + z^2}$ and (a) $\mathbf{A} = \mathbf{i}x + \mathbf{j}y + \mathbf{k}z$; (b) $\mathbf{A} = \mathbf{i}(x/r) + \mathbf{j}(y/r) + \mathbf{k}(z/r)$; (c) $\mathbf{A} = \mathbf{r}r^n$, where $\mathbf{r} = \mathbf{i}x + \mathbf{j}y + \mathbf{k}z$.

3. If $\mathbf{a}$ is a constant vector and $\mathbf{r} = \mathbf{i}x + \mathbf{j}y + \mathbf{k}z$, show that (a) $\nabla(\mathbf{a}\cdot\mathbf{r}) = \mathbf{a}$; (b) $\nabla \times (\mathbf{a} \times \mathbf{r}) = 2\mathbf{a}$.

4. Let a rigid body rotate with constant angular velocity $\mathbf{\Omega}$ about some axis through a point O in the body. If $\mathbf{r}$ is the position vector of a point $P(x,y,z)$ relative to a set of axes fixed at O, the velocity $\mathbf{v}$ of P is $\mathbf{v} = \mathbf{v}_0 + \mathbf{\Omega} \times \mathbf{r}$, where $\mathbf{v}_0$ is the velocity of O relative to some reference frame fixed in space (cf. Chap. 4, Sec. 8). Show that curl $\mathbf{v} = 2\mathbf{\Omega}$, so that the angular velocity $\mathbf{\Omega}$ at any instant of time is equal to one-half the curl of the velocity field. Note that the velocity $\mathbf{v}_0$ is independent of the coordinates (x,y,z) of points in the body.

5. Show from geometric considerations that the angle $d\theta$ subtended at the origin by an element ds of a curve is $d\theta = (\mathbf{n}\cdot\mathbf{r}/r^2)\,ds$.

6. A solid angle ω subtended by a surface σ is measured by the area subtended by the angle on a unit sphere S with center at the vertex of ω. Show that

$$\omega = -\int_\sigma \mathbf{n}\cdot\nabla\frac{1}{r}\,d\sigma$$

where $\mathbf{r}$ is the position vector of points on σ measured from the vertex of ω and $\mathbf{n}$ is the unit normal to σ. *Hint:* Apply the divergence theorem to a volume formed by the bundle of rays issuing from the solid angle and by the areas cut out by these rays on S and on σ.

7. Referring to Prob. 6, show from geometric considerations that

$$d\omega = \frac{\mathbf{n}\cdot\mathbf{r}}{r^3}\,d\sigma.$$

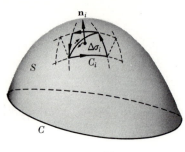

FIGURE 14

9. Stokes' Formula. Line Integrals in Space. The Stokes formula is basic to many developments in mathematical analysis and applied mathematics. The derivation of this formula in an invariant form follows the pattern used to deduce the divergence theorem in Sec. 7.

Let R be a three-dimensional region in which a continuously differentiable vector function $\mathbf{v}(P)$ is given. We consider a piecewise-smooth, open, two-sided surface S in the interior of R, and assume that the edge of S is a simple closed curve C (Fig. 14). The Stokes formula states that

$$\int_S \mathbf{n} \cdot \operatorname{curl} \mathbf{v} \, d\sigma = \int_C \mathbf{v} \cdot d\mathbf{r} \tag{9-1}$$

where $\mathbf{n}$ is a unit normal to S and the line integral over C is evaluated in the direction determined by the chosen positive orientation of $\mathbf{n}$.

We subdivide S into k approximately planar elements of area $\Delta\sigma_i$, each bounded by a simple contour C_i (say triangular) (see Fig. 14). Then formula (8-7) with ν replaced by $\mathbf{n}_i$ and A by $\Delta\sigma_i$ when applied to the element bounded by C_i yields [cf. (7-3)]

$$\mathbf{n}_i \cdot \operatorname{curl} \mathbf{v}(P_i) \, \Delta\sigma_i = \int_{C_i} \mathbf{v} \cdot d\mathbf{r} + \epsilon_i \, \Delta\sigma_i. \tag{9-2}$$

On summing these expressions over the entire surface S we get

$$\sum_{i=1}^{k} \mathbf{n}_i \cdot \operatorname{curl} \mathbf{v}(P_i) \, \Delta\sigma_i = \sum_{i=1}^{k} \int_{C_i} \mathbf{v} \cdot d\mathbf{r} + \sum_{i=1}^{k} \epsilon_i \, \Delta\sigma_i. \tag{9-3}$$

But the line integrals in (9-3) when summed over the common boundaries of adjacent elements cancel out, since such boundaries are traversed twice in opposite directions. The surviving terms yield the line integral $\int_C \mathbf{v} \cdot d\mathbf{r}$ over the boundary C.

If the greatest linear dimensions of the $\Delta\sigma_i$ tend to zero, the number k of elements $\Delta\sigma_i$ increases indefinitely, and the sum on the left becomes the surface integral $\int_S \mathbf{n} \cdot \operatorname{curl} \mathbf{v} \, d\sigma$. The sum $\sum_{i=1}^{k} \epsilon_i \, \Delta\sigma_i$ tends to zero as in the discussion of (7-3).[1] Thus, formula (9-1) is correct. It should be noted that, once a positive direction for the normal $\mathbf{n}$ has been agreed upon, the positive direction of description of the contours C_i, and hence of C, is determined by the right-hand-screw convention.

In cartesian coordinates, as follows from (8-11), formula (9-1) reads

$$\iint_S \left[\left(\frac{\partial v_z}{\partial y} - \frac{\partial v_y}{\partial z} \right) \cos(x,n) + \left(\frac{\partial v_x}{\partial z} - \frac{\partial v_z}{\partial x} \right) \cos(y,n) + \left(\frac{\partial v_y}{\partial x} - \frac{\partial v_x}{\partial y} \right) \cos(z,n) \, d\sigma \right]$$

$$= \int_C (v_x \, dx + v_y \, dy + v_z \, dz) \tag{9-4}$$

since $\mathbf{v} = \mathbf{i}v_x + \mathbf{j}v_y + \mathbf{k}v_z, \quad \mathbf{n} = \mathbf{i} \cos(x,n) + \mathbf{j} \cos(y,n) + \mathbf{k} \cos(z,n)$

and $d\mathbf{r} = \mathbf{i} \, dx + \mathbf{j} \, dy + \mathbf{k} \, dz.$

We observe that Stokes' formula (9-4) reduces to Green's theorem in the plane [Chap. 5, formula (21-6)] if C is a simple closed curve bounding a plane region R and

[1] That is, if all $|\epsilon_i|$ are less than ϵ, this sum is less than $\epsilon\sigma$, where σ is the area of S.

$v_x = M(x,y)$, $v_y = N(x,y)$, $v_z = 0$. This suggests the use of Stokes' formula in generalizing to three dimensions the theorems of Sec. 22 in Chap. 5.

We have shown in Sec. 4 that, if $\mathbf{v}(P) = \mathbf{i}v_x + \mathbf{j}v_y + \mathbf{k}v_z$ is continuously differentiable in a simply or multiply connected region R and if the line integral

$$\int_C \mathbf{v} \cdot d\mathbf{r} = \int_{(x_0,y_0,z_0)}^{(x,y,z)} (v_x\,dx + v_y\,dy + v_z\,dz) \tag{9-5}$$

is independent of the path, this integral defines a single-valued function $u(x,y,z)$ such that its total differential $du = v_x\,dx + v_y\,dy + v_z\,dz$. Moreover, the functions v_x, v_y, v_z satisfy Eqs. (4-7) and (4-8),

$$\frac{\partial v_x}{\partial y} - \frac{\partial v_y}{\partial x} = 0 \qquad \frac{\partial v_z}{\partial y} - \frac{\partial v_y}{\partial z} = 0 \qquad \frac{\partial v_x}{\partial z} - \frac{\partial v_z}{\partial x} = 0 \tag{9-6}$$

at each point of R.

But the left-hand members of Eqs. (9-6) are precisely the components of curl $\mathbf{v}$ in formula (8-11). Accordingly, if (9-5) is independent of the path and $\mathbf{v}(P)$ is continuously differentiable in R, then curl $\mathbf{v} = \mathbf{0}$ at all points of R. But the satisfaction of Eqs. (9-6) at all points of R does not ensure that the line integral (9-5) is independent of the path[1] in every connected region R. However, if R is restricted to be a simply connected region and if we suppose that curl $\mathbf{v} = \mathbf{0}$ throughout such a region, we conclude at once from Stokes' formula (9-1) that the line integral $\int_C \mathbf{v} \cdot d\mathbf{r} = \int_C (v_x\,dx + v_y\,dy + v_z\,dz) = 0$ over every simple closed path C in R. To say that this line integral vanishes over every such closed path C is equivalent to saying that it is independent of the path. (See Sec. 4.)

We summarize these conclusions in the following:

*THEOREM. If $\mathbf{v}(P)$ is continuously differentiable in a simply connected region R bounded by a closed surface S and curl $\mathbf{v} = \mathbf{0}$ at each point of R, then the line integral $\int_C \mathbf{v} \cdot d\mathbf{r} = 0$ over every simple closed path C in R and (consequently) a single-valued scalar function $u(P)$, such that $\mathbf{v} = \nabla u$, exists in R. Conversely, if $\int_C \mathbf{v} \cdot d\mathbf{r} = 0$ for every simple closed path in **any** (connected) region R then curl $\mathbf{v} \equiv \mathbf{0}$ in R.*

Example. Evaluate $\int_S \mathbf{n} \cdot$ curl $\mathbf{v}\,d\sigma$ over the surface $z = +\sqrt{a^2 - x^2 - y^2}$ if $\mathbf{v} = \mathbf{i}2y - \mathbf{j}x + \mathbf{k}z$. The surface in this example is a hemisphere of radius a, and it is clear that $\mathbf{n} = \mathbf{i}(x/r) + \mathbf{j}(y/r) + \mathbf{k}(z/r)$, where $\mathbf{r} = \mathbf{i}x + \mathbf{j}y + \mathbf{k}z$ is the position vector for points on the hemisphere. We readily check that curl $\mathbf{v} = -3\mathbf{k}$. Hence

$$\int_S \mathbf{n} \cdot \text{curl } \mathbf{v}\,d\sigma = -3 \int_S \frac{z}{a}\,d\sigma.$$

This integral can be easily evaluated by noting that (Fig. 15) $d\sigma = \sec \gamma\,dx\,dy$, where γ is the angle between the normal $\mathbf{n}$ and the positive direction of the z axis (cf. Chap. 5, Sec. 18). But from Fig. 15, $\sec \gamma = \sec \theta = a/z$, so that

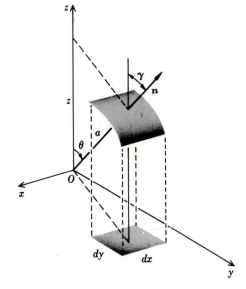

FIGURE 15

---

[1] We have already demonstrated this by the example of Sec. 23 in Chap. 4. In the two-dimensional case the condition curl $\mathbf{v} = \mathbf{0}$ reduces to $\partial M/\partial y - \partial N/\partial x = 0$, where $M \equiv v_x$, $N \equiv v_y$.

$$-3 \int_S \frac{z}{a} \, d\sigma = -3 \iint_A dx \, dy = -3\pi a^2 \tag{a}$$

since the region of integration A is a circle of radius a. The reader will check this result by taking $d\sigma = a^2 \sin \theta \, d\theta \, d\phi$ as the element of area of the surface of the sphere in spherical coordinates.

To obtain the result (a) from Stokes' formula (9-1) we compute $\int_C \mathbf{v} \cdot d\mathbf{r}$, where C is the boundary of the circle $x^2 + y^2 = a^2$. Since $d\mathbf{r} = \mathbf{i} \, dx + \mathbf{j} \, dy + \mathbf{k} \, dz$, we have

$$\int_C \mathbf{v} \cdot d\mathbf{r} = \int_C (2y \, dx - x \, dy + z \, dz). \tag{b}$$

But along C we have $dz = 0$, and the equation of C can be taken in the form

$$x = a \cos \phi \qquad y = a \sin \phi \qquad 0 \le \phi \le 2\pi.$$

We thus get for (b)

$$\int_C \mathbf{v} \cdot d\mathbf{r} = -\int_0^{2\pi} a^2(2 \sin^2 \phi + \cos^2 \phi) \, d\phi = -3\pi a^2.$$

PROBLEMS

1. If $\mathbf{v} = \mathbf{i}y + \mathbf{j}z + \mathbf{k}x$ and S is the surface of the paraboloid $z = 1 - x^2 - y^2$, $z \ge 0$, compute $\int_S \mathbf{n} \cdot \operatorname{curl} \mathbf{v} \, d\sigma$.

2. What is the value of the surface integral $\int_S \mathbf{n} \cdot \operatorname{curl} \mathbf{v} \, d\sigma$ if $\mathbf{v} = \mathbf{i}y^2 + \mathbf{j}xy + \mathbf{k}xz$ and S is the hemisphere $x^2 + y^2 + z^2 = 1$, $z \ge 0$? Evaluate this integral directly and by Stokes' formula.

3. Compute $\int_C \mathbf{v} \cdot d\mathbf{r}$ if $\mathbf{v} = \mathbf{i}(x^2 + y^2) + \mathbf{j}(x^2 + z^2) + \mathbf{k}y$ and C is the circle $x^2 + y^2 = 4$ in the plane $z = 0$.

4. Verify Stokes' formula if $\mathbf{v} = \mathbf{i}y^2 + \mathbf{j}xy - \mathbf{k}xz$ and S is the hemisphere $z = \sqrt{a^2 - x^2 - y^2}$.

5. Evaluate directly and by Stokes' formula the following integrals:

(a) $\int_C (x^2y^3 \, dx + dy + z \, dz)$, if C is the circle $x^2 + y^2 = a^2$, $z = 0$;

(b) $\int_C (y^2 \, dx + z^2 \, dy + x^2 \, dz)$, if C is the triangle with the vertices at $(0,0,0)$, $(0,a,0)$, $(0,0,a)$;

(c) $\int_C [(y^2 + 2xz^2 - 1) \, dx + 2xy \, dy + 2x^2z \, dz]$, if C is the ellipse $x^2/a^2 + y^2/b^2 = 1$, $z = 0$.

6. Compute the circulation of the vector field $\mathbf{v} = \mathbf{i}x^2y + \mathbf{j} + \mathbf{k}z$ along the circle $x^2 + y^2 = a^2$, $z = 0$. Check your result by Stokes' formula.

7. The following line integrals are independent of the path in suitable regions. Evaluate them.

(a) $\int_{(x_0,y_0,z_0)}^{(x,y,z)} \dfrac{dx + dy + dz}{x + y + z}$ 　　 (b) $\int_{(x_0,y_0,z_0)}^{(x,y,z)} \dfrac{x \, dx + y \, dy + z \, dz}{\sqrt{x^2 + y^2 + z^2}}$

(c) $\int_{(x_0,y_0,z_0)}^{(x,y,z)} (3x^2 + 3y - 1) \, dx + (z^2 + 3x) \, dy + (2yz + 1) \, dz$.

Hint: Use formulas (4-5).

ILLUSTRATIONS AND APPLICATIONS

10. Solenoidal and Irrotational Fields. Let a continuously differentiable vector function $\mathbf{v}(P)$ be specified in a region R. If $\operatorname{curl} \mathbf{v} = \mathbf{0}$ at every point of R, we say that $\mathbf{v}(P)$ is an *irrotational vector field*. If $\mathbf{v}(P)$ is such that $\operatorname{div} \mathbf{v} = 0$, the field is said to be *solenoidal*. The importance of solenoidal and irrotational vectors in applications derives from the fact that every continuously differentiable vector function $\mathbf{v}(P)$ defined in a bounded simply connected region R can be expressed as the sum of two vector functions, one of which is solenoidal and the other irrotational. We do not prove this fact here because it depends on demonstrating the existence of solutions of certain partial differential equations,[1] and it would carry us too far in the study of potential theory. Accordingly, we limit our discussion to two basic theorems concerned with solenoidal and irrotational vector fields.

The first of these theorems is a restatement of the theorem established in the preceding section.

THEOREM I. *A necessary and sufficient condition that a continuously differentiable vector field* $\mathbf{v}(P)$ *be irrotational in a simply connected region R is that* $\mathbf{v} = \nabla u$, *where u is a single-valued scalar function with continuous second derivatives.*

A function $u(P)$ whose gradient $\nabla u = \mathbf{v}$ is called the *potential* of the vector field $\mathbf{v}(P)$. For this reason an irrotational vector field is sometimes called a *potential field.*[2]

When $\mathbf{v}(P)$ satisfies the conditions of Theorem I, the potential $u(P)$ is determined by the formula $u(P) = \int_{P_0}^{P} \mathbf{v} \cdot d\mathbf{r}$.

THEOREM II. *A continuously differentiable vector function* $\mathbf{v}(P)$ *specified in a simply connected region R is solenoidal if, and only if, it is equal to the curl of some vector function* $\mathbf{w}$ *with continuous second derivatives in R.*

Let us suppose, first, that $\mathbf{v} = \operatorname{curl} \mathbf{w}$. Then

$$\operatorname{div} \mathbf{v} = \operatorname{div} \operatorname{curl} \mathbf{w} \equiv 0$$

as follows from a simple calculation[3] making use of formulas (6-6) and (8-11). Conversely, if $\operatorname{div} \mathbf{v} = 0$, we show that a vector $\mathbf{w}$ can be so constructed that $\mathbf{v} = \operatorname{curl} \mathbf{w}$. It suffices to show that the system of equations $\operatorname{curl} \mathbf{w} = \mathbf{v}$, or

$$\frac{\partial w_z}{\partial y} - \frac{\partial w_y}{\partial z} = v_x \qquad \frac{\partial w_x}{\partial z} - \frac{\partial w_z}{\partial x} = v_y \qquad \frac{\partial w_y}{\partial x} - \frac{\partial w_x}{\partial y} = v_z \qquad (10\text{-}1)$$

has a solution for w_x, w_y, w_z whenever

$$\frac{\partial v_x}{\partial x} + \frac{\partial v_y}{\partial y} + \frac{\partial v_z}{\partial z} = 0. \qquad (10\text{-}2)$$

We show how to construct one such solution in rectangular domains.[4] If we take $w_x \equiv 0$, the second and third of Eqs. (10-1) require that

$$\frac{\partial w_z}{\partial x} = -v_y(x,y,z) \qquad \frac{\partial w_y}{\partial x} = v_z(x,y,z). \qquad (10\text{-}3)$$

On integrating (10-3) with respect to x and treating y and z as constants, we get

$$w_z = -\int_{x_0}^{x} v_y(x,y,z)\, dx + \phi(y,z)$$

$$w_y = \int_{x_0}^{x} v_z(x,y,z)\, dx + \psi(y,z) \qquad (10\text{-}4)$$

---

[1] See Prob. 6. A discussion of the system of equations in question is contained in M. Mason and W. Weaver, "The Electromagnetic Field," pp. 352–365, University of Chicago Press, Chicago, 1932.

[2] The study of such fields is in the province of *potential theory.*

[3] See also Prob. 1b of Sec. 8.

[4] This limitation on the shape of the region is dictated by the use of the fundamental theorem of integral calculus in deriving the solution in the form (10-7).

where ϕ and ψ are arbitrary differentiable functions of y and z. If we insert these solutions in the first of Eqs. (10-1), we get

$$v_x = -\int_{x_0}^{x} \left(\frac{\partial v_y}{\partial y} + \frac{\partial v_z}{\partial z} \right) dx + \frac{\partial \phi}{\partial y} - \frac{\partial \psi}{\partial z}. \tag{10-5}$$

But from (10-2),

$$\frac{\partial v_y}{\partial y} + \frac{\partial v_z}{\partial z} = - \frac{\partial v_x}{\partial x}$$

so that (10-5) yields

$$v_x = \int_{x_0}^{x} \frac{\partial v_x}{\partial x} dx + \frac{\partial \phi}{\partial y} - \frac{\partial \psi}{\partial z}$$

$$= v_x(x,y,z) - v_x(x_0,y,z) + \frac{\partial \phi}{\partial y} - \frac{\partial \psi}{\partial z}.$$

This equation can be satisfied by taking $\psi \equiv 0$ and

$$\phi(y,z) = \int_{y_0}^{y} v_x(x_0,y,z) \, dy. \tag{10-6}$$

Thus, one solution of the system (10-1) and (10-2) is

$$w_x = 0$$

$$w_y = \int_{x_0}^{x} v_z(x,y,z) \, dx \tag{10-7}$$

$$w_z = -\int_{x_0}^{x} v_y(x,y,z) \, dx + \int_{y_0}^{y} v_x(x_0,y,z) \, dy.$$

The proof clearly indicates that $\mathbf{w}$ is not unique. Indeed, if we take $\mathbf{w}$ with components given by (10-7) and add to it ∇u, where u is an arbitrary scalar function with continuous second derivatives, then

$$\text{curl } (\mathbf{w} + \nabla u) = \text{curl } \mathbf{w}$$

inasmuch as curl $\nabla u \equiv \mathbf{0}$.[1]

We remark in conclusion that, whenever the divergence and curl of a continuously differentiable vector function $\mathbf{v}$ are specified in a region R and the normal component of $\mathbf{v}$ is known over the surface bounding a simply connected region R' in the interior of R, there is just one vector function $\mathbf{v}$ satisfying these conditions in R'. This uniqueness theorem is important in many applications. The reader may prove it by following suggestions given in Prob. 7 below.

PROBLEMS

1. Show that $\mathbf{v} = \mathbf{i}2xyz + \mathbf{j}x^2z + \mathbf{k}x^2y$ is irrotational, and find $u(x,y,z)$ such that $\mathbf{v} = \nabla u$.

2. Show that $\mathbf{v} = \mathbf{i}(z - y) + \mathbf{j}(x - z) + \mathbf{k}(y - x)$ is solenoidal, and find $\mathbf{w}(x,y,z)$ such that $\mathbf{v} = \text{curl } \mathbf{w}$.

3. Is $\mathbf{v} = \mathbf{i}(y^2 + 2xz^2 - 1) + \mathbf{j}2xy + \mathbf{k}2x^2z$ irrotational? If so, find u such that $\mathbf{v} = \nabla u$.

4. Is $\mathbf{v} = \mathbf{i}(x^3z - 2xyz) + \mathbf{j}(xy - 3x^2yz) + \mathbf{k}(yz^2 - xz)$ solenoidal? If so, find a $\mathbf{w}$ such that $\mathbf{v} = \text{curl } \mathbf{w}$.

5. Prove that $\mathbf{v} = r^n\mathbf{r}$, where $\mathbf{r} = \mathbf{i}x + \mathbf{j}y + \mathbf{k}z$, is irrotational. Is it solenoidal?

6. Let $\mathbf{w} = \mathbf{u} + \mathbf{v}$, where $\mathbf{u}$ is irrotational and $\mathbf{v}$ solenoidal in a given suitably restricted region R.

---

[1] Conversely, if curl $\mathbf{w}_1 = \mathbf{v}$, then curl $(\mathbf{w}_1 - \mathbf{w}) = \mathbf{0}$ and $\mathbf{w}_1 - \mathbf{w} = \nabla u$ by Theorem I. Thus every solution $\mathbf{w}_1$ is representable in the form $\mathbf{w} + \nabla u$, where $\mathbf{w}$ is the particular solution found in the text.

Then there exists a vector $\mathbf{q}$ such that $\mathbf{v} = \text{curl } \mathbf{q}$ and a scalar ϕ such that $\mathbf{u} = \nabla\phi$. Show that ϕ and $\mathbf{q}$ satisfy the following partial differential equations:

$$\nabla^2\phi = \text{div } \mathbf{w} \qquad \nabla \text{ div } \mathbf{q} - \nabla^2\mathbf{q} = \text{curl } \mathbf{w}.$$

7. If $\mathbf{v}$ is a continuously differentiable vector function defined in a regular simply connected region R bounded by the surface S and if

$$\text{curl } \mathbf{v} = \mathbf{f}(x,y,z) \qquad \text{div } \mathbf{v} = g(x,y,z)$$

in R and $\mathbf{v}\cdot\mathbf{n} = h(x,y,z)$ on S, show that $\mathbf{v}$ is uniquely determined in R by these conditions.

Outline of solution: Assume that there are two such vectors, $\mathbf{v} = \mathbf{v}_1$ and $\mathbf{v} = \mathbf{v}_2$. With $\mathbf{w} = \mathbf{v}_1 - \mathbf{v}_2$, show that there is a u such that $\mathbf{w} = \nabla u$, and deduce $\nabla^2 u = 0$. By applying the divergence theorem to the vector $u\nabla u$, show that $\int_R (\nabla u)\cdot(\nabla u)\, d\tau = 0$. Since $(\nabla u)\cdot(\nabla u) \geq 0$, this integral can vanish only if $\nabla u \equiv 0$.

8. Prove that if the field $\mathbf{v}$ is both potential and solenoidal the potential u satisfies Laplace's equation $\nabla^2 u = 0$.

9. If $\mathbf{F} = \mathbf{r}f(r)$ is a central force field (that is, $\mathbf{F}$ is directed toward a fixed point O and depends only on the distance r from O), where $f(r)$ is continuously differentiable, the field $\mathbf{F}$ is potential and the potential $u = \int_{r_0}^r rf(r)\, dr$. Verify this.

10. Let $\mathbf{F}$ in Prob. 9 be $\mathbf{F} = -\mathbf{r}/r^3$, $r \neq 0$. Show that $u = 1/r$, and verify that $\nabla^2 u = 0$.

---

11. Curvilinear Coordinates.

The chief advantage of formulating relations among geometrical and physical quantities in the form of vector equations is that the relations so stated are valid in all coordinate systems. Only when one comes to consider a special problem involving numerical computations does it prove desirable to translate vector equations into the language of special coordinate systems that seem best adapted to the problem at hand. For example, in analyzing vibrations of clamped rectangular membranes, it is usually advantageous to express the displacement vec-

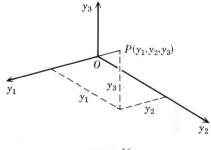

FIGURE 16

tor in cartesian coordinates. In the study of heat flow in a sphere, the geometry of the situation suggests the use of spherical coordinates, while problems concerned with the flow of currents in cylindrical conductors may indicate the use of cylindrical or bipolar coordinates. All these coordinate systems are but special cases of the general curvilinear coordinate system which we proceed to describe.

Let us refer a given region R of space to a set of orthogonal cartesian axes y_1, y_2, y_3. We denote the coordinates of any point P in R by (y_1,y_2,y_3) (Fig. 16) instead of the familiar labels (x,y,z). A set of functional relations

$$x_1 = x_1(y_1,y_2,y_3) \qquad x_2 = x_2(y_1,y_2,y_3) \qquad x_3 = x_3(y_1,y_2,y_3) \qquad (11\text{-}1)$$

connecting the variables y_1, y_2, y_3 with three new variables x_1, x_2, x_3, is said to represent a *transformation of coordinates*. We shall suppose that the functions $x_i(y_1,y_2,y_3)$ $(i = 1,2,3)$ are single-valued and are continuously differentiable at all points of the region R and that Eqs. (11-1) can be solved for the y_i to yield the *inverse transformation*

$$y_1 = y_1(x_1,x_2,x_3) \qquad y_2 = y_2(x_1,x_2,x_3) \qquad y_3 = y_3(x_1,x_2,x_3) \qquad (11\text{-}2)$$

in which the functions $y_i(x_1,x_2,x_3)$ are single-valued and continuously differentiable with respect to the variables x_i. The transformations (11-1) and (11-2) with these properties establish a one-to-one correspondence between the triplets of values (y_1,y_2,y_3) and

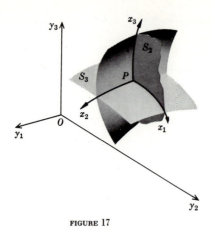

FIGURE 17

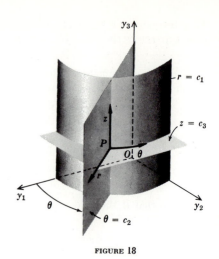

FIGURE 18

(x_1,x_2,x_3). We shall term the triplet of values (x_1,x_2,x_3), corresponding to a given point $P(y_1,y_2,y_3)$, the *curvilinear coordinates* of P and shall say that Eqs. (11-1) define a *curvilinear coordinate system* x_1,x_2,x_3. The reason for this terminology is the following: If we set in (11-1) $x_1 = c_1$ (a constant), the equation

$$x_1(y_1,y_2,y_3) = c_1 \tag{11-3}$$

represents a certain surface S_1.   Similarly, equations

$$x_2(y_1,y_2,y_3) = c_2 \tag{11-4}$$

and
$$x_3(y_1,y_2,y_3) = c_3 \tag{11-5}$$

represent surfaces S_2 and S_3. These surfaces, shown in Fig. 17, intersect at the point P, whose cartesian coordinates (y_1,y_2,y_3) can be obtained by solving Eqs. (11-3) to (11-5) for the y_i.

The surfaces S_i are called *coordinate surfaces*, and their intersections pair by pair are *coordinate lines* x_1, x_2, x_3. Thus, the x_1 coordinate line is the line of intersection of the surfaces $x_2 = c_2$ and $x_3 = c_3$. Along this line the only variable that changes is x_1, since $x_2 = c_2$ and $x_3 = c_3$ along the line x_1. Similarly, along the x_2 coordinate line the only variable that changes is x_2, while along the x_3 line the only variable that changes is x_3.

A very special case of the set of Eqs. (11-1) is

$$x_1 = y_1 \qquad x_2 = y_2 \qquad x_3 = y_3. \tag{11-6}$$

If we set $x_i = c_i$ ($i = 1,2,3$) in (11-6), we get three planes $y_i = c_i$ perpendicular to the y coordinate axes. These planes intersect at the point (c_1,c_2,c_3). The coordinate surfaces in this case are planes, and their intersections pair by pair are straight lines parallel to the coordinate axes.

As a more interesting example, consider a transformation

$$y_1 = r \cos \theta \qquad y_2 = r \sin \theta \qquad y_3 = z \tag{11-7}$$

which is of the form (11-2) if we set $x_1 = r$, $x_2 = \theta$, $x_3 = z$. The inverse of (11-7) is

$$r = +\sqrt{y_1{}^2 + y_2{}^2} \qquad \theta = \tan^{-1}\frac{y_2}{y_1} \qquad z = y_3 \tag{11-8}$$

and it is single-valued if we take $0 \leq \theta < 2\pi$ and $r > 0$. The surface $r = c_1$ is a circular cylinder $y_1{}^2 + y_2{}^2 = c_1{}^2$ whose axis coincides with the y_3 axis (Fig. 18).   The surface $\theta = c_2$ is the plane $y_2 = (\tan c_2)y_1$ containing the y_3 axis, while the surface $z = c_3$ is the plane $y_3 = c_3$ perpendicular to the y_3 axis.   The r, θ, and z coordinate lines are shown in Fig. 18, and we recognize that the curvilinear coordinate system r, θ, z is the familiar system of cylindrical coordinates.

As a final example, consider the transformation

$$y_1 = \rho \sin \theta \cos \phi \qquad y_2 = \rho \sin \theta \sin \phi \qquad y_3 = \rho \cos \theta \tag{11-9}$$

with the inverse

$$\rho = \sqrt{y_1{}^2 + y_2{}^2 + y_3{}^2} \qquad \theta = \tan^{-1}\frac{\sqrt{y_1{}^2 + y_2{}^2}}{y_3} \qquad \phi = \tan^{-1}\frac{y_2}{y_1} \tag{11-10}$$

which is single-valued if we suppose that $\rho > 0$, $0 < \theta < \pi$, $0 \leq \phi < 2\pi$.

The transformation defines a spherical system of coordinates.   The coordinate surfaces $\rho =$ const, $\theta =$ const, and $\phi =$ const are, respectively, spheres, cones, and planes, shown in Fig. 19.   The coordinate lines are the meridians, the lines of parallels, and the radial lines.

PROBLEMS

1. Discuss the curvilinear coordinates determined by $y_1 = x_1 + x_2 + x_3$, $y_2 = x_1 - x_2 + x_3$, $y_3 = 2x_1 + x_2 - x_3$.

2. Show by geometry that the coordinate lines in cylindrical and spherical coordinate systems intersect at right angles.

---

12. Metric Coefficients.

In this section we introduce an abridged notation which will enable us to write many formulas compactly and without loss of clarity.   Thus, we shall write the set of three equations of transformation (11-1) in the form

$$x_i = x_i(y_1, y_2, y_3) \qquad i = 1, 2, 3 \tag{12-1}$$

and their inverse (11-2) as

$$y_i = y_i(x_1, x_2, x_3). \tag{12-2}$$

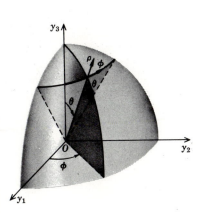

FIGURE 19

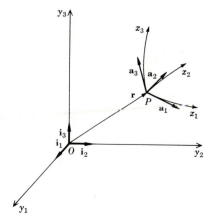

FIGURE 20

Throughout this section we shall suppose that the Latin indices i, j, k have the range of values 1, 2, 3.

If $P(y_1,y_2,y_3)$ is any point referred to a set of cartesian axes y (Fig. 20), its position vector $\mathbf{r}$ can be written in the form

$$\mathbf{r} = \mathbf{i}_1 y_1 + \mathbf{i}_2 y_2 + \mathbf{i}_3 y_3 \tag{12-3}$$

where the $\mathbf{i}_1$, $\mathbf{i}_2$, $\mathbf{i}_3$ are the unit base vectors, which in Chap. 4 we denoted by $\mathbf{i}$, $\mathbf{j}$, $\mathbf{k}$.

The square of the element of arc ds along a curve C has the form

$$(ds)^2 = (dy_1)^2 + (dy_2)^2 + (dy_3)^2 \tag{12-4}$$

and since

$$d\mathbf{r} = \mathbf{i}_1\, dy_1 + \mathbf{i}_2\, dy_2 + \mathbf{i}_3\, dy_3 \tag{12-5}$$

we can write (12-4) as a scalar product

$$(ds)^2 = \sum_{i=1}^{3} dy_i\, dy_i = d\mathbf{r}\cdot d\mathbf{r}. \tag{12-6}$$

If we replace the y_i in (12-3) by their values in terms of the x's with the aid of (12-2), $\mathbf{r}$ becomes a function of the variables x_i and we can write

$$d\mathbf{r} = \frac{\partial \mathbf{r}}{\partial x_1}\, dx_1 + \frac{\partial \mathbf{r}}{\partial x_2}\, dx_2 + \frac{\partial \mathbf{r}}{\partial x_3}\, dx_3 \tag{12-7}$$

$$\equiv \sum_{i=1}^{3} \frac{\partial \mathbf{r}}{\partial x_i}\, dx_i.$$

Now, the symbol

$$\frac{\partial \mathbf{r}(x_1,x_2,x_3)}{\partial x_i}$$

denotes the derivative of $\mathbf{r}$ with respect to a particular variable x_i ($i = 1,2,3$) when the remaining variables are held fast. Thus, if we fix the variables x_2 and x_3 by setting $x_2 = c_2$ and $x_3 = c_3$, $\mathbf{r}$ becomes a function of x_1 alone, and hence the terminus of $\mathbf{r}$ is constrained to move along the x_1 coordinate line in the x coordinate system determined by Eqs. (12-1). Consequently, the vector

$$\frac{\partial \mathbf{r}}{\partial x_1} = \lim_{\Delta x_1 \to 0} \frac{\Delta \mathbf{r}}{\Delta x_1}$$

is tangent to the coordinate line x_1. Similarly, we conclude that the vectors $\partial \mathbf{r}/\partial x_2$ and $\partial \mathbf{r}/\partial x_3$ are tangent to the x_2 and x_3 coordinate lines, respectively (Fig. 20). If we denote these vectors by $\mathbf{a}_i$, so that

$$\mathbf{a}_i = \frac{\partial \mathbf{r}}{\partial x_i} \tag{12-8}$$

we can write (12-7) as

$$d\mathbf{r} = \sum_{i=1}^{3} \mathbf{a}_i\, dx_i \tag{12-9}$$

and hence Eq. (12-6) assumes the form

$$(ds)^2 = \left(\sum_{i=1}^{3} \mathbf{a}_i\, dx_i\right) \cdot \left(\sum_{i=1}^{3} \mathbf{a}_i\, dx_i\right). \tag{12-10}$$

On expanding the scalar product in (12-10), we see that the formula can be written as

$$(ds)^2 = \sum_{i=1}^{3}\sum_{j=1}^{3} \mathbf{a}_i\cdot\mathbf{a}_j\, dx_i\, dx_j$$

and hence, with g_{ij} defined by

$$\mathbf{a}_i \cdot \mathbf{a}_j \equiv g_{ij} \qquad (12\text{-}11)$$

we can write it as

$$(ds)^2 = \sum_{i=1}^{3} \sum_{j=1}^{3} g_{ij}\, dx_i\, dx_j. \qquad (12\text{-}12)$$

In expanded form this reads

$$(ds)^2 = g_{11}(dx_1)^2 + g_{12}\, dx_1\, dx_2 + g_{13}\, dx_1\, dx_3$$
$$+ g_{21}\, dx_2\, dx_1 + g_{22}(dx_2)^2 + g_{23}\, dx_2\, dx_3$$
$$+ g_{31}\, dx_3\, dx_1 + g_{32}\, dx_3\, dx_2 + g_{33}(dx_3)^2. \qquad (12\text{-}13)$$

Since $\mathbf{a}_i \cdot \mathbf{a}_j = \mathbf{a}_j \cdot \mathbf{a}_i$, we see from the definition (12-11) that $g_{ij} = g_{ji}$. Thus the *quadratic differential form* (12-13) is symmetric.

For reasons that will appear presently, the coefficients g_{ij} in this quadratic form are called *metric coefficients*. We shall see that they can be computed directly from Eqs. (12-2) without first calculating the vectors $\mathbf{a}_i$.

The vectors $\mathbf{a}_i$, which were found to be tangent to the coordinate lines x_i at a given point P, are called *base vectors* in the curvilinear coordinate system x. Any vector $\mathbf{A}$ with the origin at P can be resolved into components A_1, A_2, A_3 along the directions of the vectors $\mathbf{a}_1$, $\mathbf{a}_2$, $\mathbf{a}_3$ (Fig. 21). Thus, the base vectors $\mathbf{a}_i$ play the same role in the system x as the base vectors $\mathbf{i}_1$, $\mathbf{i}_2$, $\mathbf{i}_3$ do in the cartesian system y. It should be noted, however, that, while the magnitudes and directions of cartesian base vectors are fixed, the vectors $\mathbf{a}_i$, in general, vary from point to point in space.

From the definition (12-11) we see on setting $i = j = 1$ that the length of $\mathbf{a}_1$ is $|\mathbf{a}_1| = \sqrt{g_{11}}$. Similarly, $|\mathbf{a}_2| = \sqrt{g_{22}}$ and $|\mathbf{a}_3| = \sqrt{g_{33}}$. These vectors are orthogonal if, and only if,

$$g_{12} = g_{21} = \mathbf{a}_1 \cdot \mathbf{a}_2 = 0 \qquad g_{31} = g_{13} = \mathbf{a}_1 \cdot \mathbf{a}_3 = 0 \qquad g_{23} = g_{32} = \mathbf{a}_2 \cdot \mathbf{a}_3 = 0.$$

A curvilinear coordinate system for which these relations hold is called *orthogonal*, and we note that in an orthogonal system the quadratic form (12-13) has the structure

$$(ds)^2 = g_{11}(dx_1)^2 + g_{22}(dx_2)^2 + g_{33}(dx_3)^2. \qquad (12\text{-}14)$$

To get at the meaning of the coefficients g_{11}, g_{22}, and g_{33}, we note that, when an element of arc ds is directed along the x_1 coordinate line, $dx_2 = dx_3 = 0$, since along the x_1 line x_2 and x_3 do not vary. Thus, (12-14) gives in this case

$$(ds_1)^2 = g_{11}(dx_1)^2$$

so that

$$ds_1 = \sqrt{g_{11}}\, dx_1. \qquad (12\text{-}15)$$

Thus, the length of the arc element ds_1 along the x_1 coordinate line is obtained by multiplying the differential of x_1 by $\sqrt{g_{11}}$. Similarly we find that the differentials of arc ds_i along the x_2 and x_3 coordinate lines are

$$ds_2 = \sqrt{g_{22}}\, dx_2 \qquad ds_3 = \sqrt{g_{33}}\, dx_3. \qquad (12\text{-}16)$$

Since the ds_i and the dx_i are real, we conclude that $g_{11} > 0$, $g_{22} > 0$, $g_{33} > 0$. In orthogonal

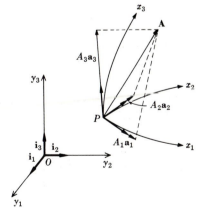

FIGURE 21

cartesian coordinates, $(ds)^2$ is given by the formula (12-4), and hence in such a system $g_{11} = g_{22} = g_{33} = 1$.

An element of volume $d\tau$ in general curvilinear coordinates is defined as the volume of the parallelepiped (see Chap. 4, Sec. 6)

$$d\tau = |\mathbf{a}_1 \cdot \mathbf{a}_2 \times \mathbf{a}_3|\ dx_1\ dx_2\ dx_3 \tag{12-17}$$

constructed on the base vectors $\mathbf{a}_i$. If the system is orthogonal, (12-17) reduces to

$$d\tau = \sqrt{g_{11}g_{22}g_{33}}\ dx_1\ dx_2\ dx_3 \tag{12-18}$$

as is immediately obvious from (12-15) and (12-16).

When a curvilinear coordinate system x is determined by equations of the form (12-1), we can write the inverse transformation (12-2) as

$$y_k = y_k(x_1, x_2, x_3) \tag{12-19}$$

and deduce the metric coefficients g_{ij} as follows: On differentiating Eqs. (12-19) with respect to x_i we get

$$dy_k = \sum_{i=1}^{3} \frac{\partial y_k}{\partial x_i}\ dx_i. \tag{12-20}$$

But in cartesian coordinates

$$ds^2 = \sum_{k=1}^{3} dy_k\ dy_k$$

and the substitution from (12-20) in this formula yields[1]

$$ds^2 = \sum_{k=1}^{3}\left[\sum_{i=1}^{3} \frac{\partial y_k}{\partial x_i}\ dx_i \sum_{j=1}^{3} \frac{\partial y_k}{\partial x_j}\ dx_j\right] = \sum_{i=1}^{3} \sum_{j=1}^{3}\left(\sum_{k=1}^{3} \frac{\partial y_k}{\partial x_i} \frac{\partial y_k}{\partial x_j}\right) dx_i\ dx_j. \tag{12-21}$$

On comparing (12-21) with (12-12), we see that

$$g_{ij} = \sum_{k=1}^{3} \frac{\partial y_k}{\partial x_i} \frac{\partial y_k}{\partial x_j} \qquad i, j = 1, 2, 3. \tag{12-22}$$

This is the desired formula for the calculation of metric coefficients.

To illustrate the use of (12-22), consider a coordinate system defined by Eqs. (11-7), which we write in the form

$$y_1 = x_1 \cos x_2 \qquad y_2 = x_1 \sin x_2 \qquad y_3 = x_3$$

to agree with the notation used in this section. From (12-22) we have

$$g_{11} = \left(\frac{\partial y_1}{\partial x_1}\right)^2 + \left(\frac{\partial y_2}{\partial x_1}\right)^2 + \left(\frac{\partial y_3}{\partial x_1}\right)^2 = \cos^2 x_2 + \sin^2 x_2 + 0 = 1$$

$$g_{22} = \left(\frac{\partial y_1}{\partial x_2}\right)^2 + \left(\frac{\partial y_2}{\partial x_2}\right)^2 + \left(\frac{\partial y_3}{\partial x_2}\right)^2 = x_1{}^2 \sin^2 x_2 + x_1{}^2 \cos x_2{}^2 + 0 = x_1{}^2$$

$$g_{33} = \left(\frac{\partial y_1}{\partial x_3}\right)^2 + \left(\frac{\partial y_2}{\partial x_3}\right)^2 + \left(\frac{\partial y_3}{\partial x_3}\right)^2 = 0 + 0 + 1 = 1$$

$$g_{12} = \frac{\partial y_1}{\partial x_1} \frac{\partial y_1}{\partial x_2} + \frac{\partial y_2}{\partial x_1} \frac{\partial y_2}{\partial x_2} + \frac{\partial y_3}{\partial x_1} \frac{\partial y_3}{\partial x_2} = \cos x_2(-x_1 \sin x_2) + \sin x_2(x_1 \cos x_2) + 0 = 0.$$

---

[1] Note that the summation index can be changed at will so that

$$\sum_{i=1}^{3} \frac{\partial y_k}{\partial x_i}\ dx_i \equiv \sum_{j=1}^{3} \frac{\partial y_k}{\partial x_j}\ dx_j.$$

We find in the same way that $g_{23} = g_{13} = 0$. Hence the system under consideration is orthogonal. The expression for ds^2 is

$$ds^2 = \sum_{i=1}^{3} \sum_{j=1}^{3} g_{ij} \, dx_i \, dx_j = (dx_1)^2 + x_1^2 (dx_2)^2 + (dx_3)^2$$

which is a familiar formula for the square of the arc element in cylindrical coordinates if we recall that $x_1 = r$, $x_2 = \theta$, $x_3 = z$. Since this system is orthogonal, the element of volume is given by (12-18), which in our case yields

$$d\tau = r \, dr \, d\theta \, dz.$$

Example. Obtain expressions for the elements of arc and volume in the coordinate system x defined by

$$y_1 = x_1 + x_2 + x_3 \qquad y_2 = x_1 - x_2 - x_3 \qquad y_3 = 2x_1 + x_2 - x_3 \qquad (12\text{-}23)$$

and discuss the system.

On making use of formula (12-22), we find as in the preceding illustration that

$$g_{11} = 6 \qquad g_{22} = 3 \qquad g_{33} = 3 \qquad g_{12} = 2 \qquad g_{23} = 1 \qquad g_{13} = -2.$$

Hence

$$ds^2 = 6(dx_1)^2 + 4 \, dx_1 \, dx_2 - 4 \, dx_1 \, dx_3 + 2 \, dx_2 \, dx_3 + 3(dx_2)^2 + 3(dx_3)^2.$$

The system is clearly not orthogonal, and to compute $d\tau$ we shall make use of formula (12-17). Now

$$\mathbf{r} = \mathbf{i}_1 y_1 + \mathbf{i}_2 y_2 + \mathbf{i}_3 y_3$$
$$= \mathbf{i}_1(x_1 + x_2 + x_3) + \mathbf{i}_2(x_1 - x_2 - x_3) + \mathbf{i}_3(2x_1 + x_2 - x_3)$$

and hence the base vectors $\mathbf{a}_i = \partial \mathbf{r}/\partial x_i$ are

$$\mathbf{a}_1 = \mathbf{i}_1 + \mathbf{i}_2 + 2\mathbf{i}_3 \qquad \mathbf{a}_2 = \mathbf{i}_1 - \mathbf{i}_2 + \mathbf{i}_3 \qquad \mathbf{a}_3 = \mathbf{i}_1 - \mathbf{i}_2 - \mathbf{i}_3.$$

Thus,

$$d\tau = |\mathbf{a}_1 \cdot \mathbf{a}_2 \times \mathbf{a}_3| \, dx_1 \, dx_2 \, dx_3 = \begin{vmatrix} 1 & 1 & 2 \\ 1 & -1 & 1 \\ 1 & -1 & -1 \end{vmatrix} dx_1 \, dx_2 \, dx_3 = 4 \, dx_1 \, dx_2 \, dx_3.$$

On solving (12-23) for the x_i, we get

$$x_1 = \tfrac{1}{2}y_1 + \tfrac{1}{2}y_2 \qquad x_2 = -\tfrac{1}{4}y_1 - \tfrac{3}{4}y_2 + \tfrac{1}{2}y_3 \qquad x_3 = \tfrac{3}{4}y_1 + \tfrac{1}{4}y_2 - \tfrac{1}{2}y_3.$$

The coordinate surfaces $x_i = c_i$ are planes, and the coordinate lines x_i are therefore straight lines.

The system in this example is a special case of an *affine coordinate system* determined by the transformation

$$y_i = a_{i1}x_1 + a_{i2}x_2 + a_{i3}x_3 \qquad i = 1, 2, 3 \qquad (12\text{-}24)$$

in which the a_{ij} are constants. Affine transformations (12-24) occur in the study of elastic deformations, in dynamics of rigid bodies, and in many other branches of mathematical physics.

PROBLEMS

1. Discuss in the manner of the preceding example a coordinate system x determined by

$$y_1 = \frac{1}{\sqrt{6}} x_1 + \frac{2}{\sqrt{6}} x_2 + \frac{1}{\sqrt{6}} x_3 \qquad y_2 = \frac{1}{\sqrt{2}} x_1 - \frac{1}{\sqrt{3}} x_2 + \frac{1}{\sqrt{3}} x_3 \qquad y_3 = \frac{1}{\sqrt{2}} x_1 - \frac{1}{\sqrt{2}} x_3.$$

2. Compute the metric coefficients appropriate to a spherical coordinate system defined by Eqs. (11-9), and thus show that $(ds)^2 = (d\rho)^2 + \rho^2(d\theta)^2 + \rho^2 \sin^2 \theta (d\phi)^2$ and $d\tau = \rho^2 \sin \theta \, d\rho \, d\theta \, d\phi$.

3. If $\mathbf{R} = \mathbf{i}x + \mathbf{j}y + \mathbf{k}z$ is the position vector of a moving point $P(x,y,z)$ in cartesian coordinates, show that the *unit* base vectors $\mathbf{e}_r$, $\mathbf{e}_\theta$, $\mathbf{e}_z$ in cylindrical coordinates (r,θ,z) [see (11-7)] are

$$\mathbf{e}_r = \mathbf{i}\cos\theta + \mathbf{j}\sin\theta \qquad \mathbf{e}_\theta = -\mathbf{i}\sin\theta + \mathbf{j}\cos\theta \qquad \mathbf{e}_z = \mathbf{k}.$$

Show that $\mathbf{R} = r\mathbf{e}_r + z\mathbf{e}_z$, compute $d\mathbf{R}/dt$ and $d^2\mathbf{R}/dt^2$, and thus show that the velocity $\mathbf{v}$ and the acceleration $\mathbf{a}$ of the point P are

$$\mathbf{v} = \frac{dr}{dt}\,\mathbf{e}_r + r\,\frac{d\theta}{dt}\,\mathbf{e}_\theta + \frac{dz}{dt}\,\mathbf{e}_z$$

$$\mathbf{a} = \left[\frac{d^2r}{dt^2} - r\left(\frac{d\theta}{dt}\right)^2\right]\mathbf{e}_r + \frac{1}{r}\frac{d}{dt}\left(r^2\frac{d\theta}{dt}\right)\mathbf{e}_\theta + \frac{d^2z}{dt^2}\,\mathbf{e}_z.$$

4. If $\mathbf{R} = \mathbf{i}x + \mathbf{j}y + \mathbf{k}z$ is the position vector of $P(x,y,z)$ in cartesian coordinates, show that the *unit* base vectors $\mathbf{e}_\rho$, $\mathbf{e}_\theta$, $\mathbf{e}_\phi$ in spherical coordinates defined by Eqs. (11-9) are

$$\mathbf{e}_\rho = \mathbf{i}\sin\theta\cos\phi + \mathbf{j}\sin\theta\sin\phi + \mathbf{k}\cos\theta$$
$$\mathbf{e}_\theta = \mathbf{i}\cos\theta\cos\phi + \mathbf{j}\cos\theta\sin\phi - \mathbf{k}\sin\theta$$
$$\mathbf{e}_\phi = -\mathbf{i}\sin\phi + \mathbf{j}\cos\phi.$$

5. If the position vector $\mathbf{R}$ of a moving point P in spherical coordinates is written as $\mathbf{R} = \rho\mathbf{e}_\rho$, where $\mathbf{e}_\rho$ is the unit vector in the direction of the increasing coordinate ρ, use the results of Prob. 4 to show that

$$\mathbf{v} = \frac{d\mathbf{R}}{dt} = \frac{d\rho}{dt}\,\mathbf{e}_\rho + \rho\,\frac{d\theta}{dt}\,\mathbf{e}_\theta + \rho\sin\theta\,\frac{d\phi}{dt}\,\mathbf{e}_\phi.$$

---

13. Gradient, Divergence, and Curl in Orthogonal Curvilinear Coordinates.

In this section we record the expressions for the gradient, divergence, curl, and Laplacian in orthogonal curvilinear coordinates. These can be obtained from the definitions (2-6), (6-4), and (8-4) in a manner so similar to that used to obtain formulas valid in cartesian coordinates that we dispense with the details of calculations.

As in Sec. 11, we suppose that a transformation

$$y_i = y_i(x_1, x_2, x_3) \qquad i = 1, 2, 3$$

wherein the variables y_i are cartesian, defines a curvilinear coordinate system x. We suppose that the coordinates x_i are orthogonal so that the quadratic differential form (12-13) has the structure

$$(ds)^2 = g_{11}(dx_1)^2 + g_{22}(dx_2)^2 + g_{33}(dx_3)^2.$$

We denote the *unit base vectors* along the x_i coordinate lines by $\mathbf{e}_1$, $\mathbf{e}_2$, $\mathbf{e}_3$ and represent a vector $\mathbf{v}\,(P)$ in the form

$$\mathbf{v} = \mathbf{e}_1 v_1 + \mathbf{e}_2 v_2 + \mathbf{e}_3 v_3. \qquad (13\text{-}1)$$

The volume element $d\tau$ formed by the coordinate surfaces $x_i = $ const and $x_i + dx_i = $ const (Fig. 22) has the shape of a rectangular parallelepiped with edges[1] $ds_i = \sqrt{g_{ii}}\,dx_i$. Hence the areas $d\sigma_{ij}$ of its faces are

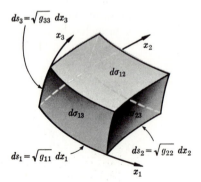

FIGURE 22

[1] See Sec. 12.

$$d\sigma_{12} = \sqrt{g_{11}g_{22}}\,dx_1\,dx_2 \qquad d\sigma_{13} = \sqrt{g_{11}g_{33}}\,dx_1\,dx_3 \qquad d\sigma_{23} = \sqrt{g_{22}g_{33}}\,dx_2\,dx_3 \qquad (13\text{-}2)$$

and its volume $d\tau$ is

$$d\tau = \sqrt{g_{11}g_{22}g_{33}}\,dx_1\,dx_2\,dx_3. \qquad (13\text{-}3)$$

To compute div $\mathbf{v}$ we calculate the flux $\int_S \mathbf{n}\cdot\mathbf{v}\,d\sigma$ over the surface of the volume element $d\tau$ and divide it by its volume (13-3). A calculation like that performed in Sec. 6 yields the result

$$\text{div } \mathbf{v} = \frac{1}{h_1 h_2 h_3}\left[\frac{\partial(v_1 h_2 h_3)}{\partial x_1} + \frac{\partial(v_2 h_1 h_3)}{\partial x_2} + \frac{\partial(v_3 h_1 h_2)}{\partial x_3}\right] \qquad (13\text{-}4)$$

where $h_i \equiv \sqrt{g_{ii}}$.

A similar but slightly longer computation also yields the formula

$$\text{curl } \mathbf{v} = \mathbf{e}_1 \frac{1}{h_3 h_2}\left[\frac{\partial(h_3 v_3)}{\partial x_2} - \frac{\partial(h_2 v_2)}{\partial x_3}\right]$$

$$+ \mathbf{e}_2 \frac{1}{h_1 h_3}\left[\frac{\partial(h_1 v_1)}{\partial x_3} - \frac{\partial(h_3 v_3)}{\partial x_1}\right] + \mathbf{e}_3 \frac{1}{h_1 h_2}\left[\frac{\partial(h_2 v_2)}{\partial x_1} - \frac{\partial(h_1 v_1)}{\partial x_2}\right] \qquad (13\text{-}5)$$

which can be written more compactly as

$$\text{curl } \mathbf{v} = \frac{1}{h_1 h_2 h_3}\begin{vmatrix} h_1 \mathbf{e}_1 & h_2 \mathbf{e}_2 & h_3 \mathbf{e}_3 \\ \dfrac{\partial}{\partial x_1} & \dfrac{\partial}{\partial x_2} & \dfrac{\partial}{\partial x_3} \\ h_1 v_1 & h_2 v_2 & h_3 v_3 \end{vmatrix}. \qquad (13\text{-}6)$$

Finally, the formula for the gradient of a scalar $u(x_1,x_2,x_3)$, as follows from (8-14), is[1]

$$\nabla u = \frac{\mathbf{e}_1}{h_1}\frac{\partial u}{\partial x_1} + \frac{\mathbf{e}_2}{h_2}\frac{\partial u}{\partial x_2} + \frac{\mathbf{e}_3}{h_3}\frac{\partial u}{\partial x_3}. \qquad (13\text{-}7)$$

Inasmuch as div $\nabla u = \nabla \cdot u$, it is easy to check that the substitution of $\mathbf{v} = \nabla u$ in (13-4) yields

$$\nabla^2 u = \frac{1}{h_1 h_2 h_3}\left[\frac{\partial}{\partial x_1}\left(\frac{h_2 h_3}{h_1}\frac{\partial u}{\partial x_1}\right) + \frac{\partial}{\partial x_2}\left(\frac{h_1 h_3}{h_2}\frac{\partial u}{\partial x_2}\right) + \frac{\partial}{\partial x_3}\left(\frac{h_1 h_2}{h_3}\frac{\partial u}{\partial x_3}\right)\right]. \qquad (13\text{-}8)$$

In cylindrical coordinates defined by the transformation

$$x = r\cos\theta \qquad y = r\sin\theta \qquad z = z$$

the metric coefficients are[2]

$$g_{11} = 1 \qquad g_{22} = r^2 \qquad g_{33} = 1$$

so that

$$h_1 = 1 \qquad h_2 = r \qquad h_3 = 1.$$

Accordingly, formulas (13-4) and (13-8) yield

$$\text{div } \mathbf{v} = \frac{1}{r}\frac{\partial(r v_r)}{\partial r} + \frac{1}{r}\frac{\partial v_\theta}{\partial \theta} + \frac{\partial v_z}{\partial z}$$

$$\nabla^2 u = \frac{1}{r}\frac{\partial(r(\partial u/\partial r))}{\partial r} + \frac{1}{r^2}\frac{\partial^2 u}{\partial \theta^2} + \frac{\partial^2 u}{\partial z^2}$$

where

$$\mathbf{v} = \mathbf{r}_1 v_r + \boldsymbol{\theta}_1 v_\theta + \mathbf{k} v_z$$

$\mathbf{r}_1$, $\boldsymbol{\theta}_1$, $\mathbf{k}$ being unit vectors in the direction of increasing r, θ, and z (Fig. 18).

[1] We use the symbol ∇u to mean grad u in curvilinear coordinates as well as in cartesian.
[2] See Sec. 12.

In spherical coordinates determined by

$$x = \rho \sin \theta \cos \phi \qquad y = \rho \sin \theta \sin \phi \qquad z = \rho \cos \theta$$

$h_1 = 1$, $h_2 = \rho$, $h_3 = \rho \sin \theta$, as follows from Prob. 2 in Sec. 12. On making use of (13-4) and (13-8) we find that in spherical coordinates

$$\operatorname{div} \mathbf{v} = \frac{1}{\rho^2} \frac{\partial(\rho^2 v_\rho)}{\partial \rho} + \frac{1}{\rho \sin \theta} \frac{\partial(\sin \theta \, v_\theta)}{\partial \theta} + \frac{1}{\rho \sin \theta} \frac{\partial v_\phi}{\partial \phi}$$

$$\nabla^2 u = \frac{1}{\rho^2} \frac{\partial(\rho^2(\partial u/\partial \rho))}{\partial \rho} + \frac{1}{\rho^2 \sin \theta} \frac{\partial(\sin \theta(\partial u/\partial \theta))}{\partial \theta} + \frac{1}{\rho^2 \sin^2 \theta} \frac{\partial^2 u}{\partial \phi^2}$$

where
$$\mathbf{v} = \boldsymbol{\rho}_1 v_\rho + \boldsymbol{\theta}_1 v_\theta + \boldsymbol{\phi}_1 v_\phi$$

and $\boldsymbol{\rho}_1$, $\boldsymbol{\theta}_1$, $\boldsymbol{\phi}_1$ are the unit vectors in the direction of increasing coordinate lines shown in Fig. 19.

─── **PROBLEMS**

1. Write the expressions for ∇u in spherical and cylindrical coordinates.

2. What is the form of ∇^2 in parabolic coordinates (u,v,ϕ) for which

$$(ds)^2 = (u^2 + v^2)[(du)^2 + (dv)^2] + u^2 v^2 (d\phi)^2$$

3. The force $\mathbf{F}$ per unit charge due to a dipole of constant strength p is

$$\mathbf{F} = \mathbf{r}_1 \left(\frac{2p \cos \theta}{r^3} \right) + \boldsymbol{\theta}_1 \left(\frac{p \sin \theta}{r^3} \right)$$

where r, θ are polar coordinates. Compute div $\mathbf{F}$ and curl $\mathbf{F}$.

───

14. Conservative Force Fields.   In the concluding sections of this chapter we illustrate the use of vector analysis in the treatment of several problems drawn from mechanics, hydrodynamics, and the theory of heat flow in solids.

When a particle of matter is displaced along a piecewise-smooth path C in a given field of force $\mathbf{F}$, the work W expended in moving it is determined by the integral

$$W = \int_C \mathbf{F} \cdot d\mathbf{r}. \tag{14-1}$$

The integral (14-1), in general, will have different values for different paths joining the same two points in the force field. If (14-1) is independent of the path, the field $\mathbf{F}$ is said to be *conservative*.

We show next that the force field determined by Newton's inverse-square law of attraction is conservative[1] in a suitable region. According to Newton's law a particle of mass m located at a point P is acted on by a force $\mathbf{F}$ whose magnitude is proportional to m and inversely proportional to the square of the distance r from P to the center of attraction O. Thus,

$$\mathbf{F} = -\frac{km}{r^2} \mathbf{r}_1 \qquad r \neq 0 \tag{14-2}$$

where $\mathbf{r}_1$ is the unit vector directed from O to P. The positive constant k is determined experimentally; it clearly depends on the choice of units of measure of $\mathbf{F}$. Physically the law (14-2) represents the force of attraction of the mass m at P by a unit mass located at O.

─────────

[1] A similar discussion applies to electrostatic force fields determined by Coulomb's law, since the mathematical structures of Newton's and Coulomb's laws are identical.

If we rewrite (14-2) in the form

$$\mathbf{F} = -\frac{km}{r^3}\mathbf{r} \qquad (14\text{-}3)$$

where $\mathbf{r} = r\mathbf{r}_1$, and insert it in the work integral (14-1), we get for the work done in displacing the particle from P_1 to P_2 along the path C, not passing through the origin $\mathbf{r} = \mathbf{0}$,

$$W = \int_C -\frac{km}{r^3}\,\mathbf{r}\cdot d\mathbf{r}. \qquad (14\text{-}4)$$

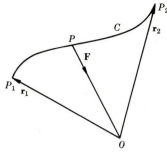

But $\mathbf{r}\cdot d\mathbf{r} = \tfrac{1}{2}d(\mathbf{r}\cdot\mathbf{r}) = r\,dr$, so that we can write (14-4) as

$$W = \int_C -\frac{km}{r^2}\,dr = \int_C km\,d\left(\frac{1}{r}\right) = km\left[\frac{1}{r}\right]_{P_1}^{P_2}. \qquad (14\text{-}5)$$

The integral (14-5) is clearly independent of the path joining P_1 and P_2 in any regular region that excludes the point $\mathbf{r} = \mathbf{0}$ and if we denote[1] $\mathbf{r}(P_2)$ by $\mathbf{r}_2$ and $\mathbf{r}(P_1)$ by $\mathbf{r}_1$ (Fig. 23), we can write

$$W = km\left(\frac{1}{r_2} - \frac{1}{r_1}\right).$$

The scalar function

$$u \equiv \frac{km}{r} \qquad (14\text{-}6)$$

appearing in (14-5) is known as the *gravitational potential*. It is easy to check that

$$\nabla u = \nabla\left(\frac{km}{r}\right) = -\frac{km}{r^3}\mathbf{r} = \mathbf{F}. \qquad (14\text{-}7)$$

The function $u(P)$ in (14-6) is continuous at all points except when $r = 0$, and since div $\nabla u = \nabla^2 u$, we readily find that the gravitational potential (14-6) satisfies Laplace's equation

$$\nabla^2 u = 0$$

except when $r = 0$.

The gravitational potential at a point P due to a continuous distribution of mass of density ρ is defined by the integral

$$u(P) = \int_\tau \frac{k\rho\,d\tau}{r} \qquad (14\text{-}8)$$

where r is the distance from the element of mass $dm = \rho\,d\tau$ to the point P.

The force of attraction of the unit mass located at P by the body is determined by the formula $\mathbf{F} = \nabla u$.

The study of the properties of the scalar function $u(P)$ defined by (14-8) is in the province of potential theory, and we shall encounter it once more in Probs. 6 and 7 of Sec. 19 and in Chap. 7.

Example. Let us compute the gravitational potential $u(P)$ of a thin homogeneous spherical shell of radius a at a point P whose distance from the center of the shell is R (Fig. 24).

The potential at P can be computed by summing potentials of the ring-shaped elements of matter bounded by the cones with the semivertical angles θ and $\theta + d\theta$. The area of the zone intercepted by these cones is $2\pi a \sin\theta\, a\, d\theta$, so that

[1] The position vector $\mathbf{r}_1$ is not to be confused with the unit vector in (14-2).

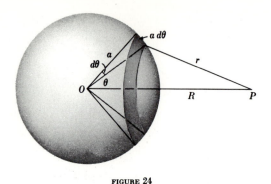

FIGURE 24

FIGURE 25

$$u(P) = \int_0^\pi \frac{k\rho 2\pi a^2 \sin\theta \, d\theta}{r} \tag{14-9}$$

where ρ is the mass per unit area of the shell.

From the cosine law of trigonometry

$$r = \sqrt{a^2 + R^2 - 2aR\cos\theta}$$

and we can write (14-9) as

$$u(P) = 2\pi k\rho a^2 \int_0^\pi \frac{\sin\theta \, d\theta}{\sqrt{a^2 + R^2 - 2aR\cos\theta}}$$

$$= \frac{2\pi k\rho a}{R} [\sqrt{(R+a)^2} - \sqrt{(R-a)^2}] \quad \text{if } R > a$$

$$= \frac{2\pi k\rho a}{R} [\sqrt{(R+a)^2} - \sqrt{(a-R)^2}] \quad \text{if } R < a.$$

If P is outside the shell, $R > a$, and we have the result

$$u(P) = \frac{4\pi k\rho a^2}{R} \equiv \frac{kM}{R} \tag{14-10}$$

where $M = 4\pi a^2 \rho$ is the mass of the shell.

When P is inside the shell, $R < a$, and we get

$$u(P) = 4\pi k\rho a \tag{14-11}$$

which is a constant.

The result (14-10) can be stated as a theorem.

THEOREM.   The potential (and hence the force of attraction $\mathbf{F} = \nabla u$) produced by a thin spherical shell at a point exterior to the shell is the same as if the mass of the shell were concentrated at its center.

The potential due to a solid sphere of constant density ρ at a point outside the sphere can be deduced at once from (14-10) by supposing the sphere to consist of thin concentric shells. We conclude that this potential has the same form as (14-10) with M replaced by the mass of the sphere. Accordingly, the force of attraction produced by a solid homogeneous sphere on a unit mass at a point P outside the sphere has the magnitude kM/R^2. This force is directed toward the center of the sphere.

From (14-11) we see that the force of attraction at a point inside the shell is zero.

The integral (14-8) becomes improper if P is within the solid, for in that case

$$r = \sqrt{(x - \xi)^2 + (y - \eta)^2 + (z - \zeta)^2}$$

becomes zero when the integration variables (ξ,η,ζ) coincide with the coordinates (x,y,z) of P. However, the concepts of potential and gravitational attraction are shown in Sec. 19 to have a meaning even when P is a point in the interior of a homogeneous solid.[1]

15. Steady Flow of Fluids. Let C be an arc of a piecewise-smooth curve in the xy plane over which a sheet of homogeneous fluid of depth 1 is flowing. The lines of flow of the fluid particles are indicated in Fig. 25 by curved arrows, and we suppose that the flow pattern is identical in all planes parallel to the xy plane. A flow of this sort is called two-dimensional.

The problem is to determine the amount of fluid that crosses C per unit time. We denote by $\mathbf{v}$ the velocity of the fluid particles at a point P on C. Assuming $\mathbf{v}$ to be continuously differentiable, we compute the volume dV of fluid crossing an element $d\mathbf{r}$ of C per unit time. Since the depth of the fluid is 1, this volume is equal to the volume of the parallelepiped

$$dV = \mathbf{k}\cdot\mathbf{v}\times d\mathbf{r}$$

where $\mathbf{k}$ is the unit vector perpendicular to the xy plane. The volume V crossing C per unit time, therefore, is

$$V = \int_C \mathbf{k}\cdot\mathbf{v}\times d\mathbf{r}.$$

But by Chap. 4, Eq. (6-2),

$$\mathbf{k}\cdot\mathbf{v}\times d\mathbf{r} = \begin{vmatrix} 0 & 0 & 1 \\ v_x & v_y & 0 \\ dx & dy & 0 \end{vmatrix} = v_x\,dy - v_y\,dx$$

since $\mathbf{v} = \mathbf{i}v_x + \mathbf{j}v_y$ and $d\mathbf{r} = \mathbf{i}\,dx + \mathbf{j}\,dy$.

Accordingly,

$$V = \int_C (v_x\,dy - v_y\,dx). \tag{15-1}$$

If C is a simple closed curve and the fluid is incompressible, the net amount of fluid crossing C is zero, because as much fluid enters the region bounded by C as leaves it. Thus a steady flow of an incompressible fluid is characterized by the equation

$$\int_C (v_x\,dy - v_y\,dx) = 0 \tag{15-2}$$

where the integral is evaluated over any simple closed curve C not enclosing the points at which the fluid is generated or absorbed. But Eq. (15-2) implies that $-v_y\,dx + v_x\,dy$ is a total differential $d\Psi(x,y)$ of the function

$$\Psi(x,y) = \int_{(x_0,y_0)}^{(x,y)} (-v_y\,dx + v_x\,dy). \tag{15-3}$$

Moreover, by (4-5), $$\frac{\partial\Psi}{\partial x} = -v_y \qquad \frac{\partial\Psi}{\partial y} = v_x \tag{15-4}$$

and v_x and v_y satisfy the condition

$$\frac{\partial(-v_y)}{\partial y} = \frac{\partial v_x}{\partial x} \tag{15-5}$$

throughout the region R in which (15-2) holds. Equation (15-5) is a consequence of (15-2); it states, in effect, that there is no fluid created or destroyed in the region R. For this reason it is called the *equation of continuity*. Since

---

[1] See in this connection Chap. 2, Sec. 12, and I. S. Sokolnikoff, "Tensor Analysis," 2d ed., sec. 93, John Wiley & Sons, Inc., New York, 1964, where it is shown that the potential u satisfies Poisson's equation $\nabla^2 u = -4\pi\rho$.

$$\text{div } \mathbf{v} = \frac{\partial v_x}{\partial x} + \frac{\partial v_y}{\partial y}$$

we can write (15-5) in vector form as

$$\text{div } \mathbf{v} = 0 \tag{15-6}$$

which is consistent with the meaning attached to the symbol div $\mathbf{v}$ in Sec. 6.

The function $\Psi(x,y)$ defined by (15-3) is the *stream function*, and the tracks of the particles of fluid, or *streamlines*, are determined by the equation $\Psi(x,y) = \text{const.}$ The velocity field satisfying (15-6), we recall, is said to be solenoidal. If the flow $\mathbf{v}$ is irrotational, then curl $\mathbf{v} = \mathbf{0}$ and there exists a scalar function $\Phi(x,y)$ such that[1]

$$\mathbf{v} = \nabla\Phi \tag{15-7}$$

or
$$v_x = \frac{\partial \Phi}{\partial x} \qquad v_y = \frac{\partial \Phi}{\partial y}. \tag{15-8}$$

The function $\Phi(x,y)$ determined by the integral

$$\Phi(x,y) = \int_{(x_0,y_0)}^{(x,y)} (v_x\, dx + v_y\, dy) \equiv \int_{P_0}^{P} \mathbf{v} \cdot d\mathbf{r}$$

is called the *velocity potential* because of the relations (15-8). We emphasize the fact that the condition for the existence of $\Phi(x,y)$ is curl $\mathbf{v} = \mathbf{0}$, or in scalar form,

$$\frac{\partial v_x}{\partial y} = \frac{\partial v_y}{\partial x}. \tag{15-9}$$

If the flow is both irrotational and solenoidal, the relations (15-4) and (15-8) hold and we conclude that

$$\frac{\partial \Phi}{\partial x} = \frac{\partial \Psi}{\partial y} \qquad \frac{\partial \Phi}{\partial y} = -\frac{\partial \Psi}{\partial x} \qquad \text{in } R. \tag{15-10}$$

These are the celebrated *Cauchy-Riemann equations* which we shall encounter again in Chap. 8.

Furthermore, if div $\mathbf{v} = 0$ and $\mathbf{v}$ is given by (15-7), we see that

$$\text{div } \nabla\Phi = \nabla^2\Phi = 0. \tag{15-11}$$

Thus, the velocity potential Φ satisfies Laplace's equation throughout any region containing no sources or sinks.

On differentiating the first of Eqs. (15-10) with respect to y and the second with respect to x and on equating $\partial^2\Phi/\partial x\partial y$ to $\partial^2\Phi/\partial y\partial x$, we find that the stream function $\Psi(x,y)$ also satisfies the equation

$$\nabla^2\Psi = 0.$$

The practical importance of these results is stressed in Chaps. 7 and 8.

The foregoing considerations can be extended to the three-dimensional flows as indicated in Sec. 17.

--- **PROBLEMS**

1. Show that the gravitational field determined by (14-2) is both solenoidal and irrotational except at $(0,0,0)$.

2. Show that the velocity field

$$\mathbf{v} = \mathbf{i}\,\frac{x}{x^2 + y^2} + \mathbf{j}\,\frac{y}{x^2 + y^2}$$

[1] See Theorem I in Sec. 10.

is solenoidal in any region that does not contain the origin $(0,0)$. Is it irrotational? Verify that the velocity potential $\Phi = \log r = \frac{1}{2} \log (x^2 + y^2)$ and the stream function $\Psi = \tan^{-1}(y/x) = \theta$. Compute the circulation around a circular path enclosing the origin, and thus obtain a physical interpretation of results.

3. Discuss a two-dimensional flow for which the velocity potential $\Phi = cx$. What is the stream function Ψ for this flow? Plot the curves $\Phi = \text{const}$ and $\Psi = \text{const}$.

4. Discuss a two-dimensional flow for which the stream function is $\Psi = 2xy$. Find the velocity potential Φ, and sketch the curves $\Phi = \text{const}$ and $\Psi = \text{const}$.

5. If $\mathbf{v}$ and $\mathbf{w}$ are irrotational vector fields, show that $\mathbf{v} \times \mathbf{w}$ is solenoidal.

6. Show that the streamlines are orthogonal to the lines $\Phi = \text{const}$.

7. Show that, when the three-dimensional flow $\mathbf{v}$ is irrotational, the streamlines satisfy the equations

$$\frac{dx}{v_x} = \frac{dy}{v_y} = \frac{dz}{v_z}.$$

8. If the velocity potential of the two-dimensional flow is $\Phi = x^2 - y^2$, find $\mathbf{v}$ and obtain the equations of the streamlines. Is this flow solenoidal? Is it irrotational?

9. Show with the aid of the Cauchy-Riemann equations that, when the stream function $\Psi(x,y)$ is given, the velocity potential Φ is determined by

$$\Phi(x,y) = \int_{(x_0,y_0)}^{(x,y)} \left(\frac{\partial \Psi}{\partial y} \, dx - \frac{\partial \Psi}{\partial x} \, dy \right).$$

10. Use the result given in the preceding problem to calculate $\Phi(x,y)$ if (a) $\Psi = x^3 - 3xy^2$; (b) $\Psi = -y/(x^2 + y^2)$, $x \neq 0$, $y \neq 0$.

---

16. Equation of Heat Flow.[1]

The following derivation of the Fourier equation of heat flow illustrates admirably the use of the divergence theorem in mathematical physics.

It is known from empirical results that heat will flow from points at higher temperatures to those at lower temperatures. At any point the rate of decrease of temperature varies with the direction, and it is generally assumed that the amount of heat ΔH crossing an element of surface $\Delta\sigma$ in Δt sec is proportional to the greatest rate of decrease of the temperature u; that is,

$$\Delta H \doteq k \, \Delta\sigma \, \Delta t \left| \frac{du}{dn} \right|.$$

Define the vector $\mathbf{q}$, representing the flow of heat, by the formula

$$\mathbf{q} = -k \, \nabla u \tag{16-1}$$

where k is a constant of proportionality known as the thermal conductivity of a substance. [The units of k are cal/(cm-sec °C).] The negative sign is chosen in the definition because heat flows from points of higher temperature to those of lower, and the vector ∇u is directed normally to the level surface $u = \text{const}$ in the direction of increasing u.

Then the total amount of heat H flowing out in Δt sec from an arbitrary region T bounded by a closed surface S is

$$H \doteq -\Delta t \int_S k \frac{du}{dn} \, d\sigma = \Delta t \int_S \mathbf{q} \cdot \mathbf{n} \, d\sigma \tag{16-2}$$

since $\mathbf{q} \cdot \mathbf{n} = -k \, du/dn$ by (16-1).

---

[1] We assume throughout this and the following section that all functions under consideration are sufficiently smooth to permit the use of the divergence and Stokes' theorems.

On the other hand, the amount of heat lost by the body T can be calculated as follows: In order to increase the temperature of a volume element by $\Delta u°$, one must supply an amount of heat that is proportional to the increase in temperature and to the mass of the volume element.   Hence

$$\Delta H \doteq c \, \Delta u \, \rho \, \Delta\tau \doteq c \, \frac{\partial u}{\partial t} \, \Delta t \, \rho \, \Delta\tau$$

where c is the specific heat of the substance $[\text{cal}/(\text{g }°\text{C})]$ and ρ is its density.   Therefore, the total loss of heat from the volume τ in Δt sec is

$$H \doteq -\Delta t \int_T \frac{\partial u}{\partial t} \, c\rho \, d\tau. \tag{16-3}$$

Equating (16-2) and (16-3) gives

$$\int_T \mathbf{q}\cdot\mathbf{n} \, d\sigma = -\int_T \frac{\partial u}{\partial t} \, c\rho \, d\tau. \tag{16-4}$$

Applying the divergence theorem to the left-hand member of (16-4) yields

$$\int_T \operatorname{div} \mathbf{q} \, d\tau = -\int_T \frac{\partial u}{\partial t} \, c\rho \, d\tau$$

and since $\mathbf{q} = -k\nabla u$, the foregoing equation assumes the form

$$\int_T \left[\operatorname{div}(-k\nabla u) + c\rho \, \frac{\partial u}{\partial t} \right] d\tau = 0. \tag{16-5}$$

Now, if k is a constant,

$$\operatorname{div}(k\nabla u) = k\nabla^2 u$$

and (16-5) becomes

$$\int_T \left(-k\nabla^2 u + c\rho \, \frac{\partial u}{\partial t} \right) d\tau = 0. \tag{16-6}$$

Since this integral must vanish for an arbitrary region T and the integrand is a continuous function, it follows that the integrand must be equal to zero, for if such were not the case, T could be so chosen as to be a region throughout which the integrand has constant sign.   But if the integrand had one sign throughout this region, the integral would have the same sign and would not vanish as required by (16-6).

Therefore, $-k\nabla^2 u + c\rho\,\partial u/\partial t = 0$,

or

$$\frac{\partial u}{\partial t} = h^2 \nabla^2 u \tag{16-7}$$

where $h^2 \equiv k/c\rho$.   Equation (16-7) was developed by Fourier in 1822 and is of basic importance in the study of heat conduction in solids.   A similar equation occurs in the study of current flow in conductors and in problems dealing with diffusion in liquids and gases.

It follows from (16-7) that a steady distribution of temperatures is characterized by the solution of Laplace's equation

$$\nabla^2 u = 0.$$

It was assumed in this derivation that the body is free from sources and sinks.   If there are sources of heat continuously distributed within T, it is necessary to add to the right-hand member of (16-3) the integral

$$\int_T f(x,y,z,t) \, d\tau$$

where $f(x,y,z,t)$ is a function representing the strengths of the sources.   The reader will show that in this case one is led to the equation

$$\frac{\partial u}{\partial t} = h^2 \nabla^2 u + \frac{f}{c\rho}$$

provided that the thermal conductivity of the substance is constant. Thus the presence of sources leads to a nonhomogeneous partial differential equation.

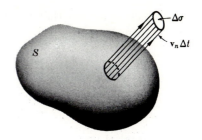

FIGURE 26

17. Equations of Hydrodynamics. Consider a region of space containing a fluid, and let $\mathbf{v}$ denote the velocity of a typical particle of the fluid. The amount Q of fluid crossing an arbitrary closed surface S in the region can be calculated by determining the flow across a typical element $\Delta\sigma$ of the surface S. A particle of fluid is displaced by Δt sec through a distance $\mathbf{v}\,\Delta t$, and since only the component of the vector $\mathbf{v}$ normal to the element $\Delta\sigma$ contributes to the flow across this element, the amount ΔQ of the fluid crossing $\Delta\sigma$ is

$$\Delta Q \doteq \rho \mathbf{v}\cdot\mathbf{n}\,\Delta\sigma\,\Delta t$$

where ρ is the density of the fluid (Fig. 26).

The entire amount Q of fluid flowing out of the region T, which is bounded by S, in Δt sec is

$$Q \doteq \Delta t \int_S \rho \mathbf{v}\cdot\mathbf{n}\,d\sigma.$$

On the other hand, the quantity of the fluid originally contained in τ will have diminished by the amount

$$Q \doteq -\Delta t \int_T \frac{\partial\rho}{\partial t}\,d\tau$$

for the change in mass in Δt sec is nearly equal to $(\partial\rho/\partial t)\,\Delta t\,\Delta\tau$, and the negative sign is taken because ρ is a decreasing function of t.

Equating these two expressions for Q gives

$$\int_S \rho \mathbf{v}\cdot\mathbf{n}\,d\sigma = -\int_T \frac{\partial\rho}{\partial t}\,d\tau \tag{17-1}$$

and the application of the divergence theorem to the left-hand member of this equation yields

$$\int_T \operatorname{div}(\rho\mathbf{v})\,d\tau = -\int_T \frac{\partial\rho}{\partial t}\,d\tau$$

or

$$\int_T \left[\operatorname{div}(\rho\mathbf{v}) + \frac{\partial\rho}{\partial t}\right] d\tau = 0.$$

Since the integrand is continuous and the region T is arbitrary, one can conclude that

$$\frac{\partial\rho}{\partial t} + \operatorname{div}(\rho\mathbf{v}) = 0. \tag{17-2}$$

This is the basic equation of hydrodynamics, known as the *equation of continuity*. It merely expresses the law of conservation of matter.

It has been assumed that there are no sources or sinks within the region occupied by the fluid. If matter is created at the rate $k\rho(x,y,z,t)$, the right-hand member of (17-1) should include a term that accounts for the increase of mass per second due to such sources, namely,

$$\int_T k\rho \, d\tau.$$

In this event the equation of continuity reads

$$\frac{\partial \rho}{\partial t} + \text{div} \, (\rho \mathbf{v}) = k\rho.$$

The constant of proportionality k is sometimes called the growth factor.

The density $\rho(x,y,z,t)$ of the fluid at the location (x,y,z) of the fluid particle depends on t explicitly and on x,y,z implicitly, since the particle coordinates change with time as the particle is displaced. Thus,

$$\frac{d\rho}{dt} = \frac{\partial \rho}{\partial t} + \frac{\partial \rho}{\partial x}\frac{dx}{dt} + \frac{\partial \rho}{\partial y}\frac{dy}{dt} + \frac{\partial \rho}{\partial z}\frac{dz}{dt}. \qquad (17\text{-}3)$$

In this equation, $d\rho/dt$ means the rate of change of density as one moves with the fluid, whereas $\partial \rho/\partial t$ is the rate of change of density at a fixed point.

Upon noting that

$$\mathbf{v} = \mathbf{i}\frac{dx}{dt} + \mathbf{j}\frac{dy}{dt} + \mathbf{k}\frac{dz}{dt}$$

and

$$\nabla \rho = \mathbf{i}\frac{\partial \rho}{\partial x} + \mathbf{j}\frac{\partial \rho}{\partial y} + \mathbf{k}\frac{\partial \rho}{\partial z}$$

we can write the formula (17-3) as

$$\frac{d\rho}{dt} = \frac{\partial \rho}{\partial t} + \mathbf{v}\cdot\nabla \rho. \qquad (17\text{-}4)$$

Substituting from (17-2) in (17-4) gives

$$\frac{d\rho}{dt} = -\,\text{div} \, (\rho \mathbf{v}) + \mathbf{v}\cdot\nabla \rho. \qquad (17\text{-}5)$$

But div $(\rho \mathbf{v}) = \mathbf{v}\cdot\nabla \rho + \rho \, \text{div} \, \mathbf{v}$ (see Prob. 3b in Sec. 6), so that (17-5) becomes

$$\frac{d\rho}{dt} = -\rho \, \text{div} \, \mathbf{v}$$

or

$$\text{div} \, \mathbf{v} = -\,\frac{1}{\rho}\frac{d\rho}{dt}. \qquad (17\text{-}6)$$

It is clear from (17-6) that div $\mathbf{v}$ is equal to the relative rate of change of the density ρ at any point of the fluid. Therefore, if the fluid is incompressible, the velocity field is characterized by the equation

$$\text{div} \, \mathbf{v} = 0. \qquad (17\text{-}7)$$

If the flow of fluid is irrotational, then curl $\mathbf{v} = \mathbf{0}$, and one is assured that there exists a scalar function Φ such that $\mathbf{v} = \nabla\Phi$. Substituting this in (17-7) gives the differential equation to be satisfied by Φ, namely,

$$\nabla^2\Phi \equiv \frac{\partial^2\Phi}{\partial x^2} + \frac{\partial^2\Phi}{\partial y^2} + \frac{\partial^2\Phi}{\partial z^2} = 0. \qquad (17\text{-}8)$$

The function Φ is called the *velocity potential*. A similar result was obtained in Sec. 15 for the two-dimensional flow.

If the fluid is ideal, that is, such that the force due to pressure on any surface element is always directed normally to that surface element, one can easily derive Euler's equa-

tions of hydrodynamics.   Denote the pressure at any point of the fluid by p; then the
force acting on a surface element $\Delta\sigma$ is $-p\mathbf{n}\,\Delta\sigma$, and the resultant force acting on an
arbitrary closed surface S is $-\int_S p\mathbf{n}\,d\sigma$.   The negative sign is chosen because the force
due to pressure acts in the direction of the interior normal, whereas $\mathbf{n}$ denotes the unit
exterior normal.

Let the body force, per unit mass, acting on the masses contained within the region
T be $\mathbf{F}$; then the resultant of the body forces is $\int_T \mathbf{F}\rho\,d\tau$.   Hence, the resultant $\mathbf{R}$ of the
body and surface forces is

$$\mathbf{R} = \int_T \mathbf{F}\rho\,d\tau - \int_S p\mathbf{n}\,d\sigma$$

$$= \int_T \mathbf{F}\rho\,d\tau - \int_T \nabla p\,d\tau \tag{17-9}$$

where the last step is obtained by making use of (8-13).

From Newton's law of motion, the resultant force is equal to

$$\mathbf{R} = \int_T \rho\,\frac{d^2\mathbf{r}}{dt^2}\,d\tau \tag{17-10}$$

where $\mathbf{r} = \mathbf{i}x + \mathbf{j}y + \mathbf{k}z$ is the position vector of the masses relative to the origin of
cartesian coordinates.   It follows from (17-9) and (17-10) that

$$\int_T \left(\mathbf{F}\rho - \nabla p - \rho\,\frac{d^2\mathbf{r}}{dt^2}\right) d\tau = 0$$

and since the volume element is arbitrary and the integrand is continuous,

$$\rho\,\frac{d^2\mathbf{r}}{dt^2} = \mathbf{F}\rho - \nabla p. \tag{17-11}$$

This is the desired equation in vector form, and it is basic in hydro- and aerodynamical
applications.

In books on hydrodynamics, the cartesian components of the velocity vector $d\mathbf{r}/dt$
are usually denoted by u, v, and w, so that

$$\frac{d\mathbf{r}}{dt} = \mathbf{i}u + \mathbf{j}v + \mathbf{k}w = \mathbf{i}\,\frac{dx}{dt} + \mathbf{j}\,\frac{dy}{dt} + \mathbf{k}\,\frac{dz}{dt}.$$

Since u, v, and w are functions of the coordinates of the point (x,y,z) and of the time t, it
follows that

$$\frac{d^2\mathbf{r}}{dt^2} = \mathbf{i}\left(\frac{\partial u}{\partial t} + \frac{\partial u}{\partial x}\frac{dx}{dt} + \frac{\partial u}{\partial y}\frac{dy}{dt} + \frac{\partial u}{\partial z}\frac{dz}{dt}\right)$$

$$+ \mathbf{j}\left(\frac{\partial v}{\partial t} + \frac{\partial v}{\partial x}\frac{dx}{dt} + \frac{\partial v}{\partial y}\frac{dy}{dt} + \frac{\partial v}{\partial z}\frac{dz}{dt}\right)$$

$$+ \mathbf{k}\left(\frac{\partial w}{\partial t} + \frac{\partial w}{\partial x}\frac{dx}{dt} + \frac{\partial w}{\partial y}\frac{dy}{dt} + \frac{\partial w}{\partial z}\frac{dz}{dt}\right).$$

Substituting this expression in (17-11) and setting $\mathbf{F} = \mathbf{i}F_x + \mathbf{j}F_y + \mathbf{k}F_z$ lead to three
scalar equations, which are associated with the name of Euler:

$$\frac{\partial u}{\partial t} + \frac{\partial u}{\partial x}u + \frac{\partial u}{\partial y}v + \frac{\partial u}{\partial z}w = F_x - \frac{1}{\rho}\frac{\partial p}{\partial x}$$

$$\frac{\partial v}{\partial t} + \frac{\partial v}{\partial x}u + \frac{\partial v}{\partial y}v + \frac{\partial v}{\partial z}w = F_y - \frac{1}{\rho}\frac{\partial p}{\partial y} \tag{17-12}$$

$$\frac{\partial w}{\partial t} + \frac{\partial w}{\partial x}u + \frac{\partial w}{\partial y}v + \frac{\partial w}{\partial z}w = F_z - \frac{1}{\rho}\frac{\partial p}{\partial z}.$$

It is possible to show with the aid of these equations (and by making some simplifying assumptions) that the propagation of sound is governed approximately by the wave equation

$$\frac{\partial^2 s}{\partial t^2} = a^2 \nabla^2 s.$$

In this equation, a is the velocity of sound and s is related to the density ρ of the medium by the formula $s = (\rho/\rho_0) - 1$, where ρ_0 is the density of the medium at rest.

18. Differentiation of Volume and Surface Integrals with Respect to a Parameter. In Chap. 5, Sec. 14, we deduced a rule for differentiating the integral $\int_{u_0(\alpha)}^{u_1(\alpha)} f(x,\alpha) \, dx$, $u_1(\alpha) \leq x \leq u_2(\alpha)$, with respect to a parameter α that appears both in the integrand and in the limits of the integral. In several branches of mechanics of continua one encounters volume, surface, and line integrals in which the integrand $u(P,t)$ contains a parameter t, where t is time. If the points P in the region of integration also depend on t, as they do when the region is in motion, the differentiation of integrals with respect to t must take account not only of the fact that $u(P,t)$ is a function of t but also that the region of integration depends on t.

Thus, consider a volume integral

$$\int_T u(P,t) \, d\tau \tag{18-1}$$

where $u(P,t)$ is a continuous scalar function in a simply connected region T. If the position of the points P in T depends on t, the region of integration ordinarily changes with t. When the region is fixed, so that T is independent of t, and $u(P,t)$ and $\partial u/\partial t$ are continuous in T for all values of t under consideration, it is true that

$$\frac{d}{dt} \int_T u(P,t) \, d\tau = \int_T \frac{\partial u(P,t)}{\partial t} \, d\tau \tag{18-2}$$

when the integrand on the right is continuous.

But if the position of points in T depends on t, the right-hand member of (18-2) should include terms that account for the change of the region of integration with t. The situation here is precisely the same as that we have already encountered in Chap. 5, Sec. 14, in connection with the integral $\int_{u_0(\alpha)}^{u_1(\alpha)} f(x,\alpha) \, d\alpha$. [See formula (14-7) of Chap. 5.] To determine the expression that should be added to (18-2) when the region T changes with t, consider the surface S bounding T at a certain time t. We suppose that S is such that the divergence theorem is applicable, and consider an element ΔS of S, with the surface area $\Delta\sigma$. In a small interval of time $(t, t + \Delta t)$ the points P of ΔS sweep out a region of space whose volume

$$\Delta\tau \doteq (\mathbf{v} \cdot \mathbf{n}) \, \Delta t \, \Delta\sigma \tag{18-3}$$

where $\mathbf{v}$ is the velocity of P and $\mathbf{n}$ is the exterior unit normal to the small element ΔS (cf. Fig. 26).

If we multiply (18-3) by $u(P,t)$ and sum such products over the entire surface S we get

$$\Delta I \equiv \Delta t \sum_S u(P,t)(\mathbf{v} \cdot \mathbf{n}) \, \Delta\sigma$$

which, on dividing by Δt and passing to the limit, yields formally

$$\frac{dI}{dt} = \int_S u(P,t)(\mathbf{v} \cdot \mathbf{n}) \, d\sigma. \tag{18-4}$$

This is the term that should be added to the right-hand member of (18-2) when the region of integration is in motion. Thus,

$$\frac{d}{dt} \int_T u(P,t)\, d\tau = \int_T \frac{\partial u(P,t)}{\partial t}\, d\tau + \int_S u(P,t)(\mathbf{v}\cdot\mathbf{n})\, d\tau. \tag{18-5}$$

But $u(\mathbf{v}\cdot\mathbf{n}) = (u\mathbf{v})\cdot\mathbf{n}$, so that the surface integral in (18-5) is

$$\int_S (u\mathbf{v})\cdot\mathbf{n}\, d\sigma = \int_T \operatorname{div}(u\mathbf{v})\, d\tau$$

by (7-4). Accordingly, (18-5) can be written as

$$\frac{d}{dt} \int_T u(P,t)\, d\tau = \int_T \left[\frac{\partial u(P,t)}{\partial t} + \operatorname{div}(u\mathbf{v}) \right] d\tau. \tag{18-6}$$

Formula (18-6) can be cast in a different form by recalling that $\operatorname{div}(u\mathbf{v}) = u \operatorname{div}\mathbf{v} + \mathbf{v}\cdot\nabla u$ and rewriting (18-6) as

$$\frac{d}{dt} \int_T u(P,t)\, d\tau = \int_T \left(\frac{\partial u}{\partial t} + \mathbf{v}\cdot\nabla u + u \operatorname{div}\mathbf{v} \right) d\tau. \tag{18-7}$$

Now, in cartesian coordinates,

$$\mathbf{v} = \mathbf{i}v_x + \mathbf{j}v_y + \mathbf{k}v_z = \mathbf{i}\frac{dx}{dt} + \mathbf{j}\frac{dy}{dt} + \mathbf{k}\frac{dz}{dt}$$

$$\nabla u = \mathbf{i}\frac{\partial u}{\partial x} + \mathbf{j}\frac{\partial u}{\partial y} + \mathbf{k}\frac{\partial u}{\partial z}$$

so that

$$\mathbf{v}\cdot\nabla u = \frac{\partial u}{\partial x}\frac{dx}{dt} + \frac{\partial u}{\partial y}\frac{dy}{dt} + \frac{\partial u}{\partial z}\frac{dz}{dt}$$

and

$$\frac{du}{dt} = \frac{\partial u}{\partial t} + \frac{\partial u}{\partial x}\frac{dx}{dt} + \frac{\partial u}{\partial y}\frac{dy}{dt} + \frac{\partial u}{\partial z}\frac{dz}{dt} = \frac{\partial u}{\partial t} + \mathbf{v}\cdot\nabla u.$$

We thus see that the sum of the first two terms in the integrand in the right-hand member of (18-7) is du/dt, so that

$$\frac{d}{dt} \int_T u(P,t)\, d\tau = \int_T \left(\frac{du}{dt} + u \operatorname{div}\mathbf{v} \right) d\tau. \tag{18-8}$$

The derivative of the surface integral $\int_S \mathbf{A}\cdot\mathbf{n}\, d\sigma$, with respect to the parameter t, where $\mathbf{A} = \mathbf{A}(P,t)$ is a vector field specified in a closed region T bounded by the surface S with fixed directions of the normals $\mathbf{n}$, can be deduced at once from (18-7). For,

$$\int_S \mathbf{A}\cdot\mathbf{n}\, d\sigma = \int_T \operatorname{div}\mathbf{A}\, d\tau$$

by the divergence theorem, and

$$\frac{d}{dt} \int_S \mathbf{A}\cdot\mathbf{n}\, d\sigma = \frac{d}{dt} \int_T \operatorname{div}\mathbf{A}\, d\tau. \tag{18-9}$$

Hence, on applying (18-6),

$$\frac{d}{dt} \int_T \operatorname{div}\mathbf{A}\, d\tau = \int_T \left[\frac{\partial(\operatorname{div}\mathbf{A})}{\partial t} + \operatorname{div}(\mathbf{v}\operatorname{div}\mathbf{A}) \right] d\tau$$

$$= \int_T \left[\operatorname{div}\left(\frac{\partial \mathbf{A}}{\partial t}\right) + \operatorname{div}(\mathbf{v}\operatorname{div}\mathbf{A}) \right] d\tau$$

$$= \int_T \operatorname{div}\left(\frac{\partial \mathbf{A}}{\partial t} + \mathbf{v}\operatorname{div}\mathbf{A}\right) d\tau. \tag{18-10}$$

An application of the divergence theorem to the volume integral in the right-hand member of (18-10) gives

$$\frac{d}{dt} \int_T \operatorname{div} \mathbf{A} \, d\tau = \int_S \mathbf{n} \cdot \left(\frac{\partial \mathbf{A}}{\partial t} + \mathbf{v} \operatorname{div} \mathbf{A} \right) d\sigma$$

$$= \int_S \left[\frac{\partial (\mathbf{A} \cdot \mathbf{n})}{\partial t} + \mathbf{v} \cdot \mathbf{n} \operatorname{div} \mathbf{A} \right] d\sigma$$

so that the relation (18-9) becomes

$$\frac{d}{dt} \int_S \mathbf{A} \cdot \mathbf{n} \, d\sigma = \int_S \left[\frac{\partial (\mathbf{A} \cdot \mathbf{n})}{\partial t} + \mathbf{v} \cdot \mathbf{n} \operatorname{div} \mathbf{A} \right] d\sigma. \qquad (18\text{-}11)$$

The corresponding formula for the derivative of the line integral $\int_C \mathbf{A} \cdot d\mathbf{r}$ is given in Prob. 1.

PROBLEMS

1. If C is a simple closed curve and $\mathbf{r} = \mathbf{r}(P,t)$ is the position vector of the point P on C at time t, show that

$$\frac{d}{dt} \int_C \mathbf{A} \cdot d\mathbf{r} = \int_C \frac{\partial \mathbf{A}}{\partial t} \cdot d\mathbf{r} + \int_C \operatorname{curl} (\mathbf{A} \times \mathbf{v}) \cdot d\mathbf{r}$$

where $\mathbf{A}(P,t)$ is a continuously differentiable vector field to which Stokes' formula is applicable. *Hint:* Use Stokes' formula and apply (18-11) to the surface integral.

2. Let $u(\alpha,\beta,\gamma) = \int_{T-R_0} f(\alpha,\beta,\gamma;x,y,z) \, d\tau$, where $T - R_0$ is an infinite region exterior to the sphere R_0 with the center at (α,β,γ). Assume that the differentiation under the integral sign is justified. Show that

$$\frac{\partial u}{\partial \alpha} = \int_{T-R_0} \frac{\partial f}{\partial \alpha} \, d\tau + \int_{S_0} f \cos (n,\alpha) \, d\sigma$$

where $\mathbf{n}$ is the interior unit normal to the surface S_0 of R_0.

3. Conclude from Prob. 2 that

$$\nabla u = \int_{T-R_0} \nabla f \, d\tau + \int_{S_0} f \mathbf{n} \, d\sigma$$

where $\nabla \equiv \mathbf{i} \partial/\partial \alpha + \mathbf{j} \partial/\partial \beta + \mathbf{k} \partial/\partial \gamma$.

---

19. Improper Multiple Integrals. The integrands of the surface and volume integrals considered thus far were assumed to be either continuous or, at worst, piecewise-continuous functions. In some applications it is necessary to consider integrals of functions that become infinite at one or more points of the region of integration. Such integrals are called *improper*, and to attach a meaning to them we follow a procedure similar to that used to define the convergence of an improper integral $\int_a^b f(x) \, dx$ of a function of one variable x.

We consider first a scalar function $f(P)$ which is continuous in a regular region T except at one point P_0, in the neighborhood of which $f(P)$ becomes infinite. We delete from T a small regular region T_0, so that P_0 is contained within T_0, and place no restrictions on the shape of T_0. The function $f(P)$ is, then, continuous in the region $T - T_0$ (shown shaded in Fig. 27), and the integral

$$\int_{T-T_0} f(P) \, d\tau \qquad (19\text{-}1)$$

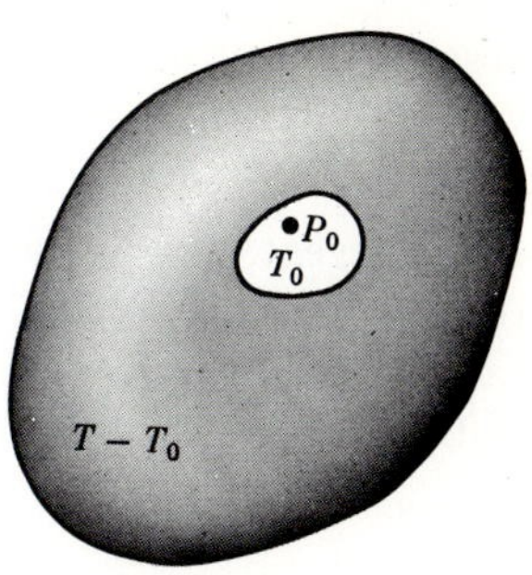

has a definite value for every choice of the region T_0.
Now, if T_0 is allowed to shrink toward P_0 in an arbitrary manner, by making the maximum diameter d of
T_0 approach zero, the region function defined by (19-1)
may have a definite limit I. We then write

$$\lim_{d \to 0} \int_{T-T_0} f(P) \, d\tau = \int_T f(P) \, d\tau = I \quad (19\text{-}2)$$

and say that the improper integral $\int_T f(P) \, d\tau$ *converges*
to the value I. If the limit of (19-1) does not exist,
the improper integral $\int_T f(P) \, d\tau$ is said to *diverge*, and
no value is attached to it.

FIGURE 27

Since no restrictions are placed on the choice of
T_0 and on the manner in which this region is allowed
to shrink toward P_0, the evaluation of the limit in (19-2) can often be simplified by
choosing a *particular* sequence of regions, such as cubes or spheres, which may be suited
to the geometry of the particular problem in hand. It is then necessary to show that
the same limit would have been obtained by use of any other *arbitrary* sequence of
regions.

A discussion of this matter is quite easy if $f(P) \geq 0$ throughout T. Under this hypothesis we shall show that *if the limit* (19-2) *exists when T_0 are restricted to be spheres
centered at P_0 then the limit also exists, and has the same value, when the sequence $\{T_0\}$ is
unrestricted.*

Indeed, since P_0 is interior to T_0 we can find a sphere S_0 centered at P_0 and contained
in T_0. Also, if the maximum diameter of T_0 is d, then T_0 is contained in a sphere S_d of
radius d centered at P_0. When d is so small that S_d is contained in the large region T,
the fact that $f(P) \geq 0$ gives

$$\int_{T-S_0} f(P) \, d\tau \geq \int_{T-T_0} f(P) \, d\tau \geq \int_{T-S_d} f(P) \, d\tau.$$

As $d \to 0$, the first and third integrals tend to the limit I by hypothesis; hence the
second does too.

Instead of having $f(P) \geq 0$ throughout T it would suffice to have $f(P) \geq 0$ only throughout
some neighborhood R of P_0. Thus, if d is so small that T_0 is contained in R, then

$$\int_{T-T_0} f(P) \, d\tau = \int_{T-R} f(P) \, d\tau + \int_{R-T_0} f(P) \, d\tau.$$

The first integral on the right is independent of T_0, and, since $f(P) \geq 0$ in R, the second tends
to a unique limit as in the previous discussion.

When $f(P) \geq 0$ and when the regions T_0 form a sequence of concentric spheres of
radius r, the integral on the left of (19-2) increases as r decreases. If the integral remains
bounded, we conclude from the fundamental principle of Chap. 1, Sec. 3, that the limit
exists. This observation leads to a comparison test for integrals analogous to the corresponding test for series given in Chap. 1, Sec. 5.

Similar remarks apply if $f(P) \leq 0$, since the minus sign can be taken outside the
integral (cf. Prob. 5). However, extension of the results to functions that change sign
in every neighborhood of P_0 requires the concept of *absolute convergence*. Assuming that
$f(P)$ is continuous except at P_0, the improper integral

$$\int_T f(P) \, d\tau$$

is said to be *absolutely convergent* if the integral

$$\int_T |f(P)| \, d\tau \tag{19-3}$$

is convergent. By writing $f(P) = |f(P)| + [f(P) - |f(P)|]$ and applying the preceding results to each term, we conclude that an *absolutely convergent improper integral is convergent.* This theorem is particularly convenient in applications because the convergence of the integral (19-3) can be tested by use of restricted regions T_0, such as spheres.

Example 1.   Consider the volume integral

$$I = \int_T \frac{d\tau}{r^n} \qquad n > 0 \tag{19-4}$$

where T is a sphere of radius a with center at P_0 and $r = \overline{P_0 P}$ is the distance from the variable point P in T to P_0.

The scalar $f(P) = 1/r^n$ becomes infinite in the neighborhood of P_0, but since $1/r^n > 0$ throughout T, the integral (19-4) can be evaluated by deleting the point P_0 from T by a sphere R of radius δ, with center at P_0, and by evaluating [see (19-2)],

$$\lim_{\delta \to 0} \int_{T-R} \frac{d\tau}{r^n}. \tag{19-5}$$

The integral in (19-5) can be computed easily by taking the origin of the spherical coordinates at P_0. On noting Example 2 of Sec. 17 in Chap. 5, we get

$$\int_{T-R} \frac{d\tau}{r^n} = \int_0^{2\pi} d\phi \int_0^\pi d\theta \int_\delta^a \frac{\sin\theta}{\rho^{n-2}} \, d\rho$$

$$= \frac{4\pi}{3-n}\left(\frac{1}{a^{n-3}} - \frac{1}{\delta^{n-3}}\right) \qquad \text{if } n \neq 3$$

$$= 4\pi(\log a - \log \delta) \qquad \text{if } n = 3.$$

These results show that the limit of (19-5) exists only if $n < 3$ and that the integral (19-4) converges to the value

$$I = \frac{4\pi}{3-n}\frac{1}{a^{n-3}} \qquad \text{if } n < 3. \tag{19-6}$$

Consider next a function $f(P)$ that is continuous at every point of some regular region T except at P_0, where it becomes infinite. If $f(P)$ satisfies the condition $|f(P)| \leq M/r^n$, where r is the distance between P and P_0 and M and n are positive constants, then

$$\int_T |f(P)| \, d\tau \leq M \int_T \frac{d\tau}{r^n}. \tag{19-7}$$

But we have shown in Example 1 that the integral in the right-hand member of (19-7) converges if $n < 3$. Accordingly, the integral $\int_T f(P) \, d\tau$ converges absolutely whenever $|f(P)| \leq M/r^n$, $n < 3$.

This result provides a useful criterion for absolute convergence of many improper integrals:

TEST FOR ABSOLUTE CONVERGENCE.   *If $f(P)$ is continuous in a regular region T except at a point P_0 where $f(P)$ is unbounded, then a sufficient condition for absolute convergence of the improper integral $\int_T f(P) \, d\tau$ is that there exist positive constants M and $n < 3$ such that $r^n|f(P)| < M$, where r is the distance between P_0 and P.*

Example 2.   In suitable units of measure the components of the force of attraction (see Sec. 14) produced by the unit mass at $P(x,y,z)$ on an element of mass $dm = \rho \, d\tau$ located at $P_0(\xi,\eta,\zeta)$ are

$$-\frac{dm}{r^2}\frac{x-\xi}{r} \qquad -\frac{dm}{r^2}\frac{y-\eta}{r} \qquad -\frac{dm}{r^2}\frac{z-\zeta}{r}$$

where

$$r = \sqrt{(x-\xi)^2 + (y-\eta)^2 + (z-\zeta)^2}.$$

If T is a regular region occupied by the body with piecewise-continuous density $\rho(\xi,\eta,\zeta)$, the x component of the force of attraction produced by the body on the unit mass at $P(x,y,z)$ is

$$F_x = -\iiint\limits_{T} \frac{\rho(\xi,\eta,\zeta)}{r^2}\frac{x-\xi}{r}\,d\xi\,d\eta\,d\zeta. \tag{19-8}$$

This integral is improper when the point $P(x,y,z)$ is in the interior of the body, since the integrand of (19-8) is unbounded in the neighborhood of P. But $\rho(\xi,\eta,\zeta)$ is bounded in T, and T is a finite region; hence there exists a constant M such that

$$\left| \frac{\rho}{r^2}\frac{x-\xi}{r} \right| \le \frac{M}{r^2}. \tag{19-9}$$

Since the exponent of r in the right-hand member of (19-9) is 2, the test for absolute convergence guarantees that the integral (19-8) is absolutely convergent. Similar considerations apply to the components F_y and F_z so that the force of attraction, $\mathbf{F}$, is defined at all interior points of the body. When T is a sphere of constant density ρ the calculations show (see Prob. 7) that F is proportional to the distance of P from the center of the sphere.

The discussion of this section can be extended in obvious ways to apply to improper integrals whose integrands become infinite at a finite number of points either in the interior or in the boundary of the region T. Also, the specialization of the main results of this section to improper integrals $\int_\Sigma f(P)\,d\sigma$, where Σ is a plane regular region, is immediate. A test for absolute convergence of such integrals is formulated in Prob. 1.

PROBLEMS

1. (*Test for Absolute Convergence of Double Integrals.*) If $f(P)$ is continuous in a plane regular region Σ, except at the point P_0 where $f(P)$ becomes infinite, then $\int_\Sigma f(P)\,d\sigma$ converges absolutely whenever $|f(P)| \le M/r^n$ with $n < 2$, M being a positive constant and r the distance from P to P_0. Prove it.

2. Apply the test in Prob. 1 to show that $\displaystyle\iint\limits_{\Sigma} \frac{x-\xi}{r^2}\,d\xi\,d\eta$, where $r^2 = (x-\xi)^2 + (y-\eta)^2$ is absolutely convergent in the square $|\xi| \le 1$, $|\eta| \le 1$.

3. If Σ is the disk $x^2 + y^2 \le 1$ and $r = \sqrt{x^2+y^2}$, determine which of the following integrals converge:

(a) $\displaystyle\int_\Sigma \log\frac{1}{r}\,d\sigma$ (b) $\displaystyle\int_\Sigma \frac{xy}{r^n}\,d\sigma$ (c) $\displaystyle\int_\Sigma \frac{x^2y^2}{r^6}\,d\sigma$ (d) $\displaystyle\int_\Sigma \frac{1}{r}\log r\,d\sigma$

(e) $\displaystyle\int_0^{2\pi}\int_0^1 \frac{\sin\theta}{r^{1/2}}\,dr\,d\theta$ (f) $\displaystyle\int_\Sigma \frac{d\sigma}{\sqrt{x+y}}$ (g) $\displaystyle\int_\Sigma \frac{d\sigma}{(x^2+y^2)^n}.$

4. If T is the sphere $x^2 + y^2 + z^2 \le 1$, and $r^2 = x^2 + y^2 + z^2$, show that

(a) $\displaystyle\int_T \frac{x^2}{r^n}\,d\tau = \frac{4\pi}{3(5-n)}$ if $n < 5$ (b) $\displaystyle\int_T \frac{x^2+y^2}{r^n}\,d\tau = \frac{2}{3}\frac{4\pi}{5-n}$ if $n < 5$.

5. Show that the operation of forming an improper integral is *linear*, that is,

$$\int_T [\alpha f(P) + \beta g(P)]\,d\tau = \alpha\int_T f(P)\,d\tau + \beta\int_T g(P)\,d\tau$$

provided α and β are constant and the integrals on the right exist. (This property was used in the discussion of absolute convergence.) *Hint:* The linearity holds when T is replaced by $T - T_0$ since the integrals are not improper. Now pass to the limit as the maximum diameter of T_0 tends to 0.

6. Let T be a sphere of radius a with its center at the origin. If the density ρ of the sphere is constant, find the gravitational potential $u(P) = \rho \int_T \dfrac{d\tau}{r}$. Refer to Sec. 14 and conclude that $u(P)$ is continuous in all space. Show by differentiating the expression for potential in the interior of the sphere that $\nabla^2 u = -4\pi\rho$. (See, in this connection, Chap. 7, Sec. 21.) *Hint:* To simplify calculations, choose P to be the point $(0,0,z)$ on the z axis, and use spherical coordinates (R,θ,ϕ) to show that

$$u(P) = \rho \int_0^{2\pi} d\phi \int_0^{\pi} d\theta \int_0^a \frac{R^2 \sin\theta}{r}\, dR = 2\pi\rho\left(a^2 - \frac{z^2}{3}\right) \tag{a}$$

where
$$r^2 = R^2 + z^2 - 2Rz\cos\theta \qquad \text{by the cosine law of trigonometry.} \tag{b}$$

To evaluate (a), integrate first with respect to θ to obtain $\displaystyle\int_0^{\pi} \frac{\sin\theta\, d\theta}{r}$ and change the variable of integration θ to r with the aid of (b). When R and ϕ are fixed, $(\sin\theta\, d\theta)/r = dr/Rz$, so that

$$\int_0^{\pi} \frac{\sin\theta}{r}\, d\theta = \int_{z-R}^{z+R} \frac{dr}{Rz} = \frac{2}{z} \qquad \text{if } z > R$$

$$= \int_{R-z}^{R+z} \frac{dr}{Rz} = \frac{2}{R} \qquad \text{if } z < R.$$

Thus, when P is within the sphere,

$$u(P) = \rho \int_0^{2\pi} d\phi \left(\int_0^z \frac{2R^2\, dR}{z} + \int_z^a \frac{2R^2\, dR}{R} \right) = 2\pi\rho\left(a^2 - \frac{z^2}{3}\right).$$

If the point P is not chosen on the z axis, then $u(x,y,z) = 2\pi\rho[a^2 - \tfrac{1}{3}(x^2 + y^2 + z^2)]$.

7. It was shown in Prob. 6 that the gravitational potential in the interior of a solid sphere of constant density ρ is $u(P) = 2\pi\rho(a^2 - r^2/3)$, where $r^2 = x^2 + y^2 + z^2$ and a is the radius of the sphere. Use the formula $\nabla u = \mathbf{F}$ (cf. Sec. 14) to show that in the interior of the sphere $\mathbf{F} = -\tfrac{4}{3}\pi\rho\mathbf{r}$, where $\mathbf{r} = \mathbf{i}x + \mathbf{j}y + \mathbf{k}z$. Since outside the sphere $\mathbf{F} = -\tfrac{4}{3}(\pi a^3\rho/r^3)\mathbf{r}$, conclude that the gravitational force $\mathbf{F}$ is continuous in all space. This result is correct, but we have not justified the use of the formula $\nabla u = \mathbf{F}$ for points in the interior of the solid. This requires the examination of the derivatives of $u(P)$ defined by (14-8).

8. Let $f(P)$, $g(P)$, and $h(P)$ be three functions continuous except at P_0, as in the text, and suppose

$$f(P) \leq g(P) \leq h(P)$$

holds throughout T. Show that the improper integral of g exists if the improper integrals of f and h exist and, furthermore,

$$\int_T f(P)\, d\tau \leq \int_T g(P)\, d\tau \leq \int_T h(P)\, d\tau.$$

The functions need not be positive. Hint: Proceed as in the proof of Chap. 1, Sec. 5, Theorem II, using the result of Prob. 5.

9. Deduce from Prob. 8 that an absolutely convergent improper integral is convergent. *Hint:* $-|f(P)| \leq f(P) \leq |f(P)|$.

Partial differential equations

7

435

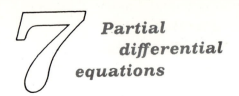

7 Partial differential equations

WHEN A PHYSICAL SYSTEM depends on more than one variable, a general description of its behavior often leads to an equation containing partial derivatives. These partial differential equations arise in such diverse subjects as meteorology, electromagnetic theory, heat transfer, nuclear physics, and elasticity, to name just a few. In contrast to the situation for ordinary differential equations, it will be seen that now the general solution is seldom sought. The main objective, rather, is to find the particular solution that satisfies the determinative conditions (the initial values and boundary values) of the problem in hand.

The first part of this chapter is concerned with a vibration problem in which the role of initial and boundary conditions is particularly easy to visualize. The concepts of linearity, superposition, wave velocity, and propagation along characteristics are introduced here. The second part presents an important method of getting series expansions, known as the method of separation of variables. The third part gives some significant integral formulas associated with the names of Fourier, Laplace, Poisson, Green, Neumann, and Helmholtz. The chapter concludes with a brief introduction to the classification of differential equations and its bearing on the question whether a problem is, or is not, well posed.

THE VIBRATING STRING

1. Arbitrary Functions. One-dimensional Waves. A *partial differential equation of order n* is an equation containing partial derivatives of order n but no higher derivatives. For example, each of the three equations

$$\frac{\partial^2 u}{\partial t^2} = a^2 \frac{\partial^2 u}{\partial x^2} \qquad \frac{\partial u}{\partial t} = \alpha^2 \frac{\partial^2 u}{\partial x^2} \qquad \frac{\partial^2 u}{\partial x^2} + \frac{\partial^2 u}{\partial y^2} + \frac{\partial^2 u}{\partial z^2} = 0$$

is a partial differential equation of order 2. In this chapter we often use the subscript notation for derivatives, so that the foregoing can be written more briefly as

$$u_{tt} = a^2 u_{xx} \qquad u_t = \alpha^2 u_{xx} \qquad u_{xx} + u_{yy} + u_{zz} = 0. \tag{1-1}$$

An n times differentiable[1] function that satisfies a given nth-order partial differential equation in a given region is called a *solution* of the equation. For example, the function

$$u = \cos x \cos at \tag{1-2}$$

---

[1] This seemingly superfluous assumption can be justified, in part, by a requirement of invariance with respect to changes of coordinates. However, its main purpose is to allow free use of the techniques of Chap. 5.

436

is a solution of the first Eq. (1-1) in the whole (x,t) plane, because (1-2) gives

$$u_x = -\sin x \cos at \qquad u_t = \cos x(-a \sin at)$$
$$u_{xx} = -\cos x \cos at \qquad u_{tt} = \cos x(-a^2 \cos at) = a^2 u_{xx}.$$

The reader will recall that the general solution of an ordinary differential equation contains arbitrary constants; for example, the general solution of $y'' + y = 0$ is

$$y = c_1 \sin t + c_2 \cos t$$

which has the arbitrary constants c_1 and c_2. We shall see that many important partial differential equations have solutions which contain arbitrary functions and, conversely, the elimination of arbitrary functions from a given expression often leads to a partial differential equation.

As an illustration of this fact, let

$$u = f(x + y) \tag{1-3}$$

where f is an arbitrary differentiable function. If the argument of f is denoted by $s = x + y$, then

$$u = f(x + y) = f(s)$$

and the chain rule gives

$$u_x = \frac{\partial u}{\partial x} = \frac{df}{ds}\frac{\partial s}{\partial x} = \frac{df}{ds} \cdot 1 = f'(s).$$

Similarly, $u_y = f'(s)$, and hence u satisfies

$$u_x = u_y \tag{1-4}$$

for any and all choices of the differentiable function f. In Prob. 9 it is shown, conversely, that every solution of (1-4) has the form (1-3).

For an example containing two arbitrary functions, let

$$U = f_1(r) + f_2(s) \qquad f_1 \text{ and } f_2 \text{ differentiable} \tag{1-5}$$

where r and s are the independent variables. Then $U_r = f_1'(r)$, and hence

$$U_{rs} = 0. \tag{1-6}$$

Conversely, (1-6) asserts that the derivative of U_r with respect to s is 0. If this holds for all values of r and s we conclude that U_r is independent of s, or, in other words,

$$U_r = h(r) \qquad \text{a function of } r \text{ only.} \tag{1-7}$$

If we write $f_1(r) = \int h(r)\,dr$, then Eq. (1-7) yields

$$\frac{\partial}{\partial r}[U - f_1(r)] = 0$$

so that $U - f_1(r) = f_2(s)$, a function of s only. Thus, U has the form (1-5).

An important example of the elimination of arbitrary functions arises from the situation shown in Fig. 1. If t is time, it is seen that $f_1(x - at)$ represents a wave form which propagates in the positive x direction with velocity a and with no change in shape, that is, with no *dispersion*. In a similar manner, $f_2(x + at)$ represents a wave form which propagates in the opposite direction with velocity a. The most general one-dimensional wave without dispersion is a superposition of two such, namely,

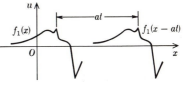

FIGURE 1

$$u = f_1(x - at) + f_2(x + at). \tag{1-8}$$

Suppose, now, that $u(x,t)$ is given by (1-8), with f_1 and f_2 twice differentiable. If we set

$$x - at = r \qquad x + at = s \tag{1-9}$$

then $u = f_1(r) + f_2(s)$, and by the chain rule

$$u_x = \frac{\partial u}{\partial x} = \frac{\partial (f_1 + f_2)}{\partial r}\frac{\partial r}{\partial x} + \frac{\partial (f_1 + f_2)}{\partial s}\frac{\partial s}{\partial x} = f_1'(x - at) + f_2'(x + at).$$

The reader may verify similarly that

$$u_t = f_1'(x - at)(-a) + f_2'(x + at)(a).$$

Differentiating again gives

$$u_{xx} = f_1''(x - at) + f_2''(x + at) \qquad u_{tt} = f_1''(x - at)(-a)^2 + f_2''(x + at)(a)^2$$

and hence u satisfies the partial differential equation

$$u_{tt} = a^2 u_{xx}. \tag{1-10}$$

We show conversely that every solution $u(x,t)$ of $u_{tt} = a^2 u_{xx}$ has the form (1-8) and thus represents the superposition of two waves propagating with velocity a. The substitution (1-9) gives

$$u(x,t) = U(r,s)$$

so that, by using the chain rule as in the previous discussion,

$$u_x = U_r + U_s \qquad u_t = -aU_r + aU_s.$$

Differentiating again yields[1]

$$u_{xx} = U_{rr} + 2U_{rs} + U_{ss} \qquad u_{tt} = a^2 U_{rr} - 2a^2 U_{rs} + a^2 U_{ss}.$$

If we substitute these values into the equation $u_{tt} = a^2 u_{xx}$ the result is $U_{rs} = 0$. As we have already seen, this ensures that U has the form $f_1(r) + f_2(s)$, and hence u has the form (1-8).

Equation (1-10) is satisfied by the most general twice-differentiable wave motion with velocity a, and, conversely, every solution represents such a motion. For this reason (1-10) is called the one-dimensional *wave equation*. Together with its analogs in higher dimensions, the wave equation is an important aid in the theory of vibration.

Although detailed study of the wave equation occupies several sections of this chapter, one of its main properties can be mentioned now. We shall show that, *if u and v are solutions, the same is true of $u + v$.* Indeed, the fact that u and v are solutions gives the representation

$$u = f_1(x - at) + f_2(x + at) \qquad v = g_1(x - at) + g_2(x + at).$$

Hence $u + v$ has the form $F_1(x - at) + F_2(x + at)$, where $F_1 = f_1 + g_1$ and $F_2 = f_2 + g_2$. Since functions of this form satisfy the wave equation, we conclude that $u + v$ satisfies it. A similar argument shows that cu is also a solution for any constant c and, more generally, $\alpha u + \beta v$ is a solution for any constants α and β. This useful fact is known as the *superposition principle*. Further discussion is given in the following examples and problems.

Example 1.   (*Standing Waves.*)   The motion given by

$$f_1(x - at) = A \sin k(x - at) \qquad A, k \text{ const} \tag{1-11}$$

---

[1] We use the fact that $U_{rs} = U_{sr}$ when U is twice differentiable.   See A. E. Taylor, "Advanced Calculus," p. 221, Ginn and Company, Boston, 1955.

represents a sine wave of amplitude A and wave-length $\lambda = 2\pi/k$, moving to the right with velocity a. The *period* T is the time required for the wave to progress a distance equal to one wavelength, so that $\lambda = aT$ or

$$T = \frac{\lambda}{a} = \frac{2\pi}{ka}.$$

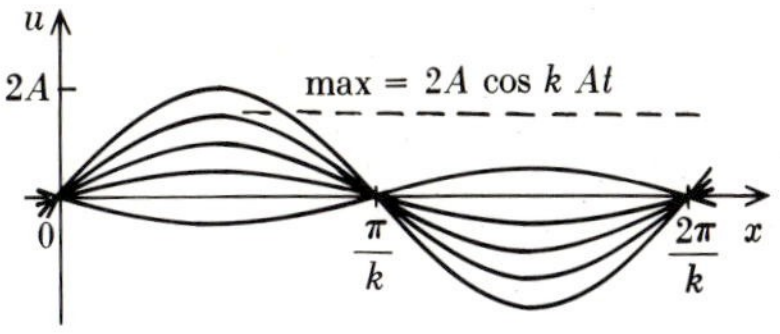

FIGURE 2

Similarly, a motion described by

$$f_2(x + at) = A \sin k(x + at) \qquad (1\text{-}12)$$

represents a sine wave, of the same amplitude and period, moving with velocity a to the left. The superposition of (1-11) and (1-12) gives

$$u = A \sin k(x - at) + A \sin k(x + at)$$

which becomes

$$u = (2A \cos kat) \sin kx \qquad (1\text{-}13)$$

when we recall the trigonometric identities

$$\sin k(x \pm at) = \sin kx \cos kat \pm \cos kx \sin kat.$$

The expression (1-13) may be regarded as a sinusoid $\sin kx$ whose amplitude $2A \cos kat$ varies with the time t in a simple harmonic manner. Several curves of (1-13) are sketched in Fig. 2 for various values of t. The points $n\pi/k$ remain fixed throughout the motion and are called *nodes*. Although the result was obtained by superposing two traveling waves, the wave form (1-13) does not appear to travel either to the left or the right, and (1-13) is said to represent a *standing wave*.

The number f of oscillations or *cycles* made by the wave per unit time is called the *frequency*. From the definition of the period T, it follows that $f = 1/T$.

Example 2.   As explained in Chap. 3, an operator $\mathbf{T}$ is *linear* if

$$\mathbf{T}(\alpha u + \beta v) = \alpha \mathbf{T}u + \beta \mathbf{T}v$$

for all constants α and β and all functions u and v for which $\mathbf{T}u$ and $\mathbf{T}v$ are meaningful. Show that the operator $\mathbf{T}$ defined by

$$\mathbf{T}w = w_{tt} - a^2 w_{xx} \qquad (1\text{-}14)$$

is linear.

The choice $w = \alpha u + \beta v$ gives

$$\mathbf{T}(\alpha u + \beta v) = (\alpha u + \beta v)_{tt} - a^2(\alpha u + \beta v)_{xx} = \alpha u_{tt} + \beta v_{tt} - a^2(\alpha u_{xx} + \beta v_{xx})$$

since differentiation is a linear operation. If we collect the terms involving α and those involving β, the right-hand side is recognized as $\alpha \mathbf{T}u + \beta \mathbf{T}v$, and that is the desired result.

In Chap. 3 it was shown that the superposition principle holds for all equations $\mathbf{T}u = 0$, where $\mathbf{T}$ is linear. Since the operator (1-14) is linear, the superposition principle for the wave equation follows as a special case.

PROBLEMS

1. Review the derivation of the chain rule in Chap. 5, Sec. 5, and do Probs. 1, 2, 3, 5, and 11 of that section.

2. If $u = f(y/x)$ with f differentiable, show that $xu_x + yu_y = 0$ for $x \neq 0$.

3. Show by direct differentiation that $u = \sin kx \sin kat$ satisfies the one-dimensional wave equation for every choice of the constant k, and express this function in the form (1-8).

4. (a) By computing u_x, u_y, and u_{xy}, obtain a second-order partial differential equation for $u = f_1(x)f_2(y)$. (b) Show that your result is equivalent to $(\log u)_{xy} = 0$ in general, and explain.

5. For many functions the chain rule applies even when the argument is complex.   Assuming this, show that

$$u = f_1(x + iy) + f_2(x - iy) \qquad i^2 = -1$$

satisfies *Laplace's equation* $u_{xx} + u_{yy} = 0$.

6. Let $f(x + iy) = u(x,y) + iv(x,y)$, where u and v are real.   Using the chain rule, show that u and v satisfy the *Cauchy-Riemann equations*

$$u_x = v_y \qquad u_y = -v_x.$$

7. (*a*) Read the discussion of linear operators and superposition in Chap. 3, Sec. 2.   (*b*) Read the discussion of nonhomogeneous equations and the principle of the complementary function in Chap. 3, Sec. 6.

8. Show that operators $\mathbf{T}$ defined by the following equations are linear:

$$\mathbf{T}u = \alpha^2 u_{xx} - u_t \qquad \mathbf{T}u = u_{xx} + u_{yy} + u_{zz} \qquad \mathbf{T}u = u_x - u_y \qquad \mathbf{T}u = xu_x + y^2 u_y + xyu.$$

9. If $u_x = u_y$, let $s = x + y$, so that $u(x,y) = U(x,s)$.   By the chain rule, show that $u_x = u_y$ implies $U_x = 0$.   Hence $U = f(s)$, and $u = f(x + y)$.

10. A homogeneous function of degree n that satisfies $u_{xx} + u_{yy} + u_{zz} = 0$ is called a *harmonic*. Given $r = \sqrt{x^2 + y^2 + z^2}$, verify that each of the following is a harmonic, and find its degree:

$$xy, \quad x^2 - y^2, \quad x, \quad 2z^2 - x^2 - y^2, \quad \tan^{-1}\frac{y}{x}, \quad z\tan^{-1}\frac{y}{x}, \quad \frac{1}{r}, \quad \frac{1}{r}\tan^{-1}\frac{y}{x}, \quad \log\frac{r+z}{r-z}, \quad \frac{x}{r^3}.$$

Hint: Since $r^2 = x^2 + y^2 + z^2$, it follows that $rr_x = x$.   Thus get r_{xx}, and use symmetry to get other derivatives.

11. As illustrated by the discussion of the text and by Prob. 9, some partial differential equations can be treated as ordinary differential equations, in which the constants of integration depend on the variables not involved in the differentiation.   Thus solve the following equations for $u(x,y)$:

$$u_x = 2x \qquad u_y = x \qquad u_{xx} = 0 \qquad u_{xx} = y \qquad u_x = 2u \qquad u_{yy} = u \qquad u_{xx} + u = x + y.$$

---

2. Transverse Oscillation of a Flexible String.

Consider an elastic string under tension, as shown in Fig. 3.   For the present the string is assumed to be of infinite length or, at any rate, so long that the effect of the end supports can be neglected for the portion of string under observation.   Also, the string is flexible, so that there is no resistance to bending.   This means that the tension force is directed tangentially to the curve formed by the string at any instant.

A complete analysis of the problem of a vibrating string requires knowledge of the nonlinear mechanics of deformable media and is somewhat out of place here.   Hence, instead of discussing the string from the point of view of nonlinear mechanics and elasticity theory, we shall assume a particular type of motion in which the string lies in a vertical plane at all times and the motion in that plane is straight up and down, not sideways. The behavior is described by giving the vertical displacement $u(x,t)$ at a distance x from some suitable origin at time t.   It will be found that this assumption of *transverse oscillation* is consistent with Newton's laws and leads to a differential equation with many desirable properties.   The

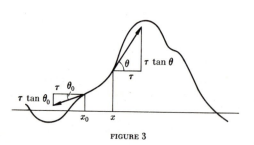

FIGURE 3

method is applicable to the general problem in which both transverse and longitudinal oscillations are present, though this aspect of the subject is not discussed here.

When the string is at rest in the horizontal position the density is denoted by $\rho = \rho(x)$, so that $\rho \, dx$ gives the differential of mass between x and $x + dx$. By hypothesis there is no sideways movement; hence the mass originally between x and $x + dx$ remains between x and $x + dx$ throughout the motion. Thus, $\rho \, dx$ describes this mass for all t, even though ρ is independent of t. The corresponding differential of momentum is $u_t \rho \, dx$, since $u_t \equiv \partial u / \partial t$ is the vertical velocity, and the total momentum of the part of the string between x_0 and x is therefore

$$M = \int_{x_0}^{x} \rho u_t \, dx. \tag{2-1}$$

We shall obtain the equation of motion by applying Newton's law

$$\frac{d}{dt} (\text{momentum}) = \text{force}$$

to the portion of the string between x_0 and x. If there is enough continuity so that (2-1) can be differentiated under the integral sign, the rate of change of momentum is

$$\frac{\partial M}{\partial t} = \int_{x_0}^{x} \rho u_{tt} \, dx \tag{2-2}$$

and it remains only to compute the force.

For the particular type of motion being considered here the horizontal component of the tensile force has a simpler behavior than the tension $\bar{\tau}$ itself. We denote this horizontal component by $\tau = \tau(x,t)$. Since there is no horizontal acceleration of the string between x_0 and x, the horizontal components of tensile force at the two ends must balance. We conclude that $\tau(x,t) = \tau(x_0,t)$ or, in other words, τ *is independent of x.*

According to Fig. 3 the vertical component of the tensile force is $\tau \tan \theta$, where θ is the angle between the tangent to the string and the x axis. Since $\tan \theta$ is the slope of the curve, $\tan \theta = du/dx$ at any given value of t. More explicitly,

$$\tan \theta = \frac{\partial u}{\partial x} \equiv u_x.$$

The net vertical component of force due to tension is therefore

$$\tau u_x \Big|_{x_0}^{x} = \int_{x_0}^{x} \frac{\partial}{\partial x} (\tau u_x) \, dx = \int_{x_0}^{x} \tau u_{xx} \, dx$$

where the second equality follows from the fact that τ is independent of x.

If in addition to the tensile force there is a distributed load $F_1 = F_1(x,t)$ the corresponding vertical force on the segment of string between x_0 and x is

$$\int_{x_0}^{x} F_1 \, dx.$$

The result of equating the rate of change of momentum (2-2) with the total force as given above is

$$\int_{x_0}^{x} \rho u_{tt} \, dx = \int_{x_0}^{x} \tau u_{xx} \, dx + \int_{x_0}^{x} F_1 \, dx.$$

Differentiation with respect to the upper limit x now produces the desired differential equation,

$$\rho u_{tt} = \tau u_{xx} + F_1$$

provided the integrands are continuous. This in turn may be written

$$u_{tt} = a^2 u_{xx} + F(x,t) \qquad \text{where } a = \sqrt{\frac{\tau}{\rho}} \qquad F = \frac{F_1}{\rho}. \tag{2-3}$$

According to our analysis the horizontal density function ρ depends on x alone, not on t, and the horizontal tensile force τ depends on t alone, not on x. For small oscillations of a uniform string under great tension both τ and ρ are practically constant, and this case is considered in the sequel. We refer to the constant τ as the *tension* and to ρ as the *density*, since these quantities actually are the tension and density when the string is at rest. Further discussion can be found in the problems at the end of this section.

When the force function $F(x,t)$ is zero, the vibrations of the string are termed *free vibrations*. By (2-3) the equation for free vibrations is

$$u_{tt} = a^2 u_{xx} \tag{2-4}$$

and hence the solution has the form (1-8). According to the discussion in Sec. 1, the motion can always be regarded as a superposition of two waves moving with velocity

$$a = \sqrt{\frac{\tau}{\rho}} \tag{2-5}$$

in opposite directions. Later we shall determine the precise form of these waves by considering the initial state of the string, that is, the state at $t = 0$, together with the conditions at the end points, $x = 0$ and $x = l$.

Inasmuch as the constant a in (2-4) involves only the ratio τ/ρ, two strings may behave similarly even if made of different materials. For example, a string with density 2ρ under tension 2τ behaves like a string with density ρ and tension τ, since both yield the same value for a. An equivalence of two different physical systems such as this is sometimes called a *principle of similitude*. The study of similitude belongs to an interesting branch of mathematical physics known as *dimensional analysis*. Although a general development[1] will not be given here, we shall describe the underlying idea as it applies to (2-5).

Equation (2-5) relates three quantities a, τ, and ρ which are expressed in different physical units. The units for τ can be found from Newton's law, since τ is a force. Hence in the mks system

$$a = \left[\frac{\text{meters}}{\text{second}}\right] \qquad \rho = \left[\frac{\text{kilograms}}{\text{meter}}\right] \qquad \tau = \left[\frac{\text{kilogram-meters}}{(\text{second})^2}\right] \tag{2-6}$$

where brackets are used to indicate that the measuring unit, rather than the value, is being described. The value of a for use in (2-5) is the *number* of such measuring units, that is, the number of meters per second, and similarly for ρ and τ.

If we decide to measure lengths in centimeters rather than in meters, the value of a will be increased by a factor 100. In other words, $100a$ cm per sec is the same as a m per sec. Similarly τ will be multiplied by 100, but ρ will be divided by 100, since the length unit for ρ in (2-6) occurs in the denominator. (Indeed, ρ kg per m is clearly the same as 0.01ρ kg per cm.) Hence, when a string has a wave velocity a, density ρ, and tension τ in the old system (2-6), the same string has velocity, density, and tension

$$100a, \qquad \frac{\rho}{100}, \qquad 100\tau \tag{2-7}$$

in the new system. Substituting into (2-5) yields

$$100a = \sqrt{\frac{100\tau}{\rho/100}}$$

[1] The reader is referred to P. W. Bridgman, "Dimensional Analysis," Yale University Press, New Haven, Conn., 1931, and S. Drobot, The Foundations of Dimensional Analysis, *Studia Math.*, 14:84–99 (1954).

which is consistent with (2-5), as it should be. One does not get a contradictory result by measuring all lengths in centimeters rather than in meters.

To change meters into centimeters, the unit of length is divided by 100. More generally, one might divide the unit by an arbitrary positive constant α. The new values of a, ρ, and τ would be, respectively,

$$\alpha a, \qquad \frac{\rho}{\alpha}, \qquad \alpha \tau \tag{2-8}$$

[compare (2-7)]. Similar changes may be made in the units of mass or of time. Equation (2-5) remains self-consistent under such changes, as the reader can verify. The question arises: Is (2-5) the *only* functional relationship

$$a = f(\rho, \tau) \tag{2-9}$$

which is consistent under such changes? If so, then we would have a proof of the functional relation (2-5), assuming merely that there is a functional relation of *some* kind.

To investigate this possibility, suppose (2-9) holds where f is an unknown function and where a, τ, and ρ stand for the numbers of their respective units of measurement in (2-6). If the unit of length is divided by α, then (2-8) gives

$$\alpha a = f\left(\frac{\rho}{\alpha}, \alpha \tau\right)$$

upon substitution into (2-9). Since α is arbitrary we may choose $\alpha = \rho$ to find

$$\rho \alpha = f(1, \rho \tau). \tag{2-10}$$

If we now divide the unit of mass by β, the value of a is unchanged but ρ and τ become $\beta \rho$ and $\beta \tau$, respectively [see (2-6)]. Substituting into (2-10) yields

$$\beta \rho a = f(1, \beta^2 \rho \tau).$$

Upon choosing $\beta^2 = (\rho \tau)^{-1}$, we get

$$(\rho \tau)^{-\frac{1}{2}} \rho a = f(1,1)$$

so that

$$a = c \sqrt{\frac{\tau}{\rho}} \tag{2-11}$$

where $c = f(1,1)$ is a *constant*, independent of a, ρ, and τ.

Finally, if we divide the unit of time by γ, the new values of a, ρ, and τ are given by (2-6) as a/γ, ρ, and τ/γ^2. Substituting into (2-11) gives

$$\frac{a}{\gamma} = c \sqrt{\frac{\tau/\gamma^2}{\rho}}$$

which reduces to (2-11) again. Thus, no new information is obtained by changing the unit of time, and the constant c in (2-11) cannot be found by dimensional analysis. But in the limiting case of small oscillations, the partial differential equation (2-4) is valid, and (2-5) shows that $c \to 1$.

PROBLEMS

1. The displacement of a certain string is

$$u(x,t) = f_1(x - at) + f_2(x + at).$$

What is the physical meaning of the condition $u(0,t) \equiv 0$? If $u(0,t) \equiv 0$, express f_1 in terms of f_2, and thus deduce

$$u(x,t) = f_2(x + at) - f_2(-x + at).$$

2. (a) Find $f(x - at)$ when $f(x) = (1 + x^2)^{-1}$; when $f(x) = \sin kx$; when $f(x) = e^x$. In each case, compute also $f'(x - at)$. *Hint:* Substitute $x - at$ for x in the expressions for $f(x)$. (b) If $u(x,t) = f(x - at) + f(x + at)$, find $u(0,t)$, $u(x,0)$, and $u(l,1/a)$ for each $f(x)$ in part a of this problem.

3. Let $\tilde{\tau}$ denote the actual tension of the string, measured tangentially, and let $\tilde{\rho}$ denote the actual linear density referred to the arc length along the curved string. Show that $\tilde{\tau}$ and $\tilde{\rho}$ are related to the parameters τ and ρ of the text by

$$\tilde{\tau} = \tau(1 + u_x^2)^{\frac12} \qquad \tilde{\rho} = \rho(1 + u_x^2)^{-\frac12}.$$

Hence $\tilde{\tau}\tilde{\rho} = \tau\rho$. Also $\tau \doteq \tilde{\tau}$ and $\rho \doteq \tilde{\rho}$ if $\frac12 u_x^2$ is small compared with unity. *Hint:* $\tilde{\tau} = \tau \sec\theta$, and $\tilde{\rho}\,ds = \rho\,dx$, where s is the arc along the string.

4. In deriving the equation for vibration of a string stretched between fixed supports it is often assumed that the actual tension $\tilde{\tau}$ of Prob. 3 is constant, as well as that the motion is vertical. Show that these assumptions are mutually contradictory. *Hint:* Since the motion is vertical, τ depends on t alone, not on x. Hence Prob. 3 shows that u_x is independent of x. If the displacement is 0 at the fixed ends, we get $u \equiv 0$, which contradicts the fact that the string is vibrating.

5. In deriving the equation for vibration of a string stretched between two fixed supports, it is often assumed that the actual density $\tilde{\rho}$ of Prob. 3 is constant. Show that the string cannot oscillate under this assumption. *Hint:* The length must increase if it oscillates, but the total mass between supports does not change.

6. (*a*) In many derivations involving small oscillations the fact that

$$\theta \doteq \tan\theta = u_x$$

is used to conclude that $\theta_x \doteq u_{xx}$. By considering $\theta = 0.001 \sin 10{,}000x$ and $\phi = 0$, show that $\theta \doteq \phi$ does not necessarily give $\theta_x \doteq \phi_x$. (*b*) In Chap. 2, Sec. 15, a nonlinear ordinary differential equation is approximated by a linear equation, and it is proved that the solutions also approximate each other. Read this discussion. (The corresponding problem for partial differential equations is, as a rule, vastly more difficult.)

7. If we walk along with a wave propagating on a string we can imagine that the string is being pulled past us over a pulley of radius $1/\kappa$, where $\kappa \equiv \lim \Delta\theta/\Delta s$ equals the radius of curvature of the curve formed by the string. (*a*) Denoting the tangential speed by $\tilde{v}$, equate the magnitude of normal force on an element of arc Δs due to tension with that due to centrifugal force. Thus obtain the approximate equalities

$$\tilde{\tau}\,\Delta\theta \doteq (\tilde{\rho}\,\Delta s)\kappa v^2 \doteq \tilde{\rho}\tilde{v}^2\,\Delta\theta$$

in the notation of Prob. 3. Hence $\tilde{v} = \sqrt{\tilde{\tau}/\tilde{\rho}}$ is the condition for equilibrium. (*b*) If the x measure of the speed, tension, and density are, respectively, $v = \tilde{v}\cos\theta$, $\tau = \tilde{\tau}\cos\theta$, $\rho = \tilde{\rho}\sec\theta$, show that the result of part a is $v = \sqrt{\tau/\rho}$. (Hence the assumption of constant $\tilde{\rho}$ and $\tilde{\tau}$ implies that $\tilde{v}$ is constant, while the assumption of constant ρ and τ, as in the text, implies that $v \equiv a$ is constant. In both cases the oscillations need not be small.)

The following problems presuppose some knowledge of elasticity theory and can be omitted without loss of continuity.

8. Show that the small longitudinal vibrations of a uniform long rod satisfy the differential equation $\rho u_{tt} = E u_{xx}$, where u is the displacement of a point originally at a distance x from the end of the rod, E is the modulus of elasticity, and ρ is the density. *Hint:* From the definition of Young's modulus E, the force on a cross-sectional area q at a distance x units from the end of the rod is $Eq(\partial u/\partial x)$, since $\partial u/\partial x$ is the extension per unit length. On the other hand, the force on an element of the rod of length Δx is approximately $\rho q\,\Delta x\,\partial^2 u/\partial t^2$.

9. If the rod of Prob. 8 is made of steel for which $E = 22 \times 10^8$ g per cm^2 and whose specific gravity is 7.8, show that the velocity of propagation of sound in steel is nearly 5.3×10^5 cm per sec, which is about sixteen times as great as the velocity of sound in air. Note that in the cgs system E must be expressed in dynes per square centimeter.

10. Obtain the equation $\rho\theta_{xx} = G\theta_{tt}$ for small angular displacement θ of a uniform shaft performing torsional oscillations. Here ρ is the density and G is the rigidity modulus.

11. Show that the differential equation of small transverse vibrations of an elastic rod carrying a load of $p(x)$ lb per unit length is

$$EI \frac{\partial^4 y}{\partial x^4} = p(x) - m \frac{\partial^2 y}{\partial t^2}$$

where E is the modulus of elasticity, I is the moment of inertia of a cross-sectional area of rod about a horizontal transverse axis through the center of gravity, and m is the mass per unit length. *Hint:* For small deflections the bending moment M about a horizontal transverse axis at a distance x from the end of the rod is given by the Euler formula $M = EI \, d^2y/dx^2$, and the shearing load $p(x)$ is given by $d^2M/dx^2 = p(x)$.

---

3. Initial Conditions.   In the previous section the wave equation

$$u_{tt} = a^2 u_{xx} \qquad a = \text{const} \tag{3-1}$$

was derived for small displacements of a uniform flexible string.   According to Sec. 1 the general solution of (3-1) is

$$u(x,t) = f_1(x - at) + f_2(x + at) \tag{3-2}$$

where f_1 and f_2 are any twice-differentiable functions.   We shall now see that these functions can be determined from the initial conditions, that is, from the conditions at time $t = 0$.

CASE I.   Initial Impulse 0.   Assume that the string is released from rest and that the initial shape is given by a known function $f(x)$.   (Such a situation arises when the string is plucked, as in a harpsichord.)   In symbols,

$$u(x,0) = f(x) \qquad u_t(x,0) = 0 \qquad |x| < \infty \tag{3-3}$$

where the second equation (3-3) expresses the fact that the vertical velocity $\partial u / \partial t$ is initially 0 for each point x of the string.   By (3-2) we get

$$u_t(x,t) = -a f_1'(x - at) + a f_2'(x + at) \tag{3-4}$$

upon using the chain rule as in Sec. 1.   Since $u_t(x,0) = 0$, Eq. (3-4) gives

$$f_1'(x) = f_2'(x)$$

after dividing by a.   It follows that $f_2(x) = f_1(x) + c$, where c is constant.   Using this equality with x replaced by $x + at$, we see that (3-2) may be written

$$u(x,t) = f_1(x - at) + f_1(x + at) + c. \tag{3-5}$$

This step is sometimes puzzling when encountered for the first time; namely, from

$$f_2(x) = f_1(x) + c$$

how can we deduce $f_2(x + at) = f_1(x + at) + c$?   The conclusion follows because the first equation holds *for all values of* x.   One cannot simply set $x = x + at$, because that would lead to $at = 0$. But one can reason as follows: $f_2(x) = f_1(x) + c$ for all x.   Hence $f_2(s) = f_1(s) + c$ for all s, and the choice $s = x + at$ yields the desired result.

So far we have used only the second initial condition (3-3).   To ensure the first condition, $u(x,0) = f(x)$, we set $t = 0$ in (3-5) and equate the result to $f(x)$; thus

$$f_1(x) + f_1(x) + c = f(x).$$

It follows that $f_1(x) = \tfrac{1}{2} f(x) - \tfrac{1}{2} c$, and substituting into (3-5) gives the final answer:

$$u(x,t) = \tfrac{1}{2} f(x - at) + \tfrac{1}{2} f(x + at). \tag{3-6}$$

The displacement $u(x,t)$ in (3-6) is the sum of two waves, each of the form $\frac{1}{2}f(x)$, which travel in opposite directions with the velocity a. Initially (that is, for $t = 0$) these waves coincide, but with the passage of time they diverge, the wave $\frac{1}{2}f(x - at)$ moving to the right and the other to the left. In particular, if the waves are of finite extent, any given point of the string is at rest in the initial position after the passage of both waves. The situation is illustrated schematically in Fig. 4 when $f(x)$ is a triangular wave on $(-k,k)$.

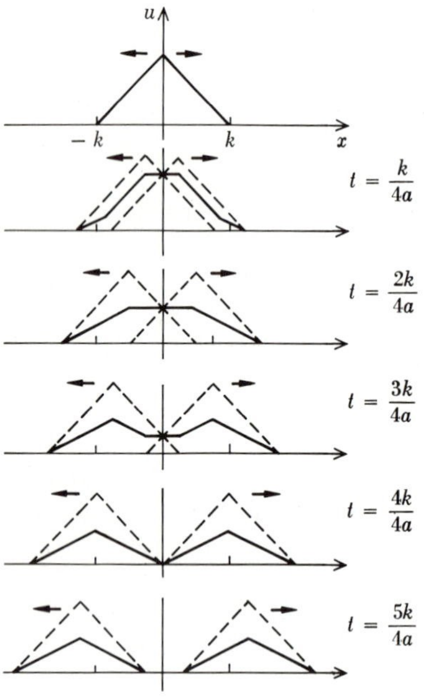

It will be observed that the wave illustrated in Fig. 4 does not satisfy our conditions of differentiability at points where the graph has corners. Hence, at these points the wave equation does not hold. However, it is possible to enlarge the scope of the term "solution" so that functions of this kind are allowed. One way of doing this is to regard any expression (3-2) as a *generalized solution* of the wave equation, whether or not f_1 and f_2 are differentiable. Another method is to regard a function as a generalized solution if it can be approximated with arbitrary precision by smooth solutions, that is, by solutions in the sense of Sec. 1. A third method is to replace the differential equation by the integrated form given in the analysis preceding (2-3).

The concept of generalized solution is important in the modern theory of differential equations and is useful in practice because it allows many examples, such as that in Fig. 4, that would otherwise have to be excluded. Since the problem of the vibrating string serves admirably to illustrate the distinction between smooth and generalized solutions, this distinction is mentioned from time to time in Secs. 3 to 6. Each of the above formulations of "generalized solution" requires a different hypothesis, but the first formulation (being the simplest) is the one we shall use.

FIGURE 4

CASE II. Initial Displacement 0. Suppose, next, that the initial displacement is 0 but that the initial velocity is not 0. (Such a situation arises when the string is struck, as in a piano.) If the initial velocity is $g(x)$ at point x of the string, the initial conditions are now

$$u(x,0) = 0 \qquad u_t(x,0) = g(x). \tag{3-7}$$

The first equation (3-7) gives

$$f_1(x) + f_2(x) = 0$$

when we recall (3-2), so that $f_2(x) = -f_1(x)$ for all values of x. Using this equality with x replaced by $x + at$, we see that (3-2) may be written

$$u(x,t) = f_1(x - at) - f_1(x + at). \tag{3-8}$$

Differentiating (3-8) with respect to t and setting $t = 0$ yield

$$u_t(x,0) = -af'_1(x) - af'_1(x) = g(x)$$

when we use the second condition (3-7).   For integrable g it follows that

$$f_1(x) = -\frac{1}{2a} \int_0^x g(s)\ ds + c \tag{3-9}$$

where c is constant, and hence (3-8) gives the final answer

$$u(x,t) = -\frac{1}{2a} \int_0^{x-at} g(s)\ ds + \frac{1}{2a} \int_0^{x+at} g(s)\ ds.$$

The result can be expressed more compactly as

$$u(x,t) = \frac{1}{2a} \int_{x-at}^{x+at} g(s)\ ds. \tag{3-10}$$

Equation (3-8), like (3-6), represents a superposition of two waves traveling in opposite directions.   Here, however, the shapes of the waves are determined by $f_1(x)$ and $-f_1(x)$, which are mirror images of each other in the x axis.   Moreover, the shapes are not found directly by the initial condition but are obtained through the integration (3-9). For this reason the waves may be of infinite extent even when the initial impulse $u_t(x,0) = g(x)$ is confined to a finite portion $-k < x < k$ of the string.   Indeed, for such a choice of $g(x)$ formula (3-10) shows that any given point x of the string eventually suffers a permanent displacement

$$\frac{1}{2a} \int_{-k}^{k} g(s)\ ds. \tag{3-11}$$

This is the case because, when $at > k + |x|$, the interval $(x - at, x + at)$ contains the interval $(-k,k)$.   Inasmuch as $g(x) = 0$ outside the interval $(-k,k)$, the integral (3-10) is then equal to (3-11).   Since each given point of the string eventually moves the same distance (3-11), the part of the string that is again at rest forms a straight line parallel to the original string. It is most interesting that this happens *regardless of the choice* of $g(x)$, provided only that $g(x) = 0$ outside some finite interval.   Graphical illustration is given in Fig. 5 for the case $g(x) = 1$ on $(-k,k)$.   Since $g(x)$ is not continuous, this example also requires the concept of "generalized solution."

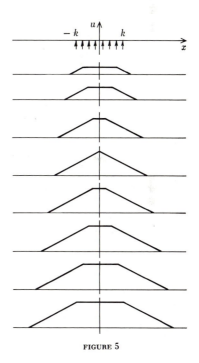

FIGURE 5

CASE III.   *Arbitrary Initial Conditions.*   Suppose, now, that both the initial displacement and the initial velocity are given by arbitrary continuous functions of x, so that the initial conditions are

$$u(x,0) = f(x) \qquad u_t(x,0) = g(x). \tag{3-12}$$

This problem can be solved by superposition of the two solutions previously obtained. Indeed, let $v(x,t)$ and $w(x,t)$ satisfy the wave equation (3-1) and the respective initial conditions

$$v(x,0) = f(x) \qquad v_t(x,0) = 0$$
$$w(x,0) = 0 \qquad w_t(x,0) = g(x). \tag{3-13}$$

Then the function

$$u(x,t) = v(x,t) + w(x,t) \tag{3-14}$$

satisfies the wave equation because v and w do, and addition of the relations (3-13) shows that u satisfies (3-12). Since the wave equation was solved in the previous discussion subject to initial conditions of the type (3-13), addition of the two solutions obtained formerly gives the solution desired now. That is,

$$u(x,t) = \tfrac{1}{2}f(x - at) + \tfrac{1}{2}f(x + at) + \frac{1}{2a} \int_{x-at}^{x+at} g(s)\, ds. \tag{3-15}$$

The expression (3-15) is known as *d'Alembert's formula;* it satisfies (3-1) and (3-12) and hence gives the motion of a string subjected to arbitrary initial displacements and velocities. This expression is meaningful as a generalized solution if f is defined and g is integrable. It is valid as a smooth solution if f is twice differentiable and g is differentiable.

The superposition method yields a solution of the problem but does not establish the uniqueness of that solution. We shall now show that every solution of (3-1) and (3-12) can be represented in the form $v + w$, with v and w as in (3-13). Since v and w were already shown to be unique in the discussion of Cases I and II, it will follow that u is also unique.

Let $u(x,t)$ be a solution of the wave equation (3-1) which satisfies the initial conditions (3-12). Let $v(x,t)$ be the unique solution of (3-1) satisfying the first conditions (3-13). Then the function $w(x,t)$ defined by

$$w(x,t) = u(x,t) - v(x,t)$$

satisfies (3-1) and the second set of initial conditions (3-13), as the reader can verify. It follows that w is uniquely determined and hence $u(x,t)$ is also uniquely determined. Because of uniqueness, (3-15) describes *the behavior of the string.* Without uniqueness, we could say only that (3-15) describes *a possible behavior of the string.*

If we introduce the *integral mean* operator M_t, where

$$M_t f = \frac{1}{2at} \int_{x-at}^{x+at} f(s)\, ds \qquad M_t g = \frac{1}{2at} \int_{x-at}^{x+at} g(s)\, ds$$

then d'Alembert's formula becomes

$$u(x,t) = \frac{\partial}{\partial t}(t M_t f) + (t M_t g). \tag{3-16}$$

Indeed, the second term of (3-16) is clearly the same as that in (3-15), and equality of the first terms for continuous f follows from the equation

$$\frac{\partial}{\partial t} \int_0^\alpha f(s)\, ds = \left[\frac{\partial}{\partial \alpha} \int_0^\alpha f(s)\, ds\right] \frac{\partial \alpha}{\partial t} = f(\alpha)\frac{\partial \alpha}{\partial t}$$

with $\alpha = x + at$ or $\alpha = x - at$, respectively (cf. Chap. 5, Sec. 14). Formulas similar to (3-16) in higher dimensions are useful in the deeper study of partial differential equations, though the limitations of space do not permit us to discuss the subject here.[1]

Example. Obtain the displacement of a semi-infinite string which is fixed at $x = 0$ and has initial displacement and velocity

$$u(x,0) = f(x) \qquad u_t(x,0) = g(x) \qquad x \geq 0. \tag{3-17}$$

Assume f and g continuous, with $f(0) = g(0) = 0$.

---

[1] See, for example, F. John, "Plane Waves and Spherical Means Applied to Partial Differential Equations," Tract 2, Interscience Publishers, Inc., New York, 1955.

We imagine an infinite string for which the initial conditions in the interval $(0,\infty)$ coincide with (3-17) and in the interval $(-\infty,0)$ are determined by

$$u(x,0) = -f(|x|) \qquad u_t(x,0) = -g(|x|) \qquad x \leq 0. \tag{3-18}$$

The point $x = 0$ of an infinite string, moving in accord with (3-17) and (3-18), will obviously be at rest, and the behavior of the infinite string for $x \geq 0$ will be identical with that of the semi-infinite string. Hence the (generalized) solution is given by d'Alembert's formula, with initial conditions as in (3-17) and (3-18).

The condition $u(0,t) = 0$ is an example of a *boundary condition*. Further discussion of boundary conditions is given in Sec. 5, where it is shown that the result obtained here is unique.

PROBLEMS

1. The displacement of a string is given by the traveling wave $u(x,t) = \sin(x - at)$. What are the initial displacement and velocity? Verify, by actual substitution into (3-15), that your initial values yield the correct result, $u(x,t) = \sin(x - at)$.

2. For a freely vibrating string the initial displacement and velocity are, respectively, $\sin x$ and $\cos 2x$. Find the displacement and velocity of the point $x = 0$ when $t = \pi$. *Hint:* First find $u(x,t)$ from (3-15).

3. A freely vibrating string was subjected to an initial displacement $6 \cos 5x$ and initial velocity 0. One second later it is found that the point $x = 0$ is displaced three units from the equilibrium position; that is, $u(0,1) = 3$. What can you say about the velocity of propagation for waves on this string?

4. The initial velocity of a freely vibrating string is xe^{-x^2}. For what choice of the initial displacement (if any) does the resulting motion represent a traveling wave traveling in the positive x direction? *Hint:* It is desired that $u(x,t) = f_1(x - at)$. Determine f_1 from the initial velocity, and then determine the initial displacement from f_1.

5. Solve Prob. 4 with the words "velocity" and "displacement" interchanged.

6. A stretched infinite string is struck so that its segment $-1 \leq x \leq 1$ is given an initial velocity 1. Use (3-15) to find the displacement, and sketch the displacement curves for $t = 1/a$ and $t = 2/a$.

7. The initial displacement and velocity of a semi-infinite string are $u(x,0) = \sin x$, $u_t(x,0) = 0$, $0 \leq x < \infty$. Find $u(x,t)$ for $t > 0$. Also find $u(x,t)$ if $u(x,0) = 0$, $u_t(x,0) = -2a \sin x$, $0 \leq x < \infty$.

---

4. Characteristics. A physical interpretation may be given not only by plotting $u(x,t)$ versus x for a succession of values of t, but also by considering the xt plane. Each point of the xt plane represents a definite position on the string at a definite time t. If we take $t = 0$ to be the present time, the half planes $t < 0$ and $t > 0$ give the past and future, respectively.

Although it is not appropriate to permit $t < 0$ when the string is plucked or struck at $t = 0$, it is appropriate if the string has been in motion for some time and the initial conditions are determined by high-speed photography. We could then take the view that we are trying to ascertain the past history of the string by observations on the present.

Since the speed of propagation is a, the disturbance at (x,t) will reach a point (x_0,t_0) given by

$$\frac{x - x_0}{t - t_0} = a \qquad \text{or} \qquad \frac{x - x_0}{t - t_0} = -a \tag{4-1}$$

for the direct wave $f_1(x - at)$ and the opposite wave $f_2(x + at)$, respectively. Equations (4-1) may be written

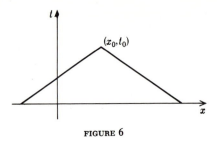

FIGURE 6

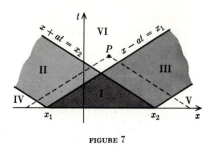

FIGURE 7

$$x - at = x_0 - at_0 \qquad x + at = x_0 + at_0. \tag{4-2}$$

If we draw the two lines (4-2) through the point (x_0,t_0), as shown in Fig. 6, their intersection with the x axis (that is, $t = 0$) gives those points on the string for which the initial condition contributes to the disturbance at (x_0,t_0). The lines (4-2) are called the *characteristics* of the partial differential equation (3-1).

Along the first line (4-2), $x - at$ is constant, and hence $f_1(x - at)$ is constant. Thus, the deflection due to the direct wave is the same at all points of the first characteristic (4-2). The second line serves the same purpose for the opposite wave, and we can say, briefly, that the disturbance travels along the characteristics.

If the initial disturbance is confined to some interval (x_1,x_2), we have the situation shown in Fig. 7. The xt plane is divided by the characteristics into six regions. In region I the points receive the disturbance from both waves, in II only from the opposite wave, and in III only from the direct wave. The points in IV and V are too far away to receive any disturbance at the corresponding times, and the points in VI are at rest because both waves have passed. That is, if P is a point in the region VI, the characteristics through P (shown dashed in the figure) intersect the x axis outside the interval (x_1,x_2). Hence the initial displacement at these points is zero, and we need consider the initial impulse only. Since the characteristics intersect outside the interval (x_1,x_2), the displacement at P due to the initial impulse is given by the constant value (3-11).

We have seen that the initial conditions determine both the direct wave and the opposite wave at each point on the x axis where these conditions are given. Since the disturbance propagates along the characteristics the following theorem is suggested:

THEOREM I. Let u and u_t be given on the interval (x_1,x_2) in Fig. 8, and suppose $u_{tt} = a^2 u_{xx}$. Then $u(x,t)$ is uniquely determined in the shaded region but is not determined at any other point.

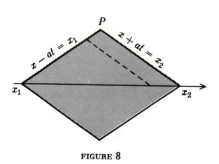

FIGURE 8

Both the initial displacement and the initial velocity have to be specified in Theorem I, just as one would expect intuitively. It is a remarkable fact that the *displacement alone* (without the velocity) will determine the solution, provided this displacement is given along two intersecting characteristics in the xt plane. Indeed, let $u(x,t)$ be given along (x_1,P) in Fig. 8. Since the direct wave $f_1(x - at)$ is constant on (x_1,P), we can ascertain the shape of the reverse wave $f_2(x + at)$ along (x_1,P).

This, in turn, gives $f_2(x + at)$ along (x_1, x_2), because the disturbance $f_2(x + at)$ propagates from (x_1, x_2) to (x_1, P) along the characteristics parallel to (x_2, P) (see the dashed line in the figure). In just the same way, when $u(x, t)$ is given on (x_2, P), we can determine the shape of the direct wave $f_1(x - at)$ on (x_1, x_2). Thus, we are led to the following theorem:

THEOREM II. Let u be specified along the two intersecting characteristics (x_1, P) and (x_2, P) in Fig. 8, and suppose that $u_{tt} = a^2 u_{xx}$. Then $u(x, t)$ is uniquely determined in the shaded region but is not uniquely determined at any other point.

Theorems I and II are the fundamental existence and uniqueness theorem for the wave equation, deduced here by physical considerations. A mathematical proof of the same results is given in Sec. 25.

In conclusion, we remark that there is an interesting connection between the theory of characteristics and the *operational solution* of differential equations. This aspect of the subject is not essential for understanding of the sequel but is developed briefly in the following problems.

--- **PROBLEMS**

1. Read the discussion of operator notation in Chap. 3, Sec. 11.

2. Operators $D_x{}^n$, $D_y{}^n$, $D_t{}^n$ are defined by

$$D_x{}^n \equiv \frac{\partial^n}{\partial x^n} \qquad D_y{}^n \equiv \frac{\partial^n}{\partial y^n} \qquad D_t{}^n \equiv \frac{\partial^n}{\partial t^n}$$

and we agree also, for example, that

$$(\alpha D_x + \beta D_y)u \equiv \alpha D_x u + \beta D_y u \equiv \alpha u_x + \beta u_y.$$

Develop the algebraic properties of these operators, after the manner of Chap. 3, Sec. 11.

3. (a) Show that $(\alpha D_x + \beta D_y)u = 0$ has a solution $u = f(\alpha y - \beta x)$, if α and β are constant and f is differentiable. (b) The lines $\alpha y - \beta x = $ const are the *characteristics* for the equation $(\alpha D_x + \beta D_y)u = 0$. Show that the characteristics satisfy $\alpha\,dy - \beta\,dx = 0$.

4. Discuss the relation between $(D_t{}^2 - a^2 D_x{}^2)u = 0$ and $dx^2 - a^2\,dt^2 = 0$.

5. (a) If m_i are constant, use the result of Prob. 3 to solve each of the equations

$$(D_x - m_1 D_y)u = 0 \qquad (D_x - m_2 D_y)u = 0.$$

(b) Show that both solutions obtained in part a satisfy

$$(D_x - m_1 D_y)(D_x - m_2 D_y)u = 0. \qquad (4\text{-}3)$$

Hint: Since m_i are constant, Eq. (4-3) may also be written

$$(D_x - m_2 D_y)(D_x - m_1 D_y)u = 0.$$

(c) Deduce that a solution of (4-3), containing two arbitrary functions, is

$$u = F_1(y + m_1 x) + F_2(y + m_2 x) \qquad m_1 \neq m_2$$
$$u = F_1(y + m_1 x) + x F_2(y + m_1 x) \qquad m_1 = m_2.$$

The lines $y + m_1 x = $ const and $y + m_2 x = $ const are characteristics. *Hint:* Since the equation is linear and homogeneous, the sum of the two solutions in part b is again a solution. The result for $m_1 = m_2$ may be verified by direct substitution.

[Similar results hold in general. The solution of

$$(D_x - m D_y)^r u = 0$$

can be shown to be

$$u = F_1(y + mx) + x F_2(y + mx) + \cdots + x^{r-1} F_r(y + mx)$$

and the solution for several such factors is obtained by addition. The process gives the "general solution" in that the number of arbitrary functions equals the order of the equation.]

6. Solve the wave equation by the method of Prob. 5. *Hint:* $(D_t + aD_x)(D_t - aD_x)u = 0$.

7. The fourth-order equation

$$\frac{\partial^4 u}{\partial x^4} + 2\frac{\partial^4 u}{\partial x^2\, \partial y^2} + \frac{\partial^4 u}{\partial y^4} = 0$$

occurs in the study of elastic plates. Show that the general solution is

$$u = F_1(y - ix) + xF_2(y - ix) + F_3(y + ix) + xF_4(y + ix).$$

Hint: The equation may be written $(D_x{}^4 + 2D_x{}^2D_y{}^2 + D_y{}^4)u = 0$ so that the decomposition into linear factors gives

$$(D_x + iD_y)(D_x + iD_y)(D_x - iD_y)(D_x - iD_y)u = 0.$$

8. As in Prob. 5, solve $u_{xx} + u_{yy} = 0$, $u_{xx} + u_{xy} = 2u_{yy}$, $u_{xx} + u_{yy} = 2u_{xy}$.

9. Consider the equation $u_{xx} + 4u_{xy} - 5u_{yy} = f(x,y)$.

(a) By the method of Prob. 5, obtain a general solution when $f = 0$.
(b) By assuming $u = cy^4$, where c is a constant to be determined, obtain a solution when $f = y^2$.
(c) Similarly, obtain a particular solution when $f = x$.
(d) By addition of the results (a), (b), and (c), obtain a general solution when $f = y^2 + x$.

10. As in Prob. 9, obtain a general solution:

$$2z_{xx} + z_{xy} - z_{yy} = 1 \qquad z_{xx} - a^2z_{yy} = x^2 \qquad z_{xx} + 3z_{xy} + 2z_{yy} = x + y.$$

---

5. Boundary Conditions.   We now suppose that the freely vibrating string is not infinite but is stretched between two points of support (Fig. 3). When the supports are on the x axis and do not move, the situation is described by

$$u(0,t) = 0 \qquad u(l,t) = 0 \qquad \text{for all } t. \tag{5-1}$$

These are called *boundary conditions*, because they refer to the boundary points of the interval $(0,l)$ in which our physical problem is defined. Although the boundary conditions obviously do not determine the motion uniquely, they do enable us to establish some of the most interesting and important properties of the motion.

Physically, one would expect the string to act like an infinite string until the disturbance created by the ends reaches the point of observation. In terms of Fig. 7, the ends $x = 0$ and $x = l$ have no effect in the region I provided the points $x = 0$ and $x = l$ lie outside the interval (x_1,x_2). When the disturbance reaches an end point, however, it is reflected, and the reflected wave must eventually be taken into account.

Because the end point is fixed the incident and reflected waves have the algebraic sum 0 at the end point, and hence there is a 180° phase shift; that is, a wave of the type $f_2(x + at)$ becomes a wave of the type $-f_2(-x + at)$ upon reflection at $x = 0$ (see Fig. 9). The change of sign in f_2 expresses the phase shift, and the change of sign in x indicates that the reflected wave

$$g(x - at) \equiv -f_2(-x + at)$$

propagates in the opposite direction.

When the wave is reflected again at $x = l$, we get another minus sign in each case, and hence the original wave $f_2(x + at)$ is restored (Fig. 9). Since the velocity is a and the length of the round-trip path is $2l$, the time for a round trip is

$$\text{period of vibration} = \frac{2l}{a}. \tag{5-2}$$

In terms of $f_2(at)$ the periodicity condition means that

$$f_2(at) = f_2\left[a\left(t + \frac{2l}{a}\right)\right] = f_2(at + 2l).$$

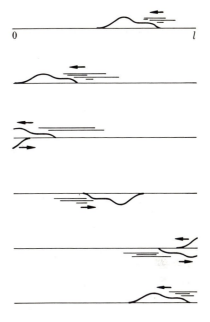

Similar remarks apply to $f_1(x)$, and hence we expect that both $f_1(x)$ and $f_2(x)$ will be periodic functions[1] with period $2l$.

To discuss the boundary conditions mathematically, let us think of the finite string as being in reality an infinite string which vibrates in such a way that the points $x = 0$ and $x = l$ remain fixed. The formula

$$u(x,t) = f_1(x - at) + f_2(x + at) \qquad (5\text{-}3)$$

holds for all solutions of the wave equation. Letting $x = 0$ gives

$$0 = f_1(-at) + f_2(at)$$

when we use the first boundary condition $u(0,t) = 0$. This shows that $f_2(s) = -f_1(-s)$ for all s, and hence (5-3) becomes

$$u(x,t) = f_1(x - at) - f_1(-x - at). \qquad (5\text{-}4)$$

FIGURE 9

Thus the effect of the boundary condition at $x = 0$ is to reduce the number of arbitrary functions from two to one.

The second boundary condition applied to (5-4) gives

$$0 = f_1(l - at) - f_1(-l - at)$$

or, if we set $s = -l - at$,

$$0 = f_1(s + 2l) - f_1(s). \qquad (5\text{-}5)$$

Since t is arbitrary, so is s, and hence $f_1(x)$ has period $2l$. (This agrees with the surmise we had formed on physical grounds.) In view of (5-4), we can summarize our result as follows:

THEOREM I. *Suppose an infinite string vibrates freely in such a way that the points $x = 0$ and $x = l$ remain fixed. Then the displacement $u(x,t)$ is periodic both in space and in time. The two periods are, respectively, $2l$ for x and $2l/a$ for t if a is the velocity of propagation.*

Hence if a string is stretched between two fixed points, the free vibrations are periodic no matter what the initial conditions may be on $(0,l)$. Since a periodic vibration is generally perceived as musical, this fact is of great importance for the development of musical instruments.

Theorem I asserts that the motion will repeat after a time $2l/a$. Hence if the minimum t period of a vibrating string is determined by observation, that minimum period will not be longer than $2l/a$. It may be shorter, however. For instance, the function

$$u(x,t) = \sin\frac{2\pi x}{l}\cos\frac{2\pi at}{l}$$

[1] A function $f(x)$ has period p if $f(x + p) \equiv f(x)$, where p is a nonzero constant.

satisfies (5-1) and the wave equation and hence represents free vibrations of a string of length l. But the minimum t period of this function is l/a rather than $2l/a$. The shorter period is explained by the fact that $u(\tfrac{1}{2}l,t) \equiv 0$; that is, the center of the string is a node. The center does not move, and the string acts like two strings of length $l/2$ placed end to end. We shall now show that there is always at least one node if the period is smaller than that given by Theorem I.

THEOREM II. If the string considered in Theorem I has an x period $2p$ or t period $2p/a$, where $0 < p < l$, then the point $x = p$ must be a node.

Suppose, first, that the x period is $2p$. Then, in particular,

$$u(p,t) = u(-p,t). \tag{5-6}$$

On the other hand, (5-4) gives $u(x,t) = -u(-x,t)$; hence

$$u(p,t) = -u(-p,t). \tag{5-7}$$

By addition of (5-6) and (5-7) we get $u(p,t) = 0$, which shows that $x = p$ is a node.

Suppose, next, that the t period is $2p/a$. The equation

$$u\left(x, t + \frac{2p}{a}\right) = u(x,t)$$

combines with (5-4) to give

$$f_1(x - at - 2p) - f_1(-x - at - 2p) = f_1(x - at) - f_1(-x - at).$$

If we let $x + at = 0$ and $x - at - 2p = s$, the equation reduces, after rearrangement, to

$$f_1(s + 2p) - f_1(s) = c \tag{5-8}$$

where $c = f_1(0) - f_1(-2p)$ is constant. Equation (5-8) shows that $f_1(s)$ increases by the amount c whenever s increases by $2p$. If $c \neq 0$, it follows that $|f_1(s)|$ is unbounded. However, $f_1(s)$ has period $2l$ by (5-5), hence is bounded, and this shows that $c = 0$ in (5-8). The choice $s = -p - at$ in (5-8) with $c = 0$ leads to the desired result:

$$u(p,t) = f_1(p - at) - f_1(-p - at) = 0.$$

To illustrate the use of Theorem II, suppose a 2-in.-diameter steel cable 100 ft long is observed to vibrate without nodes at the rate of two complete cycles per second. According to Theorem I the t period is $2l/a$ or possibly less. But Theorem II shows that the period is not less, since the motion was observed to have no fixed points. Hence

$$\frac{1}{2} = 2\frac{100}{a}$$

which gives $a = 400$ fps. This is the velocity with which waves are propagated along the cable. Since the density of steel is about 480 lb per ft³, the weight of 1 ft of cable is

$$480\pi(\tfrac{1}{12})^2 1 = (\tfrac{10}{3})\pi \doteq 10 \text{ lb per ft.}$$

This gives $\rho = \tfrac{10}{32}$ slug for the linear density, and hence the tension is

$$T = a^2\rho = (400)^2(\tfrac{10}{32}) = 50{,}000 \text{ lb.}$$

PROBLEMS

1. An infinite string vibrates freely in such a way that the two points $x = 0$ and $x = l$ remain fixed; that is, $u(0,t) = u(l,t) = 0$. Are any other points of the string necessarily fixed? Which ones?

2. Suppose a freely vibrating string of length l has just one node at $x = p$ between 0 and l. Show that the node must be at the midpoint. (The analogous result for n nodes is also true.) *Hint:*

If $p < l/2$, apply Prob. 1 to the two points $x = 0$, $x = p$. If $p > l/2$, apply Prob. 1 to the two points $x = p$, $x = l$.

3. A cable of length l ft is made of a material with density d lb per ft^3. It is found that the cable makes 10 complete oscillations in t sec. Show that the cross-sectional stress is

$$\sigma = 0.087d \left(\frac{l}{t}\right)^2 \qquad \text{psi}$$

provided the oscillations do not have a node between 0 and l. How would the result change if the midpoint remains fixed during the observed oscillations but no other point remains fixed?

4. Let $h(t)$ be a given function of t. (a) What is the physical meaning of the boundary condition $u(l,t) = h(t)$? (b) Describe a physical problem that would lead to the boundary conditions $u(x,0) = 0$, $u[x,h(t)] = 0$.

---

6. Initial and Boundary Conditions.

We shall now consider the free vibrations of a string satisfying the boundary conditions

$$u(0,t) = 0 \qquad u(l,t) = 0 \qquad \text{for all } t \tag{6-1}$$

together with the initial conditions

$$u(x,0) = f(x) \qquad u_t(x,0) = g(x) \qquad \text{for } 0 < x < l. \tag{6-2}$$

As in the preceding section we regard the finite string as being an infinite string with nodes $x = 0$, $x = l$. According to (5-4) and (5-5), the boundary conditions give

$$u(x,t) = f_1(x - at) - f_1(-x - at) \tag{6-3}$$

where $f_1(x)$ has period $2l$, and, conversely, (6-3) ensures (6-1). The initial conditions (6-2) are prescribed on $(0,l)$ for the infinite string, and our task is to assign initial conditions outside the interval $(0,l)$ in such a way that the solution has the form (6-3).

Denoting the unknown initial conditions for the infinite string by $f_0(x)$ and $g_0(x)$, we have

$$f_0(x) = f(x) \qquad g_0(x) = g(x) \qquad 0 < x < l \tag{6-4}$$

because the infinite string is to agree with the finite string on $(0,l)$. Upon setting $t = 0$ in (6-3) we get

$$f_0(x) = f_1(x) - f_1(-x).$$

Similarly, differentiating (6-3) with respect to t and putting $t = 0$ give

$$g_0(x) = -af_1'(x) + af_1'(-x).$$

These expressions show that[1]

$$f_0(x) \text{ and } g_0(x) \text{ are odd functions.} \tag{6-5}$$

Hence, f_0 and g_0 are determined on $(-l,l)$ by their values on $(0,l)$.

Finally, since $f_0(x)$ and $g_0(x)$ are expressed in terms of the function $f_1(x)$, which has period $2l$, we see that

$$f_0(x) \text{ and } g_0(x) \text{ have period } 2l. \tag{6-6}$$

Thus, f_0 and g_0 are known everywhere as soon as they are known on $(-l,l)$.

According to (3-15), the solution is

$$u(x,t) = \tfrac{1}{2}f_0(x - at) + \tfrac{1}{2}f_0(x + at) + \frac{1}{2a} \int_{x-at}^{x+at} g_0(s)\, ds. \tag{6-7}$$

---

[1] A function $\phi(x)$ is *even* if $\phi(-x) \equiv \phi(x)$ and *odd* if $\phi(-x) \equiv -\phi(x)$. An analytical and graphical discussion of such functions is given in Chap. 1, Sec. 20.

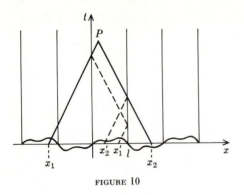

FIGURE 10

If $f_0(x)$ and $g_0(x)$ in (6-7) are determined by (6-4) to (6-6) and f_0'' and g_0' exist, it is easily verified that this function $u(x,t)$ satisfies the wave equation, the initial conditions (6-2), and the boundary conditions (6-1). Thus, (6-7) is a simple and explicit expression for the motion of a vibrating string with fixed end points. It may be noted that (6-7) is meaningful as a generalized solution under the hypothesis that f_0 is defined and g_0 is integrable.

The correspondence between the finite string and the infinite string leads to an interesting geometric construction for getting the disturbance at any point P of the strip $0 < x < l$ in the xt plane (Fig. 10). For the infinite string the disturbance at P is found by drawing characteristics as in Sec. 4 (see solid lines in Fig. 10). Since the initial conditions for the infinite string are obtained from those for the finite string by (6-4) to (6-6), the same result may be found by following the dashed lines in Fig. 10. To take account of (6-5), however, we must introduce a changed sign upon each reflection at the boundary. The disturbance at P arises from the initial disturbance at x_1' and x_2', subject to the above-mentioned convention regarding sign. This reflection of the characteristics in the boundary lines $x = 0$, $x = l$ is analogous to the reflection of waves at the end points of the string.

The procedure illustrated in Fig. 10 is an example of the *method of images*, so called because the initial conditions for the infinite string are obtained from those for the finite string by forming repeated mirror images in the lines $t = 0$, $x = 0$, and $x = l$.

Example.   Discuss the free oscillations of a string of length l which satisfies the initial conditions

$$u(x,0) = f_n \sin \frac{n\pi x}{l} \qquad u_t(x,0) = 0$$

where n is an integer and f_n is constant.

Since $\sin n\pi x/l$ is odd and has period $2l$, we may take

$$f_0(x) = f_n \sin \frac{n\pi x}{l} \qquad g_0(x) = 0$$

as initial conditions for the associated infinite string. Equation (6-7) now yields the solution

$$u(x,t) = \frac{f_n}{2} \sin \frac{n\pi(x - at)}{l} + \frac{f_n}{2} \sin \frac{n\pi(x + at)}{l} = f_n \sin \frac{n\pi x}{l} \cos \frac{n\pi at}{l}. \qquad (6\text{-}8)$$

If the initial displacement is given by a Fourier series, so that[1]

$$u(x,0) = f(x) = \Sigma f_n \sin \frac{n\pi x}{l} \qquad u_t(x,0) = 0$$

superposition of the corresponding solutions (6-8) suggests that

$$u(x,t) = \Sigma f_n \sin \frac{n\pi x}{l} \cos \frac{n\pi at}{l}. \qquad (6\text{-}9)$$

It is possible to show that this is a smooth solution under the hypothesis leading to (6-7) as a smooth solution.[2]

---

[1] Throughout Chap. 7 we use Σ as an abbreviation for $\displaystyle\sum_{n=1}^{\infty}$.

[2] This is true because the function defined by the series coincides with that given by d'Alembert's method. It may be mentioned that the conditions for (6-7) to be a *generalized solution* are too weak to allow the use of Fourier series.

By choosing $f_0(x) = 0$, $g_0(x) = g_n \sin n\pi x/l$ in (6-7), the solution satisfying

$$u(x,0) = 0 \qquad u_t(x,0) = g(x) = \Sigma g_n \sin \frac{n\pi x}{l}$$

is found to be

$$u(x,t) = \Sigma \frac{g_n l}{n\pi a} \sin \frac{n\pi x}{l} \sin \frac{n\pi at}{l}. \tag{6-10}$$

Superposition of (6-9) and (6-10) yields the general *Fourier-series solution* of the wave equation satisfying (6-1) and (6-2). The result can be expressed explicitly in terms of $f(x)$ and $g(x)$ by means of the Euler-Fourier formulas

$$f_n = \frac{2}{l} \int_0^l f(x) \sin \frac{n\pi x}{l} \, dx \qquad g_n = \frac{2}{l} \int_0^l g(x) \sin \frac{n\pi x}{l} \, dx.$$

Because of convergence questions the Fourier-series solution is harder to verify than (6-7), and it is hopelessly inferior to (6-7) for numerical computation. But Fourier series have great usefulness in that they apply to many problems in which the preceding methods fail. Examples are given for the vibrating string in the next section and for other physical systems in the sections to follow.

——————————————————————————————————— **PROBLEMS**

1. Review the discussion of sigma notation in Chap. 1, Sec. 2, and do Probs. 1 and 2 of that section. Also review the discussion of Fourier series in Chap. 1, Secs. 18, 20, 21, and 26.

2. Show that the expression (6-8) satisfies the appropriate (a) differential equation, (b) initial conditions, and (c) boundary conditions.

3. The initial displacement of a freely vibrating string of length l is

$$f(x) = \frac{2bx}{l} \qquad 0 < x < \tfrac{1}{2}l; \qquad f(x) = 2b - \frac{2bx}{l} \qquad \tfrac{1}{2}l < x < l$$

and the initial velocity is $g(x) = 0$. (a) Sketch $f(x)$ and $f_0(x)$. (b) Using (6-7) and your sketch, find the displacement of the midpoint of the string when $at = l/4$.

4. (a) Express $f(x)$ in Prob. 3 as a Fourier sine series. (b) By (a) and (6-9) show that the displacement of the string in Prob. 3 is

$$u(x,t) = \frac{8b}{\pi^2} \left(\frac{1}{1^2} \sin \frac{\pi x}{l} \cos \frac{\pi at}{l} - \frac{1}{3^2} \sin \frac{3\pi x}{l} \cos \frac{3\pi at}{l} + \cdots \right).$$

(c) Obtain an infinite-series representation for the displacement of the midpoint when $at = l/4$.

5. By d'Alembert's method, find the displacement $u(x,t)$ of a freely vibrating string that satisfies the initial conditions $u(x,0) = f(x)$, $u_t(x,0) = 0$, $x > 0$, and the boundary condition $u(0,t) = h(t)$. Interpret physically. [Cf. discussion following (5-3).]

——————————————————————————————————

7. Damped Oscillations. Forced Oscillations and Resonance. The foregoing discussion was concerned with free vibrations, so that $F(x,t) = 0$ in (2-3). It was indicated that the displacement $u(x,t)$ is always periodic in time and hence the amplitude remains constant. But, in fact, the oscillations gradually die down when a string is vibrating in air, and this behavior is to be analyzed next.

The reason for the decrease in amplitude is that the air resists the motion of an object moving through it. When there is no relative velocity, there is no resistance; when there is high velocity, there is high resistance. If the resistance is assumed proportional to the velocity, we have

$$F(x,t) = -2bu_t(x,t) \qquad b > 0, \text{ const} \tag{7-1}$$

in (2-3).   The minus sign is used because the force resists the motion and so is directed opposite to the velocity.   Our partial differential equation is now

$$u_{tt} - a^2 u_{xx} = -2bu_t \tag{7-2}$$

and the solutions of (7-2) for $b > 0$ represent the damped oscillations of the string. As before, one has the initial and boundary conditions

$$u(x,0) = f(x) \qquad u_t(x,0) = g(x) \tag{7-3}$$

$$u(0,t) = 0 \qquad u(l,t) = 0. \tag{7-4}$$

Equation (7-2) cannot be solved by the method of the preceding sections but can be solved by Fourier series.   Thus, since the solution $u(x,t)$ is a twice-differentiable function of x for each t, we may expand $u(x,t)$ in a Fourier sine series

$$u(x,t) = \sum b_n(t) \sin \frac{n\pi x}{l} \qquad 0 < x < l. \tag{7-5}$$

A sine series is chosen rather than a cosine series because such a series automatically satisfies the boundary conditions (7-4).   To satisfy the initial conditions we require

$$f(x) = \sum b_n(0) \sin \frac{n\pi x}{l} \qquad g(x) = \sum b_n'(0) \sin \frac{n\pi x}{l}. \tag{7-6}$$

These relations show that $b_n(0)$ and $b_n'(0)$ must be the Fourier coefficients of $f(x)$ and $g(x)$.   That is, if

$$f_n = \frac{2}{l} \int_0^l f(x) \sin \frac{n\pi x}{l}\, dx \qquad g_n = \frac{2}{l} \int_0^l g(x) \sin \frac{n\pi x}{l}\, dx \tag{7-7}$$

then multiplying (7-6) by $\sin n\pi x/l$ and integrating from 0 to l yield

$$b_n(0) = f_n \qquad b_n'(0) = g_n. \tag{7-8}$$

We must still satisfy the differential equation.   Upon formally substituting the series (7-5) into (7-2) we get

$$\sum b_n'' \sin \frac{n\pi x}{l} + a^2 \sum b_n \left(\frac{n\pi}{l}\right)^2 \sin \frac{n\pi x}{l} = -2b \sum b_n' \sin \frac{n\pi x}{l}$$

which gives a set of ordinary differential equations

$$b_n'' + 2bb_n' + \left(\frac{n\pi a}{l}\right)^2 b_n = 0 \tag{7-9}$$

when the coefficients of $\sin n\pi x/l$ on the left and right are equated.

In Chap. 3, Sec. 5, the equation

$$y'' + 2by' + a^2 y = 0$$

is solved subject to initial conditions of the type (7-8).   Making appropriate changes of notation in that solution we obtain $b_n(t)$, and substitution into (7-5) yields the final answer,

$$u(x,t) = e^{-bt} \sum \left[f_n \cos \omega_n t + (g_n + bf_n) \frac{\sin \omega_n t}{\omega_n} \right] \sin \frac{n\pi x}{l}$$

where

$$(\omega_n l)^2 = (n\pi a)^2 - b^2 l^2. \tag{7-10}$$

Conditions (7-2) to (7-4) are satisfied if the term-by-term differentiation is legitimate (cf. Chap. 1, Sec. 26, Theorem III).   When $b = 0$, the solution agrees with the sum of (6-9) and (6-10), as it should.   According to (7-10), the damping reduces the frequency of the corresponding terms in the series for undamped vibrations.   If $b < \pi a/l$, all the terms are oscillatory, and they have the same damping factor e^{-bt}.   But for larger values of b the first few terms may have ω_n pure imaginary. The corresponding trigonometric functions become hyperbolic functions, and the terms in question are not oscillatory. If $\omega_n = 0$, which may happen in this latter case, we replace $(\sin \omega_n t)/\omega_n$ by its limit t.

Sometimes the force function $F(x,t)$ does not involve the unknown displacement u, as in (7-1), but is determined independently. If the string is initially at rest, the corresponding mathematical problem is

$$u_{tt} - a^2 u_{xx} = F(x,t) \qquad u(x,0) = u_t(x,0) = 0 \qquad u(0,t) = u(l,t) = 0. \quad (7\text{-}11)$$

The general solution of this system by means of Fourier series is developed in the problems following Sec. 13. Just now we prefer to illustrate the important phenomenon of *resonance*. For this purpose, it is sufficient to take

$$F(x,t) = (\alpha \sin \omega t + \beta \cos \omega t) \sin \frac{n\pi x}{l} \qquad (7\text{-}12)$$

where α, β, and ω are constant.

Substitution of $F(x,t)$ and the Fourier series (7-5) into the differential equation gives

$$\sum b_n'' \sin \frac{n\pi x}{l} + \sum \omega_n^2 b_n \sin \frac{n\pi x}{l} = (\alpha \sin \omega t + \beta \cos \omega t) \sin \frac{n\pi x}{l}$$

where $\omega_n = n\pi a/l$. [Compare (7-10), with $b = 0$.] If we equate coefficients of $\sin k\pi x/l$ on the left and right we get, for $k = n$,

$$b_n'' + \omega_n^2 b_n = \alpha \sin \omega t + \beta \cos \omega t. \qquad (7\text{-}13)$$

This equation is to be solved subject to the initial conditions

$$b_n(0) = b_n'(0) = 0$$

which result from the initial conditions in (7-11). For $k \neq n$ the right-hand member of the equation for b_k is 0, and hence $b_k = 0$.

If $\omega^2 \neq \omega_n^2$, the solutions of (7-13) are all bounded, but if $\omega = \omega_n$, the particular integral involves the functions $t \sin \omega t$, $t \cos \omega t$ which increase indefinitely with t. Hence in that case the term

$$b_n(t) \sin \frac{n\pi x}{l} \qquad (7\text{-}14)$$

in the Fourier series for $u(x,t)$ becomes strongly emphasized as t increases, and we say, briefly, that the oscillation (7-14) is *resonant*.

A physical explanation is readily given in terms of the results of Sec. 5. Thus, the condition $\omega = \omega_n$ can be written as

$$\frac{2\pi}{\omega} = \frac{2l}{na}.$$

This asserts that the period of $F(x,t)$ in (7-12) is equal to the period for free oscillations of a string of length l/n. And l/n is precisely the distance between nodes for a vibration of the type (7-14).

The common feature of the problems considered here is that $F \neq 0$ and the solution involves Fourier series. We conclude with an example that shows how the case $F \neq 0$ can be attacked by d'Alembert's method (Sec. 3).

Example. A cord stretched between the fixed points $x = 0$ and $x = l$ is initially supported so that it forms a horizontal straight line. Discuss the oscillations when the support is suddenly removed.

The force function $F_1(x,t)$ in Sec. 2 is $-g\rho$, and hence the partial differential equation is

$$u_{tt} - a^2 u_{xx} = -g \qquad (7\text{-}15)$$

while the boundary and initial conditions are

$$u(0,t) = u(l,t) = 0 \qquad u(x,0) = u_t(x,0) = 0. \qquad (7\text{-}16)$$

If we succeed in finding a particular solution $u = v(x)$ of (7-15) which satisfies the boundary conditions

$$v(0) = v(l) = 0 \qquad (7\text{-}17)$$

the solution can be written as

$$u(x,t) = w(x,t) + v(x) \qquad (7\text{-}18)$$

where, as follows from (7-15) and (7-16), $w(x,t)$ satisfies

$$w_{tt} - a^2 w_{xx} = 0 \qquad w(0,t) = 0 \qquad w(l,t) = 0 \qquad w(x,0) = -v(x) \qquad w_t(x,0) = 0. \quad (7\text{-}19)$$

Since the desired particular solution $v(x)$ is to be independent of t, the choice $u(x,t) = v(x)$ in (7-15) yields $a^2 v'' = g$, so that

$$v(x) = -\frac{g}{2a^2} x(l - x) \qquad (7\text{-}20)$$

when the integration constants are determined so as to satisfy (7-17). This particular solution corresponds to the equilibrium position of the string under gravity. The value of w can now be written with the aid of (6-7) as

$$w(x,t) = \tfrac{1}{2} f_0(x - at) + \tfrac{1}{2} f_0(x + at)$$

where $f_0(x)$ is odd, has period $2l$, and is defined for $0 < x < l$ by

$$f_0(x) = -v(x) = \frac{g}{2a^2} x(l - x).$$

The required solution is $u = v + w$.

By interpreting $f_0(x)$, $f_0(x - at)$, and $f_0(x + at)$ graphically one finds that $w(x,t)$ is largest on $0 < x < l$ when $at = 0, 2l, 4l, \ldots$ and then $w(x,t) = f_0(x)$. Similarly, $w(x,t)$ is least when $at = l, 3l, 5l, \ldots$ and then $w(x,t) = -f_0(x)$. It follows that the cord oscillates between the horizontal position $u = 0$ and the position $u = 2v(x)$ in which each point is twice as low as the equilibrium position (7-20). The period is $2l/a$.

PROBLEMS

1. A string of length l vibrating in air satisfies the initial conditions $u(x,0) = f_1 \sin \pi x/l$, $u_t(x,0) = 0$. Show that the displacement of the midpoint can be written in the form

$$u(\tfrac{1}{2}l,t) = Ae^{-bt} \cos (\omega t + \phi) \qquad A, \omega, \phi \text{ const.}$$

2. Referring to Prob. 1, sketch the curves $y = \pm Ae^{-bt}$ and $y = u(\tfrac{1}{2}l,t)$ in a single neat diagram. Thus describe an experimental procedure for determining b. (When the oscillations are rapid and b is small, one can speak of the mean amplitude at a given time t. If the amplitude is A_0 at time t_0 and A_1 at time $t_0 + \tau$, the reader can verify that

$$b = \tau^{-1} \log \frac{A_0}{A_1}.$$

Since A_0 and A_1 can be found by placing a scale behind the oscillating string, this gives a method for comparing the viscosity of gases.)

3. A horizontal cable 100 ft long sags 5 ft when at rest under gravity. If the cable is disturbed so that it oscillates without nodes, what is the frequency of the oscillations? *Hint:* See (7-20).

4. A string of length l is subjected to a force $F(x,t) = \sin \omega t \sin \pi x/l$, where ω is constant. Find the displacement $u(x,t)$ if the string was initially at rest in the equilibrium position.

5. Show that the equilibrium shape of a string under a force $F(x)$ is described by

$$u = \frac{x}{a^2 l} \int_0^l \int_0^\xi F(s) \; ds \; d\xi - \frac{1}{a^2} \int_0^x \int_0^\xi F(s) \; ds \; d\xi.$$

6. Show that the function

$$v(x,t) = -\frac{1}{4a^2} \int_0^{x+at} \int_0^{x-at} F\left[\frac{1}{2} (r+s), \frac{1}{2a} (r-s)\right] ds \; dr$$

satisfies $v_{tt} - a^2 v_{xx} = F(x,t)$. *Hint:* Let $x + at = r$, $x - at = s$, $v(x,t) = V(r,s)$. Then, as in Sec. 1, $-4a^2 V_{rs} = F(x,t)$.

7. If $v(x,t)$ is the function obtained in Prob. 6, let $w(x,t)$ be determined by

$$w_{tt} - a^2 w_{xx} = 0 \qquad w(x,0) = f(x) - v(x,0) \qquad w_t(x,0) = g(x) - v_t(x,0)$$

(cf. Sec. 3). Then $u = v + w$ satisfies

$$u_{tt} - a^2 u_{xx} = F(x,t) \qquad u(x,0) = f(x) \qquad u_t(x,0) = g(x).$$

8. Solve the example in the text by means of Fourier series.

SOLUTION BY SERIES

8. Heat Flow in One Dimension. The foregoing discussion of the vibrating string enabled us to survey some typical problems in the field of partial differential equations and to illustrate a number of important methods. Among these is the method of infinite series, which will now be explored more fully and used in a variety of applications. We begin with a problem from the theory of heat conduction.

Consider a section cut from an insulated, uniform bar by two parallel planes Δx units apart (Fig. 11), and suppose that the temperature of one of the planes is u while that of the second plane is $u + \Delta u$. It is known from experiment that heat flows from the plane at higher temperature to that at the lower, the amount of heat flowing per unit area per second being approximately

$$\text{rate of flow} \doteq -k \frac{\Delta u}{\Delta x}. \tag{8-1}$$

Here k is a constant called the *thermal conductivity* of the material; its dimensions in the cgs system are cal/(cm-sec °C). In the limit as $\Delta x \to 0$, Eq. (8-1) can be regarded as an exact equality, so that

$$\text{rate of flow} = -k u_x. \tag{8-2}$$

On the other hand, if c is the heat capacity of the medium and ρ its density, the amount of heat in the section from x to $x + \Delta x$ is

$$(c\rho A \; \Delta x)\bar{u} \tag{8-3}$$

where A is the cross-sectional area and where $\bar{u}$ is the mean value of u over the interval $(x, x + \Delta x)$. For a time interval $(t, t + \Delta t)$ the increase of heat in the section $(x, x + \Delta x)$ can be computed from (8-3) and also from (8-2).

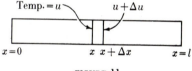

FIGURE 11

The computation yields[1]

$$c\rho A\, \Delta x\, \bar{u}(x, t + \Delta t) - c\rho A\, \Delta x\, \bar{u}(x,t) = kA\, \Delta t\, \overline{u_x}(x + \Delta x,\, t) - kA\, \Delta t\, \overline{u_x}(x,t)$$

where $\overline{u_x}$ is the mean value of u_x in the time interval $(t,\, t + \Delta t)$. Dividing by $c\rho A\, \Delta x\, \Delta t$ we obtain

$$\frac{\bar{u}(x,\, t + \Delta t) - \bar{u}(x,t)}{\Delta t} = \frac{k}{c\rho} \frac{\overline{u_x}(x + \Delta x,\, t) - \overline{u_x}(x,t)}{\Delta x} \tag{8-4}$$

and letting $\Delta x \to 0$, $\Delta t \to 0$ now gives

$$u_t = \alpha^2 u_{xx} \qquad \alpha^2 = \frac{k}{c\rho} \tag{8-5}$$

if we recall the definition of partial derivative. The fact that (8-4) involves mean values causes no trouble when u_t and u_{xx} are continuous; see Prob. 8.

We shall now solve (8-5) under the assumption that the initial temperature is a prescribed function $f(x)$,

$$u(x,0) = f(x) \qquad 0 < x < l \tag{8-6}$$

which can be represented by a convergent Fourier series. The ends of the bar are assumed to have the temperature zero:

$$u(0,t) = u(l,t) = 0 \qquad t \geq 0. \tag{8-7}$$

Since u_{xx} must exist if u satisfies (8-5), we know that $u(x,t)$ has a Fourier series in x for each fixed $t > 0$:

$$u(x,t) = \sum b_n(t) \sin \frac{n\pi x}{l}. \tag{8-8}$$

Here, a sine series is chosen because such a series automatically satisfies the requirement (8-7). Proceeding tentatively, assume that (8-8) can be differentiated term by term to give

$$\sum b_n'(t) \sin \frac{n\pi x}{l} = \alpha^2 \sum b_n(t) \left(\frac{n\pi}{l}\right)^2 \left(-\sin \frac{n\pi x}{l}\right) \tag{8-9}$$

upon substitution into (8-5). Equation (8-9) is satisfied if the coefficients of $\sin n\pi x/l$ on each side are equated:

$$b_n' = -\left(\frac{\alpha n\pi}{l}\right)^2 b_n.$$

Upon integration this gives

$$b_n(t) = c_n e^{-(\alpha n\pi/l)^2 t}$$

where the c_n are constant, and hence (8-8) becomes

$$u(x,t) = \sum c_n e^{-(\alpha n\pi/l)^2 t} \sin \frac{n\pi x}{l}. \tag{8-10}$$

The initial condition (8-6) yields

$$f(x) = u(x,0) = \sum c_n \sin \frac{n\pi x}{l}. \tag{8-11}$$

Since the Fourier series for $f(x)$ converges to $f(x)$ by hypothesis, Eq. (8-11) is assured if c_n are the Fourier coefficients,

$$c_n = \frac{2}{l} \int_0^l f(x) \sin \frac{n\pi x}{l}\, dx.$$

---

[1] It is supposed that no heat is generated within the material and that k, ρ, and c are constant over the relevant range of temperatures. If ρ is measured in grams per cubic centimeter, the dimensions of c are cal/(g °C).

The only questionable step in the foregoing discussion was the term-by-term differentiation, but this step can now be justified. Differentiating (8-10) term by term actually gives

$$u_{xx} = -\sum c_n e^{-(\alpha n\pi/l)^2 t}\left(\frac{n\pi}{l}\right)^2 \sin\frac{n\pi x}{l} \qquad u_t = -\sum c_n e^{-(\alpha n\pi/l)^2 t}\left(\frac{\alpha n\pi}{l}\right)^2 \sin\frac{n\pi x}{l} \qquad (8\text{-}12)$$

because the series (8-12) are *uniformly convergent* when $t \geq \delta > 0$. (See Chap. 1, Sec. 7, Theorem IV. The uniform convergence follows from the convergence of

$$\Sigma n^2 e^{-(\alpha n\pi/l)^2 \delta}$$

since the Fourier coefficients c_n are bounded.) Hence, (8-10) is a solution of the problem. We cannot yet say that (8-10) is *the* solution, because there might be another solution for which the term-by-term differentiation is not permissible. A uniqueness theorem is established, however, in Sec. 24.

Because of the exponential factors the series (8-10) is rapidly convergent and affords a useful means of computing the temperature. By contrast, the series obtained in Sec. 6 for solutions of the wave equation converges no better than the series for the initial values $f(x)$ and $g(x)$. The physical significance of this difference in the two cases is discussed in Sec. 27.

Example 1. Find the steady-state temperature of a uniform bar. It is required that $u(x,t)$ be independent of t, whence by (8-5)

$$\alpha^2 u_{xx} = u_t = 0.$$

Hence, $u = c_0 + c_1 x$, where c_0 and c_1 are constant. If the temperatures at the ends are u_0 and u_1, respectively, we can determine the constants and thus obtain the formula

$$u(x,t) = u_0 + \frac{x}{l}(u_1 - u_0). \qquad (8\text{-}13)$$

The rate of heat flow is given by (8-2) and (8-13) as

$$-k\frac{u_1 - u_0}{l} \qquad (8\text{-}14)$$

and, hence, (8-1) holds without approximation in the steady state.

Example 2. A rod of length 5 has the end $x = 0$ at $0°$, the end $x = 5$ at $10°$, and the initial temperature is $f(x)$. Find the temperature distribution.

If $v(x,t)$ is the unknown temperature at point x and time t, we let

$$u = v - 2x \qquad (8\text{-}15)$$

where $2x$ is the steady-state temperature determined from (8-13). Then $\alpha^2 u_{xx} = u_t$, $u(x,0) = f(x) - 2x$, $u(0,t) = u(5,t) = 0$. Hence u is given by (8-10), where the c_n's are the Fourier coefficients of $f(x) - 2x$. When we have found u, Eq. (8-15) gives v.

We have noted that the value $2x$ introduced in (8-15) is the steady-state temperature as determined by Example 1. The same method enables us to replace any constant boundary conditions by the homogeneous conditions $u(0,t) = u(l,t) = 0$. That is, if the unknown temperature $v(x,t)$ satisfies

$$v(0,t) = v_0 \qquad v(l,t) = v_1 \qquad v_0 \text{ and } v_1 \text{ const} \qquad (8\text{-}16)$$

we define u to be the difference between v and the steady-state temperature:

$$u(x,t) = v(x,t) - \left[v_0 + \frac{x}{l}(v_1 - v_0)\right].$$

Then $u(0,t) = u(l,t) = 0$, and hence u can be determined by the method of the text. A similar use of the steady-state solution was made in the example in Sec. 7.

PROBLEMS

1. Compute the loss of heat per day per square meter of a large concrete wall whose thickness is 25 cm if one face is kept at 0°C and the other at 30°C.   Use $k = 0.002$, and assume steady-state conditions.   *Hint:* The wall can be thought to be composed of bars 25 cm long perpendicular to the wall faces.   By symmetry, no heat flows through the sides of these bars in the steady state, and so (8-14) can be applied.

2. An insulated metal rod 1 m long has its ends kept at 0°C and its initial temperature is 50°C. What is the temperature in the middle of the rod at any subsequent time?   Use $k = 1.02, c = 0.06$, and $\rho = 9.6$.

3. Let the rod of Prob. 2 have one of its ends kept at 0°C and the other at 10°C.   If the initial temperature of the rod is 50°C, find the temperature of the rod at any later time.   *Hint:* See Example 2.

4. An insulated bar with unit cross-sectional area has its ends kept at temperature 0, and the initial temperature is $f(x) = c_n \sin n\pi x/l$, where c_n is constant and n is an integer.   (a) Show that the amount of heat present in the bar initially is $2lc\rho c_n/n\pi$ if n is odd and 0 if n is even. (b) Show that the net rate of flow out of the bar across the ends is $2kc_n(n\pi/l)e^{-(\alpha n\pi/l)^2 t}$ when n is odd and 0 when n is even.   *Hint:* The rate of flow out of the bar at the end $x = 0$ is $+ku_x$, not $-ku_x$.   (c) How much heat flows out of the bar in the time from $t = 0$ to t?   Evaluate as $t \to \infty$, compare (a), and explain.

5. By addition of the results in Prob. 4, obtain similar results for the bar with arbitrary initial temperature $f(x)$.

6. The equation for heat flow in a rod that loses heat by radiation into a surrounding medium of temperature 0 is

$$v_t = \alpha^2 v_{xx} - hv$$

where α and h are constant.   Show that the substitution $v = ue^{-ht}$ gives $u_t = \alpha^2 u_{xx}$.   How are the initial and boundary conditions in the original equation related to those in the new one?

7. (a) The equation for heat flow in an insulated rod in which heat is generated at a constant rate is

$$v_t = \alpha^2 v_{xx} + \sigma$$

where σ is a constant measuring the strength of the source (e.g., the rod carries a constant electric current).   By setting $v = u + cx(l - x)$, where c is a suitable constant, transform this problem into $u_t = \alpha^2 u_{xx}$ with the same boundary conditions at $x = 0$ and $x = l$.   (b) Referring to Prob. 6, discuss a similar reduction of the problem in which there are both radiation and a source function $\sigma(x)$.   *Hint:* $v = u + \phi(x)$.

8. Let $\phi = \phi(x,y)$ be a function of the two variables x and y, with ϕ and ϕ_y continuous, and let $\bar{\phi}$ denote the mean value with respect to x.   Thus,

$$\bar{\phi}(x,y) = \frac{1}{\Delta x} \int_x^{x+\Delta x} \phi(\xi,y)\, d\xi \qquad \bar{\phi}(x, y + \Delta y) = \frac{1}{\Delta x} \int_x^{x+\Delta x} \phi(\xi, y + \Delta y)\, d\xi.$$

By the mean-value theorem

$$\phi(\xi, y + \Delta y) - \phi(\xi,y) = \Delta y \phi_y(\xi,\eta) \qquad y < \eta < y + \Delta y$$

obtain the formula

$$\frac{\bar{\phi}(x, y + \Delta y) - \bar{\phi}(x,y)}{\Delta y} = \frac{1}{\Delta x} \int_x^{x+\Delta x} \phi_y(\xi,\eta)\, d\xi.$$

Using continuity of ϕ_y, deduce that the expression is as close as desired to $\phi_y(x,y)$, provided only Δx and Δy are sufficiently small.   Hence the limit is $\phi_y(x,y)$.   [The result of this problem allows a simple derivation of many partial differential equations without use of the integral-transformation theorems of Chap. 6.   For example, taking $\phi = u$ or $\phi = u_x$, we get (8-5) from (8-4).]

9. Flow of Electricity in a Cable. In Chaps. 2 and 3 it was found that an ordinary differential equation can represent the behavior of many different physical systems. The same applies to partial differential equations. As an illustration, we shall consider the flow of electricity in linear conductors (such as telephone wires or submarine cables) in which the current may leak to ground. It will be found, for certain values of the parameters, that the equation is identical with that obtained for the vibrating string in Sec. 2. For other values, the equation agrees with the equation of heat conduction obtained in the preceding section.

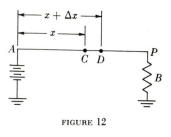

FIGURE 12

Let a long, imperfectly insulated cable carry an electric current whose source is at A (Fig. 12). The current flows to the receiving end at P through the load B and returns through the ground. It is assumed that the leaks occur along the entire length of the cable because of imperfections in the insulating sheath. Let the distance, measured along the length of the cable, be denoted by x; then the emf V (volts) and the current I (amperes) are functions of x and t. The resistance of the cable will be denoted by R (ohms per mile), and the conductance from sheath to ground by G (mhos per mile). It is known that the cable acts as an electrostatic capacitor, and the capacitance of the cable to ground per unit length is assumed to be C (farads per mile); the inductance per mile will be denoted by L (henrys per mile).

Consider an element CD of the cable of length Δx. If the emf is V at C and $V + \Delta V$ at D, the change in voltage across the element Δx is produced by the resistance and the inductance drops, so that one can write

$$\Delta V \doteq -\left(IR\,\Delta x + \frac{\partial I}{\partial t}\,L\,\Delta x \right).$$

The negative sign signifies that the voltage is a decreasing function of x. Dividing through by Δx and passing to the limit as $\Delta x \to 0$ give the equation for the voltage:

$$\frac{\partial V}{\partial x} = -IR - L\frac{\partial I}{\partial t}. \tag{9-1}$$

The decrease in current, on the other hand, is due to the leakage and the action of the cable as a capacitor. Hence, the drop in current ΔI across the element Δx of the cable is

$$\Delta I \doteq -VG\,\Delta x - \frac{\partial V}{\partial t}\,C\,\Delta x$$

so that

$$\frac{\partial I}{\partial x} = -VG - C\frac{\partial V}{\partial t}. \tag{9-2}$$

Equations (9-1) and (9-2) are simultaneous partial differential equations for the voltage and current. The voltage V can be eliminated from these equations by differentiating (9-2) with respect to x to obtain

$$I_{xx} = -V_x G - CV_{tx}.$$

Substituting for V_x from (9-1) gives

$$I_{xx} = IRG + LGI_t - CV_{tx}$$

from which V_{tx} can be eliminated by using the expression for V_{xt} obtained by differentiating (9-1). Thus one is led to

$$I_{xx} - LCI_{tt} = (LG + RC)I_t + IRG. \tag{9-3}$$

A similar calculation shows that (9-3) is also satisfied by the voltage V.

When the cable is lossless, $R = G = 0$. Equation (9-3) and the corresponding equation for V are, then,

$$I_{xx} = LCI_{tt} \qquad V_{xx} = LCV_{tt}. \tag{9-4}$$

Comparing the equation for wave motion (Sec. 1), we see that the cable propagates electromagnetic waves with velocity

$$a = (LC)^{-\frac{1}{2}}.$$

Equation (9-4) is appropriate if the frequency is high and the loss is low.

For an audio-frequency submarine cable it is more appropriate to take $G = L = 0$. The equations are then

$$I_{xx} = RCI_t \qquad V_{xx} = RCV_t. \tag{9-5}$$

Instead of representing waves, the propagation of V and I is now identical with the flow of heat in rods. Comparing with (8-5) gives

$$\alpha = (RC)^{-\frac{1}{2}}.$$

The solutions of (9-4) are oscillatory whereas those of (9-5) are not. Another difference between the two cases is seen when we consider the appropriate initial conditions. For (9-4), the discussion of Sec. 3 shows that we must specify both the initial value $I(x,0)$ and the initial derivative $I_t(x,0)$, whereas for (9-5) the initial value alone is needed.

Distinctions of this kind are so important that they form a basis for the classification of partial differential equations. Equation (9-4) is an example of a *hyperbolic* equation, and (9-5) is *parabolic*. The equation

$$u_{xx} + u_{yy} = 0$$

considered in Secs. 11 and 12 has still another behavior and is called *elliptic*. Discussion of these matters from a more general point of view forms the main topic of Secs. 24 to 28.

Example.   Consider a submarine cable l miles in length, and let the voltage at the source A, under steady-state conditions, be 12 volts and at the receiving end R be 6 volts. At a certain instant $t = 0$, the receiving end is grounded, so that its potential is reduced to zero, but the potential at the source is maintained at its constant value of 12 volts. Determine the current and voltage in the line subsequent to the grounding of the receiving end.

It is required to find V in (9-5) subject to the boundary conditions

$$V(0,t) = 12 \qquad V(l,t) = 0 \qquad t \geq 0. \tag{9-6}$$

The initial condition is

$$V(x,0) = 12 - 6\frac{x}{l} \tag{9-7}$$

since the steady-state solution of (9-5) is a linear function of x (Sec. 8, Example 1).

The voltage $V(x,t)$ subsequent to the grounding can be thought of as being made up of a steady-state[1] voltage $V_S(x)$ and a transient voltage $V_T(x,t)$ which decreases rapidly with time. Thus,

$$V(x,t) = V_S(x) + V_T(x,t). \tag{9-8}$$

Since $V_S(x)$ is linear, its value is given by the boundary conditions as

$$V_S(x) = 12 - 12\frac{x}{l}. \tag{9-9}$$

Equations (9-6) and (9-7) now yield

$$V_T(0,t) = V_T(l,t) = 0 \qquad V_T(x,0) = \frac{6x}{l}.$$

---

[1] Compare Sec. 8, Example 2.

Since V_T satisfies (9-5), we can use the solution of the heat equation (8-10) with $\alpha^2 = 1/RC$. The result is

$$V_T(x,t) = \sum_{n=1}^{\infty} \left(\frac{2}{l} \int_0^l \frac{6}{l} x_1 \sin \frac{n\pi x_1}{l} \, dx_1 \right) e^{-(1/RC)(n\pi/l)^2 t} \sin \frac{n\pi x}{l}.$$

The function V is now given by (9-8).

─── **PROBLEMS**

1. By using (9-1) with $L = 0$, find I in the example.

2. Find the emf in the cable whose length is 100 miles and whose characteristics are as follows: $R = 0.3$ ohm per mile, $C = 0.08$ μf per mile, $L = 0$, $G = 0$. If the voltage at the source is 6 volts and at the terminal end 2 volts, what is the voltage after the terminal end has been suddenly grounded?   [Use (9-5).]

3. Using (9-5), find the current in a cable 1,000 miles long, whose potential at the source, under steady-state conditions, is 1,200 volts and at the terminal end is 1,100 volts. What is the current in the cable after the terminal end has been suddenly grounded? Use $R = 2$ ohms per mile and $C = 3 \cdot 10^{-7}$ farad per mile.

4. Show that V satisfies the same equation as that obtained for I.

5. Discuss the steady-state solution of (9-3).

6. Does (9-3) have solutions $I(t)$ depending on t alone?   If so, find them, and discuss their physical significance.

7. Referring to Prob. 8 of the preceding section, carry out the derivation of (9-1) and (9-2) without using approximate equalities.

───

10. Other Boundary Conditions.   Separation of Variables.   In Sec. 8 the differential equation

$$u_t = \alpha^2 u_{xx} \tag{10-1}$$

was obtained for the temperature $u(x,t)$ of an insulated bar at point x and time t. The initial condition was

$$u(x,0) = f(x) \qquad 0 < x < l \tag{10-2}$$

and the ends were held at constant temperature.

If, instead, the ends are *insulated*, the boundary conditions are

$$u_x(0,t) = 0 \qquad u_x(l,t) = 0. \tag{10-3}$$

Equations (10-3) are appropriate because by (8-2) they state that the rate of flow across the ends is zero. We shall now consider the problem posed by (10-1) to (10-3).

The boundary conditions (10-3) are satisfied automatically if we express $u(x,t)$ as a cosine series:

$$u(x,t) = \tfrac{1}{2}a_0(t) + \sum a_n(t) \cos \frac{n\pi x}{l}. \tag{10-4}$$

Thus, u_x in (10-4) is a sine series (assuming that one can differentiate term by term), and we have already noted that the sine series vanishes at $x = 0$ and l.

Formally substituting (10-4) into the differential equation (10-1) gives

$$\tfrac{1}{2}a_0' = 0 \qquad a_n' = -\alpha^2 \left(\frac{n\pi}{l} \right)^2 a_n. \tag{10-5}$$

Solving (10-5) and substituting into (10-4), we find

$$u(x,t) = \tfrac{1}{2}c_0 + \sum c_n e^{-(n\pi\alpha/l)^2 t} \cos \frac{n\pi x}{l} \qquad (10\text{-}6)$$

where the c_n's are constant. The initial condition (10-2) shows that the c_n's are the Fourier cosine coefficients,

$$c_n = \frac{2}{l} \int_0^l f(x) \cos \frac{n\pi x}{l}\, dx \qquad (10\text{-}7)$$

and the problem is solved.[1]

We shall now solve this same problem by an important method known as *separation of variables*. It will prove interesting to compare the various stages of the solution with the answer (10-6).

The desired solution (10-6) is a sum of terms each of which has the form

$$X(x)T(t). \qquad (10\text{-}8)$$

In the method of separating variables the idea is to construct functions of the form (10-8) which satisfy the differential equation and the boundary conditions. By superposition of these functions, one then satisfies the initial conditions. The fact that there is a solution of the type (10-6) gives good reason for expecting the method to succeed.

Substituting $u = X(x)T(t)$ into (10-1) yields

$$XT' = \alpha^2 X''T$$

where the prime denotes differentiation with respect to the appropriate variable. Dividing by XT we get

$$\frac{T'}{T} = \alpha^2 \frac{X''}{X}. \qquad (10\text{-}9)$$

The variables x and t in (10-9) are *separated*, in that the left side is a function of t alone and the right side is a function of x alone. It follows that each side must be constant, independent of both x and t. A brief investigation of the effect of changing sign in (10-10) shows that XT can satisfy (10-3) only if the constant is zero or a negative number $-p^2$. Thus,

$$\frac{T'}{T} = -p^2 \qquad \alpha^2 \frac{X''}{X} = -p^2. \qquad (10\text{-}10)$$

Independent solutions of (10-10) are[2]

$$T = e^{-p^2 t} \qquad X = \cos \frac{p}{\alpha} x \qquad X = \sin \frac{p}{\alpha} x. \qquad (10\text{-}11)$$

The boundary condition $u_x(0,t) = 0$ for $u = XT$ requires that $X'(0) = 0$, and hence the appropriate choice of X in (10-11) is

$$X = \cos \frac{p}{\alpha} x. \qquad (10\text{-}12)$$

Similarly, the condition $u_x(l,t) = 0$ gives $X'(l) = 0$, so that

$$p = \frac{n\pi\alpha}{l} \qquad (10\text{-}13)$$

[1] The solution can be verified, if desired, as in the previous section. Uniqueness can be deduced by the method of Sec. 21, Prob. 5, and also by the method of Sec. 24, though the details are not given here.

[2] It is suggested that the reader compare XT at this and subsequent stages with the general term of (10-6).

where n is an integer.   By (10-11) to (10-13) we see that the function

$$T(t)X(x) = e^{-(n\pi a/l)^2 t} \cos \frac{n\pi x}{l} \qquad (10\text{-}14)$$

satisfies the differential equation and the boundary conditions.   To satisfy the initial conditions we form a superposition of terms (10-14).   The resulting series is precisely the series (10-6), and the solution is completed as before.

The merit of the separation method is that it produced the functions $\cos (n\pi x/l)$ by direct consideration of the differential equation.   If some other functions had been more appropriate, the method would have produced those other functions instead.   This fact will now be illustrated by an example.

According to Newton's law of cooling, a body radiates heat at a rate proportional to the difference between the temperature u of the radiating body and the temperature u_0 of the surrounding medium.   Thus, if our insulated rod of length l has the end $x = 0$ maintained at temperature 0 while the other end radiates into a medium of temperature $u_0 = 0$, the corresponding boundary conditions are

$$u(0,t) = 0 \qquad u_x(l,t) = -hu(l,t) \qquad (10\text{-}15)$$

where h is constant.   [The second condition (10-15) states that the rate of flow $-ku_x$ is proportional to $u(l,t) - 0$, and this agrees with Newton's law.]   If $h = 0$, there is no radiation and we have the condition for an insulated end as discussed previously.   But if $h > 0$, which we now assume, the problem is essentially different from those considered hitherto.   The difference results from the fact that (10-15) cannot be satisfied in any simple way by an ordinary Fourier series.

Actually, as we show next, the appropriate functions for the problem (10-1), (10-2), and (10-15) are not $\sin (n\pi x/l)$ or $\cos (n\pi x/l)$ but are $\sin \beta_n x$, where the β_n's are the positive roots of the transcendental equation

$$\beta \cos \beta l = -h \sin \beta l. \qquad (10\text{-}16)$$

(Cf. Prob. 10.)   Although one could hardly expect to discover the sequence $\sin \beta_n x$ by *a priori* considerations, it is produced automatically by the method of separating variables.   The solution to the problem is found to be

$$u(x,t) = \Sigma c_n e^{-\alpha^2 \beta_n^2 t} \sin \beta_n x \qquad (10\text{-}17)$$

where c_n is given in terms of the initial values by

$$c_n = \frac{\int_0^l f(x) \sin \beta_n x \, dx}{\int_0^l \sin^2 \beta_n x \, dx}. \qquad (10\text{-}18)$$

To obtain this solution by separating variables, observe that the substitution $u = X(x)T(t)$ leads to functions of the type (10-11), exactly as in the former case.   Here, however, the condition $u(0,t) = 0$ gives $X(0) = 0$, so that we require the sine rather than the cosine.   The resulting expression

$$T(t)X(x) = e^{-p^2 t} \sin \frac{p}{\alpha} x$$

becomes $\qquad\qquad\qquad e^{-\alpha^2 \beta^2 t} \sin \beta x \qquad (10\text{-}19)$

if we set $p = \alpha\beta$, and this form will be more convenient for our purposes.   The function (10-19) satisfies (10-1) and the first boundary condition (10-15) for all values of the constant β.   To satisfy the second condition (10-15) we must choose β so that

$$e^{-\alpha^2 \beta^2 t} \beta \cos \beta l = -h e^{-\alpha^2 \beta^2 t} \sin \beta l \qquad (10\text{-}20)$$

and this leads to (10-16). The resulting functions

$$e^{-\alpha^2\beta_n^2 t} \sin \beta_n x$$

satisfy both boundary conditions (10-15) and also satisfy the differential equation (10-1). If a suitable superposition (10-17) is found to satisfy the initial condition, our problem will be solved.

Setting $t = 0$ in (10-17) gives

$$f(x) = \Sigma c_n \sin \beta_n x. \tag{10-21}$$

By Chap. 3, Sec. 22, the functions $\sin \beta_n x$ are orthogonal on $(0,l)$, and hence the c_n's are given by (10-18). The solution can be verified by the method of Sec. 8 if $f(x)$ admits an expansion (10-21). Since an analog of Dirichlet's theorem holds for the sequence, $\sin \beta_n x$, this is not a serious restriction on $f(x)$.

———————————————————————————————— PROBLEMS

1. (a) Review the discussion of orthogonal functions in Chap. 1, Sec. 23, and in Chap. 3, Sec. 22. (b) Review the theory of mean convergence in Chap. 1, Sec. 24.

2. (a) If $f(x) \equiv g(t)$, where x and t are independent variables, show that $f(x)$ and $g(t)$ are constant. *Hint:* Let $t = t_0$, a fixed value. (b) Attempt to satisfy the conditions (10-3) by choosing a positive constant $+p^2$ instead of $-p^2$ in (10-10).

3. By using the functions (10-11), solve $u_t = \alpha^2 u_{xx}$, $u(0,t) = u(l,t) = 0$, $u(x,0) = f(x)$.

4. (a) Describe a physical situation which would lead to

$$u_t = \alpha^2 u_{xx} \qquad u(0,t) = u_x(l,t) = 0 \qquad u(x,0) = f(x).$$

(b) Solve by separating variables [cf. (10-11)]. (c) Verify that your result agrees with (10-16) to (10-18) for $h = 0$.

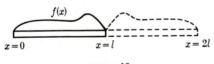

FIGURE 13

5. Solve Prob. 4 by the method of images. *Outline of Solution:* Consider a rod of length $2l$ with ends at temperature 0. Let the initial temperature $f_0(x)$ agree with $f(x)$ on $(0,l)$, and let $f_0(x)$ be symmetric about $x = l$ (Fig. 13). By symmetry, no heat flows across the center, and hence the left half of the long rod behaves like the rod of Prob. 4. The temperature $u_0(x,t)$ for the long rod can be found from (8-10).

6. The vertical displacement $u(x,t)$ of a vibrating string with fixed end points satisfies

$$u_{tt} = a^2 u_{xx} \qquad u(0,t) = u(l,t) = 0.$$

By setting $u(x,t) = X(x)T(t)$ and separating variables, obtain solutions of the form

$$\sin \frac{n\pi a t}{l} \sin \frac{n\pi x}{l} \qquad \text{and} \qquad \cos \frac{n\pi a t}{l} \sin \frac{n\pi x}{l}.$$

7. In Prob. 6, express $u(x,t)$ as an infinite series if $u(x,0) = f(x)$, $u_t(x,0) = 0$.

8. In Prob. 6, express $u(x,t)$ as an infinite series if $u(x,0) = 0$, $u_t(x,0) = g(x)$.

9. According to Prob. 11 of Sec. 2, the small transverse oscillations of an unloaded elastic rod satisfy an equation of the form

$$y_{tt} + a^2 y_{xxxx} = 0$$

where a is constant. (a) By means of separating variables, find a solution of the form $X(x)T(t)$ containing seven arbitrary constants. (b) Formulate appropriate boundary conditions for a beam clamped at one end; clamped at both ends; and pivoted at both ends. Then obtain corresponding solutions by (a). *Hint:* At the pivoted end, $u_{xx} = 0$; at a free end, $u_{xxx} = u_{xx} = 0$. (c) What initial conditions do the particular solutions in part b satisfy? (*d*) Discuss the problem of satisfying arbitrary initial conditions by infinite series.

10. Discuss the graphical solution of $\beta \cos \beta l = -h \sin \beta l$ by plotting $y = \tan \beta l$ and the straight line $y = -\beta/h$. (Cf. Chap. 10, Sec. 2, Example 2.)

---

11. Heat Flow in a Solid. By a procedure similar to that of Sec. 8 one can establish the equation

$$u_t = \alpha^2(u_{xx} + u_{yy} + u_{zz}) \qquad \alpha^2 \equiv \frac{k}{c\rho} \tag{11-1}$$

for the temperature[1] $u = u(x,y,z,t)$ in a uniform solid at time t. This is the three-dimensional form of the equation

$$u_t = \alpha^2 u_{xx} \tag{11-2}$$

obtained previously for heat conduction in a rod. The state of the solid at time $t = 0$ gives the initial condition; the state of the surface for $t > 0$ gives the boundary condition. For instance, if the surface radiates according to Newton's law, the boundary condition is

$$-k\frac{\partial u}{\partial n} = e(u - u_0) \tag{11-3}$$

where u_0 is the temperature of the surrounding medium, e the *emissivity*, and $\partial u/\partial n$ the derivative in the direction of the outward normal. When $e = 0$, Eq. (11-3) means that the body is insulated.

Sometimes there is so much symmetry that u in (11-1) does not depend on y or z. In this case (11-1) is the same as (11-2), since the terms u_{yy} and u_{zz} in (11-1) are zero, and the analysis of Secs. 8 to 10 can be applied without change.

As a specific illustration, consider a uniform plate extending from the plane $x = 0$ to the plane $x = d$. Let $u = u_0$ on the surface $x = 0$ and $u = u_1$ on the surface $x = d$, where u_0 and u_1 are constant. If the plate is infinite, or if the edges are far away from the points being considered, the symmetry suggests that u depends on x only and, hence, that (11-2) holds. The steady-state temperature is then given by Example 1 of Sec. 8, as

$$u = u_0 + \frac{x}{d}(u_1 - u_0).$$

Since the rate of flow is $-ku_x$, the amount of heat Q flowing across the area A in t sec is

$$Q = ktA \frac{u_0 - u_1}{d}.$$

If the flow of heat is steady, so that u is independent of time, $u_t = 0$ and (11-1) reduces to

$$u_{xx} + u_{yy} + u_{zz} = 0. \tag{11-4}$$

This is known as *Laplace's equation;* it occurs in a variety of physical problems. The corresponding two-dimensional form is

$$u_{xx} + u_{yy} = 0 \qquad u = u(x,y). \tag{11-5}$$

---

[1] See also the derivation in Chap. 6, Sec. 16. A similar equation governs diffusion and the drying of porous solids, with u equal to the concentration of the diffusing substance. Because of this analogy many problems on diffusion and heat conduction are mathematically indistinguishable. The constant α^2 in (11-1) is often called the *diffusivity*.

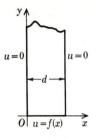

FIGURE 14

To illustrate the use of (11-5) we shall discuss the steady-state temperature in an infinitely long metal strip of width d (see Fig. 14). If the sides of the strip have the temperature zero and the bottom edge has the temperature $f(x)$, the boundary conditions are

$$u(0,y) = 0 \qquad u(d,y) = 0 \qquad u(x,0) = f(x). \qquad (11\text{-}6)$$

We assume besides that (11-5) holds for $0 < x < d, y > 0$.

It is a surprising fact that these conditions do not suffice to determine the temperature.[1] However, one expects the temperature to approach zero as one moves away from the bottom edge, so that

$$\lim_{y \to \infty} u(x,y) = 0 \qquad\qquad \text{uniformly in } x. \qquad (11\text{-}7)$$

If this condition is required, the solution can be shown to be unique (see Sec. 24).

Although the problem can be solved very simply by Fourier series, we prefer to show how the desired functions are generated by the method of separating variables. The choice $u = X(x)Y(y)$ in (11-5) gives

$$\frac{X''}{X} = -\frac{Y''}{Y} \qquad (11\text{-}8)$$

after dividing by XY. Since the variables in (11-8) are separated, each side is a constant. The boundary conditions applied to XY show (after some calculation) that the constant must be a negative number $-p^2$, and hence (11-8) gives

$$\frac{X''}{X} = -p^2 \qquad \frac{Y''}{Y} = p^2.$$

Since $(-p)^2 = p^2$, we can assume that $p > 0$ with no loss of generality.

Linearly independent solutions of these equations are, respectively,

$$\cos px, \ \sin px \qquad \text{and} \qquad e^{py}, \ e^{-py}.$$

Since $u(0,y) = 0$ requires that $X(0) = 0$, we reject the cosine, and in view of (11-7) we reject the solution e^{py}. Hence the function XY takes the form

$$XY = e^{-py} \sin px. \qquad (11\text{-}9)$$

The boundary condition $u(d,y) = 0$ gives $p = n\pi/d$, where n is an integer. Forming a linear combination of the resulting solutions (11-9) we get

$$u(x,y) = \sum c_n e^{-(n\pi/d)y} \sin \frac{n\pi x}{d} \qquad (11\text{-}10)$$

and the condition $u(x,0) = f(x)$ now shows that the c_n's are the Fourier coefficients

$$c_n = \frac{2}{d} \int_0^d f(x) \sin \frac{n\pi x}{d} \, dx.$$

The solution can be verified, if desired, as in Sec. 8.

The foregoing derivation obscures an important point which will now be discussed more fully. Although the solutions

$$e^{py} \qquad \text{and} \qquad e^{-py}$$

can be chosen for equation $Y'' = p^2 Y$, these are not the only possibilities. Another pair of independent solutions, for example, is

---

[1] The trouble is that the other end of the strip must be taken into account even though it is infinitely far away. This purpose is served by (11-7).

$$\cosh py \quad \text{and} \quad \sinh py.$$

If, now, we try to decide which of these functions satisfies (11-7) it will be found that neither one does.

What is really involved is the following: The *general* solution of $Y'' = p^2 Y$ is

$$Y = ae^{py} + be^{-py}$$

where a and b are constant. By (11-7) we get $a = 0$, and hence $Y = be^{-py}$. The reader can verify that if

$$Y = a_0 \cosh py + b_0 \sinh py$$

the condition (11-7) will give $a_0 + b_0 = 0$, and again Y is a multiple of e^{-py}. Similar remarks apply to the construction of $X(x)$ and to the derivation of (10-14).

Just as in the case of the rod, this problem involving a strip can be given a three-dimensional interpretation. That is, the strip need not be thin provided there is no variation of temperature across its thickness. By letting the thickness approach infinity, we get a semi-infinite plate. (In Fig. 14 the plate extends infinitely far toward and away from the reader; the area outlined in the figure is the *cross section* of the plate, not a frontal view.) The boundary-value problem for the plate is

$$u_{xx} + u_{yy} + u_{zz} = 0 \qquad 0 < x < d, \, y > 0, \, -\infty < z < \infty \qquad (11\text{-}11)$$

$$u(0,y,z) = 0 \qquad u(d,y,z) = 0 \qquad u(x,0,z) = f(x) \qquad (11\text{-}12)$$

$$\lim_{y \to \infty} u(x,y,z) = 0 \qquad \text{uniformly in } x \text{ and } z. \qquad (11\text{-}13)$$

If we assume u independent of z, the resulting problem is the same as that formerly considered; hence it has the solution (11-10).

The fact that $u(x,y,z)$ is independent of z does not follow from the physical symmetry but requires the condition (11-13). Indeed, the function

$$u = \sin \frac{\pi x}{d} \sin \frac{\pi y}{d} \, e^{\sqrt{2} \, \pi z/d}$$

satisfies (11-11) and (11-12) with $f(x) = 0$ and yet depends on z. Reduction of the dimension by omitting a variable is really an application of *uniqueness*. If we verify that (11-10) satisfies the problem (11-11) to (11-13) *and that the problem has no other solution*, then it is true that u must be independent of z.

─── **PROBLEMS**

1. A refrigerator door is 10 cm thick and has the outside dimensions 60 by 100 cm. If the temperature inside the refrigerator is $-10°C$ and outside is $20°C$, and if $k = 0.0002$, find the gain of heat per day across the door by assuming the flow of heat to be of the same nature as that across an infinite plate.

2. If $f(x) = 1$ and $d = \pi$, show that (11-10) gives

$$u = \frac{4}{\pi} (e^{-y} \sin x + \tfrac{1}{3}e^{-3y} \sin 3x + \tfrac{1}{5}e^{-5y} \sin 5x + \cdots).$$

3. In Prob. 2, compute the temperature at the following points: $(\pi/2,1)$, $(\pi/3,2)$, $(\pi/4,10)$.

4. Derive (11-10) by assuming a Fourier series

$$u(x,y) = \sum b_n(y) \sin \frac{n\pi x}{d}.$$

5. A semi-infinite plate 10 cm in thickness has its faces kept at 0°C and its base kept at 100°C. What is the steady-state temperature at any point of the plate?

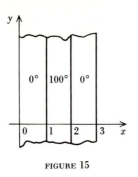

FIGURE 15

6. The faces of an infinite slab 10 cm thick are kept at temperature 0°C.   If the initial temperature of the slab is 100°C, what is the state of the temperature at any subsequent time?

7. A large rectangular iron plate (Fig. 15) is heated throughout to 100°C and is placed in contact with and between two like plates each at 0°C.   The outer faces of these outside plates are maintained at 0°C.   Find the temperature of the inner faces of the two plates and the temperature at the midpoint of the inner plate 10 sec after the plates have been put together.   Given: $\alpha = 0.2$ cgs unit. *Hint:* The boundary and initial conditions are

$$u(0,t) = 0 \qquad u(3,t) = 0 \qquad u(x,0) = f(x)$$

where $f(x) = 0$ for $0 < x < 1$ and $2 < x < 3$ but $f(x) = 100$ for $1 < x < 2$.

---

12. The Dirichlet Problem.   The Laplace equation

$$u_{xx} + u_{yy} + u_{zz} = 0 \tag{12-1}$$

was obtained in Sec. 11 for steady-state heat flow.   We shall show how the same equation arises in electrostatics and gravitation.[1]

It is a consequence of Coulomb's law that the potential due to a point charge q at (x_1,y_1,z_1) is

$$u = \frac{q}{r} \qquad \text{taking } u = 0 \text{ at } r = \infty \tag{12-2}$$

where r is the distance from the charge to the point (x,y,z) at which u is computed.   Thus,

$$r^2 = (x - x_1)^2 + (y - y_1)^2 + (z - z_1)^2 \qquad r > 0. \tag{12-3}$$

The potential due to a distribution of n point charges q_i is given by addition,

$$u = \sum_{i=1}^{n} \frac{q_i}{r_i}, \tag{12-4}$$

and the potential due to a distribution of continuous charge of density ρ in a body τ can be obtained from an expression like (12-4) by passing to the limit.

It is easily shown that $1/r$ satisfies Laplace's equation (12-1), and hence the same is true of u in (12-4) provided no r_i is zero.   This latter condition means that there is no charge at the point of observation.   One would expect, therefore, that the potential due to a continuous charge distribution will also satisfy (12-1) if there is no charge at the point of observation.   This is actually the case, and that is the reason why Laplace's equation plays such a prominent role in electrostatics.   Although a more sophisticated treatment may be given, it all comes down to the same thing; namely, $1/r$ satisfies (12-1), and the potential is given by some sort of superposition process applied to $1/r$.

Since the gravitational potential satisfies (12-2) (where q is the mass of the attracting mass point), the study of gravitation also leads to Laplace's equation.   In view of its many applications, the Laplace equation is profitably regarded as a field of study in its own right.   Such a study leads the way to a branch of analysis known as *potential theory*.

An important problem in potential theory is the *Dirichlet problem*, which can be stated as follows: Suppose given a body τ in (x,y,z) space, together with assigned values $f(x,y,z)$ on the surface of τ.   Find a function u which satisfies Laplace's equation in τ

[1] A more complete discussion is given in Chap. 6, Sec. 14.   The relation of Laplace's equation and fluid flow is developed in Chap. 6, Secs. 15 and 17, and in Chap. 8, Sec. 19.

and is equal to $f(x,y,z)$ on the surface. The foregoing discussion gives a number of physical interpretations. For instance, if u is temperature, the Dirichlet problem is to find the steady-state temperature in a uniform solid when the temperature on the surface is given. But if u is the electrostatic potential, the problem is to find the potential inside a closed surface when the potential on the surface is known. Interpretations in terms of diffusion, fluid flow, and gravitation can also be given.

Since solutions of Laplace's equation are often called *harmonic functions*, Dirichlet's problem can be stated as follows: Find a function which is harmonic in a given region and assumes preassigned values on the boundary. In two dimensions a harmonic function $u(x,y)$ satisfies

$$u_{xx} + u_{yy} = 0. \tag{12-5}$$

The region in Dirichlet's problem is now a plane region, and its boundary is a curve. The physical interpretation refers to phenomena in a thin plane sheet, or it refers to three-dimensional phenomena which show no dependence on z. The latter condition is to be expected when there is cylindrical symmetry, that is, when all planes $z = $ const exhibit the same geometry and boundary conditions.

We shall now solve the Dirichlet problem for a circle. It turns out that the problem is greatly simplified by use of polar coordinates appropriate to the circular symmetry. With

$$x = r \cos \theta \qquad y = r \sin \theta \qquad u(x,y) \equiv U(r,\theta)$$

an elementary calculation shows that (12-5) becomes

$$(rU_r)_r + \frac{1}{r} U_{\theta\theta} = 0 \tag{12-6}$$

(see Prob. 3). The boundary condition can be expressed as

$$U(R,\theta) = f(\theta) \tag{12-7}$$

where $f(\theta)$ is a known function of θ and R is the radius of the circle.

For each value of r it is clear that U has period 2π in θ, since u is single-valued, and therefore U has a Fourier series

$$U(r,\theta) = \frac{a_0(r)}{2} + \sum [a_n(r) \cos n\theta + b_n(r) \sin n\theta]. \tag{12-8}$$

Proceeding tentatively, we substitute (12-8) into (12-6) to obtain

$$\left(\frac{ra_0'}{2}\right)' + \sum [(ra_n')' \cos n\theta + (rb_n')' \sin n\theta] - \frac{1}{r} \sum (a_n n^2 \cos n\theta + b_n n^2 \sin n\theta) = 0.$$

Since the coefficients of $\cos n\theta$ and of $\sin n\theta$ must vanish,

$$(ra_n')' = \frac{1}{r} n^2 a_n \qquad (rb_n')' = \frac{1}{r} n^2 b_n.$$

These equations are both of form

$$r(ry')' = n^2 y$$

which is readily solved by the method of Chap. 3, Sec. 21. Specifically, the substitution $y = r^a$ gives

$$r(ar^a)' = n^2 r^a$$

whence $a = \pm n$. Since $a_n(r)$ and $b_n(r)$ must be finite at $r = 0$, the minus sign is excluded, and

$$a_n(r) = a_n r^n \qquad b_n(r) = b_n r^n$$

where a_n and b_n are constant.   Hence by (12-8)

$$U(r,\theta) = \frac{a_0}{2} + \sum (a_n r^n \cos n\theta + b_n r^n \sin n\theta). \tag{12-9}$$

Putting $r = R$ and using the boundary condition (12-7) give

$$f(\theta) = \frac{a_0}{2} + \sum (a_n R^n \cos n\theta + b_n R^n \sin n\theta). \tag{12-10}$$

If $f(\theta)$ has a convergent Fourier series, the validity of (12-10) is ensured by choosing $a_n R^n$ and $b_n R^n$ to be the Fourier coefficients of f:

$$a_n R^n = \frac{1}{\pi} \int_{-\pi}^{\pi} f(\phi) \cos n\phi \, d\phi \qquad b_n R^n = \frac{1}{\pi} \int_{-\pi}^{\pi} f(\phi) \sin n\phi \, d\phi. \tag{12-11}$$

The problem is now solved, but a simpler form can be found as follows: Substituting (12-11) into (12-9) gives

$$U(r,\theta) = \frac{1}{\pi} \int_{-\pi}^{\pi} \left[\frac{1}{2} + \sum \left(\frac{r}{R} \right)^n \cos n(\theta - \phi) \right] f(\phi) \, d\phi \tag{12-12}$$

when we note that

$$\cos n\theta \cos n\phi + \sin n\theta \sin n\phi = \cos (n\theta - n\phi)$$

and interchange the order of summation and integration.   The series in brackets in (12-12) can be summed as in Chap. 1, Sec. 17, Prob. 6.   The result is the *Poisson formula for a circle*[1]

$$U(r,\theta) = \frac{1}{2\pi} \int_{-\pi}^{\pi} \frac{R^2 - r^2}{R^2 - 2rR \cos (\theta - \phi) + r^2} f(\phi) \, d\phi. \tag{12-13}$$

If $f(\phi)$ is sectionally continuous, one can differentiate under the integral sign for $r < R$ to find that (12-6) holds.   Also, it is shown in Sec. 20, Prob. 3, that (12-13) gives

$$\lim_{r \to R-} U(r,\theta) = f(\theta) \tag{12-14}$$

provided f is continuous at θ.   Hence (12-13) is a solution.   In view of the derivation, it is remarkable that (12-14) holds even when the Fourier series for f does not converge to f.

The expression (12-13) gives the steady-state temperature of a thin uniform insulated disk in terms of the temperature at the boundary.   Or (12-13) can be interpreted as giving the temperature in a circular cylinder when the temperature of the surface is $f(\theta)$ independent of z.   On the other hand, the formula also gives the electrostatic potential in terms of its values on the boundary, and so on.

Example.   Let $u(x,y)$ be harmonic in a plane region, and let C be a circle contained entirely in the region.   Show that the value of u at the center of C is the average of the values on the circumference.

Without loss of generality we can take the center to be at the origin.   Equation (12-13) then gives, with $r = 0$,

$$u(0,0) = U(0,\theta) = \frac{1}{2\pi} \int_{-\pi}^{\pi} f(\phi) \, d\phi. \tag{12-15}$$

Since $f(\phi)$ stands for the values of u on the boundary, this is the required result.

[1] Another derivation is given in Chap. 8, Sec. 21.

PROBLEMS

1. Using $rr_x = x - x_1$, verify that $1/r$ in (12-2) satisfies the Laplace equation (12-1).   Thus obtain a harmonic function $u = c_0 + c_1/r$ that assumes prescribed constant values u_0 for $r = r_0$ and u_1 for $r = r_1$, where $r_1 > r_0 > 0$.   Interpret the result in terms of the electrostatic potential in a region between two concentric conducting spheres and also in terms of the steady-state temperature of a conducting spherical shell.

2. (*a*) Show that $\log r$ satisfies the two-dimensional Laplace equation (12-5).   Thus obtain the steady-state temperature in a circular ring of inner radius r_0 and outer radius r_1, given that the temperatures of the bounding circles are the constants u_0 and u_1, respectively.   *Hint:* Try $u = c_1 + c_2 \log r$.   (*b*) By interpreting the foregoing result in three dimensions, discuss the steady-state lateral heat loss from an insulated electrical cable with conductor at temperature u_0, if the cable is in a surrounding liquid of temperature u_1.   In particular, compute the rate of lateral heat loss per unit length, and verify that it is the same across the inner and outer surface of the insulation.   *Hint:* The rate of lateral heat flow is determined by $-ku_r$, where k is the thermal conductivity of the insulation.

3. If $u(x,y) = U(r,\theta)$ with $x = r\cos\theta$, $y = r\sin\theta$, show that $u_{xx} + u_{yy} = r^{-1}(rU_r)_r + r^{-2}U_{\theta\theta}$.   *Hint:* $U_r = u_x\cos\theta + u_y\sin\theta$, $U_\theta = u_x(-r\sin\theta) + u_y(r\cos\theta)$.   Similarly, compute $(rU_r)_r$ and $(U_\theta)_\theta$.

4. Derive (12-10) by considering $U = R(r)\Theta(\theta)$ and separating variables.

5. Give two physical interpretations of the following Dirichlet problem for a semicircle, where $u(x,y) = U(r,\theta)$ as in (12-6):

$$u_{xx} + u_{yy} = 0 \qquad x^2 + y^2 < 1, y > 0$$
$$U(1,\theta) = g(\theta) \qquad 0 \le \theta \le \pi; \qquad U(r,0) = U(r,\pi) = 0 \qquad 0 \le r \le 1.$$

6. Solve Prob. 5 by the method of images.   *Hint:* For $0 < \theta < \pi$, define $f(\theta) = g(\theta)$, $f(-\theta) = -g(\theta)$ and use (12-13).

7. Obtain a formula analogous to (12-13) for the region $r > R$.   (Assume that $|U(r,\theta)|$ is bounded as $r \to \infty$ and, hence, that positive values of n in the discussion of the text may be rejected.)

8. Interpret the result of Prob. 7 physically in terms of an infinite metal plate with a hole whose edges have a prescribed temperature.

9. Show that the Laplace equation $u_{xx} + u_{yy} = 0$ is *invariant*, that is, retains its form, under the following transformations:

(*a*) Translation and uniform stretching: $\tilde{x} = cx + h$, $\tilde{y} = cy + k$
(*b*) Rotation: $\tilde{x} = x\cos\theta - y\sin\theta$, $\tilde{y} = x\sin\theta + y\cos\theta$
(*c*) Inversion: $\tilde{x} = x(x^2 + y^2)^{-1}$, $\tilde{y} = y(x^2 + y^2)^{-1}$.
Hint: Set $\tilde{u}(\tilde{x},\tilde{y}) = u(x,y)$, and use the chain rule.

10. The reader sufficiently familiar with three-dimensional geometry will extend Prob. 9 to three dimensions.   (It is advisable to set $x = x_1, y = x_2, z = x_3$ and use summation signs.   "Rotation" is to be replaced by the *orthogonal transformation* discussed in Chap. 4, Sec. 19.)

---

13. Spherical Symmetry.   Legendre Functions.   Let it be required to determine the steady-state temperature in a uniform solid sphere of radius unity when one half of the surface is kept at the constant temperature 0°C and the other half at the constant temperature 1°C.   By the discussion of Sec. 11, the temperature u within the sphere satisfies Laplace's equation

$$u_{xx} + u_{yy} + u_{zz} = 0. \tag{13-1}$$

Symmetry suggests the use of spherical coordinates (r,θ,ϕ) with origin at the center of the given unit sphere (Fig. 16).   Since

$$x = r \sin \theta \cos \phi \qquad y = r \sin \theta \sin \phi \qquad z = r \cos \theta$$

Laplace's equation can be shown to be[1]

$$r(rU)_{rr} + (U_\theta \sin \theta)_\theta \csc \theta + U_{\phi\phi} \csc^2 \theta = 0 \qquad (13\text{-}2)$$

where $u(x,y,z) = U(r,\theta,\phi)$.   If the plane separating the unequally heated hemispheres is the xy plane, the symmetry suggests that U will be independent of ϕ, so that (13-2) becomes

$$r(rU)_{rr} + (U_\theta \sin \theta)_\theta \csc \theta = 0. \qquad (13\text{-}3)$$

The boundary conditions are

$$
\begin{aligned}
u = 1 &\qquad \text{for } 0 < \theta < \tfrac{1}{2}\pi &&\qquad \text{when } r = 1 \\
u = 0 &\qquad \text{for } \tfrac{1}{2}\pi < \theta < \pi &&\qquad \text{when } r = 1.
\end{aligned}
\qquad (13\text{-}4)
$$

We shall use the method of separating variables.   Substituting the form

$$U = R(r)\Theta(\theta)$$

into (13-3) gives two ordinary differential equations,

$$r(rR)'' - \alpha R = 0 \qquad (\Theta' \sin \theta)' \csc \theta + \alpha\Theta = 0 \qquad (13\text{-}5)$$

where α is an arbitrary constant.   The first of these equations can be solved by assuming that $R = r^m$ as in Chap. 3, Sec. 21.   One obtains the linearly independent solutions

$$R = r^m \qquad R = r^{-(m+1)}$$

where m satisfies the quadratic equation

$$m(m + 1) = \alpha. \qquad (13\text{-}6)$$

Changing the independent variable in (13-5) from θ to x by means of

$$x = \cos \theta \qquad \Theta(\theta) = P(x)$$

and replacing α by the expression (13-6), we get Legendre's equation

$$(1 - x^2)P'' - 2xP' + m(m + 1)P = 0 \qquad {}' = \frac{d}{dx}. \qquad (13\text{-}7)$$

When m is a nonnegative integer, a solution of (13-7) is the Legendre polynomial $P_m(x) = P_m(\cos \theta)$.   Thus, one is led to consider solutions of (13-3) which have the form

$$r^m P_m(\cos \theta) \qquad \text{or} \qquad r^{-(m+1)} P_m(\cos \theta).$$

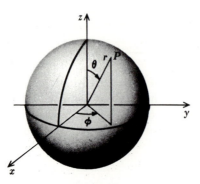

The second of these expressions is rejected because it becomes infinite as $r \to 0$, and we attempt to build up the desired solution u by forming a series

$$u = \sum_{m=0}^{\infty} A_m r^m P_m(\cos \theta). \qquad (13\text{-}8)$$

Each term of this series satisfies (13-3).
When $r = 1$, Eq. (13-8) becomes

$$u = \sum_{m=0}^{\infty} A_m P_m(\cos \theta) \qquad r = 1 \qquad (13\text{-}9)$$

[1] See Chap. 6, Sec. 13, or proceed as in Prob. 3 of the preceding section.

and if it is possible to choose the constants A_m in such a way that (13-9) satisfies the boundary condition (13-4), then (13-8) will be a solution of the problem. Since $x = \cos\theta$, the boundary condition requires

$$F(x) = \sum_{m=0}^{\infty} A_m P_m(x) \tag{13-10}$$

where $F(x) = 0$ for $-1 < x < 0$, and $F(x) = 1$ for $0 < x < 1$. As suggested by Prob. 6, the expansion (13-10) is possible for suitably restricted functions $F(x)$. According to Chap. 1, Sec. 23, and Chap. 3, Sec. 22, the coefficients are given by

$$A_m = (m + \tfrac{1}{2}) \int_{-1}^{1} F(x) P_m(x)\, dx.$$

By means of this formula, the solution is found to be

$$u = \tfrac{1}{2} + \tfrac{3}{4}r P_1(\cos\theta) - \tfrac{7}{16}r^3 P_3(\cos\theta) + \tfrac{11}{32}r^5 P_5(\cos\theta) - \cdots.$$

It is possible to establish that (13-8) is actually a solution, though the demonstration requires a detailed knowledge of Legendre functions. The uniqueness theorem established in Sec. 24 shows that there is no other solution; hence the foregoing procedure can be justified. In particular it was permissible to take m as a nonnegative integer and to use the polynomial solution of (13-7) rather than one of the infinite-series solutions.

PROBLEMS AND REVIEW PROBLEMS

1. As an infinite series, express the steady-state temperature in a circular plate of radius a which has one half of its circumference at $0°C$ and the other half at $100°C$.

2. By (12-13), find the temperature of the plate considered in Prob. 1.

3. By separating variables in polar coordinates, find the steady-state temperature in a semicircular plate of radius a if the bounding diameter is kept at the temperature $0°C$ and the circumference is kept at the temperature $100°C$.

4. Interpret the following Dirichlet problem physically, and solve:
$$u_{xx} + u_{yy} = 0 \qquad 0 < x < 1, 0 < y < 1$$
$$u(0,y) = u(1,y) = u(x,0) = 0 \qquad u(x,1) = f(x).$$

5. Derive (13-5) from (13-3).

6. The nth Legendre polynomial has the precise degree n. (a) Using this fact, show that every polynomial $P(x)$ of degree n can be written
$$P(x) = c_0 P_0(x) + c_1 P_1(x) + \cdots + c_n P_n(x)$$
where the c_k are constant. (Such an expression is called a *linear combination* of Legendre polynomials.) *Hint:* Choose c_n so that
$$P(x) - c_n P_n(x)$$
has degree at most $n - 1$, and use mathematical induction. (b) From (a) deduce that, if $f(x)$ can be approximated by a polynomial, it can be approximated in the same sense and with the same accuracy by a linear combination of Legendre polynomials. In particular, if the mean-square error
$$\int_{-1}^{1} |f(x) - P(x)|^2\, dx$$
can be made arbitrarily small by a polynomial $P(x)$, the mean-square error can be made arbitrarily small by means of the partial sums of the Legendre-Fourier series. *Hint:* See Chap. 1, Sec. 24.

7. The oscillations of a vibrating string subject to a forcing term are described by
$$u_{tt} - a^2 u_{xx} = F(x,t) \qquad u(0,t) = u(l,t) = 0.$$

(a) Assuming that $u(x,t)$ and $F(x,t)$ have Fourier-series expansions

$$u(x,t) = \sum b_n(t) \sin \frac{n\pi x}{l} \qquad F(x,t) = \sum B_n(t) \sin \frac{n\pi x}{l}$$

substitute and equate coefficients to find that

$$b_n'' + \omega_n^2 b_n = B_n \qquad \text{where } l\omega_n = n\pi a.$$

(b) If the string is initially at rest, deduce $b_n(0) = b_n'(0) = 0$, and hence by Chap. 3, Eq. (18-6),

$$\omega_n b_n(t) = \int_0^t B_n(\tau) \sin \omega_n(t - \tau) \, d\tau.$$

(c) Obtain $B_n(t)$ from $F(x,t)$ by the Euler-Fourier formula, and thus get

$$u(x,t) = \sum \frac{2}{n\pi a} \sin \frac{n\pi x}{l} \int_0^t \int_0^l \sin \frac{n\pi \xi}{l} \sin \frac{n\pi a}{l} (t - \tau) F(\xi,\tau) \, d\xi \, d\tau.$$

8. In the preceding problem, suppose $F(x,t) = a(x) \sin \omega t + b(x) \cos \omega t$. By the Euler-Fourier formula, find $B_n(t)$, and show that

$$b_n'' + \omega_n^2 b_n = \alpha \sin \omega t + \beta \cos \omega t$$

where α and β are constant.

9. Show how to reduce the general case with initial conditions

$$u(x,0) = f(x) \qquad u_t(x,0) = g(x)$$

to the case of null initial conditions considered in Prob. 7 above. *Hint:* Let $u = v + w$, where v satisfies the problem with $f = g = 0$, and w satisfies the problem with $F = 0$. The value of w is given in Sec. 6.

10. Discuss Prob. 7 if the oscillations are damped, so that the differential equation is

$$u_{tt} + 2bu_t - a^2 u_{xx} = F(x,t).$$

---

14. The Rectangular Membrane. Double Fourier Series.   Let a flexible elastic membrane be stretched over a horizontal, plane, bounding curve (Fig. 17). To explain what is meant by the *tension*, we consider the force $\Delta \mathbf{F}$ exerted by the membrane on one side of a small straight slit of length Δs. The membrane is said to be under *uniform tension* τ if this force is directed perpendicular to the slit in the plane of the membrane and has magnitude $\tau \, \Delta s$ independent of the location and orientation of the slit. A similar definition applies when the membrane does not lie in a plane or the tension is

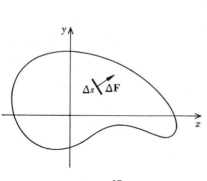

FIGURE 17

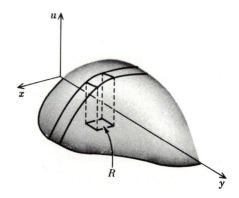

FIGURE 18

not uniform, except that we must let $\Delta s \to 0$. The role taken by the plane of the membrane in the first case is now taken by the tangent plane at the point in question.

We shall obtain a differential equation for *transverse vibration* of the membrane, in which every point moves toward or away from the plane of the bounding curve, not sideways. If the coordinate system is chosen so that the boundary curve lies in the (x,y) plane, the behavior is described by the vertical displacement $u(x,y,t)$. A differential equation for u is obtained by applying Newton's law to the portion of membrane lying above the rectangle R defined by

$$x_0 < x < x_1 \qquad y_0 < y < y_1$$

as shown in Fig. 18.

In the discussion of the vibrating string it was found desirable to use the horizontal component of the tension as primary variable, and we introduce a horizontal tension function τ to play the same role for the membrane. The precise meaning of τ is that the horizontal components of force on the element of membrane over the rectangle R are

$$H_x = \int_C \tau \, dy \qquad H_y = -\int_C \tau \, dx$$

where C is the boundary of the rectangle in the (x,y) plane. As in the discussion of the string, the fact that there is no horizontal acceleration enables us to conclude that τ is a function of t only, not of (x,y).

The vertical component of force is

$$\int_C (\tau u_x \, dy - \tau u_y \, dx) = \int_{x_0}^{x_1} \int_{y_0}^{y_1} (\tau u_{xx} + \tau u_{yy}) \, dx \, dy.$$

Here the first expression follows from the fact that the tension force is directed tangentially to the membrane, for a flexible membrane, and from the fact that u_x and u_y denote tangents of appropriate angles. The second expression follows from Green's theorem, assuming continuity.

If $\rho(x,y)$ is the density of the plane membrane at rest, the same function describes the density when it is oscillating, since the motion is wholly vertical. However, the mass is given by integration with respect to an element of area in the (x,y) plane, rather than an element of surface area of the membrane. The total momentum is obtained similarly, by integrating ρu_t.

If we equate the rate of change of momentum to the total force, the result is

$$\int_{x_0}^{x_1} \int_{y_0}^{y_1} \rho u_{tt} \, dx \, dy = \int_{x_0}^{x_1} \int_{y_0}^{y_1} (\tau u_{xx} + \tau u_{yy}) \, dx \, dy + \int_{x_0}^{x_1} \int_{y_0}^{y_1} F_1 \, dx \, dy$$

where $F_1(x,y,t)$ describes the vertical load due to pressure, gravity, air resistance, and so on. Differentiation with respect to x_1 and y_1 produces the differential equation

$$\rho u_{tt} = \tau(u_{xx} + u_{yy}) + F_1(x,y,t)$$

when the integrands are continuous. A somewhat simpler form is obtained by setting $F_1 = \rho F$ and dividing by ρ.

It has been noted that τ is independent of (x,y) and ρ is independent of t. The oscillations of a *uniform membrane*, by definition, are oscillations in which both τ and ρ are constant.

According to the foregoing discussion, *the free oscillations of a uniform membrane are described by*

$$u_{tt} = \gamma^2(u_{xx} + u_{yy}) \qquad (14\text{-}1)$$

where $\gamma = \sqrt{\tau/\rho}$ is constant. For a rectangular membrane, which is considered now, the boundary conditions are (Fig. 19)

$$u = 0 \qquad \text{for } x = 0 \qquad x = a \qquad y = 0 \qquad y = b. \qquad (14\text{-}2)$$

We also specify the initial displacement and initial velocity,

$$u(x,y,0) = f(x,y) \qquad u_t(x,y,0) = g(x,y). \qquad (14\text{-}3)$$

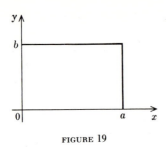

FIGURE 19

The assumption that

$$u = X(x)Y(y)T(t)$$

in (14-1) yields

$$\gamma^2 \left(\frac{X''}{X} + \frac{Y''}{Y} \right) = \frac{T''}{T} \tag{14-4}$$

upon division by XYT. Since the variables are separated, the terms in (14-4) are constant. It can be shown that these constants are negative, so that we may write

$$\frac{X''}{X} = -p^2 \qquad \frac{Y''}{Y} = -q^2 \qquad \frac{T''}{T} = -\omega^2$$

with $\gamma^2(p^2 + q^2) = \omega^2$ by (14-4).

Since $X'' + p^2 X = 0$, the function $X(x)$ is a linear combination of $\sin px$ and $\cos px$. The cosine is rejected because the condition $u = 0$ at $x = 0$ gives $X(0) = 0$, and we must have $p = m\pi/a$, where m is an integer, because the condition $u = 0$ at $x = a$ gives $X(a) = 0$. In just the same way it is found that

$$Y = \sin qy$$

where $q = n\pi/b$ for an integer n. Thus, the desired oscillation has the form

$$\sin \frac{m\pi x}{a} \sin \frac{n\pi y}{b} (A \cos \omega_{mn}t + B \sin \omega_{mn}t) \tag{14-5}$$

where A and B are constant and where $\omega_{mn} = \omega$ is given by

$$\omega_{mn}{}^2 = \left(\frac{\pi m \gamma}{a} \right)^2 + \left(\frac{\pi n \gamma}{b} \right)^2.$$

The functions (14-5) satisfy the differential equation and the boundary condition. To satisfy the initial conditions (14-3) we try a superposition, using different constants A and B for each choice of m and n:

$$u(x,y,t) = \sum_{m,n=1}^{\infty} (A_{mn} \cos \omega_{mn}t + B_{mn} \sin \omega_{mn}t) \sin \frac{m\pi x}{a} \sin \frac{n\pi y}{b}. \tag{14-6}$$

Since the initial displacement is $f(x,y)$, we must determine A_{mn} so that

$$f(x,y) = \sum_{m,n=1}^{\infty} A_{mn} \sin \frac{m\pi x}{a} \sin \frac{n\pi y}{b}.$$

Multiplying this *double Fourier series* by $\sin (m\pi x/a) \sin (n\pi y/b)$ and integrating over the rectangle give the formula

$$A_{mn} = \frac{4}{ab} \int_0^a \int_0^b f(x,y) \sin \frac{m\pi x}{a} \sin \frac{n\pi y}{b} \, dx \, dy$$

just as in the corresponding discussion for single Fourier series (Chap. 1, Sec. 18). Similarly, differentiating (14-6) with respect to t and setting $t = 0$ give

$$B_{mn} = \frac{4}{ab\omega_{mn}} \int_0^a \int_0^b g(x,y) \sin \frac{m\pi x}{a} \sin \frac{n\pi y}{b} \, dx \, dy$$

when we use the second initial condition (14-3).

The general term of the series (14-6) is a periodic function of time with period $2\pi/\omega_{mn}$. The corresponding frequencies

$$\frac{\omega_{mn}}{2\pi} = \frac{\gamma}{2}\left[\left(\frac{m}{a}\right)^2 + \left(\frac{n}{b}\right)^2\right]^{1/2} \quad \text{cps} \tag{14-7}$$

are called *characteristic frequencies*, and the associated oscillations (14-5) are called *modes*. The *fundamental mode* is the mode of lowest frequency, obtained by setting $m = n = 1$.

Similar terminology applies to the vibrating string (Secs. 2 to 6). If the length of the string is a and the equation of motion is $u_{tt} = \gamma\, u_{xx}$ the characteristic frequencies may be written in the form

$$\frac{\omega_m}{2\pi} = \frac{\gamma}{2}\left[\left(\frac{m}{a}\right)^2\right]^{1/2} \tag{14-8}$$

analogous to (14-7). The modes are described by

$$u = \sin\frac{m\pi x}{a}\,(A\cos\omega_m t + B\sin\omega_m t)$$

and the *fundamental* is the mode obtained for $m = 1$. The three-dimensional analog of (14-7) and (14-8) is discussed in Prob. 2.

In Sec. 7 it was shown for the vibrating string that the *characteristic* frequencies agree with the *resonant* frequencies, and a similar behavior is found for vibration phenomena in general. It is also true in general that the vibration can be expressed as a superposition of individual modes. This fact is illustrated by (14-6) and by the Fourier-series solution for the vibrating string.

The behavior of the vibrating membrane differs from that of the string in one respect. For each characteristic frequency of vibration of the string the corresponding mode is such that the string is divided into equal parts by the nodes whose positions are fixed. When a membrane oscillates with a given characteristic frequency, there are also points on the membrane which remain at rest. Such points form *nodal lines*. The position and the shape of the nodal lines, however, need not be the same for a given frequency, i.e., for a given characteristic frequency there may be more than one mode.

As an illustration, consider a rectangular membrane with $a = b$. The frequency equation (14-7) then yields

$$\omega_{mn} = \frac{\gamma\pi}{a}\sqrt{m^2 + n^2} = \alpha\sqrt{m^2 + n^2} \text{ where } \alpha \equiv \frac{\gamma\pi}{a}. \tag{14-9}$$

For $m = n = 1$, we get from (14-6) the fundamental mode

$$u_{11} = (A_{11}\cos\omega_{11}t + B_{11}\sin\omega_{11}t)\sin\frac{\pi x}{a}\sin\frac{\pi y}{a}$$

where $\omega_{11} = \alpha\sqrt{2}$. Since $u_{11} = 0$ for all t only when $x = 0, y = 0, x = a, y = a$, there are no nodal lines in the interior of the membrane for this frequency. If we take $m = 1, n = 2$ and $m = 2, n = 1$, we get two modes:

$$u_{12} = (A_{12}\cos\omega_{12}t + B_{12}\sin\omega_{12}t)\sin\frac{\pi x}{a}\sin\frac{2\pi y}{a}$$

$$u_{21} = (A_{21}\cos\omega_{21}t + B_{21}\sin\omega_{21}t)\sin\frac{2\pi x}{a}\sin\frac{\pi y}{a} \tag{14-10}$$

with the same frequency, since $\omega_{21} = \omega_{12} = \alpha\sqrt{5}$. For $y = a/2$, $u_{12} \equiv 0$ and for $x = a/2$, $u_{21} \equiv 0$. These nodal lines are shown in Fig. 20. By forming linear combinations of the modes in (14-10) we can get oscillations with the same frequency but with different nodal lines. Thus, if we take $A_{12} = A_{21} = 0$ and form $u_{12} + u_{21}$, we get

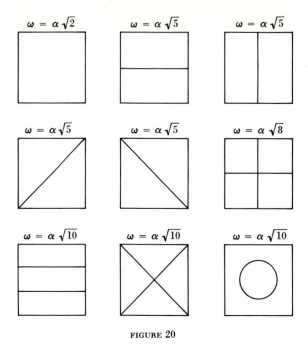

FIGURE 20

$$u_{12} + u_{21} = \sin \omega_{12} t \left(B_{12} \sin \frac{\pi x}{a} \sin \frac{2\pi y}{a} + B_{21} \sin \frac{2\pi x}{a} \sin \frac{\pi y}{a} \right)$$

$$= (\sin \omega_{12} t) 2 \sin \frac{\pi x}{a} \sin \frac{\pi y}{a} \left(B_{12} \cos \frac{\pi y}{a} + B_{21} \cos \frac{\pi x}{a} \right).$$

For this oscillation the nodal lines in the interior of the membrane are determined by

$$B_{12} \cos \frac{\pi y}{a} + B_{21} \cos \frac{\pi x}{a} = 0. \tag{14-11}$$

Equation (14-11) for $B_{12} = B_{21}$ yields the nodal line $x + y = a$ and for $B_{12} = -B_{21}$, the line $x - y = 0$ (see Fig. 20). Different nodal lines can be obtained by forming different linear combinations of the modes (14-10).

The reader will show that for $m = n = 2$, all oscillations have the same nodal lines $x = a/2$, $y = a/2$, while infinitely many different nodal lines can be obtained by forming different linear combinations of the modes u_{13} and u_{31}. A few of these are shown in Fig. 20.

Since the nodal lines may be regarded as the boundaries of new membranes contained in the original one, the character of oscillation of membranes of different shapes can be deduced from the examination of nodal lines (see Prob. 3).

Nodal lines can be observed experimentally by sprinkling a fine powder on the vibrating membrane.

─── **PROBLEMS**

1. Suppose the initial conditions for the rectangular membrane considered in the text are

$$u(x,y,0) = 0.1 \sin \frac{\pi x}{a} \sin \frac{\pi y}{b} \qquad u_t(x,y,0) = 0.$$

(a) What is the frequency of the oscillation? (b) What is the maximum speed attained by the midpoint of the membrane?

2. Analysis of a microwave resonant cavity leads to the equation $u_{tt} = \gamma^2 (u_{xx} + u_{yy} + u_{zz})$ with the boundary condition $u = 0$ or $\partial u / \partial n = 0$ on suitable portions of the planes

$$x = 0 \qquad x = a \qquad y = 0 \qquad y = b \qquad z = 0 \qquad z = c$$

(see Fig. 21). By assuming $u = XYZT$, show that the characteristic frequencies are

$$\frac{\omega_{mnp}}{2\pi} = \frac{\gamma}{2}\left[\left(\frac{m}{a}\right)^2 + \left(\frac{n}{b}\right)^2 + \left(\frac{p}{c}\right)^2\right]^{\frac{1}{2}}$$

where m, n, and p are integers.

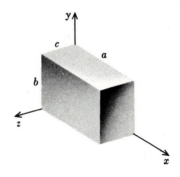

FIGURE 21

3. (a) Sketch the nodal lines for the oscillation (14-5). (b) By considering the nodal line for the oscillation which arises by adding the modes $m = 1$, $n = 2$ and $m = 2$, $n = 1$, obtain one solution for the problem of an isosceles right-triangular membrane.

15. The Circular Membrane. Bessel Functions.

To discuss the oscillations of a circular membrane with fixed edges we introduce cylindrical coordinates (r,θ,u). With

$$u(x,y,t) = U(r,\theta,t) \tag{15-1}$$

the equation of motion (14-1) takes the form (cf. Sec. 12, Prob. 3)

$$U_{tt} = \gamma^2[U_{rr} + r^{-1}U_r + r^{-2}U_{\theta\theta}]. \tag{15-2}$$

If the boundary is the circle $r = a$, then the boundary condition is

$$U(a,\theta,t) = 0. \tag{15-3}$$

To make the problem definite we also introduce initial conditions

$$U(r,\theta,0) = f(r) \qquad U_t(r,\theta,0) = 0 \tag{15-4}$$

which state, respectively, that the initial shape of the membrane is given by $f(r)$ and that the initial velocity is zero.

Since the initial shape is independent of θ, the solution presumably involves r and t only. Thus, we consider expressions of the form $R(r)T(t)$ when applying the separation method. Substituting into (15-2) gives

$$\frac{1}{T}\frac{d^2T}{dt^2} = \gamma^2\left(\frac{1}{R}\frac{d^2R}{dr^2} + \frac{1}{rR}\frac{dR}{dr}\right) \tag{15-5}$$

after division by RT. Since the left-hand member of (15-5) depends on t alone and the right-hand member on r alone, each side must be constant. It can be shown that the constant is not positive and hence may be written as $-\omega^2$. Thus (15-5) leads to

$$T'' + \omega^2 T = 0 \tag{15-6}$$

$$R'' + r^{-1}R' + k^2R = 0 \tag{15-7}$$

where $k = \omega/\gamma$.

Equation (15-6) is the familiar equation for simple harmonic motion, and Eq. (15-7) can be reduced to the Bessel equation by the substitution $x = kr$. Hence (15-7) has a solution

$$R = J_0(x) = J_0(kr).$$

The other solutions of (15-7) are rejected because they become infinite at $r = 0$, and we are led to the functions

$$J_0(kr) \sin \omega t \qquad \text{or} \qquad J_0(kr) \cos \omega t.$$

Since $u_t = 0$ when $t = 0$, we reject the solution involving the sine. The boundary

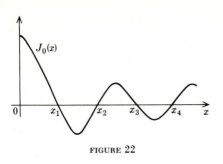

FIGURE 22

condition (15-3) applied to our elementary solution RT now gives

$$J_0(ka) \cos \omega t = 0$$

for all t. This requires that ka be a root of the equation $J_0(x) = 0$ (see Fig. 22). If the positive roots of $J_0(x)$ are denoted by x_n, the appropriate choices of k are given by $k_n = x_n/a$. Since $\omega = k\gamma$, our elementary solutions have the form

$$RT = J_0(k_n r) \cos k_n \gamma t.$$

These functions satisfy the differential equation (15-2), the boundary condition (15-3), and the second initial condition (15-4). To satisfy the first initial condition we try to represent U as a linear combination of such terms:

$$U = \sum_{n=1}^{\infty} A_n J_0(k_n r) \cos k_n \gamma t. \tag{15-8}$$

When $t = 0$, the initial condition requires that

$$f(r) = \sum_{n=1}^{\infty} A_n J_0(k_n r).$$

By Chap. 1, Secs. 23 and 24, and Chap. 3, Sec. 22, the coefficients of the Fourier-Bessel series are given by

$$A_n = \frac{2}{a^2 [J_0'(k_n a)]^2} \int_0^a f(r) J_0(k_n r) r \, dr. \tag{15-9}$$

In the terminology of the preceding section, the solution (15-8) is expressed by means of the modes. The characteristic frequencies are

$$\frac{\omega_n}{2\pi} = \frac{k_n \gamma}{2\pi} = \frac{x_n \gamma}{2\pi a}$$

and the fundamental is described by $J_0(k_1 r) \cos k_1 \gamma t$.

PROBLEMS

1. The oscillations of a cylindrical resonant cavity satisfy

$$u_{tt} = \gamma^2 (u_{xx} + u_{yy} + u_{zz}) \qquad 0 \le r < a, \, 0 < z < b$$

with boundary condition $u = 0$ on the curved surface, $u_z = 0$ on the plane ends. Obtain solutions of the form $R(r)Z(z)T(t)$ for this problem.

2. Find the distribution of temperature in a long cylinder whose surface is kept at the constant temperature zero and whose initial temperature in the interior is unity.

3. An elastic membrane subject to uniform gas pressure satisfies the equation

$$u_{tt} + p = \gamma^2 (u_{xx} + u_{yy})$$

where p is a constant depending on the pressure. If the membrane is circular, show how to reduce this problem to a problem of the type solved in the text. *Hint:* Consider the function

$$U(r,\theta,t) = u - \frac{p}{4\gamma^2} (r^2 - a^2).$$

4. The steady-state temperature in a semi-infinite circular cylinder satisfies
$$r^2u_{rr} + ru_r + u_{\theta\theta} + r^2u_{zz} = 0 \qquad u(a,z) = 0 \qquad u(r,z) \to 0 \qquad \text{as } z \to \infty.$$
By separation of variables, obtain solutions of form
$$u(r,z) = J_0(k_n r)e^{-k_n z}.$$
Express the solution satisfying $u(r,0) = 100$ as an infinite series, following the pattern of the text.

5. The transient heat flow in a circular cyclinder leads to the problem
$$ru_t = \alpha^2(ru_{rr} + u_r) \qquad u(a,t) = 0 \qquad u(r,t) \to 0 \qquad \text{as } t \to \infty.$$
Obtain solutions of the form
$$u(r,t) = e^{-\alpha^2 k^2 t}J_0(kr) \qquad k = k_n$$
by separating variables, and express the solution satisfying $u(r,0) = f(r)$ as an infinite series, following the pattern of the text.

6. The small oscillations of a flexible chain hanging under its own weight satisfy the approximate equation
$$\rho u_{tt} = (Tu_x)_x$$
where ρ is the constant density, l is the length, and $T = \rho g(l - x)$. By setting $l - x = z^2$, $u(x,t) = U(z,t)$, and separating variables, obtain solutions of the form
$$U(z,t) = J_0(kz)(c_1 \sin \omega t + c_2 \cos \omega t).$$

How are k and ω related? If the displacement at the upper end $x = 0$ is 0, what are the values of k? Discuss the use of infinite series to satisfy appropriate initial conditions.

SOLUTION BY INTEGRALS

16. The Fourier Transform.   Plane Waves.   For many partial differential equations the desired solution can be expressed as an integral involving the initial or boundary values.   This possibility was illustrated by formula (3-16) for displacement of a vibrating string and by the solution of the Dirichlet problem given in (12-13).   We shall now describe a systematic method of obtaining integral formulas.

The function $g(s)$ defined for real s by

$$\mathbf{T}f = \lim_{a \to \infty} \frac{1}{\sqrt{2\pi}} \int_{-a}^{a} e^{-ixs}f(x)\ dx = g(s) \tag{16-1}$$

is called the *Fourier transform* of $f(x)$; the operator $\mathbf{T}$ is called the *Fourier transform operator*.   The inverse operator $\mathbf{T}^{-1}$ is obtained by changing the sign of i, so that the foregoing equation may also be written

$$\mathbf{T}^{-1}g = \lim_{a \to \infty} \frac{1}{\sqrt{2\pi}} \int_{-a}^{a} e^{ixs}g(s)\ ds = f(x). \tag{16-2}$$

When such is the case, the symbol $\mathbf{T}$ satisfies the easily remembered equations

$$\mathbf{T}\mathbf{T}^{-1}f = f \qquad \mathbf{T}^{-1}\mathbf{T}f = f. \tag{16-3}$$

If the limits in (16-1) and (16-2) are regarded in the sense of mean convergence (Chap. 1, Sec. 22), and if the integrals are regarded as Lebesgue integrals (Appendix B), then (16-1) gives (16-2) and (16-2) gives (16-1), provided either of the integrals

$$\int_{-\infty}^{\infty} |f(x)|^2\ dx \qquad \text{or} \qquad \int_{-\infty}^{\infty} |g(s)|^2\ ds \tag{16-4}$$

is finite.[1] Both integrals (16-4) then have the same value. In many physical problems the common value represents the total power or energy present in the system.

To illustrate the use of the Fourier transform, we shall solve the problem

$$u_t = \alpha^2 u_{xx} \qquad t > 0, \; -\infty < x < \infty \tag{16-5}$$

$$u(x,0) = f(x) \qquad -\infty < x < \infty \tag{16-6}$$

$$u(x,t) \to 0 \qquad \text{as } t \to \infty. \tag{16-7}$$

Physically, this system describes the temperature $u(x,t)$ of an infinitely long bar at point x and time t when the initial temperature $u(x,0)$ is known.

In the method of undetermined coefficients (Chap. 3) it was found that many ordinary differential equations can be solved by a trial solution of the form e^{px}, where p is constant. A similar procedure often succeeds for partial differential equations. In the present case, the trial solution $u = e^{px+qt}$ with p and q constant leads to

$$q e^{px+qt} = \alpha^2 p^2 e^{px+qt}$$

when substituted into (16-5). Hence $q = \alpha^2 p^2$, and the trial solution is

$$e^{px+\alpha^2 p^2 t}.$$

We choose p^2 negative because of (16-7). Thus $p = is$, where s is real, and the trial solution is now

$$e^{isx-\alpha^2 s^2 t} = e^{isx} e^{-\alpha^2 s^2 t}. \tag{16-8}$$

We shall satisfy the initial condition (16-6) by forming a linear combination[2] of solutions (16-8). Thus

$$\frac{1}{\sqrt{2\pi}} e^{isx} e^{-\alpha^2 s^2 t} g(s)$$

is a solution of (16-5) no matter what value $g(s)$ may have, and the integral

$$u(x,t) = \frac{1}{\sqrt{2\pi}} \int_{-\infty}^{\infty} e^{isx} e^{-\alpha^2 s^2 t} g(s) \, ds$$

is also a solution, provided we can differentiate under the integral sign in computing (16-5). By (16-2) the latter expression can be written

$$u(x,t) = \mathbf{T}^{-1} e^{-\alpha^2 s^2 t} g(s). \tag{16-9}$$

Setting $t = 0$ and using the initial condition (16-6) give

$$f(x) = \mathbf{T}^{-1} g(s)$$

so that $\mathbf{T}f = \mathbf{T}\mathbf{T}^{-1}g = g$. Substituting into (16-9) we get the final answer,

$$u(x,t) = \mathbf{T}^{-1} e^{-\alpha^2 s^2 t} \mathbf{T}f. \tag{16-10}$$

This is an explicit formula for the temperature $u(x,t)$ in terms of the initial temperature $f(x)$.

As another example, we shall solve the Dirichlet problem for a half plane. Several physical interpretations were given in Sec. 12; the mathematical formulation is

$$u_{xx} + u_{yy} = 0 \qquad y > 0, \; -\infty < x < \infty \tag{16-11}$$

$$u(x,0) = f(x) \qquad -\infty < x < \infty \tag{16-12}$$

$$u(x,y) \to 0 \qquad \text{as } y \to \infty. \tag{16-13}$$

---

[1] This important theorem, known as *Plancherel's theorem*, is proved in E. C. Titchmarsh, "Introduction to the Theory of Fourier Integrals," chap. 3, Oxford University Press, London, 1937. For a heuristic discussion of the relation between (16-1) and (16-2) see Chap. 1, Secs. 21 and 22.

[2] This procedure is analogous to the formation of Fourier series in the method of separating variables.

The function e^{px+qy} satisfies (16-11) if $p^2 + q^2 = 0$. We choose q real and negative because of (16-13), and hence p is pure imaginary, $p = is$. The trial solution is now

$$e^{isx-|s|y}$$

when we note that $q^2 = s^2$ and that q is negative. This function satisfies (16-11) and (16-13). To satisfy (16-12) we form a linear combination as in the previous example; thus,

$$u(x,y) = \frac{1}{\sqrt{2\pi}} \int_{-\infty}^{\infty} e^{isx}e^{-|s|y}g(s)\,ds \equiv \mathbf{T}^{-1}e^{-|s|y}g.$$

Setting $y = 0$ we get $f = \mathbf{T}^{-1}g$ by (16-12). Hence $g = \mathbf{T}f$, and

$$u(x,y) = \mathbf{T}^{-1}e^{-|s|y}\mathbf{T}f. \tag{16-14}$$

If $|\mathbf{T}f|$ is bounded or if either expression in (16-4) is finite, the integrals obtained by differentiation are uniformly convergent for $y \geq y_0 > 0$, and the solution is easily justified. A similar remark applies to (16-10) for $t \geq t_0 > 0$.

The Fourier transform can be used to solve the two-dimensional wave equation

$$\gamma^2(u_{xx} + u_{yy}) = u_{tt} \qquad \gamma = \text{const} \tag{16-15}$$

and the result has an interesting physical interpretation. We suppose that the time dependence is harmonic, so that

$$u(x,y,t) = U(x,y)e^{-i\omega t} \tag{16-16}$$

where ω is constant. Substituting in (16-15) gives the *scalar wave equation*

$$U_{xx} + U_{yy} + k^2 U = 0 \qquad k = \frac{\omega}{\gamma}. \tag{16-17}$$

This equation will now be solved in the half plane $y > 0$ subject to the additional conditions

$$U(x,0) = f(x) \qquad U(x,y) \to 0 \qquad \text{as } y \to \infty. \tag{16-18}$$

Physically, the solution describes the radiation field of an antenna[1] when the aperture illumination is $f(x)$ (see Fig. 23).

By substituting the function e^{izs+qy} into (16-17) we obtain solutions

$$e^{isx}e^{\pm iy\sqrt{k^2-s^2}} \tag{16-19}$$

Because of the second condition (16-18) the coefficient of y in (16-19) has a negative real part when s is large; we shall indicate this by dropping the minus sign. Forming a linear combination of expressions (16-19) as before,

$$U(x,y) = \frac{1}{\sqrt{2\pi}} \int_{-\infty}^{\infty} e^{isx}e^{iy\sqrt{k^2-s^2}}g(s)\,ds = \mathbf{T}^{-1}e^{iy\sqrt{k^2-s^2}}g.$$

For $y = 0$ the first condition (16-18) yields $g = \mathbf{T}f$, and hence

$$U(x,y) = \mathbf{T}^{-1}e^{iy\sqrt{k^2-s^2}}\mathbf{T}f. \tag{16-20}$$

To interpret the solution physically, we have

$$u(x,y,t) = \mathbf{T}^{-1}e^{-i\omega t+iy\sqrt{k^2-s^2}}g(s)$$

by combining (16-16) and (16-20), with $\mathbf{T}f = g(s)$. Writing it in full,

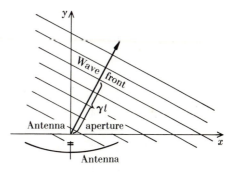

FIGURE 23

---

[1] Formulation of (16-17) in this context and discussion of the conditions for a unique solution can be found in treatises on electromagnetic theory.

$$u(x,y,t) = \frac{1}{\sqrt{2\pi}} \int_{-\infty}^{\infty} e^{i(xs+y\sqrt{k^2-s^2}-\omega t)}g(s)\ ds. \tag{16-21}$$

For simplicity we shall suppose that $g(s) = 0$ when $|s| > k$. The limits $(-\infty,\infty)$ of the integral can then be replaced by $(-k,k)$. If we now introduce a variable θ,

$$s = k \sin \theta \qquad \sqrt{k^2 - s^2} = k \cos \theta \tag{16-22}$$

we get

$$u(x,y,t) = \frac{1}{\sqrt{2\pi}} \int_{-\pi/2}^{\pi/2} e^{i(kx \sin \theta + ky \cos \theta - \omega t)}g(k \sin \theta)k \cos \theta\ d\theta.$$

This formula expresses the solution as a superposition of functions of the form

$$e^{i(kx \sin \theta + ky \cos \theta - \omega t)}. \tag{16-23}$$

In Prob. 6 it is seen that (16-23) represents a plane wave traveling with velocity γ in the direction θ (Fig. 23). Hence, *the Fourier transform procedure gives the plane-wave expansion of the antenna field.* The amplitude of the wave moving in a given direction θ is

$$g(k \sin \theta)k \cos \theta\ d\theta$$

where $g(s)$ is the Fourier transform of the aperture illumination.

PROBLEMS

1. If $f(x) = 1$ for $-c \le x \le c$ and $f(x) = 0$ elsewhere, show that

$$\mathbf{T}f = \sqrt{\frac{2}{\pi}} \frac{\sin cs}{s}.$$

Write (16-10), (16-14), and (16-20) explicitly for this case.

2. As in Secs. 2 and 3, the problem of waves on an infinite string is

$$u_{tt} = a^2 u_{xx} \qquad u(x,0) = f(x) \qquad u_t(x,0) = g(x) \qquad t > 0,\ -\infty < x < \infty.$$

(*a*) By the trial solution e^{isx+qt}, obtain the two expressions

$$e^{isx}e^{iast} \qquad \text{and} \qquad e^{isx}e^{-iast}$$

and explain how a general linear combination leads to the trial function

$$u(x,t) = \mathbf{T}^{-1}e^{iast}g_1(s) + \mathbf{T}^{-1}e^{-iast}g_2(s).$$

(*b*) Assuming that differentiation under the integrals is legitimate, evaluate u_t in part a, and by putting $t = 0$ obtain

$$f = \mathbf{T}^{-1}g_1 + \mathbf{T}^{-1}g_2 \qquad g = \mathbf{T}^{-1}iasg_1 - \mathbf{T}^{-1}iasg_2.$$

Operate on the above equations with $\mathbf{T}$ to get g_1 and g_2 in terms of $\mathbf{T}f$ and $\mathbf{T}g$ and thus obtain

$$u(x,t) = \mathbf{T}^{-1} \cos ast\ \mathbf{T}f + \mathbf{T}^{-1} \frac{\sin ast}{as}\ \mathbf{T}g.$$

3. According to Sec. 7 the equation for damped motion of waves on a string is

$$u_{tt} - a^2 u_{xx} = -2bu_t.$$

(*a*) Obtain a family of solutions of the type

$$u(x,t) = \mathbf{T}^{-1}e^{-bt}e^{t(b^2-a^2s^2)^{1/2}}g_1(s) + \mathbf{T}^{-1}e^{-bt}e^{-t(b^2-a^2s^2)^{1/2}}g_2(s)$$

by starting with $u = e^{isx+qt}$ and forming a linear combination. (*b*) Formulate appropriate initial conditions and use them to determine g_1 and g_2. Thus get $u(x,t)$, and compare Prob. 2 when $b \to 0$.

4. The displacement $u(x,t)$ of a long, stiff rod satisfies

$$EI \frac{\partial^4 u}{\partial x^4} = f(x,t) \qquad f = \text{force}$$

when the mass is negligible (cf. Sec. 2, Prob. 11). Let $U(s,t)$ be the transform of u with respect to the variable x, and let $F(s,t)$ be the transform of f. Neglecting convergence questions, show that $EIs^4U = F$, and thus obtain the solution in the form

$$u(x,t) = (EI)^{-1}\mathbf{T}^{-1}s^{-4}\mathbf{T}f.$$

　　Hint: Write the expressions

$$u(x,t) = \mathbf{T}^{-1}U(s,t) \qquad f(x,t) = \mathbf{T}^{-1}F(s,t)$$

in full, and substitute into the differential equation.

5. If the mass per unit length of the rod in the preceding example is m, the equation of motion is

$$EI \frac{\partial^4 u}{\partial x^4} + m \frac{\partial^2 u}{\partial t^2} = f(x,t).$$

Show that $u = \mathbf{T}^{-1}U$, where $U = U(s,t)$ satisfies the ordinary differential equation

$$m \frac{d^2 U}{dt^2} + s^4 EIU = F(s,t).$$

6. The *wave front* is the locus on which (16-23) is constant. Show that the wave front is given by an equation of the form

$$(x - x_0) \sin \theta + (y - y_0) \cos \theta = \gamma t$$

where $\gamma = \omega/k$ and x_0 and y_0 are constant. Deduce that the wave fronts are parallel straight lines whose common perpendicular makes an angle θ with the y axis. Also, the distance from the fixed point (x_0,y_0) to the wave front is γt, so that the velocity of propagation is γ.

7. Show that the three-dimensional wave equation

$$u_{tt} = \gamma^2(u_{xx} + u_{yy} + u_{zz}) \qquad \gamma = \text{const}$$

has solutions $u = f(ax + by + cz - \gamma t)$, where $a^2 + b^2 + c^2 = 1$. Interpret as in Prob. 6.

8. In spherical coordinates (r,θ,ϕ), let u be a function of r alone. Show that the wave equation of Prob. 7 reduces to

$$(ru)_{tt} = \gamma^2(ru)_{rr}.$$

Deduce that $ru = f(r - \gamma t) + g(r + \gamma t)$, and interpret as a superposition of *spherical waves.* *Hint:* See Secs. 13 and 1.

9. Show that $f(x \cos \theta + y \sin \theta + \gamma t)$ satisfies the two-dimensional wave equation

$$\gamma^2(u_{xx} + u_{yy}) = u_{tt}$$

if θ is constant and f is twice differentiable. Conclude that

$$u(x,y,t) = \int_0^{2\pi} f(x \cos \theta + y \sin \theta \pm \gamma t)g(\theta) \, d\theta$$

is also a solution, in general, and interpret as a superposition of plane waves. How are the waves with $+\gamma$ related to those with $-\gamma$?

10. It is shown in the theory of the Bessel function that

$$e^{-i\omega t}J_0(\omega r) = \frac{1}{2\pi} \int_0^{2\pi} e^{i\omega[r \cos (\theta - \phi) - t]}d\theta$$

where r is the polar coordinate; $x = r \cos \phi, y = r \sin \phi$. Interpret this formula as a superposition of plane waves. (The resulting wave has rotational symmetry and is called a *cylindrical wave.* Compare Sec. 15.)

---

17. The Convolution Theorem. The *convolution* $f * g$ of two functions f and g is defined by

$$f * g = \lim_{a \to \infty} \frac{1}{\sqrt{2\pi}} \int_{-a}^a f(\xi)g(x - \xi) \, d\xi = \frac{1}{\sqrt{2\pi}} \int_{-\infty}^{\infty} f(\xi)g(x - \xi) \, d\xi \qquad (17\text{-}1)$$

if the limit exists either in the ordinary sense or in the sense of mean convergence. The importance of the operation (17-1) rests on the following theorem:

CONVOLUTION THEOREM.   Let f, g, $|f|^2$, and $|g|^2$ be integrable, and let all infinite integrals be interpreted in the sense of mean convergence.   Then the product of the transforms equals the transform of the convolution:

$$\mathbf{T}(f * g) = (\mathbf{T}f)(\mathbf{T}g). \tag{17-2}$$

Although a proof in this degree of generality requires knowledge of Lebesgue integration and mean convergence, the result can be established with ease under somewhat more stringent hypotheses. For example, if the functions are sectionally continuous and the integrals are absolutely convergent, the verification involves little more than a change in order of integration. Since the details of this procedure can be found in Chap. 3, Sec. 18, they are not repeated here.

By means of the convolution theorem some of the foregoing results can be greatly simplified. Taking the transform of the formula (16-10) with respect to x gives

$$\mathbf{T}u = e^{-\alpha^2 s^2 t}\mathbf{T}f \tag{17-3}$$

for the temperature $u(x,t)$ of a rod when the initial temperature is $f(x)$. By consulting a table of Fourier transforms or by using the result of Prob. 6,

$$e^{-\alpha^2 s^2 t} = \mathbf{T}g(x) \qquad \text{where } g(x) = (2\alpha^2 t)^{-\frac{1}{2}}e^{-x^2/(4\alpha^2 t)}.$$

Hence, the result (17-3) may be written

$$\mathbf{T}u = (\mathbf{T}f)(\mathbf{T}g) = \mathbf{T}(f * g)$$

when we recall (17-2). Taking the inverse transform now yields

$$u(x,t) = f * g = (4\pi\alpha^2 t)^{-\frac{1}{2}} \int_{-\infty}^{\infty} f(\xi)e^{-(x-\xi)^2/(4\alpha^2 t)} \, d\xi. \tag{17-4}$$

The advantage of this formula is that it involves only a single integration whereas (16-10) requires two integrations. Since the integral is rapidly convergent when t is not too large, (17-4) is well suited for numerical computation.

To obtain a physical interpretation of (17-4), let the rod have the initial temperature zero except for a short piece on the interval $(x_0 - \epsilon, x_0 + \epsilon)$ (see Fig. 24). If Q cal of heat is uniformly distributed over this element of the rod, the corresponding initial temperature f is given by

$$Q = 2\epsilon c\rho f \qquad x_0 - \epsilon < x < x_0 + \epsilon$$

where c is the heat capacity and ρ the linear density. By (17-4) the resulting temperature at point x and time t is

$$u(x,t) = \frac{Q}{c\rho(4\pi\alpha^2 t)^{\frac{1}{2}}} \frac{1}{2\epsilon} \int_{x_0-\epsilon}^{x_0+\epsilon} e^{-(x-\xi)^2/(4\alpha^2 t)} \, d\xi.$$

Letting $\epsilon \to 0$ and using the mean-value theorem we get

$$\frac{Q}{c\rho(4\pi\alpha^2 t)^{\frac{1}{2}}} e^{-(x-x_0)^2/(4\alpha^2 t)}. \tag{17-5}$$

This gives the temperature distribution for an instantaneous source of strength Q at the point x_0. Now, Eq. (17-4) *represents the temperature*

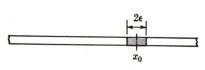

2ϵ

x_0

FIGURE 24

in the general case as a superposition of such sources. The source at $x = \xi$ has the strength

$$Q = c\rho f(\xi) \, d\xi.$$

As seen in Sec. 19, this physical interpretation enables us to solve a variety of problems in heat flow with the greatest ease.

The convolution theorem can also be applied to problems in potential theory. For example, if u is harmonic for $y > 0$ and satisfies

$$u(x,0) = f(x) \qquad u(x,y) \to 0 \qquad \text{as } y \to \infty$$

we shall show that u is given by the *Poisson formula for a half plane:*

$$u(x,y) = \frac{y}{\pi} \int_{-\infty}^{\infty} \frac{f(\xi)}{(x - \xi)^2 + y^2} \, d\xi. \tag{17-6}$$

Since this problem is the same as that in (16-11) to (16-13), the solution is given by (16-14). Taking the transform of (16-14),

$$\mathbf{T}u = e^{-|s|y}\mathbf{T}f. \tag{17-7}$$

The convolution theorem can be applied if we express $e^{-|s|y}$ as a Fourier transform. To this end we compute the inverse transform

$$\mathbf{T}^{-1}e^{-|s|y} = \frac{1}{\sqrt{2\pi}} \int_0^{\infty} e^{isx}e^{-sy} \, ds + \frac{1}{\sqrt{2\pi}} \int_{-\infty}^0 e^{isx}e^{sy} \, ds$$

$$= \frac{1}{\sqrt{2\pi}} \left(\frac{1}{y - ix} + \frac{1}{y + ix} \right) = \frac{1}{\sqrt{2\pi}} \frac{2y}{x^2 + y^2}. \tag{17-8}$$

This shows that $e^{-|s|y} = \mathbf{T}g$, where g is the function (17-8). The convolution theorem applied to (17-7) now gives $u = f * g$, and that is the desired result.

Although the derivation assumes that $\mathbf{T}f$ exists, formulas (17-4) and (17-6) are valid for much larger classes of functions. For example, the first two of the functions

$$f(x) = 1 \qquad f(x) = 1 + |x|^{3/4} \qquad f(x) = x^{100} \qquad f(x) = e^{x^{1.9}}$$

can be used in (17-6), and they can all be used in (17-4). Nevertheless, none of these functions has a Fourier transform.[1]

PROBLEMS

1. Let $f(x) = 1$ for $x > 0$ and $f(x) = 0$ for $x < 0$. Show that the Poisson formula for a half plane gives

$$u(x,y) = \frac{1}{2} + \frac{1}{\pi} \tan^{-1} \frac{x}{y}$$

and verify that the expected differential equation and boundary conditions are satisfied. Does $f(x)$ have a Fourier transform?

2. Obtain the temperature distribution for a rod extending from $x = 0$ to $x = \infty$ if the initial temperature is $f(x)$ and the end $x = 0$ is insulated. *Hint:* Consider a rod extending from $-\infty$ to ∞, with initial temperature $f_0(x)$ defined by

$$f_0(x) \text{ even} \qquad f_0(x) = f(x) \qquad \text{for } x > 0.$$

Compare Prob. 5 of Sec. 10.

---

[1] See discussion of the corresponding phenomenon for Laplace transforms in Chap. 3, Sec. 18. The method used there is not applicable here, but our solutions can be verified directly under appropriate conditions. Examples of such verification are given in the problems following Sec. 20.

3. By taking $f_0(x)$ odd in Prob. 2, find the temperature distribution when the end $x = 0$ is not insulated but is kept at the temperature zero.

4. A rod extending from $x = 0$ to $x = l$ has the initial temperature distribution $f(x)$. By regarding this rod as part of an infinite rod with the initial temperature $f_0(x)$, find the temperature $u(x,t)$ when (a) both ends are insulated and (b) both ends are kept at the temperature zero. *Hint:* Let $f_0(x)$ have period $2l$. This method of satisfying boundary conditions was used for the vibrating-string problem in Sec. 6.

5. According to Prob. 2 of Sec. 16, the formula

$$\mathbf{T}u = \frac{\sin asl}{as}\mathbf{T}g$$

holds for the string subject to $u(x,0) = 0$, $u_t(x,0) = g(x)$. By means of the convolution theorem, obtain d'Alembert's formula for this case (see Sec. 3). *Hint:* Use Sec. 16, Prob. 1, with $c = at$.

6. Let $I(x) = \mathbf{T}^{-1}e^{-cs^2}$, where c is constant. (a) Differentiate and integrate the result by parts to obtain

$$\frac{dI}{dx} = -\frac{x}{2c}I \qquad \text{hence } I(x) = I(0)e^{-x^2/4c}.$$

(b) Using Parseval's equality (Chap. 1, Sec. 22), get $[I(0)]^2 = 1/2c$. Thus deduce the formula

$$\mathbf{T}^{-1}e^{-cs^2} = (2c)^{-\frac{1}{2}}e^{-x^2/4c}.$$

In particular, conclude that $e^{-x^2/2}$ and $e^{-s^2/2}$ are transforms of each other.
(c) Taking $x = 0$, deduce that

$$\int_{-\infty}^{\infty} e^{-cs^2}\,ds = \sqrt{\frac{\pi}{c}}.$$

7. (a) Verify that each of Eqs. (16-10), (16-14), and (16-20) has the general form

$$u(x,y) = \mathbf{T}^{-1}[\Lambda(s)]^y\mathbf{T}u(x,0)$$

except perhaps for notation. What is the value of $\log \Lambda$? (b) Assuming needed convergence properties in part a, show that direct computation of $u(x, a + b)$ from $u(x,0)$ yields the same result as computation of $u(x,a)$ from $u(x,0)$ followed by computation of $u(x, a + b)$ from $u(x,a)$, when $a + b > a > 0$. (c) If $[\Lambda(s)]^y = \mathbf{T}\lambda(x,y)$, where the transform is with respect to x, the convolution theorem gives

$$u(x,y) = \lambda(x,y) * u(x,0) = \frac{1}{\sqrt{2\pi}} \int_{-\infty}^{\infty} \lambda(x - \xi, y)u(\xi,0)\,d\xi$$

and under mild assumptions the result (b) is equivalent to

$$\lambda(x,a) * \lambda(x,b) = \lambda(x, a + b).$$

For the function (17-8), show that this equation reduces to

$$\frac{1}{\pi} \int_{-\infty}^{\infty} \frac{a}{\xi^2 + a^2} \frac{b}{(x - \xi)^2 + b^2}\,d\xi = \frac{a + b}{x^2 + (a + b)^2}$$

and decide whether it is true. (This problem illustrates a so-called "semigroup property" that is important in the modern theory of partial differential equations.)

---

18. The Laplace Transform.

The Fourier transform is closely related to the Laplace transform introduced in Chap. 3, and the convolution theorems for the two transforms are also similar. We shall show how the Laplace transform is used in the solution of partial differential equations. It is assumed here that the reader is familiar with the Laplace transform and its application to ordinary differential equations, but such knowledge is not needed elsewhere in this chapter.

The main features of the method are illustrated by the solution of

$$u_t = \alpha^2 u_{xx} \qquad u(0,t) = f(t) \qquad u(x,0) = 0 \qquad t \geq 0, \, x \geq 0.$$

This problem concerns the temperature of a semi-infinite rod that has one end at temperature $f(t)$ and has initial temperature 0. Temporarily we assume that $f(t)$ is chosen so that u and u_t are continuous and satisfy inequalities of the form

$$|u(x,t)| \leq M e^{a(x+t)} \qquad |u_t(x,t)| \leq M e^{a(x+t)} \tag{18-1}$$

where M and a are constant. In particular, u is *admissible* in the sense of Chap. 3, Sec. 12, and we can introduce its Laplace transform

$$U(x,s) = \mathbf{L}u(x,t) = \int_0^\infty e^{-st} u(x,t) \, dt.$$

Here s is the usual transform variable and x is a parameter expressing the fact that U, like u, depends on x. The function U is called the *time transform* of u, because the integration is with respect to t.

By Chap. 3, Sec. 13, one has $\mathbf{L}y' = s\mathbf{L}y - y(0)$, where the prime denotes differentiation with respect to the variable of integration used in computing $\mathbf{L}y$. If we apply this to u and use partial differentiation to emphasize that x is held fast we get

$$\mathbf{L}u_t = s\mathbf{L}u - u(x,0) = sU(x,s).$$

The second equality follows from the initial condition $u(x,0) = 0$ and from the definition of U.

By (18-1) the integrals implied in $\mathbf{L}$ are uniformly convergent for large s and hence

$$\mathbf{L}\frac{\partial^2 u}{\partial x^2} = \frac{\partial^2}{\partial x^2} \mathbf{L}u = \frac{\partial^2 U}{\partial x^2}.$$

The transform of $u_t = \alpha^2 u_{xx}$ thus gives

$$sU = \alpha^2 U_{xx}.$$

This is a partial differential equation for U, but since the differentiation is with respect to x only, it can be solved as an ordinary differential equation of form $sU = \alpha^2 U''$, where s and α are constant. The result is

$$U = c_1 e^{x\sqrt{s}/\alpha} + c_2 e^{-x\sqrt{s}/\alpha}$$

where c_1 and c_2 are constant as far as the differentiation variable x is concerned.

The condition (18-1) gives an estimate for the growth of $U(x,s)$ which shows that $c_1 = 0$. Also the transform of the equation

$$u(0,t) = f(t)$$

gives $U(0,s) = F(s)$, where $F = \mathbf{L}f$, and hence $c_2 = F(s)$. We conclude that

$$U(x,s) = F(s)e^{-x\sqrt{s}/}$$

or, since $\mathbf{L}^{-1}U = u$ and $F = \mathbf{L}f$,

$$u(x,t) = \mathbf{L}^{-1}e^{-x\sqrt{s}/\alpha}\mathbf{L}f. \tag{18-2}$$

Under mild conditions the inverse transform $\mathbf{L}^{-1}$ can be expressed by an integral[1] much

---

[1] For the benefit of readers familiar with complex integration, the formula is

$$f(t) = \frac{1}{2\pi i} \int_{c-i\infty}^{c+i\infty} e^{st} F(s) \, ds \qquad c > 0 \text{ const.}$$

The subject has a large literature, and we cite only H. S. Carslaw and J. C. Jaeger, "Operational Methods in Applied Mathematics," Oxford University Press, Fair Lawn, N.J., 1945, and R. V. Churchill "Operational Mathematics," 2d ed., McGraw-Hill Book Company, New York, 1958.

like that used for $\mathbf{T}^{-1}$; hence (18-2) is a solution involving two integrations. A more satisfactory form can be obtained by the convolution theorem, however, as shown next.

By entry 6a of the table of transforms (see inside back cover)

$$e^{-x\sqrt{s}/\alpha} = \mathbf{L}g \qquad \text{where } g(t) = \frac{x}{2\alpha\pi^{1/2}t^{3/2}} e^{-x^2/(4\alpha^2t)}$$

so that (18-2) is $\mathbf{L}u = (\mathbf{L}f)(\mathbf{L}g)$. The convolution theorem of Chap. 3, Sec. 18, therefore gives

$$u(x,t) = \frac{x}{2\alpha\sqrt{\pi}} \int_0^t \frac{f(t_1)}{(t-t_1)^{3/2}} e^{-x^2/[4\alpha^2(t-t_1)]} \, dt_1.$$

Now that the solution is obtained, it can be verified under more general conditions than those used in the derivation. (Cf. problems following Sec. 20.) This is the way the transform is actually used in practice. One gets the solution by formal manipulation, without justifying each step, and at the end one justifies the solution on its own merits.

———————————————————————————————— **PROBLEMS**

1. Review Chap. 3, Secs. 12, 13, 16, and 18.

2. (a) In (18-2) if $f(t) = u_0$, a constant, obtain the formulas

$$U(x,s) = u_0 s^{-1} e^{-x\sqrt{s}/\alpha} \qquad u(x,t) = u_0 \int_0^t g(t) \, dt$$

by entries 6a and 12a of the table of transforms. (b) By a change of variable, get

$$u(x,t) = u_0 \operatorname{erf}\left(\frac{x}{2\alpha\sqrt{t}}\right) \qquad \text{where erf } z = \frac{2}{\sqrt{\pi}} \int_0^z e^{-\xi^2} \, d\xi.$$

(c) Verify the differential equation, and using erf $\infty = 1$, verify the initial and boundary conditions. [The function erf x is called the *error function*, and erfc $x = 1 - \operatorname{erf} x$ is the *complementary error function*. Thus

$$\operatorname{erf} x = 1 - \operatorname{erfc} x = 2\Phi(\sqrt{2}\, x)$$

where Φ is the function tabulated in Appendix C.]

3. A rod extending from $x = 0$ to $x = l$ has 0 initial temperature, the end $x = 0$ is insulated, and the end $x = l$ is maintained at the constant temperature u_0. If the temperature satisfies $u_t = \alpha^2 u_{xx}$, formulate appropriate initial, boundary, and growth conditions, and show that the time transform $U = \mathbf{L}u$ satisfies

$$U(x,s) = \frac{u_0 \cosh \sqrt{s}\, x/\alpha}{s \cosh \sqrt{s}\, l/\alpha}.$$

4. By the Laplace transform, solve

$$u_t = \alpha^2 u_{xx} \qquad x > 0, t > 0 \qquad u(0,t) = 0 \qquad u(x,0) = u_1$$

where u_1 is constant.

5. The temperature in a long radiating rod satisfies

$$u_t = \alpha^2 u_{xx} - hu \qquad x > 0, t > 0 \qquad u(x,0) = 0 \qquad u(0,t) = f(t).$$

Under mild additional assumptions, obtain the formula

$$U(x,s) = F(s)e^{-x\sqrt{s+h}/\alpha} \text{ where } F = \mathbf{L}f$$

for the time transform of u. By entries 6a and 9b of the table of transforms, get $u = f * g$, where

$$g(t) = \frac{x}{2\alpha\sqrt{\pi t^3}} \exp\left(-ht - \frac{x^2}{4\alpha^2 t}\right) \qquad \text{with } \exp z \equiv e^z.$$

6. The small oscillations of a semi-infinite beam satisfy

$$u_{tt} + k^2 u_{xxxx} = 0 \qquad x > 0, t > 0 \qquad u(0,t) = f(t) \qquad u_x(0,t) = 0$$

where k is constant.   Under suitable conditions, obtain the formula

$$u(x,t) = \mathbf{L}^{-1} e^{-x\sqrt{s/2k}} \cos x \sqrt{\frac{s}{2k}} \, \mathbf{L}f.$$

7. Discuss the application of the Laplace transform to the following vibrating-string problems:

(a) The semi-infinite string in the example in Sec. 3;
(b) The string with fixed ends and sinusoidal initial conditions in the example in Sec. 6;
(c) The string with damping considered in Sec. 7;
(d) The string under gravity in the example in Sec. 7;
(e) The general problem of forced oscillations (7-11).

8. A semi-infinite medium is initially at temperature 0 and is heated by radiation from a medium of constant temperature u_0 at $x = 0$.

(a) Given that the differential equation and boundary condition are

$$u_t = \alpha^2 u_{xx} \qquad x > 0, t > 0 \qquad u_x = h(u - u_0) \qquad x = 0$$

show that the Laplace transform with respect to t satisfies

$$U(x,s) = h\alpha u_0 \frac{e^{-x\sqrt{s}/\alpha}}{s(\sqrt{s} + h\alpha)}.$$

(b) Thus show, in the notation of Prob. 2, that

$$u = u_0 \operatorname{erfc} \frac{x}{2\alpha\sqrt{t}} - u_0 e^{hx + (h\alpha)^2 t} \operatorname{erfc}\left[\frac{x}{2\alpha\sqrt{t}} + h\alpha\sqrt{t}\right].$$

9. (a) Given $\alpha^2 U'' - \beta^2 U = \frac{1}{2}\delta(x)$, $x > 0$, suppose $U(x)$ is even, continuous, and tends to 0 as $x \to \infty$.   By taking $U(0) = 0$, $U'(0) = c$, where c is a suitable constant, and using the Laplace transform, obtain

$$U(x) = (2\alpha\beta)^{-1} e^{-(\beta/\alpha)|x|}.$$

(b) The singular solution for heat flow satisfies

$$u_t = \alpha^2 u_{xx} \qquad u(x,0) = \frac{1}{2}\delta(x) \qquad x > 0, t > 0.$$

Obtain an equation similar to that in part a for $U = \mathbf{L}u$; thus get U, and

$$u(x,t) = \frac{1}{2\alpha(\pi t)^{\frac{1}{2}}} e^{-x^2/(4\alpha^2 t)}.$$

10. The small longitudinal displacements of a uniform bar subject to a force $f(t)$ at one end satisfy

$$\rho u_{tt} = E u_{xx} \qquad 0 < x < l, t > 0 \qquad u(x,0) = u_t(x,0) = u(0,t) = 0$$

as well as $E u_x(l,t) = f(t)$, where E is Young's modulus and ρ the density.

(a) If $F = \mathbf{L}f$, obtain the formula

$$u(l,t) = \frac{1}{\sqrt{\rho E}} \mathbf{L}^{-1} \frac{\tanh \sqrt{\rho/E}\, sl}{s} \mathbf{L}f.$$

(b) Show that $u(l,t)$ follows a square wave if $f(t)$ is the impulse function $\delta(t)$.   *Hint:* See Chap. 3, Eq. (16-5).

11. A circular cylinder of radius r_0 has its surface at the temperature $f(t)$.   If the initial temperature is 0, the temperature $u(r,t)$ satisfies

$$u_t = \alpha^2(u_{rr} + r^{-1}u_r) \qquad 0 \leq r < r_0, t > 0$$

together with the given initial and boundary conditions. Show that

$$u(r,t) = \mathbf{L}^{-1} \frac{I_0(\sqrt{sr/\alpha})}{I_0(\sqrt{sr_0/\alpha})} \mathbf{L}f$$

in general, where $I_0(z) = J_0(iz)$ is the Bessel function with imaginary argument.

---

19. The Source Functions for Heat Flow.   According to (17-5) the function

$$\frac{Q}{\rho c (4\pi \alpha^2 t)^{\frac{1}{2}}} e^{-r^2/4\alpha^2 t} \qquad r^2 = x^2 \tag{19-1}$$

represents the temperature distribution due to an instantaneous source of strength Q at the origin.   Equation (19-1) applies to the one-dimensional heat equation $\alpha^2 u_{xx} = u_t$. The corresponding result for two dimensions is

$$\frac{Q}{\rho c (4\pi \alpha^2 t)} e^{-r^2/4\alpha^2 t} \qquad r^2 = x^2 + y^2 \tag{19-2}$$

and for three dimensions it is

$$\frac{Q}{\rho c (4\pi \alpha^2 t)^{\frac{3}{2}}} e^{-r^2/4\alpha^2 t} \qquad r^2 = x^2 + y^2 + z^2. \tag{19-3}$$

In these formulas r is *the distance from the source to the point of observation* and t is *the length of time that has elapsed* since the heat was released.   The value of ρ is, respectively, the linear, surface, or volume density.

The functions (19-1) to (19-3) are solutions, respectively, of

$$\alpha^2 u_{xx} = u_t \qquad \alpha^2 (u_{xx} + u_{yy}) = u_t \qquad \alpha^2 (u_{xx} + u_{yy} + u_{zz}) = u_t.$$

Also they give the limit 0 as $t \to 0$ through positive values, provided $r \neq 0$.   Hence the initial temperature distribution is concentrated entirely at the origin.   By integrating over the whole space it can be shown in each case that the total amount of heat present is Q when $t > 0$.   For these reasons, the physical interpretation as a *point source of strength* Q is fully justified.

The expressions (19-1) to (19-3) indicate that heat travels with infinite speed.   Even if r is large, we get a *positive* temperature for each positive t, no matter how small, but the initial temperature was zero.   By contrast, the disturbance associated with the wave equation travels with finite speed (cf. Secs. 2 and 4.)

To illustrate the use of (19-3), let us find $u(x,y,z,t)$ when the initial temperature

$$u(x_1,y_1,z_1,0) = f(x_1,y_1,z_1)$$

is given at each point (x_1,y_1,z_1) of space.   Instead of this distribution we introduce a source of strength

$$Q = c\rho f(x_1,y_1,z_1)\, dx_1\, dy_1\, dz_1 \tag{19-4}$$

at (x_1,y_1,z_1).   The temperature at point (x,y,z) and time t due to *one* such source is given by (19-3), with Q as in (19-4) and with r the distance from (x,y,z) to the source:

$$r^2 = (x_1 - x)^2 + (y_1 - y)^2 + (z_1 - z)^2.$$

The temperature at (x,y,z) due to *all* the sources is given by superposition:

$$u(x,y,z,t) = (4\pi \alpha^2 t)^{-\frac{3}{2}} \int_{-\infty}^{\infty} \int_{-\infty}^{\infty} \int_{-\infty}^{\infty} e^{-r^2/4\alpha^2 t} f(x_1,y_1,z_1)\, dx_1\, dy_1\, dz_1. \tag{19-5}$$

As another illustration, we shall find the temperature due to a point source which emits heat continually. Let $Q(t)$ represent the strength of the source, so that the amount of heat emitted in time interval $(t_1, t_1 + dt_1)$ is approximately $Q(t_1)\,dt_1$. The heat at the present time t due to the source at time t_1 is

$$\frac{Q(t_1)\,dt_1}{\rho c[4\pi\alpha^2(t - t_1)]^{3/2}}\, e^{-r^2/[4\alpha^2(t-t_1)]}$$

when we recall that t in (19-3) stands, not for the time, but for the *elapsed* time. Adding the contributions from the source at all values of t_1 prior to t gives

$$u = \frac{1}{\rho c}\int_{-\infty}^{t} \frac{1}{[4\pi\alpha^2(t - t_1)]^{3/2}}\, e^{-r^2/[4\alpha^2(t-t_1)]}Q(t_1)\,dt_1. \tag{19-6}$$

If $Q(t)$ is a constant Q, the integral can be evaluated explicitly by the change of variable

$$s^2 = \frac{r^2}{4\alpha^2(t - t_1)}.$$

The result is
$$u = \frac{Q}{4\pi\alpha^2\rho c}\frac{1}{r}. \tag{19-7}$$

This represents the temperature due to a continuous uniform source of heat at a distance r from the point of observation. Since the conditions are steady state, the solution satisfies Laplace's equation. (Compare Sec. 12, where the function $1/r$ was obtained in connection with electrostatics and gravitation.)

Example. A line contact is pressed against the plane $x = 0$ with constant normal force F per unit length, the coefficient of friction being a constant μ. At time $t = 0$ it starts to slide in a direction perpendicular to its length with constant velocity v (see Fig. 25). Obtain the temperature in the medium $x \leq 0$, assuming this temperature to have been zero initially and neglecting heat loss at the surface $x = 0$.

This problem arises in the theory of milling, leather glazing, and lathe turning. To solve it, let the line contact be initially coincident with the z axis, so that its height at time t_1 is $y = vt_1$. The heat generated by friction per unit length is $F\mu\,dy$, and hence the heat generated per unit length in the time interval dt_1 is

$$Q = F\mu v\,dt_1.$$

Using this value of Q in the result (19-2) we obtain

$$\frac{F\mu v\,dt_1}{4\rho c\alpha^2\pi(t - t_1)}\, e^{-[x^2 + (y - vt_1)^2]/4\alpha^2(t-t_1)}$$

for the contribution, at the point (x,y,z) and at the present time t, due to motion of the line contact in the time interval $(t_1, t_1 + dt_1)$. [The reader is reminded that t in (19-2) is the elapsed time and $x^2 + y^2$ in (19-2) is the square of the distance from the point of observation to the line source.] Superposition yields the final answer:

$$u(x,y,z,t) = \frac{F\mu v}{4\rho c\alpha^2\pi}\int_0^t \frac{e^{-[x^2 + (y - vt_1)^2]/4\alpha^2(t-t_1)}}{t - t_1}\,dt_1.$$

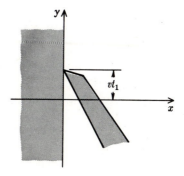

FIGURE 25

---

1. Show that (19-2) can be obtained by integrating (19-3) with respect to z and (19-1) by integrating (19-2) with respect to y.   Interpret physically.

2. What initial- or boundary-value problem is solved by (19-5)?   By (19-6)?

3. By use of (19-2), solve the initial-value problem

$$\alpha^2(u_{xx} + u_{yy}) = u_t \qquad u(x,y,0) = f(x,y).$$

4. Find the temperature distribution $u(x,y,z,t)$ for $x > 0$ due to a time-dependent distribution $f(y,z,t)$ on the plane $x = 0$.   Take the initial temperature as zero for $x > 0$.

5. State and solve the two-dimensional analog of Prob. 4; the one-dimensional analog.

---

20. A Singular Integral Formula.   The source functions considered in the preceding section have a *singularity* at $r = t = 0$, because they become infinite when $r = 0$ and[1] $t \to 0+$.   We shall now obtain a general integral formula containing a function with a singularity.   In the following sections this formula is used in the study of several partial differential equations.

The reader is assumed to be familiar with the integral theorems of vector analysis as presented in Chap. 6, but for convenience the main results are summarized here.   In the *divergence theorem*

$$\int_\tau (\nabla \cdot \mathbf{A}) \, d\tau = \int_\sigma \mathbf{A}_n \, d\sigma \qquad \begin{cases} d\tau = \text{volume element} \\ d\sigma = \text{surface element} \end{cases} \tag{20-1}$$

where $\mathbf{A}$ is a continuously differentiable vector function defined in a region τ bounded by the surface σ, the choice $\mathbf{A} = u\nabla v$ yields *Green's first identity*

$$\int_\tau [u\nabla^2 v + (\nabla u \cdot \nabla v)] \, d\tau = \int_\sigma u \frac{\partial v}{\partial n} \, d\sigma \tag{20-2}$$

when we recall that $(\nabla v)_n = \partial v/\partial n$, the normal derivative.   Writing (20-2) with u and v interchanged and subtracting give *Green's symmetric identity*

$$\int_\tau (u\nabla^2 v - v\nabla^2 u) \, d\tau = \int_\sigma \left(u \frac{\partial v}{\partial n} - v \frac{\partial u}{\partial n} \right) d\sigma. \tag{20-3}$$

The conditions for validity of these identities are discussed in Chap. 6.   For our present purposes we need an appropriate form of (20-3) when v does not satisfy the continuity conditions there required.

To this end, we show that

$$\lim_{a \to 0} \int_{\sigma_1} \frac{f(Q)}{a^c} \, d\sigma = \begin{cases} 4\pi f(P) & \text{for } c = 2 \\ 0 & \text{for } c < 2 \end{cases} \tag{20-4}$$

where f is continuous, c is constant, and the region of integration σ_1 is the surface of a sphere of radius a centered at P.   The variable point Q ranges over the surface, and $d\sigma = d\sigma_Q$.

The integral (20-4) can be written

$$\int_{\sigma_1} \frac{f(Q)}{a^c} \, d\sigma = \int_{\sigma_1} \frac{f(P)}{a^c} \, d\sigma + \int_{\sigma_1} \frac{f(Q) - f(P)}{a^c} \, d\sigma = I_1 + I_2.$$

Since the area of the sphere is $4\pi a^2$, we have, as $a \to 0$,

---

[1] The notation $t \to 0+$ indicates that $t \to 0$ through positive values.

$$I_1 = \frac{f(P)}{a^c} \int_{\sigma_1} d\sigma = \frac{f(P)}{a^c} 4\pi a^2 \to \begin{cases} 4\pi f(P) & \text{for } c = 2 \\ 0 & \text{for } c < 2. \end{cases}$$

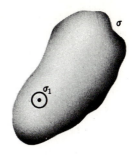

Since a surface integral does not exceed the area of the surface times the maximum value of the integrand, we have

$$|I_2| \leq 4\pi a^2 \frac{\max |f(Q) - f(P)|}{a^c} \leq 4\pi \max |f(Q) - f(P)|.$$

If f is continuous, this tends to zero as $a \to 0$, and (20-4) follows.

Let us now apply (20-3) to the function

$$v = w + \frac{1}{r} \qquad r = r(P,Q) \tag{20-5}$$

where w is twice continuously differentiable and where r is the distance from a fixed point P to the variable point of integration, Q. The region of integration is to be the region inside a given closed surface σ and outside a small sphere σ_1 of radius a centered at P (see Fig. 26). In this region $r \neq 0$, and (20-3) can be applied without hesitation.

According to (20-5) we have

$$\nabla^2 v = \nabla^2 w + \nabla^2 \frac{1}{r} = \nabla^2 w. \tag{20-6}$$

On σ_1 the outward normal n is directed along the radius into the sphere, so that

$$\frac{\partial}{\partial n} \frac{1}{r} = -\frac{\partial}{\partial r} \frac{1}{r} = \frac{1}{r^2} = \frac{1}{a^2}.$$

Since $r = a$ on σ_1, the foregoing equation and (20-5) give

$$\frac{\partial v}{\partial n} = \frac{\partial w}{\partial n} + \frac{1}{a^2} \qquad v = w + \frac{1}{a} \qquad \text{on } \sigma_1. \tag{20-7}$$

The surface integral in (20-3) can be written as an integral over σ plus an integral over σ_1. By inspection of (20-7) the integral over σ_1 is

$$\int_{\sigma_1} \left[u\left(\frac{\partial w}{\partial n} + \frac{1}{a^2}\right) - \left(w + \frac{1}{a}\right) \frac{\partial u}{\partial n} \right] d\sigma. \tag{20-8}$$

This becomes $4\pi u(P)$ as $a \to 0$, in view of (20-4). If we use this result and (20-6) in (20-3), we obtain the desired formula

$$4\pi u(P) = \int_\tau (u\nabla^2 w - v\nabla^2 u) \, d\tau + \int_\sigma \left(v\frac{\partial u}{\partial n} - u\frac{\partial v}{\partial n} \right) d\sigma \tag{20-9}$$

upon letting $a \to 0$. When P is exterior to τ, the same formula is valid, except that $4\pi u(P)$ must be replaced by 0.

Since the volume integral in (20-9) is taken over the whole region τ, it includes the point P at which $v = \infty$. The meaning of the integral is clear from the derivation, but we shall show directly that a singularity of the type $1/r$ in a volume integral causes no convergence difficulties. If τ_1 is the interior of the sphere with surface σ_1, we have

$$\int_{\tau_1} \frac{1}{r} \, d\tau = \int_0^a \frac{1}{r} 4\pi r^2 \, dr = 2\pi a^2.$$

This is clearly finite and in fact tends to zero as $a \to 0$.

In looking back over the analysis, the reader will note that the contribution associated with the singularity function $1/r$ comes from the immediate neighborhood of point P, where $r = 0$. Since u is continuous, u is nearly equal to $u(P)$ throughout this neighborhood, and that fact gives rise to the term $4\pi u(P)$ in (20-9). As suggested by the following problems, many formulas of applied mathematics have this same general structure.

PROBLEMS

1. Let $k(x,y)$ be a nonnegative integrable function of x for $-a \leq x \leq a$, where a is a positive constant or $a = \infty$. For each fixed δ satisfying $0 < \delta \leq a$, suppose

$$\lim_{y \to 0+} \int_{-\delta}^{\delta} k(x,y) \, dx = 1.$$

If $h(x)$ is bounded and integrable for $|x| \leq a$, and continuous at $x = 0$, prove that

$$\lim_{y \to 0+} \int_{-a}^{a} k(x,y)h(x) \, dx = h(0).$$

Outline of solution: Let the integral be written in the form

$$\int_{-\delta}^{\delta} k(x,y)h(x) \, dx + \left(\int_{-a}^{a} - \int_{-\delta}^{\delta} \right) k(x,y)h(x) \, dx.$$

If δ is so small that $|h(x) - h(0)| < \epsilon$, the first integral is between

$$\int_{-\delta}^{\delta} k(x,y)[h(0) - \epsilon] \, dx \qquad \text{and} \qquad \int_{-\delta}^{\delta} k(x,y)[h(0) + \epsilon] \, dx.$$

For sufficiently small positive y this is between $h(0) - 2\epsilon$ and $h(0) + 2\epsilon$. The second integral does not exceed the value it would have when $h(x)$ is replaced by M, where $|h(x)| \leq M$. This tends to 0 as $y \to 0+$, since both integrals tend to M.

2. By setting $x = x_0$ and then changing the variable of integration, show that the Poisson formula (17-6) can be written

$$u(x_0,y) = \int_{-\infty}^{\infty} k(x,y)f(x_0 + x) \, dx \qquad \text{where } k(x,y) = \frac{y}{\pi(x^2 + y^2)}.$$

Deduce that $u(x_0,y) \to f(x_0)$ as $y \to 0+$, provided f is bounded and integrable on $(-\infty,\infty)$ and continuous at x_0. *Hint:* Use Prob. 1 with $a = \infty$ and $h(x) = f(x_0 + x)$.

3. By setting $\phi - \theta = \psi$ in the Poisson formula (12-13) and using the periodicity of the integrand, show that the formula can be written

$$U(r,\theta) = \int_{-\pi}^{\pi} k(\psi,r)f(\psi + \theta) \, d\psi \qquad \text{where } k(\psi,r) = \frac{R^2 - r^2}{2\pi(R^2 - 2Rr \cos \psi + r^2)}.$$

By the series representation (12-12) for $k(\psi,r)$, show that

$$\int_{-\pi}^{\pi} k(\psi,r) \, d\psi = 1 \qquad 0 \leq r < R.$$

Deduce that $k(\psi,r)$ has the properties needed for application of Prob. 1 with $x = \psi$, $y = R - r$, and $h(x) = f(x + \theta)$. *Hint:* The integral of k from δ to π tends to 0 as $r \to R$.

4. After the manner of the preceding problems, discuss the singular integral for heat flow, Eq. (17-4).

---

21. The Poisson Equation.   If u has continuous second derivatives, the Laplacian $\nabla^2 u$ is a continuous function of position. We shall denote this function by $-4\pi\rho(x,y,z)$, so that

$$\nabla^2 u = -4\pi\rho. \tag{21-1}$$

The choice $v = 1/r$, $w = 0$ in (20-9) now yields the *Poisson formula*

$$u(P) = \int_\tau \frac{\rho}{r}\, d\tau + \frac{1}{4\pi} \int_\sigma \left(\frac{1}{r}\frac{\partial u}{\partial n} - u\frac{\partial}{\partial n}\frac{1}{r} \right) d\sigma \qquad (21\text{-}2)$$

when we divide by 4π. As before, $r = r(P,Q)$ is the distance from P to the variable point of integration Q. The formula (21-2) holds for every function having continuous second derivatives in τ and on its boundary.[1]

We now change our viewpoint. Instead of starting with u and defining ρ by (21-1), we suppose that ρ is given in advance. Equation (21-1) is now a partial differential equation for the unknown function u; it is called the *Poisson equation*. The foregoing considerations show that, if u satisfies the Poisson equation, u is given by the Poisson formula. The interest of the formula is that it yields the values of u throughout the interior of τ in terms of u and $\partial u/\partial n$ on the surface only.

For a physical interpretation, let u be the electrostatic potential due to a charge distribution of density ρ. The fact that the potential satisfies Poisson's equation is established in treatises on electrostatics,[2] so that this interpretation is consistent with (21-1). Since q/r represents the potential due to a charge q at a distance r from the point of observation, the term

$$\frac{1}{r}\,(\rho\, d\tau) \qquad \text{where } r = r(P,Q) \text{ and } \rho = \rho(Q)$$

represents the potential at P due to the charges within the volume element $d\tau$ at Q. Hence the first term of (21-2),

$$\int_\tau \frac{1}{r}\,(\rho\, d\tau)$$

represents the potential at P due to charges within the body τ. Similarly the second term in (21-2),

$$\int_\sigma \frac{1}{r}\left(\frac{1}{4\pi}\frac{\partial u}{\partial n}\, d\sigma \right)$$

represents the potential at P due to a surface-charge distribution on the surface σ.

To interpret the term

$$\int_\sigma \left(\frac{\partial}{\partial n}\frac{1}{r} \right)\left(-\frac{u}{4\pi}\, d\sigma \right) \qquad (21\text{-}3)$$

in (21-2), we consider the configuration shown in Fig. 27. Here, a charge $-q$ is introduced at the point Q on the surface σ and a charge $+q$ at a distance Δn along the outward normal $\mathbf{n}$ to σ. The distance from $-q$ to P is r, and the distance from q to P is denoted by r_1. If we take

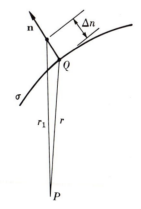

FIGURE 27

[1] Provided the region is simple enough to permit the use of (20-9). This condition on τ is hereby postulated once for all.
[2] The case $\rho = 0$ in (21-1) is discussed in Chap. 6, Sec. 14. A detailed analysis of the conditions under which Poisson's equation holds may be found in O. D. Kellogg, "Foundations of Potential Theory," p. 156, Springer-Verlag OHG, Berlin, 1929.

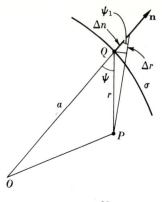

FIGURE 28

$q = m/\Delta n$, where m is constant, the potential at P is

$$q\frac{1}{r_1} - q\frac{1}{r} = q\,\Delta\frac{1}{r} = m\frac{\Delta(1/r)}{\Delta n} \to m\frac{\partial}{\partial n}\frac{1}{r}$$

as $\Delta n \to 0$. [That the limit is $m\,\partial(1/r)/\partial n$ follows from the *definition* of normal derivative, without calculation.] The limiting configuration of Fig. 27 is called a *dipole*; the constant m is called the *moment* of the dipole. We have thus found the desired interpretation of (21-3); namely, (21-3) represents the potential due to a surface distribution of dipoles having the moments $-u\,d\sigma/4\pi$. A surface distribution of dipoles such as this is called a *double layer*.

Since the volume integral in (21-2) is extended only over τ, it does not take account of the charges outside τ. That purpose is served by the surface integral in (21-2). From this viewpoint (21-2) shows that *the charges outside τ can be replaced by a suitable surface charge and double layer on σ*, without changing the potential within τ. If τ increases beyond all bounds, the limiting value of the surface integral can be thought to represent the influence of the charges at infinity.

In many important problems there are no charges at infinity, so that the limiting value of the surface integral is zero. To investigate this possibility, let σ be a large sphere of radius a centered at the origin O. By inspection of the differential triangle in Fig. 28,

$$\Delta r \sim \Delta n \cos \psi_1 \sim \Delta n \cos \psi \qquad \text{as } \Delta n \to 0$$

where ψ is the angle between OQ and PQ. The definition of normal derivative leads to

$$\frac{\partial r}{\partial n} = \lim_{\Delta n \to 0} \frac{\Delta r}{\Delta n} = \cos \psi \tag{21-4}$$

and hence

$$\frac{\partial}{\partial n}\frac{1}{r} = -\frac{1}{r^2}\frac{\partial r}{\partial n} = -\frac{\cos \psi}{r^2}. \tag{21-5}$$

If P is fixed and $a \to \infty$, it is easily seen that $r \sim a$ uniformly with respect to the point Q on σ. Hence $1/r$ has the order of magnitude $1/a$, and by (21-5) the normal derivative has the order of magnitude $1/a^2$. Now, the surface integral in (21-2) does not exceed $4\pi a^2$ times the maximum of the integrand. By the foregoing remarks, the integral therefore tends to zero as $a \to \infty$ if

$$a \max\left|\frac{\partial u}{\partial n}\right| \to 0 \qquad \text{and} \qquad \max|u| \to 0 \qquad \text{as } a \to \infty. \tag{21-6}$$

In this case (21-2) leads to the simple formula

$$u = \int \frac{\rho}{r}\,d\tau \qquad \text{integrated over all space.} \tag{21-7}$$

Referred to spherical coordinates[1] (r,θ,ϕ), the normal derivative $\partial u/\partial n$ in (21-6) is the radial derivative $\partial u/\partial r$, and $a = r$. Thus (21-6) is equivalent to

$$\lim_{r \to \infty} r\frac{\partial u}{\partial r} = 0 \qquad \lim_{r \to \infty} u = 0 \qquad \text{uniformly in } \theta \text{ and } \phi. \tag{21-8}$$

[1] The r in (21-8) has no relation to the $r = r(P,Q)$ that appears elsewhere in this discussion.

By substituting (21-5) into the integral and regrouping terms one finds that (21-8) may be replaced by the weaker condition

$$\lim_{r \to \infty} \left(r \frac{\partial u}{\partial r} + u \right) = 0 \qquad \lim_{r \to \infty} \frac{u}{r} = 0 \qquad \text{uniformly in } \theta \text{ and } \phi. \tag{21-9}$$

That is, if u satisfies (21-1) and (21-9), u can be represented in the form (21-7). When $\rho = 0$ outside a bounded region, an altogether different procedure[1] shows that the second condition (21-8) also suffices.

Example. If the region τ is a sphere of radius r_0 centered at P, every solution of Poisson's equation satisfies

$$u(P) = \frac{1}{4\pi r_0^2} \int_\sigma u \, d\sigma + \int_\tau \left(\frac{1}{r} - \frac{1}{r_0} \right) \rho \, d\tau. \tag{21-10}$$

Here $r = r(P,Q)$ is the distance from P to the variable point of integration Q. To prove (21-10) we choose $w = -1/r_0$ in (20-5) and note that on σ

$$v = 0 \qquad \frac{\partial v}{\partial n} = \frac{\partial v}{\partial r} = -\frac{1}{r^2} = -\frac{1}{r_0^2}.$$

The desired result follows at once from (20-9). The special case $\rho = 0$ in (21-10) yields the following:

AVERAGE-VALUE THEOREM: *If a function is harmonic throughout a sphere, its value at the center of the sphere equals the average of the values on the surface.*

This fact is of central importance in the study of harmonic functions.

The merit of taking $w = -1/r_0$ in (20-5) is that then $v = 0$ on σ. Hence the term involving $\partial u/\partial n$ in the surface integral (20-9) drops out. The possibility of making such a choice of v will be systematically exploited in Sec. 23.

PROBLEMS

1. If u is harmonic, show that the choice $u = v$ in (20-2) gives

$$\int_\tau (u_x^2 + u_y^2 + u_z^2) \, d\tau = \int_\sigma u \frac{\partial u}{\partial n} \, d\sigma.$$

2. Show that a solution of Poisson's equation in a closed bounded surface σ is wholly determined by its boundary values and that it is determined, apart from an additive constant, by the boundary values of the normal derivative. *Hint:* If u_1 and u_2 are two solutions, apply Prob. 1 to $u = u_1 - u_2$ and then use the result of Prob. 4. (This problem and similar ones given elsewhere refer to uniqueness only, not existence.)

3. Let u be harmonic in a region τ, and suppose u assumes its maximum value u_0 at an interior point P. Show that $u = u_0$ throughout every sphere contained in τ and centered at P. *Hint:* If $M(u)$ denotes the mean value on the surface of such a sphere, then $u_0 = M(u)$ and hence $M(u_0 - u) = 0$. Now use Prob. 4.

4. Let $f(Q)$ be continuous and nonnegative in a region τ. If $\int_\tau f \, d\tau = 0$, show that $f \equiv 0$. Similarly for surface integrals. *Hint:* If $f = \epsilon > 0$ at an interior point P, by continuity $f \geq \epsilon/2$ throughout some sphere τ_1 of radius $\delta > 0$ centered at P. But this gives

$$\int_\tau f \, d\tau \geq \int_{\tau_1} f \, d\tau \geq \int_{\tau_1} (\epsilon/2) \, d\tau > 0.$$

5. For $t > 0$, let $u_t = \alpha^2 \nabla^2 u$ in a bounded solid τ. If the conditions are such that Green's identity (20-2) can be applied and such that time differentiation and space integration can be interchanged, deduce that

---

[1] See H. B. Phillips, "Vector Analysis," p. 158, John Wiley & Sons, Inc., New York, 1933.

$$\frac{1}{2\alpha^2} \frac{\partial}{\partial t} \int u^2 \, d\tau = -\int |\nabla u|^2 \, d\tau - \int u \frac{\partial u}{\partial n} \, d\sigma.$$

If $u \, \partial u/\partial n \geq 0$ at every point of the boundary, conclude that $\int u^2 \, d\tau$ is a nonincreasing function of t and hence, by Prob. 4, $u = 0$ for $t > 0$ if $u = 0$ for $t = 0$.

6. Suppose a function $u(x,y,z,t)$ satisfies a boundary condition

$$\alpha(x,y,z,t)u - \beta(x,y,z,t) \frac{\partial u}{\partial n} = 0$$

on σ, where $\alpha \geq 0$, $\beta \geq 0$, and α and β do not vanish simultaneously. Show that $u \, \partial u/\partial n \geq 0$ and hence Prob. 5 applies. By setting $u = u_1 - u_2$ and using linearity, deduce a uniqueness theorem that generalizes the result of Prob. 2.

---

22. The Helmholtz Equation.

The *Helmholtz equation*

$$\nabla^2 u + k^2 u = 0 \tag{22-1}$$

is obtained by separating variables in the wave equation (Sec. 14) or by requiring harmonic time dependence (Sec. 16). A brief calculation[1] shows that (22-1) has the solution e^{ikr}/r, where r is the distance from a fixed point P to a variable point Q. If we set

$$v = \frac{e^{ikr}}{r} \qquad w = v - \frac{1}{r} \tag{22-2}$$

it follows that $\nabla^2 w = \nabla^2 v = -k^2 v$ for $r \neq 0$. Hence

$$u\nabla^2 w - v\nabla^2 u = u(-k^2 v) - v(-k^2 u) = 0 \tag{22-3}$$

provided $r \neq 0$ and provided u satisfies (22-1). Substituting (22-2) into (20-9) with due regard to (22-3) now yields the *Helmholtz formula*

$$u(P) = \frac{1}{4\pi} \int_\sigma \left[\frac{e^{ikr}}{r} \frac{\partial u}{\partial n} - u \frac{\partial}{\partial n} \left(\frac{e^{ikr}}{r} \right) \right] d\sigma. \tag{22-4}$$

This expresses the solution u of (22-1) as an integral involving the boundary values of u and $\partial u/\partial n$.

Sometimes the region τ is bounded and (22-1) holds at points exterior to τ. To see if (22-4) remains valid in this case, we construct a sphere τ_1 centered at the origin and having a radius a so large that τ is contained entirely within τ_1 (Fig. 29). Formula (22-4) applied to the region between τ and the surface σ_1 of the sphere gives

$$4\pi u(P) = \int_\sigma \left[\frac{e^{ikr}}{r} \frac{\partial u}{\partial n} - u \frac{\partial}{\partial n} \left(\frac{e^{ikr}}{r} \right) \right] d\sigma$$

$$+ \int_\sigma \left[\frac{e^{ikr}}{r} \frac{\partial u}{\partial n} - u \frac{\partial}{\partial n} \left(\frac{e^{ikr}}{r} \right) \right] d\sigma.$$

In the first integral the outer normal for the region of integration is the inner normal for τ. With this understanding we see that (22-4) holds in the present case, provided the integral over σ_1 tends to zero as $a \to \infty$.

To investigate the behavior as $a \to \infty$, note that on σ_1

$$\frac{\partial}{\partial n} \frac{e^{ikr}}{r} = \frac{\partial}{\partial r} \frac{e^{ikr}}{r} \frac{\partial r}{\partial n} = \frac{e^{ikr}}{r} \left(ik - \frac{1}{r} \right) \cos \psi$$

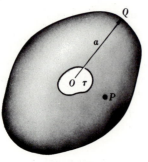

FIGURE 29

---

[1] Let the Laplacian be referred to spherical coordinates with origin at P.

by (21-4). Hence the integral over σ_1 becomes, after rearrangement,

$$\int_{\sigma_1} \frac{e^{ikr}}{r} \left[\frac{\partial u}{\partial n} - iku + iku(1 - \cos \psi) + \frac{1}{r} u \cos \psi \right] d\sigma. \tag{22-5}$$

As $a \to \infty$ with P fixed, we have $a \sim r$. Also, the law of cosines applied to Fig. 28 shows that $a^2(1 - \cos \psi)$ remains bounded. Hence, the integral will tend to 0 provided

$$a \max \left| \frac{\partial u}{\partial n} - iku \right| \to 0 \qquad \text{and} \qquad \max |u| \to 0 \qquad \text{as } a \to \infty.$$

This assumes k real, so that $|e^{ikr}| = 1$.

In spherical coordinates (r,θ,ϕ) the result of the foregoing discussion may be summarized as follows: *Formula* (22-4) *applies to the exterior of the bounded region* τ *provided*

$$r \left(\frac{\partial u}{\partial r} - iku \right) \to 0 \qquad and \qquad u \to 0 \tag{22-6}$$

as $r \to \infty$, *uniformly in* θ *and* ϕ. In just the same way it is found that (21-2) applies to the exterior of the bounded region τ provided (22-6) holds with $k = 0$.

Equation (22-6) with $k = 0$ is sometimes called the *Dirichlet condition*. It is the same as (21-8), and so means that there are no charges at infinity. Equation (22-6) with $k \neq 0$ is called the *Sommerfeld radiation condition;* it means that there are no sources of radiation at infinity.

Although (22-6) is the form usually given, it is unnecessarily restrictive. A more careful analysis of (22-5) shows that (22-6) may be replaced by the weaker condition

$$r \left(\frac{\partial u}{\partial r} - iku + \frac{u}{r} \right) \to 0 \qquad \text{and} \qquad \frac{u}{r} \to 0 \tag{22-7}$$

which reduces to (21-9) when $k = 0$.

─── **PROBLEMS**

1. (*a*) Obtain the Helmholtz equation by assuming $v = u(x,y,z) T(t)$ in the heat equation $v_t = \nabla^2 v$.
(*b*) Discuss the Helmholtz formula as applied to the cube $0 < x < \pi$, $0 < y < \pi$, $0 < z < \pi$ with $u = \sin x$, $u = \cos x$, $u = \sin x \cos y$ or $u = \sin x \sin y$ (four problems).

2. A value $\lambda = k^2$ such that the problem

$$\nabla^2 u + k^2 u = 0 \text{ in } \tau \qquad u = 0 \text{ on } \sigma$$

has a solution $u \not\equiv 0$ is called a *characteristic value*, and the corresponding solution u is a *characteristic function*. Let u be a characteristic function, and let v be any twice-differentiable function that vanishes on σ. By use of Green's identity (20-3), show that

$$\int_\tau u(\nabla^2 v + k^2 v) \, d\tau = 0.$$

3. In the preceding problem, deduce that u or $\nabla^2 v + k^2 v$ must vanish at some point of τ. Hence, if $u > 0$ and $v > 0$ in τ, show that $k^2 = -v^{-1} \nabla^2 v$ holds at some point of τ and therefore $\min (-v^{-1} \nabla^2 v) \leq k^2 \leq \max (-v^{-1} \nabla^2 v)$ where the max and min are over τ. (Thus any smooth function v that is positive in τ and 0 on the boundary furnishes both a lower and an upper bound for the characteristic value associated with a positive characteristic function u.)

4. Referring to Prob. 2, let k^2 be a characteristic value associated with a sphere of radius a, and assume that the corresponding characteristic function is positive. By applying Prob. 3 with $v = a^2 - x^2 - y^2 - z^2$ show that $ka \geq \sqrt{6}$.

5. In the notation of Prob. 2, suppose λ is a characteristic value associated with a positive function u, and μ is a characteristic value associated with a positive function v. Using Prob. 3, prove that $\lambda = \mu$.

23. The Functions of Green and Neumann.   The Laplace equation can be obtained by setting $\rho = 0$ in Poisson's equation (21-1) or by setting $k = 0$ in the Helmholtz equation (22-1).   The corresponding integral formulas, (21-2) and (22-4), both reduce to

$$u(P) = \frac{1}{4\pi} \int_\sigma \left(\frac{1}{r} \frac{\partial u}{\partial n} - u \frac{\partial}{\partial n} \frac{1}{r} \right) d\sigma. \tag{23-1}$$

This expresses every harmonic function in τ as an integral involving the boundary values and the boundary values of the normal derivative.   However, a harmonic function is determined by the *boundary values alone*, without any reference to the normal derivative.[1]   We shall now obtain a formula, similar to (23-1), in which $\partial u/\partial n$ is not present.

Such a formula can be found by an appropriate choice of v in (20-9).   Since $\nabla^2 u = 0$, the volume integral in (20-9) will drop out if

$$\nabla^2 w = 0 \qquad \text{throughout } \tau. \tag{23-2}$$

And the term involving $\partial u/\partial n$ in (20-9) will drop out if $v = 0$ on σ.   By (20-5), that condition is equivalent to

$$w = -\frac{1}{r} \qquad \text{on } \sigma. \tag{23-3}$$

Evidently, (23-2) and (23-3) determine w uniquely.   Since the function $r = r(P,Q)$ involves the fixed point P, the boundary condition (23-3) makes w, and hence v, depend on P.   The function v obtained in this way is called *Green's function* and is denoted by $G(P,Q)$.   Thus,

$$G(P,Q) = v = w + \frac{1}{r} \tag{23-4}$$

where $r = r(P,Q)$ and where w satisfies (23-2) and (23-3).   The formula (20-9) now yields

$$u(P) = -\frac{1}{4\pi} \int_\sigma u \frac{\partial v}{\partial n} d\sigma = -\frac{1}{4\pi} \int_\sigma u \frac{\partial G}{\partial n} d\sigma \tag{23-5}$$

with G given by (23-4).   The differentiation and integration in (23-5) involve Q, while P remains fixed.

What we have shown is the following: Let u satisfy

$$\nabla^2 u = 0 \text{ in } \tau \qquad u = f \text{ on } \sigma. \tag{23-6}$$

If the region τ has the Green function G, then

$$u(P) = -\frac{1}{4\pi} \int_\sigma f(Q) \frac{\partial G\,(P,Q)}{\partial n} d\sigma. \tag{23-7}$$

When a continuous function f is given in advance, it can be shown, conversely, that the function u in (23-7) satisfies (23-6).   In other words, *formula (23-7) solves the Dirichlet problem.*   The general Dirichlet problem is thus reduced to the special Dirichlet problems[2] that have to be solved in constructing Green's function.

To interpret Green's function physically, let a unit charge be placed at the point P interior to a closed, grounded conducting surface σ.   Since P is the only charge present, the potential has the form $v = w + 1/r$, where $\nabla^2 w = 0$.   Since the conductor σ is a grounded equipotential, $v = 0$ on σ, and hence v agrees with the v in the foregoing discussion.   *Thus, $G(P,Q)$ is the potential at Q due to a unit charge at P in the grounded conducting surface σ.*   Because of this interpretation the existence of Green's function is very plausible on physical grounds.[3]

---

[1] See Sec. 21, Prob. 2.

[2] One problem for each choice of P.

[3] A proof of the existence for regions likely to be met in practice is given in Kellogg, *op. cit.*, chap. 11. Uniqueness follows from Sec. 22, Prob. 2, and also from the results of Sec. 24.

The physical interpretation not only suggests that $G(P,Q)$ exists but gives a method of finding it in many cases. As an illustration, we shall construct Green's function for the half space $z > 0$ (Fig. 30). Let a charge $q = 1$ be placed at P and a charge $q = -1$ at P_1, the mirror image of P in the plane $z = 0$. By symmetry the potential $v = 0$ when $z = 0$, and hence v is Green's function. If r is the distance from P to Q and r_1 the distance from P_1 to Q, the potential is

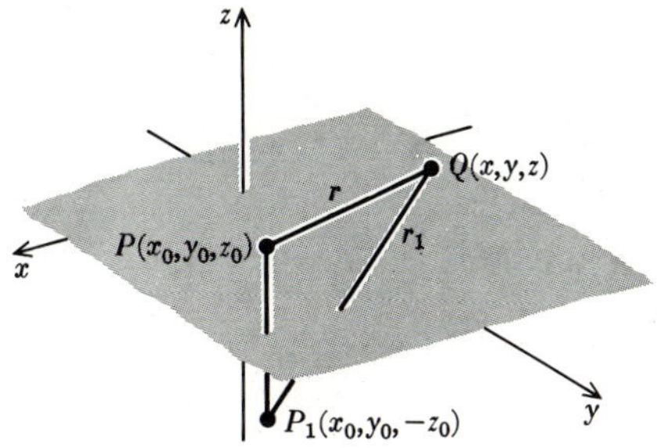

FIGURE 30

$$G(P,Q) = \frac{1}{r} - \frac{1}{r_1}.$$

As in the derivation of (21-5),

$$\frac{\partial G}{\partial n} = -\frac{2 \cos \psi}{r^2} \qquad \text{on } z = 0$$

where ψ is the angle between PQ and the normal to $z = 0$. Substituting into (23-7) yields the *Poisson formula for a half space*,

$$u(x,y,z) = \frac{1}{2\pi} \int_\sigma f \frac{\cos \psi}{r^2} \, d\sigma = \frac{z}{2\pi} \int_{-\infty}^{\infty} \int_{-\infty}^{\infty} \frac{f(x_1,y_1)}{[(x-x_1)^2 + (y-y_1)^2 + z^2]^{3/2}} \, dx_1 \, dy_1.$$

This formula represents a harmonic function for $z > 0$, which reduces to $f(x,y)$ when $z = 0$.

In terms of heat flow, the Dirichlet problem is to compute the steady-state temperature in a solid when the temperature on the surface is known. Sometimes the rate of heat flow across the surface is prescribed rather than the temperature. The problem that arises in this way is called the *Neumann problem;* it leads to the equations

$$\nabla^2 u = 0 \text{ in } \tau \qquad \frac{\partial u}{\partial n} = g \text{ on } \sigma. \tag{23-8}$$

If (23-8) is to have a solution, g must be restricted so that the rate of flow into τ equals the rate out; otherwise, a steady-state temperature cannot be expected. It is clear physically that the appropriate condition is

$$\int_\sigma g \, d\sigma = 0 \tag{23-9}$$

and, indeed, the choice $v = 1$ in (20-3) shows that (23-9) follows from (23-8).

When g satisfies (23-9), the problem is still not well posed because it has infinitely many solutions. That is, (23-8) involves the derivatives only, so that u can be altered by an additive constant. To make the solution unique, we require that

$$\int_\sigma u \, d\sigma = 0. \tag{23-10}$$

Although other conditions could be used, this one leads to simple formulas. Properly stated, the Neumann problem is to solve (23-8) when (23-9) and (23-10) hold.

By means of (20-9) we can develop a *Neumann function* $N(P,Q)$ analogous to the Green function $G(P,Q)$ of the foregoing paragraphs. As before, the condition

$$\nabla^2 w = 0 \qquad \text{throughout } \tau \tag{23-11}$$

makes the volume integral (20-9) drop out. To get rid of the surface integral involving u, we require that $\partial v/\partial n$ be constant on σ, and we recall (23-10). Since $v = w + 1/r$, this requirement is

$$\frac{\partial w}{\partial n} = -\frac{\partial}{\partial n} \frac{1}{r} + \text{const.} \tag{23-12}$$

To make w unique, we require also that

$$\int_\sigma w\, d\sigma = 0. \tag{23-13}$$

The Neumann function is

$$N(P,Q) \equiv v = w + \frac{1}{r} \tag{23-14}$$

where w satisfies (23-11) to (23-13). The solution u of (23-8) can be expressed in the form

$$u(P) = \frac{1}{4\pi}\int_\sigma gv\, d\sigma = \frac{1}{4\pi}\int_\sigma g(Q)N(P,Q)\, d\sigma. \tag{23-15}$$

When g is given in advance, it can be shown, conversely, that the function (23-15) satisfies (23-8). Hence, if we solve the particular Neumann problems involved in the construction of $N(P,Q)$, we can solve the general Neumann problem for the region.

Physically, the Neumann function represents the heat flow due to a source of strength 4π at P when the heat flows out at a uniform rate across the boundary. This shows that the condition

$$\frac{\partial v}{\partial n} = 0 \qquad \text{on } \sigma \tag{23-16}$$

analogous to (23-3) cannot be required in general; when the region is bounded, (23-16) violates the principle of conservation of heat. For *unbounded* regions, (23-16) is possible, as we see by considering the *Neumann function for a half space*,

$$N(P,Q) = \frac{1}{r} + \frac{1}{r_1}. \tag{23-17}$$

It is left for the reader to verify that (23-17) satisfies (23-16) on the plane $z = 0$ and to solve the Neumann problem.

PROBLEMS

1. In Fig. 30, let $\mathbf{P}$, $\mathbf{P}_1$, and $\mathbf{Q}$ be vectors from the origin to P, P_1, and Q, respectively, so that

$$r = |\mathbf{P} - \mathbf{Q}| \qquad r_1 = |\mathbf{P}_1 - \mathbf{Q}| \qquad \mathbf{P}_1 = \mathbf{P} - 2(\mathbf{P}\cdot\mathbf{k})\mathbf{k}$$

where $\mathbf{k}$ is the unit vector directed along the positive z axis. By expanding

$$r_1{}^2 = (\mathbf{P}_1 - \mathbf{Q})\cdot(\mathbf{P}_1 - \mathbf{Q})$$

show that

$$G(P,Q) = \frac{1}{|\mathbf{P} - \mathbf{Q}|} - \frac{1}{[|\mathbf{P} - \mathbf{Q}|^2 + 2(\mathbf{P}\cdot\mathbf{k})(\mathbf{Q}\cdot\mathbf{k})]^{1/2}}.$$

Conclude that $G(P,Q) = G(Q,P)$, and interpret this symmetry property in terms of the electrostatic potential.

2. Obtain a representation similar to that of Prob. 1 for the Neumann function associated with a half space, and deduce $N(P,Q) = N(Q,P)$.

3. In this problem, we construct Green's function for a sphere of radius R by putting a unit charge at the point P in the sphere and a charge of suitable magnitude $-q$ at the *inverse point*, P_1 (see Fig. 31). If vectors from the center O to P, P_1, and Q are $\mathbf{P}$, $\mathbf{P}_1$, and $\mathbf{Q}$, respectively, the inverse point, by definition, is given by

$$\mathbf{P}_1 = R^2|\mathbf{P}|^{-2}\mathbf{P}.$$

Thus O, P, and P_1 are collinear, and $|\mathbf{P}|\,|\mathbf{P}_1| = R^2$. The proposed formula for Green's function is

$$G(P,Q) = \frac{1}{|\mathbf{P} - \mathbf{Q}|} - \frac{q}{|\mathbf{P}_1 - \mathbf{Q}|}.$$

By writing the condition $G(P,Q) = 0$ in the form

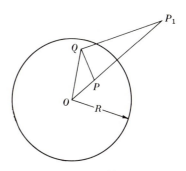

$$q^2(\mathbf{P} - \mathbf{Q})\cdot(\mathbf{P} - \mathbf{Q}) = (\mathbf{P}_1 - \mathbf{Q})\cdot(\mathbf{P}_1 - \mathbf{Q})$$

expanding, and using the definition of $\mathbf{P}_1$, show that $G(P,Q) = 0$ holds for all $|\mathbf{Q}| = R$ if, and only if, $|\mathbf{P}|q = R$. Conclude that Green's function is

$$G(P,Q) = \frac{1}{|\mathbf{P} - \mathbf{Q}|} - \frac{R|\mathbf{P}|}{|R^2\mathbf{P} - |\mathbf{P}|^2\mathbf{Q}|}.$$

4. In the preceding problem, show that $G(P,Q) = G(Q,P)$.

5. Show that the normal derivative of Green's function in Prob. 3 is

$$\frac{\partial G}{\partial n} = \frac{|\mathbf{P}|^2 - R^2}{R|\mathbf{P} - \mathbf{Q}|^{3/2}}.$$

Thus obtain *Poisson's formula*

$$u(P) = \frac{1}{4\pi R}\int_\sigma \frac{R^2 - |\mathbf{P}|^2}{|\mathbf{P} - \mathbf{Q}|^{3/2}}f(Q)\,d\sigma$$

for a function harmonic in the sphere $|\mathbf{P}| < R$ and assuming continuous boundary values $f(Q)$. *Hint:* After fixing P and the boundary point Q where the value of $\partial G/\partial n$ is to be verified, choose coordinates so that the origin is O, the x axis is along OQ, and the (x,y) plane is the plane OPQ. Then

$$\mathbf{P} = \mathbf{i}x_0 + \mathbf{j}y_0 \qquad \mathbf{Q} = \mathbf{i}x$$

and the normal derivative is the derivative with respect to x at $x = R$. Express $G(P,Q)$ in terms of x_0, y_0 and x, differentiate, and set $x = R$.

6. In the preceding problem, let $u(P)$ be harmonic in the sphere $|\mathbf{P}| \le R$, let $f(Q) = u(Q)$, and suppose $u(Q) \ge 0$ for $|\mathbf{Q}| = R$. By using the inequality

$$R + |\mathbf{P}| \ge |\mathbf{P} - \mathbf{Q}| \ge R - |\mathbf{P}|$$

for $|\mathbf{Q}| = R$, and the average-value theorem

$$u(0) = \frac{1}{4\pi R^2}\int u(Q)\,d\sigma$$

deduce the *Harnack inequality*,

$$\frac{R(R - |\mathbf{P}|)}{(R + |\mathbf{P}|)^2}\,u(0) \le u(P) \le \frac{R(R + |\mathbf{P}|)}{(R - |\mathbf{P}|)^2}\,u(0).$$

7. By letting $R \to \infty$ in the preceding problem, deduce that, if u is positive and harmonic everywhere, u is constant.

8. Write the two-dimensional Poisson formula (12-13) in a form similar to that in Prob. 5, and establish analogs of Probs. 6 and 7 in two dimensions.

9. Let the singular integral formula (20-9) be applied with $u = G(P_1,Q)$ and $v = G(P_2,Q)$, where G is Green's function and P_1 and P_2 are two distinct points of the region. By reviewing the derivation of the formula, show that the value of the integral is $4\pi G(P_1,P_2) - 4\pi G(P_2,P_1)$. On the other hand, the integral is 0 by properties of Green's function. Hence, changing notation, $G(P,Q) = G(Q,P)$.

10. As in Prob. 9, show that $N(P,Q) = N(Q,P)$.

ELLIPTIC PARABOLIC, AND HYPERBOLIC EQUATIONS

24. Classification and Stability. If a, b, and c are real continuous functions of x and y, and if H is a continuous function of the indicated arguments, the partial differential equation

$$az_{xx} + 2bz_{xy} + cz_{yy} = H(x,y,z,z_x,z_y) \tag{24-1}$$

includes many equations of mathematical physics.   It is convenient to classify equations of the type (24-1) according to the sign of the *discriminant*, $b^2 - ac$.   When $b^2 - ac < 0$, the equation is said to be *elliptic;* when $b^2 - ac = 0$, it is *parabolic;* and when $b^2 - ac > 0$, it is *hyperbolic*.   This nomenclature is suggested by analogy with the conic

$$ax^2 + 2bxy + cy^2 = H \qquad a, b, c, H \text{ const}$$

which is an ellipse, a parabola, or a hyperbola according to the sign of $b^2 - ac$.

As typical illustrations, the reader can verify that

$$u_{xx} + u_{yy} = 0 \qquad u_{xx} = u_y \qquad u_{xx} - u_{yy} = 0 \tag{24-2}$$

are elliptic, parabolic, and hyperbolic equations, respectively.   The first of these is Laplace's equation; the second is the equation for heat flow; the third describes the motion of waves on a string.[1]   The general equation (24-1) in the elliptic, parabolic, or hyperbolic case has much in common with the corresponding Eq. (24-2), and that is the reason why the classification is important.   We now discuss conditions for the unique determination of u.

CASE I.   Elliptic Equation.   Physical considerations suggest that a solution of Laplace's equation is wholly determined by its boundary values.   That is, if u_1 and u_2 satisfy

$$u_{xx} + u_{yy} = 0 \tag{24-3}$$

in a bounded region τ, and if $u_1 = u_2$ on the boundary, $u_1 = u_2$ in τ.   A mathematical proof is readily given, assuming that the function $u = u_1 - u_2$ is continuous in τ and on the boundary.

Without loss of generality, let the region τ lie between the lines $x = 0$ and $x = 1$ (so that $\cos x \neq 0$ in τ).   If v is defined by $u = v \cos x$ a short calculation shows that (24-3) yields

$$v_{xx} + v_{yy} - 2v_x \tan x - v = 0. \tag{24-4}$$

Suppose that $v > 0$ at an interior point P_0.   Then v assumes a positive maximum at an interior point P_1 (since $v = 0$ on the boundary and v is continuous).   At P_1 we have

$$v > 0 \qquad v_x = 0 \qquad v_{xx} \leq 0 \qquad v_{yy} \leq 0$$

and hence (24-4) cannot hold.   This contradiction shows that $v \leq 0$ throughout the region.   Similarly, $v \geq 0$, and hence $v \equiv 0$.   It follows that $u \equiv 0$, as was to be proved.

The same method can be used in three dimensions; the only change is that (24-4) has an extra term v_{zz}.   A decidedly less elementary proof (which applies also to the Neumann problem) was given in Sec. 21, Prob. 2.

What we have actually shown is that, if the harmonic function u satisfies $u \leq 0$ on the boundary, the same inequality holds at interior points.   Considering $u - m$ instead of u yields the following significant result:

MAXIMUM PRINCIPLE.   Let u be harmonic in a bounded region τ and let m be a constant.   If $u \leq m$ throughout the boundary of τ, then $u \leq m$ throughout τ.

Under mild restrictions on the function H, a maximum principle can be established for the general elliptic equation[2] (24-1).

[1] Let $y = \alpha^2 t$ in (8-5) and $y = at$ in (2-4).
[2] See H. Bateman, "Partial Differential Equations of Mathematical Physics," p. 135, Cambridge University Press, London, 1932.

CASE II.　Parabolic Equation.　Let u be the temperature of a thin rod extending from $x = 0$ to $x = l$.　With $y = \alpha^2 t$ the equation of heat conduction is

$$u_{xx} = u_y \qquad 0 < x < l \qquad 0 < y < \infty. \qquad (24\text{-}5)$$

As typical initial and boundary conditions, we assume that

$$u(x,0) = f(x) \qquad u(0,y) = g(y) \qquad u(l,y) = h(y). \qquad (24\text{-}6)$$

FIGURE 32

These conditions give the initial temperature and the temperature of the two ends.　In the xy plane, (24-6) specifies the value of u on the boundary of a certain semi-infinite rectangle (Fig. 32).　The physical interpretation suggests that u is uniquely determined within the rectangle, if the solution u exists, and we shall now show that this is, in general, the case.

Let u_1 and u_2 satisfy (24-5) and (24-6).　The function $u = u_1 - u_2$ then satisfies (24-5) and (24-6) with $f = g = h = 0$.　For simplicity, we shall suppose that u is continuous and bounded in the region of Fig. 32 and on its boundary, though these conditions could be weakened.

If v is defined by $u = ve^y$, substitution in (24-5) yields

$$v_{xx} = v_y + v. \qquad (24\text{-}7)$$

Suppose that $v = v_0 > 0$ at some point P_0 of the rectangle in Fig. 32.　We know that $v = 0$ on the three sides of this rectangle, and since u is bounded, the equation $v = e^{-y}u$ shows that $v < v_0$ if y is large enough.　It follows that v assumes a positive maximum at some interior point P_1. At P_1,

$$v_y = 0 \qquad v_{xx} \leq 0 \qquad v > 0$$

and hence (24-7) cannot hold.　This contradiction shows that $v \leq 0$, everywhere.　Similarly, $v \geq 0$, and hence $v \equiv 0$.　It follows that $u_1 \equiv u_2$.

The method of proving this *uniqueness theorem* leads, just as in the foregoing discussion, to the following:

MAXIMUM PRINCIPLE.　If $u \leq m$ on the boundary of the rectangle in Fig. 32, then $u \leq m$ throughout the rectangle.

By an extension of the foregoing reasoning one can establish the stability of the solution with respect to small changes in the conditions of the problem.　Instead of confining attention to Laplace's equation or the heat equation, we consider a broad class of nonlinear equations and boundary conditions that includes many cases of practical interest.　For this purpose the Laplacian is denoted by Δu, the normal derivative by u_ν, and the general point of the region by P.　Thus,

$$\Delta u = u_{xx} + u_{yy} + u_{zz} \qquad u_\nu = \frac{\partial u}{\partial n} \qquad P = (x,y,z).$$

Besides u, we introduce another function v to be compared with u.　It is assumed that the region τ is bounded and that $u - v$ is continuous in the closed region consisting of τ together with its boundary σ.　Our theorem then reads as follows:

STABILITY THEOREM.　Let a differential operator $\mathbf{D}$ and boundary operator $\mathbf{B}$ be defined for P in τ and P in σ, respectively, by

$$\mathbf{D}u = u - F(P, \text{grad } u, \Delta u), \qquad \mathbf{B}u = u - G(P,u_\nu).$$

Suppose $F[P, \text{grad } u(P), s]$ and $G(P,s)$ are nondecreasing functions of s for all relevant P. Then, if ϵ and δ are any constants,

$$|\mathbf{D}u - \mathbf{D}v| \leq \epsilon \qquad and \qquad |\mathbf{B}u - \mathbf{B}v| \leq \delta \qquad implies \quad |u - v| \leq \max(\epsilon,\delta).$$

The inequality involving **D** is assumed to hold throughout τ, while that involving **B** holds throughout σ. The conclusion holds in both τ and σ.

Before giving the proof, we discuss the significance of this result. If $w = u$ and $w = v$ are both exact solutions of the problem

$$\mathbf{D}w = f(P) \text{ in } \tau \qquad \mathbf{B}w = g(P) \text{ in } \sigma \tag{24-8}$$

the hypothesis holds with $\epsilon = \delta = 0$, and the conclusion gives $u = v$. This is a uniqueness theorem.

Suppose, next, that v is an unknown exact solution and u is a known approximation. Then (24-8) holds for $w = v$, but u satisfies conditions of the type

$$\mathbf{D}u = f(P) + \epsilon(P) \qquad \mathbf{B}u = g(P) + \delta(P)$$

with error terms $\epsilon(P)$ and $\delta(P)$ because u is only approximate. If $|\epsilon(P)| \leq \epsilon$ and $|\delta(P)| \leq \delta$, the stability theorem gives an upper bound for $|u - v|$. Thus the *solution error* is estimated in terms of the *equation errors* and *boundary errors*. Such estimates are useful in numerical analysis, as is evident.

If ϵ and δ approach zero, the theorem indicates that max $|u - v|$ also approaches zero. The approximate solution tends uniformly to the exact solution as the equation error and boundary error approach zero, and we say, briefly, that the exact solution is *stable*. In any physical problem the boundary data and the forcing term in the differential equation are known only with limited precision. Hence, *the condition of stability is essential if the problem makes sense physically*. This observation sheds light on the need for classification. It is possible to show that boundary conditions that lead to stable problems for elliptic equations do not as a rule lead to stable problems for parabolic or hyperbolic equations, and conversely.

The proof of the stability theorem is simple. Assuming that the conclusion does not hold, suppose $u - v$ assumes a maximum of value $m > \max(\epsilon, \delta)$ in τ or on σ. If the maximum occurs in τ, the conditions

$$\Delta u \leq \Delta v \qquad \text{grad } u = \text{grad } v \qquad u - v > \epsilon$$

lead to $\mathbf{D}u - \mathbf{D}v > \epsilon$, a contradiction. Similarly, if the maximum occurs on σ, the conditions

$$u_\nu \leq v_\nu \qquad u - v > \delta$$

give $\mathbf{B}u - \mathbf{B}v > \delta$. A contradiction is also obtained if $v - u$ has a maximum of value $m > \max(\epsilon, \delta)$, and this completes the proof.

It should be observed that the hypothesis about monotony of F is needed only for the known solution u, not for v. Also the precise definition of the *normal* is not important. The result holds for a region with edges and corners, such as a cube, if a suitable "inner normal" is arbitrarily defined at each point where u_ν occurs in $\mathbf{B}u$.

The stability theorem applies to the altered form of both the Laplace equation and the heat equation, that is, it applies to (24-4) and to (24-7). This fact leads to a corresponding stability theorem for the original equations. Similar results apply to general elliptic and parabolic equations and to Neumann problems,[1] though they are not presented here.

Proof of the stability theorem was based on the condition

$$\mathbf{D}u \leq \mathbf{D}v + \epsilon$$

which is an example of a *differential inequality*. The subject of differential inequalities and their bearing on nonlinear problems has a vast literature and could easily be expanded

[1] See, for example, R. M. Redheffer, Elementary Remarks on Problems of Mixed Type, *J. Math. Phys.*, March, 1964.

into a chapter as long as this one. Our purpose here is to point out the existence of this theory and its relevance to uniqueness, numerical estimation, and stability.

PROBLEMS

1. For what values of the constant k is $u_{xx} + ku_{xy} = u_{yy} = 0$ elliptic? Parabolic? Hyperbolic?

2. In what regions of the xy plane are the following elliptic? Hyperbolic?

$$u_{xx} + yu_{yy} = 0 \qquad (1 + y)u_{xx} + 2xu_{xy} + (1 - y)u_{yy} = u_x \qquad u_{xy} = (x^2 + y^2)(u_{xx} + u_{yy}).$$

3. Show that the solution of the elliptic equation $u_{xx} + u_{yy} = -ku$ is not always uniquely determined by the boundary values. *Hint:* Let the region be the square $0 \le x \le \pi$, $0 \le y \le \pi$, and separate variables. For a physical interpretation, see Sec. 14.

4. A *characteristic value* for a region τ is a constant λ such that the problem

$$u_{xx} + u_{yy} + u_{zz} + \lambda u = 0 \text{ in } \tau \qquad u = 0 \text{ on the boundary}$$

has a solution other than the trivial solution $u = 0$. Show that a characteristic value is always positive. *Hint:* If $u \not\equiv 0$, then u has a positive maximum or a negative minimum at some interior point P.

5. The semi-infinite strip $0 < x < \pi, y > 0$ has its edges kept at the constant temperature $u = 0$, whereas its end $y = 0$ is kept at the temperature $u = \sin x$. In the steady state the temperature u satisfies $u_{xx} + u_{yy} = 0$, and also

$$u(0,y) = 0 \qquad u(\pi,y) = 0 \qquad u(x,0) = \sin x.$$

(*a*) By the method of separating variables, obtain infinitely many distinct solutions to this problem. (*b*) Show that only one of these solutions satisfies $\lim_{y \to \infty} u(x,y) = 0$ uniformly in x. (*c*) If condition (*b*) is imposed, show that the problem has, in fact, only one solution. *Hint:* Use the maximum principle.

6. For all (x,y) a function $u(x,y)$ satisfies

$$\Delta u + m|\text{grad } u|^2 \ge 0 \qquad u \le m$$

where m is constant. Show that u is constant. *Hint:* If $u(P) < u(Q)$, consider $u - c \log \log r$, where c is a small positive constant. [A function such that $\Delta u \ge 0$ is said to be *subharmonic*. The result of this problem is a sharpened form of a theorem known as *Liouville's theorem for subharmonic functions;* see also Sec. 23, Prob. 7. Instead of $u \le m$ it would suffice to have $u \le \epsilon(r)|\log \log r|$, where $\epsilon(r) \to 0$ as $r \to \infty$.]

---

25. Further Discussion of Uniqueness. We now consider the hyperbolic equation

$$u_{xx} - u_{yy} = 0. \tag{25-1}$$

Since solutions of (25-1) do not satisfy the maximum principle, the foregoing methods cannot be used here.

CASE IIIa. Hyperbolic Equation, First Problem. Let the value and normal derivative of u in (25-1) be given on an interval (a,b) of the x axis (Fig. 33). Thus,

$$u(x,0) = f(x) \qquad u_y(x,0) = g(x) \qquad a < x < b.$$

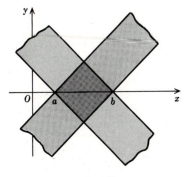

FIGURE 33

If $G(x) = \int_a^x g(s)\,ds$, d'Alembert's formula (3-15) yields an expression

$$2u(x,y) = f(x + y) + G(x + y) + f(x - y) - G(x - y) \tag{25-2}$$

which will now be used to discuss the uniqueness of u.

By hypothesis, $f(x)$ and $G(x)$ are determined for $a < x < b$ but not outside this interval.   Hence

$$
\begin{aligned}
f(x + y) + G(x + y) \text{ is determined for } a < x + y < b\\
f(x - y) - G(x - y) \text{ is determined for } a < x - y < b
\end{aligned}
\tag{25-3}
$$

but not elsewhere.   In the xy plane, the loci

$$a < x + y < b \qquad a < x - y < b$$

represent two strips, bounded by the two pairs of lines

$$x + y = a \qquad x + y = b \qquad \text{and} \qquad x - y = a \qquad x - y = b \tag{25-4}$$

(Fig. 33).   Both expressions (25-3), and hence u in (25-2), are uniquely determined in the intersection of these strips, but only there.   This shows that the region of determinacy is the doubly shaded region in the figure.

Similar behavior is found for the general hyperbolic equation, the role of the lines (25-4) being taken by the *characteristics* introduced in Sec. 28.   It is often possible to express u by an integral formula involving the initial value and normal derivative.   The method requires construction of the *Riemann function*,[1] which is in some respects analogous to the Green function of Sec. 23.

CASE IIIb.   Hyperbolic Equation, Second Problem.   The equation $u_{xx} = u_{yy}$ has the general solution

$$u(x,y) = f_1(x - y) + f_2(x + y) \tag{25-5}$$

as was shown in Sec. 1.   If u is given on two adjacent sides of the rectangle in Fig. 34, we shall use (25-5) to show that u is determined in the whole rectangle.   This corresponds to the result of Sec. 4, Theorem II, while the foregoing result corresponds to that of Sec. 4, Theorem I.

Choose a point R in the rectangle, and draw RQ and RS parallel to the sides of the rectangle, as in the figure.   With P the apex of the rectangle, $x - y$ is constant on PQ and $x - y$ is also constant on SR.   Hence the same is true of $f_1(x - y)$:

$$f_1(x - y) = \alpha \text{ at } P \text{ and } Q \qquad f_1(x - y) = \beta \text{ at } R \text{ and } S.$$

Similarly,

$$f_2(x + y) = \gamma \text{ at } P \text{ and } S \qquad f_2(x + y) = \delta \text{ at } Q \text{ and } R.$$

By using these values in (25-5) we can verify the identity

$$u(R) = u(Q) + u(S) - u(P). \tag{25-6}$$

This shows that u is determined by the data at every point in the rectangle and at no point outside the rectangle.

By a procedure known as *Picard's method*, the problem just discussed can often be

---

[1] See A. G. Webster, "Partial Differential Equations of Mathematical Physics," p. 248, B. G. Teubner Verlagsgesellschaft, mbH, Stuttgart, 1927.

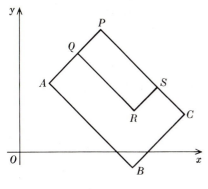

FIGURE 34

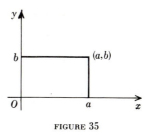

FIGURE 35

solved for the general hyperbolic equation (24-1).   It is indicated in Sec. 28 that the equation can be reduced, in general, to the form

$$u_{xy} = H(x,y,u,u_x,u_y) \tag{25-7}$$

and we shall suppose that this has been done.   For simplicity we assume the homogeneous boundary conditions

$$u(x,0) = 0 \qquad u(0,y) = 0 \qquad 0 \le x \le a, 0 \le y \le b. \tag{25-8}$$

Thus, $u = 0$ on two adjacent sides of the rectangle in Fig. 35.   The conditions (25-8) enable us to write (25-7) in the form

$$u(x,y) = \int_0^x \int_0^y H(x_1,y_1,u,u_x,u_y)\, dx_1\, dy_1$$

where the arguments of u, u_x, and u_y in the integral are x_1 and y_1.

Picard's method consists of choosing a first approximation $u^{(0)}$, evaluating the integral, and using the result as the second approximation $u^{(1)}$.   A similar process yields $u^{(2)}$, and so on.   If $u^{(n)}$ is the nth approximation, the next approximation is

$$u^{(n+1)}(x,y) = \int_0^x \int_0^y H(x_1,y_1,u^{(n)},u_x^{(n)},u_y^{(n)})\, dx_1\, dy_1. \tag{25-9}$$

Subject to mild restrictions on H it can be shown[1] that the solution is given by

$$u(x,y) = \lim_{n \to \infty} u^{(n)}(x,y).$$

As an illustration, let the equation be $u_{xy} = 1 + u$.   By (25-9),

$$u^{(n+1)} = \int_0^x \int_0^y [1 + u^{(n)}]\, dx_1\, dy_1 = xy + \int_0^x \int_0^y u^{(n)}\, dx_1\, dy_1.$$

Starting with $u^{(0)} = 0$, we get $u^{(1)} = xy$, $u^{(2)} = xy + (xy)^2/4$,

$$u^{(3)} = xy + \int_0^x \int_0^y \left[x_1 y_1 + \frac{(x_1 y_1)^2}{4} \right] dx_1\, dy_1 = xy + \frac{(xy)^2}{(2!)^2} + \frac{(xy)^3}{(3!)^2}$$

and so on.   Evidently, the process gives

$$u(x,y) = \sum_{n=1}^{\infty} \frac{(xy)^n}{(n!)^2}.$$

That this is a solution can be verified by actual substitution.

[1] R. Courant and D. Hilbert, "Methoden der mathematischen Physik," vol. II, p. 317, Springer-Verlag OHG, Berlin, 1937.

1. By means of Picard's method, solve the three equations

$$u_{xy} = 1 \qquad u_{xy} = u \qquad u_{xy} = 3 - 2u$$

subject to null boundary conditions, as in the text. Check your work by linearity.

2. Discuss the stability of solutions of the problem considered in Case IIIa of the text by use of the explicit solution given in Sec. 3.

3. Discuss the stability of solutions of the problem considered in Case IIIb by use of the explicit solution (25-6).

---

26. The Associated Difference Equations. Let h be a positive number and $u = u(x,y)$ a function of x and y. The *difference operators* Δ_x and Δ_y are defined by

$$\Delta_x u = \frac{u(x + h, y) - u(x,y)}{h} \qquad \Delta_y u = \frac{u(x, y + h) - u(x,y)}{h}. \tag{26-1}$$

Passing to the limit as $h \to 0$, we get the partial derivatives; that is,

$$\lim_{h \to 0} \Delta_x u = u_x \qquad \lim_{h \to 0} \Delta_y u = u_y \tag{26-2}$$

when the limits exist. If the *second differences* are defined by

$$\Delta_{xx} u = \frac{u(x + h, y) - 2u(x,y) + u(x - h, y)}{h^2}$$

$$\Delta_{yy} u = \frac{u(x, y + h) - 2u(x,y) + u(x, y - h)}{h^2} \tag{26-3}$$

it can be shown, in general, that

$$\lim_{h \to 0} \Delta_{xx} u = u_{xx} \qquad \lim_{h \to 0} \Delta_{yy} u = u_{yy}$$

(see Prob. 1). Hence the three *difference equations*

$$\Delta_{xx} u - \Delta_{yy} u = 0 \qquad \Delta_{xx} u = \Delta_y u \qquad \Delta_{xx} u + \Delta_{yy} u = 0$$

as $h \to 0$ become the respective differential equations

$$u_{xx} - u_{yy} = 0 \qquad u_{xx} = u_y \qquad u_{xx} + u_{yy} = 0.$$

The correspondence of difference equations and differential equations is important because there are numerical methods of solving the former which are especially adapted to high-speed computers (cf. Chap. 10, Sec. 19). As $h \to 0$, the solution of the difference equation generally tends to the solution of the corresponding differential equation. This fact gives a means of numerical approximation which has been extensively exploited. Because of space limitations we shall consider merely the determinacy of the solutions, our objective being to clarify further the distinction among elliptic, parabolic, and hyperbolic equations.

CASE I. Elliptic Equation. Using (26-3) the reader can verify that

$$\Delta_{xx} u + \Delta_{yy} u = 0 \tag{26-4}$$

can be written in the form

$$u(x,y) = \tfrac{1}{4}[u(x + h, y) + u(x - h, y) + u(x, y + h) + u(x, y - h)]. \tag{26-5}$$

This equation gives a relation between the five values of u at the five neighboring lattice points[1] illustrated in Fig. 36; in fact, *the value at the central lattice point is the arithmetic*

[1] That is, points of the form (mh,nh) with integers m and n.

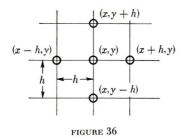

FIGURE 36

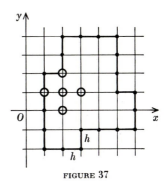

FIGURE 37

mean of the values at the four neighbors.   The corresponding property for Laplace's equation is the average-value theorem given in the example of Sec. 21.

To state the Dirichlet problem for the difference equation (26-4) we say that a point is *interior* to a region if its four nearest neighbors are points of the region.   A *boundary point* is a point for which at least one nearest neighbor belongs to the region and at least one does not.   For instance, the points • in Fig. 37 are boundary points.   In the *Dirichlet problem* a function u satisfying (26-4) is given at every boundary point, and it is required to find u at the interior points.   We shall now establish both the existence and the uniqueness of the solution.   The region is assumed bounded, so that the number of interior points is a finite number n.

Suppose, then, that u is known at every boundary point of a given region (Fig. 37). If we write Eq. (26-5) for each interior point (x,y), we obtain a system of n linear equations in n unknowns.   It will be seen presently that the determinant of this system is not zero, and hence there is one, and only one, solution.   On the other hand, if the values are not prescribed at *every* boundary point, there are always more unknowns than equations, and the solution is not determined uniquely.   These properties are analogous to those obtained previously for the Laplace equation.

To show that the determinant is not zero, we shall analyze the special case in which the boundary values are zero.   In this case the system of linear equations obtained by writing (26-4) at each interior point is homogeneous.   *If the determinant is zero*, the system will have a solution other than the trivial solution $u \equiv 0$.   Without loss of generality we can suppose that this nontrivial solution u is positive at some point.

Let the maximum value of u over all the lattice points be denoted by $m > 0$, and let P be a point where $u = m$.   Evidently P cannot be on the boundary, since $u > 0$ at P.   Hence, P is interior, and the value of u at P is the average of the values at the four neighbors.   Now $u \leq m$ at these neighbors, since m is the maximum.   If $u < m$ at any neighbor, the average is $<m$, so that $u(P) < m$.   This contradiction shows that $u = m$ at the four neighbors of P.

We can now repeat the process, starting with one of these four neighbors instead of P. Proceeding in this way we find that $u = m$ at every lattice point.   But that is impossible, since $u = 0 \neq m$ on the boundary.   Hence the assumption that the determinant was zero led to a contradiction.

PROBLEMS

1. (a) Show that $\Delta_{xx}u(x,y) = \Delta_x[\Delta_x u(x - h, y)]$.   (b) If u has a Taylor-series expansion about the point (x,y), show that $\Delta_{xx}u \to u_{xx}$ as $h \to 0$.   *Hint:* Use the first six terms of $u(x + h, y + k) = a + bh + ck + \cdots$.

2. Suppose $\Delta_{xx}u + \Delta_{yy}u = 0$, and suppose u is known for $x = 0$ and for $x = h$ $(y = h, 2h, 3h, \ldots)$. In what region of the xy plane is u determined?   *Hint:* See Fig. 36.

3. Let $\Delta_{xx}u + \Delta_{yy}u = 0$, and suppose $u(0,y) = 1$ for all y, $u(2h,y) = 2$ for all y, $u(h,0) = u(h,4h) = 0$. Find $u(h,2h)$.

4. This problem and the following ones concern functions defined at the lattice points of a finite region, as in the text.

Show that if the function $u(x,y)$ attains its maximum at an interior point $P = (x_0,y_0)$ then at P

$$\Delta_x u \leq 0 \qquad \Delta_y u \leq 0 \qquad \Delta_{xx}u \leq 0 \qquad \Delta_{yy}u \leq 0.$$

Discuss conditions for strict inequality.

5. If a function u satisfies $\Delta_{xx}u + \Delta_{yy}u \geq 0$ at every interior point, show that the maximum is taken on at the boundary. *Hint:* See Prob. 4.

6. (a) In the preceding problem, discuss restrictions on the region that enable us to conclude $u \equiv m$, if u assumes its maximum value m at an interior point. (b) Show by examples that some restriction is needed.

7. Let a difference operator **D** and boundary operator **B** be defined at lattice points $P = (x,y)$ by

$$\mathbf{D}u = u - F(P,\Delta u) \qquad \mathbf{B}u = u - G(P,u_\nu).$$

Here $\Delta u = \Delta_{xx}u + \Delta_{yy}u$ evaluated at the interior point P, and

$$u_\nu(P) = u(P) - u(P_i)$$

where P_i is an interior point adjacent to the boundary point P. If $F(P,s)$ and $G(P,s)$ are monotone nondecreasing functions of s for all relevant P show, as in Sec. 24, that

$$|\mathbf{D}u - \mathbf{D}v| \leq \epsilon \qquad \text{and} \qquad |\mathbf{B}u - \mathbf{B}v| \leq \delta \qquad \text{implies } |u - v| \leq \max(\epsilon,\delta).$$

---

27. Further Discussion of Difference Equations.

According to the foregoing discussion, the elliptic case leads to a set of simultaneous equations for determination of the unknown function u. In the parabolic and hyperbolic cases, as we shall now see, the values of u can be obtained successively.

CASE II.   Parabolic Equation. By (26-1) and (26-3) the equation

$$\Delta_{xx}u = \Delta_y u \tag{27-1}$$

takes the form

$$u(x + h, y) - (2 - h)u(x,y) + u(x - h, y) = hu(x, y + h). \tag{27-2}$$

This shows that, if u is known at the three collinear points in Fig. 38, u can be found at the fourth point.

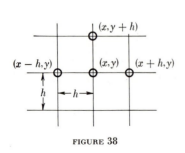

FIGURE 38

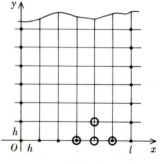

FIGURE 39

By analogy with the problem of heat flow discussed in Sec. 24, let u be given at the points • on the boundary of the semi-infinite rectangle in Fig. 39. Referring to Fig. 38, we see that u can be found at the lattice points with $y = h$ in the rectangle. Repetition gives u for $y = 2h$, and so on. Thus, u is determined throughout the rectangle, just as in the case of Fig. 32. The process works equally well when the rod is infinite and u is given at all the lattice points on the x axis.

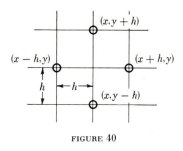

FIGURE 40

Because the pattern of Fig. 38 points upward, so to speak, it is impossible to proceed in the *negative y* direction when the rod is infinite. The very first step leads to a system of infinitely many equations in infinitely many unknowns. Inasmuch as $y = \alpha^2 t$, where t is time, this fact expresses the irreversibility of thermodynamic processes.

Further insight into the one-directional character of t is given by (8-10) and (17-4). In general, these expressions are infinitely differentiable for $t > 0$ but divergent for $t < 0$. This behavior of the heat equation contrasts to that of the wave equation. As we have repeatedly observed and will see again in the sequel, the latter is meaningful for negative t.

CASE IIIa.   Hyperbolic Equation, First Problem.   Writing the equation

$$\Delta_{xx}u - \Delta_{yy}u = 0 \tag{27-3}$$

in the form

$$u(x + h, y) + u(x - h, y) - u(x, y + h) - u(x, y - h) = 0 \tag{27-4}$$

we see that the corresponding pattern is that shown in Fig. 40. If u is given at any three of the four lattice points, (27-4) gives u at the fourth point. Inasmuch as the pattern is symmetric, one can proceed in the positive y direction and in the negative y direction with equal ease.

To discuss the analog of the initial-value problem (Sec. 25, Case IIIa), let u and $\Delta_y u$ be given in an interval of lattice points on the x axis. This is equivalent to specifying u itself on two adjacent rows of lattice points, as indicated by the black dots in Fig. 41. Considering Fig. 40 in conjunction with Fig. 41, we see that the region of determination for u consists of the lattice points in the square. The analogy with Fig. 33 is evident.

CASE IIIb.   Hyperbolic Equation, Second Problem.   If u is given on two adjacent sides of a rectangle as shown in Fig. 42, we can apply Fig. 40, starting at P. It will be found

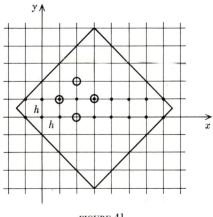

FIGURE 41

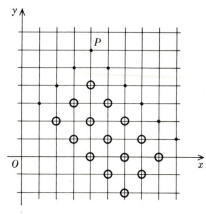

FIGURE 42

that u is determined at the indicated points O, and at no others. This behavior corresponds to that found in Sec. 25, Case IIIb.

As indicated by the problems at the end of Sec. 26, stability and uniqueness for difference equations can be studied by methods similar to those used for differential equations. A more difficult problem concerns the question whether the solutions of the difference equation tend to solutions of the differential equation as $h \to 0$. It can be stated that this happens, in general, for the Laplace and wave equations, with the difference schemes used here. For the heat equation, however, instead of the uniform lattice of Figs. 38 and 39 one must use a lattice in which the vertical step is of the order of h^2 rather than h as $h \to 0$. A discussion of convergence under this hypothesis can be found in Chap. 9, Sec. 10.

─── **PROBLEMS**

1. In Fig. 39, let $u = 0$ on the vertical rows of points •, and $u = 1$ on the horizontal row •, for $0 < x < l$. Assuming $h = 1$ in (27-2), find $u(h,6h)$.

2. In Fig. 41, let $u = 1, 0, 0, 2, 0, 0, 3, 0, 0$ on the bottom row of points • (in order), and let $\Delta_y u = 0$ at these points. Find $u(3h,5h)$.

3. Let $u(P) = 0$ in Fig. 42. Find the value of u at the opposite corner if $u = 0$ at the points • on the left of P, and $u = 2$ at the points • on the right of P.

───

28. Characteristics and Discontinuities. Canonical Form.   The function $u = f(x - at)$ represents a wave propagating in the positive x direction with velocity a. If $f(x)$ has the form shown[1] in Fig. 43, the motion exhibits a *wave front* (Fig. 44) whose locus can be found by setting the argument of f equal to c:

$$x - at = c.$$

In the xt plane, we recognize that this equation describes a characteristic of the wave equation.

Discontinuities of this general type arise in many investigations, ranging from the theory of the cracking of glass to the theory of supersonic flight. Since the subject has an important bearing on the classification of partial differential equations, a brief development is given here. We consider the equation

$$au_{xx} + 2bu_{xy} + cu_{yy} = H(x,y,u,u_x,u_y) \tag{28-1}$$

under the hypothesis that a, b, c are continuous functions of (x,y), with $a \neq 0$, and that H is a continuous function of its five arguments.

To introduce a discontinuity of the desired type, let u_1 and u_2 be two solutions valid throughout the neighborhood of a smooth curve C in the (x,y) plane, and suppose the corresponding surfaces are tangent along C. This means that the values and first derivatives of u_1 and u_2 agree along C. However, it is explicitly assumed that the second derivatives do not all agree on C.

FIGURE 43   FIGURE 44

[1] The intent is $f(x) = 0$ for $x > c$, $f'(c) = 0$, $f''(c+) - f''(c-) \neq 0$.

A single function, u, is now defined by setting $u = u_1$ on one side of C, and $u = u_2$ on the other side. Thus, u satisfies the partial differential equation (except on C) and has continuous first derivatives in a neighborhood of C, but it has a discontinuity in one or more of the second derivatives on C.

Our objective is to describe the curves C that can be the locus of a discontinuity of this kind. We shall use the symbol (u) to denote $u_1 - u_2$ at points (x,y) where u_1 and u_2 are defined, and similarly for higher derivatives. As an illustration,

$$(u_x) = u_{1x} - u_{2x} = (u)_x. \tag{28-2}$$

If (u) is evaluated on C it can be interpreted, also, as describing the *jump* in the function u when C is crossed. This is true because u has the value u_1 on one side of C and u_2 on the other side, so that the jump is $u_1 - u_2$. Similar remarks apply to the derivatives. Our hypothesis, then, is that

$$(u) = (u_x) = (u_y) = 0 \qquad \text{on } C \tag{28-3}$$

but that at least one quantity (u_{xx}), (u_{xy}), or (u_{yy}) is not zero on C.

Differentiating the relation $(u_x) = 0$ with respect to x by the chain rule yields

$$(u_{xx}) + (u_{xy})y' = 0 \qquad \text{on } C \tag{28-4}$$

when we recall that y is a function of x on C and take due account of (28-2). Similarly, differentiating $(u_y) = 0$ with respect to x yields

$$(u_{yx}) + (u_{yy})y' = 0 \qquad \text{on } C. \tag{28-5}$$

Taking the ( ) of the partial differential equation (28-1), we get

$$a(u_{xx}) + 2b(u_{xy}) + c(u_{yy}) = (H) = 0 \qquad \text{on } C \tag{28-6}$$

when H is continuous. The fact that $(H) = 0$ follows from (28-3) if we recall that H does not involve the higher derivatives of u.

Equations (28-4) to (28-6) are three linear homogeneous equations in the three unknowns (u_{xx}), $(u_{xy}) = (u_{yx})$, and (u_{yy}). By hypothesis not all these unknowns are zero; hence, the coefficient determinant must vanish:

$$\begin{vmatrix} 1 & y' & 0 \\ 0 & 1 & y' \\ a & 2b & c \end{vmatrix} = 0. \tag{28-7}$$

Expansion of this determinant yields the equation

$$a\,dy^2 - 2b\,dy\,dx + c\,dx^2 = 0 \tag{28-8}$$

after multiplication by dx^2. This is called the *characteristic equation* of the partial differential equation (28-1), and the graphs of its solutions are called *characteristic curves*. We conclude that C must be a characteristic curve, if C admits a discontinuity of the foregoing type. Conversely, if C is characteristic, the determinant (28-7) is zero, the related homogeneous equations have a nontrivial solution, and in general a discontinuity is possible.

For example, when (28-1) is the wave equation

$$a^2 u_{xx} - u_{tt} = 0$$

the differential equation (28-8) is

$$a^2\,dt^2 - dx^2 = 0.$$

Since this reduces to $dx/dt = \pm a$, the characteristics are the straight lines

$$x - at = c \qquad x + at = c.$$

This shows that the present definition of characteristic curves is consistent with that of Sec. 4 and also agrees with the remarks made in connection with Fig. 43.

It was found in Sec. 1 that the change of variable

$$r = x - at \qquad s = x + at \qquad u(x,t) = U(r,s)$$

reduces the wave equation to the form $U_{rs} = 0$, and we shall presently see that a similar result holds for (28-1).

Setting $dy = p\,dx$ in (28-8) and solving the resulting quadratic give

$$p = a^{-1}(b + \sqrt{b^2 - ac}) \qquad \text{or} \qquad p = a^{-1}(b - \sqrt{b^2 - ac}). \tag{28-9}$$

Since $p = dy/dx$, these are ordinary differential equations of the first order and can be solved by the methods of Chap. 2. If $b^2 > ac$, the roots p are real and unequal, and there are two families of characteristic curves. This is the hyperbolic case. In the parabolic case, $b^2 = ac$, the roots are equal, and there is only one family. In the elliptic case, $b^2 < ac$, and the characteristic curves are imaginary.

Since the locus is a characteristic, we conclude that a discontinuity of the type considered here may arise on two families of curves for hyperbolic equations, it may arise on one family for parabolic equations, and it cannot arise for elliptic equations. For example, the equations of fluid flow are elliptic at velocities less than the velocity of sound in the fluid. But at velocities exceeding the velocity of sound the equations become hyperbolic, and the fact that a discontinuity is now possible permits the formation of a *shock wave*.

In conclusion, we mention a remarkable connection between the characteristic curves and a process of simplifying (28-1) which is known as the *reduction to canonical form*. Equation (28-1) is said to be in *canonical form* if it has one of the three forms

$$u_{xx} + u_{yy} = H \qquad u_{xx} = H \qquad u_{xy} = H$$

where H is a function of x, y, u, u_x, and u_y. The reduction is achieved[1] by use of the characteristic curves written as relations

$$X(x,y) = c \qquad Y(x,y) = c \tag{28-10}$$

where c is constant.

CASE I. Elliptic Equation. When $b^2 - ac < 0$, the two values of p in (28-9) are conjugate complex. Hence the same is true of X and Y in (28-10); that is,

$$X = r(x,y) + is(x,y) \qquad Y = r(x,y) - is(x,y)$$

where r and s are real. In this case the reduction can be achieved by choosing r and s as new independent variables. If $u(x,y) = U(r,s)$, Eq. (28-1) gives an equation for U in which the second derivatives occur as $U_{rr} + U_{ss}$.

CASE II. Parabolic Equation. When $b^2 - ac = 0$ the two values of p in (28-9) are real and equal. Hence the same is true of X and Y in (28-10). In this case the reduction can be achieved by the change of variable

$$r = X(x,y) \qquad s = \text{any function independent of } X.$$

The second derivatives of U now occur as U_{ss}.

---

[1] A proof may be found in Webster, *op. cit.*, p. 242. We do not give the general proof here because, when applied to any specific example, the procedure is automatically justified in the course of the analysis.

CASE III. Hyperbolic Equation. When $b^2 - ac > 0$, the roots (28-9) are real and unequal, and the same is true of X and Y. The reduction is achieved by taking

$$r = X(x,y) \qquad s = Y(x,y)$$

as new independent variables. The second derivatives of U occur only as U_{rs}.

The procedure in each of the three cases is illustrated by the following example.

Example 1.   Reduce the equation

$$u_{xx} - kxu_{xy} + 4x^2 u_{yy} = 0 \tag{28-11}$$

to canonical form when $k = 0, 4,$ or 5.

According to (28-9),

$$p = -\tfrac{1}{2}kx \pm (\tfrac{1}{4}k^2 x^2 - 4x^2)^{\frac{1}{2}}. \tag{28-12}$$

When $k = 0$, this gives $p = \pm 2ix$. The equations $y' = \pm 2ix$ have the solutions

$$y - ix^2 = c \qquad y + ix^2 = c$$

where c is constant. Taking real and imaginary parts,

$$r = y \qquad s = x^2.$$

With $u(x,y) = U(r,s)$, the derivatives are

$$u_{xx} = 4x^2 U_{ss} + 2U_s \qquad u_{xy} = 2x U_{sr} \qquad u_{yy} = U_{rr}$$

and substitution into (28-11) with $k = 0$ gives the canonical form

$$U_{rr} + U_{ss} \equiv -(2x^2)^{-1} U_s = -(2s)^{-1} U_s.$$

When $k = 4$, the two roots (28-12) are both $p = -2x$. Solving this differential equation, we see that (28-10) is

$$y + x^2 = c \qquad y + x^2 = c.$$

Since $y + x^2$ and y are independent, we can take

$$r = y + x^2 \qquad s = y.$$

It is left for the reader to show that the canonical form is

$$U_{ss} = -(2x^2)^{-1} U_r = -(2r - 2s)^{-1} U_r. \tag{28-13}$$

Finally, the case $k = 5$ leads to two distinct real roots $p = -x, p = -4x$. Setting $p = dy/dx$ and solving,

$$y + \tfrac{1}{2}x^2 = c \qquad y + 2x^2 = c.$$

The change of variable

$$r = y + \tfrac{1}{2}x^2 \qquad s = y + 2x^2$$

now leads to the canonical form

$$U_{rs} = (6s - 6r)^{-1}(U_r + 4U_s). \tag{28-14}$$

Example 2. Fundamental Solutions.   A *fundamental solution* of a partial differential equation is a solution of the form $f[\lambda(x,y)]$, when λ is a *fixed* function and f an *arbitrary* function. For example, the equation $u_{xx} = u_{yy}$ has the fundamental solutions

$$f_1(x - y) \qquad \text{and} \qquad f_2(x + y) \tag{28-15}$$

in which $\lambda = x - y$ and $\lambda = x + y$, respectively. We shall now see that *if* (28-1) *has the fundamental solution* $f[\lambda(x,y)]$ *then the curves*

$$\lambda(x,y) = c \qquad c \text{ const}$$

are characteristics. For proof, it suffices to choose the arbitrary function f so that $f''(x)$ is continuous except at $x = c$. Then the function $u = f[\lambda(x,y)]$ has a discontinuity of the type previously considered on the locus $\lambda(x,y) = c$, and the desired result follows.

This explains why the techniques used in Secs. 1 to 6 to study the wave equation are not applicable to Laplace's equation. Namely, d'Alembert's method is based on the fundamental solutions (28-15), and the Laplace equation, being elliptic, has no such solutions.

PROBLEMS

1. Derive (28-13) and (28-14).

2. Describe the behavior of the characteristics of (9-3) as LC varies from zero to infinity.

3. Reduce to canonical form $3u_{xy} = u_{xx} + 2u_{yy}$, $2u_{xy} = u_{xx} + u_{yy}$, $2u_{xy} = u_{xx} + 2u_{yy}$.

Complex variable

Analytic aspects

Geometric aspects

Applications

8
Complex
variable

THIS CHAPTER contains a concise presentation of the rudiments of complex-variable theory with an indication of its many uses in the solution of important problems of physics and engineering. This theory, with roots in potential theory and hydrodynamics, is among the most fertile and beautiful of mathematical creations. Its unfolding left a deep imprint on the whole of mathematics and on several branches of mathematical physics. To an applied mathematician this theory is a veritable mine of effective tools for the solution of important problems in heat conduction, elasticity, hydrodynamics, and the flow of electric currents.

ANALYTIC ASPECTS

1. Complex Numbers.[1]   The analysis in the preceding chapters was concerned principally with functions of real variables, that is, such variables as can be represented graphically by points on a number axis, say the x axis of the cartesian coordinate system. The reader is familiar with the fact that calculation of the zeros of the function $f(x) = ax^2 + bx + c$, when the discriminant $b^2 - 4ac$ is negative, necessitates the introduction of complex numbers of the form $u + iv$, where u and v are real numbers and i is a number such that $i^2 = -1$.

A number of the form $u + iv$ can be represented by a point in a plane referred to a pair of orthogonal x and y axes if it is agreed that the number u represents the abscissa and v the ordinate of the point (Fig. 1). No confusion is likely to arise if the point (u,v), associated with the number $u + iv$, is labeled simply $u + iv$. It is clear that the point (u,v) can be located by the terminus of a vector z whose origin is at the origin O of the coordinate system. In this manner a one-to-one correspondence is established between the totality of vectors in the xy plane and the complex numbers. The vector z may be thought to represent the resultant of two vectors, one of which is of magnitude u and directed along the x axis and the other of magnitude v and directed along the y axis. Thus,

$$z = u + iv$$

where u is spoken of as the real part of the complex number z and v as the imaginary part. Therefore,

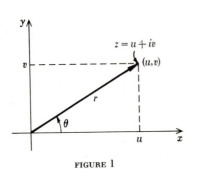

FIGURE 1

[1] An alternative formulation of the concept of complex numbers is presented in Sec. 15 of Chap. 1.

if the points of the plane are referred to a pair of coordinate axes, one can establish a correspondence between the pair of real numbers (u,v) and a single complex number $u + iv$. In this case the xy plane is called the *plane of a complex variable*, the x axis is called the *real axis*, and the y axis is called the *imaginary axis*.

If v vanishes, then

$$z = u + 0 \cdot i = u$$

is a number corresponding to some point on the real axis. Accordingly, this mode of representation of complex numbers (due to Gauss and Argand) includes as a special case the usual way of representing real numbers on the number axis.

The equality of two complex numbers,

$$a + ib = c + id$$

is interpreted to be equivalent to the two equations

$$a = c \quad \text{and} \quad b = d.$$

In particular, $a + ib = 0$ is true if, and only if, $a = 0$ and $b = 0$.

If the polar coordinates of the point (u,v) (Fig. 1) are (r,θ), then

$$u = r \cos \theta \quad \text{and} \quad v = r \sin \theta$$

so that

$$r = \sqrt{u^2 + v^2} \quad \text{and} \quad \theta = \tan^{-1} \frac{v}{u}.$$

The number r is called the *modulus*, or *absolute value*, and θ is called the *argument*, or *phase angle*, of the complex number $z = u + iv$. It is clear that the argument of a complex number is not unique, and if one writes it as $\theta + 2k\pi$, where $0 \leq \theta < 2\pi$ and $k = 0, \pm 1, \pm 2, \ldots$, then θ is called the *principal argument* of z. The modulus of the complex number z is frequently denoted by using absolute-value signs, so that

$$r = |z| = |u + iv| = \sqrt{u^2 + v^2}$$

and the argument θ is denoted by the symbol $\theta = \arg z$. Thus, $z = r(\cos \theta + i \sin \theta)$.

The reader is assumed to be familiar with the fundamental algebraic operations on complex numbers, and these will not be entered upon in detail here. It should be recalled that (cf. Chap. 1, Sec. 15)

$$z_1 + z_2 = (x_1 + iy_1) + (x_2 + iy_2) = (x_1 + x_2) + i(y_1 + y_2)$$
$$z_1 \cdot z_2 = (x_1 + iy_1)(x_2 + iy_2) = (x_1 x_2 - y_1 y_2) + i(x_1 y_2 + x_2 y_1)$$

$$\frac{z_1}{z_2} = \frac{x_1 + iy_1}{x_2 + iy_2} = \frac{x_1 x_2 + y_1 y_2}{x_2^2 + y_2^2} + i \frac{x_2 y_1 - x_1 y_2}{x_2^2 + y_2^2} \quad \text{if } z_2 \neq 0.$$

On representing complex numbers z_1 and z_2 by vectors, we can see at once from Fig. 2 that they obey the familiar "parallelogram law of addition" formulated in Chap. 4.

From elementary geometric considerations we deduce that[1]

$$|z_1 + z_2| \leq |z_1| + |z_2| \qquad (1\text{-}1)$$

that is, *the modulus of the sum of two complex numbers is less than or equal to the sum of the moduli*. This follows at once from Fig. 2 on recalling that the sum of two sides of a triangle is not less than the third side.

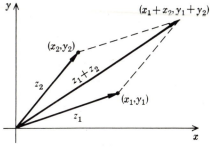

FIGURE 2

[1] See also the problems in Chap. 1, Sec. 15.

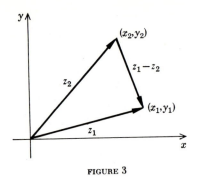

FIGURE 3

Also,

$$|z_1 + z_2| \geq |z_1| - |z_2| \tag{1-2}$$

that is, *the modulus of the sum is greater than or equal to the difference of the moduli.* This follows from the fact that the length of one side of a triangle is not less than the difference of two other sides.

Equations (1-1) and (1-2) yield a useful inequality,

$$|z_1| - |z_2| \leq |z_1 - z_2| \leq |z_1| + |z_2| \tag{1-3}$$

indicated in Fig. 3.

When calculations are carried out with complex numbers, the notion of the *conjugate complex number* is useful. We define the conjugate $\bar{z}$ of the number $z = x + iy$ by the formula

$$\bar{z} = x - iy.$$

The application of the rules for addition, multiplication, and division of complex numbers yields the following theorems:

(a)
$$\overline{z_1 + z_2} = \bar{z}_1 + \bar{z}_2 \tag{1-4}$$

or, in words, *the conjugate of the sum of two complex numbers is equal to the sum of the conjugates;*

(b)
$$\overline{z_1 z_2} = \bar{z}_1 \bar{z}_2 \tag{1-5}$$

that is, *the conjugate of the product is equal to the product of the conjugates;*

(c)
$$\overline{\left(\frac{z_1}{z_2}\right)} = \frac{\bar{z}_1}{\bar{z}_2} \tag{1-6}$$

or *the conjugate of the quotient is equal to the quotient of the conjugates.*

We note that if $\bar{z} = z$ then z is real.

The geometric interpretation of multiplication and division of complex numbers follows readily from polar representation of complex numbers. Thus,

$$\begin{aligned} z_1 z_2 &= r_1(\cos \theta_1 + i \sin \theta_1) r_2(\cos \theta_2 + i \sin \theta_2) \\ &= r_1 r_2 [\cos (\theta_1 + \theta_2) + i \sin (\theta_1 + \theta_2)]. \end{aligned} \tag{1-7}$$

That is, *the modulus of the product is equal to the product of the moduli and the argument of the product is equal to the sum of the arguments.*

Also,

$$\frac{z_1}{z_2} = \frac{r_1(\cos \theta_1 + i \sin \theta_1)}{r_2(\cos \theta_2 + i \sin \theta_2)} = \frac{r_1}{r_2} [\cos (\theta_1 - \theta_2) + i \sin (\theta_1 - \theta_2)] \tag{1-8}$$

as follows on multiplying the numerator and denominator in (1-8) by $\cos \theta_2 - i \sin \theta_2$. Thus, *the modulus of the quotient is the quotient of the moduli and the argument of the quotient is obtained by subtracting the argument of the denominator from that of the numerator.*

On extending formula (1-7) to the product of n complex numbers

$$z_k = r_k(\cos \theta_k + i \sin \theta_k) \qquad k = 1, 2, \ldots, n$$

we get

$$z_1 z_2 \ldots z_n = r_1 r_2 \ldots r_n [\cos (\theta_1 + \theta_2 + \cdots + \theta_n) + i \sin (\theta_1 + \theta_2 + \cdots + \theta_n)]$$

and, in particular, if all z's are equal,

$$z^n = [r(\cos\theta + i\sin\theta)]^n = r^n(\cos n\theta + i\sin n\theta). \tag{1-9}$$

Formula (1-9) is known as the *de Moivre formula*, and we have shown that it is valid for any positive integer n. We can show that it is also valid for negative and fractional values of n.

Indeed, from (1-8) we deduce that

$$\frac{1}{z} = \frac{\cos 0 + i\sin 0}{r(\cos\theta + i\sin\theta)} = \frac{1}{r}\left[\cos(-\theta) + i\sin(-\theta)\right]$$

and since (1-9) is known to hold for positive integers n,

$$\left(\frac{1}{z}\right)^n = z^{-n} = \left(\frac{1}{r}\right)^n\left[\cos(-\theta) + i\sin(-\theta)\right]^n$$

$$= r^{-n}[\cos(-n\theta) + i\sin(-n\theta)].$$

This establishes the result (1-9) for negative integers n.

To prove the validity of (1-9) for fractional values of n, it suffices to show that it holds when the integer n is replaced by $1/n$, for on raising the result to an integral power m, we obtain the desired formula for fractional exponents.

Let

$$w \equiv z^{1/n} = \sqrt[n]{z} \tag{1-10}$$

so that w is a solution of equation

$$w^n = z. \tag{1-11}$$

On introducing polar representations,

$$w = R(\cos\varphi + i\sin\varphi) \qquad z = r(\cos\theta + i\sin\theta) \tag{1-12}$$

where θ is the principal argument of z, we can write (1-11) with the aid of (1-9) as

$$w^n = R^n(\cos n\varphi + i\sin n\varphi) = r(\cos\theta + i\sin\theta).$$

We conclude from this that

$$R^n = r \qquad n\varphi = \theta \pm 2k\pi \qquad k = 0, 1, 2, \ldots$$

and thus

$$R = \sqrt[n]{r} \qquad \varphi = \frac{\theta \pm 2k\pi}{n} \qquad k = 0, 1, 2, \ldots.$$

Hence, from (1-12),

$$w = \sqrt[n]{r}\left(\cos\frac{\theta \pm 2k\pi}{n} + i\sin\frac{\theta \pm 2k\pi}{n}\right)$$

and, on recalling (1-10), we see that

$$z^{1/n} = [r(\cos\theta + i\sin\theta)]^{1/n} = r^{1/n}\left(\cos\frac{\theta \pm 2k\pi}{n} + i\sin\frac{\theta \pm 2k\pi}{n}\right). \tag{1-13}$$

Since $\cos(\theta \pm 2k\pi)/n$ and $\sin(\theta \pm 2k\pi)/n$ have the same values for two integers k differing by a multiple of n, the formula (1-13) yields just n distinct values for $\sqrt[n]{z}$, namely,

$$\sqrt[n]{z} = r^{1/n}\left(\cos\frac{\theta + 2k\pi}{n} + i\sin\frac{\theta + 2k\pi}{n}\right) \qquad k = 0, 1, 2, \ldots, n-1. \tag{1-14}$$

The validity of formula (1-9) for fractional values of n follows directly from (1-14) upon raising $z^{1/n}$ to an integral power m.

We illustrate the use of formula (1-14) by two examples.

Example 1. Compute $\sqrt[n]{1}$. In this case $z = 1$ and its principal argument $\theta = 0$. Formula (1-14) then yields

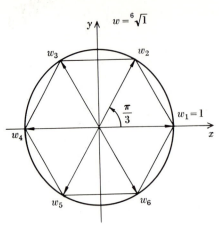

FIGURE 4

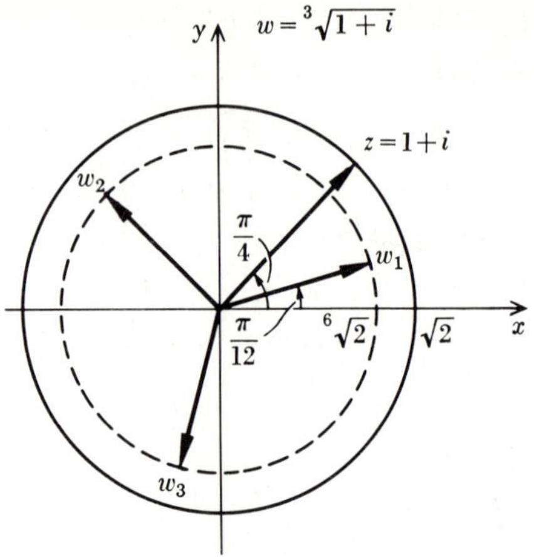

FIGURE 5

$$\sqrt[n]{1} = \cos \frac{2k\pi}{n} + i \sin \frac{2k\pi}{n} \qquad k = 0, 1, \ldots, n - 1.$$

If we plot these n roots of unity, we see that they coincide with the vertices of a regular polygon of n sides inscribed in the unit circle, with one vertex of the polygon at $z = 1$. Figure 4 shows this for $n = 6$.

Example 2. Find all roots of $\sqrt[3]{1 + i}$. Since $1 + i = \sqrt{2}\,[\cos(\pi/4) + i \sin(\pi/4)]$, formula (1-14) gives

$$\sqrt[3]{1 + i} = \sqrt[6]{2}\left[\cos \frac{(\pi/4) + 2k\pi}{3} + i \sin \frac{(\pi/4) + 2k\pi}{3} \right] \qquad k = 0, 1, 2.$$

Thus the desired roots are

$$w_1 = \sqrt[6]{2}(\cos \tfrac{1}{12}\pi + i \sin \tfrac{1}{12}\pi) \qquad w_2 = \sqrt[6]{2}(\cos \tfrac{3}{4}\pi + i \sin \tfrac{3}{4}\pi)$$

$$w_3 = \sqrt[6]{2}(\cos \tfrac{17}{12}\pi + i \sin \tfrac{17}{12}\pi).$$

These roots are represented in Fig. 5.

The reader unskilled in simple calculations involving complex numbers is urged to work out the representative problems in the following list before proceeding to the next section. The symbols Re (z) and Im (z) used in some problems in this list denote, respectively, the real and imaginary parts of a complex number z.

──────────────────────────────────── **PROBLEMS**

1. Find the moduli and principal arguments of the following numbers, and represent the number graphically:

(a) $1 + i\sqrt{3}$;   (b) $2 + 2i$;   (c) -2;   (d) i^3;   (e) $\dfrac{1}{1 + i}$;

(f) $\dfrac{1 + i}{1 - i}$;   (g) $\dfrac{3}{(i - \sqrt{3})^2}$;   (h) $(1 - i)^4$.

2. Write the following complex numbers in the form $a + bi$:

(a) $(1 - \sqrt{3}\,i)^3$;   (b) $\dfrac{(1 + i)^2}{1 - i}$;   (c) $\dfrac{2 - \sqrt{3}\,i}{1 + i}$;   (d) $\dfrac{1 + i}{2 - 3i}$;   (e) $\dfrac{3 + 2i}{3i}$.

3. Find the cubes of the following numbers:

(a) 1;   (b) $\tfrac{1}{2}(-1 + i\sqrt{3})$;   (c) $\tfrac{1}{2}(-1 - i\sqrt{3})$.

4. Find the cube roots of i, and represent them graphically.

5. Find all solutions of the equation $z^4 + 1 = 0$.

6. Verify that $z^2 - 2z + 2 = 0$ has the roots $z = 1 \pm i$.

7. Compute and represent graphically the following numbers:
(a) $\sqrt[3]{1}$;　　(b) $\sqrt[4]{1}$;　　(c) $\sqrt[5]{1}$;　　(d) $\sqrt[6]{1}$.

8. Find all the fifth roots of $1 + i$, and represent them graphically.

9. Use de Moivre's formula $[r(\cos \theta + i \sin \theta)]^n = r^n(\cos n\theta + i \sin n\theta)$ to obtain $\cos 2\theta = \cos^2 \theta - \sin^2 \theta$ and $\sin 2\theta = 2 \sin \theta \cos \theta$.

10. Write the following numbers in the form $a + bi$:

(a) $\sqrt{i}$;　　(b) $\sqrt{1 - i}$;　　(c) $\dfrac{1}{\sqrt{1 + i}}$;　　(d) $\dfrac{\sqrt{i}}{i}$;　　(e) $\sqrt{-i}$.

11. Prove that (a) $z_1 + z_2 = z_2 + z_1$;　　(b) $z_1 z_2 = z_2 z_1$;　　(c) $z_1(z_2 + z_3) = z_1 z_2 + z_1 z_3$.

12. Show that if $z_1 z_2 = 0$ then $z_1 = 0$ or $z_2 = 0$.

13. Prove formulas (1-4) to (1-6).

14. Find $|z|$, $\bar{z}$, Re (z), and Im (z) for the following:

(a) $z = 1 - 2i$;　　(b) $z = 3 + 4i$;　　(c) $z = \dfrac{1 - 2i}{3 + 4i}$.

15. Show that (a) $\overline{iz} = -i\bar{z}$; (b) $\overline{|z^3|} = |\bar{z}|^3$.

16. What is the locus of points for which
(a) $|z| = 1$?　　(b) $|z| < 1$?　　(c) $|z| \geq 1$?　　(d) $|z - 1| - |z + 1| = 1$?　　Hint: $|z| = \sqrt{x^2 + y^2}$.

17. If $z = x + iy$, what is the locus of points for which
(a) Re $(z) \geq 1$?　　(b) Im $(z) > 1$?　　(c) Re $(z^2) = 1$?

18. If $z = x + iy$, describe the loci:

(a) $|z - 1| = 2$;　　(b) $\dfrac{1}{|z|} = $ const;　　(c) $\left|\dfrac{z - 1}{z + 1}\right| = $ const.

19. Under what conditions does one have the relation
(a) $|z_1 + z_2| = |z_1| + |z_2|$?　　(b) $|z_1 + z_2| = |z_1| - |z_2|$?

20. If $z = x + iy$, write the following in the form $u + iv$:

(a) z^2;　　(b) $\dfrac{1}{z}$;　　(c) $\dfrac{1}{1 - z}$;　　(d) $z^2 + z - 1$;　　(e) $\dfrac{1}{z + i}$.

21. Show that $|z_1 + z_2|^2 + |z_1 - z_2|^2 = 2(|z_1|^2 + |z_2|^2)$.

---

2. Functions of a Complex Variable.　　A complex quantity $z = x + iy$ in which x and y are real variables is called a *complex variable*. We shall speak of the plane in which the variable z is represented as the z plane. If in some region of this plane for each $z = x + iy$ a complex number $w = u + iv$ is determined, we say that w is a function of z and write

$$w = u + iv = f(z).$$

Thus,　　　　　　　$$w = x^2 - y^2 + i2xy = (x + iy)^2 = z^2$$

is a function of z defined throughout the z plane.   Also,

$$w = u + iv = x - iy = \bar{z}$$

is a function of z.   In fact, every expression of the form $u(x,y) + iv(x,y)$ in which u and v are real functions of x and y is a function of z, since $x = \frac{1}{2}(z + \bar{z})$ and $y = (1/2i)(z - \bar{z})$ are functions of z.

A complex function $w = f(z)$ is *single-valued* if for each z in a given region of the z plane there is determined only one value of w.   If more than one value of w corresponds to z, the "function" $w = f(z)$ is *multiple-valued*.   Thus

$$w = x^2 - y^2 + i2xy = z^2$$

and

$$w = x^2 - y^2 - i2xy = \bar{z}^2$$

are single-valued functions of z.   The equation $w = \sqrt{z}$ for each $z \neq 0$ determines two complex numbers, for on setting $z = r(\cos\theta + i\sin\theta)$ and recalling formula (1-14) we get

$$w = r^{1/2}\left(\cos\frac{\theta + 2k\pi}{2} + i\sin\frac{\theta + 2k\pi}{2}\right) \qquad k = 0, 1$$

so that

$$w_1 = r^{1/2}\left(\cos\frac{\theta}{2} + i\sin\frac{\theta}{2}\right)$$

$$w_2 = r^{1/2}\left[\cos\left(\frac{\theta}{2} + \pi\right) + i\sin\left(\frac{\theta}{2} + \pi\right)\right].$$

Thus $w = \sqrt{z}$ is not single-valued and hence is not a function in the accepted sense of the term.

In order to get a function, one can consider the equation $w^2 = z$ in a *suitably restricted region* of the z plane, and add the further requirement that the admissible solution be continuous in that region.   It is then found that $w^2 = z$ defines two single-valued functions each of which can be treated by the usual methods of calculus.   These single-valued functions are called *branches* of $\sqrt{z}$, and $\sqrt{z}$ is called a multivalued function.[1]

Of course, $w = f(z)$ may be defined by different formulas in different regions of the plane, or it may not be defined at all in certain regions.

In dealing with regions of the z plane we shall distinguish interior points from those that lie on the boundaries of the region.   A characteristic property of the interior points is that about each interior point P one can draw a circle with its center at P and with nonzero radius r so small that the circle contains only points that belong to the region. The points on the boundary of the region are not interior because every circle with the boundary point as its center includes points that do not belong to the region.

A region consisting only of interior points is said to be *open*.   An example of such a region is the circular region whose points z satisfy the condition $|z| < R$.   When the boundary of the region is included in the region, the region is called *closed*.   An example of a closed region is the region consisting of points z such that $|z| \leq R$.

If every point of the region is at a finite distance from the origin, the region is said to be *finite* or *bounded*.   Thus all points of the bounded region lie within a circle $|z| = R$ if the radius R is chosen sufficiently large.   The region consisting of all points in the z plane is *unbounded*, and so is the region consisting of the points satisfying the condition $|z| \geq 1$.

[1] A more satisfactory method of treating multivalued functions is based on the concept of the *Riemann surface*, but the limitations of space do not permit an adequate discussion of these matters here.

PROBLEMS

1. Express the following functions in the form $u(x,y) + iv(x,y)$:

(a) $z^2 - z + 1$:   (b) $\dfrac{1}{z}$;   (c) $\dfrac{1}{z-1}$;   (d) $\dfrac{z-i}{z+i}$;   (e) $\dfrac{1}{z^2+i}$;   (f) $z + \frac{1}{2}(z - \bar{z})$;   (g) $\dfrac{1}{z\bar{z}}$;

(h) $\dfrac{z^2 - 2z + 1}{z + 2}$;   (i) $|z|$;   (j) $\left|\dfrac{z+1}{z-1}\right|$;   (k) $|z| - z^{-2}$;   (l) $|z + \bar{z}|^2$.

2. Describe the regions in the z plane defined by the following conditions:
(a) Re $(z) < 3$; (b) Im $(z) \geq 1$; (c) $|z| \geq 1$; (d) $1 \leq |z| < 2$; (e) $|z - 1| < 1$; (f) $|z - z_0| \leq 1$;
(g) $|z + i| > 2$; (h) Re $(z) \geq -1$; (i) $|1/z| < 2$; (j) $|z + 1|/|z - 1| = 1$.

---

3. Elementary Complex Functions.   In Sec. 1 we defined the operations of addition, multiplication, division, and root extraction for complex numbers. These suffice to determine, for any z, values of such algebraic expressions as

$$w = \frac{a_0 z^m + a_1 z^{m-1} + \cdots + a}{b_0 z^n + b_1 z^{n-1} + \cdots + b}$$

in which the powers m and n may be integers or fractions. However, they do not provide direct means for defining the complex counterparts of the real elementary transcendental[1] functions e^x, sin x, log x, tan^{-1} x, etc. A useful definition of a complex function such as e^z, for example, must specialize to e^x when z assumes real values. Also, it is desirable to preserve the familiar law of exponents $e^{z_1} e^{z_2} = e^{z_1 + z_2}$.

A definitive formula for e^z that fulfills these criteria is

$$e^z = e^{x+iy} = e^x(\cos y + i \sin y). \tag{3-1}$$

Moreover, as we shall presently see, it suggests sensible definitions for all the other elementary transcendental functions. We note first that for $x = 0$ the definition (3-1) yields

$$e^{iy} = \cos y + i \sin y. \tag{3-2}$$

On replacing y by $-y$ we get

$$e^{-iy} = \cos y - i \sin y. \tag{3-3}$$

Adding and subtracting (3-2) and (3-3) we get the *Euler formulas*

$$\cos y = \frac{1}{2}(e^{iy} + e^{-iy}) \qquad \sin y = \frac{1}{2i}(e^{iy} - e^{-iy}). \tag{3-4}$$

These formulas suggest that we define the trigonometric functions of z as follows:

$$\cos z = \frac{1}{2}(e^{iz} + e^{-iz}) \qquad \sin z = \frac{1}{2i}(e^{iz} - e^{-iz}) \qquad \tan z = \frac{\sin z}{\cos z}$$

$$\cot z = \frac{1}{\tan z} \qquad \sec z = \frac{1}{\cos z} \qquad \csc z = \frac{1}{\sin z}. \tag{3-5}$$

Using these definitions it is easy to check[2] that *all* the familiar formulas of analytic trigonometry remain valid when real arguments are replaced by the complex ones. For example,

---

[1] A variable w satisfying the equation $P(z,w) = 0$, where P is a polynomial in z and w, is called an *algebraic* function of z. A function that is not algebraic is called *transcendental*. The trigonometric and logarithmic functions and their inverses are called *elementary transcendental functions*.
[2] See Prob. 1 at the end of this section. Also, see alternative definitions of e^z, sin z, and cos z in Secs. 16 and 17 of Chap. 1.

$$\sin^2 z + \cos^2 z = 1$$
$$\sin (z_1 + z_2) = \sin z_1 \cos z_2 + \cos z_1 \sin z_2$$

and so on.

The logarithm of a complex number z is defined in the same way as in real-variable analysis. Thus,

$$w = \log z \qquad (3\text{-}6)$$

means that

$$z = e^w \qquad (3\text{-}7)$$

where e is the base of natural logarithms. Setting $w = u + iv$ in (3-7) gives

$$z = e^{u+iv} = e^u(\cos v + i \sin v) \qquad (3\text{-}8)$$

by (3-1). On the other hand, we can write z as

$$z = x + iy = r(\cos \theta + i \sin \theta)$$

so that (3-8) gives

$$r(\cos \theta + i \sin \theta) = e^u(\cos v + i \sin v).$$

It follows from this that

$$e^u = r \qquad v = \theta + 2k\pi \qquad k = 0, \pm 1, \pm 2, \dots . \qquad (3\text{-}9)$$

Since u and v are real, we conclude from (3-9) that $u = \text{Log } r$, where the symbol Log is used to denote the logarithm encountered in real-variable theory. We can thus write (3-6) in the form

$$w = u + iv = \log z = \text{Log } r + (\theta + 2k\pi)i \qquad (3\text{-}10)$$

or

$$\log z = \tfrac{1}{2} \text{Log } (x^2 + y^2) + i \tan^{-1} \frac{y}{x} \qquad (3\text{-}11)$$

since $r = \sqrt{x^2 + y^2}$ and $\theta + 2k\pi = \tan^{-1} (y/x)$.

Thus $\log z$ has infinitely many values corresponding to the different choices of the arguments θ of z. Setting $k = 0$ in (3-10) and assuming that $0 \leq \theta < 2\pi$, we get a single-valued function

$$\text{Log } z = \text{Log } r + \theta i \qquad 0 \leq \theta < 2\pi$$

which is called the *principal value* of $\log z$. If z is real and positive, the principal value of $\log z$ equals Log r.

The definition (3-10) serves to define complex and irrational powers c of the variable z by the formula

$$z^c = e^{c \log z} \qquad (3\text{-}12)$$

which, by (3-6) and (3-7), is equivalent to the statement that $\log z^c = c \log z$. Inasmuch as $\log z$ is infinitely many-valued, it follows that z^c, in general, is an infinitely many-valued function.[1] The hyperbolic functions of z are defined by the formulas

$$\sinh z = \tfrac{1}{2}(e^z - e^{-z}) \qquad \cosh z = \tfrac{1}{2}(e^z + e^{-z}) \qquad \tanh z = \frac{\sinh z}{\cosh z}$$

$$\text{sech } z = \frac{1}{\cosh z} \qquad \text{csch } z = \frac{1}{\sinh z}. \qquad (3\text{-}13)$$

These functions are clearly single-valued. The inverse trigonometric and inverse hyperbolic functions are defined in the same way as in real-variable analysis, and they are multiple-valued.[2]

Example 1. Compute e^{1-i}. On setting $x = 1$ and $y = -1$ in the formula (3-1) we get

$$e^{1-i} = e[\cos (-1) + i \sin (-1)] = e(\cos 1 - i \sin 1).$$

[1] Note, however, that z^c is single-valued when c is an integer.
[2] See Probs. 7 and 8.

Since $\cos 1 = 0.5403$, $\sin 1 = 0.8415$, and $e = 2.718$,

$$e^{1-i} = 2.718(0.5403 - i0.8415) = 1.469 - i2.287$$

to three decimal places.

Example 2.   Compute $\sin (1 - i)$.   Since

$$\sin z = \frac{1}{2i} (e^{iz} - e^{-iz}) \qquad \text{and} \qquad z = 1 - i$$

we have

$$\sin (1 - i) = \frac{1}{2i} (e^{i+1} - e^{-i-1})$$

$$= \frac{1}{2i} \{e(\cos 1 + i \sin 1) - e^{-1}[\cos (-1) + i \sin (-1)]\}$$

$$= \frac{e - e^{-1}}{2i} \cos 1 + \frac{e + e^{-1}}{2} \sin 1.$$

We can obtain the same result by making use of the addition formulas of trigonometry. Thus,

$$\sin (1 - i) = \sin 1 \cos (-i) + \cos 1 \sin (-i)$$
$$= \sin 1 \cos i - \cos 1 \sin i.$$

But by (3-5)

$$\cos i = \tfrac{1}{2}(e^{-1} + e^{1}) \qquad \sin i = \frac{1}{2i} (e^{-1} - e^{1}).$$

Substitution in the foregoing formula yields the result obtained from the definition of $\sin z$.

Example 3.   Compute $\log (1 + i)$.   Since $1 + i = \sqrt{2}[\cos (\pi/4) + i \sin (\pi/4)]$,

$$\log (1 + i) = \text{Log } \sqrt{2} + \left(\frac{\pi}{4} + 2k\pi\right) i \qquad k = 0, \pm 1, \pm 2, \ldots$$

by (3-10).   The principal value is got by setting $k = 0$.

Example 4.   Compute 2^i.   By (3-12), $2^i = e^{i \log 2}$.   But $\log 2 = \text{Log } 2 + i2\pi k$.   Hence

$$2^i = e^{i \text{ Log } 2 - 2\pi k} \qquad k = 0, \pm 1, \pm 2, \ldots.$$

Example 5.   Compute i^i.   By (3-12), $i^i = e^{i \log i}$.   But $\log i = \text{Log } 1 + i[(\pi/2) + 2k\pi] = i[(\pi/2) + 2k\pi]$, and hence

$$i^i = e^{-(\pi/2 + 2k\pi)} \qquad k = 0, \pm 1, \pm 2, \ldots.$$

Example 6.   Find all solutions of the equation $\cos z - 2 = 0$.
We have $\cos z = 2$, which gives, successively,

$$\frac{e^{iz} + e^{-zi}}{2} = 2 \qquad e^{iz} + e^{-iz} = 4 \qquad e^{2iz} - 4e^{iz} + 1 = 0.$$

Solving for e^{iz},

$$e^{iz} = \frac{4 \pm \sqrt{16 - 4}}{2} = 2 \pm \sqrt{3}.$$

Hence

$$iz = \log (2 \pm \sqrt{3})$$

and

$$z = \frac{1}{i} \log (2 \pm \sqrt{3}).$$

Since $\log (2 \pm \sqrt{3})$ is infinitely many-valued, there are infinitely many values of z, namely, $z = -i \text{ Log } (2 \pm \sqrt{3}) + 2k\pi$, $k = 0, \pm 1, \pm 2, \ldots$.

---

1. Verify the following: (a) $e^{z_1}e^{z_2} = e^{z_1+z_2}$; (b) $\sin^2 z + \cos^2 z = 1$; (c) $\cos(z_1 + z_2) = \cos z_1 \cos z_2 - \sin z_1 \sin z_2$; (d) $\cos iz = \cosh z$; (e) $\sin iz = i \sinh z$.

2. If a and b are real integers, show that $(re^{i\theta})^{a+bi} = r^a e^{-b\theta}[\cos(a\theta + b \operatorname{Log} r) + i(\sin a\theta + b \operatorname{Log} r)]$.

3. Compute and represent the numbers graphically: (a) $\cos(2 + i)$; (b) 1^i; (c) $(1 + i)^i$; (d) 2^{1+i}; (e) i^{-i}; (f) $\log e^i$.

4. Express in the form $a + bi$, where a and b are real: (a) $1/(z - 1)$; (b) $1/(z^2 + i)$; (c) $\sin(1 + i)$; (d) e^{z^2}; (e) $e^{1/z}$; (f) $\sinh(iz)$; (g) $\cosh z + \cos(iz)$.

5. Find the principal values and represent the numbers graphically: (a) $\log(-4)$; (b) $\log(5i)$; (c) $\log(1 + i)$; (d) $\log i$; (e) i^i; (f) e^{1+i}; (g) $\sin 2i$.

6. Find all solutions of the following equations: (a) $e^z + 1 = 0$; (b) $\sin z - 2 = 0$; (c) $\cos^{-1} z = 2$; (d) $\cos z - 1 = 0$; (e) $\log z = \pi i$; (f) $\sinh z = -i$.

7. The inverse functions are defined as solutions of the equation $z = f(w)$ for w in terms of z. Thus, $w = \sin^{-1} z$ if $z = \sin w = (e^{iw} - e^{-iw})/2i$. Obtain e^{iw} in this example by solving the equation $e^{2iw} - 2ize^{iw} - 1 = 0$. The result is $e^{iw} = iz \pm \sqrt{1 - z^2}$. Hence $w = \sin^{-1} z = -i \log(iz \pm \sqrt{1 - z^2})$. Show in the same way that

$$\tan^{-1} z = \frac{i}{2} \log \frac{i + z}{i - z} \quad \text{and} \quad \cos^{-1} z = -i \log(z \pm \sqrt{z^2 - 1}).$$

8. Refer to Prob. 7 and show that:

(a) $\sinh^{-1} z = \log(z + \sqrt{z^2 + 1})$;   (b) $\cosh^{-1} z = \log(z + \sqrt{z^2 - 1})$;

(c) $\tanh^{-1} z = \dfrac{1}{2} \log \dfrac{1 + z}{1 - z}$.

9. For complex numbers a, b, c, in what sense and in what circumstances is it true that $(ab)^c = a^c b^c$? *Hint:* Take $\log(ab)^c$ and examine $\arg(ab)$.

10. Show that the equations $\cot z = \pm i$ and $\tan z = \pm i$ have no solutions.

---

4. Analytic Functions of a Complex Variable. We say that a point $z = x + iy$ approaches a fixed point $z_0 = x_0 + iy_0$ if $x \to x_0$ and $y \to y_0$. Let $f(z)$ be a *single-valued* function defined in some neighborhood of the point $z = z_0$. By the *neighborhood* of z_0 we mean the set of *all points in a sufficiently small circular region with center at z_0*. As $z \to z_0$, the function $f(z)$ may tend to a definite value w_0. We say, then, that the limit of $f(z)$ as z approaches z_0 is w_0 and write

$$\lim_{z \to z_0} f(z) = w_0.$$

In particular, if $f(z_0) = w_0$, we say that $f(z)$ is *continuous* at $z = z_0$.

It is not difficult to prove that, if $f(z) = u(x,y) + iv(x,y)$ is continuous at $z_0 = x_0 + iy_0$, its real and imaginary parts u and v are continuous functions at (x_0, y_0), and conversely.

Let $w = f(z)$ be continuous at every point of some region in the z plane. The complex quantities w and z can be represented on separate complex planes, called the w and z planes. The relationship $w = f(z)$ sets up a correspondence between the points (x,y) in the z plane and the points (u,v) in the w plane (see Figs. 6 and 7), so that the corresponding points (u,v) fill some region R' in the w plane.

If $z_0 = x_0 + iy_0$ and $z = z_0 + \Delta z$ are two points in the z plane with $\Delta z = \Delta x + i \Delta y$, the corresponding points in the w plane are $w_0 = u_0 + iv_0$ and $w = w_0 + \Delta w$, where

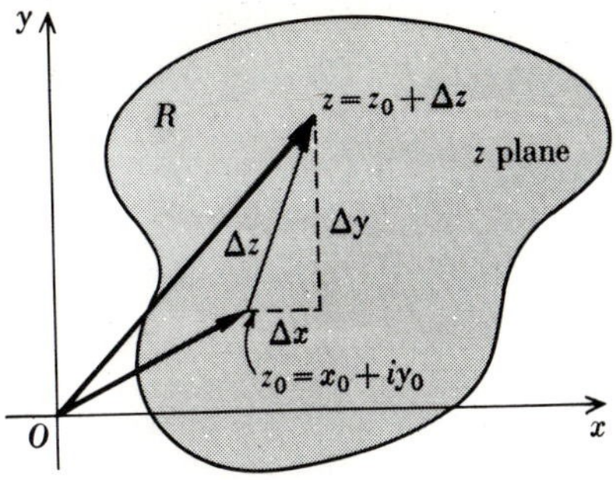

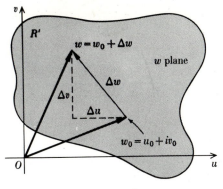

FIGURE 6

FIGURE 7

$\Delta w \equiv \Delta u + i\,\Delta v$. The change Δw in the value of $w_0 = f(z_0)$ corresponding to the increment Δz in z_0 is

$$\Delta w = f(z_0 + \Delta z) - f(z_0)$$

and we define the derivative dw/dz [or $f'(z)$] by a familiar formula

$$f'(z_0) = \lim_{\Delta z \to 0} \frac{\Delta w}{\Delta z} = \lim_{\Delta z \to 0} \frac{f(z_0 + \Delta z) - f(z_0)}{\Delta z}. \tag{4-1}$$

It is most important to note that in this formula $z = z_0 + \Delta z$ can assume any position in the neighborhood of z_0, throughout which $f(z)$ is defined, and Δz can approach zero along any one of the infinitely many paths joining z with z_0. Hence, if the derivative $f'(z_0)$ is to have a unique value, the limit in (4-1) must be independent of the way in which Δz is made to approach zero. This restriction greatly narrows down the class of complex functions that possess derivatives.

For example, if

$$w = z\bar{z}$$

on replacing z by $z + \Delta z$ and $\bar{z}$ by $\bar{z} + \overline{\Delta z}$, we get

$$w + \Delta w = (z + \Delta z)(\bar{z} + \overline{\Delta z}) = z\bar{z} + \bar{z}\,\Delta z + z\,\overline{\Delta z} + \overline{\Delta z}\,\Delta z.$$

Hence

$$\Delta w = \bar{z}\,\Delta z + z\,\overline{\Delta z} + \overline{\Delta z}\,\Delta z$$

and

$$\frac{\Delta w}{\Delta z} = \bar{z} + z\,\frac{\overline{\Delta z}}{\Delta z} + \overline{\Delta z}. \tag{4-2}$$

We show next that this quotient, in general, has no unique limit as Δz is made to approach zero along different paths. Since $z = x + iy$,

$$\Delta z = \Delta x + i\,\Delta y \qquad \overline{\Delta z} = \Delta x - i\,\Delta y$$

and we can write (4-2) as

$$\frac{\Delta w}{\Delta z} = x - iy + (x + iy)\frac{\Delta x - i\,\Delta y}{\Delta x + i\,\Delta y} + \Delta x - i\,\Delta y. \tag{4-3}$$

If we now let Δz in (4-3) approach zero along the path QRP (Fig. 8), so that first $QR = \Delta y \to 0$ and then $PR = \Delta x \to 0$, we get

$$\lim_{\Delta z \to 0} \frac{\Delta w}{\Delta z} = 2x.$$

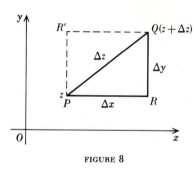

FIGURE 8

But if we take the path $QR'P$ and first allow $QR' = \Delta x \to 0$ and then $R'P = \Delta y \to 0$, we obtain

$$\lim_{\Delta z \to 0} \frac{\Delta w}{\Delta z} = -2iy.$$

Except for $x = y = 0$, these limits are distinct, and hence $w = z\bar{z}$ has no derivative except possibly at $z = 0$. As a matter of fact, it is possible to show that this function has a derivative (whose value is zero) only at the point $z = 0$.

On the other hand, if we consider

$$w = z^2$$

then $w + \Delta w = (z + \Delta z)^2 = z^2 + 2z\,\Delta z + (\Delta z)^2$, so that

$$\frac{\Delta w}{\Delta z} = \frac{2z\,\Delta z + (\Delta z)^2}{\Delta z} = 2z + \Delta z.$$

The limit of this quotient as $\Delta z \to 0$ is invariably $2z$, whatever may be the path along which $\Delta z \to 0$. In this example the derivative exists, and its value is $2z$.

We obtain next a set of conditions that real and imaginary parts of

$$w = f(z) \equiv u(x,y) + iv(x,y)$$

must fulfill if $f(z)$ is to have a unique derivative at a given point $z = x + iy$. Since $\Delta w = \Delta u + i\,\Delta v$ and $\Delta z = \Delta x + i\,\Delta y$, we get from (4-1)

$$f'(z) = \lim_{\Delta z \to 0} \frac{\Delta u + i\,\Delta v}{\Delta z}$$

$$= \lim_{\substack{\Delta x \to 0 \\ \Delta y \to 0}} \frac{\Delta u + i\,\Delta v}{\Delta x + i\,\Delta y}. \tag{4-4}$$

Now, if we let $\Delta z \to 0$ by first allowing $\Delta y \to 0$ and then $\Delta x \to 0$, we get from (4-4)

$$f'(z) = \lim_{\Delta x \to 0} \left(\frac{\Delta u}{\Delta x} + i\frac{\Delta v}{\Delta x} \right) \equiv \frac{\partial u}{\partial x} + i\frac{\partial v}{\partial x}. \tag{4-5}$$

If, on the other hand, we compute the limit in (4-4) by making first $\Delta x \to 0$ and then $\Delta y \to 0$, we obtain

$$f'(z) = \lim_{\Delta y \to 0} \left(\frac{1}{i}\frac{\Delta u}{\Delta y} + \frac{\Delta v}{\Delta y} \right) \equiv \frac{\partial v}{\partial y} - i\frac{\partial u}{\partial y}. \tag{4-6}$$

Hence, if the derivatives in (4-5) and (4-6) are to have identical values at a given point z for these two particular modes of approach of Δz to zero, we must have

$$\frac{\partial u}{\partial x} = \frac{\partial v}{\partial y} \qquad \frac{\partial v}{\partial x} = -\frac{\partial u}{\partial y}. \tag{4-7}$$

Equations (4-7) are known as the *Cauchy-Riemann equations*, and the foregoing calculation shows that they constitute *necessary conditions* for the existence of a unique derivative of $f(z) = u(x,y) + iv(x,y)$ at $z = x + iy$. As we shall presently see, these conditions also turn out to be *sufficient* if it is further assumed that the partial derivatives in (4-7) are continuous at the point (x,y).

Complex functions that have derivatives only at isolated points in the z plane are of minor interest in applications in comparison with those that have derivatives throughout the neighborhood of the given point. A function $f(z)$ that has a derivative $f'(z)$ at a given point $z = z_0$ and *at every point in the neighborhood* of z_0 is called *analytic* (or *holomorphic*) *at the point* $z = z_0$. The points of the region at which $f(z)$ ceases being analytic are called *singular points* of $f(z)$. The following theorem provides both necessary and sufficient conditions for the complex function $f(z)$ to be analytic at $z = z_0$.

THEOREM. *A necessary and sufficient condition for $f(z) = u(x,y) + iv(x,y)$ to be analytic at $z_0 = x_0 + iy_0$ is that $u(x,y)$ and $v(x,y)$ together with their first partial derivatives be continuous and satisfy Eqs. (4-7) throughout the neighborhood of (x_0,y_0).*

The necessity of conditions (4-7) for the existence of the derivative was established above. To show that the satisfaction of conditions (4-7) guarantees the existence of $f'(z)$ at all points of the neighborhood of $z_0 = x_0 + iy_0$, we note that the functions $u(x,y)$ and $v(x,y)$ are *differentiable* whenever their first partial derivatives are continuous. (See Chap. 5, Sec. 4.) Hence, for any pair of points (x,y) and $(x + \Delta y, y + \Delta y)$ in the given neighborhood, we can write

$$\Delta u = u(x + \Delta x, y + \Delta y) - u(x,y) = \frac{\partial u}{\partial x} \Delta x + \frac{\partial u}{\partial y} \Delta y + \epsilon_1 \Delta x + \epsilon_2 \Delta y$$

$$\Delta v = v(x + \Delta x, y + \Delta y) - v(x,y) = \frac{\partial v}{\partial x} \Delta x + \frac{\partial v}{\partial y} \Delta y + \epsilon_3 \Delta x + \epsilon_4 \Delta y$$

where $\epsilon_1, \epsilon_2, \epsilon_3, \epsilon_4$ approach zero as Δx and Δy approach zero. On setting $\Delta u + i \Delta v = \Delta w$ and $\Delta x + i \Delta y = \Delta z$, we obtain

$$\Delta w = \frac{\partial u}{\partial x} \Delta x + \frac{\partial u}{\partial y} \Delta y + i \left(\frac{\partial v}{\partial x} \Delta x + \frac{\partial v}{\partial y} \Delta y \right) + (\epsilon_1 \Delta x + \epsilon_2 \Delta y) + i(\epsilon_3 \Delta x + \epsilon_4 \Delta y).$$

This expression for the increment Δw of $w = f(z)$ can be written with the aid of Eqs. (4-7) in the form

$$\Delta w = \frac{\partial u}{\partial x} (\Delta x + i \Delta y) + i \frac{\partial v}{\partial x} (\Delta x + i \Delta y) + (\epsilon_1 + i\epsilon_3) \Delta x + (\epsilon_2 + i\epsilon_4) \Delta y. \qquad (4\text{-}8)$$

Since $f'(z) \equiv \lim_{\Delta z \to 0} \Delta w / \Delta z = \lim_{\substack{\Delta x \to 0 \\ \Delta y \to 0}} \Delta w / (\Delta x + i \Delta y)$, we get, after dividing both members of (4-8) by $\Delta x + i \Delta y$ and letting Δx and Δy tend to zero,

$$f'(z) = \frac{\partial u}{\partial x} + i \frac{\partial v}{\partial x} \qquad (4\text{-}9)$$

because the expressions

$$\left| (\epsilon_1 + i\epsilon_3) \frac{\Delta x}{\Delta x + i \Delta y} \right| \leq |\epsilon_1 + i\epsilon_3| \qquad \text{and} \qquad \left| (\epsilon_2 + i\epsilon_4) \frac{\Delta y}{\Delta x + i \Delta y} \right| \leq |\epsilon_2 + i\epsilon_4|$$

tend to zero as Δx and Δy approach zero. Thus, at every point of the neighborhood of z_0 the derivative of $f(z)$ is given by the formula[1] (4-9).

Calculations that have led to the rules for the differentiation of composite and inverse real functions (see Chap. 5, Secs. 5 to 7) now yield the formulas

$$\frac{dF(w)}{dz} = \frac{dF}{dw} \frac{dw}{dz}$$

and

$$\frac{dz}{dw} = \frac{1}{dw/dz} \qquad \text{if } \frac{dw}{dz} \neq 0$$

whenever $F(w)$ and $w(z)$ are analytic functions.

---

[1] The fact that the analyticity of $f(z)$ implies the continuity of the partial derivatives in (4-9) can be deduced from Cauchy's theorem in Sec. 5.

It is easy to show that familiar rules for differentiating sums, products, and quotients of real functions remain valid for analytic functions. Also, the formulas for differentiating elementary complex functions, defined in Sec. 3, are identical with the corresponding formulas in the calculus of real variables. We give a derivation of several such formulas in the following examples.[1]

Example 1.   Show that $de^z/dz = e^z$.

If $w = e^z = e^{x+iy}$, the definition (3-1) yields

$$w = u + iv = e^x(\cos y + i \sin y).$$

Here, $u = e^x \cos y$, $v = e^x \sin y$, and it follows that

$$\frac{\partial u}{\partial x} = e^x \cos y \qquad \frac{\partial u}{\partial y} = -e^x \sin y \qquad \frac{\partial v}{\partial x} = e^x \sin y \qquad \frac{\partial v}{\partial y} = e^x \cos y.$$

Since Eqs. (4-7) are satisfied and the partial derivatives are continuous, dw/dz can be calculated with the aid of either (4-5) or (4-6). Then,

$$\frac{dw}{dz} = e^x \cos y + ie^x \sin y = e^x(\cos y + i \sin y) = e^z.$$

Example 2.   Show that $(d \log z)/dz = 1/z$ if $z \neq 0$.

The function $w = \log z$, as noted in Sec. 3, is multiple-valued. However, any branch of this function got by fixing the value of k in (3-10) is single-valued, and the application of Cauchy-Riemann equations (4-7) to it shows that it is an analytic function except at $z = 0$. On fixing k we get from $w = \log z$ a single-valued function $z = e^w$ whose derivative with respect to w, by Example 1, is

$$\frac{dz}{dw} = e^w = z.$$

Hence

$$\frac{dw}{dz} = \frac{d \log z}{dz} = \frac{1}{z} \qquad \text{if } z \neq 0.$$

The point $z = 0$ is a singular point of $w = \log z$, since the derivative at that point ceases to exist.

Example 3.   Show that $dz^n/dz = nz^{n-1}$ for all values of n (real or complex).

If $w = z^n$, then

$$\log w = n \log z.$$

On differentiating this with respect to z, we get

$$\frac{1}{w} \frac{dw}{dz} = \frac{n}{z}.$$

Hence

$$\frac{dw}{dz} = n \frac{w}{z} = nz^{n-1}.$$

since $w = z^n$. This derivative ceases to exist at $z = 0$ if $n < 1$.

PROBLEMS

1. Show that

(a) $\dfrac{d}{dz}(f_1 \pm f_2) = f_1'(z) \pm f_2'(z)$;   (b) $\dfrac{d}{dz}(f_1 f_2) = f_1 f_2' + f_2 f_1'$;

(c) $\dfrac{d}{dz}\left(\dfrac{f_1}{f_2}\right) = \dfrac{f_2 f_1' - f_1 f_2'}{(f_2)^2}$;   (d) $\dfrac{d}{dz}\{f_1[f_2(z)]\} = \dfrac{df_1}{df_2}\dfrac{df_2}{dz}$

whenever f_1 and f_2 are analytic functions.

[1] See also the problems at the end of this section.

2. Show that

(a) $\dfrac{d(\cos z)}{dz} = -\sin z;$ (b) $\dfrac{d(\sin z)}{dz} = \cos z;$ (c) $\dfrac{d(\tan z)}{dz} = \sec^2 z;$

(d) $\dfrac{d(\tan^{-1} z)}{dz} = \dfrac{1}{1+z^2};$ (e) $\dfrac{d(\sinh z)}{dz} = \cosh z;$ (f) $\dfrac{da^z}{dz} = a^z \log a.$

3. Determine where each of the following functions fails to be analytic: (a) $z^2 + 2z;$ (b) $z/(z+1);$ (c) $1/z + (z-1)^2;$ (d) $\tan z;$ (e) $1/[(z-1)(z+1)];$ (f) $z\bar{z};$ (g) $e^{\bar{z}};$ (h) $x^2 - y^2 - 2ixy;$ (i) $x/(x^2+y^2) + iy/(x^2+y^2);$ (j) $|z|;$ (k) $\tan^{-1} z.$

4. Show that if $z = r(\cos \theta + i \sin \theta)$ and $f(z) = u(r,\theta) + iv(r,\theta)$ then the Cauchy-Riemann equations read

$$\frac{\partial u}{\partial r} = \frac{1}{r} \frac{\partial v}{\partial \theta} \qquad \frac{\partial v}{\partial r} = -\frac{1}{r} \frac{\partial u}{\partial \theta}.$$

5. Show with the aid of the theorem of this section that the following functions of z are not analytic at any point: (a) $\bar{z};$ (b) $\bar{z} - z^2;$ (c) $x^2 - ixy^2;$ (d) $e^x(\cos y - i \sin y);$ (e) $ze^{\bar{z}};$ (f) $|z|^2;$ (g) $|z| - \sin z.$

6. A function $f(z)$ that is analytic at all points of the z plane is called *entire*. Show that the following functions of z are entire: (a) $ze^z;$ (b) $e^{-z}(\cos y - i \sin y);$ (c) $x^2 - y^2 + i2xy;$ (d) $\cosh z;$ (e) $e^x(x \cos y - y \sin y) + ie^x(y \cos y + x \sin y).$

7. If $f(z) = \sqrt{r}\,[\cos (\theta/2) + i \sin (\theta/2)],$ $r > 0,$ $0 < \theta < 2\pi,$ show that $f(z)$ is analytic at all points except $z = 0$ and along the positive real axis.   Note Prob. 4.

8. If w is an analytic function of z in some region, show that $\sin w,$ $\cos w,$ and e^w are analytic in that region.   What are their derivatives?

9. Compute $dF(w)/dz$ if: (a) $F(w) = e^w,$ $w = z^2;$ (b) $F(w) = 1/w^2,$ $w = e^z;$ (c) $F(w) = w^2 - \log w,$ $w = e^z;$ (d) $F(w) = \tan^{-1} w,$ $w = 1 + z^2;$ (e) $F(w) = \sin w,$ $w = 1 + z^2.$

10. (a) If $u + iv$ is analytic, show that $v - iu$ is also analytic. (b) If $f(z) = u + iv \neq$ const, show that $f(\bar{z}) = u - iv$ is not analytic. (c) If in some region $f(z)$ and $\overline{f(z)}$ are analytic, show that $f(z) = $ const. (d) If $f(z) \neq$ const is analytic, show that $|f(z)|$ is not analytic. (e) If $f(z) \neq$ const is analytic, show that $|f(\bar{z})|$ is not analytic.

---

5. Integration of Complex Functions.   Cauchy's Integral Theorem.   Before proceeding to the study of integration of complex functions, the reader should review the definitions and theorems on line integrals in Secs. 19 to 23 of Chap. 5.   Throughout this chapter the term "simple closed curve" is used to mean simple closed path, so that corners are allowed.   It is assumed once for all that the regions with which we deal are connected and their boundaries form simple closed paths.

We define the integral $\int_C f(z)\, dz$ of a complex function $f(z) = u(x,y) + iv(x,y)$ along a path C in terms of real line integrals as follows:

$$\int_C f(z)\, dz = \int_C (u + iv)(dx + i\, dy)$$

$$= \int_C (u\, dx - v\, dy) + i \int_C (v\, dx + u\, dy). \tag{5-1}$$

The integral in (5-1) can also be defined in a manner of Chap. 5, Sec. 19, by the formula

$$\int_C f(z)\, dz \equiv \lim_{\substack{n \to \infty \\ \max|z_i - z_{i-1}| \to 0}} \sum_{i=1}^{n} f(\zeta_i)(z_i - z_{i-1}). \tag{5-2}$$

It is supposed that the curve C is divided into n segments by points z_i and that ζ_i is some point of the ith segment.   The limit is then computed as the number of segments is allowed to increase indefinitely in such a way that the length of the largest segment tends to zero.   The fact that the definitions (5-1) and (5-2) are equivalent follows from consideration of Sec. 19 in Chap. 5.

As an illustration of the use of formula (5-1), consider the integral

$$\int_C \bar{z}^2 \, dz \tag{5-3}$$

where the path C is a straight line joining the points $z = 0$ and $z = 1 + 2i$.   Since $\bar{z}^2 = (x - iy)^2 = x^2 - y^2 - i2xy$, we get, on substituting $u = x^2 - y^2$, $v = -2xy$ in (5-1),

$$\int_C \bar{z}^2 \, dz = \int_C \left[(x^2 - y^2) \, dx + 2xy \, dy \right] + i \int_C \left[-2xy \, dx + (x^2 - y^2) \, dy \right]. \tag{5-4}$$

But the cartesian equation of C is $y = 2x$, and hence (5-4) can be reduced to the evaluation of two definite integrals:

$$\int \bar{z}^2 \, dz = \int_0^1 5x^2 \, dx + i \int_0^1 - 10x^2 \, dx = \tfrac{5}{3} - i^{1}\tfrac{0}{3}.$$

The value of the integral (5-4) depends on the path C joining the given points $z = 0$, $z = 1 + 2i$, for according to Sec. 22 of Chap. 5 a necessary and sufficient condition that the line integral

$$\int_C M \, dx + N \, dy \tag{5-5}$$

be independent of the path in a simply connected region R is that

$$\frac{\partial M}{\partial y} = \frac{\partial N}{\partial x} \tag{5-6}$$

throughout R provided that $M(x,y)$, $N(x,y)$, and their partial derivatives in (5-6) are continuous functions throughout the closed region.   It is readily checked that Eq. (5-6) is not satisfied by the functions appearing in the line integrals in (5-4).

If, however, $f(z) = u + iv$ in (5-1) is an analytic function in a given region R, the Cauchy-Riemann equations (4-7) demand that

$$\frac{\partial u}{\partial x} = \frac{\partial v}{\partial y} \qquad \frac{\partial v}{\partial x} = -\frac{\partial u}{\partial y} \qquad \text{in } R. \tag{5-7}$$

Reference to (5-6) shows that these conditions are precisely those that ensure the independence of the path of the line integrals in (5-1), provided that the partial derivatives in (5-7) are continuous functions in the given simply connected region R.   Thus, if we suppose that $f(z)$ is analytic in the given simply connected region and $f'(z)$ is continuous there, the integral

$$\int_C f(z) \, dz$$

is independent of the path joining any pair of points in the region.   If the path C is closed, the value of this integral is zero.   We thus have a theorem, first deduced by Cauchy, which is of cardinal importance in the study of analytic functions.   Although the foregoing proof assumes the continuity of $f'(z)$ on C, the theorem can actually be established[1] under the sole hypothesis that $f(z)$ is continuous in a closed simply connected region $R + C$ and $f'(z)$ exists at each interior point of the region.   We state it in this strong form.

[1] See M. Heins, "Selected Topics in the Classical Theory of Functions of a Complex Variable," p. 135, Holt, Rinehart and Winston, Inc.; New York, 1962.

CAUCHY'S INTEGRAL THEOREM.   If $f(z)$ is continuous in a closed simply con-nected region $R + C$ and analytic within the simple closed curve C, then $\int_C f(z)\, dz = 0$.

We conclude this section by deducing, from definition (5-2), a useful inequality furnishing an upper bound for the value of the complex integral $\int_C f(z)\, dz$.   Inasmuch as the modulus of the sum of complex numbers is never greater than the sum of the moduli, calculations based on (5-2) yield

$$\left| \int_C f(z)\, dz \right| \le \int_C |f(z)| \cdot |dz|.$$

Now, if the modulus $|f(z)|$ of $f(z)$ along C does not exceed in value some positive number M, then

$$\left| \int_C f(z)\, dz \right| \le M \int_C |dz| = M \int_C |dx + i\, dy| = M \int_C ds = ML \qquad (5\text{-}8)$$

where L is the length of C.

As an illustration of the use of the inequality (5-8) we apply it to deduce an upper bound for the integral (5-3).   The modulus of $\bar{z}^2$ takes its maximum at the point $z = 1 + 2i$.   Hence we can take M in (5-3) as $|1 + 2i|^2 = 5$, and (5-8) then yields

$$\left| \int_C \bar{z}^2\, dz \right| \le 5\sqrt{5}$$

inasmuch as $L = \sqrt{5}$ for the rectilinear path in (5-3).

───────────────────────────────────── **PROBLEMS**

1. Find the value of the integral $\int_C z^2\, dz$ along the rectilinear path joining the points $z = 0$ and $z = 2 + i$.   Show that this integral is independent of the path.

2. Find the value of the integral $\int_C \bar{z}\, dz$ along the rectilinear path $y = x$ joining the points $(0,0)$ and $(1,1)$ and also along the parabola $y = x^2$ joining the same points.

3. Show that the integral $\int_C \bar{z}\, dz$ evaluated over the path $|z| = 1$ in a counterclockwise direction yields $2\pi i$.   Note that $z = e^{i\theta}$ and $\bar{z} = e^{-i\theta}$ along the path $|z| = 1$.

4. Find the value of the integral $\int_{-1}^{1} \dfrac{z-1}{z}\, dz$, where the path is the upper half of the circle $|z| = 1$.   Calculate the value of this integral over the lower half of the circle $|z| = 1$.

5. Show that $\int_C (1 + z^2)\, dz$ is independent of the path C, and evaluate this integral when C is the boundary of the square with vertices at the points $z = 0$, $z = 1$, $z = 1 + i$, and $z = i$.

6. What is the value of the integral $\int_C e^{iz}\, dz$, where C is the boundary of the square in Prob. 5?

7. Find the value of the integral $\int_0^{\pi i} e^z\, dz$ over any path joining $z = 0$ and $z = \pi i$.

8. Use formula (5-8) to show that:

(a) $\left| \int_i^{2+i} z^2\, dz \right| < 10$;   (b) $\left| \int_i^{2+i} \dfrac{dz}{z^2} \right| < 2$;   (c) $\left| \int_{-i}^{i} (x^2 + iy^2)\, dz \right| < 2$;

(d) $\left| \int_i^{2+i} \dfrac{\sin z}{z} \right| < 2 \cosh 1$

where the integration paths are straight lines joining the points appearing in the limits of these integrals.

9. Show that $\int_C^{-4} z\,dz = 0$, where C is the circle $|z| = 1$. Does this result follow from Cauchy's integral theorem?

10. Verify Cauchy's integral theorem for the integral in Prob. 5 when the path C is the circle $|z| = 1$.

---

6. Cauchy's Integral Theorem for Multiply Connected Regions.   In establishing Cauchy's integral theorem in the preceding section we assumed that the region bounded by the curve C is simply connected. It is easy to extend this theorem to multiply connected domains in the manner of Sec. 23 in Chap. 5. Thus consider, for definiteness, a doubly connected region (Fig. 9) bounded by simple closed curves C_1 and C_2, where C_2 lies entirely within C_1. We assume that $f(z)$ is analytic in the region exterior to C_2 and interior to C_1 and analytic on C_2 and C_1. The requirement of analyticity on C_1 and C_2 implies that the function $f(z)$ is analytic in an extended region (indicated by the dashed curves K_1 and K_2) that contains the curves C_1 and C_2.

If some point A of the curve C_1 is joined to a point B of C_2 by a crosscut AB, the region becomes simply connected and the theorem of Cauchy is applicable. Integrating in the positive direction[1] gives

$$\oint_{APA} f(z)\,dz + \int_{AB} f(z)\,dz + \oint_{BQB} f(z)\,dz + \int_{BA} f(z)\,dz = 0 \qquad (6\text{-}1)$$

where the subscripts on the integrals indicate the directions of integration along C_1, the crosscut AB, and C_2. Since the second and the fourth integrals in (6-1) are calculated over the same path in opposite directions, their sum is zero and one has

$$\oint_{C_1} f(z)\,dz + \oint_{C_2} f(z)\,dz = 0 \qquad (6\text{-}2)$$

where the integral along C_1 is traversed in the counterclockwise direction and that along C_2 in the clockwise direction. Changing the order of integration in the second integral in (6-2) gives

$$\oint_{C_1} f(z)\,dz = \oint_{C_2} f(z)\,dz. \qquad (6\text{-}3)$$

We see that the values of the integral of $f(z)$ over two different paths C_1 and C_2 are equal, but they need not be zero inasmuch as $f(z)$ may not be analytic at every point of the region bounded by C_2. But whatever may be the value of the integral over the path C_2, it is the same as its value over the path C_1. An important *principle of the deformation of contours* follows at once from this observation: *The integral of an analytic function over any simple closed curve C_1 has the same value over any other such curve C_2 into which C_1 can be continuously deformed without passing over singular points of $f(z)$.*

We shall see that this principle will enable us to simplify the computation of integrals of analytic functions.

The foregoing results can be extended in an obvious way (cf. Chap. 5, Sec. 23) to yield the following theorem:

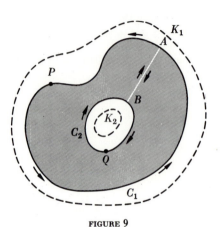

FIGURE 9

[1] That is, the direction such that a person walking along the boundary of the region finds the region to his left.

THEOREM.   If $f(z)$ is analytic in a closed multiply connected region bounded by the exterior simple closed curve C and the interior simple closed curves $C_1, C_2, \ldots, C_n$, then the integral over the exterior curve C is equal to the sum of the integrals over the interior curves provided that the integration over all the contours is performed in the same direction.

It should be noted that the requirement of analyticity of $f(z)$ in the *closed* region implied that $f(z)$ be analytic on all contours forming its boundary.   However, from the statement of Cauchy's integral theorem in the strong form in Sec. 5, one concludes that it suffices to assume that $f(z)$ is continuous in a closed region and analytic in its interior.

Before considering applications of the theorem of this section to specific problems, we deduce an important result which will enable us to compute many integrals by a method that is vastly simpler than that developed in Sec. 5.

7. The Fundamental Theorem of Integral Calculus.   Let $f(z)$ be analytic in a simply connected region R (Fig. 10), and let C be a curve joining two points P_0 and P of the region determined by the complex numbers z_0 and z.   We consider the integral

$$\int_{z_0}^{z} f(z)\, dz \qquad (7\text{-}1)$$

along C.   Since $f(z)$ is analytic, the integral (7-1) is independent of the path, and its value is completely determined by the choice of z_0 and z.   If z_0 is fixed, the integral (7-1) defines a function

$$F(z) = \int_{z_0}^{z} f(z)\, dz \qquad (7\text{-}2)$$

for every choice of z in R.

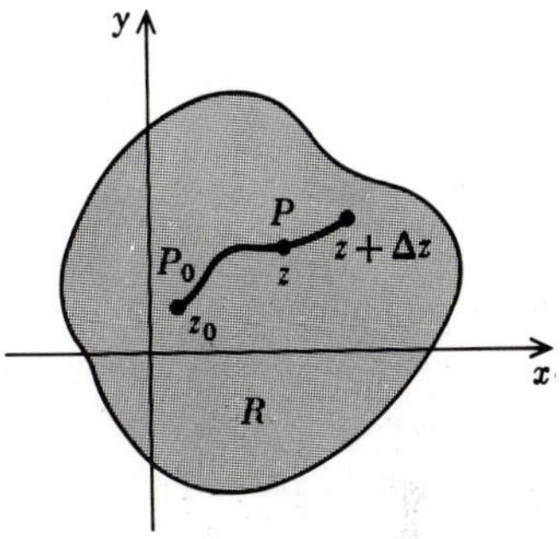

To emphasize the fact that the integration variable z plays a distinct role from the variable z appearing in the upper limit of the integral, we can rewrite (7-2) as

FIGURE 10

$$F(z) = \int_{z_0}^{z} f(\zeta)\, d\zeta. \qquad (7\text{-}3)$$

We prove next that $F(z)$ is an analytic function and, moreover, its derivative at any point z has the value of the function in the integrand at that point.   That is,

$$F'(z) = f(z).$$

We can use (7-3) to compute the difference quotient

$$\frac{F(z + \Delta z) - F(z)}{\Delta z} = \frac{1}{\Delta z}\left[\int_{z_0}^{z+\Delta z} f(\zeta)\, d\zeta - \int_{z_0}^{z} f(\zeta)\, d\zeta\right]$$

$$= \frac{1}{\Delta z}\left[\int_{z_0}^{z} f(\zeta)\, d\zeta + \int_{z}^{z+\Delta z} f(\zeta)\, d\zeta - \int_{z_0}^{z} f(\zeta)\, d\zeta\right]$$

$$= \frac{1}{\Delta z}\int_{z}^{z+\Delta z} f(\zeta)\, d\zeta \qquad (7\text{-}4)$$

and rewrite (7-4) by adding and subtracting $f(z)$ in the integrand:

$$\frac{F(z + \Delta z) - F(z)}{\Delta z} = \frac{1}{\Delta z}\int_{z}^{z+\Delta z} [f(\zeta) - f(z) + f(z)]\, d\zeta$$

$$= \frac{1}{\Delta z}\left[f(z)\int_{z}^{z+\Delta z} d\zeta\right] + \frac{1}{\Delta z}\int_{z}^{z+\Delta z} [f(\zeta) - f(z)]\, d\zeta.$$

But $\int_z^{z+\Delta z} d\zeta = \Delta z$, so that

$$\frac{F(z + \Delta z) - F(z)}{\Delta z} = f(z) + \frac{1}{\Delta z} \int_z^{z+\Delta z} [f(\zeta) - f(z)] \, d\zeta. \tag{7-5}$$

Now if

$$\lim_{\Delta z \to 0} \frac{1}{\Delta z} \int_z^{z+\Delta z} [f(\zeta) - f(z)] \, d\zeta = 0 \tag{7-6}$$

it would follow from (7-5) that $F'(z) = f(z)$. The fact that the limit in (7-6) is, indeed, zero follows at once from the estimate (5-8), for if $M = \max |f(\zeta) - f(z)|$ on the path joining z and Δz, then

$$\left| \frac{1}{\Delta z} \int_z^{z+\Delta z} [f(\zeta) - f(z)] \, d\zeta \right| \leq M.$$

But since $f(z)$ is continuous, $M \to 0$ as $\Delta z \to 0$.

Any function $F_1(z)$ such that $F_1'(z) = f(z)$ is called a *primitive* or an *indefinite integral* of $f(z)$. As in real calculus, it is easy to prove that if $F_1(z)$ and $F_2(z)$ are any two indefinite integrals of $f(z)$ they can differ only by a constant.[1]

Hence, if $F_1(z)$ is an indefinite integral of $f(z)$, it follows that

$$F(z) = \int_{z_0}^z f(z) \, dz = F_1(z) + C.$$

To evaluate C, set $z = z_0$; then, since $\int_{z_0}^{z_0} f(z) \, dz = 0$, $C = -F_1(z_0)$. Thus

$$F(z) = \int_{z_0}^z f(z) \, dz = F_1(z) - F_1(z_0). \tag{7-7}$$

The statement embodied in (7-7) establishes the connection between line and indefinite integrals and is called the *fundamental theorem of integral calculus* because of its importance in the evaluation of line integrals. It states that *the value of the line integral of an analytic function is equal to the difference in the values of any primitive at the end points of the path of integration.* [Cf. formula (20-5) in Chap. 5.]

Example 1. As an illustration of the use of formula (7-7), consider the evaluation of

$$\int_C z^2 \, dz \tag{7-8}$$

along some path C joining $z = 0$ and $z = 2 + i$. Inasmuch as $f(z) = z^2$ is analytic throughout the finite z plane, the integral (7-8) is independent of the path. Moreover, since $F(z) = \frac{1}{3}z^3$ is an indefinite integral for $f(z) = z^2$, we can write

$$\int_0^{2+i} z^2 \, dz = \frac{1}{3}z^3 \Big|_0^{2+i} = \frac{1}{3}(2 + i)^3.$$

The reader should contrast this computation with calculations required for solving this in Prob. 1 of Sec. 5.

Example 2. Evaluate $\int_C e^z \, dz$ over some path C joining $z = 0$ and $z = \pi i$. Since e^z is analytic, we get at once from (7-7)

$$\int_0^{\pi i} e^z \, dz = e^z \Big|_0^{\pi i} = e^{\pi i} - 1 = -2.$$

We indicate the nature of required calculations if this integral were to be computed by the method of Sec. 5. We first separate the integrand into real and imaginary parts,

$$e^z = e^{x+iy} = e^x \cos y + i e^x \sin y$$

---

[1] Proof: Since $F_1'(z) = F_2'(z) = f(z)$, it is evident that

$$F_1'(z) - F_2'(z) = \frac{d(F_1 - F_2)}{dz} \equiv \frac{dG}{dz} = 0.$$

But if $dG/dz = 0$, it means that $G'(z) = (\partial u/\partial x) + i(\partial v/\partial x) = (\partial v/\partial y) - i(\partial u/\partial y) = 0$, so that $\partial u/\partial x = \partial v/\partial x = \partial u/\partial y = \partial v/\partial y = 0$, and thus u and v do not depend on x and y.

and form two real line integrals

$$\int_C e^z \, dz = \int_C (e^x \cos y + ie^x \sin y)(dx + i \, dy)$$

$$= \int_C (e^x \cos y \, dx - e^x \sin y \, dy) + i \int_C (e^x \sin y \, dx + e^x \cos y \, dy).$$

Since these line integrals are independent of the path, they may be evaluated over any convenient path joining the points $(0,0)$ and $(0,\pi)$ corresponding to $z = 0$ and $z = \pi i$. The result of such calculations would yield -2, as the reader can verify.

Example 3.   Discuss the integral $\int_C (z - a)^m \, dz$, where m is an integer and a is a constant.

The function $f(z) = (z - a)^m$ is obviously analytic at all points of the z plane as long as m is a positive integer. If $m < 0$, we write $m = -n$ and consider

$$f(z) = \frac{1}{(z - a)^n} \tag{7-9}$$

where n is a positive integer.

To evaluate $\int_{z_0}^z (z - a)^m \, dz$ for $m > 0$, we note that

$$F(z) = \frac{(z - a)^{m+1}}{m + 1}$$

is an indefinite integral for $f(z) = (z - a)^m$. Accordingly

$$\int_{z_0}^z (z - a)^m \, dz = \frac{(z - a)^{m+1}}{m + 1} \bigg|_{z_0}^z. \tag{7-10}$$

If, in particular, the path C is closed, so that the limits in (7-10) coincide, we conclude that the value of the integral is zero. This result also follows from Cauchy's theorem, since $f(z) = (z - a)^m$ is analytic for all values of z when $m > 0$.

We consider next the integral

$$\int_C \frac{dz}{(z - a)^n} \qquad n > 0. \tag{7-11}$$

and note first that, if the path C passes through the point $z = a$, the integrand becomes meaningless at $z = a$. In this book[1] we shall not consider in detail integrals over those paths that go through singular points of the integrands, but special types of such integrals will occur in Sec. 22.

If C is a closed path and a is not in the region R enclosed by C, the integrand in (7-11) is analytic in the closed region R. Hence, by Cauchy's theorem the value is zero. If, however, a lies in R, Cauchy's theorem does not apply, since $f(z) = 1/(z - a)^n$ ceases being analytic at $z = a$. The integral (7-11) can, of course, be evaluated by the method of Sec. 5 once the equation of C is specified. However, it is wise to simplify calculations by making use of the principle of deformation of contours. This principle shows that, when $z = a$ is in the interior of C,

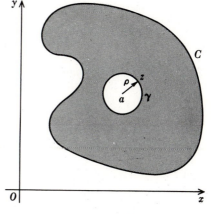

$$\oint_C \frac{dz}{(z - a)^n} = \oint_\gamma \frac{dz}{(z - a)^n}$$

where γ is a circle with center at a and with radius ρ so small that γ lies within C (Fig. 11). But the integral over γ is easily evaluated. Setting $z - a = \rho e^{i\theta}$, we get $dz = \rho e^{i\theta} i \, d\theta$ on observing that ρ is constant on γ. Hence

FIGURE 11

---

[1] When $z = a$ lies on the path of integration, the integral in (7-11) is an *improper* complex integral, and it calls for special considerations analogous to those required to treat improper real integrals. Certain types of improper complex integrals are of interest in applications. See, for example, N. I. Muskhelishvili, "Singular Integral Equations," Erven P. Noordhoff, NV, Groningen, Netherlands, 1953.

$$\oint_C \frac{dz}{(z-a)^n} = \oint_\gamma \frac{\rho e^{i\theta} i\, d\theta}{\rho^n e^{in\theta}} = \frac{i}{\rho^{n-1}} \int_0^{2\pi} e^{(1-n)\theta i}\, d\theta = \frac{i}{\rho^{n-1}} \frac{e^{(1-n)\theta i}}{i(1-n)}\bigg|_0^{2\pi} = 0 \qquad \text{if } n \neq 1. \quad (7\text{-}12)$$

If $n = 1$, we get

$$\int_C \frac{dz}{z-a} = i \int_0^{2\pi} d\theta = 2\pi i. \qquad (7\text{-}13)$$

In evaluating the integral (7-12), we noted that the integrand $e^{(1-n)\theta i}\, d\theta$, for $n \neq 1$, is the differential of $e^{(1-n)\theta i}/i(1-n)$, and we made use of the fundamental theorem of integral calculus.

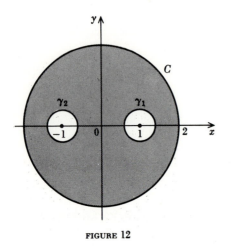

FIGURE 12

Example 4.   Evaluate the integral $\int_C \frac{dz}{z^2 - 1}$, where C is the circle $x^2 + y^2 = 4$.

The function

$$f(z) = \frac{1}{z^2 - 1} \equiv \frac{1}{(z-1)(z+1)} \qquad (7\text{-}14)$$

has two singular points $z = 1$ and $z = -1$, both of which lie within the given circle $|z| \leq 2$ (Fig. 12). If we delete these points from the circular region C by circles γ_1 and γ_2 of sufficiently small radii, $f(z)$ will be analytic in the triply connected domain exterior to γ_1 and γ_2 and interior to C. Then Cauchy's theorem for multiply connected domains permits us to write

$$\int_C f(z)\, dz = \int_{\gamma_1} f(z)\, dz + \int_{\gamma_2} f(z)\, dz. \qquad (7\text{-}15)$$

The integrals in the right-hand member in (7-15) are readily evaluated.   Since

$$\frac{1}{(z-1)(z+1)} \equiv \frac{1}{2}\frac{1}{z-1} - \frac{1}{2}\frac{1}{z+1}$$

we get

$$\int_{\gamma_1} \frac{1}{(z-1)(z+1)}\, dz = \frac{1}{2}\int_{\gamma_1} \frac{dz}{z-1} - \frac{1}{2}\int_{\gamma_1} \frac{dz}{z+1}. \qquad (7\text{-}16)$$

If the radius of γ_1 is such that γ_1 contains within it $z = +1$ but not $z = -1$, then

$$\int_{\gamma_1} \frac{dz}{z-1} = 2\pi i$$

by (7-13), and

$$\int_{\gamma_1} \frac{dz}{z+1} = 0$$

by Cauchy's integral theorem, for $1/(z+1)$ has no singularities within γ_1.   Thus, the first integral on the right in (7-15) has the value πi.   An entirely similar calculation shows

$$\int_{\gamma_2} \frac{1}{(z-1)(z+1)}\, dz = -\pi i.$$

Therefore,

$$\int_C \frac{1}{(z-1)(z+1)}\, dz = (\pi i) + (-\pi i) = 0$$

even though the integrand is not analytic in the region $|z| \leq 2$.

PROBLEMS

1. Show that $\int_{z_0}^z z\, dz = \frac{1}{2}(z^2 - z_0^2)$ for all paths joining z_0 with z.

2. Evaluate the integral $\int_C (z-a)^{-1}\, dz$, where C is a simple closed curve and a is interior to C, by expressing it as a sum of two real line integrals over C.   *Hint:* Set $z - a = \rho e^{\theta i}$; then $dz = e^{\theta i}(d\rho + i\rho\, d\theta)$.

3. Evaluate $\int_C z^{-2}\, dz$, where the path C is the upper half of the unit circle whose center is at the origin. What is the value of this integral if the path is the lower half of the circle?

4. Evaluate $\int_C z^{-1}\, dz$, where C is the path of Prob. 3.

5. Evaluate $\int_C (z^2 - 2z + 1)\, dz$, where C is the circle $x^2 + y^2 = 2$.

6. Evaluate the integral $\int_C \dfrac{z+1}{z^2}\, dz$, where C is a path enclosing the origin.

7. What is the value of the integral $\int_C (1 + z^2)^{-1}\, dz$, where C is the circle $x^2 + y^2 = 9$?

8. Discuss Prob. 7 by noting that

$$\frac{1}{1 + z^2} = \frac{1}{2i}\left(\frac{1}{z - i} - \frac{1}{z + i}\right)$$

and evaluating the integrals over the unit circles whose centers are at $z = i$ and $z = -i$. Note the theorem of Sec. 6.

9. Show that the integrals (a) $\int_C \dfrac{z\, dz}{z - 2}$, (b) $\int_C \sin z\, dz$, (c) $\int_C ze^z\, dz$, (d) $\int_C z^{-2}\, dz$ vanish if C is the unit circle $|z| = 1$.

10. Evaluate the integral $\int_C \dfrac{1}{1 - z^2}\, dz$ along the following paths C: (a) $|z| = \frac{1}{2}$; (b) $|z| = 2$; (c) $|z - 1| = 1$; (d) $|z + 1| = 1$. *Hint:* Decompose the integrand into partial fractions as in Prob. 8.

11. Use formula (7-7) to show that (a) $\int_0^{\pi i} e^z\, dz = -2$; (b) $\int_\pi^{\pi i} \sin 2z\, dz = \frac{1}{2}(1 - \cosh 2\pi)$; (c) $\int_0^i ze^{-z^2}\, dz = \frac{1}{2}(1 - e)$.

12. Show that $\int_C \dfrac{dz}{z^2(z^2 + 4)} = 0$, where C is the circle $|z| = 1$. *Hint:* Decompose the integrand into partial fractions.

---

8. Cauchy's Integral Formula. In this section we deduce with the aid of the strong form of Cauchy's theorem the remarkable fact that every function $f(z)$ continuous in a closed region $R + C$ and analytic in its interior is completely determined in the region R when the values of $f(z)$ are specified on its simple closed boundary C.

Let $f(z)$ be continuous in a simply connected region $R + C$ and analytic at all interior points of this region. If a is an interior point of R, the function

$$\frac{f(z)}{z - a} \tag{8-1}$$

is continuous in $R + C$ and analytic in R with the possible exception of the point $z = a$. If this point is excluded from the region by enclosing it in a circle γ of radius ρ and with the center at a (Fig. 11), then (8-1) will surely be analytic in the region exterior to γ and interior to C.

It follows, then, from (6-3) that

$$\int_C \frac{f(z)}{z - a}\, dz = \int_\gamma \frac{f(z)}{z - a}\, dz \tag{8-2}$$

where the paths C and γ are described in the same sense. Now the integral in the right-hand member of (8-2) can be written as

$$\int_\gamma \frac{f(z)}{z - a}\, dz = \int_\gamma \frac{f(z) - f(a)}{z - a}\, dz + f(a) \int_\gamma \frac{dz}{z - a}. \tag{8-3}$$

But by (7-13)
$$\int_\gamma \frac{dz}{z-a} = 2\pi i \qquad (8\text{-}4)$$

and we shall show next that the first integral on the right in (8-3) has the value zero. Indeed, if we take $z - a = \rho e^{i\theta}$, then, as long as z is on γ, $dz = i\rho e^{i\theta}\, d\theta$, and therefore

$$\int_\gamma \frac{f(z) - f(a)}{z - a}\, dz = i \int_\gamma [f(z) - f(a)]\, d\theta. \qquad (8\text{-}5)$$

Let the maximum of $|f(z) - f(a)|$ be M; then by (5-8)

$$\left| \int_\gamma \frac{f(z) - f(a)}{z - a}\, dz \right| \le M \int_0^{2\pi} d\theta = 2\pi M. \qquad (8\text{-}6)$$

The radius ρ is arbitrary, and if we make it sufficiently small, then $\max |f(z) - f(a)|$ can be made as small as we wish, since $f(z)$ is a continuous function. Accordingly, $M \to 0$ as $\rho \to 0$. On the other hand, from the principle of deformation of contours, the value of the integral (8-6) is *independent* of the radius ρ. Since $M \to 0$ when $\rho \to 0$, we conclude that the value of the integral (8-5) is zero.

Accordingly, (8-3), together with (8-4), gives the result

$$\int_C \frac{f(z)}{z - a}\, dz = 2\pi i f(a). \qquad (8\text{-}7)$$

We recall that the point a is any interior point of the region R bounded by C and z is the variable of integration on the contour C. If we denote the variable of integration by ζ and let z be any interior point, we can rewrite formula (8-7) as

$$f(z) = \frac{1}{2\pi i} \int_C \frac{f(\zeta)\, d\zeta}{\zeta - z}. \qquad (8\text{-}8)$$

Formula (8-8) permits us to calculate the value of $f(z)$ at any interior point from specified boundary values $f(\zeta)$ on the contour C. It is known as *Cauchy's integral formula*. This formula can be extended in the manner[1] of Sec. 6 to multiply connected domains bounded by the exterior contour C_0 and m interior contours $C_1, C_2, \ldots, C_m$. The integration in (8-8) is then performed in the clockwise sense over the interior contours and counterclockwise over the exterior contour C_0.

It is not difficult to show with the aid of formula (8-8) that an analytic function $f(z)$ has not only continuous first derivatives in the region but also derivatives of all orders. Thus an analytic function can be differentiated infinitely many times.

In fact, if we consider an integral of *Cauchy's type*,

$$F(z) = \frac{1}{2\pi i} \int_C \frac{f(\zeta)}{\zeta - z}\, d\zeta \qquad (8\text{-}9)$$

where $f(\zeta)$ is any continuous (not necessarily analytic) complex function defined on a simple closed curve C, this integral defines an analytic function $F(z)$ in the interior of C. To show this we merely have to prove that $F(z)$ has a derivative at every point of the region R bounded by C. We form the difference quotient with the aid of (8-9) and get

$$\begin{aligned}
F'(z) &= \lim_{\Delta z \to 0} \frac{F(z + \Delta z) - F(z)}{\Delta z} \\
&= \lim_{\Delta z \to 0} \frac{1}{\Delta z} \left[\frac{1}{2\pi i} \int_C \frac{f(\zeta)\, d\zeta}{\zeta - (z + \Delta z)} - \frac{1}{2\pi i} \int_C \frac{f(\zeta)\, d\zeta}{\zeta - z} \right] \\
&= \lim_{\Delta z \to 0} \left[\frac{1}{2\pi i} \int_C \frac{f(\zeta)\, d\zeta}{(\zeta - z - \Delta z)(\zeta - z)} \right].
\end{aligned}$$

[1] As in Chap. 5, Sec. 23, the term *contour* means a simple closed curve.

On taking the limit as $\Delta z \to 0$ under the integral sign, which is legitimate if $f(\zeta)$ is continuous, we get

$$F'(z) = \frac{1}{2\pi i} \int_C \frac{f(\zeta)}{(\zeta - z)^2} \, d\zeta.$$

Continuing in the same way, we find

$$F''(z) = \frac{2!}{2\pi i} \int_C \frac{f(\zeta)}{(\zeta - z)^3} \, d\zeta$$

$$\cdot \quad \cdot \quad \cdot \quad \cdot \quad \cdot \quad \cdot \quad \cdot \quad \cdot \quad \cdot \quad \cdot \quad \cdot \quad \cdot \quad \cdot$$

$$F^{(n)}(z) = \frac{n!}{2\pi i} \int_C \frac{f(\zeta)}{(\zeta - z)^{n+1}} \, d\zeta.$$

We have thus shown that $F(z)$ defined by (8-9) has derivatives of all orders even when nothing is said about the relation of the values of $F(z)$ on the boundary C to the function $f(\zeta)$ appearing in the integrand. In the special case when $f(\zeta) = F(\zeta)$, we have a formula for the nth derivative of the analytic function $f(z)$ at any interior point of R in terms of the values of $f(z)$ on C:

$$f^{(n)}(z) = \frac{n!}{2\pi i} \int_C \frac{f(\zeta)}{(\zeta - z)^{n+1}} \, d\zeta \qquad n = 0, 1, 2, \ldots . \qquad (8\text{-}10)$$

We conclude this section by noting some important consequences of formula (8-7). Let the path C be the circle $|z - a| = \rho$ with center at $z = a$ and with radius ρ. Suppose that the maximum value of the modulus of $f(z)$ on this circle is M; then by (5-8)

$$|f(a)| \leq \frac{1}{2\pi} \frac{M}{\rho} 2\pi\rho = M.$$

This result is independent of the radius ρ. Consequently $|f(z)|$ at the center a of the circle is not greater than its maximum value on the boundary. Using this result one can prove that if $f(z)$ is analytic in a given region R bounded by a curve C, and if M is the maximum value of $|f(z)|$ on C, then $|f(z)| < M$ at each interior point of R unless $|f(z)| = M$ throughout the region. This result is known as the *maximum modulus theorem*.[1] The fact that $|f(z)| \leq M$ follows from Chap. 7, Sec. 24, if we note that $\log |f(z)|$ is harmonic.

Example 1. Find the value of the integral $\int_C \frac{\sin z}{z} \, dz$ if C is the ellipse $x^2 + 4y^2 = 1$.

Since $\sin z$ is analytic in the region bounded by C, formula (8-7) yields, upon setting $f(z) = \sin z$ and $a = 0$,

$$\int_C \frac{\sin z}{z} \, dz = 2\pi i(\sin 0) = 0.$$

Example 2. Evaluate the integral $\int_C \frac{e^{-z}}{z+1} \, dz$ over the circular path $|z| = 2$.

The point $z = -1$ lies within the given circle, and since e^{-z} is analytic within C, formula (8-7) yields

$$\int_C \frac{e^{-z}}{z+1} \, dz = 2\pi i e^{-z} \Big|_{z=-1} = e 2\pi i.$$

Example 3. Find the value of the integral

$$\int_C \frac{\tan z}{[z - (\pi/4)]^2} \, dz$$

where C is the circle $|z| = 1$.

[1] See proof, for example, in E. C. Titchmarsh, "The Theory of Functions," 2d ed., p. 164, Oxford University Press, London, 1939.

The point $z = \pi/4$ lies within C, and we note that $\tan z$ is analytic for $|z| \leq 1$. From (8-10)

$$f'(a) = \frac{1}{2\pi i} \int_C \frac{f(z)}{(z-a)^2}\, dz.$$

Hence

$$\int_C \frac{\tan z}{[z-(\pi/4)]^2}\, dz = 2\pi i \left(\frac{d \tan z}{dz}\right)_{z=\pi/4} = 2\pi i \sec^2 \frac{\pi}{4} = 4\pi i.$$

PROBLEMS

1. If $f(z) = \int_C \dfrac{3\zeta^2 + 7\zeta + 1}{\zeta - z}\, d\zeta$, where C is the circle of radius 2 about the origin, find the value of $f(1 - i)$.

2. Apply Cauchy's integral formula to Prob. 7 of Sec. 7. Use the integrand in the form given in Prob. 8 of Sec. 7.

3. Evaluate the following integrals over the closed path C formed by the lines $x = \pm 1$, $y = \pm 1$:

(a) $\displaystyle\int_C \frac{\sin z}{z}\, dz$; (b) $\displaystyle\int_C \frac{\cos z}{z}\, dz$; (c) $\displaystyle\int_C \frac{e^z}{z - \frac{1}{2}i}\, dz$; (d) $\displaystyle\int_C (\sin z + e^z)\, dz$; (e) $\displaystyle\int_C \frac{\cosh z}{z}\, dz$.

4. Evaluate with the aid of Cauchy's integral formula

$$\int_C \frac{3\zeta^2 + \zeta}{\zeta^2 - 1}\, d\zeta$$

where C is the circle $|\zeta| = 2$. *Hint:* Decompose the integrand into partial fractions.

5. What is the value of the integral of Prob. 4 when evaluated over the circle $|\zeta - 1| = 1$? *Hint:* Note that $(3\zeta^2 + \zeta)/(\zeta + 1)$ is analytic for $|\zeta - 1| \leq 1$.

6. Evaluate $\displaystyle\int_C \frac{3z^2 + 2z - 1}{z}\, dz$, where C is the circle $|z| = 1$.

7. Can $|f(z)|$ assume a minimum value at an interior point of a region within which $f(z)$ is analytic? Consider $f(z) = z$.

8. Can $|f(z)|$ assume a nonzero minimum at an interior point of a region within which $f(z)$ is analytic? *Hint:* Consider $1/f(z)$.

9. Evaluate the integral $\displaystyle\int_C \frac{dz}{z^2(z^2 + 4)}$, where C is the circle $|z + 2i| = 1$. What is the value of this integral if (a) C is the circle $|z + 2i| = 1$; (b) C is the circle $|z| = 1$? *Hint:* See Example 3.

10. If C is the circle $|z| = 2$, show with the aid of formula (8-10) that (a) $\displaystyle\int_C \frac{z + z^2}{(z-1)^2}\, dz = 6\pi i$;

(b) $\displaystyle\int_C \frac{e^z}{z^3}\, dz = \pi i$; (c) $\displaystyle\int_C \frac{\sin z}{z^2}\, dz = 2\pi i$; (d) $\displaystyle\int \frac{e^{-z}\sin z}{z^2}\, dz = 2\pi i$; (e) $\displaystyle\int_C \frac{\sin z}{z^n}\, dz = \frac{(-1)^{n+1}}{(2n-1)!} 2\pi i$, if n is a positive integer.

11. If C is a circle of radius $\rho > 0$ with its center at $z = a$, show that $|f^{(n)}(a)| \leq \dfrac{n!M}{\rho^n}$, where M is the maximum of $|f(\zeta)|$ on C in formula (8-10).

12. Use the inequality $|f'(a)| \leq M/\rho$ (see Prob. 11) to prove Liouville's theorem: *If $f(z)$ is analytic and $f(z)$ is bounded for all finite values of z, then $f(z)$ is a constant.* *Hint:* Conclude from the inequality that $f'(z) = 0$ for all finite values of z.

---

9. Harmonic Functions.

We saw in the preceding section that a function analytic at a given point of the region has derivatives of all orders at that point. It follows from this that the real and imaginary parts of an analytic function $f(z) = u + iv$ have partial derivatives of all orders throughout the region where $f(z)$ is analytic, for by (4-5) and (4-6)

$$f'(z) = \frac{\partial u}{\partial x} + i\frac{\partial v}{\partial x} = \frac{\partial v}{\partial y} - i\frac{\partial u}{\partial y}$$

and since $f'(z)$ is also analytic,

$$f''(z) = \frac{\partial^2 u}{\partial x^2} + i\frac{\partial^2 v}{\partial x^2} = \frac{\partial^2 v}{\partial x\,\partial y} - i\frac{\partial^2 u}{\partial x\,\partial y} = -\frac{\partial^2 u}{\partial y^2} - i\frac{\partial^2 v}{\partial y^2}.$$

The fact that $f''(z)$ is analytic enables us to differentiate again to obtain the third partial derivatives, and so on.

Since the first and third expressions for $f''(z)$ must agree, we conclude that

$$\frac{\partial^2 u}{\partial x^2} + \frac{\partial^2 u}{\partial y^2} = 0 \qquad \frac{\partial^2 v}{\partial x^2} + \frac{\partial^2 v}{\partial y^2} = 0.$$

Any real function $u(x,y)$ with continuous second partial derivatives that satisfies Laplace's equation in a given region is called *harmonic* in that region. Thus the real and imaginary parts of a function analytic in the region R are harmonic functions. Two harmonic functions $u(x,y)$, $v(x,y)$ such that $u + iv$ is an analytic function $f(z)$ are said to be *conjugate harmonics*. We shall show next that, if one harmonic function is given, its conjugate harmonic can be determined to within a constant of integration. Let $u(x,y)$ be given in R. Then if $v(x,y)$ is a conjugate harmonic, these functions satisfy the Cauchy-Riemann equations

$$\frac{\partial u}{\partial x} = \frac{\partial v}{\partial y} \qquad \frac{\partial u}{\partial y} = -\frac{\partial v}{\partial x}. \tag{9-1}$$

Hence
$$dv = \frac{\partial v}{\partial x}\,dx + \frac{\partial v}{\partial y}\,dy = -\frac{\partial u}{\partial y}\,dx + \frac{\partial u}{\partial x}\,dy$$

and, since $\partial u/\partial x$ and $\partial u/\partial y$ are known from $u(x,y)$, we have

$$v(x,y) = \int_{(x_0,y_0)}^{(x,y)} \left(-\frac{\partial u}{\partial y}\,dx + \frac{\partial u}{\partial x}\,dy \right) \tag{9-2}$$

where the integral can be evaluated over any path joining an arbitrary point (x_0,y_0) of R with (x,y). Since the value of the line integral (9-2) depends on the choice of (x_0,y_0), it is clear that $v(x,y)$ is determined only to within an arbitrary constant. The integral is independent of the path inasmuch as

$$\frac{\partial}{\partial y}\left(-\frac{\partial u}{\partial y} \right) = \frac{\partial}{\partial x}\left(\frac{\partial u}{\partial x} \right)$$

and this equation is true because $u(x,y)$ is harmonic. It should be noted that, when the region R is not simply connected, the function $v(x,y)$ may turn out to be multiple-valued.[1]

The connection of analytic functions with Laplace's equation is one of the principal reasons for the importance of the theory of functions of complex variables in applied mathematics.

In the preceding section we noted the maximum modulus theorem for analytic functions. In Chap. 7, Sec. 24, the corresponding result was established for harmonic functions, namely, the *maximum values of harmonic functions are invariably assumed on*

[1] See Chap. 5, Sec. 23.

the boundary of the region.   We now indicate an interesting connection between these two maximum principles.

Let u be harmonic in the region R whose boundary is C.   If v is a conjugate harmonic, then $u + iv$ is an analytic function, and therefore the function

$$e^{u+iv} = e^u(\cos v + i \sin v)$$

is also analytic.   But the maximum of $|e^{u+iv}| \equiv e^u$ is assumed on the boundary C of R by the maximum modulus theorem.   Since e^u takes on its maximum on the boundary C, $u(x,y)$ must assume its maximum on C.

Example.   The function $u = x^2 - y^2$ is harmonic in every region.   Obtain a conjugate harmonic v.

Inserting u in the formula (9-2) yields

$$v(x,y) = \int_{(x_0,y_0)}^{(x,y)} (2y\,dx + 2x\,dy) = 2\int_{(x_0,y_0)}^{(x,y)} d(xy) = 2xy + c$$

where $c = -2x_0y_0$.

In this problem the integrand is so simple that we wrote its total differential by inspection. In a more complicated case it may prove more expedient to evaluate the integral over some convenient path rather than express the integrand as a total differential.

An alternative procedure is to obtain the conjugate harmonic function from Eqs. (9-1) by partial integration.   (Cf. Chap. 2, Sec. 4.)

Since $u = x^2 - y^2$, $\partial u/\partial x = 2x$, $\partial u/\partial y = -2y$, and Eqs. (9-1) demand that

$$\frac{\partial v}{\partial x} = 2y \qquad \frac{\partial v}{\partial y} = 2x.$$

The integration of the first of these equations with respect to x yields $v(x,y) = 2yx + g(y)$, where $g(y)$ is a differentiable function of y alone.   The substitution of $v(x,y)$ in the second equation requires that $2x + g'(y) = 2x$, and hence $g'(y) = 0$, or $g(y) = $ constant.

PROBLEMS

1. Prove that $v = 3x^2y - y^3$ is harmonic, and find a conjugate harmonic u.

2. Find an analytic function $f(z) = u + iv$ if (a) $u = x$; (b) $u = \cosh y \cos x$; (c) $u = x/(x^2 + y^2)$; (d) $u = e^x \cos y$; (e) $u = \log \sqrt{x^2 + y^2}$.

3. Prove: (a) If $f(z) = u(x,y) + iv(x,y)$ is analytic in R and $u(x,y) = $ const in R, then $f(z) = $ const in R.   *Hint:* Use Eqs. (9-1).   (b) If $f(z)$ is analytic in R and $f'(z) = 0$ in R, then $f(z) = $ const in R.

4. The functions $u_1 = x^2 - y^2$ and $u_2 = 3x^2y - y^3$ are defined in the disk $x^2 + y^2 \leq 1$.   At what points of this disk do u_1 and u_2 assume their maximum values?

10. Taylor's Series.   In this section we are concerned with the power-series representation of analytic functions.   The reader is advised to review Secs. 10, 11, and 16 of Chap. 1 dealing with the properties of power series.

Here we recall that when the power series $\sum_{k=0}^{\infty} a_k z^k$ converges for $z = z_1$ it converges absolutely and uniformly in every closed circular region $|z| \leq r$, where $r < |z_1|$.   A circle of radius r such that $\Sigma a_k z^k$ converges for $|z| < r$ and diverges for every $|z| > r$ is called the *circle of convergence*, and the number r is the *radius of convergence*. The radius of convergence can frequently be determined with the aid of the ratio test. Thus

$$r = \lim_{n \to \infty} \left| \frac{a_{n-1}}{a_n} \right|$$

whenever this limit exists.[1]

Example. The series $\displaystyle\sum_{n=1}^{\infty} \frac{(-1)^n z^n}{n}$ has the radius of convergence $r = 1$, since

$$\lim_{n \to \infty} \left| \frac{a_{n-1}}{a_n} \right| = \lim_{n \to \infty} \frac{n}{n-1} = 1.$$

The series $\displaystyle\sum_{n=0}^{\infty} n! z^n$ converges only for $z = 0$, since in this case

$$\lim_{n \to \infty} \left| \frac{a_{n-1}}{a_n} \right| = \lim_{n \to \infty} \frac{1}{n} = 0.$$

On the other hand, the series $\displaystyle\sum_{n=0}^{\infty} \frac{z^n}{n!}$ converges for all values of z, since

$$\lim_{n \to \infty} \left| \frac{a_{n-1}}{a_n} \right| = \lim_{n \to \infty} \frac{n!}{(n-1)!} = \infty.$$

We saw in Chap. 1, Sec. 11, that, with every real function $f(x)$ having derivatives of all orders at a given point $x = a$, we can associate the power series

$$\sum_{n=0}^{\infty} a_n (x - a)^n$$

with $a_n = f^{(n)}(a)/n!$ which usually converges to $f(x)$ in some interval about the point $x = a$. However, the existence of infinitely many derivatives at $x = a$ does not ensure the convergence of the series $\Sigma a_n (x - a)^n$ to $f(x)$. To ensure convergence, the remainder in the Taylor formula (11-1) of Chap. 1 must approach zero.

Inasmuch as every function $f(z)$ which is analytic at $z = a$ has infinitely many derivatives at that point, we can write the series

$$\sum_{n=0}^{\infty} \frac{f^{(n)}(a)}{n!} (z - a)^n$$

which converges in some circular region $|z - a| \leq r$. The question is: Does such a series invariably converge to $f(z)$?

We prove next (in contradistinction to the situation with the corresponding real series) that analytic functions can always be represented by power series.

Let $f(z)$ be analytic in some region R, and let C be a circle lying wholly in R and having its center at a. If z is any point interior to C (Fig. 13),

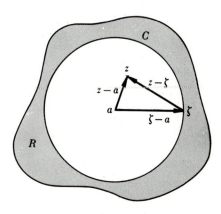

FIGURE 13

[1] See Chap. 1, Sec. 10.

it follows from Cauchy's integral formula that

$$f(z) = \frac{1}{2\pi i} \int_C \frac{f(\zeta)}{\zeta - z} \, d\zeta$$

$$\equiv \frac{1}{2\pi i} \int_C \frac{f(\zeta)}{\zeta - a} \left[\frac{1}{1 - (z-a)/(\zeta - a)} \right] d\zeta. \tag{10-1}$$

But by long division

$$\frac{1}{1-t} = 1 + t + t^2 + \cdots + t^{n-1} + \frac{t^n}{1-t}$$

and substituting this expression with $t = (z-a)/(\zeta - a)$ in (10-1) leads to

$$f(z) = \frac{1}{2\pi i} \left[\int_C \frac{f(\zeta)}{\zeta - a} \, d\zeta + (z-a) \int_C \frac{f(\zeta)}{(\zeta - a)^2} \, d\zeta + \cdots \right.$$

$$\left. + (z-a)^{n-1} \int_C \frac{f(\zeta)}{(\zeta - a)^n} \, d\zeta \right] + R_n$$

where
$$R_n = \frac{(z-a)^n}{2\pi i} \int_C \frac{f(\zeta)}{(\zeta - a)^n (\zeta - z)} \, d\zeta.$$

Making use of (8-10) gives

$$f(z) = f(a) + f'(a)(z-a) + \frac{f''(a)}{2!}(z-a)^2 + \cdots + \frac{f^{(n-1)}(a)}{(n-1)!}(z-a)^{n-1} + R_n. \tag{10-2}$$

By taking n sufficiently large, the modulus of R_n may be made as small as desired. To show this, let the maximum value of $|f(\zeta)|$ on C be M, the radius of the circle C be r, and the modulus of $z - a$ be ρ. Then $|\zeta - z| \geq r - \rho$, as shown in Fig. 13, and

$$|R_n| = \frac{|z - a|^n}{2\pi} \left| \int_C \frac{f(\zeta)}{(\zeta - a)^n (\zeta - z)} \, d\zeta \right|$$

$$\leq \frac{\rho^n}{2\pi} \frac{M 2\pi r}{r^n (r - \rho)} = \frac{Mr}{r - \rho} \left(\frac{\rho}{r} \right)^n.$$

Since $\rho/r < 1$, it follows that $\lim_{n \to \infty} |R_n| = 0$ for every z interior to C. Thus, one can write the infinite series

$$f(z) = f(a) + f'(a)(z-a) + \frac{f''(a)}{2!}(z-a)^2 + \cdots + \frac{f^{(n)}(a)}{n!}(z-a)^n + \cdots \tag{10-3}$$

which converges to $f(z)$ at every point z interior to the circle $|z - a| = r$. The series (10-3) is the Taylor series of $f(z)$ expanded about the point $z = a$. As in Chap. 1, Sec. 11, one can prove that the representation (10-3) is unique.

The radius of convergence of the series (10-3) is equal to the distance from $z = a$ to the nearest singular point $z = z_0$ of $f(z)$. Thus, $r = |z_0 - a|$. The function $f(z)$ may, of course, be analytic at some point $z = b$ outside the circle of convergence C_0 of (10-3) (Fig. 14).

If $f(z)$ is analytic at $z = b$, we can generate the Taylor series

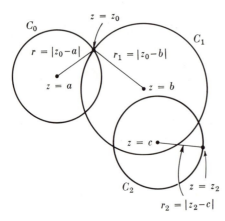

FIGURE 14

$$\sum_{n=0}^{\infty} \frac{f^{(n)}(b)}{n!} (z - b)^n \tag{10-4}$$

which converges to $f(z)$ in some circle C_1, whose radius $r_1 = |z_1 - b|$ is equal to the distance from $z = b$ to the nearest singular point $z = z_1$ of $f(z)$. If this point happens to be $z = z_0$, we have the situation shown in Fig. 14, where the circles C_0 and C_1 overlap. In the overlapping region the series (10-3) and (10-4) converge to $f(z)$. If $z = c$ is a point in the interior of the circle C_1, then $f(z)$ is surely analytic at $z = c$ and we can form the series

$$\sum_{n=0}^{\infty} \frac{f^{(n)}(c)}{n!} (z - c)^n \tag{10-5}$$

which converges to $f(z)$ in some circle C_2, whose radius $r_2 = |z_2 - c|$. Here again $z = z_2$ is the singular point of $f(z)$ nearest $z = c$. If $z = z_2$ is outside the circle C_1, the circles C_1 and C_2 will overlap, and in the overlapping region both series (10-4) and (10-5) converge to $f(z)$. In this manner the entire region where $f(z)$ is analytic can be covered by circular regions each of which is associated with some Taylor's series representation of $f(z)$. The singular points of $f(z)$ will lie on the boundaries of the circles of convergence of such series.

The situation just described is closely related to the problem of continuing an analytic function, originally specified in a given region, to a larger region. Thus, suppose that the given series $S_1 = \Sigma a_n (z - b)^n$ converges in some circle C_1 determined by $|z - b| < r_1$. At every interior point of the circle C_1, the series defines an analytic function, say $f_1(z)$, and if $z = c$ is such a point, the series $S_2 \equiv \Sigma [f_1^{(n)}(c)/n!](z - c)^n$ will converge in some circle C_2 and thus define an analytic function $f_2(z)$ in the interior of C_2. If the circles C_1 and C_2 overlap, as in Fig. 14, then in the common region $f_1(z) = f_2(z)$. It is natural to speak of $f_2(z)$ as being an *analytic continuation* of $f_1(z)$ from the region $|z - b| < r_1$ to the region $|z - c| < r_2$. Such extensions of the regions of definition of analytic functions are of considerable importance in some applications of the analytic-function theory to the solution of boundary-value problems.[1]

PROBLEMS

1. Expand $f(z) = 1/(1 - z)$ in Taylor's series about (a) $z = 0$, (b) $z = -1$, (c) $z = i$, and draw the circles of convergence for each of the series. What relation do the radii of convergence of these series bear to the distance from the point $z = 1$ to the point about which the series expansion is obtained?

2. Expand $f(z) = \text{Log } z$ in the Taylor series about $z = 1$, and determine the radius of convergence.

3. Obtain the Taylor expansion about $z = 0$ for the following functions, and determine the radii of convergence of the resulting series: (a) e^z; (b) $\sin z$; (c) $\cos z$; (d) $\text{Log }(1 + z)$; (e) $\cosh z$; (f) $\cos^2 z$. *Hint:* $\cos^2 z = \frac{1}{2}(1 + \cos 2z)$.

4. Expand $f(z) = \sinh z$ in Taylor's series about the point $z = \pi i$, and determine the radius of convergence of the resulting series.

5. Discuss the validity of the expansion $(1 + z)^m = 1 + mz + [m(m - 1)/2!]z^2 + \cdots$ for arbitrary values of m.

6. Verify the expansions:

(a) $\dfrac{1}{z^2} = \displaystyle\sum_{n=0}^{\infty} (n + 1)(z + 1)^n$ for $|z + 1| < 1$;   (b) $e^z = e \displaystyle\sum_{n=0}^{\infty} \dfrac{(z - 1)^n}{n!}$ for $|z| < \infty$.

---

[1] See also Secs. 16 and 17 of Chap. 1.

7. Show that the series $\sum_{1}^{\infty} z^n/n^2$ converges absolutely at every point of its circle of convergence $|z| = 1$ and thus defines a function $f(z)$ in the disk $|z| \leq 1$. Is $f(z)$ analytic in this disk?

8. Expand $f(z) = (z+2)^{-2}$ in the power series in $z - 1$ and find the radius of convergence of the series. *Hint:* Note that $(z+2)^{-2} = [3 + (z-1)]^{-2} = \frac{1}{9}[1 + \frac{1}{3}(z-1)]^{-2}$ and make use of the binomial theorem in Prob. 5.

9. Deduce the series for $f(z) = \tan^{-1} z$ by integrating term by term the power series in z for $f'(z)$. What is the region of convergence of the series $\sum_{n=0}^{\infty} (-1)^n z^{2n+1}/(2n+1)$?

10. Show that: (a) $\mathrm{Log} \dfrac{1+z}{1-z} = 2 \sum_{1}^{\infty} z^{2n-1}/(2n-1),\ |z| < 1$ [*Hint:* $\mathrm{Log}\,(1+z) = \sum_{1}^{\infty} (-1)^{n+1} z^n/n,$ $|z| < 1$]; (b) $\sinh z = \sum_{1}^{\infty} z^{2n-1}/(2n-1)!$; (c) $\cosh z = \sum_{0}^{\infty} z^{2n}/(2n)!$.

---

11. Laurent's Expansion.   We have just shown that a function $f(z)$ which is analytic at a given point a can be represented in the neighborhood of that point in a power series. Moreover, this series represents $f(z)$ in the interior of the circular region centered at a and whose radius is equal to the distance of a from the nearest singular point of $f(z)$. In this section we prove a more general theorem due to a French mathematician P. H. M. Laurent.

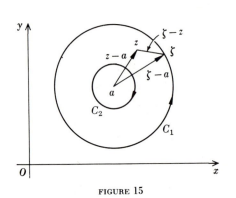

FIGURE 15

LAURENT'S THEOREM. A function $f(z)$ analytic in the interior and on the boundary of the circular ring determined by $|z - a| = R_1$ and $|z - a| = R_2$, with $R_2 < R_1$ (Fig. 15), can be represented at every interior point of the ring in the form

$$f(z) = \sum_{n=0}^{\infty} a_n(z-a)^n + \sum_{n=1}^{\infty} \frac{a_{-n}}{(z-a)^n} \tag{11-1}$$

where

$$a_n = \frac{1}{2\pi i} \oint_{C_1} \frac{f(\zeta)}{(\zeta - a)^{n+1}}\, d\zeta \qquad n = 0, 1, 2, \ldots \tag{11-2}$$

$$a_{-n} = \frac{1}{2\pi i} \oint_{C_2} \frac{f(\zeta)}{(\zeta - a)^{-n+1}}\, d\zeta \qquad n = 1, 2, \ldots \tag{11-3}$$

C_1 *and* C_2 *being the boundaries of the ring.*

To prove the theorem, we recall that Cauchy's formula (8-8), when applied to the circular ring, enables us to write

$$f(z) = \frac{1}{2\pi i} \oint_{C_1} \frac{f(\zeta)}{\zeta - z}\, d\zeta - \frac{1}{2\pi i} \oint_{C_2} \frac{f(\zeta)\, d\zeta}{\zeta - z} \tag{11-4}$$

where z is any point in the interior of the ring.

We show next that the integrals in the right-hand member of (11-4) can be represented by the series appearing in (11-1). We begin with the integral over C_1 and note that if ζ is on C_1 and z is in the ring then

$$\frac{1}{\zeta - z} \equiv \frac{1}{\zeta - a}\frac{1}{1 - (z-a)/(\zeta-a)} = \frac{1}{\zeta - a}\sum_{n=0}^{\infty}\frac{(z-a)^n}{(\zeta-a)^n}$$

since[1] $|z-a|/|\zeta-a| < 1$. Thus,

$$\frac{1}{\zeta - z} = \sum_{n=0}^{\infty}\frac{(z-a)^n}{(\zeta-a)^{n+1}} \tag{11-5}$$

and hence

$$\frac{1}{2\pi i}\oint_{C_1}\frac{f(\zeta)\,d\zeta}{\zeta - z}\,d\zeta = \frac{1}{2\pi i}\oint_{C_1}\sum_{n=0}^{\infty}\frac{f(\zeta)(z-a)^n}{(\zeta-a)^{n+1}}\,d\zeta.$$

Since integration of the series term by term can be justified as in the discussion of (10-1), we can write

$$\frac{1}{2\pi i}\oint_{C_1}\frac{f(\zeta)\,d\zeta}{\zeta - z}\,d\zeta = \frac{1}{2\pi i}\sum_{n=0}^{\infty}(z-a)^n\oint_{C_1}\frac{f(\zeta)}{(\zeta-a)^{n+1}}\,d\zeta \equiv \sum_{n=0}^{\infty}a_n(z-a)^n$$

where we define a_n by the formula (11-2). This establishes the equality of the first terms in the right-hand members of (11-1) and (11-4).

We consider next the second integral in (11-4). If ζ is on C_2, then

$$\frac{1}{\zeta - z} = -\frac{1}{z-a}\frac{1}{1 - (\zeta-a)/(z-a)} = -\sum_{n=0}^{\infty}\frac{(\zeta-a)^n}{(z-a)^{n+1}}$$

since $|\zeta-a|/|z-a| < 1$ in this case. Hence

$$\frac{1}{2\pi i}\oint_{C_2}\frac{f(\zeta)}{\zeta - z}\,d\zeta = -\frac{1}{2\pi i}\oint_{C_2}\sum_{n=0}^{\infty}\frac{f(\zeta)(\zeta-a)^n}{(z-a)^{n+1}}\,d\zeta$$

and the integration of the series term by term now yields

$$\frac{1}{2\pi i}\oint_{C_2}\frac{f(\zeta)}{\zeta - z}\,d\zeta = -\frac{1}{2\pi i}\sum_{n=0}^{\infty}\frac{1}{(z-a)^{n+1}}\oint_{C_2}f(\zeta)(\zeta-a)^n\,d\zeta \equiv -\sum_{n=1}^{\infty}\frac{a_{-n}}{(z-a)^n}$$

where a_{-n} is given by (11-3). This establishes the equality of the second terms in the right-hand members of (11-1) and (11-4), and the theorem is proved.

We note that, if $f(z)$ is also analytic in the interior of the circle C_2, the integrand in (11-3) is an analytic function and hence $a_{-n} = 0$ by Cauchy's integral theorem. In this case, (11-1) reduces to the Taylor series, since

$$a_n = \frac{1}{2\pi i}\oint_{C_1}\frac{f(\zeta)}{(\zeta-a)^{n+1}}\,d\zeta = \frac{f^{(n)}(a)}{n!}$$

by (8-10).

We can write the series (11-1) more compactly as

$$f(z) = \sum_{n=-\infty}^{\infty}a_n(z-a)^n \tag{11-6}$$

Note that $\dfrac{1}{1-t} = \displaystyle\sum_{n=0}^{\infty}t^n$ if $|t| < 1$.

where the a_n can be computed from the formula

$$a_n = \frac{1}{2\pi i} \oint_\Gamma \frac{f(\zeta)}{(\zeta - a)^{n+1}} \, d\zeta \qquad n = 0, \pm 1, \pm 2, \ldots \tag{11-7}$$

and Γ is any simple closed path[1] which lies in the ring and encloses C_2.
The representation of $f(z)$ in the series (11-6) is unique, for let

$$f(z) = \sum_{n=-\infty}^{\infty} b_n(z - a)^n \tag{11-8}$$

be another representation of $f(z)$ in the same ring. We shall show that the coefficients b_n in (11-8) are given by the same formula (11-7). We first note that the series $\sum_0^\infty b_n(z - a)^n$ and $\sum_1^\infty b_{-n}(z - a)^{-n}$ are uniformly convergent in the ring of Fig. 15. Accordingly, the series (11-8) can be integrated term by term along any simple closed path Γ that lies in the ring and encloses C_2. We also note that

$$\int_\Gamma \frac{dz}{(z - a)^k} = 0 \qquad \text{if } k \neq 1$$
$$= 2\pi i \qquad \text{if } k = 1 \tag{11-9}$$

by formulas (7-12) and (7-13).

We multiply both members of (11-8) by $\frac{1}{2\pi i} \frac{dz}{(z - a)^{m+1}}$ and obtain, on integrating over Γ,

$$\frac{1}{2\pi i} \int_\Gamma \frac{f(z)}{(z - a)^{m+1}} \, dz = \frac{1}{2\pi i} \sum_{n=-\infty}^{\infty} \int_\Gamma \frac{b_n}{(z - a)^{m-n+1}} \, dz. \tag{11-10}$$

By (11-9) every integral in the right-hand member of (11-10) vanishes except when $m - n + 1 = 1$, that is, when $n = m$. We thus obtain

$$\frac{1}{2\pi i} \int_\Gamma \frac{f(z)}{(z - a)^{m+1}} \, dz = b_m \qquad m = 0, \pm 1, \pm 2, \ldots$$

which is identical with formula (11-7).

The fact that the representation of $f(z)$ in the given ring by Laurent's series (11-6) is unique frequently enables one to deduce the series (11-6) without evaluating the coefficients a_n by formula (11-7).

For example, let $f(z) = e^z/z^2$, and let it be required to obtain the expansion

$$\frac{e^z}{z^2} = \sum_{n=-\infty}^{\infty} a_n z^n.$$

Since $e^z = 1 + z + (z^2/2!) + \cdots + (z^n/n!) + \cdots$, we have for any $z \neq 0$

$$\frac{e^z}{z^2} = \frac{1}{z^2} + \frac{1}{z} + \frac{1}{2!} + \cdots + \frac{z^{n-2}}{n!} + \cdots.$$

This is a Laurent expansion about the origin; hence it is the Laurent expansion about the origin.

The Laurent expansion for $e^{1/z}$, valid for all $|z| > 0$, can be obtained from the series $e^u = 1 + u + (u^2/2!) + \cdots$ by letting $u = 1/z$.

As another illustration, consider

$$f(z) = \frac{z}{(z - 1)(z - 3)}. \tag{11-11}$$

[1] Recall that the integrals (11-2) and (11-3) have the same values when calculated over any path Γ into which C_1 and C_2 may be deformed without leaving the ring.

This function has two singular points: $z = 1$ and $z = 3$. To obtain the Laurent series $\sum_{n=-\infty}^{\infty} a_n(z-1)^n$ valid in the neighborhood of $z = 1$, we can proceed as follows: Set $\phi(z) \equiv z/(z-3)$, and expand $\phi(z)$ in Taylor's series about $z = 1$. The result is

$$\frac{z}{z-3} = -\frac{1}{2} - 3 \sum_{n=1}^{\infty} \frac{(z-1)^n}{2^{n+1}}. \tag{11-12}$$

Since $z = 3$ is a singular point of $\phi(z)$, we conclude that (11-12) converges as long as $|z-1| < 2$. On multiplying this series by $1/(z-1)$, we get

$$\frac{z}{(z-1)(z-3)} = -\frac{1}{2(z-1)} - 3 \sum_{n=1}^{\infty} \frac{(z-1)^{n-1}}{2^{n+1}}$$

which is valid for $0 < |z-1| < 2$.

To obtain the expansion of $f(z)$ in (11-11) about $z = 3$, we set $\phi(z) = z/(z-1)$, expand it in Taylor's series about $z = 3$, and multiply the result by $1/(z-3)$.

The expansion for $f(z)$ in (11-11) valid for $|z| > 3$ can be deduced as follows: We decompose $f(z)$ into partial fractions and find

$$\frac{z}{(z-1)(z-3)} = \frac{-\frac{1}{2}}{z-1} + \frac{\frac{3}{2}}{z-3}. \tag{11-13}$$

But

$$\frac{1}{z-1} = \frac{1}{z}\frac{1}{1-(1/z)} = \frac{1}{z}\left(1 + \frac{1}{z} + \frac{1}{z^2} + \cdots\right) \qquad \text{for } |z| > 1$$

and

$$\frac{1}{z-3} = \frac{1}{z}\frac{1}{1-(3/z)} = \frac{1}{z}\left(1 + \frac{3}{z} + \frac{3^2}{z^2} + \cdots\right) \qquad \text{for } |z| > 3.$$

Substitution of these series in (11-13) yields the desired expansion.

The reader may find it instructive to obtain the same expansion by writing

$$f(z) = \frac{z}{(z-1)(z-3)} = \frac{1}{z}\frac{1}{1-(1/z)}\frac{1}{1-(3/z)} \tag{11-14}$$

and forming the product of the appropriate series for the factors in the right-hand member of (11-14).

PROBLEMS

1. (a) Read the discussion of Taylor's series in Chap. 1, Sec. 11, and study the example of that section; (b) expand $f(z)$ in (11-11) by setting $t = z - 1$, $t = z - 3$, $t = 1/z$ and expressing the results in powers of t.

2. Obtain Laurent's expansions for $f(z) = 1/[z(1-z)^2]$: (a) about $z = 0$, (b) about $z = 1$.

3. Obtain Laurent's expansion for e^{-1/z^2} valid for $|z| > 0$.

4. Expand in Laurent's series about $z = 1$: (a) $(z-1)^2$; (b) $1/(z-1)^2$; (c) $(z-1)^2 + [1/(z-1)^2]$.

5. Obtain Laurent's expansion for $f(z) = 1/[(z-1)(z-2)]$ valid in the following regions: (a) $|z-1| < 1$; (b) $|z| > 2$; (c) $1 < |z| < 2$. Note that in parts b and c the desired expansions have the forms $\sum_{n=-\infty}^{\infty} a_n z^n$. *Hint:* Show that

$$f(z) \equiv \frac{1}{z-2} - \frac{1}{z-1} \qquad \text{and} \qquad \frac{1}{z-1} = \frac{1}{z}\sum_{n=0}^{\infty}\frac{1}{z^n} \qquad \text{for } |z| > 1$$

$$\frac{1}{z-2} = -\frac{1}{2}\sum_{n=0}^{\infty}\left(\frac{z}{2}\right)^n \quad \text{for } |z| < 2 \qquad \frac{1}{z-2} = \frac{1}{z}\sum_{n=0}^{\infty}\left(\frac{2}{z}\right)^n \quad \text{for } |z| > 2.$$

6. Show that $f(z) = 1/[z^2(1 - z)]$ has the following expansions:

(a) $\displaystyle\sum_{n=0}^{\infty} z^{n-2}$, valid for $0 < |z| < 1$ (b) $\displaystyle\sum_{n=0}^{\infty} \frac{-1}{z^{n+3}}$, valid for $|z| > 1$.

7. As in Prob. 1 or as in the text show that

(a) $z^2 e^{-z} = z^2 + z + \dfrac{1}{2} + \displaystyle\sum_{n=1}^{\infty} \frac{1}{(n+2)!z^n}$ if $z > 0$;   (b) $z^{-4}e^{-z^2} = \displaystyle\sum_{n=0}^{\infty} \frac{1}{n!z^{2n+4}}$ if $z > 0$;

(c) $z^{-4}(1 + z^4)^{-1} = \displaystyle\sum_{n=0}^{\infty} (-1)^n z^{4n-4}$ if $0 < z < 1$;   (d) $(1 - z^4)^{-1} = \displaystyle\sum_{n=0}^{\infty} z^{4n}$ if $|z| < 1$;

(e) $e^z(z - 1)^{-2} = e \displaystyle\sum_{n=0}^{\infty} \frac{(z-1)^{n-2}}{n!}$ if $|z - 1| > 0$;   (f) $z^{-1}(4 - z)^{-1} = \displaystyle\sum_{n=0}^{\infty} \frac{z^{n-1}}{4^{n-1}}$ if $0 < |z| < 4$;

(g) $(1 - z^2)^{-1} = \displaystyle\sum_{n=0}^{\infty} \frac{(-1)^{n+1}}{2^{n+1}} (z - 1)^{n-1}$ if $0 < |z - 1| < 2$;

(h) $(1 - z^2)^{-1} = -\displaystyle\sum_{n=0}^{\infty} \frac{(-2)^n}{(z-1)^{n+2}}$ if $|z - 1| > 2$.

8. Expand $f(z) = (1 - z^4)^{-1}$ in Laurent's series about $z = -1$. *Hint:* Write $f(z)$ as the sum of four simple partial fractions and represent each fraction in powers of $z + 1$.

---

12. Singular Points. Residues. If $z = a$ is a singular point of an analytic function $f(z)$ and the neighborhood of $z = a$ contains no other singular points of $f(z)$, the singularity at $z = a$ is said to be *isolated*.

Thus, $f(z) = 1/z$ has an isolated singular point $z = 0$ because the region $|z| = \rho > 0$ contains no singular points other than $z = 0$ within it.   The function

$$f(z) = \frac{z - 1}{z(z^2 + 1)}$$

has three isolated singular points: $z = 0$, $z = i$, $z = -i$.   The function

$$f(z) = e^{1/(z^2 - 1)}$$

has two isolated singular points: $z = 1$ and $z = -1$.   Not all singular points of analytic functions are isolated, however.   For example,

$$f(z) = \frac{1}{\sin\,(1/z)} \tag{12-1}$$

has a singularity whenever $z = \pm(1/k\pi)$, $k = 1, 2, \ldots$.   These singular points are isolated.   But (12-1) also has a singular point $z = 0$, which is not isolated, for, no matter how small the radius ρ of the circle $|z| = \rho$ may be, this circle contains infinitely many singular points $z = \pm(1/k\pi)$ in its interior.

If $z = a$ is an isolated singular point of $f(z)$, then in the neighborhood of $z = a$ the function $f(z)$ can be represented by the Laurent series

$$f(z) = \sum_{n=0}^{\infty} a_n(z - a)^n + \sum_{n=1}^{\infty} \frac{a_{-n}}{(z - a)^n}. \tag{12-2}$$

Some coefficients in (12-2) may vanish, and there are two nontrivial cases that present themselves:

1. The expansion (12-2) contains at most a finite number m of terms with negative powers of $z - a$, so that (12-2) reads

$$f(z) = \sum_{n=0}^{\infty} a_n(z - a)^n + \frac{a_{-1}}{z - a} + \frac{a_{-2}}{(z - a)^2} + \cdots + \frac{a_{-m}}{(z - a)^m}. \qquad (12\text{-}3)$$

2. The expansion (12-2) contains infinitely many terms with negative powers of $z - a$.

The type of singularity at $z = a$ characterized by the representation (12-3) is called a *pole of order m*. A pole of order 1 is also called a *simple* pole. When the expansion (12-2) has infinitely many terms with negative powers of $z - a$, the point $z = a$ is called an *essential singular point* of $f(z)$. We shall see in Sec. 14 that the behavior of a function in the neighborhood of a pole differs radically from that at an essential singular point.

We note from (12-3) that, whenever $f(z)$ has a pole of order m, one can define a function[1]

$$\phi(z) = (z - a)^m f(z) \qquad z \neq a$$

$$\phi(a) = a_{-m}$$

which is analytic at $z = a$, but the function $(z - a)^{m-1}f(z)$ is not analytic at $z = a$. This property is used sometimes to define a pole of order m.

The coefficient a_{-1} in the Laurent representation (12-2) of $f(z)$ in the neighborhood of an isolated singular point $z = a$ plays an important role in the evaluation of integrals of analytic functions. This coefficient is called the *residue* of $f(z)$ at $z = a$.

When the singularity at $z = a$ is a *pole* of order m, the residue at a can be determined without deducing the Laurent expansion. Thus, on multiplying (12-3) by $(z - a)^m$, we get

$$\phi(z) \equiv (z - a)^m f(z)$$

$$= a_{-m} + a_{-m+1}(z - a) + \cdots + a_{-1}(z - a)^{m-1} + a_0(z - a)^m + \cdots \qquad (12\text{-}4)$$

where $a_{-m} \neq 0$. Since this is a power-series representation of $\phi(z)$, the coefficient a_{-1} in it must be the coefficient of the term $(z - a)^{m-1}$ in the Taylor expansion of $\phi(z)$ about $z = a$. Thus

$$a_{-1} = \frac{1}{(m - 1)!} \frac{d^{m-1}[(z - a)^m f(z)]}{dz^{m-1}}\bigg|_{z=a}. \qquad (12\text{-}5)$$

We formulate this result as a useful theorem:

THEOREM. *If* $\phi(z) = (z - a)^m f(z)$ *is analytic at* $z = a$ *and* $\phi(a) \neq 0$, *then* $f(z)$ *has a pole of order m at* $z = a$ *with the residue given by* (12-5).

As a special case of this theorem we note that, when the pole at $z = a$ is simple, the residue at a is given by the formula

$$a_{-1} = \lim_{z \to a} f(z)(z - a). \qquad (12\text{-}6)$$

Example 1.   Obtain the residues at the singular points of $f(z) = (1 + z)/[z(2 - z)]$. This function has a simple pole at $z = 0$ inasmuch as

[1] When $z = a$, the function $\phi(z)$ assumes the indeterminate form $0/0$. We agree to define $\phi(a) = \lim\limits_{z \to a} \phi(z)$.

$$\phi(z) = z\frac{1+z}{z(2-z)} = \frac{1+z}{2-z}$$

is analytic and does not vanish at $z = 0$.   Also

$$\phi(z) = (z-2)\frac{1+z}{z(2-z)} = -\frac{z+1}{z}$$

is analytic at $z = 2$ and does not vanish for $z = 2$.   Hence $f(z)$ also has a simple pole at $z = 2$.

The residues at these points can therefore be computed with the aid of the formula (12-6). We find that the residue at $z = 0$ is $\frac{1}{2}$ and at $z = 2$ it is $-\frac{3}{2}$.

Example 2.   The function

$$f(z) = \frac{e^z}{z^2+1} = \frac{e^z}{(z+i)(z-i)}$$

obviously has simple poles at $z = -i$ and $z = i$.   Therefore the residue at $z = i$ is

$$a_{-1} = \lim_{z \to i} (z-i)\frac{e^z}{(z+i)(z-i)} = \lim_{z \to i}\frac{e^z}{z+i} = \frac{e^i}{2i}.$$

Similarly, the residue at $z = -i$ is found to be $-e^{-i}/2i$.

Example 3.   The function $f(z) = 1/[z(z+1)^2]$ has a simple pole at $z = 0$, since

$$\phi(z) = z\frac{1}{z(z+1)^2} = \frac{1}{(z+1)^2}$$

is analytic at $z = 0$ and $\phi(0) \neq 0$.   Therefore, the residue at $z = 0$ is

$$a_{-1} = \lim_{z \to 0}\frac{1}{(z+1)^2} = 1.$$

The singularity of $f(z)$ at $z = -1$ is a pole of order 2, since

$$\phi(z) = (z+1)^2\frac{1}{z(1+z)^2} = \frac{1}{z}$$

is analytic at $z = -1$ and $\phi(-1) = -1$.   We can therefore compute the residue at $z = -1$ with the aid of (12-5).   We get

$$a_{-1} = \frac{1}{1!}\frac{d}{dz}\left(\frac{1}{z}\right)_{z=-1} = -\frac{1}{z^2}\bigg|_{z=-1} = -1.$$

Example 4.   The function $(\sin z)/z^4$ has a pole of order 3 at $z = 0$ as the reader can easily check with the aid of the theorem of this section.   Hence the residue at $z = 0$ can be computed by using formula (12-5).   It is simpler, however, in this case, to write the Laurent expansion in the neighborhood of $z = 0$ and obtain the residue from it.

Since $\sin z = z - (z^3/3!) + (z^5/5!) - \cdots$,

$$\frac{\sin z}{z^4} = \frac{1}{z^3} - \frac{1}{3!}\frac{1}{z} + \frac{z}{5!} - \cdots \qquad \text{for } |z| > 0.$$

It is clear from this that the singularity at $z = 0$ is a pole of order 3 with the residue $-1/3!$.

Example 5.   The function

$$f(z) = \cos\frac{1}{z-1}$$

has an isolated singular point at $z = 1$.   This point, however, is not a pole, for on noting that

$$\cos u = 1 - \frac{u^2}{2!} + \frac{u^4}{4!} - \cdots$$

we conclude by the substitution $u = 1/(z-1)$ that for $|z-1| > 0$

$$\cos\frac{1}{z-1} = 1 - \frac{1}{2!(z-1)^2} + \frac{1}{4!(z-1)^4} - \cdots.$$

This is the Laurent expansion about $z = 1$. Since it has infinitely many negative powers of $z - 1$, the point $z = 1$ is an essential singular point. Inasmuch as the term $(z - 1)^{-1}$ does not appear in the expansion, the residue a_{-1} at $z = 1$ is zero.

Throughout the discussion of analytic functions we restricted our consideration to single-valued functions. Whenever multiple-valued functions occurred, we confined our attention to their single-valued branches. Such branches, as we have seen, may have derivatives in certain regions of the z plane and thus are analytic in these regions. For example, $f(z) = \sqrt{z} = r^{\frac{1}{2}}[\cos(\theta + 2k\pi)/2 + i \sin(\theta + 2k\pi)/2]$, $k = 0, 1$, has two single-valued branches, each of which has a derivative $f'(z) = 1/(2\sqrt{z})$. Thus, each of these branches is analytic except when $z = 0$. Again, $f(z) = \log z = \text{Log } r + (\theta + 2k\pi)i$, $k = 0, \pm1, \pm2, \ldots$ has infinitely many branches, each of which has a derivative $f'(z) = 1/z$, and so $z = 0$ is the only point at which the branches of $\log z$ cease to be analytic. However, the singular points in these examples are not isolated for, as we shall see in Secs. 16 and 17, each branch of $\sqrt{z}$ and $\log z$ is discontinuous in the neighborhood of its singular point. Points at which the values of the branches of a multiple-valued function $f(z)$ become equal (or infinite) are called the *branch points*, and at such points the derivatives of $f(z)$ cease to exist. As just noted, the branch points are not isolated singular points of $f(z)$, and in the neighborhood of a branch point, $f(z)$ cannot be represented by the Laurent series.

PROBLEMS

1. Obtain the Laurent expansions in the neighborhood of the singular points of the following functions, and thus obtain the residues:

(a) $\dfrac{\cos z}{z^3}$;   (b) e^{-1/z^2};   (c) $\dfrac{1}{1 - z}$;   (d) $\dfrac{z}{1 - z}$;   (e) $\dfrac{e^{-z}}{z^2}$;   (f) $z^2 e^{1/z}$;   (g) $\dfrac{1 - e^{2z}}{z^4}$;   (h) $\dfrac{e^z}{(z - 1)^2}$;

(i) $\dfrac{1}{1 - z^2}$;   (j) $\dfrac{1}{z(z - 1)}$;   (k) $\dfrac{1}{e^z - 1}$;   (l) $\dfrac{e^{1/z}}{z^2}$;   (m) $\dfrac{1}{(z^2 - 1)^2}$;   (n) $\dfrac{z^2}{1 - z^4}$;   (o) $\dfrac{z^2 - 1}{(z - 2)^2}$.

2. Whenever possible, determine the residues at the poles of the functions in Prob. 1 by means of formula (12-5).

3. Obtain the residues in Examples 1, 2, and 3 of this section by deducing appropriate Laurent's series.

4. Prove the following theorem: If $f(z) = g(z)/h(z)$ is the quotient of two functions analytic at $z = a$ such that $g(a) \neq 0$, $h(a) = 0$, and $h'(a) \neq 0$, then $f(z)$ has a simple pole at $z = a$ with the residue $g(a)/h'(a)$. *Hint:* Examine the quotient of the Taylor expansions of $g(z)$ and $h(z)$ about $z = a$.

5. Use the theorem of Prob. 4 to show that $f(z) = \cot z = \cos z/\sin z$ has simple poles at $z = \pm k\pi$, $k = 0, 1, 2, \ldots$.

6. Note that $f(z) = 1/(2 - z) + 1/(z - 1)$ has the Laurent expansion

$$f(z) = \sum_{n=0}^{\infty} \frac{1}{2^{n+1}} z^n + \sum_{n=1}^{\infty} \frac{1}{z^n}$$

valid in the ring $1 < |z| < 2$. This expansion has the term $1/z$. Does it follow that $z = 0$ is a singular point of $f(z)$ with the residue equal to 1?

7. Find the residues of the following functions:

(a) $\tan z$ at $z = \pi/2$; (b) $\csc^2 z$ at $z = 0$ [*Hint:* $\sin^2 z = \frac{1}{2}(1 - \cos 2z)$]; (c) $1/(z^2 \sin z)$ at $z = 0$; (d) $\csc z$ at $z = \pm n\pi$, $n = 0, 1, 2, \ldots$; (e) $e^z \csc z$ at $z = \pm n\pi$, $n = 0, 1, 2, \ldots$.

13. Residue Theorem.   Let $f(z)$ be analytic in the given closed region R bounded by a simple closed curve C, except at the isolated singular points $z = z_1$, $z = z_2$, ..., $z = z_m$. If these points z_k are enclosed by circles Γ_k $(k = 1, 2, \ldots, m)$, so that $f(z)$ is analytic in the multiply connected region bounded by C and the Γ_k, we know that

$$\oint_C f(z) \, dz = \oint_{\Gamma_1} f(z) \, dz + \oint_{\Gamma_2} f(z) \, dz + \cdots + \oint_{\Gamma_m} f(z) \, dz. \tag{13-1}$$

But from (11-7), on setting $n = -1$, we see that

$$(a_{-1})_k = \frac{1}{2\pi i} \oint_{\Gamma_k} f(z) \, dz \tag{13-2}$$

where $(a_{-1})_k$ is the residue of $f(z)$ at $z = z_k$. We can thus write (13-1) in the form

$$\oint_C f(z) \, dz = 2\pi i \sum_{k=1}^{m} (a_{-1})_k. \tag{13-3}$$

The result embodied in this formula is known as the

RESIDUE THEOREM.   The integral of $f(z)$ over a contour C containing within it only isolated singular points of $f(z)$ is equal to $2\pi i$ times the sum of the residues at these points, whenever $f(z)$ is analytic in the closed region $R + C$.

Inasmuch as the residues of $f(z)$, as demonstrated in the preceding section, can often be easily calculated, we see that formula (13-3) provides a simple means for evaluating integrals of analytic functions with isolated singularities.

Example 1.   Evaluate $\int_C \dfrac{1+z}{z(2-z)} \, dz$, where C is the circle $|z| = 1$.

The only singular point of the integrand enclosed by C is $z = 0$. In Example 1 of Sec. 12 we saw that the residue of the integrand at $z = 0$ is $\frac{1}{2}$. Hence the value of the integral is $(2\pi i)\frac{1}{2} = \pi i$. The value of this integral over any path C enclosing $z = 0$ and $z = 2$ is $2\pi i(\frac{1}{2} - \frac{3}{2}) = -2\pi i$, since the residues at these points are $\frac{1}{2}$ and $-\frac{3}{2}$.

Example 2.   Evaluate $\int_C \dfrac{e^z}{z^2 + 1} \, dz$ over the circular path $|z| = 2$.

The residues of the integrand at $z = i$ and $z = -i$ were computed in Example 2 of Sec. 12. Hence the value of the integral is

$$2\pi i \left(\frac{e^i}{2i} - \frac{e^{-i}}{2i} \right) = 2\pi i \sin 1.$$

Example 3.   Evaluate $\int_C \cos \left(\dfrac{1}{z-1} \right) dz$.

We saw in Example 5 of Sec. 12 that $z = 1$ is an essential singular point with the residue zero. Hence the value of the integral is zero for every closed path C which does not pass through $z = 1$. If $z = 1$ lies on C, the integral is improper and other means have to be employed to determine its value.

PROBLEMS

1. Use results of Prob. 1 in Sec. 12 to obtain values of the following integrals, where C is the circle $|z| = 2$:

(a) $\int_C \dfrac{\cos z}{z^3} \, dz$; (b) $\int_C \dfrac{z \, dz}{1 - z}$; (c) $\int_C z^2 e^{1/z} \, dz$; (d) $\int_C \dfrac{1 - e^{2z}}{z^4} \, dz$; (e) $\int_C \dfrac{e^z}{(z-1)^2} \, dz$; (f) $\int_C \dfrac{1}{1 - z^2} \, dz$;

(g) $\int_C \dfrac{1}{z(z-1)} \, dz$; (h) $\int_C \dfrac{z^2}{1 - z^4} \, dz$.

2. Determine the residues of $f(z) = (z - 2)/z(z - 1)$ at $z = 0$ and $z = 1$, and thus evaluate the integral $\int_C \dfrac{z - 2}{z(z - 1)}\, dz$, where C is the circle $|z| = 2$.

3. Evaluate the integrals $\int_{C_i} \dfrac{z + 1}{z^2 - 2z}\, dz$ $(i = 1, 2)$, where C_1 is the circle $|z| = 1$ and C_2 is the circle $|z| = 3$.

4. Find the value of $\int_C \dfrac{z + 1}{(z - 2)^2}\, dz$, where (a) C is the circle $|z| = 1$; (b) C is the circle $|z| = 3$.

5. If C is the circle $|z| = 1$, evaluate: (a) $\int_C e^z \csc z\, dz$; (b) $\int_C \cot z\, dz$; (c) $\int_C \csc z\, dz$.

6. If C is the circle $|z| = 2$, show that $\quad (a)$ $\int_C \tan z\, dz = -4\pi i$; $\quad (b)$ $\int_C z e^{1/z}\, dz = \pi i$;

(c) $\int_C \dfrac{\sinh z}{z^4}\, dz = \dfrac{2\pi i}{3!}$; $\quad (d)$ $\int_C \dfrac{dz}{\sinh 2z} = -\pi i$; $\quad (e)$ $\int_C \dfrac{dz}{z \sin z} = 0$; $\quad (f)$ $\int_C \dfrac{dz}{z^2 \sin z} = \dfrac{\pi i}{3}$.

---

14. Behavior of $f(z)$ at Poles and Essential Singular Points.   From Laurent's representation (12-3) of $f(z)$ in the neighborhood of a pole $z = a$, we easily conclude that $|f(z)|$ becomes infinite as $z \to a$.   The behavior of $|f(z)|$ with an essential singularity at $z = a$ is different because the expansion (12-2) has infinitely many terms with negative powers of $z - a$.   While it is true that in this case $|f(z)|$ is also unbounded as $z \to a$, the function $|f(z)|$ oscillates as $z \to a$.   Indeed, it was shown by E. Picard that, in the neighborhood of an essential singular point, $f(z)$ assumes any preassigned value, with the possible exception of one value, infinitely many times.   A discussion of this would carry us too far in the study of analytic functions, and we merely illustrate this behavior by an example.   Since

$$e^{1/z} = 1 + \frac{1}{z} + \frac{1}{2! z^2} + \cdots \qquad |z| > 0$$

$f(z) = e^{1/z}$ has an essential singular point at $z = 0$.   We show that if A is any complex number not zero there are infinitely many values of z in the neighborhood of $z = 0$ such that

$$e^{1/z} = A \tag{14-1}$$

for on taking the logarithm of (14-1) we get infinitely many solutions

$$z = \frac{1}{\mathrm{Log}\,|A| + i(\phi + 2k\pi)} \qquad k = 0, \pm 1, \pm 2, \ldots$$

where ϕ is the principal argument of A.

GEOMETRIC ASPECTS

15. Geometric Representation.   The usefulness of graphical representation of real-valued functional relationships in the familiar three-dimensional space is too obvious to require emphasis.   The customary mode of representing real functions by curves and surfaces fails, however, when one encounters functions of more than two independent variables.   Thus, a relationship $u = f(x,y,z)$ containing three independent real variables x, y, z requires a four-dimensional space for geometric representation.   Similar difficulties arise when one attempts to represent graphically complex functions $w = f(z)$, with $z = x + iy$.   To each pair of values (x,y) there correspond two values (u,v) in $w = u + iv$, and in order to plot a quadruplet of real values (u,v,x,y) a four-dimensional space is needed.

However, a different mode of visualizing the relationship $w = f(z)$ which utilizes two separate complex planes for the representation of z and w is possible. The relationship $w = f(z)$ then establishes a connection between the points of a given region R in the z plane and another region R' determined by $w = f(z)$ in the w plane.

On separating $w = f(z)$ into real and imaginary parts one obtains two real functions

$$u = u(x,y) \qquad v = v(x,y) \tag{15-1}$$

which can be viewed as the equations of a transformation that maps a specified set of points in the xy plane into another set of points (u,v) in the uv plane.

We turn now to this mode of studying complex functions.

Example 1.   Let $w = z + a$, where $a = h + ik$ is a complex constant.

We set $w = u + iv$, $z = x + iy$, and get

$$u + iv = x + iy + h + ik$$
$$= (x + h) + i(y + k).$$

Hence
$$u = x + h \qquad v = y + k. \tag{15-2}$$

Formulas (15-2) are the familiar equations defining a *translation*, and the relationship $w = z + a$ can be visualized as representing a rigid displacement of points in the z plane, where each point is moved h units in the direction of the x axis and k units in the direction of the y axis.

Example 2.   To study the function $w = az$, where a is a constant, it is convenient to use polar coordinates.

We set $z = re^{i\theta}$, $w = \rho e^{i\phi}$, $a = Ae^{i\alpha}$ and get

$$\rho e^{i\phi} = Are^{i(\alpha+\theta)}.$$

Hence
$$\rho = Ar \qquad \phi = \alpha + \theta. \tag{15-3}$$

We see from (15-3) that the modulus of w is got by multiplying the modulus of z by A. Also the argument ϕ of w is got by adding a constant angle α to the argument θ of z. We can visualize the transformation (15-3) as representing a stretching in the ratio $A:1$ accompanied by a rotation through an angle α. A square with the center at the origin in the z plane is thus deformed into a square, a circle of radius R is transformed into a circle of radius AR, and more generally any figure is transformed into a similar figure enlarged by the factor A. If $A = 1$, we have a pure rotation through an angle α.

The same conclusions can be reached (but less readily) by setting $w = u + iv$, $z = x + iy$, $a = a_1 + ia_2$ and by deducing from $w = az$ the transformation $u = a_1x - a_2y$, $v = a_2x + a_1y$, in cartesian coordinates.

Example 3.   To study the relationship $w = 1/z$, $z \neq 0$, we again use polar coordinates. On setting $w = \rho e^{i\phi}$, $z = re^{i\theta}$, we get $\rho e^{i\phi} = (1/r)e^{-i\theta}$, so that

$$\rho = \frac{1}{r} \qquad \phi = -\theta. \tag{15-4}$$

It is clear from (15-4) that the unit circle $|z| = 1$ is transformed into the unit circle $|w| = 1$ in the w plane. Since $\phi = -\theta$, the corresponding points on these circles are obtained by reflection in the axis of reals (Fig. 16). As the point A traces out the circle $|z| = 1$ in the clockwise direction, the corresponding point A' in the w plane traces out the circle $|w| = 1$

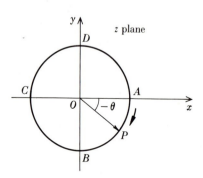

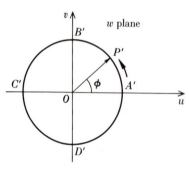

FIGURE 16

in the counterclockwise direction.   Points in the interior of $|z| = 1$ are mapped into points in the exterior of $|w| = 1$, except that the transformation of the point $z = 0$ is not defined by $w = 1/z$. Points in the neighborhood of $z = 0$ map into points at a great distance from the origin of the w plane, since $\rho = 1/r$.   To complete the correspondence of points, we can introduce a new point $w = \infty$ as the correspondent of $z = 0$.   The point $w = \infty$ is called the *point at infinity*.   If we consider the inverse transformation $z = 1/w$, we see that $w = 0$ corresponds to $z = \infty$.

The reader can show that the equations of transformation defined by $w = 1/z$ in cartesian coordinates have the form

$$u = \frac{x}{x^2 + y^2} \qquad v = -\frac{y}{x^2 + y^2}$$

with the inverse $\qquad\qquad x = \frac{u}{u^2 + v^2} \qquad y = -\frac{v}{u^2 + v^2}.$ (15-5)

PROBLEMS

1. Discuss the transformations defined by (*a*) $w = (1 + i)z$; (*b*) $w = 1/(z - 1)$; (*c*) $w = i/z$; (*d*) $w = az + b$.

2. Show that every circle in the z plane is mapped by $w = 1/z$ into a circle in the w plane, if the straight lines are considered as the limiting cases of circles.   *Hint:* Write the general equation of the family of circles, $a(x^2 + y^2) + bx + cy + d = 0$, which includes the family of straight lines when $a = 0$.   Substitute from (15-5) for x and y, and obtain the equation of the corresponding curves in the w plane.   Note particularly the case when $d = 0$, which corresponds to the family of circles passing through the origin of the z plane.

3. Show that the *bilinear transformation*

$$w = \frac{az + b}{cz + d} \qquad \text{with } ad - bc \neq 0$$

can be decomposed into successive transformations $z' = cz + d$, $z'' = 1/z'$, $w = (a/c) + [(bc - ad)/c]z''$, which are the type studied in Examples 1, 2, and 3.   Then conclude (see Prob. 2) that a bilinear transformation transforms circles into circles.   Discuss the case when $ad - bc = 0$.

4. Find the map of the line $y - x + 1 = 0$ by the transformation $w = 1/z$, and sketch.   Also, find the map of $y - x = 0$.

5. Refer to Prob. 3 and discuss the map of the region $|z| \leq 1$ by (*a*) $w = (1 + i)z$; (*b*) $w = i/z$; (*c*) $w = 1/(z - 1)$.   Note that in part *c* the point $z = 1$ maps into the point at infinity in the w plane.

6. Show that the transformation $w = (i - z)/(i + z)$ maps the x axis of the z plane onto a circle $|w| = 1$, and points in the half plane $y > 0$ onto points in $|w| < 1$.

7. Show that the transformation $w = (z - 1)/(z + 1)$ maps the half plane $x \geq 0$ on the unit circle $|w| \leq 1$.   Locate the maps of the points $z = 0$, $z = i$, $z = -i$, and $z = \infty$.

---

16. Functions $w = z^n$ and $z = \sqrt[n]{w}$.   Let us study next the mapping determined by the function

$$w = z^2.$$ (16-1)

If we set $z = re^{i\theta}$ and $w = \rho e^{i\phi}$, we get $\rho e^{i\phi} = r^2 e^{i2\theta}$, so that

$$\rho = r^2 \qquad \phi = 2\theta.$$ (16-2)

It is clear from (16-2) that the upper half of the z plane maps into the whole w plane, for when z is in the upper half plane, the range of variation of θ is $0 \leq \theta < \pi$.   Since

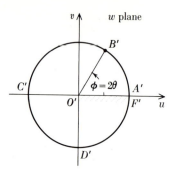

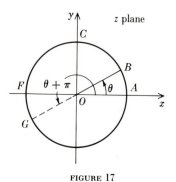

FIGURE 17

$\phi = 2\theta$, we see that the arguments of the corresponding points in the w plane vary from 0 to 2π. Points on the upper half of the circle $|z| = r$ map into the entire circle $|w| = r^2$ (Fig. 17). The half ray OA in the z plane maps into the half ray $O'A'$ in the w plane. A radial line OB, making an angle θ with the x axis, goes over into a radial line $O'B'$, making an angle $\phi = 2\theta$ with the u axis. The interior of the quadrant OAC of the circle $|z| = 1$ maps into the interior of the semicircle $|w| = 1$ in the upper half of the w plane with the boundary ABC going over into the boundary $A'B'C'$. The segment OF of the negative real axis in the z plane maps into the segment $O'F'$ along the positive u axis. To distinguish points on the positive u axis that correspond to points on the ray OA from those on OF, we can imagine that the w plane is slit along the positive u axis and suppose that the points corresponding to OA lie on the upper bank of the slit $O'A'$ and that those corresponding to OF lie on the lower bank $O'F'$.

The transformation of points determined by (16-1) can be visualized as a fanwise stretching of the upper half of the z plane in which the sector OAB opens into a sector $O'A'B'$ and the half circle $OACF$ is deformed into the whole circle $|w| = 1$. The semicircles of radius r in the z plane go over into full circles of radius $\rho = r^2$ in the w plane. Points in the lower half of the circle $|z| = 1$ map into the whole circle $|w| = 1$, inasmuch as the replacement of θ by $\theta + \pi$ in (16-2) yields $\phi = 2\theta + 2\pi$. Thus, two distinct points B and G with the arguments θ and $\theta + \pi$ in the z plane correspond to one and the same point B' in the w plane.

This is to be expected, since, on solving (16-1) for z, we get

$$z = \sqrt{w} \tag{16-3}$$

which is a double-valued function. If we set $w = \rho e^{i\phi}$ in (16-3), we get two values

$$z = \sqrt{\rho}\, e^{i(\phi/2)} \qquad z = \sqrt{\rho}\, e^{i[(\phi/2)+\pi]} = -\sqrt{\rho}\, e^{i(\phi/2)}. \tag{16-4}$$

For points along the u axis, the argument $\phi = 0$. Points on the upper bank of the slit $O'A'$ in Fig. 17 correspond to $z = \sqrt{\rho}$, and those of the lower bank $O'F'$ to $z = -\sqrt{\rho}$. Thus, along the slit, $z = \sqrt{w}$ is a discontinuous function unless $\rho = 0$.

Example 1.   Some further insight into the character of mapping by means of (16-1) can be gained by studying the maps of lines $u = $ const, $v = $ const. If we set $z = x + iy$ in (16-1), we find

$$u = x^2 - y^2 \qquad v = 2xy$$

so that the lines $u = $ const, $v = $ const map into orthogonal hyperbolas $x^2 - y^2 = $ const, $2xy = $ const. Some of these are shown in Fig. 18, in which the corresponding points are labeled by like letters.

The function

$$w = z^n \qquad n \text{ a positive integer} \tag{16-5}$$

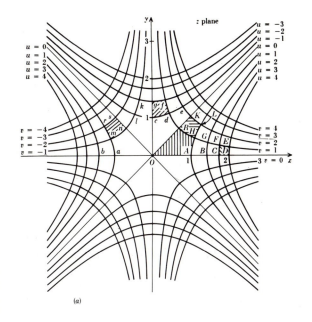

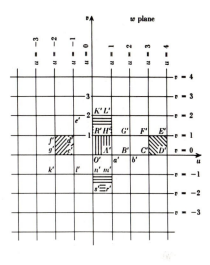

(a)

(b)

FIGURE 18

can be studied in the same way. On setting $z = re^{i\theta}$, $w = \rho e^{i\phi}$ we find

$$\rho = r^n \qquad \phi = n\theta. \tag{16-6}$$

This time a wedge of angle $2\pi/n$ in the z plane (Fig. 19) maps into the whole of the w plane, and a circular arc ACB of radius R goes over into a full circle $|w| = R^n$. An adjacent wedge OBD of angle $2\pi/n$ also maps into the whole w plane. If we divide the z plane into a set of n adjoining wedges, each of angle $2\pi/n$, the entire z plane will be mapped into the w plane n times.

Corresponding to a given point $w \neq 0$, there will be n values of z determined by the n roots [see Eq. (1-13)].

$$z = \sqrt[n]{w} = \rho^{1/n} \left[\cos\left(\frac{\phi}{n} + \frac{2\pi k}{n}\right) + i \sin\left(\frac{\phi}{n} + \frac{2\pi k}{n}\right) \right]$$

with $k = 0, 1, \ldots, n - 1$. Each of these roots lies in one of the wedges into which the z plane is divided.

Example 2. If we take $n = 3$ in (16-5), the wedge $0 \leq \arg z \leq \pi/6$ in the z plane is mapped by $w = z^3$ in a one-to-one way onto the upper half of the w plane (Fig. 20).

In many applications it proves useful to map a given simply connected region either on the half plane or on a unit circle (see Sec. 21). Now, the bilinear transformation $\zeta = (i - w)/(i + w)$ (cf. Prob. 6 of Sec. 15) maps the half plane $w \geq 0$ on the disk $|\zeta| \leq 1$, as

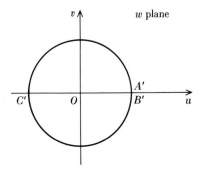

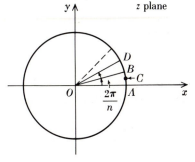

FIGURE 19

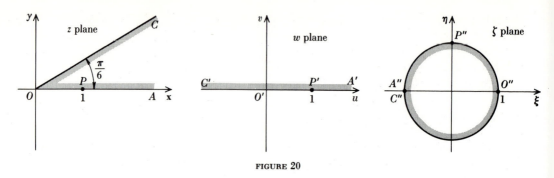

FIGURE 20

shown in Fig. 20.   Accordingly, the points of the wedge $0 \leq z \leq \pi/6$ in the z plane are mapped directly on the unit disk in the ζ plane by $\zeta = (i - z^3)/(i + z^3)$.

--- PROBLEMS

1. Follow the procedure of Example 2 to construct the map of the wedge $0 \leq \arg z \leq \pi/4$ on the unit circle $|\zeta| \leq 1$.

2. Show that $w = i(z - 1)/(z + 1)$ maps the half plane $x \geq 0$ on the circle $|w| \leq 1$.   How does this map differ from that in Prob. 7 of Sec. 15?

3. Show that $w = z + 1/z$ maps the semicircular region $|z| \leq 1$, Im $z > 0$, on the lower half ($v \leq 0$) of the w plane.

4. The transformation $w = z^2$ maps the first quadrant of the z plane on the upper half of the w plane.   Construct, in the manner of Example 2, an analytic function $\zeta = f(z)$ that maps this quadrant on the disk $|\zeta| \leq 1$.   Examine the location of the corresponding points on the boundaries of the regions.

5. Show that $w = a(z + 1/z)$, $a > 0$, maps the circles $|z| =$ const into confocal ellipses, and the radial lines $\arg z = \theta =$ const into confocal hyperbolas.

---

17. The Functions $w = e^z$ and $z = \log w$.   If we set $w = u + iv$ and $z = x + iy$ in

$$w = e^z \tag{17-1}$$

we get $$u + iv = e^{x+iy} = e^x(\cos y + i \sin y).$$

Hence $$u = e^x \cos y \qquad v = e^x \sin y. \tag{17-2}$$

It follows from these equations that

$$u^2 + v^2 = e^{2x} \qquad \frac{v}{u} = \tan y. \tag{17-3}$$

Accordingly, the lines $x =$ const map into the circles $u^2 + v^2 =$ const in the w plane, and the lines $y =$ const map into the radial lines $v/u =$ const.
    Since

$$e^{z+2k\pi i} = e^z e^{2k\pi i} = e^z \qquad k = 0, \pm 1, \pm 2, \ldots \tag{17-4}$$

we see that $w = e^z$ has an imaginary period[1] $2\pi i$.   Hence, if the z plane is divided into horizontal strips of width 2π, with the initial strip determined by $0 \leq y \leq 2\pi$ (Fig. 21),

[1] As for real functions, $f(z)$ is said to be periodic of period a if $f(z + a) = f(z)$.

the relations (17-4) ensure that the behavior of $w = e^z$ in every strip $2k\pi \leq y \leq 2(k+1)\pi$ ($k = \pm1, \pm2, \ldots$) is identical with that in the initial strip. Consequently, we can confine our attention to the behavior of $w = e^z$ in the initial strip $0 \leq y \leq 2\pi$ in which the mapping is one-to-one.

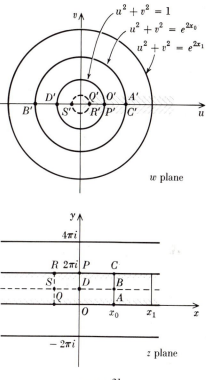

w plane

A segment AC of a straight line $x = x_0$ in the initial strip maps by (17-3) into a circle $u^2 + v^2 = e^{2x_0}$. The points $A(x_0,0)$, $C(x_0,2\pi)$ correspond to the same point $u = e^{x_0}$, $v = 0$ on the u axis. The segment OP of the y axis maps into the unit circle $u^2 + v^2 = 1$, since along OP $x = 0$; the half strip $x > 0$, $0 \leq y \leq 2\pi$ maps into the region $|w| > 1$. If $x < 0$, a segment such as QR in Fig. 21 maps into a circle whose radius is less than 1. The half strip $x < 0$, $0 \leq y \leq 2\pi$ goes into the interior of the circle $|w| = 1$. Points on the lines $y = 0$, $y = 2\pi$, forming the boundaries of the strip, map into points on the positive u axis. If we slit the w plane along the positive u axis, the points on the upper bank of the slit correspond to points on the line $y = 0$ and those on the lower bank to points on $y = 2\pi$. The interior of the rectangle $OACP$ in Fig. 21 corresponds to the interior of the ring between the circles $u^2 + v^2 = 1$ and $u^2 + v^2 = e^{2x_0}$.

z plane

FIGURE 21

We further note that a point moving along the x axis away from the origin O in the positive direction has for its image a point in the w plane that moves in the positive direction along the u axis away from the image O' on the unit circle. A point moving *away* from O in the direction of the negative x axis has for its image a point moving from O' *toward* the origin of the w plane.

If we consider some definite point w_0 in the w plane, the equation

$$w_0 = e^z \tag{17-5}$$

has for its solution

$$z = \log w_0 = \mathrm{Log}\,|w_0| + i(\phi_0 + 2k\pi) \qquad k = 0, \pm1, \pm2, \ldots \tag{17-6}$$

where ϕ_0 is the principal argument of w_0. All these values of z differ only by the imaginary part, and there is just one solution of (17-5) in each strip $2k\pi \leq y \leq 2(k+1)\pi$. The function $z = \log w$ is, therefore, infinitely many-valued. If we restrict our attention to the slit w plane so that the argument ϕ of w lies between 0 and 2π, the mapping from the w plane to the z plane will be single-valued with just one image of $\log w$ in the fundamental strip $0 \leq y \leq 2\pi$ of the z plane.

To study the map of $w = \log z$ we interchange the roles of the z and w planes in the foregoing discussion. We remark in conclusion that inasmuch as all trigonometric functions of z are defined in terms of e^z, a study of the mapping properties of such functions is reducible to the study of mapping by $w = e^{az}$.

PROBLEMS

1. Show that the semi-infinite rectangular strip $x \leq 0$, $0 \leq y \leq \pi$ in the z plane is mapped by $w = e^z$ onto the region $|w| \leq 1$, Im $w \geq 0$.

2. Since $w = \sin z = (e^{iz} - e^{-iz})/2i$, and e^z is a periodic function, $w = \sin z$ is periodic. What is its period? Consider a semi-infinite strip $0 \leq x \leq \pi/2$, $y \geq 0$, and show that it is mapped by $w = \sin z$ onto the first quadrant of the w plane. Indicate the location of the corresponding points on the boundaries of the regions.

3. Show that the rectangular region $-\pi/2 \leq x \leq \pi/2$, $0 \leq y \leq a$ is mapped by $w = \sin z$ in a semielliptical region $(u/\cosh a)^2 + (v/\sinh a)^2 \leq 1$, $v \geq 0$.

4. Note that $\cos z = \sin (z - \pi/2)$, and conclude that the maps of a given region by $w = \cos z$ and $w = \sin z$ are identical save for a translation. Also, since $\cosh z = \cos iz$, and the mapping by $\zeta = iz$ rotates the points of the z plane through an angle $\pi/2$, how are the maps by $w = \cosh z$ and $w = \cos z$ related? Sketch the maps of the rectangular region in Prob. 3 by $w = \cos z$ and $w = \cosh z$.

5. Show that the semi-infinite strip $x \geq 0$, $0 \leq y \leq \pi$ is mapped by $w = \cosh z$ onto the upper half of the w plane in such a way that $z = 0$ corresponds to $w = 1$, $z = \pi i$ corresponds to $w = -1$, and the boundary of the strip maps into the u axis.

---

18. Conformal Maps. We noted in Sec. 15 that the relationship $w = f(z)$ can be viewed as a mapping that sets up a correspondence between the points of the z and w planes. If $w = f(z)$ is analytic in some region R of the z plane, and if C is a curve in R, there is a remarkable connection between C and its image C' in the corresponding region R' in the w plane (Fig. 22). Consider a pair of points z and $z + \Delta z$ on C, and

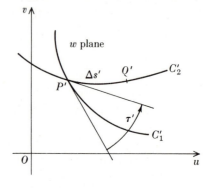

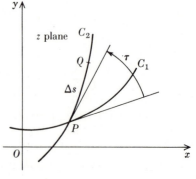

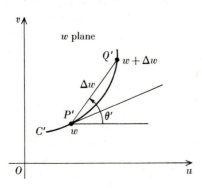

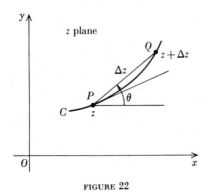

FIGURE 22

FIGURE 23

let the arc length between them be $\Delta s = PQ$. The corresponding points in the region R' are denoted by w and $w + \Delta w$, and the arc length between them by $\Delta s' = P'Q'$. Since the ratio of the arc lengths has the same limit as the ratio of the lengths of the corresponding chords,

$$\lim_{\Delta z \to 0} \frac{\Delta s'}{\Delta s} = \lim_{\Delta z \to 0} \frac{|\Delta w|}{|\Delta z|} = \lim_{\Delta z \to 0} \left|\frac{\Delta w}{\Delta z}\right| = \left|\frac{dw}{dz}\right|. \tag{18-1}$$

We shall exclude from consideration those points of R at which $dw/dz = 0$ because at such points the correspondence of values of z and w ceases to be one to one.[1]

Formula (18-1) shows that an element of arc through P, on being transformed to the w plane, suffers a change in length such that the magnification ratio is equal to the modulus of dw/dz at P. *This ratio is the same for all curves passing through P*, but ordinarily it varies from point to point in the z plane, since $|dw/dz|$ need not have the same value at all points of the z plane.

We shall see next that the argument of dw/dz determines the orientation of the element of arc $\Delta s'$ relative to Δs. The argument θ of Δz (Fig. 22) is the angle made by the chord PQ with the positive direction of the x axis, while the argument θ' of Δw is the angle made by the corresponding chord $P'Q'$ with the u axis.

Hence, the difference between the angles θ' and θ is equal to

$$\arg \Delta w - \arg \Delta z = \arg \frac{\Delta w}{\Delta z}$$

since the difference of the arguments of two complex numbers is equal to the argument of their quotient. As $\Delta z \to 0$, the vectors Δz and Δw tend to coincide with the tangents to C at P and C' at P', respectively, and hence $\arg dw/dz$ is the angle of rotation of the element of arc $\Delta s'$ relative to Δs. It follows immediately from this statement that if C_1 and C_2 are two curves that intersect at P at an angle τ (Fig. 23) the corresponding curves C_1' and C_2' in the w plane also intersect at an angle τ, for the tangents to these curves are rotated through the same angle. Thus, the sense of the angle is preserved.

A transformation that preserves angles is called *conformal*, and thus one can state the following theorem:

THEOREM. *The mapping performed by an analytic function $f(z)$ is conformal at all points of the z plane where $f'(z) \neq 0$.*

The angle-preserving property of the transformation by analytic functions has many important physical applications. We shall indicate several of these in the remaining sections of this chapter, and we merely note here that a number of results deducible analytically from Chap. 6, Sec. 15, follow directly from geometric considerations.

For example, if an incompressible fluid with a velocity potential $\Phi(x,y)$ flows over a plane (so that $v_x = \partial\Phi/\partial x$, $v_y = \partial\Phi/\partial y$), then it is known[2] that the streamlines $\Psi(x,y) = $ const are directed at right angles to the equipotential curves $\Phi(x,y) = $ const.

The orthogonality of the curves $\Phi = $ const and $\Psi = $ const in the z plane follows at once from the conformal properties of transformations by analytic functions. It was shown[3] that the functions Φ and Ψ satisfy the Cauchy-Riemann equations. One can

---

[1] If $dw/dz = f'(z) = 0$ at some point P of R, then $dz/dw = 1/f'(z)$ is not defined at the corresponding point P' for the inverse function $z = F(w)$. Thus $F(w)$ is not analytic at P'. Indeed, it can be shown that a condition for the existence of a unique differentiable solution of $w = f(z)$ at the point $z = z_0$ is precisely $f'(z_0) \neq 0$. See Prob. 4.

[2] See Chap. 6, Sec. 15, and particularly Prob. 6 of that section.

[3] See Eq. (15-10) in Chap. 6.

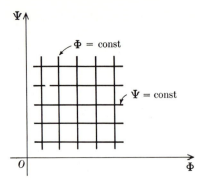

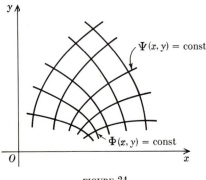

FIGURE 24

therefore assert that Φ and Ψ are the real and imaginary parts, respectively, of some analytic function $w = f(z)$; that is,

$$f(z) = \Phi(x,y) + i\Psi(x,y).$$

But the curves $\Phi = $ const and $\Psi = $ const represent a net of orthogonal lines (Fig. 24) parallel to the coordinate axes in the w plane, and they are transformed by the analytic function $w = \Phi(x,y) + i\Psi(x,y)$ into a net of orthogonal curves in the z plane.

We saw in Sec. 9 that the real and imaginary parts of every analytic function $f(z) = u(x,y) + iv(x,y)$ are harmonic: that is, they satisfy Laplace's equation in the region where $f(z)$ is analytic. Since solutions of Laplace's equation are demanded in numerous practical problems, analytic functions serve as a useful apparatus for producing such solutions. For example, if we take

$$w = u + iv = \sin z = \sin (x + iy)$$

then $u + iv = \sin x \cos iy + \cos x \sin iy$

$$= \sin x \cosh y + i \cos x \sinh y.$$

The harmonic functions $u = \sin x \cosh y$, $v = \cos x \sinh y$ are of special interest in deducing solutions of Laplace's equation in rectangular regions.[1]

Further importance of conformal transformation by analytic functions derives from the fact that a harmonic function remains harmonic when subjected to such a transformation. If a function $\phi(u,v)$ satisfies Laplace's equation

$$\frac{\partial^2 \phi}{\partial u^2} + \frac{\partial^2 \phi}{\partial v^2} = 0 \tag{18-2}$$

in some region R' of the uv plane, ϕ still satisfies Laplace's equation, in the appropriate region R of the xy plane, when the variables u, v in $\phi(u,v)$ are related to x, y by an analytic function

$$w = u + iv = f(z). \tag{18-3}$$

To see this, construct an analytic function

$$F(w) = \phi(u,v) + i\psi(u,v) \tag{18-4}$$

by calculating the conjugate $\psi(u,v)$ of the harmonic function $\phi(u,v)$.

The substitution from (18-3) in (18-4) yields

$$F[f(z)] = \Phi(x,y) + i\Psi(x,y) \tag{18-5}$$

which is analytic in the region R of the xy plane into which the region R' is mapped by (18-3). The function $\Phi(x,y)$, being the real part of the analytic function $F[f(z)]$, is harmonic.

This property of the transformation of harmonic functions by means of analytic functions is of the utmost practical importance. Suppose that we are required to find a solution $\phi(u,v)$ of Laplace's equation (18-2) such that, on the boundary C' of some

[1] See, for example, Sec. 20.

complicated region R' in the uv plane, $\phi(u,v)$ assumes specified values. If it should prove possible to find a function $w = f(z)$ that maps the region R' conformally into some simple region R (a circle, for example) in the z plane, it may be relatively easy to determine the transform $\Phi(x,y)$ of $\phi(u,v)$ in the region R with proper values of Φ on the boundary C.

If $\Phi(x,y)$ is so determined, the function $\phi(u,v)$ can be obtained by replacing the variables in $\Phi(x,y)$ by their values in terms of u and v. It is a remarkable fact, first discovered by Riemann, that every simply connected region R' (with more than one boundary point) can be mapped conformally onto the unit circle $|z| \leq 1$ in such a way that the boundary C' corresponds to the circular boundary $|z| = 1$.

We shall sketch this mode of solution of the Dirichlet problem in Sec. 21.

PROBLEMS

1. Obtain solutions of Laplace's equation from (a) $w = \cos z$; (b) $w = e^z$; (c) $w = z^3$; (d) $w = \operatorname{Log} z$; (e) $w = 1/z$.

2. Construct the conjugate harmonic functions $v(x,y)$ for the following functions:

(a) $u = \cos x \cosh y$; (b) $u = e^x \cos y$; (c) $u = y + e^x \cos y$; (d) $u = \cosh x \cos y$.

3. Examine the mapping by $w = z^2$ and $w = z^3$ at $z = 0$. Is it conformal at $z = 0$? Examine the behavior of the maps of rays issuing from $z = 0$. What are the ratios of magnification of the arc elements at $z = 1$, $z = 1 + i$, $z = i$?

4. Show that the condition $dw/dz = f'(z_0) \neq 0$ at a given point $z_0 = x_0 + iy_0$ implies that $w = f(z)$ has a unique differentiable solution $z = F(w)$ in some neighborhood of the point $w_0 = u_0 + iv_0$. *Hint:* $|f'(z)|^2 = |u_x + iv_x|^2 = u_x^2 + v_x^2$, where the subscripts denote partial derivatives. Use the Cauchy-Riemann equations to show that $|f'(z)|^2 = u_x v_y - u_y v_x$, which is the Jacobian $J(u,v/x,y)$. Refer to Chap. 5, Sec. 9, and assume that $f(z)$ is analytic.

5. Determine points in the z plane at which the mapping by the following functions is not conformal: (a) $w = z^2 + z$; (b) $w = z + 1/z$, $z \neq 0$; (c) $w = e^{-z^2}$; (d) $w = \sin z$; (e) $w = e^z$.

6. Show that $\Psi(x,y) \equiv e^{-y} \sin x$ is harmonic at all points of the xy plane. Let x and y be changed by the transformation $z = w^2$, that is, $x = u^2 - v^2$, $y = 2uv$. Show that $\psi(u,v)$, the transform of $\Psi(x,y)$, is harmonic. Find the conjugate harmonic $\Phi(x,y)$ and $F(z) = \Phi + i\Psi$.

7. Show that $\Phi(x,y) = x/(x^2 + y^2) - x - 1$ is harmonic except at $x = 0$, $y = 0$. Change the variables in $\Phi(x,y)$ by the transformation $w = \log z$ (cf. Prob. 6) and verify that the transform $\phi(u,v)$ of $\Phi(x,y)$ is harmonic.

APPLICATIONS

19. Steady Flow of Ideal Fluids. We discussed the flow of nonviscous incompressible fluids in Sec. 15 of Chap. 6, where we introduced the concept of the velocity potential $\Phi(x,y)$ and the stream function $\Psi(x,y)$. These functions were shown to be related by the Cauchy-Riemann equations

$$\frac{\partial \Phi}{\partial x} = \frac{\partial \Psi}{\partial y} \qquad \frac{\partial \Phi}{\partial y} = -\frac{\partial \Psi}{\partial x}. \tag{19-1}$$

It follows from (19-1) that

$$F(z) = \Phi(x,y) + i\Psi(x,y) \tag{19-2}$$

is an analytic function of a complex variable $z = x + iy$.   We shall call $F(z)$ the *complex potential* and show that its derivative is related simply to the velocity vector $\mathbf{v} = \nabla\Phi$ of the fluid particles.

By (4-5),

$$\frac{dF}{dz} = \frac{\partial\Phi}{\partial x} + i\frac{\partial\Psi}{\partial x} \tag{19-3}$$

and, since $\mathbf{v} = \nabla\Phi$, so that

$$v_x = \frac{\partial\Phi}{\partial x} \qquad v_y = \frac{\partial\Phi}{\partial y} = -\frac{\partial\Psi}{\partial x}$$

we can write (19-3) in the form

$$\frac{dF}{dz} = v_x - iv_y. \tag{19-4}$$

We shall see in Sec. 21 that, because of the simplicity of the complex-variable theory in comparison with the theory of real functions, it is often simpler to calculate the complex potential $F(z)$ than it is to determine either of the real functions $\Phi(x,y)$ or $\Psi(x,y)$. This determination depends on certain so-called *boundary conditions*, which are now to be described.   We first recall[1] that, since $\mathbf{v} = \nabla\Phi$ is orthogonal to the curves $\Phi(x,y) =$ const and these curves are orthogonal to the curves $\Psi(x,y) =$ const, the vector $\mathbf{v}$ is tangent to the curves $\Psi(x,y) =$ const.   Hence these curves, called *streamlines*, are the paths of the fluid particles.   When a sheet of fluid flows past an impenetrable obstacle C, the fluid particles must flow along the obstacle, and hence the boundary C must coincide with one of the streamlines.   Thus the equation of one of the streamlines, say

$$\Psi(x,y) = k \tag{19-5}$$

must coincide with the equation of the boundary C.

To determine $\Psi(x,y)$ we must then seek a solution of Laplace's equation

$$\nabla^2\Psi(x,y) = 0 \tag{19-6}$$

in the region exterior to the obstacle, which is such that on the boundary Ψ takes on a constant value.

This suggests an indirect mode of solution of the steady-fluid-flow problems.   One examines the shapes of curves $\Psi(x,y) =$ const for various harmonic functions $\Psi(x,y)$, and if a particular curve $\Psi(x,y) = k$ coincides with the boundary C of an obstacle of special technical interest, the function $\Psi(x,y)$ solves a special problem.

It follows from these remarks that any streamline $\Psi(x,y) =$ const can be regarded as a rigid boundary of some obstacle.

Instead of determining the stream function $\Psi(x,y)$, we can equally well determine a harmonic function $\Phi(x,y)$ which on the boundary C satisfies the condition

$$\frac{d\Phi}{dn} = 0 \tag{19-7}$$

where $\mathbf{n}$ is the unit normal to C, for the statement that the obstacle is rigid implies that the normal component v_n of $\mathbf{v}$ must vanish along C, since no particles of fluid can cross C.   But $v_n = \mathbf{n}\cdot\mathbf{v}$, and since $\mathbf{v} = \nabla\Phi$ and $d\Phi/dn = \mathbf{n}\cdot\nabla\Phi = v_n$, we see that (19-7) must hold on C.

It should be noted that we have assumed in the foregoing that there are no sources or sinks in the region and that the fluid is incompressible.   Moreover, the flow is irrotational; hence $\Phi(x,y)$ and $\Psi(x,y)$ are single-valued functions.   These considerations can be extended to the more general situation in which circulation is present.   However, as

[1] See Chap. 6, Sec. 2, and Sec. 18 of this chapter.

we shall see from the following examples, the complex potential $F(z)$ is then no longer single-valued.

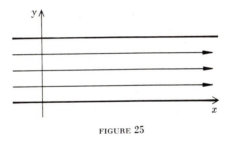

FIGURE 25

Example 1.   The simplest example of an irrotational flow is furnished by the function $F(z) = cz \equiv \Phi + i\Psi$, where c is a real constant. Since $z = x + iy$, we have $\Phi = cx$, $\Psi = cy$, and thus the curves $\Psi = $ const are straight lines parallel to the x axis. The formula (19-4) for the velocity of the fluid yields $v_x = c$, $v_y = 0$, so that the flow is parallel to the x axis. Since div $\mathbf{v} = 0$ and curl $\mathbf{v} = 0$, there are no sources or sinks in the region and the flow is irrotational.

The flow corresponding to the complex potential $F(z) = cz$ can be interpreted as a steady two-dimensional flow between two parallel plates or as a flow between two infinite parallel plates in three-dimensional space (Fig. 25).

Example 2.   As a more interesting example, consider the complex potential $\Phi + i\Psi = az^2, a > 0$. On setting $z = x + iy$, we find $\Phi = a(x^2 - y^2)$, $\Psi = 2axy$, and hence the components of the velocity $\mathbf{v}$ are $v_x = 2ax$, $v_y = -2ay$. The streamlines, determined by $\Psi = 2axy = $ const, are rectangular hyperbolas (Fig. 26). Since along the boundary of the obstacle the streamlines coincide with the boundary, the flow in Fig. 26 can be visualized as taking place in a channel formed by the hyperbola $xy = $ const and by the positive coordinate axes. The speed of flow $|\mathbf{v}| = 2a\sqrt{x^2 + y^2}$ is proportional to the distance of the fluid particle from the origin. Hence the speed is a maximum in the widest part of the channel. The amount of fluid crossing a line segment from the origin to any point in the channel is determined by the value of $\Psi(x,y)$ at that point. [See Eq. (15-3) in Chap. 6, and Example 3, below.]

To determine the flow pattern around the corner making an angle of 45°, we consider the potential $F(z) = \Phi + i\Psi = az^4, a > 0$. We find, as above, $\Phi(x,y) = a(x^4 + y^4 - 6x^2y^2)$, $\Psi(x,y) = 4axy(x^2 - y^2)$, $v_x = \partial\Phi/\partial x = 4a(x^3 - 3xy^2)$, $v_y = \partial\Phi/\partial y = 4a(y^3 - 3x^2y)$. The equations of the streamlines are $\Psi = 4axy(x^2 - y^2) = $ const. If the constant is zero, we get the streamlines $x = 0$, $y = 0$, $y = \pm x$. When the region in the first quadrant bounded by $y = x$ and $y = 0$ is regarded as the boundary of the channel, the curves $\Psi(x,y) = $ const, determine the flow shown in Fig. 27.

To facilitate plotting the streamlines, write $F(z)$ in polar form by setting $z = re^{i\theta}$, and find $\Psi = ar^4 \sin 4\theta$. The magnitude of $\mathbf{v}$ is $|\mathbf{v}| = |dF/dz| = |4az^3| = 4ar^3$, and we see that the speed of the fluid particles is proportional to the cube of the distance of the particle from the origin.

Example 3.   Let us investigate next the flow pattern determined by $F(z) = a \log z = u + iv$, $z = re^{i\theta}$, where a is a real constant.

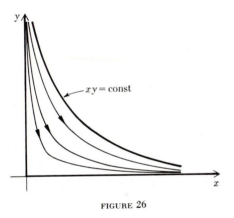

FIGURE 26

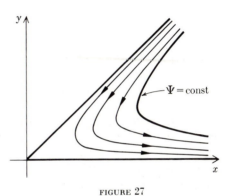

FIGURE 27

If we consider the branch of this multiple-valued function for which $0 \leq \theta \leq 2\pi$, we get $F(z) = a(\text{Log } r + i\theta)$, so that $u = a \text{ Log } r$, $v = a\theta$, $0 \leq \theta < 2\pi$.

If we set $\Psi = a\theta$, the streamlines $\Psi = \text{const}$ are the radial lines and the curves $\Phi = \text{const}$ are circles $a \text{ Log } r = \text{const}$ (Fig. 28). By Eq. (15-1) of Chap. 6, the amount of the fluid crossing per second any closed curve C is

$$V = \int_C (v_x \, dy - v_y \, dx) = \int_C \left(\frac{\partial \Psi}{\partial x} \, dx + \frac{\partial \Psi}{\partial y} \, dy \right) = \int_C d\Psi.$$

But $\Psi = a\theta$, so that

$$V = a \int_C d\theta.$$

This integral vanishes for any simple closed path that does not enclose the origin. If the origin $z = 0$ is within C, then $V = 2\pi a$. Hence, for $a > 0$, the flow is outward, and we have a source of strength $2\pi a$ at the origin. For $a < 0$, we have a sink of the same strength. Thus, div $\mathbf{v} = 0$ at all points except $z = 0$.

The circulation J is given by the integral (see Chap. 6, Sec. 8)

$$J = \int_C (v_x \, dx + v_y \, dy) = \int_C d\Phi$$

and, since $\Phi = a \text{ Log } r$, $J = 0$ and the flow is irrotational.

If, however, we take $\Phi = a\theta$ and $\Psi = a \text{ Log } r$, the roles of the curves $\Phi = \text{const}$ and $\Psi = \text{const}$ in the preceding discussion are interchanged. We thus conclude that for this flow the circulation $J = 2\pi a$ if C encloses the origin. This corresponds to the situation described as a *point vortex* at the origin.

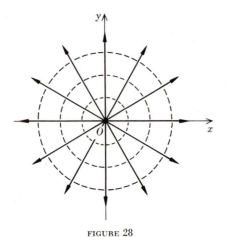

FIGURE 28

Example 4.   As our final example on fluid flow we take

$$F(z) = c \left(z + \frac{a^2}{z} \right) = \Phi + i\Psi \qquad c > 0, \, a^2 > 0. \tag{19-8}$$

If we set $z = re^{i\theta}$ in (19-8), we easily find that

$$\Phi = c \left(r + \frac{a^2}{r} \right) \cos \theta \qquad \Psi = c \left(r - \frac{a^2}{r} \right) \sin \theta.$$

For $r = a$, we have $\Psi = 0$, and hence the boundary of the circle $r = a$ is a streamline. The pattern of streamlines is shown in Fig. 29 by the solid lines, and the curves $\Phi = \text{const}$ are indicated by the dashed lines. This flow pattern corresponds to a flow around a circular cylinder. The velocity components are determined from

$$F'(z) = v_x - iv_y = c \left(1 - \frac{a^2}{z^2} \right).$$

It is easy to verify that div $\mathbf{v} = 0$ and curl $\mathbf{v} = \mathbf{0}$, so that the flow is irrotational. The points for which $v_x = v_y = 0$ are $z = \pm a$. These are called the *stagnation points*.

At the stagnation points, $F'(z) = 0$; hence the mapping by $w = F(z)$ ceases being conformal at $z = \pm a$. The region in the upper half of the xy plane bounded by the semicircle in Fig. 29 and by the intervals $-\infty < x < -a$, $a < x < \infty$ of the x axis is mapped by $w = F(z)$ onto the real axis in the w plane. For large values of $|z|$ the flow is nearly uniform with $v_x = c$. Inasmuch as the boundary consisting of the semicircle and of the above mentioned parts of the x axis coincides with a streamline, the flow shown in Fig. 29 can also be regarded as taking place in a wide channel,

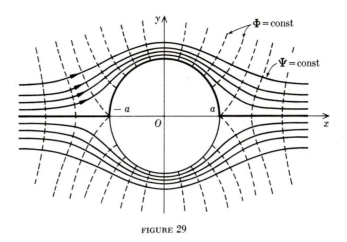

FIGURE 29

one of whose boundaries is indicated by the heavy line in Fig. 29 and the other is parallel to the x axis at a great distance from the x axis.

The reader will find it of interest to study the function

$$\Phi + i\Psi = c\left(z + \frac{a^2}{z}\right) - ic' \log z \qquad a > 0,\ c > 0$$

for which $\Psi = $ const when $|z| = a$. The function $\Psi(x,y)$ represents a flow around a circular cylinder $r = a$ with the circulation $2\pi c'$.

PROBLEMS

1. Deduce from the boundary condition (19-7) that $d\Psi/ds = 0$ along C, so that $\Psi = $ const on C. *Hint:* Note that $d\Phi/dn = (\partial\Phi/\partial x)(dx/dn) + (\partial\Phi/\partial y)(dy/dn)$. Make use of (19-1), and observe that $dx/dn = dy/ds$, $dy/dn = -dx/ds$ on C.

2. Show that $w = F(z) = cz^{\frac{2}{3}}$ maps the positive x axis and the negative y axis onto the real axis of the w plane. Conclude that this function is suitable for the study of the fluid flow around a 270° corner. *Hint:* Use polar coordinates.

3. Consider the complex potential $w = F(z) = a \sin z$. Show that $w = F(z)$ maps the semi-infinite strip $-\pi/2 \leq x \leq \pi/2$, $y \geq 0$ on the upper half of the w plane. Make use of $F(z)$ to study the fluid flow inside this strip. Determine the equation of the streamlines, and compute the velocity $\mathbf{v}$.

4. Discuss the flow with the potential

$$F(z) = c\left(ze^{i\alpha} + \frac{a^2 e^{-i\alpha}}{z}\right) \qquad c > 0,\ \alpha > 0,\ a^2 > 0.$$

Hint: See Example 4 and note that the velocity of the flow at a great distance is $\mathbf{v} = ce^{i/\alpha}$.

5. Discuss the flow with the potential

$$F(z) = c_1\left(z + \frac{a^2}{z}\right) - ic_2 \log z \qquad a^2 > 0,\ c_1 > 0,\ c_2 > 0.$$

Specialize your results by taking $c_2 = 0$, and determine the velocity at $z = \pm a$, $z = \pm 2a$, $z = \pm 3a$, and on the boundary $|z| = a$. *Hint:* Show that $\Psi = c_1(r - a^2/r)\sin\theta - c_2 \log r$.

---

20. The Method of Conjugate Functions.

We observed in the preceding section that every analytic function $F(z) = u(x,y) + iv(x,y)$ can be associated with some flow

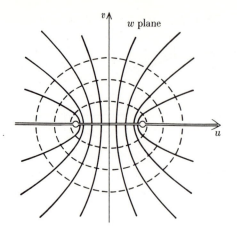

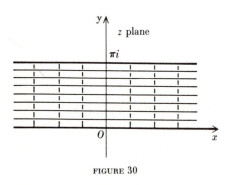

FIGURE 30

pattern of an incompressible fluid. In fact, every such function determines two flow patterns, since either of the harmonic functions $u(x,y)$, $v(x,y)$ can be regarded as determining the streamlines.

The boundary-value problem concerned with the determination of the solution of Laplace's equation $\nabla^2 u = 0$ in the interior of a given region R, which is such that u assumes preassigned values on the boundary C of R, occurs virtually in every branch of mathematical physics. The following examples contain a discussion of two functions which further illustrate the use of the analytic functions in solving important physical and engineering problems.

1. *The Transformation* $w = \cosh z$. Here $w = \frac{1}{2}(e^z + e^{-z}) = \cosh z$. But,

$$u + iv = \cosh (x + iy)$$
$$= \cosh x \cosh iy + \sinh x \sinh iy$$
$$= \cosh x \cos y + i \sinh x \sin y$$

so that $u = \cosh x \cos y$, $v = \sinh x \sin y$, or

$$\frac{u^2}{\cosh^2 x} + \frac{v^2}{\sinh^2 x} = 1 \qquad \frac{u^2}{\cos^2 y} - \frac{v^2}{\sin^2 y} = 1.$$

This transformation is shown in Fig. 30, and it may be used to obtain the electrostatic field due to an elliptic cylinder, the electrostatic field due to a charged plane from which a strip has been removed, the circulation of liquid around an elliptic cylinder, the flow of liquid through a slit in a plane, etc.

The transformation from the z plane to the w plane may be described geometrically as follows: Consider the horizontal strip of the z plane between the lines $y = 0$ and $y = \pi$, and think of these lines as being broken and pivoted at the points where $x = 0$. Rotate the strip 90° counterclockwise, and at the same time fold each of the broken lines $y = 0$ and $y = \pi$ back on itself, the strip thus being doubly "fanned out" so as to cover the entire w plane.

It is interesting to note that this same transformation $w = \cosh z$ can be used to solve a hydrodynamic problem of a different sort. When liquid seeps through a porous soil, it is found that the component in any direction of the velocity of the liquid is proportional to the negative pressure gradient in that same direction. Thus, in a problem of two-dimensional flow the velocity components (u,v) are

$$u = -k \frac{\partial p}{\partial x} \qquad v = -k \frac{\partial p}{\partial y}.$$

If these values are inserted in the equation of continuity, namely, in the equation

$$\frac{\partial u}{\partial x} + \frac{\partial v}{\partial y} = 0$$

the result is

$$\nabla^2 p \equiv \frac{\partial^2 p}{\partial x^2} + \frac{\partial^2 p}{\partial y^2} = 0.$$

Suppose, then, one considers the problem of the seepage flow under a gravity dam which rests on material that permits such seepage. One seeks (see Fig. 31) a function p that satisfies Laplace's equation and that satisfies certain boundary conditions on the surface of the ground. That is, the pressure must be uniform on the surface of the ground upstream from the heel of the dam and zero on the surface of the ground downstream from the toe of the dam. If we choose a system of cartesian coordinates u, v with origin at the midpoint of the base of the dam (Fig. 31) and u axis on the surface of the ground, it is easily checked that the function $p(u,v) = p_0 y(u,v)/\pi$, where

$$w = u + iv = a \cosh (x + iy)$$

satisfies the demands of the problem. In fact, it was seen in the study of the transformation $w = \cosh z$ that the line $y = \pi$ of the z plane folds up to produce the portion to the left of $u = -1$ of the u axis in the w plane and the line $y = 0$ of the z plane folds up to produce the portion to the right of $u = +1$ of the u axis. The introduction of

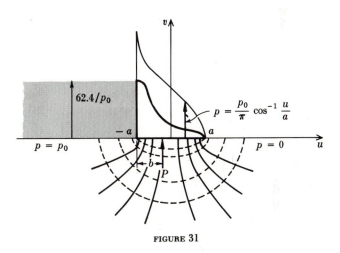

FIGURE 31

the factor a in the transformation merely makes the width of the base of the dam $2a$ rather than 2. These remarks show that $p(u,v)$ reduces to the constant π on the surface of the ground upstream from the heel of the dam. If the head above the dam is such as to produce a hydrostatic pressure p_0, one merely has to set

$$p(u,v) = \frac{p_0 y(u,v)}{\pi}.$$

One can now find the distribution of uplift pressure across the base of the dam. In fact, the base of the dam is the representation, in the uv plane, of the line $x = 0$, $0 \leq y \leq \pi$, of the xy plane. Hence, on the base of the dam the equations $u = a \cosh x \cos y$, $v = a \sinh x \sin y$ reduce to $u = a \cos y$, $v = 0$, so that $p(u,0) = (p_0/\pi) \cos^{-1} (u/a)$.

This curve is drawn in the figure. The total uplift force (per foot of dam) is

$$P = \frac{p_0}{\pi} \int_{-a}^{+a} \cos^{-1} \frac{u}{a} \, du = p_0 a$$

which is what the uplift pressure would be if the entire base of the dam were subjected to a head just one-half of the head above the dam or if the pressure decreased uniformly

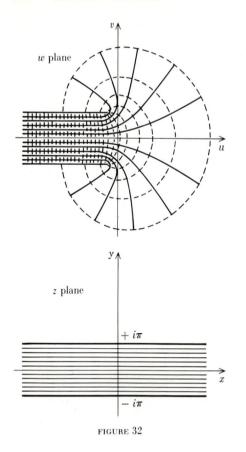

w plane

z plane

$+ i\pi$

$- i\pi$

FIGURE 32

(linearly) from the static head p_0 at the heel to the value zero at the toe. The point of application of the resultant uplift is easily calculated to be at a distance $b = 3a/4$ from the heel of the dam.[1]

2. *The Transformation* $w = z + e^z$. One has $u + iv = x + iy + e^{x+iy} = x + iy + e^x(\cos y + i \sin y)$,

so that $u = x + e^x \cos y$ $v = y + e^x \sin y$.

This transformation is shown in Fig. 32. If one considers the portion of the z plane between the lines $y = \pm\pi$, the portion of the strip to the right of $x = -1$ is to be "fanned out" by rotating the portion of $y = +1$ (to the right of $x = -1$) counterclockwise and the portion of $y = -1$ (to the right of $x = -1$) clockwise until each line is folded back on itself. This transformation gives the electrostatic field at the edge of a parallel-plate capacitor, the flow of liquid out of a channel into an open sea, etc.

21. The Problem of Dirichlet. The procedure for reducing solutions of physical problems described in the preceding section is indirect. It depends on the examination of various harmonic functions that satisfy the boundary conditions appearing in specific physical situations.

In this section we outline a general procedure for constructing harmonic functions which assume preassigned boundary values. Thus, let it be required to determine a solution of Laplace's equation

$$\nabla^2\Phi(x,y) = 0 \tag{21-1}$$

which on the boundary C of a given simply connected region R assumes preassigned continuous values

$$\Phi = \phi(s). \tag{21-2}$$

The variable s in (21-2) may be thought to be the arc parameter s measured along C from some fixed point.

The boundary-value problem characterized by Eqs. (21-1) and (21-2) is known as the *Dirichlet problem*, and it can be shown that the solution of it exists and is unique whenever the boundary C is sufficiently smooth.

We first outline a solution of this problem for the case when the region R is the unit circle $|z| \leq 1$ and later indicate how this solution can be generalized to yield a solution of the Dirichlet problem for an arbitrary simply connected region with the aid of conformal mapping.

Thus, let it be required to construct in the interior of the circle $|z| \leq 1$ a harmonic

---

[1] Some material in Secs. 18 to 20 is taken by permission from a lecture by Dr. Warren Weaver printed in the October, 1932, issue of the *American Mathematical Monthly*.

function $\Phi(x,y)$ such that on its boundary γ (Fig. 33)

$$\Phi(x,y) = f(\theta) \qquad (21\text{-}3)$$

where $f(\theta)$ is a specified continuous function of the polar angle θ.

Instead of determining $\Phi(x,y)$, it proves more convenient to determine an analytic function

$$F(z) = \Phi(x,y) + i\Psi(x,y) \qquad |z| < 1 \qquad (21\text{-}4)$$

whose real part takes on preassigned values (21-3) on $|z| = 1$, and then compute $\Phi(x,y)$ by separating $F(z)$ into its real and imaginary parts. Now, since[1] $F(z) + \overline{F(z)} = 2\Phi(x,y)$, we can write the boundary condition (21-3) in the form

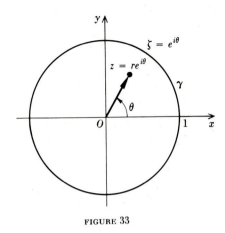

FIGURE 33

$$F(\zeta) + \overline{F(\zeta)} = 2f(\theta) \qquad (21\text{-}5)$$

where $\zeta = e^{i\theta}$ represents the values of $z = re^{i\theta}$ on the boundary γ.

An important integral formula for the calculation of the function $F(z)$ can be obtained as follows: Let z be any fixed point in the interior of the circle $|z| < 1$, and consider the function

$$\mathfrak{F}(\zeta) = \frac{F(\zeta)}{1 - \bar{z}\zeta}. \qquad (21\text{-}6)$$

Inasmuch as $F(\zeta)$ is continuous in the closed circle $|\zeta| \leq 1$ and $F(\zeta)$ is analytic for $|\zeta| < 1$, the same properties are enjoyed by $\mathfrak{F}(\zeta)$, for the denominator $1 - \bar{z}\zeta$ in (21-6) never vanishes if $|z| < 1$. Hence Cauchy's integral theorem is applicable to $\mathfrak{F}(\zeta)$ and

$$\frac{1}{2\pi} \int_\gamma \frac{F(\zeta)}{1 - \bar{z}\zeta} \frac{d\zeta}{i\zeta} = \mathfrak{F}(0) = F(0). \qquad (21\text{-}7)$$

On γ we have $\zeta = e^{i\theta}$, and hence $d\zeta/i\zeta$ is real. Taking the conjugate of (21-7) therefore gives

$$\frac{1}{2\pi} \int_\gamma \frac{\overline{F(\zeta)}}{1 - z\bar{\zeta}} \frac{d\zeta}{i\zeta} = \frac{1}{2\pi i} \int_\gamma \frac{\overline{F(\zeta)}}{\zeta - z} d\zeta = \overline{F(0)}$$

where the second equality follows because $\zeta\bar{\zeta} = |\zeta|^2 = 1$ on γ. Thus,

$$\frac{1}{2\pi i} \int_\gamma \frac{\overline{F(\zeta)}}{\zeta - z} = \overline{F(0)} \qquad (21\text{-}8)$$

Also,

$$\frac{1}{2\pi i} \int_\gamma \frac{F(\zeta)}{\zeta - z} d\zeta = F(z) \qquad (21\text{-}9)$$

by Cauchy's integral formula. By adding the relations (21-8) and (21-9) we obtain

$$F(z) = \frac{1}{2\pi i} \int_\gamma \frac{F(\zeta) + \overline{F(\zeta)}}{\zeta - z} d\zeta - \overline{F(0)}$$

and, since the numerator of the integrand in this expression is $2f(\theta)$ by (21-5), we have

$$F(z) = \frac{1}{\pi i} \int_\gamma \frac{f(\theta)}{\zeta - z} d\zeta - a_0 + ib_0 \qquad (21\text{-}10)$$

where we wrote $\overline{F(0)} \equiv a_0 - ib_0$, so that $F(0) = a_0 + ib_0$.

[1] We use bars to denote the conjugate values, so that $\overline{F(z)} = \Phi(x,y) - i\Psi(x,y)$.

The important formula (21-10) gives the general solution of the Dirichlet problem for the unit circle $|z| \leq 1$ and is called the *Schwarz formula*. The real part of $F(z)$ gives the desired solution $\Phi(x,y)$ of the Dirichlet problem. It is clear from (21-10) that the real part of $F(z)$ is determined uniquely if the value of the real constant a_0 can be determined. But, since $F(0) = a_0 + ib_0$, we get, on setting $z = 0$ in (21-10),

$$a_0 + ib_0 = \frac{1}{\pi i} \int_\gamma \frac{f(\theta)}{\zeta} \, d\zeta - a_0 + ib_0$$

and we conclude that

$$a_0 = \frac{1}{2\pi i} \int_\gamma \frac{f(\theta) \, d\zeta}{\zeta} = \frac{1}{2\pi} \int_0^{2\pi} f(\theta) \, d\theta \tag{21-11}$$

where in the last step we recalled that $d\zeta/\zeta = i \, d\theta$. Thus, the value of a_0 in (21-10) is given by (21-11).

The imaginary part of $F(z)$, that is, the function $\Psi(x,y)$ conjugate to the harmonic function $\Phi(x,y)$, is determined by (21-10) only to within an arbitrary constant b_0. This is in accord with the conclusion already reached in Sec. 9 [see formula (9-2)].

As an illustration of the use of the Schwarz formula, let us determine a harmonic function $\Phi(x,y)$ in the unit circle $|\zeta| \leq 1$, such that on the boundary $|\zeta| = 1$, $\Phi(x,y) = f(\theta) = \cos 2\theta - \sin \theta$.

Since

$$\cos 2\theta = \tfrac{1}{2}(e^{2i\theta} + e^{-2i\theta}) = \tfrac{1}{2}(\zeta^2 + \zeta^{-2})$$

$$\sin \theta = \frac{1}{2i}(e^{i\theta} - e^{-i\theta}) = \frac{1}{2i}(\zeta - \zeta^{-1})$$

we can write

$$f(\theta) = \tfrac{1}{2}\left(\zeta^2 + \frac{1}{\zeta^2}\right) - \frac{1}{2i}\left(\zeta - \frac{1}{\zeta}\right).$$

On inserting this expression in (21-10) we get

$$F(z) = \frac{1}{2\pi i}\left[\int_\gamma \frac{\zeta^2}{\zeta - z} d\zeta + \int_\gamma \frac{1}{\zeta^2(\zeta - z)} d\zeta + \int_\gamma \frac{i\zeta}{\zeta - z} d\zeta - \int_\gamma \frac{i \, d\zeta}{\zeta - z}\right] - a_0 + ib_0$$

which can be easily evaluated by the residue theorem.

But[1]

$$\int_\gamma \frac{1}{\zeta^n(\zeta - z)} d\zeta = 0 \qquad \text{for } n = 1, 2, 3, \ldots$$

and we conclude from Cauchy's integral formula that

$$F(z) = z^2 + iz - a_0 + ib_0.$$

To determine a_0 we apply formula (21-11) and find

$$a_0 = \frac{1}{2\pi} \int_0^{2\pi} (\cos 2\theta - \sin \theta) \, d\theta = 0.$$

Thus, $\Phi(x,y) = \operatorname{Re} F(z) = \operatorname{Re}(z^2 + iz) = x^2 - y^2 - y$

is the desired solution of the Dirichlet problem we have set out to solve.

We indicate next how the Dirichlet problem for an arbitrary simply connected region R can be solved when the function

$$w = w(z) \tag{21-12}$$

mapping the region R in the complex w plane conformally onto the circle $|z| \leq 1$ is known. Let $w = u + iv$; then the desired harmonic function $\Phi(u,v)$, assuming the prescribed values

$$\Phi(u,v) = \phi(s) \tag{21-13}$$

[1] The residues of $1/\zeta^n(\zeta - z)$ at $\zeta = z$ and $\zeta = 0$ are z^{-n} and $-z^{-n}$, respectively.

on the boundary C of R, is the real part of some analytic function

$$\mathfrak{F}(w) \equiv \Phi(u,v) + i\Psi(u,v). \tag{21-14}$$

On substituting in $\mathfrak{F}(w)$ from (21-12), we get $\mathfrak{F}[w(z)] \equiv F(z)$, which is analytic in the circle $|z| \leq 1$.

The values of the real part of $F(z)$ on the boundary γ of the unit circle are known, since the values of $\Phi(u,v)$ on the boundary C are specified by (21-13) and the points on C are mapped into points on γ by (21-12). We can thus write the boundary condition (21-5) in the form

$$\Phi = f(\theta) \qquad \text{on } \gamma.$$

The substitution of $f(\theta)$ in formula (21-10) then yields $F(z)$. To obtain the desired function $\Phi(u,v)$, we must calculate the real part of $\mathfrak{F}(w)$, which can be determined from $F(z)$ by expressing z in terms of w with the aid of (21-12).

It is clear that the solution of the problem of Dirichlet for an arbitrary simply connected domain hinges on the construction of a suitable mapping function (21-12). The fact that such a function exists is guaranteed by Riemann's theorem mentioned in the concluding paragraphs of Sec. 18. During recent years considerable attention has been given to the problem of developing effective methods for constructing conformal maps for simply connected domains.[1] A formula for conformal mapping of a polygonal region on the unit circle (or alternatively, on the upper half of the complex plane) has been supplied[2] by H. A. Schwarz (1843–1921) and E. B. Christoffel (1829–1900).

The theory outlined in this section can be extended to provide useful solutions of some broad classes of problems in the theory of elasticity.[3]

PROBLEMS

1. Use formula (21-10) to compute harmonic functions $\Phi(x,y)$ in the circular region $x^2 + y^2 = 1$, which assume on its boundary the following values: (a) $\Phi = x^2 + y^2$, (b) $\Phi = x^2 - y^2$, (c) $\Phi = \cos^3 \theta$, where θ is the polar angle   *Hint:* Note that $x = \frac{1}{2}(\zeta + \bar{\zeta})$, $y = (\frac{1}{2}i)(\zeta - \bar{\zeta})$ and that on the boundary of the unit circle $\bar{\zeta} = 1/\zeta$.

2. Set $z = re^{i\phi} \equiv r(\cos \phi + i \sin \phi)$, $\zeta = e^{i\theta} \equiv \cos \theta + i \sin \theta$ in (21-10); take account of (21-11); and show that the real part Φ of $F(z)$ is

$$\Phi(r,\phi) = \frac{1}{2\pi} \int_0^{2\pi} \frac{(1 - r^2)f(\theta)\, d\theta}{1 - 2r \cos (\theta - \phi) + r^2}.$$

This formula, giving the values of harmonic function Φ at every interior point (r,ϕ) of the unit circle in terms of the assigned boundary values $f(\theta)$, is known as *Poisson's integral formula.* (Cf. Chap. 7, Sec. 12.) Because of the difficulty of evaluating real integrals, this formula is generally less useful than the Schwarz formula (21-10).

3. Refer to Prob. 6 in Chap. 1, Sec. 17, and show that the integrand of the Poisson integral in Prob 2, just above, can be written

[1] There is a vast literature on this subject, and we cite only a book by L. V. Kantorovich and V. I. Krylov, "Approximate Methods of Higher Analysis," Erven P. Noordhoff, NV, Groningen, Netherlands, 1958, containing a comprehensive survey of the problem in chap. 5. A useful catalog of mapping functions is contained in the "Dictionary of Conformal Representation," Dover Publications, Inc., New York, 1952, compiled by H. Kober.

[2] This formula is contained in most books on complex-variable theory. See, for example, R. V. Churchill, "Complex Variables and Applications," 2d ed., chap. 10, McGraw-Hill Book Company, New York, 1960.

[3] See I. S. Sokolnikoff, "Mathematical Theory of Elasticity," 2d ed., McGraw-Hill Book Company, New York, 1956.

$$\frac{1 - r^2}{1 - 2r \cos(\theta - \phi) + r^2} = 1 + 2 \sum_{n=1}^{\infty} r^n \cos n(\theta - \phi) \qquad 0 \le r < 1.$$

Use this result to show that

$$\Phi(r,\phi) = \int_0^{2\pi} \frac{(1 - r^2)f(\theta)\, d\theta}{1 - 2r \cos(\theta - \phi) + r^2} = \frac{a_0}{2} + \sum_{n=1}^{\infty} r^n (a_n \cos n\phi + b_n \sin n\phi) \qquad 0 \le r \le 1$$

where

$$a_n = \frac{1}{\pi} \int_0^{2\pi} f(\theta) \cos n\theta\, d\theta \qquad b_n = \frac{1}{\pi} \int_0^{2\pi} f(\theta) \sin n\theta\, d\theta$$

are the Fourier coefficients for $f(\theta)$, $0 \le \theta \le 2\pi$. See in this connection Chap. 7, Sec. 12.

4. Use the series representation of $\Phi(r,\phi)$ in Prob. 3 to solve the following boundary-value problem:

$$\begin{aligned}
\nabla^2 \Phi(r,\phi) &= 0 & &\text{in the disk } r < 1,\ 0 \le \phi \le 2\pi \\
\Phi(1,\theta) &= \theta & &\text{if } 0 < \theta < \pi \\
&= -\theta & &\text{if } \pi < \theta < 2\pi.
\end{aligned}$$

$$\left[\text{Ans.: } \Phi(r,\phi) = \frac{\pi}{2} - \frac{4}{\pi} \sum_{n=1}^{\infty} r^{2n-1} \frac{\cos(2n-1)\theta}{(2n-1)^2}. \right]$$

5. Use Prob. 3 to show that

$$\Phi(r,\phi) = \frac{4}{\pi} \sum_{n=1}^{\infty} \frac{r^{2n-1}}{2n-1} \sin(2n-1)\phi \qquad 0 \le r \le 1$$

if $\Phi(1,\theta) = 1$, $0 < \theta < \pi$, and $\Phi(1,\theta) = -1$, $\pi < \theta < 2\pi$. What kind of heat-conduction problem is solved by $\Phi(r,\phi)$?

6. Show that the Schwarz formula (21-10) can be written as $F(z) = \dfrac{1}{2\pi i} \displaystyle\int_\gamma f(\theta) \dfrac{\zeta + z}{\zeta - z} \dfrac{d\zeta}{\zeta} + ib_0$.

---

22. Evaluation of Real Integrals by the Residue Theorem.

Formula (21–10) and the problems in Sec. 21 suggest the use of contour integration of complex functions in the calculation of certain real integrals.

Thus, consider a real integral

$$\int_0^{2\pi} F(\sin \theta, \cos \theta)\, d\theta \tag{22-1}$$

in which F is the quotient of two polynomials in $\sin \theta$ and $\cos \theta$. The evaluation of such integrals, as we shall presently see, can be reduced to the calculation of the integral of a rational function of z along the unit circle $|z| = 1$. Since rational functions have no singularities other than poles, the residue theorem (13-3) provides a simple means for evaluating integrals of the form (22-1).

We set $z = e^{i\theta}$, so that $dz = e^{i\theta} i\, d\theta$

or

$$d\theta = \frac{dz}{iz} \tag{22-2}$$

and we recall Euler's formulas,

$$\cos \theta = \frac{z + z^{-1}}{2} \qquad \sin \theta = \frac{z - z^{-1}}{2i}. \tag{22-3}$$

On inserting from (22-2) and (22-3) in (22-1) we get the integral $\int_C R(z)\, dz$ in which $R(z)$ is a rational function of z and C is the circular path $|z| = 1$. If the sum of the residues of $R(z)$ at the po es within the circle $|z| < 1$ is denoted by Σr, the residue theorem yields $\int_C R(z)\,dz = 2\pi i\,\Sigma r$, so that

$$\int_0^{2\pi} F(\sin\theta,\cos\theta)\,d\theta = 2\pi i\Sigma r. \tag{22-4}$$

Example 1.   As a specific illustration of this method of calculating integrals of the type (22-1), consider

$$I = \int_0^{2\pi} \frac{d\theta}{1 + \alpha\sin\theta} \qquad 0 < \alpha < 1. \tag{22-5}$$

On making substitutions in (22-5) from (22-2) and (22-3), we get the integral

$$I = \int_C \frac{dz}{iz[1 + \alpha(z - z^{-1})/2i]} = \frac{2}{\alpha}\int_C \frac{dz}{z^2 + (2i/\alpha)z - 1} \tag{22-6}$$

where C is the circular path $|z| = 1$.

Since the roots of $z^2 + (2i/\alpha)z - 1 = 0$ are

$$z_1 = -\frac{i}{\alpha}(1 - \sqrt{1 - \alpha^2}) \qquad z_2 = -\frac{i}{\alpha}(1 + \sqrt{1 - \alpha^2}) \tag{22-7}$$

we can write (22-6) as

$$I = \frac{2}{\alpha}\int_C \frac{dz}{(z - z_1)(z - z_2)}. \tag{22-8}$$

But it is clear from (22-7) that for $0 < \alpha < 1$ we have $|z_1| < 1$ and $|z_2| > 1$, so that only one pole $z = z_1$ of the integrand $R(z) = [1/(z - z_1)(z - z_2)]$ lies within the unit circle.   The residue of $R(z)$ at $z = z_1$, by (12-6), is $r = \lim_{z\to z_1} R(z)(z - z_1) = 1/(z_1 - z_2)$ which, on noting (22-7), yields $r = \alpha/2i\sqrt{1 - \alpha^2}$.

By the residue theorem, the value of (22-8), which is the same as that of the integral (22-5), is

$$I = \frac{2}{\alpha}2\pi ir = \frac{2\pi}{\sqrt{1 - \alpha^2}}.$$

The reader can verify by the same method, or by setting $\theta = \varphi - \pi/2$, that

$$\int_0^{2\pi} \frac{d\theta}{1 + \alpha\cos\theta} = \int_0^{2\pi} \frac{d\theta}{1 + \alpha\sin\theta} = \frac{2\pi}{\sqrt{1 - \alpha^2}} \qquad 0 < \alpha < 1. \tag{22-9}$$

The infinite integral

$$\int_{-\infty}^{\infty} \frac{f(x)}{g(x)}\,dx \tag{22-10}$$

in which $f(x)$ and $g(x)$ are polynomials in x, can also be evaluated by calculating the residues.   It should be noted that the integral (22-10) converges if, and only if,[1] $g(x) = 0$ has no real roots and the degree of $g(x)$ is at least 2 greater than that of $f(x)$.

Now, consider the complex rational function

$$R(z) = \frac{f(z)}{g(z)} \tag{22-11}$$

which, obviously, assumes along the real axis the same values as the integrand in (22-10). By hypothesis, $g(z) = 0$ has no real roots; hence no poles of $R(z)$ lie on the real axis. We form the integral

$$\int_C R(z)\,dz = \int_C \frac{f(z)}{g(z)}\,dz$$

where the path C is the boundary of the semicircular region in the upper half of the

[1] This follows directly from the usual tests on convergence of improper integrals.   See Chap. 1, Sec. 4.

z plane shown in Fig. 34. Since all roots of $g(z)$ lie at a finite distance from the origin, we can take the radius R of the semicircle C_R so great that all poles of $R(z) = f(z)/g(z)$, in the *upper half* of the z plane, lie within the semicircle. If the sum of the residues at these poles is Σr, the residue theorem yields

$$\int_C \frac{f(z)}{g(z)} = \int_{-R}^{R} \frac{f(x)}{g(x)}\, dx + \int_{C_R} \frac{f(z)}{g(z)}\, dz = 2\pi i \Sigma r. \qquad (22\text{-}12)$$

We show next that when the degree of $g(z)$ is at least 2 greater than that of $f(z)$, the integral $\int_{C_R} \dfrac{f(z)}{g(z)}\, dz \to 0$ as $R \to \infty$, so that formula (22-12) then yields

$$\int_{-\infty}^{\infty} \frac{f(x)}{g(x)}\, dx = 2\pi i \Sigma r. \qquad (22\text{-}13)$$

For proof, set $z = Re^{i\theta}$ in $R(z) = f(z)/g(z)$, and note that

$$\left| \frac{f(z)}{g(z)} \right| \le \frac{M}{R^2} \qquad M \text{ const}$$

when R is sufficiently large. Hence, by (5-8)

$$\left| \int_{C_R} \frac{f(z)}{g(z)}\, dz \right| \le \int_{C_R} \left| \frac{M}{R^2} \right| |dz| = \frac{M}{R^2}\, \pi R = \frac{M\pi}{R}$$

from which it follows that the integral over C_R tends to zero as $R \to \infty$. Thus under the stated restrictions on $f(z)$ and $g(z)$ ensuring the convergence of (22-10), formula (22-13) is true.

An improper integral like (22-10) should be understood in the sense

$$\lim_{\substack{R_1 \to \infty \\ R_2 \to \infty}} \int_{-R_1}^{R_2} \frac{f(x)}{g(x)}\, dx \qquad (22\text{-}14)$$

where R_1 and R_2 approach infinity in any manner. However, the method of calculation indicated in the text actually gives

$$\lim_{R \to \infty} \int_{-R}^{R} \frac{f(x)}{g(x)}\, dx \qquad (22\text{-}14a)$$

so that $R_1 = R_2$ in (22-14). The expression (22-14a) is termed the *Cauchy principal value* of (22-14). If (22-14) exists (as in the case considered in the text), then obviously (22-14a) exists and has the same value. But (22-14a) may exist when (22-14) does not; for example, take $f(x) = x$, $g(x) = 1 + x^2$.

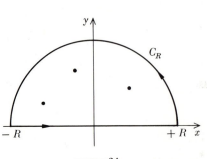

FIGURE 34

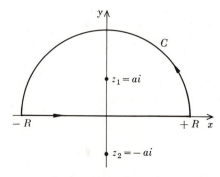

FIGURE 35

Example 2.   To illustrate the use of formula (22-13), consider an elementary integral,

$$\int_{-\infty}^{\infty} \frac{dx}{1+x^2}.$$

Here

$$R(z) = \frac{1}{1+z^2} \equiv \frac{1}{(z+i)(z-i)}$$

so that the only singularity of $R(z)$ in the upper half plane is the simple pole at $z = i$.   Since the residue of $R(z)$ at $z = i$ is $1/2i$, formula (22-13) yields

$$\int_{-\infty}^{\infty} \frac{dx}{1+x^2} = 2\pi i \frac{1}{2i} = \pi.$$

The essential considerations that have led to formula (22-13) are:

1. The integral over the semicircular boundary C_R in (22-12) approaches zero as $R \to \infty$.

2. The singularities of the integrand in the upper half of the z plane are isolated and are at a finite distance from the origin.

3. There are no singular points on the real axis.

Clearly, the same procedure can be used to evaluate integrals of the form

$$\int_{-\infty}^{\infty} F(x)\, dx$$

by computing $\int_C F(z)\, dz$ as long as the integrand $F(z)$ satisfies conditions 1, 2, and 3. Occasionally, a slight modification of the procedure outlined above can be used when $|F(z)|$ is not sufficiently small in the upper half of the z plane, so that condition 1 is not fulfilled by $F(z)$.   We illustrate this in the following example.

Example 3.   Evaluate

$$\int_{-\infty}^{\infty} \frac{\cos x}{a^2 + x^2}\, dx \qquad a > 0.$$

If we take $F(z) = (\cos z)/(a^2 + z^2)$, the method outlined above cannot be applied directly, since $|\cos z| = \tfrac{1}{2}|e^{iz} + e^{-iz}|$ becomes infinite when $z \to \infty$ along the y axis.   However, since $\cos x$ is the real part of e^{ix}, we can write

$$\int_{-\infty}^{\infty} \frac{\cos x}{a^2 + x^2}\, dx = \mathrm{Re} \int_{-\infty}^{\infty} \frac{e^{ix}}{a^2 + x^2}\, dx \tag{22-15}$$

where Re stands for the "real part of."

Now, if we take

$$F(z) = \frac{e^{iz}}{a^2 + z^2} \tag{22-16}$$

then $|e^{iz}| = |e^{i(x+iy)}| = |e^{-y}| \le 1$ if $y \ge 0$.   Thus, $F(z)$ in (22-16) is bounded in the upper half of the z plane, and there is no difficulty in showing that $\int_{C_R} F(z)\, dz \to 0$ as $R \to \infty$.   Moreover, $F(z)$ in (22-16) has only two singular points, which are poles at $z_1 = ia$ and $z_2 = -ia$.   Only one of these, $z_1 = ia$, lies in the upper half plane.   Accordingly,

$$\int_C \frac{e^{iz}}{a^2 + z^2}\, dz = 2\pi i \times \text{residue at } z_1 \tag{22-17}$$

if C is the boundary of the semicircle (Fig. 35) and R is sufficiently large to include the point $z = z_1$.

Now the residue of (22-16) at $z = ia$ is $e^{-a}/2ai$, and since $\int_{C_R} F(z)\, dz \to 0$ as $R \to \infty$, we conclude from (22-17) that

$$\int_{-\infty}^{\infty} \frac{e^{iz}\, dz}{a^2 + z^2}\, dz = \int_{-\infty}^{\infty} \frac{e^{ix}\, dx}{a^2 + x^2} = 2\pi i \frac{e^{-a}}{2ai} = \frac{\pi e^{-a}}{a}.$$

This result is real, and hence the integral in (22-15) is

$$\int_{-\infty}^{\infty} \frac{\cos x}{a^2 + x^2}\, dx = \frac{\pi e^{-a}}{a}. \tag{22-18}$$

Inasmuch as the integrand in (22-18) is even, we conclude that

$$\int_0^{\infty} \frac{\cos x}{a^2 + x^2}\, dx = \frac{\pi e^{-a}}{2a}.$$

PROBLEMS

1. Use relations (22-2) and (22-3) to write the integrals $\int_0^{2\pi} \frac{d\theta}{\frac{5}{4} + \sin \theta}$ and $\int_0^{2\pi} \frac{d\theta}{5 + 3 \cos \theta}$ in the form (22-4), and evaluate the resulting integrals by the residue theorem. Check your calculations by formula (22-9).

2. Show that

$$\int_0^{2\pi} \frac{d\theta}{(1 + \alpha \cos \theta)^2} = \frac{2\pi}{(1 - \alpha^2)^{3/2}} \qquad 0 < \alpha < 1.$$

3. Show that:

(a) $\int_{-\infty}^{\infty} \frac{x^2\, dx}{(1 + x^2)^3} = \frac{\pi}{8}$ (b) $\int_{-\infty}^{\infty} \frac{\sin \alpha x}{a^2 + x^2}\, dx = 0$ $\alpha > 0,\ a^2 > 0.$

4. Referring to Prob. 2, show that

$$\int_0^{\pi} \frac{d\theta}{(a + \cos \theta)^2} = \frac{\pi a}{(a^2 - 1)^{3/2}} \qquad \text{if } a > 1.$$

5. Show that:

(a) $\int_0^{\infty} \frac{dx}{1 + x^4} = \frac{1}{2}\int_{-\infty}^{\infty} \frac{dx}{1 + x^4} = \frac{\pi\sqrt{2}}{4}$ (b) $\int_0^{\infty} \frac{x^2\, dx}{1 + x^6} = \frac{\pi}{6}$

(c) $\int_0^{\infty} \frac{\cos ax}{1 + x^2}\, dx = \frac{\pi}{2} e^{-a}$ if $a \geq 0$ (d) $\int_{-\infty}^{\infty} \frac{x \sin ax}{4 + x^4}\, dx = \frac{\pi}{2} e^{-a} \sin a$ $a > 0.$

6. Show that

$$\int_0^{\infty} \frac{\cos x\, dx}{(1 + x^2)^2} = \frac{\pi}{2} e^{-1}.$$

7. Show that

$$\int_{-\infty}^{\infty} \frac{dx}{(1 + x^2)^n} = \frac{\pi}{2^{2n-2}} \frac{(2n - 2)!}{[(n - 1)!]^2}$$

if n is a positive integer. *Hint:* The residue of $(1 + z^2)^{-n}$ at $z = i$ is

$$\frac{-n(n + 1) \cdots (2n - 2)}{(n - 1)! 2^{2n-1}} i.$$

One way of seeing this is to let $t = z - i$, so that $(1 + z^2)^{-n} = (it)^{-n}2^{-n}(1 - \frac{1}{2}it)^{-n}$. The coefficient of $1/t$ is easily found by use of the binomial theorem.

8. Show that:

(a) $\int_0^{2\pi} \frac{d\theta}{1 - 2\alpha \cos \theta + \alpha^2} = \frac{2\pi}{1 - \alpha^2}$ $0 < \alpha < 1;$

(b) $\int_0^{2\pi} \frac{d\theta}{1 - 2\alpha \sin \theta + \alpha^2} = \frac{2\pi}{1 - \alpha^2}$ $\alpha^2 < 1;$ (c) $\int_0^{2\pi} \frac{d\theta}{2 + \cos \theta} = \frac{2\pi}{\sqrt{3}};$

(d) $\int_{-\pi}^{\pi} \frac{d\theta}{1 + \sin^2 \theta} = \sqrt{2}\pi;$ (e) $\int_0^{2\pi} \sin^{2n} \theta\, d\theta = 2\pi \frac{(2n)!}{(2^n n!)^2},$ $n = 0, 1, 2, \ldots;$

(f) $\int_0^{\infty} \frac{x \sin kx}{x^2 + a^2}\, dx = \frac{\pi}{2} e^{-ka}$ $a > 0,\ k > 0;$ (g) $\int_{-\infty}^{\infty} \frac{x^4\, dx}{(x^2 + 1)^2(x^2 + 4)} = \frac{5\pi}{18}.$

Probability 9

Fundamentals of probability theory

Probability and relative frequency

Concepts used in statistics

Probability

"... la théorie des probabilités n'est que le bon sens confirmé par le calcul."—Laplace.

THERE IS NO PART of mathematics that is more intimately connected with everyday experiences than the theory of probability, and recent developments in mathematical physics have emphasized the importance of this theory in every branch of science. Knowledge of probability is required in such diverse fields as quantum mechanics, kinetic theory, the design of experiments, and the interpretation of data. Operations analysis applies probability methods to questions in traffic control, allocation of equipment, and the theory of strategy. Cybernetics, another field of recent origin, uses the theory to analyze problems in communication and control. In this chapter on probability the reader is introduced to some of the ideas that make the subject so useful.

FUNDAMENTALS OF PROBABILITY THEORY

1. A Definition of Probability.   The idea of chance enters into everyday conversation: "It will probably rain tomorrow," "There may be a letter for me at the office," "I probably won't get double six on the next throw." It is often possible to assign a numerical measure to the notion of *probability* which these statements illustrate. Such a measure, however, must take account of the speaker's state of knowledge. For instance, in the second statement the mailman may know that a letter is there, since he put it there himself. His measure of probability and mine are therefore not the same. Probability for me is based on my knowledge, and probability for him is based on his.

From this viewpoint (which is one of several possible viewpoints) probability is a measure of knowledge.   In simple cases the state of knowledge can be accounted for, and probability can be defined as follows: We agree to regard two events as *equally likely* if our knowledge is such that we have no reason to expect one rather than the other.   For example, a 4 or a 6 is equally likely when a true die is tossed; heads and tails are equally likely in a toss of a symmetric coin; the ace of hearts and the ace of spades are equally likely to be drawn from a shuffled deck.

In the latter example how shall we measure the probability that the card drawn will in fact be the ace of hearts? We say that there is "one chance out of 52" and define the probability, accordingly, to be $\frac{1}{52}$. If it is required only that the card be an ace, common sense suggests that the probability should be four times as great, for there are four aces, equally likely, and only one ace of hearts. Reasoning in this way, we are led to the following definition:

DEFINITION.   Suppose there are n mutually exclusive, exhaustive, and equally likely cases.   If m of these are favorable to an event A, then the probability of A is m/n.

The term *mutually exclusive* means that two cases cannot both happen at once; the term *exhaustive* means that all possible cases are enumerated in the n cases. There is seldom difficulty in seeing that these conditions are satisfied, but careful analysis is sometimes needed to make sure that the cases are equally likely. For example, let two coins be tossed, and consider the probability that they both show heads. We might reason that the total number of cases is three, namely, two heads, a head and a tail, or two tails. Since only one case is favorable, the probability is $\frac{1}{3}$. Now, this reasoning is incorrect. It is true that there are three cases, but these cases are not equally likely. The case of a head and a tail is twice as likely as the others, since it can be realized with a head on the first coin or with a head on the second coin. The reader can verify that there are four equally likely cases and that the required probability is $\frac{1}{4}$.

If an event is certain to happen, its probability is 1, since all cases are favorable. On the other hand, if an event is certain not to happen its probability is zero, since no case is favorable. By means of the definition the reader may also verify the important equation

$$q = 1 - p$$

where p is the probability that an event happens and q the probability that it fails to happen.

Since one must begin somewhere, it is impossible to define everything, and every mathematical theory contains some undefined terms. These terms should be so simple that they are easily understood and also so simple that they are not readily defined in terms of anything simpler. The notion "equally likely" is an example of such a term; it was explained and illustrated in the foregoing discussion but not defined. By examples, we shall show how the notion is used.

Example 1. If a pair of dice is thrown, what is the probability that a total of 8 shows?

The first die can fall in 6 ways, and for each of these the second can also fall in 6 ways. The total number of ways is

$$6 + 6 + 6 + 6 + 6 + 6 = 6 \cdot 6 = 36$$

and these are equally likely in this problem. A sum of 8 can be obtained in 5 ways, namely, as

$$2 + 6, \qquad 6 + 2, \qquad 3 + 5, \qquad 5 + 3, \qquad 4 + 4$$

and hence the desired probability is $\frac{5}{36}$.

This computation of the total number of cases illustrates an important principle of combinatory analysis: *If one thing can be done in n different ways and another thing can be done in m different ways, both things can be done together or in succession in mn different ways.*

Example 2. In a well-shuffled deck, what is the probability that the top four cards are, respectively, the ace, 2, 3, and 4 of hearts?

To find the number of equally likely cases we consider the various possibilities for the top four cards. The first card may be any one of 52; for each determination of that card there remain 51 possibilities for the next; and so on. Repeated use of the principle mentioned at the end of Example 1 gives

$$52 \cdot 51 \cdot 50 \cdot 49$$

for the total number of cases. Since only one case is favorable, the desired probability is the reciprocal of this.

When r things are dealt into r numbered spaces from a stack of n distinct things, any particular arrangement of the objects is called "a permutation of n things r at a time." If the total number of such permutations is denoted by $_nP_r$, the foregoing reasoning yields the important formula

$$_nP_r = n(n - 1)(n - 2) \ldots (n - r + 1).$$

Example 3.   If a hand of four cards is dealt from a shuffled deck, what is the probability that the hand consists of the ace, 2, 3, and 4 of hearts?

The difference between this example and the preceding is that now the order is not relevant. Let C denote the number of distinct four-card hands, not counting order. Then the number of distinct four-card hands when the order is counted is

$$C \cdot {}_4P_4,$$

since each hand of four cards admits ${}_4P_4$ different orderings of its members. On the other hand, the number of distinct four-card hands when order is counted is also equal to ${}_{52}P_4$ by Example 2. We have, therefore,

$$C \cdot {}_4P_4 = {}_{52}P_4$$

so that $C = \dfrac{{}_{52}P_4}{{}_4P_4} = \dfrac{52 \cdot 51 \cdot 50 \cdot 49}{4 \cdot 3 \cdot 2 \cdot 1} = \dfrac{52!}{4!48!}.$

The desired probability is the reciprocal.

When r things are taken from a stack of n things, the groups so obtained are called "combinations of n things r at a time." If the number of such combinations is denoted by ${}_nC_r$, the above reasoning gives the important formula

$${}_nC_r = \frac{{}_nP_r}{{}_rP_r} = \frac{n!}{r!(n-r)!}.$$

In this formula the arrangement of members in a group is not considered. As in the case of poker hands, two groups are counted as distinct only if they have different compositions.

Example 4.   What is the probability of drawing 4 white, 3 black, and 2 red balls from an urn containing 10 white, 4 black, and 3 red balls?

We suppose that the balls are not replaced. The number of ways to get 9 balls from the 17 is ${}_{17}C_9$. The number of ways to get 4 white from the 10 white is ${}_{10}C_4$. The 3 black balls can be chosen in ${}_4C_3$ ways, and the 2 red ones in ${}_3C_2$ ways. The number of favorable cases is found by multiplication (cf. Example 1), so that the desired probability is

$$\frac{{}_{10}C_4 \cdot {}_4C_3 \cdot {}_3C_2}{{}_{17}C_9} = \frac{252}{2,431}.$$

FIGURE 1

Example 5.   If a number x is chosen at random on the interval $0 \le x < 1$, what is the probability that $\frac{1}{7} \le x < \frac{3}{7}$?

We imagine the unit interval divided into 7 segments each of length $\frac{1}{7}$ (Fig. 1). Since the point may be in any one of these there are 7 cases, and the phrase "at random" ensures that these cases are equally likely. Since only 2 cases are favorable, the desired probability is $\frac{2}{7}$.

PROBLEMS

1. What is the probability that the sum of 7 appears in a single throw with two dice? What is the probability of the sum of 11? Show that 7 is the most probable throw.

2. An urn contains 20 balls: 10 white, 7 black, and 3 red. What is the probability that a ball drawn at random is red? White? Black? If 2 balls are drawn, what is the probability that both are white? If 10 balls are drawn, what is the probability that 5 are white, 2 black, and 3 red?

3. "If three coins are tossed, some pair is sure to come down alike. The chance that the third coin fell the same way as that pair is $\frac{1}{2}$; hence the probability that all three fall alike is $\frac{1}{2}$." What (if anything) is wrong with this argument? What is the probability that three coins will fall alike?

4. What is the probability that a five-card hand at poker consists of four kings and an odd card? Five spades? A sequence in the same suit, such as 2, 3, 4, 5, 6 of hearts?

5. In how many ways can you seat eight persons at a table? Arrange eight children in a ring to dance around a Maypole? Make a bracelet of eight different beads on a loop of string?

6. The seats in a concert hall are arranged in an $m \times n$ rectangle, the side m being parallel to the stage. What is the chance that a ticket bought at random will be for a seat in back? On the side? Somewhere on the outside rows of the rectangle?

7. (*a*) How many straight lines can be drawn through seven points in such a way that each line passes through two of the points? Assume that no three points are collinear. (*b*) How many different connections must a telephone exchange be able to set up, if the exchange serves 10,000 subscribers?

8. Two dice are tossed. (*a*) What is the probability that the first die shows 2? (*b*) Suppose you are given the additional information that the total shown by both dice is 9. What is now the probability that the first die shows 2? (*c*) If no information is given, what is the probability that the total shown is 3? (*d*) If it is known that the first die gave 2, what is now the probability that the total is 3? (Assume that the various numbers on the second die are equally likely no matter what is known about the first die.)

---

2. Sample Space. The equally likely cases associated with the definition of probability represent the possible outcomes of an experiment. For instance, the 36 equally likely cases associated with a pair of dice are the 36 ways the dice may fall. Similarly if three coins are tossed, there are eight equally likely cases corresponding to the eight possible outcomes of that experiment. The set of all possible outcomes is called a *sample space;* the "points" of the sample space are events. This notion of sample space is meaningful even when the events are not equally likely and even if there are infinitely many possible outcomes. For technical reasons, however, the events composing the sample space are required to be mutually exclusive. In tossing a die the events "an even number shows" and "6 shows" are not suitable for one and the same sample space.

A finite sample space is one which has only a finite number of points. In such a space let the points (that is, events) have respective probabilities

$$p_1, p_2, \ldots, p_n$$

with
$$p_1 + p_2 + \cdots + p_n = 1.$$

Suppose the first m sample points, and only those, are favorable to another event A. Then we define the probability of A to be

$$p(A) = p_1 + p_2 + \cdots + p_m \tag{2-1}$$

(and similarly if some other set of sample points is in question). Thus, the points of the sample space are weighted according to their probabilities.

The reader should observe that this definition is consistent with that of the foregoing section: If each point of the sample space has the same probability $1/n$, the result (2-1) becomes

$$p(A) = \frac{1}{n} + \frac{1}{n} + \cdots + \frac{1}{n} = \frac{m}{n}.$$

Sample spaces with constant probability are called *uniform.*

For an example of a nonuniform sample space, consider the following experiment:

Four coins are tossed, and we are interested in the number of heads. An appropriate sample space is composed of the events

no heads, one head, two heads, three heads, four heads

with respective probabilities, or weights,

$$\tfrac{1}{16}, \tfrac{4}{16}, \tfrac{6}{16}, \tfrac{4}{16}, \tfrac{1}{16}.$$

These values are found by counting cases, as follows. The 4 coins can fall in 2^4, or 16, ways. They give *no heads* in only one case, namely, when they all fall tails, and hence the required probability is $\tfrac{1}{16}$. To obtain 1 head there are 4 cases: heads on the first coin or on the second coin, and so on. This gives $\tfrac{4}{16}$. For 2 heads, the 2 coins giving heads can be any 2 of the 4 coins. Since there are $_4C_2 = 6$ ways to choose 2 coins out of 4, there are 6 cases favorable to the event, *two heads*. The probability, then, is $\tfrac{6}{16}$. The other entries are found in the same way, or by symmetry.

To illustrate the use of this sample space, let us find the probability of getting at least two heads. Since the last three points of the sample space, and only those, are favorable to this event, the required probability is

$$\tfrac{6}{16} + \tfrac{4}{16} + \tfrac{1}{16} = \tfrac{11}{16}.$$

Again, the probability that there is an odd number of heads is

$$\tfrac{4}{16} + \tfrac{4}{16} = \tfrac{1}{2}$$

since that event corresponds to the second and fourth point. On the other hand, this sample space does not give the probability that the third coin will fall heads, although the underlying uniform space tells us that the probability is $\tfrac{1}{2}$.

Additional information concerning the experimental situation is apt to change the sample space. For example, if a toss of a die is known to have given an even number, the probabilities of 1, 3, and 5 are changed from $\tfrac{1}{6}$ to 0. This question is discussed in Examples 2 and 3.

When two sample spaces are constructed for a given experiment by the procedure of the text, it can be shown that they are consistent; that is, they give the same probability for any event to which they both apply. This fact is illustrated in the problems, though we do not give a formal proof.

The notion of sample space enables us to define probability even when there is no underlying set of equally likely cases. Suppose we are given n events and a corresponding set of nonnegative numbers p_i such that $p_1 + p_2 + \cdots + p_n = 1$. The events are said to form a *sample space*, the numbers p_i are called *probabilities*, and the probability of various associated events is defined by addition, as in the text. This abstract idea can be extended to sets of very general type, the role of the numbers p_i being taken by a so-called *measure* on the set. With such an approach, probability theory is included in a branch of mathematics known as the theory of measure.[1] A sample space defined with the help of arbitrary numbers p_i is considered in Example 1.

Example 1.   A loaded die has probabilities $p_1, p_2, p_3, p_4, p_5, p_6$ of giving the respective values 1, 2, 3, 4, 5, 6. What is the meaning of the condition $p_1 + p_2 + \cdots + p_6 = 1$? If this condition is satisfied, find the probability that a single toss will give either a 4 or a 6.

The condition means that one of the stated alternatives will certainly happen; for instance, the die does not land on edge. From a more abstract viewpoint, the condition means simply that the given events and probabilities form a sample space. When that is the case, the probability of getting 4 or 6 is $p_4 + p_6$ by definition.

The assumption that "the probabilities are p_i" is an example of a *statistical hypothesis*. It is an important task of statistical theory to test the validity of such hypotheses by examining the consequences.

[1] See Appendix B.

The reader should notice that the values p_i were not given, and could hardly be given, by considering "equally likely cases." They may be estimated, however, by repeatedly tossing the die. When p_1 is the probability of the ace, it can be shown that the proportion of aces actually observed, in a large number of tosses, is likely to be close to p_1. If there are n tosses, and if m aces are observed, this proportion m/n is called the *relative frequency*. The connection between probability and relative frequency is discussed in Secs. 8 to 10.

Example 2. Two coins are tossed. Suppose a reliable witness tells us "at least one coin showed heads." What effect does this have on the uniform sample space?

The uniform sample space had the following appearance before we received the extra information:

Event	HH	TH	HT	TT
Probability	$\frac{1}{4}$	$\frac{1}{4}$	$\frac{1}{4}$	$\frac{1}{4}$

The new information assures us that the last event is ruled out but gives no indication concerning which of the other three may have occurred. Since these three events were equally likely to begin with, they are considered to be equally likely in the new situation. (That is not a theorem, but an *axiom* of probability theory.) The new sample space, therefore, is

Event	HH	TH	HT	TT
Probability	$\frac{1}{3}$	$\frac{1}{3}$	$\frac{1}{3}$	0

Example 3. The tossing of two coins can be described by the following sample space:

Event	no heads	one head	two heads
Probability	$\frac{1}{4}$	$\frac{1}{2}$	$\frac{1}{4}$

What happens to this sample space if we know that at least one coin showed heads but have no other special information?

The first event is ruled out, but we are not told which of the remaining ones occurred. It is an axiom of probability theory that the relative probabilities of the remaining events remain unchanged in a situation such as this. Since the event "one head" is twice as likely as "two heads" in the original space, the same is assumed in the new one. The new sample space is therefore

Event	no heads	one head	two heads
Probability	0	$\frac{2}{3}$	$\frac{1}{3}$

(remember that the probabilities must add up to 1). The reader should check that this result is consistent with that of Example 2.

If the events $E_1, E_2, \ldots, E_k$ of the sample space are the ones favorable to A and have probabilities $p_1, p_2, \ldots, p_k$, the information that A happened gives a new sample space with events $E_1, E_2, \ldots, E_k$ only. The probabilities on that new sample space are

$$cp_1 \qquad cp_2 \qquad \cdots \qquad cp_k$$

where c is a constant so chosen that the sum is 1:

$$c = \frac{1}{p_1 + p_2 + \cdots + p_k}.$$

This is the general assertion that is illustrated in Examples 2 and 3.

1. A coin is tossed three times. Construct a uniform sample space for this experiment (that is, make a table showing the eight possible outcomes HHH, HHT, ... and their respective probabilities $\frac{1}{8}$, $\frac{1}{8}$, ...). According to your sample space, what is the probability of at least one H? At most one H? A run of exactly two H's in succession? A run of at least two H's in succession? H appearing before T? H appearing for the first time in the second toss? The sequence THT? The sequence TTT?

2. In Prob. 1, suppose you are concerned only with the *number* of H's. Construct an appropriate sample space (that is, make a table showing the four possible outcomes: no H's, one H, ... with their respective probabilities). Decide which questions in Prob. 1 can be answered on the basis of this sample space, answer them, and verify the agreement with your answers to Prob. 1.

3. The following argument is attributed to Leibniz: "A total of 12 with 2 dice is just as likely as a total of 11. For, 12 can materialize in just one way, namely, by getting 6 on one die and 6 on the other; and 11 can also materialize in just one way, namely, by getting 6 on one die and 5 on the other." Using the notion of sample space, explain what is wrong with Leibniz' conclusion. (With the uniform sample space, 11 can materialize in two ways. On the other hand, if we choose a sample space in which the event "6 on one die and 5 on the other" is a single point, the weight of this point is different from that of the point "6 on one die and 6 on the other.")

4. The following is due to d'Alembert: "If we want to get at least one head with two tosses of a coin, heads on the first toss makes the second toss unnecessary. So there are three cases, H, TH, and TT, of which two are favorable to heads. Hence the probability of heads is $\frac{2}{3}$." Discuss, with reference to the uniform sample space and also with reference to the sample space which has only the three points H, TH, TT. (Ambiguities such as this and the preceding can cause serious errors in practice if the notion of sample space is not well understood. In fact, one of the reasons for defining the sample space is to avoid this kind of difficulty.)

5. What happens to the uniform space associated with a pair of dice, if we are told that the total shown is 7?

6. Four coins are tossed. A reliable witness tells us that there are at least as many heads as tails. What is the most probable number of heads, and what is its probability? *Suggestion:* Use the sample space given in the text.

7. A coin is tossed three times. If we know that a sequence of two tails in a row did not occur, what is the probability that a sequence of three heads in a row did occur? *Suggestion:* Use the uniform sample space.

8. (*Review.*) A cylinder lock contains 5 tumblers, each of which may receive 10 different adjustments. How many locks can be thus designed?

9. (*Review.*) The letters of the word "tailor" are written on cards and shuffled. Four cards are drawn in order. What is the probability that the letters drawn spell, in order, the word "oral"? What is the probability that the four letters o, r, a, l are drawn in some order?

3. The Theorems of Total and Compound Probability. Statements about probability are often given an abbreviated notation. If A and B are events, AB means the event "A and B"; that is, AB happens only when both A and B happen. For example, if two cards are drawn in succession without replacing, suppose A is the event "the first draw gives a king" and B is the event "the second draw gives an ace." Then AB happens if we get a king on the first draw followed by an ace on the second.

It is customary to write $p(A)$ for "the probability of the event A." In the foregoing example, $p(A) = \frac{4}{52}$, since there are 4 kings among the 52 cards. If nothing is known about the results of the first draw, then $p(B) = \frac{4}{52}$ also.

To see this, note that the total number of cases is $52 \cdot 51$, since there are 52 ways to get the first card and, when that card is chosen, 51 ways remain to get the second. To count the cases favorable to B, observe that the ace obtained on the second draw may be any one of the 4 aces. For each choice of this ace there remain 51 possibilities for the first card. The number of favorable cases, then, is $4 \cdot 51$, and hence

$$p(B) = \frac{4 \cdot 51}{52 \cdot 51} = \frac{4}{52}. \tag{3-1}$$

Sometimes two events A and B are so related that the information that A happened changes the probability of B. To deal with this situation it is customary to write $p_A(B)$ for "the probability of B, given A." In the example cited previously,

$$p_A(B) = \frac{4}{51} \tag{3-2}$$

(for if A happened, the first draw gave a king, and hence the 4 aces are to be found among the remaining 51 cards). On the other hand, when A is the event "the first draw gives an ace" and B, as before, is the event "the second draw gives an ace," then $p_A(B) = \frac{3}{51}$ (since now only 3 aces remain when A happens). Both values for $p_A(B)$ are different from $p(B)$, the probability of ace on the second draw when nothing is said about the first draw.

In this notation *the theorem of compound probability* takes the following form:

THEOREM I. *If A and B are any events, then*

$$p(AB) = p(A)p_A(B). \tag{3-3}$$

Informally, "the probability that A and B happen is the probability that A happens times the probability that B then happens." A proof is easily given by considering equally likely cases. Let n_a, n_b, and n_{ab} denote the numbers of cases favorable to A, B, and AB, respectively. Then

$$p(AB) = \frac{n_{ab}}{n} = \frac{n_a}{n} \frac{n_{ab}}{n_a}.$$

Now, n_a/n is $p(A)$ by definition. After A has happened, the only possible cases are the n_a cases favorable to A. Of these, there are n_{ab} cases favorable to B. Since the n_a cases are to be considered equally likely, the quotient n_{ab}/n_a represents the probability of B when it is known that A happened, and this gives (3-3).

To illustrate the theorem (3-3), let us find the probability of drawing 2 aces in succession from a pack of 52 cards. The probability of an ace on the first trial is $\frac{4}{52}$. After the first ace has been drawn, the probability of drawing another ace from the remaining 51 cards is $\frac{3}{51}$, so that the probability of 2 aces is

$$\frac{4}{52} \cdot \frac{3}{51} = \frac{1}{221}.$$

This assumes that the first card is not replaced. When it is replaced, the reader will find that the desired probability is

$$\frac{4}{52} \cdot \frac{4}{52} = \frac{1}{169}.$$

For another illustration of the theorem (3-3), let us find the probability of drawing a white and a black ball in succession from an urn containing 30 black balls and 20 white balls. Here the probability of drawing a white ball is $\frac{20}{50}$. After a white ball is drawn, the probability of drawing a black ball is $^{30}/_{49}$. Hence the probability of drawing a white ball and a black ball in the order stated is

$$p = {}^{20}/_{50} \cdot {}^{30}/_{49} = {}^{12}/_{49}.$$

The events A and B are said to be *independent* if the information that A happened does not influence the probability of B. Hence for such events $p_A(B) = p(B)$, and the theorem of compound probability takes the form

$$p(AB) = p(A)p(B) \qquad \text{for independent events.} \tag{3-4}$$

For instance, let a coin and a die be tossed, and let A be the event "head shows" while B is the event "4 shows." These events are independent, and hence the probability that heads and 4 both appear is

$$p(AB) = p(A)p(B) = (\tfrac{1}{2})(\tfrac{1}{6}) = \tfrac{1}{12}.$$

The result (3-4) is readily extended to any number of independent events $A, B, C, \ldots$.

Besides the theorem of compound probability, there is a second fundamental relationship, known as the *theorem of total probability*. If A and B are two events, $A + B$ is defined to be the event "A or B or both." For instance, let A be the event "a number greater than 3 shows" while B is the event "an even number shows" in a toss of a die. Then $A + B$ happens if the die gives 2, 4, 5, or 6. In this notation the theorem of total probability reads as follows:

THEOREM II. When A and B are any events, then

$$p(A + B) = p(A) + p(B) - p(AB). \tag{3-5}$$

We can represent the statement (3-5) diagrammatically by the intersecting point sets A and B shown in Fig. 2.

Referring to the definition of probability by equally likely cases, suppose the numbers of cases favorable to A, B, AB, and $A + B$ are denoted by

$$n_a \qquad n_b \qquad n_{ab} \qquad n_{a+b}$$

respectively. To find the number favorable to $A + B$, it will not do simply to add n_a and n_b, for the cases favorable to both A and B are counted twice in this addition. To take account of that, we must subtract n_{ab}, thus:

$$n_{a+b} = n_a + n_b - n_{ab}.$$

Dividing by n, the total number of cases, gives

$$\frac{n_{a+b}}{n} = \frac{n_a}{n} + \frac{n_b}{n} - \frac{n_{ab}}{n}$$

which is equivalent to (3-5).

To illustrate the theorem, let us find the probability that at least one die gives 4, when two dice are tossed. The probability that both give 4 is $\tfrac{1}{36}$. The probability

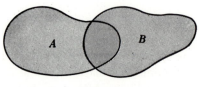

FIGURE 2

FIGURE 3

that the first gives 4 is $\frac{1}{6}$, and similarly for the second.   Hence the probability that at least one gives 4 is

$$p(A + B) = \frac{1}{6} + \frac{1}{6} - \frac{1}{36} = \frac{11}{36}. \tag{3-6}$$

This is consistent with the result given by counting cases.   Specifically, there are five cases with a 4 on the first and a number other than 4 on the second, there are five cases with 4 on the second and a number other than 4 on the first, and there is one case with 4 on both.   The number of favorable cases is therefore $5 + 5 + 1 = 11$ so that (3-6) follows.

For mutually exclusive events, that is, for events A, B which cannot both happen, $p(AB) = 0$.   Hence the theorem of total probability takes the form

$$p(A + B) = p(A) + p(B) \qquad \text{for mutually exclusive events.} \tag{3-7}$$

The statement (3-7) can be depicted by the nonintersecting point sets in Fig. 3.

For example, in a toss of a die let A be the event "4 shows" while B is the event "5 shows."   Since these events are mutually exclusive, the probability of getting either 4 or 5 is

$$p(A + B) = p(A) + p(B) = \frac{1}{6} + \frac{1}{6} = \frac{1}{3}.$$

A result similar to (3-7) applies to any number of mutually exclusive events A, B, C, ...

The foregoing analysis, by counting cases, establishes the theorems of total and compound probability for uniform sample spaces only.   Actually the results are valid for arbitrary sample spaces, as indicated in Sec. 12, Probs. 8 and 9.   Here we prefer to show how the theorems are used in practice.

Example 1.   The probability that Peter will solve a problem is p_1, and the probability that Paul will solve it is p_2.   What is the probability that the problem will be solved if Peter and Paul work independently?

The probability that both solve it is $p_1 p_2$, by the theorem of compound probability, (3-3).   Hence the probability that at least one solves it is

$$p_1 + p_2 - p_1 p_2 \tag{3-8}$$

by the theorem of total probability, (3-5).

Example 2.   Solve Example 1 by finding the probability that both fail.

Peter's probability to fail is $1 - p_1$, and Paul's probability to fail is $1 - p_2$.   The probability that both fail is

$$(1 - p_1)(1 - p_2)$$

and the probability of the contrary event, that at least one succeeds, is

$$1 - (1 - p_1)(1 - p_2). \tag{3-9}$$

The consistency of (3-8) and (3-9) is easily verified.

Example 3.   A bag contains 10 white balls and 15 black balls.   Two balls are drawn in succession.   What is the probability that one of them is black and the other is white?

The mutually exclusive events in this problem are (a) drawing a white ball on the first trial and a black ball on the second, (b) drawing a black ball on the first trial and a white on the second.   The probability of (a) is $\frac{10}{25} \cdot \frac{15}{24}$ and that of (b) is $\frac{15}{25} \cdot \frac{10}{24}$, so that the probability of either (a) or (b) is

$$\tfrac{10}{25} \cdot \tfrac{15}{24} + \tfrac{15}{25} \cdot \tfrac{10}{24} = \tfrac{1}{2}.$$

Example 4.   How often must a pair of dice be tossed to make it more likely than not that double 6 appears at least once?

The probability that double 6 does not appear on a given toss is $\frac{35}{36}$, no matter what is known about the preceding tosses.   Repeated use of the theorem of compound probability gives

$$\left(\tfrac{35}{36}\right)^n$$

for the probability that double 6 does not appear in any of n tosses. It is desired to choose n in such a way that this probability is less than $\frac{1}{2}$. Thus,

$$(\tfrac{35}{36})^n < \tfrac{1}{2}.$$

Taking the logarithm gives

$$n \log \tfrac{35}{36} < -\log 2$$

or

$$n > \frac{\log 2}{\log \tfrac{36}{35}} = 24.6.$$

Thus 25 tosses suffice, but 24 do not.

Example 5.   Peter and Paul take turns tossing a pair of dice. The first to get a throw of 7 wins. If Peter starts the game, how much better are his chances of winning than Paul's?

This problem is different from any we have considered hitherto, in that there are infinitely many possibilities; namely, Peter may win on his first throw, or on his second throw, or on his third throw, and so forth. To apply the preceding theory, we simply consider the probability that Peter wins in n throws and take the limit as $n \to \infty$. A wide variety of questions involving infinitely many outcomes may be dealt with in a similar manner.

The probability of 7 is $\frac{1}{6}$, and the probability of not getting 7 is $\frac{5}{6}$. Hence the probability that Peter wins on his first throw is $\frac{1}{6}$. The probability that Peter wins on his second throw is $(\frac{5}{6})^2(\frac{1}{6})$ (since Peter's first throw and Paul's first throw must be other than 7 but Peter's second throw must be 7). Peter's probability of winning on his third throw is $(\frac{5}{6})^4(\frac{1}{6})$, and so on.

By the theorem of total probability the probability that Peter wins is

$$\tfrac{1}{6} + (\tfrac{1}{6})(\tfrac{5}{6})^2 + (\tfrac{1}{6})(\tfrac{5}{6})^4 + \cdots = (\tfrac{1}{6})(1 + r + r^2 + \cdots) \qquad \text{where } r = (\tfrac{5}{6})^2$$

$$= \frac{1}{6}\frac{1}{1-r} = \frac{1}{6}\frac{1}{1-\tfrac{25}{31}} = \frac{6}{11}. \tag{3-10}$$

A similar procedure shows that Paul's chance of winning is $\frac{5}{11}$, or one can reason as follows: The probability that 7 does not occur in n trials is $(\frac{5}{6})^n$. Since the limit is zero, the probability of an eternal game is zero, and Peter or Paul is sure to win. Thus, Paul's chance is

$$1 - \tfrac{6}{11} = \tfrac{5}{11}.$$

PROBLEMS

1. What is the probability that 5 cards dealt from a pack of 52 cards are all of the same suit?

2. Five coins are tossed simultaneously. What is the probability that at least one of them shows a head? All show heads?

3. The probability that Paul will be alive 10 years hence is $\frac{5}{8}$ and that John will be alive is $\frac{3}{4}$. What is the probability that both Paul and John will be dead 10 years hence? Paul alive and John dead? John alive and Paul dead?

4. One purse contains three silver and seven gold coins; another purse contains four silver and eight gold coins. A purse is chosen at random, and a coin is drawn from it. What is the probability that it is a gold coin?

5. Paul and Peter are alternately throwing a pair of dice. The first man to throw a doublet is to win. If Paul throws first, what is his chance of winning on his first throw? What is the probability that Paul fails and Peter wins on his first throw?

6. How many times must a die be thrown in order that the probability that the ace appear at least once shall be greater than $\frac{1}{2}$?

7. Twenty tickets are numbered from 1 to 20, and one of them is drawn at random. What is the probability that the number is a multiple of 5 or 7? A multiple of 3 or 5?

Note that, in solving the second part of this problem, it is incorrect to reason as follows: The number of tickets bearing numerals that are multiples of 3 is 6, and the number of multiples of 5 is 4. Hence the probability that the number drawn is either a multiple of 3 or of 5 is $\frac{6}{20} + \frac{4}{20} = \frac{1}{2}$. Why is this reasoning incorrect?

8. A card is chosen at random from each of five decks. What is the probability that all are face cards? Would the probability be larger or smaller if all five cards were taken from one deck, without replacing?

9. Answer the two questions in Prob. 8 if the desired hand is 1, 2, 3, 4, 5 of clubs; if the desired hand is to have at least two aces but is otherwise unrestricted.

10. Each of two radio tubes has probability p of burning out during the first 100 hr of use. If both are put into service at the same time, what is the probability that at least one of them is still good after 100 hr? Generalize to n tubes. If $p = 0.1$, how many tubes are needed to give a probability > 0.99 that at least one is good after 100 hr?

11. The dose of a well-known cold remedy is "two capsules and one tablet." The capsules and tablets come, in proper proportions, in one bottle. If one dumps out three pills what is the chance of having the correct dose? Neglect difference in size and shape of capsules and tablets, and assume that the number of pills in the bottle is very large.

12. (*Review.*) Solve Prob. 11 when there are n pills.

---

4. Random Variables and Expectation. A process is *random* if it is impossible to predict the final state from the initial state (as, for example, in a toss of a coin or a die). Associated with a random process there may be certain numerically valued variables which themselves have a random character. For instance, if x denotes the number obtained by tossing a die, x is a variable that assumes the values

$$1, 2, 3, 4, 5, 6$$

corresponding to the six events: 1 shows, 2 shows, and so forth. The respective probabilities are

$$\tfrac{1}{6}, \tfrac{1}{6}, \tfrac{1}{6}, \tfrac{1}{6}, \tfrac{1}{6}, \tfrac{1}{6}.$$

Again, if x is the number of heads obtained when three coins are tossed, x is a variable that assumes the values

$$0, 1, 1, 1, 2, 2, 2, 3 \qquad (4\text{-}1)$$

corresponding to the various ways the coins may fall. For instance, $x = 2$ corresponds to each of the three events HHT, HTH, THH.

Similarly, if a gambler stakes d dollars on a game, the amount he wins assumes the values

$$d, -d$$

in correspondence with the events "he wins the game" and "he loses the game." If his probability of winning is p, the respective probabilities of $x = d$ and $x = -d$ are

$$p, 1 - p.$$

In each of these examples x has a definite value corresponding to each event of the sample space. We say, briefly, that x is a *random variable*. In general, x is a *random variable* if x assumes prescribed values

$$x_1, x_2, x_3, \ldots, x_n \qquad (4\text{-}2)$$

at the first, second, third, ..., nth event of the sample space, n being the total number of events in the space.

An important concept associated with a random variable is its *expectation* or *expected value* $E(x)$. Let the probabilities of the n events comprising the sample space be

$$p_1, p_2, p_3, \ldots, p_n$$

in correspondence with (4-2). Then the *expected value* is defined to be

$$E(x) = p_1 x_1 + p_2 x_2 + p_3 x_3 + \cdots + p_n x_n. \tag{4-3}$$

For example, if x is the number obtained in a toss of a die, x assumes the values 1, 2, 3, ... with corresponding probabilities $p_i = \frac{1}{6}$. Hence

$$E(x) = \tfrac{1}{6} \cdot 1 + \tfrac{1}{6} \cdot 2 + \tfrac{1}{6} \cdot 3 + \tfrac{1}{6} \cdot 4 + \tfrac{1}{6} \cdot 5 + \tfrac{1}{6} \cdot 6 = \tfrac{7}{2}.$$

Similarly, if x is the number of heads obtained when three coins are tossed, (4-1) and (4-3) give

$$E(x) = \tfrac{0}{8} + \tfrac{1}{8} + \tfrac{1}{8} + \tfrac{1}{8} + \tfrac{2}{8} + \tfrac{2}{8} + \tfrac{2}{8} + \tfrac{3}{8} = \tfrac{3}{2}$$

when we note that $p_i = \frac{1}{8}$ in this case.

By grouping terms we can write the above sum in the form

$$E(x) = \tfrac{1}{8} \cdot 0 + \tfrac{3}{8} \cdot 1 + \tfrac{3}{8} \cdot 2 + \tfrac{1}{8} \cdot 3.$$

The factors

$$0, 1, 2, 3$$

represent the *numerically distinct* values of x, and the factors

$$\tfrac{1}{8}, \tfrac{3}{8}, \tfrac{3}{8}, \tfrac{1}{8}$$

represent the probabilities corresponding to these distinct values. For example, $\frac{3}{8}$ is the probability of two heads when three coins are tossed, and hence $\frac{3}{8}$ is the probability that $x = 2$. A similar grouping of terms can be applied to the general definition (4-3) and yields the following useful theorem:

THEOREM. *The expectation $E(x)$ is given by*

$$E(x) = P_1 x_1 + P_2 x_2 + \cdots + P_r x_r$$

where $x_1, x_2, \ldots, x_r$ are the numerically distinct values of x and where P_i is the probability that $x = x_i$.

According to this result, $E(x)$ is an average over the possible values of x, weighted by their probabilities.

Since $\Sigma p_i = 1$ the expectation $E(x)$ in the original definition can be interpreted as the center of mass,

$$E(x) = \frac{p_1 x_1 + p_2 x_2 + \cdots + p_n x_n}{p_1 + p_2 + \cdots + p_n}.$$

For equally likely x_i the result reduces to the *arithmetic mean*

$$E(x) = \frac{1}{n}(x_1 + x_2 + \cdots + x_n) \qquad \text{if each } p_i = \frac{1}{n}.$$

Thus, $E(x)$ is a measure of the location of x; it is a *typical value*. The following sections show that, if sufficiently many observations of the variable x are made, the mean of those observations will almost certainly be close to $E(x)$. In this sense, $E(x)$ represents the average value attained by x in the long run.

Example 1. From an urn containing a white and b black balls, a ball is drawn at random and set aside. What is the expected number of white balls left in the urn?

Let x be the number of white balls left. If a white ball is drawn, $x = a - 1$, whereas if a black ball is drawn, then $x = a$. Hence

$$E(x) = \frac{a}{a+b}(a-1) + \frac{b}{a+b}a = a - \frac{a}{a+b} = a - p$$

where p is the probability of drawing a white ball.

Example 2. Peter plays two games, in each of which his probability of winning is 0.4. If he loses he loses his stake, but if he wins he gets double his stake. How should he divide his total fortune between the two games to maximize his expected gain?

Suppose he stakes $\$a$ on the first game and $\$b$ on the second, where $a + b = c$ is his total fortune. The following table describes the different possibilities:

Event	Win Win	Win Lose	Lose Win	Lose Lose
Probability	0.16	0.24	0.24	0.36
Gain	$a + b$	$a - b$	$-a + b$	$-a - b$

The expected gain is, therefore,

$$(0.16)(a+b) + (0.24)(a-b) + (0.24)(-a+b) + (0.36)(-a-b).$$

This reduces to $(-0.2)(a+b) = -0.2c$, which is independent of the manner of subdivision.

PROBLEMS

1. A bent coin has probability p of giving heads and probability $q = 1 - p$ of giving tails. Let x be a random variable representing the number of heads when the coin is tossed three times; x is defined on a sample space consisting of the eight events HHH, HHT, ... with associated probabilities $p^3, p^2q, \ldots$. (a) Make a table giving the eight values of x associated with the eight sample points and their respective probabilities. (b) Make a second table giving the four distinct values of x and their probabilities. (c) Compute the expectation $E(x)$ from your table (a) and also from your table (b).

2. Peter tosses a coin and a die. He gets $\$1$ if an even number shows together with heads, he loses $\$5$ if an even number shows together with tails, and he gets $\$3$ if an odd number shows. (a) Make a table of the 12 possible events 1H, 2H, ..., 6H, 1T, ..., 6T, together with their probabilities and the value of Peter's gain, x. Thus compute $E(x)$. (b) Make a table involving the four events (even, H), (even, T), (odd, H), (odd, T) and use this to compute $E(x)$. (c) Make a table of the distinct values $1, -5, 3$, with their respective probabilities, and use this to compute $E(x)$ again.

3. The sum shown in the toss of a pair of dice is 9. What is the expected number of points shown by the first die? *Hint:* Construct an appropriate sample space by the method of Sec. 2.

4. Let x denote the sum obtained by a toss of a pair of dice. (a) Make a 6×6 table showing the values of x corresponding to the 36 equally likely ways the dice can fall. Thus get $E(x)$. (b) Obtain $E(x)$ by using the theorem in this section. (c) Let y denote the number of points shown on the first die, and z the number of points shown by the second, so that $x = y + z$. Is it true that $E(x) = E(y) + E(z)$?

5. A coin is tossed repeatedly. What is the expected number of the toss at which heads first appear? *Hint:* Let x be the number of the toss at which heads first appear. Then x has the values $1, 2, 3, \ldots$ with respective probabilities $\frac{1}{2}, \frac{1}{4}, \frac{1}{8}, \ldots$. By differentiating the geometric series for $(1-r)^{-1}$, or by the binomial theorem, show that $\Sigma n r^n = r/(1-r)^2$ for any r such that $|r| < 1$.

6. (*St. Petersburg paradox.*) Peter tosses a true coin repeatedly. He agrees to pay Paul \$2 if heads appear on the first toss, \$4 if heads appear on the second toss but not on the first, and generally \$2^n if heads appear for the first time at the nth toss. How much should Paul pay for the privilege of participating in this game; i.e., what is Paul's expectation?

7. In the preceding problem, suppose Peter has only a finite fortune \$$N$, so that Paul gets \$$N$ instead of \$2^n whenever $2^n \geq N$. What is Paul's expectation now? Discuss the resolution of the paradox by assigning numerical values such as N = \$100, \$1,000, \$1,000,000.

8. A and B cut a deck of cards. Each wager \$5, and the \$10 goes to the man who cuts the higher card. A cuts and obtains the 10 of diamonds. What is B's expectation? Count aces as high. Suits in order of increasing rank are clubs, diamonds, hearts, spades.

---

5. Discrete Distributions.   When the values x_i of a random variable are distinct, the associated probabilities p_i may be written in the form

$$p_i = f(x_i).$$

Since the x_i are supposed to be all the possible values of x, we must have

$$\Sigma f(x_i) = 1 \tag{5-1}$$

just as in the preceding section $\Sigma p_i = 1$. Also $f(x) \geq 0$, because $f(x)$ is a probability.

For example, let x be the number of heads obtained when four coins are tossed. If the value $x = 0, 1, 2, 3,$ or 4 is given, the probability to assume that value is determined by the table

$x =$	0	1	2	3	4
$f(x) =$	$\frac{1}{16}$	$\frac{4}{16}$	$\frac{6}{16}$	$\frac{4}{16}$	$\frac{1}{16}$

The function $f(x)$ is called the *frequency function* because, as shown in Sec. 12, the value $f(x_i)$ is proportional to the *expected frequency* of the event $x = x_i$, in a given number of observations.

Since the values x_i are distinct, the events $x = x_1$ and $x = x_2$ are mutually exclusive. Hence, by total probability, the probability of $x = x_1$ or $x = x_2$ is

$$f(x_1) + f(x_2). \tag{5-2}$$

In just the same way the probability of $x = x_1$, or $x_2, \ldots,$ or x_k is

$$\sum_{i=1}^{k} f(x_i). \tag{5-3}$$

For example, the probability of getting an even number of heads in a toss of four coins is

$$f(0) + f(2) + f(4) = \frac{1}{16} + \frac{6}{16} + \frac{1}{16} = \frac{1}{2}$$

by addition of appropriate entries in the preceding table. The result $\frac{1}{2}$ agrees with the result obtained by symmetry considerations, as it should.

It is often desirable to consider the probability that x will not exceed a given value. If $x_1, x_2, \ldots, x_k$ are the values of x_i that do not exceed t, the probability that $x \leq t$ is given by the sum (5-3). That is, the event "$x \leq t$" is equivalent to the event "$x = x_1$, or $x = x_2, \ldots,$ or $x = x_k$." We write

$$F(t) = \sum_{x_i \leq t} f(x_i) \tag{5-4}$$

for summation over the values of x_i that do not exceed t.   The function $F(t)$ thus obtained is called the *distribution function;* it gives the probability that $x \leq t$.   When t is so small that no x_i satisfies $x_i \leq t$, the sum (5-4) has no terms, and $F(t) = 0$ for such t. When t is so large that every x_i satisfies $x_i \leq t$, then the sum (5-4) includes every x_i. In this case (5-1) gives $F(t) = 1$.

For example, if x is the number of heads obtained when four coins are tossed, the distribution function is described by the following table:

t	$t < 0$	$0 \leq t < 1$	$1 \leq t < 2$	$2 \leq t < 3$	$3 \leq t < 4$	$4 \leq t$
$F(t)$	0	$\tfrac{1}{16}$	$\tfrac{5}{16}$	$\tfrac{11}{16}$	$\tfrac{15}{16}$	1

These entries are obtained by adding the values of $f(x)$ which were found previously. For instance, $\tfrac{11}{16}$ corresponds to the interval $2 \leq t < 3$ because

$$\sum_{x_i \leq t} f(x_i) = f(0) + f(1) + f(2) = \tfrac{1}{16} + \tfrac{4}{16} + \tfrac{6}{16} = \tfrac{11}{16}.$$

The value $\tfrac{11}{16}$ is the probability of getting at most two heads when four coins are tossed.

The variables x considered so far in this chapter are called *discrete variables* because they assume isolated values only.   For instance, the number of heads obtained when several coins are tossed is an integer 0, 1, 2, 3, . . . (and cannot fill up an interval).   The distribution of such a variable is called a *discrete distribution;* it is defined for all values of x, not only for the discrete set of possible values x_k.   One may also think of the frequency function as being defined for all x, taking $f(x) = 0$ for values x other than the x_k.   (For example, the probability of getting 3.2 heads is zero.)   The fact is that $f(x)$ may be defined in any *arbitrary* fashion for values other than the x_k, provided some care is taken in the interpretation of the results.

Graphical representation of the functions $f(x)$ and $F(x)$ is given in Figs. 4 to 6. Figure 4 is valid as a probability for all x.   The relationship of $f(x)$ and $F(x)$ is clarified, however, if $f(x)$ is modified as shown in Fig. 5.   Here, the value of $f(x)$ at any integer m is used for $f(x)$ in the interval of length 1 centered about m.   The resulting step function still gives the probability that $x = x_k$, provided x_k is an integer.   The advantage of redefining $f(x)$ in this fashion rests upon the following property, which is easily verified: *If t is an integer, $F(t)$ is the area under the curve of Fig. 5 up to the value $x = t + \frac{1}{2}$.*   For instance, the area up to the value $x = 2\frac{1}{2}$ is found to be

$$f(0) + f(1) + f(2) = F(2)$$

by adding the areas of the shaded rectangles.   When the values x_k are equally spaced, similar considerations apply to any distribution and frequency functions F and f.   For

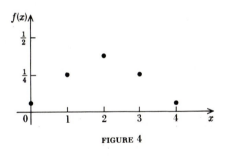

FIGURE 4

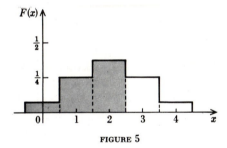

FIGURE 5

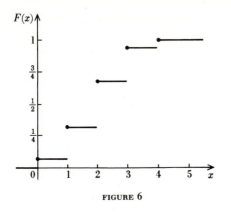

FIGURE 6

unit spacing, that is, for $x_{k+1} - x_k = 1$, Eq. (5-1) expresses the fact that the area under the curve is 1.

Actually, it is possible to describe the relationship of f and F directly, without introduction of the intermediate curve (Fig. 5). The description involves the so-called *Stieltjes integral*, which is now to be defined. Let $F(t)$ be a nondecreasing function on an interval $a \leq t \leq b$, and let $\phi(t)$ be continuous. Choose a set of points $t_1, t_2, \ldots, t_n$ on the interval $[a,b]$, and choose intermediate values ξ_k,

$$t_k \leq \xi_k \leq t_{k+1}.$$

As the subdivision given by the t_k's is made finer and finer, in such a way that

$$\max |t_{k+1} - t_k| \to 0$$

it can be shown that the expression

$$\Sigma\phi(\xi_k)[F(t_{k+1}) - F(t_k)]$$

tends to a limit (independent of the manner of subdivision and of the points ξ_k). The limit is called the *Stieltjes integral of ϕ with respect to F* and is written

$$\int_a^b \phi(t)\, dF(t).$$

When $F(t)$ is a discrete distribution corresponding to x_k and $f(x)$, the function $F(t)$ has a jump of value $f(x_k)$ at each value x_k but is constant between those values. Hence the differences

$$F(t_{k+1}) - F(t_k)$$

behave much like the function exemplified in Fig. 4. They assume the value $f(x_k)$ if the interval (t_k, t_{k+1}) contains a single point x_k, and they assume the value 0 if the interval contains no point x_k. The relationship of f and F is now described by the equation

$$F(t) = \int_{-\infty}^t dF(x)$$

where the integral is a Stieltjes integral. Although we have not defined a differential dF, we may think of $dF(x)$ as being equivalent to the frequency function $f(x)$ in the sense described above.

Example 1.   In terms of the distribution function, express the probability that $a < x \leq b$, where a and b are two numbers with $a < b$.

The event "$x \leq b$" can materialize in the mutually exclusive forms

$$x \leq a \quad \text{or} \quad a < x \leq b.$$

Hence, by total probability,

$$\Pr (x \leq b) = \Pr (x \leq a) + \Pr (a < x \leq b)$$

where Pr means "the probability that." This yields the desired expression

$$\Pr (a < x \leq b) = F(b) - F(a) \tag{5-5}$$

when we recall that the distribution function $F(t)$ satisfies

$$\Pr (x \leq b) = F(b) \qquad \Pr (x \leq a) = F(a).$$

Example 2.   In terms of the frequency function $f(x)$ the expectation of any variable $y = g(x)$ is

$$E(y) = \Sigma y_k f(x_k) \qquad y_k = g(x_k). \tag{5-6}$$

To establish this, we consider the variable y to be defined on a sample space whose points are the n events

$$x = x_1, \quad x = x_2, \quad \ldots, \quad x = x_n.$$

The probability of the event "$x = x_k$" is $f(x_k)$; the value of y corresponding to the event "$x = x_k$" is $y_k = g(x_k)$. Hence, (5-6) follows from the definition of expectation.

─── **PROBLEMS**

1. Review the discussion of sigma notation in Chap. 1, Sec. 2, and do Probs. 1 and 2 of that section.

2. Suppose a coin is tossed five times. What is the probability that this experiment will yield 0, 1, 2, 3, 4, 5 heads? If x is the number of heads, make a table representing the frequency function $f(x)$. Plot $f(x)$ and also the step-function modification (see Figs. 4 and 5).

3. In Prob. 2, make a table and also a graph for the distribution function $F(t)$.

4. If a coin is tossed five times, find the probability that the number of heads x satisfies $1 < x \le 4$ by use of (a) the frequency function $f(x)$ computed in Prob. 2, (b) the distribution function $F(t)$ computed in Prob. 3, (c) the step-function graph obtained in Prob. 2, with reference to an appropriate area under the curve.

5. With $f(x)$ as in Fig. 4, find $E(y)$ when $y = x$, $y = x + 1$, $y = 2x$, $y = 2x + 1$, and $y = x^2$.

6. Verify that the equation $E(x) = \Sigma x_k f(x_k)$ is consistent with (4-3) and with Example 2.

7. Let x be the number shown in a toss of a pair of dice. Make a table of the 12 possible values of x and their probabilities. Thus construct a graph of the frequency function $f(x)$ and of the distribution function $F(t)$. According to your graph, what is the probability that $x < 6$? Is this the same as the probability that $x > 6$?

8. Construct a frequency function and also a distribution function for the number of aces, x, in a three-card hand drawn at random from a full deck. What is $E(x)$?

9. (*Review.*) A student writes an examination consisting of 10 questions. The probability that his answer to any one question is correct is $\frac{1}{2}$. What is the probability that he answers all 10 correctly? Two correctly and eight incorrectly? Questions 3 and 7 correctly and the remaining questions incorrectly? At least two questions correctly?

10. (*Review.*) A throws three coins; B throws two coins. What is the probability that A will throw more heads than B?

───

6. Continuous Distributions. Since measurements are made only to a certain number of significant figures, the variables that arise as the result of an experiment are discrete. For example, if the diameter of a shaft is measured to the nearest 0.01 in., the measurement is a variable that assumes only isolated values, such as 3.21, 3.22, 3.23, . . . in. However, the underlying process that the measurements describe is often continuous, and even if it is discrete, it can generally be represented by continuous variables within the experimental error. Continuous variables are easier to handle analytically than discrete variables and are often used in practice.

Let a point be chosen at random on the interval $0 \le x \le 1$. How shall we measure the probabilities associated with that event? If the interval (0,1) is divided into a number of subintervals, each of length $\Delta x = 0.1$, the point x is equally likely to be in any of these subintervals (Fig. 7). The probability

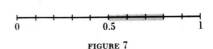

FIGURE 7

that[1] $0.5 < x < 0.8$, for example, is 0.3, since there are three favorable cases. The probability that $0.52 < x < 0.84$ is found to be $0.84 - 0.52 = 0.32$ when we divide the interval into 100 parts, and so on. This reasoning shows that the probability for x to be in a given subinterval of $(0,1)$ is the length of that subinterval. If Pr stands for "the probability that," then

$$\Pr(a < x < b) = b - a \qquad 0 \le a \le b \le 1. \tag{6-1}$$

When (6-1) holds, the variable x is said to be *uniformly distributed* on the interval $0 \le x \le 1$. Since the expression (6-1) can be written

$$\Pr(a < x < b) = \int_a^b dx = \int_a^b 1\, dx \tag{6-2}$$

we speak of the *probability density*, which in this case is unity.

More generally, a variable may be distributed with an arbitrary density $f(x)$. For such a variable the expression $f(t)\,\Delta t$ measures, approximately, the probability that x is on the interval

$$t < x < t + \Delta t.$$

An exact expression for the probability that x is on a given interval (a,b) is

$$\Pr(a < x < b) = \int_a^b f(x)\, dx. \tag{6-3}$$

This relation is illustrated in Fig. 8.

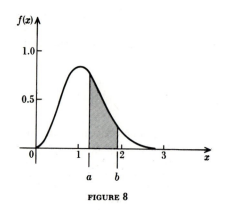

FIGURE 8

The notation (6-3) is customary in probability and statistics, but one should observe that the symbol x is used in two different senses. On the left, x is a random variable; on the right, x is the variable of integration. The result could have been written

$$\Pr(a < x < b) = \int_a^b f(\xi)\, d\xi$$

for example; see the discussion of dummy variables in Chap. 1, Sec. 2.

Since probabilities are nonnegative, $f(x) \ge 0$. If it is assumed that x always has *some* finite value, we conclude also that

$$\int_{-\infty}^{\infty} f(x)\, dx = 1. \tag{6-4}$$

This expresses the fact that $\Pr(-\infty < x < \infty) = 1$. Any nonnegative function satisfying (6-4) can be regarded as a probability density, and, conversely, a function is considered to be a probability density only if it is nonnegative and satisfies (6-4).

As a specific illustration, let

$$f(x) = 0 \quad \text{for } x < 0 \qquad f(x) = e^{-x} \quad \text{for } x \ge 0. \tag{6-5}$$

It is easily verified that (6-4) holds and hence $f(x)$ is a probability density. The probability that $x < 3$, for example, is

$$\int_{-\infty}^{3} f(x)\, dx = \int_0^3 e^{-x}\, dx = 1 - e^{-3}.$$

When $f(x)$ is a probability density, the function

$$F(t) = \int_{-\infty}^{t} f(x)\, dx \tag{6-6}$$

[1] In this section it will not matter whether the intervals include their end points or not. Thus, $\Pr(a \le x \le b) = \Pr(a < x < b)$.

is called the *distribution function*. According to our previous definitions, $F(t)$ represents the probability that x is on the interval $(-\infty, t)$; in other words,

$$F(t) = \Pr\,(x < t).$$

If $f(x)$ is continuous, then (6-6) gives $F'(t) = f(x)$, and one can speak of a *probability differential*

$$dF(t) = f(t)\,dt. \tag{6-7}$$

For example, the distribution function associated with (6-5) is

$$F(t) = 0 \qquad \text{for } t < 0 \qquad F(t) = 1 - e^{-t} \qquad \text{for } t \geq 0$$

and hence (6-7) holds for $t \neq 0$.

There is no difficulty in extending the concept of mathematical expectation to continuous variables. As the reader will recall, the expectation $E(x)$ associated with the frequency function $f(x)$ is

$$E(x) = \Sigma x_i f(x_i).$$

A similar definition is used in the continuous case, namely,

$$E(x) = \int_{-\infty}^{\infty} xf(x)\,dx = \frac{\int_{-\infty}^{\infty} xf(x)\,dx}{\int_{-\infty}^{\infty} f(x)\,dx}.$$

The latter expression follows from (6-4); it shows that $E(x)$ is the x coordinate of the center of mass for the area bounded by the curve $y = f(x)$ and the x axis. More generally, the expected value of any function $y = g(x)$ is

$$\int_{-\infty}^{\infty} yf(x)\,dx \qquad y = g(x) \tag{6-8}$$

by analogy to the result of Example 2 in Sec. 5.

We conclude this discussion by mentioning three density functions that often arise in applications, viz.,

$$e^{-\mu x}\,\frac{(\mu x)^r}{r!} \qquad \frac{1}{\sqrt{2\pi}\,\sigma}\,e^{-\frac{1}{2}[(x-\mu)/\sigma]^2} \qquad 4a\,\sqrt{\frac{a}{\pi}}\,x^2 e^{-ax^2}$$

for $x > 0$, $-\infty < x < \infty$, and $x > 0$, respectively. The first of these is the *Poisson density*, discussed in Sec. 11; the second is the *Gauss density* discussed in Secs. 9, 10, and 12; and the third is the *Maxwell-Boltzmann density*. This is important in the theory of gases, though not discussed in detail here. In each case, x is the random variable and μ, r, a, σ are constant parameters. For example, in the Maxwell-Boltzmann density, x is the speed of a gas molecule and $a = m/2kT$, where m is the mass, T is the absolute temperature, and k is a constant called the *Boltzmann constant*. A graph of the Boltzmann density for $a = 1$ is given in Fig. 8, and the Gaussian density with $\sigma = 1$, $\mu = 0$ is plotted in Fig. 10.

Example 1. A variable x is said to be uniformly distributed on (a,b) if $f(x)$ is constant on (a,b) and zero outside (a,b). Find $f(x)$ in this case.

Denoting the constant by c, we have

$$\int_{-\infty}^{\infty} f(x)\,dx = \int_a^b c\,dx = c(b - a) = 1$$

by (6-4). Solving for c yields

$$f(x) = \frac{1}{b - a} \qquad a < x < b; \qquad f(x) = 0 \qquad \text{elsewhere.}$$

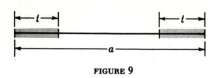

FIGURE 9

Example 2.   A stick of length a is broken at random into two pieces. Find the distribution function $F(s)$ for the length s of the shorter piece. From this, find the probability density $f(x)$ for the length l of the shorter piece.

Evidently $0 \leq s \leq a/2$ in every case.   For any t between 0 and $a/2$ we have $s < t$ if, and only if, x is on one of the intervals $(0,t)$ or $(a-t, a)$ (see Fig. 9).   The probability of that is $2t/a$, since x is uniformly distributed, and hence

$$F(t) = 2ta^{-1} \quad \text{for } 0 \leq t \leq \tfrac{1}{2}a.$$

If $t < 0$ then $F(t) = 0$, and if $t > a/2$ then $F(t) = 1$.   This is true because the probability that $s < 0$ is 0, and the probability that $s < a/2$ is 1.

By differentiation, the probability density $f(t) = F'(t)$ is

$$f(t) = 2a^{-1} \quad 0 < t < \tfrac{1}{2}a$$

and $f(t) = 0$ for $t < 0$ or $t > \tfrac{1}{2}a$.   The differentiation fails for $t = 0$ or $t = \tfrac{1}{2}a$, but it does not matter how the density is defined at these isolated points.

Example 3.   The edge x of a cube is uniformly distributed on $(0,2)$.   What is the probability that the volume should exceed its expected value?

By the result of Example 1 we have $f(x) = \tfrac{1}{2}$ for $0 < x < 2$ and $f(x) = 0$ elsewhere.   The volume is $v = x^3$, and hence

$$E(v) = \int_{-\infty}^{\infty} x^3 f(x)\, dx = \int_0^2 x^3\, \tfrac{1}{2}\, dx = 2.$$

This is the expected value of the volume.   The volume exceeds this value if, and only if, $x > \sqrt[3]{2}$.   The probability of that is

$$\int_{\sqrt[3]{2}}^{\infty} f(x)\, dx = \int_{\sqrt[3]{2}}^2 \tfrac{1}{2}\, dx = 1 - \tfrac{1}{2}\sqrt[3]{2}.$$

Example 4.   The *normal density function* is the special case $\sigma = 1$, $\mu = 0$ of the Gauss density, namely,

$$f(x) = \frac{1}{\sqrt{2\pi}}\, e^{-x^2/2}.$$

(See Fig. 10.)   If x is normally distributed, find the probability that $-1 < x < 2$ by means of the table of the function

$$\Phi(x) = \frac{1}{\sqrt{2\pi}} \int_0^x e^{-t^2/2}\, dt \qquad (6\text{-}9)$$

given in Appendix C.

The desired probability can be written

$$\frac{1}{\sqrt{2\pi}} \int_{-1}^0 e^{-t^2/2}\, dt + \frac{1}{\sqrt{2\pi}} \int_0^2 e^{-t^2/2}\, dt.$$

Since the graph of $f(x)$ is symmetric about $x = 0$, the integral from -1 to 0 is the same as the integral from 0 to 1.   Hence, the result is

$$\Phi(1) + \Phi(2) = 0.3413 + 0.4472 = 0.7885.$$

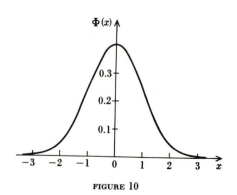

FIGURE 10

PROBLEMS

1. A function $f(x)$ is defined as follows for $0 \leq x \leq 1$, and as 0 elsewhere:

$$1, \quad 2x, \quad 5x^4, \quad \tfrac{3}{2} - x, \quad e - e^x, \quad \frac{\pi}{2}\sin \pi x, \quad 2\sin^2 \pi x.$$

In each case: (*a*) Verify that $f(x)$ is a density function; (*b*) find the probability that $x > \tfrac{1}{2}$; (*c*) find $E(x)$; (*d*) find the distribution function $F(t)$; and (*e*) use (*d*) to check (*b*).

2. A probability density is defined by $f(x) = 3x^2$ for $0 < x < 1$ and $f(x) = 0$ elsewhere. Find $E(x)$ and $E(x^2)$. Find the distribution function $F(x)$, and from this obtain a value m such that x is just as likely as not to exceed m. (The value m is called the *median* of x.)

3. The radius of a sphere is uniformly distributed on (0,1). Find the expected value of the volume. What is the probability that the volume exceeds half its maximum value? Answer the same questions for the surface.

4. A stick of length a is broken at random into two parts. What is the expected length of the shorter part? *Hint:* Use Example 2.

5. Find $E(|x|)$ if x is normally distributed (cf. Example 4).

6. Determine for what value of the constant c each of the following is a density function on $(-\infty,\infty)$:

$$ce^{-|x|}, \quad \frac{c}{1 + x^2}, \quad cxe^{-x^2}.$$

After c is properly chosen, find the corresponding distribution functions $F(t)$, and find the probability that x is positive.

7. Express $F(t)$, $E(x)$, and $E(x^2)$ for the Poisson, Gauss, and Maxwell-Boltzmann densities in integral form, but do not evaluate.

8. Show analytically and graphically that the function $\Phi(x)$ of (6-9) satisfies $\Phi(-x) = -\Phi(x)$. *Hint:* Choose $\tau = -t$ as the new variable of integration.

9. If x is normally distributed, find the probability of each of the following, as in Example 4: $0 < x < 1$; $-2 < x < 0$; $-0.5 < x < 1.5$; $1.1 < x < 2.23$; $|x| < 1$; $|x| < 2$; $|x| < 3$.

10. If the surface area of a cube is uniformly distributed on (0,6), what is the expected value of the edge? *Hint:* Let x denote the area and express the edge in terms of x.

11. If the volume of a cube is uniformly distributed on (0,1), what is the expected value of the edge? Of the surface area? What is the probability that the edge exceeds $\tfrac{1}{2}$? That the area exceeds 3?

12. In Example 2, show that the length of the longer piece is uniformly distributed on $(a/2,a)$.

13. If x has the probability density $f(x)$, show that the *theorem of total probability* is valid in the form

$$\text{Pr } (a < x < c) = \text{Pr } (a < x < b) + \text{Pr } (b < x < c)$$

for $a < b < c$. [The result is an immediate consequence of properties of integrals. It is also possible to start with the theorem of total probability, and deduce from this that probability can be represented as a Stieltjes integral (Sec. 5). Mild continuity conditions then give the representation (6-3).]

PROBABILITY AND RELATIVE FREQUENCY

7. Independent Trials. It often happens that the probability of an event cannot be determined by counting cases or by other a priori considerations. Sometimes the determination is impossible in principle; for instance, one cannot compute the probabilities associated with a loaded die or the probability that a given radio tube will fail in the first hundred hours' use. Sometimes the determination is theoretically possible but impractical. For instance, by examining every nail in a 100-lb keg one could find the probability that a nail selected at random will be defective, but this is not a useful method.

In many such cases an estimate for the probability can be obtained by repeated trials (or by inspecting a suitable *sample*, in the terminology of statistics). In the case of a biased coin, for example, if 10 tosses give 7 heads, 100 tosses give 73 heads, and 1,000 tosses give 690 heads, it appears that the probability of heads is *probably close* to 0.7. The two italicized words express a reservation that is always present in conclusions such as this.

The figures 7, 73, 690 in the above discussion represent the frequency of heads; the ratios

$$7/10, \ 73/100, \ 690/1,000$$

give the relative frequency in 10, 100, or 1,000 trials. More generally, *if an event occurs m times in n trials, the relative frequency is m/n*.

The trials in a sequence of trials are said to be *independent* if the probabilities associated with a given trial do not depend on the results of preceding trials. For example, the probability of heads on a given toss of a symmetric coin is $\frac{1}{2}$ no matter what is known about the results of previous tosses. But if we try to get an ace by drawing cards one at a time without replacing, the trials are dependent. In this case, the probability of an ace in a given trial depends on the number of aces that may have been drawn previously.

When an event has constant probability p of success, the probability of m successes in n independent trials may be computed as follows: A sequence of m successes and $n - m$ failures is represented by a sequence of m letters S and $n - m$ letters F:

$$SSFFSS \cdots SF. \tag{7-1}$$

Since the trials are independent, the probability of any one such sequence is

$$ppqqpp \cdots pq = p^m q^{n-m} \tag{7-2}$$

where $q = 1 - p$. To obtain the number of favorable sequences, observe that a sequence is determined as soon as the positions of the m letters S are fixed. The m places for these letters S can be chosen from the n places in $_nC_m$ ways, and hence the required probability is

$$p^m q^{n-m} + p^m q^{n-m} + \cdots + p^m q^{n-m} = {}_nC_m p^m q^{n-m} \tag{7-3}$$

by the theorem of total probability.

Alternatively, the reader may imagine a sample space in which each event consists of a sequence (7-1), with associated measure (7-2). Then (7-3) represents the sum of the measures of those points favorable to the event: *m successes.*

Replacing m by x gives

$$B(x) = {}_nC_x p^x q^{n-x} \equiv \frac{n!}{x!(n-x)!} p^x (1-p)^{n-x} \tag{7-4}$$

for the probability of exactly x successes in n independent trials with constant probability p. The associated distribution function is

$$F(t) = {}_nC_0 q^n + {}_nC_1 pq^{n-1} + \cdots + {}_nC_t p^t q^{n-t} \tag{7-5}$$

for integral values of t. This expression gives the probability of getting at most t successes in n trials.

Because of its connection with the binomial theorem (Prob. 6), the function $B(x)$ is called the *binomial frequency function*, $F(t)$ in (7-5) is the *binomial distribution*, and the statement that $B(x)$ gives the probability of x successes in n independent trials is called the *binomial law of probability*. Since many statistical studies involve repeated trials, the binomial law has great practical importance.

To illustrate the use of the formula (7-4) let it be required to find the probability that the ace will appear exactly 4 times in the course of 10 throws of a die. Here $p = \frac{1}{6}$, $q = \frac{5}{6}$, $n = 10$, $x = 4$. Hence the probability is

$$B(4) = \frac{10!}{4!6!}\left(\frac{1}{6}\right)^4\left(\frac{5}{6}\right)^6 = 0.05427.$$

If we define $x = 1$ when there is success and $x = 0$ when there is failure, the definition of expectation gives

$$E(x) = p \cdot 1 + q \cdot 0 = p.$$

Thus, the expected number of successes in a single trial is p. According to a general theorem established in Sec. 12, the expected number of successes in n trials must be

$$p + p + \cdots + p = np.$$

In other words, $E(x) = np$ (7-6)

where x is the number of successes in n trials. A direct proof of this important fact is given in Prob. 6.

For most distributions there is no special relation between the *expected* value and the *most probable* value, but for the binomial distribution they happen to be almost equal. Equation (7-4) yields

$$\frac{B(x + 1)}{B(x)} = \frac{(n - x)p}{(x + 1)q}$$

after slight simplification. Hence $B(x)$ is an increasing function of the integer x if, and only if,

$$\frac{(n - x)p}{(x + 1)q} > 1.$$

The latter inequality is the same as

$$(n - x)p > (x + 1)q$$

which reduces to $np > x + q$, since $p + q = 1$. We have shown, then, that $B(x + 1) > B(x)$ as long as $x < np - q$ but $B(x + 1) \le B(x)$ thereafter. Since $q < 1$, this establishes that $B(x)$ is maximum for a value of x that is within 1 of the value $x = np$. Further discussion of the function $B(x)$ is given in the following sections.

Example 1.   Ten tosses of a suspected die gave the result 1, 1, 1, 6, 1, 1, 3, 1, 1, 4. What is the probability of at least this many aces if the die is true?

The event "at least seven aces" can materialize in four mutually exclusive ways: 7 aces, 8 aces, 9 aces, 10 aces. By total probability (or by use of the distribution function) the required answer is found to be

$$B(7) + B(8) + B(9) + B(10) = {}_{10}C_7(\tfrac{1}{6})^7(\tfrac{5}{6})^3 + {}_{10}C_8(\tfrac{1}{6})^8(\tfrac{5}{6})^2 + {}_{10}C_9(\tfrac{1}{6})^9(\tfrac{5}{6}) + {}_{10}C_{10}(\tfrac{1}{6})^{10}$$

when we take $p = \frac{1}{6}$, $n = 10$. This reduces to 0.00027, approximately. Because the observed result has such small probability, one would reject the hypothesis "$p = \frac{1}{6}$" unless there is some other evidence in its favor.

Example 2.   In Example 1, let p be the unknown probability of the ace in a toss of the die. (*a*) For what value of p does the expected number of aces agree with the observed number? (*b*) For what value of p is the probability of the observed result a maximum?

Since $E(x) = np$ by (7-6), the observed and expected numbers agree when $p = x/n$, that is, when $p = 0.7$. The estimate for p given by $p = x/n$ is called an *unbiased* estimate, because $E(x/n) = p$.

For part b, the probability of getting seven aces and three other numbers is

$$p^7 q^3 \qquad \text{or} \qquad {}_{10}C_7 p^7 q^3 \text{ where } q = 1 - p$$

depending upon whether the order is considered or not. In either case the probability is maximum when $p^7(1 - p)^3$ is maximum. This, in turn, is maximum when

$$\log p^7(1 - p)^3 = 7 \log p + 3 \log (1 - p)$$

is maximum. Differentiation gives

$$\frac{7}{p} - \frac{3}{1 - p} = 0$$

or $p = 0.7$. An estimate for p such as this, which maximizes the probability of the observed result, is called a *maximum-likelihood* estimate.

PROBLEMS

1. When five coins are tossed, what is the probability of exactly two heads? At least two heads? What is the expected number of heads? The most probable number of heads?

2. If five dice are tossed simultaneously, what is the probability that (a) exactly three of them turn the ace up? (b) At least three turn the ace up?

3. If the probability that a man aged 60 will live to be 70 is 0.65, what is the probability that out of 10 men now 60 at least 7 will live to be 70?

4. A man is promised \$1 for each ace in excess of 1 that appears in 6 consecutive throws of a die. What is the value of his expectation?

5. A bag contains 20 black balls and 15 white balls. What is the chance that at least 4 in a sample of 5 balls are black?

6. (a) By use of the binomial theorem, show that

$$(q + pt)^n = B(0) + B(1)t + B(2)t^2 + \cdots + B(n)t^n.$$

(b) Interpret the identity that arises when $t = 1$. (c) Differentiate with respect to t, and interpret the identity that then arises for $t = 1$. [The function $(q + pt)^n$ is called the *generating function* of the sequence $\{B(x)\}$.]

7. (a) One hundred light bulbs were tested for 500 hr, at the end of which time 57 bulbs had failed. Obtain an unbiased estimate and also a maximum-likelihood estimate for the probability of failure in 500 hr. (b) Are these two estimates of p always equal for the binomial distribution? *Hint:* In part b, compare the result of maximizing $p^m q^{n-m}$ with respect to p and the result of choosing p so that $E(x) = m$, where m is the number of observed successes.

8. In a certain agricultural experiment, the probability that a plant will have yellow flowers is $\frac{3}{4}$. If 10,000 plants are grown, what is the probability that the number with yellow flowers will be between 7,400 and 7,600? (To appreciate later developments, observe that your answer, which should be indicated only, is difficult to compute.)

---

8. An Illustration. Some interesting conclusions concerning the binomial law are suggested by an example that presents many features of the general case. Consider a purse in which are placed two silver and three gold coins, and let it be required to find the probability of drawing exactly x silver coins in n trials, the coin being replaced after each drawing. The probability of exactly x successes in n trials is given by (7-4) where

p, the probability of drawing a silver coin in a single trial, is $\frac{2}{5}$. If the number of drawings is taken as $n = 5, 10,$ or 30, the respective frequency functions $B(x)$ are

$$B(x) = {}_5C_x(\tfrac{2}{5})^x(\tfrac{3}{5})^{5-x} \qquad B(x) = {}_{10}C_x(\tfrac{2}{5})^x(\tfrac{3}{5})^{10-x} \qquad B(x) = {}_{30}C_x(\tfrac{2}{5})^x(\tfrac{3}{5})^{30-x}.$$

By use of these expressions one can compute the values of $B(x)$ to any desired accuracy. The result of such a computation to four places of decimals is presented in the accompanying tables. In the third table the entry 0.0000 is made for $0 \leq x \leq 2$ and for $x \geq 23$ because in these cases $B(x)$ was found to be less than 0.00005. For example, the probability of drawing exactly 23 silver coins in 30 trials is

$$B(23) = {}_{30}C_{23}(\tfrac{2}{5})^{23}(\tfrac{3}{5})^7 = 0.000040128.$$

The reader can verify that the most probable values of x are exactly equal to np (and not merely within 1 of np). This behavior is always found when np is an integer.

PROBABILITY OF EXACTLY x SUCCESSES IN 5 TRIALS

x	$B(x)$	x	$B(x)$
0	0.0778	3	0.2304
1	0.2592	4	0.0768
2	0.3456	5	0.0102

PROBABILITY OF EXACTLY x SUCCESSES IN 10 TRIALS

x	$B(x)$	x	$B(x)$	x	$B(x)$
0	0.0060	4	0.2508	8	0.0106
1	0.0403	5	0.2007	9	0.0016
2	0.1209	6	0.1115	10	0.0001
3	0.2150	7	0.0425		

PROBABILITY OF EXACTLY x SUCCESSES IN 30 TRIALS

x	$B(x)$	x	$B(x)$	x	$B(x)$
≤ 2	0.0000	9	0.0823	16	0.0489
3	0.0003	10	0.1152	17	0.0269
4	0.0012	11	0.1396	18	0.0129
5	0.0041	12	0.1474	19	0.0054
6	0.0115	13	0.1360	20	0.0020
7	0.0263	14	0.1100	21	0.0006
8	0.0505	15	0.0783	22	0.0002
				≥ 23	0.0000

The values given in the tables are presented graphically in Fig. 11 after the manner described in Sec. 5. Each curve has the general shape predicted by the theory of the preceding section, but the figure shows also how the shape changes as we proceed from one curve to another. The numerical area under each curve is 1, although the curves become broader and flatter as n increases. In particular the maximum (that is, the probability of the most probable value) decreases as n increases. This is just what one would expect intuitively. (For instance, one could easily get 2 heads in 4 tosses of a coin, but one would be surprised to get exactly 500,001 heads in 1,000,002 tosses.) The

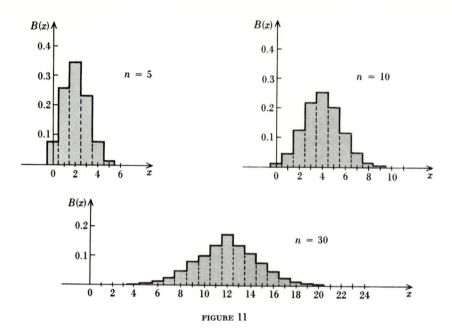

FIGURE 11

fact that the curves become broader indicates that the values of x experience a wider spread when there are more trials, and this, too, one would expect. Naturally, the curves ought to get broader if the maximum is to decrease while the area remains equal to 1.

The foregoing discussion is concerned with the frequency of success in n trials. The results are very different if, instead, one considers the *relative* frequency x/n. The distribution for the variable x/n is presented graphically in Fig. 12. These curves were obtained from the preceding by the change of scale indicated on the axes, and hence the area is still 1. Instead of becoming broader, these curves become narrower as n increases. The relative frequency x/n tends to cluster about its expected value p as n gets large. This is the reason why the relative frequency can be used to estimate an unknown probability.

The behavior suggested by this example can be summarized as follows: When the number of trials n becomes large, the absolute deviation from the expected value

$$|x - np| = |x - E(x)|$$

is likely also to be large, but the relative deviation

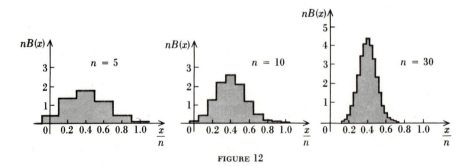

FIGURE 12

$$\left|\frac{x-np}{n}\right| = \left|\frac{x}{n} - p\right| = \left|\frac{x}{n} - E\left(\frac{x}{n}\right)\right|$$

is likely to be small.   It will be seen in Sec. 9 that the first expression is usually of the order $\sqrt{n}$ and the second, of order $1/\sqrt{n}$; compare Prob. 3.

PROBLEMS

1. Plot a distribution curve like that of Fig. 11 for the probability of x successes in four trials when $p = \frac{2}{5}$.   Shade the area corresponding to the event $1 < x \leq 3$, and find the probability of this event.

2. For $p = \frac{2}{5}$, plot the probability of the most probable number of successes versus n.   (Take points at $n = 1, 2, 3, 4, 5, 10, 30$ only; cf. Prob. 1 and accompanying tables.)   On the same figure, plot $1/\sqrt{2\pi npq}$ versus n.   (It is shown in Sec. 9 that the probability is asymptotic to $1/\sqrt{2\pi npq}$ when n is large.   This expression approaches zero as $n \to \infty$, even though we are considering the *most probable* value.)

3. Using the tables and your numerical values in Prob. 2, plot $\sqrt{n}\,B(x)$ versus $(x - np)/\sqrt{n}$ for $p = \frac{2}{5}$ and for $n = 3, 4, 5, 10$.   Use the same scale in each case.   Formulate a conjecture concerning the behavior as $n \to \infty$, and test your conjecture by plotting five well-chosen points on the curve corresponding to $n = 30$.

4. (*Review.*)   A batch of 1,000 lamps is 5 per cent bad.   If five lamps are tested, what is the probability that no defective lamp appears?   What is the chance that the test batch of five lamps is 40 per cent defective?

5. (*Review.*)   An urn contains a white and b black balls, with $a + b = N$.   What is the probability of drawing α white and β black in $n = \alpha + \beta$ draws (a) if the balls are drawn successively without replacement; (b) if the balls are drawn successively and returned after each draw? (c) What if the balls are drawn all at once, rather than successively?

9. The Laplace-de Moivre Limit Theorem.   Numerical computation of the binomial distribution is difficult when n is large.   In this section an approximate formula is obtained when n and np are both large.   In Sec. 11 a formula is found when n is large but np is not large.   These approximations, together with the exact formula when n is moderate, cover all values.   Since the cases $p = 0$ or $p = 1$ are trivial, we assume $0 < p < 1$.

In the expression

$$B(r) \doteq \frac{n!}{r!(n-r)!}\,p^r q^{n-r} \tag{9-1}$$

for the probability of r successes in n independent trials, suppose that r, n, and $n - r$ are large enough to permit the use of Stirling's formula

$$n! \sim n^n e^{-n}\sqrt{2\pi n} \tag{9-2}$$

that was established in Chap. 1, Sec. 14.   Replacing $r!$, $n!$, and $(n - r)!$ by their Stirling approximations gives, after simplification,

$$B(r) \doteq \left(\frac{np}{r}\right)^r \left(\frac{nq}{n-r}\right)^{n-r} \sqrt{\frac{n}{2\pi r(n-r)}}.$$

The subsequent algebra is somewhat easier if r and $n - r$ occur in the numerator rather than in the denominator.   One can accomplish this by transposing the radical to the left and changing the sign of the exponents on the right.   The result is the formula

$$B(r)A \doteq \left(\frac{r}{np}\right)^{-r} \left(\frac{n-r}{nq}\right)^{-(n-r)} \qquad \text{where } A = \sqrt{\frac{2\pi r(n-r)}{n}}. \tag{9-3}$$

We now let δ denote the deviation of the number of successes, r, from its expected value np.   Thus,

$$\delta = r - np \qquad r = np + \delta \qquad n - r = nq - \delta \tag{9-4}$$

where the third relation follows from the second, since $p + q = 1$.   Substitution into (9-3) gives

$$B(r)A \doteq \left(1 + \frac{\delta}{np}\right)^{-(np+\delta)} \left(1 - \frac{\delta}{nq}\right)^{-(nq-\delta)}$$

and

$$A = \sqrt{2\pi npq \left(1 + \frac{\delta}{np}\right)\left(1 - \frac{\delta}{nq}\right)}.$$

Then, $$\log B(r)A \doteq -(np + \delta) \log\left(1 + \frac{\delta}{np}\right) - (nq - \delta) \log\left(1 - \frac{\delta}{nq}\right).$$

Assuming $|\delta| < npq$, so that

$$\left|\frac{\delta}{np}\right| < 1 \qquad \text{and} \qquad \left|\frac{\delta}{nq}\right| < 1$$

permits one to write the two convergent series (see Chap. 1, Sec. 10)

$$\log\left(1 + \frac{\delta}{np}\right) = \frac{\delta}{np} - \frac{\delta^2}{2n^2p^2} + \frac{\delta^3}{3n^3p^3} - \cdots$$

and $$\log\left(1 - \frac{\delta}{nq}\right) = -\frac{\delta}{nq} - \frac{\delta^2}{2n^2q^2} - \frac{\delta^3}{3n^3q^3} - \cdots.$$

Hence, $$\log B(r)A \doteq -\frac{\delta^2}{2npq} - \frac{\delta^3(p^2 - q^2)}{2 \cdot 3n^2p^2q^2} - \frac{\delta^4(p^3 + q^3)}{3 \cdot 4n^3p^3q^3} - \cdots.$$

Now, if $|\delta|$ is so small in comparison with npq that one can neglect all terms in this expansion beyond the first and can replace A by $\sqrt{2\pi npq}$, there results the approximate formula

$$B(r) \doteq \frac{1}{\sqrt{2\pi npq}}\, e^{-\delta^2/2npq} \tag{9-5}$$

which bears the name of *Laplace's*, or the *normal, approximation*.   With $\sigma = \sqrt{npq}$, Eq. (9-5) becomes

$$B(r) \doteq \frac{1}{\sqrt{2\pi}\,\sigma}\, e^{-\delta^2/2\sigma^2}. \tag{9-6}$$

The equality is asymptotic; that is, the ratio of the two sides tends to 1 as $n \to \infty$.   A comparison of $B(r)$ with the normal approximation is given in Fig. 13.

The main usefulness of this result is to compute the probability

$$\sum_{r=r_1}^{r_2} B(r) \tag{9-7}$$

that the number of successes is between the given limits r_1 and r_2.   Equation (9-6) shows

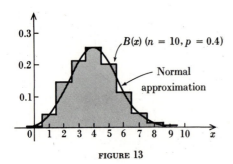

FIGURE 13

that the sum (9-7) may be approximated by a sum

$$\sum \frac{1}{\sqrt{2\pi}\,\sigma}\, e^{-\delta^2/2\sigma^2} \tag{9-8}$$

over appropriate values of δ. Since $\delta = r - np$, the difference between successive values of δ is 1, and hence if we let $t = \delta/\sigma$, the difference between successive values of t is $\Delta t = 1/\sigma$. Thus (9-8) becomes a sum over t,

$$\sum \frac{1}{\sqrt{2\pi}}\, e^{-t^2/2}\,\Delta t. \tag{9-9}$$

As $\Delta t \to 0$, the expression (9-9) approaches an integral, which may be evaluated in terms of the function

$$\Phi(t) = \int_0^t \frac{1}{\sqrt{2\pi}}\, e^{-t^2/2}\, dt \tag{9-10}$$

tabulated in Appendix C. These considerations yield the following fundamental result:

LAPLACE-DE MOIVRE LIMIT THEOREM. Let x be the number of successes in n independent trials with constant probability p. Then the probability of the inequality

$$t_1 \le \frac{x - np}{\sqrt{npq}} \le t_2 \tag{9-11}$$

approaches the limit

$$\frac{1}{\sqrt{2\pi}} \int_{t_1}^{t_2} e^{-t^2/2}\, dt = \Phi(t_2) - \Phi(t_1) \tag{9-12}$$

as $n \to \infty$.

To complete the proof one must show that the error in passing from (9-7) to (9-8) is small for large n even when the number of terms in the sum is large.[1] It is known that a better approximation is given by

$$\Phi\left(t_2 + \frac{1}{2\sigma}\right) - \Phi\left(t_1 - \frac{1}{2\sigma}\right) \text{ where } \sigma = \sqrt{npq} \tag{9-13}$$

although the improvement is not important when n is large.

To illustrate the theorem, let us find the probability that the number of aces will be between 80 and 110 when a true die is tossed 600 times. Here $n = 600$, $p = \frac{1}{6}$, $q = \frac{5}{6}$, and x varies from 80 to 110. Hence

$$t_1 = \frac{80 - 100}{\sqrt{(100)(\frac{5}{6})}} = -2.19 \quad \text{and} \quad t_2 = \frac{110 - 100}{\sqrt{(100)(\frac{5}{6})}} = 1.09.$$

The table gives $\Phi(t_2) = \Phi(1.09) = 0.362$, and similarly

$$\Phi(-2.19) = -\Phi(2.19) = -0.486.$$

The required probability is, approximately,

$$0.362 - (-0.486) = 0.848.$$

Example 1.   In the notation of the text, show that the probability $P_{\max}$ of the most probable value of r satisfies

$$P_{\max} \sim \frac{1}{\sqrt{2\pi npq}} \text{ as } n \to \infty. \tag{9-14}$$

[1] A more detailed analysis, taking due account of this question, is given in William Feller, "An Introduction to Probability Theory and Its Applications," pp. 164–172, John Wiley & Sons, Inc., New York, 1959. An expression for the error in the approximation is derived in J. V. Uspensky, "Introduction to Mathematical Probability," p. 129, McGraw-Hill Book Company, New York, 1937.

In Sec. 7 it was found that the most probable value is of the form $r = np + \theta$, $|\theta| < 1$. For this value of r we have $\delta = r - np = \theta$, and hence Eq. (9-5) shows that the associated probability is asymptotic to

$$\frac{1}{\sqrt{2\pi npq}}\, e^{-\theta^2/2npq}.$$

As $n \to \infty$, the exponential tends to 1, since θ^2 is bounded, and this yields (9-14).

Example 2.   In an agricultural experiment Mendelian theory yields a probability $p = \frac{1}{4}$ that any given plant should have blue flowers. Out of 10,000 plants it was found that 2,578 had blue flowers. Does this result contradict the theory?

According to theory, there should have been 2,500 plants with blue flowers; that is, the expected number is $np = 2,500$. There were, in fact, 78 more than this. We have to decide if this excess is too large to be attributed to chance.

Let us find the probability that the excess will be 78 or more if the hypothesis $p = \frac{1}{4}$ is indeed correct. The inequality

$$78 \le x - np \tag{9-15}$$

becomes
$$1.801 \le \frac{x - np}{\sqrt{npq}} < \infty \tag{9-16}$$

when divided by $\sqrt{npq} = 43.3$. According to the table, the probability of (9-16) is

$$\Phi(\infty) - \Phi(1.801) = 0.500 - 0.464 = 0.036.$$

Now, in a statistical test it is customary to reject the hypothesis if the hypothesis makes the probability of the observed result less than a fixed quantity α determined beforehand. The value α (which is called the *significance level* of the test) is often taken to be 0.05. Since our probability 0.036 is less than 0.05, the experimental outcome is considered too unlikely to be attributed to chance, and we reject the hypothesis "$p = \frac{1}{4}$." In this sense, the experiment contradicts Mendelian theory.

We now give another analysis that leads to the opposite conclusion. Instead of saying "the excess was 78," one could just as well say, "the discrepancy was 78," meaning

$$|x - np| = 78. \tag{9-17}$$

Both statements are equally valid descriptions of the experimental outcome. The probability of

$$|x - np| < 78$$

is found, as above, to be

$$\Phi(1.801) - \Phi(-1.801) = 0.928$$

and hence the probability of the contrary event is

$$1 - 0.928 = 0.072.$$

Since $0.072 > 0.05$, a discrepancy of "78 or more" is sufficiently probable to be attributed to chance (if, as before, our significance level is 0.05). Hence the hypothesis is not contradicted by the experiment.[1]

It requires statistical methods of considerable subtlety to decide between competing tests of a hypothesis such as the foregoing.[2] These methods show that the first procedure is appropriate for testing the hypothesis "$p = \frac{1}{4}$" against the alternative "$p > \frac{1}{4}$" whereas the second is appropriate for testing the hypothesis against the alternative "$p \ne \frac{1}{4}$."

[1] When the probability exceeds the significance level, as in this case, the hypothesis is not thereby *proved* but it is considered to have withstood the experimental test.
[2] A readable account of the subject is given in P. G. Hoel, "Introduction to Mathematical Statistics," chap. 9, John Wiley & Sons, Inc., New York, 1962.

1. Two dice are tossed 1,000 times. What is, approximately, the probability of getting a sum of 4 the most probable number of times? 500 times? (Use a table of exponentials.)

2. A true coin is to be tossed 1,600 times, and it is desired to find the probability that the number of heads x will satisfy $780 \leq x \leq 830$. (a) Show that this inequality is equivalent to

$$-1 \leq \frac{x - np}{\sqrt{npq}} \leq 1.5.$$

(b) Express the probability of the latter inequality in terms of Φ by means of the normal law. (c) Using the table, evaluate the probability.

3. By means of the normal law, obtain an approximate numerical answer to Prob. 8 of Sec. 7.

4. A machine has a probability $p = 0.01$ of producing a defective bottle. In one day's run, out of 10,000 bottles, 120 were defective. Find the approximate probability of at least this many defectives if the machine is running as usual.

5. A suspected die gave only 960 aces in 6,000 tosses. If the die is true, (a) what is the probability of getting at most 960 aces in 6,000 tosses? (b) What is the probability of getting a discrepancy $|x - np|$ of "40 or more"? (c) At a significance level of 0.05, does either calculation indicate that the die is loaded?

6. In the formula for $B(r)$, suppose $n!$ and $(n - r)!$ are replaced by their Stirling approximations, but $r!$ is not so replaced. Show that the result is the approximation

$$B(r) \doteq \frac{n^r e^{-r}}{r![1 - (r/n)]^{n-r+\frac{1}{2}}} p^r (1 - p)^{n-r}.$$

This leads to the *Poisson law* discussed n Sec. 11.

10. The Law of Large Numbers. Since $\Sigma B(r) = 1$ for each value of n, it is natural to expect, by the foregoing analysis, that

$$\frac{1}{\sqrt{2\pi}} \int_{-\infty}^{\infty} e^{-t^2/2} \, dt = 1. \tag{10-1}$$

For a direct proof of (10-1), define I by

$$I = \int_{-\infty}^{\infty} e^{-x^2/2} \, dx = \int_{-\infty}^{\infty} e^{-y^2/2} \, dy. \tag{10-2}$$

Then multiplication of the two expressions (10-2) yields

$$I^2 = \int_{-\infty}^{\infty} \int_{-\infty}^{\infty} e^{-(x^2+y^2)/2} \, dx \, dy \tag{10-3}$$

after changing to a double integral. In polar coordinates,

$$I^2 = \int_0^{2\pi} \int_0^{\infty} e^{-r^2/2} r \, dr \, d\theta = 2\pi \int_0^{\infty} e^{-r^2/2} \, d\frac{r^2}{2} = 2\pi \tag{10-4}$$

so that $I = \sqrt{2\pi}$, and (10-1) follows. The transformations are justified by the fact that (10-3) is an *absolutely convergent* double integral.

Equation (10-1) shows that the function

$$\frac{1}{\sqrt{2\pi}} \int_{-\infty}^{t} e^{-x^2/2} \, dx = \Phi(t) + \frac{1}{2}$$

is a distribution function; it is called the *normal distribution*. The theorem of the preceding section asserts that the variable δ/σ is approximately normally distributed when n is large. This fact will now be used to establish the following fundamental result, which is a special case of the *law of large numbers:*

THEOREM. *Let x be the number of successes in n independent trials with constant probability p. If ϵ is any positive number, the probability of the inequality*

$$\left|\frac{x}{n} - p\right| < \epsilon \tag{10-5}$$

tends to 1 as $n \to \infty$.

In other words, the *relative frequency* of the event is almost sure to be close to the *probability* of the event when the number of trials is large. For proof, write the inequality (10-5) in the form

$$-\epsilon < \frac{x - np}{n} < \epsilon$$

which becomes

$$-\epsilon \sqrt{\frac{n}{pq}} < \frac{x - np}{\sqrt{npq}} < \epsilon \sqrt{\frac{n}{pq}} \tag{10-6}$$

when multiplied by $\sqrt{n/pq}$. Given any number t_0 (no matter how large), we can choose n so that $\epsilon\sqrt{n/pq} > t_0$. In this case the probability of the inequality (10-6) is at least equal to the probability of

$$-t_0 < \frac{x - np}{\sqrt{npq}} < t_0. \tag{10-7}$$

As $n \to \infty$, the latter probability tends to

$$\frac{1}{\sqrt{2\pi}} \int_{-t_0}^{t_0} e^{-t^2/2}\, dt \tag{10-8}$$

by (9-12). Since t_0 is as large as we please, Eq. (10-1) shows that the integral (10-8) is as close to 1 as we please, and this completes the proof.

The theorem was established first by James Bernoulli (1654–1705) after 20 years of effort. The law of large numbers lies at the basis of all attempts to estimate a probability experimentally, and it affords a philosophical justification for such attempts. In fact, some developments of the subject *define* probability in terms of relative frequency, by the formula $p = \lim (x/n)$ as $n \to \infty$, and rely on the law of large numbers to ensure that the limit exists.

The theorem makes possible some interesting computational procedures, known as *Monte Carlo methods.* Although the method is not to be discussed at length here, we sketch an example that illustrates some of the main features. Suppose a man walks in a straight line, taking a step of length h ft every s sec (see Fig. 14). Each step is equally likely to be to the right or to the left, without regard to the preceding steps. Assuming that x is a multiple of h and t is a multiple of s, it is required to find the probability that the man is x ft from his starting point at time t.

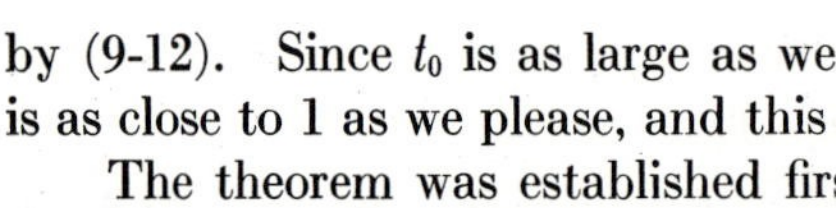

Let $U(x,t)$ represent the probability in question; that is, $U(x,t)$ is the probability of the man's being at point x at time t if he was at point $x = 0$ at time $t = 0$. Now, he can arrive at point x at time $t + s$ in two ways.

FIGURE 14

Either he was at point $x + h$ at time t and took a step to the left, or he was at point $x - h$ at time t and took a step to the right. The probability of being at $x + h$ at time t is $U(x + h, t)$ by the definition of U, and the probability of a step to the left is $\frac{1}{2}$ by hypothesis. Hence the probability of both events is

$$\tfrac{1}{2}U(x + h, t)$$

by compound probability. In just the same way, the probability of being at $x - h$ and then stepping to the right is

$$\tfrac{1}{2}U(x - h, t).$$

By total probability, the probability of getting to the point x at time $t + s$ is the sum, and we are thus led to a *difference equation* for U,

$$U(x, t + s) = \tfrac{1}{2}U(x + h, t) + \tfrac{1}{2}U(x - h, t). \tag{10-9}$$

The boundary conditions are

$$U(x,0) = 0 \quad \text{for } x \neq 0 \quad \sum_x U(x,t) = 1 \tag{10-9a}$$

which express the fact that he is sure to be at the origin when $t = 0$ and sure to be at *some* point x for all t.

To apply the Monte Carlo method to this problem, we make a large number of actual random walks experimentally. The number of times we arrive at point x at time t gives an estimate for the probability $U(x,t)$ by virtue of the law of large numbers. Hence, the calculation yields an approximate solution of the problem (10-9) without any direct use of (10-9). In practice, the "random walks" are made on a computing machine by reference to a set of random numbers. Similar methods apply to difference equations of much greater complexity than (10-9).

For readers familiar with the theory of heat conduction the foregoing example yields an interesting interpretation of the normal law. Subtracting $U(x,t)$ from both sides of (10-9) and dividing by s give

$$\frac{U(x, t + s) - U(x,t)}{s} = \frac{h^2}{2s} \frac{U(x + h, t) - 2U(x,t) + U(x - h, t)}{h^2}.$$

If we set $s = h^2$ and let $h \to 0$, this becomes, formally,

$$\frac{\partial U}{\partial t} = \frac{1}{2} \frac{\partial^2 U}{\partial x^2} \tag{10-10}$$

by Chap. 7, Sec. 26. The boundary condition is

$$U(x,0) = 0 \quad \text{for } x \neq 0 \quad \text{and} \quad \int_{-\infty}^{\infty} U(x,t)\, dx = 1.$$

Since these are the conditions for an instantaneous source of heat at the origin, a solution is given by Chap. 7, Sec. 17, namely,

$$U(x,t) = \frac{1}{\sqrt{2\pi t}}\, e^{-x^2/2t}. \tag{10-11}$$

Now, in the random walk the probability of a steps to the right and b to the left is given approximately by the normal approximation (9-5); it turns out to be

$$\sqrt{\frac{2}{\pi(a + b)}}\; e^{-(a-b)^2/2(a+b)}. \tag{10-12}$$

If the man arrives at point x at time t, he makes t/s steps altogether and x/h more steps to the right than to the left:

$$a + b = \frac{t}{s} \qquad a - b = \frac{x}{h}.$$

Substitution in (10-12) and setting $s = h^2$ yield

$$U(x,t) \doteq \frac{1}{\sqrt{2\pi t}} e^{-x^2/2t}(2h) \tag{10-13}$$

for the probability. Here $2h$ is the distance between possible values of x when t is fixed, and hence the coefficient of $2h$ may be regarded as a probability density   The condition "h small" means simply that the number of steps is large, so that the normal law is applicable.   The analogy between (10-11) and (10-13) is evident.

Since heat is due to random motion of the molecules, the analogy of the random-walk problem with the problem of heat flow has a physical basis as well as the mathematical basis outlined here.   However, the method of random walks also applies to partial differential equations that, apparently, have nothing to do with probability.

Example.   A true coin is tossed repeatedly.   It is desired to have a probability of 0.99 that the relative frequency of heads shall be within 1 per cent of the probability of heads.   How many times must the coin be tossed?

If the coin is tossed n times, the desired inequality is

$$\left| \frac{x/n - p}{p} \right| \le 0.01 \tag{10-14}$$

which is the same as

$$-0.01 \sqrt{\frac{p}{q}} \, n \le \frac{x - np}{\sqrt{npq}} \le 0.01 \sqrt{\frac{p}{q}} \, n. \tag{10-15}$$

Setting the probability of (10-15) equal to 0.99 and noting that $p = q$, we get

$$0.99 = \Phi(0.01\sqrt{n}) - \Phi(-0.01\sqrt{n}) = 2\Phi(0.01\sqrt{n})$$

by the normal approximation.   The table gives

$$0.01\sqrt{n} = 2.58$$

so that $n = 67,000$ approximately.   The fact that a problem such as this will always yield a finite value for n is the essential content of the law of large numbers.   Applying the law of large numbers in another fashion, we can interpret the result more or less as follows: If the whole coin-tossing experiment is repeated a great many times, in about 99 per cent of these experiments the inequality (10-14) will be verified.

PROBLEMS

1. In the example of the text, how many times must the coin be tossed to make the probability 0.95 that the relative frequency is within 5 per cent of the probability?

2. On an average a certain student is able to solve 60 per cent of the problems assigned to him. If an examination contains eight problems and a minimum of five problems is required for passing, what is the student's chance of passing?   *Hint:* Because of the law of large numbers, you may take the statement about the student's average performance to mean: "His probability of solving any given problem is 0.6."

3. If Paul hits a target 80 times out of 100, on an average, and John hits it 90 times out of 100, what is the probability that at least one of them hits the target when they shoot simultaneously?

4. If on an average in a shipment of 10 cases of certain goods 1 case is damaged, what is the probability that out of 5 cases expected at least 4 will not be damaged?

5. (*Review.*)   In 600 throws of a die, what is the probability that 6 will occur more than 90 times and less than 110?

11. The Poisson Law.   In the problem of repeated trials it may happen that p is too small to permit the use of the normal approximation even though n is large.   A different approximation, which is called the *Poisson law* or the *law of small numbers*, is now obtained for this case.   We shall require the approximate formula

$$(1 - ch)^{1/h} \doteq e^{-c} \tag{11-1}$$

which holds for any constant c as $h \to 0$.   Indeed, the logarithm of the expression on the left is

$$\frac{\log (1 - ch)}{h}.$$

This tends to $-c$ as $h \to 0$ by the series for the logarithm, or by l'Hospital's rule.   Since the logarithm tends to $-c$, the expression tends to e^{-c}.

To obtain the Poisson approximation, we replace $n!$ and $(n - r)!$ by their Stirling approximations in the formula $B(r)$ for the probability of r successes in n trials [formula (9-1)].   By a calculation similar to that used for (9-3) the result is

$$B(r) \doteq \frac{n^r e^{-r}}{r![1 - (r/n)]^{n-r+\frac{1}{2}}} p^r (1 - p)^r.$$

Since the expected value of r is np, we assume that r is small compared with n.   The approximation (11-1) with $h = 1/n$ and $c = r$ then gives

$$\left(1 - \frac{r}{n}\right)^{n-r+\frac{1}{2}} \doteq \left(1 - \frac{r}{n}\right)^n \doteq e^{-r}.$$

Similarly, since p is small,

$$(1 - p)^{n-r} \doteq (1 - p)^n = [(1 - p)^{1/p}]^{np} \doteq e^{-np}.$$

Substitution into the expression for $B(r)$ yields the *law of small numbers*,

$$B(r) \doteq \frac{(np)^r}{r!} e^{-np} \qquad n \text{ large, } np \text{ moderate.} \tag{11-2}$$

The result can be written

$$B(r) \doteq \frac{\mu^r}{r!} e^{-\mu} \tag{11-3}$$

where $\mu = np$ is the expected number of successes.

As an illustration, let us compute the probability that the ace of spades will be drawn from a deck of cards at least once in 104 trials.   This problem can be solved with the aid of the exact law (7-4) as follows: The probability that the ace will not be drawn in the 104 trials is

$$B(0) = {}_{104}C_0 (\tfrac{1}{52})^0 (\tfrac{51}{52})^{104} = 0.133$$

and the probability that the ace will be drawn at least once is $1 - 0.133 = 0.867$.   On the other hand, Poisson's law (11-2) gives for the probability of failure to draw the ace

$$B(0) = \frac{(104 \cdot \tfrac{1}{52})^0}{0!} e^{-104/52} = e^{-2}.$$

Hence the probability of drawing at least one ace of spades is $1 - e^{-2} = 0.865$.

The Poisson law has a significance far beyond its connection with the binomial distribution.   Suppose points x_i are distributed at random on the x axis in such a fashion that the following assumptions are valid:

1. The probability that a given number of points is in a given interval depends only on the length of that interval (and not on any information one may have about the points in adjacent intervals).

2. If $P(\Delta x)$ is the probability of 2 or more points in an interval of length Δx, then $P(\Delta x)/\Delta x \to 0$ as $\Delta x \to 0$.

3. If $P_1(\Delta x)$ is the probability of 1 point in an interval of length Δx, then $P_1(\Delta x)/\Delta x \to k$, a constant, as $\Delta x \to 0$.

In this case the probability $P_n(x)$ of n points in an interval of length x satisfies the Poisson law

$$P_n(x) = e^{-kx}\frac{(kx)^n}{n!}. \tag{11-4}$$

To prove this result, consider an interval $(0, x + \Delta x)$ of length $x + \Delta x$. We can have n points in this interval in three mutually exclusive ways. Either there are n points in x and none in Δx, or there are $n - 1$ in x and 1 in Δx, or there are fewer than $n - 1$ in x and at least 2 in Δx. The probability of this last alternative may be written $\epsilon \Delta x$, where $\epsilon \to 0$ with Δx, in view of assumption 2.

Thus, by total and compound probability,

$$P_n(x + \Delta x) = P_n(x)P_0(\Delta x) + P_{n-1}(x)P_1(\Delta x) + \epsilon \Delta x.$$

Subtracting $P_n(x)$ from both sides and dividing by Δx give

$$\frac{P_n(x + \Delta x) - P_n(x)}{\Delta x} = P_n(x)\frac{P_0(\Delta x) - 1}{\Delta x} + P_{n-1}(x)\frac{P_1(\Delta x)}{\Delta x} + \epsilon. \tag{11-5}$$

Since there must be no point, 1 point, or more than 1 point in an interval of length Δx, we have

$$P_0(\Delta x) + P_1(\Delta x) + P(\Delta x) = 1$$

which gives

$$\frac{P_0(\Delta x) - 1}{\Delta x} = -\frac{P_1(\Delta x)}{\Delta x} - \frac{P(\Delta x)}{\Delta x}. \tag{11-6}$$

Taking the limit $\Delta x \to 0$, we obtain $-k$ in (11-6), and hence taking the limit in (11-5) gives

$$\frac{d}{dx}P_n(x) = -kP_n(x) + kP_{n-1}(x) \qquad n \geq 1. \tag{11-7}$$

For $n = 0$ the term $P_{n-1}(x)$ is to be replaced by zero, so that

$$\frac{d}{dx}P_0(x) = -kP_0(x).$$

This separable differential equation yields

$$P_0(x) = ce^{-kx} = e^{-kx}$$

where the constant $c = 1$ since $P_0(0) = 1$; that is, an interval of zero length is sure to contain no points. (This follows from assumption 2.)

Substituting $P_0(x)$ in the relation (11-7) for $n = 1$ we get

$$\frac{d}{dx}P_1(x) = -kP_1(x) + ke^{-kx}$$

which yields $P_1(x) = e^{-kx}(kx)$. Proceeding step by step or using mathematical induction, we obtain (11-4).

The following are some of the phenomena which satisfy the assumptions 1 to 3 quite accurately and which, accordingly, obey a Poisson law: the distribution of blood cells on a microscope slide, the distribution of automobiles on a highway, the distribution of starting times for telephone calls, the clicks of a Geiger counter, the arrival times for customers at a theater ticket office. The first two examples are spatial distributions, while the last three refer to distributions in time.

It should be noted that none of these examples is quite the same as the problem of repeated trials with which we began our discussion. Nevertheless, probability agrees with relative frequency, in the sense that, if the theoretical distribution is a Poisson distribution, the observed distribution is likely to be close to Poisson when the number of relevant observations is large. This fact is the basis of many statistical techniques in particle physics, medicine, and biology.

Similar remarks apply to the normal law. Like the Poisson law, it has a scope vastly greater than the problem of repeated trials would lead one to expect, and, like the Poisson law, it is experimentally observed in those situations in which it is theoretically applicable.

Example 1. Show that the constant k in the Poisson law (11-4) represents the expected number of points in a unit interval. Since the probability of n points in a unit interval is

$$P_n(1) = e^{-k}\frac{k^n}{n!}$$

the expected number is

$$E(n) = \sum_{n=1}^{\infty} e^{-k}\frac{k^n}{n!}n = e^{-k}k\sum\frac{k^{n-1}}{(n-1)!} = e^{-k}ke^k = k.$$

Example 2. Suppose it is known that, in a large city, an average of two persons die daily of tuberculosis. What is the probability that x persons will die on any given day, if the deaths follow a Poisson distribution?

In this case the expected number of deaths is $\mu = 2$, so that

$$P(x) = \frac{2^x}{x!}e^{-2}.$$

The formula gives the following table:

x	0	1	2	3	4	5
$P(x)$	0.135	0.271	0.271	0.180	0.090	0.036

If conditions of population density and immunity remain constant, the actual number of deaths is likely to follow this pattern, in the long run. It is possible to use the observed data to test the hypothesis that the distribution is Poisson and thus to test whether the conditions are constant. The theory of such tests is in the province of mathematical statistics.

PROBLEMS

1. By use of the Poisson law, compute the probability of (a) just one ace in 6 tosses of a die, (b) just one double ace in 36 tosses of a pair of dice. Compare the binomial law for cases (a) and (b). Which of the two cases satisfies the assumptions of the text more exactly?

2. The probability is 0.0025 that a nail chosen at random from the output of a certain machine will be defective. What is the probability that a keg of 1,000 nails made by the machine will have at most 3 defective nails? *Hint:* The keg has "at most 3" if it has 0, 1, 2, or 3 exactly. Use the Poisson approximation.

3. In Prob. 2 it is desired to have a probability of at least 0.95 that the keg has at least 1,000 good nails. How many nails should the manufacturer put into the keg? *Hint:* If he puts in $n = 1,000 + m$ nails, he wants a probability 0.95 that the number of defective nails will be at most m. Use the Poisson law, taking $np \doteq 1,000p = 2.5$.

4. On a certain one-way highway it is proposed to install a traffic signal which has a 60-sec red interval but a long green interval. The speed of the cars may be taken as 30 mph, and the expected number is 10 cars per mile of highway. Neglecting any effects of slowing down, find the probability that just n cars will be obliged to stop when the light is red. What is the probability that at most 5 cars must stop? What is the expected number that must stop? *Hint:* Assume that the cars are distributed according to the law (11-4), and see Example 1.

5. A certain circuit can transmit three telephone calls simultaneously. The expected number of incoming calls is one per minute, and each call lasts 3 min. What is the probability of getting a busy signal? *Hint:* You will find the line busy if three calls or more have come in during the preceding 3-min interval. Use (11-4).

6. (*Review.*) How many times must one throw a die to have the probability exceed 0.99 that the ratio of the number of times 6 appears to the total number of throws deviates from $\frac{1}{6}$ by less than 0.02? What does this problem illustrate?

CONCEPTS USED IN STATISTICS

12. Expectation of a Sum. Joint Distributions. Let x_i be the r distinct values of a random variable x, and let y_j be the s distinct values of another random variable y. The *sum* $x + y$ is a random variable which is defined to be $x_i + y_j$ when $x = x_i$ and $y = y_j$. Thus, $x + y$ is defined on a sample space whose points consist of the rs events

$$x = x_i \quad \text{and} \quad y = y_j \tag{12-1}$$

for $i = 1, 2, \ldots, r$ and $j = 1, 2, \ldots, s$. One of the most important theorems in probability theory concerns sums of variables and reads as follows:

THEOREM. The expectation of the sum of two random variables is equal to the sum of the expectations, or, in symbols,

$$E(x + y) = E(x) + E(y). \tag{12-2}$$

To prove Eq. (12-2) let p_{ij} be the probability that simultaneously $x = x_i$ and $y = y_j$. Thus, p_{ij} is the probability of the event (12-1). The definition of expectation yields

$$E(x + y) = \Sigma \Sigma p_{ij}(x_i + y_j) \tag{12-3}$$

since $x_i + y_j$ is the value of $x + y$ which corresponds to the event (12-1). By rearrangement,

$$E(x + y) = \sum_i x_i \left(\sum_j p_{ij} \right) + \sum_j y_j \left(\sum_i p_{ij} \right). \tag{12-4}$$

Now, $\displaystyle\sum_j p_{ij}$ represents the probability of

$$(x = x_i, y = y_1), \quad (x = x_i, y = y_2), \quad \ldots, \quad \text{or} \quad (x = x_i, y = y_s).$$

Hence, it represents[1] the probability P_i that $x = x_i$. The theorem of Sec. 4 now gives

$$\sum_i x_i \left(\sum_j p_{ij} \right) = \sum_i x_i P_i = E(x)$$

and, similarly,

$$\sum_j y_j \left(\sum_i p_{ij} \right) = E(y).$$

This completes the proof. By repetition, a similar statement is obtained for any number of random variables; e.g., applying the above theorem to the two variables $(x + y)$ and z gives

$$E(x + y + z) = E[(x + y) + z] = E(x + y) + E(z) = E(x) + E(y) + E(z).$$

[1] This shows that $\Sigma \Sigma p_{ij} = \Sigma P_i = 1$ and therefore that the events (12-1) actually form a sample space.

As an illustration we shall find the expected number of heads when n coins are tossed. Let $X_i = 1$ if the ith coin shows heads and $X_i = 0$ otherwise. Then, for each i,

$$E(X_i) = \tfrac{1}{2} \cdot 1 + \tfrac{1}{2} \cdot 0 = \tfrac{1}{2}.$$

(The reader is cautioned that $X_1, X_2, \ldots$ are distinct variables here, not the different values x_i of a single variable.)   The number of heads m is

$$m = X_1 + X_2 + \cdots + X_n$$

and hence

$$\begin{aligned}
E(m) &= E(X_1 + X_2 + \cdots + X_n) \\
&= E(X_1) + E(X_2) + \cdots + E(X_n) \\
&= \frac{1}{2} + \frac{1}{2} + \cdots + \frac{1}{2} = \frac{n}{2}.
\end{aligned}$$

The method of this illustration can be used to find the expected number of successes in n trials when the probability p_i varies from one trial to the next.   If $X_i = 1$ when there is success in the ith trial and $X_i = 0$ otherwise, then

$$E(X_i) = p_i \cdot 1 + (1 - p_i) \cdot 0 = p_i \qquad i = 1, 2, \ldots, n.$$

The total number of successes is $m = X_1 + X_2 + \cdots + X_n$, and hence

$$E(m) = p_1 + p_2 + \cdots + p_n.$$

When all the probabilities $p_i = p$, a constant, the result is $E(m) = np$, as stated in Sec. 7.

The proof of the theorem in this section depends on the theorem of Sec. 4, which gives the expectation in terms of the *numerically distinct* values of the variable.   We now present an alternative approach based directly on the concept of sample space.

Let x be defined on a sample space containing n events, a_i, and let y be defined on another space containing the m events, b_j.   It is supposed that $x = x_i$ if the event a_i happens, and $y = y_j$ if b_j happens.   The variable $x + y$ is defined on a sample space whose mn events e_{ij} happen when, and only when, a_i and b_j both happen.   The value of $x + y$ corresponding to the event e_{ij} is defined to be $x_i + y_j$.   If p_{ij} is the probability of e_{ij}, the definition of expectation gives (12-3), which may be written in the form (12-4) as before.   Since the events of the sample space $\{b_i\}$ are mutually exclusive the sum $\sum\limits_{j} p_{ij}$ represents the probability of

$$a_i \text{ and } b_1 \qquad \text{or} \qquad a_i \text{ and } b_2 \qquad \ldots \qquad \text{or} \qquad a_i \text{ and } b_m.$$

Hence it represents p_i, the probability of a_i.   The first term in (12-4) is therefore $E(x)$ and, similarly, the second term is $E(y)$.

The sums

$$\sum_{j} p_{ij} \qquad \text{and} \qquad \sum_{i} p_{ij} \tag{12-5}$$

are called the *marginal probabilities* of a_i and b_j, respectively.   In statistical theory it is customary to start with the larger sample space $\{e_{ij}\}$ and to define the probabilities on the smaller spaces $\{a_i\}$ and $\{b_j\}$ by means of (12-5).   The above theorem is then valid automatically.

So far, we have confined attention to discrete variables.   The results can be extended to continuous variables if the concept of distribution function is appropriately generalized.

A function $f(x,y)$ is the probability density (or the *joint probability density*) for (x,y) if the probability that (x,y) is in any given region R of the xy plane is

$$\Pr\left[(x,y) \text{ in } R\right] = \iint_R f(x,y)\, dx\, dy.$$

The *distribution function*

$$F(s,t) = \int_{-\infty}^{s} \int_{-\infty}^{t} f(x,y) \, dx \, dy \tag{12-6}$$

gives the probability that $x < s$ and $y < t$. Since probabilities are nonnegative, $f(x,y) \geq 0$, and since the variables must have some finite value,

$$\int_{-\infty}^{\infty} \int_{-\infty}^{\infty} f(x,y) \, dx \, dy = 1. \tag{12-7}$$

Any nonnegative function $f(x,y)$ that satisfies (12-7) can be regarded as a joint probability density; conversely, a function is a joint probability density only if it is nonnegative and satisfies (12-7).

For example, if $f(x,y) = 1/A$ in a region R of area A and $f(x,y) = 0$ elsewhere, it is easily verified that (12-7) holds. The probability that (x,y) is in a subregion R_1 contained in R is

$$\iint_{R_1} f(x,y) \, dx \, dy = \iint_{R_1} \frac{1}{A} \, dx \, dy = \frac{A_1}{A}$$

where A_1 is the area of R_1. The variable (x,y) is then said to be *uniformly distributed in R*.

The functions

$$f_1(x) = \int_{-\infty}^{\infty} f(x,y) \, dy \qquad f_2(y) = \int_{-\infty}^{\infty} f(x,y) \, dx \tag{12-8}$$

represent, respectively, the density associated with x when nothing is said about y, and the density associated with y when nothing is said about x. They are called the *marginal densities*, and the corresponding distributions are the *marginal distributions*. In terms of the marginal densities, the definition of Sec. 6 gives

$$E(x) = \int_{-\infty}^{\infty} xf_1(x) \, dx \qquad E(y) = \int_{-\infty}^{\infty} yf_2(y) \, dy. \tag{12-9}$$

We shall use these relations to obtain the above theorem for continuous variables, assuming absolute convergence of the infinite integrals.

By definition, the expectation of any function $g(x,y)$ is

$$E(g) = \int_{-\infty}^{\infty} \int_{-\infty}^{\infty} g(x,y)f(x,y) \, dx \, dy. \tag{12-10}$$

In particular,

$$E(x+y) = \int_{-\infty}^{\infty} \int_{-\infty}^{\infty} (x+y)f(x,y) \, dx \, dy$$

$$= \int_{-\infty}^{\infty} x \left[\int_{-\infty}^{\infty} f(x,y) \, dy \right] dx + \int_{-\infty}^{\infty} y \left[\int_{-\infty}^{\infty} f(x,y) \, dx \right] dy.$$

Comparison with (12-8) and (12-9) gives $E(x) + E(y)$, as desired.

Example 1. Suppose x is a discrete random variable with the density function $f(x)$. If n observations of the value of x are made, how often should we expect $x = x_i$?

Let $X_k = 1$ if $x = x_i$ at the kth observation and $X_k = 0$ otherwise. The number of times $x = x_i$ is

$$m = X_1 + X_2 + \cdots + X_n.$$

Since the definition of expectation gives

$$E(X_k) = 1 \cdot f(x_i) + 0[1 - f(x_i)] = f(x_i)$$

we have

$$E(m) = nf(x_i).$$

Thus, the frequency function $f(x_i)$ is proportional to the expected frequency of the event $x = x_i$ in a fixed number of observations. This agrees with the interpretation given in Sec. 5.

Example 2.   A deck of cards is thoroughly shuffled. There is a *coincidence* if a card has the same position after shuffling as it had before (e.g., if it is the fourth from the top both times). Find the expected number of coincidences.

Let $X_i = 1$ if the *i*th card is in the same position before and after shuffling, and let $X_i = 0$ otherwise. Then

$$E(X_i) = \tfrac{1}{52}\cdot 1 + \tfrac{51}{52}\cdot 0 = \tfrac{1}{52}.$$

Since the number of coincidences is ΣX_i, its expectation is

$$E(X_1) + E(X_2) + \cdots + E(X_{52}) = 1.$$

Example 3.   A stick of length a is broken at random, and the longer piece is again broken. What is the probability that the three segments can form a triangle?

Let l be the length of the longer piece. If this piece is broken at a point x, the three segments are $a - l$, x, $l - x$. The condition for a triangle is that the sum of any two segments shall exceed the third:

$$a - l + x > l - x \qquad a - x > x \qquad l > a - l.$$

Since $l > a/2$ automatically, these conditions reduce to

$$l - \frac{a}{2} < x < \frac{a}{2}. \tag{12-11}$$

It is a conceptual aid to use the theorems of total and compound probability in the following manner: The probability that l is on the interval $(l, l + dl)$ is

$$\frac{2}{a}\, dl$$

as in Sec. 6. Example 2.   After l is chosen, the probability that x satisfies (12-11) is

$$\int_{l-a/2}^{a/2} \frac{1}{l}\, dx = \frac{a - l}{l}$$

since x is uniformly distributed on $(0,l)$. The probability of both these events is the product

$$\frac{a - l}{l}\frac{2}{a}\, dl$$

by compound probability, and total probability now gives the final answer:

$$\int_{a/2}^{a} \frac{a - l}{l}\frac{2}{a}\, dl = 2 \log 2 - 1.$$

Example 4.   *Buffon's Needle Problem.* A needle of length a is dropped on a board which is covered with parallel lines spaced a distance $b > a$ (Fig. 15). What is the probability that the needle intersects one of the lines?

We assume that the variables x and θ of the figure are uniformly distributed, x being the distance from the center to the nearest line. There is intersection if, and only if, $|(a/2) \cos \theta| > x$. For fixed θ, the probability of this is

$$\frac{|(a/2) \cos \theta|}{b/2} = \frac{a|\cos \theta|}{b}$$

since x is uniformly distributed on $(0,b/2)$. Using total and compound probability as in Example 3, we obtain the final answer:

$$\int_{0}^{2\pi} \frac{a|\cos \theta|}{b} \frac{d\theta}{2\pi} = \frac{a}{b}\frac{1}{2\pi} 4 \int_{0}^{\pi/2} \cos \theta\, d\theta = \frac{2a}{\pi b}.$$

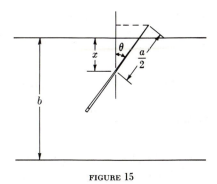

FIGURE 15

Examples 3 and 4 show how problems involving bivariate distributions can be reduced to a succession of problems involving a single variable. This technique is widely used in mathematical statistics.

PROBLEMS

1. A bent coin has probability p of giving heads and probability $q = 1 - p$ of giving tails. Make a table of the eight events HHH, HHT, . . . ; their respective probabilities; and the values of x, y, xy, and $x + y$, where x is the number of heads and y of tails in three tosses. Thus compute $E(x + y)$ and $E(xy)$ directly from the definition of expectation, and verify the theorem in this section. Is it true that $E(xy) = E(x)E(y)$?

2. Construct and solve a version of Prob. 1 based on the four distinct values of x and their respective probabilities.

3. Peter turns up the cards one at a time from a 52-card deck, and Paul tries to guess what the cards are. Find the expected number of correct guesses (*a*) when Paul calls out at random, perhaps repeating himself, (*b*) when Paul calls off the 52 cards, naming each one just once, (*c*) when Paul calls out "ace of spades" each time. (Assume that Paul has no actual insight into the behavior of the cards.)

4. In Prob. 3, suppose Peter tells Paul what the card was immediately after Paul guesses. Paul has the good sense not to call any of those cards, since he knows they have been set aside. What is the expected number of correct guesses now? *Hint:* Let $X_i = 1$ if the ith guess is correct, $X_i = 0$ otherwise. $E(X_i) = ?$ The expected number of correct guesses will be found to be approximately $\log_e 52$.

5. The probability density for bullets hitting a target is given by

$$f(x,y) = \frac{1}{2\pi\sigma_x\sigma_y} e^{-\frac{1}{2}\left[\left(\frac{x-m_x}{\sigma_x}\right)^2 + \left(\frac{y-m_y}{\sigma_y}\right)^2\right]}$$

where σ_x, σ_y, m_x, m_y are constant. Sketch the curves of constant density in the xy plane. What kind of curves are they?

6. Two points are chosen at random on a line of length a. What is the probability that the three segments can form a triangle?

7. Cards numbered consecutively from 1 to n are put in a box and then drawn out at random one at a time, without replacing. The number on the card is plotted versus the trial number at which it was drawn, and the points so obtained are joined by straight lines to give a polygonal curve. What is the expected number of maxima in that curve? *Hint:* Let $X_i = 1$ if there is a maximum at the point corresponding to the ith drawing, and $X_i = 0$ otherwise. For $2 \leq i \leq n - 1$, there are six possible arrangements of the three relevant points, and two arrangements have a maximum in the center. Thus, $E(X_i) = \frac{1}{3}$. However, if $i = 1$ or $i = n$, so that the point is an end point, then $E(X_i) = \frac{1}{2}$. (This problem is due to the great French mathematician Henri Poincaré. It shows in a striking way how the theorem in this section can be applied to dependent variables.)

8. (*Review.*) Let the events e_i of a sample space be numbered so that $e_1, \ldots, e_j$ are favorable to an event A alone, $e_{j+1}, \ldots, e_k$ are favorable to both A and another event B, while $e_{k+1}, \ldots, e_m$ are favorable to B alone. Show that the theorem of total probability is equivalent to the identity

$$p_1 + \cdots + p_m = (p_1 + \cdots + p_j + p_{j+1} + \cdots + p_k)$$

$$+ (p_{j+1} + \cdots + p_k + p_{k+1} + \cdots + p_m) - (p_{j+1} + \cdots + p_k).$$

9. (*Review.*)   In the notation of the preceding problem, deduce the theorem of compound probability from the identity

$$p_{j+1} + \cdots + p_k = (p_1 + \cdots + p_k)\left(\frac{p_{j+1}}{p_1 + \cdots + p_k} + \cdots + \frac{p_k}{p_1 + \cdots + p_k}\right).$$

Hint: Recall that the sample points favorable to B have the same relative weights after A happened as before.

---

13. Variance, Covariance, and Correlation.   Two random variables x and y are said to be *independent* if the event $x = x_i$ and the event $y = y_j$ are independent events for each choice of x_i in the range of x and each y_j in the range of y. In other words, knowledge that y has a particular value must not influence the probabilities associated with x. The numbers shown on two successive tosses of a die are independent in this sense. On the other hand, the number of heads in the first three tosses and in the first four tosses of a coin are dependent variables.

The *product* xy of two random variables is a random variable which equals $x_i y_j$ when $x = x_i$ and $y = y_j$. Although it is not usually true that the expectation of a product is the product of the expectations, this is the case when the variables are independent. In symbols,

$$E(xy) = E(x)E(y) \qquad x, y \text{ independent.} \tag{13-1}$$

The proof is simple. If p_i is the probability that $x = x_i$, and if q_j is the probability that $y = y_j$, then the assumed independence gives $p_i q_j$ for the probability that simultaneously $x = x_i$ and $y = y_j$. Hence

$$E(xy) = \Sigma\Sigma p_i q_j x_i y_j = (\Sigma p_i x_i)(\Sigma q_j y_j) = E(x)E(y).$$

When a discussion involves several variables $x, y, \ldots$, it is convenient to denote expectations by the letter μ, with a subscript to indicate the variable. Thus, we write

$$E(x) = \mu_x \qquad E(y) = \mu_y$$

and so on. For example, (13-1) in this notation takes the form

$$\mu_{xy} = \mu_x \mu_y \qquad x, y \text{ independent.} \tag{13-2}$$

To measure the deviation of a variable from its expected value μ, one introduces a quantity σ defined by[1]

$$\sigma = \sqrt{E(x - \mu)^2} \qquad \text{or} \qquad \sigma^2 = E(x - \mu)^2. \tag{13-3}$$

The expression σ is called the *standard deviation*, and its square σ^2 is called the *variance*. As for μ, here, too, it is customary to use a subscript when several variables have to be distinguished. For example,

$$\sigma_x^2 = E(x - \mu_x)^2 \qquad \sigma_y^2 = E(y - \mu_y)^2.$$

To illustrate the calculation of a variance by means of the definition, let x denote the number of heads obtained when three coins are tossed. Since $\mu = E(x) = \tfrac{3}{2}$ we have the following table:

$x =$	0	1	2	3
$x - \mu =$	$-\tfrac{3}{2}$	$-\tfrac{1}{2}$	$\tfrac{1}{2}$	$\tfrac{3}{2}$
$(x - \mu)^2 =$	$\tfrac{9}{4}$	$\tfrac{1}{4}$	$\tfrac{1}{4}$	$\tfrac{9}{4}$
Probability $p_i =$	$\tfrac{1}{8}$	$\tfrac{3}{8}$	$\tfrac{3}{8}$	$\tfrac{1}{8}$

---

[1] The intent is $E[(x - \mu)^2]$, not $[E(x - \mu)]^2$.

The definition of expectation now gives

$$\sigma^2 = E(x - \mu)^2 = \tfrac{1}{8}\cdot\tfrac{9}{4} + \tfrac{3}{8}\cdot\tfrac{1}{4} + \tfrac{3}{8}\cdot\tfrac{1}{4} + \tfrac{1}{8}\cdot\tfrac{9}{4} = \tfrac{3}{4}.$$

If $E(x) = \mu_x$ and $E(y) = \mu_y$, the quantity

$$\sigma_{xy}^2 = E(x - \mu_x)(y - \mu_y) \tag{13-4}$$

is called the *covariance* of x and y. The covariance is a generalization of the variance, in that the special case $y = x$ gives

$$\sigma_{xx}^2 = E(x - \mu_x)(x - \mu_x) = E(x - \mu_x)^2 = \sigma_x^2.$$

As an illustration, let us compute σ_{xy}^2 when x is the number of heads obtained on the first two tosses and y the number obtained altogether in three tosses of an unbiased coin. Here $\mu_x = 1$, $\mu_y = \tfrac{3}{2}$, so that we have the following table:

Event	HHH	HHT	HTH	HTT	THH	THT	TTH	TTT
$x - \mu_x$	1	1	0	0	0	0	-1	-1
$y - \mu_y$	$\tfrac{3}{2}$	$\tfrac{1}{2}$	$\tfrac{1}{2}$	$-\tfrac{1}{2}$	$\tfrac{1}{2}$	$-\tfrac{1}{2}$	$-\tfrac{1}{2}$	$-\tfrac{3}{2}$
Product	$\tfrac{3}{2}$	$\tfrac{1}{2}$	0	0	0	0	$\tfrac{1}{2}$	$\tfrac{3}{2}$

Since the associated probabilities are $\tfrac{1}{8}$, we take $\tfrac{1}{8}$ times the sum of the entries in the last row to get

$$\sigma_{xy}^2 = \tfrac{1}{2}. \tag{13-5}$$

We shall now obtain an expression for σ_{xy} which is often more useful than (13-4). Expanding the product in (13-4) gives

$$\sigma_{xy}^2 = E(xy - y\mu_x - x\mu_y + \mu_x\mu_y)$$
$$= E(xy) - E(y)\mu_x - E(x)\mu_y + \mu_x\mu_y.$$

Upon recalling that $E(x) = \mu_x$ and $E(y) = \mu_y$ we get

$$\sigma_{xy}^2 = E(xy) - E(x)E(y) = \mu_{xy} - \mu_x\mu_y \tag{13-6}$$

which is the required formula.

To apply this formula to the preceding example, we construct the following table:

Event	HHH	HHT	HTH	HTT	THH	THT	TTH	TTT
x	2	2	1	1	1	1	0	0
y	3	2	2	1	2	1	1	0
xy	6	4	2	1	2	1	0	0

Taking $\tfrac{1}{8}$ times the sum of the last entries gives $E(xy) = 2$, and hence by (13-6)

$$\sigma_{xy}^2 = 2 - (1)(\tfrac{3}{2}) = \tfrac{1}{2}.$$

The special case $x = y$ in (13-6) gives an alternative form[1] of (13-3), namely,

$$\sigma^2 = E(x^2) - \mu^2 = E(x^2) - [E(x)]^2. \tag{13-7}$$

As an illustration the reader may apply this formula to the preceding example to obtain

$$\sigma_x^2 = \tfrac{3}{2} - (1)^2 = \tfrac{1}{2} \qquad \sigma_y^2 = 3 - (\tfrac{3}{2})^2 = \tfrac{3}{4}. \tag{13-8}$$

If the variables x and y are independent, (13-2) and (13-6) give $\sigma_{xy} = 0$. Hence

[1] Note that σ^2 gives the moment of inertia of the area under the distribution curve $y = f(x)$ about the line $x = \mu$ which passes through the center of mass. From this viewpoint (13-7) is the familiar formula for moment of inertia after a change of rotational axes.

when $\sigma_{xy} \neq 0$, the variables must be related. A quantitative measure of the strength of the relationship is given by the *correlation coefficient* ρ:

$$\rho = \frac{\sigma_{xy}^2}{\sigma_x \sigma_y}. \tag{13-9}$$

For example, in the foregoing illustration, (13-5) and (13-8) yield

$$\rho = \frac{\frac{1}{2}}{\sqrt{\frac{1}{2}} \sqrt{\frac{3}{4}}} = \frac{1}{3} \sqrt{6} = 0.816. \tag{13-10}$$

Thus, if two variables x and y have a correlation coefficient $\rho = 0.8$, they are about as strongly related as are the numbers of heads on the first two tosses and on the first three tosses of an unbiased coin.

The correlation coefficient has the value 1 if $y = x$, and, as we have already observed, $\rho = 0$ when x and y are unrelated. Moreover, ρ does not change if x and y are each multiplied by a constant factor. Thus, if the correlation coefficient indicates a certain strength of relationship for x and y, it will give the same strength of relationship for $2x$ and $3y$. Similarly, ρ is unaffected by addition of a constant; for instance, $x - 2$ and $y - 3$ have the same ρ as x and y.

In spite of having these desirable properties, ρ is not always a reliable measure of dependence, and many statistical studies have led to erroneous conclusions through an incorrect interpretation of correlation. It is quite possible to have the variables so strongly related that y is a function of x and yet $\rho = 0$. Before a correlation coefficient can be used with confidence, one must know something about the underlying probability distribution.

The variables x and y are said to have a *bivariate normal distribution* when

$$f(x,y) = e^{(ax^2 + 2bxy + cy^2 + dx + ey + f)} \qquad a, b, \ldots \text{const.}$$

In this important case the theory of correlation has been fully developed, and it is found[1] that ρ actually does measure the strength of the relationship between x and y.

Example.   A variable x is said to be "normally distributed with mean μ and variance σ^2" when its density function is

$$f(x) = \frac{1}{\sqrt{2\pi}\,\sigma} e^{-\frac{1}{2}\left(\frac{x-\mu}{\sigma}\right)^2} \qquad \mu, \sigma \text{ const.}$$

Show that the mean is indeed μ and the variance σ^2.

By the definition of expectation,

$$E(x - \mu) = \frac{1}{\sqrt{2\pi}\,\sigma} \int_{-\infty}^{\infty} (x-\mu) e^{-\frac{1}{2}\left(\frac{x-\mu}{\sigma}\right)^2} dx = \frac{\sigma}{\sqrt{2\pi}} \int_{-\infty}^{\infty} t e^{-\frac{1}{2}t^2} dt = 0$$

when we set $t = (x - \mu)/\sigma$. Hence $E(x - \mu) = 0$, which gives $E(x) = \mu$. The same change of variable leads to

$$E(x - \mu)^2 = \frac{1}{\sqrt{2\pi}} \sigma^2 \int_{-\infty}^{\infty} t^2 e^{-\frac{1}{2}t^2} dt = \sigma^2$$

as we see upon integrating by parts and using (10-1). Since $\mu = E(x)$, the latter result $E(x - \mu)^2$ is the variance by definition.

PROBLEMS

1. Compute σ^2 if x is uniformly distributed on the interval $0 \leq x \leq 1$.

2. Let x be the number on top and y the number on the bottom in a toss of a true die. Compute $E(x)$, $E(y)$, $E(xy)$, and the covariance. Does your work indicate that the variables are dependent? Find the correlation coefficient.

[1] See Hoel, *op. cit.*, chap. 8.

3. Three coins are tossed. Let x be the number of heads shown by the first coin, whereas y is the number of heads shown by all the coins. Compute the correlation coefficient. Your result should be smaller than the value (13-10). Why?

4. Two continuous variables x and y are said to be *independent* if their joint density $f(x,y)$ has the form $f(x)g(y)$. Verify the theorem of compound probability for independent events, in the form

$$\Pr (a < x < b, c < y < d) = \Pr (a < x < b) \Pr (c < y < d).$$

Also verify the equation $E(xy) = E(x)E(y)$, assuming convergence of the relevant integrals.

5. Referring to (12-9) and (12-10), express the variance, covariance, and correlation coefficient in integral form, when x and y are continuous variables with joint density $f(x,y)$. By using the result of Prob. 4, show that the covariance is 0 if the variables are independent.

6. (*Chebychev's lemma.*) Let t be a random variable which does not assume negative values, and let $E(t) = \tau$. Prove that the probability of the inequality $t \leq \tau h$ is at least $1 - 1/h$, for every positive constant h. *Outline of solution:* The probability of the contrary event, $t > \tau h$, is

$$\int_{\tau h}^{\infty} f(t)\, dt \leq \int_{\tau h}^{\infty} \frac{t}{\tau h} f(t)\, dt \leq \frac{1}{\tau h} \int_0^{\infty} tf(t)\, dt = \frac{1}{h}.$$

7. (*Chebychev's inequality.*) Let x be a random variable with mean μ and variance σ^2. By Prob. 6 with $t = (x - \mu)^2$, prove that the probability of the inequality $|x - \mu| \leq \sqrt{h}\, \sigma$ is at least $1 - 1/h$, for every positive constant h.

8. Plot the probability of the inequality $|x - \mu| \leq k\sigma$ as a function of k when x is normally distributed with mean μ and variance σ. On the same axes, plot the minimum probability of this inequality as given by Prob. 7. (Specific distributions such as the Gauss, Poisson, or Maxwell-Boltzmann distributions depend on a few parameters. A method of estimation which does not assume a specific type of distribution function is called a *nonparametric method*. Since nonparametric methods assume less about the random process, they give less information.)

---

14. Arithmetic Means. The Theory of Errors. In many applications one does not consider a single value alone, but one obtains a mean of a large number of values. For example, if m denotes the measured value of the length of a rod one would make several measurements $m_1, m_2, \ldots, m_n$ and use the *arithmetic mean*

$$\overline{m} = \frac{m_1 + m_2 + \cdots + m_n}{n}.$$

If the true value of the length is v, the errors in measurement are

$$x_i = m_i - v.$$

By adding and dividing by n we get

$$\overline{x} = \overline{m} - v \tag{14-1}$$

where $\overline{x}$ is the mean of the x_i:

$$\overline{x} = \frac{x_1 + x_2 + \cdots + x_n}{n}. \tag{14-2}$$

According to (14-1), the *mean of the errors is equal to the error of the mean*. It is likely to be smaller than the error in a single measurement, because positive and negative errors tend to cancel in Σx_i.

So far we have taken the view that the x_i's are independent observations of a single random variable x, so that $\overline{x}$ is the *observed mean* or the *sample mean*. However, we can also take the view that the x_i are independent variables[1] and that $\bar{x}$ is another random variable whose value is the mean of the values of x_i. With this view it is possible to get a precise description of the improvement in accuracy as the number of measurements increases. The result is:

THEOREM.   If the variables x_i are independent, if they have the same expectation $E(x_i) = \mu$ and the same variance σ_x^2, then

$$\sigma_{\bar{x}} = \frac{\sigma_x}{\sqrt{n}}.$$

Here σ_x denotes the standard deviation of the variable x, and $\sigma_{\bar{x}}$ denotes the standard deviation of $\bar{x}$.   To prove the theorem, observe that

$$E(x_1 + \cdots + x_n) = E(x_1) + \cdots + E(x_n) = n\mu.$$

The variance of $x_1 + \cdots + x_n$ is therefore

$$E(x_1 + \cdots + x_n - n\mu)^2$$

which can be written

$$E[(x_1 - \mu) + (x_2 - \mu) + \cdots + (x_n - \mu)]^2.$$

Expanding the term in brackets we obtain

$$E\left[\sum_i (x_i - \mu)^2 + \sum_{i \neq j} (x_i - \mu)(x_j - \mu) \right]. \qquad (14\text{-}3)$$

Since the variables are independent, the covariance of x_i and x_j is zero for $i \neq j$; that is,

$$E(x_i - \mu)(x_j - \mu) = 0.$$

Also the definition of σ_x gives

$$\sigma_x^2 = E(x_i - \mu)^2.$$

Hence, taking the expectation in (14-3) yields

$$E(x_1 + \cdots + x_n - n\mu)^2 = n\sigma_x^2.$$

Dividing by n^2 we have

$$E\left(\frac{x_1 + \cdots + x_n}{n} - \mu \right)^2 = \frac{\sigma_x^2}{n}$$

which gives the conclusion upon taking the square root.

The intuitive meaning of this result is approximately as follows: Suppose a single measurement varies over an interval of length l about the true value, so that l measures the scatter or spread. Then the mean of n independent measurements will have a spread of the order of $l/\sqrt{n}$ about the true value. This shows that the improvement of accuracy due to cancellation of positive and negative errors is of the order of $\sqrt{n}$, where n is the number of measurements. The most important consideration justifying the analysis in practice is that systematic errors must be eliminated.

The foregoing conclusions are independent of the density function $f(x)$ that governs the statistical distribution of the errors. However, in a great variety of cases the errors have a Gaussian density

$$f(x) = \frac{h}{\sqrt{\pi}} e^{-h^2 x^2}. \qquad (14\text{-}4)$$

This result, known as the *Gaussian law of error*, states that the variable $\sqrt{2}\, hx$ is normally distributed. Specifically, the probability of

---

[1] The use of lower-case letters to denote variables is customary in statistical literature.   However, the x_i here are analogous to the X_i of Sec. 12, not to the x_i of Secs. 4 and 5.

$$t_1 < \sqrt{2}\,hx < t_2 \tag{14-5}$$

is given by integrating (14-4), and the change of variable $t = \sqrt{2}\,hx$ shows that the probability is

$$\frac{1}{\sqrt{2\pi}} \int_{t_1}^{t_2} e^{-t^2/2}\,dt = \Phi(t_2) - \Phi(t_1). \tag{14-6}$$

The constant h measures the accuracy of the observer and is known as the *precision constant*. The particular error that has probability $\frac{1}{2}$ to be exceeded in magnitude is called the *probable error;* it is found to be

$$\text{Probable error} = \frac{0.4769}{h}$$

by use of (14-5), (14-6), and Appendix C. Another interpretation of h is afforded by the *mean absolute error*

$$E(|x|) = \int_{-\infty}^{\infty} |x| f(x)\,dx = \frac{0.5642}{h}$$

and still a third interpretation is given by the *mean-square error*

$$E(x^2) = \int_{-\infty}^{\infty} x^2 f(x)\,dx = \frac{0.5000}{h^2}.$$

For the Gaussian distribution the theory of arithmetic means can be given a very precise formulation. It is found that $\bar{x}$ *has a Gaussian distribution with precision constant* $h\sqrt{n}$, *whenever the independent measurements* x_i *have Gaussian distributions with precision constant* h. Thus, if the inequality $|x| < \alpha$ has probability p, the inequality $|\bar{x}| < \alpha/\sqrt{n}$ has the same probability p.

The proof is omitted because it involves a tedious evaluation of multiple integrals.[1] However, the essential meaning of the result is that the "scatter" or "spread" for $\bar{x}$ is $1/\sqrt{n}$ times as great as the corresponding spread for x. When interpreted in this fashion the property follows from the above theorem.

In modern statistical theory, the Gaussian law of error is sometimes simply postulated, and sometimes it is deduced from a generalized version of the Laplace-de Moivre law known as the *central-limit theorem*. Roughly speaking, the central-limit theorem asserts that a suitably normalized sum of independent random variables is likely to be normally distributed, as the number of variables grows beyond all bounds. It is not necessary that the variables all have the same distribution function or even that they be wholly independent. When errors of measurement are due to a sum of many nearly independent random effects, one would expect the Gaussian law of error as a consequence of the central-limit theorem.

Gauss' own derivation proceeded along quite different lines. From considerations of a general nature, he deduced that the best value must be the arithmetic mean of the measurements, and he surmised that this "best value" must also equal that particular true value that would maximize the probability of the observed results. In other words, he assumed that the arithmetic mean furnishes a maximum-likelihood estimate. Under mild continuity conditions it follows necessarily that the distribution must be Gaussian. This aspect of the subject is not essential for understanding modern statistics but is developed in the following problems because of its independent interest.

[1] See J. V. Uspensky, "Introduction to Mathematical Probability," chap. 13, McGraw-Hill Book Company, 1937, for a direct verification. An indirect method based on the theory of moments is given in Hoel, *op. cit.*, sec. 6.5. See also M. E. Munroe, "The Theory of Probability," pp. 91–96, McGraw-Hill Book Company, New York, 1951.

Example.   Discuss the variance associated with the binomial distribution.

Here we let $x_i = 1$ if there is success at the ith trial in a set of independent trials with probability p, and $x_i = 0$ otherwise.   For each variable x_i we have $x_i{}^2 = x_i$ and hence

$$E(x_i{}^2) = E(x_i) = p\cdot 1 + q\cdot 0 = p.$$

By (13-7) the corresponding variance is

$$\sigma_x{}^2 = p - p^2 = p(1 - p) = pq$$

and the theorem now gives

$$\sigma_{\bar{x}} = \sqrt{\frac{pq}{n}}.$$

For these variables x_i the mean $\bar{x}$ is the relative frequency m/n, where m is the number of successes. We have, therefore,

$$\left[E\left(\frac{m}{n} - p\right)^2\right]^{1/2} = \sqrt{\frac{pq}{n}} \tag{14-7}$$

which shows that the relative frequency m/n is likely to be close to p when n is large.   The corresponding result for a general variable x is based on the above theorem; it leads to assertions concerning $|E(x) - \bar{x}|$ which are similar to the theorem established in Sec. 10 but of greater scope.

Multiplying (14-7) through by n we get

$$[E(m - np)^2]^{1/2} = \sqrt{npq}.$$

This gives an interpretation for the quantity $\sqrt{npq}$ that arose in connection with the normal law (Sec. 10); namely, $\sqrt{npq}$ *is the standard deviation of the number of successes m.*

PROBLEMS

1. In a certain experiment that satisfies the conditions of the Gauss law, the probable error is 0.01.   A measurement m_1 is about to be made.   What is the probability that the interval $(m_1 - 0.02, m_1 + 0.02)$ will contain the true value v?   *Hint:* First find h, then note that the stated result happens if, and only if, $|x_1| < 0.02$.

2. In Prob. 1, let m be the mean of 100 measurements that are about to be made.   What is the probability that the interval $(m - 0.02, m + 0.02)$ will contain the true value v?   *Hint:* The precision constant of the mean is $h\sqrt{n} = 10h$.

3. In Prob. 1, how many measurements must be made to have a probability at least 0.99 that the mean of those measurements will be within 0.02 of the true value v?

4. (a) Show that the sum of the squares of the errors $\Sigma(m_i - v)^2$ is least if the true value v happens to be the arithmetic mean of the measurements m_i.   (b) Deduce that the arithmetic mean $\overline{m}$ is a maximum-likelihood estimate for v when there are n independent measurements each satisfying (14-4).   *Hint:* It is required to choose v so that

$$f(x_1, x_2, \ldots, x_n) = f(x_1)f(x_2) \ldots f(x_n) = \left(\frac{h}{\sqrt{\pi}}\right)^n e^{-h^2 \Sigma x_i{}^2}$$

is maximum.   Use the result (a).

5. Show that the expression $|x - x_1| + |x - x_2| + \cdots + |x - x_n|$ is least if x is chosen so that just as many x_i satisfy $x_i \geq x$ as satisfy $x_i \leq x$.   (Any such choice of x is called a *median* of the x_i.) *Hint:* The expression is an increasing function of x if more than half the x_i satisfy $x_i < x$.

6. Let m_1 and m_2 be two independent measurements of a quantity m (such as the mass of an electron, for example).   It is desired to find a best estimate $m \doteq \theta(m_1, m_2)$ based on the values m_i. Discuss the justification for assuming the three properties

$$\theta(m_1 + \alpha, m_2 + \alpha) = \theta(m_1, m_2) + \alpha \qquad \theta(\beta m_1, \beta m_2) = \beta\theta(m_1, m_2) \qquad \theta(m_1, m_2) = \theta(m_2, m_1).$$

7. Show that any function θ having the properties described in Prob. 6 must be the arithmetic mean, $\theta(m_1,m_2) = \frac{1}{2}(m_1 + m_2)$. *Hint:* Regarding m_1 and m_2 as fixed, choose $\alpha = -m_2$, and substitute in the equation involving β. The result is

$$\beta\theta(m_1,m_2) = \beta m_2 + \theta(\beta m_1 - \beta m_2, 0).$$

By an appropriate choice of β, deduce that θ is a linear function.

8. For three independent measurements with the same density function $f(x)$ the joint density has the form

$$f(x_1,x_2,x_3) = f(x_1)f(x_2)f(x_3)$$

where $x_i = m_i - v$ are the errors, as in the text. Suppose this expression is maximum, as a function of v, when $v = \frac{1}{3}(m_1 + m_2 + m_3)$, and suppose f is positive and differentiable. Show that the function $F(x) = f'(x)/f(x)$ satisfies

$$F(x_1) + F(x_2) + F(x_3) = 0 \qquad \text{whenever } x_1 + x_2 + x_3 = 0.$$

Hint: At a point where the expression $\log f(m_1 - v) + \log f(m_2 - v) + \log f(m_3 - v)$ is maximum, as a function of v, its derivative with respect to v is 0.

9. By appropriate choice of the x_i in the conclusion of Prob. 8, deduce successively $F(0) = 0$, $F(x_3) = -F(-x_3)$, and

$$F(x_1 + x_2) = F(x_1) + F(x_2).$$

Assuming that F is differentiable, obtain $F'(x_1) = F'(x_2)$ by the chain rule. Hence $F'(x) = c$ is constant, hence $F(x) = cx$, and

$$f(x) = Ke^{\frac{1}{2}cx^2}.$$

Determine the sign of c and the value of the constant K by using the fact that $f(x)$ is a density function, and thus get the Gauss law of error.

10. Let x_i be independent random variables with the same expectation μ and variance σ^2. For any positive constant ϵ show that the probability of the inequality $|\bar{x} - \mu| \leq \epsilon$ is at least

$$1 - \frac{\sigma^2}{\epsilon^2 n}.$$

Hence, the probability becomes arbitrarily near to 1 when n is sufficiently large. *Hint:* Apply the Chebychev inequality (Sec. 13, Prob. 7) to the variable $\bar{x}$, with h chosen so that $\sqrt{h}\sigma_{\bar{x}} = \epsilon$.

---

15. Estimation of the Variance. If $x_1, x_2, \ldots, x_n$ are n independent observations of a variable x, the *sample variance* is defined by

$$s^2 = \frac{1}{n} \sum (x_i - \bar{x})^2 = \overline{(x_i - \bar{x})^2}. \tag{15-1}$$

Unlike the theoretical variance σ^2, the sample variance is computed from the observations and so is actually available. It will be seen, now, that s^2 can be used to estimate σ^2.
We have

$$E(ns^2) = \Sigma E(x_i - \bar{x})^2 = \Sigma E[(x_i - \mu) - (\bar{x} - \mu)]^2$$

$$= \Sigma[E(x_i - \mu)^2 - 2E(x_i - \mu)(\bar{x} - \mu) + E(\bar{x} - \mu)^2]. \tag{15-2}$$

Now, $E(x_i - \mu)^2 = \sigma^2$ by definition, and $E(\bar{x} - \mu)^2 = \sigma^2/n$ by Sec. 14. For the middle term in (15-2) we get

$$E(x_i - \mu)(\bar{x} - \mu) = \frac{1}{n} E(x_i - \mu)(x_1 + \cdots + x_i + \cdots + x_n - \mu n)$$

$$= \frac{1}{n} E(x_i - \mu)(x_i - \mu + \cdots) = \frac{1}{n} E(x_i - \mu)^2 = \frac{1}{n} \sigma^2$$

when we note that the terms not written explicitly are independent of x_i. That is, for $i \neq j$, Eq. (13-1) gives

$$E[(x_i - \mu)x_j] = E(x_i - \mu)E(x_j) = 0 \cdot \mu = 0.$$

Substituting into (15-2) yields the important formula

$$E(ns^2) = (n - 1)\sigma^2. \tag{15-3}$$

If (15-3) is divided by n, we get

$$E[\overline{(x_i - \bar{x})^2}] = \frac{n - 1}{n}\sigma^2 \tag{15-4}$$

upon recalling (15-1). On the other hand, the definition of σ^2 gives

$$E[\overline{(x_i - \mu)^2}] = \sigma^2. \tag{15-5}$$

It is not surprising that (15-4) gives a smaller value than (15-5), inasmuch as the choice $\mu = \bar{x}$ is the value of μ that *minimizes* (15-5) (cf. Prob. 4 in Sec. 14). The fact that (15-4) should be smaller is especially clear when there is only one measurement, x_1. In this case the formula gives zero, because $x_1 = \bar{x}$.

The foregoing remarks indicate that s^2 is not a suitable estimate of σ^2; it has a tendency to be too small. But if we divide (15-3) by $n - 1$ for $n \geq 2$, we get

$$E\left(\frac{n}{n - 1}s^2\right) = \sigma^2$$

which gives the following theorem:

THEOREM. Let $x_1, x_2, \ldots, x_n$ be n independent observations of a variable x, with $n \geq 2$. If s^2 is the sample variance, then the quantity

$$\hat{\sigma}^2 = \frac{n}{n - 1}s^2 \tag{15-6}$$

is an unbiased estimate of σ^2. That is, $E(\hat{\sigma}^2) = \sigma^2$.

To illustrate the use of the theorem, let $m_1 = 12$, $m_2 = 8$, $m_3 = 13$ be three measurements of an unknown quantity whose true value is v. The errors in the measurement are $x_i = m_i - v$, but since

$$x_i - \bar{x} = m_i - v - \bar{m} + v = m_i - \bar{m} \tag{15-7}$$

we can compute s^2 without knowing v. By (15-1) and (15-7),

$$ns^2 = \Sigma(x_i - \bar{x})^2 = \Sigma(m_i - \bar{m})^2.$$

In this example $\bar{m} = 11$, so that

$$ns^2 = (1)^2 + (-3)^2 + (2)^2 = 14.$$

Hence an estimate for σ^2 is

$$\hat{\sigma}^2 = \frac{ns^2}{n - 1} = \frac{14}{2} = 7.$$

According to Sec. 14 the precision constant h is estimated as $h \doteq 1/(\sqrt{2}\,\hat{\sigma}) = 1/\sqrt{14} = 0.27$. In statistics it is shown how one can determine the reliability of an estimate such as this, though we do not pursue the subject here.[1]

[1] See Hoel, *op. cit.*, chap. 11.

1. A certain experiment gave the measurements 17, 21, 20, 18, 14. Obtain an unbiased estimate for the variance of a single measurement, and, from this, estimate the precision constant.

2. If the precision constant in Prob. 1 can be assumed exactly equal to your estimate of it, (*a*) what is the probability that the next measurement will be within 0.5 of the true value? (*b*) How many measurements must you make if you want a probability 0.95 that the mean of those measurements will be within 0.1 of the true value?

3. In a certain measuring routine the cost of equipment and materials is negligible but the time required is proportional to the number of measurements. Give a rational method of adjusting the salaries of two observers whose working speeds are s_1 and s_2 if the precision constants of their measurements are h_1 and h_2. *Hint:* Consider the number of measurements each must make to attain equal reliability in the respective arithmetic means.

4. Discuss Prob. 3 if the cost of equipment is proportional to the length of time it is used and the cost of material is proportional to the number of measurements.

REVIEW PROBLEMS

5. In how many ways may 10 carnations be put into 5 vases? (This problem is much harder than it looks.)

6. Show that, if 23 people are chosen at random, it is more likely than not that at least two of them have the same birthday. *Hint:* Find the probability that no two have the same birthday, and compute its logarithm by using $\log (1 - h) \doteq -h$ for $h \leq {}^{23}\!/_{365}$.

7. If a coin is unsymmetric but just as likely to be biased toward heads as tails, prove that the chance of getting n heads in n throws is greater than for a symmetric coin, when $n \geq 2$.

8. What is the expected number of tosses of a coin needed to get three heads?

9. What is the probable error in $\log x$ if the probable error in x is E and the true value of x is $x_0 > 0$?

10. In an elimination tournament starting with $N = 2^n$ players, the players pair off at random, the winners pair off again, and so on. The players are all of unequal strengths, and the stronger of two contestants invariably wins. Show that the probability that the second prize goes to the second-best player is $\frac{1}{2}N/(N - 1)$.

Solution of equations

**Interpolation.   Empirical formulas.
Least squares**

**Numerical integration of
differential equations**

649

Numerical analysis

THE PRINCIPAL CONCERN of numerical analysis is with the construction of effective methods for the calculation of unknowns entering in the formulation of a given problem. Since every formulation of a practical problem involves assumptions and approximations, it is senseless to seek unknowns to a higher precision than is warranted by the data. A simple and perhaps crude technique giving the desired values within specified limits of tolerance is often preferred to an involved method capable of yielding an arbitrary degree of accuracy.

In recent years the growth of numerical analysis has been accelerated by the demands of science and technology for numerical solutions of many pressing problems. High-speed computing machines produced for coping with such problems are certain to open new vistas in science and leave a profound imprint in all fields of human activity.

It is the object of this chapter to present the rudiments of numerical analysis essential to all concerned with the processing of numerical data. Inasmuch as the understanding of principles must precede the acquisition of computing skills, the emphasis in the following sections is placed on basic ideas and general methods rather than on special techniques useful in solving this or that problem. Among topics included here are the determination of real roots of algebraic and transcendental equations, the basic method for solving systems of linear equations, the elements of interpolation theory, and its bearing on curve fitting and numerical solution of differential equations.

SOLUTION OF EQUATIONS

1. Graphical Methods. Geometric considerations usually are a useful guide in the construction of analytic methods of solution of practical problems. This is particularly true in the problem of determination of numerical values of the roots of algebraic and transcendental equations.[1]

If $F(x)$ is a real continuous function, the equation

$$F(x) = 0 \qquad (1\text{-}1)$$

may have real roots. The approximate values of such roots can be determined by graphing the function $y = F(x)$ and reading from the graph the values of x for which $y = 0$. This familiar procedure for graphical determination of real roots can frequently be simplified by rewriting (1-1) in the form

$$f(x) = g(x). \qquad (1\text{-}2)$$

---

[1] A polynomial equation $x^n + a_1 x^{n-1} + \cdots + a_n = 0$ is called an *algebraic* equation. An equation $F(x) = 0$ which is not reducible to an algebraic equation is called *transcendental*. Thus, $\tan x - x = 0$ is a transcendental equation, and so is $e^x + 2 \cos x = 0$.

The abscissas of points of intersection of the curves $y = f(x)$ and $y = g(x)$ will obviously be the roots of (1-2).

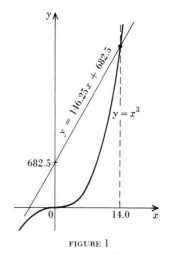

Thus, an approximate value of the real root of

$$F(x) \equiv x^3 - 146.25x - 682.5 = 0$$

can be found by graphing the function

$$y = x^3 - 146.25x - 682.5.$$

It is simpler, however, to plot the cubic

$$y = x^3$$

and the straight line (Fig. 1)

$$y = 146.25x + 682.5$$

and read off from the graph the abscissa of their point of intersection.

FIGURE 1

The general cubic equation

$$z^3 + az^2 + bz + c = 0 \tag{1-3}$$

can always be reduced to the form

$$x^3 + px + q = 0 \tag{1-4}$$

by making the substitution $z = x + h$ in (1-3) and determining h so that the coefficient of x^2 in the resulting equation is zero. Then the approximate values of real roots of Eq. (1-4) can be obtained by plotting the cubic $y = x^3$ and straight line $y = -px - q$ and by reading off the values of x that correspond to the points of intersection. If the roots of (1-4) are x_i, the roots of the original equation (1-3) are $z_i = x_i + h$.

Similarly, every quartic equation

$$z^4 + az^3 + bz^2 + cz + d = 0 \tag{1-5}$$

can be reduced by the substitution $z = x + h$ to the form

$$x^4 + px^2 + qx + r = 0 \tag{1-6}$$

in which the x^3 term is lacking. It is easy to verify that Eq. (1-6) is equivalent to the system of simultaneous equations

$$y = x^2 \qquad (x - x_0)^2 + (y - y_0)^2 = R^2 \tag{1-7}$$

where $$x_0 = \frac{-q}{2} \qquad y_0 = \frac{1 - p}{2} \qquad \text{and} \qquad R^2 = x_0{}^2 + y_0{}^2 - r.$$

Accordingly, the approximate values of the real roots of Eq. (1-6) can be determined graphically by finding the abscissas of the points of intersection of the parabola and circle defined by Eqs. (1-7).

An obvious disadvantage of graphical methods is that they require plotting curves on a large scale when a high degree of accuracy is desired. To avoid this, one obtains more precise values by applying one of the several methods of successive approximation discussed in Secs. 2 and 3. All these methods require that the desired root be first isolated; that is, they call for the determination of an interval that contains just the root in question and no others. If $F(x)$ is a continuous function, and if for a certain pair of real values $x = x_1$, $x = x_2$, the signs of $F(x_1)$ and $F(x_2)$ are opposite, it is obvious that $F(x) = 0$ has at least one real root in the interval (x_1, x_2). When $F(x)$ is monotone in the interval (x_1, x_2), and this will be the case if $F'(x)$ does not change sign in (x_1, x_2), then

$F(x) = 0$ has just one root in (x_1,x_2).   If there are several roots in (x_1,x_2), this interval is usually narrowed down by a succession of judicious trials until an interval that contains just the desired root is obtained.   For efficient application of the successive-approximation methods it is desirable that this interval be as small as possible.

We note in passing that no general methods are available for the exact determination of the roots of transcendental equations.   Also, there are no algebraic formulas for the solution of general algebraic equations of degree higher than 4.   The so-called "Cardan" and "Ferrari" solutions of the cubic and quartic equations require the calculation of cube roots of quantities which themselves are square roots.   Generally it is simpler to obtain the desired approximations by methods described in the following sections than to make use of Cardan's formulas.[1]

PROBLEMS

1. Find graphically, correct to one decimal, the real roots of:

(a) $2^x - x^2 = 0$; (b) $x^4 - x - 1 = 0$; (c) $x^5 - x - 0.5 = 0$; (d) $e^x + x = 0$; (e) $\tan x - x = 0$, $\pi < x < 3\pi/2$; (f) $x^3 - 3x + 1 = 0$; (g) $4x = 7 \sin x$.

Isolate the roots (that is, for each root find an interval that contains that root and no others).

2. A sphere 2 ft in diameter is made of wood whose specific gravity is $\frac{2}{3}$.   Find to one-decimal accuracy the depth h to which the sphere sinks in water.   *Hint:* The volume of a spherical segment is $\pi h^2(r - h/3)$.   The volume of the submerged segment is equal to the volume of displaced water, which must weigh as much as the sphere.   If water weighs 62.5 lb per ft³,

$$\pi h^2 \left(r - \frac{h}{3} \right) 62.5 = \frac{4}{3} \pi r^3 \cdot \frac{2}{3} \cdot 62.5$$

and, since $r = 1$, $h^3 - 3h^2 + \frac{8}{3} = 0$.

3. Make the substitution $z = x + h$ in $4z^3 + 6z^2 - 7z + 4 = 0$, and determine h so that the resulting equation takes the form of (1-4).   Obtain graphically (to one decimal) the real root of the given equation.

4. With the aid of the system (1-7) obtain graphically, to one decimal, the roots of $x^4 - 3x^2 - 8x - 29 = 0$.

5. Find graphically, to one decimal, the real roots of (a) $x^4 + x^2 - 2x - 2 = 0$; (b) $z^3 - 2z^2 + 3z - 5 = 0$; (c) $e^{-2x} + e^x - 4 = 0$.   *Hints:* In (a) use the system (1-7); in (b) reduce to the form (1-4); in (c) set $e^x = y$ and solve for y.

---

2. Simple Iterative Methods.   When real roots of Eq. (1-1) have been isolated, there are many methods for computing them to any degree of accuracy.   These all depend on the application of some iterative formula which furnishes values of the succeeding approximations from the preceding ones.   The nature of restrictions imposed on the function $F(x)$ in the equation

$$F(x) = 0 \tag{2-1}$$

in the two basic iterative methods discussed here is obvious from the description of the

---

[1] A numerical determination of the roots of algebraic equations is frequently accomplished by some method of synthetic division (such as Horner's method) or by the root-squaring method (Graeffe's method).   These special methods are discussed in many books.   See, for example, F. B. Hildebrand, "Introduction to Numerical Analysis," McGraw-Hill Book Company, New York, 1956.   The methods of Secs. 2 and 3 of this chapter apply to many types of equations and are generally adequate for the determination of real roots.

methods. The simplest of these is the method of *linear interpolation*, also known as the *method of false position*.

Let the root x_0 of (2-1) be isolated between x_1 and x_2. Then, in the interval (x_1, x_2), the graph of $y = F(x)$ may have the appearance shown in Fig. 2. If the points P_1 and P_2 in Fig. 2 are joined by a straight line, it will cut the x axis at some point x_3, which usually is closer to the root x_0 than either x_1 or x_2. But from similar triangles,

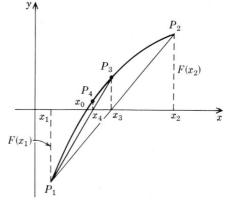

FIGURE 2

$$\frac{x_3 - x_1}{-F(x_1)} = \frac{x_2 - x_3}{F(x_2)} \qquad (2\text{-}2)$$

and on solving for x_3 we get

$$x_3 = \frac{x_1 F(x_2) - x_2 F(x_1)}{F(x_2) - F(x_1)}$$

which can also be written as

$$x_3 = x_2 - F(x_2) \frac{x_2 - x_1}{F(x_2) - F(x_1)}. \qquad (2\text{-}3)$$

To obtain an approximation x_4, we can determine the x intercept of the straight line joining the point P_1 in Fig. 2 with the point P_3, and, in general, determine the succeeding approximations from the recursion formula,

$$x_{n+1} = x_n - F(x_n) \frac{x_n - x_1}{F(x_n) - F(x_1)} \qquad n = 2, 3, 4, \ldots \qquad (2\text{-}4)$$

If $F'(x)$ and $F''(x)$ do not change signs in the interval (x_1, x_2), so that $F(x)$ is monotone and has no points of inflection in $(x_1 x_2)$, the sequence (2-4) can be shown[1] to converge to x_0 provided that x_1 is chosen so that $F(x_1) F''(x_1) > 0$.

If the fixed point x_1 in (2-4) is replaced by the variable point x_{n-1}, we get the sequence

$$x_{n+1} = x_n - F(x_n) \frac{x_n - x_{n-1}}{F(x_n) - F(x_{n-1})} \qquad n = 2, 3, \ldots \qquad (2\text{-}5)$$

which converges more rapidly to x_0 than the sequence (2-4). The method of solution of equations based on the formula (2-5) is called the *method of secants*. This method is thoroughly discussed in the cited monograph by A. M. Ostrowski.

Another useful iterative method is based on rewriting (2-1) in the form

$$f(x) = g(x). \qquad (2\text{-}6)$$

Now, if the real roots of

$$f(x) = c$$

can be determined for every real c, we can proceed as follows: Let x_1 be an approximate value of the root x_0 of (2-1). This, of course, is also an approximate root of (2-6), since (2-1) and (2-6) are equivalent equations. On setting $x = x_1$ in the right-hand member of (2-6) we get the equation

$$f(x) = g(x_1) \qquad (2\text{-}7)$$

[1] These conditions are sufficient but not necessary for the sequence (2-4) to converge to x_0. We refer the reader for proof to A. M. Ostrowski, "Solutions of Equations and Systems of Equations," Academic Press Inc., New York, 1960. Ostrowski's book contains a careful analysis of several important iterative methods, including estimates of errors.

which by hypothesis we can solve. If the solution of (2-7) is x_2, we obtain, on setting $x = x_2$ in the right-hand member of (2-6),

$$f(x) = g(x_2). \tag{2-8}$$

The solution x_3 of (2-8) is called the third approximation, and, in general, the nth approximation x_n is determined by solving

$$f(x) = g(x_{n-1}). \tag{2-9}$$

From the geometric interpretations of this procedure, which we give next, it will be seen that the sequence $x_1, x_2, \ldots, x_n, \ldots$ converges to the root x_0 of (2-1) if, in the interval of length $2\,|x_1 - x_0|$ centered at x_0, we have

$$|f'(x)| > |g'(x)| \tag{2-10}$$

and the derivatives are bounded.

Suppose, first, that the slopes of the curves

$$y = f(x) \qquad y = g(x) \tag{2-11}$$

in the interval (x_0, x_1) (Fig. 3) have the same sign and satisfy (2-10). When $x = x_1$ is taken as the first approximation to x_0, Eq. (2-7) yields the second approximation x_2, which

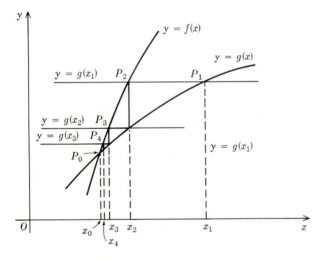

FIGURE 3

corresponds to the abscissa of the point of intersection P_2 of the straight line $y = g(x_1)$ with $y = f(x)$. Equation (2-8) gives x_3, which is the abscissa of the point of intersection P_3 of the straight line $y = g(x_2)$ with $y = f(x)$, and so on. The sequence $x_1, x_2, x_3, \ldots$ obviously converges to x_0.

The situation when the slopes of the curves (2-11) satisfy (2-10) but are opposite in sign is illustrated in Fig. 4. The value x_2 determined by solving (2-7) is the abscissa of the point of intersection P_2 of $y = f(x)$ with $y = g(x_1)$. It lies on the opposite side of the root from x_1. The third approximation x_3 is the abscissa of the intersection of $y = g(x_2)$ with $y = f(x)$, and it lies on the same side as x_1 but nearer to x_0. In Fig. 3 the approach to the intersection P_0 is along a staircase path, while in Fig. 4 it is along a spiral. In

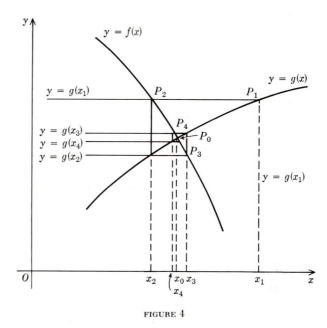

FIGURE 4

either case, the rapidity of convergence[1] depends on the nature of the functions $f(x)$ and $g(x)$.

To simplify calculations one ordinarily reduces the given equation $F(x) = 0$ to the form

$$x = g(x) \qquad (2\text{-}12)$$

so that $f(x)$ in (2-6) is equal to x. This can be accomplished by replacing the equation $F(x) = 0$ by an equivalent equation

$$x = x - kF(x)$$

where the number $k \neq 0$ is chosen so that (2-10) holds; in other words, so that

$$\frac{d}{dx}[x - kF(x)] = 1 - kF'(x)$$

is numerically small in the neighborhood of the unique root x_0.

Example 1.   Determine the approximate values of the real roots of

$$e^x - 4x = 0. \qquad (2\text{-}13)$$

The real roots of this equation are the abscissas of the points of intersection of the curves $y = e^x$ and $y = 4x$ shown in Fig. 5. It appears that the smaller of the roots, x_0, lies in the vicinity of $x = 0.3$. The larger root, ξ_0, is close to $x = 2.1$. Since for $x = x_0$ the slope of $y = 4x$ is greater than that of $y = e^x$, we write (2-13) in the form

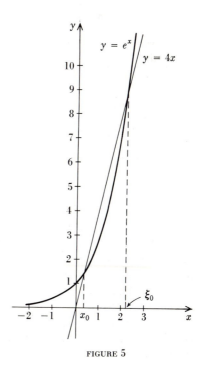

FIGURE 5

---

[1] Some criteria for the speed of convergence are given in Ostrowski, *op. cit.*, and in Hildebrand, *op. cit.*

$$x = \tfrac{1}{4}e^x$$

so that in the notation of Eq. (2-6)

$$f(x) = x \qquad \text{and} \qquad g(x) = \tfrac{1}{4}e^x.$$

The sequence of approximations x_n according to (2-9) is thus determined from

$$x_{n+1} = \tfrac{1}{4}e^{x_n} \qquad n = 1, 2, \dots \dots \tag{2-14}$$

If we take $x_1 = 0.3$, we get[1] from (2-14)

$$x_2 = \tfrac{1}{4}e^{0.3} = \tfrac{1}{4}(1.34986) = 0.3374 \qquad x_5 = \tfrac{1}{4}e^{x_4} = \tfrac{1}{4}(1.42603) = 0.3565$$

$$x_3 = \tfrac{1}{4}e^{x_2} = \tfrac{1}{4}(1.40130) = 0.3503 \qquad x_6 = \tfrac{1}{4}e^{x_5} = \tfrac{1}{4}(1.42832) = 0.3571$$

$$x_4 = \tfrac{1}{4}e^{x_3} = \tfrac{1}{4}(1.41949) = 0.3549 \qquad x_7 = \tfrac{1}{4}e^{x_6} = \tfrac{1}{4}(1.42917) = 0.3573.$$

If only three-decimal-place accuracy is required, the values obtained indicate that the computations can be terminated at this stage.

To obtain the second root we note that, at $x = \xi_0$, the slope of $y = 4x$ is less than that of $y = e^x$. If we write (2-13) in the form $e^x = 4x$ or $x = \log 4x$ so that $f(x) = x$ and $g(x) = \log 4x$, the condition (2-10) is satisfied at $x = \xi_0$.

The desired sequence $\{x_n\}$ is now given by

$$x_{n+1} = \log 4x_n \qquad n = 1, 2, \dots$$

and we can take $x_1 = 2.1$.

Using tables of natural logarithms[2] we find

$$x_2 = \log 4x_1 = \log 8.4 = 2.12823 \qquad x_6 = \log 4x_5 = \log 8.6030 = 2.15211$$

$$x_3 = \log 4x_2 = \log 8.5129 = 2.14158 \qquad x_7 = \log 4x_6 = \log 8.6084 = 2.15273$$

$$x_4 = \log 4x_3 = \log 8.5663 = 2.14783 \qquad x_8 = \log 4x_7 = \log 8.6109 = 2.15303$$

$$x_5 = \log 4x_4 = \log 8.5913 = 2.15075 \qquad x_9 = \log 4x_8 = \log 8.6121 = 2.15316.$$

The value of the root ξ_0, correct to three decimals, is 2.153. We do not give a discussion of the errors in the approximations obtained by such calculations because a rigorous analysis of errors in the iterative procedures is fairly involved.[3]

Example 2. Find an approximate value of the real root of

$$x - \tan x = 0 \tag{2-15}$$

near $x = 3\pi/2$.

From the graphs of

$$y = x \qquad \text{and} \qquad y = \tan x$$

in Fig. 6, it appears that Eq. (2-15) has just one real root in each of the intervals $(2n - 1)\pi/2 < x < (2n + 1)\pi/2$, where $n = 0, \pm 1, \pm 2, \dots$

It is convenient to rewrite (2-15) in the form

$$x = \tan^{-1} x$$

so that in the notation of (2-6) $f(x) = x$ and $g(x) = \tan^{-1} x$. This choice assures that the condition (2-10) is satisfied at the root x_0.

The sequence of approximations this time is given by

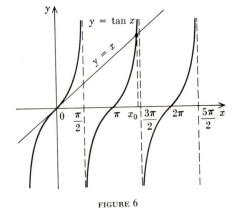

FIGURE 6

[1] In performing these calculations it is convenient to use tables such as "Tables of Exponential Functions," National Bureau of Standards, Washington, D.C., 1951.

[2] For example, "Tables of Natural Logarithms," National Bureau of Standards, Washington, D.C., 1941.

[3] See Ostrowski, *op. cit.*, and A. S. Householder, "Principles of Numerical Analysis," McGraw-Hill Book Company, New York, 1953.

$$x_{n+1} = \tan^{-1} x_n \qquad n = 1, 2, \ldots.$$

On taking $x_1 = 3\pi/2 = 4.7124$ radians, we find

$$x_2 = \tan^{-1} 4.7124 = 4.5033$$

$$x_3 = \tan^{-1} 4.5033 = 4.4938$$

$$x_4 = \tan^{-1} 4.4938 = 4.4935$$

which suggest that the root x_0, correct to three decimals, is 4.493.

These examples indicate that when Eq. (2-1) is written in the form

$$x = g(x)$$

and $|g'(x)| \leq M < 1$ in the interval of length $2|x_1 - x_0|$ centered at x_0, the recursion formula giving the desired approximating sequence is

$$x_{n+1} = g(x_n) \qquad n = 1, 2, \ldots. \tag{2-16}$$

The iteration procedure in (2-16) can also be applied to determine the real roots of the system of two equations in two unknowns,

$$F(x,y) = 0 \qquad G(x,y) = 0. \tag{2-17}$$

If this system has a real solution (x_0,y_0) and (x_1,y_1) is an approximate solution of (2-17), the set of successive approximations to (x_0,y_0) may be determined by transforming (2-17) to an equivalent system

$$x = f(x,y) \qquad y = g(x,y). \tag{2-18}$$

If $f(x,y)$ and $g(x,y)$ are such that

$$|f_x(x,y)| + |g_x(x,y)| \leq M < 1 \qquad |f_y(x,y)| + |g_y(x,y)| \leq M < 1 \tag{2-19}$$

in some neighborhood R of (x_1,y_1) containing (x_0,y_0), we construct the sequences

$$x_{n+1} = f(x_n,y_n) \qquad y_{n+1} = g(x_n,y_n) \qquad n = 1, 2, \ldots. \tag{2-20}$$

When all (x_n,y_n) belong to R, the sequences (2-20) converge to the root (x_0,y_0) of (2-17).

PROBLEMS

1. Use all methods of this section to obtain, correct to two decimals, the values of the real roots in Probs. 1 and 2 of Sec. 1.

2. Find in the manner of the examples of this section the real roots of $x^5 - x - 0.2 = 0$, correct to three decimals.

3. Find, correct to two decimals, the real root of $x = \frac{1}{4} \cos x$ by formula (2-16). Obtain the first approximation x_1 graphically.

4. Determine the initial approximations x_1 to the roots of $x \tanh x = 1$ graphically, and use (2-16) to determine the roots, correct to three decimals.

5. Apply formulas (2-4) and (2-5) to obtain, correct to two decimals, the root of $x \log x - 1 = 0$ in the neighborhood of $x_1 = 2$.

---

3. Newton's Method.   The successive terms in the approximating sequence in the method of false position (see Fig. 2) are determined by the intersection of the secant line with the x axis.   Newton proposed constructing an approximating sequence determined by the intersection with the x axis of the tangent line to the curve $y = F(x)$.

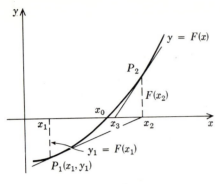

FIGURE 7

Thus let the root $x = x_0$ of

$$F(x) = 0 \qquad (3\text{-}1)$$

lie in the vicinity of $x = x_1$ (see Fig. 7). The equation of the tangent line to $y = F(x)$ at $P_1(x_1, y_1)$ is

$$y - F(x_1) = F'(x_1)(x - x_1). \qquad (3\text{-}2)$$

If the curve $y = F(x)$ has the appearance shown in Fig. 7, the tangent line (3-2) cuts the x axis at x_2, which is a better approximation to the root than x_1. To determine x_2 we set $y = 0$ and find

$$x_2 = x_1 - \frac{F(x_1)}{F'(x_1)}$$

if $F'(x_1) \neq 0$. Having determined x_2, we find in the same way that the tangent to $y = F(x)$ at $P_2[x_2, F(x_2)]$ intersects the axis at

$$x_3 = x_2 - \frac{F(x_2)}{F'(x_2)}$$

and, in general,

$$x_{n+1} = x_n - \frac{F(x_n)}{F'(x_n)} \qquad n = 1, 2, \ldots. \qquad (3\text{-}3)$$

The geometric considerations indicate[1] that when $y = F(x)$ is a monotone increasing or decreasing function in the interval (x_1, x_2) [so that $F'(x)$ does not change sign] and when there is no point of inflection in this interval [so that $F''(x)$ does not change sign] and $F(x_1)F''(x_1) > 0$ the sequence (3-3) converges to the root x_0.

The situations corresponding to the cases when there is a point of inflection or a horizontal tangent to $y = F(x)$ in the vicinity of the root are illustrated in Figs. 8 and 9. It is clear from these figures that in these cases the sequence (3-3) need not converge to x_0. Thus, before applying Newton's method, one should examine the behavior of $F'(x)$ and $F''(x)$ in the vicinity of the root.

Example. Find the angle subtended at the center of a circle by an arc whose length is double the length of its chord.

Let the arc BCA (Fig. 10) be of length $2BA$. If the angle subtended by this arc at the center of the circle is $2x$ radians, the arc $BCA = 2xr$ while $BA = 2r \sin x$, r being the radius of the circle.

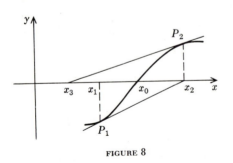

FIGURE 8

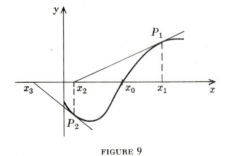

FIGURE 9

[1] For analytic criteria on the rapidity of convergence of Newton's method, see Ostrowski, *op. cit.* See also E. L. Stiefel, "An Introduction to Numerical Analysis," chap. 4, Academic Press Inc., New York, 1963.

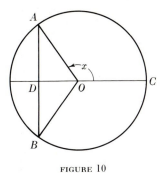

FIGURE 10

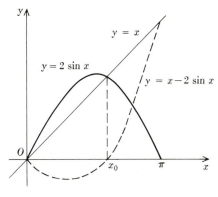

FIGURE 11

Our problem requires that

$$2xr = 4r \sin x \qquad \text{or} \qquad x - 2 \sin x = 0. \tag{3-4}$$

On graphing the functions $y = x$ and $y = 2 \sin x$ (Fig. 11), we see that they intersect at $x = 0$ and at $x = 1.88$ radians, approximately. We reject the trivial solution $x = 0$. Since $y = x - 2 \sin x$ is obviously monotone increasing and has no point of inflection near the root x_0, we can apply formula (3-3) with $x_1 = 1.88$. We find

$$x_2 = x_1 - \frac{x_1 - 2 \sin x_1}{1 - 2 \cos x_1} = 1.88 - \frac{1.88 - 2 \sin 1.88}{1 - 2 \cos 1.88} = 1.896.$$

The third approximation is

$$x_3 = x_2 - \frac{x_2 - 2 \sin x_2}{1 - 2 \cos x_2} = 1.896 - \frac{1.896 - 2 \sin 1.896}{1 - 2 \cos 1.896} = 1.8955$$

which is nearly the same as x_2. The angle subtended by the arc BCA, as given by this approximation, is 3.7910 radians.

Newton's method can be extended to the determination of real roots of the system of equations

$$F(x,y) = 0 \qquad G(x,y) = 0. \tag{3-5}$$

If the desired root (x_0,y_0) lies in the neighborhood of the point (x_1,y_1) and the Jacobian $J = \partial(F,G)/\partial(x,y) \neq 0$ in this neighborhood, an approximation (x_2,y_2) to (x_0,y_0) is $x_2 = x_1 + h_1$, $y_2 = y_1 + k_1$, where h_1 and k_1 satisfy the linear equations

$$\begin{aligned} F(x_1,y_1) + h_1 F_x(x_1,y_1) + k_1 F_y(x_1,y_1) &= 0 \\ G(x_1,y_1) + h_1 G_x(x_1,y_1) + k_1 G_y(x_1,y_1) &= 0. \end{aligned} \tag{3-6}$$

The initial point (x_1,y_1) is determined graphically by examining intersections of the curves in (3-5), and the successive approximations

$$x_{n+1} = x_n + h_n \qquad y_{n+1} = y_n + k_n \qquad n = 1, 2, \ldots$$

are determined from the system (3-6) in which the subscripts 1 are replaced by n.

PROBLEMS

1. Calculate by Newton's method, to two decimals, the roots in Examples 1 and 2 of Sec. 2.

2. Solve by Newton's method, correct to two decimals, Prob. 2 of Sec. 1.

3. Find by Newton's method, to three decimals, the angle subtended at the center of a circle by a chord which cuts off a segment whose area is one-fourth that of the circle.

4. Find by Newton's method, to three decimal places, the real roots of the following equations:
(a) $x - \cos x = 0$; (b) $x + e^x = 0$; (c) $x^4 - x - 1 = 0$; (d) $x^3 - 25 = 0$; (e) $x^5 - x - 0.2 = 0$.

5. Find by Newton's method, to three decimals, the root of the equation in Prob. 5 of Sec. 2.

6. Find by Newton's method, correct to two decimals, the roots of $2^x - 4x = 0$.

7. Find by Newton's method, correct to two decimals, the roots of the system $x^3 - y = 0$, $x^2 + y^2 - 1 = 0$. Determine the first approximations (x_1,y_1) graphically.

---

4. Systems of Linear Equations. The Gauss Reduction. No doubt the reader is familiar with Cramer's rule for solving systems of n linear equations in n unknowns by determinants as summarized in Appendix A of this book. Although Cramer's rule is important in numerous theoretical considerations, it is of questionable practical value when the given system contains more than two unknowns. Usually it is easier to obtain solutions by some process of elimination of unknowns. The simplest practical method for solving systems of linear equations, based on the idea of elimination, is the Gauss reduction method. Its several variants form the basis for most techniques used in the solutions of large systems of equations.[1]

The idea of the method is simple. Let it be required to solve a system of n linear equations

$$\begin{aligned}
a_{11}x_1 + a_{12}x_2 + \cdots + a_{1n}x_n &= c_1 \\
a_{21}x_1 + a_{22}x_2 + \cdots + a_{2n}x_n &= c_2 \\
\cdot\quad \cdot\quad \cdot\quad \cdot\quad \cdot\quad \cdot\quad \cdot\quad \cdot\quad \cdot\quad \cdot\quad \cdot\quad \cdot & \\
a_{n1}x_1 + a_{n2}x_2 + \cdots + a_{nn}x_n &= c_n
\end{aligned} \tag{4-1}$$

in n unknowns x_i. We divide the first equation in (4-1) by a_{11}, solve for x_1, and use the result to eliminate x_1 in the other equations. The resulting system of $n-1$ equations in $x_2, \ldots, x_n$ is treated in the same way; that is, we divide the first of these equations by the coefficient of x_2 and use the result to eliminate x_2 from the remaining equations. After continuing the process n times we obtain an equivalent system

$$\begin{aligned}
x_1 + a'_{12}x_2 + a'_{13}x_3 + \cdots + a'_{1n}x_n &= c'_1 \\
x_2 + a'_{23}x_3 + \cdots + a'_{2n}x_n &= c'_2 \\
\cdot\quad \cdot\quad \cdot\quad \cdot\quad \cdot\quad \cdot\quad \cdot\quad \cdot\quad \cdot\quad \cdot\quad \cdot\quad \cdot & \\
x_{n-1} + a'_{n-1,n}x_n &= c'_{n-1} \\
x_n &= c'_n
\end{aligned} \tag{4-2}$$

provided the given system has a unique solution. The substitution $x_n = c'_n$ in the preceding equation in the set (4-2) yields the value of x_{n-1}, and by working backward we obtain in succession the values of $x_{n-2}, x_{n-3}, \ldots, x_1$.

In practice the Gauss reduction can be performed in the manner indicated in the following example.

Example. Solve the system

$$\begin{aligned}
2.843x_1 - 1.326x_2 + 9.841x_3 &= 5.643 \\
8.673x_1 + 1.295x_2 - 3.215x_3 &= 3.124 \\
0.173x_1 - 7.724x_2 + 2.832x_3 &= 1.694
\end{aligned} \tag{4-3}$$

by the method of Gauss' reduction.

---

[1] Among such variants are the Crout and the Gauss-Jordan reductions, described in Hildebrand, *op. cit.*, and in many other books on numerical methods. In the following discussion, if the coefficient of x_r in the rth equation vanishes, it is necessary to renumber the variables or equations. Cf. Appendix A.

On dividing each equation in (4-3) by the coefficients of x_1 in that equation, we get

$$x_1 - 0.46641x_2 + 3.4615x_3 = 1.9849$$
$$x_1 + 0.14931x_2 - 0.37069x_3 = 0.36020 \qquad \text{(4-4)}$$
$$x_1 - 44.647x_2 + 16.370x_3 = 9.7919.$$

The subtraction of the second equation in (4-4) from the first and the third gives

$$-0.61572x_2 + 3.8322x_3 = 1.6247$$
$$-44.796x_2 + 16.741x_3 = 9.4317$$

and, on dividing these by the coefficients of x_2, we find

$$x_2 - 6.2239x_3 = -2.6387$$
$$x_2 - 0.37372x_3 = -0.21055. \qquad \text{(4-5)}$$

Subtracting the second equation from the first in (4-5) yields

$$-5.8502x_3 = -2.4282 \qquad \text{(4-6)}$$

so that $$x_3 = 0.41506.$$

The reduced system consists of the first equations in (4-4) and (4-5) and Eq. (4-6). It is

$$x_1 - 0.46641x_2 + 3.4615x_3 = 1.9849$$
$$x_2 - 6.2239x_3 = -2.6387 \qquad \text{(4-7)}$$
$$x_3 = 0.41506.$$

The substitution of the value of x_3 from the last into the second equation of (4-7) gives

$$x_2 = -2.6387 + 6.2239(0.41506) = -0.055408$$

and the first reduced equation finally yields

$$x_1 = 1.9849 + 0.46641(-0.055408) - 3.4615(0.41506) = 0.52232.$$

There are numerous modifications of the procedure just indicated, some of which are adapted for computations on desk calculators while others are more suitable for high-speed electronic computers.

————————————————————————— **PROBLEM**

Use Cramer's rule and also apply the Gauss reduction to solve the following systems:

(a) $2x + y + 3z = 2$
$\ 3x - 2y - z = 1$
$\ \ \ x - y + z = -1$

(b) $2x_1 + x_2 + 3x_3 + x_4 = -2$
$\ 5x_1 + 3x_2 - x_3 - x_4 = 1$
$\ \ x_1 - 2x_2 + 4x_3 + 3x_4 = 4$
$\ \ \ \ 3x_1 - x_2 + x_3 = 2$

(c) $1.329x_1 + 1.415x_2 - 2.291x_3 = 0.532$
$\ \ \ \ \ \ \ \ \ 1.395x_2 - 0.531x_3 = 1.211$
$\ 1.001x_1 + 2.093x_3 = 0.556.$

————————————————————————————————

5. An Iterative Method for Systems of Linear Equations.   An error resulting from approximating a given number by n digits, when its exact decimal representation requires more than n digits, is called a *round-off error*.   Except for such errors, the Gauss reduction method explained in the preceding section is exact.   When the determinant of the system (4-1) is different from zero, it yields the desired solution after a finite number of steps.   However, successive steps leading to an equivalent triangular system (4-2) may prove laborious and ill-adapted to machine calculations.   For this reason, a variety of iterative methods, which in theory require an infinite number of steps to obtain an exact solution, have been devised.

One of these methods, generally attributed to Gauss and L. Seidel, is based on the use of the iterative formula (2-16). The convergence of any iterative method obviously depends on the character of the system under consideration.

In many cases the system (4-1) can be rewritten so that in the ith equation the coefficient a_{ii} of the unknown x_i is numerically large compared with other coefficients. That is to say, the coefficients along the diagonal of the system (4-1) dominate the other coefficients. In this event, by solving the ith equation for x_i, we can rewrite such a system (4-1) in the form

$$x_1 = \frac{1}{a_{11}}\,(c_1 - a_{12}x_2 - a_{13}x_3 - \cdots - a_{1n}x_n)$$

$$x_2 = \frac{1}{a_{22}}\,(c_2 - a_{21}x_1 - a_{23}x_3 - \cdots - a_{2n}x_n)$$

$$\cdot \quad \cdot \quad \cdot \quad \cdot \quad \cdot \quad \cdot \quad \cdot \quad \cdot \quad \cdot \quad \cdot \quad \cdot \quad \cdot \quad \cdot \quad \cdot \quad \cdot$$ (5-1)

$$x_n = \frac{1}{a_{nn}}\,(c_n - a_{n1}x_1 - a_{n2}x_2 - \cdots - a_{n,n-1}x_{n-1}).$$

If we set $x_1 = x_2 = \cdots = x_n = 0$ in the right-hand members of (5-1), we obtain

$$x_i^{(1)} = \frac{c_i}{a_{ii}} \qquad i = 1, 2, \ldots, n$$ (5-2)

which is called the *first approximation* to the solution of (5-1).

The substitution of this first approximation in the right-hand members of (5-1) yields the *second approximation* $x_i^{(2)}$, and so on. The cycle is then repeated with the expectation that the values $x_i^{(k)}$ after the kth iteration are not substantially altered by further iterations.[1]

In practice the iteration process described above is usually modified by taking as the first approximation $x_1^{(1)}$ the value of x_1 obtained from the first equation in (5-1) by setting $x_2 = x_3 = \cdots = x_n = 0$. Using this value in the second equation in place of x_1 and setting $x_3 = x_4 = \cdots = x_n = 0$, one obtains the approximation $x_2^{(1)}$. To obtain $x_3^{(1)}$ one inserts for x_1 and x_2 the values $x_1^{(1)}$ and $x_2^{(1)}$ in the third equation and sets $x_4 = x_5 = \cdots = x_n = 0$. Finally, to get the value of $x_n^{(1)}$ one uses previously found values $x_1^{(1)}, \ldots, x_{n-1}^{1}$ in the last equation of the system (5-1). This process is repeated to obtain approximations of higher orders.

This particular choice of approximations usually improves the rapidity of convergence of the process. We illustrate it by an example.

Example. The system (4-3) can be rewritten in the form

$$8.673x_1 + 1.295x_2 - 3.215x_3 = 3.124$$
$$0.173x_1 - 7.724x_2 + 2.832x_3 = 1.694$$ (5-3)
$$2.843x_1 - 1.326x_2 + 9.841x_3 = 5.643$$

in which the diagonal coefficients dominate.

---

[1] There are several criteria for convergence of this process which generally are not easy to verify. See Hildebrand, *op. cit.*, for a brief discussion.

We next write (5-3) in the form (5-1) and get

$$x_1 = \frac{1}{8.673}(3.124 - 1.295x_2 + 3.215x_3)$$

$$x_2 = \frac{1}{-7.724}(1.694 - 0.173x_1 - 2.832x_3) \tag{5-4}$$

$$x_3 = \frac{1}{9.841}(5.643 - 2.843x_1 + 1.326x_2).$$

To obtain $x_1^{(1)}$ we set $x_2 = x_3 = 0$ in the first equation in (5-4) and find

$$x_1^{(1)} = \frac{3.124}{8.673} = 0.36020.$$

Inserting this value for x_1 and setting $x_3 = 0$ in the second equation in (5-4), we get

$$x_2^{(1)} = -0.21125.$$

Finally, $x_3^{(1)} = 0.44089$ is obtained by using the values $x_1^{(1)}$ and $x_2^{(1)}$ in place of x_1 and x_2 in the third of Eqs. (5-4).

A repetition of the process yields second, third, and higher approximations. These are recorded in the table:

k	1	2	3	4	5	6	7
$x_1^{(k)}$	0.36020	0.55517	0.51780	0.52312	0.52220	0.52235	0.52233
$x_2^{(k)}$	-0.21125	-0.04523	-0.05852	-0.05501	-0.05550	-0.05543	-0.05543
$x_3^{(k)}$	0.44089	0.40694	0.41594	0.41488	0.41508	0.41505	0.41505

A comparison with the values found in Sec. 4 by the Gauss reduction method shows that in this problem six iterations were necessary to get four-decimal accuracy.

INTERPOLATION. EMPIRICAL FORMULAS. LEAST SQUARES

6. Differences. One of the problems connected with the analysis of experimental data concerns the representation of such data by analytic formulas. Thus, we may wish to represent, either exactly or approximately, a set of observed values (x_i, y_i) by some relationship of the form $y = f(x)$. In such analysis the concept of *differences* is important.

We consider a set of pairs of values (x_i, y_i), where $i = 0, 1, \ldots, n-1$, which can be represented by points in the xy plane. The differences between successive pairs of ordinates y_{i+1} and y_i we call the *first forward differences* of the y's and we denote them by Δy_i. Thus,

$$\Delta y_i = y_{i+1} - y_i \qquad i = 0, 1, 2, \ldots, n-1. \tag{6-1}$$

The *second forward differences* are defined by

$$\Delta^2 y_i = \Delta y_{i+1} - \Delta y_i$$

and, in general, the kth *forward differences* are

$$\Delta^k y_i = \Delta^{k-1} y_{i+1} - \Delta^{k-1} y_i. \tag{6-2}$$

These differences are usually represented in a tabular form, as in Table 1. The quantities in each column represent the differences between the quantities in the preceding column. These are usually placed midway between the quantities being subtracted,

TABLE 1

x	y	Δy	$\Delta^2 y$	$\Delta^3 y$	$\Delta^4 y$
x_0	y_0				
		Δy_0			
x_1	y_1		$\Delta^2 y_0$		
		Δy_1		$\Delta^3 y_0$	
x_2	y_2		$\Delta^2 y_1$		$\Delta^4 y_0$
		Δy_2		$\Delta^3 y_1$	
x_3	y_3		$\Delta^2 y_2$		
		Δy_3			
x_4	y_4				
x_{n-1}	y_{n-1}				
		Δy_{n-1}			
x_n	y_n				

so that the forward differences with like subscripts lie along the downward-slanting diagonals in the table.

We note that if the rth differences $\Delta^r y_i$ are constant then all differences of order higher than r are zero. In practice, the differences can be deemed to be zero if they involve only the last significant digit of the data. (For a numerical illustration, see the example in the next section.) Now, it follows from (6-1) and (6-2) that

$$y_1 = y_0 + \Delta y_0$$
$$y_2 = y_1 + \Delta y_1 = (y_0 + \Delta y_0) + (\Delta y_0 + \Delta^2 y_0) = y_0 + 2\Delta y_0 + \Delta^2 y_0$$
$$y_3 = y_2 + \Delta y_2 = (y_0 + 2\Delta y_0 + \Delta^2 y_0) + (\Delta y_1 + \Delta^2 y_1)$$
$$= (y_0 + 2\Delta y_0 + \Delta^2 y_0) + (\Delta y_0 + \Delta^2 y_0 + \Delta^2 y_0 + \Delta^3 y_0)$$
$$= y_0 + 3\Delta y_0 + 3\Delta^2 y_0 + \Delta^3 y_0.$$

These results can be written symbolically as

$$y_1 = (1 + \Delta)y_0 \qquad y_2 = (1 + \Delta)^2 y_0 \qquad y_3 = (1 + \Delta)^3 y_0$$

in which $(1 + \Delta)^k$ is an operator on y_0 with the exponent on the Δ indicating the order of the difference. The difference operator Δ is analogous to the differential operator D introduced in Chap. 3, Sec. 11.

We easily establish by induction that

$$y_k = (1 + \Delta)^k y_0 \qquad k = 1, 2, \ldots \tag{6-3}$$

or, in the expanded form,

$$y_k = y_0 + k\,\Delta y_0 + \frac{k(k-1)}{2!}\,\Delta^2 y_0 + \frac{k(k-1)(k-2)}{3!}\,\Delta^3 y_0 + \cdots. \tag{6-4}$$

Formula (6-4) enables one to represent every value y_k in terms of y_0 and the forward differences $\Delta y_0, \Delta^2 y_0, \ldots$.

A similar formula can be derived by starting with the values of the y's at the end of Table 1 and forming the *backward* differences defined as follows: The *first backward differences* ∇y_i are

$$\nabla y_i = y_i - y_{i-1}. \tag{6-5}$$

The *second backward differences* $\nabla^2 y_i$ are defined by

$$\nabla^2 y_i = \nabla y_i - \nabla y_{i-1} \tag{6-6}$$

and, in general, the kth *backward differences* $\nabla^k y_i$ are

$$\nabla^k y_i = \nabla^{k-1} y_i - \nabla^{k-1} y_{i-1}. \tag{6-7}$$

A table of backward differences is indicated in Table 2, where the differences $\nabla^k y_i$ with a fixed subscript i along the diagonals slanting up.

TABLE 2

x	y	∇y	$\nabla^2 y$	$\nabla^3 y$	$\nabla^4 y$
x_0	y_0				
x_{n-4}	y_{n-4}				
x_{n-3}	y_{n-3}	∇y_{n-3}	$\nabla^2 y_{n-2}$		
x_{n-2}	y_{n-2}	∇y_{n-2}	$\nabla^2 y_{n-1}$	$\nabla^3 y_{n-1}$	
x_{n-1}	y_{n-1}	∇y_{n-1}	$\nabla^2 y_n$	$\nabla^3 y_n$	$\nabla^4 y_n$
x_n	y_n	∇y_n			

Now, from (6-5) to (6-7) we deduce that

$$\nabla^2 y_n = \nabla y_n - \nabla y_{n-1} = y_n - 2y_{n-1} + y_{n-2}$$
$$\nabla^3 y_n = \nabla^2 y_n - \nabla^2 y_{n-1} = y_n - 3y_{n-1} + 3y_{n-2} - y_{n-3}$$

and, in general,

$$\nabla^k y_n = \nabla^{k-1} y_n - \nabla^{k-1} y_{n-1} = \sum_{r=0}^{k} (-1)^r \binom{k}{r} y_{n-r} \tag{6-8}$$

where

$$\binom{k}{r} = \frac{k(k-1)(k-2)\ldots(k-r+1)}{r!} \tag{6-9}$$

is the binomial coefficient of x^r in the expansion of $(1+x)^k$.

By using (6-8) successively in the definitions of backward differences we find

$$y_{n-1} = y_n - \nabla y_n \equiv (1 - \nabla)y_n$$
$$y_{n-2} = y_n - 2\nabla y_n + \nabla^2 y_n \equiv (1 - \nabla)^2 y_n$$

and, in general,

$$y_{n-k} = (1 - \nabla)^k y_n \tag{6-10}$$

where ∇ is the backward-difference operator. Formula (6-10) when expanded reads

$$y_{n-k} = y_n - k\nabla y_n + \frac{k(k-1)}{2!} \nabla^2 y_n - \frac{k(k-1)(k-2)}{3!} \nabla^3 y_n + \cdots. \tag{6-11}$$

It shows that any value of y in Table 2 can be expressed in terms of y_n and backward differences $\nabla^k y_n$.

We shall use formulas (6-4) and (6-11) to derive certain interpolation formulas and to deduce some formulas for numerical integration.

── **PROBLEMS**

1. Compute the forward and backward differences for the following set of data:

x	1	2	3	4	5	6	7	8
y	2.105	2.808	3.614	4.604	5.857	7.451	9.467	11.985

2. Write expressions for the y_k, $k = 1, 2, \ldots$, in Prob. 1 by using (6-4) and (6-11).

──

7. Polynomial Representation of Data.   Unless a statement to the contrary is made, we shall suppose henceforth that the values x_i in a given set of data (x_i, y_i), where $i = 0, 1, 2, \ldots, n$, are equally spaced.   If the spacing interval is h, then

$$x_1 = x_0 + h, \qquad x_2 = x_0 + 2h, \qquad \ldots, \qquad x_n = x_0 + nh.$$

We pose the problem of representing the data by some formula $y = f(x)$, which for $x = x_0 + kh$ yields $y_k = f(x_0 + kh)$.   We shall frequently write f_k for y_k.

We observed in the preceding section that, whenever the rth differences of the y's are constant, all differences of order higher than r vanish.   In this event formula (6-4) yields, for $r \leq k$,

$$y_k = y_0 + \binom{k}{1} \Delta y_0 + \binom{k}{2} \Delta^2 y_0 + \cdots + \binom{k}{r} \Delta^r y_0 \tag{7-1}$$

where the binomial coefficients $\binom{k}{r}$ are defined by

$$\binom{k}{r} = \frac{k(k-1)(k-2)\ldots(k-r+1)}{r!}. \tag{7-2}$$

Since the x_i are spaced h units apart,

$$x_k = x_0 + kh \qquad k = 1, 2, \ldots, n$$

so that

$$k = \frac{x_k - x_0}{h}. \tag{7-3}$$

The expression (7-2) is a polynomial of degree r in k.   Therefore, on substituting in (7-1) for k from (7-3), we obtain a polynomial of degree r in x_k.   When like powers of x_k are collected, (7-1) takes the form

$$y_k = a_0 + a_1 x_k + a_2 x_k^2 + \cdots + a_r x_k^r. \tag{7-4}$$

Accordingly, the polynomial in x,

$$y(x) = a_0 + a_1 x + a_2 x^2 + \cdots + a_r x^r \tag{7-5}$$

assumes the values y_k when we set $x = x_k$.   Thus, when the rth differences of the y_k are constant and the x_k are equally spaced, the polynomial (7-5) represents these data exactly.

It is easy to prove a converse to the effect that the rth differences of the polynomial (7-5) are constant.   It would suffice to show that the first difference $\Delta y(x) = y(x + h) -$

$y(x)$ formed with the aid of (7-5) is a polynomial of degree $r - 1$, for, if differencing a polynomial once reduces its degree by 1, r successive differencings would yield a polynomial of degree 0, that is, a constant.[1]

When rth differences in a given set of data are not constant but differ from one another by negligible amounts, the polynomial (7-5) represents the data approximately.

Example.   The set of data and the forward differences tabulated below suggest that these data can be represented by a cubic polynomial $y = a_0 + a_1x + a_2x^2 + a_3x^3$ if two-decimal accuracy is sufficient.

x	y	Δy	$\Delta^2 y$	$\Delta^3 y$	$\Delta^4 y$
1	2.105				
		0.703			
2	2.808		0.103		
		0.806		0.081	
3	3.614		0.184		-0.002
		0.990		0.079	
4	4.604		0.263		-0.001
		1.253		0.078	
5	5.857		0.341		$+0.003$
		1.594		0.081	
6	7.451		0.422		-0.001
		2.016		0.080	
7	9.467		0.502		
		2.518			
8	11.985				

The coefficients a_i in this polynomial can be determined by using (7-1) with $r = 3$ and

$$y_0 = 2.105 \qquad \Delta y_0 = 0.703 \qquad \Delta^2 y_0 = 0.103 \qquad \Delta^3 y_0 = 0.081.$$

The result is

$$y_k = 2.105 + 0.703k + 0.052k(k - 1) + 0.014k(k - 1)(k - 2).$$

The a_i are determined by setting $k = x - 1$ and collecting powers of x.

─────────────────────────────── **PROBLEMS**

1. Given the table

x	19	20	21	22	23	24	25
y	81.00	90.25	100.00	110.25	121.00	132.25	144.00

compute second forward differences, and represent the data by $y = a_0 + a_1x + a_2x^2$. Determine a_0, a_1, a_2 so that the polynomial passes through (a) the first three points and (b) the last three points.

2. Discuss the calculation of the y_k in Prob. 1 from (6-4) and (6-11).

3. Construct the difference tables for (a) $y(x) = 1 + x + x^2 + x^3$, $x = 0, 1, 2, 3, 4$; (b) $y(x) = 1 - x - 2x^2 + x^3$, $x = 1, 3, 5, 7, 9, 11$. Verify that all fourth-order differences vanish.

─────────────────────────────

[1] We leave it to the reader to show that $\Delta y(x)$ is, indeed, a polynomial of degree $r - 1$. The result is analogous to the theorem that the derivative of a polynomial of degree r is a polynomial of degree $r - 1$. The expression $\Delta y = y(x + h) - y(x)$ save for the factor $1/h$ is the difference quotient used in defining the derivative.

8. Newton's Interpolation Formulas.

When the data (x_i, y_i), where $i = 0, 1, 2, \ldots, n$, are presented in tabular form, an infinite number of analytic relations $y = f(x)$ can be devised such that $y_i = f(x_i)$ either exactly or approximately. Once a suitable form of $f(x)$ is determined, the formula $y = f(x)$ can be used to calculate the ordinates y for x's not appearing in the table; that is, the formula can be used for interpolation or extrapolation.

The simplest of such formulas is a linear relationship based on the assumption that the values of y in the interval (x_i, x_{i+1}) can be represented by

$$y = y_i + \frac{y_{i+1} - y_i}{x_{i+1} - x_i}(x - x_i). \tag{8-1}$$

Formula (8-1) is precisely that used in estimating the values of such tabulated functions as logarithms by the process of "interpolation by proportional parts."

More accurate interpolation formulas are based on the assumption that the desired value of y can be computed from a polynomial

$$y = a_0 + a_1 x + a_2 x^2 + \cdots + a_m x^m \tag{8-2}$$

in which $m + 1$ coefficients a_i are so chosen that $m + 1$ pairs of tabulated values (x_i, y_i) satisfy (8-2) exactly.[1]

In the preceding section we saw that, when the data are represented by a polynomial of degree m, all forward differences of order higher than m vanish. Accordingly, formula (6-4) yields

$$y_k = y_0 + k\,\Delta y_0 + \frac{k(k-1)}{2!}\Delta^2 y_0 + \cdots + \frac{k(k-1)\ldots(k-m+1)}{m!}\Delta^m y_0 \tag{8-3}$$

and, since the x_i are equally spaced, $x_k = x_0 + kh$, so that $k = (x_k - x_0)/h$. On inserting this value of k in (8-3) we get

$$y_k = y_0 + \frac{x_k - x_0}{h}\Delta y_0 + \frac{(x_k - x_0)(x_k - x_0 - h)}{2!h^2}\Delta^2 y_0 + \cdots$$

$$+ \frac{(x_k - x_0)(x_k - x_0 - h)\cdots(x_k - x_0 - mh + h)}{m!h^m}\Delta^m y_0. \tag{8-4}$$

This relation is satisfied by $m + 1$ pairs of the tabulated values. If we assume that the value of y corresponding to an arbitrary x can be obtained from (8-4) by replacing x_k by x, we get the formula

$$y(x) = y_0 + \frac{x - x_0}{h}\Delta y_0 + \frac{(x - x_0)(x - x_0 - h)}{2!h^2}\Delta^2 y_0 + \cdots$$

$$+ \frac{(x - x_0)(x - x_0 - h)\cdots(x - x_0 - mh + h)}{m!h^m}\Delta^m y_0 \tag{8-5}$$

known as *Newton's forward-difference interpolation formula.* This formula can, of course, be used for either interpolation or extrapolation.

By replacing $(x - x_0)/h$ by a dimensionless variable X which represents the distance of x from x_0 in units of h, we get from (8-5)

$$y_X = y_0 + X\,\Delta y_0 + \frac{X(X-1)}{2!}\Delta^2 y_0 + \cdots + \frac{X(X-1)\cdots(X-m+1)}{m!}\Delta^m y_0 \tag{8-6}$$

where $X = (x - x_0)/h$ and $y_X = y(x_0 + hX) = y(x)$.

[1] These $m + 1$ pairs may include the entire set of given values (x_i, y_i), or they may be a subset so chosen that $|x - x_i|$ is as small as possible.

A similar calculation based on the use of (6-11) yields *Newton's backward-difference interpolation formula*

$$y_{n+X} = y_n + X \nabla y_n + \frac{X(X+1)}{2!} \nabla^2 y_n + \cdots$$

$$+ \frac{X(X+1) \cdots (X+m-1)}{m!} \nabla^m y_n \quad (8\text{-}7)$$

where $$X = \frac{x - x_n}{h} \quad \text{so that } x = x_n + hX$$

and $$y_{n+X} = y(x_n + hX) = y(x).$$

When the data cannot be represented by a polynomial, the right-hand members of (8-6) and (8-7) are infinite series involving differences of all orders.

Formulas (8-6) and (8-7) can be used to compute derivatives of tabulated functions. Thus, on differentiating successively (8-5) with respect to x and setting $x = x_0$ in the result, we get

$$y'(x_0) = \frac{1}{h} \left(\Delta y_0 - \tfrac{1}{2} \Delta^2 y_0 + \tfrac{1}{3} \Delta^3 y_0 - \tfrac{1}{4} \Delta^4 y_0 + \tfrac{1}{5} \Delta^5 y_0 - \cdots \right)$$

$$y''(x_0) = \frac{1}{h^2} \left(\Delta^2 y_0 - \Delta^3 y_0 + \tfrac{11}{12} \Delta^4 y_0 - \tfrac{5}{6} \Delta^5 y_0 + \cdots \right)$$

$$(8\text{-}8)$$

$$y'''(x_0) = \frac{1}{h^3} \left(\Delta^3 y_0 - \tfrac{3}{2} \Delta^4 y_0 + \tfrac{7}{4} \Delta^5 y_0 - \cdots \right)$$

$$y^{\text{iv}}(x_0) = \frac{1}{h^4} \left(\Delta^4 y_0 - 2 \Delta^5 y_0 + \cdots \right).$$

Formulas (8-8) should be used with caution because, even when $y = f(x)$ is well represented by the polynomial $P(x)$, the derivatives of $f(x)$ may differ significantly from those of $P(x)$.

Example. Using the data given in the example in Sec. 7, determine an approximate value for the y corresponding to $x = 2.2$.

First, let y be determined by using only the two neighboring observed values (hence, $m = 1$). Then, $x_0 = 2$, $y_0 = 2.808$, $\Delta y_0 = 0.806$, and $X = (2.2 - 2)/1 = 0.2$. Hence,

$$y = 2.808 + 0.2(0.806) = 2.969$$

which has been reduced to three decimal places because the observed data are not given more accurately. This is simply a straight-line interpolation by proportional parts.

If the three nearest values are chosen, $m = 2$, $x_0 = 1$, $y_0 = 2.105$, $\Delta y_0 = 0.703$, $\Delta^2 y_0 = 0.103$, and $X = 2.2 - 1 = 1.2$. Then,

$$y = 2.105 + 1.2(0.703) + \frac{(1.2)(0.2)}{2!}(0.103) = 2.961$$

correct to three decimal places.

If the four nearest values are chosen, $m = 3$, $x_0 = 1$, $y_0 = 2.105$, $\Delta y_0 = 0.703$, $\Delta^2 y_0 = 0.103$, $\Delta^3 y_0 = 0.081$, and $X = 1.2$. Therefore,

$$y = 2.105 + 1.2(0.703) + \frac{(1.2)(0.2)}{2}(0.103) + \frac{(1.2)(0.2)(-0.8)}{6}(0.081) = 2.958$$

correct to three decimal places.

1. Compute with the aid of formulas (8-6) and (8-7) the approximate values of y corresponding to $x = 5.5$ from the data of the example in Sec. 7. Use two and three neighboring values.

2. Extrapolate the value of y for $x = 8.2$ from the data in the example of Sec. 7 with the aid of (a) formula (8-6), (b) formula (8-7). Use $m = 2$.

3. Compute $y'(1)$ and $y''(1)$ from the data of the example of Sec. 7 with the aid of (8-8).

4. Given the table for $y(x)$

x	0	1	2
y	0.4384	0.4540	0.4695

determine $y(0.25)$ from (8-6) with $m = 2$ by taking $X = \frac{1}{4}$. Compare it with the value when $m = 1$ and $m = 3$.

5. Obtain, with the aid of (8-6), the interpolation polynomial for the function $y(x)$ given by the table

x	0	1	2	3	4
y	1	4	15	40	85

What is $y(1.5)$?

---

9. Lagrange's Interpolation Formula. The interpolation formulas developed in the preceding section apply only when the given set of x_i is an arithmetic progression. If this is not the case, some other type of formula must be applied.

As in Sec. 8, select the $m + 1$ pairs of observed values for which $|x - x_i|$ is as small as possible, and denote them by (x_i, y_i), where $i = 0, 1, 2, \ldots, m$. Let the mth-degree polynomials $P_k(x)$, where $k = 0, 1, 2, \ldots, m$, be defined by

$$P_k(x) = \frac{(x - x_0)(x - x_1) \cdots (x - x_m)}{x - x_k} \equiv \prod_{\substack{i=0 \\ i \neq k}}^{m} (x - x_i). \qquad (9\text{-}1)$$

Then, the coefficients A_k of the equation

$$y = \sum_{k=0}^{m} A_k P_k(x)$$

can be determined so that this equation is satisfied by each of the $m + 1$ pairs of observed values (x_i, y_i). If $x = x_k$, then

$$A_k = \frac{y_k}{P_k(x_k)}$$

since $P_k(x_i) = 0$ if $i \neq k$. Therefore,

$$y = \sum_{k=0}^{m} \frac{y_k P_k(x)}{P_k(x_k)} \qquad (9\text{-}2)$$

is the equation of the mth-degree polynomial which passes through the $m + 1$ points whose coordinates are (x_i, y_i).   If x is chosen as any value in the range of the x_i, (9-2) determines an approximate value for the corresponding y.

Equation (9-2) is known as *Lagrange's interpolation formula*.   Obviously, it can be applied when the x_i are in arithmetic progression but (8-5) is preferable in that it requires less tedious calculation.   Since only one mth-degree polynomial can be passed through $m + 1$ distinct points, it follows that (8-5), or its equivalent (8-6), and (9-2) are merely different forms of the same equation and will furnish the same value for y.

Example.   Using the data

v	10	15	22.5	33.75	50.625	75.937
p	0.300	0.675	1.519	3.417	7.689	17.300

apply Lagrange's formula to find the value of p corresponding to $v = 21$.

If the two neighboring pairs of observed values are chosen so that $m = 1$,

$$p = 0.675 \frac{21 - 22.5}{15 - 22.5} + 1.519 \frac{21 - 15}{22.5 - 15} = 1.350$$

correct to three decimal places.

If the three nearest values are chosen so that $m = 2$, then p equals

$$0.3 \frac{(21 - 15)(21 - 22.5)}{(10 - 15)(10 - 22.5)} + 0.675 \frac{(21 - 10)(21 - 22.5)}{(15 - 10)(15 - 22.5)} + 1.519 \frac{(21 - 10)(21 - 15)}{(22.5 - 10)(22.5 - 15)}.$$

Correct to three decimal places, this reduces to $p = 1.323$.

PROBLEMS

1. Using the data of the example in this section, find an approximate value for p when $v = 30$. Use $m = 1$ and $m = 2$.

2. Use $m = 1, 2,$ and 3 in formula (8-6) to find an approximate value of θ when $t = 2.3$, given

t	0	1	2	3	4	5	6	7	8
θ	60.00	51.66	44.46	38.28	32.94	28.32	24.42	21.06	18.06

3. Given the data

x	0.16	0.4	1.0	2.5	6.25	15.625
y	2	2.210	2.421	2.661	2.929	3.222

find an approximate value of y corresponding to $x = 2$.   Use formula (9-2) with $m = 1$ and $m = 2$.

4. Given the data

C	19	20	21	22	23	24	25
H	81.00	90.25	100.00	110.25	121.00	132.25	144.00

find an approximate value of H when $C = 21.6$.   Use formulas (8-6) and (8-7) with $m = 1, 2,$ and 3.

5. Obtain the Lagrange polynomial (9-2) that fits the data

x	-2	0	3	4
y	25	1	20	-23

and find $y(2)$.

---

10. Empirical Formulas. A given set of discrete data can be represented analytically in infinitely many ways. Such analytic representations are called *empirical formulas*, and the choice of the functional form for an empirical formula ordinarily depends on the use to be made of the formula. Thus, if a given set of data is to be represented by a function $f(x)$ which enters in the differential equation

$$L(u) = f(x)$$

the form of $f(x)$ may well depend on the ease with which this equation can be solved. For some types of differential operators L it may be wise to take $f(x)$ as an algebraic polynomial, in others as an exponential, and so on. Because of the commonness of algebraic and trigonometric polynomials in applications, we confine our discussion of empirical formulas primarily to these two types.

The first step usually taken by an experimenter in appraising a set of observed values (x_i, y_i) is to plot them on some coordinate paper and draw a curve through the plotted points. If the points (x_i, y_i), when plotted on a rectangular coordinate paper, lie approximately on a straight line, he assumes that the equation $y = mx + b$ represents the relationship. To determine the constants m and b, the slope and the y intercept may be read off the graph or they may be calculated by solving two linear equations for m and b got by substituting the coordinates of two judiciously chosen points on $y = mx + b$.

If the plot of points on a logarithmic coordinate paper indicates that they lie on a straight line, the desired relationship has the form

$$y = ax^m$$

for, on taking logarithms, we get

$$\log y = \log a + m \log x$$

and if coordinate axes X, Y are marked so that $\log y = Y$ and $\log x = X$, we get a linear equation

$$Y = \log a + mX.$$

Again the constants a and m can be either read off the graph or computed by solving a pair of linear equations for m and $\log a$.

Similarly, the data can be represented by an exponential function

$$y = a10^{mx}$$

if the values (x_i, y_i) when plotted on a semilogarithmic paper fall on a straight line, for on taking logarithms to the base 10 we get

$$\log y = \log a + mx$$

which is linear in $\log y$ and x.

When none of these simple functional relationships fits the data, one may determine, with the aid of Sec. 7, if the data can be fitted by a polynomial. It should be stressed, however, that ordinarily the choice of an empirical formula is governed by whatever uses are to be made of it.   Once a formula is chosen, the parameters entering in it (such as the coefficients in the polynomial representation) can be determined by imposing some criterion for the goodness of fit of the data by the chosen function.   The method of least squares, presented in the next section, provides one of the most commonly used of such criteria.

PROBLEMS

1. Plot the following data on a rectangular, logarithmic, or semilogarithmic paper to determine the approximate functional relationships between y and x.

(a)

x	3	4	5	6	7	8	9	10	11	12
y	5	5.6	6	6.4	7	7.5	8.2	8.6	9	9.5

(b)

x	1	2	3	4	5	6	7	8	9
y	2.5	3.5	4.3	5	5.6	6.2	6.6	7.1	7.5

(c)

x	1	2	3	4	5	6	7	8
y	0.5	0.8	1.2	1.9	3	4.8	7.5	11.9

2. Verify that the data in Probs. 2, 3, and 4 in Sec. 9 may be approximated by the following types of functions: $\theta = a10^{mt}$, $y = ax^m$, $H = a_0 + a_1C + a_2C^2$, respectively.   Determine the parameters graphically or analytically.

11. The Method of Least Squares.   We saw in Sec. 7 that the $m + 1$ coefficients in the polynomial

$$y = a_0 + a_1x + \cdots + a_mx^m \tag{11-1}$$

can always be determined so that a given set of $m + 1$ points (x_i, y_i), where the x's are unequal, lies on the curve (11-1).   When the x_i are equally spaced, the desired polynomial is determined by the formula (8-5) and, in the more general case, by (9-2).

When the number of points is large, the degree m of the polynomial (11-1) is high, and an attempt to represent the data exactly by (11-1) not only is laborious but may be foolish, for the experimental data invariably contain observational errors and it may be more sensible to represent the data approximately by some function $y = f(x)$ which contains a few unknown parameters.   These parameters can then be determined so that the curve $y = f(x)$ fits the data in "the best possible way."   The criteria as to what constitutes "the best possible way" are, of course, arbitrary.

For example, we may attempt to fit the set of plotted points in Fig. 12 by the straight line

$$y = a_1 + a_2x$$

and choose the parameters a_1 and a_2 so that the

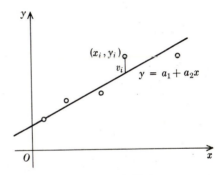

FIGURE 12

sum of the squares of the vertical deviations of the plotted points from this line is as small as possible.

More generally, if we choose to represent a set of data (x_i, y_i), where $i = 1, 2, \ldots, n$, by some relationship $y = f(x)$, containing r unknown parameters $a_1, a_2, \ldots, a_r$, and form the deviations (or the *residuals*, as they are also called)

$$v_i = f(x_i) - y_i \tag{11-2}$$

the sum of the squares of the deviations

$$S \equiv \sum_{i=1}^{n} v_i^2 = \sum_{i=1}^{n} [f(x_i) - y_i]^2 \tag{11-3}$$

is clearly a function of $a_1, a_2, \ldots, a_r$. We can then determine the a's so that S is a minimum.

Now, if $S(a_1, a_2, \ldots, a_r)$ is a minimum, then at the point in question

$$\frac{\partial S}{\partial a_1} = 0, \qquad \frac{\partial S}{\partial a_2} = 0, \qquad \ldots, \qquad \frac{\partial S}{\partial a_r} = 0. \tag{11-4}$$

The set of r equations (11-4), called *normal equations*, serves to determine the r unknown a's in $y = f(x)$. This particular criterion of the "best fit" of data is known as the *principle of least squares*, and the method of determining the unknown parameters with its aid is called the *method of least squares*. It was introduced and fully developed by Gauss[1] when he was a youth of seventeen!

We indicate the construction of the normal equations first by supposing that $y = f(x)$ is a linear function

$$y = a_1 + a_2 x. \tag{11-5}$$

The residuals (11-2) for (11-5) are $v_i = (a_1 + a_2 x_i) - y_i$, so that $S = \sum_{i=1}^{n} v_i^2$ is

$$S = (a_1 + a_2 x_1 - y_1)^2 + (a_1 + a_2 x_2 - y_2)^2 + \cdots + (a_1 + a_2 x_n - y_n)^2.$$

On differentiating S with respect to a_1 and a_2, we deduce two equations:

$$\frac{\partial S}{\partial a_1} = 2(a_1 + a_2 x_1 - y_1) + 2(a_1 + a_2 x_2 - y_2) + \cdots + 2(a_1 + a_2 x_n - y_n) = 0$$

$$\frac{\partial S}{\partial a_2} = 2x_1(a_1 + a_2 x_1 - y_1) + 2x_2(a_1 + a_2 x_2 - y_2) + \cdots + 2x_n(a_1 + a_2 x_n - y_n) = 0.$$

If we divide out the factor 2 and collect the coefficients of a_1 and a_2, we get

$$na_1 + \left(\sum_{i=1}^{n} x_i\right) a_2 = \sum_{i=1}^{n} y_i$$

$$\left(\sum_{i=1}^{n} x_i\right) a_1 + \left(\sum_{i=1}^{n} x_i^2\right) a_2 = \sum_{i=1}^{n} x_i y_i. \tag{11-6}$$

These equations can be easily solved for a_1 and a_2, since the determinant in (11-6) does not vanish. This follows from the Schwarz inequality in Chap. 4, Sec. 14.

[1] The criterion of least squares plays a fundamental role in the approximation of a suitably restricted function $f(x)$ by a linear combination of orthogonal functions. As is shown in Chap. 1, Sec. 24, the partial sums of Fourier series give the best fit in the sense of least squares. It should be noted, however, that the polynomials giving the best fit to $f(x)$ in the sense of least squares, in general, are *not* the partial sums of Maclaurin's or Taylor's series for $f(x)$. See in this connection Stiefel, *op. cit.*, sec. 3.

Example 1.   We illustrate the use of Eqs. (11-6) by calculating the coefficients in $y = a_1 + a_2 x$ to fit the following data:

x	1	2	3	4
y	1.7	1.8	2.3	3.2

In this case $n = 4$, and since

$$\sum_{i=1}^{4} x_i = 1 + 2 + 3 + 4 = 10 \qquad \sum_{i=1}^{4} y_i = 1.7 + 1.8 + 2.3 + 3.2 = 9$$

$$\sum_{i=1}^{4} x_i^2 = 1 + 4 + 9 + 16 = 30 \qquad \sum_{i=1}^{4} x_i y_i = 1.7 + 2(1.8) + 3(2 \ 3) + 4(3.2) = 25$$

the system (11-6) reads

$$4a_1 + 10a_2 = 9$$
$$10a_1 + 30a_2 = 25.$$

Solving for a_1 and a_2, we get $a_1 = 1$, $a_2 = \frac{1}{2}$, so that the desired straight line fitting the data in the sense of least squares is $y = 1 + \frac{1}{2}x$.

We suppose next that $y = f(x)$ is a polynomial

$$y = a_1 + a_2 x + a_3 x^2 + \cdots + a_r x^{r-1} = \sum_{j=1}^{r} a_j x^{j-1}. \tag{11-7}$$

The residuals v_i this time are

$$v_i = \sum_{j=1}^{r} a_j x_i^{j-1} - y_i. \tag{11-8}$$

Since $$S = \sum_{i=1}^{n} v_i^2$$

Eqs. (11-4) can be written as

$$\frac{\partial S}{\partial a_k} = 2 \sum_{i=1}^{n} v_i \frac{\partial v_i}{\partial a_k} = 0 \qquad k = 1, 2, \ldots, r. \tag{11-9}$$

From (11-8), $$\frac{\partial v_i}{\partial a_k} = x_i^{k-1}$$

so that, on dividing out the factor 2, we can write the normal equations (11-9) as

$$\sum_{i=1}^{n} v_i x_i^{k-1} = 0. \tag{11-10}$$

The substitution from (11-8) in (11-10) yields

$$\sum_{i=1}^{n} \left(\sum_{j=1}^{r} a_j x_i^{j-1} - y_i \right) x_i^{k-1} = 0$$

and on collecting the coefficients of the a_j we get a set of r linear equations

$$\sum_{j=1}^{r} \left(\sum_{i=1}^{n} x_i^{j+k-2} \right) a_j = \sum_{i=1}^{n} x_i^{k-1} y_i \qquad k = 1, 2, \ldots, r \tag{11-11}$$

for $a_1, a_2, \ldots, a_r$.

We illustrate the use of these equations by two examples.

Example 2.   Let the data in Example 1 be fitted by $y = a_1 + a_2x + a_3x^2$.   Then

$$v_i = a_1 + a_2x_i + a_3x_i{}^2 - y_i$$

and

$$\frac{\partial v_i}{\partial a_1} = 1, \qquad \frac{\partial v_i}{\partial a_2} = x_i, \qquad \frac{\partial v_i}{\partial a_3} = x_i{}^2.$$

The normal equations

$$\sum_{i=1}^{4} v_i \frac{\partial v_i}{\partial a_k} = 0 \qquad k = 1, 2, 3$$

are

$$\sum_{i=1}^{4} (a_1 + a_2x_i + a_3x_i{}^2 - y_i) \cdot 1 = 0$$

$$\sum_{i=1}^{4} (a_1 + a_2x_i + a_3x_i{}^2 - y_i)x_i = 0 \qquad \text{and} \qquad \sum_{i=1}^{4} (a_1 + a_2x_i + a_3x_i{}^2 - y_i)x_i{}^2 = 0.$$

If the coefficients of the a_j are collected and the normal equations put in the form (11-11), one obtains the three equations

$$4a_1 + \left(\sum_{i=1}^{4} x_i\right) a_2 + \left(\sum_{i=1}^{4} x_i{}^2\right) a_3 = \sum_{i=1}^{4} y_i$$

$$\left(\sum_{i=1}^{4} x_i\right) a_1 + \left(\sum_{i=1}^{4} x_i{}^2\right) a_2 + \left(\sum_{i=1}^{4} x_i{}^3\right) a_3 = \sum_{i=1}^{4} x_iy_i$$

$$\left(\sum_{i=1}^{4} x_i{}^2\right) a_1 + \left(\sum_{i=1}^{4} x_i{}^3\right) a_2 + \left(\sum_{i=1}^{4} x_i{}^4\right) a_3 = \sum_{i=1}^{4} x_i{}^2y_i.$$

Now,

$$\sum_{i=1}^{4} x_i = 1 + 2 + 3 + 4 = 10 \qquad \sum_{i=1}^{4} x_i{}^2 = 1 + 4 + 9 + 16 = 30$$

$$\sum_{i=1}^{4} x_iy_i = 1.7 + 3.6 + 6.9 + 12.8 = 25, \text{ etc.}$$

The equations become

$$4a_1 + 10a_2 + 30a_3 = 9$$
$$10a_1 + 30a_2 + 100a_3 = 25$$
$$30a_1 + 100a_2 + 354a_3 = 80.8$$

and the solutions are $a_1 = 2$, $a_2 = -0.5$, $a_3 = 0.2$.

Example 3.   Let us apply the method of least squares to fit the data

x	1	2	3	4	5	6	7	8
y	2.105	2.808	3.614	4.604	5.857	7.451	9.467	11.985

by the polynomial $y = a_1 + a_2x + a_3x^2 + a_4x^3$.

In this case $n = 8$ and Eqs. (11-11) yield four normal equations obtained by setting $k = 1, 2, 3, 4$. They are

$$8a_1 + \left(\sum_{i=1}^{8} x_i\right) a_2 + \left(\sum_{i=1}^{8} x_i{}^2\right) a_3 + \left(\sum_{i=1}^{8} x_i{}^3\right) a_4 = \sum_{i=1}^{8} y_i$$

$$\left(\sum_{i=1}^{8} x_i\right) a_1 + \left(\sum_{i=1}^{8} x_i{}^2\right) a_2 + \left(\sum_{i=1}^{8} x_i{}^3\right) a_3 + \left(\sum_{i=1}^{8} x_i{}^4\right) a_4 = \sum_{i=1}^{8} x_i y_i$$

$$\left(\sum_{i=1}^{8} x_i{}^2\right) a_1 + \left(\sum_{i=1}^{8} x_i{}^3\right) a_2 + \left(\sum_{i=1}^{8} x_i{}^4\right) a_3 + \left(\sum_{i=1}^{8} x_i{}^5\right) a_4 = \sum_{i=1}^{8} x_i{}^2 y_i$$

$$\left(\sum_{i=1}^{8} x_i{}^3\right) a_1 + \left(\sum_{i=1}^{8} x_i{}^4\right) a_2 + \left(\sum_{i=1}^{8} x_i{}^5\right) a_3 + \left(\sum_{i=1}^{8} x_i{}^6\right) a_4 = \sum_{i=1}^{8} x_i{}^3 y_i .$$

From the form of the coefficients of the a_k, it is seen that it is convenient to make a table of the powers of the x_i and to form the sums $\Sigma x_i{}^j$ and $\Sigma x_i{}^l y_i$ before attempting to write the equations in explicit form.

x_i	$x_i{}^2$	$x_i{}^3$	$x_i{}^4$	$x_i{}^5$	$x_i{}^6$
1	1	1	1	1	1
2	4	8	16	32	64
3	9	27	81	243	729
4	16	64	256	1,024	4,096
5	25	125	625	3,125	15,625
6	36	216	1,296	7,776	46,656
7	49	343	2,401	16,807	117,649
8	64	512	4,096	32,768	262,144
$\Sigma x_i{}^j$ 36	204	1,296	8,772	61,776	446,964

x_i	y_i	$x_i y_i$	$x_i{}^2 y_i$	$x_i{}^3 y_i$
1	2.105	2.105	2.105	2.105
2	2.808	5.616	11.232	22.464
3	3.614	10.842	32.526	97.578
4	4.604	18.416	73.664	294.656
5	5.857	29.285	146.425	732.125
6	7.451	44.706	268.236	1,609.416
7	9.467	66.269	463.883	3,247.181
8	11.985	95.880	767.040	6,136.320
$\Sigma x_i{}^l y_i$	47.891	273.119	1,765.111	12,141.845

When the values given in the tables are inserted, the normal equations become

$$8a_1 + 36a_2 + 204a_3 + 1.296a_4 = 47.891$$
$$36a_1 + 204a_2 + 1,296a_3 + 8,772a_4 = 273.119$$
$$204a_1 + 1,296a_2 + 8,772a_3 + 61,776a_4 = 1,765.111$$
$$1,296a_1 + 8,772a_2 + 61,776a_3 + 446,964a_4 = 12,141.845.$$

The solutions are

$$a_1 = 1.426 \qquad a_2 = 0.693 \qquad a_3 = -0.028 \qquad a_4 = 0.013.$$

Therefore, the equation, as determined by the method of least squares, is

$$y = 1.426 + 0.693x - 0.028x^2 + 0.013x^3.$$

The normal equations (11-11), corresponding to the polynomial representation of data, are linear in the coefficients a_i. They need not be linear in the unknown parameters if the function $y = f(x)$ is not a polynomial in x. In this event the solution of the system (11-4) may prove difficult, and one may be obliged to seek an approximate solution by eplacing the exact residuals (11-2) by approximate residuals which are linear in the ınknowns. This is accomplished by expanding $y = f(x)$, treated as a function of $a_1, a_2, \ldots, a_r$, in Taylor's series in terms of $a_i - \bar{a}_i \equiv \Delta a_i$, where the $\bar{a}_i$ are approximate values of the a_i. The values of $\bar{a}_i$ may be obtained by graphical means or by solving any r of the equations $y_i = f(x_i)$. The expansion gives

$$y = f(x,a_1, \ldots ,a_r) \equiv f(x, \bar{a}_1 + \Delta a_1, \ldots ,\bar{a}_r + \Delta a_r)$$

$$= f(x,\bar{a}_1, \ldots ,\bar{a}_r) + \sum_{k=1}^{r} \frac{\partial f}{\partial \bar{a}_k} \Delta a_k + \frac{1}{2!} \sum_{j,k=1}^{r} \frac{\partial^2 f}{\partial \bar{a}_j\, \partial \bar{a}_k} \Delta a_j\, \Delta a_k + \cdots \qquad (11\text{-}12)$$

where
$$\frac{\partial f}{\partial \bar{a}_k} \equiv \frac{\partial f}{\partial a_k}\bigg|_{a_k = \bar{a}_k}, \qquad \frac{\partial^2 f}{\partial \bar{a}_j\, \partial \bar{a}_k} \equiv \frac{\partial^2 f}{\partial a_j\, \partial a_k}\bigg|_{\substack{a_j = \bar{a}_j \\ a_k = \bar{a}_k}}, \text{ etc.}$$

Assuming that the $\bar{a}_i$ are chosen so that the Δa_i are small, the terms of degree higher than the first can be neglected and (11-12) becomes

$$y = f(x,\bar{a}_1, \ldots ,\bar{a}_r) + \sum_{k=1}^{r} \frac{\partial f}{\partial \bar{a}_k} \Delta a_k.$$

The n observation equations are then replaced by the n approximate equations

$$\bar{y}_i = f(x_i,\bar{a}_i, \ldots ,\bar{a}_r) + \sum_{k=1}^{r} \frac{\partial f}{\partial \bar{a}_k} \Delta a_k. \qquad (11\text{-}13)$$

If (11-13) is used, the residuals v_i will be linear in the Δa_k, and hence the resulting conditions, which become

$$\frac{\partial S}{\partial(\Delta a_k)} = 0 \qquad k = 1, 2, \ldots, r \qquad (11\text{-}14)$$

also will be linear in the Δa_k. Equations (11-14) are called the normal equations in this case.

We illustrate the use of Eqs. (11-14) in Example 4.

Example 4.   We seek to determine the constant k and a in the formula $\theta = ka^t$ chosen to represent the following data:

t	1	2	3	4
θ	51.66	44.46	38.28	32.94

The determination of k and a in this problem can be reduced to the solution of two linear equations, for if we write $\theta = ka^t$ in the form

$$\theta = k10^{bt}$$

then, on taking logarithms to the base 10, we get

$$\log \theta = \log k + bt.$$

Setting $\log \theta = y$ and $\log k = K$, we get

$$y = K + bt \tag{11-15}$$

which is linear in K and b. These constants can be determined by the procedure described above, which leads to the solution of a pair of linear equations. However, the approximation obtained by this means does not give an approximation to the *original* equation in the sense of least squares.

To illustrate the use of formulas (11-12) to (11-14), we follow a more laborious route which gives an approximation to the original equation.

When the values recorded in the table are plotted on semilogarithmic paper, it is found that $k = 60$ and $a = 10^{-0.065} = 0.86$, approximately. This suggests using $k_0 = 60$ and $a_0 = 0.9$ as the first approximations. The first two terms of the expansion in Taylor's series in terms of $\Delta k = k - 60$ and $\Delta a = a - 0.9$ are

$$\theta = 60(0.9)^t + \left(\frac{\partial \theta}{\partial k}\right)_{\substack{k=60 \\ a=0.9}} \Delta k + \left(\frac{\partial \theta}{\partial a}\right)_{\substack{k=60 \\ a=0.9}} \Delta a$$

$$= 60(0.9)^t + (0.9)^t \, \Delta k + 60t(0.9)^{t-1} \, \Delta a.$$

If the values (t_i, θ_i) are substituted in this equation, four equations result, namely,

$$\theta_i = 60(0.9)^{t_i} + (0.9)^{t_i} \, \Delta k + 60t_i(0.9)^{t_i-1} \, \Delta a \qquad i = 1, 2, 3, 4.$$

The problem of obtaining from these four equations the values of Δk and Δa, which furnish the desired values of θ_i, is precisely the same as in the case in which the original equation is linear in its constants. The residual equations are

$$v_i = (0.9)^{t_i} \, \Delta k + 60t_i(0.9)^{t_i-1} \, \Delta a + 60(0.9)^{t_i} - \theta_i \qquad i = 1, 2, 3, 4.$$

Therefore

$$S \equiv \sum_{i=1}^{4} v_i{}^2 = \sum_{i=1}^{4} [(0.9)^{t_i} \, \Delta k + 60t_i(0.9)^{t_i-1} \, \Delta a + 60(0.9)^{t_i} - \theta_i]^2$$

and the normal equations

$$\frac{\partial S}{\partial(\Delta k)} = 0 \qquad \text{and} \qquad \frac{\partial S}{\partial(\Delta a)} = 0$$

become

$$2\sum_{i=1}^{4} [0.9^{t_i} \, \Delta k + 60t_i(0.9)^{t_i-1} \, \Delta a + 60(0.9)^{t_i} - \theta_i]0.9^{t_i} = 0$$

and

$$2\sum_{i=1}^{4} [0.9^{t_i} \, \Delta k + 60t_i(0.9)^{t_i-1} \, \Delta a + 60(0.9)^{t_i} - \theta_i]60t_i(0.9)^{t_i-1} = 0.$$

When these equations are written in the form

$$p \, \Delta k + q \, \Delta a = r$$

with all common factors divided out, they are

$$\sum_{i=1}^{4} (0.9)^{2t_i} \, \Delta k + 60 \sum_{i=1}^{4} t_i(0.9)^{2t_i-1} \, \Delta a = \sum_{i=1}^{4} \theta_i(0.9)^{t_i} - 60 \sum_{i=1}^{4} (0.9)^{2t_i}$$

and

$$\sum_{i=1}^{4} t_i(0.9)^{2t_i-1} \, \Delta k + 60 \sum_{i=1}^{4} t_i{}^2(0.9)^{2t_i-2} \, \Delta a = \sum_{i=1}^{4} \theta_i t_i(0.9)^{t_i-1} - 60 \sum_{i=1}^{4} t_i(0.9)^{2t_i-1}.$$

As in Example 3, the coefficients are computed most conveniently by the use of a table.

t_i	1	2	3	4	*Totals*
$(0.9)^{t_i}$	0.9	0.81	0.729	0.6561	
$(0.9)^{2t_i}$	0.81	0.6561	0.531441	0.43046721	2.42800821
$t_i(0.9)^{2t_i-1}$	0.9	1.458	1.77147	1.9131876	6.0426576
$t_i^2(0.9)^{2t_i-2}$	1	3.24	5.9049	8.503056	18.647956
$(\theta_i)(0.9)^{t_i}$	46.494	36.0126	27.90612	21.611934	132.024654
$(\theta_i t_i)(0.9)^{t_i-1}$	51.66	80.028	93.0204	96.05304	320.76144

Substituting the values of the sums from the table gives

$$2.42800821\,\Delta k + 362.559456\,\Delta a = 132.024654 - 145.6804926$$

and

$$6.0426576\,\Delta k + 1{,}118.87736\,\Delta a = 320.76144 - 362.559456.$$

Reducing all the numbers to four decimal places gives the following equations to solve for Δk and Δa:

$$2.4280\,\Delta k + 362.5595\,\Delta a = -13.6558$$
$$6.0427\,\Delta k + 1{,}118.8774\,\Delta a = -41.7980.$$

The solutions are

$$\Delta k = -0.238 \quad \text{and} \quad \Delta a = -0.036.$$

Hence, the required equation is

$$\theta = 59.762(0.864)^t.$$

PROBLEMS

1. Apply the method of least squares to find the constants in $y = a_1 + a_2 x + a_3 x^2$ to fit the data

x	1	2	3	4	5	6
y	3.13	3.76	6.94	12.62	20.86	31.53

2. Determine by the method of least squares the constants a and n in $p = av^n$ to fit the following data by writing the equation in the form $\log p = n \log v + \log a$.

v	10	15	22.5	33.7	50.6	75.9
p	0.300	0.675	1.519	3.417	7.689	17.300

Hint: Set $\log p = y$, $\log v = x$, and determine the constants in the resulting linear equation.

3. Compare the result of Example 4 with the calculation of the constants in (11-15).

---

12. Harmonic Analysis.   The problem of representing a suitable periodic function in a trigonometric series was considered in some detail in Chap. 1. In this section we give a brief discussion of the problem of fitting a finite trigonometric sum to a set of observed values (x_i, y_i).   Let the set of observed values

$$(x_0, y_0),\ (x_1, y_1),\ \ldots,\ (x_{2n-1}, y_{2n-1}),\ (x_{2n}, y_{2n}),\ \ldots$$

be such that the values of y start repeating with y_{2n} (that is, $y_{2n} = y_0$, $y_{2n+1} = y_1$, etc.). It will be assumed that the x_i are equally spaced, that $x_0 = 0$, and that $x_{2n} = 2\pi$.   [If $x_0 \neq 0$ and the period is c instead of 2π, the variable can be changed by setting

$$\theta_i = \frac{2\pi}{c}\,(x_i - x_0).$$

The discussion would then be carried through for θ_i and y_i in place of the x_i and y_i used below.]   Under these assumptions,

$$x_i = i\,\frac{2\pi}{2n} = \frac{i\pi}{n}.$$

The trigonometric polynomial

$$y = A_0 + \sum_{k=1}^{n} A_k \cos kx + \sum_{k=1}^{n-1} B_k \sin kx \tag{12-1}$$

contains the $2n$ unknown constants

$$A_0,\ A_1,\ A_2,\ \ldots,\ A_n,\ B_1,\ B_2,\ \ldots,\ B_{n-1},$$

which can be determined so that (12-1) will pass through the $2n$ given points (x_i, y_i) by solving the $2n$ simultaneous equations

$$y_i = A_0 + \sum_{k=1}^{n} A_k \cos kx_i + \sum_{k=1}^{n-1} B_k \sin kx_i \qquad i = 0, 1, 2, \ldots, 2n - 1.$$

Since $x_i = i\pi/n$, these equations become

$$y_i = A_0 + \sum_{k=1}^{n} A_k \cos \frac{ik\pi}{n} + \sum_{k=1}^{n-1} B_k \sin \frac{ik\pi}{n} \qquad i = 0, 1, 2, \ldots, 2n - 1. \tag{12-2}$$

The solution of Eqs. (12-2) is much simplified by means of a scheme somewhat similar to that used in determining the Fourier coefficients.   Multiplying both sides of each equation by the coefficient of A_0 (that is, by unity) and adding the results give

$$\sum_{i=0}^{2n-1} y_i = 2nA_0 + \sum_{k=1}^{n}\left(\sum_{i=0}^{2n-1} \cos \frac{ik\pi}{n} \right) A_k + \sum_{k=1}^{n-1}\left(\sum_{i=0}^{2n-1} \sin \frac{ik\pi}{n} \right) B_k.$$

It can be established that (cf. Chap. 1. Sec. 17, Example 2)

$$\sum_{i=0}^{2n-1} \cos \frac{ik\pi}{n} = 0 \qquad k = 1, 2, \ldots, n$$

and

$$\sum_{i=0}^{2n-1} \sin \frac{ik\pi}{n} = 0 \qquad k = 1, 2, \ldots, n - 1.$$

Therefore,

$$2nA_0 = \sum_{i=0}^{2n-1} y_i. \tag{12-3}$$

Multiplying both sides of each equation in (12-2) by the coefficient of A_j in it, and adding the results, gives

$$\sum_{i=0}^{2n-1} y_i \cos \frac{ij\pi}{n} = \sum_{k=1}^{n}\left(\sum_{i=0}^{2n-1} \cos \frac{ik\pi}{n} \cos \frac{ij\pi}{n} \right) A_k + \sum_{k=1}^{n-1}\left(\sum_{i=0}^{2n-1} \sin \frac{ik\pi}{n} \cos \frac{ij\pi}{n} \right) B_k$$

for $j = 1, 2, \ldots, n - 1$. But

$$\sum_{i=0}^{2n-1} \cos \frac{ik\pi}{n} \cos \frac{ij\pi}{n} = 0 \qquad \text{if } k \neq j$$

$$= n \qquad \text{if } k = j$$

and

$$\sum_{i=0}^{2n-1} \sin \frac{ik\pi}{n} \cos \frac{ij\pi}{n} = 0$$

for all values of k.   Therefore,

$$nA_j = \sum_{i=0}^{2n-1} y_i \cos \frac{ij\pi}{n} \qquad j = 1, 2, \ldots, n - 1. \tag{12-4}$$

To determine the coefficient of A_n the procedure is precisely the same, but

$$\sum_{i=0}^{2n-1} \cos \frac{ik\pi}{n} \cos i\pi = 0 \qquad \text{if } k \neq n$$

$$= 2n \qquad \text{if } k = n.$$

Hence,

$$2nA_n = \sum_{i=0}^{2n-1} y_i \cos i\pi. \tag{12-5}$$

Similarly, on multiplying both sides of each equation of (12-2) by the coefficient of B_k in it and adding, one finds that

$$nB_j = \sum_{i=0}^{2n-1} y_i \sin \frac{ij\pi}{n} \qquad j = 1, 2, \ldots, n - 1. \tag{12-6}$$

Equations (12-3) to (12-6) give the constants in (12-1). A compact schematic arrangement is often used to simplify the labor of evaluating these constants.   It will be illustrated in the so-called "6-ordinate" case, that is, when $2n = 6$. The method is based on the equations that determine the constants, together with relations such as

$$\sin \frac{\pi}{n} = \sin \frac{(n-1)\pi}{n} = -\sin \frac{(n+1)\pi}{n} = -\sin \frac{(2n-1)\pi}{n}$$

$$\cos \frac{\pi}{n} = -\cos \frac{(n-1)\pi}{n} = -\cos \frac{(n+1)\pi}{n} = \cos \frac{(2n-1)\pi}{n}.$$

Six-ordinate Scheme.   Here, $2n = 6$; the given points are (x_i, y_i), where $x_i = i\pi/3$ $(i = 0, 1, 2, 3, 4, 5)$; and Eq. (12-1) becomes

$$y = A_0 + A_1 \cos x + A_2 \cos 2x + A_3 \cos 3x + B_1 \sin x + B_2 \sin 2x.$$

Make the following table of definitions:

	y_0	y_1	y_2	v_0	v_1	w_0	w_1
	y_3	y_4	y_5		v_2		w_2
Sum	v_0	v_1	v_2	p_0	p_1	r_0	r_1
Difference	w_0	w_1	w_2		q_1		s_1

It can be checked easily that Eqs. (12-3) to (12-6), with $n = 3$, become

$$6A_0 = p_0 + p_1 \qquad 3A_1 = r_0 + \tfrac{1}{2}s_1 \qquad 3A_2 = p_0 - \tfrac{1}{2}p_1$$

$$6A_3 = r_0 - s_1 \qquad 3B_1 = \frac{\sqrt{3}}{2} r_1 \qquad 3B_2 = \frac{\sqrt{3}}{2} q_1.$$

Example.   In particular, suppose that the given points are

x	0	$\dfrac{\pi}{3}$	$\dfrac{2\pi}{3}$	π	$\dfrac{4\pi}{3}$	$\dfrac{5\pi}{3}$	2π
y	1.0	1.4	1.9	1.7	1.5	1.2	1.0

Upon using these values of y in the table of definitions above,

$$
\begin{array}{ccc}
1.0 & 1.4 & 1.9 \\
1.7 & 1.5 & 1.2 \\
\hline
v_0 = 2.7 & v_1 = 2.9 & v_2 = 3.1 \\
w_0 = -0.7 & w_1 = -0.1 & w_2 = 0.7
\end{array}
$$

$$
\begin{array}{cccc}
2.7 & 2.9 & -0.7 & -0.1 \\
 & 3.1 & & 0.7 \\
\hline
p_0 = 2.7 & p_1 = 6.0 & r_0 = -0.7 & r_1 = 0.6 \\
 & q_1 = -0.2 & & s_1 = -0.8
\end{array}
$$

Therefore, the equations determining the values of the constants are

$$6A_0 = \quad 2.7 + 6.0 = \quad 8.7 \qquad \text{and} \qquad A_0 = \quad 1.45$$

$$3A_1 = -0.7 - 0.4 = -1.1 \qquad \text{and} \qquad A_1 = -0.37$$

$$3A_2 = \quad 2.7 - 3.0 = -0.3 \qquad \text{and} \qquad A_2 = -0.10$$

$$6A_3 = -0.7 + 0.8 = \quad 0.1 \qquad \text{and} \qquad A_3 = \quad 0.02$$

$$3B_1 = \frac{\sqrt{3}}{2}(0.6) \quad = \quad 0.3\sqrt{3} \qquad \text{and} \qquad B_1 = \quad 0.17$$

$$3B_2 = \frac{\sqrt{3}}{2}(-0.2) = -0.1\sqrt{3} \qquad \text{and} \qquad B_2 = -0.06.$$

Hence, the curve of type (12-1) that fits the given data is

$$y = 1.45 - 0.37 \cos x - 0.10 \cos 2x + 0.02 \cos 3x + 0.17 \sin x - 0.06 \sin 2x.$$

A convenient check upon the computations is furnished by the relations

$$A_0 + A_1 + A_2 + A_3 = y_0 \qquad \text{and} \qquad B_1 + B_2 = \frac{\sqrt{3}}{3}(y_1 - y_5).$$

Substituting the values found above in the left-hand members gives

$$1.45 - 0.37 - 0.10 + 0.02 = 1.0 \qquad \text{and} \qquad 0.17 - 0.06 = 0.11$$

which check with the values of the right-hand members.

Similar tables can be constructed for 8-ordinates, 12-ordinates, etc.

PROBLEMS

1. Use the 6-ordinate scheme to fit a curve of the type (12-1) to the data in the following table:

x	0	$\dfrac{\pi}{3}$	$\dfrac{2\pi}{3}$	π	$\dfrac{4\pi}{3}$	$\dfrac{5\pi}{3}$	2π
y	0.8	0.6	0.4	0.7	0.9	1.1	0.8

2. Make a suitable change of variable, and apply the 6-ordinate scheme to the data given in the table:

x	0	$\dfrac{\pi}{6}$	$\dfrac{\pi}{3}$	$\dfrac{\pi}{2}$	$\dfrac{2\pi}{3}$	$\dfrac{5\pi}{6}$	π
y	0.6	0.9	1.3	1.0	0.8	0.5	0.6

NUMERICAL INTEGRATION OF DIFFERENTIAL EQUATIONS

13. Numerical Integration.   The reader is familiar with the interpretation of the definite integral $\int_b^a f(x)\,dx$ as the area under the curve $y = f(x)$ between the ordinates $x = a$ and $x = b$. This interpretation underlies the construction of formulas for numerical integration contained in this section.

It will be recalled that, if the function $f(x)$ is such that its indefinite integral can be obtained in closed form, the fundamental theorem of integral calculus provides an easy means for evaluating the definite integral. However, when $f(x)$ does not have an indefinite integral expressible in terms of known functions, or when the values of $f(x)$ are given in tabular form, formulas for numerical integration are generally used to obtain an approximate value of the integral.[1]

Formulas for numerical integration, or *mechanical quadrature*, are obtained by replacing the function $f(x)$ specified at a given number of points in the interval (a,b) by a polynomial (8-5) or (9-2), depending on whether the values of x are equally or unequally spaced.

If the values of $y = f(x)$ are known at $m + 1$ points x_i, where $i = 0, 1, 2, \ldots, m$, which are spaced h units apart, an approximate value of the integral $\int_{x_0}^{x_m} f(x)\,dx$ can be computed by substituting in the integrand an approximate polynomial representation of $y = f(x)$ given by (8-5) or, equivalently, (8-6). We thus get for equally spaced values x_i

$$\int_0^m y\,dX = \int_0^m \left[y_0 + X\,\Delta y_0 + \frac{X(X-1)}{2!}\,\Delta^2 y_0 \right.$$
$$\left. + \cdots + \frac{X(X-1)\cdots(X-m+1)}{m!}\,\Delta^m y_0 \right] dX \quad (13\text{-}1)$$

where X is the dimensionless variable defined by

$$X = \frac{x - x_0}{h} \tag{13-2}$$

The upper limit of integration in (13-1) is m because $X = m$ when

$$x = x_m = x_0 + mh. \tag{13-3}$$

If $m = 1$, formula (13-1) yields

$$\int_0^1 y\,dX = \int_0^1 (y_0 + X\,\Delta y_0)\,dX = y_0 + \frac{\Delta y_0}{2} = y_0 + \frac{y_1 - y_0}{2} = \frac{1}{2}(y_0 + y_1).$$

But from (13-2) $dX = dx/h$, and on recalling (13-3), we see that this formula can be written as

$$\int_{x_0}^{x_1} y\,dx = \frac{h}{2}(y_0 + y_1). \tag{13-4}$$

---

[1] Another method, depending on power series, is discussed in Chap. 1.

Since y_0 is the ordinate of $y = f(x)$ at $x = x_0$ and y_1 is the ordinate at $x = x_1$, the right-hand member in (13-4) represents the area of the first trapezoid shown in Fig. 13. The choice of $m = 1$ in the calculations leading to (13-4) corresponds to replacing $y = f(x)$ in the interval (x_0, x_1) by the straight line through (x_0, y_0) and (x_1, y_1).

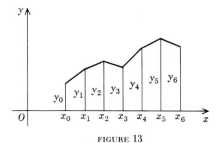

FIGURE 13

The successive application of (13-4) to intervals (x_1, x_2), (x_2, x_3), . . . , (x_{n-1}, x_n) yields

$$\int_{x_0}^{x_n} y \, dx = \int_{x_0}^{x_1} y \, dx + \int_{x_1}^{x_2} y \, dx + \cdots + \int_{x_{n-1}}^{x_n} y \, dx$$

$$= \frac{h}{2}(y_0 + y_1) + \frac{h}{2}(y_1 + y_2) + \cdots + \frac{h}{2}(y_{n-1} + y_n)$$

$$= \frac{h}{2}(y_0 + 2y_1 + 2y_2 + \cdots + 2y_{n-1} + y_n). \tag{13-5}$$

Formula (13-5) is known as the *trapezoidal rule*, for it gives the value of the sum of the areas of the n trapezoids whose bases are the ordinates $y_0, y_1, y_2, \ldots, y_n$. Figure 13 shows the six trapezoids in the case of $n = 6$.

If $m = 2$, (13-1) becomes

$$\int_0^2 y \, dX = \int_0^2 \left[y_0 + X \, \Delta y_0 + \frac{(X^2 - X)}{2} \Delta^2 y_0 \right] dX$$

$$= 2y_0 + 2 \, \Delta y_0 + \tfrac{1}{2}(\tfrac{8}{3} - 2) \, \Delta^2 y_0$$

$$= 2y_0 + 2(y_1 - y_0) + \tfrac{1}{3}(y_2 - 2y_1 + y_0)$$

$$= \tfrac{1}{3}y_0 + \tfrac{4}{3}y_1 + \tfrac{1}{3}y_2$$

or

$$\int_{x_0}^{x_2} y \, dx = \frac{h}{3}(y_0 + 4y_1 + y_2). \tag{13-6}$$

Suppose that there are $n + 1$ pairs of given values, where n is even. If these $n + 1$ pairs are divided into the groups of three pairs with abscissas x_{2i}, x_{2i+1}, x_{2i+2}, where $i = 0, 1, \ldots, (n-2)/2$, then (13-6) can be applied to each group. Hence,

$$\int_{x_0}^{x_n} y \, dx = \int_{x_0}^{x_2} y \, dx + \int_{x_2}^{x_4} y \, dx + \cdots + \int_{x_{n-2}}^{x_n} y \, dx$$

$$= \frac{h}{3}(y_0 + 4y_1 + y_2) + \frac{h}{3}(y_2 + 4y_3 + y_4)$$

$$+ \cdots + \frac{h}{3}(y_{n-2} + 4y_{n-1} + y_n)$$

$$= \frac{h}{3}[y_0 + y_n + 4(y_1 + y_3 + \cdots + y_{n-1})$$

$$+ 2(y_2 + y_4 + \cdots + y_{n-2})]. \tag{13-7}$$

Formula (13-7) is known as *Simpson's rule* with $m = 2$. Interpreted geometrically, it gives the value of the sum of the areas under the second-degree parabolas that have been passed through the points (x_{2i}, y_{2i}), (x_{2i+1}, y_{2i+1}), and (x_{2i+2}, y_{2i+2}), where $i = 0, 1, 2,$

$\ldots, (n - 2)/2$. The absolute error E in the approximation (13-7) can be shown to satisfy[1]

$$E \leq \frac{h^4}{180} (x_n - x_0)M$$

where $M = \max |f^{\text{iv}}(x)|$ in $[x_0, x_n]$.

If $m = 3$, (13-1) states that

$$\int_0^3 y \, dX = \int_0^3 \left(y_0 + X \, \Delta y_0 + \frac{X^2 - X}{2} \Delta^2 y_0 + \frac{X^3 - 3X^2 + 2X}{6} \Delta^3 y_0 \right) dX$$

$$= 3y_0 + \tfrac{9}{2} \Delta y_0 + (\tfrac{9}{2} - \tfrac{9}{4}) \Delta^2 y_0 + (\tfrac{27}{8} - \tfrac{9}{2} + \tfrac{3}{2}) \Delta^3 y_0$$

$$= 3y_0 + \tfrac{9}{2} (y_1 - y_0) + \tfrac{9}{4} (y_2 - 2y_1 + y_0)$$

$$+ \tfrac{3}{8}(y_3 - 3y_2 + 3y_1 - y_0)$$

$$= \tfrac{3}{8}(y_0 + 3y_1 + 3y_2 + y_3)$$

or

$$\int_{x_0}^{x_3} y \, dx = \frac{3h}{8} (y_0 + 3y_1 + 3y_2 + y_3). \tag{13-8}$$

If $n + 1$ pairs of values are given, and if n is a multiple of 3, then (13-8) can be applied successively to groups of four pairs of values to give

$$\int_{x_0}^{x_n} y \, dx = \frac{3h}{8} [y_0 + y_n + 3(y_1 + y_2 + y_4 + y_5 + \cdots + y_{n-2} + y_{n-1})$$

$$+ 2(y_3 + y_6 + \cdots + y_{n-3})]. \tag{13-9}$$

Formula (13-9) is called Simpson's rule with $m = 3$. It is not encountered as frequently as (13-5) or (13-7). Other formulas for numerical integration can be derived by setting $m = 4, 5, \ldots$ in (13-1), but the three given here are sufficient for ordinary purposes. In most cases, better results are obtained by securing a large number of observed or computed values, so that h will be small, and using (13-5) or (13-7).

Example 1. Using the data given in the example of Sec. 7, find an approximate value for $\int_1^7 y \, dx$. The trapezoidal rule (13-5) gives

$$\int_1^7 y \, dx = \tfrac{1}{2}(2.105 + 5.616 + 7.228 + 9.208 + 11.714 + 14.902 + 9.467) = 30.120.$$

Using (13-7) gives

$$\int_1^7 y \, dx = \tfrac{1}{3}[2.105 + 9.467 + 4(2.808 + 4.604 + 7.451) + 2(3.614 + 5.857)] = 29.989.$$

Using (13-9) gives

$$\int_1^7 y \, dx = \tfrac{3}{8}[2.105 + 9.467 + 3(2.808 + 3.614 + 5.857 + 7.451) + 2(4.604)] = 29.989.$$

If numerical integration is to be used in a problem in which the form of $f(x)$ is known, the set of values (x_i, y_i) can usually be chosen so that the x_i form an arithmetic progression and one of the formulas deduced above can be applied. Even if it is expedient to choose values closer together for some parts of the range than for other parts, these formulas can be applied successively, with appropriate values of h, to those sets of values for which the x_i form an arithmetic progression. However, if the set of given values was

[1] See J. F. Steffensen, "Interpolation," Chelsea Publishing Company, New York, 1950. A method of improving the accuracy without increasing n is given in P. Franklin, "Methods of Advanced Calculus," pp. 181–185, McGraw-Hill Book Company, New York, 1944.

obtained by observation, it is frequently convenient to use a formula that does not require that the x_i form an arithmetic progression.

Suppose that a set of pairs of observed values (x_i, y_i), where $i = 0, 1, 2, \ldots, m$, is given. The points (x_i, y_i) all lie on the curve whose equation is given by (9-2). The area under this curve between $x = x_0$ and $x = x_m$ is an approximation to the value of $\int_{x_0}^{x_m} y \, dx$. The area under the curve (9-2) is

$$\int_{x_0}^{x_m} y \, dx = \sum_{k=0}^{m} \frac{y_k}{P_k(x_k)} \int_{x_0}^{x_m} P_k(x) \, dx \tag{13-10}$$

in which the expressions for the $P_k(x)$ are given by (9-1).

If $m = 1$, (9-1) and (13-10) give

$$\int_{x_0}^{x_1} y \, dx = \frac{y_0}{x_0 - x_1} \int_{x_0}^{x} (x - x_1) \, dx + \frac{y_1}{x_1 - x_0} \int_{x_0}^{x_1} (x - x_0) \, dx$$

$$= \frac{x_1 - x_0}{2} (y_0 + y_1). \tag{13-11}$$

Formula (13-11) is identical with (13-4), as would be expected, but the formula corresponding to (13-5) is

$$\int_{x_0}^{x_n} y \, dx = \tfrac{1}{2}[(x_1 - x_0)(y_0 + y_1) + (x_2 - x_1)(y_1 + y_2)$$
$$+ \cdots + (x_n - x_{n-1})(y_{n-1} + y_n)]. \tag{13-12}$$

If $m = 2$, (13-10) becomes

$$\int_{x_0}^{x_2} y \, dx = \frac{y_0}{P_0(x_0)} \int_{x_0}^{x_2} (x - x_1)(x - x_2) \, dx$$

$$+ \frac{y_1}{P_1(x_1)} \int_{x_0}^{x_2} (x - x_0)(x - x_2) \, dx + \frac{y_2}{P_2(x_2)} \int_{x_0}^{x_2} (x - x_0)(x - x_1) \, dx$$

$$= \frac{y_0}{P_0(x_0)} \left[\frac{x_2^3 - x_0^3}{3} - \frac{(x_1 + x_2)(x_2^2 - x_0^2)}{2} + x_1 x_2 (x_2 - x_0) \right]$$

$$+ \frac{y_1}{P_1(x_1)} \left[\frac{x_2^3 - x_0^3}{3} - \frac{(x_0 + x_2)(x_2^2 - x_0^2)}{2} + x_0 x_2 (x_2 - x_0) \right]$$

$$+ \frac{y_2}{P_2(x_2)} \left[\frac{x_2^3 - x_0^3}{3} - \frac{(x_0 + x_1)(x_2^2 - x_0^2)}{2} + x_0 x_1 (x_2 - x_0) \right]$$

$$= \frac{(x_2 - x_0)^2}{6} \left[\frac{y_0}{P_0(x_0)} (3x_1 - 2x_0 - x_2) + \frac{y_1}{P_1(x_1)} (x_0 - x_2) \right.$$

$$\left. + \frac{y_2}{P_2(x_2)} (2x_2 + x_0 - 3x_1) \right]. \tag{13-13}$$

Formula (13-13) reduces to (13-6) when $x_1 - x_0 = x_2 - x_1 = h$. The formula that corresponds to (13-7) is too long and complicated to be of practical importance, and hence it is omitted here. It is simpler to apply (13-13) successively to groups of three values and then add the results.

Example 2. Using the data given in Prob. 3 of Sec. 9, find an approximate value of $\int_{0.16}^{6.25} y \, dx$. Equation (13-12) determines

$$\int_{0.16}^{6.25} y \, dx = \tfrac{1}{2}[0.24(4.210) + 0.6(4.631) + 1.5(5.082) + 3.75(5.590)] = 16.187.$$

Applying (13-13) successively to the first three values and to the last three values gives

$$\int_{0.16}^{6.25} y\,dx = \frac{(0.84)^2}{6}\left[\frac{2(1.2 - 0.32 - 1)}{(-0.24)(-0.84)} + \frac{2.210(-0.84)}{(0.24)(-0.6)} + \frac{2.421(2 + 0.16 - 1.2)}{(0.84)(0.6)}\right]$$

$$+ \frac{(5.25)^2}{6}\left[\frac{2.421(7.5 - 2 - 6.25)}{(-1.5)(-5.25)} + \frac{2.661(-5.25)}{(1.5)(-3.75)}\right.$$

$$\left. + \frac{2.929(12.5 + 1 - 7.5)}{(5.25)(3.75)}\right] = 17.194.$$

PROBLEMS

1. Determine the values of $\int_1^7 y\,dx$ by applying (13-5) and (13-7) to the following data:

x	1	2	3	4	5	6	7
y	2.157	3.519	4.198	4.539	4.708	4.792	4.835

2. Apply formula (13-12) to compute $\int_{10}^{50.625} p\,dv$ from the data of the example in Sec. 9.

3. Work the preceding problem by applying (13-13).

4. Apply formulas (13-5) and (13-7) to compute $\int_{19}^{25} H\,dC$ from the data of Prob. 4 in Sec. 9.

5. Find approximate values of $\int_0^6 \sqrt{4 + x^3}\,dx$ by applying formulas (13-5) and (13-7) with $x_m = m$, $m = 0, 1, 2, \ldots, 6$.

14. Euler's Polygonal Curves.

As noted in Chaps. 2 and 3, the methods available for the exact solution of differential equations apply only to a few, principally linear, types of differential equations. Many equations arising in applications are not solvable by such methods, and one is obliged to devise techniques for the determination of approximate solutions.

We begin with the consideration of the first-order equation

$$y' = f(x,y) \tag{14-1}$$

and seek its solution $y = y(x)$ taking on a prescribed value $y_0 = y(x_0)$ at $x = x_0$.

At each point of the region where $f(x,y)$ is continuous, Eq. (14-1) determines the slope of the integral curve passing through that point. The equation of the tangent line at the point (x_0,y_0) to the integral curve $y = y(x)$ is

$$y - y_0 = f(x_0,y_0)(x - x_0). \tag{14-2}$$

If we advance along this line a short distance to a point (x_1,y_1), we can compute from (14-1) the value $y'(x_1) = f(x_1,y_1)$ which, in general, will not be equal to the slope of $y = y(x)$ at $x = x_1$, because the point (x_1,y_1) ordinarily will not lie on the integral curve $y = y(x)$. But if (x_1,y_1) is close to (x_0,y_0), the slope of the integral curve at $x = x_1$ will not differ much from $f(x_1,y_1)$. To put it differently, the linear function (14-2) approximates the solution of (14-1) in the neighborhood of the point (x_0,y_0). (See Chap. 2, Sec. 14.)

We consider next the straight line through (x_1,y_1) with the slope $f(x_1,y_1)$ and proceed along it a small distance to a point (x_2,y_2). At (x_2,y_2) we draw another straight line with the slope $f(x_2,y_2)$ and advance along it to a point (x_3,y_3). By continuing this construction we obtain a polygon consisting of short straight-line segments joining the points (x_0,y_0), (x_1,y_1), (x_2,y_2), . . . , (x_n,y_n). The polygonal curve so obtained is called *Euler's polygon*. This polygonal curve can be expected to approximate the integral curve reasonably well when the points (x_i,y_i) are not too far apart and the end point (x_n,y_n) is not too far away from (x_0,y_0).

The end points of the segments forming Euler's polygon clearly satisfy [cf. (14-2)]

$$
\begin{aligned}
y_1 - y_0 &= f(x_0,y_0)(x_1 - x_0) \\
y_2 - y_1 &= f(x_1,y_1)(x_2 - x_1) \\
&\cdot\ \cdot\ \cdot\ \cdot\ \cdot\ \cdot\ \cdot\ \cdot\ \cdot\ \cdot \\
y_n - y_{n-1} &= f(x_{n-1},y_{n-1})(x_n - x_{n-1})
\end{aligned}
\tag{14-3}
$$

and if each interval $x_i - x_{i-1}$ is of length h, we can write (14-3) as

$$
y_{m+1} = y_m + f(x_m,y_m)h \qquad m = 0, 1, 2, \ldots, n - 1.
\tag{14-4}
$$

The recursion formula (14-4) enables one to compute successively the approximate values of the ordinates of the integral curve $y = y(x)$ at $x_k = x_0 + kh$, where $k = 1, 2, \ldots, n$. It may suffice for rough calculations if the spacing interval h is small and m not too large.[1]

A more accurate formula can be obtained by constructing, instead of the chain of rectilinear segments, a chain made up of parabolic segments. Thus, we can draw through (x_1,y_1) a parabola

$$
y = a_0 + a_1(x - x_1) + a_2(x - x_1)^2
\tag{14-5}
$$

which at $x = x_0$ has the slope $f(x_0,y_0)$ and at $x = x_1$ the slope $f(x_1,y_1)$. A simple calculation of the constants in (14-5) yields

$$
y_2 = y_1 + \{y'(x_1) + \tfrac{1}{2}[y'(x_1) - y'(x_0)]\}h.
\tag{14-6}
$$

This formula serves to determine y_2 if $y_1 = y(x_1)$, $y'(x_1)$, and the difference $\nabla y_1' \equiv y'(x_1) - y'(x_0)$ are known. Now, if we suppose that the solution $y(x)$ can be represented by Taylor's formula

$$
y(x) = y_0 + y'(x_0)(x - x_0) + \tfrac{1}{2}y''(x_0)(x - x_0)^2 + \cdots + R_n
\tag{14-7}
$$

we can calculate the needed quantities in the right-hand member of (14-6).

The coefficients in (14-7) can be calculated from (14-1) whenever $f(x,y)$ has a sufficient number of partial derivatives, for on setting (x_0,y_0) in (14-1), we get $y'(x_0) = f(x_0,y_0)$. Differentiating (14-1) with respect to x yields

$$
y''(x) = f_x(x,y) + f_y(x,y)y'(x)
\tag{14-8}
$$

and substituting $x = x_0$, $y = y_0$ in (14-8) gives

$$
y''(x_0) = f_x(x_0,y_0) + f_y(x_0,y_0)y'(x_0).
$$

By differentiating (14-8), we obtain $y'''(x)$, and so on. The value of R_n in (14-7) in general cannot be computed, but by neglecting it we get an approximate value of $y(x)$.

Once the coefficients in (14-7) are determined, we use (14-7) to compute $y(x_1) = y_1$. The value of $y'(x_1)$ is then determined by (14-1), since $y'(x_1) = f(x_1,y_1)$. The substitution in (14-6) then yields y_2.

Having computed y_2, we can advance another step and compute y_3 from [cf. (14-6)]

$$
y_3 = y_2 + \{y'(x_2) + \tfrac{1}{2}[y'(x_2) - y'(x_1)]\}h.
$$

This requires calculating $y'(x_2)$ from (14-1).

---

[1] For a comprehensive treatment of Euler's, Adams', and other numerical methods of solution of ordinary differential equations see Peter Henrici, "Discrete Variable Methods in Ordinary Differential Equations," John Wiley & Sons, Inc., New York, 1962.

The general recursion formula, based on the parabolic approximation, is

$$y_{m+1} = y_m + [y'(x_m) + \tfrac{1}{2} \nabla y'(x_m)]h \tag{14-9}$$

where $\nabla y'(x_m) = y'(x_m) - y'(x_{m-1})$.

More elaborate recursion formulas can be constructed by using polynomials of higher degree instead of (14-5). Such formulas lie at the basis of the Adams method of integration of differential equations discussed in the next section.

--- **PROBLEMS**

1. Construct a polygonal approximation, in the interval $(-1,1)$, to the solution of $y' = \tfrac{1}{2}xy$ which is such that $y(0) = 1$. Take the spacing interval $h = 0.2$. Also obtain the exact solution, and plot it on the same sheet of paper.

2. Determine the coefficients in (14-5), and thus deduce (14-6).

3. Use the equation in Prob. 1 to illustrate the calculation of y_2 from formula (14-6). Also obtain y_3. Take $x_0 = 0$, $y_0 = 1$, $x_1 = 0.2$, $x_2 = 0.4$.

15. The Adams Method.

We extend the considerations of the preceding section by developing a step-by-step procedure for computing an approximate solution of

$$y' = f(x,y) \tag{15-1}$$

taking on a prescribed value y_0 at $x = x_0$. The ordinates y_m, approximating the ordinates of the integral curve $y = y(x)$ at $x = x_m$, will be determined for equally spaced values of x, so that $x_m = x_0 + hm$, where $m = 0, 1, 2, \ldots$. Thus our approximate solution will appear in a tabulated form for a discrete set of values of x.

By the fundamental theorem of integral calculus,

$$\int_{x_m}^{x_{m+1}} y'(x)\,dx \equiv \int_{x_m}^{x_{m+1}} \left(\frac{dy}{dx}\right) dx = y(x_m + h) - y(x_m)$$

so that

$$y_{m+1} = y_m + \int_{x_m}^{x_m+h} y'(x)\,dx \tag{15-2}$$

where $y_{m+1} \equiv y(x_m + h)$ and $y_m \equiv y(x_m)$.

Now, if the variable x in the integral of (15-2) is replaced by

$$x = x_m + hX \tag{15-3}$$

where X is a new dimensionless variable, (15-2) becomes[1]

$$y_{m+1} = y_m + h \int_0^1 y'(x_m + hX)\,dX. \tag{15-4}$$

But we saw in Sec. 8 that when a function $y'(x)$ is approximated by a polynomial of degree n taking on the values $y'_m, y'_{m-1}, \ldots, y'_{m-n}$ at $x = x_m, x_{m-1}, \ldots, x_{m-n}$, then [cf. (8-7)]

$$y'(x_m + hX) = y'_m + X \nabla y'_m + \frac{X(X+1)}{2!} \nabla^2 y'_m + \cdots$$

$$+ \frac{X(X+1) \cdots (X+n-1)}{n!} \nabla^n y'_m. \tag{15-5}$$

If we insert (15-5) in (15-4) and carry out simple integrations, we find that

[1] By (15-3) $dx = h\,dX$, and at the limits $x = x_m$ and $x = x_m + h$, the values of X are $X = 0$ and $X = 1$.

we obtain a polygon consisting of short straight-line segments joining the points (x_0,y_0), (x_1,y_1), (x_2,y_2), . . . , (x_n,y_n). The polygonal curve so obtained is called *Euler's polygon*. This polygonal curve can be expected to approximate the integral curve reasonably well when the points (x_i,y_i) are not too far apart and the end point (x_n,y_n) is not too far away from (x_0,y_0).

The end points of the segments forming Euler's polygon clearly satisfy [cf. (14-2)]

$$\begin{aligned}
y_1 - y_0 &= f(x_0,y_0)(x_1 - x_0) \\
y_2 - y_1 &= f(x_1,y_1)(x_2 - x_1) \\
&\cdots\cdots\cdots\cdots\cdots\cdots \\
y_n - y_{n-1} &= f(x_{n-1},y_{n-1})(x_n - x_{n-1})
\end{aligned} \tag{14-3}$$

and if each interval $x_i - x_{i-1}$ is of length h, we can write (14-3) as

$$y_{m+1} = y_m + f(x_m,y_m)h \qquad m = 0, 1, 2, \ldots, n - 1. \tag{14-4}$$

The recursion formula (14-4) enables one to compute successively the approximate values of the ordinates of the integral curve $y = y(x)$ at $x_k = x_0 + kh$, where $k = 1, 2, \ldots, n$. It may suffice for rough calculations if the spacing interval h is small and m not too large.[1]

A more accurate formula can be obtained by constructing, instead of the chain of rectilinear segments, a chain made up of parabolic segments. Thus, we can draw through (x_1,y_1) a parabola

$$y = a_0 + a_1(x - x_1) + a_2(x - x_1)^2 \tag{14-5}$$

which at $x = x_0$ has the slope $f(x_0,y_0)$ and at $x = x_1$ the slope $f(x_1,y_1)$. A simple calculation of the constants in (14-5) yields

$$y_2 = y_1 + \{y'(x_1) + \tfrac{1}{2}[y'(x_1) - y'(x_0)]\}h. \tag{14-6}$$

This formula serves to determine y_2 if $y_1 = y(x_1)$, $y'(x_1)$, and the difference $\nabla y_1' \equiv y'(x_1) - y'(x_0)$ are known. Now, if we suppose that the solution $y(x)$ can be represented by Taylor's formula

$$y(x) = y_0 + y'(x_0)(x - x_0) + \tfrac{1}{2}y''(x_0)(x - x_0)^2 + \cdots + R_n \tag{14-7}$$

we can calculate the needed quantities in the right-hand member of (14-6).

The coefficients in (14-7) can be calculated from (14-1) whenever $f(x,y)$ has a sufficient number of partial derivatives, for on setting (x_0,y_0) in (14-1), we get $y'(x_0) = f(x_0,y_0)$. Differentiating (14-1) with respect to x yields

$$y''(x) = f_x(x,y) + f_y(x,y)y'(x) \tag{14-8}$$

and substituting $x = x_0$, $y = y_0$ in (14-8) gives

$$y''(x_0) = f_x(x_0,y_0) + f_y(x_0,y_0)y'(x_0).$$

By differentiating (14-8), we obtain $y'''(x)$, and so on. The value of R_n in (14-7) in general cannot be computed, but by neglecting it we get an approximate value of $y(x)$.

Once the coefficients in (14-7) are determined, we use (14-7) to compute $y(x_1) = y_1$. The value of $y'(x_1)$ is then determined by (14-1), since $y'(x_1) = f(x_1,y_1)$. The substitution in (14-6) then yields y_2.

Having computed y_2, we can advance another step and compute y_3 from [cf. (14-6)]

$$y_3 = y_2 + \{y'(x_2) + \tfrac{1}{2}[y'(x_2) - y'(x_1)]\}h.$$

This requires calculating $y'(x_2)$ from (14-1).

---

[1] For a comprehensive treatment of Euler's, Adams', and other numerical methods of solution of ordinary differential equations see Peter Henrici, "Discrete Variable Methods in Ordinary Differential Equations," John Wiley & Sons, Inc., New York, 1962.

The general recursion formula, based on the parabolic approximation, is

$$y_{m+1} = y_m + [y'(x_m) + \tfrac{1}{2}\nabla y'(x_m)]h \qquad (14\text{-}9)$$

where $\nabla y'(x_m) = y'(x_m) - y'(x_{m-1})$.

More elaborate recursion formulas can be constructed by using polynomials of higher degree instead of (14-5). Such formulas lie at the basis of the Adams method of integration of differential equations discussed in the next section.

── **PROBLEMS**

1. Construct a polygonal approximation, in the interval $(-1,1)$, to the solution of $y' = \tfrac{1}{2}xy$ which is such that $y(0) = 1$. Take the spacing interval $h = 0.2$. Also obtain the exact solution, and plot it on the same sheet of paper.

2. Determine the coefficients in (14-5), and thus deduce (14-6).

3. Use the equation in Prob. 1 to illustrate the calculation of y_2 from formula (14-6). Also obtain y_3. Take $x_0 = 0$, $y_0 = 1$, $x_1 = 0.2$, $x_2 = 0.4$.

──

15. The Adams Method.   We extend the considerations of the preceding section by developing a step-by-step procedure for computing an approximate solution of

$$y' = f(x,y) \qquad (15\text{-}1)$$

taking on a prescribed value y_0 at $x = x_0$. The ordinates y_m, approximating the ordinates of the integral curve $y = y(x)$ at $x = x_m$, will be determined for equally spaced values of x, so that $x_m = x_0 + hm$, where $m = 0, 1, 2, \ldots$. Thus our approximate solution will appear in a tabulated form for a discrete set of values of x.

By the fundamental theorem of integral calculus,

$$\int_{x_m}^{x_{m+1}} y'(x)\,dx \equiv \int_{x_m}^{x_{m+1}} \left(\frac{dy}{dx}\right) dx = y(x_m + h) - y(x_m)$$

so that

$$y_{m+1} = y_m + \int_{x_m}^{x_m + h} y'(x)\,dx \qquad (15\text{-}2)$$

where $y_{m+1} \equiv y(x_m + h)$ and $y_m \equiv y(x_m)$.

Now, if the variable x in the integral of (15-2) is replaced by

$$x = x_m + hX \qquad (15\text{-}3)$$

where X is a new dimensionless variable, (15-2) becomes[1]

$$y_{m+1} = y_m + h \int_0^1 y'(x_m + hX)\,dX. \qquad (15\text{-}4)$$

But we saw in Sec. 8 that when a function $y'(x)$ is approximated by a polynomial of degree n taking on the values $y'_m, y'_{m-1}, \ldots, y'_{m-n}$ at $x = x_m, x_{m-1}, \ldots, x_{m-n}$, then [cf. (8-7)]

$$y'(x_m + hX) = y'_m + X\,\nabla y'_m + \frac{X(X+1)}{2!}\nabla^2 y'_m + \cdots$$

$$+ \frac{X(X+1)\cdots(X+n-1)}{n!}\nabla^n y'_m. \qquad (15\text{-}5)$$

If we insert (15-5) in (15-4) and carry out simple integrations, we find that

[1] By (15-3) $dx = h\,dX$, and at the limits $x = x_m$ and $x = x_m + h$, the values of X are $X = 0$ and $X = 1$.

$$y_{m+1} = y_m + h(y'_m + \tfrac{1}{2} \nabla y'_m + \tfrac{5}{12} \nabla^2 y'_m + \tfrac{3}{8} \nabla^3 y'_m + {}^{251}\!/_{720} \nabla^4 y'_m$$
$$+ \cdots + a_n \nabla^n y'_m) \qquad (15\text{-}6)$$

where
$$a_n = \int_0^1 \frac{X(X+1) \cdots (X+n-1)}{n!} \, dX. \qquad (15\text{-}7)$$

Formula (15-6) enables us to compute the ordinate y_{m+1} if we know y_m, y'_m, and the backward differences $\nabla^k y'_m$. When the $\nabla y'_m$ vanish, (15-6) reduces to (14-4), and when the $\nabla^2 y'_m$ vanish, we get (14-9). As was the case with (14-9), the values of y_m, y'_m and the $\nabla^k y'_m$ in (15-6) are not available to us at the start. They must be computed by some means before (15-6) can be used to evaluate y_{m+1}. The number of the $\nabla^k y_m$ depends on the degree n of the polynomial chosen to approximate $y'(x)$. Once we agree on the value of n, we can compute y_m, y'_m and the requisite number of the $\nabla^k y'_m$ with the aid of Taylor's representation of the solution $y = y(x)$, as was done in Sec. 14.

We illustrate the procedure in detail in the following example.

Example. Use Adams' method to obtain, in the interval $(0,1)$, an approximate solution of

$$y' = y + x \qquad (15\text{-}8)$$

taking on the value $y_0 = 1$ at $x = 0$.

Let us subdivide the interval $(0,1)$ into subintervals of length $h = 0.1$, so that

$$x_k = x_0 + kh = 0.1k \qquad k = 0, 1, 2, \ldots, 10.$$

Furthermore, let us agree to retain in (15-6) the differences of y'_m up to and including those of order 3. This corresponds to approximating $y'(x)$ in (15-2) by a polynomial of degree 3.

To compute y_{m+1} from (15-6) we need y_m, y'_m, $\nabla y'_m$, $\nabla^2 y'_m$, and $\nabla^3 y'_m$. The calculation of the third differences $\nabla^3 y'_m$ requires at least four values y'_m, y'_{m-1}, y'_{m-2}, y'_{m-3}, as is obvious from the following table.

$$
\begin{array}{ccccc}
\vdots & \vdots & \vdots & \vdots & \\
y'_{m-3} & & & & \\
& \nabla y'_{m-2} & & & \\
y'_{m-2} & & \nabla^2 y'_{m-1} & & \\
& \nabla y'_{m-1} & & \nabla^3 y'_m & \\
y'_{m-1} & & \nabla^2 y'_m & & \\
& \nabla y'_m & & & \\
y'_m & & & &
\end{array}
$$

If we determine y'_0, y'_1, y'_2, y'_3, we shall be in a position to fill in the values in this table with $m = 3$ and then proceed to determine y_4 from (15-6).

Since $y_0 = 1$ for $x_0 = 0$, Eq. (15-8) yields

$$y'_0 = 1. \qquad (15\text{-}9)$$

To compute y_1, y_2, and y_3 we use Taylor's series

$$y(x) = y_0 + y'_0(x - x_0) + \frac{y''_0}{2!}(x - x_0)^2 + \frac{y'''_0}{3!}(x - x_0)^3 + \cdots \qquad (15\text{-}10)$$

with $y_0 = 1$ and $x_0 = 0$. The coefficients in (15-10) can be calculated from (15-8). Differentiating (15-8), we get

$$y''(x) = y'(x) + 1 \qquad (15\text{-}11)$$

and on setting $x = 0$ and recalling that $y'(0) = y'_0 = 1$, we get $y''(0) = 2$. Successive differentiations of (15-11) give

$$y'''(x) = y''(x), \qquad y^{IV}(x) = y'''(x), \qquad \ldots, \qquad y^{(n)}(x) = y^{(n-1)}(x) \qquad (15\text{-}12)$$

and since $y''(0) = 2$, we get from (15-12)

$$y'''(0) = 2, \qquad y^{IV}(0) = 2, \qquad \ldots, \qquad y^{(n)}(0) = 2.$$

Accordingly, (15-10) becomes
$$y(x) = 1 + x + x^2 + \frac{x^3}{3} + \frac{x^4}{3 \cdot 4} + \frac{x^5}{3 \cdot 4 \cdot 5} + \cdots.$$

Setting $x = 0.1$, we get
$$y_1 = 1 + 0.1 + (0.1)^2 + \frac{(0.1)^3}{3} + \frac{(0.1)^4}{3 \cdot 4} + \frac{(0.1)^5}{3 \cdot 4 \cdot 5} + \cdots = 1.1103.$$

In the same way, using $x = 0.2$ and $x = 0.3$, we obtain
$$y_2 = 1.2428 \qquad y_3 = 1.3997.$$

The desired values of y_1', y_2', and y_3' can now be computed from (15-8). We find that
$$y_1' = y_1 + x_1 = 1.1103 + 0.1 = 1.2103$$
$$y_2' = y_2 + x_2 = 1.2428 + 0.2 = 1.4428$$
$$y_3' = y_3 + x_3 = 1.3997 + 0.3 = 1.6997.$$

We can now proceed to construct the table of differences shown below.

x	y	y'	$\nabla y'$	$\nabla^2 y'$	$\nabla^3 y'$
0	1.0000	1.0000			
			0.2103		
0.1	1.1103	1.2103		0.0222	
			0.2325		0.0022
0.2	1.2428	1.4428		0.0244	
			0.2569		0.0026
0.3	1.3997	1.6997		0.0270	
			0.2839		
0.4	1.5836	1.9836			
0.5	1.7974				

The substitution from this table in (15-6), with $m = 3$ and $n = 3$, yields
$$y_4 = 1.3997 + 0.1[1.6997 + \tfrac{1}{2}(0.2569) + \tfrac{5}{12}(0.0244) + \tfrac{3}{8}(0.0022)] = 1.5836.$$

This value is recorded in the table for $x = 0.4$.

To compute y_5 we must extend the table, since formula (15-6) requires the knowledge of y_4' and assorted differences of y_4'. By (15-8)
$$y_4' = y_4 + x_4 = 1.5836 + 0.4 = 1.9836.$$

The calculated values (recorded below the heavy line in the table) can now be used in (15-6), with $m = 4$, $n = 3$, to compute y_5. We have
$$y_5 = 1.5836 + 0.1[1.9836 + \tfrac{1}{2}(0.2839) + \tfrac{5}{12}(0.0270) + \tfrac{3}{8}(0.0026)] = 1.7974.$$

This value is recorded in the table for $x = 0.5$.

We leave it to the reader to make further extensions in the table required for the calculation of $y_6, y_7, \ldots, y_{10}$.

PROBLEMS

1. Complete the table in the example by computing $y_6, y_7, \ldots, y_{10}$.

2. Since (15-8) is a linear equation, its solution satisfying the condition $y(0) = 1$ is easily found to be $y = 2e^x - x - 1$. Compare the exact and the approximate values $y_1, y_2, \ldots, y_{10}$.

3. Apply the Adams method to obtain an approximate solution of $y' = y$ with $y(0) = 1$. Use $h = 0.1$, and compute $y(0.3)$, $y(0.4)$, $y(0.5)$, and $y(0.6)$ from (15-6) with $n = 2$. Compare with the exact solution.

4. Use $h = 0.1$ and (15-6) with $n = 3$ to find an approximate value of $y(-0.6)$ for the integral curve of $y' = x^2 + y^2$ through $(-1,0)$.

5. Apply the Adams method to obtain an approximate solution of $y' = -x + y$ with $y(0) = 1.500$. Use $h = 0.25$ and compute $y(1.00)$, $y(1.25)$, $y(1.5)$.

---

16. Equations of Higher Order.   Systems of Equations.   The methods of Secs. 14 and 15 can be extended to obtain numerical solutions of equations of higher order.   Thus, the second-order equation

$$y'' = f(x,y,y')   \tag{16-1}$$

with initial conditions

$$y(x_0) = y_0   \qquad y'(x_0) = y_0'   \tag{16-2}$$

can be written as a system of two equations of first order by setting

$$y' = z.   \tag{16-3}$$

The substitution in (16-1) from (16-3) then yields the second equation

$$z' = f(x,y,z).   \tag{16-4}$$

In indicating the extension we shall consider, instead of the system (16-3) and (16-4), a more general system

$$\begin{aligned} y' &= f_1(x,y,z) \\ z' &= f_2(x,y,z) \end{aligned}   \tag{16-5}$$

with initial conditions

$$y(x_0) = y_0   \qquad z(x_0) = z_0.   \tag{16-6}$$

When solutions of the system (16-5) can be expanded in Taylor's series

$$\begin{aligned} y(x) &= y(x_0) + y'(x_0)(x - x_0) + \frac{y''(x_0)}{2!}(x - x_0)^2 + \cdots \\[2mm] z(x) &= z(x_0) + z'(x_0)(x - x_0) + \frac{z''(x_0)}{2!}(x - x_0)^2 + \cdots \end{aligned}   \tag{16-7}$$

the coefficients in (16-7) can be computed by differentiating Eqs. (16-5) successively as was done in Secs. 14 and 15.

The construction of Euler's polygonal approximation also follows the pattern of Sec. 14.   Thus, the equation of the straight line through (x_0,y_0,z_0) tangent to the integral curve of the system (16-5) is[1]

$$\begin{aligned} y - y_0 &= f_1(x_0,y_0,z_0)(x - x_0) \\ z - z_0 &= f_2(x_0,y_0,z_0)(x - x_0). \end{aligned}   \tag{16-8}$$

When abscissas are spaced uniformly h units apart,

$$x_1 = x_0 + h, \qquad x_2 = x_0 + 2h, \qquad \ldots, \qquad x_k = x_0 + kh$$

and from (16-8) it follows that the approximate solutions at $x_1, x_2, \ldots$ are

$$\begin{aligned} y_1 &= y_0 + f_1(x_0,y_0,z_0)h \\ z_1 &= z_0 + f_2(x_0,y_0,z_0)h \\ y_2 &= y_1 + f_1(x_1,y_1,z_1)h \\ z_2 &= z_1 + f_2(x_1,y_1,z_1)h \\ &\cdot \ \cdot \ \cdot \ \cdot \ \cdot \ \cdot \ \cdot \ \cdot \\ &\cdot \ \cdot \ \cdot \ \cdot \ \cdot \ \cdot \ \cdot \ \cdot \\ y_{k+1} &= y_k + f_1(x_k,y_k,z_k)h \\ z_{k+1} &= z_k + f_2(x_k,y_k,z_k)h. \end{aligned}$$

[1] The integral curve of the system (16-5) is, in general, a space curve, to which the tangent line is determined by the intersection of the planes (16-8).

If, instead of approximating the solution in each interval by a linear function, we make use of the polynomial approximations in the manner of Sec. 15, we obtain

$$y_{m+1} = y_m + h(y_m' + \tfrac{1}{2} \nabla y_m' + \tfrac{5}{12} \nabla^2 y_m' + \cdots + a_n \nabla^n y_m')$$
$$z_{m+1} = z_m + h(z_m' + \tfrac{1}{2} \nabla z_m' + \tfrac{5}{12} \nabla^2 z_m' + \cdots + a_n \nabla^n y_m')$$

$$(16\text{-}9)$$

with a_n determined by (15-7).

In computing y_{m+1} and z_{m+1} from (16-9), we must first obtain the values of y_m, z_m, y_m', z_m' and the required differences, as was done in Sec. 15.

Example.   Obtain the solution of the system

$$y' = x + z$$
$$z' = 1 + y$$

$$(16\text{-}10)$$

in the form (16-7), which is such that

$$y(0) = -1 \qquad z(0) = 1.$$

$$(16\text{-}11)$$

On setting $x_0 = 0$ in (16-7) we get

$$y(x) = y(0) + y'(0)x + \frac{1}{2!} y''(0)x^2 + \cdots$$

$$z(x) = z(0) + z'(0)x + \frac{1}{2!} z''(0)x^2 + \cdots$$

$$(16\text{-}12)$$

the coefficients in which can be computed by differentiating (16-10) and noting (16-11).   We obtain from (16-10)

$$y''(x) = 1 + z'(x) \qquad z''(x) = y'(x)$$
$$y'''(x) = z''(x) \qquad z'''(x) = y''(x)$$
$$\cdots \cdots \cdots \cdots$$
$$y^{(n)}(x) = z^{(n-1)}(x) \qquad z^{(n)}(x) = y^{(n-1)}(x).$$

$$(16\text{-}13)$$

The substitution from (16-11) in (16-10) yields $y'(0) = 1$, $z'(0) = 1 - 1 = 0$, and making use of these values in (16-13) we find

$$y''(0) = 1 + 0 = 1 \qquad z''(0) = y'(0) = 1$$
$$y'''(0) = z''(0) = 1 \qquad z'''(0) = y''(0) = 1$$
$$\cdots \cdots \cdots \cdots \qquad \cdots \cdots \cdots \cdots$$
$$y^{(n)}(0) = 1 \qquad z^{(n)}(0) = 1.$$

Accordingly, (16-12) yields

$$y(x) = -1 + x + \frac{x^2}{2!} + \frac{x^3}{3!} + \cdots$$

$$z(x) = 1 + 0x + \frac{x^2}{2!} + \frac{x^3}{3!} + \cdots.$$

$$(16\text{-}14)$$

By eliminating z from the system (16-10) we see that it is equivalent to the second-order equation

$$y'' - y = 2$$

with $y(0) = -1$, $y'(0) = 1$.   Its solution is readily found to be

$$y = e^x - 2$$

$$(16\text{-}15)$$

and from the first of Eqs. (16-10) we conclude that

$$z = e^x - x.$$

$$(16\text{-}16)$$

The Maclaurin expansions of these solutions are precisely (16-14).

It may be instructive to compute the polygonal approximations to the solution of (16-10) at $x = 0.2$ and $x = 0.4$.

On setting the differences in (16-9) equal to zero, we get

$$y_{m+1} = y_m + hy_m' \qquad z_{m+1} = z_m + hz_m'.$$

$$(16\text{-}17)$$

Now, if we take $x_1 = 0.2$, so that $h = 0.2$, we obtain from (16-17)

$$y_1 = y(0.2) = -1 + (0.2)1 = -0.8$$
$$z_1 = z(0.2) = 1 + (0.2)0 = 1$$

since $y_0' = 1$ and $z_0' = 0$.

The exact solution (16-15) and (16-16) yields

$$y(0.2) = e^{0.2} - 2 = -0.7786$$
$$z(0.2) = e^{0.2} - 0.2 = 1.0214.$$

Using $y_1 = -0.8$, $z_1 = 1$ in (16-17), we obtain

$$y_2 = y(0.4) = y_1 + 0.2y_1'$$
$$z_2 = z(0.4) = z_1 + 0.2z_1'. \tag{16-18}$$

The values of y_1' and z_1' can be calculated from (16-10) by setting $x = 0.2$, $z = 1$, and $y = -0.8$. We find that

$$y_1' = 0.2 + 1 = 1.2 \qquad z_1' = 1 + y_1 = 1 - 0.8 = 0.2$$

and then (16-18) yields

$$y_2 = -0.8 + (0.2)(1.2) = -0.56$$
$$z_2 = 1 + (0.2)(0.2) = 1.04$$

while the corresponding exact values are

$$y(0.4) = e^{0.4} - 2 = -0.5082$$
$$z(0.4) = e^{0.4} - 0.4 = 1.0918.$$

The reader is advised to obtain more accurate polygonal approximations by taking the interval $h = 0.1$ and to compare the polygonal approximations with the values given by (16-9) in which the differences of order higher than 1 are set equal to zero.

_____ **PROBLEMS**

1. Obtain from (16-7) a fourth-degree polynomial approximation to the solution of

$$y' = e^x + z \qquad z' = e^{-x} + y$$

with $y(0) = 0$ and $z(0) = 0$.

2. Use a polygonal approximation to compute y_1, y_2, y_3, y_4 for the system in Prob. 1 by taking $x_1 = 0.1$, $x_2 = 0.2$, $x_3 = 0.3$, $x_4 = 0.4$.

3. Use a polygonal approximation to compute y_1, y_2, y_3, corresponding to $x_1 = 0.1$, $x_2 = 0.2$, $x_3 = 0.3$ for $y'' - y^2 = x$, with initial conditions $y(0) = 1$, $y'(0) = 0$. *Hint:* Set $y' = z$, and consider the system $y' = z$, $z' = x + y^2$ with $y(0) = 1$ and $z(0) = 0$.

4. Obtain the solution for Prob. 3 in Maclaurin's series.

5. Solve the system in Prob. 3 by the Adams method. Retain only the second differences in (16-9), and use the result of Prob. 4 to start the iteration.

_____

17. Boundary-value Problems. In many physical problems solutions of the second- and higher-order differential equations that satisfy preassigned conditions at more than one point of the interval are required. A simple example of this occurs in the study of deflections of a beam supported at several points. Problems of this sort are termed *boundary-value problems* to distinguish them from *initial-value problems* in which the conditions on solutions are imposed only at one point.

An important feature of the boundary-value problems is that their solutions (if they exist at all) need not be unique. When the general solution of the differential equation can be obtained, the conditions imposed on solutions of the boundary-value problem can usually be met by determining the values of arbitrary constants in the general solution[1] so that the specified conditions are satisfied. However, general solutions of differential equations can rarely be written down, and one is obliged to seek solutions of boundary-value problems by numerical methods. The methods available for numerical solution of initial-value problems require that the integral curve be uniquely determined at the starting point and thus do not apply to problems in which solutions must satisfy specified conditions at more than one point. To solve a boundary-value problem numerically one must employ laborious trial-and-error procedures utilizing the solutions of suitable initial-value problems.

We outline briefly the procedure commonly followed in solving a two-point boundary-value problem for the second-order differential equation.[2]

Let it be required to determine a solution of

$$y'' = f(x,y,y') \tag{17-1}$$

which assumes at the end points of the interval $a \le x \le b$ the values

$$y(a) = A \qquad y(b) = B. \tag{17-2}$$

Now, if in addition to the value $y(a) = A$ we specify the slope $y'(a)$ at $x = a$, the solution of (17-1) is uniquely determined,[3] but this solution will satisfy the condition $y(b) = B$ only for some value of the slope $y'(a)$ which is not known. Physical or geometric considerations may suggest an approximate value of the slope, say $y'(a) = C$, which is such that the integral curve of (17-1) satisfying the conditions

$$y(a) = A \qquad y'(a) = C \tag{17-3}$$

also satisfies the condition $y(b) = B$.

The procedure used in solving the boundary-value problem consists in actually constructing the solution $y = y(x)$ satisfying the conditions (17-3) and computing the value of $y(x)$ at $x = b$. If it is tolerably near B, we have the desired approximate solution of the boundary-value problem. If not, we choose another value of the slope $y'(a)$ and try again. The procedure is clearly laborious and far from being elegant.

18. Characteristic-value Problems. Closely associated with boundary-value problems are *characteristic-value problems*. These are generally concerned with solutions of the two-point boundary-value problems for differential equations containing parameters.

A simple instance of the characteristic-value problem occurs in the study of small vibrations of an elastic string of finite length. (Cf. Chap. 7.) When the initial shape and initial velocity of the string are specified, its subsequent displacement $u(x,t)$ is determined by solving the equation

$$\frac{\partial^2 u}{\partial t^2} = a^2 \frac{\partial^2 u}{\partial x^2} \tag{18-1}$$

where a is a physical constant. If the string is of length l and its ends are fixed at $x = 0$ and $x = l$, the solution of (18-1) must satisfy the end conditions

$$u(0,t) = 0 \qquad u(l,t) = 0. \tag{18-2}$$

---

[1] This was the procedure followed in solving the boundary-value problems in Chap. 3, Sec. 10.

[2] For a more detailed discussion of such problems see W. E. Milne, "Numerical Solutions of Differential Equations," chap. 7, John Wiley & Sons, Inc., New York, 1952, and Henrici, *op. cit.*, chap. 7. Both these books contain extensive bibliographical references.

[3] We suppose that $f(x,y,y')$ is such that the initial-value problem has a unique solution and also such that the boundary-value problem in (17-1) and (17-2) has a solution.

When we attempt to obtain solutions of (18-1) by the method of separation of variables,[1] that is, by assuming that $u(x,t)$ is expressible in the form

$$u(x,t) = y(x)T(t) \tag{18-3}$$

where $y(x)$ is a function of x alone and $T(t)$ is a function of t alone, we are led to a pair of ordinary differential equations

$$\frac{d^2y}{dx^2} + \lambda^2 y = 0 \qquad \frac{d^2T}{dt^2} + a^2\lambda^2 T = 0 \tag{18-4}$$

where λ is a constant. This constant must be chosen so that the end conditions (18-2) are satisfied.

From the assumed form of solution (18-3) and from (18-2) it follows that the solutions of (18-4) must be such that

$$y(0) = 0 \qquad y(l) = 0. \tag{18-5}$$

We thus have a two-point boundary-value problem for Eq. (18-4) with the end conditions (18-5).

The determination of suitable solutions this time is very simple because the general solution of the first equation in (18-4) is

$$y = c_1 \cos \lambda x + c_2 \sin \lambda x. \tag{18-6}$$

If we impose the conditions (18-5) on (18-6) and reject the trivial solution $y \equiv 0$, we find infinitely many solutions

$$y = c_2 \sin \lambda x \tag{18-7}$$

where

$$\lambda = \frac{k\pi}{l} \qquad k = 1, 2, \ldots. \tag{18-8}$$

The values of λ in (18-8) are called the *characteristic values* of the boundary-value problem of (18-4) and (18-5), and the solutions (18-7) with appropriate λ's are *characteristic functions* of this problem.[2]

The simplicity of the characteristic-value problem defined by (18-4) and (18-5) masks some important features of the general problem. These features become clearer if we consider the determination of small vibrations of an elastic rod of variable cross section. In this case the separation of variables in the appropriate partial differential equation leads to the equation

$$\frac{d^2}{dx^2}\left[p(x)\frac{d^2y}{dx^2}\right] - \lambda q(x)y = 0 \tag{18-9}$$

in which $p(x)$ and $q(x)$ are known functions and λ an unknown constant. If the rod is of length l, with the end points at $x = 0$ and $x = l$, the solutions of (18-9) must satisfy suitable conditions determined by the mode of fixing the ends. If the end $x = 0$ is clamped, then $y(0) = y'(0) = 0$; if it is simply supported, then $y(0) = y''(0) = 0$; if it is free, then $y''(0) = y'''(0) = 0$. Similar conditions are imposed at the end $x = l$.

For definiteness, we suppose that the ends of the rod are free (a ship floating at sea). We then seek a solution of (18-9) such that

$$y''(0) = y'''(0) = 0 \qquad y''(l) = y'''(l) = 0. \tag{18-10}$$

Since (18-9) is a linear equation, its general solution is the sum of four linearly independent solutions

---

[1] See Chap. 7, Sec. 10.

[2] The terms *eigenvalue* and *eigenfunction* are used by some writers to mean "characteristic value" and "characteristic function," respectively. These stem from German words *Eigenwert* and *Eigenfunktion*.

$$y(x,\lambda) = c_1 y_1(x,\lambda) + c_2 y_2(x,\lambda) + c_3 y_3(x,\lambda) + c_4 y_4(x,\lambda) \qquad (18\text{-}11)$$

where λ is the parameter appearing in (18-9) and the c_i are arbitrary constants. On imposing the end conditions (18-10) on (18-11) we get a system of four equations:

$$c_1 y_1''(0,\lambda) + c_2 y_2''(0,\lambda) + c_3 y_3''(0,\lambda) + c_4 y_4''(0,\lambda) = 0$$
$$c_1 y_1'''(0,\lambda) + c_2 y_2'''(0,\lambda) + c_3 y_3'''(0,\lambda) + c_4 y_4'''(0,\lambda) = 0$$
$$c_1 y_1''(l,\lambda) + c_2 y_2''(l,\lambda) + c_3 y_3''(l,\lambda) + c_4 y_4''(l,\lambda) = 0$$
$$c_1 y_1'''(l,\lambda) + c_2 y_2'''(l,\lambda) + c_3 y_3'''(l,\lambda) + c_4 y_4'''(l,\lambda) = 0.$$

This system of four linear equations in the unknowns c_i will have a nontrivial solution if, and only if, the determinant $D(\lambda)$ of the coefficients of the c's is zero.[1] The equation

$$D(\lambda) = 0 \qquad (18\text{-}12)$$

is the *characteristic equation*, and its solutions are the *characteristic values* of the problem. In general, (18-12) is a transcendental equation, and its solution poses many vexing problems.[2] Usually it is solved by numerical methods. Because of the importance of characteristic equations in analyzing the behavior of dynamical systems, they have been studied extensively and there is a vast literature on the subject of numerical determination of characteristic values.[3]

19. Method of Finite Differences.   We conclude this chapter with a brief description of the most commonly used method for solving boundary-value problems in partial differential equations, known as the *method of finite differences*. In this method the differential equation is replaced by an approximating difference equation, and the continuous region in which the solution is desired by a set of discrete points. This permits one to reduce the problem to the solution of a system of algebraic equations, which may involve hundreds of unknowns. Ordinarily, some iterative technique has to be devised to solve such systems, and high-speed electronic computers have been developed largely because of the need for coping with problems of this sort.[4]

The main disadvantage of all numerical techniques is that they give numerical values for unknown functions at a set of discrete points instead of the analytic expressions defined over the initial region R. Of course, when the boundary-value data are determined by measurements at a finite set of points of R, the difference-equations methods may be the best mode of attack on the problem. Any analytic technique would require fitting curves to the discontinuous data.

We proceed to the outline of the general procedure followed in reducing the given analytic boundary-value problem to a problem in difference equations. For definiteness let the region R be bounded by a simple closed curve C. We seek to determine the function $u(x,y)$ satisfying a given differential equation in R. From the definition of partial derivatives it follows that

$$\frac{\partial u}{\partial x} = \lim_{h \to 0} \frac{u(x+h,\,y) - u(x,y)}{h}.$$

Also, if the second partial derivatives are continuous one can show that

---

[1] See Appendix A.

[2] An instance of a simple transcendental characteristic equation appears in Chap. 7, Eq. (10-16), in which the parameter is denoted by β. See also Chap. 2, Sec. 16, and Chap. 3, Sec. 15, where $P(\lambda) = 0$ is an algebraic equation. A concise presentation of characteristic-value problems is contained in Stiefel, *op. cit.*

[3] For bibliography see Henrici, *op. cit.*, Milne, *op. cit.*, and Hildebrand, *op. cit.*

[4] One of the best available treatments of partial differential equations by finite-difference methods is contained in G. E. Forsythe and W. Wasow, "Finite Difference Methods for Partial Differential Equations," John Wiley & Sons, Inc., New York, 1960.

$$\frac{\partial^2 u}{\partial x^2} = \lim_{h \to 0} \frac{u(x+h,\,y) - 2u(x,y) + u(x-h,\,y)}{h^2}$$

$$\frac{\partial^2 u}{\partial x\,\partial y} = \lim_{\substack{h \to 0 \\ k \to 0}} \frac{u(x+h,\,y+k) - u(x+h,\,y) - u(x,\,y+k) + u(x,y)}{hk}$$

and so on.

For small values of h and k the partial derivatives are nearly equal to the difference quotients appearing in the right-hand members of these formulas. If one replaces derivatives in the given differential equation by difference quotients, there results a difference equation that is a good approximation to the given equation when h and k are small.

Thus, to Laplace's equation

$$\nabla^2 u \equiv \frac{\partial^2 u}{\partial x^2} + \frac{\partial^2 u}{\partial y^2} = 0$$

there corresponds the difference equation

$$\Delta_{xx} u + \Delta_{yy} u = 0$$

where

$$\Delta_{xx} u \equiv \frac{1}{h^2} [u(x+h,\,y) - 2u(x,y) + u(x-h,\,y)]$$

$$\Delta_{yy} u \equiv \frac{1}{h^2} [u(x,\,y+h) - 2u(x,y) + u(x,\,y-h)].$$

In a difference equation the values of $u(x,y)$ are related at a set of discrete points determined by the choices of h and k. Ordinarily these points are chosen so that they form a square net[1] with specified mesh size h.

The usual procedure is to cover the region R by a net consisting of two sets of mutually orthogonal lines a distance h apart (Fig. 14) and mark off a polygonal contour C' so that it approximates sufficiently closely the boundary C. The domain R' in which the solution of the difference equation is sought is formed by the lattice points of the net contained within C'. The assigned boundary values on C are then transferred in some manner to the lattice points on C'. When the lattice points on C' do not coincide with points on C, the desired values can be got by interpolation.[2]

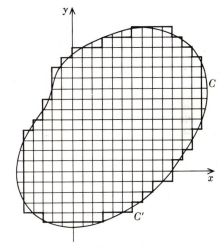

FIGURE 14

One then seeks a solution of the difference equation that satisfies the boundary conditions imposed at the lattice points on C'. Usually, this leads to a consideration of a system of a large number of algebraic equations in many unknowns.

[1] Rectangular, polygonal, and curvilinear nets are also used. See, for example, D. Y. Panov, "Handbook on Numerical Solution of Partial Differential Equations," Moscow, 1951, which contains a good account of the difference-equations techniques. A German translation of this book was published by Akademie-Verlag, Berlin, 1955. See also Appendix to S. Timoshenko and J. N. Goodier, "Theory of Elasticity," 2d ed., McGraw-Hill Book Company, New York, 1951.

[2] See, for example, Milne, *op. cit.*, Forsythe and Wasow, *op. cit.* Further discussion of difference equations is given in Chap. 7, Secs. 26 and 27, and in Chap. 9, Sec. 10.

The literature on finite-difference methods is extensive. An illustration of the use of the method of finite differences in solving a boundary-value problem in Laplace's equation is included in I. S. Sokolnikoff, "Mathematical Theory of Elasticity," sec. 124, McGraw-Hill Book Company, New York, 1956, which contains further references.

Appendixes

A Determinants

1. The Definition and Properties of Determinants. A determinant of the first order consists of a single element a and has the value a. A determinant of the second order contains four elements in a 2×2 square array and has the value

$$\begin{vmatrix} a_1 & a_2 \\ b_1 & b_2 \end{vmatrix} = a_1b_2 - a_2b_1. \tag{1-1}$$

A determinant of third order is similarly defined, in terms of second-order determinants:

$$\begin{vmatrix} a_1 & a_2 & a_3 \\ b_1 & b_2 & b_3 \\ c_1 & c_2 & c_3 \end{vmatrix} = a_1 \begin{vmatrix} b_2 & b_3 \\ c_2 & c_3 \end{vmatrix} - a_2 \begin{vmatrix} b_1 & b_3 \\ c_1 & c_3 \end{vmatrix} + a_3 \begin{vmatrix} b_1 & b_2 \\ c_1 & c_2 \end{vmatrix}. \tag{1-2}$$

By analogy, a determinant of order n consists of a square $n \times n$ array of elements a_{ij}

$$\begin{vmatrix} a_{11} & a_{12} & \cdots & a_{1n} \\ a_{21} & a_{22} & \cdots & a_{2n} \\ \cdot & \cdot & \cdots & \cdot \\ a_{n1} & a_{n2} & \cdots & a_{nn} \end{vmatrix}$$

to which a numerical value is assigned as follows: Denoting the determinant by D, let the elements in the first row be a_{1i}, and let M_{1i} be the determinant of order $n-1$ formed when the first row and ith column of D are deleted. Then, by definition,

$$D = a_{11}M_{11} - a_{12}M_{12} + \cdots + (-1)^{1+n}a_{1n}M_{1n}. \tag{1-3}$$

The definition is inductive; a determinant of order n is defined in terms of those having order $n-1$.

The expansion (1-3) is termed a *Laplace development* of the determinant on elements of the first row. The determinant M_{1i} is called the *minor* of the element a_{1i}; the signed determinant $(-1)^{1+i}M_{1i}$ is the *cofactor* of a_{1i}. More generally, the determinant M_{ij} formed when the ith row and jth column are deleted is the *minor* of the element a_{ij} in this row and column. The signed determinant

$$A_{ij} = (-1)^{i+j}M_{ij} \tag{1-4}$$

is the *cofactor* of a_{ij}. It is a fundamental theorem that a *determinant may be evaluated by a Laplace development on any row or column;* in other words,

$$\begin{vmatrix} a_{11} & a_{12} & \cdots & a_{1n} \\ a_{21} & a_{22} & \cdots & a_{2n} \\ \cdot & \cdot & \cdots & \cdot \\ a_{n1} & a_{n2} & \cdots & a_{nn} \end{vmatrix} = \sum_{i=1}^{n} a_{ij}A_{ij} = \sum_{j=1}^{n} a_{ij}A_{ij}. \tag{1-5}$$

The proof may be given by induction directly or may be based on the following considerations, which are also established by induction. The expansion of an nth-order determinant is a sum of the $n!$ terms $(-1)^k a_{k_1 1} a_{k_2 2} \cdots a_{k_n n}$, where $k_1, k_2, \ldots, k_n$ are the numbers $1, 2, \ldots, n$ in some order. The integer k is defined as the number of *inversions of order* of the subscripts $k_1, k_2, \ldots, k_n$ from the normal order $1, 2, \ldots, n$ where a particular arrangement is said to have k inversions of order if it is necessary to make k successive interchanges of adjacent elements in order to make the arrangement assume the normal order. There are $n!$ terms, since there are $n!$ permutations of the n first subscripts, and each term contains as a factor one, and only one, element from each row and one, and only one, element from each column.

For example, consider the third-order determinant

$$D = \begin{vmatrix} a_{11} & a_{12} & a_{13} \\ a_{21} & a_{22} & a_{23} \\ a_{31} & a_{32} & a_{33} \end{vmatrix}.$$

The six terms of the expansion are, apart from sign,

$$a_{11}a_{22}a_{33} \qquad a_{11}a_{32}a_{23} \qquad a_{21}a_{12}a_{33} \qquad a_{21}a_{32}a_{13} \qquad a_{31}a_{12}a_{23} \qquad a_{31}a_{22}a_{13}.$$

The first term, in which the first subscripts have the normal order, is called the *diagonal term*, and its sign is positive. In the second term the arrangement 132 requires the interchange of 2 and 3 to make it assume the normal order; therefore $k = 1$, and the term has a negative sign. Similarly, the third term has a negative sign. The fourth term has a positive sign, for the arrangement 231 requires the interchange of 3 and 1 followed by the interchange of 2 and 1 to assume the normal order. Similarly, the fifth term has a positive sign. In the sixth term, it is necessary to make three interchanges (3 and 2, 3 and 1, and 2 and 1) in order to arrive at the normal order; hence, this term will have a negative sign. It follows that

$$D = a_{11}a_{22}a_{33} - a_{11}a_{32}a_{23} - a_{21}a_{12}a_{33} + a_{21}a_{32}a_{13} + a_{31}a_{12}a_{23} - a_{31}a_{22}a_{13}.$$

The main result of this discussion is that a determinant is the sum of all the $n!$ products that can be formed by taking *exactly one element from each row and each column* and multiplying by 1 or -1 according to a definite rule.

Example 1. By a development on the second column, evaluate

$$D = \begin{vmatrix} 1 & 0 & -1 & 2 \\ 6 & 0 & 4 & 3 \\ 4 & 7 & 0 & 2 \\ 2 & 0 & 2 & 3 \end{vmatrix}.$$

The $(-1)^{i+j}$ rule for determining sign means that the sign of the minors alternates as one proceeds from one element to an adjacent one in the same row or column, and the sign starts with $+$ in the upper left-hand corner. Thus,

$$D = 0(-M_{12}) + 0(M_{22}) + 7(-M_{32}) + 0(M_{42}).$$

Crossing out the row and column containing 7 gives the determinant M_{32}, whence

$$D = -7M_{32} = (-7)\begin{vmatrix} 1 & -1 & 2 \\ 6 & 4 & 3 \\ 2 & 2 & 3 \end{vmatrix} = (-7)\left[(1)\begin{vmatrix} 4 & 3 \\ 2 & 3 \end{vmatrix} - (-1)\begin{vmatrix} 6 & 3 \\ 2 & 3 \end{vmatrix} + (2)\begin{vmatrix} 6 & 4 \\ 2 & 2 \end{vmatrix} \right]$$

$$= (-7)[(1)(6) - (-1)(12) + 2(4)] = -182.$$

Example 2. The following determinant is said to be in *diagonal form*. Show that its value is $abcd$ no matter what elements are put in place of the asterisks:

$$D = \begin{vmatrix} a & * & * & * \\ 0 & b & * & * \\ 0 & 0 & c & * \\ 0 & 0 & 0 & d \end{vmatrix}.$$

Successive Laplace developments on first columns give

$$D = a \begin{vmatrix} b & * & * \\ 0 & c & * \\ 0 & 0 & d \end{vmatrix} = ab \begin{vmatrix} c & * \\ 0 & d \end{vmatrix} = abcd.$$

Evidently, a similar result is true in general.

Example 3.   If the elements are differentiable functions of t, show that

$$\frac{d}{dt} \begin{vmatrix} a_1 & a_2 & a_3 \\ b_1 & b_2 & b_3 \\ c_1 & c_2 & c_3 \end{vmatrix} = \begin{vmatrix} a_1' & a_2' & a_3' \\ b_1 & b_2 & b_3 \\ c_1 & c_2 & c_3 \end{vmatrix} + \begin{vmatrix} a_1 & a_2 & a_3 \\ b_1' & b_2' & b_3' \\ c_1 & c_2 & c_3 \end{vmatrix} + \begin{vmatrix} a_1 & a_2 & a_3 \\ b_1 & b_2 & b_3 \\ c_1' & c_2' & c_3' \end{vmatrix}.$$

A typical term in the expansion is $\pm a_i b_j c_k$.   Differentiating gives

$$\pm (a_i b_j c_k)' = \pm a_i' b_j c_k \pm a_i b_j' c_k \pm a_i b_j c_k'$$

and the sum on i, j, k of these three types of terms yields the expanded form of the three determinants on the right.   A corresponding result for determinants of order n is proved in the same way.

The fundamental theorem (1-5) leads to some important properties of determinants that are now enumerated.

1. *If each element in a row or column of a determinant is zero, the determinant is zero.*

2. *If each element in a row or column is multiplied by m, the determinant is multiplied by m.*

3. *If each element of a row or column is a sum of two terms, the determinant equals the sum of the two corresponding determinants; for example,*

$$\begin{vmatrix} a_1 & b_1 & c_1 \\ a+\alpha & b+\beta & c+\gamma \\ a_2 & b_2 & c_2 \end{vmatrix} = \begin{vmatrix} a_1 & b_1 & c_1 \\ a & b & c \\ a_2 & b_2 & c_2 \end{vmatrix} + \begin{vmatrix} a_1 & b_1 & c_1 \\ \alpha & \beta & \gamma \\ a_2 & b_2 & c_2 \end{vmatrix}. \tag{1-6}$$

These three results become obvious when we make a Laplace development on the row or column in question.   In (1-6), for example, let A, B, C be the cofactors of the elements in the second row.   The determinant is

$$(a + \alpha)A + (b + \beta)B + (c + \gamma)C$$

which equals $(aA + bB + cC) + (\alpha A + \beta B + \gamma C)$.   This, in turn, is the sum of the expansions of the two determinants on the right of (1-6).   The proof for $n \times n$ determinants is very similar and should be supplied by the reader.

4. *If two rows or two columns are proportional, the determinant is zero.*

5. *If two rows or two columns are interchanged, the determinant changes sign.*

6. *If rows and columns are interchanged, the determinant is unaltered.*

The properties 4, 5, and 6 are easily verified for 2×2 determinants and then proved in general by mathematical induction.   To obtain item 6, for example, expand the original determinant on elements of the first row and the new one on elements of the first column.   The theorem for order n then follows from the theorem for order $n - 1$. As an illustration we have

$$\begin{vmatrix} a_1 & b_1 & c_1 \\ a_2 & b_2 & c_2 \\ a_3 & b_3 & c_3 \end{vmatrix} = a_1 \begin{vmatrix} b_2 & c_2 \\ b_3 & c_3 \end{vmatrix} - a_2 \begin{vmatrix} b_1 & c_1 \\ b_3 & c_3 \end{vmatrix} + a_3 \begin{vmatrix} b_1 & c_1 \\ b_2 & c_2 \end{vmatrix} \tag{1-7}$$

which coincides with the expansion (1-2) when we interchange rows and columns of the second-order determinants on the right-hand side of (1-7).

7. *The value of a determinant is unaltered if a multiple of one row (or column) is added to another.*

8. *If the cofactors for one row (or column) are combined with the elements of another, as in (1-8), the resulting sum is zero:*

$$\sum_{i=1}^{n} a_{ij}A_{ik} = 0 \qquad \sum_{i=1}^{n} a_{ji}A_{ki} = 0 \qquad k \neq j. \tag{1-8}$$

These results follow from those already established. To illustrate the proof of item 7, we have

$$\begin{vmatrix} a_1 & b_1 + mc_1 & c_1 \\ a_2 & b_2 + mc_2 & c_2 \\ a_3 & b_3 + mc_3 & c_3 \end{vmatrix} = \begin{vmatrix} a_1 & b_1 & c_1 \\ a_2 & b_2 & c_2 \\ a_3 & b_3 & c_3 \end{vmatrix} + \begin{vmatrix} a_1 & mc_1 & c_1 \\ a_2 & mc_2 & c_2 \\ a_3 & mc_3 & c_3 \end{vmatrix} \tag{1-9}$$

by 3; and the second determinant on the right of (1-9) is zero by 4. The reader should extend the proof to $n \times n$ determinants.

The result 8 follows from 4. Thus, the first expression (1-8) is the expansion of the determinant that arises when the row

$$a_{1k}, a_{2k}, \ldots, a_{nk}$$

is replaced by $a_{1j}, a_{2j}, \ldots, a_{nk}$, and hence it is the expansion of a determinant with two equal rows.

9. *If two determinants A and B of order n are given and a new determinant C is formed, the element in the ith row and jth column of which is obtained by multiplying each element in the ith row of A by the corresponding element in the jth row of B and adding the products thus formed, then $C = AB$.*

Thus, if the elements of A and B are denoted by a_{ij} and b_{ij}, respectively, the element c_{ij} in the ith row and jth column of the product determinant C is

$$c_{ij} = a_{i1}b_{j1} + a_{i2}b_{j2} + \cdots + a_{in}b_{jn}. \tag{1-10}$$

The validity of rule 9 for determinants of order n follows from considerations similar to those we give next for the case when $n = 2$.

If the determinants A and B are of second order, formula (1-10) states that their product C is

$$C = \begin{vmatrix} a_{11}b_{11} + a_{12}b_{12} & a_{11}b_{21} + a_{12}b_{22} \\ a_{21}b_{11} + a_{22}b_{12} & a_{21}b_{21} + a_{22}b_{22} \end{vmatrix}. \tag{1-11}$$

Since the elements in (1-11) are binomials, we can write C by using property 3 as the sum of four determinants:

$$C = \begin{vmatrix} a_{11}b_{11} & a_{11}b_{21} \\ a_{21}b_{11} & a_{21}b_{21} \end{vmatrix} + \begin{vmatrix} a_{11}b_{11} & a_{12}b_{22} \\ a_{21}b_{11} & a_{22}b_{22} \end{vmatrix} + \begin{vmatrix} a_{12}b_{12} & a_{11}b_{21} \\ a_{22}b_{12} & a_{21}b_{21} \end{vmatrix} + \begin{vmatrix} a_{12}b_{12} & a_{12}b_{22} \\ a_{22}b_{12} & a_{22}b_{22} \end{vmatrix}.$$

On factoring out the elements b_{11} and b_{21} in the first determinant, we obtain a determinant with two like rows, and hence its value is zero. Similar remarks apply to the fourth determinant. The second determinant, on factoring out b_{11} and b_{22}, yields $b_{11}b_{22}A$, while the third has the value $-b_{12}b_{21}A$. Thus,

$$C = A(b_{11}b_{22} - b_{12}b_{21}) = AB.$$

If a similar procedure is applied to determinants of order n, the expression C given by rule 9 appears as a sum of a large number of determinants whose elements have the form $a_{ik}b_{jk}$. On factoring out the b_{jk} we find that each determinant in this sum has the form

$$K(\text{determinant of } a\text{'s}) \ (\text{product of some } b\text{'s})$$

where $K = 0, 1,$ or -1. The sum therefore gives

$$C = AQ(b)$$

where A is the determinant of the a's and $Q(b)$ denotes a polynomial in the b's.

In the same way, or by symmetry,

$$C = BP(a)$$

where B is the determinant of the b's and $P(a)$ is a polynomial in the a's. These formulas show that $C = 0$ if $A = 0$ or $B = 0$, so that the desired result $C = AB$ is verified in that case.

If neither A nor B is 0, the equation $BP(a) = AQ(b)$ can be divided by AB to give

$$\frac{P(a)}{A} = \frac{Q(b)}{B}.$$

Since the left side is a function of the a's alone while the right side is a function of the b's alone we conclude that each side must be constant.[1] This gives

$$C = (\text{const}) \ AB.$$

We can evaluate the constant by considering the special case in which A and B have 1 on the main diagonal and 0 elsewhere. Then $A = B = C = 1$; hence the constant is 1, and rule 9 follows.

Since the value of a determinant is unchanged when its rows and columns are interchanged, there are four ways in which the determinant C may be written. Thus, if we interchange the rows and columns of B, the elements c_{ij} in C will be given by (1-10) in which the subscripts on the b's are interchanged.[2]

Example 4.   Without expanding, show that

$$\begin{vmatrix} 1 & x_1 & x_1^2 \\ 1 & x_2 & x_2^2 \\ 1 & x_3 & x_3^2 \end{vmatrix} = (x_1 - x_2)(x_3 - x_2)(x_1 - x_3).$$

The determinant is a polynomial in x_1, and it vanishes when $x_1 = x_2$, since the first two rows are then proportional. Hence it is divisible by $x_1 - x_2$. Similarly, it is divisible by $x_3 - x_2$ and $x_1 - x_3$. It therefore equals

$$E(x_1 - x_2)(x_3 - x_2)(x_1 - x_3)$$

for some polynomial E. Since the determinant is of degree 3 in x_1, x_2, x_3, we must have $E = \text{const}$, and comparing coefficients of $x_2 x_3^2$ shows that $E = 1$.

Example 5.   Write the product of the determinants

$$A = \begin{vmatrix} 1 & 2 & 1 \\ 3 & 0 & 1 \\ 0 & 2 & 1 \end{vmatrix} \quad \text{and} \quad B = \begin{vmatrix} -1 & 4 & 2 \\ 2 & -1 & 3 \\ 0 & 2 & -1 \end{vmatrix}$$

as a single determinant of third order.

Using rule 9, we find

$$AB = \begin{vmatrix} -1+8+2 & 2-2+3 & 0+4-1 \\ -3+0+2 & 6-0+3 & 0+0-1 \\ 0+8+2 & 0-2+3 & 0+4-1 \end{vmatrix} = \begin{vmatrix} 9 & 3 & 3 \\ -1 & 9 & -1 \\ 10 & 1 & 3 \end{vmatrix}.$$

To check the result, we find on expanding the determinants A, B, and AB that $A = -2$, $B = 21$, and $AB = -42$.

[1] Compare Chap. 7, Sec. 10, Prob. 2. We are indebted to a student, Joseph Motzkin, for this elegant proof of rule 9.

[2] Cf. Chap. 4, Eq. (15-3).

Example 6.   Show that a trigonometric polynomial

$$y = a_1 \sin x + a_2 \sin 2x + a_3 \sin 3x \qquad (1\text{-}12)$$

passing through three assigned points (x_i, y_i) is given, in general, by

$$\begin{vmatrix} y & \sin x & \sin 2x & \sin 3x \\ y_1 & \sin x_1 & \sin 2x_1 & \sin 3x_1 \\ y_2 & \sin x_2 & \sin 2x_2 & \sin 3x_2 \\ y_3 & \sin x_3 & \sin 2x_3 & \sin 3x_3 \end{vmatrix} = 0.$$

Expanding on the first row gives

$$c_1 y + c_2 \sin x + c_3 \sin 2x + c_4 \sin 3x = 0$$

where c_i are the appropriate cofactors.   Hence y has the form (1-12), if $c_1 \neq 0$.   Moreover, when $x = x_i$ and $y = y_i$ the equation is true, since two rows of the determinant are then equal.

PROBLEMS

1. By a Laplace development on the first row, evaluate

$$\begin{vmatrix} 1 & 2 & 3 \\ 3 & 1 & 2 \\ 2 & 3 & 1 \end{vmatrix}, \quad \begin{vmatrix} -1 & 2 & 2 \\ -3 & 6 & 6 \\ 5 & 7 & 9 \end{vmatrix}, \quad \begin{vmatrix} 1 & 0 & 0 \\ 0 & 0 & 1 \\ 0 & 1 & 0 \end{vmatrix}, \quad \begin{vmatrix} 1 & 2 & 3 \\ 4 & 5 & 6 \\ 7 & 8 & 9 \end{vmatrix}.$$

2. Evaluate the determinants in Prob. 1 by a Laplace development on (*a*) the first column, (*b*) the second row.

3. Evaluate this determinant by development on

(*a*) The first column
(*b*) The second row
(*c*) The first row
(*d*) The third column

$$\begin{vmatrix} 1 & -1 & 1 & -1 \\ 0 & 1 & -1 & 1 \\ 0 & 0 & 1 & -1 \\ 1 & 0 & 0 & 1 \end{vmatrix}.$$

4. Show that

$$\begin{vmatrix} x_1 & 1 \\ x_2 & 1 \end{vmatrix}, \quad \frac{1}{2} \begin{vmatrix} x_1 & y_1 & 1 \\ x_2 & y_2 & 1 \\ x_3 & y_3 & 1 \end{vmatrix}$$

represent, respectively, the (signed) length of the segment (x_1, x_2) and the area of the triangle with vertices (x_i, y_i).

5. Evaluate, using some of the properties 1 to 7:

$$\begin{vmatrix} x & 1 & 1 \\ 1 & x & 1 \\ 1 & 1 & x \end{vmatrix}, \quad \begin{vmatrix} y+z & x & x \\ y & x+z & y \\ z & z & x+y \end{vmatrix}, \quad \begin{vmatrix} 0 & -a & -b \\ a & 0 & -c \\ b & c & 0 \end{vmatrix}.$$

Hint: In the last determinant, interchange rows and columns.

6. Write as determinants of third order the product of the first determinant in Prob. 1 by the second and third determinants.

7. Using determinants, find a, b, c if $y = a + b \cos x + c \cos 2x$ passes through $(0,0)$, $(\pi/2, 1)$, $(\pi, -2)$.

8. (*a*) Find a cubic containing the points $(0,1)$, $(1,-1)$, $(3,4)$, $(-1,0)$.   *Hint:* Consider a determinant with top row y, 1, x, x^2, x^3.   (*b*) Write the equation of a polynomial of degree n whose graph contains $n + 1$ assigned points (x_i, y_i).

2. Cramer's Rule.   Consider the set of simultaneous equations

$$
\begin{aligned}
a_1x + b_1y + c_1z &= d_1 \\
a_2x + b_2y + c_2z &= d_2 \\
a_3x + b_3y + c_3z &= d_3.
\end{aligned}
\tag{2-1}
$$

Now, by rules 2 and 7 of the preceding section,

$$
x\begin{vmatrix} a_1 & b_1 & c_1 \\ a_2 & b_2 & c_2 \\ a_3 & b_3 & c_3 \end{vmatrix} = \begin{vmatrix} a_1x & b_1 & c_1 \\ a_2x & b_2 & c_2 \\ a_3x & b_3 & c_3 \end{vmatrix} = \begin{vmatrix} a_1x + b_1y + c_1z & b_1 & c_1 \\ a_2x + b_2y + c_2z & b_2 & c_2 \\ a_3x + b_3y + c_3z & b_3 & c_3 \end{vmatrix}.
$$

Hence, if x satisfies (2-1), it is necessary that

$$
x\begin{vmatrix} a_1 & b_1 & c_1 \\ a_2 & b_2 & c_2 \\ a_3 & b_3 & c_3 \end{vmatrix} = \begin{vmatrix} d_1 & b_1 & c_1 \\ d_2 & b_2 & c_2 \\ d_3 & b_3 & c_3 \end{vmatrix}.
\tag{2-2}
$$

The determinant on the left of (2-2) is termed the *coefficient determinant* of the system (2-1); we denote it by D.   Equation (2-2) and the corresponding relations for y and z may then be written

$$
xD = \begin{vmatrix} d_1 & b_1 & c_1 \\ d_2 & b_2 & c_2 \\ d_3 & b_3 & c_3 \end{vmatrix} \qquad yD = \begin{vmatrix} a_1 & d_1 & c_1 \\ a_2 & d_2 & c_2 \\ a_3 & d_3 & c_3 \end{vmatrix} \qquad zD = \begin{vmatrix} a_1 & b_1 & d_1 \\ a_2 & b_2 & d_2 \\ a_3 & b_3 & d_3 \end{vmatrix}.
\tag{2-3}
$$

If $D \neq 0$, we may divide by D to express x, y, and z as quotients of two determinants.

To show that these values of x, y, and z actually satisfy the system (2-1), substitute into (2-1) and multiply through by D.   The equations become

$$
a_k\begin{vmatrix} d_1 & b_1 & c_1 \\ d_2 & b_2 & c_2 \\ d_3 & b_3 & c_3 \end{vmatrix} + b_k\begin{vmatrix} a_1 & d_1 & c_1 \\ a_2 & d_2 & c_2 \\ a_3 & d_3 & c_3 \end{vmatrix} + c_k\begin{vmatrix} a_1 & b_1 & d_1 \\ a_2 & b_2 & d_2 \\ a_3 & b_3 & d_3 \end{vmatrix} = d_k\begin{vmatrix} a_1 & b_1 & c_1 \\ a_2 & b_2 & c_2 \\ a_3 & b_3 & c_3 \end{vmatrix}
$$

with $k = 1$, 2, or 3, respectively.   Now, the determinant

$$
\begin{vmatrix} a_k & b_k & c_k & d_k \\ a_1 & b_1 & c_1 & d_1 \\ a_2 & b_2 & c_2 & d_2 \\ a_3 & b_3 & c_3 & d_3 \end{vmatrix}
$$

is zero because two rows are equal, and it yields the desired relation when expanded on elements of the first row (use rule 5 of the preceding section).

The foregoing method applies to n equations in n unknowns, and yields:

CRAMER'S RULE: Let[1]

$$
\begin{aligned}
a_{11}x_1 + a_{12}x_1 + \cdots + a_{1n}x_n &= k_1 \\
a_{21}x_1 + a_{22}x_2 + \cdots + a_{2n}x_n &= k_2 \\
\cdots\cdots\cdots\cdots\cdots\cdots\cdots& \\
a_{n1}x_1 + a_{n2}x_2 + \cdots + a_{nn}x_n &= k_n
\end{aligned}
\tag{2-4}
$$

be a system of n equations in the n unknowns x_i such that the coefficient determinant D is not zero.   The system (2-4) has a unique solution $x_i = D_i/D$, where D_i is the determinant formed by replacing the elements $a_{1i}, a_{2i}, \ldots, a_{ni}$ of the ith column of D by $k_1, k_2, \ldots, k_n$, respectively.

Consider the *homogeneous system* that arises from (2-4) when the right-hand members are replaced by zero.   This system obviously has a solution $x_1 = x_2 = \cdots = x_n = 0$,

[1] A compact derivation of this rule is given in Chap. 4, Sec. 15.

the *trivial solution.* If the coefficient determinant is not zero, the solution is unique by Cramer's rule. Hence *a homogeneous system can have a nontrivial solution only if the coefficient determinant is zero.* One can prove, conversely, that there is always a nontrivial solution of the homogeneous equations if the determinant is zero.

The rectangular array

$$\begin{pmatrix} a_1 & b_1 & c_1 & d_1 \\ a_2 & b_2 & c_2 & d_2 \\ a_3 & b_3 & c_3 & d_3 \end{pmatrix} \tag{2-5}$$

is termed the *augmented matrix* of the system (2-1). By striking out one or another column of the matrix (2-5), we are led to the square arrays

$$\begin{pmatrix} b_1 & c_1 & d_1 \\ b_2 & c_2 & d_2 \\ b_3 & c_3 & d_3 \end{pmatrix}, \quad \begin{pmatrix} a_1 & c_1 & d_1 \\ a_2 & c_2 & d_2 \\ a_3 & c_3 & d_3 \end{pmatrix}, \quad \begin{pmatrix} a_1 & b_1 & d_1 \\ a_2 & b_2 & d_2 \\ a_3 & b_3 & d_3 \end{pmatrix}, \quad \begin{pmatrix} a_1 & b_1 & c_1 \\ a_2 & b_2 & c_2 \\ a_3 & b_3 & c_3 \end{pmatrix}.$$

Since these arrays are square, they have corresponding determinants. Now (2-3) shows that all these determinants must be zero if $D = 0$ and if the system (2-1) actually has a solution. In other words, if $D = 0$ but a third-order determinant formed from (2-5) is not zero, the system (2-1) is inconsistent.

The foregoing results are included in a general theory of linear systems, which is now discussed. An $m \times n$ *matrix* is a system of mn quantities a_{ij}, called *elements*, arranged in m rows and n columns. The array is customarily enclosed in parentheses, thus:

$$A \equiv \begin{pmatrix} a_{11} & a_{12} & \cdots & a_{1n} \\ a_{21} & a_{22} & \cdots & a_{2n} \\ & \cdot \cdot \cdot \cdot \cdot \cdot \\ a_{m1} & a_{m2} & \cdots & a_{mn} \end{pmatrix}.$$

If $m = n$, then A is the *coefficient matrix* of the system (2-4); the *augmented matrix* is obtained by adjoining a column with elements (in order) $k_1, k_2, \ldots, k_n$. If the matrix is square, one can form *the determinant of the matrix*, a determinant whose elements have the same arrangement as those of the matrix. From any matrix, smaller matrices can be formed by striking out some of the rows and columns. Certain of these smaller matrices are square, and their determinants are called *determinants of the matrix*. A matrix A is said to be of *rank r* if there is at least one r-rowed determinant of A that is not zero, whereas all determinants of A of order higher than r are zero or nonexistent. (The latter alternative arises if r equals the smaller of the two numbers m and n.) The rank is zero if all elements are zero. With these preliminaries we can state the following:

FUNDAMENTAL THEOREM. Suppose we are given a set of m linear equations in n unknowns. Let the rank of the coefficient matrix be r, and let the rank of the augmented matrix be r'. If r' > r, the equations have no solution. If r' = r = n, there is one, and only one, solution. If r' = r < n, we may give arbitrary values to n − r of the unknowns and express the others in terms of these.

When $r = r'$, the common value is called the *rank* of the system. The r unknowns which are expressed in terms of the others must be associated with some nonvanishing determinant of order r, and any such set of r unknowns can be used.

To illustrate the use of the theorem, we consider a set of n homogeneous linear equations in n unknowns. Since the right-hand members are 0 the augmented matrix cannot have a rank greater than that of the coefficient matrix. Hence, $r' = r$. If the coefficient determinant is zero, then $r \leq n - 1$ and, by the fundamental theorem, at least one unknown can be prescribed at will.   Since this arbitrary value can be chosen different from 0, we obtain the following result, already mentioned above:

COROLLARY.   A homogeneous system of n equations in n unknowns has a nontrivial solution if, and only if, the coefficient determinant is zero.

Proof of the corollary hinges on proof of the fundamental theorem, which is outlined in the next section.   Here we prefer to give some simple examples.

Example 1.   By Cramer's rule, find x and y, given

$$\begin{aligned} 3x + y + 2z &= 3 \\ 2x - 3y - z &= -3 \\ x + 2y + z &= 4. \end{aligned} \tag{2-6}$$

The coefficient determinant D is found to be 8, so that

$$8x = \begin{vmatrix} 3 & 1 & 2 \\ -3 & -3 & -1 \\ 4 & 2 & 1 \end{vmatrix} = 8 \qquad 8y = \begin{vmatrix} 3 & 3 & 2 \\ 2 & -3 & -1 \\ 1 & 4 & 1 \end{vmatrix} = 16.$$

Thus, $x = 1$, $y = 2$.   If z is desired, one can find it from the third equation (2-6):

$$z = 4 - x - 2y = 4 - 1 - 4 = -1.$$

Example 2.   For what values of λ do the equations

$$\begin{aligned} ax + by &= \lambda x \\ cx + dy &= \lambda y \end{aligned} \tag{2-7}$$

have a solution other than $x = y = 0$?
Transposing to the left gives the homogeneous system

$$\begin{aligned} (a - \lambda)x + by &= 0 \\ cx + (d - \lambda)y &= 0. \end{aligned}$$

For a nontrivial solution, the coefficient determinant must vanish:

$$\begin{vmatrix} a - \lambda & b \\ c & d - \lambda \end{vmatrix} \equiv \lambda^2 - (a + d)\lambda + (ad - bc) = 0. \tag{2-8}$$

Thus, (2-7) has a nontrivial solution if, and only if, λ satisfies the quadratic (2-8).

Example 3.   If the equations

$$\begin{aligned} ax^2 + bx + c &= 0 \\ \alpha x^2 + \beta x + \gamma &= 0 \end{aligned} \tag{2-9}$$

have a common root, show that

$$\begin{vmatrix} a & b & c & 0 \\ 0 & a & b & c \\ \alpha & \beta & \gamma & 0 \\ 0 & \alpha & \beta & \gamma \end{vmatrix} = 0. \tag{2-10}$$

If x is the common root, (2-9) gives

$$\begin{aligned} ax^3 + bx^2 + cx &= 0 \\ ax^2 + bx + c &= 0 \\ \alpha x^3 + \beta x^2 + \gamma x &= 0 \\ \alpha x^2 + \beta x + \gamma &= 0. \end{aligned} \tag{2-11}$$

The system (2-11) is a homogeneous linear system in the "unknowns" x^3, x^2, x, 1, of which one (namely, 1) is not zero.   Hence the coefficient determinant vanishes, and (2-10) follows.

Example 4.   Give a necessary condition that these three lines be concurrent:

$$\begin{aligned} ax + by + c &= 0 \\ a_1 x + b_1 y + c_1 &= 0 \\ a_2 x + b_2 y + c_2 &= 0. \end{aligned} \tag{2-12}$$

Let (x,y) be the point at which the three lines meet. With this particular choice of x and y, the three equations (2-12) are satisfied simultaneously. Now, these equations may be regarded as simultaneous equations in three unknowns x, y, and 1, one of which (namely, 1) is not zero. Hence the coefficient determinant vanishes:

$$\begin{vmatrix} a & b & c \\ a_1 & b_1 & c_1 \\ a_2 & b_2 & c_2 \end{vmatrix} = 0.$$

(The condition is also sufficient if no two of the three lines are parallel. The reader should observe the duality between points and lines that is illustrated by this and the following example.)

Example 5.   Find a necessary and sufficient condition that the three points (x,y), (x_1,y_1), (x_2,y_2) lie on a line.

If the equation of the line is

$$ax + by + c = 0 \tag{2-13}$$

we have, besides (2-13),

$$ax_1 + by_1 + c = 0$$
$$ax_2 + by_2 + c = 0.$$

These equations may be regarded as a system in the unknowns a, b, c, which cannot all vanish if (2-13) represents a line. Hence the coefficient determinant must vanish:

$$\begin{vmatrix} x & y & 1 \\ x_1 & y_1 & 1 \\ x_2 & y_2 & 1 \end{vmatrix} = 0. \tag{2-14}$$

Conversely (2-14) ensures that the system has a nontrivial solution a, b, c. Compare Prob. 4 in Sec. 1.

Example 6.   Show that the following equations are consistent if, and only if, $k = 9$:

$$\begin{aligned} 2x + 3y &= 1 \\ x - 2y &= 4 \\ 4x - y &= k. \end{aligned} \tag{2-15}$$

The coefficient matrix has rank 2, and hence the equations are consistent if, and only if, the augmented matrix also has rank 2. This entails

$$\begin{vmatrix} 2 & 3 & 1 \\ 1 & -2 & 4 \\ 4 & -1 & k \end{vmatrix} = 0$$

which yields $1(7) - 4(-14) + k(-7) = 0$ or $k = 9$. The same result is found if we regard (2-15) as a system in the three unknowns x, y, k and solve by Cramer's rule. The reader should obtain the result of Examples 3 and 4 by considering the augmented matrix, as in the present example.

PROBLEMS

1. Solve, by Cramer's rule, the systems:

(a)
$$\begin{aligned} x + 2y + 3z &= 3 \\ 2x - y + z &= 6 \\ 3x + y - z &= 4; \end{aligned}$$

(b)
$$\begin{aligned} 2x + y + 3z &= 2 \\ 3x - 2y - 2z &= 1 \\ x - y + z &= -1; \end{aligned}$$

(c)
$$\begin{aligned} x + 2y &= 1 \\ 2x - y - 2z &= 3 \\ -x + y + 3z &= 2; \end{aligned}$$

(d)
$$\begin{aligned} 2x + y + 3z + w &= -2 \\ 5x + 3y - z - w &= 1 \\ x - 2y + 4z + 3w &= 4 \\ 3x - y + z &= 2. \end{aligned}$$

2. Obtain nonzero solutions when they exist.

(a) $x + 3y - 2z = 0$
$\quad 2x - y + z = 0;$

(b) $x - 2y = 0$
$\quad 3x + y = 0$
$\quad 2x - y = 0;$

(c) $3x - 2y + z = 0$
$\quad x + 2y - 2z = 0$
$\quad 2x - y + 2z = 0;$

(d) $2x - 4y + 3z = 0$
$\quad x + 2y - 2z = 0$
$\quad 3x - 2y + z = 0;$

(e) $4x - 2y + z = 0$
$\quad 2x - y + 3z = 0$
$\quad 2x - y - 2z = 0$
$\quad 6x - 3y + 4z = 0;$

(f) $x + 2y + 2z = 0$
$\quad 3x - y + z = 0$
$\quad 2x + 3y + 2z = 0$
$\quad x + 4y - 2z = 0.$

3. Investigate the following systems and find solutions whenever the systems are consistent:

(a) $x - 2y = 3$
$\quad 2x + y = 1$
$\quad 3x - y = 4;$

(b) $2x + y - z = 1$
$\quad x - 2y + z = 3$
$\quad 4x - 3y + z = 5;$

(c) $3x + 2y = 4$
$\quad x - 3y = 1$
$\quad 2x + 5y = -1;$

(d) $2x - y + 3z = 4$
$\quad x + y - 3z = -1$
$\quad 5x - y + 3z = 7.$

4. (a) Give a necessary and sufficient condition that four points in space be coplanar. (b) Give a necessary and sufficient condition that four planes be concurrent.

5. As in Example 5, find a necessary and sufficient condition that four points lie on a circle.

6. Give a relation that the coefficients must satisfy if

$$ax^3 + bx^2 + cx + d = 0$$
$$\alpha x^2 + \beta x + \gamma = 0$$

have a common root.

7. Give a condition on the coefficients of a general cubic $f(x)$ if it has a double root. *Hint:* $f(x)$ and $f'(x)$ have a root in common.

8. The system $ax + by = c$, $\alpha x + \beta y = \gamma$ represents two lines that may intersect at one point, may be parallel, or may coincide. Discuss the system geometrically, and thus obtain all the relevant results involving rank. *Hint:* Begin by showing that the lines are parallel if, and only if, the coefficient determinant is zero.

9. An equation $ax + by + cz = d$ represents a plane, and two planes are parallel if, and only if, corresponding coefficients a, b, c are proportional:

$$a = ka_1 \qquad b = kb_1 \qquad c = kc_1.$$

(You may assume these geometric facts.) As in Prob. 8, give a complete geometric discussion of the behavior of two equations in three unknowns.

10. As in Prob. 9, discuss the general system of three equations in three unknowns.

---

3. Further Discussion of Rank. Although the fundamental theorem of the preceding section plays a role in many important investigations, it gives no clue to how one should proceed for efficient numerical computation. A particularly simple method can be based on the *Gauss reduction* introduced in Chap. 10, Sec. 4, and for the reader's convenience an independent development of the Gauss reduction is given here. It will be found that the method not only facilitates numerical computation but actually yields a proof of the fundamental theorem.

To illustrate the Gauss reduction, consider the following system of three equations in five unknowns:

$$x_1 + 2x_2 + 2x_3 - 3x_4 + 4x_5 = 1 \qquad (a)$$

$$x_1 + 3x_2 - 3x_3 - 6x_4 + 2x_5 = 2 \qquad (b)$$

$$2x_1 + 5x_2 + 2x_3 - 3x_4 + 3x_5 = 9. \qquad (c)$$

If the first equation (a) is subtracted from the second, the latter becomes

$$x_2 - 5x_3 - 3x_4 - 2x_5 = 1. \qquad (d)$$

Also, if twice (a) is subtracted from (c), then (c) becomes

$$x_2 - 2x_3 + 3x_4 - 5x_5 = 7. \qquad (e)$$

This is the first step of the Gauss reduction. It replaces the system (abc) by an algebraically equivalent system (ade), in which (de) do not involve x_1.

Proceeding to the second step, we subtract (d) from (e). The result is

$$x_3 + 2x_4 - x_5 = 2 \qquad (f)$$

after division by 3 to make the coefficient of x_3 equal 1. The original system (abc) is algebraically equivalent to (adf), in which (e) does not contain x_1, and (f) does not contain x_1 or x_2.

If arbitrary values

$$x_4 = k \qquad x_5 = l$$

are chosen, then by (f)

$$x_3 = 2 - 2k + l.$$

Substitution of x_3, x_4, and x_5 into (d) gives

$$x_2 = 11 - 7k + 7l$$

and finally, by use of these values in (a),

$$x_1 = -25 + 21k - 20l.$$

These formulas give the complete solution of the system (adf); since this system is algebraically equivalent to (abc), they also give the complete solution of (abc).

The foregoing operations on the system (abc) correspond to certain operations on the augmented matrix of this system,

$$\begin{pmatrix} 1 & 2 & 2 & -3 & 4 & 1 \\ 1 & 3 & -3 & -6 & 2 & 2 \\ 2 & 5 & 2 & -3 & 3 & 9 \end{pmatrix} \qquad (3\text{-}1)$$

The first step of the Gauss reduction amounts to subtracting the first row from the second and then subtracting twice the first row from the third. The resulting matrix is the augmented matrix for the system (ade), as the reader can verify. The second step in the Gauss reduction amounts to subtracting the second row from the third, in this new matrix, and then dividing the elements of the third row by 3. The resulting matrix

$$\begin{pmatrix} 1 & 2 & 2 & -3 & 4 & 1 \\ 0 & 1 & -5 & -3 & -2 & 1 \\ 0 & 0 & 1 & 2 & -1 & 2 \end{pmatrix} \qquad (3\text{-}2)$$

is the augmented matrix for the system (adf).

Matrix (3-2) has rank 3 because the determinant formed by the first three columns is $1 \cdot 1 \cdot 1 = 1$ by inspection. Now, the operations leading from (3-1) to (3-2) do not change the value of the determinant, except for the last operation, which divides the determinant by 3. Since the final determinant has the value 1, the original determinant, formed by the first three columns in (3-1), must have had the value 3. In particular, it is not zero; hence the original system and the reduced system have the same rank.

In the course of the foregoing illustration we multiplied an equation of the linear system by a nonzero constant, to make the coefficient of a designated unknown equal to 1. We also subtracted one equation from another. Each of these operations is called an *elementary operation* on the system or on the augmented matrix.

Quite often these two elementary operations are sufficient to complete the Gauss reduction. Sometimes, however, the process stops, because a coefficient has the value 0. For example, if the second step of the reduction leads to the system

$$x_3 + 4x_4 = 2 \qquad\qquad (g)$$

$$x_2 + 5x_3 - 2x_4 = 7 \qquad\qquad (h)$$

$$2x_2 + 2x_3 + 6x_4 = 2 \qquad\qquad (i)$$

we cannot use the first equation to eliminate x_2 from the following equations, because x_2 does not occur in the first equation.

In such a case one would simply interchange the first equation with one of the others for which the coefficient is not zero. The system (ghi) could be replaced by (hgi), for example. The operation of interchanging two equations in this way is also called an elementary operation. It corresponds to interchanging two rows of the augmented matrix.

Another exceptional case is illustrated by the system

$$x_1 + 2x_2 + 3x_3 + 4x_4 = 6$$
$$x_3 + 3x_4 = 3.$$

To keep the pattern of the foregoing illustrations, one could interchange the roles of the unknowns x_2 and x_3, so that the system appears as

$$x_1 + 3x_3 + 2x_2 + 4x_4 = 6$$
$$x_3 + 0x_2 + 3x_4 = 3.$$

A change such as this in the ordering of the unknowns is also called an elementary operation. It corresponds to an interchange of two columns in the coefficient matrix.

By use of rules 2, 3, 5, and 7 of Sec. 1 it is not difficult to show that the rank r of the coefficient matrix and the rank r' of the augmented matrix are unchanged by the four elementary operations (cf. Prob. 4). Since the Gauss reduction is achieved by a succession of elementary operations, we conclude that the *reduced system* obtained by this process has the same values of r and r' as has the original system. The reduced system has the form

$$y_1 + b_{12}y_2 + b_{13}y_3 + \cdots + b_{1n}y_n = m_1$$
$$y_2 + b_{23}y_3 + \cdots + b_{2n}y_n = m_2$$
$$\cdots\cdots\cdots\cdots\cdots\cdots \qquad\qquad (3\text{-}3)$$
$$y_r + \cdots + b_{rn}y_n = m_r$$

together with some additional relations, when $r < n$, of the form

$$0 = m_{r+1}, \qquad 0 = m_{r+2}, \qquad \ldots, \qquad 0 = m_n. \qquad\qquad (3\text{-}4)$$

These additional equations have left-hand members 0 because otherwise the rank of the coefficient matrix would be larger than r.

If any value m_{r+1}, m_{r+2}, $\ldots$, or m_n is not zero, the equations are inconsistent, and also, as is easily seen, $r' > r$. Conversely, if $r' > r$, at least one of these quantities m_k must be nonzero, and the system is inconsistent. Thus, the fundamental theorem is verified in this case.

If $r' = r$, Eqs. (3-4) are satisfied, and we need consider only (3-3). When $r = n$, the last equation gives $y_n = m_n$, the preceding equation gives y_{n-1}, and so on. Hence the solution is unique. This also agrees with the theorem.

Finally, if $r' = r$ and $n > r$, we can assign arbitrary values to the $n - r$ unknowns

$$y_{r+1}, y_{r+2}, \ldots, y_n$$

and determine y_r from the last equation of (3-3). The preceding equation gives y_{r-1}, and so on. The theorem is therefore verified in this case also.

As far as specific computation is concerned, it should be stressed that the procedure illustrated here is enormously more efficient than that of the preceding section. Given a numerical system of rank 6 with 20 equations and 30 unknowns, nobody in his senses would check the rank by examining all subdeterminants of orders 6 and 7, nor would he solve by Cramer's rule. Both problems are settled in practice by one or another variant of the Gauss reduction.

As far as theory is concerned, our objective has been not so much to *prove* the fundamental theorem as to make clear why the fundamental theorem is true. However, the argument can be developed into a proof along lines suggested by Probs. 4 to 6.[1]

PROBLEMS

1. By means of the Gauss reduction, determine the ranks (r,r') for selected systems among the 14 of Probs. 1 to 3 in Sec. 2. Also find complete solutions when they exist.

2. Test the following systems for consistency and solve if consistent when $\alpha = 0$ and $\beta = 1$; when $\alpha = 3$ and $\beta = 2$; and when $\alpha = 3$ and $\beta = -2$:

(a)
$$x + 3y + 2z + w = \beta$$
$$\alpha x + 2y + z + 2w = 0$$
$$2x - y - z + w = 2$$

(b)
$$x + y + z + w = 1$$
$$2x + 3y = 2$$
$$3x + 4y + z + w = 3$$
$$5x + 6y + 3z + \alpha w = \beta + 3.$$

3. Write a system of four linear homogeneous equations in four unknowns in the reduced form (3-3), and verify the corollary of the preceding section for this case.

4. Verify that r and r' are unchanged (a) when an equation is divided by a nonzero constant; (b) when two equations are interchanged; (c) when two equations are subtracted. *Hint:* In part c, which is the only troublesome case, let rows R and S of the augmented matrix be subtracted. By part b we can assume that R is the top row and S the second row. There is no difficulty about determinants that involve both R and S or about those that involve neither R nor S. If a determinant involves R but not S, let $D(R)$ be such a determinant, and let $D(S)$ denote the corresponding determinant obtained when R is replaced by S. Then the original system has determinants $D(R)$ and $D(S)$, and the new system has $D(R)$ and $D(S - R)$. Since $D(S - R) = D(S) - D(R)$, the new determinants are zero if both the old ones were, and at least one of the new ones is nonzero if at least one of the old ones was.

5. Show that each of the four elementary operations is reversible, so that the reduced system is algebraically equivalent to the original system.

6. For brevity, we did not give an explicit verbal formulation of the concept "reduced system." (a) Using the discussion of (abc) and (3-3) as a guide, supply the missing formulation, both in terms of linear systems and in terms of the augmented matrix. (b) Prove that a general linear system can be reduced to the form (a) by a succession of elementary operations. *Hint:* Use mathematical induction on the number of equations.

---

[1] See, in this connection, Garrett Birkhoff and Saunders MacLane, "A Survey of Higher Algebra," p. 228, The Macmillan Company, New York, 1953. An interesting discussion of linear systems from a somewhat different point of view can be found in Lowell J. Paige and J. Dean Swift, "Elements of Linear Algebra," Ginn and Company, Boston, 1961.

Comparison of the Riemann and Lebesgue integrals

1. The Riemann Integral. Let a function $f(x)$ be given on the interval $a \leq x \leq b$ (Fig. 1). To define the *Riemann integral*

$$\int_a^b f(x)\,dx \tag{1-1}$$

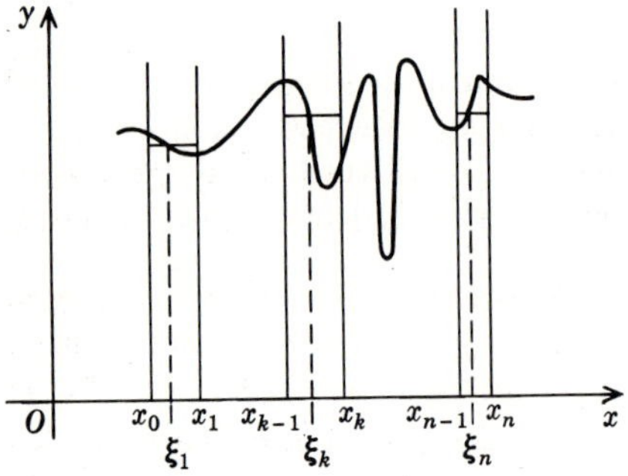

we divide the interval $[a,b]$ into smaller intervals by points x_k,

$$a = x_0 < x_1 < x_2 \cdots < x_n = b.$$

It will be desirable to consider a sequence of subdivisions which are made finer and finer by choosing more and more points x_k. The precise requirement is

$$n \to \infty \qquad \text{and} \qquad \max_k |x_k - x_{k-1}| \to 0.$$

To describe this situation we say, in brief, that the subdivision becomes *arbitrarily fine*.

Let ξ_k be an arbitrary point on the interval $[x_{k-1}, x_k]$. With $y_k = f(\xi_k)$ as shown in Fig. 1, the sum

FIGURE 1

$$s = y_1(x_1 - x_0) + y_2(x_2 - x_1) + \cdots + y_n(x_n - x_{n-1}) \tag{1-2}$$

represents a certain area that presumably approximates the area under the curve $y = f(x)$. The geometric interpretation suggests that s has a unique limit s_0, independent of the manner of subdivision, provided the subdivision becomes arbitrarily fine. When s actually has this behavior, $f(x)$ is said to be *Riemann integrable*, and the limit s_0 is called the *Riemann integral*.

The Riemann integral does not exist if $f(x)$ oscillates too violently. For example, let $a = 0$, $b = 1$ and define

$$f(x) = 2 \quad \text{for } x \text{ rational[1]}; \qquad f(x) = 3 \quad \text{for } x \text{ irrational.} \tag{1-3}$$

It is easily shown that every interval (no matter how small) contains both rational and irrational numbers, so that the graph of $f(x)$ has the appearance suggested by Fig. 2. If we choose ξ_k rational, then $f(\xi_k) = 2$ and

$$s = 2(x_1 - x_0) + 2(x_2 - x_1) + \cdots + 2(x_n - x_{n-1}) = 2(x_n - x_0) = 2$$

[1] A *rational* number is a fraction p/q, where p and q are integers. Thus $\frac{1}{3}$ and $-1\frac{7}{8}$ are rational, but $\sqrt{2}$ is not.

no matter how fine the subdivision may be.   On the other hand, if the ξ_k's are all irrational, $s = 3$.   This shows that the limit of s depends on the manner of subdivision and, hence, that the Riemann integral does not exist.   As we shall see presently, the Lebesgue integral for this function does exist and can be evaluated explicitly.

2. Measure.
The decisive idea in the Lebesgue integral is the notion of *measure*, which will now be described.   The measure of an open[1] interval $a < x < b$ is simply the length $b - a$.   If a set consists of a finite collection of such intervals (Fig. 3), the measure is the sum of the lengths.   The same definition applies when there are infinitely many intervals.   The sum of the lengths is now an infinite series, but since the terms are positive, *the sum does not depend on the order of the terms* (Chap. 1, Sec. 7, Theorem III).   Thus, the measure is well defined in this case also.

FIGURE 2

The notion of measure can be extended to still more general sets E as follows: Let I be a collection of open intervals which contains[2] E, and let $m(I)$ denote the measure of I.   We approximate E better and better by these sets I, so that $m(I)$ becomes smaller and smaller.   The smallest value for $m(I)$ that is given by this process is called the *outer measure* of E and is denoted by $m_0(E)$.

Strictly speaking, the "smallest value" need not be attained, and the precise definition of outer measure is as follows: The outer measure is the largest number c such that $m(I) \geq c$ for all sets I of the above-described type.   The number c is called the *greatest lower bound* of the numbers $m(I)$; its existence can be established by the fundamental principle quoted in Chap. 1, Sec. 3.

A collection of open intervals, such as I in the foregoing discussion, is called an *open set*.   As we have seen, outer measure is defined by considering the open sets containing E.   The points of $[a,b]$ not belonging to a given open set form a *closed set*.   By considering closed sets contained in E one can define the inner measure $m_i(E)$.   If $m_i(E) = m_0(E)$, the set E is said to be *measurable* and the common value is called the *measure* of E.

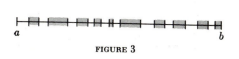

FIGURE 3

To illustrate the calculation of a measure, let the set E consist of the rational points x on $0 \leq x \leq 1$, that is, the points whose coordinate x is a rational number.   By taking first the rational numbers p/q with denominator $q = 1$, then those with $q = 2$, and so on, we see that the rational numbers can be arranged in a sequence

$$r_1, r_2, r_3, \ldots, r_n, \ldots \qquad (2\text{-}1)$$

Given $\epsilon > 0$, construct an open interval of length $\epsilon/2$ centered at r_1, an interval of length $\epsilon/2^2$ centered at r_2, and so on.   The nth interval is of length $\epsilon/2^n$ and is centered at r_n.   If I denotes the set consisting of all these open intervals, then

$$m(I) \leq \frac{\epsilon}{2} + \frac{\epsilon}{2^2} + \cdots + \frac{\epsilon}{2^n} + \cdots = \epsilon. \qquad (2\text{-}2)$$

[We have inequality rather than equality in (2-2) because some of the intervals may overlap.]

The foregoing construction shows that the outer measure of E is $\leq \epsilon$.   Since ϵ is arbitrary,

[1] An interval is *open* if the end points do not belong to the interval and *closed* if they do.   Thus $a \leq x \leq b$ is a closed interval.   The notation (a, b) and $[a, b]$ is used for open and closed intervals respectively.
[2] That is, every point of E is interior to one of the intervals belonging to the set I.

the outer measure must be zero. Because $m_i(E) \leq m_0(E)$, it follows that the inner measure is also zero, and hence $m(E) = 0$.

As a second illustration we shall find $m(E')$, where E' is the set of all irrational numbers on $[0,1]$. One of the most important properties of measure is that it is *additive;* if E and E' are two measurable sets with no point in common, then

$$m(E + E') = m(E) + m(E').$$

(We use $E + E'$ as an abbreviation for the set of all the points belonging either to E or to E'.) In the present case E is the set of rational points on $[0,1]$, and E' the set of irrational points on $[0,1]$. Evidently $E + E'$ is the set of all points on $[0,1]$, so that $m(E + E') = 1$. Since E and $E + E'$ are measurable, it can be shown that E' is also measurable and the above equation then gives

$$m(E') = 1 - m(E) = 1 - 0 = 1.$$

3. The Lebesgue Integral.   A function $y = f(x)$ is said to be *measurable* if the set of points x at which $f(x) < c$ is measurable for any and all choices of the constant c. It can be shown that the set e_k at which $y_{k-1} \leq f(x) < y_k$ is then measurable for all choices of y_{k-1} and y_k. To define the Lebesgue integral of $f(x)$, let the y axis be subdivided by points y_k, as shown in Fig. 4, and form the sum

$$\sigma = y_1 m(e_1) + y_2 m(e_2) + \cdots + y_n m(e_n).$$

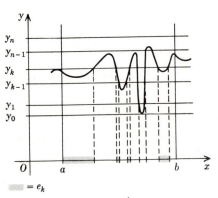

= e_k

FIGURE 4

When $f(x)$ is measurable and bounded, the sum σ has a unique limit σ_0, independent of the manner of subdivision, provided the subdivision becomes arbitrarily fine. This limit σ_0 is called the *Lebesgue integral* of $f(x)$ and is written in the form (1-1).

The most obvious difference between Riemann's definition and Lebesgue's is that in the former the x axis and in the latter the y axis is subdivided. This distinction, however, is superficial. The important fact is that Riemann's definition is based on the notion *length of an interval* whereas Lebesgue's is based on the more general notion, *measure of a set*. The intervals $x_k - x_{k-1}$ in Riemann's definition play the same role as the sets e_k in Lebesgue's.

Riemann's definition breaks down if $f(x)$ does not remain close to y_k throughout most of the intervals $[x_{k-1},x_k]$. Lebesgue's definition cannot break down in this way, because $f(x)$ is automatically close to y_k throughout the set e_k. That is why (in contrast to the former definition) the latter carries with it an assertion that the integral actually exists.

To illustrate the calculation of a Lebesgue integral we shall integrate the function (1-3) illustrated in Fig. 2. If the intervals (y_{p-1},y_p) and (y_{q-1},y_q) contain 2 and 3, respectively (Fig. 5), the sets e_p and e_q are the only ones that are not empty. Thus $m(e_k) = 0$ for $k \neq p$ or q, and the sum reduces to

$$\sigma = y_p m(e_p) + y_q m(e_q).$$

Since e_p is the set of rational points and e_q the set of irrational points, these sets have the measures 0 and 1, respectively. Hence $\sigma = y_q$. As the subdivision becomes arbitrarily fine, $y_q \to 3$ and the Lebesgue integral is found to be

$$\int_0^1 f(x) \, dx = 3. \tag{3-1}$$

It can be shown that if the Riemann integral exists the Lebesgue integral exists also and the two have the same value. On the other hand, the latter may exist when the former does not, as we have just seen. Because of its greater generality the Lebesgue integral has many desirable properties, of which we mention the following:

LEBESGUE THEOREM ON BOUNDED CONVERGENCE. Suppose $|f_n(x)| \leq M$, where M is constant; suppose $f_n(x)$ are Lebesgue integrable; and suppose $\lim f_n(x) = f(x)$ on an interval $[a,b]$. Then $f(x)$ is Lebesgue integrable, and

$$\lim \int_a^b f_n(x)\ dx = \int_a^b f(x)\ dx.$$

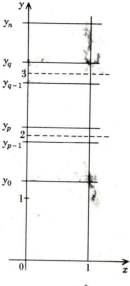

FIGURE 5

To see why the theorem fails for Riemann integrals, let $f_n(x) = 2$ at the first n rational points r_k in the sequence (2-1) and $f_n(x) = 3$ elsewhere. Then $|f_n(x)| \leq 3$, and as a Riemann or Lebesgue integral,

$$\int_0^1 f_n(x)\ dx = 3. \qquad (3\text{-}2)$$

Evidently, $\lim f_n(x) = f(x)$, where $f(x)$ is the function (1-3). Taking the limit of the expression (3-2) as $n \to \infty$, we get

$$\lim \int_0^1 f_n(x)\ dx = 3 = \int_0^1 f(x)\ dx \qquad (3\text{-}3)$$

provided the latter integral is the Lebesgue integral (3-1). Equation (3-3) does not hold for Riemann integration because, as we have seen, $f(x)$ is not Riemann integrable.

C Table of normal distribution

The entries give the values of $\Phi(x) = \dfrac{1}{\sqrt{2\pi}} \displaystyle\int_0^x e^{-t^2/2}\,dt$.   The normal distribution is

$$F(x) = \tfrac{1}{2} - \Phi(-x) \quad \text{for } x \leq 0; \qquad F(x) = \tfrac{1}{2} + \Phi(x) \text{ for } x \geq 0.$$

x	0.00	0.01	0.02	0.03	0.04	0.05	0.06	0.07	0.08	0.09
0.0	0.0000	0.0040	0.0080	0.0120	0.0160	0.0199	0.0239	0.0279	0.0319	0.0359
0.1	0.0398	0.0438	0.0478	0.0517	0.0557	0.0596	·0.0636	0.0675	0.0714	0.0753
0.2	0.0793	0.0832	0.0871	0.0910	0.0948	0.0987	0.1026	0.1064	0.1103	0.1141
0.3	0.1179	0.1217	0.1255	0.1293	0.1331	0.1368	0.1406	0.1443	0.1480	0.1517
0.4	0.1554	0.1591	0.1628	0.1664	0.1700	0.1736	0.1772	0.1808	0.1844	0.1879
0.5	0.1915	0.1950	0.1985	0.2019	0.2054	0.2088	0.2123	0.2157	0.2190	0.2224
0.6	0.2257	0.2291	0.2324	0.2357	0.2389	0.2422	0.2454	0.2486	0.2517	0.2549
0.7	0.2580	0.2611	0.2642	0.2673	0.2704	0.2734	0.2764	0.2794	0.2823	0.2852
0.8	0.2881	0.2910	0.2939	0.2967	0.2995	0.3023	0.3051	0.3078	0.3106	0.3133
0.9	0.3159	0.3186	0.3212	0.3238	0.3264	0.3289	0.3315	0.3340	0.3365	0.3389
1.0	0.3413	0.3438	0.3461	0.3485	0.3508	0.3531	0.3554	0.3577	0.3599	0.3621
1.1	0.3643	0.3665	0 3686	0.3708	0.3729	0.3749	0.3770	0.3790	0.3810	0.3830
1.2	0.3849	0.3869	0.3888	0.3907	0.3925	0.3944	0.3962	0.3980	0.3997	0.4015
1.3	0.4032	0.4049	0.4066	0.4082	0.4099	0.4115	0.4131	0.4147	0.4162	0.4177
1.4	0.4192	0.4207	0.4222	0.4236	0.4251	0.4265	0.4279	0.4292	0.4306	0.4319
1.5	0.4332	0.4345	0.4357	0.4370	0.4382	0.4394	0.4406	0.4418	0.4429	0.4441
1.6	0.4452	0.4463	0.4474	0.4484	0.4495	0.4505	0.4515	0.4525	0.4535	0.4545
1.7	0.4554	0.4564	0.4573	0.4582	0.4591	0.4599	0.4608	0.4616	0.4625	0.4633
1.8	0.4641	0.4649	0.4656	0.4664	0.4671	0.4678	0.4686	0.4693	0.4699	0.4706
1.9	0.4713	0.4719	0.4726	0.4732	0.4738	0.4744	0.4750	0.4756	0.4761	0.4767
2.0	0.4772	0.4778	0.4783	0.4788	0.4793	0.4798	0.4803	0.4808	0.4812	0.4817
2.1	0.4821	0.4826	0.4830	0.4834	0.4838	0.4842	0.4846	0.4850	0.4854	0.4857
2.2	0.4861	0.4864	0.4868	0.4871	0.4875	0.4878	0.4881	0.4884	0.4887	0.4890
2.3	0.4893	0.4896	0.4898	0.4901	0.4904	0.4906	0.4909	0.4911	0.4913	0.4916
2.4	0.4918	0.4920	0.4922	0.4925	0.4927	0.4929	0.4931	0.4932	0.4934	0.4936
2.5	0.4938	0.4940	0.4941	0.4943	0.4945	0.4946	0.4948	0.4949	0.4951	0.4952
2.6	0.4953	0.4955	0.4956	0.4957	0.4959	0.4960	0.4961	0.4962	0.4963	0.4964
2.7	0.4965	0.4966	0.4967	0.4968	0.4969	0.4970	0.4971	0.4972	0.4973	0.4974
2.8	0.4974	0.4975	0.4976	0.4977	0.4977	0.4978	0.4979	0.4979	0.4980	0.4981
2.9	0.4981	0.4982	0.4982	0.4983	0.4984	0.4984	0.4985	0.4985	0.4986	0.4986
3.0	0.4987	0.4987	0.4987	0.4988	0.4988	0.4989	0.4989	0.4989	0.4990	0.4990

x	0.0	0.2	0.4	0.6	0.8
1.0	0.3413447	0.3849303	0.4192433	0.4452007	0.4640697
2.0	0.4772499	0.4860966	0.4918025	0.4953388	0.4974449
3.0	0.4986501	0.4993129	0.4996631	0.4998409	0.4999277
4.0	0.4999683	0.4999867	0.4999946	0.4999979	0.4999992

† This table is reproduced by permission from the "Biometrica Tables for Statisticians," vol. 1, 1954, edited by E. S. Pearson and H. O. Hartley and published by the Cambridge University Press for the Biometrica Trustees.

Answers

CHAPTER 1

Section 1, Page 5

1. 3.14159.   **2.** $\frac{1}{2}(7 \times 8) = 28$.   **4.** $na + \frac{1}{2}n(n-1)d$.
11. 5, 15, 10, 20, 15, 30.   **12.** 0, 3, 6, 6, 9.

Section 2, Page 8

1. First series: $1 + 2 + 3 + \cdots + n + \cdots$.   **2.** First series: $\Sigma 1/k$.
3. 6, $\frac{7}{8}$, $\frac{49}{36}$, $\frac{71}{105}$, $\frac{439}{500}$, $\frac{133}{60}$.   **4.** Prob. 1, parts a, e, f; Prob. 2, parts c and e.

Section 3, Page 11

1. First series: $\frac{1}{2} + \frac{1}{4} + \frac{1}{6} + \cdots + 1/(2n) + \cdots$.   **2.** First series: $\Sigma(\frac{2}{3})^{k+1}$.
3. Prob. 1a; 1e for $x/\pi = $ integer; none in Prob. 2.
4. Prob. 1b; 1e for $x/\pi \neq $ integer; Prob. 2, parts a and d.
5. Prob. 1, parts a, d; Prob. 2, parts b, c, f.   **8.** Div, con, div, div.
11. First answer: If $a_n \to a$ and $b_n \to b$, then $a_n b_n \to ab$.

Section 4, Page 14

1. Div, e^{-1}, $\frac{1}{3}$, $(\log 2)^{-1}$, div.   **2.** Con, con, div, con, div.   **3.** $c < 0$, con for $c < 0$.
4. $(100.9)^{10} < n < (101)^{10}$; hence $n \doteq 10^{20}$.
5. $10 < s < 11$, about 10^{10} terms; $s = 2.15 + R$, $8.36 < R < 8.50$, about 3.5×10^{18} terms.
6. (b) Div.

Section 5, Page 17

1. Div, con, con, div, con.   **2.** Div, con, div, con, con, con.
3. Con, div, con, con, div, con.   **4.** Div, con, con, div, div, div.
5. Div, con, div, div, div, con.   **6.** (b) No. For example, $a_n = b_n = n$, $c_n = 1 - n$, $d_n = 2 - n$.

Section 6, Page 19

1. Con, div, con, con for $|x| < \sqrt{2}$, con, con, con.
2. Con, con for $c > 1$, div.   **3.** Con, con, con, div, con.

Section 7, Page 23

1. Cond con, div, abs con, abs con, abs con.
2. Abs con: $|x| < 1$, all x, $|x| < 1$, $|x| > 1$, $-\frac{4}{3} < x < 4$, $x = 0$, $|x - 2| < 1$.
   Cond con: $x = 1$, never, never, $x = -1$, $x = -\frac{4}{3}$, never, $x = 1$.
3. About 100 terms; -0.95.   **4.** (a) Abs con; (b) div; (c) cond con.   **5.** $\Sigma n 2^{1-n}$.
6. 999, 9999.   **7.** First answer: $s = 2x$ for $|x| < 1$, $s = 1$ at $x = 1$.

Section 8, Page 27

3. (a) No condition; (b) $a_n \to 0$; (c) $a_n/n \to 0$.   **6.** Yes; no; no; yes.
7. No.   **8.** Yes; no.   **9.** No; no.

Section 9, Page 30

3. Differentiated series diverges for all x.

5. Uniform con for: $|x| \leq c < \infty$, $|x| \leq c < 1$, $|x| \leq 1$, $|x| < \infty$, $|x| \leq c < 0.1$, $|2x - \pi n| \geq c > 0$ where n is the odd integer nearest to $2x/\pi$, $1 < c \leq |x| < \infty$.

6. Uniform con for: $|x| \leq c < \infty$, $|x| \leq c < 1$, $|x| \leq c < 1$, $|x| < \infty$, $|x| \leq c < 0.1$; same as Prob. 5, $|x| \geq c > 1$ or $|x| \leq c < 1$.

9. (c) About $e^{1,000-\gamma} \doteq 10^{434}$ terms.

Section 10, Page 35

1. (a) $-1 < x < 1$, $-\sqrt{2} < x < \sqrt{2}$, all x, $-3 < x < 3$, $-\sqrt[4]{3} < x < \sqrt[4]{3}$.
   (b) $-3 \leq x < 1$, $-2 \leq x \leq 0$, $1 \leq x \leq 3$, $x = 1$, $-4 < x < 0$.

3. (b) $\tan^{-1} x = \sum \dfrac{(-1)^n}{2n+1} x^{2n+1}$.   4. (d) $2.72, 0.368$.

5. $\sum (-1)^n \dfrac{x^{4n+1}}{4n+1}$, $\sum (-1)^n \dfrac{x^{5n+6}}{(n+1)(5n+6)}$, $\sum \dfrac{n+1}{3n+7} x^{3n+7}$, $\sum \dfrac{nx^{n+14}}{n+14}$, $\sum \dfrac{(-1)^n x^{8n+3}}{8n+3}$.

Section 11, Page 39

1. $-1 + 3x - 3x^2 + x^3 = (x-1)^3$, $2 + x^2 = 3 + 2(x-1) + (x-1)^2$,
   $x^4 = 1 + 4(x-1) + 6(x-1)^2 + 4(x-1)^3 + (x-1)^4$,

   $$\sum \frac{(-1)^n}{2^{n+1}} x^n = \sum \frac{(-1)^n}{3^{n+1}} (x-1)^n, \quad \sum \frac{2^n x^n}{n!} = e^2 \sum \frac{2^n (x-1)^n}{n!},$$

   $$\sum (-1)^n \frac{\pi^{2n+1}}{(2n+1)!} x^{2n+1} = \sum (-1)^{n-1} \frac{\pi^{2n+1}(x-1)^{2n+1}}{(2n+1)!},$$

   $$\cos 1 \sum (-1)^n \frac{x^{2n}}{(2n)!} + \sin 1 \sum (-1)^n \frac{x^{2n+1}}{(2n+1)!} = \sum (-1)^n \frac{(x-1)^{2n}}{(2n)!}.$$

2. (a) $e^a \sum \dfrac{(x-a)^n}{n!}$; (b) $\sum (-1)^n \dfrac{(x-1)^{n+1}}{n+1}$.

3. $\Sigma(-1)^n(x-1)^n$, $\Sigma \tfrac{1}{4}[(-3)^{-n-1} - 1](x-1)^n$, $\Sigma \tfrac{1}{8}[2(-1)^{n+1} - 1 - (-3)^{-n-1}](x-1)^n$,
   $\Sigma(-1)^{n+1}[4^{-n-1} - 2(5)^{-n-1}](x-1)^n$, $\Sigma \tfrac{1}{4}[2(-4)^{-n} + 5(-3)^{-n-1} - 1](x-1)^n$.

4. $\tfrac{1}{2}\Sigma(-1)^n \dfrac{1}{(2n)!}\left(x - \dfrac{\pi}{6}\right)^{2n} + \tfrac{1}{2}\sqrt{3}\,\Sigma(-1)^n \dfrac{1}{(2n+1)!}\left(x - \dfrac{\pi}{6}\right)^{2n+1}$

6. $\sum 3(-9)^n \dfrac{x^{2n+1}}{(2n+1)!}$, $\sum (-1)^n \dfrac{x^{2n}}{n!}$, $\sum (-1)^n \dfrac{x^{4n+2}}{(2n+1)!}$, $2\sum \dfrac{x^{2n}}{(2n)!}$, $\sum \dfrac{2x^{2n+1}}{(2n+1)!}$, $\sum (-1)^n \dfrac{x^{4n+2}}{n+1}$,

   $$\sum c_n x^n, \text{ where } c_0 = 0, c_n = \sum_{k=1}^{n} \frac{-1}{(n-k)!k}.$$

7. See Sec. 9.

8. $\sum \dfrac{1 \cdot 3 \cdot 5 \cdots (2n-1)}{2 \cdot 4 \cdot 6 \cdots 2n}\left(\dfrac{x}{1+x}\right)^{n+1}$, coefficient for $n = 0$ being 1.

Section 12, Page 41

1. $\sum \dfrac{(-1)^n x^{2n+1}}{(2n+1)n!}$, $\sum \dfrac{2x^{2n+1}}{(2n+1)(2n+1)!}$, $\sum \dfrac{(-1)^n x^{4n+3}}{(4n+3)(2n+1)!}$,

   $\sum \dfrac{(-1)^n x^{2n+1}}{(2n+1)(2n+1)!}$, $\sum \dfrac{(-1)^n x^{2n+1}}{(2n+1)(2n+2)!}$.

3. $\sum \dfrac{(-1)^n x^{n+p}}{n!(p+n)}, p > 0$; $\sum \dfrac{(-1)^n x^{n+p+2}}{(n+1)(n+p+2)}, p > -1, x \leq 1$;

   $\sum \dfrac{(-1)^n x^{4n+2p+2}}{(4n+2p+2)(2n)!}, p > -1$; $\sum \dfrac{1 \cdot 3 \cdot 5 \cdots (2n-1)}{2 \cdot 4 \cdot 6 \cdots 2n} \dfrac{x^{4n+p+1}}{4n+p+1}, p > -1$, where
   first coefficient $= 1$.

4. $\dfrac{1}{p} + \sum_{1}^{\infty} \dfrac{(q-1)(q-2) \cdots (q-n)}{n!(p+n)(-1)^n}, q > 0, p \neq 0, -1, -2, \ldots$.

Section 13, Page 44

2. $x + \tfrac{1}{3}x^3 + \tfrac{2}{15}x^5$, $1 + x + \tfrac{1}{2}x^2$, $1 + \tfrac{1}{2}x^2 + \tfrac{5}{24}x^4$, $\tfrac{1}{2} + \tfrac{1}{4}x - \tfrac{1}{48}x^3$,
   $1 - \tfrac{1}{2}x - \tfrac{1}{12}x^2$, $1 - \tfrac{1}{4}x^2 - \tfrac{1}{96}x^4$, $1 - \tfrac{1}{2}x - \tfrac{11}{24}x^2$, $x - \tfrac{1}{3}x^3 + \tfrac{1}{10}x^5$.

3. (a) 0.00167.   **4.** $\tfrac{10}{3} + \tfrac{4}{3}x^2 + \cdots$.   **5.** 3.004; 0.985; 0.545; 2.036.
6. 0.310, 0.020, -0.103, 0.94.   **7.** $|\alpha| \leq 0.24$ radian $\doteq 14°$.   **9.** $\tfrac{1}{3}x^3 - \tfrac{1}{42}x^7$.
10. 0.621, 0.493, 0.747, 0.764.

Section 14, Page 47

1. $(p/\delta e)^p$.   **5.** $c < -1, c < -\tfrac{1}{2}, c > \tfrac{1}{2}$.

Section 16, Page 53

3. $|z| < 1, |z| < 1, |z| < \infty, |z| < e^{-1}, |z| < \infty$.
4. Partial answer: $|z - 3| < \tfrac{1}{2}, z = -4, |z + 1| < 1$.
5. First case: the strip defined by $1 < x < 3, -\infty < y < \infty$.

Section 17, Page 55

2. $\dfrac{\sin (n + 1)x/2 \sin (nx/2)}{\sin (x/2)}, x \neq 0, \pm 2\pi, \pm 4\pi, \ldots$ **4.** 12π.

Section 18, Page 60

1. 0 if $m^2 \neq n^2$, 2π if $m = n = 0$, π if $m^2 = n^2 \neq 0$.

2. $\tfrac{3}{4}\pi + \displaystyle\sum_{1}^{\infty} \dfrac{(-1)^{n-1} \cos (2n - 1)x - \sin (2n - 1)x + \sin 2(2n - 1)x}{2n - 1}$.

3. $\dfrac{1}{\pi} - \dfrac{2}{\pi} \displaystyle\sum_{1}^{\infty} \dfrac{\cos 2nx}{4n^2 - 1} + \tfrac{1}{2} \sin x$.   **4.** $\dfrac{4}{\pi} \displaystyle\sum_{1}^{\infty} \dfrac{n \sin 2nx}{4n^2 - 1} + \tfrac{1}{2} \cos x$.

6. $\dfrac{\pi}{2} + 4 \displaystyle\sum_{1}^{\infty} \dfrac{\sin (2n - 1)x}{2n - 1} - \displaystyle\sum_{1}^{\infty} \dfrac{\sin 2nx}{n}$.   **7.** $\tfrac{1}{4} + \dfrac{1}{\pi} \displaystyle\sum_{1}^{\infty} \dfrac{\sin (2n - 1)x}{2n - 1} + \displaystyle\sum_{1}^{\infty} \dfrac{(-1)^n \sin nx}{n}$.

Section 19, Page 62

1. First sketch for $-\pi < x < \pi$; then extend to adjacent intervals by repetition; finally add dots indicating correct value $\tfrac{1}{2}[f(x+) + f(x-)]$ at points of discontinuity.

4. $\dfrac{a}{2\pi} + \dfrac{1}{\pi} \displaystyle\sum_{1}^{\infty} \dfrac{\sin na \cos nx}{n} + \dfrac{1}{\pi} \displaystyle\sum_{1}^{\infty} \dfrac{(1 - \cos na) \sin nx}{n}$.

Section 20, Page 66

1. Even, even, odd, even, neither, even, odd, even, odd.

4. $\dfrac{\pi}{4} + \displaystyle\sum_{1}^{\infty} (-1)^{n-1} \dfrac{\cos (2n - 1)x}{2n - 1} + 2 \displaystyle\sum_{1}^{\infty} \dfrac{\sin (n - 1)x}{2n - 1}$,

$\dfrac{\pi}{4} + 2 \displaystyle\sum_{1}^{\infty} \dfrac{(-1)^{n-1} \cos (2n - 1)x}{2n - 1}, 4 \displaystyle\sum_{1}^{\infty} \dfrac{\sin (2n - 1)x}{2n - 1}$.

11. $\tfrac{1}{2} - \tfrac{1}{2} \cos 2x, \tfrac{1}{2} + \tfrac{1}{2} \cos 2x$.

Section 21, Page 70

2. $\dfrac{4}{\pi} \displaystyle\sum_{0}^{\infty} \dfrac{\sin (2n + 1)(\pi/2)x}{2n + 1}$.   **3.** $\tfrac{1}{2} - \displaystyle\sum_{1}^{\infty} \dfrac{4}{\pi^2(2n - 1)^2} \cos (2n - 1)\pi x$.

4. $\cos \pi x$.   **5.** $f(x) = \tfrac{1}{2} + \displaystyle\sum_{1}^{\infty} \dfrac{(-1)^{n+1}}{2n - 1} \cos (2n - 1) \dfrac{x}{2}$.   **6.** $f(x) = 4 \displaystyle\sum_{1}^{\infty} \dfrac{\cos 2n\pi x}{(2n\pi)^2}$.

Section 22, Page 73

2. $e^{\alpha x} = \dfrac{\sinh \pi\alpha}{\pi\alpha} + \dfrac{\sinh \pi\alpha}{\pi^2} \displaystyle\sum_{1}^{\infty} \dfrac{(-1)^n 2\alpha \cos nx}{\alpha^2 + n^2} - \dfrac{2 \sinh \pi\alpha}{\pi^2} \displaystyle\sum_{1}^{\infty} \dfrac{(-1)^n n \sin nx}{\alpha^2 + n^2}$.

4. $\sin \beta x = \dfrac{\sin \pi \beta}{i\pi} \displaystyle\sum_1^\infty \dfrac{(-1)^n n e^{inx}}{\beta^2 - n^2}.$ **5.** $\dfrac{\sin \pi \beta}{\pi} \displaystyle\sum_1^\infty \dfrac{(-1)^n 2n \sin nx}{\beta^2 - n^2},$ $\dfrac{\sin \pi \beta}{\pi}\left(\dfrac{1}{\beta} + \displaystyle\sum_1^\infty \dfrac{(-1)^n 2\beta \cos nx}{\beta^2 - n^2}\right).$

6. Partial answer: $\dfrac{1}{2} + \displaystyle\sum_{-\infty}^\infty \dfrac{1 - \cos n\delta}{(n\delta)^2} e^{inx}.$

Section 24, Page 79

2. $a_1 = \dfrac{4}{\pi},\ a_2 = 0,\ a_3 = \dfrac{4}{3\pi}.$

Section 26, Page 86

5. $x^5 = \displaystyle\sum_1^\infty (-1)^{n+1}\left(\dfrac{2\pi^4}{n} - \dfrac{40\pi^2}{n^2} + \dfrac{240}{n^4}\right)\sin nx,$

$x^6 = \tfrac{1}{7}\pi^6 + \displaystyle\sum_1^\infty (-1)^n \left(\dfrac{12\pi^4}{n^2} - \dfrac{120\pi^2}{n^3} + \dfrac{720}{n^6}\right)\cos nx.$

6. $2 - \tfrac{1}{3}\pi^2 + \pi^4 + \displaystyle\sum_1^\infty (-1)^n \left(\dfrac{40\pi^2 - 4}{n^2} - \dfrac{240}{n^4}\right)\cos nx - \displaystyle\sum_1^\infty (-1)^n \left(\dfrac{6 + 2\pi^2}{n} - \dfrac{12}{n^3}\right)\sin nx.$

CHAPTER 2

Section 1, Page 92

1. 2, 2, 2, 1.
2. (b) $y = x^2, x^2, x^2 + 1, x^2 - 2$; (c) The differential equation makes the slopes agree at corresponding values of x; (d) Yes.
3. Yes.   **4.** Exceptional if the points are on a line $x = c$, as (1,2) and (1,3), for example.
5. (c) $y = \tfrac{1}{3}x^3 + x.$ **7.** $x^2 y = c,\ xy = c,\ xy = c,\ x^2 + 3y^2 = c,\ x^2 + 3y^2 = c.$
9. Partial answer: $v = \pm\sqrt{64 - 2gs},\ t = 8/g \doteq 0.2485$ sec.   **10.** $\tfrac{2}{3}g.$
11. Partial answer: $v = \sqrt{2gh},\ s = \tfrac{1}{2}gt^2 \sin \theta.$
13. First equation: $y''(t) + y'(t) = \sin t, -\infty < t < \infty$; or $y'' + y' = S$, where S is the function defined by $S(t) = \sin t, -\infty < t < \infty.$

Section 2, Page 97

2. $\tfrac{1}{2}\log 2.$ **3.** $2.32, 3.32, \infty$ hr.   **4.** 2 in.   **5.** $30e^{-kt}, 30k^{-1}(1 - e^{-kt}).$

7. $3\tfrac{1}{3}$ g, $3\dfrac{\log 2}{\log 3}$ years.   **9.** $200\sqrt{5}$ lb.

Section 3, Page 100

1. (a) $x = \tan (y + c),\ \sin^{-1} x - \sin^{-1} y = c$ or $x = \pm 1$ or $y = \pm 1,$
$\qquad y - 1 = ce^{x^2}(y + 1),\ 4\cos y = \sin 2x - 2x + c;$
$\qquad$ (b) $y = (1 - x)^{-1},\ y = 1 + 2x,\ \tan y = \sec x - 1 + \tan 1,\ y = \tan (2\sqrt{1 + x} - 2 + \tfrac{1}{4}\pi).$
2. (a) $3x^2 y + y^3 = c,\ y = x \sin \log (cx)$ or $y = \pm x,\ y = -x \sin^{-1} \log (cx);$
$\qquad$ (b) $x^2 - 2xy = y^2 - 2,\ 3y^3 \log y + x^3 = y^3,\ y + y \log x = x,\ y = x.$
3. 3, 2, 1, 25, 4.
5. (a) $2\tan^{-1} e^y + \log [c \tanh (x/2)] = 0$ or $x = 0,\ y = x \log cx,\ y = ce^{-2\sqrt{x/y}}$ or $y = 0,\ x = ce^{2\sqrt{y/x}}$
$\qquad$ or $y = 0;$ (b) $x^{-1} - y^{-1} = \log (cy)$ or $y = 0,\ \log (cx) + e^{-y/x} = 0,$
$\qquad\qquad y(4 - \log y^2) = \tan^2 x + c,\ \tan \tfrac{1}{2}y = c \tan \tfrac{1}{2}x.$
6. $e^x \cos (2 + y - x) = c,\ x + y + 3 = \tan (y + c),\ y^2 = x^2 \log (cx)^2,\ e^{y-x} = 2 + ce^{y+x}.$
7. $v = v_0 \tanh (gt/v_0),\ v = v_0(1 - e^{-2gs/v_0^2})^{1/2}.$ **8.** 72 sec.   **9.** About $\tfrac{1}{4}$ cfs.

Section 4, Page 104

1. (a) $e^x + x - y = c$, not exact, $x^3 y - xy^3 = c$, (b) $x^2 + y^2 = c,\ ye^x = c,\ xy(x^2 + y^2) = c,$
$\qquad$ (c) $x^2 + \sin xy = c,\ x^2 y + xy^2 + x = c$, (d) not exact, $x \cos xy^2 = c.$
2. $y^4 + 4(x^3 + y^3)^{4/3} = c,\ (x^2 + y^2)^3 + x^6 = c.$ **3.** $(x + y)e^{xy} = c.$
4. $y + (x + y) \log c(x + y) = 0,\ xe^{x^2 y} = c.$
6. $A = b,\ B = c;\ 3F = x(ax^2 + 2bxy + cy^2) + y(Ax^2 + 2Bxy + Cy^2).$
8. (a) $y = cx$; (b) $x^2 + y^2 = c^2$; (c) $c \pm x = \sqrt{1 - y^2} - \text{sech}^{-1} y$; (d) $y = ce^x$; (e) $y = \cosh x.$

Section 5, Page 107

  2. (a) $x^3 + xy = cx^2$, $ye^{1/xy} = cx$ or $xy = 0$, $(x^2 + y^2)^{1/2} - xy = c$;
      (b) $x^3 y = ce^y$, $y - 2\tan^{-1}(x/y) = c$ or $y = 0$, $ye^{2xy} = cx$;
      (c) $xe^y = cy$, $x^2 + y^2 = cy$, $\tan^{-1}(y/x) + (x^2 + y^2)^{-1/2} = c$.

  3. (a) $(1 + x^2)(y - 1)^2 = c$, $3(x^2 + 1)y = x^3 + c$,
        $xy = 2x\sin x + (2 - x^2)\cos x + c$, $y = e - c\exp(-e^x)$;
      (b) $y = 1 - 2e^{-x^2/2}$, $y = e^{-x^2}(x - 1)$, $y = \cos^2 x + 2\sin x - 2$, $4y = x^4 - 3e^{-x^4} - 1$.

  4. (a) $3x^4 y^4 - 2x^6 = c$, $y^{-2} = x + \frac{1}{2} + ce^{2x}$, $y^{-5} = \frac{5}{2}x^3 + cx^5$, $x = y\log cx$;
      (b) $y^{-1} = 1 + \log x + cx$, $y^{-2} = 1 + x^2 + ce^{x^2}$, $y^3 = 4 + ce^{-x}$, $e^{2x} = y^2(6x + c)$.

  5. $x^{-2} = y + \frac{1}{2} + ce^{2y}$, $2x = y^3 + cy$, $e^y - xy = c$, $\sec y + \tan y = ce^x \sec y$.

  6. (a) $y = \sin x + ce^x$, $1 - x = ce^{-y^2/2}$, $\sqrt{x^2 - y^2} = x\log cx$ or $y^2 = x^2$;
      (b) $x^3(1 + 4y^2)^3 = c$, $x\sin^{-1} x + \sqrt{1 - x^2} = e^y(y - 1) - \frac{1}{2}y^2 + c$, $y = ce^{x - x^2/2}$.

Section 6, Page 111

  1. (a) $(y - c_1 e^{3x})(y - c_2 e^{-x}) = 0$, $y = x + c$, $(y^2 - x^2 - c_1)(xy - c_2) = 0$,
        $(y - c_1 e^{x^4/4})(y - c_2 e^x)(y - c_3 e^{-x}) = 0$;
      (b) $x = 2p - \frac{4}{3}p^{-3} + c$, $y = p^2 - 2p^{-2}$; $x = 2\tan^{-1} p - p^{-1} + c$, $y = \log(p^3 + p)$;
        $(y - c_1 x - c_1^2)(y + \frac{1}{4}x^2 - c_2) = 0$; $x = e^p(1 + p) + 2p + c$, $y = p^2(1 + e^p) - 1$;
        $(y - 1)[(y - x - c)^2 + 4x - 4cx] = 0$.
      (c) $(y^2 - 4x)(cy - 1 - c^2 x) = 0$; $y = \sin(x + c)$; $(y - c_1 e^x)(y + x^2 - c_2) = 0$;
        $(y - e^x - c_1)(2y + x^2 - c_2) = 0$; $8x = 3\log(2p - 1) + 6p(p + 1) + c$,
        $16y = 3\log(2p - 1) + 6p(p + 1) + 8p^3 + c$, or $16y = 8x + 1$.

  2. Yes: $x = -g'(p)$, $y = g(p) - pg'(p)$ if g is differentiable.

  3. (a) $x = \sin(c_1 y + c_2)$, $y = \frac{1}{3}x^3 + c_1 x^2 + c_2$, $y = \frac{1}{2}x^2 + c_1 x^{-1} + c_2$,
        $y = (x + c_1)\log(x + c_1) - x\log x + c_2$, $ec_1 e^y = x^2 + c_2$;
      (b) $y = c_1 + c_2 x$, $y = 2c_1 \tan(c_1 x + c_2)$, $y = c_1 + c_2 e^x$, $y = c_1 \sin(2x + c_2)$,
        $y = -\log(c_1 x + c_2)$, $e^{c_1 x} + c_1 e^{c_1 x - y} = c_2$, $\pm x = \int (c + \sin y)^{-1/2}\, dy$.
      (c) $y = c_2 - (1 + x)\log c_1 x$, $y = c_2 e^{c_1 x}$, $y = \frac{2}{3}x^2 + c_1 x^{-1} + c_2$, $e^{-y} = c_2 \cos(x + c_1)$,
        $y = (x + c_1)\log(x + c_1) - 3x\log x + 2x + c_2$.

  4. $p^2 = 2x^5 + m^2 - 2$, $m^2 = 2$, $(7y + c)^2 = 8x^7$; $xp^2 = x^4 + m^2 - 1$, $m^2 = 1$, $(5y + c)^2 = 4x^5$;
      $e^{p+x} = 2(x^2 - 1) + e^{m+1}$, $m = \log 2 - 1$, $y = x(\log 2x^2 - \frac{1}{2}x - 2) + c$;
      $(xp)^2 = \frac{4}{3}(x^3 - 1) + m^2$, $m^2 = \frac{4}{3}$, $27(y + c)^2 = 16x^3$.

  5. $p^2 = 2y + y^2 + m^2$, $m^2 = 1$, $y = ce^{\pm x} - 1$; $p^2 = 2e^y + m^2 - 2$, $m^2 = 2$, $x = c \pm \sqrt{2}e^{-y/2}$;
      $p^2 = \frac{1}{2}y^4 + m^2$, $m = 0$, $y(c \pm x) = \sqrt{2}$;
      $p + 1 = (m + 1)e^{-y} + y$ or also $p = 0$ if $m = 0$, $m = -1$, $y = 1 + ce^x$;
      $p^2 = \frac{1}{2} - y + (m^2 - \frac{1}{2})e^{-2y}$, $m^2 = \frac{1}{2}$, $4y + (c \pm x)^2 = 2$.

  6. (a) Circles centered at the origin.

  7. (a) $4xy + y^4 = c$, $y\tan^{-1} x = c$, $x - y = c(1 + xy)$;
      (b) $x\sin 2y = c$, $e^{x-y} = c - x$, $x = \frac{1}{2}y - \frac{1}{4} + ce^{-2y}$, $x^{-2} = y + \frac{1}{2} + ce^{2y}$;
      (c) $y^2 = 2x^2 \log cx$, $y = e^{3x} + ce^{2x}$, $y^2 = x^2(2e^y + c)$, $y^{-2} = \dfrac{x}{3} + \dfrac{1}{18} + ce^{6x}$;
      (d) $(y - \log c_1 x)(y - x - c_2) = 0$, $y = x - \log cx$,
        $y = v\log cvx$ where $v = \sqrt{x^2 + y^2} - y$, $x + y = \tan(x + c)$.

  9. $x^2 + 2xy - y^2 - 2x + 2y = c$.   **10.** $\dfrac{du}{dv} = \dfrac{u + v}{2u + v}$, $\dfrac{3u + v}{2u + v}$, $\sin\dfrac{u - v}{u + v}$, $\left(\dfrac{u + 2v}{2u - v}\right)^3$.

  11. Let v_0 be the speed of the boat and $v_1 = rv_0$ the speed of the river. Choose coordinates so that the y axis is the left bank, the boat heads toward the origin, and the direction of flow is toward $y = +\infty$. Then $2y = x[(cx)^{-r} - (cx)^r]$, $c > 0$. When $r = 1$ the boat follows a parabolic path on which it approaches, but does not reach, a point $(0, y_0)$, $y_0 > 0$.

  12. $t = |x_0| b/(b^2 - a^2)$; almost within $\frac{1}{2}|x_0|$.

Section 7, Page 116

  1. (b) Infinitely many through the origin, one through other points; none through the origin, one through other points.
      (c) For $y = f(x)$: infinitely many through the origin, one through points where $x \neq 0$, none through points where $x = 0$ and $y \neq 0$; none through points on the x axis, one through other points.

  2. (b) Increasing for $x > 0$, $xy > 0$, $xy < 0$, $x(y - x) > 0$.
      (c) Concave up for all (x,y), $y(y^2 - x^2) > 0$, $y > 0$, $x < 0$.
      (d) At $(x,0)$ slopes are x, ∞, 0, -1; at $(0,y)$ they are 0, 0, $-\infty$, ∞.

(e) Partial answer: Slope $= 1$ on locus: $x = 1$, $x = y$, $x = -y$, $2x = y$.

(g) $y = \frac{1}{2}x^2 + c$, $x^2 - y^2 = c$, $xy = c$, $y = -x \log cx$.

3. Yes; yes.   **4.** Partial answer: $y' = $ max on circles $x^2 + y^2 = (n + \frac{1}{2})\pi$, $n = 0, 2, 4, \ldots$.

5. (d) $4(y - x) = y'(y' - 2)$.   **6.** Partial answer: $4y + (y')^2 = 4y'x$; yes.

9. Partial answer: $(x - c)^2 + y^2 = a^2$.

Section 8, Page 119

1. (a) $y = cx$, $x^2 - y^2 = c$, $x^2 + ny^2 = c$, $y^9 = cx^4$, $3x^2y - y^3 = c$, $2y + x^2 = c$;

(b) $y^2 = c - 2x$, $y = ce^{-2x}$, $y^2 = -\log cx$, $\theta = c$, $r = ce^{-\theta^2/2}$, $r = c(1 + \cos \theta)$, $r = c \cos \theta$.

5. Self-orthogonal family.   **6.** $F(x, y, -1/y') = 0$, if $\alpha = \pi/2$.

7. (a) $y^2 + 2xy - x^2 = c$; (b) $x^2 + y^2 = c(y - x)$.

8. $x = A \cos (\theta + n)[\tan (\theta/2)]^{1/n}$, $y = A \sin \theta[\tan (\theta/2)]^{1/n}$, where $n = \pm a/v_0$ and θ is a parameter.

Section 9, Page 122

2. The rate g would be thought of as $f(t)g$ instead, where $f(t) = 0$ for $0 \le t < t_0$, and $f(t) = 1$ for $t \ge t_0$. Equation (9-1) would be written: $dy/dt = wr + f(t)g - ry/G$.

3. $dy/dt = wr + kA_0e^{-kt} - ry/G$.   **5.** $(A - y)/(B - y) = (A/B)e^{kt(A-B)}$.

6. Let x represent the amount of substance dissolved after time t, A the amount present when $x = 0$, $t = 0$, and c the proportionality constant. If v is the volume of solvent and S the saturate concentration, then $dx/dt = c(A - x)(S - x/v)$, if the dissolving substance does not change the volume v.

7. (a) $I = L^{-1}e^{-(R/L)t} \int_0^t E_0(t)e^{(R/L)t} \, dt$.

Section 11, Page 127

1. $t = v_0/\mu g$, $s_0 = v_0^2/2\mu g$.   **5.** $v = v_0 + c \log 2$.

6. Equation is $dv/dt + kv(m_0 - rt)^{-1} = -g + rc(m_0 - rt)^{-1}$,

$v = A(m_0 - rt)^{k/r} + g(m_0 - rt)(r - k)^{-1} + rck^{-1}$,

$s = A[m_0^{(k+r)/r} - (m_0 - rt)^{(k+r)/r}](k + r)^{-1} + g[(m_0 - rt)^2 - m_0^2][2r(k - r)]^{-1} + rck^{-1}t$,

where $A = m_0^{-k/r}[v_0 - gm_0(r - k)^{-1} - rck^{-1}]$.

7. $y = -g(2v_0^2 \cos^2 \alpha)^{-1}x^2 + x \tan \alpha$; range is max for $\alpha = \pi/4$; region is described by $y \le v_0^2/(2g) - gx^2/(2v_0)^2$.   **9.** $\tan \theta = \tan \theta_0 + 2eE/(\omega m_0 c)$.

Section 13, Page 137

1. $EIy_{\max} = Wl^3/48$.

2. $EIy_{\max} = 5wl^4/384$, if simply supported; $EIy_{\max} = wl^4/8$, if used as a cantilever.

Section 14, Page 141

2. $2y = \log [c(1 + x)(1 - x)^{-1}]$.   **3.** $|x| < \frac{1}{2}$, $|x| < \frac{1}{2}\pi$.

5. Regions: $y^2 > 4x^2 \ne 0$, $x^2 > y^2 \ne 0$, $4y > x^2$, $y^2 > x^2 \ne 0$.

Solutions: $2cy = c^2 + 4x^2$, $y^2 + 2cx + c^2 = 0$, $y = \frac{1}{2}x^2 + cx + c^2$.

Singular solution: $y^2 = 4x^2$, $y^2 = x^2$, $4y = x^2$. The singular solution can be found by each method. Answers are not provided for the last equation because it is obtained from the second by interchanging x and y.

10. $|y - y_0| \le m|x - x_0|$.

Section 15, Page 147

1. (a) First equation: $y_1' = y_2$, $(1 - x^2)y_2' = xy_2$.   (b) Second equation: $y_1' = y_2$, $y_2' = y_1y_2$.

(c) Third equation: $y_1' = y_2$, $xy_2' = 4x - 2y_2$.

2. (b) $y = c$, $y = cx$, $x^2 + y^2 = c$, $xy = c$, $y = cx^2$, $y = x \log cx$. In the last case $r' = r$, $\theta' = -1$, hence $r = ce^{-\theta}$.

6. $xy^a = c$, saddle; $x = cy^a$, stable node; stable focus, solvable by (3-3) but best analyzed as in Fig. 15.

Section 16, Page 151

1. (a) $y_1 = c_1e^{-t} + c_2e^{3t}$, $y_2 = -2c_1e^{-t} + 2c_2e^{3t}$.

(b) $x = c_1t + c_2$, $y = c_1$; $x = c_1e^t$, $y = c_2e^t$; $x = c_1e^{it} + c_2e^{-it}$, $y = i(c_1e^{it} - c_2e^{-it})$;

$x = c_1e^{-t}$, $y = c_2e^t$; $x = c_1e^t$, $y = c_2e^{2t}$; $x = c_1e^t$, $y = c_2e^t + c_1te^t$;

$x = e^t(c_1e^{it} + c_2e^{-it})$, $y = ie^t(c_1e^{it} - c_2e^{-it})$.

The behavior as $t \to \infty$ is easily analyzed, and, except in the last case, the orbits are easily found by eliminating t. These results agree with the answers given previously. Expressions involving e^{it} can be reduced to real form by choosing $c_2 = \bar{c}_1$.

3. $x = c_1e^{at}$, $y = c_2e^{-t}$; $x = c_1e^{-at}$, $y = c_2e^{-t}$; $x = e^{-at}(c_1e^{it} + c_2e^{-it})$, $y = ie^{-at}(c_1e^{it} - c_2e^{-it})$.

Section 17, Page 155

1. $y_1 = \dfrac{x^3}{3}$, $y_3 = \dfrac{x^3}{3} + \dfrac{x^7}{63} + \dfrac{2x^{11}}{2,079} + \dfrac{x^{15}}{59,535}$.

   For the error estimate, use $m_0 = \frac{1}{2}$, $k \le 2(\max |y|) \le 2 \max |x| = \sqrt{2}$.

2. Since $|y| \le |x|$ in the given region, use $y' = x^2 + y^2$ to get $|y| \le \frac{2}{3}|x|^3$. The choice $k = 2|y|$, $m_0 = x^2$ gives the desired result. For $|x| = 0.1$ and $n = 3$ this estimate is better than that of Prob. 1 by a factor of about 70,000,000,000.

CHAPTER 3

Section 1, Page 170

2. (a) $y'' + 10y = 0$, $f = \sqrt{10}/2\pi$; (b) $y = 2\cos\sqrt{10}t$; (c) $y = \sqrt{10}\sin\sqrt{10}t$.
3. $y = (mg/k)\cos\sqrt{k/mt}$.   4. $y = y_0\cos t + y_1\sin t$.   5. $\omega = \sqrt{1/LC}$.   6. $v^2 + y^2 = 2E$.
7. $m = \omega_1^2 m_1/(\omega^2 - \omega_1^2)$, $k = \omega^2 m$.   8. about $4/\pi \doteq 1$.

Section 2, Page 172

1. $y = c_1 e^t + c_2 e^{2t}$, $y = e^t$.
2. $y = c_1 e^{-3t} + c_2 e^{-4t}$, $y = c_1 e^{-3t} + c_2 e^{-6t}$, $y = c_1 e^{-3t} + c_2 e^{-5t}$, $y = c_1 e^{-t} + c_2 e^t$.
3. $y = \frac{1}{3}(19e^{-2t} - 7e^{-5t})$, $y = \frac{1}{4}(23e^{-2t} - 7e^{-6t})$, $y = \frac{1}{5}(27e^{-2t} - 7e^{-7t})$, $y = \frac{1}{6}(11e^{3t} + 13e^{-3t})$.
4. $y = y_0 \exp[p_0(t_0 - t)]$.
5. Partial answer: linear, nonlinear, nonlinear, linear, linear, linear.
7. (a) An ellipse, $9y^2 + 4v^2 = 81$; (b) Any for which $9y_0^2 + 4y_1^2 = 81$.

Section 3, Page 175

1. $y = (1 + 2t)e^{2t}$, $y = 0$.   2. (a) $y = (c_1 + c_2 t)e^{-3t}$, $y = (c_1 + c_2 t)e^{5t}$, $y = c_1 + c_2 t$;
   (b) $y = (c_1 + c_2 t)e^t$, $y = (c_1 + c_2 t)e^{5t/2}$, $y = c_1 e^{2t} + c_2 e^{-4t/3}$, $y = e^{-t/2}(c_1 e^{kt} + c_2 e^{-kt})$ with $k = \sqrt{5}/2$.
3. 1; $t^{a+b-1}(b - a)$; -2; -1.   4. Tends to zero.   5. $v/u = \text{const.}$
9. $c_2 = \pm\omega$, c_1 and c_3 arbitrary; or $c_1 = 0$, c_2 and c_3 arbitrary.

Section 4, Page 178

3. $e^{5t}\cos t$, $e^{5t}\sin t$.
4. (a) $y = c_1 e^{-9t} + c_2 e^{6t}$, $y = c_1 e^{3t} + c_2 e^{2t}$, $y = e^{-3t}(c_1 \cos t + c_2 \sin t)$;
   (b) $y = (c_1 + c_2 t)e^t$, $y = c_1 e^{2t} + c_2 e^{-2t}$, $y = e^{-5t}(c_1 \cos 2t + c_2 \sin 2t)$;
   (c) $y = c_1 \cos 2t + c_2 \sin 2t$, $y = (c_1 + c_2 t)e^{2t}$, $y = e^{-3t/2}[c_1 \cos(\sqrt{7}t/2) + c_2 \sin(\sqrt{7}t/2)]$;
   (d) $y = e^{2t}(c_1 \cos t + c_2 \sin t)$, $y = e^t(c_1 \cos 7t + c_2 \sin 7t)$, $y = e^{-3t}(\cos\sqrt{26}t + c_2 \sin\sqrt{26}t)$.
6. (a) $y = c_1 e^{(-2+\sqrt{4-a^2})t} + c_2 e^{(-2-\sqrt{4-a^2})t}$, if $a^2 \ne 4$; $y = c_1 e^{-2t} + c_2 t e^{-2t}$, if $a^2 = 4$.
7. (a) $y = c_1 e^{(-b+\sqrt{b^2-2})t} + c_2 e^{(-b-\sqrt{b^2-2})t}$, if $b^2 \ne 2$; $y = c_1 e^{-bt} + c_2 t e^{-bt}$, if $b^2 = 2$.

Section 5, Page 182

3. $y = 10\cos\sqrt{245}t$.   4. $y = 10e^{-5t}\left(\cos\sqrt{220}t + \dfrac{5}{\sqrt{220}}\sin\sqrt{220}t\right)$; $R = 400\sqrt{245}$ dynes.

5. $V = 100\sqrt{2}e^{-500t}\cos\left(500t - \dfrac{\pi}{4}\right)$; $V = 100e^{-500\sqrt{2}t}(1 + 500\sqrt{2}t)$, $R = 500\sqrt{2}$.

6. $V = 20\sqrt{5}e^{-50,000t}(5 \sinh 10,000\sqrt{5}t + \sqrt{5}\cosh 10,000\sqrt{5}t)$.

11. $y'' + y'\log 4 + [(120\pi)^2 + (\log 2)^2]y = 0$ if t is in minutes.

Section 6, Page 185

1. (a) $y = 12t^2 - 24$, $y = 24t - 12t^2 + 4t^3$, $y = t^4$; (b) $y = e^{2t}$, $y = 12te^{2t}$, $y = 6t^2e^{2t}$.
2. (a) $y = c_1 e^{3t} + c_2 e^{2t} + \frac{1}{2}e^{4t}$, $y = (c_1 + c_2 t)e^{-t} + t - 2$, $y = c_1 e^{-3t} + c_2 e^{-2t} + \frac{1}{12}e^t$;
   (b) $y = (c_1 + c_2 t)e^t + t + 2$, $y = c_1 e^t + c_2 e^{-t} - 5t + 2$, $y = c_1 e^t + c_2 e^{-t} + e^{2t}(\frac{1}{3}t - \frac{7}{9})$;
   (c) $y = (c_1 + c_2 t)e^t + \frac{1}{6}t^3 e^t$, $y = c_1 e^{3t} + c_2 t e^{3t} + \frac{1}{2}t^2 e^{3t}$, $y = c_1 + c_2 e^{-9t} + \frac{1}{8}t$;
   (d) $y = c_1 \cos 3t + c_2 \sin 3t + (9t^2 - 18t + 7)/81$, $y = c_1 e^t + c_2 e^{-t} + \frac{1}{2}te^t$,
       $y = c_1 \cos t + c_2 \sin t + t^3 - 5t$;
   (e) $y = c_1 e^{2t} + c_2 e^{3t} - e^{2t}(t^4/4 + t^3 + 3t^2 + 6t)$, $y = c_1 e^t + c_2 te^t + e^t(\frac{1}{6}t^3 - \frac{1}{2}t^2)$,
       $y = c_1 + c_2 t + e^{-t}(48 + 18t + 6t^2 + t^3)$;
   (f) $y = c_1 e^{3t} + c_2 e^{2t} + (19 + 30t + 18t^2)/36$, $y = c_1 + c_2 e^{5t} - (6t + 15t^2 + 25t^3)/125$,
       $y = c_1 + c_2 e^{-3t} + \frac{1}{40}e^{5t}$.
3. $y = 1 - e^{-t} + t^2 - t$, $y = 0$, $y = \frac{1}{6}e^{-3t} - \frac{1}{2}e^{-t} + \frac{1}{3}$, $y = \frac{1}{8} - t/4 - t^2/4 - e^{2t}/8 + te^{2t}/2$.
4. (a) $y = -1 - t + t^2 + e^{-t}$; $y = -\frac{4}{9} + t/3 + 2e^{-t} - 5e^{-3t}/9$;
   (b) $y = \frac{1}{8}t - \frac{4}{9}$; $y = -11 + 11e^{-t} + 12t - 6t^2 + 2t^3$.

Section 7, Page 188

1. (b) $y = -2 \cos 2t + 16 \sin 2t$; $y = -16 \cos 2t - 2 \sin 2t$. Because the roots of the characteristic equation have negative real parts.

2. (a) $y = c_1 e^t + c_2 e^{2t} - (3 \sin 2t + \cos 2t)/20$, $y = c_1 \sin 2t + c_2 \cos 2t - \frac{1}{5} \cos 3t$;

   (b) $y = c_1 e^t + c_2 e^{-t/2} + 2 \sin t$, $y = c_1 e^{-2t} + c_2 e^{-3t} + 3t e^{-t} + \frac{1}{30} e^{3t}$;

   (c) $y = e^{-t}(c_1 \sin 2t + c_2 \cos 2t) + e^t(\frac{1}{20} \sin 2t - \frac{1}{10} \cos 2t)$, $y = c_1 e^{3t} + c_2 e^{-2t} + \frac{1}{102} e^{3t}(5 \sin 3t - 3 \cos 3t)$.

   (d) $y = c_1 e^{5t} + c_2 e^{-5t} + \frac{1}{10} t e^{5t} - \frac{1}{25} t^2 + \frac{2}{25} t - \frac{2}{625}$, $y = c_1 \sin t + c_2 \cos t + \frac{9}{8} \cos 3t - \sin 2t$.

   (e) $y = e^t(c_1 \sin 3t + c_2 \cos 3t) + \frac{1}{8} e^t + \frac{1}{37} (6 \cos 3t + \sin 3t)$,
   $y = c_1 \sin t + c_2 \cos t + 1 - \frac{1}{2} t \cos t - \frac{1}{5} \cos 2t$.

3. (a) $y = (\sinh t - \sin t)/2$, $y = 0$, $y = \frac{1}{4}(t \sin t - t^2 \cos t)$;

   (b) $y = -\frac{1}{4} + \frac{3}{20} e^{2t} + e^{-t}(\frac{1}{10} \cos t - \frac{1}{5} \sin t)$,
   $y = -\cos t + (t \sin t)/2 + 1$, $y = e^t/10 - (3 \cos 3t + \sin 3t)/30$.

4. $y = \frac{1}{3} \sin 2t - \frac{1}{3} \sin t - \cos t$, $y = t^2 + e^t$, $y = \frac{1}{3} (\sin t + \sin 2t)$.

5. (a) $y = [(3 - \omega^2) \cos \omega t + 4\omega \sin \omega t]/(\omega^4 + 10\omega^2 + 9)$; not stable;
   $y = [10t(3 \cos t + \sin t) - 11 \cos t - 2 \sin t]/400$;

   (b) $y = [(13 - 4\omega^2) \sin \omega t - 8\omega \cos \omega t]/[(13 - 4\omega^2)^2 + 64\omega^2]$, not stable;
   $y = [(2 - 9\omega^2) \sin \omega t - 6\omega \cos \omega t]/[(2 - 9\omega^2)^2 + 36\omega^2]$.

8. $y = [(a^2 - \omega^2) \cos \omega t + 2\omega \sin \omega t]/[(a^2 - \omega^2)^2 + 4\omega^2]$,
   $y = [(1 - \omega^2) \cos \omega t + 2b\omega \sin \omega t]/[(1 - \omega^2)^2 + 4b^2\omega^2]$, $y = \cos \omega t/(1 - \omega^2)$.

Section 8, Page 191

4. $G = [(1 - \omega^2)^2 + 4b^2\omega^2]^{-\frac{1}{2}}$, $\phi = \tan^{-1}[2b\omega/(1 - \omega^2)]$.   5. b/a, ω/a.   6. $y = (1 - \cos \omega t)/\omega^2$.

Section 9, Page 194

1. $y = c_1 + c_2 t + c_3 t^2 + c_4 t^3$, $y = c_1 + c_2 t + c_3 t^2 + c^4 t^3 + t^4 + \cos t$.

2. $y = c_1 + c_2 e^t + c_3 e^{2t} + 2t + e^{5t}$.

3. $y = c_1 + c_2 e^{-t} + e^{t/2}(c_3 \cos \frac{1}{2} t\sqrt{3} + c_2 \sin \frac{1}{2} t\sqrt{3}) + (\omega^3 \cos \omega t + \sin \omega t)/(w^7 + \omega)$,
   $y = c_1 + c_2 e^{-t} + e^{t/2}(c_3 \cos \frac{1}{2} t\sqrt{3} + c_2 \sin \frac{1}{2} t\sqrt{3}) + (\omega^3 \sin \omega t - \cos \omega t)/(w^7 + \omega)$.

4. $y = -\frac{1}{2} e^{-t} + \frac{5}{2} e^t - t$.   5. $y = e^{-2t}/3 + 17 e^t/12 + 7 e^{-t}/2 - e^{-3t}/4 - 2t^2 + 2t - 5$.

6. $y = c_1 e^t + c_2 e^{-t} + c_3 e^{-3t} + c_4 t e^{-3t}$.

7. $y = [(12 - 8\omega^2) \cos \omega t + (19\omega - \omega^3) \sin \omega t]/[(12 - 8\omega^2)^2 + (19\omega - \omega^3)^2]$; yes.

8. $y = A e^{rt}/P(r)$.

Section 10, Page 197

1. (a) $y = 0$, $y = c \sin \pi x$; (b) no solutions exist; $y = 1 - \cos x + c \sin x$.   2. $n^2\pi^2/T^2$, n integer.

Section 11, Page 200

1. $y = c_1 e^t + c_2 e^{-t/2}$, $y = c_1 e^{-t} + c_2 e^t$, $y = c_1 e^{-2t} + c_2 e^t$, $y = (c_1 + c_2 t)e^{3t}$.

2. (a) $y = c_1 + c_2 e^{-3t/2} + c_3 e^{5t}$, $y = c_1 \sin t + c_2 \cos t + e^{-t}(c_3 \cos 2t + c_4 \sin 2t)$;

   (b) $y = c_1 e^{-t} + c_2 t e^{-t} + c_3 t^2 e^{-t}$, $y = c_1 e^{-2t} + e^t(c_2 \sin \sqrt{3}t + c_3 \cos \sqrt{3}t)$;

   (c) $y = (c_1 + c_2 t)e^t + c_3$, $y = (c_1 + c_2 t + c_3 t^2)e^{-t} + c_4$;

   (d) $y = c_1 \cos kt + c_2 \sin kt + c_3 \cosh kt + c_4 \sinh kt$, $y = c_1 e^{-t} + (c_2 + c_3 t)e^{2t}$;

   (e) $y = c_1 + e^{t/2}[c_2 \cos (\sqrt{15}t/2) + c_3 \sin (\sqrt{15}t/2)] + t^2/2 + (t + e^t)/4$,
   $y = e^{t/\sqrt{2}}[c_1 \sin (\sqrt{2}t/2) + c_2 \cos (\sqrt{2}t/2)] + e^{-t/\sqrt{2}}[c_3 \sin (\sqrt{2}t/2) + c_4 \cos (\sqrt{2}t/2)] + \cos t$;

   (f) $y = c_1 e^t + c_2 e^{2t} + c_3 t e^{2t} - \frac{1}{4} t^2 - t - \frac{11}{8}$, $y = c_1 e^{-t} + c_2 e^t + c_3 e^{2t} + \frac{1}{2} + \frac{1}{2} t e^t$.

   (g) $y = c_1 + c_2 t + e^{-t}(c_3 + c_4 t)$,
   $y = c_1 e^{-t} + c_2 \cos t + c_3 \sin t + (\frac{1}{4} t - \frac{3}{8})e^t$.

   (h) $y = c_1 + c_2 t + c_3 t^2 + c_4 e^{-t} + (4 \cos 4t - \sin 4t)/1,088$,
   $y = c_1 e^{-t} + e^{-t/2}(c_2 \cos \frac{3}{2}\sqrt{3}t + c_4 \sin \frac{3}{2}\sqrt{3}t) - (6 \cos 2t + 5 \sin 2t)/13$.

   (i) $y = c_1 + c_2 t + e^t(c_3 + c_4 t) + t^2(12 + 3t + \frac{1}{2} t^2 + \frac{1}{20} t^3)$,
   $y = c_1 e^{2t} + (c_2 + c_3 t)e^t - 4 - t$.

3. (a) $y = \frac{3}{50}(-7 \cos 2t + \sin 2t)$, $y = (\frac{5}{1.014})(-17 \sin 3t - 7 \cos 3t)$;

   (b) $y = \frac{1}{442}(3 \cos t + 2 \sin t)$, $y = \frac{1}{195}(4 \sin t - 7 \cos t)$.

4. $c_1 = \frac{17}{25}$, $c_2 = -\frac{47}{400}$.

5. (a) $y = c_1 \cos t + c_2 \sin t$, $x = c_1 \sin t - c_2 \cos t$; $y = c_1 e^t + c_2 e^{-t}$, $x = c_1 e^t - c_2 e^{-t}$;
   $y = (c_1 + c_2 t)e^t$, $x = e^t(c_1 + \frac{1}{2} c_2 + c_2 t)$.

(b) $y = c_1 e^t + c_2 e^{-t} + c_3 \cos t + c_4 \sin t$, $x = c_1 e^t + c_2 e^{-t} - c_3 \cos t - c_4 \sin t$;
$\quad y = c_1 e^t - \tfrac{1}{2} e^{-t/2}[(c_2 + c_3\sqrt{3}) \cos (\sqrt{3}t/2) + (c_3 - \sqrt{3}c_2) \sin (\sqrt{3}t/2)]$,
$\quad x = c_1 e^t + [c_2 \cos (\sqrt{3}t/2) + c_3 \sin (\sqrt{3}t/2)]e^{-t/2}$;
$\quad y = c_1(1 + \sqrt{2})e^{\sqrt{2}t} + c_2(1 - \sqrt{2})e^{-\sqrt{2}t}$, $x = c_1 e^{\sqrt{2}t} + c_2 e^{-\sqrt{2}t}$.

6. $x = c_1 e^{(3 + \sqrt{17})t/2} + c_2 e^{(3 - \sqrt{17})t/2} + \tfrac{1}{2} e^t - \tfrac{1}{2}$.

7. $y = t^3 + e^t$, $y = 1 - e^{-t} - 2[e^{t/2} \sin (\sqrt{3}t/2)]/\sqrt{3}$.

Section 12, Page 204

1. $1/s$, $1/s^2$.   **2.** $b/(s^2 - b^2)$, $s/(s^2 - b^2)$.

3. $(3s^2 + 2s - 8)/s^3$, $2s/(s^2 + 9) - 8/(s^2 + 4)$,
$\quad 1/(s - 1) + 1/(s^2 - 2s + 2)$, $1/(2s) - s/[2(s^2 + 4)]$.

4. $e^{2t} + 2e^{-3t} + 3e^{-4t}$, $\tfrac{5}{2} \sin 2t + 3 \sinh 2t$, $3 \cos 4t + \tfrac{1}{4} \sin 4t$, $e^{-2t} + 4e^{-3t} \cos 2t - \tfrac{17}{2}e^{-3t} \sin 2t$.

5. $(5/s)(1 - e^{-s})$, $[1/(s + 1)](1 - e^{-(s+1)})$, $2e^{-s}/s + 3e^{-s}/s^2 + 1/s - 3/s^2$,
$\quad \pi(e^{-s} + 1)/(s^2 + \pi^2)$, $1/(1 + s)^2 - e^{-(1+s)}(s + 2)/(1 + s)^2$.

6. (a) e^{-sa}/s, $(Ae^{-sa} + Be^{-sb})/s$; (c) $[Ae^{-sa} + Be^{-sb} + Ce^{-sc}]/s$.

Section 13, Page 207

2. $y = y_0 e^{-pt}$.   **3.** $y = t^2 + t + 1$.   **4.** $y = 12t^2$, $y = 4t^3$, $y = t^4$, $y = 4t^3$, $y = 0$, $y = t^7/105$.

5. $y = y_1 t^2/2 + y_2 t^3/6$.

6. $y = e^{-2t}(\tfrac{2}{3} \cosh 2t\sqrt{3} - \tfrac{3}{2}\sqrt{3} \sinh 2t/\sqrt{3})$.

7. (a) $y = e^{-t} \cos t$; (b) $y = \sin t + \cos t$.

8. (a) $y = (e^{-2t} + 3)/4$; (b) $y = e^{-3t/2}(\cosh \tfrac{1}{2}\sqrt{5}t - \tfrac{3}{5}\sqrt{5} \sinh \tfrac{1}{2}\sqrt{5}t)$.

9. $y = e^{-7t/2}(4 \cosh \tfrac{3}{2}t + \tfrac{26}{3} \sinh \tfrac{3}{2}t) \equiv \tfrac{19}{3}e^{-2t} - \tfrac{7}{3}e^{-5t}$,
$\quad y = e^{-4t}(4 \cosh 2t + \tfrac{15}{2} \sinh 2t)$,
$\quad y = \tfrac{1}{5}e^{-9t/2}(20 \cosh \tfrac{5}{2}t + 34 \sinh \tfrac{5}{2}t)$, $y = 4 \cosh 3t - \tfrac{1}{3} \sinh 3t$.

10. (a) $y = \tfrac{4}{3}e^{t/4} \sinh (3t/4)$, $y = \sinh t$, $y = \tfrac{2}{3}e^{-t/2} \sinh \tfrac{3}{2}t$, $y = te^{3t}$;
$\quad$ (b) $y = e^{t/4}(\cosh \tfrac{3}{4}t - \tfrac{1}{3} \sinh \tfrac{3}{4}t)$, $y = \cosh t$,
$\quad y = e^{-t/2}(\cosh \tfrac{3}{2}t + \tfrac{1}{3} \sinh \tfrac{3}{2}t)$, $y = e^{3t} - 3te^{3t}$.

11. (a) First problem: $y = e^{-3t/2}[y_0 \cosh \tfrac{15}{2}t + (\tfrac{2}{15}y_1 + \tfrac{1}{5}y_0) \sinh \tfrac{15}{2}t]$
$\quad\quad\quad\quad\quad\quad\quad\quad\quad\quad\quad \equiv \tfrac{1}{15}(9y_0 + y_1)e^{6t} + \tfrac{1}{15}(6y_0 - y_1)e^{-9t}$.

12. $L(y) = [F(s) + (s + 2b)y_0 + y_1]/(s^2 + 2bs + a^2)$.

13. (a) First problem: $Ly = 6s(s^2 + 4)^{-1}(s + 1)^{-2}(s + 2)^{-1}$.

Section 14, Page 210

2. $y = \tfrac{1}{3} \sin t + \cos 2t - (\sin 2t)/6$.   **3.** $y = \tfrac{1}{6}t^3 - t + \sinh t$.

4. $y = 1 - \cos t$ for $t > 0$, $y = 0$ for $t < 0$, no.

13. $q(\omega e^{-pt} + p \sin \omega t - \omega \cos \omega t)$, $q(p \cos \omega t + \omega \sin \omega t - pe^{-pt})$,
$\quad$ where $p = (RC)^{-1}$ and $q = pE_0(p^2 + \omega^2)^{-1}$.

14. Partial answer: $\tfrac{1}{2}E_0(\sin \omega_0 t - \omega_0 t \cos \omega_0 t)$,
$\quad E_0 \omega_0 (\omega_0^2 - \omega^2)^{-1}(\omega_0 \sin \omega t - w \sin \omega_0 t)$, where $\omega_0 = (LC)^{-1/2}$.

Section 15, Page 214

1. (a) $y = 4(e^{-2t} - e^{-t})$; (b) $y = -te^{-4t}$; (c) $y = (\tfrac{6}{5})e^{-13t/2} \sinh (5t/2)$; (d) $y = -2te^{-t}$;

2. (a) $y = 4e^{-t} - 3e^{-2t}$; (b) $y = e^{-4t}(1 + t)$; (c) $y = \tfrac{2}{5}e^{-4t} + \tfrac{3}{5}e^{-9t}$; (d) $y = e^{-t} - 2te^{-t}$;

4. (a) $Lx = (s^3 - 2s + 3)/(s^4 + 3s^2 + 2)$, $Ly = (s^2 - 4s + 5)/(s^4 + 3s^2 + 2)$;
$\quad$ (b) $Lx = [s(s^2 + 4) - s + 1]/(s^2 + 4)^2$, $Ly = (s^2 + 5 - s)/(s^2 + 4)^2$;
$\quad$ (c) $Lx = [s(s^2 + 7) + 2]/(s^4 + 13s^2 + 36)$, $Ly = (s^2 + 3s + 6)/(s^4 + 13s^2 + 36)$;
$\quad$ (d) $Lx = [s(s^2 + 1) + 2s + 2]/(s^2 + 1)^2$, $Ly = (s^2 - 2s - 1)/(s^2 + 1)^2$.
$\quad$ (e) $Lx = (s^3 + s + 1)/(s^4 + 2s^2 + 2)$, $Ly = (s^2 - s + 1)/(s^4 + 2s^2 + 2)$.

5. $y = 2 + e^{-2t} - 3e^{-t} - te^{-t}$, $z = -3 - 2e^{-2t} + 5e^{-t} + 2te^{-t}$.

6. $y = z = w = e^{t/2}$.   **10.** $x = a(1 - \cos \omega t)$, $y = a(\omega t - \sin \omega t)$, where $\omega = He/m$, $a = E/\omega H$.
$\quad$ This is a cycloid generated by a circle of radius a.

Section 16, Page 219

7. $y = e^{-3t} \left[\tfrac{1}{4}t \cosh \sqrt{2}t - \dfrac{1}{4\sqrt{2}} \sinh \sqrt{2}t \right]$.

Section 17, Page 223

1. $V = (e^{-t} - e^{-3t})/2$.
2. Partial answer: $y = 1, t, t^2/2$ for $t > 0$.
3. y''', y'', y', y continuous, respectively, and higher derivatives discontinuous.   In last case, y is discontinuous.
4. All linear except $Nf = [f(1)]^2$.

Section 18, Page 226

1. $y = \int_0^t f(\tau)e^{\tau-t}\,d\tau,\ y = \int_0^t f(\tau)\sinh\,(t-\tau)\,d\tau,\ y = \int_0^t f(\tau)e^{\tau-t}\sin\,(t-\tau)\,d\tau$.

2. $y = \dfrac{1}{\beta-\alpha}\displaystyle\int_0^t f(\tau)[e^{-\alpha(t-\tau)} - e^{-\beta(t-\tau)}]\,d\tau$.

3. Rest solutions: $y = \displaystyle\int_0^t f(\tau)(t-\tau)\,d\tau,\ y = \int_0^t f(\tau)(t-\tau)e^{-\alpha(t-\tau)}\,d\tau$.

5. $y = \omega(\omega^2 - \omega_1^2)^{-1}(\omega\sin\omega_1 t - \omega_1\sin\omega t)$ if $\omega \neq \omega_1$.

10. $y = \frac{1}{2} - e^{-t} + \frac{1}{2}e^{-2t}$, for $f(t) = H(t)$; $y = \frac{1}{20}e^{3t} - \frac{1}{5}e^{-2t} - \frac{1}{4}e^{-t}$, for $f(t) = e^{3t}$; $y = -\frac{3}{4} + \frac{1}{2}t + e^{-t} - \frac{1}{4}e^{-2t}$, for $f(t) = t$.

Section 19, Page 230

2. $y = -t^3 e^{-4t}/18 + e^{-t}/81;\ y = 6e^t(\cos 2t - \sin 2t);\ y = e^{-t}\cos t + e^{-t}t^6/720$.
3. $y = (t/18)(\sin t + c_1\cos 2t + c_2\sin 2t),\ y = (t^2/96)\cos 2t,$
   $y = (t^3/240)(2\cos t + \sin t)$.

4. $y = e^{-t}\displaystyle\int_0^t e^{-\tau_1}\int_0^{\tau_1} e^{2\tau}f(\tau)\,d\tau\,d\tau_1 = e^{-t}\int_0^t (e^{\tau} - e^{2\tau-t})f(\tau)\,d\tau$.

5. $I = R^{-1}V_0 + (K - R^{-1}V_0)e^{-Rt/L} - K\cos\omega t + (RK)(\omega L)^{-1}\sin\omega t$ with $K = A\omega L/(R^2 + \omega^2 L^2)$.
6. $I = L^{-1}e^{-Rt/L}$.

7. $RLI = \displaystyle\int_0^t V(\tau)e^{-(R/L)(t-\tau)}\,d\tau = L\int_0^t V'(\tau)[1 - e^{(R/L)(t-\tau)}]\,d\tau + L[1 - e^{-(R/L)t}]V(0)$.

Section 20, Page 234

2. $y = \Sigma x^n/n! = e^x,\ y = \Sigma 2^n x^n/n! = e^{2x},\ y = -1 - x + 2\Sigma x^n/n! = 2e^x - 1 - x,$
   $y = 1,\ y = \Sigma x^{2n}/(2^n n!) = e^{x^2/2}$.

3. $y = -2 - 2x - x^2;\ y = x + \displaystyle\sum_{n=2}^{\infty} \frac{x^n}{n(n-1)}$.   4. $y = 1 + x^2 + x^4/2, k = 2$,   exact.

5. $y = \sin^{-1} x = \displaystyle\sum \frac{1\cdot 3\cdots(2n-1)}{2\cdot 4\cdots 2n}\ \frac{1}{2n+1}\ x^{2n+1}$, where first coefficient $= 1$.

6. $y = 1 - \dfrac{x^3}{3!} + \dfrac{1\cdot 4x^6}{6!} - \dfrac{1\cdot 4\cdot 7x^9}{9!} + \dfrac{1\cdot 4\cdot 7\cdot 10x^{12}}{12!} - \cdots;$

   $y = 1 + x + \dfrac{2!x^4}{4!} + \dfrac{3!x^5}{5!} + \dfrac{2!6!x^8}{4!8!} + \dfrac{3!7!x^9}{5!9!} + \cdots$.

7. $y = a_0\Sigma(-1)^n x^{2n}/(2n)! + a_1\Sigma(-1)^n x^{2n+1}/(2n+1)!,\ y = 1 + (a_0 - 1)\cos x + a_1\sin x,$

   $y = a_0 + a_1\displaystyle\sum \frac{x^{2n+1}}{2^n(2n+1)\cdot n!} + \sum \frac{2^n n!x^{2(n+1)}}{[2(n+1)]!},\ y = a_0 + a_1\sum \frac{x^{2n+1}}{2^n(2n+1)\cdot n!},$

   $y = a_0\left(1 + \dfrac{x^3}{3!} + \dfrac{1\cdot 4x^6}{6!} + \dfrac{1\cdot 4\cdot 7x^9}{9!} + \cdots\right) + a_1\left(x + \dfrac{2x^4}{4!} + \dfrac{2\cdot 5x^7}{7!} + \dfrac{2\cdot 5\cdot 8x^{10}}{10!} + \cdots\right)$

   $+ \dfrac{x^2}{2!} + \dfrac{3x^5}{5!} + \dfrac{3\cdot 6x^8}{8!} + \dfrac{3\cdot 6\cdot 9x^{11}}{11!} + \cdots$.

8. At least if $|x| < 2$, $y = c_1\displaystyle\sum \frac{x^n}{n!} + c_2(1 - x) = c_1 e^x + c_2(1 - x)$, yes.

9. (a) $y = c\displaystyle\sum x^n$.   General solution: $y = c_1\sum x^n + c_2\sum \frac{(-1)^n x^n}{n!}$;

   (b) $y = a_0\left[1 + \dfrac{(x-2)^3}{2\cdot 3} + \dfrac{(x-2)^6}{2\cdot 3\cdot 5\cdot 6} + \dfrac{(x-2)^9}{2\cdot 3\cdot 5\cdot 6\cdot 8\cdot 9} + \cdots\right]$

   $+ a_1\left[x - 2 + \dfrac{(x-2)^4}{3\cdot 4} + \dfrac{(x-2)^7}{3\cdot 4\cdot 6\cdot 7} + \dfrac{(x-2)^{10}}{3\cdot 4\cdot 6\cdot 7\cdot 9\cdot 10} + \cdots\right]$.   15. $x^2 y'' + 12y = 6xy'$.

Section 21, Page 239

1. (a) $y = c_1 x^2 + c_2 x^3$, $y = c_1 x^n + c_2 x^{-n-1}$.

2. $y = \sum \dfrac{(-1)^n x^{2n}}{(2n+1)!} = \dfrac{\sin x}{x}$, $y = 1 - \dfrac{x^2}{2^2} + \dfrac{x^4}{2^2 \cdot 4^2} - \dfrac{x^6}{2^2 \cdot 4^2 \cdot 6^2} + \cdots$.

7. $a_n n(n \pm 2p) = -a_{n-2}$.   9. (b) $y = \sqrt{2/(\pi x)}(c_1 \cos x + c_2 \sin x)$.

12. $y = x[c_1 + c_2 \log x + c_3 (\log x)^2 + \tfrac{1}{24}(\log x)^4]$.   See shift theorem of Sec. 19.

13. (a) $y = c_1 x^{-2} + c_2 x^{-1} + \tfrac{1}{2}\log x - \tfrac{3}{4}$, $y = x^{1/2}(c_1 \cos \tfrac{1}{2}\sqrt{3} \log x + c_2 \sin \tfrac{1}{2}\sqrt{3} \log x) + \tfrac{1}{8}x^2$.
    (b) $y = c_1 x^2 + c_2 x^{(5+\sqrt{21})/2} + c_3 x^{(5-\sqrt{21})/2} - \tfrac{1}{2}$, $y = c_1 x^2 + c_2 x - x[(\log x)^2/2 + \log x]$.

Section 23, Page 246

1. $y = -1/x$, $y = \log x$, $y = -\tfrac{1}{2}\cos 2x$.

2. $y = c_1 x + c_2 x[\log|1+x| - \log|1-x|]$, $y = c_1 x^{-1}\sin x + c_2 x^{-1}\cos x$, $y = c_1 x^2 + c_2 x e^x$.

3. $y = e^x[(3x - x^3)c_1 + c_2]$.   4. $a = -\tfrac{1}{2}$.

10. (a) $(1 - x^2)^{-1}$, e^{x^2}, $(1 - x^2)^{-1/2}$, $(1 - x)e^x$.

13. $y = c_1 x + c_2 x^2 + c_3 x^3$.

Section 24, Page 250

1. $W = 0$.   2. (a) All linearly dependent; (b) all linearly independent.   4. $y^{IV} = 0$.

Section 25, Page 254

1. (a) $y = (x^3 - x^2)/3 + 2x/9 - \tfrac{2}{27}$, $y = x + 2$, $y = -\sin x \log|\csc x + \cot x|$;
    (b) $y = e^x/12$, $y = \sin x$, $y = ae^{ax}\displaystyle\int x^3 e^{x^2 - ax}\,dx - ae^{-ax}\displaystyle\int x^3 e^{x^2 + ax}\,dx$, where $a^2 = 2$.

3. $y = -(x^2 \log x)/9$.   4. $y = c_1 e^x + c_2 x + x^2 + 1$.

7. (a) $y = \displaystyle\int_{x_0}^x (x - t)e^{x-t}f(t)\,dt$;
    (b) $G(x,t) = (\tfrac{1}{2}x - 1)te^{x-t}$ if $0 < t < x$, $G(x,t) = (\tfrac{1}{2}t - 1)xe^{x-t}$ if $x < t < 1$.

CHAPTER 4

Section 2, Page 261

2. $\mathbf{A} + \mathbf{B} + \mathbf{C} = 0$.   3. $\mathbf{A} = \tfrac{1}{2}(\mathbf{S} + \mathbf{D})$, $\mathbf{B} = \tfrac{1}{2}(\mathbf{S} - \mathbf{D})$.

5. (a) $\dfrac{\mathbf{A}}{|\mathbf{A}|}$; (b) $\left(\dfrac{\mathbf{A}}{|\mathbf{A}|} + \dfrac{\mathbf{B}}{|\mathbf{B}|}\right)$.   9. (a) They form a right triangle; (b) $OP = \tfrac{1}{2}(\mathbf{A} + \mathbf{B})$.

Section 3, Page 262

1. (a) $5\mathbf{j}$, $-5\mathbf{j}$;
    (b) $\mathbf{A} + \mathbf{B} = 2\mathbf{i} + 3\mathbf{j} + 4\mathbf{k}$; $(\mathbf{A} + \mathbf{B}) + \mathbf{C} = 3\mathbf{i} + 3\mathbf{j} + 3\mathbf{k}$; $\mathbf{B} + \mathbf{C} = 2\mathbf{i} + \mathbf{j}$; associative law;
    (c) $5\mathbf{i} + 10\mathbf{j} \times 15\mathbf{k}$, $-2\mathbf{i} - 4\mathbf{j} - 6\mathbf{k}$, $3\mathbf{i} + 6\mathbf{j} + 9\mathbf{k}$, $3\mathbf{i} + 6\mathbf{j} + 9\mathbf{k}$, distributive law;
    (d) $3\mathbf{i} + 6\mathbf{j} + 9\mathbf{k}$, $3\mathbf{i} + 3\mathbf{j} + 3\mathbf{k}$, $6\mathbf{i} + 9\mathbf{j} + 12\mathbf{k}$; (e) -4.

Section 4, Page 264

1. (a) $10, 2, 8$; (b) $6, 4, \mathbf{i} + 3\mathbf{j} + \mathbf{k}, 10$; (c) 12;
    (d) $\cos^{-1} 3/\sqrt{21}$; (e) $4/\sqrt{5}$; (f) $s = 4$; (g) $-\mathbf{i} - \mathbf{j} + \mathbf{k}$.

2. (b) $x = -10, y = 4$ $z = \tfrac{1}{2}$.   3. (a) No; (b) yes.   4. (a) $-\tfrac{4}{9}$.

Section 5, Page 266

1. (a) $-2\mathbf{i} + 3\mathbf{j} - 4\mathbf{k}, 5\mathbf{i} - 4\mathbf{j} + 3\mathbf{k}, 3\mathbf{i} - \mathbf{j} - \mathbf{k}, 2\mathbf{i} + 3\mathbf{j} + 3\mathbf{k}, 3\mathbf{i} - \mathbf{j} - \mathbf{k}$; (c) $\mathbf{i} + 2\mathbf{j} + 2\mathbf{k}$; (d) $\tfrac{3}{2}$.

6. $\mathbf{N} = (3\mathbf{i} + 2\mathbf{j} - 4\mathbf{k})/\sqrt{29}$, $\sqrt{121\tfrac{1}{29}}$.

7. (a) No, no; (b) yes, no; (c) yes, no; (d) no, no; (e) no, no.

9. (a) $(\mathbf{i} + \mathbf{j} + \mathbf{k})/\sqrt{3}$; (c) $(8\mathbf{i} + \mathbf{j} - 2\mathbf{k})/\sqrt{69}$.   10. (a) $\sqrt{3}/2$; (c) $\sqrt{69}/2$.

Section 6, Page 268

2. (a) 0; (b) $x = \tfrac{8}{5}$; (d) $0, 0$.

Section 7, Page 270

1. (a) $\mathbf{R}'(t) = 2\mathbf{i} + 6t\mathbf{j} + 3t^2\mathbf{k}$; (b) $\mathbf{R}'(1) = 2\mathbf{i} + 6\mathbf{j} + 3\mathbf{k}$; (c) $\mathbf{v} = 2\mathbf{i} + 6\mathbf{j} + 3\mathbf{k}$, $|\mathbf{v}| = 7$.

2. (a) $\mathbf{v} = \mathbf{R}'(t) = \mathbf{i} + \mathbf{j}\cos t - \mathbf{k}\sin t$; $|\mathbf{R}'(t)| = \sqrt{2}$; (b) $s = 2\sqrt{2}$.   9. $\mathbf{R}\cdot\mathbf{R}' \times \mathbf{R}'''$, $\mathbf{R} \times \mathbf{R}''$.

Section 8, Page 275

1. (a) $W = 0$; (b) $W = -2$; (c) $W = 4$.   2. (a) $\mathbf{T} = -\mathbf{i} + 2\mathbf{j} - 3\mathbf{k}$; (b) $\mathbf{T} = 2\mathbf{i} - 4\mathbf{j} + 4\mathbf{k}$.

3. (a) $\mathbf{v} = k\mathbf{A} \times \mathbf{B}$; (b) $\mathbf{v} = k\mathbf{C} \times (\mathbf{A} - \mathbf{B})$.   6. (a) $\mathbf{R} = \tfrac{1}{3}(2\mathbf{i} + 5\mathbf{j})$; (b) $\mathbf{R}(1,2) = \tfrac{1}{3}(4\mathbf{i} + \mathbf{j})$.

Section 9, Page 277

1. (a) $\mathbf{n} = \mathbf{i} + 2\mathbf{j} + 3\mathbf{k}$; (b) $\cos^{-1} 6/\sqrt{42}$; (c) $9/\sqrt{14}$.

2. (a) $\mathbf{i} + 2\mathbf{j} + 3\mathbf{k}$; (b) $x = 1 + t$; $y = 2t$; $z = 1 + 3t$; (c) $\sqrt{6/7}$;
(e) $\frac{3}{2}\mathbf{i} + \mathbf{j} + \frac{1}{2}\mathbf{k}$; (f) $\frac{8}{9}\mathbf{i} - \frac{2}{9}\mathbf{j} + \frac{2}{3}\mathbf{k}$.

3. (a) $\mathbf{R} = 5\mathbf{i} - 2\mathbf{j} + (-4\mathbf{i} + 4\mathbf{j} - \mathbf{k})t$; (b) $-4\mathbf{i} + 4\mathbf{j} - \mathbf{k}$;
(d) $-4x + 4y - z + c = 0$; (e) $\mathbf{R} = -3\mathbf{i} + \mathbf{k} + (-4\mathbf{i} + 4\mathbf{j} - \mathbf{k})t$.

4. (a) $D^2 = 3t^2 + 8$; (b) $t = 0$; $D = 2\sqrt{2}$.   **5.** $\mathbf{R} = (\mathbf{i} + \mathbf{j} + 3\mathbf{k})t$, $(-\frac{1}{8}, -\frac{1}{8}, -\frac{3}{8})$.

Section 10, Page 279

1. $\mathbf{n} = a\mathbf{i} + b\mathbf{j} + c\mathbf{k}$.

2. (a) $-\mathbf{i} + 6\mathbf{j} + 2\mathbf{k}$; (b) $-x + 6y + 2z = 10$; (c) $2\mathbf{i} + \mathbf{j} + 3\mathbf{k} + (-\mathbf{i} + 6\mathbf{j} + 2\mathbf{k})t = R$.

4. $\theta = \cos^{-1} 9/\sqrt{102}$.   **5.** $-16\mathbf{i} + 8\mathbf{j} + 4\mathbf{k}$.   **6.** $\mathbf{R} = (i a^2 + j b^2 + k c^2)/\pm\sqrt{a^2 + b^2 + c^2}$.

7. $x + y + z = \pm\sqrt{a^2 + b^2 + c^2}$.   **10.** $\pi/3$.

Section 11, Page 283

1. (a) $(-\mathbf{i} + \mathbf{j} - \mathbf{k})/\sqrt{3}$; (b) $-x + (y + 2) - (z - 2) = 0$; (c) $\sqrt{3} + \log(1 + \sqrt{3/2})$.

2. (a) $\mathbf{v} = 2t\mathbf{i} + 2\mathbf{j} + 2t\mathbf{k}$; $\mathbf{A} = 2\mathbf{i} + 2\mathbf{k}$; (b) $v = 2\sqrt{2t^2 + 1}$; (c) $\kappa = \dfrac{\sqrt{4t^2 + 2}}{2(2t^2 + 1)^2}$; $N = \dfrac{\mathbf{i} - 2t\mathbf{j} + \mathbf{k}}{\sqrt{4t^2 + 2}}$.

3. Let $\mathbf{R}(t) = (a_2 t^2 + a_1 t + a_0)\mathbf{i} + (b_2 t^2 + b_1 t + b_0)\mathbf{j} + (c_2 t^2 + c_1 t + c_0)\mathbf{k}$; then an equation of the plane through the plane curve is
$(b_1 c_2 - b_2 c_1)(x - a_0) + (a_2 c_1 - a_1 c_2)(y - b_0) + (b_2 a_1 - b_1 a_2)(z - c_0) = 0$.

5. (a) $\mathbf{T} = (\mathbf{i} + 2\mathbf{j})/\sqrt{5}$; $\mathbf{N} = (\mathbf{j} - 2\mathbf{i})\backslash\sqrt{5}$; (b) $\mathbf{V} = \mathbf{i} + 2\mathbf{j}$; $A = 2\mathbf{j}$;
(c) $V_t = \sqrt{5}$; $A_t = 4/\sqrt{5}$; (d) $s' = \sqrt{1 + 4t^2}$; $s'' = 4t/\sqrt{1 + 4t^2}$.

6. (a) $A_n = 2/\sqrt{5}$; (b) $\kappa = 2/(5\sqrt{5})$.

7. $\mathbf{T} = (\mathbf{i} + 2\mathbf{j} + 3\mathbf{k})/\sqrt{14}$, $\mathbf{B} = (3\mathbf{i} - 3\mathbf{j} + \mathbf{k})/\sqrt{19}$, $\mathbf{N} = (-11\mathbf{i} - 8\mathbf{j} + 9\mathbf{k})/\sqrt{266}$.

8. $\dfrac{x - 2}{1} = \dfrac{y - 4}{4} = \dfrac{z - 8}{12}$, $12x - 6y + z = 8$.   **9.** $A_t = 0$, $A_n = 2$.

10. $x - y - \sqrt{2}z = 0$.   **11.** $\kappa = \tau = (\operatorname{sech}^2 t)/(2a)$.

Section 15, Page 291

3. (a) $a_{11} a_{22} a_{33} a_{44}$; (b) $x_2 x_3^2 - x_3 x_2^2 - x_1 x_3^2 + x_3 x_1^2 + x_1 x_2^2$; (c) same as (b); (d) $a_1 b_2 c_3$.

Section 16, Page 295

3. (a) $\begin{pmatrix} 13 \\ 13 \\ 11 \end{pmatrix}$; (b) $\begin{pmatrix} 13 & 0 & 0 \\ 13 & 0 & 0 \\ 11 & 0 & 0 \end{pmatrix}$.   **5.** $2; 3; 2; 2; 3$ if $k \neq 0$, 0 if $k = 0$.

Section 17, Page 298

1. $A^{-1} = \begin{pmatrix} 10/9 & 5/9 & -4/9 \\ -8/9 & -4/9 & 5/9 \\ -5/3 & -1/3 & 2/3 \end{pmatrix}$, $B^{-1} = \begin{pmatrix} 1 & 0 & 0 \\ -2 & 1 & 0 \\ 2/3 & -1/3 & 1/3 \end{pmatrix}$.

6. $A^{-1} = \begin{pmatrix} 1/2 & 1/2 \\ -1/2 & 1/2 \end{pmatrix}$, $A' = \begin{pmatrix} 1 & 1 \\ -1 & 1 \end{pmatrix}$.

7. $A^{-1} = \begin{pmatrix} 1/\sqrt{2} & 0 & -1/\sqrt{2} \\ 0 & 1 & 0 \\ 1/\sqrt{2} & 0 & 1/\sqrt{2} \end{pmatrix}$.   **8.** $A^{-1} = \begin{pmatrix} -3 & 2 \\ 2 & -1 \end{pmatrix}$, $A' = \begin{pmatrix} 1 & 2 \\ 2 & 3 \end{pmatrix}$.

10. $AB = \begin{pmatrix} 4 & -1 \\ -1 & 0 \end{pmatrix}$, $BA = \begin{pmatrix} 1 & 2 \\ 5 & 1 \end{pmatrix}$, $(BA)^{-1} = \begin{pmatrix} -1/9 & 2/9 \\ 5/9 & -1/9 \end{pmatrix}$.

Section 20, Page 306

1. $\begin{pmatrix} -1/2 & -1/2 \\ -1/2 & -3/2 \end{pmatrix}\begin{pmatrix} 2 & 3 \\ -1 & -2 \end{pmatrix}\begin{pmatrix} -3 & 1 \\ 1 & -1 \end{pmatrix} = \begin{pmatrix} 1/2 & 0 \\ 0 & 2 \end{pmatrix}$.   **3.** $\begin{pmatrix} 1 & 1 & -1/2 \\ 0 & -1 & -1 \\ 0 & 0 & 1 \end{pmatrix}$.

4. $\begin{pmatrix} 2 & 0 & 0 \\ 0 & -6 & 0 \\ 0 & 0 & -27 \end{pmatrix}$.

CHAPTER 5

Section 2, Page 316

1. (a) Entire xy plane; (b) entire xy plane; (c) $y^2 \leq 5$;
(d) $x^2 + y^2 \neq 0$; (e) $x \neq 0$; (f) $(x - 1)^2 + y^2 \leq 1$.

2. (a) Circles $x^2 + y^2 = a^2$; (b) parabolas $y = a - x^2$, $a > 0$; (c) straight lines $y = cx$, $x \neq 0$; (d) circles $x^2 + y^2 = a^2$; (e) parabolas $y = ax^2$.

3. (a) Planes $x + y + z = \text{const}$; (b) spheres $x^2 + y^2 + z^2 = \text{const}$; (c) paraboloids of revolution $z = a(x^2 + y^2)$, $x \neq 0$, $y \neq 0$.

4. (a) Points of discontinuity on parabola $y = x^2$; (b) points on $x^2 + y^2 = 1$; (c) $x = y = 0$.

Section 3, Page 318

1. (a) $\dfrac{-y}{x^2}, \dfrac{1}{x}$; (b) $3x^2y - \dfrac{y}{x^2 + y^2}$, $x^3 + \dfrac{x}{x^2 + y^2}$; (c) $y \cos xy + 1$, $x \cos xy$; (d) $e^x \log y$, e^x/y;

   (e) $2xy + \dfrac{1}{\sqrt{1 - x^2}}$, x^2.

2. (a) $2xy - z^2$, $x^2 + z$, $y - 2xz$; (b) $yz + \dfrac{1}{x}$, $xz + \dfrac{1}{y}$, xy; (c) $\dfrac{z}{\sqrt{y^2 - x^2}}$, $\dfrac{-zx}{y\sqrt{y^2 - x^2}}$, $\sin^{-1}\dfrac{x}{y}$;

   (d) $\dfrac{x}{\sqrt{x^2 + y^2 + z^2}}$, $\dfrac{y}{\sqrt{x^2 + y^2 + z^2}}$, $\dfrac{z}{\sqrt{x^2 + y^2 + z^2}}$;

   (e) $\dfrac{-x}{(x^2 + y^2 + z^2)^{3/2}}$, $\dfrac{-y}{(x^2 + y^2 + z^2)^{3/2}}$, $\dfrac{-z}{(x^2 + y^2 + z^2)^{3/2}}$.

6. (a) 1; (b) $(2x^2 - y^2 - z^2)/(x^2 + y^2 + z^2)^{5/2}$.

Section 4, Page 322

1. $\pi/6$ ft^3.   2. 11.7 ft.   3. 0.139 ft.   4. 2,250.   5. 10.85.   6. \$8.64.
7. 0.112; 0.054.   8. 53.78; 0.93.   9. 0.003π; 0.3 per cent.   10. 1.6π; π.

Section 5, Page 325

1. (a) $e^{t^2}\left(2t \sin\dfrac{t-1}{t} + \dfrac{1}{t^2}\cos\dfrac{t-1}{t}\right)$.   2. $6/\sqrt{3}$.

3. $\dfrac{1}{x^2 + y^2}\left(x\sqrt{x^2 + y^2}\,\dfrac{\partial f}{\partial u} - y\dfrac{\partial f}{\partial v}\right)$; $\dfrac{1}{x^2 + y^2}\left(y\sqrt{x^2 + y^2}\,\dfrac{\partial f}{\partial u} + x\dfrac{\partial f}{\partial v}\right)$;

   $(x^2 + y^2)^{-1/2}\sqrt{(x^2 + y^2)\left(\dfrac{\partial f}{\partial u}\right)^2 + \left(\dfrac{\partial f}{\partial v}\right)^2}$.

6. $du = 2x\,dx + 2y\,dy = 2r\,dr$.   7. $f_u = \dfrac{e^{xy}}{u^2 + v^2}(uy + vx)$; $f_v = \dfrac{e^{xy}}{u^2 + v^2}(vy - ux)$.

10. (a) $2x$, $2(x + \tan x \sec^2 x)$; (b) $\cos\theta\dfrac{\partial V}{\partial x} + \sin\theta\dfrac{\partial V}{\partial y}$, $r\left(\cos\theta\dfrac{\partial V}{\partial y} - \sin\theta\dfrac{\partial V}{\partial x}\right)$, $\dfrac{\partial V}{\partial z}$.

Section 6, Page 328

1. $xx_0/a^2 + yy_0/b^2 = 1$.   2. $ay + bx = \sqrt{2}ab$.   3. $y' = -(9x^2y^2 + \cos y)/(6x^3y - x \sin y)$.

Section 7, Page 330

1. (a) $z_x = 3x^2y/(\cos z - 2z)$, $z_y = 3x^2/(\cos z - 2z)$; (c) $z_x = -(z^2 + 3y)/(2xz - y)$, $z_y = (z - 3x)/(2xz - y)$.

2. $z_x = -1$, $z_{xx} = -2$, $z_y = 1$, $z_{yy} = 0$, $z_{xy} = 1$.   3. $x - y + z = 1$.

5. $z_{xx} = (4 - 4z + z^2 + x^2)/(2 - z)^3$.

Section 8, Page 333

1. $u_x = (2v - x^2)/[u(v - u)]$, $v_x = (x^2 - 2u)/[u(v - u)]$.
2. $w_x = -(2v + 3u)/v^2$, $w_y = (u + v)/v^2$.   3. $u_x = (3x^2 + 4v)/(2u + 2v)$.

Section 10, Page 337

1. $\dfrac{\pi^2}{4} + \left(\dfrac{\pi^2}{4} - \dfrac{\pi}{2}\right)h + (\pi - 1)k + (\pi - 1)hk + k^2 + \dfrac{\pi^3}{48}h^3 + \dfrac{\pi^2}{8}h^2k + \left(1 + \dfrac{\pi}{4}\right)hk^2 + \frac{1}{6}k^3 + \cdots$

   where $h = x - 1$, $k = y - \pi/2$.

2. $e\left\{1 + (h + k) + \dfrac{1}{2!}[h^2 + 4hk + k^2] + \cdots\right\}$, $h = x - 1$, $k = y - 1$.

3. $1 + x + \dfrac{1}{2!}(x^2 - y^2) + \dfrac{1}{3!}(x^3 - 3xy^2) + \dfrac{1}{4!}(x^4 - 6x^2y^2 + y^4) + \cdots$

Section 11, Page 340

8. $8abc/3\sqrt{3}$.   **9.** $a/3, b/3, c/3$.

10. $\sqrt{3}P/(2\sqrt{3}+3), (\sqrt{3}+1)P/2(2\sqrt{3}+3), P/(2\sqrt{3}+3)$.

11. $l = h = \dfrac{1}{5\pi}\sqrt[3]{60\pi^2 V}, d = \sqrt{5}l$.

Section 12, Page 344

3. (a) $164°15', 90°$; (b) $105°45', 90°$.   **4.** $3a/\sqrt{2}$.   **5.** $d/\sqrt{a^2+b^2+c^2}$.

6. $x = \pm 4/\sqrt{5}, y = \pm 3/\sqrt{5}$.   **8.** $x = \Sigma m_i x_i/\Sigma m_i, y = \Sigma m_i y_i/\Sigma m_i$.

9. $I_1:I_2:I_3 = R_1^{-1}:R_2^{-1}:R_3^{-1}$.   **10.** $\sqrt{2A}, \sqrt{2A}, \sqrt{2A}$.

Section 13, Page 347

1. $x = \pm a\sqrt{a}/\sqrt{a+b}, y = \pm b\sqrt{b}/\sqrt{a+b}$.   **2.** Cube with edge $s/12$.

Section 14, Page 350

1. $\dfrac{\pi \sin(\pi\alpha/2)}{2\alpha} + \dfrac{\cos(\pi\alpha/2) - 1}{\alpha^2}$.   **2.** $\alpha\pi$.   **3.** $\alpha\left(\dfrac{\pi}{2} - \log 2\right)$.

4. $-\tan\alpha$.   **5.** $2x^2$.   **7.** $\alpha\pi(\alpha^2 - 1)^{-3/2}$.   **11.** 3π.

Section 15, Page 354

5. $(py')' - qy - f = 0$.

Section 18, Page 363

1. $\pi a^2/2$.   **2.** $4a^2(\pi/2 - 1)$.   **3.** $\frac{4}{3}a^3(\pi/2 - \frac{2}{3})$.   **4.** $8a^2$.

5. $\bar{x} = a\cos^2(\alpha/2)$.   **7.** $\sqrt{2}\, a^2/2$.

Section 22, Page 373

2. (a) $u = x^3/3 + xy^2$; (b) $u = y\sin x$; (c) $u = \frac{1}{2}(x+1)^2 + \frac{1}{2}(y+1)^2$; (d) $u = xy^2$.

CHAPTER 6

Section 2, Page 385

1. At $(1,2,3)$, $\nabla u = 2\mathbf{i} + 4\mathbf{j} + 6\mathbf{k}$; $du/dn = 2\sqrt{14}$; at $(0,1,2)$, $\nabla u = 2\mathbf{j} + 4\mathbf{k}$; $du/dn = 2\sqrt{5}$.

2. (a) $-(i\mathbf{x} + j\mathbf{y} + k\mathbf{z})(x^2 + y^2 + z^2)^{-3/2}$; (b) $2(i\mathbf{x} + j\mathbf{y} + k\mathbf{z})(x^2 + y^2 + z^2)^{-1}$.

3. $du/ds = -3/\sqrt{6}$.   **4.** $du/ds = -7/\sqrt{5}$.   **5.** $\mathbf{n} = \frac{1}{3}(\mathbf{i} - 2\mathbf{j} + 2\mathbf{k})$.

6. $\frac{1}{3}(2\mathbf{i} - 2\mathbf{j} - \mathbf{k}), \frac{1}{3}(-2\mathbf{i} + 2\mathbf{j} + \mathbf{k})$.   **8.** $du/ds = -3$.   **9.** $dv/ds = 6/\sqrt{5}$.

10. $\cos^{-1}[-9/(5\sqrt{157})]$.   **11.** $(x^2 + y^2 + z^2)^{-1}$.   **13.** Circular cones.

Section 3, Page 389

1. 0.   **2.** $(\pi^2 + 4)/8$.   **3.** (a) $\frac{5}{6}$; (b) $-\frac{1}{2}$; (c) $\frac{5}{3}$.

4. Helical path, $\pi^2/8 - 1$; rectilinear path, $\pi^2/8 - 1$.   **5.** (a) $W = -g$; (b) $W = -g$.

6. 13.   **7.** $\frac{8}{15}$.   **8.** 0.   **9.** $\frac{1}{3}$.   **10.** 0.

Section 4, Page 391

1. $\frac{3}{2}$.   **2.** $u = xy + \frac{1}{3}y^3 - \frac{1}{3}x^3$.

Section 6, Page 397

1. (a) 3; (b) $2/r$; (c) 0.   **2.** $0, 2/r$.   **5.** $5u, 0$.

6. div $\mathbf{v} = f'(r) + 2f(r)/r, r \neq 0$; (a) 4π; (b) $4\pi a^3$.

Section 7, Page 399

2. $\pi a^2 b$.   **3.** $4\pi abc$.   **4.** $4\pi a^5$.

9. (a) $12\pi a^5/5$; (b) $a^3/2$; (c) $\pi a^2 b^2/2$; (d) $a^2 b^2(2a/3 + \pi b/8)$; (e) 0.

Section 8, Page 403

2. (a) 0; (b) 0; (c) 0.

Section 9, Page 406

1. $-\pi$.   **2.** 0.   **3.** 0.   **5.** (a) $-\pi a^6/8$; (b) $-a^3$; (c) 0.   **6.** $-\pi a^6/8$.

7. (a) $\log(x + y + z) + c$; (b) $\sqrt{x^2 + y^2 + z^2} + c$; (c) $x^3 + 3xy - x + yz^2 + z + c$.

Section 10, Page 408

2. $\mathbf{w} = \mathbf{j}(xy - \frac{1}{2}x^2) + \mathbf{k}[z(x + y) - \frac{1}{2}(x^2 + y^2)]$. 　3. $u = xy^2 + x^2z^2 - x$.

4. $\mathbf{w} = \mathbf{j}(xyz^2 - \frac{1}{2}x^2z) + \mathbf{k}(-\frac{1}{2}x^2y + x^3yz)$. 　5. No.

Section 12, Page 415

2. $g_{11} = 1$, $g_{22} = \rho^2$, $g_{33} = \rho^2 \sin^2 \theta$, $g_{12} = g_{23} = g_{13} = 0$, where $\rho = x_1$; $\theta = x_2$; $\phi = x_3$.

Section 13, Page 418

1. $\nabla u = \mathbf{r}_1 \dfrac{\partial u}{\partial r} + \dfrac{\mathbf{\theta}_1}{r} \dfrac{\partial u}{\partial \theta} + \mathbf{k} \dfrac{\partial u}{\partial z}$.

2. $\nabla^2 = \dfrac{1}{(u^2 + v^2)u} \dfrac{\partial}{\partial u} + \dfrac{1}{(u^2 + v^2)v} \dfrac{\partial}{\partial v} + \dfrac{1}{u^2 + v^2} \dfrac{\partial^2}{\partial u^2} + \dfrac{1}{u^2 + v^2} \dfrac{\partial^2}{\partial v^2} + \dfrac{1}{u^2 v^2} \dfrac{\partial^2}{\partial \phi^2}$.

3. $\operatorname{div} \mathbf{F} = -3p \cos \theta / r^4$; $\operatorname{curl} \mathbf{F} = 0$.

Section 15, Page 422

2. Irrotational. 　4. $\Phi = x^2 - y^2$; hyperbolas. 　8. Irrotational and solenoidal.

10. $\Phi = y^3 - 3x^2y$; $\Phi = x(x^2 + y^2)^{-1}$.

CHAPTER 7

Section 1, Page 439

3. $u = \frac{1}{2} \cos k(x - at) - \frac{1}{2} \cos k(x + at)$. 　4. $uu_{xy} = u_x u_y$.

11. $u = x^2 + f_2(y)$, $u = xy + f_2(x)$, $u = xf_1(y) + f_2(y)$, $u = yx^2/2 + xf_1(y) + f_2(y)$, $u = f(y)e^{2x}$,

$u = f_1(x)e^y + f_2(x)e^{-y}$, $u = f_1(y) \cos x + f_2(y) \sin x + x + y$.

Section 2, Page 443

1. String is fixed at $x = 0$.

2. (a) $[1 + (x - at)^2]^{-1}$, $-2(x - at)[1 + (x - at)^2]^{-2}$; $\sin k(x - at)$, $k \cos k(x - at)$; e^{x-at}, e^{x-at};

(b) $u = [1 + (x - at)^2]^{-1} + [1 + (x + at)^2]^{-1}$, $u(0,t) = 2(1 + a^2t^2)^{-1}$, $u(x,0) = 2(1 + x^2)^{-1}$,

$u(l,1/a) = (l^2 - 2l + 2)^{-1} + (l^2 + 2l + 2)^{-1}$; $u(x,t) = \sin k(x - at) + \sin k(x + at)$,

$u(0,t) = 0$, $u(x,0) = 2 \sin kx$, $u(l,1/a) = 2 \cos k \sin kl$; $u(x,t) = e^{x-at} + e^{x+at}$, $u(0,t) = e^{at} + e^{-at}$,

$u(x,0) = 2e^x$, $u(l,1/a) = e^l(e + e^{-1})$.

Section 3, Page 449

2. $u(0,\pi) = \dfrac{1}{2a} \sin 2a\pi$; $u_t(0,\pi) = \frac{1}{2} \cos 2a\pi$. 　3. $a = \pm\dfrac{\pi}{15} + \dfrac{2n\pi}{5}, n = 0, \pm1, \pm2, \ldots$.

4. $\text{Const} + (2a)^{-1}e^{-x^2}$. 　5. $ae^{-x^2}(2x^2 - 1)$.

7. $\frac{1}{2} \sin (x + at) + \frac{1}{2} \sin (x - at)$; $\cos (x + at) - \cos (x - at)$.

Section 4, Page 451

5. (a) $f(y + m_1x)$, $f(y + m_2x)$. 　6. $f_1(x + at) + f_2(x - at)$.

8. (a) $F_1(y - ix) + F_2(y + ix)$, $F_1(y + x) + F_2(y - 2x)$, $F_1(y + x) + xF_2(y + x)$.

9. (d) $F_1(y - 5x) + F_2(y + x) - y^4/60 + x^3/6$.

10. $\dfrac{x^2}{4} + F_1(2y + x) + F_2(y - x)$; $\dfrac{x^4}{12} + F_1(y - ax) + F_2(y + ax)$;

$\frac{1}{6}x^3 + \frac{1}{12}y^3 + F_1(y - 2x) + F_2(y - x)$.

Section 5, Page 454

1. $-l, \pm2l, \pm3l, \ldots$. 　3. σ would be $0.22d(l/t)^2$.

Section 6, Page 457

3. (b) $b/2$. 　4. (a) $\dfrac{8b}{\pi^2} \sum_{n=0}^{\infty} (-1)^n \dfrac{1}{(2n + 1)^2} \sin \dfrac{(2n + 1)\pi x}{l}$; (c) $\dfrac{4\sqrt{2}b}{\pi^2} \sum_{n=0}^{\infty} \dfrac{(-1)^n}{(2n + 1)^2}$.

Section 7, Page 460

3. 0.45 oscillation per second.

4. $u(x,t) = \dfrac{1}{\omega_n^2 - \omega^2} \sin \omega t \sin \dfrac{\pi x}{l}$, if $\omega_n \neq \omega = \pi a/l$; $u(x,t) = -t \dfrac{\cos \omega t}{2\omega} \sin \dfrac{\pi x}{l}$, if $\omega_n = \omega = \pi a/l$.

Section 8, Page 464

1. 2.07×10^6 cal/(m²)(day).   **2.** $u\left(\frac{l}{2},t\right) = \frac{200}{\pi} \sum_{n=1}^{\infty} \frac{1}{n} \sin\left(\frac{n\pi}{2}\right) e^{-1.77 n^2 \pi^2 10^{-4} t}.$

3. $v(x,t) = \frac{20}{\pi} \sum_{n=1}^{\infty} \frac{5 - 4\cos n\pi}{n} e^{-1.77 n^2 \pi^2 10^{-4} t} \sin\frac{n\pi x}{100} + \frac{x}{10}.$ **4.** (c) $\frac{2kc_n l}{\pi n \alpha^2}\left[1 - e^{-(\alpha n\pi/l)^2 t}\right].$

Section 9, Page 467

1. $I(x,t) = \frac{12}{Rl} - \sum_{n=1}^{\infty} \left(\frac{2}{l}\int_0^l \frac{6}{l}\xi \sin\frac{n\pi\xi}{l}\,d\xi\right) e^{-(1/RC)(n\pi/l)^2 t} \frac{n\pi}{Rl}\cos\frac{n\pi x}{l}.$

2. $V = 6 - \frac{3x}{50} - \frac{4}{\pi}\sum_{n=1}^{\infty}\frac{(-1)^n}{n} e^{-(1{,}000/0.24)n^2\pi^2 t}\sin\frac{n\pi x}{100}.$

3. $I = 0.6 + 1.1\sum_{n=1}^{\infty}(-1)^n e^{-n^2\pi^2 t/0.6}\cos\frac{n\pi x}{1{,}000}.$ **5.** $I = c_1 e^{\sqrt{RG}x} + c_2 e^{-\sqrt{RG}x}.$

Section 10, Page 470

3. $u(x,t) = \sum c_n e^{-(\alpha n\pi/l)^2 t}\sin\frac{n\pi x}{l},\; c_n = \frac{2}{l}\int_0^l f(x)\sin\frac{n\pi x}{l}\,dx.$

4. (b) $u(x,t) = \sum c_n e^{-[\pi\alpha(2n-1)/2l]^2 t}\sin\frac{\pi}{2l}(2n-1)x;\; c_n = \frac{2}{l}\int_0^l f(x)\sin\frac{\pi}{2l}(2n-1)x\,dx.$

7. $\sum a_n\cos\frac{n\pi a t}{l}\sin\frac{n\pi x}{l};\; a_n = \frac{2}{l}\int_0^l f(x)\sin\frac{n\pi x}{l}\,dx.$

8. $\sum b_n\sin\frac{n\pi a t}{l}\sin\frac{n\pi x}{l};\; b_n = \frac{2}{n\pi a}\int_0^l g(x)\sin\frac{n\pi x}{l}\,dx.$

9. (a) $(c_1\sin\sqrt{p/a}x + c_2\cos\sqrt{p/a}x + c_3\sinh\sqrt{p/a}x + c_4\cosh\sqrt{p/a}x)(c_5\sin pt + c_6\cos pt).$

Section 11, Page 473

1. 1.5552×10^6 cal per day.   **3.** $0.44883,\ 0.14922,\ 0.00004.$

5. $u(x,y) = \frac{400}{\pi}\sum\frac{1}{2n-1} e^{-(2n-1)\pi y/10}\sin(2n-1)\frac{\pi x}{10}.$

6. $u(x,t) = \frac{400}{\pi}\sum\frac{1}{2n-1} e^{-[\alpha(2n-1)\pi/10]^2 t}\sin(2n-1)\frac{\pi x}{10}.$ **7.** $35.5,\ 41.9.$

Section 12, Page 477

1. Partial answer $u = \dfrac{u_1 r_1 - u_0 r_0}{r_1 - r_0} - \dfrac{r_1 r_0}{r}\dfrac{u_1 - u_0}{r_1 - r_0}.$

2. (a) $u = \dfrac{u_1\log(r/r_0) + u_0\log(r_1/r)}{\log r_1/r_0},$ (b) $2\pi k\dfrac{u_0 - u_1}{\log r_1/r_0}.$

6. $U(r,\theta) = \frac{1}{2\pi}\int_0^\pi\left[\frac{R^2 - r^2}{R^2 - 2Rr\cos(\theta - \phi) + r^2} - \frac{R^2 - r^2}{R^2 - 2Rr\cos(\theta + \phi) + r^2}\right]f(\phi)\,d\phi.$

7. $U(r,\theta) = \frac{1}{2\pi}\int_0^{2\pi}\frac{r^2 - R^2}{r^2 - 2rR\cos(\theta - \phi) + R^2}f(\phi)\,d\phi.$

Section 13, Page 479

1. $U = 50 + \frac{200}{\pi}\sum\frac{1}{(2n-1)a^{2n-1}}r^{2n-1}\sin(2n-1)\theta.$

2. $U = \frac{50}{\pi}\int_0^\pi\frac{a^2 - r^2}{a^2 - 2ar\cos(\theta - \phi) + r^2}\,d\phi.$

3. $U = \frac{400}{\pi}\sum\frac{1}{(2n-1)a^{2n-1}}r^{2n-1}\sin(2n-1)\theta.$

4. $u(x,y) = \sum a_n\sin\pi nx\sinh\pi ny;\; a_n = \frac{2}{\sinh\pi n}\int_0^1 f(x)\sin\pi nx\,dx.$

Section 14, Page 484

1. (a) $\dfrac{\gamma}{2}\left(\dfrac{1}{a^2} + \dfrac{1}{b^2}\right)^{\frac{1}{2}}$; (b) $\dfrac{\pi}{10}\,\gamma\left(\dfrac{1}{a^2} + \dfrac{1}{b^2}\right)^{\frac{1}{2}}$.

Section 15, Page 486

1. $J_0(k_n r)\cos\left(m\dfrac{\pi}{b}z\right)(A_{mn}\cos\omega_{mn}t + B_{mn}\sin\omega_{mn}t)$; $\omega_{mn}^2 = \gamma^2\left(k_n^2 + \dfrac{\pi^2 m^2}{b^2}\right)$.

2. $\Sigma A_n e^{-\alpha^2 k_n^2 t} J_0(k_n r)$, where $1 = \Sigma A_n J_0(k_n r)$.   4. $\Sigma A_n e^{-k_n z} J_0(k_n r)$, where $100 = \Sigma A_n J_0(k_n r)$.

5. $\Sigma A_n e^{-\alpha^2 k_n^2 t} J_0(k_n r)$, where $f(r) = \Sigma A_n J_0(k_n r)$.   6. $k^2 = 4\omega^2/g$.

Section 17, Page 493

1. No.   2. $(\pi\alpha^2 t)^{-\frac{1}{2}}\displaystyle\int_0^\infty f(\xi)e^{-(x^2+\xi^2)/(4\alpha^2 t)}\cosh\dfrac{x\xi}{2\alpha^2 t}\,d\xi$.

3. $(\pi\alpha^2 t)^{-\frac{1}{2}}\displaystyle\int_0^\infty f(\xi)e^{-(x^2+\xi^2)/(4\alpha^2 t)}\sinh\dfrac{x\xi}{2\alpha^2 t}\,d\xi$.

4. $(4\pi\alpha^2 t)^{-\frac{1}{2}}\displaystyle\int_0^l f(s)\left\{\sum_{n=-\infty}^{\infty}\left[e^{-(x-2nl-s)^2/(4\alpha^2 t)} \pm e^{-(x-2nl+s)^2/(4\alpha^2 t)}\right]\right\}ds$.

Section 18, Page 496

4. $u(x,t) = u_1\left[1 - \dfrac{x}{2a\sqrt{\pi}}\displaystyle\int_0^t \dfrac{e^{-x^2/[4a^2(t-t_1)]}}{(t-t_1)^{3/2}}\,dt_1\right]$.

Section 19, Page 500

3. $u(x,y,t) = (4\pi\alpha^2 t)^{-1}\displaystyle\int_{-\infty}^{+\infty}\int_{-\infty}^{+\infty} e^{-[(x-x_1)^2+(y-y_1)^2]/(4\alpha^2 t)}f(x_1,y_1)\,dx_1\,dy_1$.

4. $u(x,y,z,t) = \displaystyle\int_0\int_{-\infty}^{+\infty}\int_{-\infty}^{+\infty}\dfrac{f(y_1,z_1,t_1)}{[4\pi\alpha^2(t-t_1)]^{3/2}}e^{-[x^2+(y-y_1)^2+(z-z_1)^2]/[4\alpha^2(t-t_1)]}\,dy_1\,dz_1\,dt_1$.

5. $u(x,y,t) = \displaystyle\int_0^t\int_{-\infty}^{+\infty}\dfrac{f(y_1,t_1)}{4\pi\alpha^2(t-t_1)}e^{-[x^2+(y-y_1)^2]/4\alpha^2(t-t_1)}\,dy_1\,dt_1$;

$u(x,t) = \displaystyle\int_0 \dfrac{f(t_1)}{[4\pi\alpha^2(t-t_1)]^{\frac{1}{2}}}e^{-x^2/4\alpha^2(t-t_1)}\,dt_1$.

Section 24, Page 515

1. $|k| < 2$; $|k| = 2$; $|k| > 2$.   2. Elliptic for $0 < y$, $x^2 + y^2 < 1$, $x^2 + y^2 > \frac{1}{2}$.

5. (a) $[ce^y + (1-c)e^{-y}]\sin x$.

Section 25, Page 518

1. First equation $3\Sigma 2^{n-1}(-1)^{n+1}(xy)^n/(n!)^2$.

Section 26, Page 519

2. At the lattice points in the region $y \geq x$, $x > 0$; $y \geq h - x$, $x \leq 0$.   3. $\frac{9}{7}$.

Section 27, Page 522

1. 19.   2. 2.   3. 2.

Section 28, Page 526

3. $U_{rs} = 0$; $U_{rr} = 0$; $U_{rr} + U_{ss} = 0$.

CHAPTER 8

Section 1, Page 532

1. (a) 2, $\pi/3$; (b) $2\sqrt{2}$, $\pi/4$; (c) 2, π; (d) 1, $3\pi/2$; (e) $\sqrt{2}/2$, $7\pi/4$; (f) 1, $\pi/2$; (g) $\frac{3}{4}$, $\pi/3$; (h) 4, π.

2. (a) -8; (b) $-1 + i$; (c) $(2 - \sqrt{3})/2 - (2 + \sqrt{3})/2$.   3. (a) 1; (b) 1; (c) 1.

4. $\cos(\pi/6) + i\sin(\pi/6)$, $\cos(\pi/6 + 2\pi/3) + i\sin(\pi/6 + 2\pi/3)$,
$\cos(\pi/6 + 4\pi/3) + i\sin(\pi/6 + 4\pi/3)$.

5. $\pm\frac{1}{2}(2 \pm i\sqrt{2})$.   7. (a) 1, $\frac{1}{2}(-1 + i\sqrt{3})$, $\frac{1}{2}(-1 - i\sqrt{3})$; (b) 1, i, $-i$, -1.

8. $2^{1/10}e^{\pi i/20}$, $2^{1/10}e^{9\pi i/20}$, $2^{1/10}e^{17\pi i/20}$, $2^{1/10}e^{25\pi i/20}$, $2^{1/10}e^{33\pi i/20}$.

10. (a) $\pm\sqrt{2}\,(1+i)/2$; (b) $2^{1/4}e^{7\pi i/8}$, $2^{1/4}e^{11\pi i/8}$; (c) $2^{-1/4}e^{-7\pi i/8}$, $2^{-1/4}e^{-11\pi i/8}$.

14. (a) Re $(z) = 1$, Im $(z) = -2$; (b) Re $(z) = 3$, Im $(z) = 4$; (c) Re $(z) = -\frac{1}{5}$, Im $(z) = -\frac{2}{5}$.

16. (a) Circle $x^2 + y^2 = 1$; (b) disk $x^2 + y^2 < 1$; (c) infinite region $x^2 + y^2 \geq 1$;
    (d) hyperbola $12x^2 - 4y^2 = 3$.
17. (a) Region $x \geq 1$, $-\infty < y < \infty$; (b) open region $y > 1$, $-\infty < x < \infty$;
    (c) hyperbola $x^2 - y^2 = 1$.
18. (a) Circle $(x - 1)^2 + y^2 = 4$; (b) circle $x^2 + y^2 = 1/(\text{const})^2$;
    (c) circle $x^2 + y^2 - 2x(1 + c^2)/(1 - c^2) = c^2/(1 - c^2)$, $c^2 \neq 1$.
20. (a) $x^2 + y^2 - i2xy$; (b) $(x - iy)/(x^2 + y^2)$; (c) $[(1 - x) + iy]/[(1 - x)^2 + y^2]$;
    (d) $(x^2 + y^2 + x - 1) + i(2xy + y)$; (e) $[x - i(y + 1)]/[x^2 + (y + 1)^2]$.

Section 2, Page 535

1. (a) $(x^2 - y^2 - x + 1) + i(2xy - y)$; (b) $x/(x^2 + y^2) - iy/(x^2 + y^2)$;
   (c) $[(x - 1) - iy]/[(x - 1)^2 + y^2]$; (d) $(x^2 + y^2 - 1)/[x^2 + (y + 1)^2] - i2x/[x^2 + (y + 1)^2]$;
   (e) $[x^2 - y^2 - i(1 + 2xy)]/[(x^2 - y^2)^2 + (1 + 2xy)^2]$; (f) $x + i2y$; (g) $(x^2 + y^2)^{-1}$.
2. (a) Open region $x < 3$, $-\infty < y < \infty$; (b) the region $y \geq 1$, $-\infty < x < \infty$; (c) the region exterior
   to the circle of radius 1 with center at the origin and including circular boundary; (d) circular ring
   centered at the origin with interior radius 1, exterior radius 2, including the boundary of the inner
   circle; (e) open circular region with center at $(1,0)$ of radius 1; (f) closed circular region of radius 1
   with center at $z_0 = x_0 + iy_0$; (g) open region exterior to the circle of radius 2 with center at $(0,-1)$.

Section 3, Page 538

3. (a) $\frac{1}{2}(e^{-1} + e) \cos 2 + \frac{1}{2}i(e^{-1} - e) \sin 2$; (b) $e^{2k\pi}$, $k = 0, \pm1, \pm2, \cdots$;
   (c) $ei \, \text{Log} \, \sqrt{2} - (\pi/4 + 2k\pi)$; (e) $e^{\pi/2 + 2k\pi}$.
4. (a) $[(x - 1) - iy]/[(x - 1)^2 + y^2]$; (b) $[x^2 - y^2 - i(1 + 2xy)]/[(x^2 - y^2)^2 + (1 + 2xy)^2]$;
   (c) $\frac{1}{2}(e + e^{-1}) \sin 1 + \frac{1}{2}i(e - e^{-1}) \cos 1$; (d) $e^{x^2 - y^2}(\cos 2xy + i \sin 2xy)$;
   (e) $e^{x/(x^2 + y^2)}\{\cos [y/(x^2 + y^2)] - i \sin [y/(x^2 + y^2)]\}$.
5. (a) $\text{Log } 4 + \pi i$; (b) $\text{Log } 5 + i\pi/2$; (c) $\text{Log } \sqrt{2} + i(\pi/4 + 2k\pi)$; (d) $2k\pi i$; (e) $e^{-\pi/2}$;
   (f) $e(\cos 1 + i \sin 1)$; (g) $\frac{1}{2}i(e^2 - e^{-2})$.
6. (a) $(\pi + 2\pi k)i$; (b) $\pi/2 + 2\pi k - i \text{ Log } (2 \pm \sqrt{3})$; (d) $2\pi k$, $k = 0, \pm1, \pm2, \pm \cdots$.

Section 4, Page 542

3. (b) $z = -1$; (c) $z = 0$; (d) $z = \pi/2 + k\pi$, $k = 0, \pm1, \pm2, \cdots$; (e) $z = 1$, $z = -1$;
   (f), (g), (h), (i), (j) at all points; (k) $z = \pm i$.
8. $\cos w \, dw/dz$, $-\sin w \, dw/dz$, $e^w \, dw/dz$.
9. (a) $2ze^{-z^2}$; (b) $-2e^{-2z}$; (c) $4z^2e^{2z^2} - 2z$; (d) $2z/[1 + (1 + z^2)^2]$; (e) $2z \cos (1 + z^2)$.

Section 5, Page 545

1. $\frac{1}{3}(2 + 11i)$.   2. 1 along rectilinear, $1 + i/3$ along parabolic.   4. $2 + \pi i$, $2 - \pi i$.
5. 0.   6. 0.   7. (-2).

Section 7, Page 550

2. $2\pi i$.   3. 2, upper half; -2, lower half.   4. (a) $-\pi i$; (b) πi.   5. 0.   6. $2\pi i$.
7. 0.   10. (a) 0; (b) 0; (c) $-\pi i$; (d) πi.

Section 8, Page 554

1. $2\pi i(8 - 13i)$.   2. 0.   3. (a) 0; (b) $2\pi i$; (c) $2\pi i e^{i/2}$; (d) 0; (e) $2\pi i$.   4. $2\pi i$.
5. $4\pi i$.   6. $-2\pi i$.

Section 9, Page 556

1. $u = x^3 - 3xy^2$.
2. (a) $x + iy$; (b) $\cosh y \cos x - i \sinh y \sin x$; (c) $1/z$; (d) $e^x(\cos y + i \sin y)$; (e) $\log z$.
4. Max $u_1 = 1$ at $(\pm1,0)$.

Section 10, Page 559

1. (a) $1 + \sum_{n=1}^{\infty} z^n$, $|z| < 1$; (b) $\frac{1}{2}\left[1 + \sum_{n=1}^{\infty} (z + 1)^n/2^n\right]$, $|z + 1| < 2$;

(c) $\dfrac{1}{1 - i}\left[1 + \sum_{n=1}^{\infty} \left(\dfrac{z - i}{1 - i}\right)^n\right]$, $|z - i| < \sqrt{2}$.

2. $\displaystyle\sum_{n=1}^{\infty} (-1)^{n-1} \frac{(z-1)^n}{n}$, $R = 1$.

3. (a) $\displaystyle\sum_{n=0}^{\infty} \frac{z^n}{n!}$, $R = \infty$; (b) $\displaystyle\sum_{n=1}^{\infty} \frac{(-1)^{n-1}z^{2n-1}}{(2n-1)!}$, $R = \infty$; (c) $1 + \displaystyle\sum_{n=1}^{\infty} \frac{(-1)^n z^{2n}}{(2n)!}$, $R = \infty$;

(d) $\displaystyle\sum_{n=1}^{\infty} (-1)^{n-1} \frac{z^n}{n}$, $R = 1$; (e) $\displaystyle\sum_{n=0}^{\infty} z^{2n}/(2n)!$, $R = \infty$.

4. $-\displaystyle\sum_{n=1}^{\infty} (z - \pi i)^{2n-1}/(2n-1)!$, $R = \infty$.

Section 11, Page 563

2. (a) $\dfrac{1}{z} + 2 + 3z + 4z^2 + \cdots$; (b) $\dfrac{1}{(1-z)^2} + \dfrac{1}{1-z} + 1 + (1-z) + (1-z)^2 + (1-z)^3 + \cdots$.

3. $1 - \dfrac{1}{z^2} + \dfrac{1}{2!z^4} - \dfrac{1}{3!z^6} + \dfrac{1}{4!z^8} - \cdots$ **4.** Functions *are* expressed in Laurent's series

5. (a) $-(z-1)^{-1} - \displaystyle\sum_{n=0}^{\infty} (z-1)^n$; (b) $\displaystyle\sum_{n=0}^{\infty} (2^n - 1)/z^{n+1}$; (c) $-\displaystyle\sum_{n=0}^{\infty} \left(\dfrac{z^n}{2^{n+1}} + \dfrac{1}{z^{n+1}} \right)$.

Section 12, Page 567

1. (a) $+\dfrac{1}{z^3} - \dfrac{1}{2!}\dfrac{1}{z} + \dfrac{1}{4!}z - \dfrac{1}{6!}z^3 + \cdots$, residue $(-\tfrac{1}{2})$; (b) $1 - \dfrac{1}{z^2} + \dfrac{1}{2!}\dfrac{1}{z^4} - \dfrac{1}{3!}\dfrac{1}{z^6} + \cdots$, residue (0);

(c) $\dfrac{-1}{z-1}$, residue (-1); (d) $= -1 - \dfrac{1}{z-1}$, residue (-1);

(e) $\dfrac{1}{z^2} - \dfrac{1}{z} + \dfrac{1}{2!} - \dfrac{z}{3!} + \dfrac{z^2}{4!} - \cdots$, residue (-1);

(f) $z^2 + z + \dfrac{1}{2!} + \dfrac{1}{3!z} + \dfrac{1}{4!z^2} + \cdots$, residue $(\tfrac{1}{3}!)$;

(g) $-\dfrac{2}{z^3} - \dfrac{4}{2!z^2} - \dfrac{8}{3!z} - \dfrac{16}{4!} - \dfrac{32z}{5!} - \cdots$, residue $(-\tfrac{8}{3}!)$;

(h) $e\left[\dfrac{1}{(z-1)^2} + \dfrac{1}{z-1} + \dfrac{1}{2!} + \dfrac{z-1}{3!} + \dfrac{(z-1)^2}{4!} + \cdots \right]$, residue (e);

(i) at $z = 1$, residue $-\tfrac{1}{2}$, at $z = -1$, residue $\tfrac{1}{2}$;

(j) $\dfrac{-1}{z} - \displaystyle\sum_{n=0}^{\infty} z^n$, residue -1 at $z = 0$, $\dfrac{1}{z-1} + \displaystyle\sum_{n=0}^{\infty} (z-1)^n$, residue 1 at $z = 1$;

(k) residues -1 at $z = 2n\pi i$, $n = 0, 1, 2, \ldots$; (l) $\displaystyle\sum_{n=0}^{\infty} 1/(z^{n+2}n!)$, residue 0;

(m) at $z = 1$ residue $-\tfrac{1}{4}$, at $z = -1$ residue $\tfrac{1}{4}$;

(n) at $z = 1$ residue $-\tfrac{1}{4}$, at $z = -1$ residue $\tfrac{1}{4}$, at $z = i$ residue $i/4$, at $z = -i$ residue $-i/4$;

(o) at $z = 2$ residue 0.

6. No.   The expansion does not represent $f(z)$ in the neighborhood of $z = 0$.

7. (a) -1; (d) $1/\cos n\pi$; (e) $e^{\pm n\pi} \cos n\pi$.

Section 13, Page 568

1. (a) $-\pi i$; (b) $2\pi i$; (c) $2\pi i/3!$; (d) $-8\pi i/3$; (e) $2\pi i e$; (f) 0; (g) 0.   **2.** $2\pi i$.   **3.** πi; $2\pi i$.

4. (a) 0; (b) $2\pi i$.   **5.** (b) $2\pi i$; (c) $2\pi i$.

Section 15, Page 571

4. $u^2 + v^2 - u - v = 0$, $u = -v$.

Section 18, Page 579

1. (a) $\cos x \cosh y$, $\sin x \sinh y$; (b) $e^x \cos y$, $e^x \sin y$;
   (d) $\log (x^2 + y^2)^{1/2}$, $\tan^{-1} (y/x)$; (e) $x/(x^2 + y^2)$, $-y/(x^2 + y^2)$.

2. (c) $v = e^x \sin y - x$; (d) $\sinh x \sin y$.

5. (a) $z = -\frac{1}{2}$; (b) $z = \pm 1$; (c) $z = 0$; (d) $z = \pm \pi/2, \pm 3\pi/2, \pm 5\pi/2, \ldots$.

6. $\phi = e^{-2uv} \sin (u^2 - v^2), \Phi = e^{-y} \cos x, F = e^{iz}$.   7. $\phi = (e^{-u} - e^u) \cos v - 1$.

CHAPTER 9

Section 1, Page 598

1. $\frac{1}{6}$, $\frac{1}{18}$.   2. $\frac{3}{20}$, $\frac{1}{2}$, $\frac{7}{20}$; $\frac{9}{38}$; $1{,}323/46{,}189$.   3. $\frac{1}{4}$.   4. $\dfrac{48!5!}{52!}$, $\dfrac{13!47!}{8!52!}$, $36 \dfrac{47!5!}{52!}$.

5. $8!$, $7!$, $7!/2$.   6. $\dfrac{1}{n}, \dfrac{2}{m}, \dfrac{2m + 2n - 4}{mn}$.   7. (a) 21; (b) 49,995,000.

8. (a) $\frac{1}{6}$; (b) 0; (c) $\frac{1}{18}$; (d) $\frac{1}{6}$.

Section 2, Page 602

1. $\frac{7}{8}$, $\frac{1}{2}$, $\frac{1}{4}$, $\frac{3}{8}$, $\frac{1}{2}$, $\frac{1}{4}$, $\frac{1}{8}$, $\frac{1}{8}$.   2. Questions 1, 2, 8 can be answered.   6. $\frac{6}{11}$.

7. $\frac{1}{5}$.   8. 100,000.   9. $\frac{1}{360}$; $\frac{1}{15}$.

Section 3, Page 606

1. $33/16{,}660$.   2. $\frac{31}{32}$; $\frac{1}{32}$.   3. $\frac{3}{32}$; $\frac{5}{32}$; $\frac{9}{32}$.   4. $\frac{41}{60}$.   5. $\frac{1}{6}$; $\frac{5}{36}$.

6. $n > (\log 2)/(\log 6 - \log 5)$.   7. $\frac{3}{10}$; $\frac{9}{20}$.   8. $\left(\dfrac{3}{13}\right)^5 > \dfrac{12 \cdot 11 \cdots 8}{52 \cdot 51 \cdots 48}$.

9. $\dfrac{5!}{52^5} < \dfrac{5!}{52 \cdots 48}$, $\dfrac{18{,}781}{13^5} > \dfrac{270{,}840}{52 \cdot 51 \cdot 50 \cdot 49}$.   10. $1 - p^n, n = 3$.   11. $\frac{4}{9}$.

12. $\dfrac{\frac{2}{3}n(\frac{2}{3}n - 1)}{(n - 1)(n - 2)}$.

Section 4, Page 609

1. $3p$.   2. $\frac{1}{2}$.   3. $\frac{9}{2}$.   4. 7; yes.   5. 2.   6. ∞.

7. $n + 2^{-n}N \doteq 1 + \log_2 N$, where $n = [\log_2 N]$.

8. $-\$75/52$, if no money changes hands in the event of a tie.

Section 5, Page 613

2. $\frac{1}{32}$, $\frac{5}{32}$, $\frac{10}{32}$, $\frac{10}{32}$, $\frac{5}{32}$, $\frac{1}{32}$.   4. $\frac{25}{32}$.   5. 2, 3, 4, 5, 5.   7. $\frac{5}{18}$; no.

8. $E(x) \doteq 0.23$.   9. $\dfrac{1}{2^{10}}$; $\dfrac{45}{2^{10}}$; $\dfrac{1}{2^{10}}$; $1 - \dfrac{11}{2^{10}}$.   10. $\frac{1}{2}$.

Section 6, Page 617

1. (b) $\frac{1}{2}$, $\frac{3}{4}$, $\frac{31}{32}$, $\frac{3}{8}$, $e^{1/2} - e/2$, $\frac{1}{2}$, $\frac{1}{2}$; (c) $\frac{1}{2}$, $\frac{2}{3}$, $\frac{5}{6}$, $\frac{5}{12}$, $\frac{1}{2} e - 1$, $\frac{1}{2}$, $\frac{1}{2}$.

2. $0.75, 0.60, F(x) = x^3$ for $0 < x < 1, m = 0.794$.   3. $\frac{1}{3}\pi$; 0.206.   4. $\frac{1}{4}a$.   5. $\sqrt{2/\pi}$.

6. $\frac{1}{2}$; $1/\pi$; none.   9. 0.3413; 0.4772; 0.6247; 0.1228; 0.6826; 0.9544; 0.9974.   10. $\frac{2}{3}$.

11. $\frac{3}{4}$; $\frac{13}{5}$; $\frac{7}{8}$; $1 - \frac{1}{4}\sqrt{2}$.

Section 7, Page 620

1. $\frac{5}{16}$; $\frac{13}{16}$; $\frac{1}{2}$; 2 and 3.   2. (a) $125/3{,}888$; (b) $\frac{23}{648}$.

3. $(0.65)^{10} + 10(0.65)^9(0.35) + 45(0.65)^8(0.35)^2 + 120(0.65)^7(0.35)^3$.

4. $5(\frac{1}{6})^6 + 24(\frac{1}{6})^5(\frac{5}{6}) + 45(\frac{1}{6})^4(\frac{5}{6})^2 + 40(\frac{1}{6})^3(\frac{5}{6})^3 + 15(\frac{1}{6})^2(\frac{5}{6})^4 = (\frac{5}{6})^6$.   5. $741/2{,}728$.

7. $0.57, 0.57$; $\dfrac{m}{n} = \dfrac{m}{n}$.

Section 8, Page 623

1. 0.499.   4. $p \doteq (0.95)^5 \doteq 0.77$; $p \doteq 0.021$.   5. (a) $\dfrac{(_aC_\alpha)(_bC_\beta)}{_NC_n}$; (b) $_nC_\alpha \dfrac{a^\alpha b^\beta}{N^n}$; (c) same as (a).

Section 9, Page 627

1. 0.039.   2. $\phi(1.5) - \phi(-1.0) = 0.806$.   3. 0.979.

4. 0.0222.   5. (a) 0.083; (b) 0.166; (c) no.

Section 10, Page 630

1. 1,540.   2. 46,413/78,125.   3. $^{49}\!/\!_{50}$.   4. 0.91854.   5. 0.73.

Section 11, Page 633

1. (a) 0.368, 0.402; (b) 0.368, 0.373.   2. 0.758.   3. 1,005.   4. Expected number = 5.
5. 0.577.   6. About 2,300.

Section 12, Page 638

1. 3; $6pq$; no.   2. $E(XY) = 6pq$.   3. (a) 1; (b) 1; (c) 1.
4. $\frac{1}{52} + \frac{1}{51} + \frac{1}{50} + \cdots + \frac{1}{1} \doteq 6.83$.   5. Ellipses.   6. $\frac{1}{4}$.   7. $\frac{1}{3}(n+1)$.

Section 13, Page 641

1. $\frac{1}{12}$.   2. $\frac{7}{2}$, $\frac{7}{2}$, $2\frac{3}{3}$; yes; $\rho = -1$.   3. $\frac{1}{3}\sqrt{3} = 0.577$.

4. Partial answer: $\sigma_{xy}^2 = \iint xyf(x,y)\,dx\,dy - \left(\iint xf(x,y)\,dx\,dy\right)\left(\iint yf(x,y)\,dx\,dy\right)$.

8. Partial answer: 0.954 and 0.750 when $k = 2$.

Section 14, Page 645

1. 0.82.   2. 0.9992.   3. 15.

Section 15, Page 648

1. 7.5; 0.26.   2. (a) 0.145; (b) 29.   5. Partial answer: 5^{10} if carnations and vases are all different
and some vases can be left empty.

CHAPTER 10

Section 1, Page 652

1. (a) $-0.8 < x_1 < -0.7$, $x_2 = 2$, $x_3 = 4$; (b) $-0.8 < x_1 < -0.7$, $1.2 < x_2 < 1.3$;
(c) $-0.8 < x_1 < -0.7$, $-0.6 < x_2 < -0.5$, $1.0 < x_3 < 1.1$;
(d) $-0.6 < x_1 < -0.5$; (e) $4.4 < x_1 < 4.5$; (f) $-1.8, 4, 1.5$; (g) $0, \pm 0.7$.
2. $h = 1.23$.   3. $h = \frac{1}{2}$.   4. $3, -2.3$.   5. (a) $1.3, -0.7$; (b) 1.8; (c) 1.4.

Section 2, Page 657

1. Prob. 1: (a) -0.75; (b) $-0.73, 1.22$; (c) $-0.77, -0.55, 1.08$; (d) -0.57; (e) 4 49; Prob 2: 1.226.
2. $-0.942, -0.200, 1.045$.   3. 0.24.   4. ± 1.20.   5. 2.51.

Section 3, Page 659

3. 2.310 radians.
4. (a) 0.739; (b) -0.567; (c) $-0.725, 1.221$; (d) 2.924; (e) $1.045, -0.942, -0.200$.
5. 2.506.   6. 4, 0.31.   7. $0.83, 0.5; -0.83, -0.5$.

Section 4, Page 661

(a) $x = {}^{15}\!/\!_{13}$; $y = {}^{20}\!/\!_{13}$; $z = -\frac{8}{13}$; (b) $x_1 = 1$; $x_2 = -1$; $x_3 = -2$; $x_4 = 3$;
(c) $x_1 = -0.107$; $x_2 = 0.988$; $x_3 = 0.317$.

Section 7, Page 667

1. $y = 0.25x^2 - 0.50x + 0.25$.

Section 8, Page 670

1. 6.654; 6.611.   2. 9.466; 12.549.   4. 0.4423.   5. $y = 1 + x + x^2 + x^3$.

Section 9, Page 671

1. 2.784, 2.700.   2. If $\theta_0 = 60$, $\theta = 40.82, 42.52, 42.50$.   3. 2.581, 2.627.   4. 106.09.
5. $y = x^2 - 10x + 1$; $y(2) = -15$.

Section 10, Page 673

1. (a) $y = \frac{1}{2}x + \frac{7}{2}$; (b) $y = 2.5x^{0.5}$; (c) $0.3(10^{0.2x})$.

Section 11, Page 680

1. $y = 4.98 - 3.13x + 1.26x^2$.   2. $p = 0.003v^2$.   3. $K = 1.778$; $b = 1.9349$; $\theta = 60.02(0.861)^t$.

Section 12, Page 683

1. $y = 0.75 + 0.10 \cos x - 0.05 \cos 3x - 0.29 \sin x$.
2. $y = 0.85 - 0.25 \cos 2x - 0.05 \cos 4x + 0.05 \cos 6x + 0.26 \sin 2x - 0.03 \sin 4x$.

Section 13, Page 688

1. 25.252, 25.068.   2. 132.137.   3. 128.6.   4. 666.25, 666.00.   5. 39.30, 38.98.

Section 14, Page 690

1. $y(\pm 0.2) = 1; y(\pm 0.4) = 1.02; y(\pm 0.6) = 1.061; y(\pm 0.8) = 1.124; y(\pm 1.0) = 1.214$.   The corresponding exact values are 1.010, 1.041, 1.094, 1.174, 1.284.

2. $y = y_1 + y'(x_1)(x - x_1) + \dfrac{1}{2h}[y'(x_1) - y'(x_0)](x - x_1)^2$.

3. $y_1 = 1.0100; y_2 = 1.0403; y_3 = 1.0927$.

Section 15, Page 692

1. $y_6 = 2.0442; y_7 = 2.3274; y_8 = 2.6509; y_9 = 3.0190; y_{10} = 3.4363$.
3. $y(0.3) = 1.3498; y(0.4) = 1.4917; y(0.5) = 1.6485; y(0.6) = 1.8218$.
4. 0.2740.   5. $y(1.00) = 3.359; y(1.25) = 3.994; y(1.50) = 4.739$.

Section 16, Page 695

1. $y = x + x^2 + \frac{1}{6}x^3 + \frac{1}{6}x^4 + R_4; z = x + \frac{1}{2}x^3 + R_4$.   2. 0.1; 0.2205; 0.3627; 0.5281.
3. 1.01; 1.031; 1.063.
4. $y = 1 + \frac{1}{2}x^2 + \frac{1}{6}x^3 + \frac{1}{12}x^4 + \frac{1}{60}x^5 + \cdots; z = x + \frac{1}{2}x^2 + \frac{1}{3}x^3 + \frac{1}{12}x^4 + \cdots$.
5. 1.0052; 1.0215; 1.0502.

APPENDIX A

Section 1, Page 707

1. $18, 0, -1, 0$.   3. 1.
5. Obtain a partial check of your results by assigning simple numerical values to the variables.
6. Check by evaluating your determinant and comparing with the product as given by above answers to Prob. 1.
7. $y = \cos x - \cos 2x$.
8. (a) $y = 1 - \frac{5}{4}x - \frac{3}{2}x^2 + \frac{3}{4}x^3$; (b) follow the pattern (a) but do not expand.

Section 2, Page 711

1. (a) $(2, -1, 1)$; (b) $(1, \frac{3}{2}, -\frac{1}{2})$; (c) $(3, -1, 2)$; (d) $(1, -1, -2, 3)$.
2. (a) $(-k/7, 5k/7, k)$; (b) $(0, 0)$; (c) $(0, 0, 0)$; (d) $(k/4, 7k/8, k)$; (e) $(k, 2k, 0)$; (f) $(0, 0, 0)$.
3. (a) $(1, -1)$; (b) inconsistent; (c) inconsistent; (d) $(1, 3k - 2, k)$.
4. Straightforward generalization of Examples 4 and 5.
5. A determinant with ith row $x_i^2 + y_i^2$ x_i y_i 1 must vanish. If the "circle" does not include a "straight line," no subdeterminant with rows x_i y_i 1 can vanish.
6, 7. Straightforward generalization of Example 3.

Section 3, Page 714

2. (a) $r = r' = 3, x = 1, y = -3k, z = 4k, w = k; r = 2, r' = 3$, inconsistent;
$r = r' = 2, x = \frac{1}{7}(4 + k - 4l), y = -\frac{1}{7}(6 + 5k + l), z = k, w = l$;
(b) $r = r' = 3, x = -3k, y = 2k + \frac{2}{3}, z = k, w = \frac{1}{3}$;
$r = r' = 2, x = 1 - 3k - 3l, y = 2k + 2l, z = k, w = l; r = 2, r' = 3$, inconsistent.

Gravitational potential, 419
Greatest lower bound, 717
Green's function, 251, 508, 511
Green's identities, 399, 500
Green's theorem, 370, 375
  consequences of, 373

Harmonic analysis, 680
Harmonic functions, 440, 475, 555
  average value theorem for, 505
  conjugate, 555
    method of, 583
  maximum principle for, 512, 555
Harnack's inequality, 511
Heat equation, 423, 461
  boundary and initial conditions for, 467
  Fourier series solution of, 462, 472
  Fourier transform solution of, 488
  Laplace transform solution of, 495
  separation of variables in, 467, 472
Heat flow, 423, 461, 471
  connection with random walks, 629
  from source, 492, 498, 629
Heaviside's expansion theorem, 225
Heaviside's superposition principle, 227
Heaviside's unit function, 218, 227
Helix, circular, 282
  hyperbolic, 283
Helmholz equation, 506
Hermite equation, 235
Hermite polynomials, 235
  orthogonality of, 240
Hermitian forms, 308
Hermitian matrices, 308
Holomorphic function, 541
Homogeneous function, 101, 103, 324, 440
Horner's method, 652n.
Hydrodynamics, 425
  (See also Fluid flow)
Hypergeometric equation, 239, 240

Implicit functions, 326, 330
  change of variables in, 331
  defined by systems of equations, 330
  differentiation of, 328
  existence theorem for, 335
Improper integrals, 12, 430
Indicial equation, 237
Infinite series (see Series)
Integral calculus, fundamental theorem of, 347 547
Integrals, absolutely convergent, 431, 432
  of analytic functions, 543
    bounds for, 545
  of Cauchy's type, 552
  change of variables in, 354, 358
  of complex functions, 543
  along contour (see Line integrals)

Integrals, differentiation of, 347, 428, 553
  elliptic, 40, 129, 137
  evaluation of, by fundamental theorem, 347, 548
    by numerical methods, 685
    by residue theorem, 590
    by series, 39
  improper, 12, 430, 549n., 592
    principal value of, 592
  indefinite, 548
  independent of path, 368, 389
  Lebesgue, 718
  line (see Line integrals)
  multiple, 354, 358, 430
    change of variables in, 355, 359, 372
  Riemann, 716
  Stieltjes, 612
  surface, 361
  transformation of, 370, 375, 398–404, 500
Integrating factors, 94, 105
Integration, numerical, 684
Interior point, 138, 314, 534
Interpolation, 663–671
  Lagrange's formula for, 670
  linear, 653
  Newton's formulas for, 668
Interval, closed and open, 24n., 314, 717n.
  of convergence, 32
Invariants, 381, 390
Irrotational fields, 406
Isoclines, 113
Isogonal curves, 119
Isolated singular points, 564
Isoperimetric problems, 353
Iterative methods, 652–662

$J_n(x)$ (see Bessel's functions)
Jacobian, 331

$K_n(x)$ (see Bessel's functions)
Kepler's laws, 129
Kronecker delta, 288

Lagrange equation, 109
Lagrange interpolation formula, 670
Lagrange multipliers, 342, 345, 353
Laguerre equation, 235
Laguerre polynomials, 235
  orthogonality of, 244
Lambert's law of absorption, 97
Laplace-de Moivre limit theorem, 623
Laplace operator, 396, 417, 418
Laplace transform, 201–231, 494
  bilateral and unilateral, 224n.
  convolution theorem for, 224
  of Dirac's "function," 220
  existence theorem for, 203, 227